C0-ATI-147

BIOMEDICAL TECHNOLOGY AND DEVICES

SECOND EDITION

HANDBOOK SERIES FOR MECHANICAL ENGINEERING

PUBLISHED TITLES

BIOMEDICAL TECHNOLOGY AND DEVICES

SECOND EDITION

Edited by
James E. Moore Jr.
Duncan J. Maitland

CRC Press
Taylor & Francis Group
Boca Raton London New York

CRC Press is an imprint of the
Taylor & Francis Group, an **informa** business

CRC Press
Taylor & Francis Group
6000 Broken Sound Parkway NW, Suite 300
Boca Raton, FL 33487-2742

Printed on acid-free paper
Version Date: 20130516

International Standard Book Number-13: 978-1-4398-5959-9 (Hardback)

Library of Congress Cataloging-in-Publication Data

Biomedical technology and devices / editors, James E. Moore, Jr., Duncan J. Maitland. -- 2nd ed.
    p. ; cm. -- (Mechanical engineering handbook series)
    Includes bibliographical references and index.
    ISBN 978-1-4398-5959-9 (hardback : alk. paper)
    I. Moore, James E., Jr., 1964- II. Maitland, Duncan J. III. Series: Mechanical engineering handbook series.
    [DNLM: 1. Biomedical Engineering--instrumentation. 2. Biomedical Technology--instrumentation. QT 36]

R856.15.B565 2013
610'.28--dc23
                                                                        2013014920

Visit the Taylor & Francis Web site at
http://www.taylorandfrancis.com

and the CRC Press Web site at
http://www.crcpress.com

Contents

SECTION I Basic Clinical Measurements

SECTION II Imaging Techniques

SECTION III Biological Assays

SECTION IV Biomaterials and Tissue Engineering

SECTION V Interventional Disease Treatment

SECTION VI Monitoring

SECTION VII Recovery

SECTION VIII Alternative and Emerging Techniques

Preface

In an attempt to summarize the field of medical technology, a natural tendency is to begin to list the areas of medicine in which technology is being applied. If, on the other hand, one tries to name an area of medicine in which it is not being applied, it becomes clear that technology is a dominant factor in modern healthcare delivery. The domain of biomedical technology has evolved through the interaction of physicians, scientists, and engineers who have combined their skills to develop better methods for diagnosing and treating disease. Recent advances in the fields of molecular biology, biomaterials, tissue engineering, and imaging have given clinicians exciting new tools. The ever-increasing significance of computational sciences, bioinformatics, visualization, telecommunications, and robotics gives promise of new clinical devices and procedures that will revolutionize medical practice. The majority of clinicians are fascinated by, and very receptive to, new technological advances. However, the degree to which novel technologies are included in everyday practice depends on many factors. In the free-market environment of the United States, financial considerations such as potential market share, intellectual property protection, regulatory approval, and insurance reimbursement often determine the fate of an emerging technology. On the other hand, if the market demands that a new technology become available, it will find its way into clinical practice. The rapidly evolving nature of biomedical technology development makes it difficult to define exactly the field itself, much less attempt to describe its entirety in a single volume. Thanks to the expertise and special effort of the contributing authors, this handbook represents our attempt at preparing a current and, as much as possible, comprehensive professional reference source, as well as a textbook for undergraduate and graduate students in biomedical engineering programs. Yet, it should at best be considered a still photograph of a swiftly moving, constantly reshaping cloud.

Acknowledgments

A volume with such diversity in topics could not have been compiled without the enthusiastic contributions from scientists with diverse interests. Contributions from 77 authors, representing 53 universities, hospitals, and companies, in 8 countries around the world are included here. This book is a testament to their extraordinary effort and belief in the purpose of its existence.

We would also like to acknowledge the hard work and endless support extended by the staff of CRC Press, particularly Jill Jurgensen and Michael Slaughter. Dr. Moore also gratefully acknowledges the support of the Biomedical Engineering Department at Texas A&M University, where he was on the faculty for much of the preparation of this book.

Editors

James E. Moore Jr., PhD, earned his bachelor's in mechanical engineering with highest honor from the Georgia Institute of Technology, followed by an MS and a PhD from the same school and institute. Following a postdoctoral fellowship at the Swiss Institute of Technology in Lausanne, he has held faculty positions at Florida International University, Texas A&M University, and now Imperial College London. His research focuses on the biomechanics of the cardiovascular and lymphatic systems.

Duncan J. Maitland, PhD, has worked as an engineer in aerospace, national defense, and biomedical applications since 1985. He earned his BEE (electrical engineering) and MS (physics) from Cleveland State University and his PhD in biomedical engineering from Northwestern University. After his PhD, he worked at Lawrence Livermore National Laboratory for 12 years and subsequently joined the Department of Biomedical Engineering at the Texas A&M University in 2008. His research projects include endovascular interventional devices, microactuators, optical therapeutic devices, and basic device–body interactions/physics, including computational and experimental techniques.

Contributors

Elena Alegre Aguarón
Department of Biomedical Engineering
Columbia University
New York, New York

Stelios T. Andreadis
Department of Chemical and Biological
    Engineering
State University of New York at Buffalo
Buffalo, New York

Brian E. Applegate
Department of Biomedical Engineering
Texas A&M University
College Station, Texas

Kyriacos A. Athanasiou
Department of Biomedical Engineering
University of California at Davis
Davis, California

Daniel S. Beasley
Department of Audiology and Speech Pathology
The University of Memphis
Memphis, Tennessee

Christie Bergerson
Department of Biomedical Engineering
Texas A&M University
College Station, Texas

A.G. Brown
Department of Pathology
Queen Mary University of London
London, United Kingdom

Mary Beth Browning
Department of Biomedical Engineering
Texas A&M University
College Station, Texas

Shelby Buffington
Department of Biomedical Engineering
Texas A&M University
College Station, Texas

L.-C. Chen
Department of Biomedical Engineering
University of Michigan
Ann Arbor, Michigan

Rajiv Chopra
Department of Radiology
University of Texas Southwestern Medical Center
Dallas, Texas

Kevin Cleary
The Sheikh Zayed Institute for
    Pediatric Surgical Innovation
Georgetown University Medical Center
Washington, DC

F.J. Clubb, Jr.
Texas A&M Institute for Preclinical Studies
Texas A&M University
College Station, Texas

Elizabeth Cosgriff-Hernández
Department of Biomedical Engineering
Texas A&M University
College Station, Texas

Eric M. Darling
Division of Biology and Medicine
Brown University
Providence, Rhode Island

Brent DeGeorge
Department of Plastic Surgery
University of Virginia Health System
Charlottesville, Virginia

Michael D. Eggen
Medtronic, Inc.
Minneapolis, Minnesota

Carl P. Frick
Department of Mechanical Engineering
University of Wyoming
Laramie, Wyoming

Ken Gall
School of Materials Science and Engineering
Georgia Institute of Technology
Atlanta, Georgia

Thomas J. Gampper
Department of Plastic Surgery
University of Virginia Health System
Charlottesville, Virginia

Herbert Jay Gould
School of Audiology and Speech-Language
    Pathology
The University of Memphis
Memphis, Tennessee

S.E. Greenwald
Barts & the London School of Medicine and
    Dentistry
Queen Mary University of London
London, United Kingdom

Craig J. Hartley
Division of Cardiovascular Sciences
Baylor College of Medicine
Houston, Texas

Jonathan Hartman
Department of Neurosurgery
Kaiser Permanente Sacramento Medical Center
Sacramento, California

Danika Hayman
Department of Bioengineering
Imperial College London
London, United Kingdom

Jagtar Singh Heir
Department of Anesthesiology and Perioperative
    Medicine
University of Texas M.D. Anderson Cancer Center
Houston, Texas

Hermann Hinrichs
Liebniz Institute for Neurobiology
University of Magdeburg
Magdeburg, Germany

Clark T. Hung
Department of Biomedical Engineering
Columbia University
New York, New York

Kullervo Hynynen
Department of Medical Biophysics
Sunnybrook Research Institute
Toronto, Ontario, Canada

Paul A. Iaizzo
Department of Surgery
University of Minnesota
Minneapolis, Minnesota

Manish Kaushik
Department of Nephrology
Ospedale San Bortolo
and
International Renal Research Institute
Vicenza, Italy

Eiji Kawasaki
Department of Metabolism/Diabetes and
    Clinical Nutrition
Nagasaki University Hospital
Nagasaki, Japan

Spencer S. Kee
Department of Anesthesiology and Perioperative
    Medicine
University of Texas M.D. Anderson Cancer
    Center
Houston, Texas

Terri-Ann N. Kelly
Department of Biomedical Engineering
Columbia University
New York, New York

Jeong Chul Kim
Department of Nephrology
Ospedale San Bortolo
and
International Renal Research Institute
Vicenza, Italy

Tal A. Kohen
Heller Institute of Medical Research
Sheba Medical Center
Tel Hashomer, Israel

Elisa E. Konofagou
Department of Biomedical Engineering
Columbia University
Boston, Massachusetts

Gerald J. Kost
Department of Pathology and Laboratory
    Medicine
University of California at Davis
Davis, California

Jean-Pierre Lalonde
Medtronic, Inc.
Minneapolis, Minnesota

Pedro Lei
Department of Chemical and Biological
    Engineering
State University of New York at Buffalo
Buffalo, New York

Mark Lenox
Texas A&M Institute for Preclinical Studies
Texas A&M University
College Station, Texas

Robert S. Litman
Department of Pharmacology
Nova Southeastern University
Fort Lauderdale, Florida
and
The Ohio State University
Columbus, Ohio

W.R. Lloyd III
Department of Biomedical Engineering
University of Michigan
Ann Arbor, Michigan

Christopher K. Macgowan
Department of Medical Imaging
University of Toronto
Toronto, Ontario, Canada

Duncan J. Maitland
Department of Biomedical Engineering
Texas A&M University
College Station, Texas

Kristen C. Maitland
Department of Biomedical Engineering
Texas A&M University
College Station, Texas

Robert L. Mauck
Department of Orthopaedic Surgery
University of Pennsylvania
Philadelphia, Pennsylvania

Mary P. McDougall
Department of Biomedical Engineering
Texas A&M University
College Station, Texas

James E. Moore, Jr.
Department of Bioengineering
Imperial College London
London, United Kingdom

Daniel S. Moran
Heller Institute of Medical Research
Sheba Medical Center
Tel Hashomer, Israel
and
Ariel University
Ariel, Israel

Michael R. Moreno
Department of Biomedical Engineering
Texas A&M University
College Station, Texas

Lawrence J. Mulligan
Medtronic, Inc.
Minneapolis, Minnesota

Andrea D. Muschenborn
Department of Biomedical Engineering
Texas A&M University
College Station, Texas

M.-A. Mycek
Department of Biomedical Engineering
University of Michigan
Ann Arbor, Michigan

Charles C. Nguyen
School of Engineering
Catholic University of America
Washington, DC

Jesung Park
Department of Biomedical Engineering
Texas A&M Univeristy
College Station, Texas

Dejan B. Popović
Department of Biomedical Engineering
University of Belgrade
Belgrade, Serbia

Ronaldo V. Purugganan
Department of Anesthesiology
    and Perioperative Medicine
University of Texas M.D. Anderson Cancer Center
Houston, Texas

Elizabeth Rebello
Department of Anesthesiology
    and Perioperative Medicine
University of Texas M.D. Anderson Cancer Center
Houston, Texas

A. Roberts
College of Veterinary Medicine
Texas A&M University
College Station, Texas

Timothy P.L. Roberts
Department of Medical Imaging
University of Toronto
Toronto, Ontario, Canada

Jennifer N. Rodriguez
Department of Biomedical Engineering
Texas A&M University
College Station, Texas

Claudio Ronco
Department of Nephrology
Ospedale San Bortolo
and
International Renal Research Institute
Vicenza, Italy

David L. Safranski
MedShape, Inc.
Atlanta, Georgia

Ryan L. Shelton
Biomedical Engineering Department
Texas A&M University
College Station, Texas

Sebina Shrestha
Department of Biomedical Engineering
Texas A&M University
College Station, Texas

Kathryn E. Smith
MedShape, Inc.
Atlanta, Georgia

Robert Staruch
Imaging Research
Sunnybrook Research Institute
Toronto, Ontario, Canada

Nam Tran
Department of Pathology and
    Laboratory Medicine
University of California at Davis
Davis, California

Thomas D. Wang
Department of Internal Medicine
University of Michigan
Ann Arbor, Michigan

Michael D. Weil
Sirius Medicine, LLC
Fort Collins, Colorado

Yossi Weiss
Department of Health System Management
Ariel University
Ariel, Israel

R.H. Wilson
Department of Biomedical Engineering
University of Michigan
Ann Arbor, Michigan

Christopher M. Yakacki
Department of Mechanical Engineering
University of Colorado Denver
Denver, Colorado

I

Basic Clinical Measurements

Introduction: Technology as a Tool for Aiding Initial Diagnosis

From the moment we enter a doctor's office or hospital, we are confronted with equipment designed to improve the diagnosis and treatment of disease. Most of the initial clinical evaluation is aided by seemingly simple devices. However, the fact that these devices have become commonplace does not detract from the elegance of their technological development.

The widespread use of basic clinical technologies has greatly benefited patient care; body temperature, heart activity, blood pressure, and brain activity can be measured much faster and more accurately

1

than ever before. Moreover, certain patterns in the data gathered by these relatively simple devices can often be used to provide an accurate diagnosis, as in the case of cardiac arrhythmia.

In this section, the underlying technology of devices used at the front lines of clinical medicine is presented, along with basic information on their use and the potential risks they may pose to patients. Continued advances in this area will enhance the ability of primary care physicians to identify and treat disease accurately and efficiently.

1

Measuring Body Temperature

Daniel S. Moran
Ariel University

*Heller Institute, Sheba
Medical Center*

Tal A. Kohen
*Heller Institute, Sheba
Medical Center*

Yossi Weiss
Ariel University

1.1 Introduction

Knowledge of body temperature is required for a wide range of physiological and clinical studies. However, body temperature is not uniform throughout, neither in distribution within the body nor over time, and it is imperative to choose an appropriate measurement site and method. To make these decisions, basic knowledge of how the human body responds to environmental factors is required and is described below.

Several mathematical models have been developed to describe the distribution of body temperature in humans in various environments and under widely ranging physiological conditions [1]. A theoretical concept of a warm "core" surrounded by a cooler "shell" is widely used to describe the body's central region of temperature that is maintained by a variety of physiological mechanisms [2]. This core is roughly a cylindrical region along the axis of the trunk and includes the inner regions of the neck and head. While there are small differences in temperature from place to place within the core that depend on local blood supply, metabolic rate, and the temperature of neighboring tissues, a single temperature is used as a reasonable approximation of core temperature (T_c) [3].

Under neutral environmental conditions (i.e., ambient temperature of 28–30°C and 40–60% relative humidity), T_c is approximately 37°C in a resting, healthy nude subject [4]. The skin temperature of the forehead, trunk, fingers, and toes are progressively cooler, and in a cold environment, may be much lower. When sufficiently cold, the skin may freeze even if T_c falls by only a degree or two. By providing insulation, clothing can substantially raise the peripheral skin temperature and protect against hypothermia.

To properly interpret a temperature measurement at any given site, some knowledge is required concerning the physical and physiological factors which create that temperature, and how changes in those factors may affect it. Skin temperature of a resting, nude human is largely determined by the environmental temperature, relative humidity, wind speed, intensity of solar radiation, atmospheric pressure, metabolic rate, and skin blood flow [5]. This blood flow varies over a wide range depending on the body's requirements at the time. During exercise at room temperature, as body temperature rises, warm blood flows from the core to the extremities substantially increases, raising the temperature of the skin and dissipating this heat by radiation, conduction, convection, and the evaporation of sweat [6]. The relative contribution of each heat dissipation mechanism varies with physiological status and environmental conditions. At cold ambient temperatures at rest, blood flow to the skin is reduced, decreasing skin temperature and minimizing heat loss from the core [5].

Even at rest and under constant environmental conditions, body temperature rises and falls by about 0.5–0.6°C from a mean of approximately 37°C, in a circadian cycle that is lowest in the morning and highest in the late afternoon (this varies somewhat from person to person) [7]. Where, then, is the best place to measure the temperature—skin or core? It depends on the nature of what is intended: is it a clinical study of fever, or a physiological study of heat loss to the environment? For most purposes, the best measure of abnormal health and altered physiological status is the measurement of T_c. These measurements can be made sequentially during the course of an experiment lasting minutes to hours, or periodically at the same time each day to account for the circadian variation.

Body temperature is a result of the steady state between the rate of metabolic heat production plus absorption from the environment and the rate of heat dissipation. When the absorption of heat from high environmental temperature and/or high metabolic rate from exercise is greater than the body's ability to dissipate it, temperature rises and hyperthermia results. Fever, on the other hand, results when an infection or absorbed substance causes a release of cytokines from various body tissues that circulate to the brain, act on the thermoregulatory center, and raise the "set point" temperature from 37°C. Under such conditions, a 37°C body temperature is "considered" by the hypothalamus to be a hypothermic state. The body defends this new set point temperature and elevates T_c by means of an increased metabolic rate through shivering and by reduced dissipation of heat through vasoconstriction, two classic homeostatic mechanisms of hypothermia.

Hypothermia results when heat is dissipated from the body faster than it is produced. It usually occurs due to a cold environmental temperature or an increased conduction of heat by water because of water immersion. In addition, hypothermia may result from the wearing of wet clothes in cold weather, or from a reduction of metabolic heat production due to severe illness. Regardless of the mechanisms, hyperthermia, fever, and hypothermia can be fatal.

1.1.1 History of the Thermometer

Illness has long been known to lead to fever [8]. It is written in Deuteronomy, "The Lord shall smite thee with a consumption, with a fever, and with an inflammation …" [9] seventeenth century Florentine glassblowers were the first to devise quantitative instruments for measuring temperature. In some of these instruments, carefully weighted hollow glass balls were suspended in a liquid medium whose density depended on temperature. The balls in turn would float or sink depending on the local density of the medium. The temperature of the liquid depended on the ambient temperature surrounding the instrument, its distance from the outer surface of the liquid, and the convective movement of the liquid. This system was not very accurate because the temperature indicated by the glass balls depended not only on ambient temperature, density, and location, but also on the shape of the container, as this influenced convection currents.

The first pioneer in the history of the thermometer was Galileo, who prepared a "thermoscope" from a thin glass tube with one end inserted into a sealed bulb containing liquid [8]. As temperature rose and fell, the liquid volume expanded and contracted causing the level of the liquid in the tube to rise and fall. However, because one end of the tube was open to the air, the height of the liquid was influenced by atmospheric pressure in addition to temperature. This was clear to the astronomer Olaus Roemer in 1702, who recognized the importance of sealing the open end of the tube as well as the bulb. Roemer also developed a temperature scale based on two fixed points, the boiling point of water defined as 60° and the temperature of melting ice set at 7.5° [10]. In 1720, Daniel Gabriel Fahrenheit added what he considered a more "rational" scale to his thermometer assigning a value of 0° to the freezing point of a saturated solution of salt and water and setting the upper point by "placing the bulb of the thermometer in the mouth or armpit of a healthy male" [10]. Eventually, this scale was modified so that it could be conveniently calibrated by defining the freezing and boiling points of distilled water as 32°F and 212°F, respectively. Today, the most widely used scale is the centigrade scale, suggested by Celsius in 1742. Celsius originally set the boiling point of water as 0° and the freezing of ice at 100°, but these points were reset into the present format by the Danish botanist Linnaeus in 1750 [8].

In 1868, Carl Wunderlich published *Fever in Disease*, which summarized his observations of thousands of temperature measurements in patients with various diseases [11]. He found that fevers rarely exceeded 105°F and that a practical thermometer for clinical use required the measurement of only the limited temperature range of 95–105°F. For convenience, he added a small kink in the tube to prevent the mercury from returning to the bulb when the thermometer was removed from the fevered patient. His views that temperature should be measured in all patients met considerable resistance from the conservative medical community, and years passed before temperature determinations became routine in medical practice.

In a different approach to temperature measurements, nineteenth-century chemists found that esters of cholesterol formed loose structures—today called liquid crystals—that changed color as their temperature changed. Lehmann first introduced these in 1877 as a means of measuring skin temperature (T_{sk}), but others did not employ them for a century. Initially, the crystals were mixed with black pigments and adhered to a flat surface that was placed on the skin. Because surface veins provided local warm regions, they could be readily identified on the liquid crystal surface by their color, which was different from that of the rest of the (cooler) skin. Nevertheless, this method was not suitable for widespread physiological or medical use, as it had a narrow temperature range and long response time, was uncomfortable for patients, was limited to a specialized laboratory environment, and could only be used once before being thrown away.

The advent of microencapsulation technology allowed for the development of simplified devices containing microencapsulated liquid crystals embedded onto flexible surfaces. These had improved contact areas as the use of inflated airbags with transparent windows allowed them to be pressed against uneven and curved skin surfaces. Advances in chemistry led to the development of reusable liquid crystal devices, which could be produced in physiologically relevant temperature ranges, and each responding to small temperature changes. Such systems have been used to diagnose pathological conditions in the breast, lower back, and veins [12,13]. However, the liquid crystal technique was only useful when a pathology led to alterations in skin temperature, not to changes in core temperature alone.

The most widely used instruments for measuring core temperature are the mercury and alcohol thermometers. In recent years, however, the use of mercury thermometers has been declining. They have been banned from some government laboratories due to increasing concern over mercury toxicity to the environment and penalties by the U.S. Environmental Protection Agency that may result. Prohibitions may be extended to alcohol-based thermometers if investigators can show that they too carry a high risk.

Infrared (IR) thermometers inserted into the ears are increasingly being used in hospital and home-care environments because of their convenience, economic benefit, and environmental safety.

A relatively new method for measuring core temperature is the "temperature pill." This is a small, sealed sensor containing a miniaturized transmitter that emits a signal whose frequency depends upon temperature. A nearby receiver translates the frequency into temperature and records the results over time.

We will review various devices and methods to measure T_c as a diagnostic tool for illnesses and infectious diseases, with a focus on their practical incorporation to prevent heat or cold injuries during commercial activities, sporting events, military training, and combat.

An ideal thermometer must satisfy several conditions [14]:

- It must be accurate to at least 0.1° in order to monitor temperatures associated with fever, as well as the normal circadian variation in temperature.
- It must provide a true local temperature and not be influenced by changes in temperature of distant areas of the body such as in the limbs and skin, or by the environmental temperature.
- It must be stable long enough so that, after calibration, its accuracy is maintained.
- It must be small enough and of a shape appropriate for its use in the body (mouth, rectum, esophagus).
- The area of use should not influence the temperature being measured. For instance, heavy breathing usually cools a mouth thermometer.
- It must be simple to operate.

Errors preventing true measurements of the core body temperature include noncompliance with the measuring protocol, for example, inserting a rectal thermometer to an incorrect depth, an insufficient measuring time, and possible chemical incompatibility between the sensor and the surrounding medium [14].

1.2 Noninvasive Sites for T_c Measurements

1.2.1 Oral

The method of inserting a thermometer into the sublingual pocket (i.e., under the tongue) is the most widely used method for measuring body temperature in the home environment, in the clinic, and often in clinical research due to easy accessibility of the mouth. The temperature in this region is usually, but not always, close to T_c because of large arterial blood flow from a branch of the carotid artery. Furthermore, this thermometer responds rapidly with changes in T_c. Nevertheless, this measurement may be substantially affected by the following factors:

- The skin on the head or face being cooler than the rest of the body
- Exposure to cold air
- Rapid or irregular mouth-breathing patterns, especially in athletes or crying children that may increase local evaporation thus lowering the temperature
- Ingestion of hot or cold beverages
- Recent smoking
- Neurological compromise making compliance difficult, for example, by heat illness
- Temperature variations in different parts of the mouth cavity

1.2.2 Axilla

Measuring temperature at the axilla (armpit) is slower than oral temperature measurement, as it takes longer to reach equilibrium, and this method is less accurate than measurements in the rectum, mouth, or the ear [15]. In addition, the axillary temperature is usually appreciably lower than the T_c, especially in athletes. Owing to its inherent inaccuracy, it is not recommended for clinical use.

1.2.3 Tympanic Membrane

A branch of the internal carotid artery supplies blood to the thermoregulatory center in the hypothalamus region of the brain, while another branch supplies the tympanic membrane. Because a high proportion of blood flow from the heart goes to the brain, the temperature of the brain is close to the temperature of the heart and the core. A thermometer was developed to measure the temperature at this site, in part because of easy access to the ear canal. However, many studies show that this method is unreliable, especially during physical exertion in the heat, due to errors in measurement introduced by dirt and inaccurate placement of the thermometer by an unskilled person [14–18]. Furthermore, methods involving probes in physical contact with the tympanic membrane carry a small risk of perforating the membrane [7].

1.2.4 Body Surface

Skin temperature is normally measured by thermistors placed on the skin, covered by a layer of insulation and secured in place by tape. Because this temperature varies from region to region, an integrated T_{sk} is normally taken by averaging recorded temperatures in a dozen or more locations, unless a temperature is required at a specific site. This body surface temperature does not reflect T_c, but only the temperature of the skin.

A method that allows for the measurement of T_c using skin probes is "heat flux." This technique was proposed in the 1970s, but has been improved over the past decade. The zero heat flux approach involves completely insulating a small area of skin, so that when thermal equilibrium is reached, the skin has the same temperature as deeper tissue. This can be achieved passively with perfect thermal insulation or by active heating to compensate for heat loss. The forehead is one of the most suitable sites for this technique. As active heating requires a relatively large amount of energy, there is a limitation for battery-operated thermometers. Zero heat flux sensors have been commercialized in Japan, but are not available in Europe or the United States [7].

A second technique of "heat flux" analyzes the temperature gradient of a well-defined thermal bridge (double sensor) placed on a skin area without the use of active heating. T_c is determined from the measured temperatures based on a mathematical model. This approach has exhibited accuracy in clinical trials (Gunga et al. showed that the double sensor differed by −0.16°C to 0.1°C from the mean T_{rec}), but disadvantages include an initial response time of several minutes and relatively large sizes [7].

1.3 Invasive Sites for T_c Measurements

1.3.1 Rectal

The rectum contains a high density of blood vessels that carry large volumes of blood from the core and is therefore an accurate and convenient location for measuring T_c. Temperatures obtained from the rectum tend to be slightly higher than other regions, especially during leg exercise, as there is a warming effect of blood returning to the core from the metabolically active leg muscles. Conversely, in severe pathological conditions such as the hypodynamic state of shock, in which metabolic rate may fall, the rectal temperature (T_{re}) may be lower than T_c.

In particular, the rectum is the most widely used site for physiological and clinical studies of heat illnesses and for diagnostic tests, especially in infants and children. For accuracy, the thermometer must be inserted at a consistent distance into the rectum. Furthermore, the response time for determining rectal temperature is somewhat slower than that of other locations, such as the esophagus. This slow response time is a disadvantage when rapid changes in core temperature are being studied. This method also has the disadvantage of requiring a hardwire connection to a recording device, thus limiting physical activity. However, in a side-by-side comparison between rectal, aural, and axillary temperatures, it was concluded that rectal temperature monitoring offers the greatest combination of accuracy and precision [15].

1.3.2 Esophagus

A small thermistor or thermocouple attached to a thin insulated wire can be swallowed so that it remains in place within the esophagus, above the lower esophageal sphincter (entry to the stomach). This deep body location is close to the heart and aorta, and measurements at this location respond rapidly to changes in T_c.

The main reason this site is not more widely used is the difficulty of insertion, given that the esophageal thermistor is threaded through the nasal passages and down into the esophagus. Not all subjects can tolerate this procedure as it may irritate the nasal passages and be quite uncomfortable [19]. Among other issues with this method are uncertainty that the thermistor is placed correctly in the esophagus and that the subject has not recently consumed any hot or cold beverages. T_{es}, although less accurate than pulmonary artery measurements, is appropriate for use in physiological research, but has relatively little use in clinical practice.

1.3.3 Nasopharynx

Nasopharyngeal (upper throat behind the nose) temperature is often measured with an esophageal probe that is placed above the palate. The temperature is close to that of the brain/core if the probe makes good contact with the mucosa, but this method has the risk of aspiration or entrapment of the probe [7].

1.3.4 Pulmonary Artery

The pulmonary artery is a large vessel that carries blood with low oxygen content from the right ventricle of the heart to the lungs, where the blood is reoxygenated. An invasive procedure exhibiting significant medical risk is required to reach this location and therefore its use is severely limited to situations where such a procedure is justified [20]. A long, flexible catheter (narrow tube) containing a tiny thermistor is inserted into a large vein in the neck, leg, or arm, threaded toward the heart, through the vena cava, into the right atrium and ventricle and then floated up into the pulmonary artery. The large blood flow through the pulmonary artery from the core and surroundings renders this location very accurate for measuring T_c and it is therefore considered the "gold standard."

1.3.5 Urinary Bladder

This method is based on the belief that the temperature of urine is closely related to the temperature of the body. Bladder temperature is measured via a urinary catheter and it correlates well with rectal, esophageal, and pulmonary artery temperature [7]. It has been determined, however, that this temperature depends on the rate of urination, and limited accuracy is observed with low urine flow. As a result, this method is rarely used.

1.4 Instruments for T_c Measurement

Temperature is a measure of the average kinetic energy of the particles in a system. When heat is added to a system, the kinetic energy of all the atoms is increased, their electrons become more energetic, and their temperature rises. The operation of thermistors and thermocouples described below is based on this principle.

1.4.1 Digital Electronic Thermometers

Digital electronic thermometers contain one of two types of temperature-sensing elements—thermistors or thermocouples. Long, narrow flexible cables connect these devices to an electronic "black box" where temperature is displayed digitally, generally to 0.1°C. These thermometers can be used for measuring rectal, oral, or axillary temperatures, and are convenient for use in research, in the clinic, and in the home environment. Their use increased when mercury thermometers were banned in a number of countries for environmental protection and safety reasons. Both the thermistor- and thermocouple-based digital thermometers are stable, accurate, and respond rapidly to temperature changes. They are smaller than other types of temperature probes because their temperature-sensing elements are minute (<50 μm^2).

1.4.1.1 Thermistors

Thermistors contain an alloy of heavy metals (of the semiconductor type), whose atoms form microscopic crystal lattices. When this alloy is placed in an electric circuit at a constant ambient temperature, a stable voltage is maintained [21]. Electrons in semiconducting metals are easily released by only a small increase in thermal energy, that is, by an increase in temperature. When the electric circuit is established at a given voltage, freed electrons cross the semiconductor lattice [22]. The number of thermal electrons released by the alloy increases exponentially with temperature, and by Ohm's law, the electrical resistance of the lattice falls exponentially. A rise in temperature, therefore, leads to a reduced voltage, and by precalibration, this voltage measurement is converted into and displayed as temperature [14].

1.4.1.2 Thermocouples

A thermocouple contains a junction of two wires composed of different metals welded together. When placed in an electric circuit, a current will flow when the two metals are at different temperatures [16,17]. The physical basis for current flow across the junction is as follows.

In any given metal, some of the electrons present can move freely, almost as if they were in a gas. If one end of a wire is heated, more electrons will be "freed" in the warmer region. With their higher energy, these electrons carrying negative charges will undergo a net diffusion toward the cooler region. When two metals of intrinsically different physical and chemical properties are welded together, differences between the sides are displayed in the following: number of electrons "freed" at a given temperature, rate of migration of electrons down a thermal gradient, and rate of heat flow. As a result of these differences, heat and electrons will flow out of one side of the junction and be absorbed by the other. In an open circuit, the flow of electrons across the junction continues until the net negative charge created by the movement of negatively charged electrons onto one side of the junction produces a negative electromotive force (emf) that counterbalances the flow. In addition to this thermally produced emf at the junction, there is a small chemical effect due to the difference in chemical potential between the two metals.

In an open circuit, the two metals produce an emf at the junction proportional to the difference in temperature between the two sides of the junction. For most pairs of metals, this emf is in the order of microvolts per °C difference. In practice, two separate thermocouples are run simultaneously in a differential mode, with one maintained at a constant reference temperature and the other used as the temperature probe. When the latter is placed on or into warm tissue, heat transfers across the wall of the probe and its temperature approaches the local body temperature. The voltage generated by the difference in temperature between the two probes is converted into temperature by precalibration.

1.4.1.3 Response Time of Digital Electronic Thermometers

The time required for an accurate measurement can be seconds or minutes, depending on which of the two possible modes of operation is employed. In the steady-state mode, requiring several minutes, the final temperature is displayed only after the sensor has reached thermal equilibrium. In the predictive mode, the initial rate of temperature change is measured for a few seconds, following which a mathematically calculated final temperature is predicted. The prediction of the final temperature depends on precisely when the probe is first in contact with the body. While this method is faster, it is obviously not as reliable since it is an extrapolation. Moreover, this method is particularly technician-dependent as the probe must be placed accurately, and the thermometer must be activated with precision.

1.4.2 IR Radiometry

A new and popular method for measuring body temperature at home and in many clinics is the aural thermometer that measures temperature from the tympanic membrane, tympanic canal, or temporal artery.

Emission radiometry utilizes the temperature-dependent blackbody emission phenomenon [23]. Within the ear, IR radiation is emitted from all surfaces, at frequencies that are dependent on local temperature. An IR probe can be placed inside the ear canal to measure this temperature—a rapid, cost-effective, noninvasive, and particularly convenient method, compared to others. However, despite its use in certain athletic activities, the aural thermometer is generally not considered accurate enough for most research [24,25]. In one study, aural temperatures were approximately 3°F lower than rectal temperatures measured simultaneously [26].

In use, a disposable otoscope probe consisting of a flaring cylinder with a highly polished interior surface is placed into a patient's ear. This probe only collects IR radiation from the direction in which it is aimed, and the field of view "seen" by the transducer depends on both its distance from the target surface and the viewing angle. Conventional waveguides in commercial use have conical fields of view of 45° around their axes, where the radius of the field is equal to its distance from the target.

IR thermometers operate very rapidly, typically requiring less than 5 s, because they simply record radiation; they do not have to absorb heat from a tissue, which requires time for adequate heat transfer.

1.4.2.1 Tympanic Temperature

The tympanic membrane is both well supplied with arterial blood and convenient to access for measuring T_c. A disadvantage of this site, however, is the need to maneuver the probe to obtain input from the tympanic membrane, exclusively, and not from any other local tissue, so that the measurement is accurate and reproducible.

1.4.2.2 Ear Canal

The temperature within the ear canal is not uniform throughout and is influenced by ambient temperature [26]. It is typically warmest close to the tympanic membrane and coolest near the ear lobe. To utilize ear canal measurements, algorithms have been designed to provide temperatures close to T_c. Using this site has the advantage that no special maneuvers of the IR probe are required, however; a correction factor must be applied to the data to better account for the cooler temperature of the ear canal due to heat loss to the environment.

1.4.2.3 Shortcomings of IR Measurements

Despite the apparent ease of this method, numerous studies have reported difficulties in obtaining true tympanic temperatures [16–18].

These thermometers typically have a viewing angle that is too wide, resulting in a field of view that is too large to exclusively measure the temperature of the tympanic membrane. In fact, the probes from these devices may not even "see" the membrane at all because of poor user technique, or due to atypical anatomy of a particular ear canal. Research has shown that even when the thermometer is placed correctly, the temperature recorded from the tympanic membrane includes a component from an averaged temperature of the ear canal wall. As a result, temperature readings are generally lower than the actual temperature of the tympanic membrane and lower than T_c [16–18]. To compensate for these lower readings, some instruments contain an offset system for compensation. The system is calibrated from previously obtained data and shows computerized and updated temperatures from other locations (e.g., oral, rectal, axilla). One method for overcoming these difficulties is scanning, in which a series of measurements of the tympanic membrane are taken, with the maximal temperature measured considered to be that of the tympanic membrane.

There are several scientific reports on the accuracy of the aural thermometer [14–18]. For example, when the measurement technique of medical staff is inconsistent, inaccurate measurements are recorded.

The lack of standardization stemming from the use of different types of thermometers results in a wide range of readings. Cerumen (earwax) buildup on the thermometer speculum causes an inaccurate measurement when a disposable probe is not used, or the top is broken from improper use. Additional factors that can affect the accuracy include an abnormally high-temperature gradient between the interior and exterior parts of the ear canal due to wind, inadequate depth of insertion of the thermometer, and external conditions such as environmental temperature.

In summary, because of the complexity of the technique and many sources of error, aural temperature is not reliable for establishing fever or hypothermia. In 1996, Hooker and Houston [18] summarized a study comparing emergency room temperature measurements in the aural and oral cavities: "Tympanic thermometers are convenient and well accepted and do not require contact with mucous membranes. However, most authors have shown that the tympanic thermometer is very insensitive to fever and current research indicates that its sensitivity is not as good as that of oral electronic thermometers in the detection of fever." It is therefore not recommended to use these thermometers in screening for fever in the emergency room.

1.4.3 Microwave Radiometry

Although IR radiometry detects radiation in the IR range (~9–14 µm), microwave radiometry (MR) uses wavelengths in the submillimeter to centimeter range. This spectral region is quite effective in penetrating tissue, and microwaves emitted from a depth of a few centimeters can reach the exterior of the body at detectable levels [23]. With this technique, an antenna is used as a receiver as opposed to the photodetectors or pyroelectric sensors used with IR radiometry. Han et al. used MR to measure neonatal brain temperature during hypothermic treatment and observed a standard error of 0.75°C in the estimated central brain temperature [7].

The main disadvantage with radiometric measurements, in general, is that the source of interest (e.g., the carotid artery) is surrounded by tissue that emits its own radio frequency. Moran et al., however, used a "double container model" to show that the temperature of a localized source can be observed in spite of the surrounding tissue [23]. MR has the advantage of deep tissue penetration (compared to IR radiometry), temperature measurements with no toxicity, and a passive and safe device that produces real-time, accurate data and can be employed throughout a variety of situations [23]. That being said, MR has poor temporal and spatial resolution, sensitivity to interferences from other electrical sources, and potential dependence on dielectric properties rather than temperature [7].

1.4.4 Temperature Pill

Advances in microcircuitry technology have led to the development of a sophisticated "temperature pill" (T_{pill}). This vitamin pill-sized device (smaller than 0.5 in. diameter × 1 in.) is sealed with biocompatible epoxy and/or silicone and contains both a battery and an FM transmitter whose frequency depends upon the local temperature. A small receiver ($6 \times 2.5 \times 12$ cm³) located nearby picks up the signal, amplifies it, converts it to temperature after appropriate calibration, and displays/records it for later retrieval. Once the pill is swallowed, it traverses the alimentary tract and records local temperatures on a receiver located on the subject's belt. The FDA has certified this device, and other sensors/FM circuits can be added to simultaneously measure blood pressure, heart rate, pH, and local chemistry. The T_{pill} system has been found to be accurate, reliable, and comfortable compared to the T_{re} and T_{es} methods, and temperature and response time tended to fall between the two [27]. The measurements were found to be valid not only during the steady state, but also during rising and falling body temperatures [25 27]. Because of the few limitations to its use, T_{pill} is increasingly used in physiological research [28] and even in unusual clinical situations, such as in monitoring T_c of saturation divers to prevent hypothermia [29].

Not only can the pill be swallowed for short application (~1–2 days), but it can also be manufactured small enough (9 mm diameter × 35 mm) to pass through a 10-mm endoscopic tube for long-term implantation, and has been used to monitor intrauterine temperature and pressure for up to 10 months [30].

This device (pill and receiver) provides considerable advantages over previous techniques in experimental physiology because: (1) it does not require limiting the physical activity of the subject since no wire is required between the sensor and recorder; (2) it is not limited to the laboratory or hospital environment, but can be used in field settings; and (3) it may be used experimentally as well as clinically.

Disadvantages of the temperature pill include: (1) discomfort in swallowing for some individuals; (2) as the pill progresses down the gastrointestinal tract, local temperatures vary from place to place, and are subject to different influences; (3) high cost of the receiver (~$2000) and pill (~$40 for a one-time use); (4) impractical or not cost-effective for most clinical uses; and (5) inappropriate for the home environment.

1.4.5 Methods under Development

MR thermometry utilizes the fact that various temperature-dependent physical parameters affect MR signals. Recent studies (2010) show results with a precision of ±0.3°C and a mean difference between brain temperature and T_{rec} of 1.3 ± 0.4°C [7]. This technique could lead to the measurement of deep tissue temperatures during MR imaging diagnostics.

Measuring the time or frequency spectrum of backreflected ultrasound pulses can lead to an estimation of tissue temperature. This is possible due to the fact that the speed of sound depends on the temperature of the material it penetrates. In *in vivo* experiments, Seip and Ebbini found a standard deviation of 0.5°C [7].

Light absorption by tissue chromophores is significantly related to temperature. Therefore, when near-IR light (tissue penetration of a few centimeters) is employed, a spectral change of the backscattered light will depend on a temperature change. Hollis was able to determine the temperature within 0.4°C (standard error) in a tissue-like phantom [7].

Using a compartment model of a subject (i.e., its metabolism and blood flow), it appears feasible to accurately predict T_c. This technique does not directly measure temperature, but estimates the body core temperature based on simulations. Gribok et al. used an autoregressive model to predict T_c in humans during physical activity [7].

1.5 Summary

Table 1.1 summarizes the T_c measurement sites discussed in this chapter. Until as recently as 1987, esophageal temperature was considered the optimal method for measuring body core temperature in the research setting; since then, it has been rectal temperature [20]. If that could not be measured, oral temperature was considered the best alternative. Nevertheless, measuring temperature with a mercury thermometer under the tongue still remains the most popular and accepted method in the world. In many instances, when it is not possible to measure temperature orally, and body core temperature is needed, the measurement takes place in the rectum. Measuring temperature in the brain or the pulmonary artery would be the "gold standard" for temperature measurements, but it is rarely possible to do so.

In recent years, a great deal of effort has been expended in advancing sophisticated new techniques in medicine such as MR technology and biotechnology, leading to improved diagnostic capability and new approaches to therapy. At the same time, old, familiar methods are being replaced. Despite the favorable climate for developing new technology, however, there is still no ideal system for measuring body core temperature that exhibits the following properties [7]:

- Small
- Easy to use
- Comfortable
- Fast (for spot checks)

TABLE 1.1 Characteristics of T_c Measurement Sites

Location	Accuracy	Response	Invasiveness	Comfort
Oral	Medium	Fast	Low	Medium
Axilla	Low	Slow	Low	High
Tympanic membrane (noncontact/IR)	Low	Fast	Low	High
Tympanic membrane (contact)	Good	Fast	Medium	Low
Forehead	Good	Slow	Low	High
Rectal	Medium	Slow	Medium	Low
Esophagus	Good	Fast	Medium	Low
Nasopharynx	Good	Fast	Medium	Low
Pulmonary artery	Gold standard	Fast	High	Low
Urinary bladder	Limited	Fast	High	Low
Gastrointestinal (temperature pill)	Good	Medium/slow	Low	High

Source: Adapted from Imhoff, M., Muhlsteff, J., and Wartzek, T. 2011. *Biomed Tech* 56:241–57.

- Continuous
- Accurate
- Precise
- Noninvasive
- Low-energy consumption
- Affordable

References

1. Havenith, G. 2001. Human surface to mass ratio and body core temperature in exercise heat stress— A concept revisited. *J Therm Biol* 26:387–93.
2. Elizondo, R. 1989. Regulation of body temperature. In *Human Physiology*, eds. R.A. Rhodes and R.G. Pflanzer, pp. 823–40. Philadelphia, PA: Saunders College Publishing.
3. Wenger, C.B. 2002. Human adaptation to hot environments. In *Medical Aspects of Harsh Environments*, eds. K.B. Pandolf and R.E. Burr, pp. 51–86. Washington, DC: Borden Institute, Walter Reed Army Medical Center.
4. Bligh, J. and Johnson, K.G. 1973. Glossary of terms for thermal physiology. *J Appl Physiol* 35:941–61.
5. Pozos, R.S. and Danzl, D.F. 2001. Human physiological responses to cold stress and hypothermia. In *Medical Aspects of Harsh Environments*, eds. K.B. Pandolf and R.E. Burr, pp. 351–82. Washington, DC: Borden Institute, Walter Reed Army Medical Center.
6. Sawka, M.N. and Wenger, C.B. 1988. Physiological responses to acute exercise heat stress. In *Human Performance Physiology and Environmental Medicine at Terrestrial Extremes*, eds. K.B. Pandolf, M.N. Sawka, and R.R. Gonzalez, pp. 97–151. Indianapolis, IN: Benchmark Press.
7. Imhoff, M., Muhlsteff, J., and Wartzek, T. 2011. Temperature measurement. *Biomed Tech* 56:241–57.
8. Ring, E.F.J. 1988. Progress in the measurement of human body temperature. *IEEE Eng Med Biol Mag* 17:19–24.
9. *The Bible*, Deuteronomy 28:22.
10. Gough, J.B. 1996. Fahrenheit's thermometer. In *A History of the Thermometer and Its Uses in Meteorology*, ed. W.E. Knowles Middleton. Baltimore, MD: Johns Hopkins Press.
11. Wunderlich, C. 1871. *On the Temperature in Disease: A Manual of Medical Thermometry*, Transl., London: New Sydenham Society.
12. Sterns, E.E., Zee, B., SenGupta, S., and Saunders, F.W. 1996. Thermography. Its relation to pathologic characteristics, vascularity, proliferation rate, and survival of patients with invasive ductal carcinoma of the breast. *Cancer* 77:1324–8.

13. Yang, W.J. and Yang, P.P. 1992. Literature survey on biomedical applications of thermography. *Biomed Mater Eng* 2:7–18.
14. Cetas, T.C. 1997. Thermometers. In *Fever: Basic Mechanisms and Management*, ed. P.A. Mackowiak, pp. 11–34. Philadelphia, PA: Lippincott-Raven Publication.
15. Cattenaeo, C.G., Frank, S.M., Hesel, T.W., El Rahmany, H.K., Kim, L.J., and Tran, K.M. 2000. The accuracy and precision of body temperature monitoring methods during regional and general anesthesia. *Anesth Analg* 90:938–45.
16. Amoateng-Adjepong, Y., Del Mundo, J., and Manthous, C.A. 1999. Accuracy of an infrared tympanic thermometer. *Chest* 115:1002–5.
17. Briner, W.W., Jr. 1996. Tympanic membrane vs rectal temperature measurement in marathon runners. *JAMA* 276:194.
18. Hooker, E.A. and Houston, H. 1996. Screening for fever in an adult emergency department: Oral vs tympanic thermometry. *South Med J* 89:230–4.
19. Stuart, M.C., Lee, S.M.C., and Williams, W.J. 2000. Core temperature measurement during supine exercise: Esophageal, rectal and intestinal temperature. *Aviat Space Environ Med* 71:939–45.
20. Brengelmann, G.L. 1987. Dilemma of body temperature measurement. In *Man in Stressful Environments: Thermal and Work Physiology*, eds. K. Shiraki and M.K. Yousef, pp. 5–22. Springfield, IL: Thomas.
21. Childs, P.R.N., Greenwood, J.R., and Long, C.A. 2000. Review of temperature measurement. *Rev Sci Instrum* 71:2959–78.
22. Bloomfield, L.A. 2001. Thermometers and thermostats. In *How Things Work. The Physics of Everyday Life*. Indianapolis, IN: John Wiley & Sons.
23. Eliyahu, U., Heled, Y., Hoffman, J., Margaliot, M., Moran, D.S., and Rabinovitz, S. 2004. Core temperature measurement by microwave radiometry. *J Therm Biol* 29:539–42.
24. Armstrong, L.E., Maresh, C.M., Crago, A.E., Adams, R., and Roberts, W.O. 1994. Interpretation of aural temperatures during exercise, hyperthermia, and cooling therapy. *Med Exerc Nutr Health* 3:9–16.
25. Armstrong, L.E., Crago, A.E., Adams, R., Senk, J.M., and Maresh, C.M. 1994. Use of the infrared temperature scanner during triage of hyperthermic runners, *Sports Med Training Rehab* 5:1–3.
26. Ash, C.J., Cooke, J.R., McMurry, T.A., and Auner, C.R. 1992. The use of rectal temperature to monitor heat stroke. *Missouri Med* May: 288.
27. O'Brien, C., Hoyt, R.W., Buller, M.J., Castellani, J.W., and Young, A.J. 1998. Telemetry pill measurement of core temperature in humans during active heating and cooling. *Med Sci Sports Exerc* 30:468–72.
28. Coyne, M.D., Kesick, C.M., Doherty, T.J., Kolka, M.A., and Stephenson, L.A. 2000. Circadian rhythm changes in core temperature over the menstrual cycle: Method for noninvasive monitoring. *Am J Physiol Regul Integr Comp Physiol* 279: R1316–20.
29. Mekjavic, I.B., Golden, F.S., Eglin, M., and Tipton, M.J. 2001. Thermal status of saturation divers during operational dives in the North Sea. *Undersea Hyperb Med* 28:149–55.
30. NASA. 2002. Implantable Biotelemetry System for Preterm Labor and Fetal Monitoring. http://technology.arc.nasa.gov/techopps/biotelemetry.html.

2

Ultrasonic Blood Flow and Velocity Measurement

Craig J. Hartley

Baylor College of Medicine

2.1 Introduction

During the past 50 years, ultrasound has developed into a widely used research and clinical modality with its most widespread and familiar applications in noninvasive two-dimensional and color Doppler imaging.[1–4] From its earliest days, ultrasound has found nonimaging medical applications using noninvasive as well as invasive, intraoperative, implantable, and intravascular transducers and sensors to measure dimensions, displacement, velocity, and flow. Here, we will concentrate on the ultrasonic measurement of blood flow and velocity.

2.2 Ultrasound Physics

Ultrasound is usually defined as a mechanical vibration with a frequency above the range of human hearing. The frequencies (f) usually employed in medical applications are in the range between 500 kHz and 100 MHz. Acoustic signals at these frequencies can be directed and coupled into body tissues where they propagate at the speed of sound. While traveling through the various tissues, the sound waves undergo absorption, refraction, reflection, and scattering, which depend on the acoustic properties of the tissues (density, speed of sound, absorption coefficient, and homogeneity) and the changes in these properties at the tissue interfaces.[5,6] Thus, sound, which is transmitted into body tissues, undergoes a very complex series of interactions in which it can be partially passed, redirected, reflected, and/or

15

weakened by each tissue and interface through which it passes. The reflections at the interfaces return-ing to the sending transducer produce the images with which we are familiar.

The speed of sound (*c*) in water, blood, and most body tissues is approximately 1500 ± 100 m/s or 1.5 mm/μs, so at frequencies from 500 kHz to 100 MHz, the wavelength ($\lambda = c/f$) is between 0.015 and 3.0 mm.[7] The higher frequencies have shorter wavelengths and give higher resolutions, but are also attenu-ated to a greater extent and do not penetrate as far into the tissue without unacceptable loss of signal.[6] Thus, low frequencies (1–5 MHz) are used where greater penetration is required (noninvasive imaging and Doppler), and higher frequencies (5–50 MHz) are used where high resolution is required (invasive and intravascular imaging and velocimetry). Frequencies used for blood flow and velocity measurements from extravascular cuff-type transit time and Doppler probes are between 450 kHz and 20 MHz.

2.3 Ultrasonic Transducers

An important part of any ultrasound instrument is the transducer, which converts electrical energy into mechanical vibration and vice versa and defines the direction, frequency, and geometry of the sound beam. The active element is usually a piezoelectric material that ranges from single-crystal quartz, which has a high sensitivity and narrow bandwidth, to polymers which have lower sensitivity but wider band-width.[8–10] The choice of material depends on the application; one of the more common materials used in medical ultrasound is piezoelectric ceramic such as lead–zirconate–titanate (LZT or PZT).[11] Ceramics have properties that are intermediate between crystals and polymers, provide a good compromise between sensitivity and bandwidth, and are available from several suppliers in sheets with metallic electrodes (sil-ver, gold, or nickel) plated to each face.[12] The ceramic is generally fabricated in "thickness mode" where the thickness (1/2 wavelength) determines the resonant frequency which can range from a few hundred kilohertz to more than 100 MHz.[13] When properly polarized during manufacture, the piezoelectric mate-rial thins or thickens when a voltage is applied and conversely develops an electrical potential between its electrodes when subjected to a mechanical force. It can thus act as both a transmitter and a receiver of ultrasound. The sheets are cut into discs, squares, or strips for fabrication into a complete transducer consisting of the piezoelectric element or elements; acoustic matching layers; acoustic backing; acoustic focusing or diverging lenses; a holder or body consisting of metal, plastic, silicone rubber, or epoxy; lead wires; and an electrical connector. The word "transducer" is often used to refer to the piezoelectric element or to the completed device which is also referred to as a "scan head," "array," "probe," "sensor," or "crystal" depending on its configuration, shape, and application.[9] Imaging transducers are relatively complex because the sound beam must be electrically steered or mechanically directed to scan an area of interest. However, ultrasound can also be used in nonimaging applications to measure dimensions, velocity, flow, and displacement of tissues and fluids utilizing the principles outlined below. Compared to imaging, these methodologies use fairly simple transducers and signal processing, and many can produce outputs compatible with standard physiological recorders and data acquisition systems.

2.4 Transit-Time Dimension

One of the first applications of ultrasound in medicine was to measure dimensions using the transit-time principle.[14–16] If a pulse of sound transmitted by one transducer is received by a second transducer or is reflected by a target back to the same transducer, the pulse arrival time (*t*) is related to the distance (*d*) between the transducers or to the reflector by the speed of sound (*c*) as shown in Figure 2.1a and b. The equations for the one-way ($t_{1\text{way}}$) and two-way ($t_{2\text{way}}$) transit times are shown in Figure 2.1:

$$t_{1\text{way}} = \frac{d}{c} \tag{2.1}$$

$$t_{2\text{way}} = \frac{2d}{c} \tag{2.2}$$

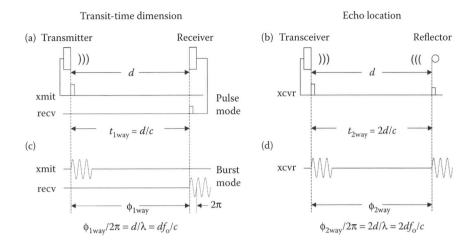

FIGURE 2.1 Drawing showing how ultrasound can be used to measure distance via transit time (a, c) or pulse echo (b, d) methods for pulse (a, b) or burst (c, d) excitation of the transmitter. Equations are shown relating one-way (1way) and two-way (2way) transit time (*t*) and phase (φ) to the distance (*d*) between the crystals or from the crystal to a reflector, where *c* is the speed of sound (~1500 m/s or 1.5 mm/μs), λ is the wavelength, and f_0 is the ultrasonic frequency.

If a flip/flop is set at transmission of the pulse and reset upon receipt of the pulse, the width provides a simple measure of the distance between the transducers updated at a typical pulse repetition frequency (PRF) of 1–10 kHz. Compared to imaging, the signal processing is very simple. This method requires that the transducers be inserted into or attached to the tissue of interest and is commonly used to measure ventricular diameters,[14,16] myocardial segment length[15,17] and wall thickness,[18,19] and arterial diameter.[20,21] With proper synchronization, several dimensions can be measured simultaneously.[14,15,22] Although the accuracy is limited by the wavelength (typically 0.3 mm at 5 MHz), the sensitivity to motion or change in dimension is on the order of 1 μm.

If a tone burst is transmitted instead of a single-cycle pulse, the phase (φ) of the received burst measured in radians with respect to the transmitted burst could also be used as a measure of distance as shown in Figure 2.1c and d. The equations for one-way and two-way phases are also shown below in terms of the wavelength (λ) and the transmitted burst frequency (f_0):

$$\frac{\phi_{1way}}{2\pi} = \frac{d}{\lambda} = \frac{df_o}{c} \tag{2.3}$$

$$\frac{\phi_{2way}}{2\pi} = \frac{2d}{\lambda} = \frac{2df_o}{c} \tag{2.4}$$

In general, pulse mode is used to measure the distance between two transducers,[14,18,22] and burst mode is used to measure the change in position or displacement of tissues with a single echo transducer.[23–27]

If the fluid and/or the target are moving, the velocity (*V*) affects the arrival time, the phase, and the frequency of the received signals as shown in Figure 2.2. In the pulse mode, the moving fluid speeds up the arrival of a pulse moving with the flow (t_{2-1}) and retards the arrival of a pulse moving against the flow (t_{1-2}). If we alternately transmit from each crystal, receive on the other, and subtract the arrival times, the difference in arrival times (Δ*t*) divided by the average arrival time (t_{avg}) is directly proportional to the velocity as shown in the following equation provided that $V \ll c$:

$$\frac{\Delta t}{t_{avg}} = \frac{2V}{c} \tag{2.5}$$

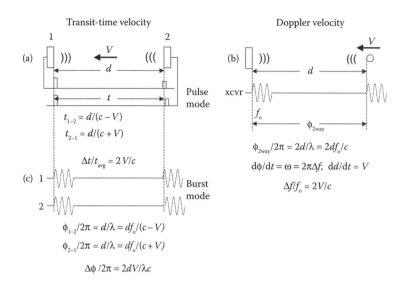

FIGURE 2.2   Drawing showing how ultrasound can be used to measure the velocity of a moving fluid or a reflecting target via transit time (a, c) or Doppler (b) methods. The fluid velocity (V) adds or subtracts from the speed of sound (c) to change the arrival time (t) or phase (ϕ) of pulses traveling with (2–1) or against (1–2) the flow. The difference in transit times (Δt) or phase ($\Delta\phi$) is proportional to the velocity. Since the phase of an echo (ϕ_{2way}) is proportional to the distance (d), the derivative of phase (Doppler frequency, Δf) is proportional to the derivative of the distance (reflector velocity, V). The equations hold only when $V \ll c$ and the velocity is in the direction of sound propagation.

In the burst mode, the relative phase of the received bursts ($\Delta\phi$) is also proportional to velocity:

$$\frac{\Delta\phi}{2\pi} = \frac{2dV}{\lambda c} \tag{2.6}$$

In the echo mode, the phase of the echo changes with each successive burst as the target moves with respect to the transducer. If we differentiate both sides of Equation 2.4 noting that the derivative of the phase is angular frequency ($\omega = 2\pi f$) and the derivative of the distance is velocity, we get an equation relating the Doppler shift frequency (Δf) to the velocity (V) of the reflector[28]:

$$\frac{\Delta f}{f_o} = \frac{2V}{c} \tag{2.7}$$

Thus, ultrasound can be used to measure either the distance or the velocity depending on the conditions, the transducer, and the signal processing applied.

2.5 Transit-Time Velocity and Flow

The differential transit-time principle was first applied to the measurement of biological flows in the 1950s[29,30] and is now in wide use in both industrial and medical applications. This method can operate with catheter-mounted transducers immersed in the fluid[31] or with extravascular or cuff-type probes[30,32] as shown in Figure 2.3. The simplest approach is to place the transducers diagonally on opposite sides of the vessel (Figure 2.3a). This requires modification to Equations 2.5 and 2.6 to account for the angle between the sound beam and the direction of flow, but with the constant angle, the difference between upstream and downstream transit times (or phase shift) is still proportional to the average velocity along the sound path as shown by the equations in Figure 2.3 and below.

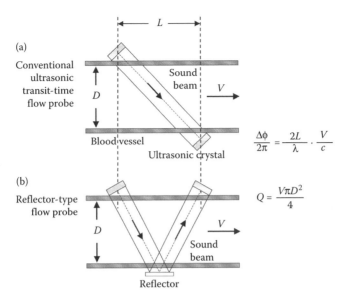

FIGURE 2.3 Ultrasonic transit-time methods for measuring blood flow through an exposed vessel using a conventional probe (a) or a reflector probe (b). The governing equation relating the difference in phase between upstream and downstream transits ($\Delta\phi$) to the average velocity (V) along the sound path is shown. In theory, if the crystals are wider than the vessel, then flow anywhere in the lumen contributes equally to the average velocity. Flow (Q) is then determined by multiplying the velocity by the cross-sectional area. Because of the angle, the sensitivity is proportional to the length (L) along the vessel between the crystals rather than to the crystal spacing.

$$\frac{\Delta\phi}{2\pi} = \frac{2LV}{\lambda c} \tag{2.8}$$

Volume flow (Q) is calculated by multiplying the average velocity across the lumen by the cross-sectional area of the vessel:

$$Q = \frac{V\pi D^2}{4} \tag{2.9}$$

To be sensitive to volume flow and independent of vessel diameter and velocity profile, the sound beam must cover the entire vessel uniformly.[33] To achieve this, the piezoelectric crystals must be at least as long as the vessel diameter. In addition, the sensitivity to flow increases with the length of the probe (L) and the ultrasonic frequency ($f_o = c/\lambda$). These requirements and the need for stable and rigid geometries and insensitivity to variations in vessel angle have led to some innovative probe configurations. The reflector probe shown in Figure 2.3b allows the two transducers to be mounted in a rigid frame, and the dual path minimizes the sensitivity to angle variations.[34] Implantable transit-time probes based on the reflector design are available from Transonic Systems, Ithaca, NY, in sizes to fit vessels from under 1 mm up to several centimeters.[35]

A simplified block diagram of a transit-time flowmeter is shown in Figure 2.4.[36] It uses the burst mode illustrated in Figure 2.2c with both crystals driven simultaneously. After the short transit time, the bursts are received at the same time, and their phases are compared and sampled. After amplifying and filtering to remove the PRF, the flow signal can be displayed and recorded. Notice that an offset must usually be added to compensate for any differences in components or transducers in the signal path. In a more practical implementation, electronic switches are included to alternately reverse the crystal connections and/or the inputs to the phase detector in an attempt to cancel any differences in the crystals or

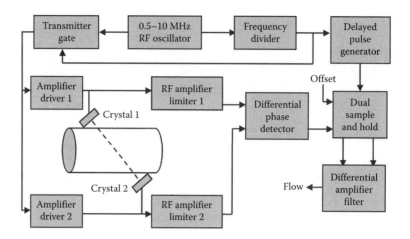

FIGURE 2.4   Simplified block diagram of one implementation of a transit-time flowmeter based on measuring the differential phase of ultrasonic bursts traveling simultaneously in opposite directions between two crystals as illustrated in Figure 2.2c. The phase is sampled during the reception of the burst by a delayed pulse following each transmission, held until the next sample, and filtered to produce an output proportional to flow. Other designs use alternate pulsing of the two crystals and switching such that the same signal path is used for each direction of measurement. This cancels or minimizes the effects of small differences in component values which would otherwise cause unacceptable offsets in measuring the very small phase shifts.

in the signal paths. This, and the careful matching of load impedances during transmission and reception, and improved transducer designs have minimized zero-drift and made the ultrasonic transit-time flowmeter a practical and widely used device.[34]

2.6 Doppler Velocity

Another method to measure blood flow with ultrasound is Doppler velocimetry.[28,37,38] As indicated in Equation 2.7, the velocity of a target can be estimated by measuring the difference in frequency between the transmitted wave and the signal reflected from the target. The difference frequency is known as the Doppler frequency and is directly proportional to the component of the velocity along the sound beam. When applied to blood flow measurement, the situation is complicated by several factors as shown in Figure 2.5: (1) the direction of blood flow is not generally in the direction of the sound beam, (2) the blood cells that reflect the sound are very small and are poor reflectors,[39–42] (3) many cells are in the sound beam or sample volume (SV) at the same time, and (4) the cells do not necessarily move at the same velocity or direction. The signals from each blood cell or reflector add together with each blood cell, contributing a signal whose amplitude and frequency vary according to its velocity, direction, and position within the SV. The practical implications of these complicating factors will be explained below.

2.7 Continuous Wave Doppler

The first Doppler velocimeters utilized continuous wave (CW) ultrasound with one transducer acting as a constant transmitter and another simultaneously as a receiver.[37,38] The transducers can be placed on a catheter inside the vessel[43–46] or more commonly on a probe or cuff outside the vessel as shown in Figure 2.5c.[38,47] From outside the vessel, the angle between each transducer and the direction of flow (θ) must be considered as shown by the Equation 2.10

$$\frac{\Delta f}{f_o} = 2\left(\frac{V}{c}\right)\cos\theta \qquad (2.10)$$

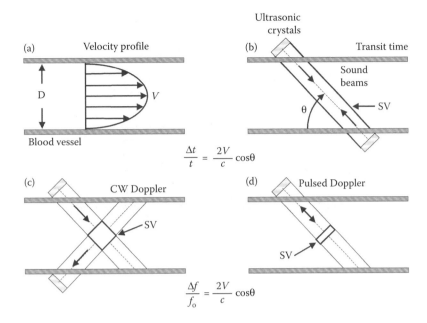

$$\frac{\Delta t}{t} = \frac{2V}{c} \cos\theta$$

$$\frac{\Delta f}{f_o} = \frac{2V}{c} \cos\theta$$

FIGURE 2.5 Ultrasonic methods for measuring blood flow in an exposed vessel: (a) an idealized velocity profile for laminar flow, (b) transit time, (c) CW Doppler, and (d) pulsed Doppler. The transit-time method (b) requires two crystals on opposite sides of the vessel, its SV includes the entire sound path between the crystals, and no reflectors are required in the fluid for operation. The CW Doppler method (c) also uses two crystals which can be on the same or opposite sides of the vessel. Its SV is the region where the transmitting and receiving sound beams cross, and its operation requires reflectors (blood cells) in the fluid. The pulsed Doppler method (d) uses a single crystal, and its SV can be controlled electronically in both length and position along the sound beam. In addition, the pulsed Doppler method can measure velocity from several SVs along the sound beam simultaneously. Normalized equations governing transit-time and Doppler methods are very similar in form. In each case, the measured parameter (differential transit-time or Doppler frequency shift) varies in proportion to the average velocity (V) in the SV.

The volume from which signals originate is often referred to as the SV. In the transit-time flowmeter, the SV consists of the area between the crystals as shown in Figure 2.5b, and the velocity is thus averaged across the entire lumen. In CW Doppler, signals are generated by any reflector in the area where the transmitting and receiving beams cross, as shown in Figure 2.5c. Because of absorption and attenuation, reflections from close targets will have higher signals than distant targets, and reflections from targets near the edges of the beams are weaker than from those near the center. Thus, the SV has an amplitude as well as a geometry, and the summing of the signals within the SV produces a weighted average due to these nonuniformities. Although the shape of the SV can be varied by controlling the beam shapes and crossing zone through sizing, angling, and focusing of the transducers, the control is limited, and the size and shape of the SV in CW Dopplers is often ill-defined.

2.8 Pulsed Doppler Velocity

Pulsed Doppler systems allow for better control of the SV by transmitting and receiving short pulses from the same transducer at different times as shown in Figure 2.5d.[28,48,49] The axial length of the SV is determined primarily by the lengths of the transmit and receive pulses, and its position along the sound beam is controlled by the time delay between transmission and reception. By controlling the beam width through focusing and sizing, the dimensions and shape of the SV can be controlled much more accurately in pulsed vs. CW Doppler systems.

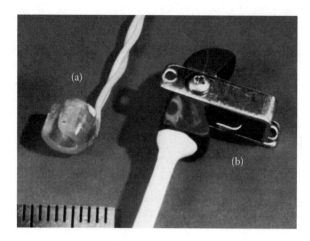

FIGURE 2.6   Photograph of a pulsed Doppler cuff-type probe (a) and a reflector-type transit-time probe (b) each sized to fit a 4-mm-diameter blood vessel. The scale is in millimeters.

Also shown in Figure 2.5a is the velocity profile across the vessel which may be parabolic as shown or much more complex. The way the SV intersects the velocity profile is extremely important in interpreting the signals from any of the ultrasonic velocimeters. Ideally, if volume flow is to be sensed, the SV should cover the entire vessel uniformly to average the entire lumen (best done with transit time but also possible with CW and pulsed Doppler methods); and if local velocity is to be sensed, the SV should be as small as possible (best done with the pulsed Doppler method). Figure 2.6 shows a photograph of a 20-MHz pulsed Doppler cuff (a) and a transit-time probe with a stainless-steel reflector (b). Both are sized to fit around a 4-mm-diameter vessel.

2.9 Doppler Signal Processing

The final Doppler signal is a summation of the signals from each reflector in the SV with the frequency determined by the reflector velocity and angle, and the amplitude determined by its position in the SV. The result is a wideband signal with its spectral content related to the velocity distribution within the SV. The task of the Doppler signal processor is to extract the information contained in the signal and to present it in a meaningful way. The available options include audio only for listening,[37] frequency-to-voltage conversion for a recorder output,[38,50] and spectral analysis and display.[47] An additional concern is whether nondirectional or directional demodulation is needed.[51,52]

A block diagram of a directional, 20-MHz pulsed Doppler velocimeter with frequency–voltage conversion is shown in Figure 2.7. A 20-MHz oscillator provides all of the timing and phase reference signals via frequency division and phase shifting. The transducer is energized at a PRF of 62.5 kHz by an 8-cycle tone burst similar to that shown in Figure 2.2b. The returning echoes are amplified and compared in-phase to two reference signals in quadrature (90° out of phase). The two-phase signals are sampled after a variable delay (which defines the location of the SV) and filtered to produce in-phase (I) and quadrature (Q) Doppler signals. These I and Q signals, when plotted on an X–Y display, show in polar coordinates the amplitude and phase of the Doppler vector which rotates at the Doppler frequency in a direction determined by the direction of flow. The signal processor must measure the frequency of rotation or angular velocity of the vector, the direction of rotation, and generate a suitable output (instantaneous, mean, peak, average, or spectrum). In complex flow regimes, there may be motion in several directions at once producing a very complex signal. The processor shown counts all X- and Y-axis (zero) crossings of the Doppler vector using the sign of the other signal to determine the direction and produces a directional display of the average frequency.[49,51]

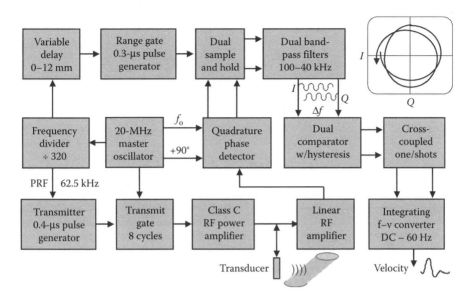

FIGURE 2.7 Block diagram of a 20-MHz pulsed Doppler velocimeter using quadrature phase detection and range gating. After sampling and filtering, the in-phase (I) and quadrature (Q) Doppler signals can be viewed as the X and Y components of a phase vector which rotates at the Doppler frequency in a direction determined by the direction of flow. The Doppler signals can be further processed to produce a directional waveform proportional to the average frequency as shown, or they can be connected to a spectrum analyzer.

The simple CW Doppler devices introduced in the early 1960s used nondirectional demodulation (producing only the X or Y component of the vector) and often contained only an audio output that was interpreted by listening to the signal. When used in research applications, a recordable and quantifiable output was required, and several methods were developed to generate a voltage proportional to the average frequency of the signal. The simplest of these is the zero-crossing counter (ZCC) that operates by counting the number of times the audio signal passes through zero in a given interval.[38,53] With a slight increase in complexity, the ZCC method can work with quadrature inputs to provide a direction-sensitive output[45,51,54] as shown in Figure 2.7. The ZCC method is simple, reliable, provides an accurate output for single-frequency or narrow-band signals with good signal-to-noise ratios and is incorporated into many commercially available CW and pulsed Doppler devices.[22,49] As an example, Figure 2.8 shows simultaneously measured arterial pressure, 10-MHz transit-time aortic flow, 20-MHz Doppler coronary flow, and 5-MHz transit-time myocardial dimension signals from a dog with implantable ultrasonic sensors. The transit-time flow probe is configured as in Figure 2.4 with burst excitation and phase detection, the segment length crystals are configured as in Figure 2.1a with pulse excitation, and the Doppler velocity signal is derived from a 20-MHz pulsed Doppler probe as in Figure 2.6a with the SV centered in the vessel where the velocity gradient is small and the spectrum tends to be narrow. However, the performance of a ZCC degrades with wideband signals and with low or marginal signal-to-noise ratios often encountered in noninvasive applications,[53] and a better signal processor is required.

From the first applications of Doppler ultrasound, it was recognized that there was valuable information in the shape of the spectrum that could be appreciated by listening to the sounds, but that was difficult to quantify or display. Early attempts at spectral analysis used swept filters,[55] banks of filters,[56] phase-locked loops,[57–59] or zero-crossing-interval histograms[60] to produce various forms of time–frequency displays using analog signal processing. The advent of digital signal processing in the 1970s enabled a wide range of additional methods for spectral analysis which continue to be improved upon.[4] The most common spectral analyzer in use today is the fast Fourier transform (FFT) which is used in various forms in most clinical Doppler devices.[59] Other methods have included autoregressive (AR),[61]

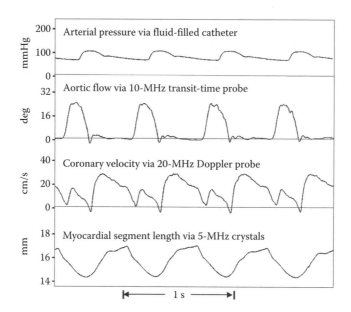

FIGURE 2.8 Multiple physiological signals from an instrumented dog. Pressure was measured with a fluid-filled catheter placed in the descending aorta and connected to an external pressure transducer, aortic flow was measured with a 10-MHz transit-time probe on the ascending aorta, coronary flow was measured with a 20-MHz pulsed Doppler probe on the left anterior descending coronary artery, and LV myocardial segment length was measured with a pair of 5-MHz transit-time crystals (sonomicrometry) imbedded into the myocardium.

time–frequency distribution,[62,63] and others too numerous to include. The FFT algorithm acquires a series of short (64–1024 points) samples of the Doppler signal upon which a spectrum is calculated and displayed. Then, after a short time delay, a new set of samples is acquired and a new spectrum is calculated either in real-time or from previously sampled data. Depending on the time resolution required, the time delay may or may not exceed the total number of samples in the FFT resulting in either overlap or complete separation of adjacent spectra. Figure 2.9 shows FFT (a) and ZCC (b) displays of the Doppler signal taken from the author's common carotid artery using a 10-MHz pulsed Doppler probe held against the neck. The velocity scale on the right is calculated from the Doppler shift on the left using Equation 2.10 with a 45° angle. The dark line on the FFT display shows the peak of the spectrum and corresponds to the maximum velocity in the SV. The ZCC signal approximates the average velocity in the SV. Note that the peaks of the spectral velocity signal are more uniform than those of the ZCC signal. We and others have found that the maximum velocity derived from the spectrum is a more robust signal that is less affected by vessel wall motion, probe motion, signal strength, noise, or slight misalignment of the probe.[62,64–66]

2.10 Multigate and Color Doppler

Pulsed Doppler devices are often used to measure the velocity distribution across a vessel, valve, or chamber during the cardiac cycle. Multiple range-gating allows velocity to be sensed at several locations or SVs along the sound beam at the same time.[28] The simplest method uses several analog processors operating in parallel to produce quadrature audio signals from each gate or depth. The complexity of parallel processing limits the number of gates typically from 8 to 80.[67–70]

Digital processing can also be used with no practical limits on the number of range gates.[69,71,72] This approach is used in color Doppler imaging devices to sense velocity over a two-dimensional region in the image.[4,73] Color Doppler instruments allow for visual interpretation of velocity patterns and

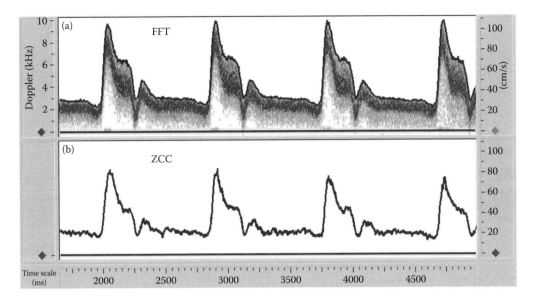

FIGURE 2.9 FFT display (a) and average zero-crossing frequency signal (b) from a common carotid artery made with a 10-MHz pulsed Doppler transducer applied to the neck. The solid line over the FFT display shows the maximum frequency calculated from the spectrum.

distributions, but quantification is difficult because the number of samples from each measurement site is limited by the need to maintain a high frame rate for the image and because of the way frequencies are mapped into colors.

2.11 Feature Extraction

The Doppler signal contains potentially valuable information about the flow field within the SV, and spectral processing is the best way to extract the maximum number of parameters. The underlying assumption is that frequency components in the Doppler spectrum are directly related to velocity components within the SV. As an example, Figure 2.10 shows intracardiac velocity signals taken noninvasively from an anesthetized mouse using a 10-MHz probe applied just below the sternum and pointed toward the heart with the SV placed in the left ventricle at a depth of 7–8 mm. Using the envelope or maximum value of the spectrum, several useful parameters relating to cardiac function can be extracted as shown. From the aortic outflow wave, these include heart rate and period, systolic time intervals, peak ejection velocity, mean velocity, rise time, peak acceleration, and area under the ejection curve (stroke distance). From the mitral filling wave, we can measure filling times, peak early filling velocity, peak late filling velocity, and areas and slopes of the waves. Thus, with proper signal processing, Doppler velocimetry can provide useful indexes of left ventricular systolic and diastolic functions[74] as well as peak and mean filling and ejection velocities.[64]

It is also possible to measure flow and velocity in peripheral arteries of animals as small as mice using transit time and pulsed Doppler ultrasound.[35,65] Figure 2.11 shows Doppler velocity signals (d) taken noninvasively from nine sites (c) in an anesthetized mouse (a) using a 20-MHz Doppler probe (b). The shape of the velocity wave in a given vessel is a function of the vascular impedance of the arteries distal to the measurement site and is often used to estimate the severity of vascular disease and stenoses.[4,75,76] It can be seen in Figure 2.11 that the upstroke of the velocity wave at each site with respect to the ECG increases with distance from the heart. By measuring the difference in arrival times and the distance between measurement sites, the pulse-wave velocity of the arteries between the sites can be calculated.[77] Pulse-wave velocity is a function of arterial stiffness and is known to increase with age, hypertension,

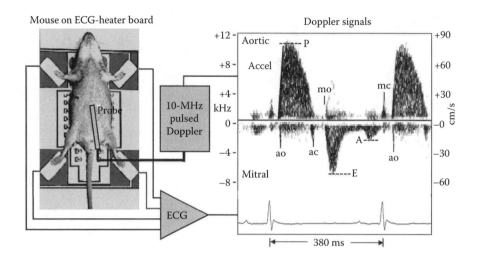

FIGURE 2.10 FFT display of cardiac Doppler signals taken noninvasively from an anesthetized mouse using a 10-MHz probe placed just below the sternum and pointed toward the heart. With the SV in the left ventricle, both inflow and outflow signals can be obtained. Labels show the opening (o) and closing (c) of the aortic (a) and mitral (m) valves, peak ejection velocity (P) and acceleration (Accel), and peak early (E), and late (A) filling velocities. From these signals, it is possible to obtain accurate timing of cardiac events such as preejection time, filling and ejection times, and isovolumic contraction and relaxation times as indexes or systolic and diastolic ventricular functions.

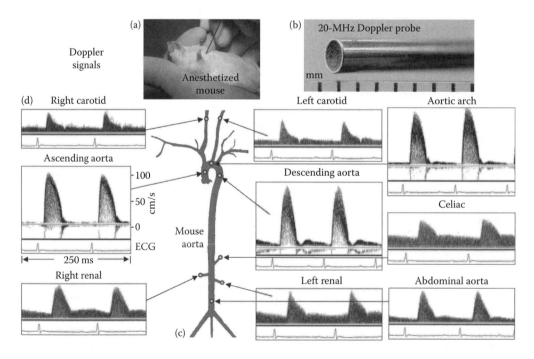

FIGURE 2.11 Doppler signals (d) from several peripheral vessels (c) in an anesthetized mouse (a) taken with the 2-mm-diameter 20-MHz probe (b). All signals were taken with the mouse supine except for the renal signals which were obtained with the mouse prone and the probe placed lateral to the spine.

and other conditions[78,79] and has been proposed as an independent risk factor for cardiovascular disease.[80] It can be measured noninvasively with Doppler ultrasound.

2.12 Converting Velocity to Volume Flow

In general, the measurement of volume flow requires knowledge of the average luminal velocity and the cross-sectional area of the vessel at the site where velocity is measured. There are several possible ways to accomplish this using ultrasound. Transit-time velocimetry can be converted to volume flow if the sound beam covers the entire vessel uniformly.[33,34] This turns out to be fairly easy to accomplish in practice, and most of the commercially available transit-time flowmeters utilize this principle.[35] CW Doppler could also be sensitive to volume flow if the sound beam covered the entire vessel uniformly, the cross-sectional area was known, and the output was related to the average frequency or first moment of the spectrum.[47,81–83] Although numerous attempts have been made, it has proved difficult to obtain uniform insonation together with an accurate measure of vessel diameter. In theory, pulsed Doppler can be used to measure the flux of blood through a surface which intersects the vessel.[84,85] This method is known as the *attenuation compensated flowmeter* and should sense volume flow independent of vessel size, orientation, or velocity profile. The method utilizes two sound beams: a broad beam which covers the entire vessel; and a smaller one that is centered in the vessel and used to estimate the attenuation along the signal path. Although the method has been proved to work in the laboratory, it has been difficult to implement practically due to problems in obtaining uniform insonation of one and only one vessel.

Pulsed Doppler can also be used to measure the velocity profile using a movable range gate or multiple range gates.[28] The point velocity measurements could then be combined with an area estimate to calculate the volume flow.[32,67,68,86] Still another approach is to measure the centerline velocity using a Doppler crystal mounted at a known angle in a rigid cuff of known diameter (Figure 2.6a), assume a parabolic velocity profile where the centerline velocity is twice the average (Figure 2.5a), and calculate volume flow based on the assumptions (Equation 2.9). It has been shown that, despite the obvious shortcomings, this method works fairly well in practice[87] and is much simpler than the algorithms using multiple range gates.

2.13 Other Applications of Doppler Velocimetry

In addition to the applications mentioned above, noninvasive Doppler ultrasound is used to estimate the degree of stenosis in peripheral vessels such as carotid and femoral arteries by alterations to the Doppler spectrum and blood flow waveforms.[4,55,76] Flow disturbances including turbulence and vorticity can also be detected and evaluated.[65,88–90] Doppler catheters can be used to assess deeper vessels such as coronary arteries to estimate the effects of stenoses on coronary blood flow and vascular reserve.[91–94] Doppler is also used to detect and quantify valvular heart disease, including insufficiency and stenosis by estimating regurgitant fraction and pressure drop.[95,96]

2.14 Artifacts and Limitations

There are numerous potential sources of error when using ultrasound to sense blood velocity or flow. Doppler and transit-time instruments measure only the component of velocity along the sound beam and provide no information about the other components of the velocity vector. Thus, some assumptions regarding the true direction of flow are required to estimate the actual velocity or flow. Usually, it is assumed that velocity is parallel to the vessel walls and that the Doppler device measures a component of this velocity according to the angle between the sound beam and the vessel axis (Figure 2.5). However, branching, curvature, tortuosity, stenosis, pulsatility, turbulence, and so on can invalidate this assumption and produce errors in the estimation of velocity. These errors are minimal with transit-time methods because the SV is large and the velocity is averaged over most of the lumen to estimate volume flow, but the errors can be significant with Doppler methods.

Stability and accuracy are concerns with both transit-time and Doppler methods. The first transit-time flowmeters had unacceptable drift and zero stability,[30] and it was this severe problem that led to the development of Doppler methods.[38] Drift is caused by geometric instability and fluid absorption by the probe and by the thermal drift in the electronic components which must measure nanosecond time differences. These problems have largely been solved by the new generation of transit-time probes and instruments, and both short- and long-term zero stability and accuracy are now acceptable. Doppler velocimetry does not rely on any inherent property of the transducer for accuracy, so the transducers can be made much simpler as shown in Figure 2.6 and are not critical to the accuracy of the measurements. The Doppler frequency shift is easy to measure and calibrate in the instrument, and zero frequency is always zero velocity. The probe either works or it does not, and most of the potential errors are due to the relationship between the measured velocity and volume flow as described above.

Pulsed Doppler signal processing involves sampling and the possibility of aliasing if the sample rate is not high enough.[97–99] In a directional Doppler velocimeter, the Doppler vector (Figure 2.7) must be sampled at least twice during each revolution for the sampled version to have the same frequency and direction as the true vector. If the sample rate is too low, the frequency is underestimated and the apparent direction of rotation is reversed.[98] Aliasing can be resolved by increasing the PRF and sampling rate, by additional signal processing,[97,98] or simply by shifting the spectral display to place the aliased signals in their proper place.[98] Aliasing is not a problem with transit-time or CW Doppler methods.

A related problem is range ambiguity resulting from having multiple pulses in flight at the same time. This is caused by high PRF and low absorption such that the echoes from one pulse are still being received when the next pulse is transmitted. Since the echoes from the two pulses overlap, the range-gate samples from two (or more) locations at the same time. This often occurs in commercial ultrasound systems when used in "high PRF mode" to avoid aliasing at high velocities. The solution is to lower the PRF if possible.

It is often desired in fluid mechanical studies to make point velocity measurements at several locations to determine the shape of the velocity profile[100,101] or to detect the presence, location, and duration of flow disturbances.[65] The pulsed Doppler method can provide this. The SV in a pulsed Doppler system has dimensions determined by the diameter, wavelength, and focus of the transducer, by the burst length, gate length, gate delay, and filtering within the instrument, and by the acoustic properties of the scattering medium. The result is a complex four-dimensional surface (x, y, z, and amplitude) over which the signals are averaged. Assuming that each red cell generates a spectral component proportional to its velocity, the spectral distribution should represent the weighted velocity distribution within the SV. However, other factors contribute to further broadening of the spectrum. These include transit time and geometric broadening[102] which are due to the limited bandwidth of the short burst, the limited time each scatterer spends in the SV, and the geometry of the beam and the transducer. The result is a Doppler spectrum that is always broader than the velocity distribution would predict.

In sensing volume flow with Doppler, it is assumed that the average frequency of the Doppler spectrum represents the average velocity in the SV (Figure 2.8). The presence of flow disturbances, turbulence, and/or vorticity can seriously affect the accuracy of this assumption by generating variable nonaxial velocity components in the SV. The spectral width and dynamics can be used to detect these effects,[88,89] but under those conditions the average frequency does not relate well to the average axial velocity.

The nature of the Doppler spectrum defies rigorous analysis by conventional means because it is dynamic and nonstationary.[4] For instance, FFT analyzers work best on long samples with stationary spectra. Frequency resolution improves with longer samples but at the expense of temporal resolution. To be useful for Doppler signals, compromises must be made. Typically, short (1–20 ms) samples are used, and a stationary condition is assumed over this short interval. But this condition is violated especially during rapid acceleration. Despite its well-known limitations, the FFT remains the standard for Doppler signal analysis.

2.15 Summary

In the last 20 years, transit-time ultrasound has supplanted the electromagnetic flowmeter as the "gold standard" for measuring blood flow in animals and in humans from extracorporeal probes placed on exposed arteries. At the same time, pulsed Doppler ultrasound has become the method of choice for high-resolution and noninvasive measurements of blood velocity and for the detection of flow disturbances secondary to cardiovascular disease. Both methods are capable of sensing flow in vessels from <0.2 mm and above in animals ranging from mice to humans and larger.

References

1. Wells, P.N.T., *Biomedical Ultrasonics*, Academic Press, New York, 1977.
2. Altobelli, S.A., Voyles, W.F., and Greene, E.R., *Cardiovascular Ultrasonic Flowmetry*, Elsevier, New York, 1985.
3. Kremkau, F.W., *Doppler Ultrasound: Principles and Instruments*, W.B. Saunders, Philadelphia, 1990.
4. Evans, D.H. et al., *Doppler Ultrasound: Physics, Instrumentation, and Clinical Applications*, John Wiley & Sons, New York, 1989.
5. Morse, P.M. and Ingard, K.U., *Theoretical Acoustics*, McGraw-Hill, New York, 1968.
6. Christensen, D.A., *Ultrasonic Bioinstrumentation*, John Wiley & Sons, New York, 1988.
7. Goldman, D.E. and Hueter, T.F., Tabular data of the velocity and absorption of high-frequency sound in mammalian tissues, *J. Acoust. Soc. Am.*, 28, 35, 1956.
8. Snook, K.A. et al., Design, fabrication, and evaluation of high frequency, single-element transducers incorporating different materials, *IEEE Trans. Ultrason. Ferroelectr. Freq. Control*, 49, 169, 2002.
9. Shung, K.K. and Zipparo, M.J., Ultrasonic transducers and arrays, *IEEE Eng. Med. Biol.*, 15, 20, 1996.
10. Ritter, T.A. et al., A 30-MHz piezo-composite ultrasound array for medical imaging applications, *IEEE Trans. Ultrason. Ferroelectr. Freq. Control*, 49, 217, 2002.
11. Kossoff, G., The effects of backing and matching on the performance of piezoelectric ceramic transducers, *IEEE Trans. Sonics Ultrason.*, SU-13, 20, 1966.
12. Desilets, C.S., Fraser, J.D., and Kino, G.S., The design of efficient broad-band piezoelectric transducers, *IEEE Trans. Sonics Ultrason.*, SU-25, 115, 1978.
13. Zipparo, M.J., Shung, K.K., and Shrout, T.R., Piezoceramics for high-frequency (20–100 MHz) single-element imaging transducers, *IEEE Trans. Ultrason. Ferroelectr. Freq. Control*, 44, 1038, 1997.
14. Rushmer, R.F., Franklin, D.L., and Ellis, R.M., Left ventricular dimensions recorded by sonocardiometry, *Circ. Res.*, 4, 684, 1956.
15. Theroux, P. et al., Regional myocardial function during acute coronary artery occlusion and its modification by pharmacologic agents in the dog, *Circ. Res.*, 35, 896, 1974.
16. Stegall, H.F. et al., A portable simple sonomicrometer, *J. Appl. Physiol.*, 23, 289, 1967.
17. Hill, R.C. et al., Perioperative assessment of segmental left ventricular function in man, *Arch. Surg.*, 115, 609, 1980.
18. Sasayama, S. et al., Dynamic changes in left ventricular wall thickness and their use in analyzing cardiac function in the conscious dog, *Am. J. Cardiol.*, 38, 870, 1976.
19. Gallagher, K.P. et al., Significance of regional wall thickening abnormalities relative to transmural myocardial perfusion in anesthetized dogs, *Circulation*, 62, 1266, 1980.
20. Pagani, M. et al., Measurements of multiple simultaneous small dimensions and study of arterial pressure dimension relations in conscious animals, *Am. J. Physiol Heart Circ. Physiol.*, 235, H610, 1978.
21. Bertram, CD., Ultrasonic transit-time system for arterial diameter measurement, *Med. Biol. Eng. Comput.*, 15, 489, 1977.
22. Hartley, C.J. et al., Synchronized pulsed Doppler blood flow and ultrasonic dimension measurement in conscious dogs, *Ultrasound Med. Biol.*, 4, 99, 1978.

23. Baker, D.W. and Simmons, V.E., Phase track techniques for detecting arterial blood vessel wall motion, *Proc. 21st ACEMB*, 8.6, 1968 (Abstract).
24. Hokanson, D.E. et al., A phase-locked echo tracking system for recording arterial diameter changes in vivo, *J. Appl. Physiol.*, 32, 728, 1972.
25. Hartley, C.J. et al., Doppler measurement of myocardial thickening with a single epicardial transducer, *Am. J. Physiol. Heart Circ. Physiol.*, 245, H1066, 1983.
26. Zhu, W. et al., Validation of a single crystal for the measurement of transmural and epicardial thickening, *Am. J. Physiol Heart. Circ. Physiol.*, 251, H1045, 1986.
27. Hartley, C.J. et al., An ultrasonic method for measuring tissue displacement: Technical details and validation for measuring myocardial thickening, *IEEE Trans. Biomed. Eng.*, 38, 735, 1991.
28. Baker, D.W., Pulsed ultrasonic Doppler blood flow sensing, *IEEE Trans. Sonics Ultrason.*, SU-17, 170, 1970.
29. Kalmus, H.P., Electronic flowmeter system, *Rev. Sci. Instrum.*, 25, 201, 1954.
30. Franklin, D.L., Baker, D.W., and Ellis, R.M., A pulsed ultrasonic flowmeter, *IRE Trans. Med. Electron.*, 6, 204, 1959.
31. Plass, K.G., A new ultrasonic flowmeter for intravascular application, *IEEE Trans. Biomed. Eng.*, BME-11, 154, 1964.
32. Keller, H.M. et al., Non-invasive measurement of velocity profiles and blood flow in the common carotid artery by pulsed Doppler ultrasound, *Stroke*, 7, 370, 1976.
33. Rader, R.D., A diameter-independent blood flow measurement technique, *Med. Instrum.*, 10, 185, 1976.
34. Drost, C.J., Vessel diameter-independent volume flow measurements using ultrasound, *Proc. 17th San Diego Biomed.Symp.*, 1978, pp. 299–302 (Abstract).
35. D'Almeida, M.S., Gaudin, C, and Lebrec, D., Validation of 1 and 2 mm transit-time ultrasound flow probes on mesenteric artery and aorta of rats, *Am. J. Physiol. Heart Circ. Physiol.*, 268, H1368, 1995.
36. Hartley, C.J., A phase detecting ultrasonic flowmeter, *Proc 25th ACEMB*, 331972 (Abstract).
37. Satomura, S., Ultrasonic Doppler method for the inspection of cardiac functions, *J. Acoust. Soc. Am.*, 29, 1181, 1957.
38. Franklin, D.L., Schlegal, W., and Rushmer, R.F., Blood flow measured by Doppler frequency shift of back-scattered ultrasound, *Science*, 134, 564, 1961.
39. Carstensen, E.L., Li, K., and Schwan, H.P., Determination of the acoustic properties of blood and its components, *J. Acoust. Soc. Am.*, 25, 286, 1953.
40 Shung, K.K., Sigelmann, R.A., and Reid, J.M., Scattering of ultrasound by blood, *IEEE Trans. Biomed. Eng.*, BME-23, 460, 1976.
41. Angelsen, B.A.J., A theoretical study of the scattering of ultrasound from blood, *IEEE Trans. Biomed. Eng.*, 27, 61, 1980.
42. Shung, K.K., Physics of blood echogenicity, *J. Cardiovasc. Ultrasonography*, 2, 401, 1983.
43. Stegall, H.F., Stone, H.L., and Bishop, VS., A catheter-tip pressure and velocity sensor, *Proc. 20th Annual Conf. on Engineering in Medicine and Biology*, 27(4), 1967 (Abstract).
44. Benchimol, A. et al., Aortic flow velocity in man during cardiac arrythmias measured with the Doppler catheter-flowmeter system, *Am. Heart J.*, 78, 649, 1969.
45. Kalmanson, D. et al., Retrograde catheterization of left heart cavities in dogs by means of an orientable directional Doppler catheter-tip flowmeter: A preliminary report, *Cardiovasc. Res.*, 6, 309, 1972.
46. Reid, J.M. et al., A new Doppler flowmeter system and its operation with catheter mounted transducers, *Cardiovascular Applications of Ultrasound*, Reneman, R.S., Ed., Elsevier, New York, 1974, pp. 183–197.
47. Brody, W.R. and Meindl, J.D., Theoretical analysis of the CW Doppler ultrasonic flowmeter, *IEEE Trans. Biomed. Eng.*, 21, 183, 1974.
48. Peronneau, P.A. et al., Theoretical and practical aspects of pulsed Doppler flowmetry: Real-time application to the measurement of instantaneous velocity profiles *in vitro* and *in vivo*, *Cardiovascular Applications of Ultrasound*, Reneman, R.S., Ed., Amsterdam, North-Holland, 1974, pp. 66–84.

49. Hartley, C.J. and Cole, J.S., An ultrasonic pulsed Doppler system for measuring blood flow in small vessels, *J. Appl. Physiol.*, 37, 626, 1974.

50. Satomura, S., Study of the flow patterns in peripheral arteries by ultrasonics, *J. Acoust. Soc. Jpn.*, 15, 151, 1959.

51. McLeod, F.D., A directional Doppler flowmeter, *Proc. 7th ICMBE*, 14, 1967 (Abstract).

52. Coghlan, B.A. and Taylor, M.G., Directional Doppler techniques for detection of blood flow velocities, *Ultrasound Med. Biol.*, 2, 181, 1976.

53. Lunt, M.J., Accuracy and limitations of the ultrasonic Doppler blood velocimeter and zero-crossing detector, *Ultrasound Med. Biol.*, 2, 1, 1975.

54. Hartley, C.J. and Cole, J.S., A single crystal ultrasonic catheter tip velocity probe, *Med. Instrum.*, 8, 241, 1974.

55. Felix, W.R. et al., Pulsed Doppler ultrasound detection of flow disturbances in arteriosclerosis, *J. Clin. Ultrasound*, 4, 275, 1976.

56. Cross, G. and Light, L.H., Direction-resolving Doppler instrument with improved rejection of tissue artifacts for transcutaneous aortovelography, *Physiol. Soc*, 5P, 1971.

57. Giddens, D.P. and Khalifa, A.M., Turbulence measurements with pulsed Doppler ultrasound employing a frequency tracking method, *Ultrasound Med. Biol.*, 8, 427, 1982.

58. Sainz, A.J., Roberts, V.C., and Pinardi, G., Phased-locked loop techniques applied to ultrasonic Doppler signal processing (blood flow measurements), *Ultrasonics*, 14, 128, 1976.

59. Brigham, E.O., *The Fast Fourier Transform*, Prentice-Hall, Englewood Cliffs, NJ, 1974.

60. Daigle, R.E. and Baker, D.W., A readout for pulsed Doppler velocity meters, *ISA Trans.*, 16, 41, 1977.

61. Kitney, R.I. and Giddens, D.P., Analysis of blood velocity waveforms by phase shift averaging and autoregressive spectral estimation, *J. Biomech. Eng.*, 105, 398, 1983.

62. Evans, D.H., Doppler signal processing, *Cardiovascular Ultrasonic Flowmetry*, Altobelli, S.A., Voyles, W.F., and Greene, E.R., Eds., Elsevier, New York, 1985, pp. 239–261.

63. Cohen, L., Time–frequency distributions—A review, *Proc. IEEE*, 77, 941, 1995.

64. Hartley, C.J., Michael, L.H., and Entman, M.L., Noninvasive measurement of ascending aortic blood velocity in mice, *Am. J. Physiol. Heart Circ. Physiol.*, 268, H499, 1995.

65. Hartley, C.J. et al., Noninvasive cardiovascular phenotyping in mice, *ILAR J.*, 43, 147, 2002.

66. Sudhir, K. et al., Measurement of volumetric coronary blood flow with a Doppler catheter: Validation in an animal model, *Am. Heart J.*, 124, 870, 1992.

67. Casty, M. and Giddens, D.P., 25 + 1 channel pulsed ultrasound Doppler velocity meter for quantitative flow measurements and turbulence analysis, *Ultrasound Med. Biol.*, 10, 161, 1984.

68. Stacey-Clear, A. and Fish, P.J., Repeatability of blood flow measurement using multichannel pulsed Doppler ultrasound, *Br. J. Radiol.*, 57, 419, 1984.

69. Hoeks, A.P.G., Reneman, R.S., and Peronneau, P.A., A multigate pulsed Doppler system with serial data processing, *IEEE Trans. Sonics Ultrason.*, 28, 242, 1981.

70. Kajiya, F. et al., Evaluation of human coronary blood flow with an 80-channel 20 MHz pulsed Doppler velocitometer and zero-cross and Fourier transform methods during cardiac surgery, *Circulation*, Suppl. III, 53, 1986.

71. Brandestini, M.A., A digital 128-channel transcutaneous blood-flowmeter, *Biomed. Tech.*, 21, 291, 1976.

72. Nowicki, A. and Reid, J.M., An infinite gate pulse Doppler, *Ultrasound Med. Biol.*, 7, 41, 1981.

73. Merritt, C.R., Doppler color flow imaging, *J. Clin. Ultrasound*, 15, 591, 1987.

74. Taffet, G.E. et al., Noninvasive indexes of cardiac systolic and diastolic function in hyperthyroid and senescent mouse, *Am. J. Physiol. Heart Circ. Physiol.*, 270, H2204, 1996.

75. Hartley, C.J. et al., Hemodynamic changes in apolipoprotein E-knockout mice, *Am. J. Physiol. Heart Circ. Physiol.*, 219, H2326, 2000.

76. Skidmore, R., Woodcock, J.P., and Wells, P.N.T., Physiological interpretation of Doppler-shift waveforms, *Ultrasound Med. Biol.*, 6, 7–10, 219–225, 227, 1980.

77. Hartley, C.J. et al., Noninvasive determination of pulse-wave velocity in mice, *Am. J. Physiol. Heart Circ. Physiol.*, 273, H494, 1997.

78. Avolio, A.P. et al., Effects of aging on arterial distensibility in populations with high and low prevalence of hypertension: Comparison between urban and rural communities in China, *Circulation*, 71, 202, 1985.
79. Nichols, W.W. and O'Rourke, M.F., *McDonald's Blood Flow in Arteries: Theoretical Experimental and Clinical Principles*, Edward Arnold, London, 1998.
80. Arnett, D.K., Evans, G.W., and Riley, W.A., Arterial stiffness: A new cardiovascular risk factor? *Am. J. Epidemiol.*, 140, 669, 1994.
81. Arts, M.G.J. and Roevros, J.M.G.J., On the instantaneous measurement of blood flow by ultrasonic means, *Med. Biol Eng.*, 10, 23, 1972.
82. Gerzberg, L. and Meindl, J.D., Mean frequency estimator with applications in ultrasonic Doppler flowmeters, *Ultrasound in Medicine*, White, D.N. and Brown, R.E., Eds., Plenum Press, New York, 1977, pp. 1173–1175.
83. Gill, R.W., Performance of the mean frequency Doppler demodulator, *Ultrasound Med. Biol.*, 5, 237, 1979.
84. Hottinger, C.F., Blood flow measurement using the attenuation-compensated volume flowmeter, *Ultrasonic Imaging*, 1, 1, 1979.
85. Hottinger, C.F. and Meindl, J.D., Unambiguous measurement of volume flow using ultrasound, *IEEE*, 63, 984, 1975.
86. Marquis, C. et al., Femoral blood flow determination with a multichannel digital pulsed Doppler: An experimental study on anaesthetized dogs, *Vasc. Surg.*, 17, 95, 1983.
87. Ishida, T. et al., Comparison of hepatic extraction of insulin and glucagon in conscious and anesthetized dogs, *J. Endocrinol.*, 112, 1098, 1983.
88. Cloutier, G., Chen, D., and Durand, L.G., Performance of time–frequency representation techniques to measure blood flow turbulence with pulsed-wave Doppler ultrasound, *Ultrasound Med. Biol.*, 27, 535, 2001.
89. Cloutier, G., Allard, M.F., and Durand, L.G., Characterization of blood flow turbulence with pulsed-wave and power Doppler ultrasound imaging, *J. Biomech. Eng.*, 118, 318, 1996.
90. Wang, Y. and Fish, P.J., Comparison of Doppler signal analysis techniques for velocity waveform, turbulence, and vortex measurement: A simulation study, *Ultrasound Med. Biol.*, 22, 635, 1996.
91. Cole, J.S. and Hartley, C.J., The pulsed Doppler coronary artery catheter: Preliminary report of a new technique for measuring rapid changes in coronary artery flow velocity in man, *Circulation*, 56, 18, 1977.
92. Wilson, R.F. et al., Transluminal, subselective measurement of coronary artery blood flow velocity and vasodilator reserve in man, *Circulation*, 72, 82, 1985.
93. Sibley, D.H. et al., Subselective measurement of coronary blood flow velocity using a steerable Doppler catheter, *J. Am. Coll. Cardiol.*, 8, 1332, 1986.
94. Hartley, C.J., Review of intracoronary Doppler catheters, *Int. J. Cardiac Imaging*, 4, 159, 1989.
95. Hatle, L., Non-invasive assessment and differentiation of left ventricular outflow obstruction with Doppler ultrasound, *Circulation*, 64, 381, 1981.
96. Hatle, L. et al., Noninvasive assessment of pressure drop in mitral stenosis by Doppler ultrasound, *Br. Heart J.*, 40, 131, 1978.
97. Tortoli, P., Valgimigli, F., and Guidi, G., Clinical evaluation of a new anti-aliasing technique for ultrasound pulsed Doppler analysis, *Ultrasound Med. Biol.*, 15, 749, 1989.
98. Hartley, C.J., Resolution of frequency aliases in pulsed Doppler velocimeters, *IEEE Trans. Sonics Ultrasonics*, SU-28, 69, 1981.
99. Bom, N., De Boo, J., and Rijsterborgh, H., On the aliasing problem in pulsed Doppler cardiac studies, *J. Clin. Ultrasound*, 12, 559, 1984.
100. Rabinovitz, R.S. et al., Fluid dynamics of the left main coronary bifurcation, *Proc. 40th ACEMB*, 154, 1987 (Abstract).
101. Vieli, A., Jenni, R., and Anliker, M. Spatial velocity distributions in the ascending aorta of healthy humans and cardiac patients, *IEEE Trans. Biomed. Eng.*, 33, 28, 1986.
102. Newhouse, V.L. et al., The dependence of ultrasound Doppler bandwidth on beam geometry, *IEEE Trans. Sonics Ultrason.*, SU-27, 50, 1980.

3

Evoked Potentials

Hermann Hinrichs
University of Magdeburg

3.1 Basic Operational Mechanisms

The ongoing background electroencephalogram (EEG) is discussed in Chapter 5. In this chapter, we discuss techniques for extracting from the ongoing EEG measures of the brain's specific reaction to peripheral sensory stimulation, which are known as *evoked potentials* (EPs). A comprehensive description of the underlying physiological mechanisms maybe found in Kandel et al. (2000). Techniques for measuring EPs can also be used to measure sensory nerve potentials that occur before the effects of peripheral stimulation are apparent in the brain. A brief overview of the main technical components necessary to record an EP is provided in Figure 3.1.

EPs can be extracted from the background EEG if subjects are exposed to repeated brief sensory stimuli. For standard neurological applications, *auditory, somatosensory*, and *visual* stimuli are the most important. The corresponding specific EPs are called AEPs (auditory evoked potentials), SEPs (somatosensory evoked potentials), and VEPs (visual evoked potentials), respectively. The most common methods of stimulation are (1) presenting auditory clicks via earphones (AEP), (2) presenting brief electrical pulses to peripheral sensory nerves such as the *median* or the *tibial nerve*, and (3) presenting inverting checkerboard patterns on a computer screen (VEP). Further modalities, involving olfactory or thermal stimuli, for example, will not be discussed here. Different subtypes of AEP and SEP can be observed with different latencies after stimulus onset. In this chapter, we focus on the earliest responses known as the *brainstem AEP* (BAEP, brainstem auditory evoked potential) and the *early SEP*. There are later AEP and SEP components, but these reflect higher processing stages and have gained less importance in everyday neurological examination. The required frequency of repeated stimulus presentation strongly varies with modality and ranges from about 1 Hz for VEPs to 2–3 Hz for SEPs, up to 10–20 Hz for BAEPs.

In contrast to the more-or-less random background EEG, EPs are, to a large extent, reproducible when peripheral stimuli are repeatedly presented. This is the reason why they can be measured, despite the fact that their amplitude is much lower than that of the background EEG. The basic technique needed to extract these small signals is to *average* (Epstein and Boor, 1988) the voltages measured in short EEG

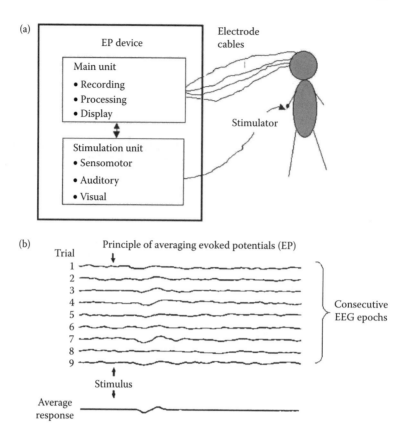

FIGURE 3.1 (a) Basic components of an EP recorder; (b) principle to extract an EP from an ongoing EEG (for details, see Figure 3.11).

epochs that are synchronized with the stimulus presentation. Assuming a stable EP pattern and a background EEG that is random with respect to the stimulation timing, such averaging will attenuate the background EEG amplitude by $1/\sqrt{n}$, where n is the number of averaged epochs. Because the EP fraction of the signal remains constant, the EP emerges from the background more and more clearly with a growing number of averaged epochs. In principle, the EP can be determined with arbitrary precision by adding sufficient epochs. However, practically there are limits (1) because of limited examination time and (2) because EP morphology is not perfectly stable due to physiological noise, habituation, and other effects. Also, as discussed below, minor signal distortions due to artifacts often cannot be fully controlled and therefore degrade EP quality and reproducibility. The number of repeated stimuli needed for a reliable EP measurement significantly depends on the modality, ranging from about 50 (VEP) up to more than 1000 (BAEP).

The EP reflects specific neurophysiological functions, in contrast to the background EEG, which provides an overview of global brain function. A normal EP indicates that both the specific sensory pathways as well as the corresponding brain areas are intact. In contrast, pathological EPs permit—to varying extents, depending on the modality—diagnostic conclusions about the origin and type of the underlying malfunction. Due to the considerable intra-individual variability of the EP, it is good practice in terms of quality control to derive at least two consecutive EP measurements and check whether they are similar. In special applications, like function monitoring in the intensive care unit, or during operations, EPs may be repeatedly acquired according to fixed or random schedules in order to monitor brain functions continuously over hours, days, or even weeks.

In contrast to the background EEG, the EP is usually recorded from only a few (typically two) electrodes placed over the specific brain areas. In addition, one or two additional electrodes are sometimes placed on the neck to record cervical SEPs. Thus, only two to four EP channels are generally recorded. The electrical activity is measured against electrodes placed somewhere else on the skull where activity related to the EP is negligible. In accordance with the organization of the relevant sensory nerve pathways, the brain responses are mainly ipsilateral (BAEP) or contralateral (VEP, SEP) to the side of stimulation. For a comprehensive examination, therefore, EP recordings with alternating sides of stimulation are needed. Alternatively, in certain situations, bilateral stimulation may be useful. Appropriate electrode sites for recording the various EP types are as follows, labeled according to the internationally standardized 10–20 positioning system (Jasper, 1958) (see Figure 3.2): (1) VEP: O1 (left) and O2 (right) vs. Fz; (2) BAEP: left and right mastoid vs. Cz; (3) SEP: left and right sensory cortex vs. Fz. For instance, for the median nerve, the appropriate positions are C3′ and C4′, which are slightly more posterior than C3 and C4. Cervical potentials may be acquired from the C2 position (not shown in Figure 3.2) on the neck, again using Fz as reference electrode.

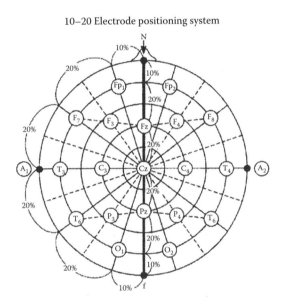

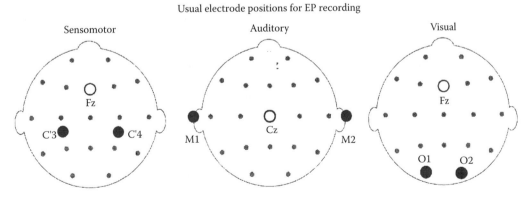

FIGURE 3.2 (Upper) The international 10–20 system (Jasper, 1958) for electrode placement and labeling. (Lower) Scalp locations (according to the 10–20 system) used for different EP modalities. M1 and M2 indicate the left and right mastoids located close to the ears.

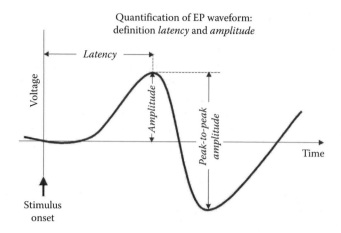

FIGURE 3.3 Definition of parameters used for EP quantification.

Details regarding the various stimulation techniques are explained below. EP recorders are usually embedded in general neurophysiological devices capable of recording and analyzing electromyographic (EMG) activity, neurographic measures (nerve velocity, etc.), and EP. All these methods share the same amplifiers, stimulation units (where applicable), and computer environment. Alternatively, there are some EEG recorders on the market capable of recording and analyzing EPs as an additional tool.

As shown symbolically in Figure 3.3, the EP elicited by the stimulus exhibits a sequence of one or more positive and negative deflections (also called *peaks* or *components*) arising from the noise floor which are usually characterized by their *latency* (i.e., the delay between stimulus onset and the peak of the deflection) and *amplitude*. The surface EP typically covers an amplitude range of 5–10 μV (VEP, SEP) or 0.5 μV (BAEP). The first clear and stable VEP component (a positive deflection, see Figure 3.4) occurs at about 100 ms (called P100 or P1) after stimulus onset, frequently preceded by a small negative deflection at 75 ms (N75), and followed by another negative deflection at about 145 ms (N145), all reflecting visual cortex activity. The latencies of the earliest cortically recorded SEP components are around 10 ms, depending on the site of stimulation. In principle, the more proximal the stimulation is, the shorter the latency. In clinical examinations, it is most usual to stimulate the median nerve. The corresponding SEP shows a pronounced negative peak in the scalp recording at about 20 ms (N20) post-stimulus, followed by a positive deflection at around 25 ms (P25). In addition, under favorable conditions (i.e., low noise), an early negative wave may be observed at 13 ms (N13). This component coincides with a strong negative peak obtained in recordings from the neck (C2 electrode site). Both peaks, therefore, represent the same neurophysiological entity arising as a post-synaptic potential in the spinal cord. Similar temporal relations can be observed with SEPs derived when stimulating arbitrary sensory nerves. BAEPs recorded at the scalp start with an early peak at around 2 ms after stimulus onset and extend up to approximately 6 ms. Five principal signal deflections can be distinguished, which are labeled I through V, in order of their occurrence. Components I and II are attributed to the acoustic nerve, whereas the three later peaks (III, IV, V) originate in the brain stem. Two additional peaks may be found in the latency range of 6–10 ms. However, these components have not gained clinical importance.

The main SEP and VEP components (N20, P25, P100) are generated in close vicinity to the relevant scalp electrode. On the contrary, the early cortical SEP component N13 as well as all BAEP components originate in regions located far away from the electrodes. They can nevertheless be measured due to volume conduction in the brain. To distinguish between these two mechanisms, the terms *near-field* and *far-field potentials* have been introduced.

Owing to the strongly differing temporal structures of the different EPs, the amplifier frequency band best suited to record them varies significantly. Reasonable bandpass settings are approximately

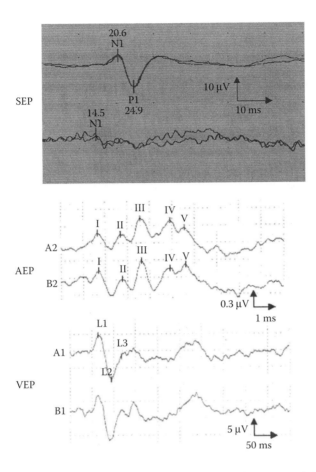

FIGURE 3.4 Examples of the various EP modalities. As usual in clinical routine, each EP is recorded and displayed twice in order to control its reproducibility. Note the different scaling of time and amplitude.

1–100 Hz (VEP), 5–1000 Hz (SEP), and 100–3000 Hz (BAEP). These settings are somewhat arbitrary and depend on individual laboratory standards rather than official recommendations. The number of epochs needed to extract an EP mainly depends on its amplitude. Typical values are 64–128 (VEP), 128–256 (SEP), and 1024–2048 (BAEP).

With respect to EP evaluation, the most important characteristics are the latencies of certain components and—to a lesser extent due to large inherent variability—their amplitudes. In addition, sometimes latency or amplitude differences between consecutive peaks may carry further diagnostic information.

The differential amplifiers used for EP recording are very much similar to EEG amplifiers, except for the larger bandwidth needed for BAEP acquisition. The same holds for considerations regarding the main physical properties of the electrodes providing the appropriate low-impedance contact between the skin and the amplifier input. However, the requirements to be able to transmit very low frequencies approaching direct current (DC) are less restrictive with EPs, because this frequency range is less important here. Therefore, the selection of electrode types is less critical with EPs than with EEG.

As in the case of EEG recording, external signal components often overlay the original EP and thus corrupt the measured signal, even after averaging. There are both technical and biological sources of these so-called *artifacts*. Among the biological artifacts picked up by the amplifiers are residual electrocardiogram (ECG), electromyogram (EMG) generated by muscles in the vicinity of the electrode, electrooculogram (EOG) resulting from eye movements, and slowly fluctuating electrode voltages

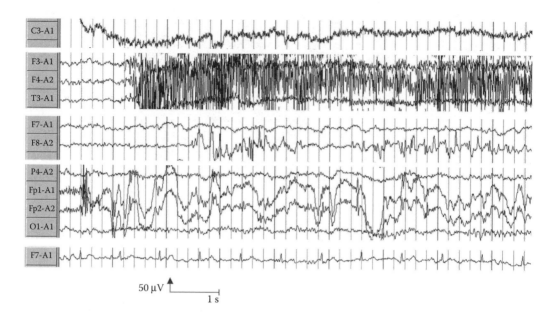

FIGURE 3.5 Examples of different types of EEG artifacts (shown only selected channels per artifact) that may affect the quality of the EP extracted from these signals: line (mains) artifact, that is, 50-Hz activity; electromyogram (EMG) acitivity generated by scalp muscles if the subject is not sufficiently relaxed; movement of electrodes on the skin; eye movements: the moving eye—as an electrical dipole—generates slowly varying electrical potentials which are picked up by the EEG electrodes; electrocardiographic (ECG) activity embedded in the ongoing EEG.

induced by transpiration, which influences the electrode's electrochemical conditions. Some examples are demonstrated in Figure 3.5. Technical artifacts have a variety of causes, including inductive nonsymmetric coupling of the 50/60 Hz line (mains) frequency, which usually occurs when electrode impedances are excessively large, fluctuating electrode voltages caused by electrode movement, distortions of the ambient electrostatic field due to movement of the patient or the technician, unstable electrical contact between the electrode and the amplifier input, and bad or poorly cleaned electrode material.

The significance of these artifacts differs depending on the type of EP. For example, low-frequency disturbances caused EOG or transpiration hardly influence BAEPs, whereas VEPs may be severely distorted by these artifacts. By contrast, under favorable conditions, a VEP may not be significantly degraded by EMG, which would, however, severely affect BAEP quality. A general rule for EP recording, therefore, is to exclude artifactual epochs as much as possible from the averaging process. Under difficult conditions, as, for instance, in intensive care units, 50- or 60-Hz line (mains) frequency artifacts sometimes cannot be avoided. Notch filters are capable of selectively suppressing these frequencies. However, these filters, although available on most devices, must be applied with great caution in the case of SEPs, because they may suppress parts of the real EP signal as well. Thus, the usage of notch filters is still debated with respect to SEP. Even with VEPs, notch filters must be applied very carefully because they may cause latency shifts.

Traditionally, during EP acquisition, the actual digitally recorded epoch as well as the emerging average potential were displayed on an analog cathode ray tube. Once acquired, the EP was written as a hard copy to photosensitive paper strips. The evaluation was completed by interactive amplitude and latency measurement, either using the hardcopy or done directly on the screen, sometimes supported by cursors. Modern EP systems are supported by digital post-processing aides, including automatic peak detection. Also, signal display now greatly benefits from modern digital displays offering better quality and larger flexibility (e.g., easy rescaling, color marking of selected signal epochs). The graphical resolution of these screens is much better than is actually required if one takes into account the signal-to-noise ratio of the EPs. In addition to the ability to produce additional hardcopy documentation, modern EP devices offer the possibility of storing the data on digital media such as magnetic disks, and so forth. Details regarding these issues are discussed in the next section.

There are now a large variety of digital post-processing methods, such as source analysis to localize the intracerebral neural generators of individual EP components from multichannel recordings, decomposition by principal component analysis, sophisticated peak detection techniques, and so forth. However, routine EP interpretation still relies almost exclusively on inspection of peak latencies and amplitudes. Only for special applications such as monitoring of long-term cerebral function are automated procedures routinely applied.

3.2 Signal Processing

In this section, the main elements of a modern fully digital EP recorder will be described. An excellent overview of EP techniques is found in a book edited by Deuschl and Eisen (1999) comprising separate chapters (written by different groups of authors) for the various EP modalities. As shown in Figure 3.6, an EP recording may be subdivided into the following stages:

- Stimulators
- Electrodes
- Analog preamplifier; safety considerations
- Analog-to-digital conversion
- Digital signal processing; filtering, averaging, and scaling
- Signal display on screen and on paper
- Long-term data storage and retrieval

3.2.1 Stimulators (see Figure 3.7)

3.2.1.1 Somatosensory Evoked Potential

The basic stimulation unit consists of an electronic circuit capable of applying short electrical impulses (50–200 μs) of predefined, yet scalable constant currents or voltages to the skin. A pair of electrodes (metal plates of a few millimeters diameter) provides the necessary low resistance electrical contact

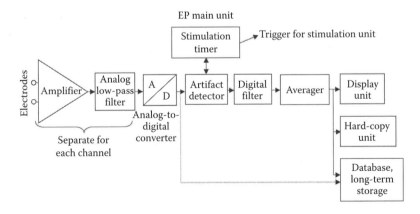

FIGURE 3.6 Technical block diagram of an EP recorder. The digital filters allow for filtering of the signal with a variable bandwidth according to the actual needs. In some designs, this feature may be replaced by switchable analog filters. Yet another solution might be to filter the averaged rather than the raw data. There is still no canonical technical concept. For later evaluation, arbitrary sets of averaged EPs can be fetched from the archive.

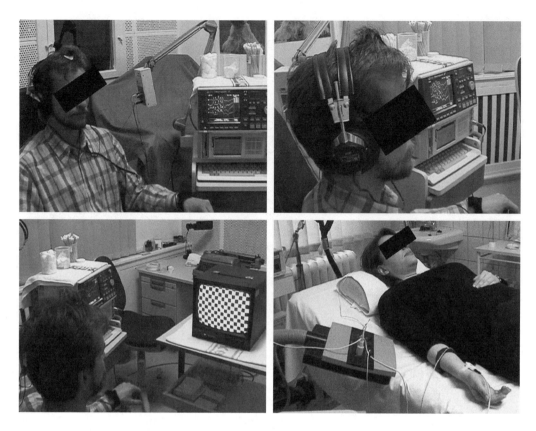

FIGURE 3.7 Recording an EP. (Upper) AEP, auditory stimulation is accomplished by a headset. (Lower left) VEP, note the monitor in front of the subject providing the flickering checkerboard pattern used for stimulation. (Lower right) SEP, note the electrode pair applied at the subjects's wrist for electrical stimulation of the median nerve. For EP recordings of other nerves, the electrodes have to be placed accordingly.

between the poles of the stimulator and the skin. Usually, the cathode is placed a few centimeters proximal to the anode, both with an orientation parallel to the sensory nerve fiber under consideration. The selection of electrode material is not critical (e.g., refined steel or silver are suitable). To enhance the electrical contact, an electrolytic paste or saline solution is brought into the gap between electrode surface and skin by various methods. Today, most stimulators feed constant currents through the closed loop during the stimulation period. Constant voltage stimuli may be used as well, although they lead to less reliable stimulation in the case of nonoptimal electrical contacts.

For the adjustment of stimulus strength, the following procedure is performed before recording. Starting from a low value, the current is usually increased until a motor response is observed. For the subsequent repeated stimulation, this current intensity is maintained or even slightly increased. In the case of pathological SEPs, it is often helpful to record the SEP also from the contralateral hemisphere. This SEP may serve as a reference for the interpretation. Owing to the electrical stimulation, a large-amplitude artifact is seen in the averaged EP at zero latency. However, due to the short pulse duration and the short time constant of the filter settings (see below), this artifact does not interfere with the EP components under consideration. Typical stimulators allow for the stimulation frequency to be varied over a wide range (typical values are 0.5–10 Hz). For safety reasons, the application unit (providing the stimuli) is electrically insulated from the other electronic circuits of the recording unit.

3.2.1.2 Brainstem Auditory Evoked Potential

Simple rectangular auditory pulses (clicks) of 100-µs duration are presented to the subject over ordinary high-quality headphones. Typical pulse generators provide both rarefaction and compression pulses, including alternating sequences. A sound pressure level of around 70 dB above sensory level (SL) is required to acquire stable BAEP, with the precise value varying between laboratories. The SL is the minimal sound pressure that is detected by the subject with a probability larger than chance. It must be determined individually for each subject before the EP acquisition starts. BAEPs are usually recorded separately for each ear. To prevent interference from ambient noise, the opposite ear is usually "neutralized" by applying white noise at a sound pressure level of approximately 30 dB below the click intensity. Typical auditory stimulators allow the stimulation frequency to be adjusted over a range of around 1–50 Hz, and the intensity to be adjusted up to 100 dB or louder.

3.2.1.3 Visual Evoked Potential

Sequentially inverting black-and-white checkerboards presented on a regular computer screen is currently the most popular method of visual stimulation. A reasonable value for the total field of view of the stimulation pattern, in terms of degrees of visual angle, is in the range of $15° \times 12°$, with about eight-squared checks per row. The pattern should be presented at the largest possible contrast ratio and with good luminance. These parameters must be carefully controlled and kept constant because the VEP latency and amplitude vary with these physical features. Normative VEP parameter values are useful only if the the luminance and contrast values are specified. Typical values for the luminance and contrast ratio are 35–50 cd/m^2 and 50–95%, respectively (Altenmüller et al., 1989; Celesia and Brigell, 1999), as measured by a photometer device. The checkerboard is typically inverted at a rate of about 1 Hz. Again, the stimulators provide a large range of possible frequency values as well as different settings of the checkerboard (number of checks per row and per column, half-field stimulation, etc.). For the stimulation of unconscious patients, there is a variant of this type of stimulator in which grids of light-emitting diodes (LEDs) are placed in a pair of goggles. Groups of these LEDs are switched on and off alternately. This stimulation can be applied even with the eyes closed. Yet another stimulation device involves repeatedly flashing bulbs, but is used only rarely. Usually each eye is stimulated separately in separate sessions, with the opposite eye being hidden by a shield. Some laboratories append a third session with binocular stimulation.

3.2.2 Recording Electrodes (see Figure 3.8)

Measuring the EP requires the setting up of a closed loop passing the current from its neuronal source through the various layers of tissue to the electronic amplifier, and back again through the head tissue to the neuronal source. Within this circuit, the coupling between skin surface and amplifier input plays a critical role: in contrast to the brain tissue, the skin itself usually exhibits a high resistance and therefore the electrode must be combined with a wet electrolyte in order to bridge this junction. The number of electrodes needed for EP recording is less than with EEG. Therefore, the standard method of electrode fixation is to use a special electrolytic paste that softly glues the electrode, and at the same time provides a good electrical contact. For special cases like long-term monitoring, the fixation of the electrodes is enhanced by the application of a collodium-acetone solution.

Among the metals used for electrodes are silver, gold, platinum, tin, and refined steel. Electrodes may consist of a plate covering an area of a few square millimeters, but thin needles of refined steel are also frequently used. The polarization observed with pure noble metal (see Chapter 5 for details) does not play a role here, because low-frequency components—which would be suppressed by polarizing electrodes—do not significantly contribute to the EP waveform. If EPs are to be recorded cortically (e.g., for presurgical evaluation, or even during operations) grids or strips with a fixed geometrical arrangement of a large

EP recording electrodes

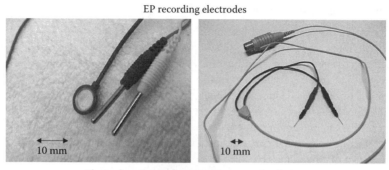

Electrode pair used for somatosensory stimulation

FIGURE 3.8 Examples of electrodes used for EP recording and stimulation. (Upper) Recording electrodes. (Lower) Pair of electrodes arranged in a fixed frame to apply electrical stimuli for somatosensory stimulation.

number of electrodes are available that allow for a multichannel recording to achieve a high spatial resolution. However, this topic is beyond the scope of this chapter. Recently, *graphite* electrodes have been introduced which have special advantages for EEG recordings during magnetic resonance tomography (MRI).

The electrolyte–electrode unit is, physically speaking, an electrochemical element similar to a battery, generating a DC voltage that is uniquely defined by the sort of metal used for the electrode. This voltage is several magnitudes larger than the EEG and EP voltage. Given identical electrode types for all recording sites, the voltage cancels out if differential amplifiers are used. Cancellation of this voltage is a prerequisite for being able to amplify the EEG by a factor of approximately 10,000 (or even more in some available devices) without blocking the amplifiers. If different electrode metals are used simultaneously, resulting in different electrode potentials, amplifier overload may result in clipped signals. Mixing of electrode types should therefore be avoided.

For a good quality recording, the impedance of the electrode–skin junction should be kept below 5 kΩ. In contrast to the EEG, this value can well be achieved under almost any conditions, given the small number of electrodes needed for EP recording. Abrading the skin may help to reach this value, as well as slightly moving the electrode. Keeping impedances as low as possible keeps electrical noise low and reduces the risk of picking up the 50- or 60-Hz line (mains) frequency. The measurement of electrode impedance is usually a special function included in the EP recorder, using the same measurement method as in EEG recorders. Details can be found in Chapter 5.

3.2.3 Amplifier

Standard low-noise operational amplifiers with high input impedance (in most cases >10 MΩ) are used to amplify the voltage differences between pairs of electrodes (i.e., differential amplifiers are used). The amplifier is split into two modules, with the first stage providing a modest gain of about 10. Before entering the second stage, the signals pass a coupling capacitor that removes potential residual high-voltage DC potentials that might occur if electrode potentials are not perfectly equal over the electrodes involved (e.g., due to unclean surfaces). The overall gain in most EP systems is at least 10,000 yielding a signal amplitude of a few volts at the amplifier's output. Due to the DC blocking capacitor between its two modules, the amplifier has a high-pass characteristic with a low-frequency cutoff that is defined by the capacitor. If several different capacitors are provided, a switch allows several cutoff frequencies to be selected. Traditionally, as with EEG recorders, the time constant of this circuit is specified in addition to the lower cutoff frequency. Typical values are 1, 0.3, 0.1, and 0.03 s, corresponding to 0.16, 0.5, 1.6, and 5 Hz, respectively. Modern digitized EP recorders provide a larger flexibility than older ones with respect to these settings. This flexibility is accomplished by providing just one high-capacity coupling capacitor in the amplifier and defining the final cut-off frequency using a digital filter (see below). The same approach is usually used to filter out higher frequencies, according to the demands of the various EP modalities: The amplifier transmits the signal with an upper cut-off frequency meeting the demands of all EP modalities. Then the signal is converted from the analog to the digital domain, with the final bandwidth being defined afterwards using digital filters.

Owing to the large-amplitude range covered by the various EP modalities, as well as the superimposed EEG, it is desirable to have different amplifier gain factors (also known as sensitivity, specified in terms of microvolts per unit). The average EEG amplitudes observed during BAEP recording are well below the values observed during VEP recording, due to the fact that the background EEG is almost perfectly suppressed by the high-pass filter applied during BAEP acquisition. On the other hand, BAEP average waveforms exhibit much smaller amplitudes compared to the VEP. Variable amplifier gain therefore helps to adjust the actual signal amplitude range to the input range of the analog-to-digital converter (ADC; see below), thereby keeping digitization noise low. Another solution to this problem, which is easier to realize with modern digital components, is to extend the ADC input range by increasing the number of bits per sample, thereby extending the input range without increasing digitization noise.

Besides providing adequate signal transmission, the amplifiers must be designed to match the safety demands of the relevant *IEC 601* norm. The main goal is to rule out the possibility that the current flowing from the amplifier backward through the tissue exceeds 100 μV, even in the case of a technical failure of the electronics. This can be achieved by placing appropriate resistors between the electrode cable and the amplifier input pins. In addition, modern EP amplifiers usually use optical transmitters to provide full electrical insulation between the front-end preamplifier and subsequent electronics, in order to prevent high voltages entering into the front end. This isolation is known as floating input, because there is no stable relationship between the absolute signal potentials within the preamplifier and the ground potential of the subsequent amplification stages.

One important characteristic of the differential amplifier is its *common mode rejection* (CMR) characteristic, which is defined by the ratio $g1/g2$, where $g1$ is the gain applied to the difference between the voltages at the two inputs and $g2$ is the gain applied to the voltage that is common to the two inputs. A large CMR is a prerequisite for efficient suppression of unwanted signal components that are common to the inputs. Modern EP amplifiers achieve a CMR of 80 dB or even better, thereby reducing common signal components by at least 1:10,000.

The intrinsic amplifier noise level significantly depends on the bandwidth. Given a setting of 1–100 Hz, as would be appropriate for VEP recording, the resulting noise will be about 0.5 μV_{eff} before amplification, whereas this level may go up to 3 μV_{eff} if the bandwidth is extended to approximately 3000 Hz, as in the case of BAEP recordings. The total noise is slightly larger because the thermal noise originating at the electrodes adds to this noise.

3.2.4 Analog-to-Digital Converter

Going from analog-to-digital signal representation (see Figure 3.9a) requires constraints to be placed on (1) the spectral bandwidth, (2) the amplitude resolution, and (3) the amplitude range.

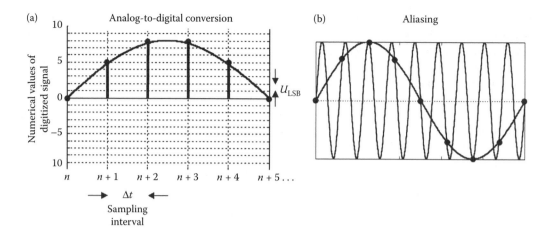

FIGURE 3.9 (a) Principle of analog-to-digital conversion. The continuous analog signal is converted into a sequence of numbers (indicated on the *y*-axis) representing the signal amplitudes at discrete points in time with a limited number of different amplitude values. The resulting temporal and amplitude resolution are labeled as Δt and U_{LSB}. The sampling rate thus is $1/\Delta t$. (b) Aliasing: A high-frequency component (thin line) mimics a low-frequency component (bold line) if sampled with a too low rate.

The sampling rate and the number of bits per sample (sometimes called *resolution*) then need to be adapted appropriately.

According to the *Shannon Nyquist theorem*, the minimum sampling rate *fs* for adequate digital representation of analog signals is 2 * *fn*, where *fn* is the highest frequency occurring in the signal. If this rule is violated (i.e., *fs* < 2 * *fn*), a component at frequency *f* > *fn* will result in a spurious frequency component of a lower frequency after digitization, with the frequency of the spurious component being given by *fn* − (*f* − *fn*). This effect is known as aliasing (see Figure 3.9b). To prevent such distortion, analog low-pass filters (antialiasing filters) with an appropriate cutoff frequency are used to suppress high frequencies before digitization. Due to the limited steepness of the roll-off characteristic of the filter, the sampling rate should be slightly higher than twice the cutoff frequency, *fc*. Given the largest EP bandwidth of 3 kHz, reasonable minimum values for *fc* and *fs* are 3 and 10 kHz, respectively. Better performance, in terms of phase linearity within the passband (i.e., below 3 kHz), can be achieved by increasing *fc* to, for instance, 6 kHz, and *fs* to 20 kHz. In technical terms, this corresponds to mild oversampling (see Figure 3.10). The advantage of this approach is that the final 3-kHz cutoff frequency can be realized using digital filters (see below). Such filters provide superior performance and better control of the overall filter characteristics, especially with respect to phase linearity, which is important because of

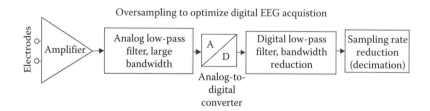

FIGURE 3.10 After passing an analog filter of large bandwidth, the signals are digitally sampled with a correspondingly high rate. A subsequent digital filter substantially reduces the bandwidth. Finally, the digital data rate is reduced accordingly. Usual reduction ratios (the *decimation factor*) are 1:4 or 1:16.

potential EP latency shifts that might arise from different phase shifts. In the case of SEPs and VEPs, the bandwidth can be adapted by changing the parameters of the digital filter accordingly.

Current commercial EP recorders provide a resolution of either 12 or 16 bits per sample. The ADC input range covered by these 4096 or 65,536 different code words must be determined by a trade-off between the amplitude range to be coded and the *least significant bit* (LSB), which determines the additional *quantization noise* generated by the ADC. The LSB is the input amplitude step (labeled as U_{LSB} in Figure 3.9a) corresponding to the numerical difference between two adjacent digital codes, and thus determines the precision of the ADC. This precision should be kept significantly below the analog noise level to prevent further degrading of the signal-to-noise ratio. In the case of EPs, the noise level as well as the amplitudes of the waveforms significantly vary with modality. This variation is best accommodated by a large ADC resolution, so that the ADC tolerates a large input range and also provides high precision, in terms of a small LSB value. A 16-bit ADC is thus superior to a 12-bit ADC. However, a 12-bit ADC may still be sufficient, if an amplifier with adaptable gain and bandwidth is provided.

3.2.5 Digital Signal Processing

Once the EEG/EP has been digitized, all basic functions of the EP recorder can be realized using numerical algorithms and data transfer operations. The main functions are: (1) subsequent *averaging* of stimulus-related epochs and (2) digital filtering.

3.2.5.1 Averaging

The stimulation unit sends a trigger signal to the recorder each time a stimulus is presented. Let us assume the set of indices $\{t_j, j = 1, 2, \ldots, nt\}$, denoting the temporal stimulation sequence. Averaging of an EP for one channel can then be formally described as

$$\mathrm{EP}_i = \frac{1}{nt} \sum_{j=1}^{nt} \mathrm{EEG}_{t_j+i}$$

where EEG_{t_j+i} is the ith sample of the t_j-th epoch and EP_i is the ith sample of the averaged EP. During the recording process, nt is continuously increased until the predefined number of epochs is reached. This process is schematically shown in Figure 3.11.

With respect to signal quality, most EP recorders automatically detect and reject gross artifacts before averaging. These simple pattern recognition algorithms mainly check the signal for unusually large amplitudes or amplitude gradients, so that minor artifacts may remain undetected. However, due to the averaging procedure, these minor artifacts will not significantly degrade the final EP waveform, with the exception of continuous small-amplitude EMG artifacts, which may significantly corrupt the EP and should be avoided.

3.2.5.2 Digital Filtering

With respect to EP recording, filtering means suppression of selected frequency ranges either toward the upper or lower frequencies (*low-pass* or *high-pass* filtering). In addition, a dedicated filter is usually provided for the selective suppression of one fixed frequency component (*notch filter*), with the aim of rejecting 50- or 60-Hz line (mains) frequency artifacts. The latter should be applied carefully in order to avoid distortions that can occur when significant signal components are affected by the notch filter. The superior performance of digital filters over analog filters—especially regarding their perfectly linear phase characteristics—as well as some typical features of filters are discussed in detail in the EEG chapter of this book. Digital filtering is usually applied continuously online immediately after digitization by the ADC. By setting the filter coefficients appropriately, the bandwidth can easily be adapted to

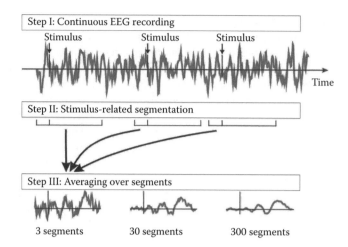

FIGURE 3.11 Principle of stimulus-related averaging. Triggered by the stimulus sequence epochs of fixed duration are extracted from the ongoing EEG. All these epochs are averaged (omitting artifacts) to obtain the low-amplitude EP embedded in the high-amplitude background EEG.

the EP modality and EEG quality. Within certain limits, filters may also be applied after averaging, for instance, in order to remove residual high-frequency noise.

3.2.6 Data Display

For routine applications, EP evaluation is still based exclusively on visual inspection. In the past, the average EP traces were written on paper strips with limited adjustability regarding the scaling of the waveform. With the advent of digital, graphical computer screens and high-resolution printers, the average waveform can be displayed with much higher quality in terms of graphic resolution. A practical example is given in Figure 3.12. The incoming raw data may also be displayed with high graphical quality and may even be separately digitally filtered for display purposes only.

Latency and amplitude measurements are greatly assisted by cursors continuously indicating the amplitude and latency values where they are positioned. Using double cursors, peak-to-peak latencies and amplitudes are also easily measured. The evaluation is further supported by automatic peak detectors. These tools work well with VEPs, but can also be used with SEPs, depending on the signal quality, but are less reliable with BAEPs. The main problem with peak detection is to distinguish real peaks from noise components. This is a serious problem, especially with BAEPs, where a certain EP component is sometimes missing (e.g., due to fusion with an adjacent component), or where the component amplitudes are close to the noise floor. During EP evaluation, it is often helpful to have another EP at hand (for instance, a previous recording of the same patient) that should be displayed simultaneously. For this purpose, most current EP systems can open additional graphic windows and display a separate EP in each window.

3.2.7 Data Storage

According to official regulations, the EP data or figures must be safely stored for 10 years or longer. The data volume per EP is in the range of a few kilobytes. Therefore, a huge number of EP datasets may be stored on current long-term mass storage devices such as optical disks or CDs. Compared to optical disks, CDs are less reliable because they are not housed in their own cartridges. The long-term use of pure magnetic media such as floppy disks and magnetic tapes is discouraged in the IFCN recommendations for digital EEG recorders (Nuwer et al., 1998), because the risk of losing data cannot be controlled

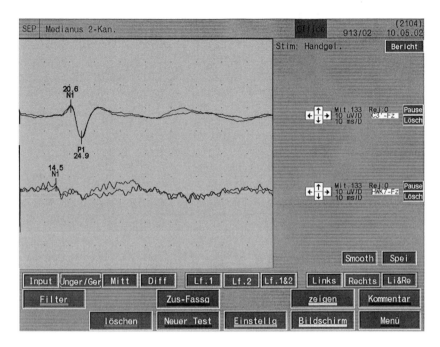

FIGURE 3.12 Screen shot taken from a recording EP device.

unless the environmental conditions of the storage room (e.g., temperature, dust, magnetic fields) are well controlled. Alternatively, only the printed EP figures may be stored. Care should be taken if a thermo-transfer printer is used for this purpose, because the resulting documents lack long-term stability.

3.2.8 Post-Processing

In routine applications, quantitative analysis techniques for EPs have not gained great importance. As mentioned before, some simple quantitative post-processing is applied in the context of long-term monitoring, for which continuous EP recordings are made over hours or days. Automatic peak detectors are helpful for documenting slowly varying latencies or amplitudes in terms of trend curves. To gain further insights into the underlying physiological processes, *source analysis* techniques (Fuchs et al., 1999; Scherg et al., 1999) are useful in order to identify the cerebral origin of certain EP components. However, this technique requires many more than two EP channels and, consequently, is mainly applied in research rather than in routine clinical applications.

3.2.9 Event-Related Potentials

Event-related potentials (ERPs) are obtained from a generalization of the EP method. ERPs reflect the specific brain electrical activity associated with cognitive tasks. The term ERP, rather than EP, is used because the relevant brain events are elicited in response to internal mental events, not just in response to external sensory stimuli (Picton et al., 2000). A simple example is the so-called *auditory oddball paradigm*, in which the subject is presented a long sequence of auditory tones of a particular frequency (e.g., 1000 Hz), which contains rarely occurring tones of a slightly deviant frequency (e.g., 1100 Hz), positioned at random within the sequence. The subject must press a button when he or she detects the deviant stimuli. If the brain potentials are selectively averaged for the frequent and rare stimuli, different waveforms result for the two stimulus classes, with a specific and reproducible difference at about

300 ms after stimulus onset. This so-called *P300* component reflects stimulus *processing* rather than mere perception; it reflects the subject's detection of stimulus deviance in comparison to a norm, and not just the physical discrepancy in frequency between the frequent and rare tones. Thus, the basic idea with ERPs is to vary the brain's response by varying the subject's cognitive processing. The difference between the corresponding average potentials is assumed to represent a brain response specific to a particular mental process.

A huge variety of ERP experiments have been designed and performed in the past 20 years (for a review, see Münte et al., 2000). Some topics that have been addressed with this technique are visual and auditory perception and attention, music perception and processing, memory, language, and emotion. Because of their millisecond resolution, ERPs are particularly suitable for revealing and contrasting the time course of different mental processes. By means of source analysis techniques (see above), the approximate spatial location of the neural generators involved in these processes can also be identified. ERPs are usually acquired as multichannel recordings (with current recorders, typically 32 or more channels), with the electrodes distributed widely over the scalp. In amplification and recording, the lower cutoff frequency has to be reduced to about 0.01 Hz (corresponding to a time constant of a few seconds), because many ERP components are characterized by spectral components in this frequency range. In certain cases, a bandpass down to DC is used, because slow shifts in brain potentials (e.g., related to cognitive stategies that extend over substantial periods) are of specific interest. Presumably, because of the more cognitive processes involved, ERPs are less reliable (i.e., less consistently evoked) than EPs, and thus more subject to noise problems. For a review of recording and publication standards for ERPs, see Picton et al. (2000).

3.3 Detection of Malfunctions

EPs are the method of choice to test the integrity of sensory pathways and, in the case of SEPs and VEPs, to test the integrity of cortical sensory processing. As a general principle, axonal lesions as well as demyelinization can be detected and distinguished using EPs. An axonal loss leads to an amplitude reduction of certain peaks, whereas demyelinization primarily causes latency shifts, and sometimes extended peak widths and amplitude reductions. In addition, damage to inhibitory or excitatory afferent fibers acting on the sensory nerve may indirectly affect the EP, leading to a slightly degraded morphology. Regarding the differentiation of normal and pathological EPs, the question of the appropriate normal reference EP arises. In the case of longer observation periods (i.e., long-term monitoring), the EP history of the patient may serve as a reference in certain situations. Another solution may be to use the EP on the side opposite to the suspect EP, evoked by the opposite-side stimulation. Sometimes, however, predetermined normal values are needed for a valid evaluation. Although a large number of articles have published possible normative values, the most appropriate way of establishing normative values is to do so in the local laboratory, because the precise values may depend on a wide variety of factors specific to the laboratory. These factors include the type of electrode, the lower and upper cutoff frequencies, and the kind of artifact rejection employed.

The remaining paragraphs describe some typical diagnostic applications of the EP method. For further reading see, for instance, Cracco and Bodis-Wollner (1986).

3.3.1 Somatosensory Evoked Potential

Here, we sketch some malfunctions that can be observed with SEPs derived from stimulation of the median or tibial nerve. In the case of peripheral axonal lesions, the early components observed at cervical electrode sites are depressed or even completely missing. Depending on the severity of the lesion, later cortical peaks may also be depressed or missing. Cervical gray matter lesions usually lead to an amplitude reduction of the early component (N13) of the median nerve SEP, whereas the corresponding cortical peaks are still normal. A spinal cord injury syndrome can be differentially diagnosed by means of tibial SEP. In the case of a complete section of the spinal cord, the early cervical as well as the cortical peaks are completely missing, whereas an incomplete section causes an amplitude reduction and/or latency shift of these components. Pathological brainstem processes (for instance, vascular lesions) leave the early component (N13 in the tibial SEP) intact, but may lead to a delayed peak latency and/or an amplitude reduction. Lesions of higher neural structures usually cause similar effects. If the somatosensory cortex is affected, the corresponding component (N19 in the median and P37 in the tibial SEP) is depressed or even missing. Median or tibial nerve SEPs can also serve as an indicator in brain-death diagnosis. A missing main cortical component (N19 in the median and P37 in the tibial SEP) on both sides, both observed with contralateral stimulation, provides evidence for brain death, given further support from additional clinical signs.

3.3.2 Brainstem Auditory Evoked Potential

Objective hearing tests rely primarily on BAEPs, because various pathogenetic mechanisms can be identified from different specific variations of some or all of the BAEP peaks. Depending on the specific peaks affected, the BAEP can especially distinguish between a malfunction of the cochlea or the primary acoustic nerve, both affecting the early components, and a brainstem lesion, leading to latency shifts or amplitude reductions of later components. Impaired hearing associated with *acusticus neurinoma* is usually characterized by delayed peaks and/or reduced amplitudes of later components. Therefore, BAEPs are valuable indicators for specific function monitoring during surgical therapy of this tumor.

3.3.3 Visual Evoked Potential

Various reasons for impaired vision can be discriminated by means of the VEP. Beyond variations of the VEP components (amplitude reduction or latency shift of the N75, P100, or N145), it is also important for diagnosing whether one or both eyes are involved. On the other hand, the VEP may be normal in the case of cortical blindness and may thus help to identify this pathomechanism. VEPs are also a valuable indicator for multiple sclerosis (MS) diagnosis, because the optic nerve already suffers from demyelinization in the early stage of the disease. Consequently, a delayed latency of P100 is consistent with MS. VEPs are also sensitive with respect to a maculo-papillar degeneration; the P100 latency is then specifically shifted to around 135 ms.

References

Altenmüller, E., Diener, H.C., and Dichgans, J., Visuell evozierte Potentiale (VEP), in *Evozierte Potentiale*, 2nd ed., Stöhr, M., Dichgans, J., Diener, H.C., and Buettner, U.W., Eds., Springer, Berlin, 1989.

Celesia, G.G. and Brigell, M.G., Recommendation standards for pattern electroretinograms and visual evoked potentials, in *Recommendations for the Practice of Clinical Neurophysiology: Guidelines of the International Federation of Clinical Neurophysiology*, Deuschl, G. and Eisen, A., Eds., Suppl. 52 to *Electroencephalogr. Clin. Neurophysiol.*, Elsevier, Amsterdam, 1999.

Cracco, R.Q. and Bodis-Wollner, I., Eds., *Evoked Potentials*, Alan R. Liss, New York, 1986.

Deuschl, G. and Eisen, A., Eds., Recommendations for the practice of clinical neurophysiology: Guidelines of the International Federation of Clinical Neurophysiology. Supplement 52 to *Electroencephalogr. Clin. Neurophysiol.*, Elsevier, Amsterdam, 1999.

Epstein, C.M. and Boor, D.R., Principles of signal analysis and averaging. *Neurol. Clin.*, 6(4), 649–656, 1988.

Fuchs, M., Wagner, M., Kohler, T., and Wischmann, H.A., Linear and nonlinear current density reconstructions, *J. Clin. Neurophysiol.*, 16(3), 267–295, 1999.

Jasper, H.H., The ten–twenty system of the International Federation, *Electroencephalogr. Clin. Neurophysiol.*, 10, 371–375, 1958.

Kandel, E.R., Schwartz, J.H., and Jessell, T.M., *Principles of Neural Science*, 4th ed., McGraw-Hill, New York, 2000.

Münte, T., Urbach, T.P., Düzel, E., and Kutas, M., Event-related brain potentials in the study of human cognition and neuropsychology, in *Handbook of Neuropsychology*, Vol. 1, Boller, F. and Grafman, J., Eds., Elsevier, Amsterdam, 2000, pp. 139–236.

Nuwer, M.R., Comi, G., Emerson, R., Fuglsang-Frederiksen, A., Guerit, J.M., Hinrichs, H., Ikeda, A., Luccas, F.J.C., and Rappelsberger, P., IFCN standards for digital recording of clinical EEG, *Electroencephalogr. Clin. Neurophysiol*, 106, 259–261, 1998.

Picton, T.W., Bentin, S., Berg, P., Donchin, E., Hillyard, S.A., Johnson, R., Jr., Miller, G.A., Ritter, W., Ruchkin, D.S., Rugg, M.D., and Taylor, M.J., Guidelines for using event-related potentials to study cognition: Recording standards and publication criteria, *Psychophysiology*, 37, 127–152, 2000.

Scherg, M., Bast, T., and Berg, P., Multiple source analysis of interictal spikes: Goals, requirements, and clinical value, *J. Clin. Neurophysiol.*, 16(3), 214–224, 1999.

Additional Recommendations and Standards

A comprehensive set of further recommendations is available via internet from the *International Federation of Clinical Neurophysiology* (IFCN) home page at http://www.ifcn.info/.

International Electrotechnical Commission, Geneva, Switzerland, IEC 601 standard "Medical electrical equipment", Part 2–26: Particular requirements for the safety of electroencephalographs (IEC 601-2-26), 1994.

International Electrotechnical Commission, Geneva, Switzerland, IEC 601 standard "Medical electrical equipment," Part 2-40: Particular requirements for the safety of electromyographs and evoked response equipment (IEC 60601-2-40), 1998.

Noachtar, S., Binnie, C., Ebersole, J., Maugière, F., Sakamoto, A., and Westmoreland, B., A glossary of terms most commonly used by clinical electroencephalographers and proposal for the report form for the EEG findings, in *Recommendations for the Practice of Clinical Neurophysiology: Guidelines of the International Federation of Clinical Neurophysiology*, Deuschl, G. and Eisen, A., Eds., Suppl. 52 to *Electroencephalogr. Clin. Neurophysiol.*, Elsevier, Amsterdam, 1999.

4

Electroencephalography

Hermann Hinrichs
University of Magdeburg

4.1 Basic Operational Mechanisms

The electrical potentials generated by the brain's neural activity can be observed at the scalp using appropriate amplification techniques. The measured signal is called the electroencephalogram (EEG). It reflects *global brain function*, rather than brain function related to the performance of specific cognitive tasks. Within the framework of everyday neurological examination, therefore, the EEG serves to provide initial information about global brain condition. For clinical examination purposes, the EEG is recorded over a period of approximately 15–20 min, with the patient sitting relaxed in a comfortable chair, keeping his or her eyes closed as illustrated in Figures 4.1 and 4.2.

The activity thus measured is called the *background EEG*, to distinguish it from activity measured while the patient is engaged in processing of specific external or internal stimuli. Advanced applications of EEG recording such as sleep staging, brain function monitoring in the intensive care unit, and so forth can lead to recording periods of several hours or even days. Additionally, with certain special patient groups, the brain's electrical signal may even be recorded from the surface of the cortex or from inside the brain, but discussion of such specialized recordings is beyond the scope of this chapter.

To serve as an index of the function of a spatially distributed neural network, the EEG must be recorded from multiple measurement positions distributed over the scalp, resulting in a number of different measured signals. For routine applications, the measurement positions are arranged according to an international standard called the 10–20 system (Jasper, 1958; see Figure 4.3). This standardized system facilitates the comparison and interpretation of EEG from different recording sessions and/or patients. The 10–20 system comprises 19 scalp electrodes and 2 ear electrodes. Accordingly, modern EEG devices record at least 21 different EEG signals, referred to as *channels*. For advanced applications, systems with 32 channels or even more (up to 512) are on the market. The American Electroencephalographic Society

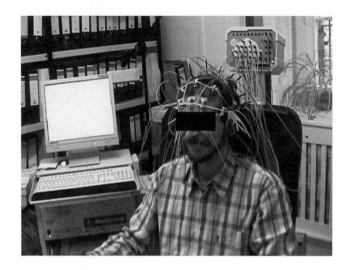

FIGURE 4.1   Recording an EEG.

(1994) has published guidelines for the nomenclature of electrode locations that goes beyond the 10–20 system of Jasper.

Figure 4.4 shows some typical normal background EEG waveforms. The fluctuations in the waveform typically cover an amplitude range of 50 µV, peak to peak. In certain cases, however, voltage fluctuations may exceed 200 µV. While the EEG contains random noise, it also contains systematic frequency components, with the main component, in most cases, being a frequency of about 10 Hz (the so-called *Alpha rhythm*). For clinical applications, EEG frequency components ranging from 0.5 to 30 Hz are typically of most interest. To achieve proper signal representation and interpretation, therefore, a recording system bandwidth of 0.5 to about 70 Hz is required. The recording of the tiny scalp voltages requires both a low-noise large-gain amplifier and a stable, low-resistance electrical contact between the amplifier and the scalp tissue. The latter can be attained using appropriate electrodes either glued or mechanically fixed on the skin. With modern components and electronics, the necessary amplifier specifications can be reached even with restrictive safety limits protecting the patient from potential risks arising from failure of the electronics or the system operator.

The voltages observed at the electrodes are actually a combination of the EEG and potentials induced by the ambient environment. Among these external potentials are the 50- or 60-Hz line (mains) frequency and electrostatic charging of the patient's body. In addition, static electrode potentials also

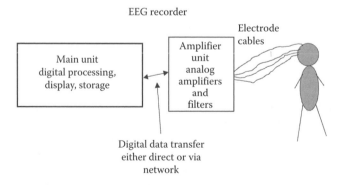

FIGURE 4.2   Basic components of an EEG device.

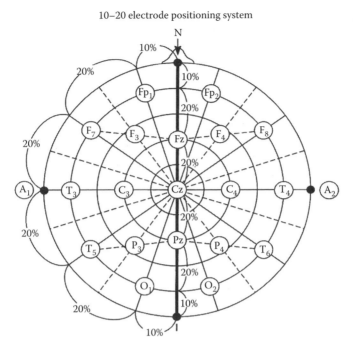

FIGURE 4.3 Schematic diagram of the international 10–20 system for electrode placement and labeling. The term "10–20" refers to the relative distances between adjacent electrodes as indicated in the figure. (Adapted from Jasper, H.H., *Electroencephalogr. Clin. Neurophysiol.*, 10, 371–375, 1958.)

contribute to the total voltage picked up at the amplifier input. The amplitude of these external potentials usually exceeds the EEG by several magnitudes. Therefore, *differential amplifiers* are employed that magnify only the differences between EEG signals picked up at pairs of electrodes, and not the absolute amplitudes (see Figure 4.5).

The external potentials therefore tend to cancel each other out because they are almost identical at all electrode sites. There are no electrically neutral points on the surface of the head, or anywhere else on the body, so there is no physiologically canonical way of combining the various electrodes into pairs. Instead, the EEG has been examined using a number of different *electrode montages*, meaning that the EEG can be recorded with different types of electrode pairings. Each type of pairing yields different measured EEG signals and thus a different view of EEG topography.

Although the set of montages used within a particular clinical EEG laboratory is typically standardized, there are still no standard montages with international acceptance across laboratories. Traditionally, there are two groups of montages that are called *bipolar* and *unipolar*. With bipolar montages, arbitrary pairs of electrodes are formed for the difference amplification process. With unipolar montages, all electrodes in a particular hemisphere are paired with one common electrode (which is usually located at the mastoid or at the ear lap) per hemisphere. The *referential* scheme is a variant of the unipolar montage in which all electrodes, regardless of hemisphere, share one common reference electrode. During recording, the reference electrode is often placed on the mastoid or earlobe. Given the use of modern digital EEG devices, however, the recorded voltage differences may be recalculated after the recording with respect to a virtual reference electrode, derived by averaging across all electrodes (a so-called *common average reference*). EEG recordings based on a referential montage (often called *reference recordings*) are preferable for this reason, because any other montage can be derived from the signals by mere recalculation.

Regarding the electrical contact between the skin and the amplifier input, the impedance should be (1) negligible with respect to the amplifier input resistance and (2) almost constant over the relevant

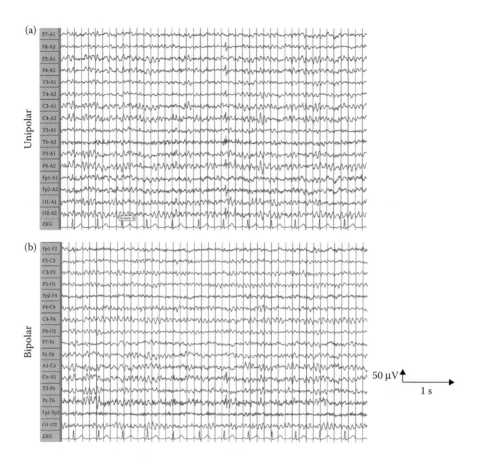

FIGURE 4.4 Example of a normal human multichannel EEG. (a) Display according to a *unipolar* recording, that is, all signals refer to the electrical potential at either the left or right ear electrode (A1 and A2, respectively). (b) The same EEG-epoch after re-referencing according to a *bipolar* montage (according to the electrode labels seen at the left margin).

frequency range, especially toward low frequencies that approach direct current (DC). To meet these requirements, electrically conducting electrodes (usually made of metal) are fixed on the skin by various techniques. An electrolyte typically consisting of paste or gel provides a low resistance between the electrode surface and the skin. As an inevitable side effect, this electrochemical system acts like an electrical battery generating a constant voltage of its own, the size of which depends on the electrode

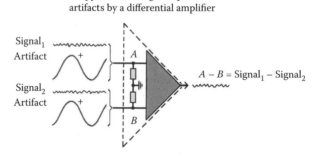

FIGURE 4.5 Principle of differential amplifier as used in EEG recorders to remove superimposed common high-voltage activity form pairs of EEG signals.

material. The difference in this constant voltage between different electrode types can be in the range of volts. Therefore, only identical electrodes must be used for EEG recording to prevent the differential amplifiers from being overloaded when fed with voltages that vary widely between pairs of electrodes.

Further external signal components are often superimposed on the EEG, corrupting the measured signal. There are both biological and technical sources of these so-called *artifacts*. Some examples are shown in Figure 4.6. Among the biological artifacts are residual electrocardiogram (ECG), electromyogram (EMG) generated by muscles in the vicinity of the electrodes, electro-oculogram (EOG) resulting from eye movement, and slowly fluctuating electrode voltages induced by transpiration, which influences the electrochemical conditions of the electrodes. Technical artifacts have a variety of causes, including inductive nonsymmetric coupling of the 50/60-Hz line (mains) frequency, which can occur when electrode impedances are too large, movement of the electrodes causing fluctuating electrode voltages, distortions of the ambient electrostatic field due to movement of the patient or the technician, unstable electrical contact between electrode and amplifier input, and bad electrode material due, for instance, to insufficient cleaning of the electrodes. Despite using differential amplifiers, these artifacts often significantly exceed the amplitude of the ongoing EEG and can sometimes completely hide it. The 50/60-Hz artifacts and the EMG can be particularly problematic in this respect. EEG recording systems are usually designed to record moderately distorted EEG without clipping, although the signal display may still exhibit clipping, depending on the graphical scaling used to display the recorded voltages.

If artifacts cannot be avoided despite taking measures such as electrode replacement, the interpretability of the EEG can sometimes be improved if low- or high-frequency components of the recorded waveform are filtered out, thereby reducing the impact of the artifact. However, unusual low- or high-frequency EEG activity may be indicative of pathology, and so it may be advantageous instead to extend the frequency band compared to the routine setting. Whichever the case, the bandwidth of EEG

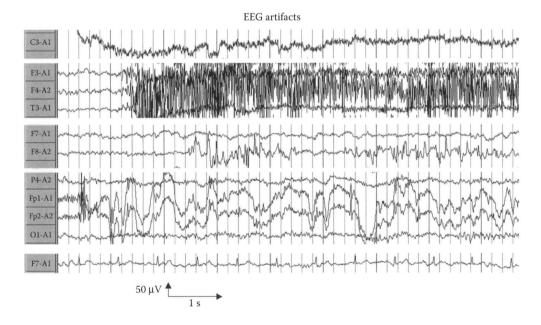

FIGURE 4.6 Examples of different types of artifacts (shown only selected channels per artifact): line (mains) artifact, that is, 50-Hz activity; electromyogram (EMG) activity generated by scalp muscles if the subject is not sufficiently relaxed; movement of electrodes on the skin; eye movements: the moving eye—as an electrical dipole—generates slowly varying electrical potentials which are picked up by the EEG electrodes; electrocardiographic (ECG) activity embedded in the ongoing EEG.

recorders can be interactively adjusted by modifying both the lower cutoff frequency f_1 (traditionally specified in terms of the corresponding *time constant*, defined as $1/2f_1$) and the upper cutoff frequency. Also, if line (mains) frequency artifacts cannot be removed by improving the electrode impedance, an appropriate notch filter helps to selectively suppress this component. From a technical point of view, adjustment of these filter specifications may be done by switching the corresponding amplifier settings or by modifying parameters of the internal digital post-processing algorithms. With modern EEG systems, digital solutions are usually preferred.

Traditionally, EEG signals were written continuously in real-time onto continuous z-folded paper strips. Since the advent of digital EEG systems, the signals are displayed on high-resolution graphic displays, both during recording and afterward for detailed evaluation. Hardcopies are only written for selected representative EEG segments as part of the physician's report. The full dataset is stored digitally on modern mass storage devices, which serve as stable long-term EEG archives.

For routine applications, the EEG evaluation is still based on a visual inspection of the signal traces, taking into account the spatial and temporal distribution of both the spectral structure and the amplitudes of the multichannel signal. The resulting report is usually a mixture of quantitative measures (for instance, the topographical area amplitude and frequency of dominant EEG activity, or the relative distribution of EEG activity over several frequency bands) and qualitative statements (for instance, the occurrence of specific signal patterns, ratings of the signals with respect to possible pathology, and comparisons to previous recordings). There are now many post-processing tools available (for instance, for spectral power analysis, pattern detection, and correlation analysis), but with respect to routine applications none of these tools has been widely accepted so far. Nevertheless, these tools have extended the applicability of EEG recordings well beyond their original application in neurological diagnosis. Among the new applications for EEG are the analysis of the action of cerebrally active drugs and the long-term monitoring of cerebral function (see below).

Regarding the technical structure of modern EEG systems, different concepts are on the market. The system may consist of either (1) a front-end signal-preamplifier and main amplifier sending the acquired data via a dedicated cable to a dedicated computer providing the further functionality needed for data storage and evaluation (stand-alone system); or (2) the same system as just described, but with the dedicated computer connected to a central dataserver via a regular network, and with the server also being used as a *reader station* (i.e., for data display, or to allow further computers to access the data for evaluation); or (3) a front-end system sending the data via network connection directly to a server displaying and storing the data and/or distributing them to other systems linked via the same network. Moreover, EEG devices are increasingly being integrated into larger IT networks combining various biomedical techniques and at the same time organizing patient data management. Finally, there is a growing tendency to extend the functionality of EEG systems to include other neurophysiological methods like evoked potentials (EP), electromyography (EMG), or neurography.

4.1.1 Photic Stimulation

One of the main applications of EEG is the diagnosis of epilepsy. Specific spike-like EEG signals, often accompanied by a subsequent slow-wave forming a so-called *spike wave complex* indicate a risk of epileptic seizure. Therefore, these patterns carry a high level of diagnostic information. However, in many patients, the spikes occur only rarely requiring an unacceptably long recording period to pick up at least a few spikes. It is known that some of these patients are sensitive with respect to repeated flashlight exposure. Under these circumstances, a seizure may be provoked thereby improving the chances of picking up a large number of epileptic EEG patterns. Therefore, EEG recorders are usually equipped with a flashlight generator, with a flash frequency tunable over a range of approximately 0.1–30 Hz. This generator may be triggered either by the EEG recorder or by an internal clock. The flash onset is recorded and displayed simultaneously with the EEG so as to be able to correlate it with EEG variations during evaluation. Besides provoking epileptic EEG activity, the photic stimulation, if presented at an appropriate frequency, sometimes provokes synchronized EEG activity at the same frequency. This phenomenon is known as *photic driving*.

4.2 Signal Processing

In this section, the main elements of modern digital EEG recorders will be described in more detail. Older analog devices that recorded the signals directly on endless paper strips will not be discussed. An excellent overview of the traditional analog techniques was provided by Cooper et al. (1981). As shown in Figure 4.7, EEG recording may be subdivided into the following logical stages:

- Electrodes
- Analog preamplifier; safety considerations
- Analog-to-digital conversion
- Digital signal processing
- Signal display on screen and on paper
- Long-term data storage and retrieval

The remainder of this section is arranged according to this list of topics.

4.2.1 Electrical Coupling of Tissue/Skin and Electronics

Measuring the EEG requires the setting up of a closed-loop passing current from its neuronal source through the various intervening layers of tissue to the electronic amplifier and back again through the head tissue to the neuronal source. Within this circuit, the coupling between skin surface and amplifier input plays a critical role; in contrast to the brain tissue, the skin itself usually exhibits a high resistance and therefore electrodes must be used in combination with a wet electrolyte to bridge this junction.

Standard electrodes consist of a metal plate covering an area of a few square millimeters. The electrolyte (saline solution) is either contained in liquid form in a pad wrapping the electrode or it is bound in a

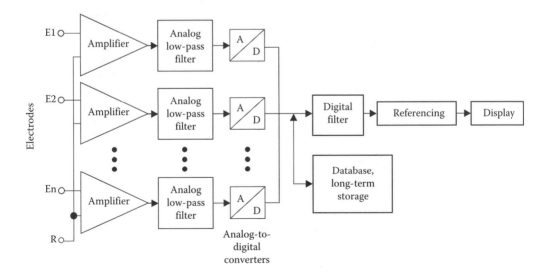

FIGURE 4.7 Technical block diagram of an EEG recorder. Rather than spending separate ADC for each channel, some manufacturers provide just one ADC scanning all channels periodically by means of an analog multiplexer located between the ADC and the low-pass filters. During recording, the incoming actual EEG is displayed on the screen. For later evaluation, arbitrary multiple sets of EEG signals can be fetched from the archive and displayed in separate windows.

gel or paste filling the gap between the electrode and the skin. Various methods of fixing the electrodes on the head surface are available: a grid of rubber strips, special caps with the electrodes being integrated into the cap textile, gluing using a collodium-acetone solution, and paste (i.e., the paste providing the electrolyte also serves as a fixation aide).

Among the metals used for electrodes are platinum, silver, gold, tin, and refined steel. Although all these materials are good conductors, they become *polarized* for biochemical reasons if a constant or only slowly varying voltage is present. This means that an inverse voltage is generated that subtracts from the real signal resulting in a decrease of the observed voltage. These electrodes are less suitable for the transmission of low-frequency EEG components (i.e., they act like high-pass filters). For this reason, silver electrodes coated with a thin silver chloride layer (Ag/AgCl electrodes) are the current standard for routine applications. The chloride layer fully prevents polarization. A special variant is made by sintering a mixed powder of Ag/AgCl onto the electrode rather than applying it as a discrete metal coating. This electrode requires less maintenance because, with the conventional Ag/AgCl electrode, the AgCl coating is easily damaged during routine use. If so, polarization occurs and the electrode needs to be recoated with a new AgCl layer. Some electrode types are shown in Figure 4.8.

Special electrodes are available for particular critical applications: for example, for EEG recordings during operations, or in the intensive care unit, if patients are unconscious, it is important to apply the electrodes quickly and stably. For these purposes, needles of refined steel are available that are pricked into the skin. In this context, the drawback of polarization is offset by ease of handling and stability of contact. For recording EEG in patients with certain types of epilepsy, *sphenoidal* electrodes consisting of either a steel, platinum, or Ag/AgCl wire are available. These are inserted through the muscle as close as possible to the anterior part of the brain's temporal lobe, permitting the recording of epileptogenic activity with greater sensitivity. Additional invasive electrode locations are often used for epilepsy diagnosis, such as the *nasopharyngeal* and *foramen* ovale electrode locations. In the context of presurgical epilepsy

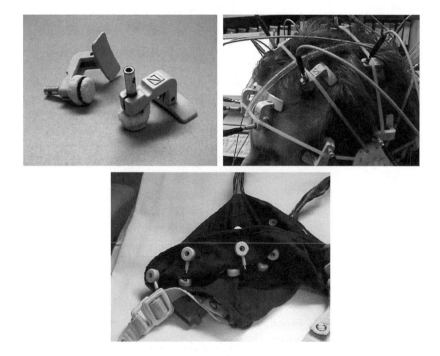

FIGURE 4.8 EEG recording electrodes. The electrodes shown in the upper row have to be applied individually using a dedicated flexible rubber frame (upper right). In contrast, electrocaps are available (lower) where all the electrodes are automatically placed at their right position once the cap is applied.

diagnosis, the recording of electrical activity at the cortex or even in deeper brain regions is sometimes required for precise localization of the epileptic focus. In these cases, recordings are made from electrodes arranged on a grid or strip, as well as multiple electrodes assembled on a single piece of wire, usually using platinum as the electrode material. Recently, graphite scalp electrodes have been introduced that have special advantages for EEG recordings made during magnetic resonance tomography (MRI).

As previously mentioned, another electrochemical effect, in addition to polarization, results from the fact that the unit composed of the electrolyte and electrode is—physically speaking—an electrochemical element like a battery, generating a DC voltage. This DC voltage is uniquely defined by the kind of electrode metal and is several orders of magnitude larger than the EEG voltage. Given identical electrode types for all recording sites, this DC voltage cancels out as long as differential amplifiers are used. Cancellation of this voltage is a prerequisite for being able to amplify the EEG by a factor of approximately 10,000 without blocking the amplifiers. However, if different electrode metals are used simultaneously (leading to different electrode potentials), the difference between the electrode potentials will not be negligible and may provoke amplifier overload, or drive the amplifier into amplitude ranges in which linearity is not guaranteed. Mixing of electrode types should therefore be avoided. Alternatively, if different electrode types cannot be avoided, the total gain may be reduced, at the cost of a reduction in signal-to-noise ratio.

For a good quality recording, the impedance of the electrode–skin junction should be kept below 5 kΩ. However, this limit is sometimes too strict for applications in which patients cannot tolerate a long preparation phase. As a rule of thumb, keeping the impedance below 20 kΩ still results in acceptable recording quality. The importance of keeping impedances as low as possible derives from the following facts:

1. The amplitude of the EEG as measured at the amplifier's input depends on the ratio of the electrode impedance and the amplifier input impedance. Osselton (1965) has shown that the relative drop of the measured EEG amplitude is $(R1 + R2)/(Rin + Rl + R2)$, where Rl and $R2$ are the impedance of the two electrodes whose voltage difference is being amplified, and Rin is the amplifier's input impedance. However, with modern EEG recorders the input impedance is usually much better than 10 MΩ, resulting in a signal loss below 1% even for electrode impedances up to 50 kΩ.

2. By the general rules of physics, a resistor R generates a thermal noise voltage Vn which is specified by the equation

$$Vn_{\text{eff}} = \sqrt{4kTBR}$$

   over the spectral bandwidth B, where k is the Boltzman constant and T the absolute temperature. With large resistances, this additional noise contributes significantly to the total measurement noise, thus degrading the signal-to-noise ratio.

3. Owing to electromagnetic and electrostatic coupling, the line (mains) frequency (50 or 60 Hz) can be coupled into the measurement setup. When the electrode impedances are large, the resulting currents generate a 50/60-Hz AC voltage at these impedances, thus causing an artifact superimposed on the EEG signal. Due to the fact that the differential amplifiers (see above) record the *difference* between two voltages picked up at two electrodes, this effect is most pronounced if the impedances of the two electrodes used for the measurement differ greatly; it does not depend so much on the absolute values of the electrode impedances.

The ability to measure electrode impedances is usually included as a special function in modern EEG recorders. For this purpose, a constant alternating current (AC; usually 10 Hz) is fed from the recording electronics to the electrodes. According to Ohm's law, the voltage observed under these conditions directly reflects the resistance. Depending on the various recorder types and suppliers, graphical or alphanumerical methods are used to display the actual resistance values for all electrodes at a glance. For practical reasons, it is advantageous to have this display at the preamplifier located close to the patient's head, rather than only at the main recording unit.

4.2.2 Amplifier

Standard low-noise operational amplifiers with high-input impedance (>10 MΩ) are used to amplify the voltage differences between pairs of electrodes. The amplifier is split into two modules, with the first stage providing a modest gain of about 10. Before entering the second stage, the signals pass a coupling capacitor that removes potential residual high-voltage DC potentials that might occur if electrode potentials are not equal over the electrodes involved (which in practice usually cannot be avoided). The overall gain in most EEG systems is on the order of 10,000–20,000, yielding an EEG amplitude of about 1 V at the amplifier's output. Due to the DC-blocking capacitor between the two modules, the amplifier has a high-pass characteristic with a low-frequency cutoff that is defined by the capacitor. Traditionally, the time constant of this circuit is specified, rather than the lower cutoff frequency. The usual values are 0.03, 0.1, 0.3 (standard), 1, and 3 s, corresponding to 5, 1.6, 0.5, 0.16, and 0.05 Hz. Short-time constants facilitate the interpretation of EEG signals when there are large superimposed low-frequency components, due either to artifacts or to pathological activity. However, if pathological activity at low frequencies needs to be evaluated with high sensitivity, for instance, for brain death diagnosis, larger time constants are required. EEG recorders therefore allow switching between different settings. On the amplifier side, this is accomplished by hardware switches that select between various capacitor values. An alternative (in many available systems, additional) approach is to change the effective time constants by modifying the coefficients of the digital filters during signal post-processing (see below).

Besides providing adequate signal transmission, the amplifiers must be designed to match the safety demands specified by the *IEC 601* standard as formulated by the International Electrotechnical Commission, Geneva, Switzerland (1994). The main goal is to rule out the possibility that the current flowing from the amplifier input through the tissue exceeds 100 μV, even in the case of a failure of the electronics. Such protection can be achieved using appropriate resistors between the electrode cable and amplifier's input pins. In addition, modern EEG amplifiers usually provide full electrical insulation (using optical transmitters) of the front-end amplifier from subsequent electronics to prevent high voltages entering into the front end. Such decoupling is known as *floating input*, because there is no stable relation between the absolute signal amplitude within the preamplifier and the ground potential of the subsequent stages.

For special applications, *DC recording* units are available that transmit the input difference signals without any frequency limitations, that is, without any DC-suppressing coupling capacitor. To avoid excessively large amplitudes that would exceed the amplifier's dynamic range, it is necessary to provide an individual DC voltage for each channel that is subtracted from the difference signal, thus compensating for the residual DC component attributable to fluctuating electrode potentials. From time to time, these recorders need to be reset interactively in order to adapt the voltage of this DC compensation signal. Alternatively, a slow voltage follower may track the fluctuating DC, adapting the compensation signal automatically. However, in a strict sense, this is no longer a pure DC recorder.

Modern EEG systems are designed as referential recorders, meaning that all electrodes are measured with respect to one common reference electrode placed somewhere on the head. Accordingly, this electrode is internally connected to the inverting input pins of all the difference amplifier channels. One important characteristic of these difference amplifiers is their *common mode rejection* (CMR) characteristics, which are defined by the ratio $g1/g2$, where $g1$ is the gain applying to the difference voltage at the two input pins and $g2$ is the gain applying to the common voltage at both input pins. A large CMR is a prerequisite for efficient suppression of noise components present at both input pins. Modern EEG amplifiers achieve a CMR of 80 dB or even better, thereby reducing common noise components by at least 1:10,000.

The intrinsic noise level of modern EEG amplifiers is about 0.5 μV_{eff} at a bandwidth of 100 Hz before amplification. Adding the noise originating at the electrodes, a total noise floor up to 0.7 μV_{eff}, corresponding to an approximate peak-to-peak level of 2–3 μV (Gaussian amplitude distribution), is realistic.

4.2.3 System Calibration

Although modern operational amplifier electronics are very stable with respect to their specifications, it is still good practice to calibrate the total system each time an EEG is recorded. For this purpose, a highly stable reference signal is provided that can be internally connected to all amplifier inputs on demand. Using a 1-Hz square wave signal (which is offered by most systems), the main features of the amplifier and all post-processing steps including filters can be checked at a glance.

4.2.4 Analog-to-Digital Converter

Conversion from analog-to-digital signal representation (see Figure 4.9a) requires constraints to be placed on (1) the spectral bandwidth, (2) the amplitude resolution, and (3) the amplitude range. The sampling rate and the number of bits/sample (sometimes called the *resolution*) then needs to be adapted appropriately. According to the *Shannon Nyquist theorem*, the minimum sampling rate fs for adequate digital representation of analog signals is $2 \times fn$, where fn is the highest frequency occurring in the signal. If this rule is violated (i.e., $fs < 2 \times fn$), a component at frequency $f > fn$ will result in a spurious frequency component of a lower frequency after digitization, with the frequency of the spurious component being given by $fn - (f - fn)$. This effect is known as *aliasing* (see Figure 4.9b). To prevent such distortion, analog low-pass filters (*antialiasing filters*) with an appropriate cutoff frequency are used to suppress high frequencies before digitization. Owing to the limited steepness of the roll-off characteristic of the filter, the sampling rate should be slightly higher than twice the cutoff frequency fc. Typical values for fc and fs are 100 and 256 Hz, respectively. This bandwidth is larger than would be necessary for the EEG alone. However, high-frequency artifacts such as electromyographic (EMG) activity sometimes can only be distinguished from EEG with the use of this larger bandwidth. Some EEG systems extend the analog bandwidth and the sampling rate even further and go down to the lower rate only

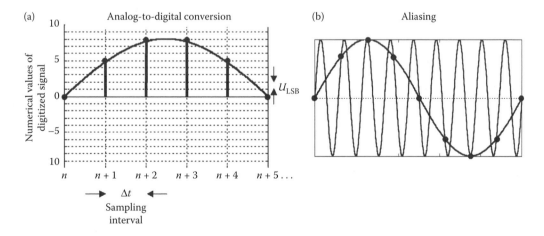

FIGURE 4.9 (a) Principle of analog-to-digital conversion. The continuous analog signal is converted into a sequence of numbers (indicated on the *y*-axis) representing the signal amplitudes at discrete points in time with a limited number of different amplitude values. The resulting temporal and amplitude resolution are labeled as Δt and U_{LSB}. The sampling rate thus is $1/\Delta t$. (b) Aliasing: A high-frequency component (dotted line) mimics a low-frequency component (straight line) if sampled with too low a rate.

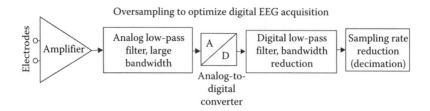

FIGURE 4.10 After passing an analog filter of large bandwidth, the signals are digitally sampled with a correspondingly high rate. A subsequent digital filter substantially reduces the bandwidth. Finally, the digital data rate is reduced accordingly. Usual reduction ratios (the *decimation factor*) are 1:4 or 1:16.

after digital filtering and subsequent down sampling (i.e., sampling rate reduction, also known as *decimation*) as illustrated in Figure 4.10. Once the signals have been digitized with a sampling rate of fs, all further digital processing occurs within the limited frequency range $0 - fn$ Hz.

Current commercial EEG recorders provide a resolution r of either 12 or 16 bits/sample. The ADC input range covered by these 4096 or 65,536 different code words must be determined by a trade-off between the amplitude range to be coded and the *least significant bit* (LSB), which determines the additional *quantization noise* generated by the ADC. The LSB is the input amplitude step (labeled as U_{LSB} in Figure 4.9b) corresponding to the numerical difference between two adjacent digital codes, and thus determines the precision of the ADC. This precision should be kept significantly below the analog noise level to prevent further degrading of the signal-to-noise ratio. Reasonable LSB values are thus well below 1 μV before amplification. The amplitude range R can then be calculated as LSB $\times 2^{r-1}$. For example, assuming an LSB of 0.5 μV and an r of 12 bits, the range is $R = 0.5 \times 4096$ μV or ± 1.024 mV. This value is sufficient for ordinary EEG signals, but may still clip large-amplitude artifacts. For instance, severe EMG artifacts occasionally exceed this limit. Also, in the case of DC recordings, this bound may be exceeded in unfavorable situations. Two ways of extending the ADC amplitude have been implemented in commercial systems: either the larger 16-bit resolution is used, extending the range by a factor of 16, or the ADC input range is arbitrarily shifted toward larger values, at the cost of precision/digitization noise. The second solution is only acceptable if low-amplitude signal components are of little interest.

4.2.5 Digital Signal Processing

Once the EEG has been digitized, all the basic functions of the EEG recorder can be realized by using numerical algorithms and data transfer operations executed either by the main processor or by dedicated signal processors. The main functions are: (1) digital filtering and (2) calculation of virtual EEG signals corresponding to various electrode montages. Filtering is sometimes necessary to facilitate visual EEG evaluation in cases where abnormal pathological or artifactual signal components at certain frequency bands overlay (and thereby hide) the background activity under consideration.

4.2.5.1 Digital Filtering

With respect to EEG recording, filtering aims at the suppression of selected frequency ranges, either toward the upper frequencies (*low-pass* filtering) or the lower frequencies (*high-pass* filtering). In addition, a dedicated filter is usually provided for the selective suppression of one fixed frequency component (*notch filter*) with the goal of rejecting line frequency artifacts (50 or 60 Hz). Ideally, a filter transmits frequency components within its *passband* with a gain of 1 (i.e., perfect transmission) and frequencies within its *stopband* with a gain of 0 (i.e., perfect suppression). However, for various reasons, real-world filters do not achieve this perfect performance. Instead, the characteristics of real-world

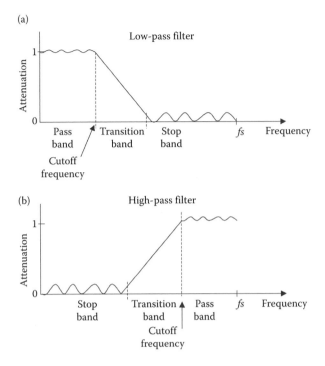

FIGURE 4.11 Schematic frequency responses of digital low-pass (a) and high-pass (b) filters. The Nyquist frequency is indicated by *fs*. The frequency characteristic repeats periodically toward higher frequencies because the corresponding signals cannot be distinguished from low-frequency components due to aliasing (see above).

filter algorithms can be described as follows (see Figure 4.11). Within the passband, the gain stays in a range $1 \pm \varepsilon$, where ε is known as the *passband ripple*, whereas throughout the stopband the gain is below $\delta \ll 1$, where δ is known as the *stopband ripple*. The frequency range between the passband and the stopband is named *transition band*, within which the gain monotonically decreases from 1 to 0. The steepness of this decreasing characteristic is specified by the *roll-off* value, usually given in terms of the logarithmic attenuation (decibel, dB) per octave. The frequencies defining the edges of the passband and stopband are also called *cutoff frequencies*.

Compared to analog filters, digital filters are advantageous for several reasons: (1) they can easily be designed to work without phase distortions (i.e., they can have a perfect linear phase characteristic); (2) changing the cutoff frequencies is accomplished by simple modification of the numerical parameters rather than changing hardware components; and (3) the filter characteristics are strictly identical for all channels, whereas different analog filters differ with respect to their exact specifications, due to variability in electronic components. Moreover, digital filters are cheaper because the only investment needed is the design, which is easily accomplished using standard tools, and only has to be done once when developing the EEG recorder. Digital filtering may be applied in real time, that is, while recording the signals, as well as later during data analysis and evaluation.

The general algorithm of a linear phase digital filter calculates the filtered signal samples as a weighted average of a limited number of subsequent samples of the input signal. The type of filter is defined by the sequence of weighting factors. For instance, if all the factors are all more or less similar, the output signal will come close to a moving average of the input signal. This filter thus resembles a low-pass filter. Alternatively, if the successive weighting factors have alternating signs, the filter will behave like a difference operator (which suppresses slowly varying signal components) and thus act like a high-pass filter.

The formal definition is as follows:

$$y_i = \sum_{j=0}^{p-1} a_j * x_{i-j}$$

where $a_j = a_{p-j}$, the filter output being designated by y_i, the filter coefficients by a_j (i.e., the weighting factors mentioned before), the filter order by p, and input data by x_i. The subscript i indicates the time step (i.e., the recording time between samples is $i * \Delta t$, with Δt specifying the sampling interval $1/fs$). From a numerical point of view, the various filter types differ only in the filter coefficients a_j and the filter order p. The latter mainly determines the computational effort needed to apply the filter. Large filter orders are necessary to realize filters with very small passband and stopband ripples, a narrow transition band and other rigid features. Therefore, the filter characteristic needs to be carefully defined in order to keep the computational demands within reasonable limits. In addition, with very narrow transition bands (resulting in steep roll-off slopes, which is optimal in terms of frequency characteristic; see previously), the filter output tends to overshoot (i.e., distortions) in the case of steep signal gradients if the order p is too low. With respect to the EEG, this aspect is especially important for notch filter design.

Although digital filters, in principle, have completely flexible adjustment of cutoff frequencies, commercial EEG recorders usually provide a few preset cutoff frequencies for the passband. Typical examples are 15, 30, and 70 Hz for the upper cutoff, and 0.053, 0.16, 0.53, and 1.6 Hz for the lower cutoff (the latter corresponding to time constants of 3, 1, 0.3, and 0.1 s, respectively).

4.2.5.2   Re-Referencing EEG Signals According to Various Electrode Montages

As previously mentioned, with older analog EEG recording devices, the electrode montage used during the recordings determined the electrode montage to be used for display and evaluation of the signals. On the contrary, with modern digital EEG recorders, the signals are stored in a form that permits subsequent reprocessing. Given that the signals have been recorded with a referential electrode montage to begin with, it is then possible to recalculate the signals to simulate any other possible electrode montage.

To illustrate how to calculate an arbitrary EEG difference signal from the raw data, let us consider how the signals resulting from a bipolar electrode montage can be computed from the signals obtained from the original referential recording. If we assume that $E1_i - R_i$ and $E2_i - R_i$ (where i denotes the sequence of time points 1, 2, etc.) are the digital representations of two referentially recorded EEG channels picked up at electrodes $E1$ and $E2$, with R denoting the reference electrode, the difference signal $E1_i - E2_i$ is easily computed as

$$E1_i - E2_i = (E1_i - R_i) - (E2_i - R_i)$$

More complicated montages comprising channels from more than just two electrodes can be derived as well by such simple numerics. Alternatively, virtual references may be generated by combining a group of electrodes. The popular *common average reference* is a special example in which the average over all recorded EEG channels is taken as the reference, thus representing a spatial average EEG. The detailed definition of this type of reference varies among EEG laboratories.

4.2.6   Data Display

For routine diagnostic applications, EEG evaluation is still based exclusively on visual inspection. In the past, the EEG traces were continuously written on continuous z-folded paper strips, with a more or less fixed temporal and amplitude scaling of 3 cm/s and 70 µV/cm, respectively, 10 s of traces recorded per page. Different scalings were available for special applications. With modern systems, the EEG is analyzed using high-resolution computer screens to display the signals. Usually, the graphical

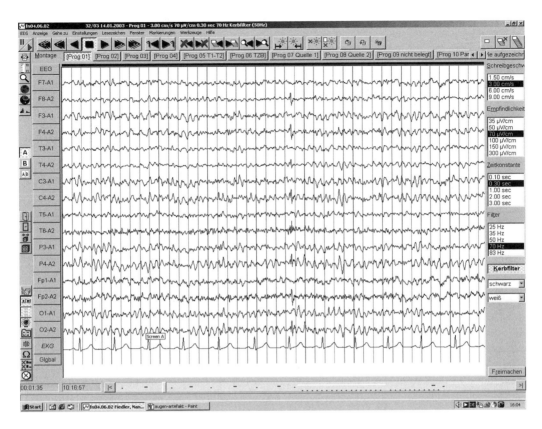

FIGURE 4.12   Screen shot taken from a recording EEG device.

resolution is better than 1024×768 pixels, thus allowing for a quality that is almost—although still not fully—comparable to the traditional paper recording. The advantages of this technique over paper recordings are obvious: signals can be displayed repeatedly at different scalings, color coding permits different signals to be distinguished, especially if the signals cross each other when there are large amplitudes, the different channels can be arbitrarily arranged on the screen, and measurement of amplitude and period duration is supported by providing interactive cursors, and so forth. In Figure 4.12, a snapshot of an EEG recorder screen is displayed.

Regarding real-time display, the operator can mark several event types (for instance, patient movement, eye movement, artifact, etc.) by pressing certain buttons or clicking the mouse. The number and meaning of different event codes can usually be defined by the user. Of course, all this marker information is stored with the raw data so as to be available during later evaluation. During EEG evaluation, it is often helpful to have another EEG at hand (for instance, a previous recording of the same patient) that can be displayed simultaneously. For this purpose, most current EEG systems can open additional graphic windows and display a separate EEG in each window. This feature may also be used to compare different sections of the same EEG during evaluation. In addition to the screen display, hardcopies may be generated from selected parts of the EEG in order to serve as traditional printed documents in an EEG evaluation report. Conventional inkjet or laser printers clearly provide sufficient resolution to print the EEG traces with high quality.

4.2.7   Amplitude Mapping

The traditional way of displaying the observed EEG is to draw the amplitude of the *n* channels over time as *n* separate traces. This method allows for an excellent temporal analysis but is less suitable for

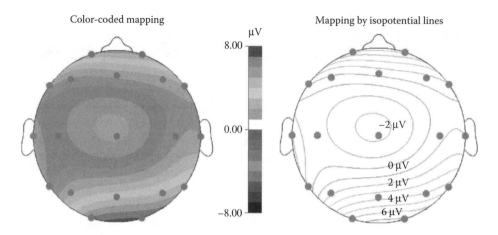

FIGURE 4.13 Display of spatial amplitude distributions applying mapping techniques. Amplitudes between electrodes (marked by dots) are estimated by means of spatial interpolation algorithms.

topographical evaluations. Therefore, a complementary method has been developed showing the amplitudes of all channels at one instant in time, or averaged across a short time interval, for the topographical region covered by the electrode set (assuming a reference montage). Figure 4.13 shows an example of such a *topographic map*. Amplitude values between electrodes are estimated from the measured values by interpolation techniques, for example, by two-dimensional spline algorithms (see Perrin et al., 1987). The amplitude values are color or grayscale coded. Alternatively, equipotential lines may be used to display the spatial amplitude distribution. A sequence of maps derived for different time points provides a concise overview of the spatiotemporal evolution of the EEG.

A crucial issue with amplitude mapping is the selection of the reference electrode. Maps derived using different reference signals (for instance, the central electrode Cz vs. a common average reference) may look completely different. Also, artifacts may lead to substantially corrupted maps, with the consequent risk of a severe misinterpretation if only the map is used for evaluation, rather than using it in conjunction with the raw underlying signals.

4.2.8 Data Storage

According to official regulations, clinical EEG data must be safely stored over 10 years or longer. The data volume per EEG session is typically about 10 MB (assuming 21 channels, a 256-Hz sampling rate, 2 bytes/sample, and a 15-min recording time). Despite the constantly growing capacity of magnetic disk storage devices, this medium is still not large enough to hold all EEG records acquired over several years. In addition, magnetic disks may "crash" and thus are not acceptable with respect to data safety. Therefore, current EEG devices are equipped with long-term storage devices. The IFCN recommendations for digital EEG recorders (Nuwer et al., 1998) encourage users to use optical disks to archive data, rather than magnetic tape media. The latter are less stable due to suboptimal mechanical and magnetic characteristics; in particular, there is a risk of erasing data due to ambient magnetic fields in the vicinity of strong electric currents. Recently, recordable CDs have become popular as storage devices. The stability of this medium may be sufficient for a 10-year period, but care must be taken to properly handle and store the CDs because they are not as well mechanically shielded as regular optical disks, which are permanently housed in individual cassettes.

As a general recommendation, the raw data should be stored on disk without any further processing. The patient information as well as the all-relevant technical parameters, plus event marks, should be kept within the same file. This guarantees that a particular patient's EEG file can be retrieved even if the patient database crashes.

4.2.9 Visual Analysis

In routine clinical EEG applications, the focus of the visual analysis is on (1) the spectral structure of the EEG, (2) the topographical distribution of the various frequency components, (3) potential hemispheric asymmetries, (4) focused abnormal activity, and (5) spikes or a complex of spikes with a subsequent slow wave, especially in the context of EEG recorded from epileptic patients. The spectral evaluation concentrates on estimating the main amplitude and frequency in four basic frequency bands: *Delta* (0.5–4 Hz), *Theta* (4–8 Hz), *Alpha* (8–13 Hz), and *Beta* (13–30 Hz). These bands are accepted with only minor variations worldwide. In most cases, the main activity of the adult EEG is observed in the Alpha band at around 10 Hz. Often one of the main problems is to discriminate artifactual activity from real EEG. Therefore, in addition to a high-quality EEG recording, it is important to get sufficient information from the technician about potential artifacts that may have been introduced during recording.

4.2.10 Quantitative Analysis

Advanced applications such as monitoring of long-term brain function, sleep staging, evaluating the effects of drugs on the EEG, and so forth are not feasible by means of visual analysis because this method does not involve statistical analysis. Moreover, visual techniques are too time-consuming and prone to error if hours of EEG need to be analyzed routinely. Therefore, a large variety of quantitative computer-based analysis techniques have been developed. A comprehensive presentation of all relevant methods exceeds the scope of this paper. Here, though, some methods are sketched that have gained some practical importance.

4.2.10.1 Spectral Analysis

This technique aims at estimating the spectral distribution of the signal power. Depending on the application, the result is just one power spectrum per channel, or a sequence of short-term spectra per channel. For further processing, the spectral data (some 100 numbers per spectrum) are usually concentrated in band-power values representing the average power and the median frequency in a small set of frequency bands. Various approaches to estimating the spectra have been proposed. Among these are (1) calculating and averaging the short-term spectra of consecutive signal segments using the fast Fourier transform (FFT); (2) parametric spectral analysis by fitting stepwise constant or time-varying linear stochastic models to the data; and (3) wavelet-based time-varying spectral analysis. A typical EEG power spectrum derived with method (1) is shown in Figure 4.14.

Spectral analysis is not restricted to the separate analysis of individual signals, but may be used to analyze the *joint* spectral power in pairs of signals, resulting in *cross-spectra*. A special normalized variant of this technique results in *coherence spectra*, which reflect the spectrally resolved correlation between pairs of signals. Coherence spectra have been used by a variety of authors to analyze functional connectivity between different brain regions.

4.2.10.2 Pattern Recognition

One of the paramount applications of the EEG is supporting the diagnosis of epilepsy. EEG spikes, and/or spikes combined with a subsequent slow wave, are highly specific in indicating a risk of epileptic seizures. In many epileptic patients, these patterns occur only very rarely or in an unusual morphology, even under photic stimulation. Hours or even days of continuous EEG recordings are sometimes required to pick up a sufficient number of spikes for a valid evaluation. Visual inspection is hardly feasible for these long-term EEG traces. To recognize these patterns, automatic procedures have been developed that detect steep slopes, sudden amplitude increases, unusual local spectral structure, and so forth. Once these patterns have been identified, a second important step is to discriminate true patterns from artifacts that frequently mimic these patterns.

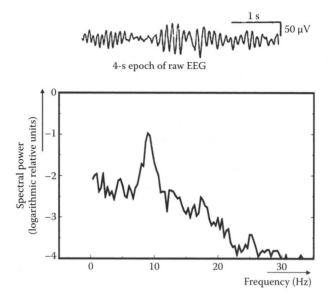

FIGURE 4.14 Example of an EEG power spectrum (representing one channel) estimated by averaging over 20 short-term spectra calculated from subsequent 4-s epochs (one example shown in the upper trace). Note the peak near 10 Hz representing the dominating 10-Hz background activity seen in most nonpathological EEG recordings.

4.2.10.3 Sleep Staging

EEG recordings are central in sleep analysis, because the main EEG activity shifts more and more toward lower frequencies with deeper sleep stages. According to rules defined by Rechtschaffen and Kales (1968) (R&K), the EEG plus some additional physiological signals (electrooculogram [EOG], EMG) are rated in 30-s steps. As a result, each 30-s period is assigned to four different sleep stages, plus the *rapid eye movement* (REM) stage. Applying this procedure, the evaluation of an 8-h sleep recording can be compressed into one comprehensive diagram showing the sequence of the 30-s sleep stages. An example of this so-called *hypnogram* is shown in Figure 4.15.

Since the introduction of the R&K rules, sleep analysis has relied on mere visual inspection, with each such analysis requiring several hours of visual inspection. Algorithms have therefore been developed for automatic analysis. The principal idea is to perform a spectral analysis and classify the low-frequency activity using a discriminant function. In addition, some coherence measures help to discriminate normal sleep activity from REM activity. Recently, neural network techniques have been successfully applied in classifying sleep activity. Nevertheless, in many laboratories, visual inspection of the signals is still the gold standard, because many automatic procedures still lack reliability, producing conclusions that differ from those of the human expert.

4.3   Detecting Malfunctions

The capacity of the EEG to provide a quick overview of global brain function is exploited for routine neurological examinations, as well as for long-term brain function monitoring during operations and in the intensive care unit. Here, some typical applications are briefly discussed. A more comprehensive description is found in Niedermeyer and Lopes da Silva (1998).

4.3.1   Detecting Focused Malfunction

In a routine neurological examination, the EEG is often used either to exclude or to prove topographically focused central functional disturbances. For this purpose, the physician looks for

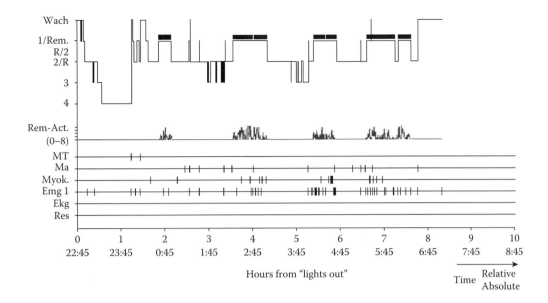

FIGURE 4.15 Example of a hypnogram (upper trace) resulting from an evaluation of a whole night recording according to the Rechtschaffen and Kales rules. Besides several channels of EEG, additional signals are included, among them the electrooculogram (EOG) and an electromyogram (EMG). Each step represents a 30-s epoch classified into one of five different sleep stages (plus wakeness). Note the almost periodic repetition of low and deep sleep stages. In the lower part of the figure, some additional results are shown like phases of rapid eye movements (REM, defining a sleep stage of its own), EMG activity, and so on.

topographically abnormal frequency distributions (especially frequencies below 8 Hz) occurring at a limited number of electrode sites, thus reflecting a limited brain area. The EEG is not only sensitive to such disturbances, but can also provide an estimate of severity of malfunction. However, conclusions as to the reason for the malfunction (which might be a tumor, an edema, a brain contusion, etc.) cannot be based on the EEG alone. A classical application for the EEG is estimation of the severity of head injuries. Also, following head injury, the course of the central nervous system function can be noninvasively monitored over hours, days, or weeks in the intensive care unit. Encephalitis is one of the rare diseases for which the EEG alone can provide strong evidence for its differential diagnosis, based on specific patterns of focused abnormalities over the temporal lobe that are almost unique to this disease.

4.3.2 Epilepsy

Two prominent issues in the context of epilepsy diagnoses are (1) to classify the type of epilepsy and (2) to identify the epileptic focus driving the pathological activity. Both issues are important for selecting an appropriate therapeutic approach. The EEG is capable of addressing both questions. In general, most epileptic patients exhibit specific EEG patterns in a restricted brain area, which can be observed in a subset of skull electrodes located near these areas. These patterns usually comprise one or several subsequent spikes (duration < 100 ms), which are sometimes followed by a slow wave (duration = a few hundred milliseconds). Several variants of this so-called *spike wave complex* are specific to particular variants of epilepsy, permitting the use of EEG for differential diagnosis.

In addition, the topographical distribution of the abnormal activity can help to localize the epileptic focus. *Source analysis* techniques may further support this localization. Given a high signal-to-noise ratio, they are able to relate the surface voltage distribution to the underlying intracerebral neural sources. One of the main problems here is to observe a sufficient number of interictal spikes within a routine EEG recording session of only 15–20 min. Therefore, long-term EEG recording techniques are available that use ambulatory devices to pick up these EEG sequences. However, in severe cases in which only surgical therapy can help, depth recordings may be required to identify the epileptic focus with sufficient precision.

4.3.3   Coma Staging

In comatose patients, both adequate therapy and prognosis depend on the coma depth and its development. The EEG is a sensitive measure for evaluating these. Again, certain morphological and topographical EEG patterns provide specific evidence of various coma stages.

4.3.4   Brain Death Diagnosis

The spontaneous EEG is one of the most sensitive measures available for brain death diagnosis. According to official German recommendations, the EEG can prove brain death if (1) there is clinical evidence for this pathological status, (2) no brain-related electrical activity can be recorded even at an increased amplifier gain (or larger-scale factor for the screen display), and (3) the low cutoff frequency is extended to 0.16 Hz and the recording time is extended to 30 min. However, artifacts obscuring the underlying brain electrical activity often preclude a valid EEG interpretation. Also, certain drug states (for instance, due to barbiturate abuse) as well as a low body temperature may simulate a brain-death EEG, and these factors must be taken into consideration before EEG recording.

4.3.5   Intraoperative Brain Function Monitoring

Intraoperative EEG monitoring mainly focuses on two fields of application: control of anesthesia and general brain function monitoring during operations in the brain. Pichlmayr and Lips (1983) and others have shown that the depth of anesthesia is significantly reflected in specific EEG patterns, which can be automatically detected and classified in real time by appropriate computer algorithms. As direct measures of brain function, in contrast to epiphenomena such as blood pressure and heart rate, these indices allow for a more sensitive and immediate control of the depth of anesthesia, thereby saving drugs and reducing side effects. In addition, the risk of unintended wake-ups is minimized.

During surgery on brain vessels (for instance, arteria carotis endarterectomy), a temporary clamping of the vessel is sometimes required. In such situations, the EEG allows one to check whether the missing oxygen supply is tolerated by the brain. Potentially impaired brain function can be detected immediately from pathological EEG signs like decreased amplitudes and/or slowing in the relevant topographical area. Monitoring of the EEG can be done visually. However, for longer-term monitoring, support from automatic computer-based procedures is advantageous. Special devices have been developed for these applications. In principle, all these algorithms analyze the time-varying spectral distribution of the EEG individually for each channel. The critical parameters extracted from this spectral analysis are plotted in terms of a trend curve, thus providing significant data reduction and noise removal. However, a human interpreter still has to make the final decision regarding the presence of pathological changes in brain activity.

References

American Electroencephalographic Society, Guideline thirteen: Guidelines for standard electrode position nomenclature, *J. Clin. Neurophysiol.*, 11, 111–113, 1994.

Cooper, R., Osselton, J.W., and Shaw, J.C., *EEG Technology*, 3rd ed., Butterworth-Heinemann, Oxford, 1981.

International Electrotechnical Commission, Geneva, Switzerland, IEC 601 standard "Medical electrical equipment," Part 2-26: Particular requirements for the safety of electroencephalographs (IEC 601-2-26), 1994.

Jasper, H.H., The ten–twenty system of the International Federation, *Electroencephalogr. Clin. Neurophysiol.*, 10, 371–375, 1958.

Niedermeyer, E. and Lopes da Silva, F., *Electroencephalography: Basic Principles, Clinical Applications, and Related Fields*, 4th ed., Williams & Wilkins, Baltimore, MD, 1998.

Nuwer, M.R., Comi, G., Emerson, R., Fuglsang-Frederiksen, A., Guerit, J.M., Hinrichs, H., Ikeda, A., Luccas, F.J.C., and Rappelsberger, P., IFCN standards for digital recording of clinical EEG, *Electroencephalogr. Clin. Neurophysiol.*, 106, 259–261, 1998.

Osselton, J.W., The influence of bipolar and unipolar connection on the net gain and discrimination of EEG amplifiers, *Am. J. EEG Technol.*, 5, 53, 1965.

Perrin, F., Pernier, J., Bertrand, O., Giard, M.H., and Echallier, J. F., Mapping of scalp potentials by surface spline interpolation, *Electroencephalogr. Clin. Neurophysiol.*, 66, 75–81, 1987.

Pichlmayr, I. and Lips, U., EEG monitoring in anesthesiology and intensive care, *Neuropsychobiology*, 10(4), 239–248, 1983.

Rechtschaffen, A. and Kales, A., *A Manual of Standardized Terminology, Techniques and Scoring System for Sleep Stages of Human Subjects*, Natl. Inst. Neurol. Dis. Blind (NIH Publ. 204), Bethesda, MD, 1968.

Additional Recommendations, Standards, and Further Reading

American Electroencephalographic Society, Guidelines for writing EEG reports, *J. Clin. Neurophysiol.*, 1, 219–222, 1984.

Chatrian, G.E., Bergamasco, B., Bricolo, A., Frost, J., and Prior, P., IFCN recommended standards for electrophysiologic monitoring in comatose and other unresponsive states, *Electroencephalogr. Clin. Neurophysiol.*, 99, 103–126, 1966.

Ebner, A., Sciarretta, C.M., Epstein, C.M., and Nuwer, M., EEG instrumentation, in *Recommendations for the Practice of Clinical Neurophysiology: Guidelines of the International Federation of Clinical Neurophysiology*, Deuschl, G. and Eisen, A., Eds., Suppl. 52 to *Electroencephalogr. Clin. Neurophysiol.*, Elsevier, Amsterdam, 1999.

Noachtar, S., Binnie, C., Ebersole, J., Maugière, R., Sakamoto, A., and Westmoreland, B., A glossary of terms most commonly used by clinical electroencephalographers and proposal for the report form for the EEG findings, in *Recommendations for the Practice of Clinical Neurophysiology: Guidelines of the International Federation of Clinical Neurophysiology*, Deuschl, G. and Eisen, A., Eds., Suppl. 52 to *Electroencephalogr. Clin. Neurophysiol.*, Elsevier, Amsterdam, 1999.

Nunez, P.L., *Electrical Fields of the Brain*, Oxford University Press, New York, 1981.

Nuwer, M.R., Quantitative EEG. I. Techniques and problems of frequency analysis and topographic mapping, *J. Clin. Neurophysiol.*, 5, 1–43, 1988.

Nuwer, M.R., The development of EEG brain mapping, *J. Clin. Neurophysiol.*, 7, 459–471, 1990.

A comprehensive set of further recommendations is available via the Internet from the *International Federation of Clinical Neurophysiology* (IFCN) home page at http://www.ifcn.info/.

5

Hearing and Audiological Assessment

Herbert Jay Gould
The University of Memphis

Daniel S. Beasley
The University of Memphis

The traditional role of audiological assessment has been twofold: to determine the effects of a hearing loss on a person's verbal communication ability and to establish the site of difficulty within the auditory system. Assessment of communication skills leads to the implementation of strategies for improving quality of life through use of assistive listening devices, hearing aids, or alternative communication modes such as sign language. Determination of the site of lesion within the auditory system assists in decisions regarding medical treatment plans for the alleviation of hearing problems. Historically, the primary tools for both of these tasks have been the pure tone audiogram and speech recognition testing.

Today, these two measures, although still important in the assessment process, have been joined by increasingly sophisticated measures designed to better pinpoint the site of lesion within the auditory system. Previously, a diagnosis of a "sensorineural" hearing loss indicated that the lesion could be located anywhere within the cochlea (sensory) through the VIIIth cranial nerve (neural). With today's advanced technology and related diagnostic techniques, audiologists can determine damage at the cellular level within the cochlea *per se*, separating inner hair cell damage from outer hair cell damage as well as identifying sites of lesion within the peripheral and central auditory nervous system.

In this chapter, an overview of the auditory system, basic acoustics, and audiologic assessment will be presented. Sources will be provided to allow the readers to expand their knowledge beyond this introductory commentary to specific areas of interest.

5.1 Structure of the Auditory System

The auditory system can be divided into four major subsystems. These subsystems include (1) an outer ear which collects and funnels the sound, (2) a middle ear which matches the air-borne sound impedance to that of the fluid-filled cochlea, (3) an inner ear which converts the energy to neural impulses, and (4) the central auditory nervous subsystems that perform complex perceptual judgments on the incoming information. Each of these subsystems contains a number of complex components that can affect an individual's hearing ability.

5.1.1 The Outer Ear

The outer ear consists of the pinna (auricle) and external auditory meatus (ear canal) as shown in Figure 5.1. The complex shape of the outer ear provides two functions.

First, the pinna serves as a collector of sound and provides resonance to the signal based on the sound's angle of incidence. This modification of the sound provides significant clues to the location of sounds in the vertical plane. The pinna then funnels the sound to the ear canal.

The canal is approximately 25 mm in length and functions as a tube closed on one end (the eardrum). This provides a resonance peaking at approximately 3500 Hz. This resonance when combined with the resonance effects of the head, torso, and pinna create an approximate 15-dB SPL [decibels re: sound pressure level (SPL)] boost in the signal level striking the tympanic membrane for frequencies in the 2000–6000-Hz region.

The second major role of the outer ear system is to provide a protective function to the middle ear system by placing the tympanic membrane (eardrum) deep inside the skull. Cerumen (earwax) helps in this protective function and moves debris outside. The small guard hairs within the ear canal point laterally and help prevent the entrance of miniscule inert and biologic matter.

5.1.2 The Middle Ear

The middle ear provides a transformer function. Sound is principally transmitted through the air to hair cells in the inner ear which are contained in a fluid bath. The difference in density between the air and fluid creates an impedance mismatch. To transmit the sound effectively to the nerve endings, the air/fluid impedance mismatch must be overcome.

The main structures of the middle ear are designed to perform this transformer function (Figure 5.2). The entry point to the middle ear is the tympanic membrane, which seals off the middle ear from the external auditory meatus. Suspended within the middle ear is the ossicular chain comprised of a group of three tiny bones: the malleus, the incus, and the stapes. The chain connects the tympanic membrane to the oval window, which is the entry point to the inner ear, or cochlea. The malleus is attached to the tympanic membrane laterally and is suspended by ligaments anteriorly and superiorly. The tensor tympani muscle projects from the cochleariform process on the medial wall of the tympanic cavity and

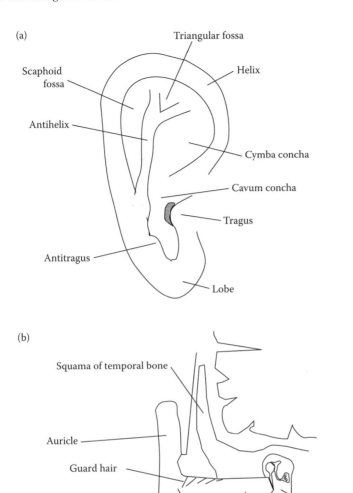

FIGURE 5.1 (a) Lateral view of the auricle; (b) coronal cut through the external and middle ear looking in an anterior-to-posterior direction. (Figure drawn by Herbert Jay Gould. With permission.)

connects to the medial aspect of the malleus. Contraction of this muscle can modify ossicular chain motion; however, this modification appears to be larger in some animals than in humans (Borg 1972).

The posterior aspect of the head of the malleus is attached to the body of the incus. The incus, through its long and lenticular processes, provides a bridge to the stapes. The stapes inserts into the oval window of the inner ear. In the head/neck region of the stapes, the stapedius muscle extends to the posterior medial wall of the middle ear at the pyramidal eminence. The stapedius muscle and the tensor tympani act to modify ossicular chain motion in the presence of loud sounds. When activated, the stapedius

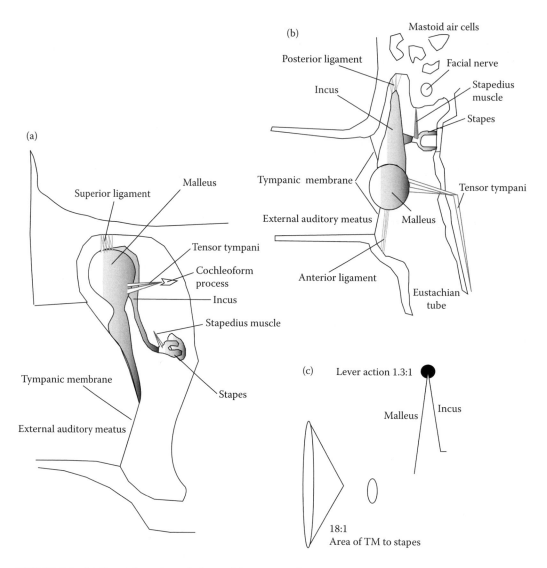

FIGURE 5.2 (a) Coronal cut through the middle ear space looking in an anterior-to-posterior direction. The Eustachian tube is not seen as it enters into the anterior wall. (b) Transverse cut through the middle ear space looking from superior to inferior. Note the axis formed through the anterior and posterior ligaments, head of malleus, and body/short process of incus. (c) The middle ear mechanism provides a boost in signal pressure through the area difference of the tympanic membrane and stapes footplate as well as the lever formed by the malleus and incus through the axis of rotation. (Figure drawn by Herbert Jay Gould. With permission.)

muscle increases the stiffness of the ossicular chain, raising the resonant frequency, and changes the axis of stapedial rotation, thereby reducing the velocity of the stapedius footplate in the oval window.

The middle ear has three components related to its function as a transforming mechanism. The area difference of the tympanic membrane to that of the stapedial footplate provides a pressure increase of approximately 18 to 1. Second, the lever action of the ossicular chain provides a 1.3 to 1 mechanical advantage. Finally, there is a minor lever action of the tympanic membrane. Overall, the pressure gain due to the function of the middle ear is approximately 26–30 dB SPL.

This gain of the middle ear is often compared with the loss associated with transmission at an air/water boundary. The loss of 30 dB makes this comparison intuitively satisfying. However, the cochlea is not an expanse of water, but a fluid in a hard-walled cavity with two relief ports, the round and oval windows. Furthermore, the entire length of the cochlea is partitioned by internal structures. Durrant and Lovrinic (1995) provide a thorough discussion of the middle ear transformer.

When considering the middle ear, it is also important to think of the sound paths to the cochlea. Although the middle ear provides a modest 20–30 dB boost to the sound pressure entering the inner ear, losses up to 60 dB are often seen in cases were the tympanic membrane and ossicular chain are disrupted. The three paths of sound to the cochlea are via the ossicular chain to the oval window, airborne directly to the round window, and finally through bone conduction. Sound enters the inner ear from different sides of the cochlear partition, resulting in necessary phase and pressure level differences between the ossicular and airborne paths. If the phase and amplitude were the same on both sides of the partition, the traveling wave deflection of the partition would be canceled and no sound would be perceived.

The arrangement of the tensor tympani and the stapedius muscle contribute to an action known as the acoustic reflex. This reflex provides a change in the middle ear impedance characteristics when the muscles are activated (Wever and Lawrence 1954). Wever and Lawrence listed four general theories for the acoustic reflex: (1) the intensity-control theory, (2) the frequency-selection theory, (3) the fixation theory, and (4) the labyrinthine pressure theory. The intensity-control theory, in a simplified form, suggests that the reflex is related to a protective function preventing damage to the ear from intense sound levels. The frequency-selection theory implies that the eardrum/middle ear mechanism frequency response is regulated by the tonus of the middle ear muscles. This theory, in its original form, is incorrect in that the muscle system does not permit fine frequency tuning in the middle ear. However, Dorman and coworkers (1987) have suggested that the broad upward frequency shift might play a role in speech perception in noise. They hypothesize that the reflex shifts the middle ear resonant frequency to slightly higher frequencies, an important element for understanding speech, while suppressing low frequencies, which are more characteristic of noise. This role is still being debated (Phillips et al. 2002). The fixation theory suggests that the middle ear muscles provide part of the supportive mechanism for the middle ear and contribute to the strength and rigidity of the ossicular chain. The final theory of labyrinthine pressure regulation has been rejected.

It is necessary for the resting middle ear pressure to be equal to that of the surrounding atmosphere for it to work optimally. If there is a pressure imbalance, the tympanic membrane will be moved from its normal rest position, resulting in a change in hearing. A passage, known as the Eustachian tube, traveling from the middle ear cavity into the nasopharyngeal cavity, regulates middle ear pressure. The Eustachian tube exits the anterior wall of the middle ear space and opens at the posterior lateral aspect of the nasopharyngeal area adjacent to the adenoid. The Eustachian tube is normally closed, opening only when the muscles raising the soft palate, at the rear of the oral cavity, are activated, such as during swallowing or yawning. The typical "stuffy ear" feeling when flying is due to the tube being unable to adjust to rapid large changes in atmospheric pressure. During childhood, as the bones of the face grow, the Eustachian tube angle changes significantly, until about 9 years of age. This change in angle results in an associated change in the vector of the muscles opening it. This, coupled with a reduction of the size of the adenoid with age, are primary factors accounting for the relatively low incidence of middle ear problems in adults, relative to that seen in early childhood.

The view that middle ear pressure regulation is solely the realm of the Eustachian tube is a simplification of a complex process that is still not fully understood. Since the *hydrops ex vacuo* theory was proposed by Politzer in the late 1800s, a significant body of work has accumulated on middle ear gas exchange (Flisberg 1967, 1970; Flisberg et al. 1963; Ingelstedt 1967; Jones 1961; Magnuson 2001; Murphy 1979; Sade 2000; Sade and Hadas 1979; Sade et al. 1976).

5.1.3 The Inner Ear

The inner ear combines the end organ for hearing and the end organ for balance. They share a common vestibular space and fluid arrangement (Figure 5.3). Due to the focus on the auditory system in this

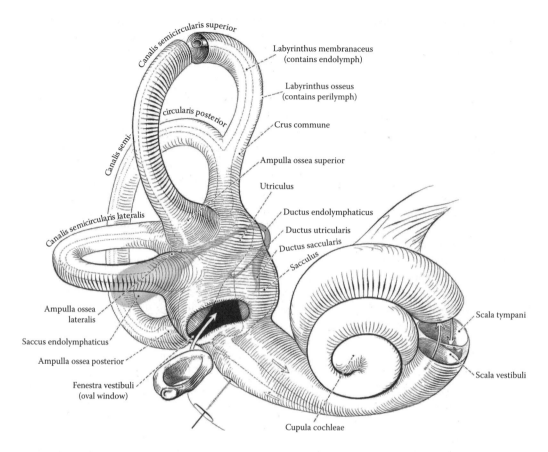

FIGURE 5.3 The stapes footplate enters the inner ear in the vestibule at the juncture of the cochlea and semicircular canals. (Drawing by Biaglio John Melloni, as shown in *Some Pathological Conditions of the Eye, Ear, and Throat*, courtesy of Abbot Laboratories, North Chicago, IL.)

chapter, a description of the vestibular system will be omitted. However, it should be noted that hearing and balance problems are often related. The reader is referred to Shepard and Telian (1996) as well as Jacobson et al. (1993) for additional information on balance disorders and assessment.

As depicted in Figure 5.3, the cochlea starts at the vestibule of the inner ear and travels as a spiral tube within the temporal bone at the base of the skull. It is divided into three fluid-filled chambers: scala vestibuli, scala media, and scala tympani (Figure 5.4). The scala vestibuli and scala tympani are joined at the apex of the spiral through an opening called the helicotrema. The scala media is formed between the other two scala by two membranous structures: Reissner's membrane and the basilar membrane. The organ of Corti lies on top of the basilar membrane and is shielded superiorly by the tectorial membrane.

The basilar membrane is critical to the function of the cochlea. The membrane varies in width and stiffness along its length and provides the initial impetus to sort sounds tonotopically (by frequency). High frequencies establish deflections at the stiff base of the basilar membrane, while lower frequencies move progressively apical (toward the apex) as the stiffness lessens. This forms the basis for the traveling wave theory of hearing which is attributed to Georg von Békésy in his text *Experiments in Hearing* (Von Békésy 1960). This theory states that signal frequency is encoded at the location of maximum deflection in the traveling wave and the signal loudness is the amplitude of that deflection. This theory was originally conceived with a passive inner ear system. However, it is now known that movement of the basilar membrane is modified by the outer hair cells in the organ of Corti (Robles and Ruggero 2001).

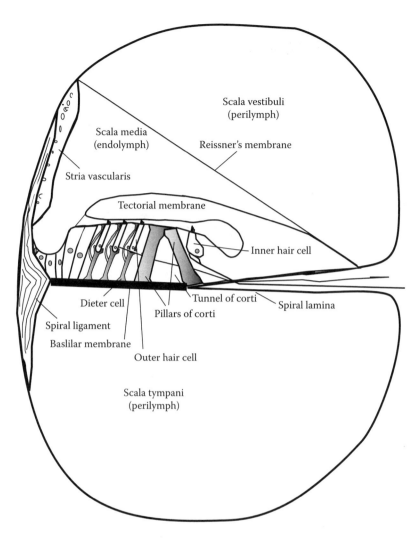

FIGURE 5.4 The cochlea is a three-chambered structure with the end organ of hearing, Organ of Corti, located in the medial chamber. The high concentration of potassium which is required for the hair cells to function is produced by the stria vascularis on the lateral wall. (Figure drawn by Herbert Jay Gould. With permission.)

The Organ of Corti contains the sensory cells for hearing, cells that provide an amplification/tuning function and supporting cells (see Figure 5.4). The sensory cells are the inner hair cells, while the outer hair cells provide amplification and tuning of incoming acoustic signals. These two cell types are located on either side of the pillars of Corti, the inner hair cells medially and the outer hair cells laterally to the pillars. The inner hair cells form a single row of closely spaced cells surrounded by supporting cells. In contrast, the outer hair cells form three rows of cells with each row separated by a space (Nuel's space). Each outer hair cell is supported by a Dieters cell. The Organ of Corti is bounded laterally by cells of Henson, Claudius, and Boettcher.

Outer hair cells are long test tube-shaped structures with cilia (hairs) located at the top. For each cell, the lateral-most cilia tips are attached to the tectorial membrane. The point of motion for the tectorial membrane is medially displaced from that for the basilar membrane on which the hair cell rides (Figure 5.5). The difference in motion between the basilar membrane and the tectorial membrane spreads the cilia bundle and opens a mechanical gate within the cilia. This gate regulates the flow of potassium into the cell. When the gate is opened, potassium enters the cell will shorten. When the

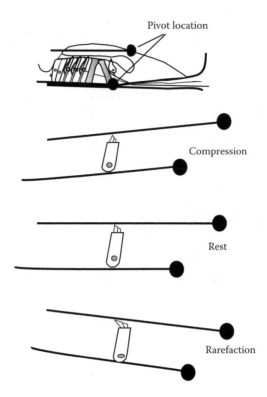

FIGURE 5.5 The basilar membrane and the tectorial membrane have different points of pivot. This difference creates a shearing action on the outer hair cells opening a mechanical gate in the cilia and allowing potassium to enter the cell. The gate is open for one-half of the cycle during rarefaction of sound pressure and is closed during the compression phase of the cycle. (Figure drawn by Herbert Jay Gould. With permission.)

gate is closed, potassium is pumped out and the cell lengthens. This expansion and contraction is due to a motor protein, prestin, located along the cell wall. The outer hair cells have been referred to as the cochlear amplifier and are essential for hearing low-intensity sounds. Damage to the outer hair cells is a major factor in many cochlear hearing problems. The active role of the outer hair cells also results in a much steeper and narrower deflection of the basilar membrane at lower intensity levels, thereby providing greater frequency resolution within the cochlea.

The inner hair cells have a gourd-like shape with a narrower apex than base. The inner hair cells are the actual sensory receptors that will initiate activity within the central nervous system in the presence of auditory stimuli. Operation of both inner and outer hair cells is dependent upon a strong concentration of potassium generated on the lateral aspect of the scala media in the stria vascularis. Potassium appears to be circulated from the stria vascularis through the hair cells and back to the stria by way of the supporting cells in the Organ of Corti. The transport mechanism is the connexin 26 protein which forms a gap junction between adjacent cells. Approximately 30% of congenital nonsyndromic deafness is now thought to be related to a failure of connexin within the inner ear, which leads to diminished potassium levels within the scala media (Kikuchi et al. 2000; Spiess et al. 2002).

5.1.4 Central Auditory Nervous System

The central auditory nervous system is comprised of both afferent (incoming) and efferent (outgoing) neuronal structures. It is probably the most complex subcortical sensory processing system in humans. The system is bilaterally represented and has multiple nuclei within the brain stem.

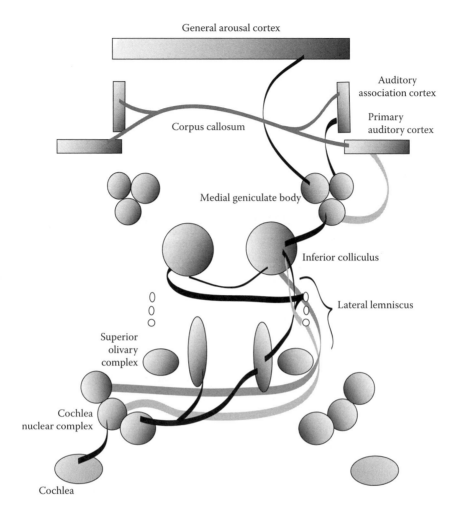

FIGURE 5.6 The ascending auditory pathway is primarily crossed. For clarity, the figure only traces one pathway from the cochlea to the cortex for low-frequency information. (Figure drawn by Herbert Jay Gould. With permission.)

The afferent system begins at the base of the inner hair cells (Figure 5.6), as the auditory portion of the VIIIth cranial nerve, and passes through a boney opening, known as the internal auditory meatus, into the cranial vault. The internal auditory meatus also serves as a passage for the vestibular (balance) portion of the VIIIth nerve, the facial nerve and a blood vessel supplying the inner ear (labyrinthine artery). Schwannomas (benign growths) are common within the internal auditory meatus, typically arising from the vestibular portion of the VIIIth cranial nerve.

The neural arrangement is different for the inner and outer hair cells. Each inner hair cell communicates with approximately 10 afferent (ascending) neurons. Each neuron, however, only communicates with a single inner air cell. This provides a significant redundancy of neural response for activation of each sensory cell. The neurons exhibit a tuning curve that is similar to that seen for the hair cell. In contrast, outer hair cells primarily receive efferent (descending) input with each neuron connecting to multiple cells over approximately ¼ turn to the cochlea.

The VIIIth cranial nerve enters the posterior lateral aspect of the brainstem and connects to the cochlear nuclear complex (CNC). In the CNC, basic timing and pitch extraction is initiated. There are several intricate neuronal feedback loops within the nuclei as well as connections from the efferent system. These connections appear to help in sound localization and noise suppression. At this point, interaction also occurs with other neural systems such as sensory input from the trigeminal nerve and from proprioceptors in the neck-and-shoulder region. This allows for coordination of head position with auditory input for sound localization.

Auditory information exits the CNC by three distinct pathways. The anterior-most pathway (anterior acoustic stria) enters another nuclear body, the superior olivary complex (SOC), bilaterally. The SOC has a major role in auditory localization. Output from the SOC is primarily to the ipsilateral (same side) lateral lemniscus. The lateral lemnisci have several nuclei and there is a shunting of some neurons across the midline at the Bundle of Probst. The intermediate and dorsal acoustic stria bypass the SOC and directly enter the lateral lemniscus, which is a major ascending pathway, primarily on the side contralateral (opposite) to the ear receiving the signal. The lemnisci enter into the inferior colliculus on the posterior aspect of the brainstem.

The inferior colliculus is a major way station in the pathway. Most neurons synapse at this point and some of the neurons cross the midline through a pathway called the anterior commissure. Several studies have suggested that processing of amplitude and frequency modulation may occur at this level (Eggermont 2001; Giraud et al. 2000; Kiren et al. 1994; Kollmeier and Koch 1994; Lorenzi et al. 1995; Muller-Preuss et al. 1994). Neurons exit the inferior colliculi through a pathway, the brachium of the inferior colliculus, and enter the medial geniculate body of the thalamus.

The medial geniculate body can be divided into three major areas: ventral, dorsal, and medial. Fibers exiting the medial geniculate body carry information to the cortical areas of the brain. The ventral medial geniculate projects to the primary auditory cortex through the internal capsule and contains about 90% of the fibers leaving the structure. The dorsal medial geniculate body projects to the auditory association cortex through the internal capsule. The medial portion of the geniculate receives input from both ventral and dorsal medial geniculate structures as well as other nuclei of the thalamus associated with other sensory systems. The cortical projections from the medial portion of the geniculate are widely distributed and are thought to be associated with alerting and arousal.

The primary auditory cortex is situated on the posterior superior surface of the temporal lobe on the transverse gyrus of Heschl. The superior temporal gyrus and Wernicke's area are also associated with auditory processing. The auditory areas in the right and left temporal lobes of the brain are connected by a bundle of neurons which passes through the corpus callosum. This connection permits the bilateral transfer of information at the cortical level.

As the auditory system leaves the primary auditory cortex, it takes two different routes to the brain's associative and executive areas. One route is posteriorly through the arcuate fasciculus. This route passes under Wernicke's area around the tip of the Sylvian fissure, then anteriorly to the frontal areas. The second route passes anteriorly into the anterior areas of the temporal lobe and through the uncinate fasciculus to the frontal lobe. These two pathways have been hypothesized as providing different processing strategies for the auditory signal (Belin and Zatorre 2000).

The complex arrangement of the auditory pathway allows information to be processed bilaterally in an efficient manner. The parallel processing of information that is initiated at the lower brainstem level is integrated at higher levels such as at the level of the inferior colliculus and the cortex. The redundancy inherent in information transfer through this maze of interactive subsystems ensures accuracy in perception and communication skills even in light of neurological insults and injury. Similarly, the intact system is capable of processing auditory information which has been severely degraded by the presence of competing signals or other interfering factors.

5.2 Sound and the Decibel

Hearing is the conscious interpretation of vibration as sound. Sound requires three key elements: (1) a source of vibratory energy, (2) a medium in which to transmit the vibration, and (3) a receiver. Sounds can be quantified based on their physical frequency, complexity, and SPL.

The frequency of sound is measured in hertz (Hz) which is the number of 360° phase shifts per second. The frequency response characteristic of the ear is significantly impacted by the anatomical and physiological characteristics of the outer and middle ear systems (Webster 1995). The range of frequencies processed by different species varies greatly. Elephants, for example, respond down into the human subsonic frequency range (Reuter et al. 1998), whereas bats can respond in the human ultrasonic range (Coles et al. 1989; Guppy and Coles 1988). The normal frequency range for humans is 20–20,000 Hz.

The complexity of sound is based on the number of frequencies present and the physical combinations of those frequencies. At the simplest level, when a single frequency is present, it is referred to as a pure tone. As additional frequencies are added, the sound becomes more complex. Complex sounds that have a regular repetition of the combined frequencies take on a tonal quality. At the other extreme, frequencies in a sound that have a truly random pattern are perceived as noise.

The frequency structure of a sound is also related to the duration of the sound and its rise and fall time characteristics. Short-duration electrical signals used to generate sounds will have a band of frequencies associated with them. This is referred to as the frequency bandwidth (BW) and is represented as

$$BW = \frac{1}{\text{Duration}}$$

The bandwidth is modified by the characteristics of the transducer which is used to convert the electrical energy into sound. Similarly, sound frequency is affected by the durational rise and fall characteristics of the signal. Sounds that come on abruptly will have broader frequency ranges than those that have slower onset times or ones that are gated (shaped) to reduce the amount of frequency spread. For example, the Blackman window is a common gating function used to reduce frequency spread in short-duration tones generated for auditory brainstem response (ABR) measurement (Gorga and Thornton 1989; Purdy and Abbas 2002; Reuter et al. 1998).

Most mammalian auditory systems are able to receive and process a broad range of energy that is measured in terms of pressure. In humans with normal hearing, a pressure of 20 µPa causes the eardrum to move approximately 1/1,000,000,000th of an inch, resulting in the perception of sound. As pressure increases, the sound is perceived as becoming increasingly louder up to a pressure level that is 10^{14} times greater than the softest sound pressure perceived. To deal with such a broad range of numbers, SPLs are expressed as a logarithmic ratio with a specific reference value.

Most sound level measurements in the environment as well as hearing aid specifications are reported in terms of SPL, referenced to 20 µPa, and measured as decibels (dB). dB SPL is represented by the formula

$$\text{dB SPL} = 20 \log \frac{\text{Pressure measured}}{\text{Pressure reference}}$$

This formula illustrates that 0 dB SPL is not the absence of sound. Sound can, and often does, fall into the negative dB range.

Figure 5.7 shows the minimum audible pressure map for the human ear (Dadson and King 1952). Such maps for other species will vary depending on the structure of their outer and middle ear systems. As can be seen, different SPLs are required at each frequency to be just audible. To account for the ears differing sensitivity by frequency, a number of different dB scales have been created. These scales are based on dB SPL but have a weighting factor for each frequency (ANSI-S1.4-1983) (Institute 1983). The linear scale with no correction is now called dBZ.

The three most common weighted scales used are dBA, dBB, and dBC. The dBA scale has its primary use in the area of noise damage-risk assessment. The dBA scale is based on a single sound level measurement that is taken across the frequency spectrum. This scale applies correction factors to the SPL value at each frequency that makes up the sound to account for the responsiveness of the human ear. The dBA

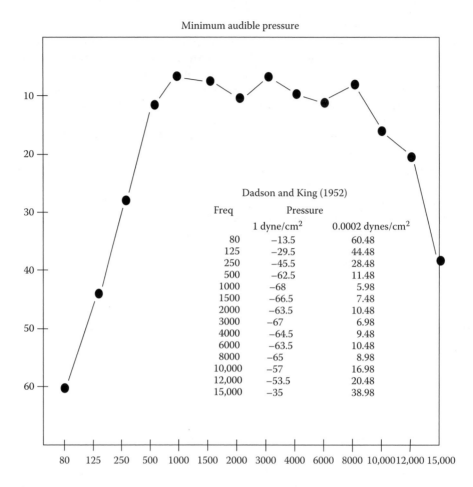

FIGURE 5.7 The minimal audible pressure curve originally determined by Dadson and King (1952). The original data as expressed in dynes/cm² have been converted by the author to the now commonly used reference of 20 μPa. (Figure drawn by Herbert Jay Gould. With permission.)

scale uses a weighting factor that is based on the inverse of a 40-phon curve, where a phon is defined as the perceived equal loudness level across frequencies. The 40-phon curve equates loudness at all audible frequencies to that of a 40-dB SPL, 1000-Hz tone.

The dBB scale is seldom used. It reflects human hearing in a manner similar to the dBA scale except that the weighting factor is based on a smoothed inverse of the 70-phon curve. The loudness function represented by the 70-phon curve is flatter than that seen for the 40-phon curve. This reflects the underlying tuning curve for hair cells which tends to flatten with increased sound pressure (Yost 2000).

The dBC scale is used when a very low, inaudible frequency is present that may distort the overall sound pressure measure being taken of sounds in the audible range. There is virtually no weighting of the response throughout the audible portion of the frequency spectrum. The reader is referred to Durrant and Lovrinic (1995), Lipscomb (1994), Yost (2000), and ANSI S1.4-1983 (Institute 1983) for expanded discussion of these measurement scales.

Audiograms are graphic representations of the basic hearing ability of humans, reflecting the threshold at which a frequency becomes just audible. Sound level for human hearing measurements on the audiogram also have had correction factors applied. The audiogram specifies hearing in terms of dB HL. The 0 dB HL level at each frequency corresponds to a different SPL value (ANSI 3.6 1996) (Institute 1996). These differences are due to the audibility differences between the frequencies as well as the physical characteristics of the transducer coupling with the ear. Specifically, each earphone and cushion arrangement used to measure hearing will alter these correction factors based on the relative amount of air space enclosed between the transducer's diaphragm and the tympanic membrane (eardrum). Because of the physical variability associated with the devices and even the testing environment, very precise calibration measurements must be made and maintained.

Decibel levels used to express values relative to an individual's hearing threshold are measured in terms of sensation level (SL). Thus, if a person had a hearing threshold of 40 dB HL and you presented a sound to them at 70 dB HL, that tone would be at 30 dB SL. That is, the SL of presentation was 30 dB greater than threshold for the signal. The SL measure is often referred to when performing audiometric measures using speech materials.

5.3 Annoyance and Damage—Risk from Noise

Sound can be annoying and may damage the ear at even moderately loud levels given sufficient time. There are a number of scales for estimating the annoyance level, likelihood of interference with speech and probability of damage to the ear. Many local communities have enacted noise control ordinances which vary from using simple subjective criteria to specific limiting objective levels of noise.

Several agencies and organizations are involved in studying noise and setting regulations regarding noise levels. These include the Environmental Protection Agency (EPA), the Occupational Safety and Health Agency (OSHA), and National Institute for Occupational Safety and Health (NIOSH) which is an institute of the Center for Disease Control (CDC), and the World Health Organization (WHO).

Evaluation of noise for annoyance or for possible damage encompasses two similar but different areas. The EPA and WHO have suggested various criteria for environmental and community noise (Berglund et al. 1999; Noise_Pollution_Clearinghouse 2012).

The preferred noise criteria (PNC) curves were created as a means to set design goals for structures and as a means of evaluating existing noise environments (Beranek et al. 1971). The curves are generated from an octave band analysis of the environmental sound from 31.5 to 8000-Hz center frequency. Each curve represents a spectrum of sound that is acceptable for various activities. For example, a PNC 10–20 is acceptable for concert halls, while a PNC of 50–60 is acceptable for large work areas where speech still needs to be heard.

While PNC curves address goals for listening environments, OSHA, NIOSH, and the ISO (International Standards Organization) have all addressed damage-risk criteria for workplace noise exposure. These standards are based on damage-risk assessment, but are also influenced by economic factors. They should be considered as guidelines to possible outcomes for the majority of individuals. However, it should be recognized that some individuals appear to be very susceptible to noise damage, while others have almost no effect from moderate amounts of noise. OSHA 1910.95(b)(2) permits 8 h of exposure at 90 dBA at a slow response setting. It reduces the permissible exposure by half for every 5 dBA increase in SPL. NIOSH recommends a maximum of 85 dBA for 8 h and a reduction in exposure time of half for every 3 dBA increase in exposure level (NIOSH 1998).

5.4 Assessing the Auditory System

Audiometric assessment of the auditory system is not pathology-specific. Its main goals are to locate the site of a disruption in the system and to determine the severity of that disruption. The traditional categories of hearing loss have been conductive, sensorineural, mixed, and functional. The sensorineural component can now be subcategorized into cochlear and retro-cochlear, and, in many instances, even finer delineations are provided within these subcategories.

The severity of a disruption in the auditory system can be viewed in several ways, including degree of loss for specific tonal frequency, loss in speech intelligibility or, more generally, the amount of disruption to everyday activities as a result of communicative difficulty. Predictions can be made from one view of the loss to another, and variations between the prediction and the measured value may point to possible sites of auditory lesions.

5.4.1 Transducers

Acoustic signal generation for hearing assessment requires the use of transducers to generate the auditory signals. Loudspeakers are used to generate signals in a sound field. When using loudspeakers in a room, care must be taken to maintain the orientation of the subject relative to the speaker as well as fixing the subject's location. Variations in location and orientation may significantly alter the signal level at the ear. Use of pure tones when testing with loudspeakers is inadvisable due to the creation of standing waves in the testing environment.

Specific hearing measurements are performed either with supra-aural headphones or with insert earphones. TDH-39 and TDH-49 supra-aural headphones mounted in MX41AR cushions traditionally have been the most common transducers used for measuring hearing. When using headphones, it is imperative that the diaphragm of the earphone be placed directly over the external auditory meatus. Without proper placement, the signal can be attenuated due to blockage by the tissue of the ear and face. It also is necessary to ensure that the external auditory meatus does not collapse under headphone pressure. A collapse of the meatus will result in elevated thresholds with significant variability, particularly in the low-frequency region. On visual inspection, ear canals that have a narrow oval appearance tend to collapse. It has been estimated that collapse of the ear canal will occur in 3–4% of the population while undergoing hearing testing using supra-aural headphones (Bess 1971; Chaiklin and McClellan 1971; Marshall and Grossman 1982).

The headphone band should be snug holding the headphone cushions securely in place in order to maximize attenuation between ears. With properly fit TDH-type headphones, a nominal 40-dB value is accepted as the amount of inter-aural attenuation. The cables from the headphone should be directed down the back of the subject to prevent extraneous noise generated by their rubbing on clothing.

In the last 10 years, the use of earphones inserted into the ear canal, such as ER-3A and ER-5A, has gained popularity. Insert earphones have significant advantages in testing. They eliminate the possibility of ear canal collapse and provide a greater inter-aural attenuation, thus reducing the need for masking the nontest ear.

Insert earphones are particularly advantageous in auditory assessment of young children. The bulky TDH headphones and cushions are uncomfortable for many young children and will often provide a poor fit. Normal attenuation between ears (inter-aural attenuation) appears to be improved as the foam ear tip of the insert earphone seals the ear canal (Sobhy and Gould 1993). Again, the cables should be directed behind the subject to prevent extraneous noise.

Insert earphones have several advantages when used to generate stimuli during electrophysiological tests such as the ABR. First, there is a lower electrical noise associated with the inserts than with the traditional (TDH 39) headphones. This means that complex and heavy metal shielding is not required. Second, the insert transducer is separated from the head by a length of the tube, creating a time delay between the electrical impulse to the earphone and the arrival of the sound at the ear. This helps isolate any remaining electrical artifact from the desired response.

When performing electrophysiological tests, care should be taken to have the transducer body located away from electrodes and the electrode cables to prevent pickup of the electrical signals that are generated. The insert transducer tube should not be altered; it provides a resonance that is calculated as part of the earphone's response characteristic.

5.5 Pure Tone Audiometry

The mainstay of audiometric measurement has been the pure tone audiogram. Pure tone audiometry provides a method for separating conductive losses located in the outer and middle ears, from sensorineural losses associated with problems in the cochlea and central nervous system. It is performed using both air-conducted signals which pass through the outer and middle ears as well as bone-conducted signals which are generated by an oscillator pressed against the mastoid or forehead. Bone conducted signals vibrate the skull, directly affecting the cochlea and bypass the outer and middle ear system. Typically, a problem in the outer or middle ear is suspected when the air conduction results are poorer than the bone conduction results. If lowered bone conduction scores are seen with the reduced air scores, a mixed loss is present, simultaneously indicating problems in both the outer/middle ear and the inner ear.

5.5.1 Method

Test instructions for pure tone audiometry should be brief, simple, and to the point to minimize error-producing confusion on the part of the test subject. Typical instructions are: "You will hear a series of sounds, some will be very soft. Raise your hand whenever you hear the sound." The use of "tones" or "whistles" are often substituted for the more generic word "sound."

The pure tone audiogram is then obtained at octave intervals from 250 to 8000 Hz. Signals are initially presented at 30 dB HL for 2 s at the frequency being tested. Longer signals increase the probability of a false-positive response. If signals have very short durations (<500 ms), then there will be inadequate sensory integration leading to higher threshold levels. The signal is lowered by 10 dB HL until no response is given, followed by raising the signal in 5-dB HL steps until a response is given. The signal is then lowered by 10 dB HL and the steps repeated. Once the subject responds at the same level, two out of three times on the ascending signal increments, threshold is obtained and the process is repeated at a new frequency.

Typically, the better hearing ear, by subject report, is measured first, commencing at 1000 Hz. After all frequencies have been measured, the test is repeated at 1000 Hz. The subject should have a threshold within 5 dB of the previous score for the measure to be considered reliable.

If the subject has a hearing loss greater than 30 dB and does not respond to the initial signal presentation, the signal level should be raised in 20 dB steps until a response is obtained. At that point, the process of bracketing the threshold begins as described above.

5.5.2 The Audiogram

The audiogram is used for recording the individual's auditory thresholds. It is constructed such that the distance of 20 dB on the *Y*-axis is equivalent to one octave on the *X*-axis. Right ear air-conducted thresholds are represented by a red circle, while left ear air-conducted thresholds are represented by a blue X. Scores between 0 and 25 dB HL are considered to be within the normal range. Figure 5.8 provides an example of an audiogram.

There are often a number of different symbols on the audiogram, including boxes, brackets [ ], and arrows, < > which represent different measures such as bone conduction testing and the use of masking. Each audiogram should have a legend listing the symbols and their interpretation.

5.5.3 Masking

Masking is the use of a noise to prevent hearing of a specified signal. It is typically employed to prevent the nontest ear from responding to a signal presented to the test ear. As mentioned earlier, inter-aural attenuation for air-conducted signals is approximately 40 dB. Therefore, a signal that is presented to one ear at 50 dB will cross the head and be received by the opposing ear at a level of 10 dB.

The crossing of signals from one ear to another can create significant problems if there are hearing differences greater than 40 dB between the ears for air-conducted signals. A 'shadow' audiogram will often be seen where the poorer ear will yield thresholds that are 40–50 dB worse than the hearing ear if masking is not appropriately applied. For bone conduction measurement, a vibrator is placed on the skull, usually over the mastoid process, resulting in an inter-aural attenuation of 0 dB. Therefore,

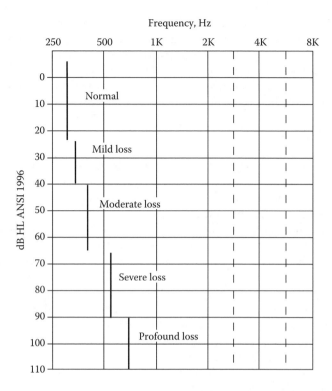

FIGURE 5.8 Audiograms are constructed with each 20 dB increase in hearing level equivalent to a one-octave increase in frequency. The common descriptive terms for hearing loss based on the pure tone average are shown with range bars to the left of the descriptor. (Figure drawn by Herbert Jay Gould. With permission.)

masking during the bone conduction test becomes critical in any test situation where there is a 10 dB or greater difference in hearing between the ears for air conduction.

The ideal masker would be a signal that is exactly the same as the test signal. However, such a signal would lead to confusion as to what to respond to by most subjects. Therefore, noise signals are employed for masking purposes. Most audiometers have different noise types to choose from when selecting a masker. The goal in choosing the most appropriate noise masker is to concentrate the maximum energy into the area of the test signal. In a flat spectrum signal, the level per cycle (LPC) can be calculated from the overall signal level (OL) and the signal bandwidth (BW):

$$LPC = OL - 10 \log_{10} BW$$

For pure tones, narrow band maskers are most efficient, whereas speech signals require wider band maskers.

A secondary problem occurs when masking. The masker has the same inter-aural attenuation as the original signal and the masking signal can cross over to the test ear. When this happens over masking has occurred. To avoid over-masking an ear, formulas have been established for calculating appropriate masking levels (Martin and Clark 2002).

Figure 5.9 graphically illustrates the masking concept. The *y*-axis represents the threshold in the test ear and the *x*-axis represents the masking applied to the nontest ear. If masking noise is inadequate, every 10 dB increase in the masker will result in a similar change in the threshold in the test ear. Once adequate masking is achieved, the test ear threshold will stabilize or plateau until the masker signal crosses to the test ear. This should be through a range of 30–40 dB of masker increase, that is, the amount of inter-aural attenuation. Once the masker crosses to the test ear, the threshold will again rise

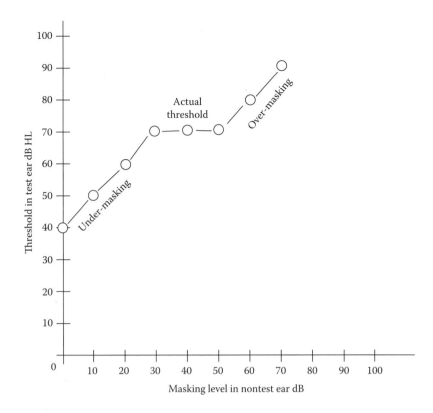

FIGURE 5.9 Under-masking occurs when the masker is too weak to prevent crossover of the test stimulus to the nontest ear. Over-masking occurs when the masker is too loud and crosses to the test ear. (Figure drawn by Herbert Jay Gould. With permission.)

in a linear fashion with increases in masker presentation level. The above procedure to find appropriate masking levels has been described as the Hood plateau method (Hood 1960).

5.5.4 Degree of Hearing Loss

Degree of hearing loss is often categorized as normal, mild, moderate, severe, and profound (Figure 5.8). This description is based on the pure tone average (PTA), which is the average threshold loss at 500, 1000, and 2000 Hz. These frequencies were chosen based on their contribution to the understanding of speech. "Normal" hearing is on average less than or equal to 20 dB HL; individuals with this degree of loss should not experience a great deal of difficulty in normal listening situations. Individuals with a mild loss have a PTA between 20 and 40 dB HL and will typically have only mild difficulty in most listening situations. They should be able to carry on a conversation in quiet at 3 ft. However, as distance or competing noise increases, their difficulty will become more apparent.

Moderate hearing loss is a PTA ranging from 40 to 70 dB HL. Individuals with a moderate loss will experience difficulty in conversation at 3 ft in quiet and will rely on visual cues to help in communication.

Severe loss is a PTA of 70–90 dB HL. Because conversational speech is generally in the 60–70 dB SPL range, a listener with a severe loss will miss most, if not all, of conversational level speech even in quiet. Individuals with moderate and severe hearing losses typically benefit from the use of assistive listening devices and hearing aids.

Profound hearing loss is a PTA of 90 dB HL or greater. Individuals with a profound loss may benefit from hearing aids but will often still have difficulty in oral communication situations. Frequency transposition aids or multichannel cochlear implants may provide significant benefit (Davis-Penn and Ross 1993; Tyler and Tye-Murray 1991).

It should be noted that, though generalizations about daily function can be made based on the PTA, there is a great deal of individual variability. Some individuals with moderate hearing loss will experience significantly more difficulty than others with severe-to-profound hearing losses.

5.6 Speech Reception Threshold

The speech reception threshold (SRT) is the signal level at which a person can correctly identify spondaic word stimuli 50% of the time. Spondaic words are two syllables in length and have equal stress on both syllables (e.g., baseball). The PTA and SRT should correlate well and be within 8 dB HL of each other for the audiogram to be considered reliable. Gaps larger than this may be seen in steeply sloping losses where there is a precipitous loss across the speech range and into the upper frequencies. In such cases, a two-frequency average at 500 and 1000 Hz will be a better correlate with the SRT.

A discrepancy between the PTA and SRT is also seen in cases of functional hearing loss. Functional loss occurs when an individual hears better than audiometric test results indicate. Individuals with functional loss may or may not be aware of this problem. Because the individual may not be aware of the functional nature of the loss, as in cases of hysterical deafness, the term pseudo-hypoacousis is now used rather than the more pejorative designation "malingering."

As hearing measurements, technology, and procedures have evolved, the SRT, although still routinely used by many audiologists, is taking on a less vital role in hearing assessment. It is often used for determining the presentation level for a speech word recognition test, though this is just as easily and accurately set using the PTA. The SRT is probably most useful in testing young children for whom the audiogram is thought to be questionable or is impossible to obtain.

5.6.1 Method

Instructions should be kept simple. The subject should be instructed to repeat the word that they hear and, if they are not sure of the word, they should be instructed to guess. A typical instruction would be: "Please read the list of words on the sheet in front of you. You will be hearing these words through the earphones. Some of the words will be very quiet. I want you to repeat the word that you hear. If you are unsure of a word make a guess."

The subject should be familiarized with the word list prior to the beginning of the procedure by either letting them read the list or presenting it at 40–60 dB SL re PTA. The familiarization list order should be random relative to the final test format. There have been several different methodologies for performing the SRT test. The procedures advocated by the American Speech-Language-Hearing Association (ASHA 1988) as well as one recommended by Martin and Dowdy (1986) appear to provide good reliability. The difference is that Martin and Dowdy (1986) use a 5-dB step rather than the 2-dB step used in the ASHA method. The procedure starts by presenting one spondee at 30-dB HL and lowering the presentation level in 10-dB steps until the first spondee is missed. Each successive spondee is presented at 5-dB increments until one is correctly identified. The level is then lowered by 10-dB and the process is repeated until three correct responses have been obtained at a given level.

5.7 Speech Recognition Testing

Speech and language recognition is a complex cognitive task requiring a relatively intact central auditory nervous system to function normally. There is a significant body of literature surrounding both the task and methods for testing the various processes that are used in recognizing speech. Research in this area can be traced back to Campbell in 1910 using nonsense syllables to assess telephone circuit performance (Mendel and Danhauer 1997).

However, the modern era of speech perception research started during World War II. During this time, research was performed on normal hearing individuals to assess battlefield communication systems. Investigators involved in this research later adapted word lists and techniques to clinical audiological practices.

An early outgrowth of this research was the development of the articulation index (AI) (Fletcher 1950; French and Steinberg 1947). The AI is a mathematical model for estimating the intelligibility of speech based on the acoustical characteristics of the stimulus and the transmission medium. The method has been standardized as ANSI 3.5-1997 (Institute 1997). This standard has been helpful in the study of communications systems.

If the audiometric threshold is considered a special case of a filter in the transmission line, the AI can be used to predict speech recognition scores based on the stimulus material and audiometric configuration (Halpin et al. 1996; Studebaker et al. 1999a,b). Deviations between the predicted value and the actual recognition score obtained for a subject provides clinical insights into cochlear function and central processing of auditory information. Unfortunately, the acoustical characteristics of each stimulus set must be precisely known for prediction to approach any degree of accuracy.

A second outgrowth of the World War II research was the establishment of the field of audiology. An initial focus of the field was in assessment and rehabilitation of returning veterans with hearing loss. Part of this assessment process required estimating the effects of a loss on speech communication. To assess speech recognition, the word lists and procedures originally used during the war for assessing communication systems were employed. As problems with these materials were uncovered, they were replaced with the recorded CID W-22 word lists (Penrod 1985). The CID W-22 word recognition test remains in common clinical practice. Another list, the Northwestern University Auditory Test number Six (NU-6), is also commonly used as a substitute for the CID W-22.

Unfortunately, there are problems in using both of these lists for assessment of communication ability. They are administered in quiet, while we live in a noisy world. They are single words, while we listen to connected discourse. They are presented in optimal conditions, while we often function in less-than-ideal conditions. Because of these and other problems, other measures have been created for assessing speech communication function.

Today, assessment of speech encompasses a wide range of materials. These materials are designed to assess everything from simple perception to more complex language-processing strategies. Stimuli include nonsense syllables, words, and sentences. The material can be presented monaurally (to one ear), or binaurally (both ears). Binaural signals may be presented diotic (same sound to both ears) or dichotic (different signals to each ear) as competing messages. It is becoming common practice to assess speech perception in noise as part of the standard assessment using tests such as the SIN (Speech In Noise Test) or QuickSIN (Killion et al. 2004).

In assessing the central auditory function, degraded speech stimuli are often used because they task the system sufficiently to identify problems in auditory processing located in the central nervous system. Popular methods of degrading speech include the use of filtering, time compression, distortion, or competing signals. The reader is referred to Lucks-Mendel and Danhauer (Mendel and Danhauer 1997) for a more thorough overview of speech perception and assessment.

5.8 Immittance Audiometry

Acoustical immittance audiometry, within limits, assesses middle ear pressure, impedance/admittance, and function of the acoustic reflex. This battery of measures provides useful clinical information on middle ear status and can provide information regarding central auditory nervous system function up to the level of the SOC.

The history of acoustic immittance measures dates to Lucae in 1867 (Feldman and Wilber 1976). He along with later researchers used mechanical acoustic impedance bridges until the development of the electromechanical bridge in the early 1960s. This development led to increased research into middle ear

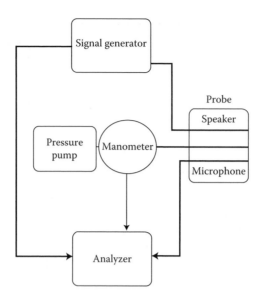

FIGURE 5.10   Block diagram of the components of a tympanometer. (Figure drawn by Herbert Jay Gould. With permission.)

function and to eventual application of these measures to clinical practices. Because early electrome-chanical bridges used either impedance or its reciprocal, admittance, some confusion in terminology developed. Now all measures are referred to as immittance, and terminology and procedures have been standardized (ANSI S3.39 1987) (Institute 1987).

The basic system for acoustic immittance is diagramed in Figure 5.10. The system generates a probe tone into a sealed ear canal. A portion of this tone's sound pressure will pass through the tympanic membrane and middle ear to the cochlea. The remaining pressure will be retained within the external ear canal. The microphone in the probe picks up the amount of retained pressure and passes it on to the analytical circuit of the bridge. The immittance is indirectly measured by the sound pressure of the probe tone in the ear canal. The more pressure transmitted through the tympanic membrane, the greater the immittance.

The immittance measure is confounded by the size of the external ear canal and the depth of inser-tion of the probe tube. Different volumes will result in different pressures. To remove this confounding variable, pressure is applied within the sealed external auditory meatus. This pressure increase tightens the tympanic membrane, thereby preventing passage of sound pressure into the middle ear system. Subtracting the pressurized measure from the un-pressurized measure provides the static immittance value for the middle ear system. Typically, a pressure of +200 mmH$_2$O, relative to ambient, is sufficient to isolate the ear canal from the middle ear.

Unfortunately, static immittance exhibits a large variation in the normal population. Both the sensi-tivity and the specificity of the measure are low but, as part of the diagnostic battery, it provides impor-tant information.

The tympanogram demonstrates acoustic immittance over a range of external auditory meatus pres-sure values. Measurements are typically acquired over a range of ±200 mmH$_2$O relative to ambient. Lower pressures, down to −400 mmH$_2$O, may be used in the case of some middle ear disorders. The resulting shape of the tympanogram is dependent on the middle ear status and the frequency of the probe tone. Tones in frequency ranges of 600 Hz and above result in multipeaked configurations that are often hard to interpret. Using a 220-Hz probe tone results in various single peaked functions which were classified by Jerger (1970). In this system, shapes are classified as A, B, or C depending on the loca-tion of the immittance peak on the tympanogram (Figure 5.11). The classification system was originally

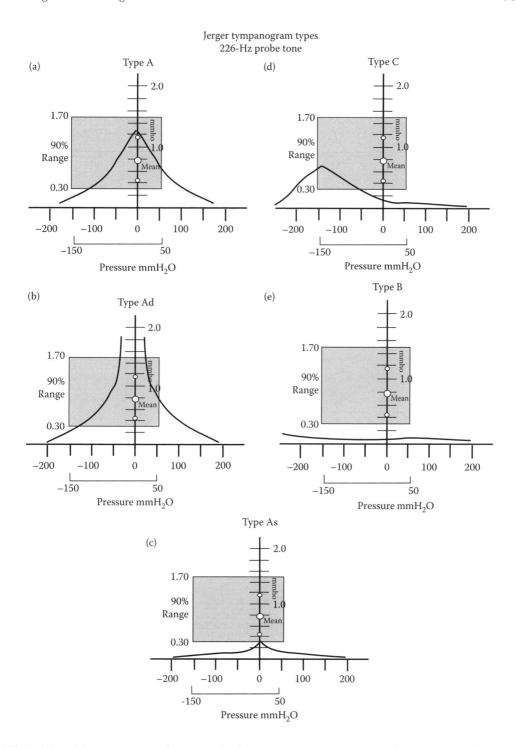

FIGURE 5.11 Tympanogram configurations. (a–c) Jerger type A tympanograms indicating a normal pressure in the middle ear space. (d) Jerger type C tympanogram normally associated with Eustachian tube failure and beginning otitis media. (e) A flat tympanogram, Jerger type B. Type B tympanograms are associated with excessive middle ear fluid or a perforated tympanic membrane. (Figure drawn by Herbert Jay Gould. With permission.)

developed during a period before immittance measures were expressed routinely in mmho's, a mho is a measure of acoustic conductance, and many impedance bridges provided data in arbitrary units. Therefore, the classification system is qualitative and is dependent on the experience and quality of the examiner's judgment. That being said, the system is in widespread clinical use.

All type A tympanograms have a peak compliance occurring within the normal pressure range, which is considered to be ambient to −100 mmH$_2$O. The lower limit can vary depending on the clinical setting and is often extended down to −150 mmH$_2$O. Occasionally, slightly positive pressures may be seen, though this is atypical.

There are two subclassifications of type A tympanograms which are based on the compliance level at the tympanogram pressure peak. The first subtype is a deep configuration (A$_d$), the pattern of which suggests an excess amount of energy flow into the middle ear system at the pressure peak. Type A$_d$ is typically associated with very flaccid tympanic membranes or with breaks in the ossicular chain. The second subtype (A$_s$) is a shallow configuration which reflects a lack of energy flow into the middle ear system. Type A$_s$ tympanograms are associated with heavily scarred tympanic membranes or with ossicular fixation, particularly at the point of stapes footplate entrance to the oval window of the inner ear.

The type C tympanogram has a pressure peak less than −100 mmH$_2$O relative to ambient. To be classified as a type C, a peak must be seen, although it may be somewhat rounded. The peak may be in the normal admittance range or it may be reduced. Type C tympanograms indicate a problem with pressure regulation in the middle ear and are typically related to a failure of the Eustachian tube. Children are particularly susceptible to Eustachian tube problems due to the angle of the tube and the position of the adenoid adjacent to the opening into the nasopharynx.

It should be noted that the pressure peak on the tympanogram is only an approximation of the true middle ear pressure. As the ear canal pressure is changed, the position of the tympanic membrane changes, either increasing or decreasing the volume of the middle ear space. Because pressure and volume are a constant according to Boyle's law, the changes in volume will introduce a related pressure change.

Type B tympanograms are flat, exhibiting no peak and little if any change in compliance across the pressure range. These tympanograms will have very low static immittance values which, combined with small to normal canal size values, indicate that there is little energy flow into the middle ear space. This condition is seen when the middle ear space has fluid in it. In contrast, a type B tympanogram with a low static immittance value and a high canal size value suggests a perforation of the tympanic membrane. In effect, the perforation opens up the volume of the middle ear space and mastoid air cell area. As the pressure freely enters the middle ear space, there is no displacement of the tympanic membrane as the canal pressure is changed.

5.8.1 Acoustic Reflex

The acoustic reflex can also be measured using immittance audiometry. The reflex is a contraction of the stapedius and tensor tympani muscles caused by a sound louder than approximately 80 dB SPL. Contraction of the muscles alters the immittance of the middle ear system. The reflex is bilateral, simultaneously affecting both ears equally.

The stimulus level required to elicit the acoustic reflex is relatively constant at approximately 80–90 dB SPL for individuals with cochlear hearing losses below 55 dB HL. This means that as hearing loss increases, there is a decrease in the SL required to elicit the response. After approximately 55 dB HL, there is a relatively linear increase in SPL required to elicit the response. The decreased SL phenomenon is related to loudness recruitment which is an abnormally fast growth in loudness for increasing SPLs and is typically indicative of abnormal cochlear function.

The immittance change is time locked to the stimulus onset and should be sustained for tones less than 2000 Hz for at least 30 s. At 500 and 1000 Hz, a decrease in the reflex value by 50%, over a 10-s period, at 5 dB SL re acoustic reflex threshold, is suggestive of an VIIIth nerve lesion.

5.9 Otoacoustic Emissions

An otoacoustic emission (OAE) is an acoustic signal that has been generated by active processes in the inner ear and is recorded from the ear canal. The emission has been linked to the expansion and contraction of outer hair cells participating in the active processes within the cochlea. Gold (1948) conjectured that an active cochlear process would produce a low-level tone emitted into the ear canal. Unfortunately, amplifier sensitivity and noise levels did not permit him to successfully record an emission. OAE, however, was successfully recorded by Kemp 30 years later (Kemp 1978). Since then there have been numerous articles on the topic. There are four types of emissions, including spontaneous, transient evoked, distortion product, and stimulus frequency. This classification scheme is based on the stimulus type used to elicit the response. All OAEs are believed to result from the movement of outer hair cells and require an intact outer and middle ear to be successfully recorded. For excellent reviews of this topic, see Robinette and Glattke (1997) and Hall (2000).

5.9.1 Spontaneous Otoacoustic Emissions

Spontaneous otoacoustic emissions (SOAEs) require no external stimulus. At present, SOAEs can be recorded in approximately 70% of ears in normal hearing individuals (Penner et al. 1993). This number has steadily risen in the past few years due to improvements in recording technology. Microphone sensitivity and noise floor levels are critical in the recording because the low SPLs of the response are easily missed or masked by outside noise. The presence of SOAEs appears to vary by racial group and gender, but age does not appear to be a factor through young adulthood (Bilger et al. 1990; Martin et al. 1990; Strickland et al. 1985; Whitehead et al. 1993). However, there may be a decreased presence of SOAEs in individuals over 60 even with normal hearing (Stover and Norton 1993).

SOAEs have narrow bandwidths and relatively stable amplitudes and frequencies (van Dijk and Wit 1990). It is common for SOAEs to be present at multiple frequencies within an ear, and these frequencies are not necessarily the same in both ears of a single subject. SOAEs are only seen in relatively normal cochlea and are not recorded if there is a hearing loss greater than 30 dB HL. It has been conjectured that there may be natural imperfections in the Organ of Corti that might cause a perturbation and reverse the traveling wave (Kemp 1986). In spite of a few anecdotal reports, SOAEs do not appear to be associated with tinnitus (Penner 1990; Penner and Glotzbach 1994).

All OAEs must be transmitted from the cochlea back through the middle ear system before being recorded. Therefore, it is important to know the effects of the middle ear system on the OAE response. Middle ear pressure changes can cause a shift in SOAE frequency of up to 50 Hz. Although these shifts are typically upward in frequency, some downward shifts have been noted. There is typically a reduction in amplitude of the response as the middle ear pressure is moved from ambient. These shifts are consistent with changes in middle ear function. Stiffening of the tympanic membrane causes an upward shift in the resonance of the middle ear system. In addition, the change in stiffness makes it more difficult for the tympanic membrane to move and results in lower amplitude deflections and reduced input to, and from, the cochlea.

5.9.2 Transient Evoked Otoacoustic Emissions

Contrary to SOAEs, which occur in a passive state, transient evoked otoacoustic emissions (TEOAEs) are elicited by brief duration stimuli. The emission was reported by Kemp who introduced the first commercially available unit for recording the response (Kemp 1978). Many of the published studies of TEOAEs have used this equipment and its default parameters.

TEOAEs are present in all normally functioning ears with thresholds better than 30 dB HL. TEOAEs provide an efficient screening tool because the emission disappears if a mild hearing loss is present. However, it is not possible to predict audiometric threshold levels from the TEOAE amplitude.

The small amplitude of the TEOAE necessitates a signal-processing paradigm to extract the response from stimulus artifact and background noise as shown in Figure 5.12. Because the TEOAE is a deterministic signal in a relatively random background noise, a signal-averaging strategy can be employed to elicit the response. Using an averaging paradigm, stimuli are presented and the data averaged until the background noise is significantly attenuated. This is similar to the strategy used for evoked potential measurements discussed elsewhere in this book. Unlike evoked potentials, both the stimulus and the response are acoustic, thereby requiring a different strategy to eliminate the stimulus artifact.

Fortunately, the largest part of the stimulus artifact occurs before the response is emitted from the cochlea back into the ear canal. This means that the response collection system can simply turn off during this time frame, or alternatively, discard the data in this time frame post hoc. Nonetheless, during the time frame of interest, there will still be a small, lower-level, stimulus artifact that may remain and obscure the data. This problem is managed as shown in the stimulus column in Figure 5.12. The stimuli are presented in chains of four with each chain presenting three stimuli of one polarity and a fourth stimulus 180° out of phase and equal in intensity to the sum of the previous three. This results in the stimulus artifact summing to zero.

The TEOAE is a nonlinear response from the cochlea. Therefore, when the polarity of the signal is reversed, the TEOAE response does not equal the sum of the previous three responses. This means that, although part of the response is canceled, a measurable portion remains in the average.

The TEOAE system also permits evaluation of the response repeatability by collecting stimuli into two separate analysis groups, labeled group A and group B in Figure 5.12. Stimulus chains, and responses, are alternately assigned to each group during the collection process. Once an average is obtained, the groups can be correlated for a statistical measure of similarity. If the responses are similar, the averages can be combined, thereby increasing the signal-to-noise ratio.

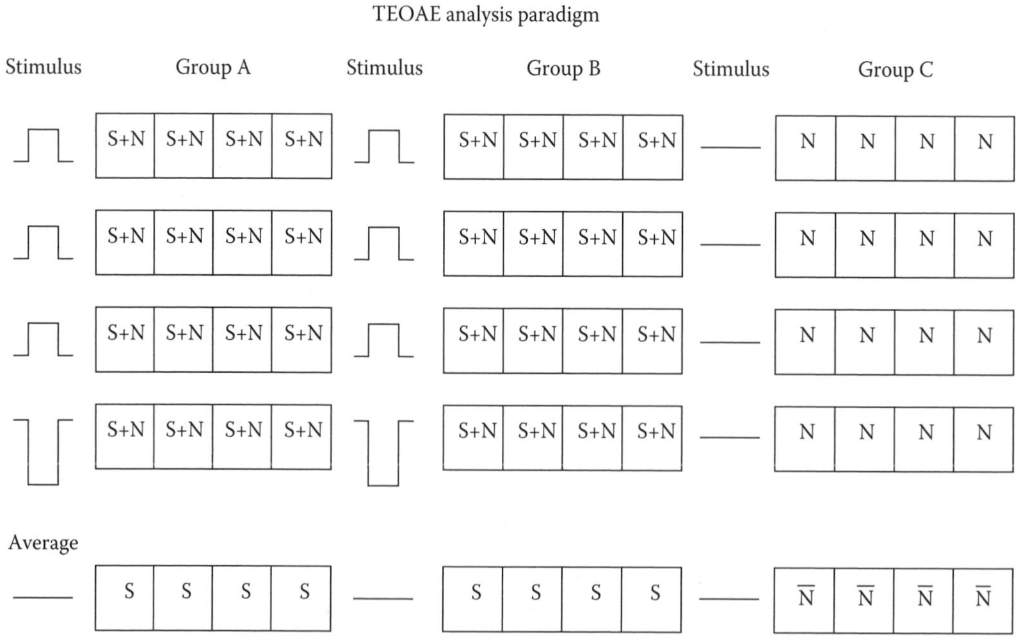

FIGURE 5.12 The paradigm for analyzing the responses during the TEOAE evaluation. Averaging within groups A and B provides for cancelation of stimulus artifact and reduction of noise. Correlation between groups A and B demonstrates the responses reliability. Comparison of the FFT from groups A and B to group C provides a measure of OAE strength compared to background noise. (Figure drawn by Herbert Jay Gould. With permission.)

If a repeatable response is observed, it is important to know the frequency content and to determine if it is consistent with the stimulus and different from the background noise. To obtain the estimate of background noise, the data collection system maintains a third buffer, group C (see Figure 5.12), which is collected during times of no stimulus presentation. To evaluate the frequency content, the system performs FFTs on the combined TEOAE response data, the noise data and finally the stimulus. The FFT should show larger amplitudes at stimulus frequencies for the TEOAE data than for the background noise data if a response is present.

The TEOAE in infants is approximately 10 dB larger than that seen in adults. Kok and coworkers reported a 78% prevalence of TEOAEs in newborns less than 36 h of age; however, this increased to 99% by 108 h of age (Kok et al. 1993). Current reports of 90% success rate, now often reported in clinical data, may reflect that the subject was screened more than once during the first 36 h period. In adults, there appears to be little variation in response amplitude with age. However, it has been found that the right ear has a slightly larger response than the left and that female responses are larger than male responses.

The TEOAE response is affected by stimuli that are presented in the contralateral ear (Moulin et al. 1993; Thornton and Slaven 1995; Veuillet et al. 1992). The reduction in response amplitude that is seen is due to input to the cochlea from the efferent auditory neurological subsystem, most likely the crossed olivocochlear bundle (Hood et al. 2003; Hurley et al. 2002). A similar effect is seen with the eighth nerve response during electrocochleography, where the N1 response (first negativity) is reduced in the presence of bilateral stimulation.

5.9.3 Distortion Product Otoacoustic Emissions

Distortion product otoacoustic emission (DPOAE) measurements were introduced by Kemp in 1979. The emission is the result of an inter-modulation between two frequencies presented to the ear. The stimulus frequencies called primaries are typically labeled $f1$ (lower frequency) and $f2$ (higher frequency). If the two primary frequencies are presented to the ear such that the $f2/f1$ ratio is 1.2 (or slightly larger, i.e., 1.22), distortion by-products are created in the inner ear. The cochlea produces a number of distortion by-products simultaneously at areas that can be mathematically determined (Gelfand 1998). However, the largest of these is at the frequency $2f1–f2$ and is known as the cubic-difference tone, as depicted in Figure 5.13.

The primary frequencies ($f1$ and $f2$) have associated presentation levels of L1 and L2. L1 and L2 presentation levels are either equal in SPL, or L2 is presented at a level lower than L1. The optimum L1/L2 level is dependent upon frequency and on the L1 absolute level (Whitehead et al. 1993, 1995). It is relatively common in clinical situations to place L2 approximately 5 dB below L1.

The DPOAE response, similar to that seen for TEOAE, is small, ranging between 5 and 15 dB SPL and, like the TEOAE, requires a processing strategy to extract it from the background noise. Unlike the TEOAE, the stimulating frequencies are distant to the DPOAE frequency response. Therefore, the stimulus artifact will not interfere with the response and there is no need to remove it from the recording.

5.9.4 Stimulus Frequency Otoacoustic Emissions

Stimulus frequency otoacoustic emissions (SFOAEs) are elicited by a frequency glide presented to the test ear. When the response is present, an out-of-phase signal will be returned into the ear canal. The recording of this response is technically difficult, and therefore SFOAEs are not currently used in clinical settings. Shera and Guinan (1999) argue that the TEOAE and SFOAE should be joined conceptually because they occur at the stimulation frequency, whereas the DPOAE occurs at a frequency that is not present in the evoking stimuli. They further argue that the underlying mechanism is different for the responses, with the TEOAE dependent on a linear coherent reflection of the signal, while the DPOAE depends on a nonlinear distortion. They suggest that, because the underlying mechanism is different,

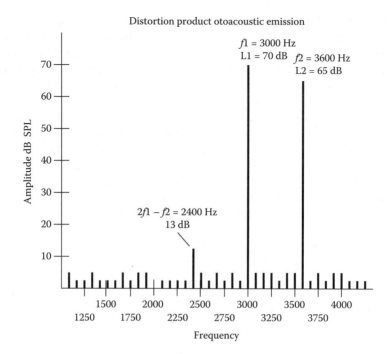

FIGURE 5.13 The distortion product OAE is generated by the interaction of two primary tones labeled $f1$ and $f2$. The interaction of the tones creates the emission which is recorded at the frequency $2f1 - f2$. (Figure drawn by Herbert Jay Gould. With permission.)

there should be differences in the manner in which the measures (TEOAE or DPOAE) react to different cochlear pathologies.

5.10 Electrophysiological Assessment of the Auditory System

Other chapters in this book have addressed how to collect evoked potentials from a wide range of sensory and cognitive systems. We have chosen to briefly address the auditory evoked potentials here because they have significant impact on audiological assessment both for determining site of an auditory impairment as well as estimating hearing ability in difficult-to-test populations.

The auditory evoked potentials are a single event emanating from an auditory stimulus that have been segmented into a series of subpotentials based on historical, technological, and physiological restraints. The primary responses used in the evaluation of the auditory system include: electrocochelography (ECochG), ABR, middle latency response (MLR), and the late auditory response (LAR). In addition, there are several other responses such as the P300, mismatched negativity (MMN), contingent negative variation (CNV), and N400 which are still in research phases as possible additions to the audiological test battery.

5.10.1 Electrocochleography

The electrocochleogram (ECochG) is generated in the cochlea and auditory portion of the VIIIth cranial nerve. It is recorded either as a trans-tympanic response with the noninverting electrode placed on the promontory of the cochlea or as an extra-tympanic recording with the noninverting electrode in the ear canal. The three components of the ECochG are the cochlear microphonic (CM), summating potential (SP), and the compound action potential (AP). The origin of the CM is the outer and inner hair cells and the electrical potential that is generated follows the stimulus waveform. The summating potential convolves with the CM and reflects the stimulus envelope. As the SP does not change polarity with changes in stimulus polarity, averaging a compression generated response to a rarefaction response will cancel the CM and yield the SP. The negative AP response occurs after the CM and SP at about 2 ms.

The ECochG was used in the 1970s and early 1980s for determination of auditory threshold (Naunton and Zerlin 1976, 1977; Schoonhoven et al. 1999). With the advent of ABR, this usage has significantly decreased in the United States, and the ECochG is now mainly used for determination of endolymphatic hydrops associated with Meniere's disease. Specifically, there is excessive endolymphatic pressure in the inner ear which results in vertigo, low-frequency hearing loss, and tinnitus. The symptoms fluctuate over time but eventually the hearing loss tends to become more severe, broader in frequency, and permanent. The increased endolymphatic pressure results in an increase in the SP amplitude relative to the AP amplitude and an increased traveling wave velocity within the cochlea.

5.10.2 Auditory Brainstem Response

The ABR also is known as the brainstem electric response (BSER) and brainstem auditory electric response (BAER). Audiologists tend to refer to the response as ABR, while neurologists tend more toward BSER or BAER. The response is most often credited to Jewett and Romano (Jewett et al. 1970). However, Sohmer and Feinmesser (1967), as part of an article on cochlear potentials, published waveforms several years earlier and only referred to them as possibly coming from the brainstem.

The ABR is a series of five waves conventionally labeled with Roman numerals. Wave I is generated by the VIIIth nerve and is therefore the same as the ECochG AP wave. There is a complex interaction of electrical signals generated at various levels of the brainstem that contribute to waves III, IV, and V. Hall (pp. 41–46) provides an excellent summary of possible ABR generator sites (Hall 2007).

The ABR has supplanted ECochG as the primary method for auditory threshold estimation in the difficult-to-test populations. It is used extensively in newborn screenings along with OAEs. The ABR has an advantage in screening in that it evaluates the VIIIth nerve and brainstem which is farther along the auditory pathway than the OAE which only evaluates outer hair cell function. ABR is the recommended technique for screening in neonatal intensive care nurseries where there is a higher likelihood of not only hearing loss, but also loss related to VIIIth nerve and other central nervous system problems. Auditory neuropathy/dysynchrony disorder occurs when the signal transduced by the cochlea is not transmitted to the brainstem (Starr et al. 1991, 1996). Individuals presenting such results have outer hair cell function, and therefore OAEs but no ABR. Such individuals function as hearing impaired or deaf, however, typically experience less benefit with amplification which is mainly targeted at outer hair cell hearing loss.

Soon after the ABR was discovered, it began to play a significant role in detecting tumors involving the VIIIth nerve and cerebellopontine angle as well as multiple sclerosis. The original benefit of ABR was that it could be used to effectively screen individuals for the likelihood of a central disorder before subjecting the patient to ionizing radiation. With modern imaging techniques, these risks are greatly reduced leaving cost and availability to be the major advantages of ABR to retro-cochlear lesion detection.

Several new techniques have appeared for analysis of brainstem responses. Two of these techniques, Stacked ABR and CHAMPs, use the derived band technique (Don et al. 1997, 2005a,b). The derived band technique collects a series of responses to clicks combined with different high-pass masked pink (shaped) noise. Subtracting various combinations of the responses, it is possible to isolate responses generated within specific frequency bands.

5.10.3 Middle Latency Response

The MLR was first reported by Geisler et al. (1958). It spans a range from about 10 to 80 ms and consists of four primary waves, namely Pa, Na, PB, and Nb. Additional waves P0 and N0 were later reported by Goldstein and coworkers (Goldstein and Rodman 1967; Mendel and Goldstein 1969).

The response has had a long and rocky history in the field of audiology. Its discovery is attributable to improvements in amplification and in signal averaging. The response was immediately assailed as a myogenic (muscle) response (Bickford et al. 1964). It was not until Harker and coworkers (1977) used a technique testing a human subject with and without succinylcholine that the majority of the response was considered to be neurogenic rather than myogenic in origin. Nevertheless, care must be taken with this response to avoid muscle artifact at high stimulus levels.

A second controversy soon enveloped the response as regards to the effects of filtering (Kavanagh and Domico 1986; Kavanagh and Franks 1989; Kavanagh et al. 1988) but this controversy has again been laid to rest with modern nonphase shift digital filters. At the time these controversies were resolved, the ABR was discovered and proved to be a highly effective measure of auditory threshold. Therefore, use of MLR never gained wide clinical acceptance as a measure of auditory threshold.

Since then the MLR has gained some acceptance as part of a test battery for auditory processing disorder (APD). The test uses multiple electrodes in a coronal chain to determine whether there is a difference in the strength of the right vs. left hemisphere (Musiek et al. 1994).

5.10.4 Late Auditory Response

The LAR is usually attributed to Pauline Davis (1939). This response was from single-trial data and was often difficult to observe in the ongoing EEG. Subsequent advances in amplification and the advent of digital computers and signal averaging permitted a much clearer and more robust response. Many attempts were initially made to use the LAR for threshold estimation. However, this cortical response is significantly affected by maturation through infancy and early childhood as well as by the subject's attention level and other behavioral factors. Ponton and coworkers (1999) demonstrated that the early changes in the late potential were not a simple shifting of peak latencies and amplitudes but reflected development of new waveforms with maturation. As the ABR proved to be a reliable tool for threshold estimation, much of the interest in the late potential died away. However, there has been a recent renewed interest in the response for its potential (1) to assess individuals with cochlear implants (Sharma et al. 2002, 2005) and (2) as a tool to assess central auditory dysfunction at a cortical level (Cone-Wesson and Wunderlich 2003; Kraus 1995; Sharma et al. 1993).

5.11 Common Pathological Conditions of the Auditory System

It would be impossible to cover all of the diseases, ototoxic effects, and traumas that could alter hearing function in a chapter such as this. Also, as stated earlier, audiology is site-specific rather than pathology-specific. Thus, various pathologies may reflect large variations in auditory function depending on how the pathology is impinging upon the auditory system. In this section, several common disease states seen in each of the auditory subsystems will be presented along with typical audiometric configurations associated with them.

5.11.1 Outer and Middle Ear Disease

In a conductive hearing loss, threshold by bone conduction is better than threshold by air conduction for a given signal. Conductive losses are associated with the external and middle ear systems that transfer the sound to the cochlea.

The external ear is relatively resilient to conductive problems and typically requires a complete blockage of the external auditory meatus before significant changes in threshold are seen. If the meatus is blocked, losses up to 60 dB with flat configurations are typical.

Pathology of the tympanic membrane (TM) results in a complex array of findings. Small perforations of the TM can result in little or no loss. Losses associated with larger perforations will vary with their extent and location. Tympanometric measures for perforations demonstrate flat tympanogram and large (>2 cm³) ear canal volume with extremely low static compliance.

Scarring of the TM, known as tympanosclerosis, is seen via otoscopy as an area of white calcification. These rarely cause significant change in thresholds unless a majority of the TM is affected. In these cases, shallow tympanograms (Type A_s) with reduced compliance may be observed. A similar finding is observed in otosclerosis, a disease discussed later in this section.

"Monomeric" tympanic membranes lack the fibrous middle layer of the TM. Mild–to-maximum conductive losses may be seen in these cases, and tympanometry yields a deep tympanogram (type A_d) with large static compliance. Similar tympanometric findings are seen for ossicular discontinuity.

Middle ear pathology also covers a broad range of structural malformations and pathological agents. These can have little to no effect on the audiogram or may appear as a maximum conductive hearing loss. The most common problem in childhood is otitis media, with or without effusion.

Otitis media often starts as a pressure imbalance between the middle ear space and the outside atmosphere due to a failure of the Eustachian tube to open properly. As air is absorbed in the middle ear space, a pressure gradient develops across the tympanic membrane which results in a "stiffness tilt" to the audiogram. Figure 5.14a demonstrates an audiogram with a larger conductive loss in the lower frequencies than in the higher frequencies. As the disease process continues, fluid enters the middle ear space, adding mass to the system. This typically results in a drop in the high-frequency region of hearing as shown in Figure 5.14b. The fluid will eventually thicken and the configuration of the hearing loss will flatten as shown in Figure 5.14c. This process can result in a maximum conductive loss of approximately 60 dB HL.

Tympanometry in otitis media will first present as a type C tympanogram with relatively good static compliance. As pressure drops in the middle ear space and fluid accumulates, the pressure peak will become more negative and present reduced static compliance values. The end result is a type B, flat, tympanogram with normal canal volume and a low static compliance value. As the condition is treated, the process reverses itself to a type C tympanogram and progresses to the normal type A.

In adulthood, otosclerosis is a well-known pathological process. Otosclerosis is the slow change in the normal bone of the stapes footplate and surrounding cochlea to a softer spongy bone. Fixation of the stapes footplate in the oval window of the cochlea results from this bone remodeling. The pathological process with eventual fixation of the stapes causes a conductive loss that is progressive over time. There will also be a small drop in bone conduction threshold at 2 kHz. This drop, known as Carhart's notch, as well as the conductive loss, can be rectified with replacement of the stapes with a prosthetic strut. The tympanogram associated with otosclerosis is typically type A_s, with an absent stapedial reflex. Occasionally a small reflex is seen, but traveling in the opposite direction from expected. This is thought to occur from action of the tensor tympani muscle. As noted previously, other ossicular fixations as well as tympanosclerosis will present with similar findings.

Ossicular discontinuity is a break in the ossicular chain, usually occurring at the incudo-stapedial joint. This can result from head trauma or from bone necrosis in long-standing cases of otitis media. If any linkage remains between the parts of the ossicular chain, such as bands of fibrous scar tissue, hearing loss can range from mild-to-maximum conductive loss. The tympanogram is type A_d and the acoustic reflex is absent. As in otosclerosis, a small reverse reflex from the tensor tympani can occasionally be observed.

5.11.2 Sensorineural Hearing Loss

Sensorineural hearing loss is defined as a loss in which there is no gap between the air-conducted and bone-conducted threshold responses to a given signal (Figure 5.15). Such losses typically reflect cochlear and, to a lesser extent, neural pathologies. Several sites have been identified within the cochlea

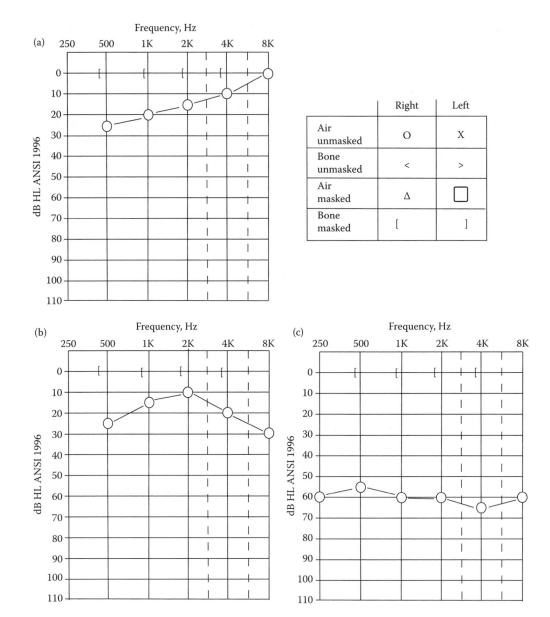

FIGURE 5.14 Air and bone conduction thresholds for the right ear demonstrating (a) a typical stiffness tilt audiogram often seen in early otitis media. (b) As fluid accumulates in the middle ear space, a mass tilt is superimposed on the stiffness tilt seen in (a). (c) Flat, maximum conductive loss is seen in otitis media once middle ear fluid is thickened. (Figure drawn by Herbert Jay Gould. With permission.)

as potential points of disruption to the auditory signal. Given today's technology and the improvements in audiometric testing procedures and materials, the term sensory-neural may now be divided into the categories of cochlear loss and neural loss.

A principal site of auditory disorder in the cochlea is the amplification system of the outer hair cells. Outer hair cells appear to be more susceptible to damage than inner hair cells and are particularly susceptible to insult from the environment. Noise and ototoxicity are two common environmental factors leading to hearing loss in adults.

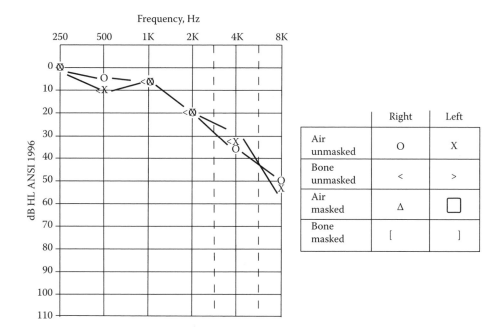

FIGURE 5.15 An air and bone conduction audiogram, demonstrating a high-frequency sensory-neural hearing loss. As the unmasked bone scores interweave with the air conduction scores, no masking is necessary. (Figure drawn by Herbert Jay Gould. With permission.)

5.11.3 Noise-Induced Hearing Loss

Noise is a major factor in modern life and is a significant contributor to outer hair cell damage. Noise-induced hearing loss has a characteristic notch in the 4000–6000-Hz region for both air- and bone-conducted scores (Figure 5.16). This notch progresses with exposure and, in older individuals, will combine with presbycusis (loss associated with aging) to render significant high-frequency impairments.

As noted earlier, OSHA and NIOSH have established noise exposure limits for the workplace. However, noise in modern day life is not limited to the workplace. Many recreational activities now include use of power tools and engines that produce high levels of potentially damaging noise (Bess and Powell 1972; Chung et al. 1981; McClymont and Simpson 1989; Molvaer and Natrud 1979; Plakke 1983; Shirreffs 1974). Properly fitting hearing protection in the form of ear muffs and ear plugs are effective in preventing noise-induced losses and should always be worn in situations where noise is present.

5.11.4 Presbycusis

Presbycusis is hearing loss due to aging and is one of the most common forms of loss. Prevalence estimates suggest that approximately 25% of the population aged 65–70 years of age experience hearing loss. This increases to approximately 40% for those over age 75. Such losses lead to difficulties in the quality of life for older individuals, and this social problem will increase as the average life span increases. There are multiple physiological etiologies for presbycusis, including outer hair cell loss, inner hair cell loss, metabolic changes associated with degeneration of the stria vascularis, and loss of neural elements (Frisina and Frisina 1997; Schuknecht and Gacek 1993). Remediation of the loss and fitting of appropriate amplification remains a challenge (Cohn 1999).

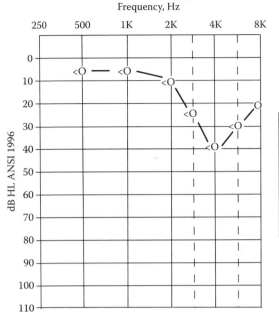

FIGURE 5.16 The audiogram demonstrates right ear thresholds seen in a noise-induced hearing loss. The notch in the 4000-Hz region, with improvement at higher frequencies is characteristic of noise damage to the cochlea. (Figure drawn by Herbert Jay Gould. With permission.)

5.11.5 Auditory Neuropathy/Dys-Synchrony

A relatively newly discovered disorder is auditory neuropathy or auditory dys-synchrony reported by Starr and colleagues (1996). This disorder is a disruption of the inner hair cell/neural junction or a conduction problem of the VIIIth nerve. Advent of OAE testing led to the discovery and diagnosis of the disorder, which is characterized by an absent ABR, absent acoustic reflexes, and strong outer hair cell function with OAEs. For individuals exhibiting auditory dys-synchrony, hearing aid use typically is not recommended. There have been some reports of improved function with cochlear implants. A cochlear implant is an electronic device that receives acoustic input, encodes it, and then transmits the coded signal to an array of electrodes that have been surgically threaded into the cochlea (Cooper 1991).

5.11.6 Meniere's Disease/Endolymphatic Hydrops

Meniere's disease is a constellation of symptoms and includes fluctuating sensory neural hearing loss, vertigo and tinnitus, and a feeling of fullness in the ear (Mateijsen et al. 2001). These symptoms appear to be from failure to regulate fluid appropriately in the scala media. The term endolymphatic hydrops is often inappropriately used in place of Meniere's disease. Hearing loss is typically unilateral (>80%) and more severe in the low-frequency region. As the disease progresses, the hearing loss configuration tends to flatten and become more severe. One common measure for Meniere's disease is electrocochleography (EcochG) (Ferraro and Tibbils 1999; Storms et al. 1996). With EcochG, the ratio of the summating potential to the action potential is evaluated and, as a rule of thumb, values larger than 0.4 are considered abnormal. This value is dependent on a number of factors including the presence of symptoms at the time of test and the depth of electrode insertion. A second measure used in assessing Meniere's is the glycerol test (Mori et al. 1985; Thomsen and Vesterhauge 1979). Pure tone and speech discrimination scores are obtained, and the patient is then given a diuretic, typically glycerol. Hearing is then measured

2 h after taking the diuretic. If either a 12-dB improvement in the PTA or a 16% improvement in speech recognition scores is observed, the results are positive for the disease.

5.11.7 Ototoxicity

Modern antibiotics, chemotherapeutics, analgesics, and environmental treatments can be toxic to the inner ear. Mycin antibiotics, for example, are known as toxic to hair cell structures in both the cochlea and vestibular systems. They can produce profound hearing loss and balance problems.

cis-Platinum is a widely used chemotherapy agent for certain cancers. Hearing loss related to *cis*-platinum ototoxicity is bilateral, typically symmetrical, progressive, and irreversible. Degree of loss is dependent on dosage, route of administration, age, previous exposure to cranial irradiation, and any previous cochlear damage (Berg et al. 1999; Nagy et al. 1999).

Salicylates (aspirin) are also ototoxic (Cazals 2000). However, the effects are reversible. High-dose levels of aspirin, as used for arthritis pain relief, can result in tinnitus and sensory neural loss. Stopping the use of aspirin normally reverses the hearing loss and relieves the tinnitus.

5.11.8 Central Disorders

The most common problems with audition in the central nervous system are generically known as acoustic nerve tumors. This term is only somewhat accurate in that many of the growths are vestibular schwannomas. Nevertheless, the effects on audition are the same: the growth disrupts normal transmission in nerve conduction from the cochlea. Often there is little or no loss of hearing for pure tones in the affected ear in the early stages. Speech recognition is often affected and, if lower than expected scores are observed relative to the PTA, a problem is suspect. A hallmark of VIIIth nerve involvement is abnormal auditory adaptation or fatigue. Tone decay procedures and Bekesy audiometry were initially used as part of the test battery for the detection of VIIIth nerve lesions. These measures were replaced by acoustic reflex decay testing and, more recently, by auditory brain stem response tests (ABR). ABR in turn has largely been supplanted by the use of magnetic resonance imaging (MRI). However, a new variant of the ABR test may significantly improve its diagnostic efficiency (Don et al. 1997).

Problems central to the auditory nerve that impinge upon the auditory system can also result in auditory dysfunction. These include, but are not limited to, multiple sclerosis, ischemic attack, vascular disorder, and neoplasms. In general, these entities have very subtle effects on audition due to the highly redundant and broadly distributed nature of the auditory nervous system. Therefore, pathology in one part of the system may be compensated for by the remaining unaffected areas. Similarly, most auditory signals themselves have a built-in redundancy. For these reasons, a general rule in behavioral assessment of the central auditory nervous system suggests both the neurological system and the acoustic signal must have a reduction in their redundant characteristics before a problem is uncovered. Therefore, behavioral auditory measures for central disorders often involve complex manipulations of the auditory signal in order to stress the auditory pathways employed in the processing of those signals.

5.12 Amplification

Although there are numerous pathological processes resulting in hearing loss, a vast majority of individuals with moderate-to-profound loss have cochlear losses which are not correctable by medical or surgical means. For these individuals, amplification and assistive listening devices typically provide significant improvement in their ability to hear and understand speech.

Hearing aids come in a number of styles and circuit designs. The device most appropriate for any individual is dependent on numerous factors such as hearing thresholds, dynamic range, speech comprehension, life style, cognitive ability, and dexterity. The fitting of a hearing aid must be part of a total rehabilitative package that will permit the person to make maximum use of any residual hearing. Some individuals adapt well to amplification, while others have significant difficulty (Alberti et al. 1984; Cox and Alexander 2000; Malinoff and Weinstein 1989; Mulrow et al. 1992).

5.12.1 Traditional Hearing Aids

There are several basic hearing aid styles in use today. The body aid style is not frequently used except for young children and where large amounts of power are required, for example, bones conduction. The body aid is similar in size to an iPod and is connected electrically with a cord to the earphone. This positions the microphone and speaker, known as a receiver in hearing aid parlance, at a distance from one another. This separation reduces the problem of feedback, which produces a loud squeal.

The behind-the-ear (BTE) aid is a small aid that is placed behind the auricle. The microphone is typically mounted at the top of the aid and the receiver situated inside the aid and coupled acoustically to the ear by a tube. Recently, the mini BTE has gained significantly in popularity. This aid is smaller than the traditional BTE and has moved the receiver from the body of the aid to the ear canal. This separation of microphone and receiver coupled with modern feedback algorithms in the aid permit open-fit ear canals. An open fitting allows for a reduction in low-frequency sound, while maintaining high-frequency information. This type of aid works well for audiometric configurations with good low-frequency hearing and sharply falling scores above 1000 Hz.

The in-the-ear (ITE) style of hearing aid fits into the ear canal with the electronics situated in the area of the concha. These small aids are popular but require manual dexterity for placement and operation of their small controls.

The final style is a completely-in-canal (CIC) aid. These aids fit down into the ear canal and take advantage of the resonant characteristics of the auricle. CIC aids require dexterity for insertion and a canal morphology that is large enough to permit placement of the electronics into the area. A newer version of this style aid uses a long life battery and can be left in place for up to 120 days.

The circuits in hearing aids come in both analog and digital form and offer varied filtering and signal compression characteristics. The digital designs offer significantly more flexibility in fitting than their analog counterparts. Signal filters in the digital designs are narrower and provide greater control in the final signal configuration. In spite of significant cost differentials favoring analog over digital aids, the market has shifted to predominantly digital configurations over the last 10 years.

Signal compression helps prevent distortion, discomfort, and damage to the cochlea through over-amplification. If a linear circuit is used, peak clipping of the signal will occur at the maximum signal output. This peak clipping creates distortion in the output signal and lowers speech intelligibility. Loud sounds which occur in our modern environment also can create a problem by overamplifying a signal, thereby creating discomfort and possible additional noise damage to the cochlea. Signal compression establishes a nonlinear amplification such that amplification for louder sounds is reduced until, at particularly high signal levels, little if any amplification is present.

Microphone arrays allowing directionality have provided a significant improvement in hearing aid design. This design on BTE and ITE aids typically permits amplification of signals located to the front of the listener, while suppressing sounds of less significance from the sides and back of the listener. Thus, background noise, a major problem for the hearing aid user, is significantly reduced. For further information on hearing aid design and fitting the reader is referred to Valente (2002) and Sandlin (2000).

Relatively new changes in hearing aid technology have resulted in hearing aids that communicate with one another as well as with other electronic devices. When an individual is fitted with two hearing aids, known as a binaural fitting, the aids have traditionally operated as independent devices. New technology now permits the devices to coordinate their output to enhance the listening environment. Similarly, a number of hearing aids now connect with Bluetooth technology to interface the user directly to the phone, television, or other Bluetooth-enabled devices.

5.12.2 Implantable Aids

There are now three types of implantable hearing aids. The most recent of these is similar to a traditional hearing aid. However, the device uses an accelerometer attached to the malleus as the input instead of a microphone. A second accelerometer is attached to drive the stapes in place of a traditional receiver. In a small clinical trial (61 subjects), it performed better than traditional hearing aids. The advantage of this device is that it makes full use of the pinna and ear canal resonances. In addition, because it is implanted, it is in operation in all environments, unlike a hearing aid which, for example, can be damaged if it gets wet. There are several potential disadvantages, including cost (up to 4 or 5 times the cost of a topline nonimplanted digital hearing aid), surgical risk, and battery replacement requiring additional surgery every 4–9 years.

A second type of implantable device is the cochlear implant. Cochlear implants have become firmly established for cases of severe-to-profound cochlear hearing loss. The implant provides a significant advantage in auditory rehabilitation for individuals with significant hearing loss and poor speech recognition. The implant is an array of electrodes which is threaded into the cochlea. The electrodes send electrical signals to the tonotopically organized VIIIth nerve receptors spread along the length of the cochlea. There is a signal processor connected to the electrode array which codes the electrical signals based on the acoustic input. This code activates combinations of electrodes to represent the incoming sounds. Implants have been successful in providing hearing function to the profoundly hearing impaired.

The cochlear implant requires that the auditory portion of the VIIIth nerve is intact and the cochlea will permit entry of the electrode array. However, some pathology, such as meningitis, can result in calcification of the cochlea, thereby preventing insertion of the electrode. Surgical intervention for other pathologies, such as neurofibromatosis, can result in sectioning of the VIIIth nerve, thereby severing the connection from the cochlea to the brain. For these cases, a new device, the brainstem implant, is currently being developed (Lenarz et al. 2001, 2002; Nevison et al. 2002; Toh and Luxford 2002). This device places the electrode array on the posterior lateral surface of the brainstem over the cnc. Initial trials with this device have shown successes similar to those seen with early single-channel cochlear implants.

References

Alberti, P. W., M. K. Pichora-Fuller et al. 1984. Aural rehabilitation in a teaching hospital: Evaluation and results. *Ann Otol Rhinol Laryngol* **93**(6 Pt 1): 589–594.

Asha, A. S. L. h. A. 1988. Guidelines for determining the threshold level for speech. *Asha* **30**: 85–89.

Belin, P. and L. Zatorre 2000. 'What', 'where', and 'how' in auditory cortex. *Nat Am Neurosci* **3**(10): 965–966.

Beranek, L., W. Blazier et al. 1971. Preferred noise criterion (PNC) curves and their application to rooms. *J Acoust Soc Am* **50**(5): 1223–1228.

Berg, A. L., J. B. Spitzer et al. 1999. Ototoxic impact of cisplatin in pediatric oncology patients. *Laryngoscope* **109**(11): 1806–1814.

Berglund, B., T. Lindvall et al. 1999. Guidelines for community noise. From http://whqlibdoc.who.int/hq/1999/a68672.pdf.

Bess, J. C. 1971. Ear canal collapse. A review. *Arch Otolaryngol* **93**(4): 408–412.

Bess, F. H. and R. L. Powell 1972. Hearing hazard from model airplanes. A study on their potential damaging effects to the auditory mechanism. *Clin Pediatr (Phila)* **11**(11): 621–624.

Bickford, R., J. Jacobson et al. 1964. Nature of averaged evoked potentials to sound and other stimuli in man. *Annu NY Acad Sci* **112**: 204–223.

Bilger, R. C., M. L. Matthies et al. 1990. Genetic implications of gender differences in the prevalence of spontaneous otoacoustic emissions. *J Speech Hear Res* **33**(3): 418–432.

Borg, E. 1972. Excitability of the acoustic m. stapedius and m. tensor tympani reflexes on the nonanesthetized rabbit. *Acta Psychol Scand* **85**(3): 374–389.

Cazals, Y. 2000. Auditory sensori-neural alterations induced by salicylate. *Prog Neurobiol* **62**(6): 583–631.

Chaiklin, J. B. and M. E. McClellan 1971. Audiometric management of collapsible ear canals. *Arch Otolaryngol* **93**(4): 397–407.

Chung, D. Y., R. P. Gannon et al. 1981. Shooting, sensorineural hearing loss, and workers' compensation. *J Occup Med* **23**(7): 481–484.

Cohn, E. S. 1999. Hearing loss with aging: Presbycusis. *Clin Geriatr Med* **15**(1): 145–161, viii.

Coles, R. B., A. Guppy et al. 1989. Frequency sensitivity and directional hearing in the gleaning bat, *Plecotus auritus* (Linnaeus 1758). *J Comp Physiol [A]* **165**(2): 269–280.

Cone-Wesson, B. and J. Wunderlich 2003. Auditory evoked potentials from the cortex: Auditory applications. *Curr Opin Otolaryngol Head Neck Surg* **11**(5): 372–377.

Cooper, H. 1991. *Practical Aspects of Audiology Cochlear Implants: A Practical Guide.* Singular, San Diego.

Cox, R. M. and G. C. Alexander 2000. Expectations about hearing aids and their relationship to fitting outcome. *J Am Acad Audiol* **11**(7): 368–382; quiz 407.

Dadson, R. S. and J. H. King 1952. A determination of the normal threshold of hearing and its relation to the standardization of audiometers. *J Laryngol Otol* **66**: 366–378.

Davis, P. 1939. Effects of acoustic stimuli on the waking human brain. *J Neurophysiol* **2**: 494–499.

Davis-Penn, W. and M. Ross 1993. Pediatric experiences with frequency transposing. *Hear Instr* **44**(4): 27–32.

Don, M., B. Kwong et al. 2005a. A diagnostic test for Meniere's disease and cochlear hydrops: Impaired high-pass noise masking of auditory brainstem responses. *Otol Neurotol* **26**(4): 711–722.

Don, M., B. Kwong et al. 2005b. The stacked ABR: A sensitive and specific screening tool for detecting small acoustic tumors. *Audiol Neuro-Otol* **10**(5): 274–290.

Don, M., A. Masuda et al. 1997. Successful detection of small acoustic tumors using the stacked derived-band auditory brain stem response amplitude. *Am J Otol* **18**(5): 608–621; discussion 682–605.

Dorman, M. F., J. M. Lindholm et al. 1987. Vowel intelligibility in the absence of the acoustic reflex: Performance-intensity characteristics. *J Acoust Soc Am* **81**(2): 562–564.

Durrant, J. D. and J. H. Lovrinic. 1995. *Bases of Hearing Science.* Williams & Wilkins, Baltimore, MD.

Eggermont, J. J. 2001. Between sound and perception: Reviewing the search for a neural code. *Hear Res* **157**(1–2): 1–42.

Feldman, A. S. and L. A. Wilber. 1976. *Acoustic Impedance and Admittance: The Measurement of Middle Ear Function.* Williams & Wilkins, Baltimore, MD.

Ferraro, J. A. and R. P. Tibbils. 1999. SP/AP area ratio in the diagnosis of Meniere's disease. *Am J Audiol* **8**(1): 21–28.

Fletcher, H., Galt, R. H. 1950. The perception of speech and its relation to telepony. *J. Acoust. Soc. Am.* **22**(2): 89–151.

Flisberg, K. 1967. Determination of the airway resistance of the eustchian tube. *Acta Otolaryngol* **63**(Suppl 224): 376–384.

Flisberg, K. 1970. The effects of vacuum on the tympanic cavity. *Otolaryngol Clin No Am* **3**: 3–13.

Flisberg, K., S. Ingelstedt et al. 1963. On middle ear pressure. *Acta Otolaryngol* **56**(Suppl 182): 43–56.

French, N. R. and J. C. Steinberg 1947. Factors governing the intelligibility of speech sounds. *J Acoust Soc Am* **19**(1): 90–119.

Frisina, D. R. and R. D. Frisina 1997. Speech recognition in noise and presbycusis: Relations to possible neural mechanisms. *Hear Res* **106**(1–2): 95–104.

Geisler, C. D., L. S. Frishkopf et al. 1958. Extracranial responses to acoustic clicks in man. *Sci News* **128**(3333): 1210–1211.

Gelfand, S. A. 1998. *Hearing, An Introduction to Psychological and Physiological Acoustics.* M. Dekker, New York.

Giraud, A. L., C. Lorenzi et al. 2000. Representation of the temporal envelope of sounds in the human brain. *J Neurophysiol* **84**(3): 1588–1598.

Gold, T. 1948. Hearing II: The physical basis of action in the cochlea. *Proc R Soc Lond Ser B: Biol Sci* **135**: 492–498.

Goldstein, R. and L. Rodman 1967. Early components of averaged evoked responses to rapidly repeated auditory stimuli. *J Speech Hearing Res* **10**: 697–705.

Gorga, M. P. and A. R. Thornton 1989. The choice of stimuli for ABR measurements. *Ear Hear* **10**(4): 217–230.

Guppy, A. and R. B. Coles 1988. Acoustical and neural aspects of hearing in the Australian gleaning bats, *Macroderma gigas* and *Nyctophilus gouldi*. *J Comp Physiol [A]* **162**(5): 653–668.

Hall, J. W. 2000. *Handbook of Otoacoustic Emissions*. Singular, San Diego, CA.

Hall, J. 2007. *New Handbook of Auditory Evoked Responses*. Pearson, Boston, MA.

Halpin, C., A. Thornton et al. 1996. The articulation index in clinical diagnosis and hearing aid fitting. *Curr Opin Otolaryngol Head Neck Surg* **4**: 325–334.

Harker, L. A., E. Hosick et al. 1977. Influence of succinylcholine on middle component auditory evoked potentials. *Arch Otolaryngol* **103**(3): 133–137.

Hood, J. 1960. The principles and practice of bone conduction audiometry. *Laryngoscope* **70**: 1211–1228.

Hood, L. J., C. I. Berlin et al. 2003. Patients with auditory neuropathy/dys-synchrony lack efferent suppression of transient evoked otoacoustic emissions. *J Am Acad Audiol* **14**(6): 302–313.

Hurley, R. M., A. Hurley et al. 2002. The effect of midline petrous apex lesions on tests of afferent and efferent auditory function. *Ear Hear* **23**(3): 224–234.

Ingelstedt, S. 1967. Mechanics of the human middle ear. *Acta Oto-Laryngo* **64**(Suppl 228): 1–57.

Institute, A. N. S. 1983. ANSI S1.4-1983 American National Standard Specification for Sound Level Meters, American Institute of Physics for the Acoustical Society of America: vii, 18.

Institute, A. N. S. 1987. ANSI S3.39-1987 American National Standard Specification for instruments to measure aural acoustic impedance and admittance (Aural accoustic immittance), American Institute of Physics for the Acoustical Society of America: vi, 21.

Institute, A. N. S. 1996. ANSI S3.6-1996 American National Standard Specification for Audiometers, Acoustical Society of America: vi, 33.

Institute, A. N. S. 1997. ANSI S3.5-1997 Methods for calculation of the Speech Intelligibility Index, Acoustical Society of America: iii, 22.

Jacobson, G. P., C. W. Newman et al. 1993. *Handbook of Balance Function Testing*. Mosby Year Book, St. Louis.

Jerger, J. 1970. Clinical experience with impedance audiometry. *Arch Otolaryngol* **92**: 311–324.

Jewett, D. L., M. N. Romano et al. 1970. Human auditory evoked potentials: Possible brain stem components detected on the scalp. *Science* **167**(3924): 1517–1518.

Jones, M. 1961. Pressure changes in the middle ear after altering the composition of contained gas. *Acta Otolaryngol* **53**: 1–11.

Kavanagh, K. T. and W. D. Domico 1986. High-pass digital filtration of the 40 Hz response and its relationship to the spectral content of the middle latency and 40 Hz responses. *Ear Hear* **7**(2): 93–99.

Kavanagh, K. T. and R. Franks 1989. Analog and digital filtering of the brain-stem auditory evoked-response. *Ann Otol Rhinol Laryngol* **98**(7): 508–514.

Kavanagh, K. T., W. D. Domico et al. 1988. Digital filtering and spectral analysis of the low intensity ABR. *Ear Hear* **9**(1): 43–47.

Kemp, D. T. 1978. Stimulated acoustic emissions from within the human auditory system. *J Acoust Soc Am* **64**(5): 1386–1391.

Kemp, D. T. 1986. Otoacoustic emissions, travelling waves and cochlear mechanisms. *Hear Res* **22**: 95–104.

Kikuchi, T., J. C. Adams et al. 2000. Potassium ion recycling pathway via gap junction systems in the mammalian cochlea and its interruption in hereditary nonsyndromic deafness. *Med Electron Microsc* **33**(2): 51–56.

Killion, M. C., P. A. Niquette et al. 2004. Development of a quick speech-in-noise test for measuring signal-to-noise ratio loss in normal-hearing and hearing-impaired listeners. *J Acoust Soc Am* **116**(4 Pt 1): 2395–2405.

Kiren, T., M. Aoyagi et al. 1994. An experimental study on the generator of amplitude-modulation following response. *Acta Otolaryngol Suppl* **511**: 28–33.

Kok, M. R., G. A. van Zanten et al. 1993. Click-evoked oto-acoustic emissions in 1036 ears of healthy newborns. *Audiology* **32**(4): 213–224.

Kollmeier, B. and R. Koch 1994. Speech enhancement based on physiological and psychoacoustical models of modulation perception and binaural interaction. *J Acoust Soc Am* **95**(3): 1593–1602.

Kraus, N., T. McGee et al. 1995. Central auditory system plasticity associated with speech discrimination training. *J Cogn Neurosci* **7**(1): 25–32.

Lenarz, T., M. Moshrefi et al. 2001. Auditory brainstem implant: part I. Auditory performance and its evolution over time. *Otol Neurotol* **22**(6): 823–833.

Lenarz, M., C. Matthies et al. 2002. Auditory brainstem implant: part II: Subjective assessment of functional outcome. *Otol Neurotol* **23**(5): 694–697.

Lipscomb, D. M. 1994. *Hearing conservation in industry, schools, and the military*. Singular Pub. Group, San Diego, CA.

Lorenzi, C., C. Micheyl et al. 1995. Neuronal correlates of perceptual amplitude-modulation detection. *Hear Res* **90**(1–2): 219–227.

Magnuson, B. 2001. Physiology of the eustachian tube and middle ear pressure regulation. *Physiology of the Ear*. A. F. a. S.-S. Jahn, J. Singular-Thomson Learning, San Diego, CA, pp. 75–99.

Malinoff, R. L. and B. E. Weinstein. 1989. Measurement of hearing aid benefit in the elderly. *Ear Hear* **10**(6): 354–356.

Marshall, L. and M. Grossman 1982. Management of ear canal collapse. *Arch Otolaryngol* **108**: 357–361.

Martin, F. N. and J. G. Clark 2002. *Introduction to Audiology*. Allyn and Bacon, Boston, MA.

Martin, F. N. and L. K. Dowdy 1986. A modified spondee threshold procedure. *J Aud Res* **26**(2): 115–119.

Martin, G. K., R. Probst et al. 1990. Otoacoustic emissions in human ears: Normative findings. *Ear Hear* **11**(2): 106–120.

Mateijsen, D. J., P. W. Van Hengel et al. 2001. Pure-tone and speech audiometry in patients with Meniere's disease. *Clin Otolaryngol* **26**(5): 379–387.

McClymont, L. G. and D. C. Simpson 1989. Noise levels and exposure patterns to do-it-yourself power tools. *J Laryngol Otol* **103**(12): 1140–1141.

Mendel, L. L. and J. L. Danhauer 1997. *Audiologic Evaluation and Management and Speech Perception Assessment*. Singular Pub. Group, San Diego, CA.

Mendel, M. and R. Goldstein 1969. The effect of test conditions on the early components of the averaged electroencephalic response. *J Speech Hear Res* **12**: 344–350.

Molvaer, O. I. and E. Natrud 1979. Ear damage due to diving. *Acta Otolaryngol Suppl* **360**: 187–189.

Mori, N., A. Asai et al. 1985. Comparison between electrocochleography and glycerol test in the diagnosis of Meniere's disease. *Scand Audiol* **14**(4): 209–213.

Moulin, A., L. Collet et al. 1993. Contralateral auditory stimulation alters acoustic distortion products in humans. *Hear Res* **65**(1–2): 193–210.

Muller-Preuss, P., C. Flachskamm et al. 1994. Neural encoding of amplitude modulation within the auditory midbrain of squirrel monkeys. *Hear Res* **80**(2): 197–208.

Mulrow, C. D., M. R. Tuley et al. 1992. Correlates of successful hearing aid use in older adults. *Ear Hear* **13**(2): 108–113.

Murphy, D. 1979. Negative pressure in the middle ear by ciliary propulsion of mucus through the eustachian tube. *Laryngoscope* **89**: 954–961.

Musiek, F., J. Baran et al. 1994. *Neuroaudiology Case Studies*. Singular Publishing Group, San Diego, CA.

Nagy, J. L., D. J. Adelstein et al. 1999. Cisplatin ototoxicity: The importance of baseline audiometry. *Am J Clin Oncol* **22**(3): 305–308.

Naunton, R. F. and S. Zerlin 1976. Basis and some diagnostic implications of electrocochleography. *Laryngoscope* **86**(4): 475–482.

Naunton, R. F. and S. Zerlin 1977. The evaluation of peripheral auditory function in infants and children. *Otolaryngol Clin North Am* **10**(1): 51–58.

Nevison, B., R. Laszig et al. 2002. Results from a European clinical investigation of the nucleus multichannel auditory brainstem implant. *Ear Hear* **23**(3): 170–183.

NIOSH 1998. *Criteria for a Recommended Standard: Occupational Noise Exposure.* Publication No. 98-126, from http://www.cdc.gov/niosh/docs/98-126/chap1.html#11.

Noise_Pollution_Clearinghouse 2012. EPA Document Collection: Noise related, from http://www.nonoise.org/epa/.

Penner, M. J. 1990. An estimate of the prevalence of tinnitus caused by spontaneous otoacoustic emissions. *Arch Otolaryngol Head Neck Surg* **116**(4): 418–423.

Penner, M. J. and L. Glotzbach 1994. Covariation of tinnitus pitch and the associated emission: A case study. *Otolaryngol Head Neck Surg* **110**(3): 304–309.

Penner, M. J., L. Glotzbach et al. 1993. Spontaneous otoacoustic emissions: Measurement and data. *Hear Res* **68**(2): 229–237.

Penrod, J. 1985. Speech discrimination testing. *Handbook of Clinical Audiology.* J. Katz. Williams & Wilkins, Baltimore, MD, pp. 235–253.

Phillips, D. P., A. Stuart et al. 2002. Re-examination of the role of the human acoustic stapedius reflex. *J Acoust Soc Am* **111**(5 Pt 1): 2200–2207.

Plakke, B. L. 1983. Noise levels of electronic arcade games: A potential hearing hazard to children. *Ear Hear* **4**(4): 202–203.

Ponton, C. W., J. J. Eggermont et al. 1999. Maturation of human central auditory system activity: Evidence from multi-channel evoked potentials. *Clin Neurophysiol* **111**: 220–236.

Purdy, S. C. and P. J. Abbas. 2002. ABR thresholds to tonebursts gated with Blackman and linear windows in adults with high-frequency sensorineural hearing loss. *Ear Hear* **23**(4): 358–368.

Reuter, T., S. Nummela et al. 1998. Elephant hearing. *J Acoust Soc Am* **104**(2 Pt 1): 1122–1123.

Robinette, M. S. and T. J. Glattke 1997. *Otoacoustsic Emissions: Clinical Applications.* Thieme, New York, NY.

Robles, L. and M. A. Ruggero 2001. Mechanics of the mammalian cochlea. *Physiol Rev* **81**(3): 1305–1352.

Sade, J. 2000. The buffering effect of middle ear negative pressure by retraction of the pars tensa. *Am J Otol* **21**(1): 20–23.

Sade, J. and Hadas, E., 1979. Prognostic evaluation of secretory otitis media as a function of mastoid pneumotisation. *Arch Otorhinolaryngol* **225**: 39–44.

Sade, J., Halevy, A., Hadas, E., and Saba, K., 1976. Clearance of middle ear effusions and middle ear pressures. *AORL* **85**(Suppl 25): 58–62.

Sandlin, R. E. 2000. *The Textbook of Hearing Aid Amplification.* Singular Thomson Learning, San Diego, CA.

Schoonhoven, R., P. J. Lamore et al. 1999. The prognostic value of electrocochleography in severely hearing-impaired infants. *Audiology* **38**(3): 141–154.

Schuknecht, H. F. and M. R. Gacek 1993. Cochlear pathology in presbycusis. *Ann Otol Rhinol Laryngol* **102**(1 Pt 2): 1–16.

Sharma, A., M. F. Dorman et al. 2002. A sensitive period for the development of the central auditory system in children with cochlear implants: Implications for age of implantation. *Ear Hear* **23**(6): 532–539.

Sharma, A., M. F. Dorman et al. 2005. The influence of a sensitive period on central auditory development in children with unilateral and bilateral cochlear implants. *Hear Res* **203**(1–2): 134–143.

Sharma, A., N. Kraus et al. 1993. Acoustic versus phonetic representation of speech as reflected by the mismatch negativity event-related potential. *Electroencephalogr Clin Neurophysiol* **88**(1): 64–71.

Shepard, N. T. and S. A. Telian. 1996. *Practical Management of the Balance Disorder Patient.* Singular Pub. Group, San Diego, CA.

Shera, C. A. and J. J. Guinan, Jr. 1999. Evoked otoacoustic emissions arise by two fundamentally different mechanisms: A taxonomy for mammalian OAEs. *J Acoust Soc Am* **105**(2 Pt 1): 782–798.

Shirreffs, J. H. 1974. Recreational noise: Implications for potential hearing loss to participants. *J Sch Health* **44**(10): 548–550.

Sobhy, O. A. and H. J. Gould 1993. Interaural attenuation using insert earphones: Electrocochleographic approach. *J Am Acad Audiol* **4**(2): 76–79.

Sohmer, H. and M. Feinmesser 1967. Cochlear action potentials recorded from the external ear in man. *Ann Otol* **76**: 427–435.

Spiess, A. C., H. Lang et al. 2002. Effects of gap junction uncoupling in the gerbil cochlea. *Laryngoscope* **112**(9): 1635–1641.

Starr, A., T. W. Picton et al. 1996. Auditory neuropathy. *Brain* **119** (Pt 3): 741–753.

Starr, A., D. McPherson et al. 1991. Absence of both auditory evoked potentials and auditory percepts dependent on timing cues. *Brain* 114: 1157–1180.

Storms, R. F., J. A. Ferraro et al. 1996. Electrocochleographic effects of ear canal pressure change in subjects with Meniere's disease. *Am J Otol* **17**(6): 874–882.

Stover, L. and S. J. Norton 1993. The effects of aging on otoacoustic emissions. *J Acoust Soc Am* **94**(5): 2670–2681.

Strickland, E. A., E. M. Burns et al. 1985. Incidence of spontaneous otoacoustic emissions in children and infants. *J Acoust Soc Am* **78**(3): 931–935.

Studebaker, G. A., G. A. Gray et al. 1999a. Prediction and statistical evaluation of speech recognition test scores. *J Am Acad Audiol* **10**(7): 355–370.

Studebaker, G. A., D. M. McDaniel et al. 1999b. Monosyllabic word recognition at higher-than-normal speech and noise levels. *J Acoust Soc Am* **105**: 2431–2444.

Thomsen, J. and S. Vesterhauge 1979. A critical evaluation of the glycerol test in Meniere's disease. *J Otolaryngol* **8**(2): 145–150.

Thornton, A. and A. Slaven 1995. The effect of stimulus rate on the contralateral suppression of transient evoked otoacoustic emissions. *Scand Audiol* **24**: 83–90.

Toh, E. H. and W. M. Luxford 2002. Cochlear and brainstem implantation. *Otolaryngol Clin North Am* **35**(2): 325–342.

Tyler, R. S. and N. Tye-Murray 1991. Cochlear implant signal-processing strategies and patient perception of speech and enviromental sounds. *Practical Aspects of Audiology: Cochlear Implants*. H. Cooper. Singular, San Diego, CA.

Valente, M. 2002. *Strategies for Selecting and Verifying Hearing Aid Fittings*. Thieme, New York, NY.

van Dijk, P. and H. P. Wit 1990. Amplitude and frequency fluctuations of spontaneous otoacoustic emissions. *J Acoust Soc Am* **88**(4): 1779–1793.

Veuillet, E., L. Collet et al. 1992. Differential effects of ear-canal pressure and contralateral acoustic stimulation on evoked otoacoustic emissions in humans. *Hear Res* **61**(1–2): 47–55.

Von Békésy, G. 1960. *Experiments in Hearing*. McGraw-Hill, New York, NY.

Webster, D. B. 1995. *Neuroscience of Communication*. Singular Pub. Group, San Diego, CA.

Wever, E. G. and M. Lawrence 1954. *Physiological Acoustics*. Princeton University Press, Princeton, NJ.

Whitehead, M. L., N. Kamal et al. 1993. Spontaneous otoacoustic emissions in different racial groups. *Scand Audiol* **22**(1): 3–10.

Whitehead, M. L., B. B. Stagner et al. 1995. Dependence of distortion-product otoacoustic emissions on primary levels in normal and impaired ears. II. Asymmetry in L1,L2 space. *J Acoust Soc Am* **97**(4): 2359–2377.

Yost, W. A. 2000. *Fundamentals of Hearing: An Introduction*. Academic Press, San Diego, CA.

II

Imaging Techniques

Contributions of Technological Developments to Imaging

Imaging the interior workings of the human body has long been an elusive ambition of healthcare providers. From the early drawings of Leonardo da Vinci, humans have aspired to visualize the body's machinery as accurately as possible. The development of x-ray imaging provided the first glimpse of a living human structure. This opened the door for visual diagnosis. The importance of such developments cannot be understated, given the degree to which humans rely on visual information.

Most imaging techniques aim to provide static images of hard and soft tissues. The use of contrast agents and multimodality imaging has greatly improved the resolution of the images obtained. However, recent technological and computational advances, along with the development of specific radiopharmaceuticals, have extended these capabilities to actual functional imaging. It is now possible to assess certain physiological processes and organ functions in real time. The variety of imaging technologies available to clinicians is growing rapidly. The development of mesoscale imaging technologies has yielded devices that can be deployed inside the body via catheters to provide direct live images without contrast agents or ionizing radiation. Extraordinary advances in all areas of imaging technology have allowed clinicians to make accurate and efficient diagnoses, often in a noninvasive fashion.

6

Magnetic Resonance Imaging

Timothy P.L.
Roberts
Children's Hospital of
Philadelphia

Christopher K.
Macgowan
University of Toronto

Mary P. McDougall
Texas A&M University

6.1 Introduction

Magnetic resonance imaging (MRI) has evolved in the last three decades to become the method of choice for the vast proportion of noninvasive cross-sectional "scanning" in diagnostic radiology. Apart from simply "looking inside the body," the success and range of MRI has built upon the variety of physiological and pathological insights available from different MRI approaches. That is to say, there is not "an MRI," but rather a broad variety of "MRI pulse sequences" that generate images whose appearance may reflect not merely tissue density (such as x-ray computed tomography [CT]), but also properties of the physicochemical microenvironment, such as fluid mobility, macromolecule (e.g., protein) content, metabolic products, blood flow and perfusion, and so on. Mastery of this range of MRI techniques provides the radiologic "artist" with a broad "palette" of physiological sensitivities to exploit or deny as the image is created.

While many textbooks are devoted to MRI, and even specific aspects thereof, the purpose of this chapter is to provide a swift overview of the MRI family of techniques, to discuss applications in the study of the brain and the heart, and to introduce the varied roles of MRI in diagnostic radiology, in preclinical drug development and in basic science.

Founded on the science of nuclear magnetic resonance (NMR), the development of techniques for spatial encoding of NMR signals during the 1970s led to the birth and subsequent explosion of MRI. The MRI scanner thus consists of similar components to an NMR spectrometer, with the addition of systems for controlling the local magnetic field that provide the ability to spatially localize the signals and thus form an image.

6.2 Hardware Components of an MRI Scanner

6.2.1 Magnet

The NMR experiment relies on placing the test sample (or in this case, patient) in a uniform and strong external magnetic field. In standard clinical practice the magnetic field strength may be as low as 0.2 T or as high as 3 T, with 1.5 T being the most established clinical field strength, and 3 T considered state-of-the art for routine clinical use. The highest field strength commercially available/supported for whole-body clinical scanners is 7 T, although limited accessibility due to cost and the multiplicity of technical challenges associated with the higher field strength currently limit the technology to preclinical research applications at specialized sites. Dedicated research systems for human imaging exist at field strengths as high as 8 T and 9.4 T, and even higher field systems at 10.5 T, 11.7 T, and 14 T are in the process of being installed or in the development stages. Smaller-bore experimental MRI systems for small animal imaging and chemical analysis exist with field strengths up to 20 T. The static magnetic field strength of an MRI scanner is commonly referred to as B_0. Essentially all systems employ superconducting technology, with liquid helium cooling.

6.2.2 Gradient Coils/Amplifiers

The feature that distinguishes the MRI scanner from the NMR spectrometer is the ability to perform spatial encoding of detected signals for subsequent image formation. At the heart of this lies a requirement to control the "local magnetic field." Over and above the static (and uniform) magnetic field, B_0, an MRI scanner uses (three) magnetic field gradient coils to impose linear variations in magnetic field in any or all of the three Cartesian axes. Such magnetic field gradients are typically not in place throughout the image acquisition process but are "switched on" as required during the playing out of the MRI pulse sequence, a set of instructions controlling the timing of RF and gradient systems. Magnetic field gradient strengths on clinical MRI scanners are of the order of mT/m (some systems currently offer the capability of 60 mT/m) and may have 40 cm or more of spatial extent; magnetic field gradient coils are associated with powerful current amplifiers, leading to systems capable of slew rates up to 200 mT/m/ms. That said, high gradient values and short switching times can lead to peripheral nerve stimulation (PNS) in patients, and American and European regulatory agencies specify safe limits on dB/dt (change in field per unit time). Interestingly, the characteristic acoustic noise of the MRI is a product of such rapid switching of magnetic field gradients in the presence of the main static magnetic field.

6.2.3 RF Coils/Transmission and Reception System

With the sample placed in the static magnetic field, B_0, the MRI experiment continues by excitation and detection of the NMR response. As with NMR, the origin of the signal used in MR imaging is the nuclear spin of the tissue constituent atoms. While NMR responses are detectable from a variety of non-zero spin nuclei (including biologically relevant nuclei such as ^{23}Na and ^{31}P), the vast majority of clinical MRI uses the ^{1}H proton as the source of signal. Not only does ^{1}H have the greatest sensitivity of all nuclei to the NMR experiment (with the exception of the essentially irrelevant tritium ion), but it must be remembered that the human body is composed of approximately 70% water (H_2O), each water molecule containing two ^1H protons in a liquid state optimal for NMR. Thus, the ^1H signal source is present at concentrations of approximately 80 M. Boltzmann statistics relates the populations of nuclear spins in the two energy states associated with the application of the external field (B_0 in this case). The source of our signal is the small excess of spins that align in the low-energy state, a number that increases with field strength. However, it is clear that, even at 1.5 T or 3 T, the NMR experiment is extremely inefficient (with an excess population in the low-energy state of only a few spins per million). Thus, the high available spin density of the ^1H in water molecules of the human body is fortuitous!

The frequency of electromagnetic energy used for excitation of the excess low-energy spins is the same as the frequency of the received signal and is governed by Equation 6.1, called the Larmor equation,

$$\omega = \gamma B, \tag{6.1}$$

where ω is the Larmor or resonant frequency, γ the gyromagnetic ratio ($\gamma/2\pi = 42.57$ MHz/T for the ^1H proton) and B the *local* magnetic field. Specifically, B is dominated by the static field B_0, but it is of critical importance to the ability to spatially localize the received signal that B also contains additional local field influences (e.g., magnetic field gradients applied). For the field strengths used for MRI, signals are generated and received at frequencies in the MHz range. Hence, a radiofrequency (RF) coil, similar to the concept of an antenna, is used to transmit RF energy into the sample (patient). Similarly, the same (or a second) coil is used to detect the weak NMR response, prior to preamplification and digitization. Most clinical scanners contain a built-in RF "body coil," but oftentimes an RF coil more tailored to the anatomy under study is used in order to increase the sensitivity of the experiment. For instance, MR imaging of the brain is done with a "head coil"; MR mammography is done with a "breast coil," and so on. In addition, it is now standard for clinical systems to come equipped with multiple receiver channels, enabling the use of multiple RF coils ("RF array coils") to receive the signal from the patient. This increases the signal-to-noise ratio of the experiment, which is typically exploited to accelerate the image acquisition time. This is of particular benefit to dynamic applications such as cardiac imaging and functional studies.

6.3 Image Encoding

While a detailed discussion of the process of image formation is beyond the scope of this chapter, an overview illustrating the main conceptual stages is worthwhile. Consider the ensemble of nuclear spins to be represented by a vector, M, initially aligned with the external magnetic field B_0. The first stage (RF excitation) involves rotating the vector M off the main axis of B_0 through an angle α. This is achieved by applying a magnetic field oscillating at the Larmor frequency via an RF coil, discussed above. The amplitude of the magnetic component of such an RF "pulse" is generally labeled B_1. The rotation or "flip" angle is predicted by the amplitude and duration of the RF pulse:

$$\alpha = \gamma B_1 t, \tag{6.2}$$

where α is the flip angle, γ the gyromagnetic ratio, and t the pulse duration. After the B_1 field (RF pulse) is discontinued, the component of the tilted vector M that is transverse to B_0 then undergoes circular motion (precesses) around the axis of B_0 at the Larmor frequency. Since the magnetic vector M carries with it a tiny magnetic field, this also rotates at that frequency. A conducting coil (our RF coil in this case) experiencing a rate of change of magnetic flux (the rotating transverse component of M in this case) has an electromotive force (EMF) induced in it according to Faraday's law of electromagnetic induction. The small voltage detectable at the ports of the RF coil constitutes our NMR signal that is amplified and digitized. In practice, as will be discussed below, the amplitude of the detected oscillating signal decays over time as the underlying spins contributing to the vector M lose phase coherence due to slightly different precession rates. Overall this damped oscillation—the NMR response—is known as the free induction decay (FID), "free" because it is no longer "forced" by the B_1 field, "induction" because of the mechanism of signal detection, and "decay" because of the signal's decreasing amplitude.

However, imaging requires that spatial information is encoded in the NMR response. How can that be achieved? In our three-dimensional spatial world, there are three mechanisms by which spatial information can be introduced into the process of magnetic resonance image formation: slice selection, frequency encoding, and phase encoding. We will deal briefly with these concepts (see Figure 6.1).

6.3.1 Slice Selection

If the RF field, B_1, is modulated in amplitude, it will cause excitation of spins resonant at a range of Larmor frequencies. If the amplitude modulation takes the form of a sinc function $(1/t)\sin(\pi t)$ envelope, then to a first approximation the frequency response profile takes the form of bandpass (i.e., a "rect function" describes the frequency content of a sinc function). A narrow band of frequencies is uniformly excited, with spins whose resonant frequency lies outside the band being unaffected and remaining aligned with the B_0 field, the nominal z-axis. (Recall that the transverse component of M is our source of signal and therefore if spins remain aligned with the B_0 field, they will not contribute to our signal.) But why would spins have different resonant frequencies? Simultaneously with the application of the sinc-modulated RF pulse, a linear magnetic field gradient is applied along one spatial axis, leading to a spatially varying local effective magnetic field and thus a spatially varying resonant frequency. If the gradient direction is the z-axis and the amplitude is G_z, the Larmor equation becomes modified to

$$\omega(z) = \gamma(B_0 + G_z \cdot z). \tag{6.3}$$

Thus, a narrow band of frequencies excited has a direct interpretation as a narrow band of z-coordinates, or a "slice." This interpretation is based on knowledge of the linear magnetic field gradient, G_z. Thus,

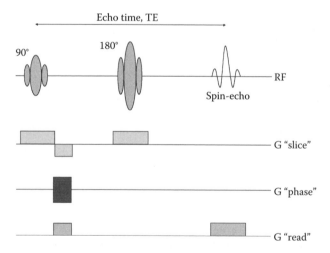

FIGURE 6.1 The timing diagram schematic for an MRI "spin-echo" pulse sequence. Essentially portraying the timing of instructions to the RF transmission system and each of the three spatial magnetic field gradient systems, this schematic representation illustrates the relative timing of 90° and 180° RF pulses (shown as sine-modulated) and subsequent spin-echo formation at the echo time, TE. It also shows (as rectangles) the application of pulses of magnetic field gradient in the slice, phase-encode and readout spatial directions. Note the ladder representation of the phase encoding gradient pulse, implying that each time the shown RF excite/data readout module is repeated a different value of the magnetic field gradient amplitude is used for phase encoding.

the application of a gradient field, G_z, has transformed the property of "frequency selectivity" into the physical property of "spatial selectivity," and we have reduced the dimensionality of the image formation problem from three to two.

6.3.2 Frequency Encoding

As described above, the fundamental NMR response is a damped oscillation at the Larmor frequency, with the decay of the signal response due to the dephasing (loss of coherence of phase) of the spins that were tipped into the transverse plane. Some of the dephasing can be overcome and the NMR response can be restored as an "echo," time shifted with respect to the initial excitation. This will be discussed further below. The NMR signal response must be collected during this echo formation, known as "readout." While the echo occurs at a constant time with respect to the initial excitation, the approach to the echo depends on the precessional frequency of the underlying spins, which is dependent on the local field to which the spins are exposed. Therefore, if the echo is acquired (sampled/digitized) with the coincident application of a magnetic field gradient in one of the two remaining spatial directions, for instance G_x, spin ensembles will be forced into formation of "lower frequency" and "higher frequency" echoes according to their spatial coordinate along this axis (e.g., x), which determines the effective local magnetic field they experience. Since these echoes are inherently summed in the echo acquisition process, some form of frequency analysis is required to resolve the frequency distribution of component echoes in the composite signal. By employing a one-dimensional Fourier transform of the echo, such a profile of spin abundance according to frequency can be determined. With knowledge of the gradient field strength and orientation, this frequency abundance profile can be interpreted as revealing the one-dimensional spin distribution in space (since low frequencies are associated with low spatial coordinates in the direction of the applied magnetic field gradient because of the lower effective local magnetic field). This process is known *as frequency-encoding.*

6.3.3 Phase Encoding

A single, slice-selected, frequency-encoded NMR response can thus be interrogated to reveal a one-dimensional "projection" of a selected slice of the object under study. It does not, however, allow construction of a two-dimensional image. In fact there is insufficient information in a single echo to allow this. Consequently many echoes must be collected by repeated application of the RF excitation/data readout module in order to encode in the second dimension. (Note: In practice, the sequence repeat time (TR) must be set to allow sufficient signal potential [recovery of spins to the low-energy state, ready to be "re-tipped"] for subsequent cycles of the module. This ultimately limits most MRI acquisition rates.) Each collected echo is distinguished by its experience in terms of the remaining spatial axis. After spins have been excited by the RF pulse and during subsequent precession in the transverse plane, a magnetic field gradient is imposed for a short duration, aligned along this third spatial dimension (y). Spins undergo irrecoverable dephasing according to their spatial coordinate in this dimension and according to the magnitude of the gradient field, G_y. This is called *phase encoding.* As the module is repeated, the magnitude of G_y is incremented (hence the ladder representation in the pulse sequence diagram). Consider a peripheral (high y-coordinate) distribution of spins—as G_y is increased, rapid dephasing occurs and spin-echo signal is strongly attenuated. Consider a more central (low y-coordinate) distribution: even increasing G_y only leads to slow dephasing, and subsequently spin-echo amplitude persists, only weakly attenuated. Thus, observing the rate of loss of spin-echo amplitude as a function of increasing G_y (phase-encoding gradient strength), we can make an interpretation in terms of spin distribution in the y-direction. This process can be thought of as gathering "independent views" along the phase-encoding axis. For increased clarity, consider a basic case: To form a 256×256 image of a slice, one would slice encode with G_z according to the description above, frequency encode each echo in the x direction with G_x while digitizing/sampling 256 points, and repeat the process 256 times, incrementing the phase-encode gradient value G_y each time and thus acquiring a unique echo ("independent view") each time. Display and reconstruction of these repeated echoes are discussed below.

6.3.4 *k*-Space

A commonly employed representation is to portray the multiple (typically 128 or 256) spin echoes in a two-dimensional grid (where the *x*-axis scales with readout time) and the *y*-axis offset is proportional to the phase-encoding gradient strength. This "*k*-space" (frequency domain) representation allows a two-dimensional Fourier transformation (FT) to yield an interpretable (space domain) image.[*] In general, the object of the MR pulse sequence is to fill "*k*-space" with data, for subsequent FT and interpretation. Thus, the pursuit of faster imaging approaches, for example, can be viewed conceptually as the pursuit of strategies to fill *k*-space faster. This has seen the evolution, beyond spin-echo imaging of gradient-recalled echo (GRE) imaging, fast spin-echo (FSE) imaging, spiral imaging, and echo planar imaging (EPI). These are discussed at length in the suggested reading, but differ fundamentally as to how *k*-space is filled—either one "line" per RF excitation pulse (as discussed above), or multiple lines per excitation (as in FSE and, ultimately, EPI) where the entirety of *k*-space may be filled with data from a single RF excitation. In that case, TR is rendered largely irrelevant, with image acquisition time being limited primarily by how fast the readout data can be digitized.

6.4 Endogenous Tissue Contrast

MRI has evolved into a powerful and, importantly, versatile tool in diagnostic radiology. This arises largely from the fact that the NMR signal that is detected is sensitive to a variety of factors in the local micro-environment of the ^{1}H protons. Initially considered sources of signal loss and image "artifacts" in many cases, many of these sensitivities have been harnessed and may be exploited as sources of "image contrast" with appropriate construction of the MRI pulse sequence to allow weighted exposure or sensitivity to such factors. Although many exist, four categories of endogenous image-contrast sources will be discussed, demonstrating the range of physiological inference that can be derived from sensitivity to such physical factors.

6.4.1 Relaxation Times

6.4.1.1 $T_2{}^*$-Weighting

After excitation, the nuclear spins can be considered to rotate, or process, in a transverse magnetization plane (i.e., transverse to the main magnetic field B_0). As discussed above, this precession rate or frequency is dictated by the Larmor equation and is dependent on the local magnetic field. We depend on controlling our local magnetic fields and thus the spatially dependent frequency of precession with our three purposefully imposed magnetic field gradients for slice, frequency, and phase-encoding. However, it is not only application of external magnetic field gradients that leads to spatially varying local magnetic fields. Inhomogeneities in the main magnetic field uniformity associated with magnet construction also contribute to varying precessional frequencies and thus dephasing of the precessing spins underlying the detected signal (free induction decay), leading to signal decay following RF excitation. The rate of loss of FID amplitude relates to the diversity of magnetic fields experienced by the excited nuclear spin population. Often approximated by an exponential decay, a single time constant $(T_2{}^*)$ can

---

[*] In fact, the axes of *k*-space (k_x and k_y) are formed by considering integrated gradient field history in the pulse sequence timing. For a constant readout gradient G_x, $k_y(t)$ can be seen to equal $\gamma G_x \cdot t$, such that the product of k_x and the *x*-coordinate will yield the phase angle of a spin population at coordinate *x*. (Note that for a constant gradient field amplitude, k_x is proportional to readout time *t*.) Similarly, the k_y coordinate is given by $\gamma G_y t_p$, where t_p is the duration of the phase-encoding gradient pulse application.

thus be used to characterize the rate of signal loss (and by inference field inhomogeneity). This is mathematically expressed as

$$S(\text{FID}) \propto -\frac{t}{T_2^*} \qquad (6.4)$$

where S is the free induction decay signal amplitude, t the time, and T_2^* the "effective transverse relaxation time constant." T_2^* typically takes values of the order of 10–100 ms in human tissues in clinical MRI scanners (1.5 T), with uniform environments associated with longer T_2^* values (more persistent FID). However, other sources of magnetic field inhomogeneity dominate the effects of field construction. Different tissue types themselves become magnetized in the presence of an external field. The degree to which this magnetization occurs is governed by the tissue property of magnetic susceptibility χ. Thus, at interfaces between different tissue types (especially tissue/bone or tissue/air), local magnetic field homogeneity is disrupted by magnetic susceptibility differences. This can lead to signal voiding (often termed "magnetic susceptibility artifact") as well as signal mismapping on image formation (accumulated phase cannot be distinguished from deliberate "phase encoding"). T_2^*-sensitivity, although imposing sensitivity to the above "artifacts," can also be exploited by deliberately manipulating the local magnetic field homogeneity of tissue, for example, by the administration of a blood-borne magnetic contrast agent, which disturbs local field uniformity wherever it travels. This can be readily visualized on images acquired with T_2^*-sensitivity and assessments of blood supply (perfusion) to a tissue can be made (see perfusion below). Furthermore, even without an exogenous contrast agent, transient changes in the balance of diamagnetic oxy-hemoglobin to paramagnetic deoxyhemoglobin, arising as a consequence of neuronal activity can be detected (due to the field-disturbing effect of the deoxyhemoglobin). This contrast mechanism (conventionally referred to as BOLD, blood oxygenation level dependent) can be seen to simply be an extension of T_2^*-weighted sensitivity in conjunction with a varying intravascular oxy/deoxy-hemoglobin balance (see fMRI below).

6.4.1.2 T_2-Weighting

However, to mitigate the effects of such spin dephasing and loss of signal in the free induction decay, a commonly employed MRI technique known as the "spin echo" was developed. After initial RF excitation and subsequent precession (with T_2^* dephasing/decay) a second RF pulse is applied to cause the nuclear spin magnetization to "flip" in the transverse plane (see Figure 6.2). The spatial location of the nuclei is not changed, just the magnetization orientation is changed. Thus, spins continue to precess at different rates. However, where this previously led to *dephasing*, after the second (180°) RF pulse, the different precession rates now lead to *rephasing*. At a time after the 180° RF pulse equal to the time between initial RF excitation and the 180° RF pulse, spins will transiently realign, leading to recovery of signal, known as the *spin echo*. The concept of a spin echo is often liked to a race in which the contestants all run at different but constant speeds (spins precess at different frequencies). A given time after the start of the race, the contestants are all told to turn around (a 180° RF pulse is applied). Assuming none of the contestants change speed, they will all return to the starting line at the same time (spins rephase to form an echo). The time between initial excitation and echo formation is known as the *echo time*, TE. However, the amplitude of the spin-echo does not retain the initial amplitude of the free induction decay. This is because the spin-echo mechanism refocuses or reverses the dephasing effects of *time-invariant* spatial magnetic field inhomogeneities (such as those caused by magnet construction that stay constant). In addition to such constant inhomogeneities, however, another mechanism exists by which local magnetic field variations, or fluctuations, occur due to the interaction of excited spins with each other. Consider the tiny magnetic field of an excited nuclear spin. As the water molecules diffuse, they may approach or depart from each other—transiently changing the effective local field each nuclear spin experiences in a random and irreversible fashion. Such irreversible transverse dephasing and subsequent signal loss

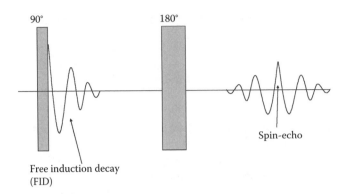

90°   180°

Free induction decay
(FID)

Spin-echo

FIGURE 6.2 The FID occurring after the initial 90° excitation exhibits rapid signal loss via T_2^*-dephasing. Application of a 180° refocusing RF pulse leads to the formation of the spin-echo whose amplitude is restored. Note the amplitude at echo time TE is nonetheless attenuated, not by T_2^* mechanisms, but by T_2 decay. The degree of such attenuation depends on the tissue property of T_2 and can be a valuable contrast mechanism in (T_2-weighted) MRI.

in the spin–echo is known as transverse or spin–spin relaxation, is a tissue-dependent property, and is characterized by an exponential time constant, T_2, analogous to T_2^*, described as

$$S(\text{spin echo}) \propto \exp\left(-\frac{\text{TE}}{T_2}\right) \tag{6.5}$$

where S is the spin-echo amplitude, TE the echo time, and T_2 the transverse relaxation time constant.

T_2-weighting (i.e., the sensitivity of the resultant MR image to irreversible T_2 effects) is thus controlled by choice of echo time, TE. Since species with long T_2 values (e.g., fluids such as cerebrospinal fluid, CSF) retain signal even at long TEs, while species with shorter T_2 values lose signal amplitude rapidly with increasing TE, choice of long TE values will introduce T_2-dependent contrast between tissues. This is generally referred to as T_2-*weighting*. The T_2-weighted image has become the mainstay of much clinical diagnostic radiology, in many cases revealing and delineating pathological tissues by hyperintensity associated with altered fluidity/water mobility. Of course, such contrast is gained at the expense of overall signal as in general signal is *lost* (just at different rates) with increasing TE. Commonly, MR images are described and evaluated in terms of a "contrast-to-noise ratio" (CNR) as opposed to the more typical signal-to-noise ratio (SNR).

6.4.1.3 T_1-Weighting

As discussed above, the process of phase encoding for image formation requires the repetition of the RF excite/data readout module to collect a set of typically 128 or 256 phase-encoded spin echoes. Each module is separated by an interval, TR. During TR, nuclear spins, having been tipped out of their equilibrium magnetic alignment with the external field B_0, begin to recover or relax longitudinally. The longitudinal relaxation time is characterized as an exponential recovery process with time constant, T_1. T_1, like T_2, being the property of the tissue and reflects the ability of the excited spin system to "give back" energy acquired during excitation. It is thus facilitated by magnetic interactions with microstructural and macromolecular entities in tissue (often referred to as the "lattice"). Thus, free fluids are associated with long T_1 times and slow recovery. As a general rule of thumb, to approach asymptotic recovery, and allow full signal potential for subsequent RF excite/data readout modules of the pulse sequence, TR should be set to ~$5T_1$. To the extent that a given TR is less than $5T_1$ for a given

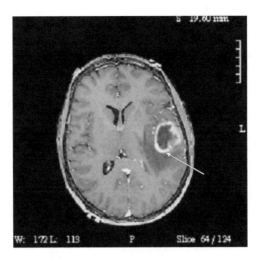

FIGURE 6.3 An axial Tl-weighted image through a left hemispheric high-grade glioma after administration of Gd-based contrast medium shows characteristic "ring enhancement" of the lesion (associated with blood–brain barrier breakdown).

species, subsequent echoes will be attenuated by a factor that depends on T_1 and images are said to be T_1-weighted, expressed as

$$S(\text{TR}) \propto \left\{1 - \exp\left(-\frac{TR}{T_1}\right)\right\} \tag{6.6}$$

T_1-weighted imaging becomes especially valuable after administration of T_1-shortening Gd-based contrast agents. To the extent that these contrast agents undergo selective accumulation, there is regional T_1-shortening and consequent hyperintensity on T_1-weighted images. This is particularly useful for identifying areas of blood–brain barrier disruption, where intravascular contrast agent accumulates in the extravascular space of a lesion such as a brain tumor (see Figure 6.3).

A critical take-home point from the above discussion of T_1- and T_2-weighting is that by deliberate manipulation of the pulse sequence parameters, specifically TR and TE, the user can control the sensitivity or weighting of the resultant image, in a manner not available to other cross-sectional imaging modalities such as CT. This represents both the power and the complexity of MRI as an imaging technique.

6.4.2 Flow

Bulk flow can also be visualized on MR images. Two phenomena exist that relate to the flow of blood water ¹H protons and their subsequent appearance: in-flow enhancement and flow-related dephasing (flow artifact). Both will be elaborated upon in the following discussion of MR angiography. However, they arise from different mechanisms. In-flow enhancement can be seen as a consequence of T_1-weighting (particularly of the so-called gradient-recalled echo images). T_1-weighting is achieved as discussed above by using TR values that are considerable shorter than $5T_1$ leading to signal suppression of longer T_1 entities. This assumes that ¹H spins experience the repeated train of RF pulses. While this is indeed the case for stationary spins, in-flowing blood may not have experienced the full train of selective excitation pulses and may thus not be "T_1-weighted" but will rather deliver its full signal potential; thus vascular structures with in-flow appear hyperintense. If flow is too slow, the blood protons nevertheless experience a train of RF pulses and indeed exhibit signal suppression. Flow-related dephasing, on the other hand, describes the accumulation of phase relative to stationary spins arising from flow during the application of gradient field pulses. While a balanced pulse sequence design will lead to no net accumulation of phase in the slice select or readout directions (but only in the phase-encode direction), spins that relocate between gradient pulses in the pulse sequence module will acquire a net phase proportional to their velocity. This may be considered an artifact (since this accumulated phase will be misinterpreted by the image reconstruction and signals will be mismapped in the phase-encode direction), or exploited as a source of contrast, by observing the phase angle (retrievable from the complex acquired data), it is possible to deduce the flow velocity, or rather deliberately employ "flow encoding."

6.4.3 Diffusion

Exploiting (again) the spatially dependent accumulation of a "phase" angle by spins experiencing a magnetic field gradient pulse, the pulsed gradient spin-echo (PGSE) approach was developed for studying molecular diffusion in the time interval between application of matched magnetic field gradient pulses. A stationary spin would experience equal-and-opposite phase angles and so would be unimpeded by such a pulsed gradient pair. However, to the extent that there is relocation of spin position (by mechanisms such as, but not limited to, diffusion) during the interval separating the gradient pulses (typically ~40 ms), rephasing and dephasing will be imbalanced and there will be a net signal loss. The degree to which an imaging sequence is made sensitive to the processes of diffusion (more correctly, "apparent diffusion," since any form of displacement will contribute to signal loss) is defined by a quantity termed the *b-value*, which for rectangular gradient pulses of duration, δ, separation, Δ, and strength, G, is given by

$$b = \gamma^2 \delta^2 G^2 (\Delta - \delta/3) \tag{6.7}$$

Additional signal loss on such diffusion-weighted images (DWIs) is approximated by a factor

$$S \propto \exp(-bD) \tag{6.8}$$

where D is the apparent diffusion coefficient (ADC). Spatial maps of the ADC can be derived from two images acquired with different b-values (typically 0 and 1000 s/mm^2).

The adoption of diffusion-weighted imaging has seen a recent explosion due to overcoming hurdles limiting ultrafast image acquisition (to freeze effects of gross motion) as well as the appreciation of the sensitivity of DWI to acute pathophysiological changes during acute ischemic stroke. In particular, the ADC has been observed to drop by as much as 50% in minutes to hours following an acute stroke, leading to a readily appreciated regional hyperintensity of the ischemic tissue on a DWI (see Figure 6.4). Furthermore, diffusion, or specifically "apparent diffusion," may exhibit directional preference. Such a phenomenon is termed *anisotropy* and can be assessed by comparing images acquired with diffusion sensitivity gradient pulses applied along different spatial axes. As such, both the magnitude and preferred direction of anisotropic diffusion can be assessed. Recent developments in the field of diffusion tensor imaging (where the diffusion process is considered as a 3×3 matrix of diffusion coefficients) has allowed the following or tracking of anisotropic structures such as white matter fibers.

6.4.4 Magnetization Transfer

Another available contrast mechanism that shows considerable utility in the study of white matter development and disorders, and also in the study of cartilage, exploits the phenomenon of magnetization transfer to probe macromolecular or microstructural entities (e.g., myelin or collagen) that

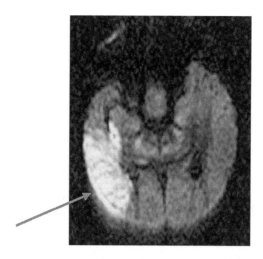

FIGURE 6.4 A diffusion weighted echo planar image (with diffusion sensitivity, b-value, of 1000 s/mm^2) shows pronounced hyperintensity in the right posterior areas (arrow), associated with acute cerebral ischemia.

interact with free water protons. A simple model describes the protons of the macromolecular structure as "bound," while water molecule protons are "free." The bound state leads to sustained spin–spin interactions leading to fast signal decay (see discussion of T_2 above) and a correspondingly broad spectral resonance. The free water protons have a much longer T_2 and corresponding narrow spectral line. Consequently, the bound pool does not contribute directly to image intensity, and only the free pool is interrogated. As such, applying "off-resonance" irradiation (~1 kHz from the nominal free water resonance) is expected to have no direct effect on the free water (narrow resonance) protons, and thus no direct effect on observed signal intensity, but would nevertheless magnetically saturate spins of the bound pool. This would not have particular significance without the phenomenon of chemical exchange in which protons from the free and bound pools may interact and indeed exchange with each other, provided the appropriate chemical environment is present (as in collagen or myelin). Such exchange leads to a decrease in the free water signal since some of the spins carry with them the saturation imposed upon them while they were in the bound state. The degree of such signal loss can be seen to reflect the density of such bound protons (and by extension, the concentration of macromolecular structures) and the opportunity for chemical exchange. While this signal loss can be substantial (~50%) in healthy white matter and/or cartilage, it decreases in demyelinating lesions and as an indicator of breakdown of the proteoglycan–collagen matrix in cartilage decay, for example, in osteoarthritis.

6.5 Contrast Agents

Despite the above-discussed plethora of endogenous tissue contrast available to the MRI experiment, with associated "exquisite soft tissue contrast" in resulting images, there are occasions when exogenous contrast media (or tracers) are nonetheless beneficial.

6.5.1 Static Enhancement

The most widespread application of contrast-enhanced MRI is the delineation of pathological tissue by virtue of increased extravasation of a blood-borne contrast agent, or magnetic dye. The most common (and clinically approved) contrast media are based on small organic chelates of the gadolinium (Gd)$^{3+}$ ion with total molecular weights of 500–1000 Da (e.g., Magnevist™, Omniscan™, and Prohance™). As water approaches the Gd^{3+} ion, longitudinal relaxation of the water ^{1}H protons is facilitated. On

T_1-weighted images this leads to elevated signal intensity and thus a regional probe of contrast agent distribution (see Figure 6.3).

6.5.2 Dynamic Enhancement: Perfusion and Permeability

The above "static" use of Gd-based contrast media does not fully exploit the opportunities of using an exogenous blood-borne tracer. Drawing on the methodologies of nuclear medicine, it is possible to dynamically image or "track" the passage of contrast media after bolus injection. Typical injected volumes are of the order of 10–20 cm³ (to achieve an overall dose of 0.1 mmol/kg BW with a 0.5 M stock solution). With power injectors delivering contrast media at a rate of up to 5 cm³/s, the bolus injection can be seen to be as short as 2–4 s. To increase sensitivity to the low blood volume in the brain, imaging is typically performed with T_2^*-weighting to track the dynamic variations in field homogeneity disturbance associated with the paramagnetic Gd ion (see Figure 6.5a). Signal intensity loss (Figure 6.5b) can then be related to change in effective transverse relaxation rate ΔR_2^*, where $R_2^* = 1/T_2^*$; this change is assumed proportional to instantaneous contrast agent concentration (Figure 6.5c,d). Interestingly, because the T_2^* mechanism is a "field effect," the influence of the contrast agent extends beyond its immediate intravascular environment, disturbing magnetic field homogeneity (and leading to signal loss) also in the neighboring parenchyma. This is the rationalization of the increased sensitivity of T_2^*-based (or "negative

(a)

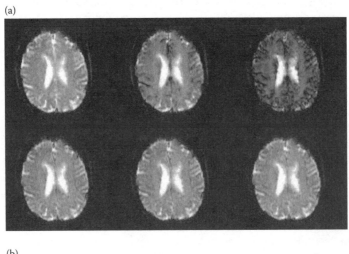

(b)

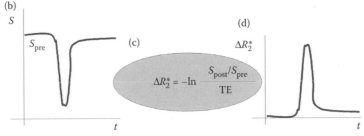

FIGURE 6.5 MR perfusion. (a) Six from a series of dynamic T_2^*-weighted images acquired during passage of a bolus of Gd-based contrast agent (temporal resolution ~1 s). In this example, the "negative enhancing" effect of the contrast agent can be seen as it passes through the brain and subsequently washes out (returning signal intensity toward pre-contrast levels). (b) Signal intensity time course can be converted via simple algebra (c) to a time course of ΔR_2^*, the change is effective transverse relaxation rate (d). This quantity is assumed proportional to instantaneous tracer concentration and can be used in kinetic modeling.

enhancing") bolus-tracking methods, compared to T_1-based ("positive enhancing") methods, which rely on close proximity of water ^{1}H protons to the contrast agent tracer. Dynamic T_1-weighted imaging, using the same contrast agent, may, however, also be used for perfusion assessment, as is typically the case in studies of, for example, myocardial perfusion. Dynamic imaging can be subsequently modeled using tracer kinetic techniques, to estimate parameters of perfusion including blood volume, blood flow, and mean transit time. Quantitation accuracy is improved by inclusion in the model of an "arterial input function" which can be simultaneously measured in the imaging and reflects the time course of the contrast agent bolus in the feeding arterial structures (and thus accommodates any delay or dispersion that has occurred since the time of injection). Such assessments of perfusion, especially in the brain, have shown tremendous clinical promise in the characterization of ischemic territories, as well as showing increasing utility in the field of oncologic imaging.

Elevated microvascular permeability is a hallmark of many malignant tumors and a correlate of the phenomenon of *angiogenesis*, the formation of new blood vessels, critical for tumor growth. Dynamic contrast-enhanced MRI can be used to estimate microvascular permeability, since initial enhancement of tissue after contrast agent administration relates predominantly to contrast agent occupying the intravascular space (fractional blood volume), but *progressive* enhancement revealed on subsequent images occurs as a consequence of contrast agent extravasation beyond the compromised blood–brain barrier and into the tumor interstitium. While multicompartment tracer kinetic modeling approaches are employed for the accurate estimation of microvascular permeability, an intuitive approach suggests that the *rate* (or slope) of such progressive enhancement relates to the rate of contrast agent leakage out of the intravascular space and thus the permeability of the vessel walls (Figure 6.6).

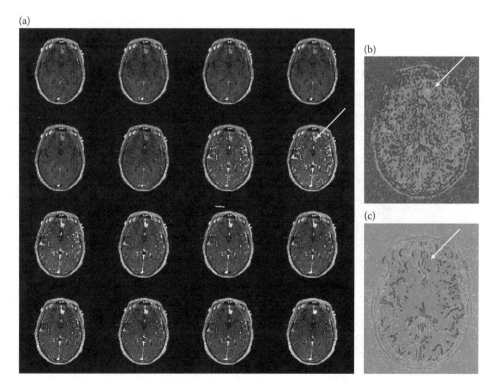

FIGURE 6.6 (**See color insert.**) MR permeability assessment. (a) Sixteen of a series of dynamic T_1-weighted images acquired during passage of Gd-based contrast agent in a patient with an intracranial tumor (arrow). Kinetic modeling of the progressive positive enhancement yields estimates of (b) fractional blood volume and (c) microvascular permeability, both of which are elevated in this tumor.

6.5.3 New Developments

A variety of novel contrast media are under development either with increased compartmental specificity (e.g., blood pool or intravascular agents) or with biochemical targeting (e.g., conjugated to antibodies specific for certain receptor expression) or even "smart" or switchable, in which the contrast medium is latent until biochemically "activated"—a reaction that leads typically to a conformational change and increased access of water to the relaxation enhancing moiety. While these developments are just beginning to be seen pre-clinically, they appear poised to advance MRI into the emerging field of molecular/metabolic imaging.

6.6 Functional Magnetic Resonance Imaging

In the last decade, the field of human brain mapping has been revolutionized by the advent and development of the technique of functional magnetic resonance imaging (fMRI) (Figure 6.7). As discussed above in the description of T_2^*, the blood oxygenation level-dependent (BOLD) contrast mechanism is used to track changes in the balance of diamagnetic oxyhemoglobin to paramagnetic deoxyhemoglobin as a consequence of the phenomenon of neurovascular coupling and the electrical activity of neurons during stimulation or task performance. In fact, so intricate are the experimental paradigms currently being explored that it is possible to claim that where once we imaged the *brain*, we may now begin to image the *mind*.

The imaging methodology underlying BOLD fMRI is essentially identical to that of a dynamic contrast-enhanced (DCE) perfusion study. Rapid images (typically using echo planar imaging, or EPI) are acquired with T_2^*-weighting during alternating periods, or single events, of various stimuli, or control conditions. The central hypothesis is that regional neuronal activation leads to regional increase in oxy- to deoxyhemoglobin and thus regional increase in signal. The rapid imaging

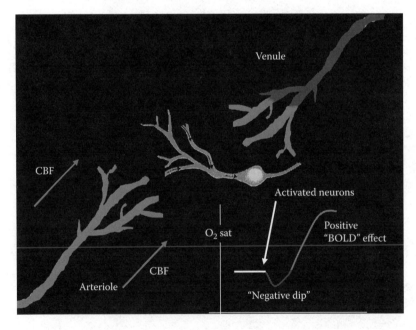

FIGURE 6.7 (See color insert.) Schematic representation of the fMRI BOLD contrast mechanism. Neuronal activation leads to an increase in regional cerebral blood flow (CBF), without a concomitant increase in tissue oxygen consumption, leading to an increase in the oxygenation of the capillary bed and postcapillary venules. This increase in oxygenation can be visualized as an increase in signal on T_2^*-weighted images, as paramagnetic deoxyhemoglobin is displaced.

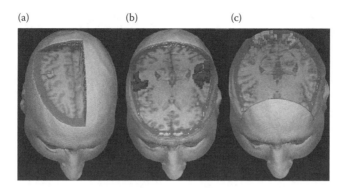

FIGURE 6.8 Human brain mapping. The BOLD fMRI experiment can be performed while stimulating various sensory (or cognitive systems) and corresponding pixels identified and highlighted on anatomic three-dimensional reconstructions. In this case, pixels of the (a) motor, (b) auditory, and (c) visual systems were identified in a single volunteer during repeated scanning with hand motor activity, auditory tone presentation, and checkerboard visual stimulation, respectively.

strategies allow this transient phenomenon to be captured for subsequent analysis. Owing to the small magnitude of the signal change, experimental paradigms conventionally require multiple repeated image acquisitions either during alternating "blocks" of stimulus and rest conditions (block design paradigms), or at specific times post-single stimulus (or cognitive) events (event-related paradigms). Subsequent statistical analysis is employed to define which of the observed signal intensity responses can be related to the parameters of the experimental model (e.g., stimulus time course). See Figure 6.8 for an illustration of this.

6.7 Magnetic Resonance Angiography and Flow Quantification

In the past, x-ray angiography and digital subtraction angiography were the standard diagnostic procedures for imaging vascular disease. The rapid advances in the past decade in three-dimensional magnetic resonance (MR) and CT angiographic methods, however, allow physicians to obtain more detailed information about vascular anatomy and physiology. Multi-detector spiral CT can provide high-resolution images of vessels quickly, but requires potentially high doses of radiation and contrast media. These may not be tolerated in a pediatric environment or in patients that need regular examination. As described in this section, MR angiography is a noninvasive alternative to CT angiography and can also provide valuable physiological information ranging from the chemical composition of tissue to the oxygenation and flow of blood.

An ideal angiographic method is able to generate a strong signal, whatever the signal is, from within a vessel while also suppressing signal from tissue outside the vessel. Three methods for achieving this using MR are explained: time-of-flight angiography, contrast-enhanced angiography, and phase-contrast (PC) angiography. The benefits and drawbacks of each method are discussed. Flow quantification using PC imaging is also explained.

6.7.1 Time-of-Flight Angiography

Time-of-flight (TOF) MR angiography relies on the saturation* of the magnetization of static tissue within a repeatedly excited volume. Tissue outside the excited volume, including blood, remains highly magnetized because it does not experience the same radio frequency (RF) excitations. Unsaturated

---

* Saturation refers to the demagnetization of tissue that results in suppression of the MR signal.

blood that flows into the saturated volume will therefore appear brighter than the surrounding tissue. This so-called "inflow effect" is the basis of all TOF angiography (Figure 6.9).

Saturation of static tissue occurs when the time between successive RF pulses does not allow for complete T_1 decay of the magnetization. As a result, the magnetization will reach a reduced steady-state value after the application of numerous RF pulses. The steady-state magnetization, M, of static tissue imaged using a conventional[*] pulse sequence depends on the T_1-decay constant of the tissue and the TR of the pulse sequence according to the following expression:

$$M = M_0 \frac{(1 - e^{-TR/T_1})\sin\alpha}{1 - e^{-TR/T_1}\cos\alpha}, \tag{6.9}$$

where α is the flip angle of the RF pulse and M_0 the maximum unsaturated magnetization of the tissue (Figure 6.10).

TOF MR angiography is designed to minimize M in static tissue while maximizing M in the moving blood. As shown in Equation 6.9, the saturation of static tissue can be increased by using a pulse sequence with a short TR or a flip angle near 90°, which makes static tissue to appear darker in the acquired MR images. Assuming an RF pulse with $\alpha = 90°$ is used, Equation 6.9 simplifies to

$$M = M_0(1 - e^{-TR/T_1}). \tag{6.10}$$

The brightness of a vessel in an angiogram depends on the fraction of a voxel refilled with magnetized blood each TR, as depicted in Figure 6.9. Given constant plug flow in a cylindrical vessel oriented perpendicularly to the imaged slice, the fractional filling, f, of a voxel depends on the slice thickness, L, the velocity of blood, v, and the TR:

$$f = \begin{cases} TR.\dfrac{v}{L} & v < L/TR \\ 1 & v \geq L/TR \end{cases}. \tag{6.11}$$

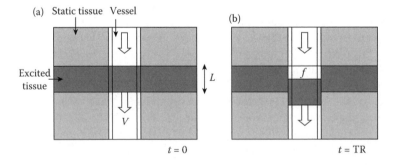

FIGURE 6.9 Explanation of two-dimensional TOF angiography. (a) A slice of thickness L positioned perpendicularly to a vessel of interest is repeatedly excited using an RF pulse with a flip angle of 90°. This leads to saturation of the signal within both the vessel and surrounding static tissue at time $t = 0$. (b) Blood flow at a constant velocity v causes the movement of saturated blood out of the excited slice and fully magnetized blood into the slice. This inflow effect results in a bright signal within the vessel, relative to the surrounding tissue, depending on the fraction f of the slice thickness refilled at time $t = TR$ when the signal is measured.

---

[*] For example, a spoiled gradient-echo pulse sequence.

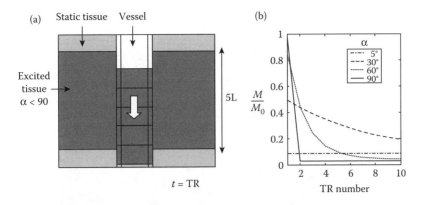

FIGURE 6.10 Explanation of three-dimensional TOF saturation effects. (a) Three-dimensional TOF angiography involves exciting thick slabs of tissue and spatially encoding all three dimensions of the resulting volume. A disadvantage of this approach is the progressive saturation of the blood as it propagates through the volume. Vessels in the resulting TOF angiogram will appear to be fading along the direction of flow. (b) Magnetization as a function of the number of applied RF pulses with $T_1 = 1200$ ms (i.e., typical of blood at 1.5 T) and TR = 40 ms, using different RF flip angles. Although the magnetization is initially large following a 90° RF pulse, the steady-state magnetization is much smaller than with the other flip angles shown. The flip angle that maximizes the steady-state signal is known as the Ernst angle, $\theta_E = \cos^{-1}(\exp(-TR/T_1))$.

Vessel intensity will therefore increase relative to the surrounding tissue as TR increases or L decreases, and reach a maximum when the entire voxel is fully replenished between successive excitations. Since TR must remain short to ensure that static tissue remains well saturated, thin slices are typically imaged to increase f. This also improves the through-plane resolution of the vessel, but results in longer scan times.

A series of contiguous parallel slices imaged using this TOF approach can be stacked to form a three-dimensional angiogram. This three-dimensional spatial information is usually displayed as a two-dimensional projection of the data (see Figure 6.11). Alternatively, a three-dimensional surface of the vessels can be displayed (see Figure 6.12). In both visualization strategies, the anatomy can be rotated in real time to assist in the interpretation of complex vascular anatomy.

Producing a spatially well-defined thin slice requires both a high-amplitude slice-selection gradient and a narrow-bandwidth RF pulse. Unfortunately, the simultaneous implementation of these two requirements leads to an increase in the duration of the excitation and adds to the minimum TR of the sequence. As shown in Equations 6.9 and 6.11, a longer TR results in a greater inflow of blood but also leads to an increase in background signal from static tissue (i.e., greater steady-state magnetization and less saturation).

Instead of collecting a series of parallel two-dimensional slices over a long period of time, three-dimensional MR data can be acquired. These data provide high spatial resolution by exciting a thicker slice and spatially encoding the through-plane dimension as well as the in-plane dimensions. This approach not only allows for a shorter TR, but also results in the progressive saturation of the blood as it travels through the thick volume and experiences multiple excitations. As the saturation of the blood approaches that of the surrounding static tissue, image contrast between the vessel and the static tissue drops. Using smaller flip angle RF pulses reduces this saturation, but also reduces the saturation of the static tissue. Methods that combine a smaller flip angle with multiple overlapping thin three-dimensional volumes have been developed to avoid the blood saturation problem. Alternatively, ECG-gated selective inversion-recovery sequences have been implemented to image vessels in large three-dimensional volumes.

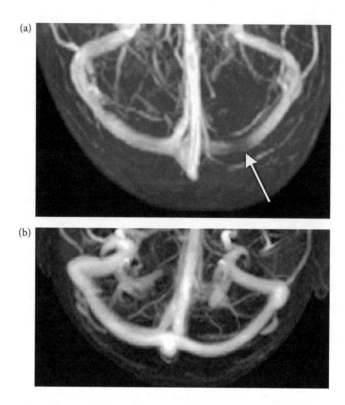

FIGURE 6.11 Comparison between (a) three-dimensional TOF and (b) three-dimensional contrast-enhanced angiography of the veins at the back of the head, displayed as (maximum intensity) projections of the three-dimensional data. The improved resolution of (b) is apparent in the clearer depiction of the small cranial vessels and boundary of the large vein (cerebral sinus). A suspected left transverse thrombosis present in the TOF MR angiogram (white arrow) is not present in the contrast-enhanced image, indicating a blood-saturation artifact in (a). (Images courtesy of Dr. M. Shroff, Hospital for Sick Children and University of Toronto, Toronto, Canada.)

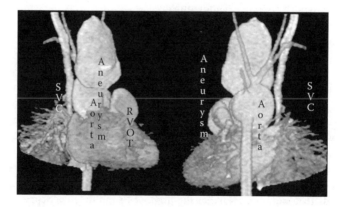

FIGURE 6.12 Two views of a surface-rendered contrast-enhanced angiogram of an aortic aneurysm. Despite the slow blood flow present in the aneurysm, the use of a contrast agent makes the aneurysm clearly visible. (Images courtesy Dr. Shi-Joon Yoo, Hospital for Sick Children and University of Toronto, Toronto, Canada.)

Another form of TOF angiography is known as arterial spin labeling (ASL). The method first "tags" blood using an inversion pulse applied to the upstream segment of a vessel. The tagged blood is then allowed to propagate through the vasculature before being imaged downstream. This process is repeated with the inversion pulse turned off, and the two images are subtracted to suppress static tissue. The resulting difference image ideally includes signal from the tagged blood only. Using different delays between blood tagging and image acquisition provides time-lapse information about the propagation of blood from the initial tagging site.

In general, TOF angiography provides a noninvasive measurement of vessel anatomy without the need for intravascular contrast agents. Despite its strengths, the dependency of TOF angiography on high flow rates limits its application in some vessel geometries. Vessels that curve into the plane of a two-dimensional acquisition, for example, will appear darker over that segment. This results from the fact that the longer blood remains within the imaging volume, the greater its saturation. Similarly, blood flow at the surface of a vessel is often slower than it is at the center. In this situation, flow at the edges of the vessel may appear darker due to increased saturation. This orientation-dependent artifact can mimic vessel narrowing similar to stenotic disease. Furthermore, small vessels and aneurysms can possess slow flow or no flow at all. The reduced inflow effect associated with these conditions limits the usefulness of TOF in such applications. Last, the long scan times of TOF angiography may preclude its use in vessels undergoing physiologic motion.

6.7.2 Contrast-Enhanced Angiography

Contrast-enhanced MR angiography involves the injection of an intravascular contrast agent to reduce the T_1 of blood. The resulting difference between the T_1 of the blood and of the surrounding tissue produces bright vessels in T_1-weighted images, with the degree of enhancement depending on the concentration of the agent and its relaxivity (change in relaxation rate per unit concentration of contrast agent, measured in s^{-1} mM^{-1}).

Magnetic resonance contrast agents are normally injected intravenously, either manually or using a power injector. Rapid time-resolved imaging during contrast administration allows the passage of this injected bolus to be seen during its first pass through the vascular system. These images are often collected as projections of thick anatomical slices; however, newly developed methods using a variety of data acquisition and signal processing tricks are becoming available for time-resolved three-dimensional contrast-enhanced MR angiograms (see Figure 6.12).

Time-resolved angiography provides useful information about the flow patterns present in complex cardiovascular diseases such as congenital heart disease. Imaging during the first pass of the contrast agent is also important for generating angiograms of the arteries without venous enhancement. Alternatively, arteries and veins can be separated by their temporal rather than spatial characteristics through a pixel-by-pixel analysis of contrast-enhancement curves.

Vessel brightness in an MR angiogram depends on the order in which k-space data are collected and on the concentration of the contrast agent throughout the acquisition. For example, if data collection begins at the center of k-space (the so-called "centric-ordered" or "elliptic centric-ordered" acquisitions) before the contrast agent has arrived, an abrupt increase in signal will occur when the contrast agent enters the imaging volume. An abrupt signal change in k-space produces ringing artifacts in the image, while the relatively brighter signal at the k-space periphery will lead to enhanced vessel edges in the angiogram. This intimate relationship between vessel appearance and acquisition timing makes the accurate estimation of contrast arrival time important to contrast-enhanced MR angiography.

Ideally, contrast-enhanced angiograms are acquired while the contrast agent is most concentrated within the vessels of interest. The transit time of the contrast agent from the injection site to the intended vessel has traditionally been estimated based on patient mass, heart rate, and experience. Recently, improved estimation methods have been introduced. One such method involves the injection of a small test bolus of contrast agent. The transit time of the test bolus is measured using rapid imaging during injection. The measured transit time is then used as the delay between the full injection of contrast agent and the start of the angiographic acquisition. An extension of this method uses real-time MR imaging to detect the arrival of the full contrast bolus, which then automatically triggers the longer angiographic acquisition without the need for a test bolus.

The major advantage of contrast-enhanced angiography is its insensitivity to specific flow conditions. As a result, the method avoids many artifacts that affect other MR angiographic methods, including pulsatile flow artifacts, vessel-orientation saturation, slow-flow saturation, and patient motion. A disadvantage of the method is the need for a venous injection that is minimally invasive but requires patient preparation and relatively expensive contrast material.

6.7.3 PC Angiography

Both TOF angiography and contrast-enhanced angiography rely on changes in the magnitude of the MR signal-to-image vessels. PC angiography instead uses differences in the phase of the signal between static and moving structures. The transverse magnetization of a structure moving through a magnetic-field gradient accumulates phase at a rate depending on the motion. This effect is the basis of phase-sensitive motion measurements and is a consequence of the Larmor frequency dependence on magnetic-field strength.

The precessional frequency, ω, of transverse magnetization moving through a magnetic-field gradient is given by

$$\omega(t) = \gamma \bar{r}(t) \cdot \bar{G}(t) + \Delta\omega, \tag{6.12}$$

where γ is the gyromagnetic ratio and $\bar{r} = (r_x, r_y, r_z)$ is the position of the magnetization with respect to the motion-encoding magnetic-field gradient $\bar{G} = (G_x, G_y, G_z)$, both of which depend on time t. $\Delta\omega$ represents all other sources of frequency variation, such as the difference in frequency between fat and water. Noting that frequency is simply the temporal derivative of phase ($\omega = d\theta/dt$) and expanding $\bar{r}(t)$ as a Taylor series produces the following expression for the phase of a moving structure gives

$$\theta(t) = \gamma \int_0^t \left(\bar{r}_0 + \bar{v}_0 \tau + \frac{\bar{a}_0}{2}\tau^2 + \cdots \right) \cdot \bar{G}(\tau) d\tau + \Delta\theta, \tag{6.13}$$

where $\bar{r}_0, \bar{v}_0,$ and $\bar{a}_0$ represent the position, velocity, and acceleration at $t = 0$, respectively, and $\Delta\theta$ represents other constant sources of phase accumulation corresponding to $\Delta\omega$.

In PC angiography, blood flow is assumed to move at a constant velocity over the measurement period, making $\bar{a}_0$ and all higher-order terms of the Taylor series zero. As a result, Equation 6.13 simplifies to

$$\theta(t) = \gamma \left(\bar{r}_0 \int_0^t \bar{G}(\tau) d\tau + \bar{v}_0 \int_0^t \bar{G}(\tau) \cdot \tau d\tau \right) + \Delta\theta. \tag{6.14}$$

If the goal of PC angiography is to highlight flowing blood based on its phase accumulation, then the first integral term must be eliminated to prevent static tissue from dominating the phase calculation. This is accomplished by using a motion-encoding gradient with a net area of zero

$$\int_0^t \bar{G}(\tau) d\tau = 0.$$

Finally, the phase errors amalgamated into the $\Delta\theta$ term are eliminated by performing two measurements, the first using a motion-encoding gradient of $\overline{G}(\tau)$ and the second using a gradient of $-\overline{G}(\tau)$. The phase difference between these two acquisitions (Figure 6.13), given by the following expression, produces a cancellation of $\Delta\theta$ and leaves

$$\theta(t) = 2\gamma \cdot \overline{v}_0 \int_0^t \overline{G}(\tau) \cdot \tau \, d\tau. \tag{6.15}$$

The brightness of blood in PC angiograms depends only on the blood velocity and does not require fully polarized blood entering the volume of interest. This allows large three-dimensional volumes to be imaged continuously. However, a limitation of PC angiography is its relatively long acquisition time compared with either TOF or contrast-enhanced angiography. The phase measurement described above provides motion sensitivity along only one dimension, which means that the creation of angiograms of complex vascular anatomy requires at least four complete sets of k-space data in order to detect flow in all three dimensions.

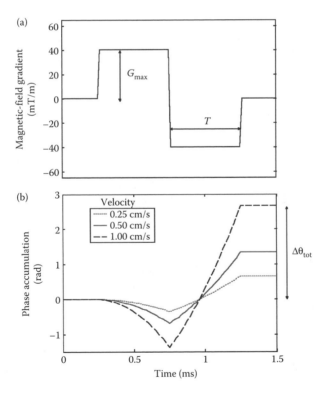

FIGURE 6.13 Phase accumulation due to flow through a magnetic-field gradient. (a) Bi-polar magnetic-field gradient applied to encode motion, with a total area of zero. (b) Phase accumulated by blood moving at a constant velocity during the application of the gradient in (a). At the end of the application of the gradient, a net phase is accumulated which is proportional to the velocity: $\Delta\theta_{tot} = \gamma v T^2 G_{max}$. Above a critical velocity, known as the aliasing velocity ($v = \pi/\gamma G_{max} T^2$), a phase greater than π radians will be accrued by the flowing blood. This results in blood moving above this velocity appearing to move in the opposite direction. Such aliasing artifacts can be avoided using a weaker or shorter motion-encoding gradient.

6.7.4 PC Flow Quantification

The same principles used to generate PC angiograms can also be used to make quantitative measurements of blood flow. From time-resolved images of blood velocity, the velocity waveform, peak velocity, or total volumetric blood flow can be calculated (see Figures 6.14 through 6.16).

PC imaging is normally performed using a single slice oriented perpendicularly to a vessel of interest. Time-resolved measurements are then made, typically with flow encoding applied parallel to the vessel only rather than in all three dimensions. In a uniform vessel, flow is directed primarily along the length of the vessel and may be encoded in this direction alone in order to reduce total scan time.

Volumetric flow measurements are useful for quantifying the total blood pumped by the heart each time it contracts, or the volume of blood provided to an organ. Volumetric flow measurements are obtained through post-processing of the time-resolved velocity measurements by calculating the average velocity over the vessel cross-section. This average velocity, multiplied by the area of the region of interest, represents the volume of blood crossing the imaged slice over that fraction of the cardiac cycle. Similar calculations carried out over the entire cardiac cycle are summed to produce the total volume per cardiac cycle.

To increase the effective temporal resolution of flow measurements, cardiac triggering is often used to combine data collected from different heartbeats. If the pattern of flow is not exactly the same following each cardiac contraction, this procedure can introduce errors to the flow measurement. These

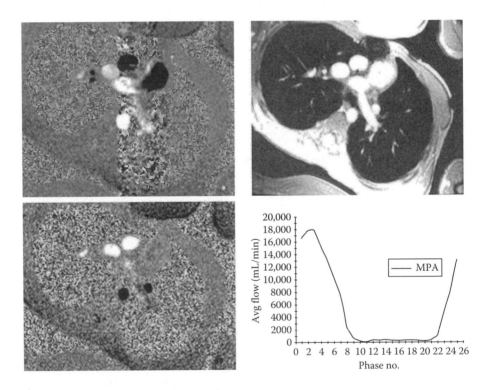

FIGURE 6.14 Phase-contrast images of flow through the main pulmonary artery (MPA). At the top right is an axial magnitude image of the chest, directly above the heart. At the left are two phase-contrast velocity images at the same location corresponding to different times in the cardiac cycle. The volume of blood passing through the MPA was measured in each velocity image to produce the graph at the bottom right. (Images courtesy Dr. Shi-Joon Yoo, Hospital for Sick Children and University of Toronto, Toronto, Canada.)

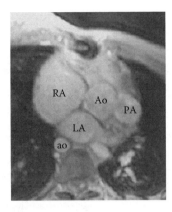
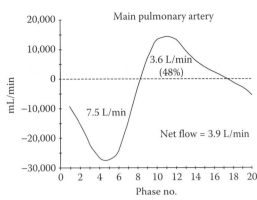

FIGURE 6.15 Pulmonary regurgitation in the pulmonary artery (PA) of a patient after surgery. Valves in the heart, when functioning properly, allow blood to flow in only one direction. As shown in the graph at the right, velocity measurements in the PA of this patient exhibit flow in both directions. In this graph, negative velocities represent flow in the proper physiologic direction toward the lungs, while positive velocities represent flow back into the heart. The total area under this curve represents the volume of blood pumped to the lungs each minute. The fraction of reversed flow (48%) provides a clinical estimation of valve dysfunction. MR is considered as the best method of evaluation of right ventricular function in conjunction with accurate simultaneous estimation of pulmonary flow. (Images courtesy Dr. Shi-Joon Yoo, Hospital for Sick Children and University of Toronto, Toronto, Canada.)

beat-to-beat changes may result from cardiac arrhythmia or changes in cardiac contraction due to respiration. Errors in the electrocardiogram triggering of the data acquisition also affect the accuracy of the data amalgamation. Triggering errors are compounded by respiration, which moves the heart and alters the received electrocardiogram signal.

A straightforward approach to eliminating this trigger error is to collect all the flow data in one heartbeat. Data from consecutive heartbeats then provide independent flow measurements that could be averaged together later. Magnetic resonance measurements of flow generally assume that the data are acquired under exactly the same conditions each heartbeat, or that beat-to-beat fluctuations, will average to zero when the data are combined, which is not strictly true. This argues for a flow measurement that can be completed in real time.

There are also immediate clinical benefits to fast MR measurements of blood flow. By reducing the total scan time of the study, patients may spend less time in the magnet. The risk of corrupted data due to patient motion is also reduced. Fast measurements reduce or eliminate the need for breath holding,* which can be difficult for cardiovascular or disoriented patients, or patients under general anesthesia. Finally, measurements with high temporal resolution allow flow under changing physiological conditions to be investigated (e.g., during stress testing or pharmaceutical treatment).

There is a limit, however, to how quickly a flow-sensitive MR image can be acquired. This limit is due primarily to the speed with which the spatial distribution of the MR signals can be spatially encoded (k-space can be filled). One way to decrease scan time and increase temporal resolution is to discard some or all spatial information. By forcing the MR signal to emanate only from a volume of interest, the need for spatial encoding is reduced. For this reason, methods have been developed using the so-called volume-selective excitations to measure motion more quickly (see Figure 6.17).

---

* Breath holding refers to the practice of collecting MR data while the patient suspends respiration. The purpose of breath holding is to eliminate respiratory motion during data acquisition, which would otherwise corrupt the data.

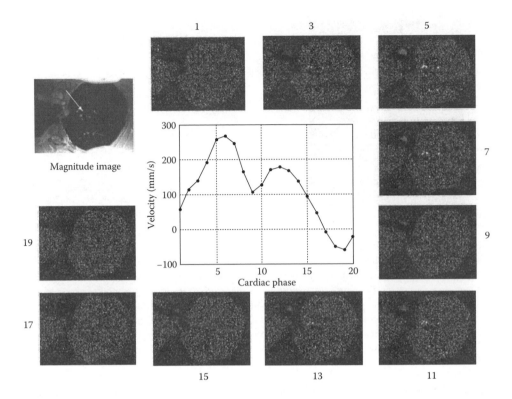

FIGURE 6.16 Phase-contrast measurements of blood velocity in a pulmonary vein. The upper left frame of this figure provides a magnitude image corresponding to an axial slice through the upper left lung of a healthy volunteer. The white arrow indicates a small pulmonary vein in which velocity was analyzed. The remaining images are phase-contrast velocity maps corresponding to the odd-numbered phases (labels) of the cardiac cycle, proceeding clockwise from the upper left. The central graph displays the velocity of the blood as measured from the phase-contrast data throughout the cardiac cycle.

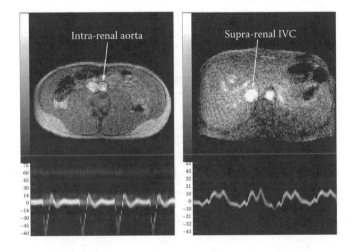

FIGURE 6.17 Real-time flow quantification using one-dimensional velocity encoding. The anatomical images at the top display the regions in which real-time flow measurements were performed (arrows to circled vessel). At the bottom are the corresponding measurements of the velocity spectra across four heartbeats, acquired at a temporal resolution of 42 ms.

6.8 MR Spectroscopy (MRS) and Spectroscopic Imaging (MRSI)

In addition to being dependent on physical factors that influence the local magnetic field, as described above, the NMR signal is also dependent on the local chemical environment of the ^1H protons. Indeed, the shielding effect of molecular electrons upon the ^1H nuclei and the subsequent characteristic "chemical shift" of the proton resonance according to its molecular environment ($\sim CH_2$ vs $\sim CH_3$, etc.) has been well known for many decades. In principle, molecules can thus be considered to have a "spectral" signature, with resonances at different frequencies (within a few parts per million [ppm], of the nominal 1H spectral peak), at intensities reflecting the abundance of protons in that environment (e.g., methyl, methylene, hydroxyl, etc.). Although a major component of structural organic chemistry, NMR spectroscopy has recently demonstrated convincing application in clinical use, especially in the area of brain tumors, Alzheimer's disease, and metabolic abnormalities of infant development. Beyond the brain, considerable utility is demonstrated in the study of prostate cancer and other branches of oncology. It seems likely that further applications will emerge, particularly as sensitivity to these small signals increases with higher field strength availability.

In the field of intracranial brain tumors, in which MRS and its imaging analog, MRSI, have demonstrated most convincing utility, the technique focuses on relative quantitation of key metabolites and metabolic products. *N*-acetyl-aspartate (NAA), an amino acid and marker of neuronal integrity, is readily visualized in healthy brain at a chemical shift of 2 ppm (relative to the nominal proton resonance; the protons of water are, in fact, found to exhibit a resonance at 4.7 ppm). Creatine is visible at 3 ppm and free choline and choline-containing compounds are observed at 3.2 ppm. Other metabolites such as myoinositol, alanine, and glutamate/glutamine are commonly reported as well, at clinical field strengths. In tumors, the link between pathology and metabolic function demonstrated by MRS is the general decrease in abundance of NAA, while the strength of the choline (Cho) resonance increases. Quantitation is beginning to enable assessment of tumor grade (aggression), and it is proving sensitive for the prediction of future sites of tumor spread or recurrence (by identifying abnormalities in the spectral signature, prior to conventional imaging-based evidence of tumor).

Recent implementation of two- and three-dimensional MRSI allows (low-resolution) for assessment of regional variations in NAA, Cr, and Cho, including generation of MRS-derived "metabolite maps," indicating the regional abundance of each metabolite, calculated by integrating intensity across the characteristic range of chemical shift frequencies associated with each metabolite (Figure 6.18). Despite the potential utility of metabolic insights gained from MR spectroscopy, slow acquisition rates and low spatial resolution ($\sim 1 \times 1 \times 1$ cm^3) have limited its application. Future promise of faster MRSI (by use of higher field strength magnets as well as multiple echo and/or parallel imaging strategies) encourages the more routine adoption of higher resolution MRSI in clinical practice.

6.9 MR Evaluation of Cardiac Function: An Example of an Integrated MRI Examination

The evaluation of cardiac function and viability remains one of the most important diagnostic procedures in Western cultures due to the high mortality associated with heart disease. To diagnose cardiac problems, a variety of imaging modalities are enlisted, including MR imaging. This serves as an illustrative example of the combined use of several MRI-based approaches each with different sensitivities in the evaluation of a clinical situation. Many similar examples of the multiple capabilities of MRI can be found in characterization of the brain and other organ systems.

MRI is known for its excellent depiction of cardiovascular anatomy, but it is also sensitive to physiological information such as blood flow, the chemical composition of tissue, and the oxygen saturation of blood. This unique wealth of information from MR imaging, known as the "one-stop shop," makes it a growing method for evaluating many cardiac problems. Given the risky and expensive procedures needed to treat cardiac dysfunction, it is essential to have accurate diagnostic information. Two

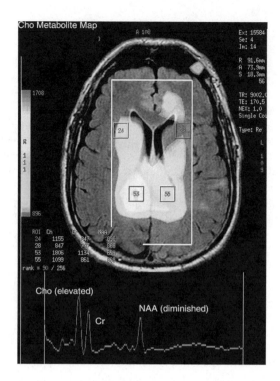

FIGURE 6.18   MR spectroscopy. Cho metabolite map from a patient with a right hemisphere glioma (hot spot on Cho color overlay). Below: representative voxel spectrum from tumor showing diminished NAA and elevated Cho.

important aspects of an MR cardiac evaluation are described below: myocardial function and coronary angiography.

6.9.1 Myocardial Evaluation by MR

As with any mechanical pump, the performance of the heart depends intimately on its motion. Imaging of the contraction and relaxation of the heart provides a qualitative measurement of the mechanical health of the tissue. When injured, regions of the myocardium* may not contract as strongly as the surrounding healthy tissue. Furthermore, an efficient cardiac contraction requires that all four chambers of the heart contract in the appropriate sequence. Abnormalities in the electrical stimulation of the heart may result in an incorrect contraction pattern, which can also be diagnosed through MR imaging and subsequent analysis of myocardial wall thickness at each stage of the cardiac cycle, with subsequent determination of the degree of wall thickening associated with diastole (Figure 6.19).

A strength of MR imaging is its ability to measure the three-dimensional anatomy of the myocardium throughout the cardiac cycle. This is accomplished by acquiring a series of parallel slices that can be stacked together to form a three-dimensional volume at each point in the cardiac cycle. From this four-dimensional data, the change in volume of the heart can be measured during contraction. A small change in the volume of the cardiac chambers is associated with myocardial dysfunction that

---

* Muscle of the heart.

(a) (b)

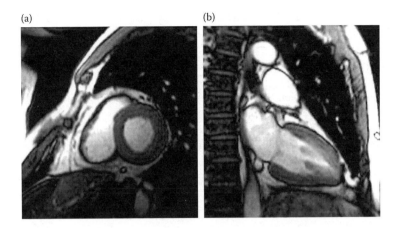

FIGURE 6.19 MR images of the heart from two standard anatomic views: (a) short-axis view and (b) long-axis view. Time-resolved images from multiple parallel slices can be combined to create four-dimensional data of the contracting heart.

may be corrected surgically or with medication. The noninvasive nature of MR imaging allows it to be performed before and after such therapy in order to monitor its success.

When a mechanical dysfunction of the myocardium is detected, the next step is to determine whether the muscle is dead or merely hibernating (or is "stunned") due to reduced blood flow, similar to having the muscles in a limb "fall asleep."

One method to measure myocardial damage with MR, known as "delayed enhancement," involves the injection of an intravascular contrast agent. This agent has been found to collect preferentially in damaged tissue several minutes after injection, making the damaged tissue appear brighter in the image. This effect is thought to result from damage to the capillaries in the affected muscle, causing them to leak contrast agent into the surrounding tissue. Conversely, rapid T_1-weighted imaging of the myocardium during the first pass of the contrast agent can provide information about myocardial blood flow, with regions receiving less blood appearing darker (less enhanced) in the MR images.

6.9.2 MR Coronary Angiography

A common source of myocardial damage and dysfunction occurs when the blood supply to this muscle gets reduced or cut off completely. This can result from blockage of the vessel feeding the myocardium, known as the coronary vessels. Because the coronary arteries are small and moving, imaging of these vessels has represented one of the greatest recent challenges to cardiovascular MR.

Imaging of the coronaries requires the synchronization of the data acquisition to both the cardiac and respiratory cycles to ensure the vessels are in the same location throughout the acquisition. While the use of a pressure-sensitive belt strapped around the chest provides a reasonable measure of respiratory position for many clinical applications, this method is too inaccurate for high-resolution imaging of coronary vessel obstruction. Instead, MR methods (typically employing an additional "navigator" echo) that measure the displacement of the diaphragm or coronaries directly during the acquisition of high-resolution images have been developed and have resulted in the increasingly routine volumetric imaging of left and right coronary arteries.

Thus, MRI is able to assess the functional consequence of myocardial injury (through focal abnormalities in contractility), the nature of the injury (degree of ischemia, based on myocardial perfusion and/or late enhancement studies) and in principle the origin of the injury (an occlusion or stenosis of the relevant coronary artery). While MRI is by no means the only technique available for cardiac imaging, the integrated study it promises offers compelling argument for its adoption.

6.10 Conclusion

It is hoped that the above discussion has offered some insights into the power and versatility of the family of MRI techniques, their synergistic use in clinical diagnosis, prognosis, and their potential role in characterizing response to interventional therapy in an increasingly physiologically specific manner. The field continues to evolve with exciting developments in hardware (e.g., higher field strength magnets, multiple RF coils, stronger and faster magnetic field gradient systems) and software (e.g., parallel imaging reconstruction algorithms). These will no doubt contribute to developments in both increasing the physiological specificity of MRI approaches and increasing the utility of MRI in a real time, or interactive manner.

Suggested Further Reading

Axel, L. and Dougherty, L., MR imaging of motion with spatial modulation of magnetization, *Radiology*, 171, 841–845, 1989.

Belliveau, J.W., Kennedy, D.N., McKinstry, R.C. et al., Functional mapping of the human visual cortex by magnetic resonance imaging, *Science*, 254(5032), 716–719, 1991.

Bryant, D.J., Payne, J.A., Firmin, D.N., and Longmore, D.B., Measurement of flow with NMR imaging using a gradient pulse and phase difference technique, *J. Comput. Assist. Tomogr.*, 8(4), 588–593, 1984.

Burstein, D., MR imaging of coronary artery flow in isolated and *in vivo* hearts, *J. Magn. Reson. Imaging*, 1, 337–346, 1991.

Duyn, J.H. and Moonen, C.T., Fast proton spectroscopic imaging of human brain using multiple spinechoes, *Magn. Reson. Med.*, 30, 409–414, 1993.

Dydak, U., Weiger, M., Pruessmann, K.P. et al., Sensitivity-encoded spectroscopic imaging, *Magn. Reson. Med.*, 46, 713–722, 2001.

Einthoven, W., Fahr, G., and De Waart, A., On the direction and manifest size of the variations of the potential in the human heart and on the influence of the position of the heart on the form of the electrocardiogram, *Am. Heart J.*, 40, 163–193, 1950.

Farb, R.I., McGregor, C., Kim, J.K., Laliberte, M., Derbyshire, J.A., Willinsky, R.A., Cooper, P.W. et al. Intracranial arteriovenous malformations: Real-time auto-triggered elliptic centric-ordered three-dimensional gadolinium-enhanced MR angiography—Initial assessment, *Radiology*, 220(1), 244–251, 2001.

Feinberg, D.A., Crooks, L., Hoenninger, J., III, Arakawa, M., and Watts, J., Pulsatile blood velocity in human arteries displayed by magnetic resonance imaging, *Radiology*, 153, 177–180, 1984.

Firmin, D.N., Nayler, G.L., Kilner, P.J., and Longmore, D.B., The application of phase shifts in NMR for flow measurement, *Magn. Reson. Med.*, 14, 230–241, 1990.

Haacke, E.M., Brown, R.W., Thompson, M.R., and Venkatesan, R., Eds., *Magnetic Resonance Imaging: Physical Principles and Sequence Design*, Wiley-Liss, New York, 1999.

Higgins, C.B., Hricak, H., and Helms, C.A., Eds., *Magnetic Resonance Imaging of the Body*, 3rd ed., Lippincott-Raven, Philadelphia, 1997.

Hu, B.S., Pauly, J.M., and Nishimura, D.G., Localized real-time velocity spectra determination, *Magn. Reson. Med.*, 30, 393–398, 1993.

Jack, C.R., Jr., Thompson, R.M., Butts, R.K., Sharbrough, F.W., Kelly, P.J., Hanson, D.P., Riederer, S.J., Ehman, R.L., Hangiandreou, N.J., and Cascino, G.D., Sensory motor cortex: Correlation of presurgical mapping with functional MR imaging and invasive cortical mapping, *Radiology*, 190(1), 85–92, 1994.

Kim, J.K., Farb, R.I., and Wright, G.A., Test bolus examination in the carotid artery at dynamic gadolinium-enhanced MR angiography, *Radiology*, 206(1), 283–289, 1998.

Korin, H.W., Felmlee, J.P., Ehman, R.L., and Riederer, SJ., Adaptive technique for three-dimensional MR imaging of moving structures, *Radiology*, 177(1), 217–221, 1990.

Kucharczyk, J., Roberts, T., Moseley, M.E., and Watson, A., Contrast-enhanced perfusion-sensitive MR imaging in the diagnosis of cerebrovascular disorders, *J. Magn. Reson. Imaging*, 3(1), 241–245, 1993a.

Kucharczyk, J., Vexler, Z.S., Roberts, T.P., Asgari, H.S., Mintorovitch, J., Derugin, N., Watson, A.D., and Moseley, M.E., Echo-planar perfusion-sensitive MR imaging of acute cerebral ischemia, *Radiology*, 188(3), 711–717, 1993b.

Kwong, K.K., Belliveau, J.W., Chesler, D.A., et al., Dynamic magnetic resonance imaging of human brain activity during primary sensory stimulation, *Proc. Natl. Acad. Sci. USA*, 89, 5675–5679, 1992.

LeBihan, D., Ed., *Diffusion and Perfusion Magnetic Resonance: Applications to Functional MRI*, Raven Press, New York, 1995.

Li, D., Haacke, E.M., Mugler, J.P., III, Berr, S., Brookeman, J.R., and Hutton, M.C., Three-dimensional time-of-flight MR angiography using selective inversion recovery RAGE with fat saturation and ECG-triggering: Application to renal arteries, *Magn. Reson. Med.*, 31(4), 414–422, 1994.

Lythgoe, M.F., Thomas, D.L., Calamante, F., Pell, G.S., Busza, A.L., King, M.D., Sotak, C.H., Williams, S.R., Ordidge, R.J., and Gadian, D.G., Acute changes in MRI diffusion, perfusion, T_1 and T_2 in a rat model of oligemia produced by partial occlusion of the middle cerebral artery, *Magn. Reson. Med.*, 44, 706–712, 2000.

Macgowan, C.K. and Wood, M.L., Motion measurements from individual MR signals using volume localization, *J. Magn. Reson. Imaging*, 9(5), 670–678, 1999.

Macgowan, C.K. and Wood, M.L., Fast measurements of the motion and velocity spectrum of blood using MR tagging, *Magn. Reson. Med.*, 45(3), 461–469, 2001.

Maki, J.H., Chenevert, T.L., and Prince, M.R., Three-dimensional contrast-enhanced MR angiography, *Magn. Reson. Imaging*, 8(6), 322–344, 1996.

Mazaheri, Y., Carroll, T., Du, J., Block, W.F., Fain, S.B., Hany, T.F., Aagaard, B.D., Strother, CM., Mistretta, C.A., and Grist, T.M., Combined time-resolved and high-spatial-resolution three-dimensional MRA using an extended adaptive acquisition, *J. Magn. Reson. Imaging*, 15(3), 291–301, 2002.

Moonen, C.T.W. and Bandettini, P.A., Eds., *Functional MRI*, Springer-Verlag, Berlin, 1999.

Mori, S., Kaufman, W.E., Pearlson, G.D., et al., *In vivo* visualization of human neural pathways by magnetic resonance imaging, *Ann. Neurol.*, 47, 412–414, 2000.

Moseley, M.E., Kucharczyk, J., Mintorovitch, J., Cohen, Y., Kurhanewicz, J., Derugin, N., Asgari, H., and Norman, D., Diffusion-weighted MR imaging of acute stroke: Correlation with T_2-weighted and magnetic susceptibility-enhanced MR imaging in cats, *Am. J. Neuroradiol.*, 11, 423–429, 1990.

Nayler, B.L., Firmin, D.N., and Longmore, D.B., Blood flow imaging by CINE magnetic resonance, *J. Comput. Tomogr.*, 10, 715–722, 1986.

Nield, L.E., Qi, X., Yoo, S.J., Valsangiacomo, E.R., Hornberger, L.K., and Wright, G.A., MRI-based blood oxygen saturation measurements in infants and children with congenital heart disease, *Pediatr. Radiol*, 32(7), 518–522, 2002.

Ogawa, S., Lee, T.M., Kay, A.R., et al., Brain magnetic resonance imaging with contrast dependent on blood oxygenation, *Proc. Natl. Acad. Sci. USA*, 87, 9868–9872, 1990.

Ogawa, S., Menon, R.S., Tank, D.W., et al., Functional brain mapping by blood oxygenation level-dependent contrast magnetic resonance imaging. A comparison of signal characteristics with a biophysical model, *Biophys. J.*, 64(3), 803–812, 1993.

Parker, D.L., Yuan, C, and Blatter, D.D., MR angiography by multiple thin slab three-dimensional acquisition, *Magn. Reson. Med.*, 17(2), 434–451, 1991.

Pohmann, R., von Kienlin, M., and Haase, A., Theoretical evaluation and comparison of fast chemical shift imaging methods, *J. Magn. Reson.*, 129, 145–160, 1997.

Pruessmann, K.P., Weiger, M., Scheidegger, M.B., and Boesiger, P., SENSE: Sensitivity encoding for fast MRI, *Magn. Reson. Med.*, 42, 952–962, 1999.

Redpath, T.W. and Norris, N.G., A new method of NMR flow imaging, *Phys. Med. Biol.*, 29(7), 891–898, 1984.

Rosen, B.R., Belliveau, J.W., Vevea, J.M., and Brady, T.J., Perfusion imaging with NMR contrast agents, *Magn. Reson. Med.*, 14(2), 249–265, 1990.

Saeed, M., Wendland, M.F., Yu, K.K., Lauerma, K., Li, H.T., Derugin, N., Cavagna, F.M., and Higgins, C.B., Identification of myocardial reperfusion with echo planar magnetic resonance imaging. Discrimination between occlusive and reperfused infarctions, *Circulation*, 90(3), 1492–1501, 1994.

Sorensen, A.G., Buonanno, F.S., Gonzalez, R.G., Schwamm, L.H., Lev, M.H., Huang-Hellinger, F.R., Reese, T.G. et al. Hyperacute stroke: Evaluation with combined multisection diffusion-weighted and hemodynamically weighted echo-planar MR imaging, *Radiology*, 199(2), 391–401, 1996.

Stark, D.D. and Bradley, W.G., Eds., *Magnetic Resonance Imaging*, 2nd ed., C.V. Mosby, St. Louis, 1992.

Stejskal, E.O. and Tanner, J.E., Spin diffusion measurements: Spin echoes in the presence of a time-dependent field gradient, *J. Chem. Phys.*, 42, 288–292, 1965.

Strecker, R., Scheffler, K., Klisch, J., Lehnhardt, S., Winterer, J., Laubenberger, J., Fischer, H., and Hennig, J., Fast functional MRA using time-resolved projection MR angiography with correlation analysis, *Magn. Reson. Med.*, 43(2), 303–309, 2000.

Stuber, M., Bornert, P., Spuentrup, E., Botnar, R.M., and Manning, W.J., Selective three-dimensional visualization of the coronary arterial lumen using arterial spin tagging, *Magn. Reson. Med.*, 47(2), 322–329, 2002.

Sussman, M.S., Stainsby, J.A., Robert, N., Merchant, N., and Wright, G.A., Variable-density adaptive imaging for high-resolution coronary artery MRI, *Magn. Reson. Med.*, 48(5), 753–764, 2002.

Turski, P.A., Korosec, F.R., Carroll, T.J., Willig, D.S., Grist, T.M., and Mistretta, C.A., Contrast-enhanced magnetic resonance angiography of the carotid bifurcation using the time-resolved imaging of contrast kinetics (TRICKS) technique, *Magn. Reson. Imaging*, 12(3), 175–181, 2001.

Van Bruggen, N. and Roberts, T.P.L., Eds., *Biomedical Imaging in Experimental Neuroscience*, CRC Press, Boca Raton, FL, 2002.

Villringer, A., Rosen, B.R., Belliveau, J.W., Ackerman, J.L., Lauffer, R.B., Buxton, R.B., Chao, Y.S., Wedeen, V.J., and Brady, T.J., Dynamic imaging with lanthanide chelates in normal brain: Contrast due to magnetic-susceptibility effects, *Magn. Reson. Med.*, 6, 164–174, 1988.

Warach, S., Chien, D., Li, W, Ronthal, M., and Edelman, R.R., Fast magnetic resonance diffusion-weighted imaging of acute human stroke, *Neurology*, 42(9), 1717–1723, 1992.

Wilke, N.M., Jerosch-Herold, M., Zenovich, A., and Stillman, A.E., Magnetic resonance first-pass myocardial perfusion imaging: Clinical validation and future applications, *J. Magn. Reson. Imaging*, 10(5), 676–685, 1999.

7

Ultrasonic Imaging

Elisa E. Konofagou
Columbia University

7.1 Fundamentals of Ultrasound

In the past 30 years, ultrasound has become a very powerful imaging modality mainly due to its unique temporal resolution, low cost, nonionizing radiation, and portability. Lately, unique features such as harmonic imaging, three-dimensional visualization, transducer micromachining, elasticity imaging, and the use of contrast agents have added to the higher quality and wider applications of diagnostic ultrasound images. In this chapter, a short overview of the fundamentals of diagnostic ultrasound and a brief summary of its many applications and methods are provided.

7.1.1 Ultrasound Echoes

Sounds with a frequency above 20 kHz are called ultrasonic since they occur at frequencies inaudible to the human ear. When emitted at short bursts, propagating through media such as water with low reflection coefficient and reflected by obstacles along their propagation path, detection of the reflection, or *echo*, of the ultrasonic wave can help localize the obstacle. This principle has been used by sonar (sound navigation and ranging) and inherently used by marine mammals, such as dolphins and whales, that allows them to localize prey, obstacles, or predators. In fact, the frequencies used for "imaging" vary significantly depending upon the application: from underwater sonar (up to 300 kHz), diagnostic ultrasound, therapeutic ultrasound, and industrial nondestructive testing (0.8–20 MHz) to acoustic microscopy (12 MHz to above 1 GHz).

7.1.2 Wave Equation

As the ultrasonic wave propagates through the tissue, its energy and momentum are both transferred to the tissue. No net transfer of mass occurs at any particular point in the medium unless this is induced by the momentum transfer. As the ultrasonic wave passes through the tissue, or medium, the peak local pressure in the medium increases. The oscillations of the particles result in harmonic pressure variations within the medium and to a pressure wave that propagates through the medium as neighboring particles move with respect to one another (Figure 7.1). The particles of the medium can move back and forth in a direction parallel (longitudinal wave) or perpendicular (transverse wave) to the traveling direction of the wave.

Let us consider the first case.

Assuming that a small volume of the medium that can be modeled as a nonviscous fluid (no shear waves can be generated) is shown in Figure 7.2, an applied force $5F$ produces a displacement of $u + \delta u$ in the z-position on the right-hand side of the small volume. A gradient of force $\partial F / \partial z$ is thus generated across the element in question, and, assuming that the element is small enough so that the measured

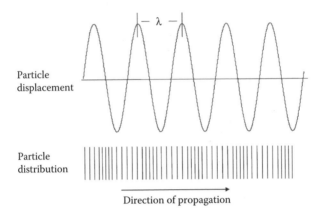

FIGURE 7.1 Particle displacement and particle distribution for a traveling longitudinal wave. The direction of propagation is from left to right. A shear wave can be created in the perpendicular direction, that is, the particles would be moving in a direction orthogonal to the direction of propagation. (Adapted from Wells, P.N.T., *Biomedical Ultrasonics*, Medical Physics Series, Academic Press, London, 1977.)

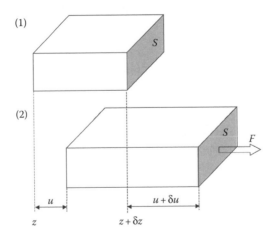

FIGURE 7.2 A small volume of the medium of impedance Z (1) at equilibrium and (2) undergoing oscillatory motion when an oscillatory force F is applied on its cross-sectional surface. (Adapted from Christensen, P.A., *Ultrasonic Bioinstrumentation*, 1st ed., 1988, Copyright Wiley-VCH Verlag GmbH & Co. KGaA.)

quantities within the medium are constant, it can be assumed as being linear, or

$$\delta F = \frac{\partial F}{\partial z} \delta z \tag{7.1}$$

and according to Hooke's law

$$F = KS \frac{\partial u}{\partial z}, \tag{7.2}$$

where K is the adiabatic bulk modulus of the liquid and S the area of the region on which the force is exerted. By taking the derivative of both sides of Equation 7.2 with respect to z and following Newton's second law, from Equation 7.1, we obtain the well-known "wave equation":

$$\frac{\partial^2 u}{\partial z^2} - \frac{1}{c^2} \frac{\partial^2 u}{\partial t^2} = 0, \tag{7.3}$$

where c is the speed of sound given by

$$c = \sqrt{\frac{K}{\rho}} = \sqrt{\frac{1}{\rho k}},$$

where ρ is the density of the medium and K the compressibility of the medium. Equation 7.3 relates the second differential of the particle displacement with respect to distance to the acceleration of a simple harmonic oscillator. Note that the average speed of sound in most soft tissues is about 1540 m/s with a total range of ±6%. For the shear wave derivation of this equation, refer to Wells[1] or Kinsler and Frey[2] among others.

The solution of the wave equation is given by a function u, where

$$u = u(ct - z). \tag{7.4}$$

An appropriate choice of function for u in Equation 7.4 is

$$u(t,z) = u_0 \exp[jk(ct - z)], \tag{7.5}$$

where j is equal to $\sqrt{-1}$ and k is the wavenumber and equal to $2\pi/\lambda$, with λ denoting the wavelength.

7.1.3 Impedance, Power, and Reflection

The pressure wave $p(t, z)$ that results from the displacement generated and given by Equation 7.5 is given by

$$p(t,z) = p_0 \exp[jk(ct - z)], \tag{7.6}$$

where p_0 is the pressure wave amplitude and j is equal to $\sqrt{-1}$. The particle speed and the resulting pressure wave are related through the following relationship:

$$u = \frac{p}{Z}, \tag{7.7}$$

where Z is the acoustic impedance defined as the ratio of the acoustic pressure wave at a point in the medium to the speed of the particle at the same point. The impedance is thus characteristic of the medium and is given by

$$Z = \rho c. \tag{7.8}$$

The acoustic wave intensity is defined as the average flow of energy through a unit area in the medium perpendicular to the direction of propagation.[2] By following that definition, the intensity can be found by[3]

$$I = \frac{p_0^2}{2Z} \tag{7.9}$$

and usually measured in units of mW/cm² in diagnostic ultrasound.

A first step into understanding the generation of ultrasound images is to follow the interaction of the propagating wave with the tissue. Thanks to the varying mean acoustic properties of tissues, a wave transmitted into the tissue will get partly reflected at areas where the properties of the tissue and, thus its impedance, are changing. These areas constitute a so-called impedance mismatch (Figure 7.3).

The reflection coefficient R of the pressure wave at an incidence angle of θ_i is given by

$$R = \frac{p_r}{p_i} = \frac{Z_2 \cos\theta_i - Z_1 \cos\theta_t}{Z_2 \cos\theta_i + Z_1 \cos\theta_t}, \tag{7.10}$$

where θ_t is the angle of the transmitted wave (Figure 7.3) also related to the incidence angle through Snell's law:

$$\lambda_2 \cos\theta_i = \lambda_1 \cos\theta_t, \tag{7.11}$$

where λ_1 and λ_2 are the wavelengths of the waves in media 1 and 2, respectively, and related to the speeds in the two media through

$$c = \lambda f, \tag{7.12}$$

where f is the frequency of the propagating wave.

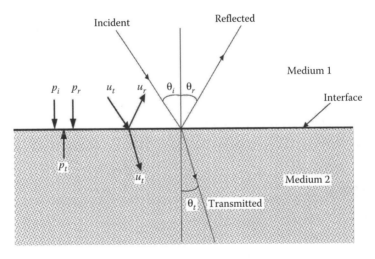

FIGURE 7.3 An incident wave at an impedance mismatch (interface). A reflected and a transmitted wave with certain velocities and pressure amplitudes are created ensuring continuity at the boundary.

As Figure 7.3 also shows, the wave impingent upon the impedance mismatch also generates a transmitted wave, that is, a wave that propagates through. The transmission coefficient is defined as

$$T = \frac{p_t}{p_i} = \frac{2Z_2 \cos\theta_i}{Z_2 \cos\theta_i + Z_1 \cos\theta_t}.$$ (7.13)

According to the parameters reported by Jensen[3] on impedance and speed of sound of air, water, and certain tissues, the reflection coefficient at a fat–air interface is equal to –99.94%, showing that virtually all of the energy incident on the interface is reflected back in tissues such as the lung. A more realistic example found in the human body is the muscle–bone interface, where the reflection coefficient is 49.25%, demonstrating the challenges encountered when using ultrasound for the investigation of bone structure. On the other hand, given the overall similar acoustic properties between different soft tissues, the reflection coefficient is too low when used to differentiate between different soft tissue structures ranging only between –0.1 and 0.1.

The values mentioned above determine both the interpretation of ultrasound images, or sonograms, and the design of transducers, as discussed in the following sections.

7.1.4 Tissue Scattering

In the previous section, the notions of reflection, transmission, and propagation were discussed in the simplistic scenario of plane wave propagation and its impingement on plane boundaries. In tissues, however, such a situation is rarely encountered. In fact, tissues are constituted by cells and groups of cells that serve as complex boundaries to the propagating wave. As the wave propagates through all these complex structures, reflected and transmitted waves are generated at each one of these interfaces depending on the local density, compressibility, and absorption of the tissue. The groups of cells are called "scatterers," as they scatter acoustic energy. The backscattered field, or what is "scattered back" to the transducer, is used to generate the ultrasound image. In fact, the backscattered echoes are usually coherent and can be used as "signatures" of tissues that are, for example, in motion or under compression, as it will be later shown.

An example of such an ultrasound image can be seen in Figure 7.4. The capsule (i.e., the outermost layer) of the prostate is shown to have a strong echo, mainly due to the high impedance mismatch between the surrounding medium, gel in this case, and the prostate capsule. However, the remaining area of the prostate is depicted as a grainy region surrounding the fluid-filled area of the urethra (dark, or low-scattering, area in the middle of the prostate). This grainy appearance is called "speckle,"

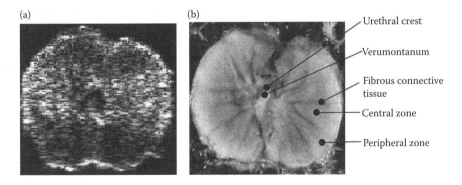

FIGURE 7.4 Sonogram of (a) an *in vitro* canine prostate and (b) its corresponding anatomy at the same plane as that scanned.

a term borrowed from the laser literature.[4] Speckle is produced by the constructive and destructive interference of the scattered signals from structures smaller than the wavelength, hence the appearance of bright and dark echoes, respectively. Thus, speckle does not necessarily relate to a particular structure in the tissue.

Given its statistical significance, the amplitude of speckle has been represented as having a Gaussian distribution with a certain mean and variance.[5] In fact, these same parameters have been used to indicate that the signal-to-noise ratio of an ultrasound image is fundamentally limited to only 1.91.[5] As a result, in the past, several authors have tried different speckle cancelation techniques[6] in an effort to increase the image quality of diagnostic ultrasound. However, speckle offers one important advantage that has rendered it vital in the current applications of ultrasound (Section 7.9). Despite it being described solely by statistics, speckle is not a random signal. As mentioned earlier, speckle is coherent, that is, it preserves its characteristics when shifting from position. Consequently, motion estimation techniques that can determine anything from blood flow to elasticity are made possible in a field that is widely known as "speckle tracking." This is further discussed in later sections of this chapter.

7.1.5 Attenuation

As the ultrasound wave propagates inside the tissue, it undergoes a loss of power dependent on the distance traveled in the tissue. Attenuation of the ultrasonic signal can be attributed to a variety of factors, such as divergence of the wavefront, reflection at planar interfaces, scattering from irregularities or point scatterers, and absorption of the wave energy.[7] In this section, we will concentrate on the latter as it is the strongest factor in soft (other than lung) tissues. In this case, the absorption of the wave's energy leads to heat increase. The actual cause of absorption is still relatively unknown but simple models have been developed to demonstrate the dependence of the resulting wave pressure amplitude decrease in conjunction with the viscosity in tissues.[8]

By not going into detail concerning the derivations of such a relationship, an explanation of the phenomenon is provided here. Let us consider a fluid with a certain viscosity that provides a certain resistance to a wave propagating through its different layers. To overcome the resistance, a certain force per unit area, or pressure, needs to be applied that is proportional to the shear viscosity of the fluid η as well as the spatial gradient of the velocity,[7] or

$$p \propto \eta \frac{\partial u}{\partial z}. \tag{7.14}$$

Equation 7.14 shows that a fluid with higher viscosity will require higher force to experience the same velocity gradient compared to a less viscous fluid. By considering Equations 7.2 and 7.14, an extra term can be added to the wave equation that includes both the viscosity and compressibility of the medium,[7] or

$$\frac{\partial^2 u}{\partial z^2} + \left(\frac{4\eta}{3} + \xi \right) k \frac{\partial^3 u}{\partial z^2 \partial t} - \frac{1}{c^2} \frac{\partial^2 u}{\partial t^2} = 0, \tag{7.15}$$

where ξ denotes the dynamic coefficient of compressional viscosity. The solution to this equation is given by

$$u(t,z) = u_0 \exp(-\alpha z) \exp[jk(ct - z)], \tag{7.16}$$

where α is the attenuation coefficient also given by (for $a \ll k$)

$$\alpha = \frac{\left(\frac{4\eta}{3} + \xi\right)k^2}{2\rho c}. \tag{7.17}$$

From Equation 7.16, the effect of attenuation on the amplitude of the wave is clearly demonstrated. An exponential decay on the envelope of the pressure wave highly dependent on the distance results from the tissue attenuation (Figure 7.5). The intensity of the wave will decrease much faster, given that from Equation 7.9

$$I(t,z) = \frac{p_0^2}{Z} \exp(-2\alpha z)\exp[2jk(ct - z)] \tag{7.18}$$

or the average intensity is equal to

$$\langle I \rangle = I_0 \exp(-2\alpha z). \tag{7.19}$$

Another important effect that tissue attenuation has on the propagating wave is a frequency shift. This is because a more complex form for the attenuation α is

$$\alpha = \beta_0 + \beta_1 f, \tag{7.20}$$

where β_0 and β_1 are the frequency-independent and frequency-dependent attenuation coefficients.[3] In fact, the frequency-dependent term is the largest source of attenuation and increases linearly with frequency. As a result, the spectrum of the received signal changes as the pulse propagates through the tissue in such a way that a shift to smaller frequencies, or downshift, occurs. In addition, the downshift is dependent on the bandwidth of the pulse propagating in the tissue and the mean frequency of a spectrum (in this case Gaussian[3]) can be given by

$$\langle f \rangle = f_0 - (\beta_1 B^2 f_0^2)z, \tag{7.21}$$

where f_0 and B denote the center frequency and bandwidth of the transducer. Thus, according to Equation 7.21, the downshift due to attenuation depends on the frequency-dependent attenuation coefficient,

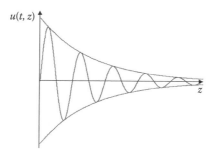

FIGURE 7.5 Attenuated wave of Figure 7.1. Note the envelope of the wave dependent on the attenuation of the medium.

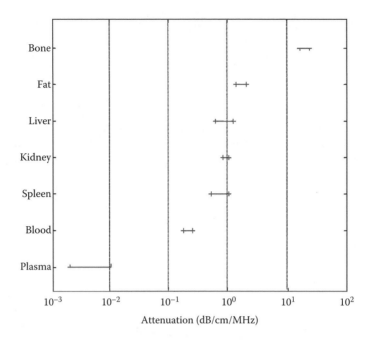

FIGURE 7.6 Attenuation values of certain fluids and soft tissues. (From Haney, M.J. and O'Brien, W.D., Jr., in *Tissue Characterization with Ultrasound*, Greenleaf, J.F., Ed., CRC Press, Boca Raton, FL, 1986, 15–55.)

the transducer center frequency, and its bandwidth. A graph showing the typical values of frequency-dependent attenuation coefficients (measured in dB/cm/MHz) is given in Figure 7.6.

7.2 Transducers

The pressure wave that was discussed in the previous section is generated using an ultrasound transducer, which is typically a piezoelectric material. "Piezoelectric" denotes the particular property of certain crystal polymers of transmitting a pressure ("piezo" means "to press" in Greek) wave generated when an electrical potential is applied across the material. Most importantly, since this piezoelectric effect is reversible, that is, a piezoelectric crystal will convert an impinging pressure wave to an electric potential, the same transducer can also be used as a receiver. Such crystalline or semicrystalline polymers are the polyvinylidene fluoride (PVDF), quartz, barium titanate, and lead zirconium titanate (PZT).

A single-element ultrasound transducer is shown in Figure 7.7. Depending upon its thickness (l) and propagation speed (c), the piezoelectric material has a resonance frequency given by

$$f_0 = \frac{c}{2l}. \tag{7.22}$$

The speed in the PZT material is around 4000 m/s, so for a 5-MHz transducer, the thickness should be 0.4 mm. The matching layer is usually coated onto the piezoelectric crystal to minimize the impedance mismatch between the crystal and the skin surface and, thus, maximize the transmission coefficient (Equation 7.13). To overcome the aforementioned impedance mismatch, the ideal impedance Z_m and thickness d_m of the matching layer are, respectively, given by

$$Z_m = \sqrt{Z_T Z} \tag{7.23}$$

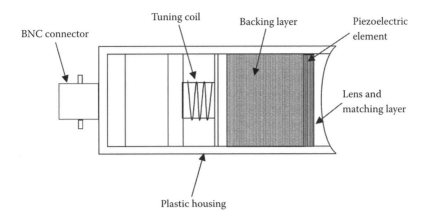

FIGURE 7.7 Typical construction of a single-element transducer. (From Jensen, J.A., *Estimation of Blood Velocities Using Ultrasound*, Cambridge University Press, Cambridge, UK, 1996.)

and

$$d_m = \frac{\lambda}{4}, \tag{7.24}$$

with Z_T denoting the transducer impedance and Z the impedance of the medium.

The backing layers behind the piezoelectric crystal are used to increase the bandwidth and the energy output. If the backing layer contains air, then the air–crystal interface yields a maximum reflection coefficient given the high impedance mismatch. Another byproduct of an air-backed crystal element is that the crystal remains relatively undamped, that is, the signal transmitted will have a low bandwidth and a longer duration. On the other hand, as will be seen in Section 7.3.3, the axial resolution of the transducer depends on the signal duration, or pulse width, transmitted. As a result, there is a trade-off between transmitted power and resolution of an ultrasound system. Depending on the application, different backing layers are therefore used. Air-backed transducers are used in continuous-wave and ultrasound therapy applications. Heavily backed transducers are utilized to obtain high resolution, for example, for high-quality imaging at the expense of lower sensitivity and reduced penetration.

For imaging purposes, an assembly of elements such as that in Figure 7.8 is usually used and called an "array" of such elements. In an array, the elements are stacked next to each other at a distance equal to less than a wavelength for the minimum interference and reduced grating lobes (Section 7.3.2). The most common are shown in Figure 7.8. The linear array has the simplest geometry. It selects the region of interest by firing elements above that region (Figure 7.8a). The beam can then be moved on a line by firing groups of adjacent elements and then the rectangular image obtained is formed by combining the signals received by all the elements. A curved array is used when the transducer is smaller than the area scanned (Figure 7.8b). A phased array can be used to change the "phase," or delay, between the fired elements and thus achieve steering of the beam. The phased array is usually the choice for cardiovascular exams, when the window between the ribs allows for a very small transducer to image the whole heart. Focusing and steering can both be achieved by changing the pulsing delays between elements (Figure 7.9).

7.3 Ultrasound Fields

In the previous section, the different priorities of resolution and depth in pulsed-wave and continuous-wave practices were described. In this section, their distinct applications are discussed.

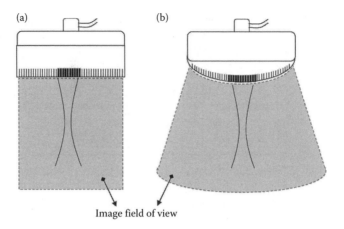

FIGURE 7.8 (a) Linear and (b) curved array transducer used for B-scan acquisition and according to the type of application.

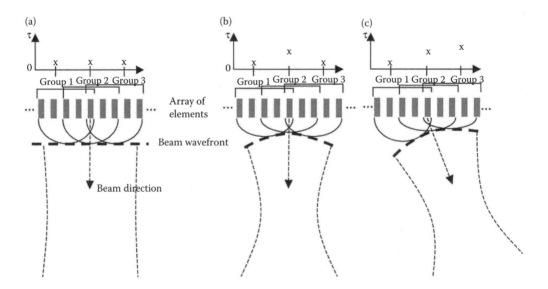

FIGURE 7.9 Electronic (a) beam forming, (b) focusing, and (c) focusing and beam steering as achieved in phased arrays. The time delay between the firings of different elements is denoted here by τ.

7.3.1 Continuous Wave

Let us consider the field produced by the simplest transducer geometry: a circular aperture single-element transducer with a radius r. It is considered as the simplest transducer geometry, since the surface of the transducer can be viewed as a vibrating piston with constant amplitude and phase. The simplest kind of wave emitted from such a transducer is a continuous plane wave, that is, where the wavefront, or the surface in which the motion is everywhere in phase, is assumed to be planar. For this to hold, the source of the wave has to be much smaller in size than the wavelength of the emitted wave. According to Huygen's principle, any wave can be considered as the sum of contributions from a particular distribution of sources, which have the suitable phase and amplitude to generate the wave in question. Therefore, the circular transducer can be assumed as being composed of several Huygen's sources all identical in size and at a uniform distribution.

In this case, a simple solution of the field distribution along the axis of symmetry of the source (Figure 7.10) can be found. The field, or beam, is found to have two distinct regions. The beam intensity along the central axis of the field $I(z)$ is given by

$$\frac{I(z)}{I_0} = \sin^2\left\{\frac{\pi}{\lambda}\left[\sqrt{r^2 + z^2} - z\right]\right\}, \tag{7.25}$$

where I_0 is the maximum intensity. The maxima and minima of the field in Equation 7.25 are found at

$$z_m = \frac{4r^2 - (2m+1)^2\lambda^2}{4(2m+1)\lambda}, \quad m = 0, 1, 2, \ldots \tag{7.26}$$

and

$$z_n = \frac{r^2 - n^2\lambda^2}{2n\lambda}, \quad n = 1, 2, 3, \ldots, \tag{7.27}$$

respectively. The last axial maximum occurs at $m = 0$ when

$$z_0 = \frac{4r^2 - \lambda^2}{4\lambda}$$

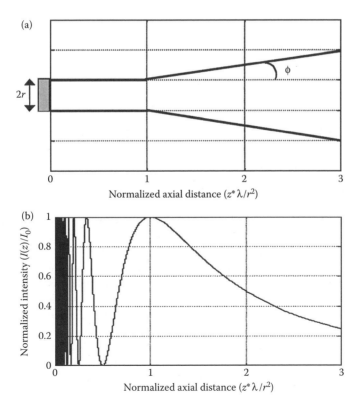

FIGURE 7.10 (a) Continuous-wave field for a circular aperture transducer of radius r and divergence angle ϕ. (b) Normalized intensity amplitude along the symmetry axis of the transducer shown on top at 5-MHz center frequency and 7-mm radius.

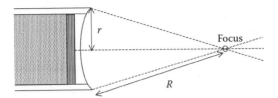

FIGURE 7.11 Field of view of a concave transducer with a radius of curvature R and a radius of aperture r.

and, if $r^2 \gg \lambda^2$,

$$z_0 = \frac{r^2}{\lambda}. \tag{7.28}$$

The region before z_0 is called the near field, or Fresnel zone, of the transducer and the region after z_0 is the far field, or Fraunhofer zone (Figure 7.10). Therefore, z_0 is called the transition point or transition distance beyond which the field becomes more uniform. The angle of divergence in the far field (Figure 7.10) is given by

$$\phi = \arcsin\left(0.61\frac{\lambda}{r}\right). \tag{7.29}$$

A slightly more complex but more applicable transducer geometry is that of a concave transducer (Figure 7.11). In that case, the variation in intensity is found by[9]

$$\frac{I(z)}{I_0} = kr^2 Jinc(kr\sin\phi), \tag{7.30}$$

where $Jinc(x) = J_1(x)/x$, $J_1(x)$ is the first-order Bessel function. The term $Jinc(kr \sin \phi)$ is also known as the *directivity* factor. The beam profile is shown in Figure 7.12.

7.3.2 Pulsed Pressure and Pulse Echo Fields

A pulsed pressure field can be generated either by spherical waves emitted by the aperture or by the intersection between spherical waves and the aperture. The latter observation is very important in treating the subsequent imaging in a systems approach.

As mentioned in Section 7.3, the reflected wave or signal is generated following the interaction of the transmitted wave with relatively small structures that cause a perturbation due to their density, compressibility, and absorption. The reflected wave can be safely assumed to be spherical in the case when the transmitted wave is backscattered and the scatterer–transducer distance is sufficiently large. In other words, when the same transducer is used for transmission and reception, the same impulse response can be used to describe the generation and reception of the scatterer field. By following a systems approach, the reflected waveform, $r(x, y, z)$, can be expressed as the convolution (denoted here by $\otimes$) of the wave field incident on the tissue structure, $p(x, y, z)$, and the impulse response, $f(x, y, z)$, that is associated with the acoustic properties of the tissue structure. That is

$$r(x,y,z) = p(x,y,z) \otimes f(x,y,z), \tag{7.31}$$

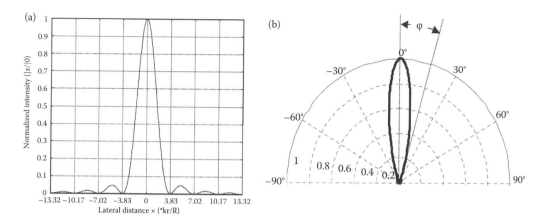

FIGURE 7.12 (a) The beam profile in the far field at distance z from the transducer of radius r. (b) The same profile in an angular plot.

where $p(x, y, z) = p_x(x) \otimes p_y(y) \otimes p_z(z)$ is also called the *point-spread function* (psf), with $p_x(x)$ denoting the pulse-echo impulse response of the system along the axis of propagation, or axial direction, and is responsible for the radio frequency (RF) content of the signal, $p_y(y)$, the pulse-echo impulse response perpendicular to the axial direction but in the same plane, or lateral direction, and determines the *beam profile*, and $p_z(z)$ the pulse-echo impulse response perpendicular to the imaging plane, or elevational (azimuthal) direction[5] (Figure 7.13). A more complex relationship for $r(x, y, z)$ can be obtained showing its time dependence, or four-dimensional characteristics; the reader is encouraged to consult Section 2.5 in Jensen.[3]

The signal of Equation 7.31 is also called the "raw signal" or "RF signal" to denote the unprocessed ultrasonic tissue response received by the system. This is not the signal displayed by the ultrasonic scanners used in clinical environments and several processing and display methods can be applied on the received signal to obtain the physical and physiological parameters of interest to the clinicians. This is thoroughly discussed in Section 7.4.

As mentioned in Section 7.2, most imaging systems use arrays for faster and more controlled scanning. In this case, the fields are calculated based on the psfs of the individual, usually identical, trans-

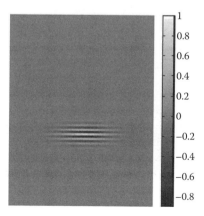

FIGURE 7.13 (**See color insert.**) A two-dimensional version of the psf $p(x, y, z)$.

ducer elements. Following Huygen's principle and assuming that the wave propagation is linear, the individual impulse responses can be added according to

$$p_a(x, y, z) = \sum_{j=0}^{N-1} p_j(x, y, z), \tag{7.32}$$

where N is the number of elements, $p_j(x, y, z)$ is the psf for each element, and $p_a(x, y, z)$ is the psf for the whole array at the tissue location of $(x, y\ z)$. If the elements are assumed to be very small, the field point far away from the array, the individual transducer elements can be assumed as having Dirac psfs and for the example of a linear array of Figure 7.8a, the amplitude of the beam profile can be found by[3]

$$\left| P_a(f) \right| = \left| \frac{k}{d} \frac{\sin\left(N \dfrac{Dn\sin\vartheta}{c} f \right)}{\sin\left(\dfrac{Dn\sin\vartheta}{c} f \right)} \right|, \tag{7.33}$$

where k is the wavenumber, d the distance of the array to the field point, D the distance between two neighboring elements in the array, and ϑ the angle of divergence.

The main lobe of the beam profile of Equation 7.33 occurs naturally at $\vartheta = 0$ but other maxima of the response are found at

$$\varphi_g = \arcsin\left(\frac{n\lambda}{D} \right), \tag{7.34}$$

where $n = 1, 2, \ldots$. For an array at 5 MHz with an element spacing equal to 0.15 mm, the angle φ_g for the next first maximum is equal to 11.8°. This means that the received signal will also be affected by scatterers that are off the image axis, in this case those positioned at 11.8° off the axis. These responses are naturally not desirable and are called *grating lobes* (Figure 7.14). Since the angle φ_g increases with a decreasing element spacing, one way to make sure they are out of the image plane (i.e., past 90° in Figure 7.14) is to ensure that the elements are separated by less than a wavelength (in the example mentioned earlier, it would mean that $D < 0.3$ mm). Typically, an element spacing equal to half a wavelength is chosen (like

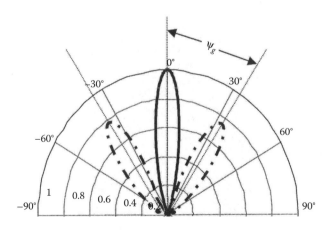

FIGURE 7.14   The beam profile from an N-element linear array.

in the example just discussed) so as to allow for steering of the beam. Steering, however, can move the grating lobes closer to the center of the beam, dependent on the steering angle.

7.3.3 Axial, Lateral Resolution, and Focal Spot Size

7.3.3.1 Axial Resolution

As explained earlier, different designs and trade-offs need to be considered depending on whether the application requires continuous-wave or pulsed-wave excitation. In the former case, a high efficiency is desired, while in the latter, high-quality imaging should result. The transducer parameter that can best express the difference between the priorities in the two cases is the quality factor, or Q, defined as

$$Q = \frac{f_0}{B},\tag{7.35}$$

where f_0 is the center frequency of the transducer while B is the bandwidth of the spectrum of the pulse at the –3 dB level (Figure 7.15). Figure 7.15 shows the difference between two extreme cases of transducer designs. Since Q is inversely proportional to the bandwidth, high Q denotes long pulsewidth, or continuous-wave, while low Q denotes pulse-wave. Low Q, therefore, also denotes high axial resolution.

Axial, or longitudinal, resolution is defined as the minimum distance between two scatterers that can be discerned by the system (located along the axis of wave propagation). When the pulse is short, the echoes successively reflected by the two scatterers can be better differentiated. When the pulse is too long, the two echoes at reception blend together, making the differentiation between two scatterers impossible. Therefore, the axial resolution (AR) is directly proportional to the Q of the transducer and in fact is given by[7]

$$AR \approx \frac{Q\lambda}{4}.\tag{7.36}$$

7.3.3.2 Lateral Resolution and Focal Spot Size

Another important imaging parameter of an ultrasound transducer is the lateral resolution, defined as the minimum distance between two scatterers that can be discerned by the system and located perpendicular to the axis of wave propagation but in the imaging plane. Compared to the axial resolution,

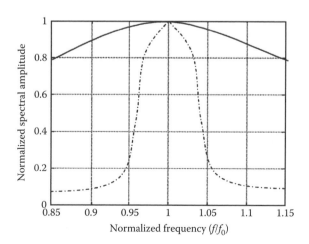

FIGURE 7.15 Two different frequency responses for $Q = 2$ (solid) and $Q = 20$ (dotted) transducers. (Adapted from Wells, P.N.T., *Biomedical Ultrasonics*, Medical Physics Series, Academic Press, London, 1977.)

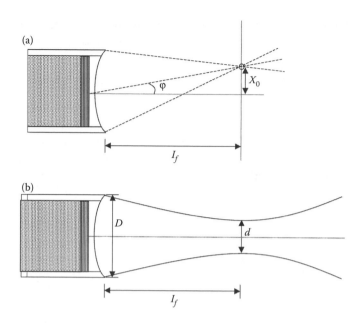

FIGURE 7.16 (a) The lens of the transducer has the property of focusing the waves at a distance x_0 from the focal point (to be compared with Figure 7.11). (b) The lens focuses the beam to a focal spot size equal to d.

the lateral resolution is defined more by the lens than the pulse characteristics. The beam width of the transducer is often too wide to ensure good definition of lateral features in scanned objects. Therefore, a lens is often used as shown in Figure 7.11. The role of the lens is to direct waves entering at a certain angle ϕ to a particular distance in the focal plane x_0.[7] The lens in Figure 7.16 has a focal length given by

$$l_f = \frac{R}{1-n}, \tag{7.37}$$

where $n = c/c_1$ and c_l is the speed of sound in the lens and the "divergence angle" of the lens is given by

$$\sin \varphi = \frac{x_0}{l_f}. \tag{7.38}$$

Therefore, if a lens is placed in front of a circular disk transducer with a field given by Equation 7.30, from Equations 7.30 and 7.38, the directivity factor becomes equal to

$$H_l(x_0) = Jinc\left(kr\frac{x_0}{l_f}\right). \tag{7.39}$$

The field pattern of this focused transducer is exactly the same as the far field of the unfocused one with the same characteristics. The focused spot is the central portion of that field and its size is equal to the diameter of that portion, which is equal to the distance d between the two first zeros of the field in Equation 7.39 given by[7]

$$d = 2.44 F\lambda, \tag{7.40}$$

where *F* is also known as the *F*-number of the transducer defined as

$$F = \frac{l_f}{D}. \tag{7.41}$$

The focal spot size (Equation 7.40) also determines the lateral resolution since two scatterers will be discernable in the lateral direction if they are separated by a distance at least equal to the focal spot size. Clearly, axial (Equation 7.36) and lateral (Equation 7.40) resolutions are determined by the wavelength, which sets their lower limits. For high-quality imaging, short wavelengths need to be employed, thus denoting high transducer frequencies that are in turn limited by the resulting higher attenuation (Equation 7.20). This constitutes the well-known resolution versus depth penetration trade-off that needs to be taken into account in the design of the optimal ultrasound system.

7.3.4 Nonlinear Effects

Until now, the tissue was assumed to respond linearly to the mechanical pressure applied by the propagating wave. This only applies to waves of very small amplitude. In fact, waves at higher amplitudes with pressures in the MPa range are routinely utilized in clinical scanners, clearly challenging the models presented until now. This means that, as the wave propagates through a nonlinear medium, it no longer has a constant speed but rather a speed dependent upon the local wave amplitudes. As a result, the propagating wave changes in shape as it propagates and the subsequent distortion introduces higher harmonics than the fundamental one initially transmitted. These higher harmonics are more affected by attenuation (Equation 7.20) and the distortion increases with frequency, the nonlinearity of the medium, lower speed of sound, and lower attenuation coefficient. Compared to what was discussed in the previous subsections, the nonlinear effects will cause the wave to be absorbed faster, the psf (Figure 7.13) to be more spatially variant, and the variability of the beam shape to be higher. There are, however, several advantages of utilizing this effect in certain applications, such as harmonic imaging and with contrast agents. These effects are further detailed in Section 7.5.

7.4 Ultrasonic Imaging

Ultrasonic imaging is usually known as *echography* or *sonography* depending on which side of the Atlantic Ocean one is scanning from. As mentioned earlier, the signal acquired by the scanner can be processed and displayed in several different fashions. In this section, the most typical and routinely used ones are discussed.

7.4.1 A-Mode

Figure 7.17 shows a block diagram of the different steps that are used to acquire, process, and display the received signal from the tissue.

7.4.1.1 Transducer Frequency

A pulse of a given duration, frequency, and bandwidth is first transmitted. As mentioned before, a trade-off between penetration (or low attenuation) and resolution exists. Therefore, the chosen frequency will depend on the application. Usually, for deeper organs such as the heart, the uterus, and the liver, the frequencies are restricted in the range of 3–5 MHz, while for more superficial structures, such as the thyroid, the breast, the testis, and applications on infants, a wider range of 4–10 MHz is applied. Finally, for ocular applications, a range of 7–15 MHz is determined by the low attenuation, low depth, and higher resolution required.

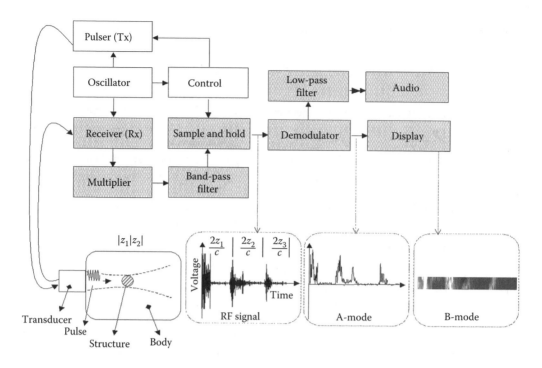

FIGURE 7.17   Block diagram of a pulsed-wave system and the resulting signal or image at three different steps.

The pulse is usually a few cycles of that frequency length (usually 3–4 cycles) so as to ensure high resolution (Equation 7.36) and is generated by the transmitter through a voltage-step sinusoidal function at a voltage amplitude (100–500 V) and a frequency equal to that of the resonance frequency of the transducer elements. For static structures, a single pulse or multiple pulses (usually used for averaging later) could be used at an arbitrary frequency. However, for moving structures, such as blood, liver, and the heart, a fundamental limit on the maximum pulse repetition frequency (PRF) is set by the maximum depth of the structure, or PRF (kHz) = $c/2D_{max}$. Typically, the PRF is in the range of 1–3 kHz.

7.4.1.2 RF Amplifier

The received signal needs to be initially amplified so as to guarantee a good signal-to-noise ratio. At the same time, the input of the amplifier should be devoid of the high-voltage pulse to protect the circuits but also to maintain its low noise and high gain. A typical dynamic range expected at the output is on the order of 70–80 dB.

7.4.1.3 Time-Gain Compensation

As mentioned in Section 7.1.5, attenuation is unavoidable as the wave travels through the medium and increases with depth. To avoid artificial darkening of deeper structures as a result, a voltage-controlled attenuator is usually employed, where a control voltage is utilized to manually adjust the system gain accordingly after reception of an initial scan. A logarithmic voltage ramp is usually applied that compensates for a mean attenuation level with depth.[6] The dynamic range becomes further reduced to 40–50 dB.

7.4.1.4 Compression Amplifier

The signals will ultimately be displayed as a grayscale on a cathode ray tube (CRT), where the dynamic range is typically only 20–30 dB. To this purpose, an amplifier with a logarithmic response is utilized.

7.4.1.5 Demodulation or Envelope Detection

Since the image is a grayscale picture, the amplitude of the signal is displayed. For this, the envelope of the RF signal needs to be calculated. This is usually achieved by using Hilbert transforms. The resulting signal is called a detected A-scan, A-line, or *A-mode scan* (A for amplitude). An example of that is shown in Figure 7.17.

7.4.2 B-Mode

When the received A-scans are spatially combined after acquisition using either a mechanically moved transducer or the previously mentioned arrays and brightness modulate the display in a two-dimensional format, the brightness or B-mode is created, which has a true image format and is by far the most widely used diagnostic ultrasound mode. By default, sonogram or echogram refers to B-mode. Figure 7.18a shows an image of the left ventricle imaged parasternally. One of the biggest advantages of ultrasound scanning is real-time scanning and this is achieved due to the shallow depth of scanning in most tissues and the high speed of sound. The frame rate is usually on the order of 30–100 Hz (while in the M-mode version, it can be as fast as the PRF itself; see below). The frame rate is limited by the number of A-mode scans acquired, N_A, and the maximum depth, that is, $PRF_{max} = (c/2D_{max})/N_A$.

7.4.3 M-Mode

Another fashion of displaying the A-scans is in function of time, especially in cases where tissue motion needs to be monitored and analyzed. In this case, only one A-scan from a particular tissue structure is displayed in brightness mode but followed in time depending on the PRF used, and is called Motion-, or M-mode scan. A depth-time display is then generated. A typical application of the M-mode display is used in the examination of heart-valve leaflets motion and Doppler displays (see later section). Figure 7.18b shows the M-mode version of the B-mode to follow the motion of the ventricle at the papillary muscle level.

7.4.4 C-Mode

The constant-depth, or C-scan, is not used as widely as the aforementioned modes, mainly due to its distinct use of scanning. Instead of relying on the acquisition of reflected echoes from the medium, the pulse is transmitted from one side of the body by a transmitter to be received on the other side at

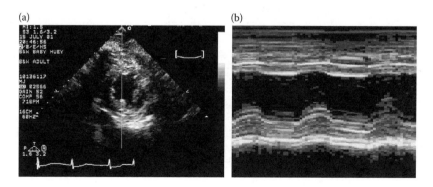

FIGURE 7.18 (a) B-scan of the left ventricle from an apical view; (b) M-mode image of the same view taken at the level of the papillary muscles shown in (a) over three cardiac cycles. (Courtesy of Scott D. Solomon, Brigham and Women's Hospital, Boston, MA.)

the same depth by a separate transducer—borrowing from the x-ray CT scan principle. As a result, the scanning motion is perpendicular to the transmitted beam. The two main applications include attenuation measurement along the transmitted beam that depends on both the absorption by the tissue structures and reflection losses at the tissue interfaces, and measurement of the acoustical index of refraction defined as $n = c_w/c$, where c_w is the speed of sound in water and c is the speed of sound in the tissue. The latter is achieved by measuring the time lapsing between pulse transmission and reception, or *time-of-flight*, to calculate the phase velocity of the tissue along the path that separates the two transducers. The C-scans have enjoyed applicability in tissues that are more superficial and relatively homogeneous so as to ensure travel of the echo through all the interfaces. Such an application is the human female breast.

7.4.5 Doppler

7.4.5.1 Doppler Equation

Let us consider the case of a scatterer, for example, a red blood cell, as shown in Figure 7.19 at a distance x from the transducer and moving with a constant velocity v and at an angle 0 with respect to the transducer beam of frequency $\omega = 2\pi f$ given by

$$u(t) = u_0 \cos(\omega t). \tag{7.42}$$

The received beam will be given by

$$u_r(t) = u_{r0} \cos\left\{\omega\left(t - \frac{2x}{c}\right)\right\}, \tag{7.43}$$

where $2x$ is the round-trip distance equal to $tv \cos \theta$ (Figure 7.19) and assuming $v \ll c$, we obtain

$$u_r(t) = u_{r0} \cos\left\{\omega\left(1 - \frac{2v\cos\theta}{c}\right)t\right\} \tag{7.44}$$

so that the frequency of the received signal is given by

$$f_r = f\left(1 - \frac{2v\cos\theta}{c}\right)$$

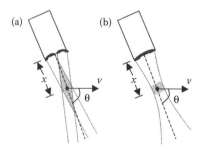

FIGURE 7.19   (a) Continuous- and (b) pulsed-wave systems.

or

$$\Delta f = -\frac{2vf\cos\theta}{c} \tag{7.45}$$

that denotes the frequency shift caused in the received signal due to velocity v of the scatterer, which can have a positive or negative value (depending on whether the flow is forward or reverse, according to the conventions used) and at an angle θ to the central axis of the transducer beam. This shift is also known as the *Doppler shift* and Equation 7.45 is known as the *Doppler equation*. According to Equation 7.45, with a frequency f on the order of 2–10 MHz and $v\cos\theta$ in the range of 0–5 m/s, the Doppler shift varies usually in the range 0–14 kHz, which is well within the human audio range and can be broadcasted through the speakers of the system (Figure 7.17).

There are several advantages and disadvantages in the formulation and use of the Doppler equation. First, the formulation is very simple and easy to remember. However, the angle is very difficult to determine, thus making the accurate measurement of the velocity cumbersome. Second, the equation describes the motion of a single scatterer. In practice, a cloud of scatterers is moving with each scatterer at a different velocity, causing a different frequency shift. A spectrum is thus generated (Figure 7.20) that is typically changing during a cardiac cycle, especially when measuring pulsatile flow in key vessels inside the body. The multidimensionality of the Doppler spectrum allows the simultaneous observation of the distribution of blood velocities inside a vessel, their time variations, and their magnitudes. This constitutes a unique feature of ultrasound systems. The spectral characteristics also depend on the geometry of the vessel and the parameters of the beam. However, qualitative analysis of the Doppler spectra is capable of determining the type of flow inside a vessel, that is, whether the flow is parabolic, turbulent, or other, by merely measuring the increase in the *bandwidth* of the Doppler spectrum, otherwise known as *spectral broadening*. The magnitude of the Doppler spectrum depends on the compliance of the vessel wall and the flow impedance.[3,6] However, stationary structures, for example, within the vessel wall, usually constitute the main noise source in the velocity measurement, since they can generate disproportionally (often 10–100 times) larger echoes, or *clutter*, than the low backscatter echoes from blood. In addition, as discussed earlier, attenuation also affects the received echo spectrum and often

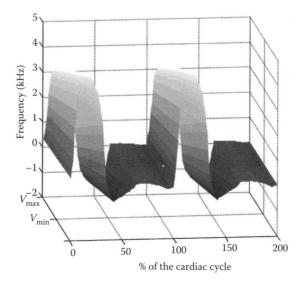

FIGURE 7.20 Simulated Doppler spectrum at different instances of two cardiac cycles.

to a much larger extent than the Doppler effect.[3] On the other hand, the coherent interference of waves in a random medium, which was earlier defined as "speckle," also contains information on the motion of the scatterers in question. The expansive field of "speckle tracking" techniques is briefly discussed in Section 7.4.5.3. Finally, the Doppler equation does not take the beam finiteness into account and, as a result, when calculating the spectrum, frequency leakage occurs.[6] The well-known trade-off between spatial and frequency resolution also holds here. The aforementioned characteristics are only valid in the case of continuous-wave systems. As mentioned earlier, most imaging systems use pulsed-wave, or pulse-echo, methods to achieve higher resolution and those are discussed in the following section.

7.4.5.2 Continuous-Wave Velocity and Flow Detector

There are two types of Doppler systems: the continuous wave and pulsed wave (Figure 7.19). The continuous-wave system, as briefly discussed in the previous subsection, is the simpler of the two. A continuous pressure wave is transmitted by one transducer element, while another receives the echoes scattered from the tissue during transmission. The frequency shifts, or differences in frequency, are measured using the A-scans and displayed and/or made audible. For those measurements, several methods have been developed over the years.[6] One technique of choice has been quadrature phase demodulation, in which the received echo (Equation 7.44) is multiplied by a quadrature signal with frequency equal to that of the transmitted wave (Equation 7.42) to yield

$$u_{rq}(t) = u_{r0} \cos\left\{\omega\left(1 - \frac{2v\cos\theta}{c}\right)t\right\}\cos(\omega t)$$

$$= \frac{u_{r0}}{2}\left[\cos\left\{\frac{2\omega v\cos\theta}{c}t\right\} + \cos\left(2\omega t - \frac{2\omega v\cos\theta}{c}t\right)\right]. \qquad (7.46)$$

A bandpass filter is used to remove the higher-frequency component (second additive term in Equation 7.46) as well as the DC component introduced by stationary echoes.[3] After filtering, only the first additive term remains that contains the information on the frequency shift and the velocity (Equation 7.45). An example of such a display in a clinical environment is shown in Figure 7.21 and is also often called "sonogram."

Since all Doppler frequencies fall in the audible human range, the Doppler spectrum can be listened to and judged by a clinician, and a very simple method that has been used in both old and current

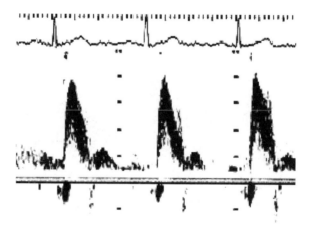

FIGURE 7.21 Continuous-wave (CW) Doppler sonogram with the corresponding EKG (top) of a normal aorta at the suprasternal notch over three cardiac cycles.

systems can provide a direct feedback regarding the blood flow characteristics. For example, a forward (reverse) flow occurs when the argument in the cosine term is positive (negative). It turns out[3] that the forward and reverse flows can be separated by shifting the phase of the Doppler signal by 90° and then adding its imaginary and real parts. The two flows are then heard over two individual speakers providing the flow direction (Figure 7.17). For more information on the subject as well as quantitative methods for flow measurements, the reader is encouraged to consult Jensen.[3]

7.4.5.3 Pulsed-Wave Velocity and Flow Detector

The previous subsection dealt with the application of the Doppler equation and the use of the Doppler spectrum for flow pattern detection and characterization. Despite its simplicity, a serious drawback of the continuous-wave approach is the lack of range, or depth, information due to the fact that the transmit and receive elements are in close proximity. An inherent limitation of the technique is that there is nonuniqueness between a Doppler spectrum and a flow profile, rendering the method questionable for the reliable differentiation between normal and pathological cases. To overcome these limitations, pulsed-wave systems were developed that allow for the investigation of flow patterns in individual vessels or individual parts of a vessel (Figure 7.19). The pulsed-wave systems combine characteristics from pulse echo imaging and CW Doppler methods. A pressure wave of a certain frequency is emitted, similar to the CW system, but at short pulses and a certain PRF. RF echoes are then received by the same transducer following each pulse and then gated with an adjustable duration. This gate allows for the selection of the depth at which the velocity will later be measured.

An important limitation of the Doppler spectrum method is that, as briefly mentioned previously, the attenuation effect can be much more dramatic than the sought-after Doppler shift. For example, in the case where the vessel is at a depth of 5 cm from the transducer, the attenuation of the medium is at 0.5 dB/cm/MHz and the relative transducer bandwidth is at 80%, a downshift of the transducer center frequency equal to 16 kHz occurs. Assuming a velocity of 0.5 m/s, the Doppler shift will be on the order of 1–2 kHz, thus completely overshadowed by the downshift due to attenuation. In other words, in a pulsed-wave system, the classic Doppler effect is not utilized.

The Doppler effect is in fact interpreted in the time domain so as to avoid the undesirable and unavoidable frequency effects from other noise sources. Except for the scaling change of the signal in the time domain (following the bandwidth change in the frequency domain), the signal also experiences a time delay as a result of the velocity. In other words, if two pulses are emitted successively, according to the amount of time lapsing between the two pulses, as the cloud of scatterers moves, the received signal will also move accordingly, away or toward the transducer. The scatterers move between pulses that are emitted at a delay inversely proportional to the PRF, or T_{PRF}. If the received signals acquired at pulses 1 and 2 are given by u_{r_1} and u_{r_2}, respectively, they can be linked through the following equation[3]:

$$u_{r_2}(t) = u_{r_1}\left(t - \frac{T_{\mathrm{PRF}}}{\alpha}\right), \tag{7.47}$$

where

$$\alpha = 1 - \frac{2v\cos\theta}{c}.$$

According to Equation 7.47, the velocity of the scatterers results in a shift of the received signal that is dependent on the velocity, beam-vessel angle, speed of sound, and the transmitted pulse delay. The so-called speckle-tracking techniques are then employed to estimate the shift resulting from the scatterer motion that is equivalent to speckle displacement. No scatterer motion, or stagnant flow, will result in the two received waveforms being identical. Assumptions to obtain Equation 7.47

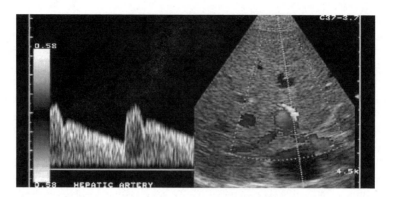

FIGURE 7.22   A triplex-mode sonogram showing the PW Doppler sonogram (left) at the level of the hepatic artery in B-scan of the liver (right). Usually, the color (shown in this example on a grayscale) here indicates the direction of the flow, for example, toward or away from the transducer. (Courtesy of CoreVision Pro, Toshiba Medical Systems, Irvine, CA.)

are that the displacement is relatively small, noise is negligible, and that transverse motion is not significant. The displacement can be safely assumed small in the case where $vT_{PRF} < b$, where b is the beamwidth.

An important limitation of a pulsed-wave Doppler system is that the frequency shift has to be at least smaller than half of the PRF. Otherwise, aliasing of the measured Doppler frequencies will occur. Since PRF (kHz) = $c/2D_{max}$ (from the pulse echo imaging section) and following Equation 7.45, the maximum velocity that can be measured without ambiguity at the maximum allowable gate depth, D_{max}, is equal to

$$v_{max} = \frac{c^2}{8 f \cos\theta D_{max}}. \tag{7.48}$$

This is otherwise known as the "range-velocity" trade-off. The inverse proportionality between velocity and center frequency and depth presents serious limitations to the pulsed Doppler systems for large-depth applications. Several techniques have been developed to maximize performance and include an adjustable gate depth according to the PRF used and a sharp cut-off smoothing filter with an adjustable cut-off frequency, also dependent on the PRF. Finally, these shortcomings can also be avoided by combining the B-scan mode with the Doppler sonogram mode, otherwise known as *Duplex mode*, or the *Triplex mode* that is like the Duplex mode but with an additional color Doppler image superimposed on the B-scan (Figure 7.22). A user-defined line of site and range gate for the pulsed Doppler is overlapped onto a B-scan image, allowing for the identification of the vascular anatomy of interest (and possible artifacts in the Doppler estimate arising as a result) and for a more reliable and quantifiable Doppler estimate through angle θ and vessel diameter determination.

7.5 Current Developments

Despite the fact that diagnostic ultrasound is an older imaging modality compared to MRI and PET, it is very intriguing to see that it continues to expand as a field and offers numerous applications. In the past decade, several leaps have been made with the advent of faster computer processors, contrast agents, the utilization of nonlinear wave propagation, signal-processing techniques, and complex transducer architecture, to name a few. In this section, a short overview is presented of the key techniques that can routinely be used on ultrasound machines in the future, if not already.

7.5.1 Contrast Agents and Harmonic Imaging

During the wave interaction with tissues, besides the linear wave propagation discussed earlier, nonlinear effects also occur, especially at higher intensities (on the order of MPa) that are routinely used by diagnostic scanners. This is because the pressure exerted by the beam can no longer be considered as negligible compared to the pressure of the medium. As a result, nonlinear waves can be generated that are dependent on the acoustic pressure, the medium characteristics, and depth through which the beam travels. The medium characteristics can be given by the B/A parameter shown in Figure 7.23 for different media. Typically, the higher the B/A characteristic, the higher the nonlinearity term in the wave equation, and the higher the distortion in the resulting wave.

One of the main problems with the standard use of ultrasound arises from high attenuation in some tissues (Figure 7.6) and thus is most prominent in the imaging of small vessels and velocity measurements, as described in the previous section. To overcome this limitation in the case of blood flow measurements, contrast agents are routinely employed. Contrast agents are typically microspheres of encapsulated gas or liquid coated by a shell, usually of albumin. Owing to the high impedance mismatch created by the gas or liquid contained, the resulting backscatter by the contrast agents is a lot higher than that of the blood echoes.

An alternative method to generate higher backscatter due to the increased impedance mismatch is based on the harmonics generated by the bubble's interaction with the ultrasonic wave. This interaction results in a vibration of the latter at a resonance frequency f_r given by[2]

$$f_r = \frac{1}{2\pi a}\sqrt{\frac{3\gamma P}{\rho}}, \tag{7.49}$$

where a is the radius of the contrast agent, γ the adiabatic constant, P the applied pressure, and ρ the density of the bubble. The bubble vibration also generates harmonics above and below the fundamental frequency, with the second harmonic possibly exceeding the first harmonic. In other words, the contrast agent introduces nonlinear backscattering properties into the medium where it lies. Several processes of filtering out undesired echoes from stationary media surrounding the region, where flow characteristics

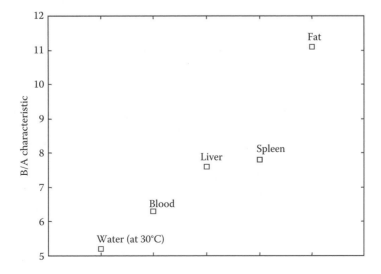

FIGURE 7.23 The B/A characteristic for different media. (From Jensen, J.A., *Estimation of Blood Velocities Using Ultrasound*, Cambridge University Press, Cambridge, UK, 1996.)

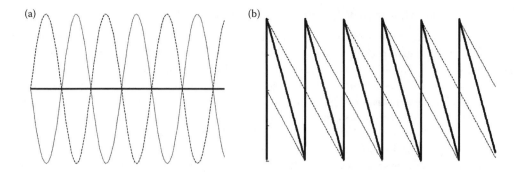

FIGURE 7.24 The pulse inversion method with (a) the transmitted pulses [initial in dashed and phase-inverted in dotted lines, respectively; summed waveform (bold solid) resulting in cancelation] and (b) the resulting potential sawtooths from the respective transmitted pulses of the left diagram. Note that the summed waveform (in bold solid) results in a sawtooth of twice the frequency.

are assessed, result in weakening of the overall signal at the fundamental frequency. Therefore, since residual harmonics will result from moving scatterers, all motion characteristics can be obtained from the higher harmonic echoes, after using a high-pass filter and filtering out the fundamental frequency spectrum that also contains the undesired stationary echoes. Another method for distilling the harmonic echo information is the more widely used phase or pulse inversion method. Figure 7.24 shows an example of how this method works. Instead of one, two pulses are sequentially transmitted with their phases reversed. Upon reception, the echoes resulting from the two pulses are summed up. A sinusoid with a particular frequency f_0 will be canceled at summation, while, for example, a sawtooth that contains a much higher amount of frequency content will have its fundamental at f_0 term removed with the extant components remaining at $2f_0$, $4f_0$, and so on.

Despite the fact that the idea of contrast agent use originated for applications in the case of blood flow, the same type of approach can be applied in the case of soft tissues as well. After being injected into the bloodstream, the contrast agents can also appear and remain on the tissues and offer the same advantages of motion detection and characterization as in the case of blood flow. An example of the contrast improvement provided by contrast agents in cardiac tissues is shown in Figure 7.25. However, it turns out that contrast agents are not always needed for the imaging of tissues at higher harmonics, especially since backscatter from tissues can be up to two orders of magnitude higher than backscatter from blood. The nonlinear wave characteristic of the tissues themselves is thus sufficient in itself to allow imaging of tissues (Figure 7.23), despite the resulting higher attenuation at those frequencies. The avoidance of patient discomfort following contrast agent injection is one of the major advantages of this approach in tissues. Imaging using the harmonic approach (whether with or without contrast agents) is generally known as *harmonic imaging*. Compared to the standard approach, harmonic imaging in tissues offers the ability to distinguish between noise and fluid-filled structures, for example, cysts and the gall bladder. In addition, harmonic imaging allows for better edge definition in structures and, therefore, is generally known to increase image clarity, mainly due to the much smaller influence of the transmitted pulse to the received spectrum. Harmonic imaging is now available in most clinical ultrasound systems. One of the main requirements for harmonic imaging is the large bandwidth of the transducer at the receiver so as to allow reception of the higher-frequency components. This is also in very good agreement with the higher-resolution requirement for imaging.

7.5.2 Elasticity Imaging

Another field that, like harmonic imaging, has emerged out of ultrasonic imaging in the last decade is *elasticity imaging*. Its premise is built on two proven facts: (1) significant differences between mechanical

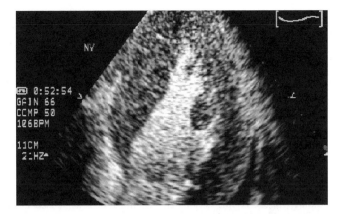

FIGURE 7.25 Long-axis echocardiogram of a left ventricle using harmonic imaging. Note the increase in resolution and image clarity as a result of the contrast agent introduction (to be compared to the echocardiograms without contrast of Figure 7.18). (Courtesy of Scott D. Solomon, Brigham and Women's Hospital, Boston, MA.)

properties of several tissue components exist and (2) that the information contained in the coherent scattering, or speckle, is sufficient to depict these differences following an external or internal mechanical stimulus. Figure 7.26 demonstrates the validity of the first fact by showing the range of elastic moduli for several different normal and pathological human breast tissues. Not only is the hardness of fat different than that of glandular tissue, but, most importantly, the hardness of normal glandular tissue is different than tumorous tissue (benign or malignant) by up to one order of magnitude. This is also the reason why palpation has been proven an infallible tool in the detection of cancer.

The second observation is based on the fact that coherent echoes can be tracked during or after the tissue in question undergoes motion and/or deformation caused by the mechanical stimulus, for example, an external vibration or a quasistatic compression. Figure 7.27 shows the general concept behind the elasticity imaging techniques and, more specifically, the example of an applied compression to detect a harder lump in a method called *elastography*. Speckle-tracking techniques are also employed here for the motion estimation. In fact, Doppler techniques, such as those used for blood velocity estimation, were initially applied to track motion during vibration (*sonoelasticity imaging* or *sonoelastography*).

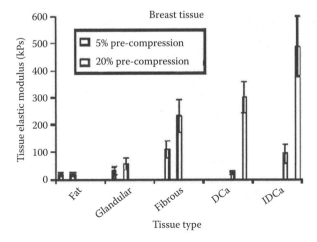

FIGURE 7.26 Elastic moduli (or stiffnesses) of normal and tumorous breast tissues. DCa, ductal carcinoma; IDCa, invasive ductal carcinoma. (Courtesy of Tom Krouskop, Baylor College of Medicine, Houston, TX.)

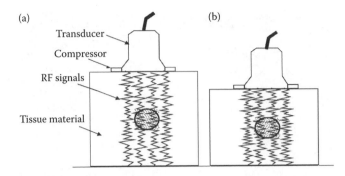

FIGURE 7.27 The principle of elastography. The tissue is insonified (a) before and (b) after a small uniform compression. In the harder tissues (e.g., the circular lesion depicted), the echoes will be less distorted than in the surrounding tissues.

Parameters such as velocity and strain are estimated and imaged in conjunction with the mechanical property of the underlying tissue. The higher the velocity or strain estimated, the softer the material and vice versa. In fact, these parameters can also be used for the recovery of the underlying tissue property but these methods have been proved cumbersome and unreliable in their numerous assumptions given the unknown stress distributions that often fail in the case of highly inhomogeneous tissues.

Examples of elastograms, or strain images, in the same *in vitro* prostate shown in Figure 7.4 (Figure 7.28) and in the case of two pathological, clinical breast cases are presented in Figure 7.29. In the case of the canine prostate, through comparison to Figure 7.4, complementary information based on the distinct mechanical responses and properties of the several anatomical structures of the prostate, such as the urethra and the peripheral zones, is provided. In the case of the breast, both benign and malignant tumors can be depicted in elastograms, whether visible on the sonogram or not. Through comparison between the sonographic and elastographic characteristics of the several tissue components, their malignancy type could be characterized.

Owing to the vast impact that these techniques could have in imaging and characterization of tissues based on their mechanical attributes, a variety of very promising methods have been recently developed, spanning from hand-held and real-time application of elastography to elastic modulus maps based on

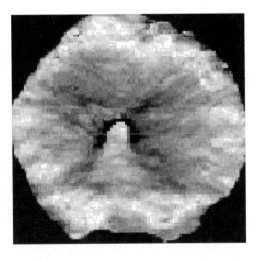

FIGURE 7.28 Prostate elastogram of the prostate in Figure 7.4. Black and white denote highest and lowest strains, respectively. (From Ophir, J. et al., *C.R. Acad. Sci. Paris*, Tome 2, Ser. IV(8), 1193–1212, 2001.)

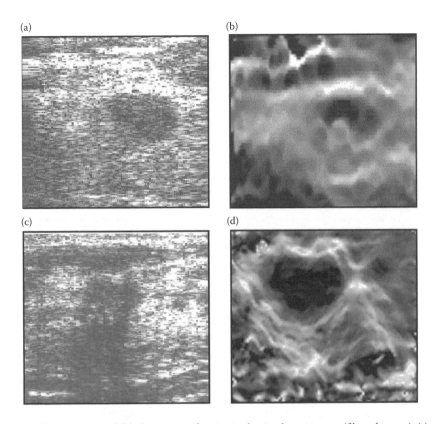

FIGURE 7.29   (a) Sonogram and (b) elastogram of an *in vivo* benign breast tumor (fibroadenoma); (c) sonogram and (d) elastogram of an *in vivo* malignant breast tumor (invasive ductal carcinoma). In these elastograms, black and white denote lowest and highest strains, respectively.

the wavelength of propagation through different tissues following an applied stimulus (*transient elastography*) and the use of the internal radiation force resulting from the pressure of the beam itself to locally displace (*remote palpation*) or vibrate (*ultrasound-stimulated vibroacoustography*), to name a few. This field has not been proved as applicable clinically as that of harmonic imaging due to its inherent, more complicated nature but all these techniques are on the brink of becoming readily applicable and very useful in conjunction with the standard, current use of ultrasonic imaging.

7.5.3 Three-Dimensional Imaging

Until now, only two-dimensional sonograms were presented and discussed. However, ultrasonic imaging has recently also expanded to three-dimensional imaging in clinical applications, especially cardiology and obstetrics. This is due to the advancement in transducer design, namely the two-dimensional arrays, where beams are generated in all three planes (Figure 7.30c), but three-dimensional images can also be achieved with sweeping a linear or phased array out of plane (Figure 7.30a and b). An example of such a sonogram is shown in Figure 7.31.

7.5.4 Acoustic Microscope

As mentioned in Section 7.3.3.1, the resolution of ultrasonic imaging systems is ultimately limited by the wavelength used. Therefore, in principle, the smaller the wavelength or the higher the frequency (Equation 7.12), the better the resolution (Equations 7.36 and 7.40). The acoustic microscope uses much

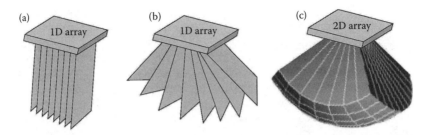

FIGURE 7.30 Three methods for three-dimensional ultrasonic imaging: (a) linear and (b) tilt mechanical scanning of standard (ID) array for the acquisition of a series of parallel two-dimensional images and three-dimensional image reconstruction; (c) two-dimensional ($N \times N$) array producing a pyramidal scan for the direct generation of three-dimensional images.

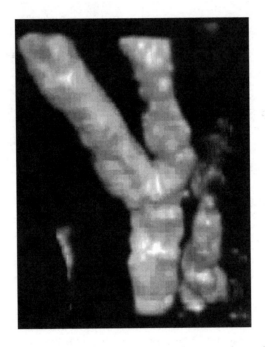

FIGURE 7.31 Three-dimensional sonogram of a carotid artery bifurcation obtained using color Doppler energy (EKG-gated). The jugular next to the carotid is also evident in these images due to the nondirectionality of the CDE. (Image obtained using Acuson (Siemens) Sequoia and courtesy of John A. Hossack (University of Virginia, Charlottesville, VA, biomedical engineering), Sandy Napel (Stanford University, Stanford, CA, radiology), and R. Brooke Jeffrey (Stanford University, Stanford, CA, radiology).)

higher frequencies than those used in the medical imaging field to achieve resolution of the order of a micron—similar resolution to that of optical microscopes. The highest frequency currently used is on the order of 1.5 GHz with a corresponding wavelength equal to 1.0 μm.

The aforementioned trade-off between resolution and penetration applies here as well. Samples, thus, cannot be insonified at depths larger than a few microns. As a result, the acoustic microscope does not have the standard ultrasonic imaging applications. However, it provides the possibility of imaging tissues at the cellular level and allows observation of cell motion and interaction among other things. In addition, the acoustic microscope can provide completely different information on the tissues compared to the optical or electron microscopes, namely it can show maps of attenuation as well as acoustic phase

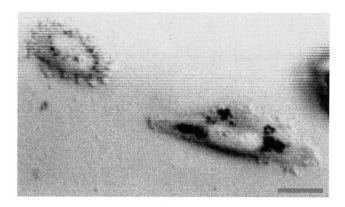

FIGURE 7.32 Acoustic microscope images obtained at 1 GHz of melanoma kidney cells. The bar is equal to 20 μm. (Courtesy of C. Miyasaka and B. R. Tittmann, Department of Engineering Science and Mechanics, Pennsylvania State University, College Park, PA.)

and impedance interfaces.[7] In other words, the acoustic microscope can depict tissue properties and characteristics that otherwise are unobtainable, even with other microscopes. An example of an acoustic microscope image is shown in Figure 7.32.

References

1. Wells, P.N.T., *Biomedical Ultrasonics*, Medical Physics Series, Academic Press, London, 1977.
2. Kinsler, L.E. and Frey, A.R., *Fundamentals of Acoustics*, 2nd ed., John Wiley & Sons, New York, 1962.
3. Jensen, J.A., *Estimation of Blood Velocities Using Ultrasound*, Cambridge University Press, Cambridge, UK, 1996.
4. Burckhardt, C.B., Speckle in ultrasound B-mode scans, *IEEE Trans. Son. Ultrason.*, SU-25, 1–6, 1978.
5. Wagner, R.F., Smith, S.W., Sandrik, J.M., and Lopez, H., Statistics of speckle in ultrasound B-scans, *IEEE Trans. Son. Ultrason.*, 30, 156–163, 1983.
6. Bamber, J.C. and Tristam, M., *Diagnostic Ultrasound*, Webb, S., Ed., IOP Publishing Ltd., London, 1988, pp. 319–386.
7. Christensen, P.A., *Ultrasonic Bioinstrumentation*, 1st ed., John Wiley & Sons, New York, 1988.
8. Morse, P.M. and Ingard, K.U., *Theoretical Acoustics*, McGraw-Hill, New York, 1968.
9. Kino, G.S., *Acoustic Waves: Devices, Imaging and Analog Signal Processing*, Prentice-Hall, Upper Saddle River, NJ, 1987.
10. Haney, M.J. and O'Brien, W.D., Jr., Temperature dependence of ultrasonic propagation in biological materials, in *Tissue Characterization with Ultrasound*, Greenleaf, J.F., Ed., CRC Press, Boca Raton, FL, 1986, pp. 15–55.
11. Ophir, J., Kallel, E., Varghese, T., Konofagou, E.E., Alam, S.K., Garra, B., Krouskop, T., and Righetti, R., Elastography, optical and acoustic imaging of acoustic media, *C.R. Acad. Sci. Paris*, Tome 2, Ser. IV(8), 1193–1212, 2001.

Emission Imaging

Mark Lenox
Texas A&M University

8.1 Introduction

Medical imaging has completely changed the way clinicians practice medicine. It provides them with a relatively noninvasive method to diagnose a wide variety of disease. Innovation in medical imaging has been proceeding at an extraordinary pace in the last 100+ years since Wilhelm Röntgen discovered x-rays in 1895 and was later awarded the first Nobel Prize in Physics. This seemingly simple discovery, ionizing radiation, spawned the entire field of diagnostic radiology. In the intervening years since Röntgen's discoveries, others have discovered many ways to noninvasively measure processes occurring inside a living subject, including the use of magnetic fields, and light. However, the largest bulk of diagnostic radiology is still performed using ionizing radiation, of which x-rays are a subset, making it the workhorse of modern diagnostic radiology.

Medical imaging techniques that use ionizing radiation fall into two categories: transmission and emission. Transmission methods apply some form of radiation from outside the body and measure the effect as it passes through. Examples of this include radiographs and computed tomography (CT) scans. This type of imaging is typically very good for the determination of anatomical structure. Emission methods rely on some kind of injected tracer or probe. These tracers circulate within the body and emit radiation, which is detected externally. Examples of this type of imaging include planar gamma cameras, single-photon emission computed tomography (SPECT), and positron emission tomography (PET). Since the tracers can be attached to molecules that are metabolically active, this type of imaging is typically very good for the determination of function and is also often referred to as functional or molecular imaging.

The actual implementation of all medical imaging modalities that are based on ionizing radiation rely on two key concepts: detection and measurement of a variety of types of radiation, and mathematics. In emission methods, we rely principally on gamma radiation for the detection and measurement portion, and the radon transform (Deans, 1983) for the mathematics. Transmission methods also rely on the radon transform for their mathematical basis, but utilize externally generated radiation. Systems that are built to provide more than one angular view of coverage around a subject and reconstruct that information into a three-dimensional (3D) volume are termed tomographic.

This chapter concentrates on the concepts and capabilities of the emission-based modalities PET and SPECT.

8.2 Radiation Detection and Measurement

The implementation of emission-based medical imaging modalities is based on the detection of electromagnetic (EM) radiation that is generated as a result of some type of nuclear decay. Principally, we think of radiation as either EM waves/photons or particles. Examples of EM radiation include gamma, x-rays, radio waves, and visible light. Particle-based radiation includes types such as alpha (charged helium nuclei) and beta (electrons).

8.2.1 Electromagnetic Radiation

Electromagnetic radiation exists in a spectrum, but can generally be classified as either ionizing or nonionizing (Figure 8.1). Ionizing radiation contains enough energy that when it interacts with matter it can knock electrons loose, thus forming a charged particle, also called an ion. The energy of a photon is measured in terms of electron volts (eV). By definition, 1 eV represents the amount of energy required to move a single electron across an electric potential of 1 V. It is approximately equivalent to 1.602E-19 joules (Knoll, 2000).

The energy (E) of a photon and the frequency (v) on the electromagnetic spectrum are related linearly with the following formula:

$$E = hv$$

where h is the Planck's constant and is equal to 4.13566733E-15 eV-s.

The energy required to ionize an atom varies for different elements, but from a radiological health and safety perspective, the generally accepted level of energy required to cause ionization is 10 eV (Allisy et al., 2011). This energy corresponds closely to the first ionization level of oxygen and hydrogen, since the body largely consists of water. This frequency is in the visible violet and ultraviolet portion of the spectrum. As the frequency is increased, the energy of the photons increases as well. As the frequency increases beyond the visible spectrum, electromagnetic waves become x-rays. As the frequency continues to increase, EM waves become classified as gamma (γ) radiation.

8.2.2 Nuclear Decay

Elements are considered stable if they do not undergo radioactive decay. The nucleus of all elements is primarily made up of a collection of positively charged particles called protons and neutrally charged particles called neutrons. An isotope of an element has the same number of protons, and therefore the same atomic number as the original element, but a different number of neutrons, and a different atomic mass. A stable isotope is in the lowest energy state for that number of protons. The addition or subtraction of particles in the nucleus can put it in a higher energy state in which it is no longer stable. The end result of this instability is that the nucleus will undergo a change that will take it back down to a lower energy state. This change results in the emission of some type of energy or particle and is termed nuclear decay. There are several different types of emissions that can occur, with the result being the emission of energy or a particle (Table 8.1).

Alpha emission emits large number of charged particles. These particles do not penetrate very far through normal tissue. This characteristic makes alpha radiation useful in depositing energy in tissue for therapeutic purposes, but renders them useless for imaging. Beta particles are electrons, similar to

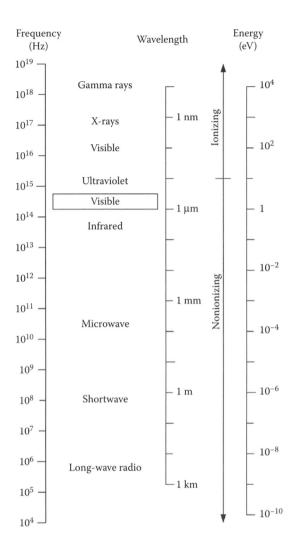

FIGURE 8.1 Electromagnetic spectrum.

alpha particles, and they do not penetrate a very significant distance in tissue. To detect changes, some percentage of photons must pass entirely through the material being measured. For this reason, emission imaging largely relies on gamma emission and indirectly, positron emission.

Gamma emissions are photons of pure energy. Depending on their energy, they can travel through significant mass unimpeded. Any gamma photon with an energy level above approximately 50 keV has a reasonable probability of penetrating entirely through a typical person, making it viable as an imaging method. For larger patients, attenuation at energies below 50 keV may not provide sufficient imaging

TABLE 8.1 Types of Radiation

Type	Energy/Particle Emitted
Alpha	Alpha particle, charged helium nucleus
Beta	Beta particle, electron
Gamma	Gamma photon, high-energy EM particle/wave
Positron	Positron, positive electron

TABLE 8.2   Common Isotopes Used as Imaging Tracers

Isotope	Energy	Type	Half-Life
O-15	1.7 MeV	Positron	122.24 s
F-18	0.64 MeV	Positron	109.77 min
C-11	0.96 MeV	Positron	20.37 min
N-13	1.2 MeV	Positron	9.97 min
Cu-64	0.65 MeV	Positron	12.8 h
Cu-62	2.93 MeV	Positron	10 min
Rb-82	3.15 MeV	Positron	1.27 min
Y-90	0.80 MeV	Positron	64 h
Tc-99m	142 keV	Gamma	6.0058 h
Ga-67	1000 keV	Gamma	3.26 days
In-111	171 keV and 245 keV	Gamma	2.804 days
I-123	159 keV	Gamma	13.22 h
Tl-201	135 keV and 167 keV	Gamma	73 h

Source: Adapted from Brookhaven National Laboratory, Chart of the Nuclides, http://www.nndc.bnl.gov/chart/, accessed May 5, 2012.

statistics. Energies above 500 keV start to suffer from lack of attenuation, the inability to collimate, and difficulty stopping causing degradation in image quality.

Positrons, being positive electrons, suffer from the same low-penetration effects as electrons; however, they are different in a significant way. When an isotope undergoes positron decay, the nucleus emits a positron and a neutrino. Thus, one of the protons becomes a neutron, the atomic number goes down by 1, but the atomic mass stays relatively the same as positrons and neutrinos do not have much mass. When the positron leaves the nucleus and travels away, it eventually encounters an electron, the two particles undergo positron annihilation and both cease to exist. In their place are two gamma photons at 511 keV each that travel away from the site of annihilation nearly 180° opposed from each other. Even though we cannot detect the positrons themselves directly outside the body, the resulting 511-keV gamma photons can eventually leave the body to be detected later. The distance that the positron travels before it encounters an electron is referred to as the mean free path. The mean free path varies depending on the energy of the positron as it is ejected from the nucleus and is among the largest single limitations to resolution in PET (Table 8.2).

Positron imaging applications rely on the annihilation photons at 511 keV and not the primary energy of the positron to image, making them suitable for a subject of any size. The acquisition energy is always 511 keV because that is the energy of the annihilation photons. Higher-energy positrons cause higher mean free path distances and increased noncolinearity of the annihilation gamma photons, resulting in reduced spatial resolution. For imaging purposes, we assume that gamma photons travel in a straight line unless scattered or entirely absorbed.

8.2.3 Detection

There are a variety of methods to detect the different types of radiation. Geiger Mueller (GM) tubes, for example, use a charged grid to directly detect ionization. From the perspective of medical imaging, scintillation detectors are the primary method.

8.2.3.1 Scintillation Detectors

In a scintillation detector, a gamma photon travels through space until it interacts with the detector material. Upon interaction, energy is deposited, mostly in the form of heat. Some energy is deposited in the valence electron shell of the material, pushing the outer shell electrons of some number of atoms into their next higher band. The higher electron shell is not stable in this configuration, and after some time, the electron exits the upper band and falls back to the original band, emitting yet another photon of energy. The energy level of the new photon depends on the bandgap between electron shells. For some elements, the bandgap happens to reflect the energy of a photon that falls in the visible light spectrum. Thus, with a single gamma photon, the energy is divided up into a shower of many lower-energy photons that can be detected and counted with a detector that is sensitive to the specific wavelength of light emitted. Table 8.3 lists some of the more common commercially used scintillators for emission medical imaging systems.

Density is of high importance in a scintillator, as the denser scintillators are able to stop higher-energy photons in a shorter distance. The light yield and decay time are also important as they contribute to photon statistics. Higher light yields generally translate to better energy resolution and better intrinsic spatial resolution. Faster decay times improve the detector performance with respect to pileup, allowing for higher acquisition speeds.

To be useful, an electronic signal must be generated that allows for electronic counting of photons that are incident on the detector. The most common method of performing this function is through the use of a photomultiplier. Solid-state detectors such as avalanche photodiodes (APDs) and silicon photomultipliers (SiPMs) could be similarly utilized, and with continued decreases in price, they will continue to see strong growth in this application. Photomultipliers work through the process of secondary electron emission. They are built as a series of plates called dynodes that are charged to high voltages. When photons of light strike the photocathode, a shower of electrons is created that are accelerated and drawn toward the first dynode. As they strike each dynode in the chain, more electron showers are created in a multiplicative effect. The total voltage across all dynodes in a photomultiplier (PMT) is typically on the order of 800–1600 V. Increasing the voltage increases the multiplicative effect, but generally decreases the life expectancy of the tube. In ordinary use, photomultipliers are very reliable and generally last the entire life of the imaging system with few exceptions (Figure 8.2).

TABLE 8.3   Common Scintillators and Their Relevant Physical Properties

Material	Density (g/cm³)	Attenuation Length (mm)	Luminosity (ph/MeV)	Decay Half-Life (ns)	Wavelength (nm)	Hygroscopic	Energy Resolution (%)
NaI	3.67	29.1	41,000	230	410	Y	5.6
LSO	7.4	11.4	25,000	40	420	N	7.9
BGO	7.1	10.4	8200	300	480	N	9.0
LuAP	8.3	10.5	12,000	18	365	N	~15
LaBr	5.3	21.3	61,000	35	370	Y	2.9

Source: Adapted from Knoll, G., *Radiation Detection and Measurement*, 3rd Ed., 2000, Copyright Wiley-VCH Verlag GmbH&Co. KGaA.

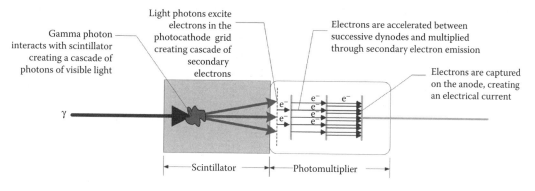

FIGURE 8.2   Scintillator detector schematic with attached light detector (photomultiplier).

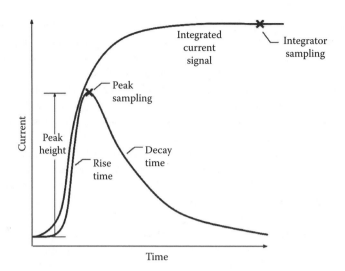

FIGURE 8.3 Electrical output from photomultiplier in response to a single incident event on the scintillator including both the real-time response and an integrated response. Peak sampling must occur at exactly the correct time, while integrated sampling can vary in time without significantly degrading the end result.

The output of the photomultiplier is an electrical current that reflects the incoming flux of light photons from the scintillator (Figure 8.3). The actual number of light photons measured at the photocathode is proportional to the energy of the incident gamma photon and the light yield of the scintillator. The shape of the time course of current is highly dependent on the physical characteristics of the scintillator as well as the incident gamma photon. Before the event occurs, there is very little current. Once the photon interaction has occurred, there is a sharp rise in visible light photon flux and thus current from the light detector. This is followed by an exponential decay as the energy in the scintillator decays off. The three characteristics that are of interest are the absolute peak height, rise time, and decay time. The peak height is proportional to the energy of the incident gamma photon and the relative light output of the scintillator. The rise time and decay times are mainly dependent on the scintillator and, to a lesser extent, on the choice of light detection method (photomultipliers, APDs, SiPM, etc.).

Since the energy of the photon is proportional to the peak height of the current, it is possible to measure the energy by simply measuring the peak height using either an analog sample and hold circuit, or a high-speed analog-to-digital converter (ADC). This technique is subject to a few problems, in that the precise timing of the peak must be determined. It should also be noted that since the energy of the incident gamma is also proportional to the total number of electrons output from the light detector, it is possible to simply integrate the current over time and sample at a later time. This technique requires more complex hardware but tends to provide a cleaner measure of total energy and is often used in commercial systems where accurate energy discrimination is important.

The characteristics of different scintillators have a large impact on the performance of the detection system. Higher-density scintillators have a higher efficiency versus high-energy photons, for example, so it is important that the scintillator be chosen appropriately for the application. Single-photon applications tend toward photon energies in the 80–200 keV range. These applications commonly utilize sodium iodide (NaI) as a scintillator, which has high light yield and is relatively inexpensive, but not particularly dense. PET applications require detection of 511 keV photons making BGO or LSO a far better choice due to their higher densities.

The light yield of the scintillator has a dramatic effect on the ability of the system to accurately characterize the incident gamma photon interaction event. Higher light yields provide a better statistical measure of the event, especially in terms of energy. Energy is represented as a histogram for evaluating

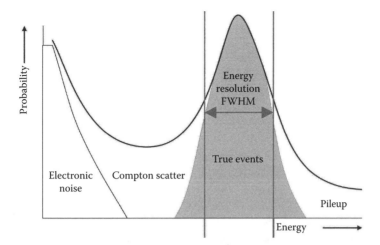

FIGURE 8.4 Energy spectrum showing true events, pileup, electronic noise, and Compton scatter. Energy performance is measured as the FWHM of the histogrammed energy peak. Where pileup and Compton scatter are minimized, the energy resolution can be measured directly instead of a Gaussian fit to the measured peak.

its quality. At the lower end of the spectrum, electronic noise and scatter come into play. The design of the trigger circuitry plays a major role in the effect of electronic noise on the energy spectrum as a whole and will be discussed in a later section on constant fraction discriminators (CFDs). Scattered events are created by photons that have otherwise interacted elsewhere and lost some energy. These events are of less informational value than events that have not been scattered because presumably at the time of the scattering event the direction was changed as well as the energy, and there is no method to determine the original direction. At the upper end of the spectrum, event pileup is the dominating factor. In this case, multiple events arrive at the same, or close to the same, time and their light photons intermingle and add together to create a single event of higher energy that can otherwise be expected. In the central portion of the spectrum are the true events or those events that impact the scintillator without scattering or pileup. Ideally, this would represent an impulse function on the energy spectrum; however, given the realities of construction of detector systems and the uncertainties thereof, the true event spectrum has a normal distribution with some amount of width. This is represented as the full-width-at-half-max (FWHM) of the energy peak (Figure 8.4).

Energy resolution of the scintillator is a statistical process that is inherent in the physical characteristics of the material. So, instead of representing the energy resolution FWHM in units of energy as is done in the case of something physical like spatial resolution, it is represented as a percentage of the peak energy:

$$\text{Energy resolution} = \frac{\text{Energy FWHM}}{\text{Peak energy}}(\%)$$

8.2.3.2 Detection Electronics

Once an incident gamma photon interacts with a scintillation detector, the output from the photomultiplier is an electrical current that is proportional to the flux of incident light photons on the photocathode grid of the light detector. The function of the detector electronics is to identify where the event occurred and measure the energy. In the case of PET detectors, the precise time of the event must be determined as well.

A very basic electronic package consists of a preamplifier, ADC, a trigger circuit, some digital logic, and memory (Figure 8.5). In this configuration, the preamplifier takes as input the output of

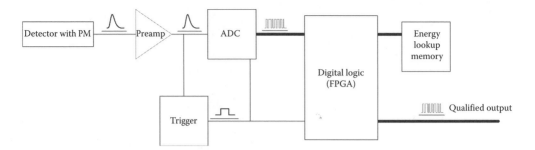

FIGURE 8.5 Basic acquisition electronics for a single channel.

the photomultiplier, applies some gain and filtering, and then sends the analog signal to the ADC and trigger circuit. The trigger determines a threshold and when that threshold is crossed, triggers the ADC and the digital logic to start recording samples. The digital logical processes those samples and determines if the pulse matches the characteristics required, and then packages the information in digital form and sends it to the next part of the processing stream.

In an electronic configuration of this type, it is possible to continuously trigger the ADC and digitally integrate the raw signal, or apply an analog integrator or shaping circuit to the preamp to sample for event energy.

8.2.3.2.1 Anger Logic

Previous examples have used a detector where a single piece of scintillator material is coupled to a single light detector. Commercial detector systems seldom use this type of configuration due to the high cost. Instead, detector systems for scintillation cameras typically function by sharing light between multiple detectors. This basic method was invented by Hal Anger in 1957, and patented by him in 1961 (Patent 3,011,057). In this approach, individual scintillator crystal blocks are put together in an array over a light guide, typically glass, with an array of light detectors behind it. A simple Anger camera consists of four photomultipliers bonded to an array of crystals (Figure 8.6).

In this design, light from a single event is distributed to all four photomultipliers in proportion to their location. The position of an incident event on the array can be computed based on the relative outputs of the four photomultipliers (Figure 8.7).

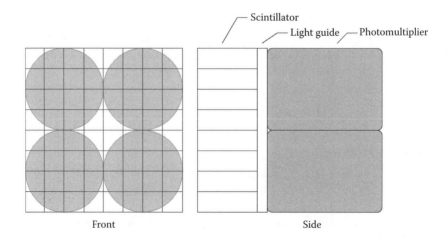

FIGURE 8.6 Simple Anger detector array with four photomultipliers in a 2 × 2 array bonded to a light guide and detectors.

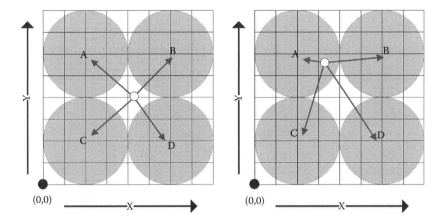

FIGURE 8.7 Anger logic physical representation. As the event occurs nearer to a particular photomultiplier, a higher proportion of the light photons is collected in that PMT and less in the others.

Thus, the total energy (E) of the event is the sum of all the signals from all four photomultipliers:

$$E = A + B + C + D$$

The ratio of the photomultiplier signals in the direction of measurement to the total energy yields the position along that direction:

$$X = \frac{B + D}{A + B + C + D}$$

$$Y = \frac{A + B}{A + B + C + D}$$

By combining electronics and photomultipliers together into arrays, a far larger number of crystals can be decoded for position and energy per acquisition channel, dramatically reducing cost. Histogramming the incoming events for a pixilated detector results in a position profile where the individual crystals can be seen (Figure 8.8).

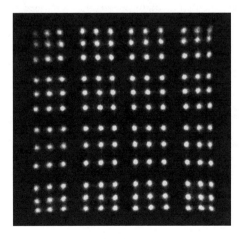

FIGURE 8.8 Position profile of a 12×12 high-resolution detector array.

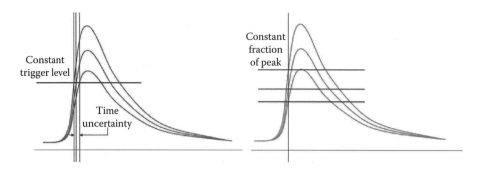

FIGURE 8.9 Constant level triggering results in variable measurements for time of arrival based on peak height. CFDs provide consistent timing independent of peak height at a specified fraction of the peak height depending on the specified attenuation and delay parameters.

8.2.3.2.2 Trigger Circuits

Single-photon applications are not particularly time-dependent, but PET relies on timing to perform physical collimation, and for these applications, precise timing is very important. The trigger circuitry is crucial to the time performance of any given system, and the CFD is a common method for PET scanners.

A level-based trigger is a simple comparator that triggers for an incoming event based purely on the level of the signal. Even though incoming events are technically the same energy, their actual recorded energy, and thus pulse height varies due to a variety of factors, including position within the array, scattering, and the accuracy of the electronics built to measure it. This variation in pulse height can cause inaccuracies in the recorded event arrival time (Figure 8.9).

CFDs when used as a trigger circuit vary the trigger level to that of a constant fraction of the peak. In this way, they always trigger on the same proportion of the peak height, measuring time of arrival more consistently.

The CFD circuit itself is relatively simple, consisting of an amplifier, time delay, and a comparator (Figure 8.10). The input signal is separately delayed and attenuated. The attenuated version is subtracted from the delayed version, resulting in an initial downward trend in the signal. As the delayed signal starts to dominate the result, the output trends upward until it crosses a zero level. The output is processed through the comparator that selects above or below zero with high gain, resulting in an inverted pulse whose upward-moving portion corresponds to the constant fraction arrival time of the pulse.

This circuit is normally implemented in analog electronics for speed purposes. It has been proposed that by using continuously sampled digital signals, a purely digital CFD could be built, but the timing performance would be limited by the ADC frequency.

8.2.4 Reconstruction

Emission imaging techniques can be either planar or tomographic in nature (Figure 8.11). Planar imaging such as that done with older gamma cameras is two-dimensional in nature and requires no

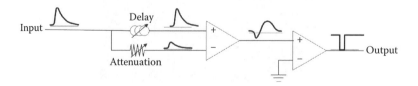

FIGURE 8.10   CFD circuit based on an amplifier, comparator, attenuator, and delay.

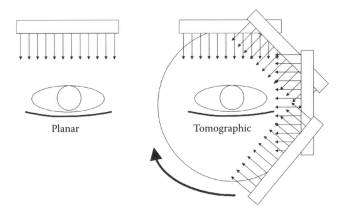

FIGURE 8.11 Planar versus tomographic imaging. A planar image is generated from a single viewpoint and has no depth. Tomographic imaging is done from multiple views. An entire 360° set of views is not required to reconstruct the 3D volume, but image quality improves with increasing numbers of views.

reconstruction to view an image. A planar view consists of a row of parallel projections projected through the image volume onto an array of detectors. 3D techniques such as SPECT and PET acquire multiple planar images of parallel projections taken from a variety of angles around the subject and require reconstruction to transform the acquired planar projections into a 3D image volume. It is also possible to build systems based on fan or cone-beam projections using different types of collimation.

The mathematics of all reconstruction techniques are fundamentally based on a simple premise. There is a 3D volume with an unknown distribution of tracer (x) that is coupled to a set of projection observations (b), through a transformation (A):

$$b = Ax$$

Typically, to determine the distribution from the observations, we would invert the transformation matrix A and solve for x directly.

$$x = A^{-1}b$$

Unfortunately, for these systems, the system transformation matrix is ill composed and does not invert easily if at all, so other mathematical methods must be employed. These methods are categorized into three main areas: analytic, algebraic, and statistical. Collectively, this is referred to as the inverse problem.

8.2.4.1 System Matrix and the Radon Transform

The system transformation matrix A relates the forward parallel projections of the image volume on the observed information. The basis for this connection is the radon transform (Deans, 1983).

By definition, the radon transform is a line integral through an image (Figure 8.12). We define a line AB at an angle θ offset from the origin a distance t as

$$x \cos \theta + y \sin \theta = t$$

Integrating the line AB over the entire length at a specific angle θ and offset t is defined as follows:

$$P_{\theta}(t) = \int_{A}^{B} f(x, y) \, ds$$

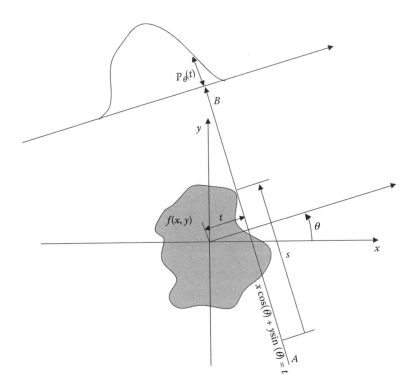

FIGURE 8.12 Radon transform is a line integral through an image $f(x, y)$.

Using a delta function to sample $P_\theta(t)$ at a particular location results in the formal definition of the two-dimensional (2D) radon transform as used in imaging applications:

$$P_\theta(t) = \iint\limits_{-\infty}^{\infty} f(x, y)\delta(x\cos\theta + y\sin\theta - t)\,dx\,dy$$

The radon transform can be used to represent the physical process of describing a distribution. As the line integral crosses through the imaging volume, it accumulates information about what it has passed through and then represents that information on some kind of detector. In CT, by applying linear attenuation, it can be used to describe density. In PET and SPECT, it can be used to describe the distribution of a tracer.

The system matrix A consists of the individual sampled forms of the radon transform created for all required angles and offsets in a given imaging area. Clearly, this matrix can be large depending on the sampling frequency chosen.

8.2.4.2 Analytic Techniques

If the camera system is carefully designed, the system matrix has a significant amount of repetitive content. By taking advantage of this, the solution to the reconstruction problem can be computed directly through analytical techniques. This limits the construction of the camera in some significant ways; however, it dramatically reduces the computational load required to reconstruct a given image volume. The principal requirements associated with analytical reconstruction methods involve uniformity of sampling and circular detector orbits. Analytical techniques are the most commonly used reconstruction methodologies due to their efficient computational requirements and generally robust nature.

The basis for analytical reconstruction is the Fourier slice theorem, which states that a one-dimensional Fourier transform of a set of parallel projections is equal to a slice of the 2D Fourier transform of the original object (Kak and Slaney, 2001). Given this connection between physical and Fourier space, it would seem possible to directly solve for a series of slices defined by a series of parallel projections. In theory, this is true; however, in practice, it is physically impossible to acquire an infinite number of projections. Thus, the sampling in frequency space becomes nonuniform. There are far more sampling points near the central, low-frequency, areas than in the higher-frequency areas, resulting in blurring of the final result. To overcome this sampling deficiency, we employ a technique known as filtered backprojection.

Filtered backprojection, also referred to as FBP, is an analytical approach based on the Fourier slice theorem that is designed to make the same informatic contribution to all points in the reconstructed volume. It is based on a two-step process consisting of a filtering step, and a backprojection step. The filtering step generates a set of filtered projections Q for a given set of raw projections S at an angle θ and offset t:

$$Q_\theta(t) = \int_{-\infty}^{\infty} S_\theta(w) \, |w| \, e^{j2\pi wt} dw$$

In this equation, the frequency response of the filter $|w|$ can be set arbitrarily. Typically, this is chosen to be a ramp filter with a response of zero at DC and steadily higher-pass response at higher frequencies. This type of filter, really a differentiator, enhances edges and provides optimal image resolution at the expense of high-frequency noise. Other filter types contribute some form of smoothing. All of the filtered projections at their various angles are then added together across the image volume through a backprojection step defined as follows:

$$f(x, y) = \int_{-\infty}^{\infty} Q_\theta(x \cos \theta + y \sin \theta) \, d\theta$$

For all their efficiency, FBP and all of the derivatives thereof suffer from a few problems. They are all susceptible to statistical noise. When there is limited dynamic range between areas with and without distribution in projection space, streak artifacts occur. This characteristic makes FBP a desirable method for reconstruction in situations where sufficient raw data statistics are present and high patient throughput is desirable. As imaging methodologies put increasingly higher value on lowering the dose of ionizing radiation delivered to the patient, analytical techniques will become less useful.

8.2.4.3 Iterative Techniques

Iterative reconstruction techniques can be used in applications where the assumptions made by analytical techniques are not possible. This allows the designer of imaging systems far greater freedom to design systems that work better with the structure they are intended to scan. They function well in a low statistics environment, which is an important feature in the reduction of delivered radiation dose to an imaging patient during the study (Moscariello et al., 2011; Winkelhener et al., 2011). Iterative techniques generally follow mathematical methodologies for solving large ill-composed matrix problems such as Gauss–Seidel, Jacobi, or conjugate gradient methods. In these approaches, a current estimated solution is forward projected to compare it against known observations, then either a difference (error) or ratio (probability) is backprojected into the estimated image to create the next iteration. The computations converge based on either an error norm for algebraic methods or a probability for statistical methods. Many differences exist regarding when updates are performed and which projections are used during each iteration in an effort to optimize the convergence speed of these otherwise computationally demanding methods.

Algebraic techniques to solve the inverse problem are based around the minimization of an error norm. In these methods, an iterative approach is used where a new estimate for the answer is generated by forward projecting the current estimate on the raw observation data, and then backprojecting the difference (error) between the raw data and the estimated projection data on the estimated image to yield a new estimated image. Basic variants include algebraic reconstruction technique (ART), simultaneous algebraic reconstruction technique (SART), and simultaneous iterative reconstruction technique (SIRT) (Kak and Slaney, 2001).

A typical example, preconditioned simultaneous iterative reconstruction technique (PSIRT) (Gregor et al., 2009), solves the weighted least-squares problem:

$$x^\star = \operatorname{argmin} \|Ax - b\|_R^2$$

PSIRT is based on the following update scheme:

$$x^{k+1} = x^k + \alpha CA^T R(b - Ax^k)$$

In this case, x is the image estimate for a given iteration, A is the system matrix, and b is the acquired projection data. The constant α is a feedback gain factor. As written, PSIRT is a Richardson iteration, and can be shown to have guaranteed convergence with $0 < \alpha < 2/\lambda_{max}$, where λ_{max} is the maximum eigenvalue of A. The R and C matrices are diagonal matrices of inverse row and column sums of the system matrix.

Algebraic techniques are very compute-intensive; however, they have several advantages over FBP. They perform well on raw data with inherent noise and converge well on normal tomographic information (Gregor et al., 2008). They are also useful if a large number of projections are not available, the sampling is nonuniform, or the ray projections are subject to bending/refraction. In addition, they are subject to a number of techniques to speed up their execution time. These include numerical techniques, such as ordered subsets (Hudson and Larkin, 1994), and physical techniques, such as parallelization (Gregor et al., 2009).

Statistical methods of image reconstruction are based on the concept of the maximization of likelihood. Examples include expectation maximization (EM) (Michel et al., 1999) and maximum a posteriori (MAP) (Lange and Fessler, 1995). Since the physics of emission imaging is based on radioactive decay, it is modeled as a Poisson process. The inclusion of statistical counting process methodologies directly into the reconstruction is thought to more accurately model the actual physics (Lange and Fessler, 1995). Statistical methods are useful in situations where the counting statistics are low.

8.3 Single-Photon Imaging

Single-photon imaging is composed of a variety of techniques that have been developed over the last 50+ years and are loosely termed nuclear medicine. These techniques include planar methods such as parallel collimated gamma cameras as well as tomographic methods such as SPECT. All that single-photon techniques require is a radioactive isotope that emits at least one gamma photon. If the tracer emits multiple photons, that information is generally lost.

8.3.1 Theory of Operation

Single-photon imaging is based primarily on low-energy gamma emitters like Technetium 99 m. In this type of radioactive decay, a single gamma photon is emitted from the nucleus and detected on a detector. As there is no way of knowing the direction that the photon is traveling when it hits the detector, physical collimation is used to restrict the flux to only those photons traveling at a particular angle to

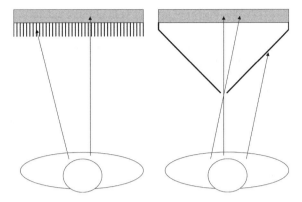

FIGURE 8.13   Flat parallel and pinhole single-photon collimators.

the detector. Multiple types of collimators are available, but the most common types fall into two categories, flat and pinhole. Flat collimators generate parallel projection images, while pinhole collimators generate cone-beam information. Either type can be viewed planar or reconstructed tomographically, but pinhole collimators are nearly always used tomographically. Multiple detector arrays and collimators can be positioned around the patient to increase sensitivity with a subsequent increase in system cost (Figure 8.13).

The typical scintillator used in low-energy single-photon cameras is NaI. It is cheap with good energy resolution and well-understood performance. NaI is hygroscopic, so it needs to be in a completely sealed container to prevent contamination by water and the destruction of the detector. Collimators are normally interchangeable except for very specialized cameras; thus, the selection of collimator can have a very large impact on performance for a given application. Very fine pitch collimators can be built that can deliver very high resolution, at the expense of sensitivity.

The intrinsic resolution of all detectors is measured by their absolute ability to localize a very small source without any sort of collimation in place (Figure 8.14). Placing a highly collimated source against the detector yields a spread of measured locations. The FWHM of this measurement yields the intrinsic resolution.

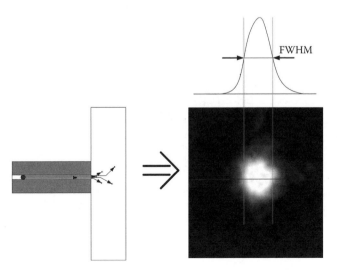

FIGURE 8.14   Detector intrinsic resolution.

There are a variety of factors that affect the intrinsic resolution of a detector, including scattering in the scintillator as well as the relative energy of the photon being measured. Higher-energy photons can travel further in the scintillator and thus spread more before being stopped. Thus, the specifications for a given detector will be determined for a particular isotope.

8.3.1.1 Parallel Collimators

Parallel collimators are the predominant collimator type used in clinical single-photon applications (Figure 8.15). They are constructed of a grid, typically of lead or tungsten. Holes tend to be either square or hexagonal.

The performance of a flat collimator is a basic function of geometry and the photon energy it is designed to perform with. Collimators built to function with higher energy must have thicker walls to stop the higher-energy photons and prevent blurring. It is possible to build a single-photon collimator to image 511-keV photons from PET tracers (and throw away the second photon), but the performance lags significantly the coincidence time collimation used in dedicated PET scanners.

The line source spatial resolution R_{ls} of a single-photon collimated detector varies as a function of the distance from the collimator:

$$R_{ls} = W_b \frac{L + d}{L}$$

The focal distance D_f is determined based on where the view lines cross:

$$D_f = \frac{W_b L}{W_a - W_b}$$

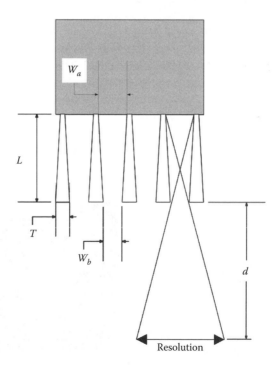

FIGURE 8.15   Flat single-photon collimator dimensions.

The sensitivity of a collimator to a uniform plane distribution is a good way to evaluate the expected overall sensitivity performance of a system. For a parallel collimator, the sensitivity as a percentage can be described as follows:

$$S = \frac{W_a^2 W_b^2}{4\pi L^2 (W_b + T)^2}$$

A cursory analysis of the equations shows the resolution of the collimator improving, getting smaller, linearly with respect to the width between the slats. Increasing the number of holes increases the relative size of the thickness of the slats with respect to the width between the slats, decreasing sensitivity as the square of the distance. This is the classic trade-off of sensitivity versus resolution seen in many imaging system types.

8.3.1.2 Pinhole Collimators

Pinhole collimators are commonly used in preclinical single-photon imaging applications. A source on one side of the shield is visible to the detector by only a small pinhole. Sensitivity is reduced as the distance between the collimator and source becomes large, but the resolution can be extremely high for pinhole collimators. It is possible to build multipinhole systems where multiple views can be acquired at the same time with the same detector position. The location of each pinhole is optimized to spread the response function over a different portion of the detector face (Figure 8.16).

The ideal sensitivity of a pinhole collimator S, the integral of the point response function, is given by the following equation (Grenier et al., 1974):

$$S = \frac{F\pi D^2}{8}(1 - \cos\theta)$$

where D is the diameter of the pinhole, θ the angle between the source and the normal vector to the hole, and F the total flux of gamma rays (counts/min/cm^2) emitting from a uniform plane source placed at an angle θ to the normal axis of the hole.

Note that the sensitivity is independent of the distance between the source and the hole as long as the source covers the entire solid angle of the hole. Furthermore, this equation assumes a negligible keel width, an idealized knife edge hole.

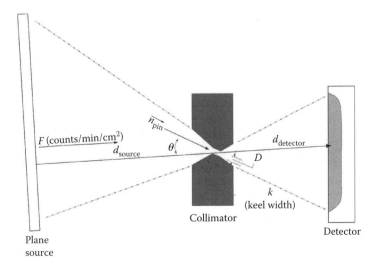

FIGURE 8.16 Pinhole collimator geometry.

The spatial resolution R of a pinhole collimator is defined as the distance between point sources before they overlap. Since pinhole collimators carry a magnification factor that is dependent on the distance between the source and the detector, this must be corrected for geometrically. The spatial resolution of a pinhole collimator with a hole of diameter D is represented by the following equation:

$$R = \left(1 + \frac{d_{source}}{d_{detector}}\right)D$$

Theoretically, pinhole collimators can have extremely high resolution, dependent only on how small the pinhole can be fabricated, improving linearly with the diameter. Taking advantage of the magnification factors can further improve the effective resolution. Of course, the same techniques will also decrease the sensitivity with the square of the diameter, thus lowering count statistics by the square of the diameter as well. Realistic imaging times limit the size of the hole when used to image living things with finite lifetimes.

Both sensitivity and resolution measurements are dependent on photon energy. Photons with energy above approximately 200 keV or pinholes with extremely thin and sharp keels can cause the effective diameter of the hole to be larger than the physical diameter (Accorsi and Metzler, 2005). The sensitivity and resolution of an entire system are the convolution of the intrinsic characteristics of the detector and the collimator chosen.

8.4 Positron Emission Tomography

PET is a nuclear medicine technique that was developed to measure *in vivo* concentrations of a wide variety of tracers. In fact, it is the selection of available positron-emitting isotopes that is one of the principal strengths of PET. Positron emitters include ^{18}F, ^{11}C, ^{15}O, and ^{13}N, the building blocks of organic chemistry. PET is characterized by high sensitivity, moderate resolution, and the ability to perform absolute quantification. Modern systems are combined with CT for anatomic registration.

8.4.1 Theory of Operation

A PET camera consists of a series of detectors that surround the patient axially. A tracer made with a positron-emitting isotope is injected into the patient and the resulting distribution is measured in three dimensions, tomographically. PET does not detect positrons directly, but instead it takes advantage of the dual photon nature of positron annihilation to measure the activity in the field of view. Each time an isotope decays, the positron emitted annihilates with an electron, and two gamma photons are detected in the ring. Each time two gammas are detected on the ring at close to the same time, they are said to be coincident, and the pair form a line of response (LOR) in the image. There is no physical collimation in a modern PET scanner like there is in single-photon imaging. All acquisition is performed fully in three dimensions (Figure 8.17).

Timing is of importance in PET system design as it is the primary collimation method for the determination of valid detected events. PET is different in this way from single-photon methods that rely on mechanical collimation. During a positron annihilation event, two gamma photons are created at exactly the same time. Through the use of accurate timing, it is possible to determine that the individually detected gammas were likely created together during an annihilation event as well as approximate their location in space. This is possible because after the annihilation event, the two gamma photons travel in collinear opposite directions toward the detectors at the same speed (Figure 8.18).

The distance traveled for each photon (γ_a and γ_b) can be determined by examining the arrival times of the two photons. It is possible for two unrelated events to trigger opposing detectors randomly, or even in multiples, resulting in falsely recorded lines of response. Better time resolution can reduce these

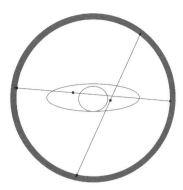

FIGURE 8.17 Simple PET scanner. The detector ring entirely surrounds the subject so that positron annihilation coincidence pairs can be detected. Covering only a portion of the ring dramatically reduces the sensitivity of the device. All detectors are open; there are no physical collimators between the subject and the detector. All events are collimated electronically, adding to the sensitivity of the system.

FIGURE 8.18 Annihilation event definition.

effects by more accurately rejecting pairs of single events that were not created at exactly the same time (Figure 8.19).

Histogramming the arrival time difference for a finite period of time for event pairs yields two distinct components. For pairs that are at the center of the field of view and are related to each other because they were created by an annihilation event, a narrow Gaussian distribution centered around a time difference of zero is formed. These types of events are said to be true coincident events. For event pairs that are not related through positron annihilation, their arrival times are a uniform random distribution. Since a time difference histogram is the convolution of the two uniform distributions, a triangular distribution is formed. These types of events are termed random. A PET scanner cannot directly determine whether or not an event is coincident or random, as it sees all events coming in as the sum of the two distributions. The sum of true coincident events and random events is termed prompt events. To differentiate between the two types from a stream of only prompts, it must do so statistically (Figure 8.20).

The timing performance of a detector system is determined from the time difference histogram of the true coincident events. By definition, it is the FWHM of the coincident histogram for a point source in the center of the field of view expressed in units of time. Placing the source off center will of course skew the histogram in the direction of the physical displacement as d_a and d_b will have changed and the photons are all traveling at the same speed, the speed of light. Use of a source other than a point source also affects the coincidence time difference histogram (TDH). Using a source that has some measurable internal volume will increase the width of the coincident TDH because of the physical variability that can exist in d_a and d_b.

For a given source, the randoms rate is related to the timing performance of the system and the singles rate that the detectors are firing at. For two detectors, operating at a count rate of S_1 and S_2, respectively, the expected rate of random events R within a given window T is given by the following formula:

$$R = 2\tau S_1 S_2$$

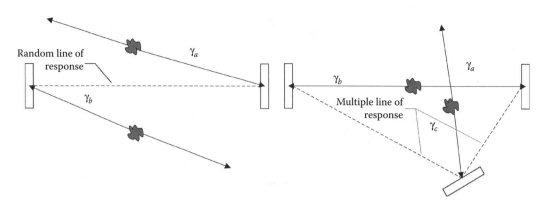

FIGURE 8.19   Random and multiple events result in false lines of response registered by the system.

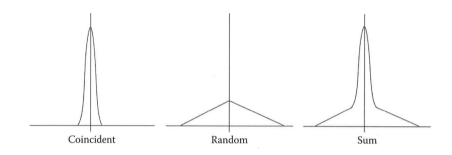

FIGURE 8.20   Time difference histograms of coincident events, random events, and the sum of the two distributions that represent what a PET scanner measures directly.

To differentiate between true coincident events and random events, two approaches exist. One approach relies on the measured single-channel count rates to generate a random count rate for each detector pair. This value is then subtracted from the in-window counts acquired to result in the actual number of true events. This approach requires significant overhead. Another approach is to estimate the randoms through a process called delayed windowing and subtract them from the total events (also called prompt events). This process can be performed online as the scan is being taken and thus provide real-time input to the operators. In this technique, a coincidence window is placed around the center of the time difference histogram. All event pairs with a time difference within the coincidence window are identified as prompt events. Prompt events are the sum of both random and true events. A second set of windows, termed random windows, each the same width as the primary window, are placed at one-half the distance (time) to the end of the entire time window. Since the random distribution is triangular, the drop is linear, and placing the random windows at that location gives them exactly half of the response that is expected in the center. The combination of both the positive and negative random windows provides an approximation of the random events within the coincidence window, which are then subtracted (Figure 8.21).

Just as improved light yield from the detector improves the energy resolution of a system, it also improves the timing. Increased photon count yields a better statistical measurement of the incoming event pulse as seen by the electronics. A steeper rise time on the pulse provides for better performance of the CFD as well. In addition, electronics and light detector characteristics in the detector (photomultipliers, APDs, and SiPMs) are all reflected in the final coincident time difference histogram for the system. The lesser the uncertainty of the timing, the narrower the coincident time difference histogram. As the time histogram narrows, a narrower coincidence window can be applied without rejecting a high

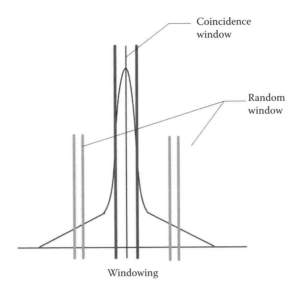

Windowing

FIGURE 8.21 Coincidence and random window definition.

proportion of true events (which would reduce sensitivity), yet rejecting more randoms. This improves the overall counting performance of the system.

In the last 20 years, a variety of commercial systems have been produced. A large contributor to improved PET system performance has been in the area of timing. Improvements in timing result in reduced random contribution (noise) in reconstructed images. This is very important in scanning systems where only a small part of the subject being scanned is actually in the field of view of the system. Contributions to randoms from activity located just outside the field of view can be dramatic. In the furthest extreme, modern PET scanners with sufficiently accurate timing can provide additional information on location of each event to the reconstruction algorithm (time-of-flight) increasing the signal-to-noise performance of the system.

System Type	Year	Timing Performance (FWHM)
BGO detectors with photomultipliers	1993	8 ns
Early LSO detectors without time-of-flight	1997	3 ns
Modern LSO time-of-flight with fast PMT	2009	500 ps
LSO with APDs (MRI-compatible)	2011	5 ns
LSO with SiPM (MRI-compatible)	Future	300 ps

8.4.2 Measures of Performance

Sensitivity is defined as the percentage of events that are correctly recognized by a system. This can be expressed as a percentage for a point source located at a particular location (usually the center), or as a volumetric efficiency that expresses the number of counts extracted per amount of activity in a specified phantom. PET scanners are identified by their high sensitivity. From this perspective, the ideal device would be a sphere. In this way, there would be a high probability that all positron annihilation events would be detected, as the solid angle would be 100%. It is difficult and expensive to build such a spherical device, so most systems are designed around a cylindrical geometry that cuts a smaller solid angle. As the length of the cylinder goes to infinity, the solid angle approaches 100% (Figure 8.22).

Computing the solid angle Ω of a cylinder can be done accurately (Verghese et al., 1972), but as long as the cylinder's axial length is significantly smaller than its radius, it can be approximated as the outer

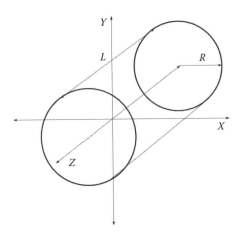

FIGURE 8.22 PET detector cylinder definition of orientation.

surface area of the cylinder (not including the ends) divided by the surface area of a sphere of the same radius:

$$\Omega_{app} = \frac{A_{cyl}}{A_{sphere}} = \frac{2\pi RL}{4\pi R^2} = \frac{L}{2R}$$

One common measure of sensitivity is the point source sensitivity measured at the center of the field of view. Sensitivity can also be represented as the probability of detecting an actual occurrence. It is thus the ratio of detected events to actual events. In the case of dual-photon imaging, this is represented by the following, where Ω is the solid angle represented as a percentage of coverage and ρ_x is the probability of interacting with a given detector if the photon actually strikes the detector:

$$S = \Omega^2 \rho_1 \rho_2$$

The resolution of a PET scanner depends on the intrinsic resolution of the detectors, the sampling frequency of the image volume, and the positron range of the isotope being imaged. In modern PET pixellated detector systems, a very high signal-to-noise ratio exists in the design of the detector itself, yielding an intrinsic detector resolution close to the size of the pixels. In fact, newer designed systems using APDs and SiPMs have a one-to-one connection between the detector crystal and light detector, making the detector size exactly equal to the intrinsic resolution. For PET cameras built in a circular geometry, a fan relationship exists that establishes the maximum spatial sampling frequency (Figure 8.23).

The sampling pitch at the center of the field of view is thus equal to half the detector pitch using basic triangular geometry:

$$s = \frac{d}{2}$$

The sampling pitch varies over the field of view, but at the center, it is the same for all possible angles, so the center is used for simplicity. Given that the Nyquist theorem provides for the complete replication of a structure if it is sampled at least double the maximum frequency in the structure, the best system resolution we could expect to get at the center, based purely on sampling, would be equal to the detector size (d).

Rotating the sampling pattern by the minimum amount, one crystal, yields the next closest pattern. Further rotation yields an interference pattern that causes a singularity at the center location making

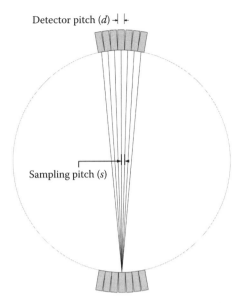

Detector pitch (*d*)

Sampling pitch (*s*)

FIGURE 8.23 Fanbeam sampling of a circular geometry PET camera.

the measurement of system resolution at that precise location suspect. System resolution measurements are often cited at the center of the field of view and offset by 1 and 10 cm for this reason.

Computing the number of lines that pass through a particular voxel by ray tracing yields a better measure of the true expected performance of a system in three dimensions. Sampling can be very nonuniform, especially if nonstandard geometries are chosen. Most PET scanners only approximate a circular or spherical geometry. Most are constructed with detector arrays that have flat faces and are assembled into a ring. This results in a structure that is actually a multisided polygon, not a circle, resulting in circular artifacts in the sampling matrix. Image reconstruction takes these nonuniformities into consideration and normalizes them out (Figure 8.24).

System resolution is also limited by limitations imposed by the depth of interaction of the photons with the detector. Unless the system is equipped with detectors that are able to sense depth of interaction (Schmand et al., 1999), or a statistical location confidence measure (Lenox et al., 2012), the detector electronics can only qualify that the event occurred in a given detector (Figure 8.25).

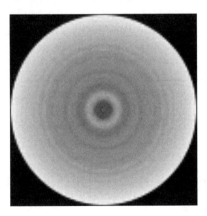

FIGURE 8.24 Sampling density of the field of view of a PET tomograph. The physical construction of most tomographs involves detectors with flat faces arranged in a ring that approximates, but does not actually equate to a perfect circle. Instead, the polygonal structure causes variation in sampling that appears as concentric rings.

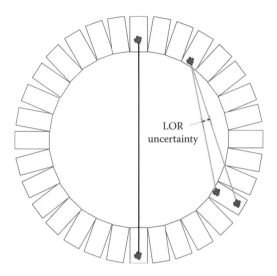

FIGURE 8.25 Depth of interaction of the photon with the detector blurs the image by introducing uncertainty in the LOR as the LOR moves further away from the center of the FOV. At the center of the FOV, the depth of interaction does not cause uncertainty.

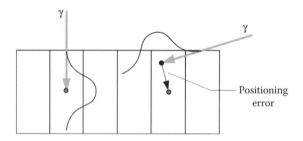

FIGURE 8.26 Interaction depth varies with incident angle.

The geometry must be determined from a preknown point for the reconstruction. This location is often determined as the interaction centroid, the point of highest probability of interaction of the crystal for a photon with a perpendicular incident angle. This assumption is not valid as the incident angle of the photon on the detector becomes more oblique. A positioning error develops that moves the measured LOR from the actual point of interaction to the centroid. This error can be corrected by modeling the point spread function (PSF) of the detector geometry and incorporating this model into the reconstruction (Panin et al., 2006) (Figure 8.26).

8.4.3 Examples

8.4.3.1 PET/CT

In the last decade, most PET scanners built were combined PET/CT devices. These combine the anatomic capabilities of a CT with the functional capabilities of PET. One current production device is the Biograph mCT (Siemens Molecular Imaging, Knoxville, USA) (Figure 8.27). This device includes a 20-cm axial field-of-view PET scanner combined with 128-slice CT. Iterative PET reconstruction algorithms include ordered subsets expectation maximization (OSEM) and a time-of-flight augmented OSEM both of which include PSF modeling to improve spatial resolution. Attenuation correction is performed through the use of CT.

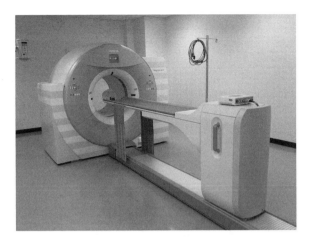

FIGURE 8.27   Siemens biograph mCT PET/CT installed at the Texas A&M Institute for Preclinical Studies. This system is made up of a 128-slice CT with time-of-flight PET.

Measured physical PET performance of the mCT (Jakoby et al., 2011) is summarized as follows:

Parameter	Value
Detector size	4 mm × 4 mm × 20 mm
Energy resolution	11.5 ± 0.2% FWHM
Time resolution	527.5 ± 4.9 ps FWHM
Spatial resolution (center of FOV)	4.1 ± 0.0 mm FWHM
Sensitivity (center of FOV)	0.96%

8.4.3.2 SPECT/CT

As with PET, the state-of-the-art single-photon SPECT systems are also often combined with CT for accurate coregistration with anatomy and estimation of attenuation correction. An example of this type of device is the Siemens Symbia SPECT/CT (Tables 8.4 and 8.5; Figure 8.28). This system combines up to 16-slice CT with two single-photon detector arrays. A variety of collimators are available that can be interchanged on the detectors.

8.5 Applications

Both PET and SPECT rely on the injection of a radiotracer to perform their function. The choice of the tracer has a significant impact on the diagnostic result. Radiopharmaceuticals have been developed for investigating a wide variety of biological processes.

PET radiopharmaceuticals are based principally on ^{18}F for several reasons. [18]F is manufactured by cyclotron bombardment and has a half-life that is sufficiently long enough to allow transport over reasonable distances. This allows economical manufacturing with centrally located cyclotron facilities. Also, the chemistry of ^{18}F, it being close in the periodic table to ^{18}O, provides for reasonably fast chemical substitution for [18]O. The faster the chemical synthesis can be done, the higher and more economical the yield. [11]C is another popular isotope used in PET, as carbon is common in most organic molecules. The disadvantage of ^{11}C is the very short half-life that makes synthesis and transport very difficult. This fact makes an on-site cyclotron necessary, a large expense. [13]N is popular when synthesized into ammonia and used as a perfusion agent, but due to the short half-life, this approach also requires an on-site cyclotron. [15]O is used principally as a perfusion agent in the form of ^{15}O-H_2O. In this form, [15]O becomes an ideal perfusion agent as H_2O is not metabolized and thus has a very linear response for measuring

TABLE 8.4 Flat Collimators Available for the Siemens Symbia SPECT/CT

Collimator	LEHS (Low Energy High Sensitivity)	LEAP (Low Energy All Purpose)	LEHR (Low Energy High Resolution)	LEUHR (Low Energy Ultra-High Resolution)	LEFB (Low Energy Fan Beam)	ME (Medium Energy)	HE (High Energy)	EHE (Extra High Energy)
Isotope/ energy	^{99m}Tc	^{99m}Tc	^{99m}Tc	^{99m}Tc	^{99m}Tc	^{67}Ga	^{131}I	^{18}F
# Holes (×1000)	28	90	148	146	64	14	8	4
Hole length (mm)	24.05	24.05	24.05	35.8	35	40.64	59.7	50.5
Septal thickness (mm)	0.36	0.2	0.16	0.13	0.16	1.14	2.0	3.4
Hole diameter (mm)	2.54	1.45	1.11	1.16	1.53	2.94	4	2.5
Sensitivity	1020 cpm/ μCi	330 cpm/ μCi	202 cpm/ μCi	100 cpm/ μCi	280 cpm/ μCi	310 cpm/ μCi	147 cpm/ μCi	185 cpm/ Ci
Geometric resolution at 10 cm (mm)	14.6	8.3	6.4	4.6	6.3	10.8	13.2	10.6
System resolution at 10 cm (mm)	15.6	9.4	7.4	6.0	7.3	12.5	13.4	19.0

TABLE 8.5 Pinhole Collimators for the Siemens Symbia SPECT/CT

	^{99m}Tc	^{123}I	^{131}I
Cone length	4 mm, 6 mm, 8 mm	4 mm, 6 mm, 8 mm	4 mm, 6 mm, 8 mm
Sensitivity at 10 cm w/4 mm	123 cpm/μCi	111 cpm/μCi	67 cpm/μCi
Sensitivity at 10 cm w/6 mm	271 cpm/μCi	243 cpm/μCi	133 cpm/μCi
Sensitivity at 10 cm w/8 mm	478 cpm/μCi	426 cpm/μCi	221 cpm/μCi
Geometric resolution at 10 cm w/4 mm	6.2 mm FWHM	6.3 mm FWHM	7.5 mm FWHM
Geometric resolution at 10 cm w/6 mm	9.3 mm FWHM	9.3 mm FWHM	10.6 mm FWHM
Geometric resolution at 10 cm w/8 mm	12.3 mm FWHM	12.4 mm FWHM	13.6 mm FWHM
System resolution at 10 cm w/4 mm	6.6 mm FWHM	6.6 mm FWHM	7.6 mm FWHM
System resolution at 10 cm w/6 mm	9.5 mm FWHM	9.5 mm FWHM	10.7 mm FWHM
System resolution at 10 cm w/8 mm	12.5 mm FWHM	12.5 mm FWHM	13.7 mm FWHM

flow, volume, and mean transit time. Again, owing to a very short half-life of just a few minutes, ^{15}O requires an on-site cyclotron.

Limitations in availability, but the high desirability of PET compounds based on isotopes other than ^{18}F, have created a need for a new type of easily installed, easily operable cyclotron. One such design is the biomarker generator built by ABT Molecular Imaging (Knoxville, TN). These systems are relatively small, capable of 7.5 MeV positive ion bombardments of 5 μA (ABT, 2012). However, when paired with very efficient microfluidic chemistry, systems are able to produce single doses of a variety of radiopharmaceuticals on demand for both ^{18}F and ^{11}C (Table 8.6 and Figure 8.29).

SPECT radiopharmaceuticals are based principally on ^{99m}Tc with a variety of others also available.

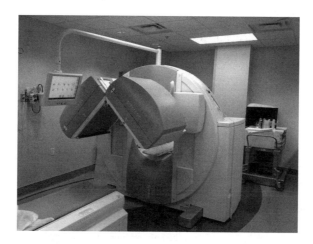

FIGURE 8.28 Siemens Symbia SPECT/CT. A 16-slice CT with dual single-photon detector arrays. Suitable for a variety of applications in both cardiology and oncology. Detector heads are positioned at 90° for cardiac imaging, but can be reconfigured depending on the application.

Radiopharmaceutical	Usage	Reference
^{123}I-Isopropyliodoamphetamine (IMP)	Perfusion	Kawamura et al. (1990)
^{99m}Tc-Hexamethylpropylamine oxime (HMPAO)	Perfusion	Alptekin et al. (2001)
^{133}Xe	Perfusion (via inhalation)	Kikuchi et al. (2001)
^{123}I-quinuclidinyl-iodo-benzilate (QNB)	Acetylcholine muscarinic antagonist (Alzheimers)	Pakrasi et al. (2007)
^{123}I-iodo-hydroxy-methoxy-N[(ethyl-pyrrolidinyl)methyl]benzamide (IBZM)	Dopamine receptor imaging for Parkinsons and schizophrenia	Plotkin et al. (2005)
^{111}In-capromab-pendetide	Tumor imaging	Seo et al. (2010)
^{111}In	White blood cell labeling	Bar-Shalom et al. (2006)
^{99m}Tc-Sestamibi	Cardiovascular	Heinle et al. (2000)
^{99m}Tc-Arcitumomab	Tumor imaging	Libutti et al. (2001)
^{99m}Tc-Votumumab	Tumor imaging	Olafsen et al. (2010)
^{99m}Tc-Hynic-octreotide	Tumor imaging	Sun et al. (2007)
^{67}Ga	Inflammation	Bar-Shalom et al. (2006)

8.5.1 Pharmacokinetics

As drugs, including tracers, move through a living thing, they exhibit a variety of characteristics. They can be transported by force through perfusion, move across cell boundaries by diffusion, and are metabolized or excreted. The quantification of their transport characteristics is termed pharmacokinetics (PK). The effect that the drug has on the body once it reaches the target location is termed pharmacodynamics (PD). The determination of the kinetic constants that characterize the PK of a particular drug have long been determined through direct sampling of various tissues in a variety of ways, usually very invasively by taking direct samples of blood and tissue. Both PET and SPECT imaging techniques are able to measure the concentration of a tracer at a particular point in time and space, in a relatively noninvasive manner. The use of PET and SPECT is a very powerful tool to characterize PK for drug discovery and development.

There are four principal components of PK, including absorption, distribution, metabolism, and excretion. Together, this is often abbreviated as ADME.

Absorption: Delivery of a given material into the circulation. This can be IV, oral, or though a skin patch, for example.

Distribution: Transport through the body, generally via perfusion in the bloodstream, but could also be through cerebrospinal fluid (CSF), or simply general diffusion.

Metabolism: Occurs at the point of use. This can be located at both the target therapeutic location and elsewhere.

Excretion: Removal of the material from the bloodstream in a nontherapeutic manner. Generally through either the liver or kidneys.

TABLE 8.6 PET Radiopharmaceuticals and Their Applications

Radiopharmaceutical	Usage	Reference
[18]F fluorodeoxyglucose (^{18}F-FDG)	PET glucose metabolism	Hamacher et al. (1986)
[18]F sodium fluoride (^{18}F-NaF)	Bone imaging	Beheshti et al. (2008)
[18]F fluorodopamine (^{18}F-DOPA)	Dopamine receptor	Goldstein et al. (1993)
[18]F fluorothymidine (^{18}F-FLT)	Cell proliferation through TK1 during DNA synthesis	Shields et al. (1998b)
[18]F fluroroethyl-ʟ-tyrosine (^{18}F-FET)	Assessment of gliomas	Weser et al. (1999)
[18]F-fluorocholine (^{18}F-FCH)	Lipid synthesis	Kwee et al. (2006)
[18]F-fluoromisonidazole (^{18}F-FMISO)	Tumor hypoxia	Bruehlmeier et al. (2004)
[18]F-fluoroazomycin arabinoside (^{18}F-FAZA)	Tumor hypoxia	Souvatzoglou et al. (2007)
[18]F-fluoro-β-estradiol (^{18}F-FES)	Quantifying ER expression in tumors	Couturier et al. (2004)
[18]F-fluorodihydrotestosterone (^{18}F-FDHT)	Prostate cancer	Larson et al. (2004)
[18]F-flumazenil (^{18}F-FMZ)	Benzodiazephine receptor (epilepsy)	Leveque et al. (2003)
[18]F-fluoropropylcarbometh-oxyiodophenylortropane (^{18}F-FP-CIT)	Dopamine transporter imaging for evaluation of Parkinson's disease	Wang et al. (2006)
[18]F-fluoroannexin V (^{18}F-annexin V)	Early-stage cellular apoptosis	Murakami et al. (2004)
[18]F-5-fluoropentyl-2-methyl-malonic acid (^{18}F-ML-10)	Cellular apoptosis	Reshef et al. (2008)
[11]CO	Red blood cell distribution	Hostetler and Burns (2002)
[11]C Raclopride	Dopamine D2 receptor antagonist	Farde et al. (1985)
[11]C-hydroxyephedrine (^{11}C-MEHD)	Accumulates in nerve terminals	Munch et al. (2000)
[11]C-acetate	Myocardial viability/fatty acid metabolism	Wolpers et al. (1997)
[11]C-methionine (^{11}C-MET)	Tumor evaluation, especially in the brain	Nuutinen et al. (1999)
[11]C-choline	Phospholipid precursor used to evaluate tumors	Hara et al. (2002)
[11]C-thymidine	Precursor for DNA synthesis used to measure rates of cell proliferation	Shields et al. (1998a)
[11]C-flumazenil (^{11}C-FMZ)	Benzodiazepine receptor used for evaluation of epilepsy	Foged et al. (1997)
[13]N-ammonia	Perfusion	Fricke et al. (2005)
[15]O-water	Perfusion	Langen et al. (2008)
[82]Rb-rubidium chloride	Perfusion	Lautamaki et al. (2009)
[68]Ga-iron hydroxide colloid	Liver and spleen function	Kumar et al. (1981)
^{68}Ga-PSMA	Prostate-specific membrane antigen	Banerjee et al. (2010)
^{68}Ga-HBED	Hepatopilary function	Eder et al. (2009)
[62]Cu-ATSM	Tumor hypoxia	Dehdashti et al. (2003)
[62]Cu-pyruvaldehyde Bix(N4-methyl) thiosemicarbazone ([62]Cu-PTSM)	Perfusion	Okazawa et al. (1995)
[124]I	Direct measurement of thyroid, can also be used to label antibodies	Phan et al. (2008)
[90]Y	Liquid brachytherapy for cancer therapeutic applications	Zhou et al. (2011)

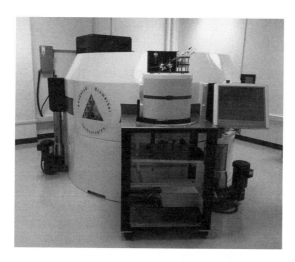

FIGURE 8.29 Single-dose biomarker generators can deliver small amounts of radiopharmaceuticals by combining a small cyclotron with microfluidic chemistry. High levels of automation allow for use with fewer staff and low infrastructure requirements. (Courtesy of ABT Molecular Imaging, Knoxville, TN.)

8.5.1.1 Compartmental Modeling

Although the body is very complex, containing a plethora of compartments and transport methods for drugs, we generally simplify the system into just a few compartments and combine the transport methods together to form simple systems that can be solved by differential equations. The basic premise of compartmental modeling is the preservation of mass. Once into the system via absorption, the material must be either metabolized or excreted or the concentration will stabilize at a particular level. The goal of PK modeling is to calculate the transport characteristics from the distributions measured because those characteristics can have a large impact on the therapeutic success or failure of a drug. A drug can only function therapeutically if the concentration is high enough for a sufficient time on the target. If clearance rates are too high, then high doses are required, and these may cause unintended side effects (Figure 8.30).

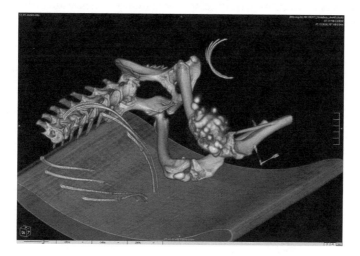

FIGURE 8.30 **(See color insert.)** PET/CT volume rendering of a drug distribution. ^{90}Y is used to treat osteosarcoma in the distal femur of a dog. (Adapted from Zhou, J. et al. 2011. *J. Nucl. Med.*, 52(Supplement 1), 129.) (Texas A&M Institute for Preclinical Studies. With permission.)

The pharmacokinetic characteristics of a drug concentration on a target are described by the concentration half-life of the drug on the target (τ). Knowing the value of τ allows the designer of a drug to specify the drug dose as well as the time duration between doses necessary to keep the concentration in the therapeutic range. In the case of radiolabeled drugs, the biological clearance half-life τ_{bio} is convolved with the decay characteristics of the chosen isotope τ_{decay} to understand the concentrations actually measured. Since the isotope is known, the τ_{decay} is known and the biological characteristics can be separated out:

$$\frac{1}{\tau_{meas}} = \frac{1}{\tau_{bio}} + \frac{1}{\tau_{decay}}$$

Determination of pharmacokinetic characteristics is done using compartmental modeling. The most basic model is a two-compartment design with a central compartment and a target compartment (Figure 8.31).

The flow in and out of each compartment is described with rate constants and the assumption that mass is conserved. The change in concentration at a particular location can be described as a first-order differential equation for most biological processes. Using this method, the change in concentration C with respect to time goes down linearly with the concentration:

$$\frac{dC}{dt} = -kC$$

Solving this equation, where C_0 is the initial concentration, t the time, and k the rate constant, yields

$$C = C_0\, e^{-kt}$$
$$\frac{C}{C_0} = e^{-kt}$$

The ratio of C/C_0 represents the fraction of the original concentration and is often used to simplify the computation of k, as it can be numerically determined from actual measurements. A further rewrite of the basic equation yields

$$\log_e \frac{C}{C_0} = -kt$$

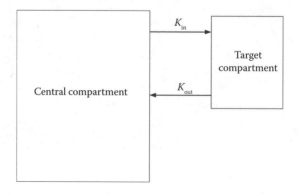

FIGURE 8.31 Basic two-compartment model.

This basic behavior results in the characteristic logarithmic model, where the concentration is often described by the half-life of the measure at a given point. This can be accommodated by rewriting the exponential in terms of powers of 2 with an appropriate change to k now reflecting the reciprocal of the half-life in units of time. The two are related as follows (Figure 8.32):

$$\log_e \frac{C}{C_0} = -kt_{\frac{1}{2}} = \log_e 0.5$$

$$\tau_{meas} = t_{\frac{1}{2}} = \frac{0.693}{k}$$

For the purposes of drug targeting, the body can be modeled with a three-compartment scheme where the blood is considered a central compartment, and the target area is considered separately. The principal biological construct behind this premise is that blood is highly perfusive, so a direct injection into the bloodstream is immediately responsive. The peripheral areas are serviced by the blood, but the transport mechanism to get a drug across the membrane into the cell is often modeled as a diffusion process. For any drug that is metabolized in the cell, a third compartment is added.

Thus, for any given compartment, it is possible to write an equation that describes the concentration of drug in that compartment as a function of time. Given an absorption function $A(t)$, which is normally modeled as a step function to simulate a syringe injecting drug into the bloodstream, by conservation of mass, the following equation describes the concentration of drug in the central compartment:

$$C_c(t) = \text{initial} + \text{absorption} - \text{loss to target} + \text{return from target} - \text{excretion}$$

For a known diffusion process, we rely on Fick's first law of diffusion, where a flux F (change in mass per time) is determined by the permeability of a cellular membrane P_{cell} and the concentration in the two

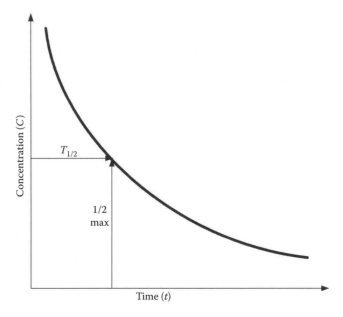

FIGURE 8.32 Concentration decays logarithmically with respect to time but rate is linear with respect to concentration.

compartments. Thus, to describe the loss from the central compartment to the target compartment, per unit area, we can describe the flow as follows:

$$F = \frac{dC}{dt} = P_{cell} (C_c - C_t)$$

Unfortunately, the transport is not always well known and may, in fact, be a combination of factors that make a precise analytic description difficult. Furthermore, even though we model the various intercompartmental connections as if they were actual physical transport, they may not really relate in the same way. This is the case for metabolic activity where a state change occurs. Thus, to estimate the combined transport method, we lump all of them together and rely on the first-order estimation previously described.

8.5.2 Cardiology

Understanding blood flow is a useful diagnostic parameter in cardiology. By definition, blood flow is determined in flow units, such as mL/min, and is normally used in conjunction with vessels. Perfusion is similar in that it also measures in units of flow, but it macroscopically describes flow over a mass volume (i.e., mL/min/g). Thus, from a diagnostic perspective, a measure of perfusion can tell the clinician how much blood is being delivered to a given volume in a given period of time. Also of interest is the blood volume and mean transit time, literally the volume of blood in a given mass of tissue that is available to move and the amount of time necessary for a unit volume to be replaced. These three parameters are related as follows:

$$F = \frac{V}{T_{mtt}}$$

Perfusion is a physiomechanical process. There are two approaches to measuring perfusion; these involve tracers that are either freely diffusible or nonfreely diffusible. A freely diffusible tracer can move in and out of the compartment based only on biophysical transport. Examples of such tracers include water or gases that are not metabolized such as xenon. A nondiffusible tracer gets to a certain location and then stops because it is metabolically changed or simply physically halted. Examples of this type of tracer would include microspheres that become stuck in capillaries once they reach the region of interest.

An injection of the tracer is given and then it travels through the bloodstream (central compartment) and is taken up in the target tissue (target compartment). The presence of the tracer can be determined through imaging means as a function of time. This is normally termed a time–activity curve, or TAC.

Considering the multicompartment model shown in Figure 8.33, if a perfusion agent is injected in a fast bolus, the transfer of the material from the central compartment to the target compartment is controlled by the rate constants k_{ct} and k_{tc}. Nonfreely diffusible agents, also called trapped perfusion agents, have negligible return, and thus k_{tc} is insignificant. This vastly simplifies the measurement as perfusion rate is simply the total accumulated material normalized by the cardiac output. Freely diffusible agents can pass in and out of the target compartment. Furthermore, the perfusion agents commonly used are generally not metabolized with the ideal perfusion agents being ^{15}O-water for PET or ^{133}Xe gas for SPECT (Figure 8.34).

The target compartment concentration, C_t, is highly dependent on the central compartment concentration, C_c. The two are related as follows, the rate of change of concentration in the target compartment is equal to the flux into the compartment minus the flux out of the compartment:

$$\frac{dC_t(t)}{dt} = k_{ct}C_c(t) - k_{tc}C_t(t)$$

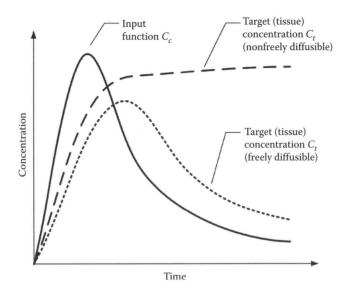

FIGURE 8.33 Time–activity curve of both freely diffusible and nonfreely diffusible tracers and their input function. The target tissue being measured is always delayed and attenuated from the input function, the extent of which is determined by the input and output rate constants.

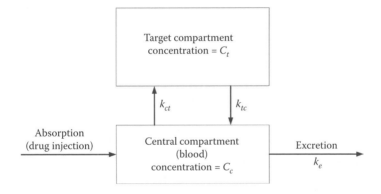

FIGURE 8.34 Simplified two-compartment model for evaluation of perfusion.

Integrating both sides of the equation yields

$$C_t(t) = \int_0^t k_{ct} C_c(t) \, dt - \int_0^t k_{tc} C_t(t) \, dt$$

The central compartment concentration ($C_c(t)$) can be measured by taking blood samples at particular times, as well as by serial imaging. The target concentration $C_t(t)$ is also measured by serial imaging. Rearranging this equation and putting it into a form that resembles a linear equation yields the following:

$$C_t(t) = k_{ct} \int_0^t C_c(t) \, dt - k_{tc} \int_0^t C_t(t) \, dt$$

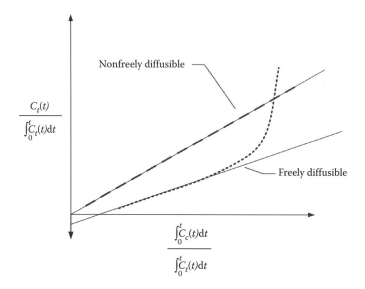

FIGURE 8.35 Characteristic graph for freely diffusible and nonfreely diffusible tracers shows a linear relationship once steady-state transients have passed in the free case. Nonfreely diffusible tracers have no transient effects.

$$\frac{C_t(t)}{\displaystyle\int_0^t C_t(t)\mathrm{d}t} = k_{ct}\frac{\displaystyle\int_0^t C_c(t)\mathrm{d}t}{\displaystyle\int_0^t C_t(t)\mathrm{d}t} - k_{tc}$$

By computing the integrals of C_t and C_c, the equation describes a line with a slope k_{ct} and with a y-intercept k_{tc}. Both constants can be estimated for steady state through linear regression for both freely and nonfreely diffusible tracers. Nonfreely diffusible tracers show their linear regression y-intercept to be zero (Figure 8.35).

Freely diffusible tracers must be analyzed after sufficient time t to provide for transient suppression (Logan, 2000). Once known, a variety of physical parameters can be estimated from the transport constants, that is, permeability through Fick's law or the perfusion dC/dt for steady state.

8.5.3 Oncology

[18]F-fluorodeoxyglucose (FDG) is an analog of glucose, and since glucose metabolism is so pervasive, it can provide considerable insights into the detection of disease, especially cancer. This is a clinically important tracer and represents a majority of all PET studies performed. Yet, ^{18}F-FDG is not a glucose, but it is a fluorinated glucose. In this particular case, the chemical changes associated with fluorination of the compound change it in such a way that it moves partially through the metabolic pathway and eventually gets stopped (Figure 8.36).

As with freely diffusible perfusion tracers, as a small molecule, ^{18}F-FDG is also freely diffusible up to the target compartment. Once inside the target compartment and available for reaction, the reaction is an enzymatic reaction governed by the Michaelis–Menton equation. This equation states that the reaction rate R for the conversion of S to P can be stated as follows:

$$R = \frac{V_m[S]}{[S] + K_m}$$

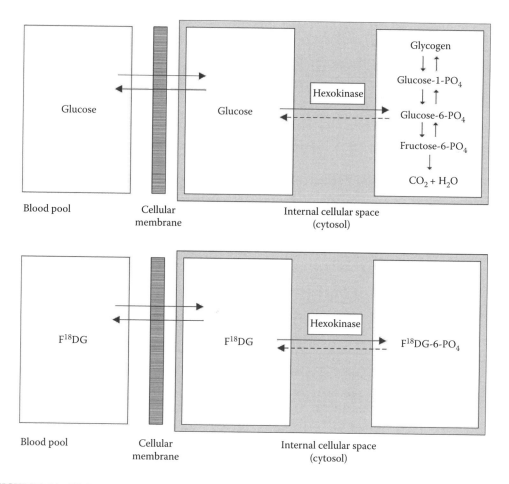

FIGURE 8.36 FDG versus normal glucose metabolism. FDG is trapped as ^{18}F-FDG-6-PO$_4$.

where V_m is the maximum rate of the reaction, S the substrate concentration, and K_m the concentration of S that causes a reaction rate of one-half the maximum.

Similar to the simpler two-compartment models used to describe perfusion tracers, a three-compartment model can be devised to describe the overall process of 18F-FDG as well as other tracers (Figure 8.37).

The rate of change of a given compartmental concentration is related through the conservation of mass as

$$\frac{dC_t(t)}{dt} = k_{ct}C_c(t) - k_{tc}C_t(t) - k_{tm}C_t(t) + k_{mt}C_m(t)$$
$$\frac{dC_m(t)}{dt} = k_{tm}C_t(t) - k_{mt}C_t(t)$$

As with perfusion, these equations can be solved and metabolic rates determined (Hawkins et al., 1992) by relying on the physiology that ^{18}F-FDG is trapped metabolically, k_{mt} is zero, and thus the equations simplify. From a practical perspective, however, to do this requires dynamic information in the form of blood concentrations and regional measurements. Practically, this takes considerable time to

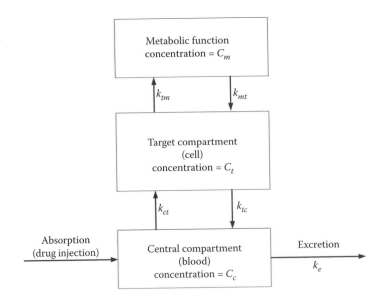

FIGURE 8.37 Three-compartment model to describe metabolically active materials.

measure as the scan must continue past the peak blood concentration to evaluate the washout phase. Typically, this information requires more than 60 min. Diagnostically, a simpler approach, the standardized uptake value (SUV) has been determined that has been shown to be similar in accuracy to kinetic modeling methods (Thie, 2004). This technique allows single time point imaging. The SUV is defined as the ratio of uptake to injected dose per body weight (Figure 8.38):

$$\text{SUV} = \frac{C_t(t)}{d_{t0}/m}$$

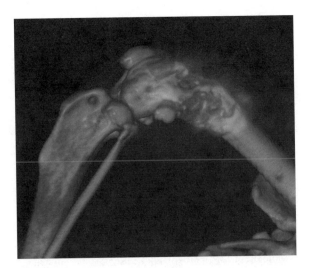

FIGURE 8.38 **(See color insert.)** PET/CT of a canine osteosarcoma. Brightly colored areas indicate high-glucose SUV and correspond physically to the bone lysis and proliferation. (Texas A&M Institute for Preclinical Studies. With permission.)

References

ABT. 2012. Advanced Biomarker Technologies, Product Brochure, Knoxville, TN.

Accorsi, R. and Metzler, S. D., 2005. Resolution-effective diameters for asymmetric-knife-edge pinhole collimators. *IEEE Trans. Med. Imaging*, 24(12), 1637–1646.

Alptekin, K. et al., 2001. Tc-99m HMPAO brain perfusion SPECT in drug-free obsessive-compulsive patients without depression. *Psychiatry Res. Neuroimaging*, 107, 51–56.

Allisy, A. et al. 1996. International Commission on Radiation Units and Measurements, Report #64—Medical Imaging—The Assessment of Image Quality.

Banerjee, S. et al., 2010. ^{68}Ga-labeled inhibitors of prostate-specific membrane antigen (PSMA) for imaging prostate cancer. *J. Med. Chem.*, 53, 5333–5341.

Bar-Shalom, R. et al., 2006. SPECT/CT using ^{67}Ga and ^{111}In-labeled leukocyte scintigraphy for diagnosis of infection. *J. Nucl. Med.*, 47, 587–594.

Beheshti, M. et al., 2008. Detection of bone metastases in patients with prostate cancer by 18F fluorocholine and 18F fluoride PET-CT: A comparative study. *Eur. J. Nucl. Med. Mol. Imaging*, 35, 1766–1774.

Brookhaven National Lab, Chart of the Nuclides, http://www.nndc.bnl.gov/chart/, accessed May 5, 2012.

Bruehlmeier, M. et al., 2004. Assessment of hypoxia and perfusion in human brain tumors using PET with 18F-fluoromisonidazole and 15O-H$_2$O. *J. Nucl. Med.*, 45, 1851–1859.

Couturier, O. et al., 2004. Fluorinated tracers for imaging cancer with positron emission tomography. *Eur. J. Nucl. Med. Mol. Imaging*, 31, 1182–1206.

Deans, S. R., 1983. *The Radon Transform and Some of Its Applications*. John Wiley & Sons, New York.

Dehdashti, F. et al., 2003. Assessing tumor hypoxia in cervical cancer by positron emission tomography with ^{60}Cu-ATSM: Relationship to therapeutic response—A preliminary report. *Int. J. Radiat. Oncol. Biol. Phys.*, 55, 1233–1238.

Eder, M. et al., 2010. ^{68}Ga-labelled recombinant antibody variants for immuno-PET imaging of solid tumours. *Eur. J. Nucl. Med. Mol. Imaging*, 37, 1397–1407.

Farde, L. et al., 1985. Substituted benzamides as ligands for visualization of dopamine receptor binding in the human brain by positron emission tomography. *Proc. Natl. Acad. Sci.*, 82, 3863–3867.

Foged, C. et al., 1997. Bromine-76 and carbon-11 labelled NNC 13-8199, metabolically stable benzodiazepine receptor agonists as radioligands for positron emission tomography (PET). *Eur. J. Nucl. Med.*, 24, 1261–1267.

Fricke, E. et al., 2005. Attenuation correction of myocardial SPECT perfusion imaging with low-dose CT: Evaluation of method by comparison with perfusion PET. *J. Nucl. Med.*, 46, 736–744.

Goldstein, D. et al., 1993. Positron emission tomographic imaging of cardiac sympathetic innervation using 6-[18F] fluorodopamine: Initial findings in humans. *J. Am. College Cardiol.*, 22(7), 1961–1971.

Gregor, J. et al. 2008. Computational analysis and improvement of SIRT. *IEEE Trans. Med. Img.*, 27(7), 918–924.

Gregor, J., Lenox, M., and Arrowwood, B., 2009. Multi-core cluster implementation of SIRT with application to cone beam micro-CT. *High Performance Medical Imaging Workshop, IEEE NSS/MIC*, October 2009.

Grenier, R. P. et al. 1974. A computerized multicrystal scintillation camera. *Inst. Nuc. Med.*, 2, 101–134.

Hamacher, K., Coenen, H. H., and Stocklin, G., 1986. Efficient stereospecific synthesis of no-carrier-added 2-[18F]-fluoro-2-deoxy-D-glucose using aminopolyether supported nucleophilic substitution. *J. Nucl. Med.*, 27, 235–238.

Hara, T. et al., 2002. Development of 18F-fluoroethylcholine for cancer imaging with PET: Synthesis, biochemistry, and prostate cancer imaging. *J. Nucl. Med.*, 43, 187–199.

Hawkins, R. A., Choi, Y., Huang, S., Messa, C., Hoh, C., and Phelps, M. E., 1992. Quantitating tumor glucose metabolism with FDG and PET. *J. Nucl. Med.*, 33(3), 339–344.

Heinle, S. et al., 2000. Assessment of myocardial perfusion by harmonic power Doppler imaging at rest and during adenosine stress: Comparison with 99m Tc-sestamibi SPECT imaging. *Circulation*, 102, 55–60.

Hostetler, E. and Burns, H. 2002. A remote-controlled high pressure reactor for radiotracer synthesis with [11C] carbon monoxide. *Nucl. Med. Biol.*, 29, 845–848.

Hudson, H. M. and Larkin, R. S., 1994. Accelerated image reconstruction using ordered subsets of projection data. *IEEE Trans. Med. Imaging*, 13(4), 601–609, 1994.

Jakoby, B. W. et al. 2011. Physical and clinical performance of the MCT time-of-flight PET/CT scanner. *Phys. Med. Biol.*, 56, 2375–2389.

Kak, A. C. and Slaney, M., 2001. *Principles of Computerized Tomographic Imaging.* Society for Industrial and Applied Mathematics.

Kawamura, M. et al., 1990. Single photon emission computed tomography (SPECT) using *N*-isopropyl-*p*-(^{123}I)iodoamphetamine (IMP) in the evaluation of patients with epileptic seizures. *Eur. J. Nucl. Med.*, 16, 285–292.

Kikuchi, K. et al., 2001. Measurement of cerebral hemodynamics with perfusion-weighted MR imaging: Comparison with pre- and post-acetazolamide 133Xe-SPECT in occlusive carotid disease. *Am. J. Neuroradiol.*, 22, 248–254.

Knoll, G., 2000. *Radiation Detection and Measurement.* 3rd Ed., John Wiley & Sons, USA.

Kumar, B. et al., 1981. Positron tomographic imaging of the liver: 68Ga iron hydroxide colloid. *Am. J. Roentgenol.*, 136, 685–690.

Kwee, S. et al., 2006. Localization of primary prostate cancer with dual-phase 18F-fluorocholine PET. *J. Nucl. Med.*, 47, 262–269.

Lange, K. and Fessler, J. A., 1995. Globally convergent algorithms for maximum *a posteriori* transmission tomography. *IEEE Trans. Image Proc.*, 4(10), 1430–1438.

Langen, A. et al., 2008. Reproducibility of tumor perfusion measurements using 15O-labeled water and PET. *J. Nucl. Med.*, 49, 1763–1768.

Larson, S. et al., 2004. Tumor localization of 16β-18F-fluoro-5α-dihydrotestosterone versus 18F-FDG in patients with progressive, metastatic prostate cancer. *J. Nucl. Med.*, 45, 366–373.

Lautamaki, R. et al., 2009. Rubidium-82 PET-CT for quantitative assessment of myocardial blood flow: Validation in canine model of coronary artery stenosis. *Eur. J. Nucl. Med. Mol. Imaging*, 36, 576–586.

Lenox, M. et al., 2012. *Real-Time Line of Response Position Confidence Measurement*, US Patent 8,148,694.

Leveque, P. et al., 2003. Quantification of human brain benzodiazepine receptors using [18F]fluoroethylflumazenil: A first report in volunteers and epileptic patients. *Eur. J. Nucl. Med. Mol. Imaging*, 30, 1630–1636.

Libutti, S. et al., 2001. A prospective study of 2-[18F]fluoro-2-deoxy-D-glucose/positron emission tomography scan, 99m Tc-labeled arcitumomab (CEA-scan), and blind second-look laparotomy for detecting colon cancer recurrence in patients with increasing carcinoembryonic antigen levels. *Ann. Surg. Oncol.*, 8(10), 779–786.

Logan, J. 2000. Graphical analysis of PET data applied to reversible and irreversible tracers. *Nuc. Med. Bio.*, 27(7), 661–670.

Michel, C., Liu, X., Sanabria, S., Lonneux, M., Sibomana, M., Bol, A., Comtat, C., Kinahan, P., Townsend, D., and Defrise, M., 1999. Weighted schemes applied to 3D-OSEM reconstruction in PET. *IEEEE NSS/MIC Conf. Record*, 3, 1152–1157.

Moscariello, R. A. et al., 2011. Coronary CT angiography: Image quality, diagnostic accuracy, and potential for radiation dose reduction using a novel iterative image reconstruction technique—comparison with traditional filtered back projection. *Eur. Radiol.*, 21, 2130–2138.

Munch, G. et al., 2000. Evaluation of sympathetic nerve terminals with [11C] epinephrine and [11C] hydroxyephedrine and positron emission tomography. *Circulation*, 101, 516–523, 1985.

Murakami, Y. et al., 2004. 18F-labelled annexin V: A PET tracer for apoptosis imaging. *Eur. J. Nucl. Med. Mol. Imaging*, 31, 469–474.

Nuutinen, J. et al., 1999. Evaluation of early response to radiotherapy in head and neck cancer measured with [^{11}C]methionine-positron emission tomography. *Radiother. Oncol.*, 52, 225–232, 1985.

Okazawa, H. et al., 1995. Clinical application of ^{62}Zn/^{62}Cu positron generator: Perfusion and plasma pool images in normal subjects. *Ann. Nucl. Med.*, 9(2), 81–87.

Olafsen, T. and Wu, A., 2010. Antibody vectors for imaging. *Semin. Nucl. Med.*, 40, 167–181.

Pankrasi, S. et al., 2007. Muscarinic acetylcholine receptor status in Alzheimer's disease assessed using (R,R) ^{123}I-QNB SPECT. *J. Neurol.*, 254, 907–913.

Panin, V. Y. et al. 2006. Fully 3-D reconstruction with system matrix derived from point source measurements. *IEEE Trans. Med. Img.*, 25(7), 907–921.

Phan, H. et al., 2008. The diagnostic value of ^{124}I-PET in patients with differentiated thyroid cancer. *Eur. J. Nucl. Med. Mol. Imaging*, 35, 958–965.

Plotkin, M. et al., 2005. Combined ^{123}I-FP-CIT and ^{123}I-IBZM SPECT for the diagnosis of parkinsonian syndromes: Study of 72 patients. *J. Neural Transm.*, 112, 677–692.

Reshef, A. et al., 2008. Molecular imaging of neurovascular cell death in experimental cerebral stroke by PET. *J. Nucl. Med.*, 49, 1520–1528.

Schmand, M. et al., 1999. HRRT, a new high resolution LSO-PET research tomography. *J. Nucl. Med.*, 40(5), 76.

Seo, Y. et al., 2010. *In vivo* tumor grading of prostate cancer using quantitative ^{111}In-capromab pendetide SPECT/CT. *J. Nucl. Med.*, 51, 31–36.

Shields, A. et al., 1998a. Carbon-11-thymidine and FDG to measure therapy response. *J. Nucl. Med.*, 39, 1757–1762.

Shields, A. et al., 1998b. Imaging proliferation *in vivo* with [F-18]FLT and positron emission tomography, *Nat. Med.*, 4(11), 1334–1336.

Souvatzoglou, M. et al., 2007. Tumour hypoxia imaging with [18F]FAZA PET in head and neck cancer patients: A pilot study. *Eur. J. Nucl. Med. Mol. Imaging*, 34, 1566–1575.

Sun, H. et al., 2007. ^{99}m Tc-HYNIC-TOC scintigraphy in evaluation of active Graves' ophthalmopathy (GO). *Endocrinology*, 31, 305–310.

Thie, J. A., 2004. Understanding the standardized uptake value, its methods, and implications for usage. *J. Nucl. Med.* 45(9), 1431–1434.

Verghese, K., Gardner, R. P., and Felder, R. M., 1972. Solid angle subtended by a circular cylinder. *Nucl. Instrum. Methods*, 101(2), 391–393.

Wang, J. et al., 2006. 18F-FP-CIT PET imaging and SPM analysis of dopamine transporters in Parkinson's disease in various Hoehn and Yahr stages. *J. Neurol.*, 254, 185–190.

Weser, H. et al., 1999. Synthesis and radiopharmacology of O-(2-[18F]fluoroethyl)-L-tyrosine for tumor imaging. *J. Nucl. Med.*, 40, 205–212.

Winkelhener, A. et al., 2011. Raw data-based iterative reconstruction in body CTA: Evaluation of radiation dose saving potential. *Eur. Radiol.*, 21, 2521–2526.

Wolpers, H. et al., 1997. Assessment of myocardial viability by use of 11C-acetate and positron emission tomography. *Circulation*, 95, 1417–1424.

Zhou, J., Lenox, M., Fossum, T., Frank, K., Simon, J., Stearns, S., Ruoff, C., and Akabani, G., 2011. Dosimetry of Y-90 liquid brachytherapy in dogs with osteosarcoma using PET/CT. *J. Nucl. Med.*, 52(Suppl 1), 129.

9

Endoscopy

Kristen C. Maitland
Texas A&M University

Thomas D. Wang
University of Michigan

9.1 Introduction

Endoscopy is a powerful medical instrument used for *in vivo* diagnosis and in the treatment of human diseases. Major developments in technology have paved the way for endoscopes to assume a key role in the practice of medicine.[1] Endoscopes have unparalleled ability to visualize lesions within internal organs with high resolution for minimally invasive medical purposes. These instruments can be inserted into natural body orifices (mouth, nose, anus, urethra) to access hollow organs, such as the oropharynx, esophagus, stomach, small intestine, pancreas and biliary ducts, colon, larynx, bronchial tree, and urinary bladder. This chapter will first present the fundamental concepts of endoscopy, including the optical, mechanical, and electronic designs required for the proper function of an endoscope.

The basic theory of endoscopic image formation, such as image detection, spatial intensity distribution, image resolution, signal to noise, and color image processing, will also be elaborated. Finally, a brief description of new emerging endoscopic technology will be introduced with reference to more detailed discussions. Numerous images are displayed to illustrate the features of each technology.

Technological innovation has significantly changed the practice of endoscopy in recent years. The first endoscopes consisted of crude, rigid tubes that provided only a limited view of easily accessible organs.[1] Recent developments have substantially upgraded the capabilities of endoscopes. Miniaturized semiconductor detectors provided a substantial improvement in image resolution along with a reduction in instrument size. Finally, ultrasound (US) imaging has been combined with endoscopy to enable visualization beyond the tissue surface. Conventional endoscopy is based on the detection of diffusely reflected white light from the tissue surface, using subtle changes in the color and shadows to reveal structural changes.[2] New advancements in optical imaging are going beyond white light and are taking full advantage of light's properties. Several technologies that are being developed for clinical use include optical coherence tomography (OCT), high-magnification endoscopy, chromoendoscopy, fluorescence imaging, and confocal endomicroscopy. Also, wireless capsule and virtual endoscopy have demonstrated the potential to collect endoscopic images with minimal discomfort to the patient. Further technological improvements promise to improve the capability of endoscopes to visualize, diagnose, and treat human diseases, and are expected to revolutionize the practice of medical endoscopy in the near future.

9.2 Basic Components of an Endoscope

In this section, an upper gastrointestinal (GI) endoscope, or esophagogastroduodenoscope (EGD), will be used to illustrate the basic components of an endoscope. An upper GI endoscope is used to visualize the esophagus, stomach, and duodenum. A schematic diagram detailing the individual components of the endoscope is shown in Figure 9.1. The distal tip contains the optics required for illuminating and collecting endoscopic images, channels for delivery of instruments, and the mechanisms for providing air, water, and suction.[2] The bending section contains a set of hinges that allow the distal tip to deflect at large angles as high as 270°. The insertion tube comprises the distal part of the endoscope and is

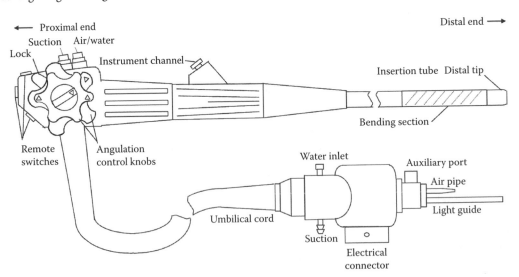

FIGURE 9.1 Basic components of a conventional medical endoscope. The insertion tube is located on the distal end, and a short bending section connects to the distal tip where the camera is located. The proximal end contains the angulation control knobs; valves for suction, air, and water; and remote switches. The instrument channel is used to deliver accessories to the distal tip.

covered with a rugged plastic. The proximal end of the endoscope contains the entry into the instrument channel, the angulation control knobs for manipulating the distal end, a lock for maintaining distal tip deflection, air/water, and suction valves, and remote switches for freezing, capturing, and storing images. An umbilical cord connects the endoscope to the video processor, and contains the light guide, electrical connectors, and conduits for air, water, and suction. The total length of the upper endoscope is about 1.5 m and can be even more longer for a colonoscope or enteroscope.

9.2.1 Distal End

The distal end of an endoscope contains the optics that form the endoscopic images. A cross-section view of a forward-viewing endoscope is shown in Figure 9.2a, and displays the location of the objective lenses, illumination lenses, air/water conduit, and detector. The corresponding end-view is shown in Figure 9.2b, and two illumination lenses are located on either side of the objective to more uniformly illuminate the image. A large instrument channel is present for delivery of instruments, removal of tissue, and suction. The objective usually contains several optical elements, such as a diverging lens, intermediate lens(es), pupil, and an achromat. The air/water channel directs either water flowing across the outer surface of the objective to clear debris or air to insufflate and expand the organ being examined. The first endoscopes used a coherent optical fiber imaging bundle to transmit the image to the proximal end of the endoscope where it could be viewed directly by the endoscopist through an eyepiece. Owing to the recent advances in semiconductor technology, the imaging bundles have now been replaced by miniaturized charge-coupled device (CCD) detectors located in the distal tip directly behind the objective lens to produce a video image.

9.2.2 Objective Lens

The objective lens of the endoscope is designed to provide a large field of view with high image resolution. Since it is very difficult to achieve both these requirements with a simple lens, endoscopes use multiple lenses to form the image. The design of these lens systems is complex and is usually done with an optical ray tracing program. The optical train of lens elements required to produce the endoscopic image is shown in Figure 9.2a. The angle between a ray of light and the normal to the objective is defined as θ. A diverging lens (negative focal length) is needed to produce a large angle θ to maximize the field of view. A pupil is located behind the objective lens to block extraneous internal reflections. An intermediate lens helps focus the image onto the detector. An achromat corrects for chromatic aberrations so that all the colors in the visible spectrum will focus onto the same plane at the detector. The focal length of the

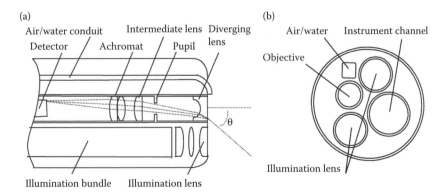

FIGURE 9.2 A detailed view of the distal tip. (a) A cross-section view shows the design of the optics, detector, and air/water conduits. (b) The end-view shows the relative location of these elements.

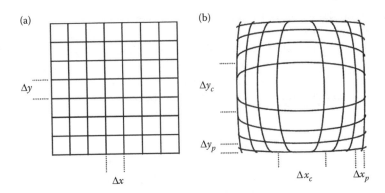

FIGURE 9.3   Effect of barrel distortion in the objective lens. (a) Image of a grid of boxes with no barrel distortion. (b) Image with barrel distortion, where the height and width Δx_c and Δy_c in the center of the image are significantly larger than that in the periphery Δx_p and Δy_p.

objective is different in the center of the image than in the periphery, and this design introduces barrel distortion into the image.[3] The effect of barrel distortion is illustrated by an image of a grid of boxes with equal dimensions. In Figure 9.3a, there is no barrel distortion present and all of the boxes throughout the grid have the same width Δx and height Δy. For an objective lens with barrel distortion, the image magnification is higher in the center of the image than in the periphery. As shown in Figure 9.3b, the boxes in the center of the grid have significantly larger height and width dimensions, Δx_c and Δy_c, than those in the periphery, Δx_p and Δy_p. Although electronic video endoscopes have largely replaced fiber optic endoscopes, the term "resel" will be used to designate the smallest resolution element in image detection to accommodate both video and older fiber optic endoscopes. A resel corresponds to either an optical fiber in an imaging bundle or a pixel (picture element) in a CCD array. The endoscopist can maximize the resolution of a region of interest by placing it in the center of the image where the density of image resels is the highest.

9.2.3 Insertion Tube

An exposed cross-section of the insertion tube is shown in Figure 9.4.[2] The outer part consists of a durable plastic covering capable of withstanding caustic bodily fluids, such as gastric acid and bile, and disinfectants for cleaning the instrument. Under the covering is a wire mesh that runs along the length

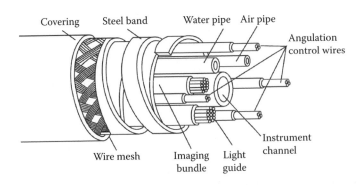

FIGURE 9.4   The contents of the insertion tube include the light guide, imaging bundle, angulation control wires, air and water pipes, and instrument channel, and are enclosed by a wire mesh and steel bands for mechanical protection.

of the tube to prevent twisting or stretching during use. Below the mesh are helically shaped steel bands that maintain the round shape and also provide mechanical protection. The contents of the insertion tube include the light guide, imaging bundle, angulation control wires, air and water pipes, and instrument channel. The tube is designed with a stiffness that varies along the length of the endoscope to facilitate its insertion into the GI tract. The distal end can be manipulated to produce large bending angles that are needed for visualization of tissue located behind the endoscope (retroflexed view) and for fine movements to traverse tortuous internal organs, such as the intestines.

9.2.4 Angulation Control

The angulation control of the distal end is performed by the left hand of the endoscopist, leaving the right hand free to hold and manipulate the insertion tube. These angulations are produced by a durable set of guide wires that deflect the distal end in four directions, termed up, down, left, and right, by convention. The angulation control knobs are connected to a sprocket that moves the guidewires connected to the distal tip. If desired, a brake can be applied to lock the position of the deflection. The second, third, and fourth fingers grip the proximal end of the instrument against the palm, leaving the left thumb to control the up/down knob and the left index finger to operate the air/water and suction valves. The right hand is used to torque and advance the insertion tube, insert accessories down the instrument channel, control the right/left knobs, and activate the remote switches.

9.2.5 Air, Water, and Suction Valves

A set of buttons and valves are located on the proximal end of the endoscope to deliver air and water or to suction intraluminal contents. When the small opening in the air/water valve is covered, air supplied by a pump within the video processor is emitted from the nozzle on the distal tip. Air is introduced into an organ, such as the colon or stomach, to expand the mucosal folds for better visualization. Air can also be removed from a distended organ for decompression. When the air/water valve is depressed, water is delivered from the pressurized water container located on the video processor and out of the nozzle on the distal tip, and over the surface of the objective lens. Water is sprayed over the outer surface of the objective to remove debris that is obstructing the view. Intraluminal contents, such as fluid or stool, can be aspirated through the endoscope to a collection bottle via an external suction pump that provides negative pressure. The umbilical cord contains the connections for the light source, water container, suction pump, and electrical cord to safely return any leakage current. A vent on the connector allows the interior of an airtight and fluid-tight endoscope to be vented before the instrument is placed in an evacuated chamber for sterilization.

9.3 Endoscopic Imaging

9.3.1 Image Illumination

A variety of light sources such as metal halide, halogen, and xenon lamps can be used to provide image illumination. These light sources can provide an incident power $P_i(\lambda_i)$ at incident wavelengths λ_i over the visible and infrared band up to 300 W or more. Because of the intense heat generated by these light sources, filters are used to block the infrared portion, allowing only visible light to be delivered at the distal tip. In addition, large heat sinks and air circulation within the light source prevent excessive heating. The light guide consists of an incoherent fiber optic bundle that has a large numerical aperture (NA) lens at the proximal end to maximize the amount of illumination light collected. There is a diverging lens at the distal end that is designed to illuminate the field of view of the objective. The optical fibers in the light guide have high NA and high transmission efficiency. Also, they are made as large as possible without compromising their flexibility and durability to produce the highest-possible packing fraction.

Recently, white light-emitting diodes (LEDs) placed at the distal tip of the endoscope have been used for illumination.[4] The use of LEDs reduces the overall system size and energy consumption.

9.3.2 Image Detection

Image detection was first performed by the eye of the endoscopist viewing the proximal end of a fiber optic imaging bundle through an eyepiece. Later, a camera is attached to the eyepiece to detect the image and display it on a video monitor. While this method is simple and effective, there have been several limitations.[5] First, if the endoscopist uses the eyepiece, no one else can view the image, and the images cannot be captured or stored. Second, a Moiré pattern is produced by the superposition of the image from the fiber optic bundle and the detector array from the camera, resulting in the appearance of an interference pattern artifact. Third, the fiber optic imaging bundle is inefficient in collecting light, requiring high-power light sources. Finally, the number of optical fibers packed into the bundle, determining the image resolution, is limited by instrument size and stiffness. Also, the individual fibers may break over time, creating gaps in the image. In modern video endoscopes, a miniaturized CCD array located directly behind the objective lens detects the image and eliminates many of these shortcomings.

9.3.3 Video Endoscopy

The CCD array is a semiconductor detector made from silicon that is sensitive to light.[6,7] The surface of a CCD is divided into a two-dimensional array of pixels that may total over a million in number. There is a thin strip of dead space in between the pixels, as shown in Figure 9.5a. The distance from the center of the pixel to the edge of the active area is denoted by r_i, and the distance to the middle of the dead zone is denoted by r_o. The packing fraction f_p for a CCD array is given by

$$f_p = \left(\frac{r_i}{r_o} \right)^2 \tag{9.1}$$

When a photon of light becomes incident on the surface of the silicon, electrons are generated and collected in potential wells. The amount of charge collected is proportional to the intensity of light detected. These charges accumulate over a period of exposure, and then are transferred out of the potential wells

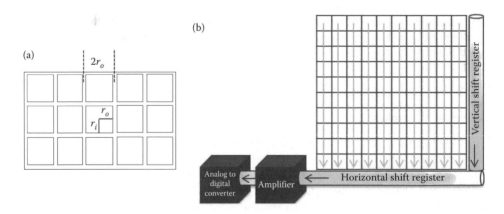

FIGURE 9.5 A CCD is an array of pixel detectors. (a) There is thin strip of dead space in between the pixels in the CCD array, and the pixel spacing determines the resolution. (b) The read-out process for the CCD array. The total charge from each pixel in a row is transferred to the pixel located in the row directly below via a vertical shift register. The charge corresponding to the pixels in the last row is transferred by a horizontal shift register to the amplifier and converted into a voltage.

through a read-out process for storage and display. The schematic of the read-out process for a CCD array of size X by Y pixels is shown in Figure 9.5b. The total charge from each pixel in a row is transferred to the pixel located in the row directly below via a vertical shift register. Hence, the charges are coupled to one another as they move across the grid. The charge corresponding to the pixels in the last row is transferred by a horizontal shift register to the amplifier and converted into a voltage. This voltage signal is then transmitted electronically through the endoscope to the video processor. CCD detectors have several important features that make them well suited for image detection in endoscopy. In addition to low power requirements, high resolution, and good color reproduction, CCDs have a sensitivity and dynamic range greater than that of photographic film by a factor of 100 and 50 times, respectively. Further, CCDs are not susceptible to mechanical vibration and shock and are not damaged by bright illumination. New developments in semiconductor technology have enabled the miniaturization of the CCD arrays. These detectors can be fabricated small enough to detect the image directly behind the objective lens, providing better light transmission efficiency and increased image resolution. Thus, endoscopes can be designed with a smaller light guide and a larger field of view. Also, the number of pixels that could be packed onto a CCD was significantly higher than that of the fibers in an imaging bundle.[8-14]

The maximum angle at which the imaging system will collect light can be as high as 130° with a wide-angle lens.[15] The range of distances d over which the image remains in focus is called the depth of field, and typically ranges from 3 mm to as high as 100 mm. The typical working distance of an endoscope is about 5–75 mm. The area of tissue imaged is called the field of view.

9.4 Endoscopic Image Formation

The objective lens of the endoscope collects light returning from the tissue and forms an image. Because of the large angles involved in the delivery and collection of light, the intensity along a cross section of an image from a uniform plane of tissue is not uniform. Instead, the intensity peaks in the center of the image and decreases significantly in the periphery. In conventional endoscopy, the returning light is diffusely reflected from the tissue surface, and the signal is intense. However, in new endoscopic techniques such as fluorescence, the returning light, generated spectroscopically, is weak. These methods require quantitative analysis, and an understanding of the light distribution throughout the image is needed. A mathematical model for endoscopic image formation will be presented.[16]

9.4.1 Spatial Distribution of Illumination Light

A set of spherical coordinates is defined at the distal end of the endoscope in Figure 9.6. The origin is located at the center of one of the illumination lenses, and has the distance variable r directed radially toward the tissue. The tissue is flat in shape, and oriented perpendicular to the endoscope at a distance d. The distance variable z defines the optic axis of the illumination lens, pointing perpendicularly through the tissue surface. The angle θ is defined between r and z. The variable ρ' represents the radial distance from the optic axis to an arbitrary region A_r in the plane of the tissue. The centers of the objective and illumination lenses are separated by a distance ρ_o. Light delivered from the illumination port has a maximum divergence angle of θ_m. The intensity of illumination light is assumed to be distributed in a Lambertian fashion, or uniformly over the surface of the illumination lens. The product of the incident power $P_i(\lambda_i)$ and the exposure time Δt determines the total energy incident onto the tissue. The incident fluence $F_i(\rho', d, \lambda_i)$, energy per unit area, valid for $0 \le \rho' \le d \tan\theta$, can be found to be as follows:

$$F_i(\rho', d, \lambda_i) = \frac{P_i(\lambda_i)\Delta t}{2\pi(1 - \cos\theta_m)d^2 \left(1 + \left(\frac{\rho'}{d}\right)^2\right)^2} \qquad (9.2)$$

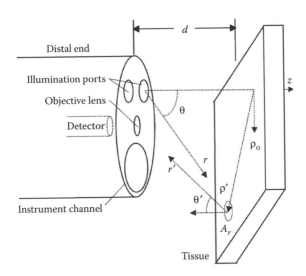

FIGURE 9.6 A set of spherical coordinates is defined at the distal end of the endoscope for the mathematical model of endoscopic image formation.

Thus, the illumination energy on the tissue surface is not uniform but it is maximum at the center of the image ($\rho' = 0$) and falls off toward the periphery ($\rho' = d \tan \theta$). The fluence produced by the other illumination port is similar.

9.4.2 Light Capture Efficiency

For either diffuse reflectance or fluorescence, the returning light is proportional to the incident light by a tissue factor ε_t. The total number of resels is designated as N_r. For a fiber optic endoscope, N_r is the total number of fibers in the imaging bundle, and for a video endoscope, N_r is the product of the number of pixels X in each row and Y in each column of the CCD array. The average area A_r viewed per resel can be approximated by the ratio of the total area imaged by the endoscope and the total number of resels N_r

$$A_r = \frac{\pi (d \tan \theta_m)^2}{N_r} \tag{9.3}$$

As with the light from the illumination lens, the distribution of the light returning from the tissue region A_r is assumed to be Lambertian. At a distance r', the radiated energy can be taken to be uniformly distributed over the surface of a sphere whose area is given by $4\pi r'^2$. A fraction of the returning light, radiating isotropically toward the endoscope, is collected by the objective lens, which has a radius r_L and an area of πr_L^2. Thus, the fraction of the returning light that is collected is found from the ratio of the surface area of the lens with that of the sphere of radius r'.

9.4.3 Spatial Distribution of Returning Light

After the returning light is collected, it passes through the lenses in the objective with a transmission efficiency of T_o and incurs losses due to the packing fraction of the detector f_p. The approximate number of photons collected per resel can be determined from the collected light energy using the energy per photon of hc/λ_r as a conversion factor. The designation h is Planck's constant, c is the speed of light, and λ_r is the wavelength of the returning light. The total number of signal photons collected per resel, $N_s(\rho', d, \lambda_i, \lambda_r)$, valid for $0 \leq \rho' \leq d \tan \theta$, can be found to be as follows:

$$N_s(\rho', d, \lambda_i, \lambda_r) = \frac{K}{d^2\left[1+\left(\dfrac{\rho'-\rho_o}{d}\right)^2\right]^2\left[1+\left(\dfrac{\rho'}{d}\right)^2\right]^2}, K = \frac{\lambda_r T_o f_p}{hc} \frac{r_L^2 \varepsilon_t \tan^2\theta_m P_i(\lambda_i)\Delta t}{8(1-\cos\theta_m)N_r} \quad (9.4)$$

As in the case of illumination, the spatial distribution of the returning light is not uniform, but is maximum at the center of the image ($\rho' = 0$) and falls off in the periphery ($\rho' = d\tan\theta$). The CCD detects the signal photons N_s, and they are converted into photoelectrons N_p with an efficiency η.

9.4.4 Experimental Validation of Model

The spatial distribution and distance dependence of an endoscopically collected image has been confirmed with fluorescence using a video colonoscope with the following objective parameters of $\theta_m = 60°$, $r_L = 1.25$ mm, and $\rho_o = 6$ mm.[16] An argon laser with excitation wavelengths $\lambda_i = 351$ and 364 nm was used to excite autofluorescence over the spectral bandwidth from 400 to 700 nm from a flat specimen of resected colonic mucosa. In Figure 9.7a, the spatial distribution of fluorescence intensity collected is compared with that of the model, as predicted from Equation 9.4, for a distance of $d = 20$ mm, which is a typical operating distance. All values were ratioed by the peak intensity for normalization. As shown in Equation 9.4, the spatial profile of the detected image is not uniform but peaks near the center and falls off with ρ'. This nonuniform spatial distribution of collected light must be taken into account in quantitative methods for evaluating endoscopic images.

Furthermore, in Figure 9.7b, the dependence of image intensity as a function of distance is shown on a log–log plot. The fluorescence intensity at $\rho' = 0$ shown for distances d ranging between 5 and 35 mm in 2.5-mm increments from experimentally collected images is compared to the model results from Equation 9.4. The intensities at each distance d were divided by the value at $d = 20$ mm for normalization. The model predicts that the number of photons, N_s, will decrease with distance as $1/d^2$ for $d \gg \rho_o$. In addition, the model shows that as d approaches 0, N_s reaches a maximum value. By taking the natural log of both sides of Equation 9.4, the results should approximate the line $\ln(N_s) = \ln(K) - 2\ln(d)$ for a $1/d^2$ dependence for large values of d. The experimental results shown in Figure 9.7b confirm the $1/d^2$ dependence for $d > 15$ mm. The corresponding distances in mm are shown in parentheses below the data points. For $d > 15$ mm, the measured values reach a maximum, as expected.

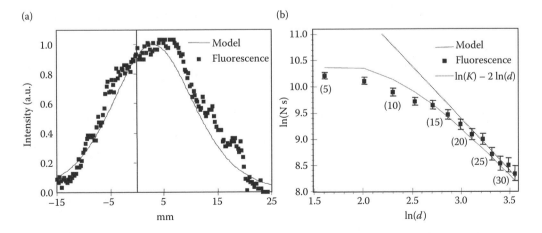

FIGURE 9.7 (a) A profile through the center of an endoscopically collected fluorescence image of colonic mucosa along at $d = 20$ mm is shown with the predicted model results. (b) The dependence of endoscopically collected fluorescence intensity over the range of distances between 5 and 35 mm is shown in a log–log plot.

9.5 Endoscopic Image Resolution

The image resolution defines the level of detail that can be clearly seen on an endoscopic image. The transverse resolution is the smallest distance at which two objects located perpendicular to the optic axis (parallel to the surface of the tissue) can be distinguished while the axial resolution defines the smallest distance at which two objects located along the optic axis (into the tissue) can be distinguished. For conventional white-light endoscopy, only the transverse resolution parameter is relevant because the light detected is reflected from the tissue surface and does not penetrate below. In other endoscopic methods where the light detected has penetrated with depth into the tissue, such as endoscopic ultrasound (EUS), OCT, and confocal endomicroscopy, the axial resolution is also relevant. As discussed earlier, because of barrel distortion in the objective lens, the image resolution varies across the field of view. Thus, endoscopes are characterized by an average rather than an absolute transverse resolution. This average transverse resolution parameter can be approximated by geometric arguments to be the ratio of the total area imaged by the endoscope, $\pi(d \tan \theta_m)$,[2] and the total number of resels, N_r. This value is the same as A_t in Equation 9.3, which is valid over the depth of focus of the objective. In Figure 9.8a, a family of curves shows the average transverse resolution of endoscopic images as a function of distance d varying from 5 to 75 mm, the typical range of working distances. The total resels N_r vary from 50,000 to 200,000, and reflect typical values found in fiber optic and video endoscopes. The total resels N_r is about 50,000 for fiber optic endoscopes and about 100,000–300,000 for conventional video endoscopes. Because more pixels can be packed into a CCD array than optical fibers in an imaging bundle, the average resolution for video endoscopes is generally better. As shown in Figure 9.8a, the average transverse resolution for both fiber optic and video endoscopes is submillimeter over the depth of focus, and improves as the distal end of the endoscope is moved closer to the tissue surface.

9.6 Endoscopic Image Signal-to-Noise Ratio

The signal-to-noise ratio (SNR) is another important imaging parameter in endoscopy. The SNR should be optimized in an endoscope so that smaller-diameter light guides can be used to illuminate large fields of view. These designs result in thinner endoscopes that are more comfortable for patient-use. The noise in an endoscopic image arises from dark current $D\Delta t$ and detector read-out noise N_{ro} in the detector, as shown in Figure 9.9. The conversion of signal photons to photoelectrons follows Poisson statistics,

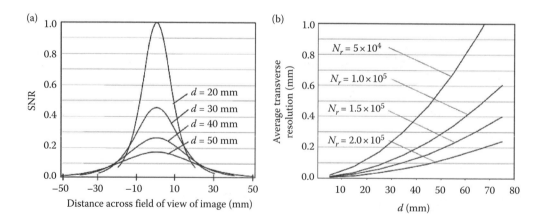

FIGURE 9.8 (a) A family of curves shows how the normalized SNR varies spatially for distances d ranging between 20 and 50 mm for a video endoscope. (b) A family of curves shows the average transverse resolution of endoscopic images as a function of distance d varying from 5 to 75 mm, the typical range of working distances, which improves as the distal end of the endoscope is moved in closer to the tissue surface.

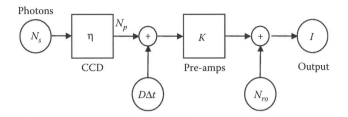

FIGURE 9.9 The noise in an endoscopic image arises from dark current $D\Delta t$ and detector read-out noise N_{ro} in the detector.

characterized by an average value $\bar{N}_p = \eta N_s$ and standard deviation $\sigma_p = \sqrt{\eta N_s}$. The expected output intensity $E(I)$ and variance σ_I^2 are given by the following:

$$E(I) = K\eta N_s \tag{9.5}$$

$$\sigma_I^2 = K^2[\eta N_s + D\Delta t] + N_{ro}^2 \tag{9.6}$$

The SNR, defined as the ratio of the expected intensity $E(I)$ and the standard deviation of the signal σ_I, is given by the following:

$$\text{SNR} = \frac{E(I)}{\sigma_I} = \frac{\eta N_s}{\sqrt{\eta N_s + D\Delta t + \left(\dfrac{N_{ro}}{K}\right)^2}} \tag{9.7}$$

An intensified CCD detector may be used for fluorescence imaging, and the number of photoelectrons ηN_s is amplified by a large gain factor to suppress the instrument noise term

$$N_s(\rho', d, \lambda_i, \lambda_r) = \frac{K}{d^2\left(1 + \left(\dfrac{\rho' - \rho_o}{d}\right)^2\right)^2 \left(1 + \left(\dfrac{\rho'}{d}\right)^2\right)^2}, \quad K = \frac{\lambda_r T_o f_p}{hc}\frac{r_L^2\,\varepsilon_t\,\tan^2\theta_m\,P_i(\lambda_i)\Delta t}{8(1 - \cos\theta_m)N_r}$$

that can be ignored in the denominator of Equation 9.7. This condition is called shot-noise limited detection, and the SNR reduces to the following:

$$\text{SNR} = \sqrt{\eta N_s} \tag{9.8}$$

A video endoscope can also be used for fluorescence imaging. In this situation, the instrument noise is usually much larger than the number of detected photoelectrons, and the photoelectron term ηN_s can be ignored in the denominator of Equation 9.7. This condition is called instrument-noise limited detection, and the SNR over the field of view can be approximated by the following:

$$\text{SNR}(\rho, d) = \frac{K\eta}{d^2\left(1 + \left(\dfrac{\rho - \rho_o}{d}\right)^2\right)^4 \sqrt{D\Delta t + \left(\dfrac{N_{ro}}{K}\right)^2}} \tag{9.9}$$

The family of curves in Figure 9.8b shows how the normalized SNR varies with ρ' for distances d ranging between 20 and 50 mm for a video endoscope.

9.7 Color-Image Processing

9.7.1 RGB Color Wheel

CCD detectors are sensitive to light over the entire visible spectrum and generally produce monochromatic (black and white) images. Thus, image-processing techniques, such as RGB sequential imaging or color filtering, are required to produce color images.[2] In the RGB sequential imaging method, a color wheel, shown in Figure 9.10a is placed in front of the light source in the video processor. The wheel contains a set of red, green, and blue filters, and rotates at approximately 30 Hz (video rate). The timing diagram shown in Figure 9.10b illustrates the read-out process. The times at which the start and end of the red filter is exposed to the light is designated as t_{rs} and t_{re}, respectively. The CCD array will collect red light over this period of exposure. The intervening space between the red and green filters on the wheel is opaque and blocks any light from exposing the CCD detector during the time between t_{re} and t_{gs}. During this period, the data in the CCD array are read out and stored in the red page of image memory. This process is repeated in an identical fashion to produce the green and blue reflectance images. The time required to produce all three images takes about 1/30 s. The red, green, and blue images located in memory are then superimposed to produce the color video image. The color images acquired with this method use the full resolution of the CCD array.

9.7.2 Color Filtering

Color images can also be produced with a colored filter placed over the CCD array rather than using a filter wheel. In this method, all of the color components are collected, at the same period of exposure; however, a loss of image resolution occurs as a trade-off. The two most common filters used are called RGB striped and mosaic. The array corresponding to an RGB striped filter, shown in Figure 9.11a, consists of three vertical filter columns, or stripes, that transmit red, green, and blue light in each column, respectively. The red filter allows reflected red light to pass and blocks green and blue. The green and blue filters work in a similar fashion. When an image is detected, the read-out process is performed as discussed before, but the columns of pixels corresponding to the red, green, and blue filters are separated into different pages of image memory. The images from the three pages in memory are then recombined for display purposes.

The array corresponding to a mosaic filter, shown in Figure 9.11b, has cyan (c), yellow (Y), and white (no filter) filters arranged in a 2 × 2 pixel pattern repeated over the array. These filters allow more of the

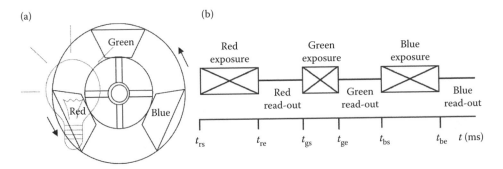

FIGURE 9.10 (a) In the RGB sequential imaging method, a color wheel is placed in front of the light source, located within the video processor. The wheel contains a set of red, green, and blue filters, and rotates at 30 cycles per second. (b) The timing diagram of the red, green, and blue exposure and read-out periods is shown.

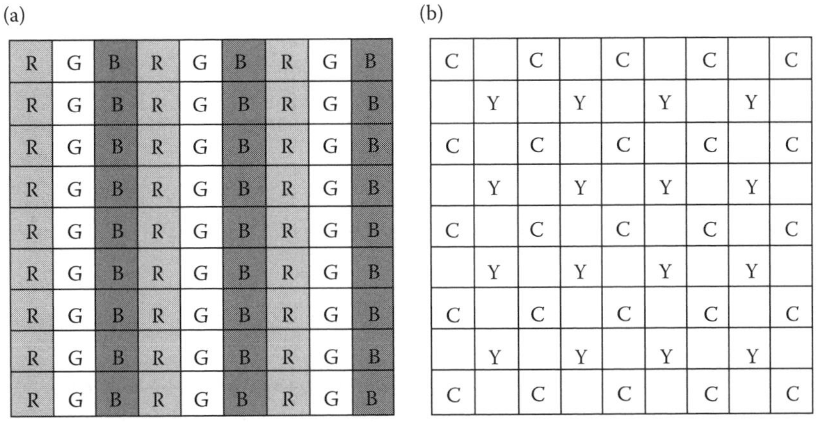

(a)

R	G	B	R	G	B	R	G	B
R	G	B	R	G	B	R	G	B
R	G	B	R	G	B	R	G	B
R	G	B	R	G	B	R	G	B
R	G	B	R	G	B	R	G	B
R	G	B	R	G	B	R	G	B
R	G	B	R	G	B	R	G	B
R	G	B	R	G	B	R	G	B
R	G	B	R	G	B	R	G	B
R	G	B	R	G	B	R	G	B

(b)

C		C		C		C		C
	Y		Y		Y		Y	
C		C		C		C		C
	Y		Y		Y		Y	
C		C		C		C		C
	Y		Y		Y		Y	
C		C		C		C		C
	Y		Y		Y		Y	
C		C		C		C		C

FIGURE 9.11 (a) A striped filter is shown with three vertical columns that transmit red, green, or blue light. (b) A mosaic filter is shown with cyan, yellow, and white (no filter) filters arranged in a 2 × 2 pixel pattern repeated over the CCD array.

light to pass than RGB striped filters, resulting in a higher SNR. The cyan filter allows blue and green light to pass, the yellow filter passes red and green light, and the white filter lets all of the light to pass. By adding or subtracting the intensity of light detected from adjacent pixels, the individual red, green, and blue components can be obtained for each group of four pixels. For example, the intensity at pixel coordinates (x, y) on the array of the blue $I_B(x, y)$ is produced from the difference between the intensity of white $I_W(x, y)$ and yellow $I_Y(x, y)$. Similarly, the intensity of red $I_R(x, y)$ is obtained from that of white $I_W(x, y)$ and cyan $I_C(x, y)$. The intensities for the red, green, and blue components of the image at pixel location (x, y) are summarized in Equation 9.10.

$$I_B(x, y) = I_W(x, y) - I_Y(x, y)$$

$$I_R(x, y) = I_W(x, y) - I_C(x, y) \tag{9.10}$$

$$I_G(x, y) = I_W(x, y) - I_B(x, y) - I_R(x, y)$$

9.8 Endoscopic Retrograde Cholangiopancreatography

Tissue structures such as the pancreatic and biliary ducts are too small to directly insert an endoscope. Endoscopes can still be used to image these structures radiographically through fluoroscopy with a procedure called endoscopic retrograde cholangiopancreatography (ERCP). In ERCP, a cannula is inserted into the papilla, the site where the pancreatic and common bile ducts join, and contrast is injected. This contrast is viewed with delivery of x-rays onto a fluoroscopy monitor. Because the papilla, located within the Ampulla of Vater, is on the side wall of the duodenum, there is not enough space for a forward-viewing endoscope to maneuver and perform the cannulation. Instead, a side-viewing endoscope, or duodenoscope, is used. The optics for a side-viewing endoscope are designed for imaging tissue located along the parallel to the axis of the endoscope. In Figure 9.12a, a cross-section view shows an illumination bundle that is curved to deliver light to the side of the endoscope. Also, the objective collects returning light coming from the side using a prism to reflect the image to the detector. Furthermore, the side-viewing endoscope has a forceps raiser located in the instrument channel that elevates the catheter during cannulation, as shown in the side-view in Figure 9.12b. Only one illumination lens is located on the distal tip. Side-viewing duodenoscopes are about 120 cm in working length, and have a variety

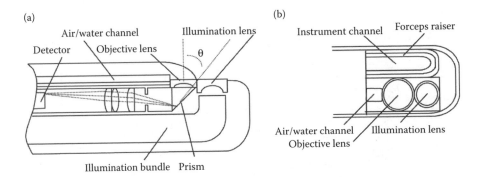

FIGURE 9.12 (a) A cross-section view of a side-viewing endoscope shows the optics packaged to view images oriented parallel to the axis endoscope rather than perpendicular. (b) A side view shows one illumination lens and a forceps raiser to help guide cannulation.

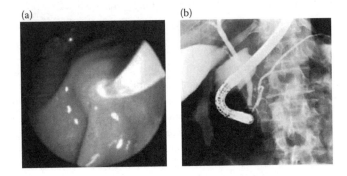

FIGURE 9.13 (a) A side-viewing duodenoscope is used to insert the cannula into the papilla. (b) A radiograph shows the position of the duodenoscope with contrast being injected into the papilla. The contrast is filling the common bile duct shown on the left and the pancreatic duct shown on the right.

of instrument channel sizes, ranging from 2.8 to 4.0 mm. The larger-diameter channels are useful for special instruments used in therapeutic biliary endoscopy, such as for endoscopic sphincterotomy, stone extraction, and stent placement. In Figure 9.13a, an image from a side-viewing duodeoscope is shown with a cannula inserted into the papilla. In Figure 9.13b, a radiograph shows the position of the duodenoscope with contrast filling the common bile duct (left) and the pancreatic duct (right).

9.9 Endoscopic Ultrasound

EUS combines endoscopy with US, using a miniaturized US transducer attached to the distal tip. EUS collects ultrasound images with the transducer placed in close proximity to the structure being imaged, such as the esophagus, unlike transcorporal ultrasound, where the sound travels through many layers of tissue.[17,18] The resolution obtained with EUS is significantly better because obstructing layers of bone and gas can be avoided and higher acoustic frequencies can be used. Further, contrast agents are increasingly being used in EUS to enhance vasculature in an organ of interest and to differentiate benign from malignant tissue.[19] Since ultrasound imaging is described in another chapter, this section will discuss only technology specific to EUS. The clinical applications of these technological advances include determining the grade and stage of invading tumors and guiding fine needle aspiration (FNA) for tumor cytology. There are two modes of scanning—radial sector and linear array—and they are often used clinically in a complementary fashion.[20–25]

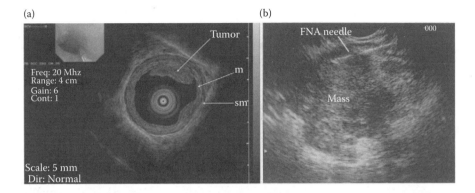

FIGURE 9.14   (a) Endoscopic ultrasonography with 20-MHz miniprobe shows thickening of the mucosa (m) and submucosa (sm) and a submucosal squamous carcinoma (Tumor) of the esophagus. (Reprinted from *Ann Thorac Surg*, **85**(1), Rampado, S. et al., Endoscopic ultrasound: Accuracy in staging superficial carcinomas of the esophagus, 251–256, Copyright 2008, with permission from Elsevier.) (b) EUS-guided FNA of a pancreatic adenocarcinoma of the body of the pancreas using a linear array echoendoscope at 7.5 MHz. (Reprinted from Macmillan Publishers Ltd., *Am J Gastroenterol*, Erickson, R.A. and A.A. Garza, Impact of endoscopic ultrasound on the management and outcome of pancreatic carcinoma, **95**(9): 2248–2254, 2000.)

9.9.1 Radial Sector Scanning

Radial sector scanning provides ultrasound images in the plane oriented perpendicular to the axis of the endoscope. A radial sector echoendoscope uses a rotating acoustic mirror that can scan images in sectors of 360° or 180°. They are available with transducer frequencies ranging from 7.5 to 30 MHz.[26] Higher frequency yields higher resolution at the expense of penetration depth. Depending on selection of transducer frequency, axial resolution down to 100 μm or penetration as deep as 100 mm can be achieved. An example of a 360° radial sector scan of the esophagus obtained with a 20-MHz endoscopic ultrasonography miniprobe is shown in Figure 9.14a. The layers of the wall, mucosa (m), and submucosa (sm) can be seen clearly.[27] A tumor is visible with signs of hypoechogenic disruption of the third layer (stage T1sm). On pathological examination, the tumor was confirmed as T1sm squamous carcinoma of the esophagus.

9.9.2 Linear Array Scanning

Linear array scanning provides ultrasound images in the plane oriented parallel to the axis, and is useful for FNA. A linear array scanning echoendoscope uses an array of transducers that provides a wedge-shaped image in sectors of 100° to 180°. They are available with 5- and 7.5-MHz transducers that have variable focal distances. The linear array can also provide color-flow mapping and Doppler imaging to locate blood flow in vessels for identifying landmarks and for assessing vascular invasion by tumors. In addition, this scanning mode provides continuous real-time visualization of a needle during EUS-guided FNA for cytology. The oblique-viewing echoendoscope has a curved linear array transducer located on the distal tip with a fine needle for aspiration protruding from the instrument channel. Figure 9.14b shows an endosonographic view of an EUS-guided FNA of a pancreatic tumor with the needle identified.[28]

9.10 Optical Coherence Tomography

OCT performs cross-sectional imaging in a manner similar to that of EUS, except the phase property of light rather than the backscattering behavior of sound is used.[29] Moreover, OCT has up to 10–100 times better axial resolution than EUS at a penetration depth of up to 1 mm. Depending on the light source, axial resolution can range between 1 and 15 μm.[30–32] OCT is a technology that is being developed to

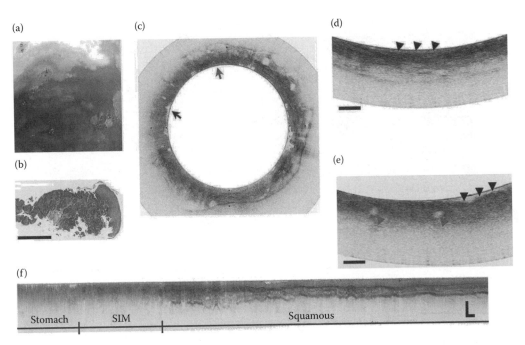

FIGURE 9.15 **(See color insert.)** Barrett's esophagus with dysplasia (a) as viewed with videoendoscopy showing patchy mucosa consistent with SIM. (b) Histology image of biopsy taken from squamocolumnar junction diagnosed with intestinal metaplasia and low-grade dysplasia. (c) Cross-sectional OFDI image showing SIM without dysplasia [arrow, top; expanded in (d)] and with high-grade dysplasia [arrow, left; expanded in (e)]. (f) Longitudinal slice highlights the transition from gastric cardia, through a 9-mm segment of SIM and finally into squamous mucosa. Scale bars: 1 mm. (Reprinted from *Gastrointest Endosc*, **68**(4), Suter, M.J. et al., Comprehensive microscopy of the esophagus in human patients with optical frequency domain imaging, 745–753, Copyright 2008, with permission from Elsevier.)

assess tissue histopathology otherwise known as optical biopsy.[33–35] OCT has demonstrated the ability for distinguishing the layers of mucosa in the GI tract with the potential to detect precancerous lesions such as dysplasia.[36–42] OCT instrumentation is described in depth in another chapter in this book; therefore, this section summarizes the endoscopic application of OCT.

OCT imaging can resolve layers of normal esophagus, which correspond to epithelium, lamina propria, muscularis mucosa, submucosa, and muscularis propria. OCT diagnostic criteria have been established for gastric cardia, specialized intestinal metaplasia (SIM), high-grade dysplasia, and intramucosal carcinoma.[36,39,41] Features include changes in the layer architecture, reflectivity, image penetration, and regularity of the layer structure. Optical frequency domain imaging (OFDI), or swept-source OCT, has enabled high-speed OCT imaging of the entire distal esophagus.[42] The OFDI instrument employs a balloon catheter to stabilize the distal optics during helical scanning. Volumetric imaging from 6 cm (z) lengths of distal esophagus were obtained in 2 min with resolutions of 20 μm (φ) × 8 μm (r) × 50 μm (z). Figure 9.15 illustrates radial and longitudinal images acquired with OFDI in a patient with Barrett's esophagus diagnosed with dysplasia at the squamocolumnar junction. Furthermore, OCT imaging has demonstrated potential for other diagnostic purposes, such as the identification of premalignant tissue in the lung,[43] ovary,[44] cervix,[45] measurement of blood flow via Doppler,[46] and the characterization of atherosclerotic plaque in coronary arteries.[47]

9.11 High-Magnification and High-Resolution Endoscopy

High-magnification endoscopy is a new technology that uses optical zoom to visualize the details of the mucosal surface and to identify malignant from benign tissue.[48,49] While most raised mucosal lesions, such as polyps, can be identified easily with a conventional endoscope, there is increased evidence for the existence of flat lesions that have malignant potential.[50,51] These lesions cannot be seen easily on white-light endoscopy. High-magnification endoscopes have an optical zoom consisting of a movable motor-driven lens in the optical train to attain image magnifications as high as ×170, whereas most standard instruments can magnify up to ×35. In the high-magnification mode, the depth of field is usually significantly reduced, and the endoscope must be placed very close to the mucosal surface. Translucent caps at the tip may be used to stabilize the instrument at the optimal focal distance. Also, the field of view decreases significantly, and the images are susceptible to motion artifacts.

In contrast to high-magnification endoscopy that uses optical zoom, high-resolution or high-definition (HD) endoscopy uses higher pixel count CCD detectors to increase resolution. Conventional video endoscopes have CCDs with 100,000 to 300,000 total resels. The transverse resolution varies with distance as shown in Figure 9.9a. High-resolution endoscopes produce images with resolutions that range from 850,000 to greater than 1 million pixels. A greater number of pixels increases resolution and enables superior electronic or digital zooming; however, image quality is lost at some point without the addition of optical zoom. High-magnification high-resolution endoscopes can provide transverse resolution as low as 10 μm. In comparison, the naked human eye has a transverse resolution of approximately 125 μm. High-magnification endoscopes are often used in combination with chromoendoscopy.

An example of high-magnification endoscopy is illustrated with images taken of neoplastic and non-neoplastic colorectal polyps.[52] Figures 9.16a and 9.16d are conventional white-light nonmagnifying endoscopic images. Figures 9.16b and 9.16e are magnifying narrow-band imaging (NBI) endoscopic images

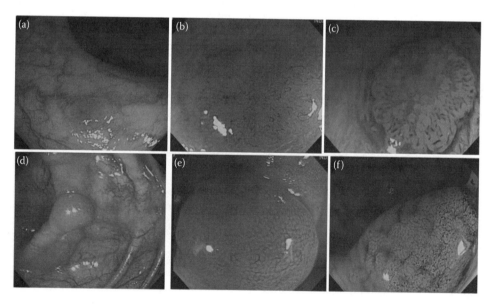

FIGURE 9.16 (See color insert.) (a) Conventional white-light nonmagnifying video colonoscopic view of a colorectal IIa neoplastic polyp. (b) Magnifying (×100 magnification) NBI view of neoplastic polyp with remarkable superficial mesh-capillary pattern. (c) Magnifying chromoendoscopy of a IIa neoplastic polyp with indigo carmine showing IIIs and IIIL pit patterns. (d) Conventional view of a Is nonneoplastic polyp. (e) Magnifying NBI view of a Is nonneoplastic polyp without a superficial mesh-capillary pattern. (f) Magnifying chromoendoscopy of a Is nonneoplastic polyp with a type II pit pattern. (Reprinted with kind permission from Springer Science+Business Media: *Int J Colorectal Dis*, Comparative study of conventional colonoscopy, magnifying chromoendoscopy, and magnifying narrow-band imaging systems in the differential diagnosis of small colonic polyps between trainee and experienced endoscopist, **24**(12), 2009, 1413–1419, Chang, C.C. et al.)

and Figures 9.16c and 9.16f are magnifying chromoendoscopy images obtained at ×100 magnification. Magnification and the methods for contrast enable visualization of the capillary-mesh and pit patterns.

9.12 Chromoendoscopy and Narrow-Band Imaging

Chromoendoscopy is an endoscopic technique that involves the application of tissue stains to highlight regions of interest in the mucosa, and is often used together with high-magnification endoscopy.[53–55] The use of these methods is becoming more widespread for detection of flat lesions with malignant potential that are not apparent on conventional endoscopy. There are two common types of stains: contrast and vital stains. Contrast stains, such as indigo carmine and cresyl violet, collect within the surface defects of the lesion and enhance the contrast in color.[56,57] These dyes are useful for identifying flat adenomas and carcinoma *in situ* in the colon and stomach. They are not absorbed by the mucosa, thus produce no systemic effects to the patient. Vital stains such as Lugol's iodine, methylene blue, and toluidine blue bind to the tissue and produce a chemical change. Lugol's solution contains iodine and potassium iodide and reacts with glycogen to produces a brown-black color.[58,59] This dye is used to enhance lesions such as squamous cell carcinoma and high-grade dysplasia of the esophagus, which are poor in glycogen. Methylene blue is absorbed by the cytosol of the epithelium of the esophagus, and the small and large intestine.[60–62] This dye is particularly useful for staining the intestinal metaplasia associated with Barrett's esophagus, which turns a bright blue color and appears granular. Toluidine blue binds to the nucleus of cells, and highlights mucosa with increased DNA synthetic activity. This dye is also used for enhancing regions of Barrett's esophagus to screen for high-grade dysplasia. An example of chromoendoscopy with application of indigo carmine to evaluate polyps in the colon is shown in Figures 9.16c and 9.16f.[52] After indigo carmine is sprayed onto the lesion, the borders of the lesion are clearly defined and visualization of the pit pattern is significantly enhanced.

NBI, also called digital or virtual chromoendoscopy, uses optical filters to control the illumination or separation of the red–green–blue channels of the CCD in detection to simulate the contrast effects of chromoendoscopy.[63,64] The contrast enhances mucosal morphology and microvasculature architecture. The source of contrast is the differential attenuation of the different wavelength bands. Shorter wavelengths (blue) are biased toward the superficial tissue, whereas longer wavelengths (red) penetrate more deeply. Additionally, enhanced absorption of blue light by hemoglobin highlights vessel architecture, as is demonstrated in Figures 9.16b and 9.16e.[52] Image-processing algorithms can further enhance visualization of mucosal architecture.

9.13 Fluorescence Imaging

The endoscopic techniques discussed previously provide images of structural changes in the tissue. New endoscopic methods collecting fluorescence can determine the biochemical and molecular changes in the tissue. Fluorescence can originate from endogenous molecules, such as aromatic amino acids, NADH, FAD, and porphyrins, termed autofluorescence.[65] Also, fluorescence can be produced by exogenously administered drugs, such as hematoporphyrin derivative and 5-aminolevulinic acid (5-ALA), called photodynamic diagnosis. Both these fluorescence methods are used to identify cancer at an early stage.

9.13.1 Autofluorescence Imaging

Early autofluorescence endoscopy through fiber optic endoscopes encountered challenges in signal collection and sensitivity. Disappointing results in studies using these instruments to detect neoplasia tempered the initial enthusiasm for autofluorescence endoscopy.[66,67] With the advantages of high-resolution video endoscopy, including increased sensitivity and resolution over fiber optic endoscopes, and integration with other contrast enhancement techniques such as NBI or chromoendoscopy, autofluorescence imaging (AFI) has reemerged for potential high-sensitivity neoplasia detection.[68] The AFI system

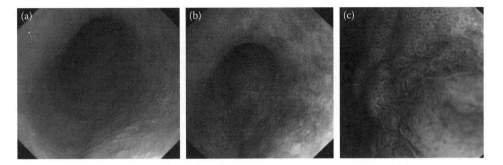

FIGURE 9.17 (**See color insert.**) Images of Barrett's esophagus obtained using the endoscopic trimodality imaging system from a patient with multifocal high-grade dysplasia. (a) Only subtle irregularities are seen under conventional white-light endoscopy. (b) Multiple dark purple spots are visible on a green background with autofluorescence imaging. (c) Detailed imaging of a suspicious area using NBI reveals irregular mucosal and vascular patterns. All autofluorescence positive areas with irregular patterns on NBI contained high-grade dysplasia. (Reprinted from *Best Pract Res Cl Ga*, **22**(4), Curvers, W.L., R. Kiesslich, and J.J.G.H.M. Bergman, Novel imaging modalities in the detection of oesophageal neoplasia, 687–720, Copyright 2008, with permission from Elsevier.)

uses two CCDs in a video endoscope—one for conventional white-light endoscopy and a second for autofluorescence. The autofluorescence image is a pseudocolored image incorporating autofluorescence from blue excitation light and either green only or green and red reflectance. AFI alone has been found to have high sensitivity to detect early neoplasia in Barrett's esophagus; however, with a high false-positive rate.[69] This may be attributed to changes in autofluorescence signal with inflammation that appear similar to neoplasia.[68] Therefore, AFI is termed a "red flag" technology for detection of suspicious lesions. The combination of AFI with NBI has been explored to reduce the false-positives with promising results.[70] Figure 9.17 demonstrates the contrast enhancement of AFI in Barrett's esophagus and the subsequent evaluation with NBI to identify high-grade dysplasia.[68] In addition to premalignant tissue in other areas of the GI tract,[71,72] AFI has also demonstrated the potential to detect lesions in the bladder,[73] head and neck,[74,75] and lung.[76,77]

9.13.2 Photodynamic Imaging

Photodynamic imaging collects fluorescence generated from externally administered agents, or prodrugs, which are internally converted and selectively taken up by cells that have higher metabolic activity.[78,79] A promising agent is 5-ALA, a porphyrin substrate in the heme biosynthetic pathway that induces the formation of protoporphyrin IX (PPIX), a compound with high fluorescence efficiency. 5-ALA has demonstrated potential to detect premalignant tissue in the bladder,[80] colon,[81] esophagus,[82] and oropharynx.[83] Excess exogenous application of 5-ALA leads to a three- to sixfold higher intracellular accumulation of PPIX in areas of high-grade dysplasia and malignant lesions compared with the surrounding mucosa. PPIX fluorescence is excited by blue light (peak of 405 nm), and the emission is red at wavelengths above 600 nm. ALA can be administered orally for systemic distribution, and the drug is cleared in about 48 h, and can produce mild skin photosensitivity and transient elevation of liver enzymes. ALA can also be administered intravesically for detection of transitional cell carcinoma (TCC) of the urinary bladder.[84] TCC is a flat urothelial lesion that has high malignant potential. These lesions are often missed on conventional endoscopy.

9.14 Confocal Endomicroscopy

Confocal endomicroscopy is a new technology that enhances magnification down to the cellular and subcellular level enabling *in vivo* microscopy during real-time endoscopy.[85–87] This instrument is generally passed through the working channel of a conventional endoscope to provide an optical biopsy of the mucosal layer. Confocal images with resolution on the order of histology can be acquired *in vivo* increasing diagnostic yield of excisional biopsies. Confocal microscopy is a laser scanning technique that uses a confocal pinhole to collect light from the focus in the sample while significantly rejecting out of focus light, resulting in a high spatial resolution in three dimensions. By scanning in two dimensions, an optically sectioned image is generated for a single focal plane. Scanning can be performed at either the proximal end of a fiber bundle[88] or the distal tip using a piezoelectric fiber scanner,[89] microelectromechanical systems (MEMS) scanning device,[90] or a technique called spectral encoding.[91] Confocal endomicroscopes can have either a fixed-depth focal plane or an axial scanning mechanism to image at varying depths. Advances in miniaturized high NA objective lenses and MEMS devices have reduced the outer diameter of the endomicroscopes.[90,92] Fluorescent agents, such as intravenous fluorescein sodium (approved by the United States Food and Drug Administration for *in vivo* use for optical imaging), cresyl violet, or topical acriflavine, can be used to provide contrast in fluorescence confocal endomicroscopy.

Confocal endomicroscopic images are acquired as optical sections parallel to the surface of the tissue down to approximately 250 μm below the surface, and typically with a field of view up to 500×500 μm, lateral resolution of 0.5–3 μm, and axial resolution of 3–10 μm.[86] Figure 9.18 shows representative *in vivo* fluorescence confocal images of colon tissue stained with fluorescein and corresponding histology images sectioned with the same orientation.[93] Confocal images in normal colon shown in Figures 9.18a and 9.18b illustrate the regular arrangement of crypts as seen in histology in Figure 9.18e and with higher magnification in Figure 9.18f. The star-shaped luminal openings of the crypts in hyperplastic tissue are evident in the confocal image in Figure 9.18c and histology in Figure 9.18g. Colitis-associated dysplasia shows distorted crypt and tissue structure in the confocal image in Figure 9.18d and histology in Figure 9.18h.

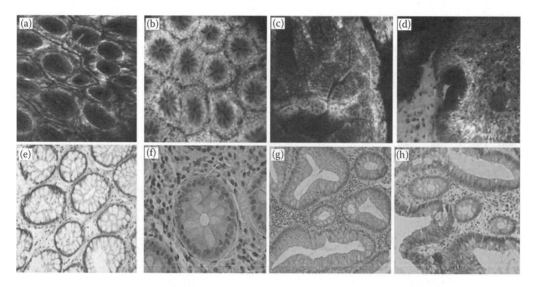

FIGURE 9.18 **(See color insert.)** *In vivo* confocal endomicroscopy images of colon tissue stained with fluorescein. Normal crypt structure imaged (a) within the lamina propria readily identifies single capillaries labeled with fluorescein (red arrow) and (b) closer to the surface with visible mucin in goblet cells presenting as dark subcellular parts (blue arrow). (e) and (f) Corresponding histology. (c) Elongated crypts with corresponding star-shaped luminal openings (red arrow) were confirmed by histology (g) as hyperplastic tissue (red arrow). (d) Ulcerative colitis-associated neoplasia shows distorted crypt architecture with inflammation and dysplastic changes. (e) Histology revealed colitis-associated dysplasia with low-grade intraepithelial neoplasia. (Reprinted by permission from Macmillan Publishers Ltd., *Nat Clin Pract Oncol*, Kiesslich, R. et al., Technology insight: Confocal laser endoscopy for *in vivo* diagnosis of colorectal cancer, **4**(8): 480–490, 2007.)

9.15 Wireless Capsule

In addition to video endoscopes, the development of miniaturized detector technology has allowed for images to be collected with a wireless capsule.[94–97] All of the optics and electronics are contained within a small capsule that can be swallowed painlessly and passed throughout the GI tract by peristalsis. There are no cables, wires, or optical fibers. This method of endoscopy is particularly suited for imaging regions of the GI tract, which cannot be easily reached by conventional endoscopes, such as the small bowel. The illumination light is provided by battery powered white LEDs. The images are collected by a low noise complementary metal oxide silicon (CMOS) detector that produce images with quality comparable to a CCD detector but with consumption of significantly less power. The images are transmitted from within the body via ultra high frequency (UHF) band radio telemetry to aerials worn around the waist and stored on a portable solid state recorder. The integration of the detector, transmitter, and LEDs was performed by application-specific integrated circuit (ASIC) design. Future developments may include locomotion in autonomous endoscopes.[98]

A schematic diagram of the capsule is shown in Figure 9.19a.[99] The capsule has dimensions of 11×26 mm. There is an optical dome that covers the four white LEDs, short focal length objective lens, and CMOS detector. Two silver oxide batteries provide all the required power for image illumination, detection, and transmission up to 6–8 h. The images are delivered via a transmitter and antenna at a rate of two frames per second. The capsule passes through the human GI tract in an average time of 80 (range of 17–280) min. Example endoscopic images collected with the wireless capsule are shown in Figure 9.19b of the mosaic pattern of the mucosa and in Figure 9.19c of ulceration in the distal jejunum in patients with celiac disease.[100] The anatomic location of the capsule within the small bowel is determined by calculating the time of travel within the small bowel relative to the total small bowel transit time and the strength of the transmitted signal. The wireless capsule has the potential for diagnostic use in occult GI bleeding.

9.16 Virtual Endoscopy

All of the endoscopic methods discussed so far require an imaging instrument to be inserted into the patient. This process can cause significant patient discomfort and distress. Virtual endoscopy has the

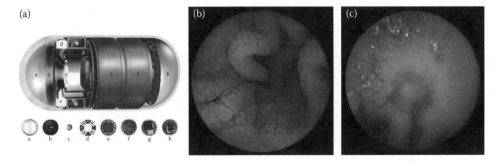

FIGURE 9.19 **(See color insert.)** Wireless capsule endoscopy device and images. (a) The M2A capsule camera consisting of a disposable plastic capsule. The contents include an optical dome [a], a lens holder [b], a short focal-length lens [c], six white light-emitting diode illumination sources [d], a CMOS chip camera [e], two silver oxide batteries [f], a UHF band radio telemetry transmitter [g], and an antenna [h]. (Reprinted by permission from Macmillan Publishers Ltd., *Nat Rev Drug Discov*, Qureshi, W.A., Current and future applications of the capsule camera, **3**(5): 447–450, 2004.) Wireless capsule endoscopy is used in complicated celiac disease to identify mucosal abnormalities such as (b) mosaic pattern of the mucosa or (c) ulceration in the distal jejunum. (Reprinted from *Gastrointest Endosc*, **62**(1), Culliford, A. et al., The value of wireless capsule endoscopy in patients with complicated celiac disease, 55–61, Copyright 2005, with permission from Elsevier.)

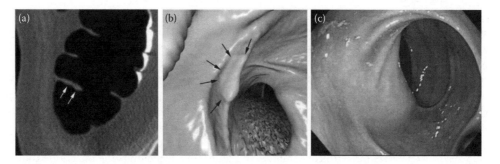

FIGURE 9.20 A 1.2-cm flat tubular adenoma with low-grade dysplasia (black and white arrows) detected on screening CT colonography. (a) Sagittal two-dimensional multiplanar reformatted image. (b) Three-dimensional endoluminal view. (c) Subsequent colonoscopy before polypectomy in a patient with multiple prior incomplete endoscopies. (Reprinted from *J Am Coll Radiol*, **6**(11), McFarland, E.G. et al., ACR Colon Cancer Committee white paper: Status of CT colonography 2009, 756–772 e4, Copyright 2009, with permission from Elsevier.)

potential to provide high-quality endoscopic images in cases where conventional endoscopy is incomplete or inadequate, such as in the presence of an obstructing lesion or patient noncompliance. Virtual endoscopy is a technology where images are acquired with a noninvasive imaging modality, such as computed tomography (CT), magnetic resonance imaging, or ultrasound, and are computer reconstructed to produce volumetric data.[101–107] This three-dimensional image is then projected with an endoluminal perspective to simulate an endoscopic image. Virtual endoscopy is most advanced with CT because of the development of the helical CT scanners that allow scans to be completed in only a few seconds. The fast image acquisition times minimize artifacts from patient respiration and motion. These scans are then postprocessed to produce a set of complementary two- and three-dimensional images.[108]

The three-dimensional reconstructions can be performed with a mathematical process called surface or volume rendering. Surface rendering involves a preprocessing step that identifies iso-intense surfaces from an endoluminal perspective and reduces the data to a set of surface triangles. Data deep to the identified surfaces are discarded to reduce the data set and to increase the processing speed. Volume rendering generates images by assigning degrees of opacity to various structures within the image based on the attenuation coefficient of each voxel (volume element). The sense of depth and distance is achieved with perspective volume rendering where an object grows larger as the observer approaches. An example of images produced for virtual colonoscopy is shown in Figure 9.20.[109] In Figure 9.20a, a 1.2-cm flat tubular adenoma with low-grade dysplasia is identified on a two-dimensional sagittal CT scan of the colon. With volume rendering, the endoluminal three-dimensional view is shown in Figure 9.20b. The virtual images of the polyp were confirmed on conventional colonoscopy performed the same day, shown in Figure 9.20c.

9.17 Summary and Conclusions

Endoscopy is a field that has undergone tremendous technological development in recent years, and new innovations promise to radically improve the capabilities of endoscopy in the near future. These technological improvements are very diverse and interdisciplinary. Advances in semiconductor technology have allowed detectors to become miniaturized, and have led to the development of video endoscopes and wireless capsules. These endoscopes can view larger areas with higher resolution. At the same time, the diameters of the insertion tubes are becoming smaller in size, which provides more patient comfort. The addition of ultrasound technology to endoscopy has allowed structural information to be collected beyond the tissue surface to depths of several millimeters and even into adjacent tissue structures. New methods are being developed to elicit even more diagnostic information from tissue. OCT and confocal endomicroscopy provide morphologic information with subcellular resolution for real-time histopa-thology. Fluorescence spectroscopy, using endogenous fluorophores and photosensitizers, reveals bio-chemical and molecular information below the tissue surface. Magnification and chromoendoscopy provide high-resolution details about the mucosal surface. The wireless capsule provides endoscopic images from previously uncharted territory such as the small bowel. Finally, virtual endoscopy provides endoluminal views without the endoscope. Thus, endoscopy is a highly technological field that promises exciting new developments in the near future.

Symbols Used in the Endoscopic Imaging Model

Symbol	Description
A_r	Area of tissue imaged per resel
c	Speed of light
d	Distance between endoscope and tissue
ε_t	Conversion efficiency of incident to return light
F_i	Incident illumination fluence
f_p	Packing fraction of detector
h	Planck's constant
Θ	Angle from normal to endoscope
Θ'	Angle from normal to tissue
θ_m	Maximum collection angle
λ_i	Wavelength of incident light
λ_r	Wavelength of returning light
η	Quantum efficiency of detector
N_r	Number of resolution elements
N_s	Number of signal photons
N_p	Number of photoelectrons
P_i	Incident illumination power
r	Radial distance between endoscope and tissue
r'	Radial distance between tissue and endoscope
r_L	Radius of objective lens
ρ_o	Distance between illumination and objective lenses
ρ'	Radial distance along tissue surface
r_i	Inner radius of resel
r_o	Outer radius of resel
T_o	Transmission efficiency of optical train
Δt	Exposure time of detector
z	Distance variable from endoscope to tissue

Glossary of Terms

Term	Definition
Aberration	Imperfection in objective lens causing loss of focus
Achromat	Lens element that corrects focus for different wavelengths
5-ALA	5-Aminolevulinic acid, agent used for photodynamic diagnosis
Angulation control	Mechanism for steering distal tip of endoscope
ASIC	Application-specific integrated circuit
Axial resolution	Ability to resolve detail with depth below tissue surface
Barrel distortion	Lens aberration that increases magnification in the center of the image
CCD	Charge-coupled device semiconductor detector

CMOS	Complementary metal oxide semiconductor
Dark current	Charge generated in detector in the absence of light
Depth of field	Range of distance over which the image remains in focus
Distal	End of endoscope furthest from endoscopist
EGD	Esophagogastroduodenoscopy, esophagus, stomach, and duodenum
ERCP	Endoscopic retrograde cholangiopancreatography
EUS	Endoscopic ultrasound
Field of view	Diameter of region imaged by endoscope
Fluorescence	Spectroscopic light generated in visible spectrum
FNA	Fine needle aspiration for tissue histology
GI	Gastroenterology, study of stomach, liver, and intestinal disease
LED	Light-emitting diode for image illumination
Moiré	Wavy effect produced by the convergence of lines or patterns
Monochromatic	One color, or black and white
Mosaic	Repeated pattern of filters for producing color image
NA	Numerical aperture equal to sine of angle θ in free air
NBI	Narrow-band imaging
Objective	Group of lenses in endoscope that forms the image
OCT	Optical coherence tomography
OFDI	Optical frequency domain imaging
Penetration depth	Distance light travels into tissue prior to attenuation
Pixel	Picture element
Photon	Individual package of light
Photoelectron	Charge produced by detected photon
Prism	Optical element for diverting image in side-viewing endoscope
Proximal	End of endoscope closest to endoscopist
Pupil	Aperture that blocks stray light
Read-out noise	Electronic noise involved in acquisition of charge from detector
Reflectance	Light diffusely scattered from tissue
Resel	Resolution element
RGB	Primary colors red, green, and blue
Shift register	Memory for reading out a row of data in CCD array
SNR	Signal-to-noise ratio
Transverse resolution	Ability to resolve detail across tissue surface
US	Ultrasound

References

1. Hopkins, H.H., The development of the modern endoscope. *Natnews*, 1980. **17**(8): 18–22.
2. Sivak, M.V., *Gastroenterologic Endoscopy*. 2000, Philadelphia, PA: W. B. Saunders Company.
3. Hecht, E., *Optics*. 4th ed. 2002, San Francisco, CA: Addison-Wesley.
4. Nishikawa, J., H. Yanai, T. Okamoto, S. Higaki, S. Hashimoto, S. Kurai, and I. Sakaida, A novel colonoscope with high color-rendering white light-emitting diodes. *Gastrointest Endosc*, 2011. **73**(3): 598–602.
5. Reedy, R.P., Coherent fiber optics test techniques. *Opt Eng*, 1980. **19**(4): 556–560.
6. Boyle, W. and G. Smith, *Charge Coupled Semiconductor Devices*. Vol. J. 49. 1970: Bell Syst. Tech.
7. Beyond, J.D.E.a.L., D.R., *Charge-Coupled Devices and Their Applications*. 1980, New York: McGraw-Hill.
8. Sivak, M.V. and D.E. Fleischer, Colonoscopy with a VideoEndoscope: Preliminary experience. *Gastrointest Endosc*, 1984. **30**(6706081): 1–5.
9. Schapiro, M., Electronic video endoscopy. A comprehensive review of the newest technology and techniques. *Pract Gastroenterol*, 1986. **10**: 8–18.

10. Demling, L. and H.J. Hagel, Video endoscopy. Fundamentals and problems. *Endoscopy*, 1985. **17**(4054060): 167–169.

11. Sivak, M.V., Video endoscopy. *Clin Gastroenterol*, 1986. **15**(3731515): 205–234.

12. Knyrim, K., Video-endoscopes in comparison with fiberscopes: Quantitative measurement of optical resolution. *Endoscopy*, 1987. **19**(04): 156.

13. Knyrim, K., H. Seidlitz, N. Vakil, F. Hagenmuller, and M. Classen, Optical performance of electronic imaging systems for the colon. *Gastroenterology*, 1989. **96**(2914640): 776–782.

14. Lux, G., K. Knyrim, R. Scheubel, and M. Classen, Electronic endoscopy—fibres or chips? *Z Gastroenterol*, 1986. **24**(7): 337–343.

15. Cotton, P.B. and C.B. Williams, *Practical Gastrointestinal Endoscopy: The Fundamentals*. 6th ed. 2008, Chichester, UK: Wiley-Blackwell.

16. Wang, T.D., G.S. Janes, Y. Wang, I. Itzkan, J. Van Dam, and M.S. Feld, Mathematical model of fluorescence endoscopic image formation. *Appl Opt*, 1998. **37**(34): 8103–8111.

17. Kremkau, F.W., Clinical benefit of higher acoustic output levels. *Ultrasound Med Biol*, 1989. **15 Suppl 1**(2672517): 69–70.

18. Goldstein, A.R., C.M., Wilson, S.R., and Charboneau, J.W., Physics of ultrasound, In *Diagnostic Ultrasound*. 1991, St. Louis: Mosby-Year Book. pp. 2–18.

19. Reddy, N.K., A.M. Ioncica, A. Saftoiu, P. Vilmann, and M.S. Bhutani, Contrast-enhanced endoscopic ultrasonography. *World J Gastroenterol*, 2011. **17**(1): 42–48.

20. Menzel, J., W. Domschke, H.J. Brambs, N. Frank, A. Hatfield, C. Nattermann, S. Odegaard, H. Seifert, K. Tamada, T.L. Tio, and E.C. Foerster, Miniprobe ultrasonography in the upper gastrointestinal tract: State of the art 1995, and prospects. *Endoscopy*, 1996. **28**(8886639): 508–513.

21. Yanai, H., M. Tada, M. Karita, and K. Okita, Diagnostic utility of 20-megahertz linear endoscopic ultrasonography in early gastric cancer. *Gastrointest Endosc*, 1996. **44**(8836713): 29–33.

22. Saitoh, Y., T. Obara, K. Einami, M. Nomura, M. Taruishi, T. Ayabe, T. Ashida, Y. Shibata, and Y. Kohgo, Efficacy of high-frequency ultrasound probes for the preoperative staging of invasion depth in flat and depressed colorectal tumors. *Gastrointest Endosc*, 1996. **44**(8836714): 34–39.

23. Chak, A., M. Canto, P.D. Stevens, C.J. Lightdale, F. Van de Mierop, G. Cooper, B.J. Pollack, and M.V. Sivak, Clinical applications of a new through-the-scope ultrasound probe: Prospective comparison with an ultrasound endoscope. *Gastrointest Endosc*, 1997. **45**(9087836): 291–295.

24. Adrain, A.L., H.-C. Ter, M.J. Cassidy, T.D. Schiano, J.-B. Liu, and L.S. Miller, High-resolution endoluminal sonography is a sensitive modality for the identification of Barrett's metaplasia. *Gastrointest Endosc*, 1997. **46**(2): 147–151.

25. Chak, A., G. Isenberg, S. Mallery, J. Van Dam, G.S. Cooper, and M.V. Sivak Jr, Prospective comparative evaluation of video US endoscope. *Gastrointest Endosc*, 1999. **49**(6): 695–699.

26. Committee, A.T., Endoscopic ultrasound probes. *Gastrointest Endosc*, 2006. **63**(6): 751–754.

27. Rampado, S., P. Bocus, G. Battaglia, A. Ruol, G. Portale, and E. Ancona, Endoscopic ultrasound: Accuracy in staging superficial carcinomas of the esophagus. *Ann Thorac Surg*, 2008. **85**(1): 251–256.

28. Erickson, R.A. and A.A. Garza, Impact of endoscopic ultrasound on the management and outcome of pancreatic carcinoma. *Am J Gastroenterol*, 2000. **95**(9): 2248–2254.

29. Huang, D., E. Swanson, C. Lin, J. Schuman, W. Stinson, W. Chang, M. Hee et al., Optical coherence tomography. *Science*, 1991. **254**(5035): 1178–1181.

30. Herz, P.R., Y. Chen, A.D. Aguirre, J.G. Fujimoto, H. Mashimo, J. Schmitt, A. Koski, J. Goodnow, and C. Petersen, Ultrahigh resolution optical biopsy with endoscopic optical coherence tomography. *Opt Express*, 2004. **12**(15): 3532–3542.

31. Tomlins, P.H. and R.K. Wang, Theory, developments and applications of optical coherence tomography. *J Phys D Appl Phys*, 2005. **38**(15): 2519–2535.

32. Fercher, A.F., W. Drexler, C.K. Hitzenberger, and T. Lasser, Optical coherence tomography—Principles and applications. *Rep Prog Phys*, 2003. **66**(2): 239–303.

33. Tearney, G.J., M.E. Brezinski, B.E. Bouma, S.A. Boppart, C. Pitris, J.F. Southern, and J.G. Fujimoto, *In vivo* endoscopic optical biopsy with optical coherence tomography. *Science*, 1997. **276**(9197265): 2037–2039.

34. Brezinski, M.E., G.J. Tearney, S.A. Boppart, E.A. Swanson, J.F. Southern, and J.G. Fujimoto, Optical biopsy with optical coherence tomography: Feasibility for surgical diagnostics. *J Surg Res*, 1997. **71**(9271275): 32–40.

35. Vakoc, B.J., D. Fukumura, R.K. Jain, and B.E. Bouma, Cancer imaging by optical coherence tomography: Preclinical progress and clinical potential. *Nat Rev Cancer*, 2012. **12**(5): 363–368.

36. Poneros, J.M., S. Brand, B.E. Bouma, G.J. Tearney, C.C. Compton, and N.S. Nishioka, Diagnosis of specialized intestinal metaplasia by optical coherence tomography. *Gastroenterology*, 2001. **120**(1): 7–12.

37. Sivak Jr, M.V., K. Kobayashi, J.A. Izatt, A.M. Rollins, R. Ung-runyawee, A. Chak, R.C.K. Wong, G.A. Isenberg, and J. Willis, High-resolution endoscopic imaging of the GI tract using optical coherence tomography. *Gastrointest Endosc*, 2000. **51**(4): 474–479.

38. Vakoc, B.J., M. Shishko, S.H. Yun, W.Y. Oh, M.J. Suter, A.E. Desjardins, J.A. Evans, N.S. Nishioka, G.J. Tearney, and B.E. Bouma, Comprehensive esophageal microscopy by using optical frequency-domain imaging (with video). *Gastrointest Endosc*, 2007. **65**(6): 898–905.

39. Evans, J.A., B.E. Bouma, J. Bressner, M. Shishkov, G.Y. Lauwers, M. Mino-Kenudson, N.S. Nishioka, and G.J. Tearney, Identifying intestinal metaplasia at the squamocolumnar junction by using optical coherence tomography. *Gastrointest Endosc*, 2007. **65**(1): 50–56.

40. Isenberg, G., M.V. Sivak, A. Chak, R.C.K. Wong, J.E. Willis, B. Wolf, D.Y. Rowland, A. Das, and A. Rollins, Accuracy of endoscopic optical coherence tomography in the detection of dysplasia in Barrett's esophagus: A prospective, double-blinded study. *Gastrointest Endosc*, 2005. **62**(6): 825–831.

41. Evans, J.A., J.M. Poneros, B.E. Bouma, J. Bressner, E.F. Halpern, M. Shishkov, G.Y. Lauwers, M. Mino-Kenudson, N.S. Nishioka, and G.J. Tearney, Optical coherence tomography to identify intramucosal carcinoma and high-grade dysplasia in Barrett's esophagus. *Clin Gastroenterol H*, 2006. **4**(1): 38–43.

42. Suter, M.J., B.J. Vakoc, P.S. Yachimski, M. Shishkov, G.Y. Lauwers, M. Mino-Kenudson, B.E. Bouma, N.S. Nishioka, and G.J. Tearney, Comprehensive microscopy of the esophagus in human patients with optical frequency domain imaging. *Gastrointest Endosc*, 2008. **68**(4): 745–753.

43. Hariri, L.P., M.B. Applegate, M. Mino-Kenudson, E.J. Mark, B.D. Medoff, A.D. Luster, B.E. Bouma, G.J. Tearney, and M.J. Suter, Volumetric optical frequency domain imaging of pulmonary pathology with precise correlation to histopathology. *Chest*, 2013. **143**(1): 64–74.

44. Hariri, L.P., G.T. Bonnema, K. Schmidt, A.M. Winkler, V. Korde, K.D. Hatch, J.R. Davis, M.A. Brewer, and J.K. Barton, Laparoscopic optical coherence tomography imaging of human ovarian cancer. *Gynecol Oncol*, 2009. **114**(2): 188–194.

45. Kuznetsova, I.A., N.D. Gladkova, V.M. Gelikonov, J.L. Belinson, N.M. Shakhova, and F.I. Feldchtein, *OCT in Gynecology Optical Coherence Tomography*, W. Drexler and J.G. Fujimoto, Editors. 2008, Berlin: Springer. pp. 1211–1240.

46. Chen, Z., T.E. Milner, X. Wang, S. Srinivas, and J.S. Nelson, Optical Doppler tomography: Imaging *in vivo* blood flow dynamics following pharmacological intervention and photodynamic therapy. *Photochem Photobiol*, 1998. **67**(9477766): 56–60.

47. Chamié, D., D. Prabhu, Z. Wang, H. Bezerra, A. Erglis, D. Wilson, A. Rollins, and M. Costa, Three-dimensional fourier-domain optical coherence tomography imaging: Advantages and future development. *Curr Cardiovasc Imag Rep*, 2012. **5**(4): 221–230.

48. Bruno, M.J., Magnification endoscopy, high resolution endoscopy, and chromoscopy; towards a better optical diagnosis. *Gut*, 2003. **52**: 7–11.

49. ASGE Technology Committee, High-resolution and high-magnification endoscopes. *Gastrointest Endosc*, 2009. **69**(3): 399–407.

50. Jaramillo, E., M. Watanabe, P. Slezak, and C. Rubio, Flat neoplastic lesions of the colon and rectum detected by high-resolution video endoscopy and chromoscopy. *Gastrointest Endosc*, 1995. **42**(2): 114–122.

51. Jaramillo, E., M. Watanabe, R. Befrits, E. Ponce de Leon, C. Rubio, and P. Slezak, Small, flat colorectal neoplasias in long-standing ulcerative colitis detected by high-resolution electronic video endoscopy. *Gastrointest Endosc*, 1996. **44**(8836711): 15–22.

52. Chang, C.C., C.R. Hsieh, H.Y. Lou, C.L. Fang, C. Tiong, J.J. Wang, I.V. Wei, S.C. Wu, J.N. Chen, and Y.H. Wang, Comparative study of conventional colonoscopy, magnifying chromoendoscopy, and magnifying narrow-band imaging systems in the differential diagnosis of small colonic polyps between trainee and experienced endoscopist. *Int J Colorectal Dis*, 2009. **24**(12): 1413–1419.

53. Kiesslich, R. and M.F. Neurath, Chromoendoscopy and other novel imaging techniques. *Gastroenterol Clin N*, 2006. **35**(3): 605–619.

54. Brown, S.R. and W. Baraza, Chromoscopy versus conventional endoscopy for the detection of polyps in the colon and rectum. *Cochrane DB Sys Rev*, 2010. 2010(10): 1–32.

55. Subramanian, V., J. Mannath, K. Ragunath, and C.J. Hawkey, Meta-analysis: The diagnostic yield of chromoendoscopy for detecting dysplasia in patients with colonic inflammatory bowel disease. *Aliment Pharm Ther*, 2011. **33**(3): 304–312.

56. Pohl, J., E. Lotterer, C. Balzer, M. Sackmann, K.D. Schmidt, L. Gossner, C. Schaab, T. Frieling, M. Medve, G. Mayer, M. Nguyen-Tat, and C. Ell, Computed virtual chromoendoscopy versus standard colonoscopy with targeted indigocarmine chromoscopy: A randomised multicentre trial. *Gut*, 2009. **58**(1): 73–78.

57. Pohl, J., A. Schneider, H. Vogell, G. Mayer, G. Kaiser, and C. Ell, Pancolonic chromoendoscopy with indigo carmine versus standard colonoscopy for detection of neoplastic lesions: A randomised two-centre trial. *Gut*, 2011. **60**(4): 485–490.

58. Sugimachi, K., K. Kitamura, K. Baba, M. Ikebe, and H. Kuwano, Endoscopic diagnosis of early carcinoma of the esophagus using Lugol's solution. *Gastrointest Endosc*, 1992. **38**(6): 657–661.

59. Yokoyama, A., T. Ohmori, H. Makuuchi, K. Maruyama, K. Okuyama, H. Takahashi, T. Yokoyama, K. Yoshino, M. Hayashida, and H. Ishii, Successful screening for early esophageal cancer in alcoholics using endoscopy and mucosa iodine staining. *Cancer*, 1995. **76**(6): 928–934.

60. Canto, M.I.F., S. Setrakian, J. Willis, A. Chak, R. Petras, N.R. Powe, and M.V. Sivak Jr, Methylene blue–directed biopsies improve detection of intestinal metaplasia and dysplasia in Barrett's esophagus. *Gastrointest Endosc*, 2000. **51**(5): 560–568.

61. Sharma, P., M. Topalovski, M.S. Mayo, and A.P. Weston, Methylene blue chromoendoscopy for detection of short-segment Barrett's esophagus. *Gastrointest Endosc*, 2001. **54**(3): 289–293.

62. Kiesslich, R., M. Hahn, G. Herrmann, and M. Jung, Screening for specialized columnar epithelium with methylene blue: Chromoendoscopy in patients with Barrett's esophagus and a normal control group. *Gastrointest Endosc*, 2001. **53**(1): 47–52.

63. Wilson, B.C., Detection and treatment of dysplasia in Barrett's esophagus: A pivotal challenge in translating biophotonics from bench to bedside. *J Biomed Opt*, 2007. **12**(5).

64. Kuiper, T. and E. Dekker, IMAGING NBI-detection and differentiation of colonic lesions. *Nat Rev Gastroentero*, 2010. **7**(3): 128–130.

65. Richards-Kortum, R. and E. Sevick-Muraca, Quantitative optical spectroscopy for tissue diagnosis. *Annu Rev Phys Chem*, 1996. **47**: 555–606.

66. Kara, M.A., M.E. Smits, W.D. Rosmolen, A.C. Bultje, F.J.W. ten Kate, P. Fockens, G.N.J. Tytgat, and J.J.G.H.M. Bergman, A randomized crossover study comparing light-induced fluorescence endoscopy with standard videoendoscopy for the detection of early neoplasia in Barrett's esophagus. *Gastrointest Endosc*, 2005. **61**(6): 671–678.

67. Borovicka, J., J. Fischer, J. Neuweiler, P. Netzer, J. Gschossmann, T. Ehmann, P. Bauerfeind, G. Dorta, U. Zurcher, J. Binek, and C. Meyenberger, Autofluorescence endoscopy in surveillance of Barrett's esophagus: A multicenter randomized trial on diagnostic efficacy. *Endoscopy*, 2006. **38**(9): 867–872.

68. Curvers, W.L., R. Kiesslich, and J.J.G.H.M. Bergman, Novel imaging modalities in the detection of oesophageal neoplasia. *Best Pract Res Cl Ga*, 2008. **22**(4): 687–720.

69. Kara, M.A., F.P. Peters, F.J.W. ten Kate, S.J. van Deventer, P. Fockens, and J.J.G.H.M. Bergman, Endoscopic video autofluorescence imaging may improve the detection of early neoplasia in patients with Barrett's esophagus. *Gastrointest Endosc*, 2005. **61**(6): 679–685.

70. Kara, M.A., F.P. Peters, P. Fockens, F.J.W. ten Kate, and J.J.G.H.M. Bergman, Endoscopic video-autofluorescence imaging followed by narrow band imaging for detecting early neoplasia in Barrett's esophagus. *Gastrointest Endosc*, 2006. **64**(2): 176–185.

71. Kuiper, T., F.J.C. van den Broek, A.H. Naber, E.J. van Soest, P. Scholten, R.C. Mallant-Hent, J. van den Brande et al., Endoscopic trimodal imaging detects colonic neoplasia as well as standard video endoscopy. *Gastroenterology*, 2011. **140**(7): 1887–1894.

72. Kobayashi, M., H. Tajiri, E. Seike, M. Shitaya, S. Tounou, M. Mine, and K. Oba, Detection of early gastric cancer by a real-time autofluorescence imaging system. *Cancer Lett*, 2001. **165**(2): 155–159.

73. Koenig, F., F.J. McGovern, H. Enquist, R. Larne, T.F. Deutsch, and K.T. Schomacker, Autofluorescence guided biopsy for the early diagnosis of bladder carcinoma. *J Urology*, 1998. **159**(6): 1871–1875.

74. Shin, D., N. Vigneswaran, A. Gillenwater, and R. Richards-Kortum, Advances in fluorescence imaging techniques to detect oral cancer and its precursors. *Future Oncol*, 2010. **6**(7): 1143–1154.

75. Žargi, M., I. Fajdiga, and L. Šmid, Autofluorescence imaging in the diagnosis of laryngeal cancer. *Eur Arch Oto-Rhino-Laryngology*, 2000. **257**(1): 17–23.

76. Kusunoki, Y., F. Imamura, H. Uda, M. Mano, and T. Horai, Early detection of lung cancer with laser-induced fluorescence endoscopy and spectrofluorometry. *Chest*, 2000. **118**(6): 1776–1782.

77. Shibuya, K., T. Fujisawa, H. Hoshino, M. Baba, Y. Saitoh, T. Iizasa, M. Suzuki, M. Otsuji, K. Hiroshima, and H. Ohwada, Fluorescence bronchoscopy in the detection of preinvasive bronchial lesions in patients with sputum cytology suspicious or positive for malignancy. *Lung Cancer*, 2001. **32**(1): 19–25.

78. Bhunchet, E., H. Hatakawa, Y. Sakai, and T. Shibata, Fluorescein electronic endoscopy: A novel method for detection of early stage gastric cancer not evident to routine endoscopy. *Gastrointest Endosc*, 2002. **55**(4): 562–571.

79. Namihisa, A., H. Miwa, H. Watanabe, O. Kobayashi, T. Ogihara, and N. Sato, A new technique: Light-induced fluorescence endoscopy in combination with pharmacoendoscopy. *Gastrointest Endosc*, 2001. **53**(3): 343–348.

80. Riedl, C.R., D. Daniltchenko, F. Koenig, R. Simak, S.A. Loening, and H. Pflueger, Fluorescence endoscopy with 5-aminolevulinic acid reduces early recurrence rate in superficial bladder cancer. *J Urology*, 2001. **165**(4): 1121–1123.

81. Eker, C., S. Montan, E. Jaramillo, K. Koizumi, C. Rubio, S. Andersson-Engels, K. Svanberg, S. Svanberg, and P. Slezak, Clinical spectral characterisation of colonic mucosal lesions using autofluorescence and delta aminolevulinic acid sensitisation. *Gut*, 1999. **44**(10075958): 511–518.

82. Endlicher, E., R. Knuechel, T. Hauser, R.-M. Szeimies, J. Schölmerich, and H. Messmann, Endoscopic fluorescence detection of low and high grade dysplasia in Barrett's oesophagus using systemic or local 5-aminolaevulinic acid sensitisation. *Gut*, 2001. **48**(3): 314–319.

83. Leunig, A., C.S. Betz, M. Mehlmann, H. Stepp, S. Arbogast, G. Grevers, and R. Baumgartner, Detection of squamous cell carcinoma of the oral cavity by imaging 5-aminolevulinic acid-induced protoporphyrin IX fluorescence. *Laryngoscope*, 2000. **110**(1): 78–83.

84. Zaak, D., M. Kriegmair, H. Stepp, H. Stepp, R. Baumgartner, R. Oberneder, Schneede, S. Corvin, D. Frimberger, R. Knüchel, and A. Hofstetter, Endoscopic detection of transitional cell carcinoma with 5-aminolevulinic acid: Results of 1012 fluorescence endoscopies. *Urology*, 2001. **57**(4): 690–694.

85. Goetz, M., A. Watson, and R. Kiesslich, Confocal laser endomicroscopy in gastrointestinal diseases. *J Biophotonics*, 2011. **4**(7–8): 498–508.

86. Jabbour, J.M., M.A. Saldua, J.N. Bixler, and K.C. Maitland, Confocal endomicroscopy: Instrumentation and medical applications. *Ann Biomed Eng*, 2012. **40**(2): 378–97.

87. Liu, J.T.C., N.O. Loewke, M.J. Mandella, R.M. Levenson, J.M. Crawford, and C.H. Contag, Point-of-care pathology with miniature microscopes. *Anal Cell Pathol*, 2011. **34**(3): 81–98.

88. Tanbakuchi, A.A., A.R. Rouse, J.A. Udovich, K.D. Hatch, and A.F. Gmitro, Clinical confocal micro-laparoscope for real-time *in vivo* optical biopsies. *J Biomed Opt*, 2009. **14**(4): 044030.

89. Ota, T., H. Fukuyama, Y. Ishihara, H. Tanaka, and T. Takamatsu, *In situ* fluorescence imaging of organs through compact scanning head for confocal laser microscopy. *J Biomed Opt*, 2005. **10**(2): 024010.

90. Piyawattanametha, W., H. Ra, M.J. Mandella, K. Loewke, T.D. Wang, G.S. Kino, O. Solgaard, and C.H. Contag, 3-D near-infrared fluorescence imaging using an MEMS-based miniature dual-axis confocal microscope. *IEEE J Sel Top Quant*, 2009. **15**(5): 1344–1350.

91. Pitris, C., B.E. Bouma, M. Shiskov, and G.J. Tearney, A GRISM-based probe for spectrally encoded confocal microscopy. *Opt Express*, 2003. **11**(2): 120–124.

92. Carlson, K., M. Chidley, K.B. Sung, M. Descour, A. Gillenwater, M. Follen, and R. Richards-Kortum, *In vivo* fiber-optic confocal reflectance microscope with an injection-molded plastic miniature objective lens. *Appl Opt*, 2005. **44**(10): 1792–1797.

93. Kiesslich, R., M. Goetz, M. Vieth, P.R. Galle, and M.F. Neurath, Technology insight: Confocal laser endoscopy for *in vivo* diagnosis of colorectal cancer. *Nat Clin Pract Oncol*, 2007. **4**(8): 480–490.

94. Gong, F., P. Swain, and T. Mills, Wireless endoscopy. *Gastrointest Endosc*, 2000. **51**(6): 725–729.

95. Iddan, G., G. Meron, A. Glukhovsky, and P. Swain, Wireless capsule endoscopy. *Nature*, 2000. **405**: 417.

96. Appleyard, M., A. Glukhovsky, and P. Swain, Wireless-capsule diagnostic endoscopy for recurrent small-bowel bleeding. *New Engl J Med*, 2001. **344**(3): 232–233.

97. Appleyard, M., Z. Fireman, A. Glukhovsky, H. Jacob, R. Shreiver, S. Kadirkamanathan, A. Lavy, S. Lewkowicz, E. Scapa, R. Shofti, P. Swain, and A. Zaretsky, A randomized trial comparing wireless capsule endoscopy with push enteroscopy for the detection of small-bowel lesions. *Gastroenterology*, 2000. **119**(6): 1431–1438.

98. Cheng, W.B., M.A. Moser, S. Kanagaratnam, and W.J. Zhang, Overview of upcoming advances in colonoscopy. *Dig Endosc*, 2012. **24**(1): 1–6.

99. Qureshi, W.A., Current and future applications of the capsule camera. *Nat Rev Drug Discov*, 2004. **3**(5): 447–450.

100. Culliford, A., J. Daly, B. Diamond, M. Rubin, and P.H.R. Green, The value of wireless capsule endoscopy in patients with complicated celiac disease. *Gastrointest Endosc*, 2005. **62**(1): 55–61.

101. Hara, A.K., C.D. Johnson, J.E. Reed, D.A. Ahlquist, H. Nelson, R.L. Ehman, and W.S. Harmsen, Reducing data size and radiation dose for CT colonography. *Am J Roentgenol*, 1997. **168**(5): 1181–4.

102. Johnson, C.D. and D.A. Ahlquist, Computed tomography colonography (virtual colonoscopy): A new method for colorectal screening. *Gut*, 1999. **44**(10026308): 301–305.

103. Vining, D.J., Virtual endoscopy: Is it reality? *Radiology*, 1996. **200**(8657938): 30–31.

104. Fenlon, H.M. and J.T. Ferrucci, Virtual colonoscopy: What will the issues be? *Am J Roentgenol*, 1997. **169**(2): 453–8.

105. Ahlquist, D.A., A.K. Hara, and C.D. Johnson, Computed tomographic colography and virtual colonoscopy. *Gastrointest Endosc Clin N Am*, 1997. **7**: 439–452.

106. Fenlon, H.M., P.D. Clarke, and J.T. Ferrucci, Virtual colonoscopy: Imaging features with colonoscopic correlation. *Am J Roentgenol*, 1998. **170**(5): 1303–9.

107. Dachman, A.H., J.K. Kuniyoshi, C.M. Boyle, Y. Samara, K.R. Hoffmann, D.T. Rubin, and I. Hanan, CT colonography with three-dimensional problem solving for detection of colonic polyps. *Am J Roentgenol*, 1998. **171**(4): 989–95.

108. Fenlon, H.M., D.P. Nunes, P.C. Schroy, M.A. Barish, P.D. Clarke, and J.T. Ferrucci, A comparison of virtual and conventional colonoscopy for the detection of colorectal polyps. *New Engl J Med*, 1999. **341**(20): 1496–1503.

109. McFarland, E.G., J.G. Fletcher, P. Pickhardt, A. Dachman, J. Yee, C.H. McCollough, M. Macari et al., ACR Colon Cancer Committee white paper: Status of CT colonography 2009. *J Am Coll Radiol*, 2009. **6**(11): 756–772 e4.

10

Optical Coherence Tomography

Ryan L. Shelton
Texas A&M University

Sebina Shrestha
Texas A&M University

Jesung Park
Texas A&M University

Brian E. Applegate
Texas A&M University

10.1 Introduction

Optical coherence tomography (OCT) is a high-resolution optical imaging modality that was invented around 1990 [1–3] and rapidly developed as a diagnostic imaging tool for retinal diseases. Its rapid development can be attributed at least in part to the fact that no other imaging modality was able to produce high-resolution cross-sectional images through the entire thickness of the retina without making contact with the eye. Carl Ziess Meditec introduced the first commercial OCT system for retinal imaging in 1995. Since then, commercial instruments for clinical imaging have been introduced for the anterior segment of the eye and more recently for intravascular imaging of the coronary arteries. In the last 10 years, a number of small- and medium-sized companies have brought to market research-grade OCT systems intended for use in biomedical and related fields. The field has matured to the point where incorporating OCT imaging as part of a research plan is as easy as incorporating some of the more classic optical imaging modalities like fluorescence microscopy.

In this chapter, we hope to provide the reader with a basic understanding of the physics behind OCT, functional extensions of OCT, and examples of clinical and preclinical applications. Given the confines of a chapter, the detail with which we cover these topics will be limited; hence, we encourage the interested reader to consult one or more of the several books [4–7] dedicated entirely to OCT to delve more deeply into the subject.

10.2 Theory

OCT fundamentally interrogates the changes in refractive index in tissue to render a morphological image. Near-infrared light is loosely focused on the tissue. At points along the path of the light in the tissue where there exists a change in the refractive index, light is reflected back. Measuring the intensity of the light returning as a function of time enables the reconstruction of a line image of the changes in refractive index along the light path in the tissue. Fortunately, structural variations in the tissue typically are accompanied by relatively large changes in refractive index, which appear as peaks in the backscattered light intensity. By moving the loosely focused light across the tissue and recording a line image at each position, tissue cross-sectional and volumetric images may be generated. The imaging depth of OCT is both tissue- and wavelength-dependent, but typically falls in the range 1–2 mm. The axial and lateral resolutions are typically in the range 1–15 and 10–30 μm, respectively.

To appreciate the underlying physics of OCT, it is illustrative to start by drawing an analogy with pulse-echo ultrasound. Here, a sound pulse is launched into the tissue and its echo is recorded as a function of time to generate a line image of the changes in acoustic impedance. Changes in acoustic impedance are typically largest at the interface of different morphological structures; hence, they generate the strongest echoes. The resolution of clinical ultrasound typically falls in the range 0.1–1 mm. Given that sound travels at ~1500 m/s in tissue, to accurately measure the time-of-flight of the sound pulse, a temporal resolution of ~70–700 ns would be needed. If we were to take the same approach with OCT and launch a short pulse of light and try to measure its time-of-flight with a detector, we would need a temporal resolution of ~3–50 fs due to the much higher velocity of light ($\sim 3 \times 10^8$ m/s). Since such detectors do not exist, an alternative approach is necessary.

To measure the time-of-flight of optical radiation, one can take advantage of the coherence properties of light and utilize interferometry. Consider a simple Michelson-type interferometer illuminated by a monochromatic laser source as depicted in Figure 10.1. The intensity of light measured at the detector (I_D) is given by

$$I_D = I_1 + I_2 + 2\sqrt{I_1 I_2}\cos(2k\Delta z) \tag{10.1}$$

where the subscripts 1 and 2 refer to the two arms of the interferometer, k is the spatial frequency (aka wavenumber $k = 2\pi/\lambda$), and Δz is the optical pathlength difference (i.e., Δz is the relative difference in the distance between the mirrors in the two arms of the interferometer and the beam splitter). If we divide Δz by the speed of light, we have the relative difference in time between light returning from the two arms of the interferometer. Essentially, we can measure the time-of-flight of light in one arm of the interferometer against the time-of-flight of light in the other arm. In other words, Δz is precisely what we would like to measure.

Now, suppose that instead of a single monochromatic light source, we use an array of n light sources with spatial frequency k_i, each separated in spatial frequency by δk with an overall intensity envelope well described by a Gaussian function with center frequency k_0 and full-width at half-maximum of Δk or

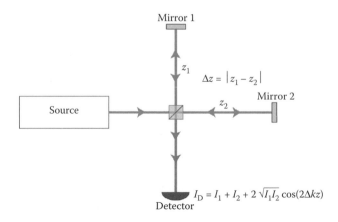

FIGURE 10.1 Simple Michelson interferometer. The arrowheads indicate the direction of light propagation.

$$I_S(k_i) = \sum_{i=1}^{n} e^{-((k_i - k_0)^2 \, 4\ln 2 / \Delta k^2)}. \tag{10.2}$$

This situation is illustrated in Figure 10.2a for $n = 50$. If we illuminate the interferometer with each of these sources sequentially and measure the intensity at the detector, we can map out Equation 10.1 as a function of k with sampling δk or

$$I_D(k_i) = \sum_{i=1}^{n} \left[I_{S,1}(k_i) + I_{S,2}(k_i) + 2\sqrt{I_{S,1}(k_i) I_{S,2}(k_i)} \cos(2k_i \Delta z) \right] \tag{10.3}$$

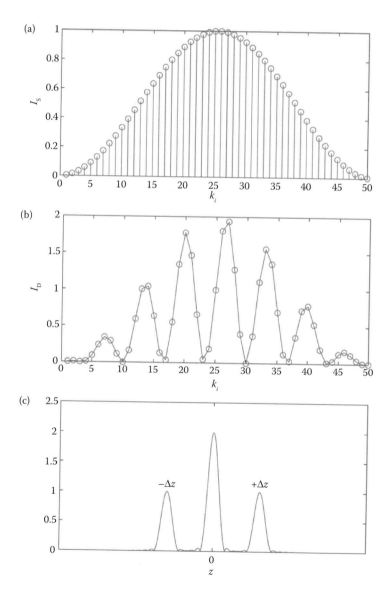

FIGURE 10.2 (a) Normalized spectrum of monochromatic sources. (b) Optical signal measured at the detector in the Michelson interferometer for each source in (a). (c) Magnitude of the discrete inverse Fourier transform of (b).

where the subscripts 1 and 2 indicated the portion of the source intensity that returns from the two arms of the interferometer. Equation 10.3 for $n = 50$ is shown in Figure 10.2b. The magnitude of the discrete inverse Fourier transform of $I_D(k_i)$ yields the spectrum in Figure 10.2c. The central peak at $z = 0$ is due to the first two terms of Equation 10.3, $I_{S,1}(k_i) + I_{S,2}(k_i)$. The peaks at $z = \pm\Delta z$ are due to the last term of Equation 10.3, $\sqrt{I_{S,1}(k_i)I_{S,2}(k_i)}\cos(2k_i\Delta z)$. If we had multiple reflectors in one of the arms of the interferometer, that is, an array of Δz, then we would have multiple peaks in the spectrum on either side of $z = 0$. The relative intensity of each peak would be proportional to the strength of the reflection. The changes in refractive index in a tissue sample constitute an array of Δz; therefore, if we replace the mirror in one arm of the interferometer with a tissue sample, we can make a depth-resolved measurement of the tissue's reflectivity.

Each measurement of I_D yields a line image of the tissue sample in depth. This line image is referred to as an A-line or A-scan. To generate a cross-sectional (B-scan) or volumetric image, the light illuminating the sample must be scanned across its surface, that is, in the x, y-plane. Two-dimensional (2D) and three-dimensional (3D) images are formed by combining the results of a large number of A-scans made at different positions in the x, y-plane. It is also sometimes useful to measure I_D at the same x, y-position, but over a period of time, to generate what are called M-scans. This mode of imaging is most typically done when trying to measure sample motion.

Since the central peak at $z = 0$ carries no information about the depth-dependent reflectivity of the sample, it is usually removed. A typical method is to directly measure the first two terms of Equation 10.3 or synthesize them from the interferometric data and subtract them before computing the discrete Fourier transform. Likewise, the mirror image at $-\Delta z$ contains no new information; hence, only the positive frequencies are typically displayed. The negative frequencies are a consequence of taking a Fourier transform of real data. A problem can arise when there are negative pathlength differences. In that case, the peaks at Δz can overlap with the peaks at $-(-\Delta z)$. In other words, it is not possible to distinguish negative pathlength differences from negative frequencies. This issue is sometimes referred to as the complex ambiguity. There are a number of methods to overcome this problem, including constructing a complex dataset [8–10] and imparting a carrier frequency to the interferometric signal to move the signal further from $z = 0$ [11,12]. However, the most typical strategy is simply ensuring that the signal only has positive pathlength differences.

Once a sample is introduced that contains multiple reflections is introduced, another interference term arises from interference between reflections at different depths in the sample. This is sometimes called the "autocorrelation" term. The pathlength differences between reflections in the sample are typically small; hence, the signal from this term usually falls close to $z = 0$. While the autocorrelation signal is typically weak because of the weak sample reflectivity, overlap with the autocorrelation term could obscure the desired signal. It can be avoided in several ways, but the most typical strategy is the same as for the complex ambiguity, that is, ensure that the minimum pathlength difference is sufficiently far from $z = 0$.

Since we are sampling Equation 10.3 at discrete k, the maximum pathlength difference we can measure is a function of the sampling frequency and is given by the Nyquist sampling theorem as $(4\delta k)^{-1}$. (Note that the extra factor of 2 arises from the fact that the light passes through the sample twice.) Additionally, the full-width at half-maximum of each peak in the Fourier transform is a function of the width of the source spectrum. Based on the Rayleigh criterion, this parameter also corresponds to the axial resolution. The axial resolution of an OCT system assuming a Gaussian source spectrum is

$$\delta z = \frac{4\ln 2}{n\Delta k} = \frac{2\ln 2\lambda_o^2}{n\pi\Delta\lambda} \tag{10.4}$$

where n is the refractive index of the sample, λ_o the center source wavelength, and $\Delta\lambda$ the full-width at half-maximum bandwidth of the source. Clearly, an increase in the width of the source spectrum improves the axial resolution of the imaging system.

The lateral resolution is governed by how tight the source is focused onto the sample. Since a line image is collected with one spectral acquisition, it is important to maintain the lateral resolution over the entire imaging depth. For a Gaussian beam intensity profile, the Rayleigh range is typically used as the governing equation for defining the focal region where twice the Rayleigh range corresponds to the desired imaging depth, that is

$$2z_R = \frac{2\pi w_o^2}{\lambda} \tag{10.5}$$

where z_R is the Rayleigh range parameter and w_o the e^{-2} radius of the beam intensity at the focus. At the edge of the Rayleigh range, the e^{-2} radius of the Gaussian beam intensity drops to $\sqrt{2}w_o$. Clearly, a compromise must be struck between the desire to minimize the lateral resolution (w_o) and to maximize the available imaging depth ($2z_R$).

A similar compromise must be made with respect to the source wavelength. The tissue scattering coefficient (μ_s) is a strong function of wavelength, becoming weaker with increasing wavelengths. A weaker scattering coefficient leads to better tissue imaging depth. Unfortunately, water absorption gets stronger with increasing wavelengths; hence, there must be a compromise between minimizing the scattering coefficient and minimizing the water absorption. For this reason, imaging around 800 nm is optimal when imaging through a significant concentration of water, as in retinal imaging in the eye where the beam must pass through the aqueous humor before reaching the retina. Applications where water absorption is less of an issue tend to use longer wavelengths, typically 1300 nm.

So far, only a hypothetical and nonexistent source consisting of an array of monochromatic light sources has been discussed. Real light sources come in two distinct designs, which also dictate how the light is detected. The first is a narrow-band laser whose spatial frequency is swept as a function of time, that is, $k(t)$. This source is very similar to our hypothetical source except that it has a finite instantaneous bandwidth (i.e., not monochromatic) and may be essentially arbitrarily sampled (i.e., not limited to 50 samples). In this case, the light in the interferometer may be detected by a photodiode and sampled with a fast digitizer to generate the desired number of samples (n in Equation 10.3). This version of OCT is commonly called swept-source OCT (SS-OCT) or optical frequency domain imaging (OFDI).

Swept lasers for SS-OCT widely employ semiconductors as gain media due to their high gain, broad bandwidth, and fast gain response [13]. Since swept source lasers perform continuous wavelength tuning in the optical cavity, different configurations of optical cavities have been developed for rapid sweeping of narrow linewidths over a wide tuning range for SS-OCT applications [14–17]. A rapid scanning intracavity filter using a polygon mirror scanner, diffraction grating, and telescope is a good example for illustrating the continuous sweep operation. By implementing the polygon-based filter into an extended-cavity semiconductor laser, high-speed swept-source lasers can be designed [14].

With real light sources and detectors comes the issue of signal roll-off (or fall-off). This is a reduction in interferometric signal as Δz gets larger. In other words, the amplitude in the Fourier transform of a given reflector will be a function of Δz. For SS-OCT, the dominant source of roll-off is the instantaneous bandwidth of the light source. To a reasonable approximation, the roll-off is equivalent to the coherence length of the laser source. A narrow instantaneous bandwidth translates into a long coherence length. For comparison, the hypothetical light source with a delta function (monochromatic) instantaneous bandwidth has an infinite coherence length. In general, longer coherence lengths are better.

The second is a broadband source that can be either continuous wave like a superluminescent light-emitting diode (SLED) or pulsed like a femtosecond laser source. Since the light source is broadband, the interferometric signal is mapped as a function of spatial frequency by detecting the light with a spectrometer. The number of samples is then governed by the number of pixels on the camera detector of the spectrometer. This version is commonly called Fourier domain OCT (FD-OCT) or spectral domain OCT (SD-OCT); however, these terms have also been used to refer to the swept-laser source version, especially in the earlier literature. In SD-OCT, the spectrometer resolution is equivalent to the instantaneous bandwidth in SS-OCT. However, even a spectrometer with infinitely narrow resolution (delta function) will have roll-off due to the integration of multiple wavelengths illuminating any given pixel on the camera. Both the sampling and spectrometer resolution are important in determining the roll-off function in SD-OCT [14,18]. In general, larger numbers of pixels (sampling) and narrower spectrometer resolution are best.

We have restricted our discussion so far to OCT based on spectral interferometry. However, most of the early research in OCT and the initial commercial instruments were based on what is called time-domain OCT (TD-OCT). In TD-OCT, Equation 10.1 is mapped as a function of Δz rather than k. This is accomplished by rapidly changing the optical pathlength in the arm of the interferometer containing the mirror, which effectively makes Δz a function of time. In 2003, it was shown by several groups [19–21] that OCT based on spectral interferometry was much more sensitive by ~20–30 dB than TD-OCT. Since then, most research and commercial instruments have switched to one of the spectral interferometry-based techniques. For a more complete discussion of the theory from the perspective of TD-OCT, the reader is encouraged to consult Ref. [4].

SS-OCT and FD-OCT have two primary benefits over TD-OCT: sensitivity and greatly increased line rate. An important secondary benefit is the lack of any moving parts in the FD-OCT system, in sharp contrast to the requirement of modulating the reference mirror in a TD-OCT setup. Capitalizing on these benefits, real-time volumetric imaging has been achieved using the SS-OCT and FD-OCT systems, which has accelerated clinical applications of OCT. These systems have been applied to a variety of clinical applications including ophthalmology [22,23], cardiology [24,25], and dermatology [26], otolaryngology [27], and pulmonary [28], and developmental biology [29–31], as well as many other areas.

10.3 Applications

Below are examples of the established clinical applications of OCT in the eye and for imaging the coronary arteries. These examples are meant to illustrate both the image quality and specific applications. Beyond the established clinical applications, there are numerous others in various stages of investigation. In the interest of brevity, we have only highlighted two, middle ear and esophagus.

10.3.1 Diagnostic Imaging of the Retina and Anterior Segment of the Eye

By far, the most prominent clinical imaging application of OCT is in the field of ophthalmology. Ophthalmic applications of OCT have garnered the largest share of research efforts and commercial interest. Recent improvements in imaging speed and resolution has spurred the development of systems capable of high-resolution volumetric imaging largely free of motion artifact.

Retinal imaging was the earliest application of OCT. Povazay et al. [22] used a high-speed FD-OCT system for 3D visualization of the retina and choroid and compared the results to a commercially available system. Augmenting their standard OCT system with adaptive optics (AO-OCT) allowed for cellular resolution to measure cone densities in normal and pathologic patients. Pathologic differences in the morphology of the retina and choroid were also investigated. Figure 10.3a–d illustrates the comparison of FD-OCT and AO-OCT when used to diagnose retinitis pigmentosa (RP).

Grulkowski et al. [32] used a high-speed CMOS camera in an FD-OCT system to acquire real-time volumes in the anterior segment and outer eye. Real-time volumetric imaging allowed for 3D visualization of the pupillary response to light stimulation, and the eye blink. Modifying the system for removal of the complex conjugate image allowed for simultaneous acquisition of the entire anterior segment. Figure 10.3e–g shows an example of detailed cross-section of anterior segment when a contact lens is on.

10.3.2 Intravascular Imaging of the Coronary Arteries

Intravascular OCT imaging of the coronary arteries has only been FDA approved since mid-2010, although it has been used clinically in Japan and Europe much longer. OCT is capable of producing

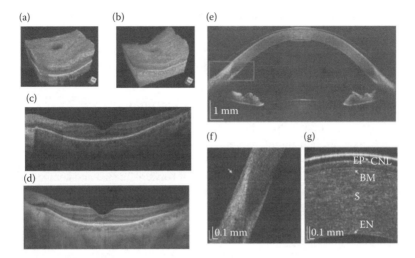

FIGURE 10.3 Application of OCT in ophthalmology. (a–d) OCT images of a patient with retinitis pigmentosa (RP): (a) 3D-OCT image of retina taken using adaptive OCT at 800 nm, which is compared to (b) 3D image acquired with high-speed 1060 nm OCT. (c, d) Cross-sectional images of (a) and (b), respectively. Layers of retinal and choroid are visible in all the images. From (b) and (d), RP was diagnosed based on the thickness of the layers, comparing that of a normal eye. Likewise, (a) and (c) depicted cellular changes, enabling measurement of cone density. (e) High-resolution and high-density cross-sectional image of the anterior segment of myopic eye with contact lens on. The red rectangles indicate location of the magnified regions shown in (f) and (g). (f) Magnification of the limbal region with the contact lens (indicated by the arrow). (g) Magnification of the apex region of the cornea. CNL, contact lens; EP, epithelium; BM, Bowman's membrane; S, stroma; EN, endothelium.

3D reconstructions of vascular walls, providing visualization of coronary plaques and stenosis for rapid assessment and diagnosis. Vulnerable plaques, plaques with a thin, fibrous cap over a necrotic core, are high risk to the patient. OCT has been shown to perform very well in visualizing these plaques, as well as provide a measurement of cap thickness. In addition, intravascular OCT is used for assessing stent placement and efficacy as in Figure 10.4. As the stent can be visualized along with

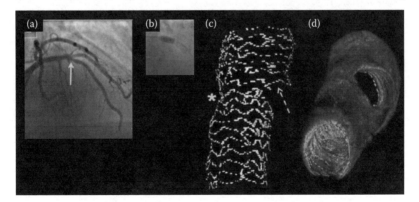

FIGURE 10.4 Application of OCT in intravascular imaging. (a) Angiographic view of lesion in the left ventricular wall with ischemia. (b) An optimal angiographic result after stent placement at the bifurcated region, shown with a yellow arrow in (a). (c) Offline 3D reconstruction of the FD-OCT images of the stent allows for visualization of its placement and indicates the location for correction. For example, the yellow asterisk shows stent distortion. (d) External 3D view of the intravasculature helps to view any compression developed due to the stent. (Adapted from Chamié, D. et al., *Three-Dimensional Fourier-Domain Optical Coherence Tomography Imaging: Advantages and Future Development*. Current Cardiovascular Imaging Reports: pp. 1–10. With permission.)

the vessel, stent fractures can be detected with OCT. Repeated long-term studies can monitor stent efficacy and prevent possible occlusions and aneurysms. Prior to the introduction of intravascular OCT, intravascular ultrasound (IVUS) was the primary tool used to generate this type of information. Improved resolution is the most significant advantage OCT has over IVUS.

10.3.3 Diagnostic Imaging of the Barrett's Esophagus

In addition to the aforementioned clinical applications, endoscopic OCT for imaging the gastrointestinal (GI) tract has gone through a number of preclinical trials. Westphal et al. [34] investigated endoscopic OCT for imaging in the lower GI tract and correlated the images directly to histological sections. Mucosal and submucosal layer with different contents of fat were imaged and compared (Figure 10.5). Endoscope pressure effects on the OCT images were also investigated and discussed. Clinical trials for

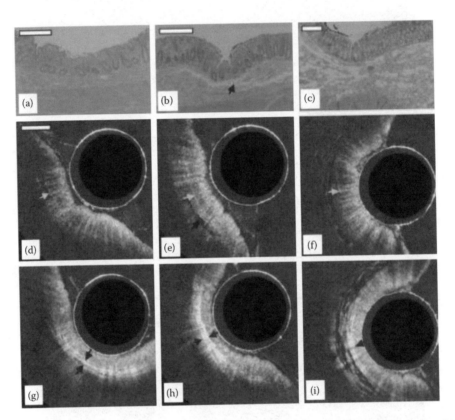

FIGURE 10.5 (See color insert.) Sequences of endoscopic OCT (EOCT) images and corresponding histology sections, with increasing adipose tissue content in GI tract submucosa (bars = 0.5 mm). (a) Histologic section with lowest submucosal fat content. (b) Histologic section with medium submucosal fat content shown with the blue arrow. (c) Histologic section with highest submucosal fat content. Different probe pressure was applied while imaging and the effect was investigated. (d) EOCT image of (a) acquired while exerting minimal pressure with probe showing border between mucosa and submucosa (yellow arrow). (e) EOCT image of (b), acquired while exerting minimal pressure with probe, showing mucosa/submucosa (yellow arrow) and adipose band seen in (b) (blue arrow). (f) EOCT image of (c), acquired while exerting minimal pressure with probe, showing border between mucosa and submucosa (yellow arrow). (g) EOCT image of (a), acquired with high-probe pressure; thickness of submucosa indicated by red arrows. (h) EOCT image of (b), acquired with high-probe pressure; thickness of submucosa indicated by red arrows. (i) EOCT image of (c), acquired with high-probe pressure; thickness of submucosa is indicated by red arrows. In (d), (e), and (f), the tissue below the mucosa/submucosa line becomes darker and darker.

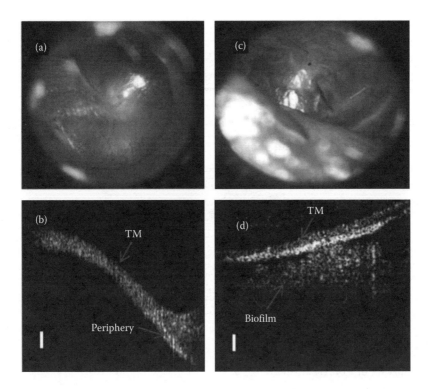

FIGURE 10.6 **(See color insert.)** (a, b) Normal human ear. (a) Video otoscope image of the tympanic membrane (TM). (b) Cross-sectional OCT image of the TM. (c, d) Chronic middle-ear infection with a thick, highly scattering biofilm that is a major source of infection. (c) Video otoscopy image showing a less translucent TM. (d) Cross-sectional OCT image showing the lateral spatial extent of biofilm. Based on the scattering properties of the TM, its thickness was measured for diagnosis. (Scale bar in (b) and (d): 100 μm.) (Adapted from Nguyen, C.T. et al., *Proceedings of the National Academy of Sciences of the United States of America*, 2012. 109(24): 9529–9534. With permission.)

endoscopic OCT of the GI tract are currently underway, investigating patients with Barrett's esophagus, which has been shown to be linked to esophageal cancer.

10.3.4 Diagnostic Imaging of the Middle Ear

OCT applications in the ear have become more prominent in recent years. In addition to developmental studies of *ex vivo* tissue in the inner and middle ear, clinical studies investigating middle ear infections have recently been demonstrated [35]. In this study by Nguyen et al., a handheld OCT probe was developed to investigate the presence of biofilms behind the tympanic membrane in normal and pathologic cases (Figure 10.6). Clinical studies have shown that chronic ear infections are strongly associated with the presence of a biofilm behind the tympanic membrane. An algorithm was developed to classify the data taken as either normal or abnormal with an associated biofilm or effusion with a sensitivity of 83% and a specificity of 98%.

10.4 Functional Extensions

10.4.1 Polarization-Sensitive Optical Coherence Tomography

Polarization-sensitive optical coherence tomography (PS-OCT) is an imaging modality that combines the advantages of OCT and scanning laser polarimetry (SLP). PS-OCT can noninvasively acquire depth-resolved reflectivity and polarization properties in turbid anisotropic media such as biological tissues.

The polarization properties including birefringence, dichroism, and optic axis orientation in PS-OCT are determined from depth-dependent variation in the polarization state of backscattered light by various polarization analysis methods such as Jones or Mueller matrices, Stokes parameters, and complex polarization ratio (CPR) [36–38]. As the polarization properties in biological tissues change under various pathological conditions, the measurement of depth-resolved polarization properties by PS-OCT is being widely investigated as a promising diagnostic technique by comparing the polarization properties between normal and abnormal tissues [39–42].

10.4.1.1 Polarization Properties of Biological Tissues

Birefringence is one of the major polarization properties in many biological tissues such as retina, cornea, skin, muscle, nerve, tendon, cartilage, and teeth. Most birefringence in biological tissues is called form-birefringence, which originates from anisotropic structures such as fibers. As the electric field of the incident light propagates into fibrous tissues, different phase velocities and refractive indices are observed with respect to the fiber axis orientation. The difference in the refractive indices generates form-birefringence in the fibrous tissues [43]. Because dichroism on the incident polarization is only observed in anisotropic media that also exhibit birefringence, it can be detected in many birefringent biological tissues. The dependence of absorbance of the electric fields on the orientation of the electric field vector of incident light creates dichroism in the birefringent tissues [44]. Practically, phase retardation from birefringence, diattenuation from dichroism, and fiber axis orientation are widely measured to investigate the variation of pathological state in the biological tissues [36,45,46].

10.4.1.2 Analysis of Polarization

Several mathematical representations and geometrical visualization are employed to describe the polarization state of light and analyze the change of the polarization state of light in complex systems containing polarizing elements [47–49].

Jones vectors and Jones matrices: The electric field of light (E) is generally expressed as horizontal ($\mathbf{E}_h$) and vertical ($\mathbf{E}_v$) polarization bases, which are mathematically described by the complex forms with the amplitudes and phases ($\mathbf{E}_h = E_h\, e^{jh}$ and $\mathbf{E}_v = E_v\, e^{jv}$). Two complex forms are written as a 2×1 matrix referred to as a Jones vector. As the Jones vector is usually normalized, the sum of the squares of two elements is unity:

$$\mathbf{E} = \begin{bmatrix} \mathbf{E}_h \\ \mathbf{E}_v \end{bmatrix} = \begin{bmatrix} E_h e^{jh} \\ E_v e^{jv} \end{bmatrix} \tag{10.6}$$

Polarization-sensitive optical elements such as polarizers and retarders can be described as a complex 2×2 matrix called Jones matrix. For instance, a linear polarizer with horizontal transmission axis has Jones matrix described by $\mathbf{J}_{lp} = \begin{bmatrix} 1 & 0 \\ 0 & 0 \end{bmatrix}$, and a quarter-wave plate (QWP) with a horizontal fast axis has a Jones matrix described by $\mathbf{J}_{qwp} = e^{-j/4} \begin{bmatrix} 1 & 0 \\ 0 & j \end{bmatrix}$. The composite effect on the polarization state of light ($\mathbf{E}_f$) propagating through n optical elements ($\mathbf{J}_n$) can be described by multiplying the Jones vector of input light by the Jones matrices for each optical element ($\mathbf{E}_f = \mathbf{J}_n\mathbf{E} = \mathbf{J}_n\mathbf{J}_{n-1} \ldots \mathbf{J}_2\mathbf{J}_1\mathbf{E}$) [48].

Stokes parameters and Mueller matrices: Jones vector and matrix representation is a very straightforward method to describe the polarization state of light, but it has a limited ability to describe partially polarized light and depolarization processes. Stokes parameters and Mueller matrix representation based on the measurement of irradiance (I_r, sometimes called "intensity") is employed to overcome this limitation. The irradiance can be written as the complex electric field components ($I_r = \mathbf{E}\mathbf{E}^*$, where * represents the complex conjugate and the angular brackets denote time averaging over an interval) [49]. Stokes parameters consist of four components (I, Q, U, and V) representing the total irradiance (I_t) and three polarization states of light. The irradiance of linear polarizers with angles 0°, 90°, +45°, and −45°

along the horizontal axis ($I_{0°}$, $I_{90°}$, $I_{+45°}$, and $I_{-45°}$) and irradiance of right and left circular polarizers (I_{rc} and I_{lc}) are generally used to calculate the Stokes parameters. Normalized Stokes parameters are defined as

$$I = I_t, \qquad Q = (I_{0°} - I_{90°})/I_t, \qquad U = (I_{+45°} - I_{-45°})/I_t, \qquad V = (I_{rc} - I_{lc})/I_t \qquad (10.7)$$

where I represents the total irradiance of light, Q the polarized state along the horizontal and vertical axes, U the polarized state along the +45° and −45°, and V the right and left circularly polarized state of light. The degree of polarization (DOP) can also be defined by

$$DOP = \sqrt{Q^2 + U^2 + V^2} \qquad (10.8)$$

DOP ranges in value from unpolarized light (DOP = 0) to totally polarized light (DOP = 1). The Stokes parameters can be written as a 4×1 real vector called Stokes vector ($\mathbf{S} = [I\ Q\ U\ V]^T$, where T is the transpose operation) to describe polarized light [47]. Polarization-sensitive optical elements in an optical system can be described as a 4×4 real matrix known as a Mueller matrix ($\mathbf{M}$). When incident polarized light is transmitted through n optical elements ($\mathbf{M}_n$), the transmitted polarized light ($\mathbf{S}_f$) can be analyzed by multiplying the Stokes vector of incident light (S) by the Mueller matrices for each optical element ($\mathbf{S}_f = \mathbf{M}_n S = \mathbf{M}_n \mathbf{M}_{n-1} \ldots \mathbf{M}_2 \mathbf{M}_1 S$) [49].

Geometrical visualization: There are two general methods to visually represent polarization states of light: CPR and Poincaré sphere [50,51]. The CPR is a 2D representation of polarized light using the ratio of two Jones vectors according to a selected polarization orthogonal basis set in the complex plane. Linearly horizontal and vertical polarization states (h, v) are selected as the orthogonal basis, and the CPR ($\mathbf{C}$) is expressed as a complex form ($\mathbf{C} = E_h/E_v = (E_h/E_v)e^{j(\theta_h - \theta_v)}$) and visualized on Cartesian complex plane. Poincaré sphere is a 3D geometrical tool to visually analyze changes to the polarization state of light in an optical system. The polarization state of light is described as a 3×1 real vector from three components of Stokes parameters ($\vec{S} = [Q\ U\ V]$), and mapped to a point of sphere with unity radius using an Q, U, and V Cartesian coordinate system. The transformation of polarized light in the optical system can be observed by checking a trajectory of the points on the Poincaré sphere. The complex plane demonstrating CPR is directly related to the Poincaré sphere used to represent Stokes vectors under a projective transformation known as stereographic projection. If the linearly horizontal and vertical polarization states (h, v) are considered, the origin on the complex plane and the point of $Q = 1$ representing the linearly horizontal state are tangent to each other. The point $Q = -1$ representing the linearly vertical state is expressed as infinity of real axis on the complex plane [50,51].

10.4.1.3 Instrumentation

Compared to standard OCT, which only acquires depth-resolved reflectivity measurement in samples, PS-OCT generally records horizontal ($|\tilde{\Gamma}_h(z)|$) and vertical ($|\tilde{\Gamma}_v(z)|$) interference fringe magnitudes, as well as relative phase ($\angle\tilde{\Gamma}_{v-h}(z)$) between horizontal and vertical magnitudes as a function of depth (z). This is accomplished by controlling polarization optics in the illumination source, sample, and reference paths of the OCT system [52,53]. Although a variety of approaches have been developed for implementing PS-OCT, we will limit our discussion to a fairly simple SS-OCT implementation using a bulk-optic Michelson interferometer. In Figure 10.7, collimated laser light passing though a linear polarizer has a pure linear horizontal input state and is divided into reference and sample paths by a nonpolarizing beam splitter (NPBS). Light in the reference path passes though a QWP at 22.5° and has equal amplitude and phase in both the horizontal and vertical components. Light in the sample path passes through another QWP at 45° to convert linear to circular polarization states of light. Reflected

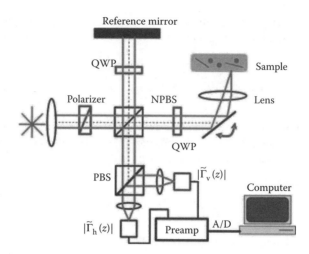

FIGURE 10.7 Schematic of basic time-domain PS-OCT instrument. QWP, quarter-wave plate; NPBS, nonpolarizing beam splitter; PBS, polarizing beam splitter.

light from the sample has an arbitrary polarization state, dependent on the polarization properties of the sample. Light returning from sample and reference paths recombines and interferes in the NPBS and is directed to the detection path. The horizontally and vertically polarized components of the beam are divided by a polarizing beam splitter (PBS), and separately detected by two detectors. Horizontal ($|\tilde{\Gamma}_h(z)|$) and vertical ($|\tilde{\Gamma}_v(z)|$) interference fringe magnitudes and relative phase ($\angle\tilde{\Gamma}_{v-h}(z)$) are acquired directly from the discrete Fourier transform. The diverse analytic methods discussed above have been developed to compute polarization properties of samples, including depth-resolved phase retardation, birefringence, diattenuation, and optic axis orientation from $|\tilde{\Gamma}_h(z)|$, $|\tilde{\Gamma}_v(z)|$, and $\angle\tilde{\Gamma}_{v-h}(z)$.

10.4.1.4 Applications of PS-OCT

Noninvasive quantification of depth-resolved polarization properties by PS-OCT is an important application in the clinical management and basic understanding of diseases. These include, but are not limited to, corneal and retinal diseases, burn injury, osteoarthritis, and dental caries [39].

OCT is currently one of the most advanced clinical imaging techniques in ophthalmology due to its ability to quantitatively measure variations in morphology ranging from the anterior to the posterior eye segment. PS-OCT has been developed as a diagnostic instrument in ophthalmology by detecting change in polarization states of several ophthalmological structures [i.e., cornea, retinal nerve fiber layer (RNFL), and retinal pigment epithelium (RPE)] [39]. Figure 10.8 shows PS-OCT images of the fovea region, including RNFL and RPE from a healthy human subject. PS-OCT can image depth-resolved polarization properties of the tissue structures of the anterior eye segment, including cornea, anterior chamber, and sclera, and discriminate the abnormal pathological variation from normal tissue structures [54,55]. The change of polarization patterns (i.e., retardation and fast axis orientation) caused from the change of lamellar structures in the cornea can be measured by PS-OCT and used as a valuable diagnostic tool for keratoconus, that is, degeneration of the corneal structure [56]. PS-OCT is particularly suited to examine correlation between thickness and polarization properties (i.e., form-birefringence) in the human RNFL *in vivo* to potentially diagnose glaucoma, the second most common cause of blindness after cataracts. Since glaucomatous damage results in thinning of RNFL and changes to the birefringence due to the loss of ganglion cells, quantitative measurements of thickness and polarization properties in peripapillary region of the RNFL can help to discriminate glaucoma from normal eyes [57,58]. PS-OCT imaging of the macular region, especially the RPE segment related to the metabolism of the photoreceptors, is another significant application in ophthalmology.

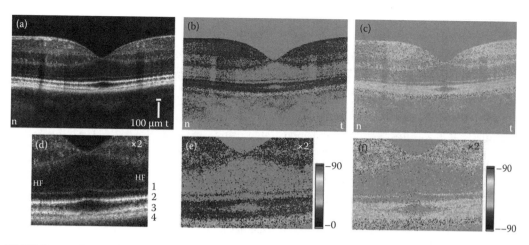

FIGURE 10.8 **(See color insert.)** PS-OCT images of the fovea region of a healthy human subject. (a) Intensity image, (b) retardation image, (c) axis orientation image corresponding to horizontal axis; 2× magnified views are shown in (d)–(f). Numbers from 1 to 4 represent the posterior retinal layers. HF, Henle's fiber layer, n, nasal; t, temporal. (Adapted from Leitgeb, R., C.K. Hitzenberger, and A.F. Fercher, *Optics Express*, 2003. **11**(8): 889–894. With permission.)

Intensity-based OCT imaging has difficulty detecting the RPE because it incorrectly identifies adjacent layers, such as Bruch's membrane or the photoreceptor layers, but the polarization-based PS-OCT images can clearly distinguish the RPE from other layers. Automated RPE segmentation based on PS-OCT images can be provided to detect the existence of drusens and measure their size. Drusens are deposits of extracellular materials beneath the RPE and are the most common indicator in diagnosing age-related macular degeneration (AMD) [59,60].

Osteoarthritis, joint inflammation from cartilage degeneration, is caused from nonuniform loss of proteoglycans associated with increased water content and increased fibrillation of the collagen network. Noninvasive, quantitative characterization of polarization properties including form-birefringence and optic axis orientation in the collagen network by PS-OCT has been investigated for both early diagnosis of osteoarthritis and *in vivo* assessment of the efficacy of novel therapeutic modalities such as transplantation, artificial matrices, and pharmaceuticals [40,61].

PS-OCT has been applied to elucidate traumatic, functional, or physiological alternations such as the severity and depth of burns and wound healing in dermatology. Collagen in burned skin tissue is denatured, and polarization properties of the collagen are changed as the temperature increases. Therefore, PS-OCT can more accurately classify skin burns by the assessment of depth as well as polarization properties [62].

Examination of dental caries lesion by PS-OCT in dentistry is another potential application. PS-OCT can detect fast depolarization processes of the incident polarized light in the area of demineralization and discriminate between healthy and carious tooth tissues by the difference in reflectivity as well as polarization. The early detection of the demineralization process in the surface of the tooth can provide *in vivo* monitoring of caries lesions and prevent further lesion progression [42,63].

10.4.2 Doppler and Phase-Sensitive OCT

Among the most prevalent functional extensions of OCT is Doppler OCT. Widely used in ultrasonography, Doppler imaging utilizes the Doppler effect to measure motion. In biological specimens, the primary source of motion inside the tissue arises from erythrocytes flowing through blood vessels. When applied to OCT, Doppler can be used to measure the flow of these erythrocytes [64–66], complementing the high-resolution, 3D structural information already obtained by a conventional OCT system.

Doppler shifts are measured at each depth in the OCT stack, allowing for 3D maps of vessel flow information *in vivo* [67,68].

Motion in the sample produces Doppler frequency shifts, which may be detected by measuring the phase of the interference between the reference and sample arms of the OCT system. The Doppler shift is directly measured by observing the phase of the interference over several A-lines at a single spatial location. The phase changes can then be used to calculate flow velocity if the angle between the vessel axis and optical axis is known and the direction of flow is not normal to the optical axis.

While DOCT was first successfully demonstrated using time-domain OCT (TD-OCT) systems [69–71], it is now almost exclusively performed using Fourier domain OCT (FD-OCT) systems [72–74]. The reason for this is due to the increased speed and sensitivity realized by the FD-OCT system, allowing real-time, *in vivo* measurements. An additional benefit of using an FD-OCT system is that the phase information is readily available through the complex Fast Fourier Transform (FFT) taken during the FD-OCT processing.

In typical DOCT, the Doppler frequency shift is defined as

$$\Delta f = \frac{\Delta \varphi}{2\pi \Delta t} \qquad (10.9)$$

where $\Delta \varphi$ is the average phase shift over repeated A-lines and Δt the time between consecutive A-line acquisitions. The range over which the Doppler shift can be accurately measured is governed by physical restraints. The minimum measurable shift is determined by the phase noise of the OCT system. This can be directly measured by looking at the phase off of an immobile specular reflector, such as a mirror. This is one area in which FD-OCT gains a significant advantage over TD-OCT due to its superior phase stability. The maximum Doppler shift that may be detected without ambiguity is the frequency corresponding to a change in phase of $\pm\pi$ radians. Larger shifts will phase wrap, resulting in a measured phase that is offset by 2π radians. Many algorithms have been developed to unwrap the phase; however, a single phase shift of 2π or larger results in a truly ambiguous signal, since the unwrapping techniques cannot reconcile whether the phase was wrapped only once, or whether it was wrapped multiple times.

Once the Doppler frequency shift has been measured, the flow velocity in the direction parallel to the optical axis (z) may be calculated by multiplying the Doppler shift by the fringe spacing, resulting in the following relation:

$$v(z) = \frac{\Delta \varphi(z) \lambda}{4\pi \Delta t} \qquad (10.10)$$

where λ is the center wavelength of the source. Since most flow in biological samples will not be parallel to the optical axis, this equation has limited use. To make it applicable to all vessel orientations, it must be noted that the measured Doppler shift corresponds to the projection of the velocity vector onto the optical axis. For this reason, the angle between the direction of flow and the optical axis must be known to obtain accurate flow velocity. The modified equation is

$$v(z) = \frac{\Delta \varphi(z) \lambda}{4\pi \Delta t \cos(\theta)} \qquad (10.11)$$

where θ is the angle between the particle flow direction and the optical axis. It is important to recognize that flow normal to the optical axis (i.e., $\theta = 90°$) cannot be evaluated.

DOCT has been successfully implemented in applications for ophthalmology, where retinal hemodynamics is related to diseases such as glaucoma [75], diabetic retinopathy [76], and AMD [77]. While other techniques such as angiography enable visualization of retinal vasculature, exogenous contrast agents such as fluorescein must be used [78]. DOCT does not require contrast agents, and thus is a less (non) invasive technique. Other noninvasive techniques for imaging retinal vasculature, such as laser speckle imaging [79] and laser Doppler imaging [80], have been developed; however, neither technique is capable of cellular-scale depth sectioning. DOCT provides high-resolution, 3D imaging of coregistered morphology and flow [81,82].

10.4.2.1 Applications of DOCT

DOCT has been used extensively for studying developmental models by imaging developing embryos [83–85]. Imaging the hemodynamics of the whole embryo along the entire developmental process provides insights into cardiac and vasculature development, which can in turn provide valuable information regarding cardiac and vascular disease [66,86,87]. Microcirculation in the brain [88] and kidney [89] has also been demonstrated.

There are also a few techniques that are similar to DOCT, in that they analyze the phase of consecutive A-lines to deduce information regarding flow in tissue. Phase variance OCT (pvOCT) [90] calculates the variance of the phase of consecutive spatial locations between multiple B-scans. Blood flow in the tissue can be mapped out using this technique, although flow velocity cannot be measured. Figure 10.9 shows a 3D dataset of human retinal vasculature imaged with pvOCT overlaid on a standard fluorescein angiography image.

Another technique derived from DOCT and used to image vasculature is optical microangiography (OMAG) [91]. OMAG is another phase-based, noninvasive method for mapping vasculature in three dimensions. OMAG is an extension of a technique termed "full-range complex FD-OCT" [92], which uses phase modulation of the OCT B-scan to suppress the complex conjugate image and utilize the entire Fourier spectral range for depth imaging. Further processing of the phase changes induced by moving scatterers yields the OMAG image. One OMAG volume can produce flow information both into and out of the plane of imaging. Figure 10.10 shows an OMAG image of a rat cortex taken with the skull intact.

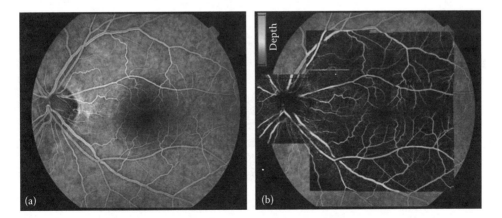

FIGURE 10.9 **(See color insert.)** Large field-of-view stitched pvOCT imaging overlaid on a fundus FA. (a) Fluorescein angiography. (b) Depth color-coded imaging with the 10 volumes. Total image acquisition time for 10 volumetric images is approximately 35 s. (Adapted from Kim, D.Y. et al., *Biomedical Optics Express*, 2011. 2(6): 1504–1513. With permission.)

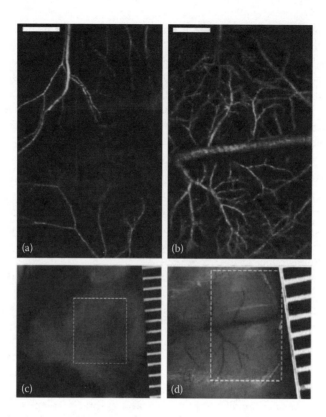

FIGURE 10.10 The head of an adult mouse with the skin and skull intact was imaged with OMAG *in vivo*. (a,b) Projection views of blood perfusion from within the skin and the brain cortex, respectively. Capillary blood flow can be seen from (b). It took ~7.5 min to acquire the 3D data to obtain (a) and (b) using the current system setup. (c) Photograph taken right after the experiments where viewing the vasculatures through the skin is impossible. (d) Photograph showing blood vessels over the cortex after the skull and the skin of the same mouse were carefully removed. The superficial major blood vessels show excellent correspondence with those in (b). The area marked with dashed white box represents 4.2×7.2 mm^2, and the scale bar indicates 1.0 mm. (Adapted from Wang, R.K.K. and S. Hurst, *Optics Express*, 2007. **15**(18): 11402–11412. With permission.)

References

1. Huang, D. et al., Optical coherence tomography. *Science*, 1991. **254**: 1178–1181.
2. Fercher, A.F., K. Mengedoht, and W. Werner, Eye-length measurement by interferometry with partially coherent light. *Optics Letters*, 1988. **13**: 186–188.
3. Fercher, A.F. et al., *In vivo* optical coherence tomography. *American Journal of Ophthalmology*, 1993. **116**(1): 113–114.
4. Bouma, B.E. and G.J. Tearney, *Handbook of Optical Coherence Tomography*, 2002, New York: Marcel Dekker. x, 741 pp., 16 p. of plates.
5. Brezinski, M.E., *Optical Coherence Tomography: Principles and Applications*, 2006, Amsterdam; Boston: Academic Press. Vol. xxvii, 599 pp.
6. Regar, E., T.G. van Leeuwen, and P.W. Serruys, *Optical Coherence Tomography in Cardiovascular Research*, 2007, Essex, United Kingdom: Informa UK Ltd.
7. Drexler, W., J.G. Fujimoto, and SpringerLink (Online service), *Optical Coherence Tomography Technology and Applications, in Biological and Medical Physics, Biomedical Engineering*, 2008, Berlin; New York: Springer. p. xxix, 1346 pp.

8. Sarunic, M.V., B.E. Applegate, and J.A. Izatt, Real-time quadrature projection complex conjugate resolved Fourier domain optical coherence tomography. *Optics Letters*, 2006. **31**: 2426–2428.

9. Choma, M.A., C.H. Yang, and J.A. Izatt, Instantaneous quadrature low-coherence interferometry with 3 × 3 fiber-optic couplers. *Optics Letters*, 2003. **28**(22): 2162–2164.

10. Sarunic, M.V. et al., Instantaneous complex conjugate resolved Fourier domain and swept-source OCT using 3 × 3 couplers. *Optics Express*, 2005. **13**: 957–967.

11. Davis, A.M., M.A. Choma, and J.A. Izatt, Heterodyne swept-source optical coherence tomography for complete complex conjugate ambiguity removal. *Journal of Biomedical Optics*, 2005. **10**(6): 064005.

12. Yun, S.H. et al., Removing the depth-degeneracy in optical frequency domain imaging with frequency shifting. *Optics Express*, 2004. **12**(20): 4822–4828.

13. Siegman, A.E., *Lasers*, 1986, Mill Valley, California: University Science Books. Vol. xxii, 1283 pp.

14. Yun, S.H. et al., High-speed spectral-domain optical coherence tomography at 1.3 μm wavelength. *Optics Express*, 2003. **11**(26): 3598–3604.

15. Choma, M.A., K. Hsu, and J.A. Izatt, Swept source optical coherence tomography using an all-fiber 1300-nm ring laser source. *Journal of Biomedical Optics*, 2005. **10**(4): 044009.

16. Flanders, D.C., *Short Cavity Tunable Laser with Mode Position Compensation*, U.S.P. Office, Editor 2002, Axsum Technologies, Inc., USA.

17. Huber, R., M. Wojtkowski, and J.G. Fujimoto, Fourier Domain Mode Locking (FDML): A new laser operating regime and applications for optical coherence tomography. *Optics Express*, 2006. **14**(8): 3225–3237.

18. Dorrer, C. et al., Spectral resolution and sampling issues in Fourier-transform spectral interferometry. *Journal of the Optical Society of America B—Optical Physics*, 2000. **17**(10): 1795–1802.

19. de Boer, J.F. et al., Improved signal-to-noise ratio in spectral-domain compared with time-domain optical coherence tomography. *Optics Letters*, 2003. **28**(21): 2067–2069.

20. Choma, M.A. et al., Sensitivity advantage of swept-source and Fourier-domain optical coherence tomography. *Optics Express*, 2003. **11**(18): 2183–2189.

21. Leitgeb, R., C.K. Hitzenberger, and A.F. Fercher, Performance of Fourier domain vs. time domain optical coherence tomography. *Optics Express*, 2003. **11**(8): 889–894.

22. Povazay, B. et al., Wide-field optical coherence tomography of the choroid in vivo. *Investigative Ophthalmology & Visual Science*, 2009. **50**(4): 1856–1863.

23. Torzicky, T. et al., High-speed retinal imaging with polarization-sensitive OCT at 1040 nm. *Optometry and Vision Science*, 2012. **89**(5): 585–592.

24. Yun, S.H. et al., Comprehensive volumetric optical microscopy in vivo. *Nature Medicine*, 2006. **12**(12): 1429–1433.

25. Tearney, G.J. et al., Three-dimensional coronary artery microscopy by intracoronary optical frequency domain imaging. *Jacc-Cardiovascular Imaging*, 2008. **1**(6): 752–761.

26. Tsai, M.T. and F.Y. Chang, Visualization of hair follicles using high-speed optical coherence tomography based on a Fourier domain mode locking laser. *Laser Physics*, 2012. **22**(4): 791–796.

27. Liu, G.J. et al., Imaging vibrating vocal folds with a high speed 1050 nm swept source OCT and ODT. *Optics Express*, 2011. **19**(12): 11880–11889.

28. Lee, S.W. et al., Quantification of airway thickness changes in smoke-inhalation injury using in-vivo 3-D endoscopic frequency-domain optical coherence tomography. *Biomedical Optics Express*, 2011. **2**(2): 243–254.

29. Luo, W. et al., Three-dimensional optical coherence tomography of the embryonic murine cardiovascular system. *Journal of Biomedical Optics*, 2006. **11**(2): 021014.

30. Davis, A.M. et al., *In vivo* spectral domain optical coherence tomography volumetric imaging and spectral Doppler velocimetry of early stage embryonic chicken heart development. *Journal of the Optical Society of America A—Optics Image Science and Vision*, 2008. **25**(12): 3134–3143.

31. Kagemann, L. et al., Repeated, noninvasive, high resolution spectral domain optical coherence tomography imaging of zebrafish embryos. *Molecular Vision*, 2008. **14**(253–255): 2157–2170.

32. Grulkowski, I. et al., Anterior segment imaging with spectral OCT system using a high-speed CMOS camera. *Optics Express*, 2009. **17**(6): 4842–4858.

33. Chamié, D. et al., *Three-Dimensional Fourier-Domain Optical Coherence Tomography Imaging: Advantages and Future Development.* Current Cardiovascular Imaging Reports, pp, 1–10.

34. Westphal, V. et al., Correlation of endoscopic optical coherence tomography with histology in the lower-GI tract. *Gastrointestinal Endoscopy*, 2005. **61**(4): 537–546.

35. Nguyen, C.T. et al., Noninvasive *in vivo* optical detection of biofilm in the human middle ear. *Proceedings of the National Academy of Sciences of the United States of America*, 2012. 109(24): 9529–9534.

36. Park, B.H. et al., Jones matrix analysis for a polarization-sensitive optical coherence tomography system using fiber-optic components. *Optics Letters*, 2004. **29**(21): 2512–2514.

37. Jiao, S.L. et al., Fiber-based polarization-sensitive Mueller matrix optical coherence tomography with continuous source polarization modulation. *Applied Optics*, 2005. **44**(26): 5463–5467.

38. Park, J. et al., Complex polarization ratio to determine polarization properties of anisotropic tissue using polarization-sensitive optical coherence tomography. *Optics Express*, 2009. **17**(16): 13402–13417.

39. Pircher, M., C.K. Hitzenberger, and U. Schmidt-Erfurth, Polarization sensitive optical coherence tomography in the human eye. *Progress in Retinal and Eye Research*, 2011. **30**(6): 431–451.

40. Matcher, S.J., A review of some recent developments in polarization-sensitive optical imaging techniques for the study of articular cartilage. *Journal of Applied Physics*, 2009. **105**(10): 102041.

41. Kaiser, M. et al., Noninvasive assessment of burn wound severity using optical technology: A review of current and future modalities. *Burns*, 2011. **37**(3): 377–386.

42. Fried, D. et al., Imaging caries lesions and lesion progression with polarization sensitive optical coherence tomography. *Journal of Biomedical Optics*, 2002. **7**(4): 618–627.

43. de Boer, J.F. and T.E. Milner, Review of polarization sensitive optical coherence tomography and Stokes vector determination. *Journal of Biomedical Optics*, 2002. **7**(3): 359–371.

44. Chipman, R.A., Polarization analysis of optical-systems. *Optical Engineering*, 1989. **28**(2): 90–99.

45. Kemp, N.J. et al., Fibre orientation contrast for depth-resolved identification of structural interfaces in birefringent tissue. *Physics in Medicine and Biology*, 2006. **51**(15): 3759–3767.

46. Yamanari, M. et al., Phase retardation measurement of retinal nerve fiber layer by polarization-sensitive spectral-domain optical coherence tomography and scanning laser polarimetry. *Journal of Biomedical Optics*, 2008. **13**(1): 014013.

47. Shurcliff, W.A. and S.S. Ballard, *Polarized Light.* 1964, Princeton, NJ: D. Van Nostrand.

48. Pedrotti, F.L. and L.S. Pedrotti, *Introduction to Optics.* 2nd ed. 1993, Englewood Cliffs, NJ: Prentice Hall.

49. Hecht, E., *Optics.* 3rd ed. 1998, Reading, MA: Addison-Wesley.

50. Azzam, R.M.A. and N.M. Bashara, *Ellipsometry and Polarized Light.* 1977, Amsterdam, New York: North-Holland Pub. Co. Vol. xvii, 529 pp.

51. Brosseau, C., *Fundamentals of Polarized Light: A Statistical Optics Approach.* 1998, New York: Wiley. Vol. xiv, 405 pp.

52. Kemp, N.J. et al., High-sensitivity determination of birefringence in turbid media with enhanced polarization-sensitive optical coherence tomography. *Journal of the Optical Society of America A— Optics Image Science and Vision*, 2005. **22**(3): 552–560.

53. Gotzinger, E., M. Pircher, and C.K. Hitzenberger, High speed spectral domain polarization sensitive optical coherence tomography of the human retina. *Optics Express*, 2005. **13**(25): 10217–10229.

54. Miyazawa, A. et al., Tissue discrimination in anterior eye using three optical parameters obtained by polarization sensitive optical coherence tomography. *Optics Express*, 2009. **17**(20): 17426–17440.

55. Fanjul-Velez, F. et al., Polarimetric analysis of the human cornea measured by polarization-sensitive optical coherence tomography. *Journal of Biomedical Optics*, 2010. **15**(5): 056004.

56. Gotzinger, E. et al., Imaging of birefringent properties of keratoconus corneas by polarization-sensitive optical coherence tomography. *Investigative Ophthalmology & Visual Science*, 2007. **48**(8): 3551–3558.

57. Townsend, K.A., G. Wollstein, and J.S. Schuman, Imaging of the retinal nerve fibre layer for glaucoma. *British Journal of Ophthalmology*, 2009. **93**(2): 139–143.

58. Elmaanaoui, B. et al., Birefringence measurement of the retinal nerve fiber layer by swept source polarization sensitive optical coherence tomography. *Optics Express*, 2011. **19**(11): 10252–10268.

59. Baumann, B. et al., Segmentation and quantification of retinal lesions in age-related macular degeneration using polarization-sensitive optical coherence tomography. *Journal of Biomedical Optics*, 2010. **15**(6): 061704.

60. Schlanitz, F.G. et al., Performance of automated Drusen detection by polarization-sensitive optical coherence tomography. *Investigative Ophthalmology & Visual Science*, 2011. **52**(7): 4571–4579.

61. Ugryumova, N. et al., Novel optical imaging technique to determine the 3-D orientation of collagen fibers in cartilage: Variable-incidence angle polarization-sensitive optical coherence tomography. *Osteoarthritis and Cartilage*, 2009. **17**(1): 33–42.

62. Pierce, M.C. et al., Collagen denaturation can be quantified in burned human skin using polarization-sensitive optical coherence tomography. *Burns*, 2004. **30**(6): 511–517.

63. Kang, H.B., C.L. Darling, and D. Fried, Nondestructive monitoring of the repair of enamel artificial lesions by an acidic remineralization model using polarization-sensitive optical coherence tomography. *Dental Materials*, 2012. **28**(5): 488–494.

64. Chen, Z. et al., Noninvasive imaging of *in vivo* blood flow velocity using optical Doppler tomography. *Optics Letters*, 1997. **22**: 1119–1121.

65. Yazdanfar, S., A.M. Rollins, and J.A. Izatt, Imaging and velocimetry of the human retinal circulation with color Doppler optical coherence tomography. *Optics Letters*, 2000. **25**(19): 1448–1450.

66. Yang, V.X. et al., High speed, wide velocity dynamic range Doppler optical coherence tomography (Part II): Imaging *in vivo* cardiac dynamics of *Xenopus laevis*. *Optics Express*, 2003. **11**(14): 1650–1658.

67. Mariampillai, A. et al., Doppler optical cardiogram gated 2D color flow imaging at 1000 fps and 4D *in vivo* visualization of embryonic heart at 45 fps on a swept source OCT system. *Optics Express*, 2007. **15**(4): 1627–1638.

68. Kehlet Barton, J. et al., Three-dimensional reconstruction of blood vessels from *in vivo* color Doppler optical coherence tomography images. *Dermatology*, 1999. **198**(4): 355–361.

69. Izatt, J.A. et al., *In vivo* bidirectional color Doppler flow imaging of picoliter blood volumes using optical coherence tomography. *Optics Letters*, 1997. **22**(18): 1439–1441.

70. Chen, Z.P. et al., Optical Doppler tomographic imaging of fluid flow velocity in highly scattering media. *Optics Letters*, 1997. **22**(1): 64–66.

71. Zhao, Y. et al., Phase-resolved optical coherence tomography and optical Doppler tomography for imaging blood flow in human skin with fast scanning speed and high velocity sensitivity. *Optics Letters*, 2000. **25**(2): 114.

72. White, B.R. et al., *In vivo* dynamic human retinal blood flow imaging using ultra-high-speed spectral domain optical Doppler tomography. *Optics Express*, 2003. **11**(25): 3490–3497.

73. Leitgeb, R.A. et al., Real-time measurement of *in vitro* flow by Fourier-domain color Doppler optical coherence tomography. *Optics Letters*, 2004. **29**(2): 171–173.

74. Zhang, J. and Z. Chen, *In vivo* blood flow imaging by a swept laser source based Fourier domain optical Doppler tomography. *Optics Express*, 2005. **13**(19): 7449–7457.

75. Flammer, J. et al., The impact of ocular blood flow in glaucoma. *Progress in Retinal and Eye Research*, 2002. **21**(4): 359–393.

76. Cunha-Vaz, J.G. et al., Studies on retinal blood flow. II. Diabetic retinopathy. *Archives of Ophthalmology*, 1978. **96**(5): 809–811.

77. Friedman, E., A hemodynamic model of the pathogenesis of age-related macular degeneration. *American Journal of Ophthalmology*, 1997. **124**(5): 677–682.

78. Kang, S.W., C.Y. Park, and D.I. Ham, The correlation between fluorescein angiographic and optical coherence tomographic features in clinically significant diabetic macular edema. *American Journal of Ophthalmology*, 2004. **137**(2): 313–322.

79. Briers, J.D. and A.F. Fercher, Retinal blood-flow visualization by means of laser speckle photography. *Investigative Ophthalmology & Visual Science*, 1982. **22**(2): 255–259.

80. Michelson, G. et al., Principle, validity, and reliability of scanning laser Doppler flowmetry. *Journal of Glaucoma*, 1996. **5**(2): 99–105.

81. Makita, S. et al., Comprehensive *in vivo* micro-vascular imaging of the human eye by dual-beam-scan Doppler optical coherence angiography. *Optics Express*, 2011. **19**(2): 1271–1283.

82. Schmoll, T., C. Kolbitsch, and R.A. Leitgeb, Ultra-high-speed volumetric tomography of human retinal blood flow. *Optics Express*, 2009. **17**(5): 4166–4176.

83. Larina, I.V. et al., Live imaging of blood flow in mammalian embryos using Doppler swept-source optical coherence tomography. *Journal of Biomedical Optics*, 2008. **13**(6): 060506.

84. Jenkins, M.W. et al., Measuring hemodynamics in the developing heart tube with four-dimensional gated Doppler optical coherence tomography. *Journal of Biomedical Optics*, 2010. **15**(6): 066022.

85. Larina, I.V. et al., Live imaging of rat embryos with Doppler swept-source optical coherence tomography. *Journal of Biomedical Optics*, 2009. **14**(5): 050506.

86. Yelbuz, T.M. et al., Optical coherence tomography—A new high-resolution imaging technology to study cardiac development in chick embryos. *Circulation*, 2002. **106**(22): 2771–2774.

87. Li, P. et al., Assessment of strain and strain rate in embryonic chick heart *in vivo* using tissue Doppler optical coherence tomography. *Physics in Medicine and Biology*, 2011. **56**(22): 7081–7092.

88. Srinivasan, V.J. et al., Quantitative cerebral blood flow with optical coherence tomography. *Optics Express*, 2010. **18**(3): 2477–2494.

89. Wierwille, J. et al., *In vivo*, label-free, three-dimensional quantitative imaging of kidney micro-circulation using Doppler optical coherence tomography. *Laboratory Investigation*, 2011. **91**(11): 1596–1604.

90. Kim, D.Y. et al., *In vivo* volumetric imaging of human retinal circulation with phase-variance optical coherence tomography. *Biomedical Optics Express*, 2011. **2**(6): 1504–1513.

91. Wang, R.K. et al., Three dimensional optical angiography. *Optics Express*, 2007. **15**(7): 4083–4097.

92. Wang, R.K.K., *In vivo* full range complex Fourier domain optical coherence tomography. *Applied Physics Letters*, 2007. **90**(5): 054103.

93. Wang, R.K.K. and S. Hurst, Mapping of cerebro-vascular blood perfusion in mice with skin and skull intact by optical micro-angiography at 1.3 μm wavelength. *Optics Express*, 2007. **15**(18): 11402–11412.

11

Biophotonics: Clinical Fluorescence Spectroscopy and Imaging

W.R. Lloyd III
University of Michigan

L.-C. Chen
University of Michigan

R.H. Wilson
University of Michigan

M.-A. Mycek
University of Michigan

11.1 Introduction

The goal of this chapter is to provide an overview of several well-known fluorescence spectroscopy and imaging techniques that have been employed in clinical settings. The chapter provides an introduction to fluorescence, starting with a discussion of fluorescence techniques and instrumentation, followed by clinical applications, and closing with clinical design considerations for those researchers designing and developing their own instrumentation. The concepts described within this chapter are also applicable to laboratory bench research and small-animal studies.

Fluorescence detection has been employed for a range of spectroscopy and imaging applications in biomedicine, owing to the high sensitivity and selectivity (Lakowicz 2006) inherent in the techniques. Applications of fluorescence sensing in biomedicine include cell sorting (Kumar and Borth 2012), cell viability assays (Darzynkiewicz et al. 1994), cell function (Lippincott-Schwartz 2011), fluorescence resonance energy transfer (FRET) (Roy et al. 2008), and DNA analysis (Ansorge 2009), including applications in both animal and human studies (Mycek and Pogue 2003). For clinical applications, tissue fluorescence is sensitive to a number of morphological and biochemical changes that occur during disease progression. Fluorescence spectroscopy and imaging techniques have shown significant potential for clinically translatable tissue diagnostics in small animals and humans (Wagnieres et al. 1997, 1998).

11.2 Fluorescence

Each photon has a characteristic energy according to its wavelength and the speed of light, as governed by the Planck–Einstein equation. If an excitation photon is absorbed by a molecule, fluorescence emission is one of the several processes that can occur. The fluorescence process is illustrated in Figure 11.1. When a photon is absorbed by a fluorescent molecule, a ground-state electron is raised to a higher-energy quantum state. The electron may then radiatively decay back to the ground quantum state, yielding a fluorescence photon of longer wavelength and less energy than the excitation photon.

In cells and tissues, fluorescent molecules can be endogenous or exogenous. Endogenous fluorophores are naturally occurring biomolecules capable of fluorescence. Exogenous fluorophores are nonnative biomolecules that can be employed as labels to increase contrast from tissues, most frequently employed during imaging applications to localize a specific target molecule.

There are advantages and drawbacks to relying on endogenous or exogenous fluorescence for studies in cells and tissues. Advantages to endogenous fluorescence sensing include label-free, nonperturbing tissue assessment. Applications employing endogenous fluorescence sensing require no additional United States Food and Drug Administration (FDA) approvals for safety of exogenous fluorescent dyes. Additionally, endogenous fluorescence often varies owing to factors related to disease progression and tissue microenvironment (Lakowicz 2006; Provenzano et al. 2009). However, endogenous fluorophores typically have lower fluorescence yields than exogenous fluorophores (Lakowicz 2006; Wagnieres et al. 1998). A number of commonly studied native tissue fluorophores are excited in the near-ultraviolet and visible regions of the electromagnetic spectrum. These include extracellular fluorophores, such as collagen, keratin, and elastin, and intracellular fluorophores, such as nicotinamide adenine dinucleotide (phosphate) (NAD(P)H) and flavin adenine dinucleotide (FAD) (Wagnieres et al. 1998).

Exogenous fluorophores are often employed as molecules that selectively target tumor cells (Kuil et al. 2010; Zhang et al. 2010a). Fluorescent dyes can be introduced into the bloodstream to map out lymph nodes (Zhang et al. 2010b) and blood vessels (Barretto et al. 2011; Kalchenko et al. 2012). Advantages to employing exogenous dyes include relatively high fluorescence yields from molecules that can be developed for specific binding, thereby illuminating a particular disease state. However, in order for an exogenous fluorophore to be approved by the FDA for *in vivo* human use, the safety of that fluorophore must first be rigorously confirmed.

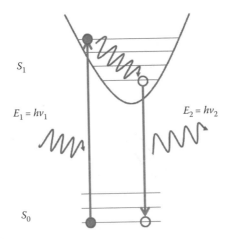

FIGURE 11.1 The fluorescence process can be described with a simplified Jablonski diagram. Fluorescence occurs when an excitation photon is absorbed by a fluorescent molecule in the ground state (S_0) resulting in an electron in a higher-energy quantum state (S_1). The excited electron decays to the lowest energy level in the higher quantum state and then radiatively relaxes back to the ground state, yielding a fluorescence emission photon with energy less than (wavelength longer than) the excitation photon.

The interrogated fluorophores and sampled tissue volume are determined by the design of the experimental set-up. The excitation wavelength determines which fluorophores in a sample will be excited. For many *in vivo* applications that use endogenous fluorescence, near-ultraviolet wavelengths are employed for excitation. The tissue volume studied depends on the wavelength and delivery method of the excitation light, the tissue optical properties at the fluorescence excitation and emission wavelengths, and the method employed to collect the fluorescent light that returns to the surface. For clinical applications, fiber-optic fluorescence spectroscopy is often employed to analyze tissue volumes in the order of 1 mm³ (Fang et al. 2011; Utzinger and Richards-Kortum 2003) or employed with depth-selective fiber-optic probes to excite specific regions of the tissue (Utzinger and Richards-Kortum 2003). Fluorescence-imaging techniques can be employed to optically interrogate thin tissue sections with resolution enabling single-molecule detection (Huang et al. 2009). With two-photon excitation microscopic fluorescence imaging, tissue layers less than 5 μm in thickness can be "optically sectioned" for imaging (Chen et al. 2012; Gareau et al. 2004; Pena et al. 2005).

11.3 Clinical Fluorescence Techniques and Instrumentation

Fluorescence methods employed for tissue diagnostics fall into two broad categories, spectroscopy and imaging, for fluorescence intensity or fluorescence lifetime sensing. This section provides an overview of the techniques, instrumentation, data acquisition, and data analysis associated with these methods.

11.3.1 Spectroscopy

Fluorescence spectroscopy measures the intensity spectrum of emitted light from a sample after interrogation by a narrow-band excitation source (Vishwanath and Ramanujam 2011). One implementation of fluorescence spectroscopy in tissues is illustrated in Figure 11.2. Here, excitation photons (at wavelength λ_{ex}) are delivered to the tissue and propagate according to the tissue optical properties in each layer (absorption coefficient μ_{ai}, scattering coefficient μ_{si}). Some of these photons will be absorbed by

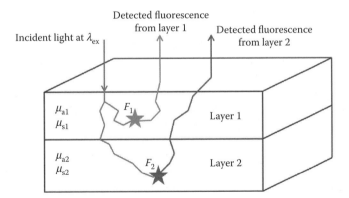

FIGURE 11.2 Fluorescence spectroscopy is characterized by excitation photons (incident light) entering a tissue at wavelength λ_{ex}, propagating according to the tissue optical properties (absorption coefficient μ_{ai}, scattering coefficient μ_{si}; i = layer, all $\mu = f(\lambda)$) at the excitation wavelength, being absorbed and isotropically reemitted as fluorescence by a fluorophore (F_1, F_2; stars in layer 1 and layer 2, respectively), and fluorescence photons propagating back to the tissue surface according to the tissue optical properties at the emission wavelength. Tissue optical properties are frequently different in each layer of a multilayered tissue, impacting the photon trajectories in each layer. Upon exiting from the tissue surface, the photons (detected fluorescence from layer 1 and layer 2) can be collected by fiber-optic probes and sent to a detector. The detected fluorescence intensity spectrum contains information about the fluorophores in each tissue layer. Measured data can be analyzed to characterize the sample and extract tissue information, including fluorophore concentrations and tissue optical properties.

fluorophores (e.g., F_1, F_2) and reemitted as fluorescence photons at a wavelength that is longer than the excitation wavelength. These fluorescence photons will then continue to propagate through the tissue, and those that arrive back at the tissue surface can be detected. Detected photons are commonly directed to a spectrograph and charge-coupled device camera (Chandra et al. 2006; Vishwanath and Ramanujam 2011) or spectrometer (Fawzy and Zeng 2008) to measure wavelength-resolved spectra. To detect photons that reached the surface at different distances from the source, fiber-optic probes with different source–detector separations can be employed.

The tissue sample volume interrogated is directly related to the experimental design: excitation/detection fiber-optic probe geometry, excitation source wavelength, and tissue optical properties (Vishwanath and Mycek 2005). For example, for fiber-optic spectroscopy studies, the interrogated sample volume will be related to the size of the optical fibers employed and their center-to-center spacing. Alternative source–detector geometries can be employed to change the sample volume interrogated (Utzinger and Richards-Kortum 2003).

11.3.2 Imaging

A fluorescence image of a tissue is a spatially resolved map of detected tissue fluorescence across a specimen. The resulting 2D (or 3D) image of the tissue is comprised of individual "pixels" of fluorescence information in two (or three) spatial dimensions. Depending on the clinical application, images can be acquired from the tissue surface (e.g., tumor-margin detection or endoscopic imaging; Pogue et al. 2010; Thong et al. 2011) or from the bulk sample (e.g., tomographic imaging; Chaudhari et al. 2009; Kumar et al. 2008). Fluorescence images can be acquired with macroscopic (e.g., wide-field whole-animal imaging; Leblond et al. 2010; McGinty et al. 2010) or microscopic (e.g., imaging of individual cells; Graves et al. 2004; Leblond et al. 2010; Skala et al. 2007) spatial resolution. Image contrast can originate from endogenous (e.g., collagen, NADH, or FAD; Chen et al. 2012; Skala et al. 2007) or exogenous (e.g., fluorescent dyes; Fujiwara et al. 2009; Kosaka et al. 2009) molecular fluorophores. Fluorescence imaging can be applied to clinical applications (e.g., human; Weissleder and Pittet 2008) or preclinical in small animals (e.g., murine tumor model; Leblond et al. 2010). Finally, fluorescence images can spatially map fluorescence intensity (Chen et al. 2012) or fluorescence lifetime (Chang et al. 2007) information (described below). Here, we will discuss fluorescence-imaging technology and its application to cancer detection.

Fluorescence-imaging systems comprise excitation light sources, light delivery and collection optics, filters, and technology for spatially resolved fluorescence detection (Leblond et al. 2010). Light sources for fluorescence excitation include broad-wavelength sources (e.g., UV or xenon arc lamps; Thong et al. 2009) or narrow-wavelength sources (e.g., lasers or LEDs; Chen et al. 2012). For a wide-field imaging system, the excitation beam is expanded before illuminating the sample (Chang et al. 2007) and the fluorescence is detected with a CCD camera (Chang et al. 2009). For a point-scanning imaging system, a scanning unit (e.g., a set of mirrors on galvanometers) is employed to translate the focused excitation beam over the imaging field (Chen et al. 2012) and fluorescence emission is detected from each scanned point with a high-sensitivity photomultiplier tube or avalanche photodiode to build-up the image pixel-by-pixel. Objective lenses can be used to enhance image resolution, especially at the microscopic level (Barretto et al. 2009). For tomographic applications, fiber-optic bundles are commonly employed to excite and detect fluorescence from different spatial locations, and the resulting data set is analyzed to create images of fluorophore distributions in tissues (Chaudhari et al. 2009; Leblond et al. 2010). Thus, fluorescence technologies can be employed to acquire images at the single cell, tissue, or whole-body levels, as shown in Figure 11.3.

11.3.3 Fluorescence Lifetime Sensing

The electronic excited-state lifetime of a fluorophore is the average time that an excited electron spends in the excited state and is typically in the order of picoseconds to nanoseconds in duration. Fluorophore

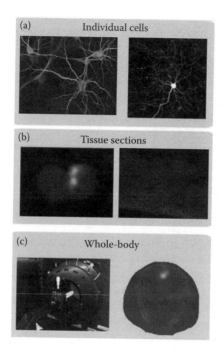

FIGURE 11.3 (**See color insert.**) Fluorescence imaging was employed for *in vitro*, *ex vivo*, and *in vivo* preclinical studies. (a) Individual cells were imaged with two-photon microscopy that shows *in vitro* mouse hippocampal neuron and glial cells. Neurons and glial cells were transfected with green fluorescent protein (GFP) (green) and red fluorescent protein (red), respectively. (Courtesy of Dr. De Koninck.) (b) Tissue sections, *ex vivo* sliced brain tissue sections, were imaged from a mouse that show glioma cells highlighted with fluorescence from GFP (green) and Protoporphyrin IX (PpIX, red). (c) Whole-body fluorescence tomography was employed to spatially map *in vivo* PpIX fluorescence from a brain tumor model in a mouse. (Reprinted from *Journal of Photochemistry and Photobiology B*, 98(1), F. Leblond et al., Preclinical whole-body fluorescence imaging: Review of instruments, methods and applications, 77–94, Copyright 2010, with permission from Elsevier, and Dr. Paul De Koninck, www.greenspine.ca.)

lifetime is an intrinsic property of each individual fluorophore (Lakowicz 2006) and is affected by fluorophore microenvironment (Lakowicz 2006), including tissue pH, oxygenation, and temperature, while being generally insensitive to fluorescence intensity-based artifacts such as optical scattering from collagen fibers and optical absorption from blood (Chorvat and Chorvatova 2009; Vishwanath and Mycek 2004). Relative fluorophore contributions can be resolved from time-resolved fluorescence measurements even if the steady-state spectra of the fluorophores overlap (Chorvat and Chorvatova 2009; Lloyd et al. 2010).

Fluorophore lifetime can be detected in either the time (Lakowicz 2006; Lloyd et al. 2010) or frequency (Boens et al. 2007; Lakowicz 2006) domains. Once limited to *in vitro* tissue studies, the first endoscopic *in vivo* fluorescence lifetime spectroscopy measurements on human tissues were made on colonic polyps in 1998, in the time domain, and showed promise as an optical diagnostic technique capable of distinguishing precancerous tissue types *in vivo* (Mycek et al. 1998). Here, we will discuss time-domain lifetime sensing instrumentation and its application for clinical diagnostics.

Instrumentation for time-domain fluorescence lifetime sensing commonly employs one of two techniques for data acquisition: (1) time-correlated single-photon counting (TCSPC) or (2) direct waveform recording (DWR; Figure 11.4). [Other time-domain technologies, including streak cameras (Glanzmann et al. 1999), do not offer significant advantages to TCSPC or DWR.] TCSPC can be considered the current gold standard for fluorescence-lifetime detection, because of its high accuracy and sensitivity to low

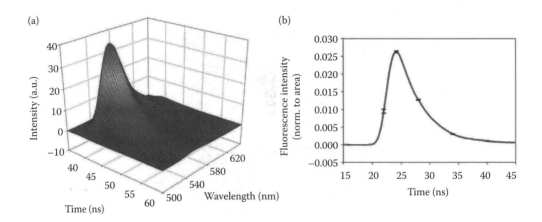

FIGURE 11.4 A clinically compatible system was developed to measure wavelength- and time-resolved fluorescence intensity simultaneously via DWR. (a) A representative fluorescence WTM is shown. The data contained within a WTM can yield wavelength-resolved data by integrating the WTM over time or time-resolved data by integrating the WTM over wavelength. (b) A time-resolved fluorescence decay curve is obtained from repeated measurements at a single emission wavelength. This decay curve provides a precise measurement of time-resolved fluorescence, as the error bars represent a standard deviation no greater than 1% of total signal at peak intensity. This high-precision data has the potential to enable quantitative analysis of subtle differences in measured fluorescence data resulting from changes in tissue parameters with disease (see data analysis and modeling section for more information). (W.R. Lloyd et al. Instrumentation to rapidly acquire fluorescence wavelength-time matrices of biological tissues. *Biomedical Optics Express* 1(2); 2010. Reprinted with permission from Optical Society of America.)

fluorophore concentrations (Birch and Imhof 2002; Chang et al. 2007; Chorvat and Chorvatova 2009; DeBeule et al. 2007; Lakowicz 2006; Urayama et al. 2003), but the relatively long data-acquisition times relative to DWR (acquisition speeds for comparable data collection are 10^5 times longer than DWR; Muretta et al. 2010) are disadvantageous for clinical applications.

DWR detection commonly employs transient digitizers coupled to photomultiplier tubes (Lloyd et al. 2010) or avalanche photodiodes (Chandra et al. 2006). One limitation of common DWR systems is lack of compatibility with high-repetition rate sources, relying instead on low-repetition rate lasers (e.g., nitrogen lasers operating at ~10 Hz). The newest developed DWR systems can operate with laser sources at repetition rates up to 25 kHz, drastically increasing data collection speeds with improved signal averaging (Lloyd et al. 2010; Muretta et al. 2010).

One difficulty in monitoring biological specimens is that endogenous fluorophores (including collagen, elastin, keratin, NAD(P)H, and FAD) have broad, overlapping fluorescence emission spectra in the wavelength range from ~300 to 800 nm (Wagnieres et al. 1998). One technique to isolate relative fluorescence contributions from individual species within a multifluorophore sample is to collect time-resolved fluorescence decays and extract fluorescence lifetimes (Chandra et al. 2006). With the addition of a monochromator or band-pass filters (Figure 11.4), fluorescence decays can be collected at multiple emission wavelengths (Chorvat and Chorvatova 2009; Lloyd et al. 2010), providing wavelength-time matrices (WTMs) for quantitative analysis.

11.3.4 Data Analysis and Modeling

The detected fluorescence signal can be mathematically modeled as a linear combination of the fluorescence spectra of the fluorophores emitting in the detected wavelength range (Chang et al. 2006; Lakowicz 2006; Vo-Dinh 2003; Volynskaya et al. 2008; Wilson et al. 2010). For a medium with N fluorophores that emit in the wavelength range of interest, this equation takes the following steady-state (Equation 11.1) and time-resolved (Equation 11.2) forms:

$$F(\lambda) = \sum_{i=1}^{N} C_i F_i(\lambda) \tag{11.1}$$

$$F(t) = \sum_{i=1}^{N} C_i F_i(t) \tag{11.2}$$

In Equations 11.1 and 11.2, $F_i(\lambda)$ ($F_i(t)$) are the fluorescence spectra (decays) from each fluorophore in the tissue, and their relative contributions to the total fluorescence $F(\lambda)$ ($F(t)$) are weighted according to the fit coefficients C_i. Fitting Equation 11.1 to the steady-state fluorescence or fitting Equation 11.2 to the time-resolved fluorescence will extract the percentage contributions $\%C_i$ of each fluorophore to the total fluorescence:

$$\%C_i = (100)\left(C_i \middle/ \sum_{i=1}^{N} C_i \right) \tag{11.3}$$

For fluorescence studies in tissues, it is important to note that in Equations 11.1 and 11.2, $F(\lambda)$ and $F(t)$ represent the native "intrinsic" fluorescence of the tissue without any distortion from tissue absorption and scattering. However, the measured steady-state fluorescence $F_{measured}(\lambda)$ reflects the "intrinsic fluorescence" that has (typically) been attenuated by different amounts at different wavelengths, in accordance with the absorption and scattering properties of the medium. These attenuation artifacts must be removed from the measured spectrum (Figure 11.5) before the data are modeled with Equation 11.1. This procedure typically employs information from concurrently measured tissue reflectance spectra and can be accomplished with photon propagation models (Fawzy and Zeng 2008; Fawzy et al. 2006; Müller et al. 2001; Volynskaya et al. 2008; Wu et al. 1993), photon–tissue interaction models (Wilson et al. 2009, 2010), or Monte-Carlo simulations (Palmer and Ramanujam 2006, 2008).

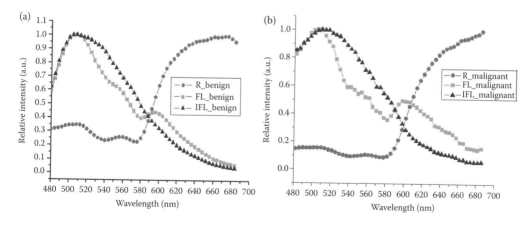

FIGURE 11.5 Correction of measured fluorescence spectra (FL, light gray squares) from benign (a) and malignant (b) human lung tissue lesions for absorption and scattering, using measured reflectance data (R, gray circles), to produce attenuation-free "intrinsic" steady-state tissue fluorescence (IFL, black triangles). (Reprinted from Y.S. Fawzy and H. Zeng. Intrinsic fluorescence spectroscopy for endoscopic detection and localization of the endobronchial cancerous lesions. *Journal of Biomedical Optics* 13; 2008. With permission from SPIE.)

By contrast, for bulk tissues, the fluorescence lifetime extracted from measured time-resolved fluorescence decays is generally free of absorption and scattering artifacts, so it can be modeled directly with Equation 11.2, without the need for attenuation correction, using techniques such as multiexponential decay (Lakowicz 2006), stretched exponential decay (Lee et al. 2001), Laguerre deconvolution (Dabir et al. 2009), and phasor analysis (Digman et al. 2008). In layered epithelial tissues, measured fluorescence decay times can reflect both tissue morphology and biochemistry, as demonstrated by Monte-Carlo simulations of time-resolved fluorescence (Vishwanath and Mycek 2004, 2005).

11.4 Clinical Applications

The sensitivity of fluorescence measurements to disease-related biochemical and morphological changes in tissue has led to promising developments in fluorescence-based diagnostic technology for a wide variety of clinical applications. These applications range from cancer diagnostics to dental caries (Thomas et al. 2010) to inflammatory diseases, such as arthritis (Werner et al. 2011) or Crohn's disease (Hundorfean et al. 2012). Potential future clinical applications include diabetes, neurological diseases, arthritis, and metabolic diseases (Kirschner 2010). The remainder of this section will be focused on cancer diagnostics.

11.4.1 Cancer Diagnostics

Fluorescence techniques have been developed and employed for clinical cancer diagnostics for nearly two decades, as chronicled in a recent review article (Liu 2011). Cancers are characterized by a number of biological and morphological tissue changes, which can be markedly different among tissue types, thus requiring specific fluorescence instrumentation and analysis algorithms for accurate classification.

A wide variety of cancers have been previously studied with fluorescence-sensing methods, including head and neck (Meier et al. 2010) [including oral (Roblyer et al. 2008), skin (DeBeule et al. 2007; Fujiwara et al. 2009; Konig et al. 2007; Rigel et al. 2010), breast (Poellinger et al. 2011; Volynskaya et al. 2008), colon (Mycek et al. 1998), brain (Butte et al. 2011), lung (Kobayashi et al. 2002; Kusunoki et al. 2000), stomach (Mayinger et al. 2004a,b), prostate (Boutet et al. 2009), cervix (Chang et al. 2006), ovarian (van Dam et al. 2011), and pancreatic (Chandra et al. 2007b, 2009; Wilson et al. 2009, 2010)]. Although some of these studies were *ex vivo*, the fluorescence technology reported is clinically translatable and the research goal is to develop *in vivo* clinical technology.

Below, we will focus on three subfields of cancer diagnostics: point-spectroscopic detection, imaging for cancer detection and tumor margin assessment, and endoscopy. Multiple optical modalities can be employed simultaneously to increase contrast between cancerous and normal tissues (Evers et al. 2012). For example, optical-reflectance technology has been used in conjunction with fluorescence instrumentation to target the absorption and scattering properties of tissues (Wilson et al. 2010). Here, devices with dual modalities will be mentioned, but the results shown are only from clinical fluorescence technology.

11.4.1.1 Point Spectroscopic Detection

Fluorescence spectroscopy measurements have the potential to detect cancers occurring in bulk tissues. Point detection measurements are typically characterized by analyzing bulk tissue volumes (~1 mm³) using fiber-optic spectroscopy. The exact volume interrogated with specific fiber-optic probe geometry and spacing will vary with optical tissue parameters and can be estimated with Monte-Carlo simulations (Vishwanath and Mycek 2004, 2005).

The first noninvasive optical assessment for pancreatic cancer using bimodal (fluorescence and reflectance) optical spectroscopy, a real-time measurement technique that is compatible with minimally invasive diagnostic procedures, was performed recently (Chandra et al. 2007a, 2010; Wilson et al. 2009, 2010). Figure 11.6 shows the average fluorescence spectra from 96 measurements, with visible differences between normal, chronic pancreatitis (inflamed), and adenocarcinoma (cancerous)

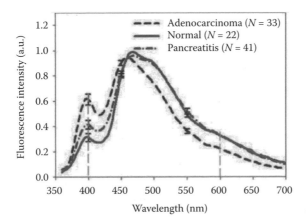

FIGURE 11.6 Mean fluorescence spectra for point detection of pancreatic cancer from freshly excised human pancreatic tissues. Visible differences were observed from normal, chronic pancreatitis (inflammation), and adenocarcinoma (cancerous) sites. Observed fluorescence differences at ~400 nm were attributed to increased extracellular collagen content, whereas differences at wavelengths >500 nm were attributed to intracellular fluorescence from NAD(P)H and FAD. Error bars represent standard error. (Reprinted from M. Chandra et al. Spectral areas and ratios classifier algorithm for pancreatic tissue classification using optical spectroscopy. *Journal of Biomedical Optics* 15(1); 2010. With permission from SPIE.)

tissues resulting from changes in tissue biochemistry and morphology (Chandra et al. 2010; Wilson et al. 2010). Primary differences were observed in the fluorescence spectra between adenocarcinoma (pancreatic cancer) and benign tissues (both normal tissue and chronic pancreatitis) at ~400 nm and at wavelengths greater than 500 nm. The differences at ~400 nm were attributed to changes in the relative tissue collagen concentration. The differences at wavelengths greater than 500 nm were attributed to relative changes in the cellular NADH and FAD concentrations. In total, 96 spectra (from 50 sites) were measured. Developed classifier algorithms resulted in a sensitivity, specificity, negative predictive value, and positive predictive value of 85%, 89%, 92%, and 80%, respectively (Chandra et al. 2010).

Point spectroscopy has also been employed to detect head-and-neck carcinomas with time-resolved laser-induced fluorescence spectroscopy, using a nitrogen laser to excite *in vivo* fluorescence primarily from collagen and NADH. Preliminary results showed that malignant tissue had lower fluorescence with the fluorescence peak intensity occurring at longer wavelengths. Average lifetimes showed significant differences (Meier et al. 2010). In another fluorescence lifetime study, a clinical system was developed to endoscopically diagnose squamous cell carcinomas in concert with white-light imaging. Results showed that the detection sensitivity for early lesions with fluorescence sensing was 1.6 times higher than that with white-light bronchoscopy alone (Gabrecht et al. 2007).

11.4.1.2 Imaging for Cancer Detection and Tumor-Margin Assessment

Fluorescence imaging has been shown to improve upon the diagnostic capabilities of standard white-light reflectance imaging of tissue surfaces (DeBeule et al. 2007; de Leeuw et al. 2009; Fujiwara et al. 2009) and therefore has potential for detecting surface epithelial cancers, which account for ~85% of all cancers (Medicine). One important application of fluorescence imaging is tumor margin detection: the ability to delineate cancer from normal tissues during biopsy and surgical excision.

Oral cancer has been a primary target for optical technology because the oral cavity is readily accessible with hand-held optical probes (Lane et al. 2006, 2012a,b; McGee et al. 2009; Pavlova et al. 2008; Poh et al. 2007, 2009; Scheer et al. 2011; Truelove 2010). This technology has begun to show clinical promise not only for cancerous lesion detection, but also for detecting precancers and other high-risk conditions (Poh et al. 2007). Figure 11.7 shows a fluorescence measurement acquired from a hand-held

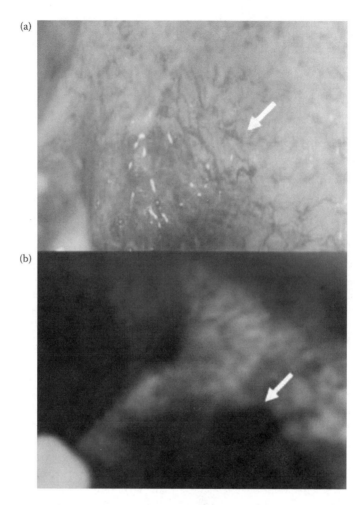

FIGURE 11.7 (**See color insert.**) A pilot study was performed to detect oral malignancy with fluorescence imaging. (a) Under white-light imaging, there is little visible contrast between normal and cancerous tissue. (b) After excitation with 400–460-nm blue light, the endogenous fluorescence image contains a noticeable dark region (white arrow). After histological analysis, this spot was confirmed to be carcinoma *in situ*. (C.F. Poh et al., Direct fluorescence visualization of clinically occult high-risk oral premalignant disease using a simple hand-held device. *Head & Neck*. 2007. 29(1). Copyright Wiley-VCH Verlag GmbH & Co. KgaA. Reprinted with permission.)

probe that illuminates the tissue with blue light from 400 to 460 nm and detects green and red (~500–700 nm) autofluorescence. Figure 11.7 shows one such suspicious lesion inside the oral cavity, where a malignancy is not visible under white-light imaging, but visible in the fluorescence image. Cancerous and suspicious tissues exhibited decreased autofluorescence relative to normal mucosa, attributed to changes in collagen biochemistry, increased hemoglobin absorption, and decreased FAD fluorescence (Lane et al. 2006). An initial pilot study with 44 patients distinguished carcinomas (severe dysplasia, carcinoma *in situ*, or invasive carcinoma) from normal mucosa with 98% sensitivity and 100% specificity (Lane et al. 2006).

Skin cancer has also been a prominent target for fluorescence-based technologies (Brancaleon et al. 2001; DeBeule et al. 2007; de Leeuw et al. 2009; Fujiwara et al. 2009). One approach employed hyperspectral fluorescence lifetime imaging to obtain spectral and temporal decay profiles after excitation from both 355 and 435–445 nm sources (DeBeule et al. 2007). Preliminary results from 23 human sites

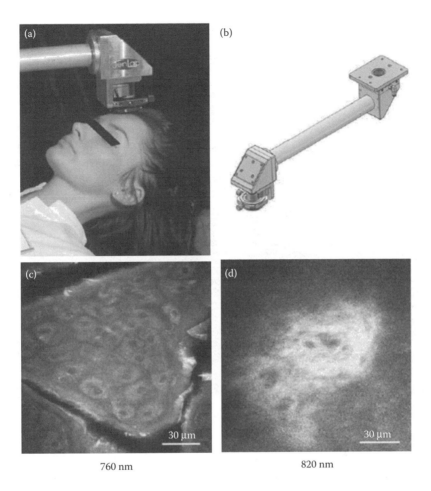

760 nm   820 nm

FIGURE 11.8 Imaging capabilities of DermaInspect system (a, b), a nonlinear optical microscopic fluorescence imaging system, are demonstrated with images from *in vivo* human skin: stratum spinosum (c) and a papilla (d). The images show morphological features similar to those visible via histopathology, demonstrating the potential of the technique for tissue characterization. (K. König et al.: Multiphoton tissue imaging using high-NA microendoscopes and flexible scan heads for clinical studies and small animal research. *Journal of Biophotonics*. 2008. 1(6). Copyright Wiley-VCH Verlag GmbH & Co. KgaA. Reprinted with permission.)

showed that the endogenous fluorescence induced by these two excitation sources enabled detection of different types of skin lesions for 355 nm (basal cell carcinoma, squamous cell carcinoma, and seborrhoeic keratosis) and 435 nm (dysplastic naevi and melanoma skin cancer) excitations. The observed changes in tissue autofluorescence were attributed to changes in NADH, FAD, collagen, and keratin concentrations and biochemistry (DeBeule et al. 2007).

Recently, multiphoton excitation fluorescence has been combined with advanced scanning techniques and gradient refractive index (GRIN) lens technology to produce a miniaturized fiber-optic probe for clinical use (Wu et al. 2009). Figure 11.8 shows the ability of a nonlinear optical microscopic fluorescence imaging system (DermaInspect, JenLab) to image different layers of the human skin (Konig et al. 2008) *in vivo*.

Surface fluorescence imaging with exogenous contrast agents has been employed for intraoperative cancer detection and visualization (Gotoh et al. 2009; van Dam et al. 2011), sentinel lymph node (SLN) mapping (Crane et al. 2011; Fujiwara et al. 2009; Miyashiro et al. 2008; Troyan et al. 2009), and

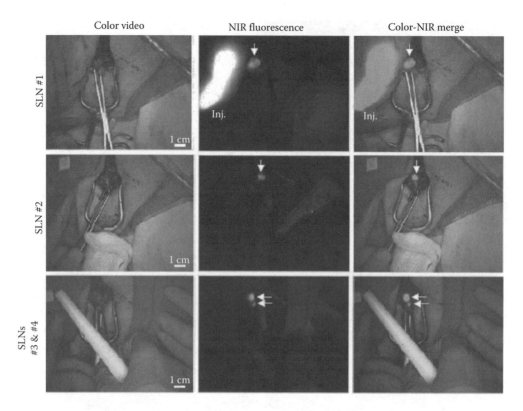

FIGURE 11.9 **(See color insert.)** Images of breast cancer SLN mapping with NIR fluorescence imaging. White-light images (left column) are shown alongside NIR fluorescence (800 nm) images (middle column) after injection (Inj.) of a fluorescent dye. False-color (green) overlays of the NIR fluorescence are shown in the right column. Four SLNs (arrows) were identified for removal from the patient, whereas only three were detected with the current "gold standard," ^{99m}Tc-sulfur colloid lymphoscintigraphy. (Reprinted with permission from Springer Science+Business Media: The Flare((Tm)) Intraoperative near-infrared fluorescence imaging system: A first-in-human clinical trial in breast cancer sentinel lymph node mapping. *Annals of Surgical Oncology*, 16(10), 2009, S.L. Troyan et al.)

angiography (Lee et al. 2010) in a clinical setting. SLN mapping is a procedure that allows clinicians to biopsy lymph nodes and identify whether cancer has spread from its site of origin. This procedure was performed clinically in 10 patients with skin cancer (Fujiwara et al. 2009), in 3 patients with gastric cancer (Miyashiro et al. 2008), in 10 patients with vulvar cancer (Crane et al. 2011), and in groups of 6 to 24 patients with breast cancer (Hundorfean et al. 2012; Lee et al. 2010; Troyan et al. 2009). Figure 11.9 shows an example of near-infrared (NIR, 700–1000 nm; Frangioni 2003) fluorescence imaging for SLN mapping using indocyanine green (ICG), an FDA-approved fluorescence imaging agent for cardiocirculatory and liver diagnostics. In the study, four SLNs (arrows) were identified during an operation from a woman with breast cancer (Troyan et al. 2009).

In addition, tumor-specific fluorescent contrast agents were imaged for 10 patients with ovarian cancer (van Dam et al. 2011) and 10 patients with hepatocellular carcinoma (Gotoh et al. 2009). In the respective studies, it was reported that contrast agents, folate-FITC (fluorescein isothiocyanate) and ICG, directly targeted the tumors and showed bright fluorescence during operations.

11.4.1.3 Endoscopy

Fluorescence technology can be incorporated into an endoscope to access internal tissues and organs. Advances in light sources, fiber bundles, miniature optics, scanning systems, and high-performance detection techniques have enabled fluorescence-imaging technology to become miniaturized into hand-held fiber probes for clinical endoscopic applications (Bao et al. 2008; Flusberg et al. 2005).

A conventional endoscope for imaging internal organs (such as the gastrointestinal tract, bladder, or cervix) consists of a white-light source, a light delivery system for illumination, a lens system for imaging, and an imaging detector (Flusberg et al. 2005). Although traditional white-light endoscopy provides spatially resolved morphological information about the tissue surface, fluorescence endoscopy provides complementary information, including localized tissue functional properties.

For example, lung cancer was studied with a commercialized laser-induced fluorescence endoscope (LIFE; Xillix Technologies and Olympus Optical; Olympus). Results from a large clinical study showed that LIFE technology increased endobronchial lesion-detection sensitivity (when compared to white-light bronchoscopy alone) by an average of 33% and demonstrated potential to more effectively detect early-stage cancers (Edell et al. 2009; Moghissi et al. 2008; Yasufuku 2010). Biochemical and morphological tissue changes highlighted as indicators of contrast were increased epithelial layer thickness and changes in relative concentrations of endogenous fluorophores, including FAD, collagen, and NADH. Tumors were shown to have increased autofluorescence at wavelengths in the low-energy (red) region of the visible spectrum, a potential indication of greater porphyrin content resulting from increased tumor hypervascularity.

In addition to the macroscopic surface fluorescence imaging described above, endoscopic techniques have been developed for microscopic fluorescence endoscopy, including confocal and nonlinear optical microscopic fluorescence endoscopy.

A confocal microscopic fluorescence endoscope illuminates tissue and detects fluorescence from a thin focal plane within a sample, employing a pinhole to eliminate out-of-focus light (Kara et al. 2007). In recent years, this technology has been employed for clinical studies, showing promise for *in vivo* applications. Nonlinear optical microscopic fluorescence endoscopy via multiphoton excitation has significant sectioning capabilities compared to confocal imaging owing to the nonlinear, two-photon absorption that is highly localized spatially and therefore does not require a pinhole for optical sectioning (Figure 11.8). Nonlinear optical imaging provides high-spatial resolution in the focal plane and little out-of-focus photobleaching (Gareau et al. 2004; Pena et al. 2005).

Barrett's esophagus, a condition resulting from gastroesophageal reflux disease, was studied with endoscopic surveillance. In a clinical study (Pohl et al. 2008), 38 patients were assessed with conventional high-resolution endoscopy, confocal microscopic fluorescence endoscopy, and subsequent tissue biopsy (Figure 11.10). As shown in Figure 11.10, confocal microscopic fluorescence endoscope detected the irregularly shaped epithelial lining, gland fusion, and irregular vasculature. The study reported a negative predictive value of more than 98%, suggesting that the fluorescence technique has the potential to reliably rule out the presence of disease.

11.5 Clinical Design Considerations

11.5.1 Regulatory Issues

Device safety is charged to the device manufacturer, both during the FDA approval process and afterwards, when monitoring the device once it is employed for clinical use. To achieve FDA approval, Class-III clinical devices (any device that is employed to support human life, prevent human harm, or present an amount of unreasonable risk of patient injury or illness) must undergo FDA premarket approval (PMA) to assess their safety and effectiveness and to designate them as an adjunct or replacement technology (FDA 2009). For a specific technology that poses potential risk to patient health, safety, or welfare, an investigational device exemption (IDE) may be required. Devices employing laser sources add additional concerns regarding photosensitivity, including radiation hazards and thermal tissue heating (Jawad et al. 2011; Niemz 2007).

The first regulation for clinical fluorescence technologies was enacted in 1990 with the Safe Medical Devices Act, shortly followed by Medical Devices Amendments in 1992. In 1997, the FDA outlined clear

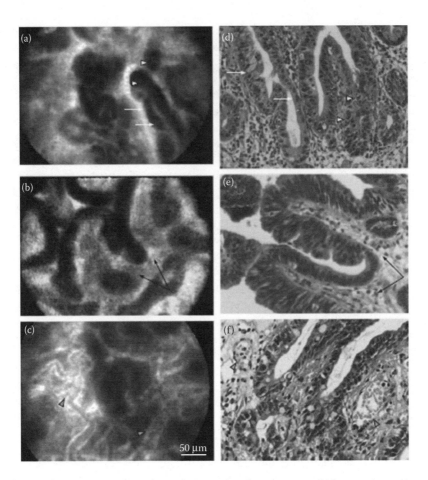

FIGURE 11.10 **(See color insert.)** Barrett's esophagus with early-stage mucosal adenocarcinoma, imaged *in vivo* with confocal microscopic fluorescence endoscopy (a–c) and examined via histopathology using hematoxylin and eosin stains (d–f). White arrows (a, d) indicate the irregularly shaped epithelial lining with variable width, white triangles (a, d) show regions of higher cell density, black arrows (b, e) indicate gland fusion, and gray arrowheads (c, f) show blood vessels that are irregularly shaped and dilated. (Reprinted from H. Pohl et al. Miniprobe confocal laser microscopy for the detection of invisible neoplasia in Patients with Barrett's oesophagus. *Gut* 57(12); 2008. With permission from BMJ Publishing Group Ltd.)

guidelines for developing fluorescence technology targeting cervical cancer (Services 1997), which can be extended to similar fluorescence technologies. These FDA requirements outline the procedure for Investigational Device Exemption/Investigational New Drug (IDE/IND) approval and define the permitted clinical uses of the technology. Four uses were permitted: (1) adjunct tool to cytology, (2) screening device after abnormal cytology, (3) localize biopsy sites, and (4) primary screening as an alternative to cytology.

Fluorescence-based clinical technologies typically indicate a significant risk device, requiring an IDE application, which must include all technology details, including the device description, a complete list of the patient-contacting materials, and software details. Preclinical testing must conform to good laboratory practices (GLP), which is general practice guideline to ensure that biocompatible materials and optical radiation levels employed will not compromise patient safety (Aldrich et al. 2012).

For laser-based devices, safety must be demonstrated according to American National Standards Institute (ANSI) standards (Delori et al. 2007). Before clinical trials begin, a local institutional review board (IRB) must review and approve the study. To gain IRB approval, technology must demonstrate

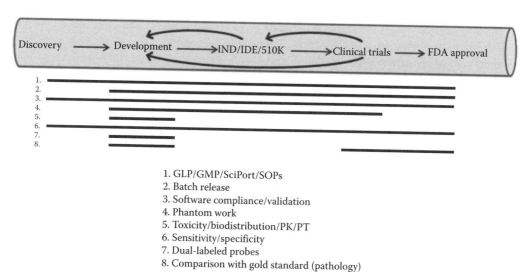

1. GLP/GMP/SciPort/SOPs
2. Batch release
3. Software compliance/validation
4. Phantom work
5. Toxicity/biodistribution/PK/PT
6. Sensitivity/specificity
7. Dual-labeled probes
8. Comparison with gold standard (pathology)

FIGURE 11.11 Schematic of the "translational pipeline" from device discovery through FDA approval. An emphasis is placed on the iterative process during device development and clinical trials. The goal of a common translational pipeline is to improve and standardize the approval process for medical imaging technologies. With faster approval times, and thereby expedited returns on investment, investors should be more apt to fund clinical-imaging technology companies developing fluorescence instrumentation. (M.B. Aldrich et al. Seeing it through: Translational validation of new medical imaging modalities. *Biomedical Optics Express* 3(4); 2012. Adapted with permission from Optical Society of America.)

two layers of safety, defined as either procedural or instrument-based barriers to protect the patient from the technology and vice versa (Marcus and Biersach 2003). The detailed requirements for performing phase I and II clinical trials are discussed in Section 11.5.3.

Also pertinent to study success is the sterilization of all materials that will contact the patient (Rutala and Weber 2004). Detailed guidelines regarding disinfection and sterilization practices for healthcare facilities were outlined in 2008 by the Healthcare Infection Control Practices Advisory Committee (HICPAC) (Rutala et al. 2008). Specific to fiber-optic probes were sterilization guidelines for steam, flash steam, and ethylene oxide sterilization. Previous studies have outlined procedures used to sterilize fiber-optic probes and endoscopes (Rutala et al. 2007; Utzinger and Richards-Kortum 2003). A new probe disinfector has been FDA approved for disinfection of transesophageal ultrasound probes, but has yet to be employed for fluorescence fiber-optic probes or endoscopes (2011).

Recently, regulatory processes for medical-imaging technology seeking FDA approval and clinical use were discussed for four separate medical imaging modalities (including NIR fluorescence and multispectral fluorescence imaging) (Aldrich et al. 2012). The study concluded that each of the four technologies was at a different stage of the translational pipeline (Figure 11.11) and that each group had taken a slightly different approach to gaining approval. This outcome suggested that a standardized pipeline would advance promising technology to a commercial stage more efficiently returning greater value from the funding provided by both private and taxpayer dollars.

11.5.2 Integration with Hospital Environment

To successfully integrate fluorescence technology into the clinic, it must possess several attributes to ensure adoption by physicians (Liu 2010).

In clinical diagnostic and surgical procedures, clinicians make real-time decisions based on the findings of the procedure (e.g., the presence or absence of cancer). Therefore, clinicians require instrumentation that immediately provides accurate and reliable diagnostics. Fluorescence technologies must significantly improve upon a physician's current tools. If diagnostics are negligibly improved, the training time and equipment cost cannot be justified by the physician or hospital. The technology must also uphold IRB safety standards (Marcus and Biersach 2003) and include back-up plans that protect the patient from potential device failure.

Fluorescence-based instrumentation for disease detection may be used by clinicians throughout a procedure. For instance, optical instrumentation for cancer detection could be employed for both point detection and tumor margin analysis. Therefore, the technology should be designed to smoothly incorporate into the current clinical protocol. To achieve this goal, the instrument should be intuitive to use and it should provide the clinician with diagnostically significant information that enables an improvement upon current methods for disease detection. This information should be easy for the clinician to interpret: for fluorescence imaging, displays should incorporate well-understood color schemes (e.g., green for go and red for stop) and false coloring for clear distinction of healthy and diseased tissue; for point detection, the instrument should clearly indicate the presence or absence of disease (e.g., H for healthy and D for diseased) on an easy-to-read interface within the clinical procedure room (Martin et al. 2008).

11.5.3 Clinical Trials

Before a new medical technology can be made commercially available, it must successfully undergo a series of clinical trials (clinicaltrials.gov). Phase-I clinical trials employ a group of 20–80 patients and are focused on evaluating the safety of the technology. Phase-II clinical trials involve a patient set of 100–300 people and are employed to test both the effectiveness and the safety of the technology. Phase-III clinical trials include 1000–3000 patients, a large enough sample size to compare the new technology to currently employed techniques and develop protocols for safe use of the technology. Phase-IV clinical trials occur after marketing of the technology has begun, and they are designed to gain a more complete understanding of the advantages and drawbacks of the new technology so that a strategy for using the technology can be optimized.

Numerous fluorescence-based techniques have undergone clinical trials. For example, from 2002 to 2009, the ability of a fluorescence-based device to detect cervical (pre) cancer was tested in a clinical trial sponsored by the M.D. Anderson Cancer Center (Houston, TX) in collaboration with the National Cancer Institute (Health 2009). In this trial, 100 patients who were undergoing colposcopy at one of five different clinical locations were also given a fluorescence-based cervical examination, and these results were employed to assess the diagnostic accuracy of the fluorescence technology. The same sponsor and collaborator conducted a larger clinical trial from 1998 to 2009 that enrolled 1070 healthy patients (Health 2011). In this trial, fluorescence and reflectance measurements were acquired from each of two sites on the cervix (one normal columnar site, one normal squamous site) and biopsies were obtained from both of these locations. The goal of this trial was to better characterize the optical spectra of normal cervical tissues, so that normal tissue, inflamed tissue, and precancerous tissue could be better distinguished from each other using fluorescence- and reflectance-based techniques. Techniques assessed in these trials were employed to distinguish diseased tissues (moderate dysplasia, severe dysplasia, or malignant) from nondiseased tissues (mild dysplasia or nondysplastic) with a sensitivity of 100%, specificity of 71%, and area under the receiver-operating characteristic (ROC) curve of 0.85 (Cantor et al. 2011).

Optical devices that employ fluorescence for oral cancer detection are currently undergoing clinical trials. The multispectral digital microscope (MDM) (Park et al. 2005; Roblyer et al. 2008) obtains wide-field fluorescence images of the oral cavity under 365, 405, and 450 nm excitation, in addition to white-light reflectance and orthogonal polarization images. The FastEEM4 (Freeberg et al. 2007) excites tissue fluorescence at multiple wavelengths to measure spectrally resolved fluorescence excitation–emission matrices (EEMs). (Previous versions of this instrument were employed in clinical trials for cervical cancer detection (Marin et al. 2006).) The PS2-Oral (Rahman et al. 2008) incorporates an optical setup for fluorescence and reflectance imaging onto a surgical headlight system. A clinical trial employing MDM, FastEEM4, and the PS2-Oral began in 2007 and is scheduled for completion in 2014 with an estimated enrollment of 200 patients (Health).

11.6 Commercialized Clinical Fluorescence Technologies

A number of clinically compatible devices that employ fluorescence for disease detection are commercially available (Administration 2012). Typically, these devices provide tissue analysis using contrast from the tissue fluorescence measured at the surface, thereby assisting with visual diagnostic examination.

The VELscope (LED Dental Inc.) is a hand-held device (8.6″ × 2.2″ × 3.4″) that illuminates the oral mucosal tissue on the inside of the mouth with blue (400–460 nm; Farah et al. 2012; VELscope 2012) light in order to detect malignant and premalignant lesions using endogenous tissue fluorescence. Normal tissue will emit fluorescent light that causes the surface of the mouth to glow a green (>480 nm; Scheer et al. 2011) color. However, cancerous and precancerous tissue will not emit a notable amount of fluorescence, so the surfaces of diseased regions will appear darker than the surrounding normal tissue. The VELscope procedure can easily be incorporated into a dental exam, as the total time required to examine the patient with the device is only about 2 min. Recently, a study involving 620 patients (Truelove et al. 2011) demonstrated that the VELscope was able to assist with detecting all 28 of the lesions that were not found during standard visual examination. However, the utility of this device is limited if the patient has inflamed tissue or if the device is interrogating a region with high pigmentation or prominent blood vessels near the tissue surface (Farah et al. 2012). Under these circumstances, the fluorescence can be attenuated by tissue optical absorption from these features and the surface may appear dark even if there is no disease present.

The Identafi 3000 (DentalEZ) also uses tissue autofluorescence (similar to the VELscope) via a violet light source (Identafi). This device was approved by the FDA in 2009 for use in assisting with visual examinations of the oral cavity to detect cancerous and precancerous lesions. It was also approved as a tool to assist with the identification of tumor margins during oral surgery. An ongoing clinical trial (Lane et al. 2012a,b) is testing the ability of the instrument to identify oral neoplasia. This trial involves more than 300 patients from two different comprehensive cancer centers that are independently researching the diagnostic accuracy of the device. The device employs 405-nm excitation to measure fluorescence images of stromal neovasculature and stromal breakdown, both associated with lesion growth. The device also employs white light for reflectance measurements and amber light (545 nm; Lane et al. 2012a) for improved detection of vasculature.

The WavSTAT optical biopsy system (SpectraScience) delivers laser light (410 nm; Ferrer-Roca 2009) onto human colon tissue via a fiber-optic probe incorporated into a pair of biopsy forceps inserted through the instrument channel of a colonoscope (Benes and Antos 2009). The detected endogenous fluorescence signal from the colon tissue is translated into a binary diagnostic result of "suspect" or "nonsuspect" for the site of interest, thereby removing the need for the examiner to have expertise in analysis or interpretation of complex fluorescence spectra. Preliminary *in vivo* studies showed promise in distinguishing hyperplastic (benign, with no malignant potential) polyps and from adenomatous (premalignant) polyps in the colon (Benes and Antos 2009).

The LUMA imaging system (SpectraScience) was approved by the FDA in 2006 for use by clinicians performing cervical cancer examinations, to determine whether other regions of the cervix should be biopsied after colposcopy (Administration 2012). The device employs 337-nm laser light for fluorescence excitation and two flash lamps for reflectance (Kendrick et al. 2007). The lamps and laser are coupled to a fiber-optic probe to deliver light to the tissue, and the system scans a region of 25 mm diameter, mapping a suspicious area of the cervix (Kendrick et al. 2007). The scan takes roughly 12 s and the probe does not contact the tissue during this process (Kendrick et al. 2007). In two randomized clinical trials, the true positive biopsy rate was found to increase by at least 25% when the LUMA device was employed along with colposcopy, and the corresponding increase in the false-positive rate was only 4% (Huh et al. 2004; Kendrick et al. 2007).

11.7 Conclusions

Fluorescence spectroscopy and imaging technologies can be applied to address a wide range of unmet clinical needs in tissue diagnostics. For successful clinical integration of these technologies, a clearly defined "translational pipeline" should be further developed to expedite the timeline from preliminary clinical trials to commercialization and adoption by the medical community. As additional fluorescence-based technologies progress through the FDA approval process in the future, the impact of fluorescence spectroscopy and imaging technologies on clinical patient care is expected to increase, directly addressing the need for cost-effective, real-time, objective, and noninvasive tissue diagnostics.

Acknowledgments

The authors were supported in part by the National Institutes of Health (R01-DE 019431 to M.-A.M.), the Wallace H. Coulter Foundation (to M.-A.M.), and the U.S. Department of Education (GAANN Fellowship to W.R.L.).

References

Administration, US Food and Drug. Medical Devices—Recently-Approved Devices. 2012. (4/1/2012). Accessed. http://www.fda.gov/MedicalDevices/ProductsandMedicalProcedures/DeviceApprovals andClearances/Recently-ApprovedDevices/default.htm.

Aldrich, M. B., M. V. Marshall, E. M. Sevick-Muraca et al. Seeing it through: Translational validation of new medical imaging modalities. *Biomedical Optics Express* 3(4); 2012: 764–76.

Ansorge, W. J. Next-generation DNA sequencing techniques. *New Biotechnology* 25(4); 2009: 195–203.

Bao, H., J. Allen, R. Pattie, R. Vance, and M. Gu. Fast handheld two-photon fluorescence microendoscope with a 475 Microm X 475 Microm field of view for *in vivo* imaging. *Optics Letters* 33(12); 2008: 1333–5.

Barretto, R. P. J., T. H. Ko, J. C. Jung et al. Time-lapse imaging of disease progression in deep brain areas using fluorescence microendoscopy. *Nature Medicine* 17(2); 2011: 223–8.

Barretto, R. P. J., B. Messerschmidt, and M. J. Schnitzer. *In vivo* fluorescence imaging with high-resolution microlenses. *Nature Methods* 6(7); 2009: 511–2.

Benes, Z. and Z. Antos. Optical biopsy system distinguishing between hyperplastic and adenomatous polyps in the colon during colonoscopy. *Anticancer Research* 29(11); 2009: 4737–9.

Birch, D. and R. Imhof. Time-domain fluorescence spectroscopy using timecorrelated single-photon counting. Ed. Lakowicz, J. Vol. 1. *Topics in Fluorescence Spectroscopy*. Springer, USA, 2002. 1–95. Print.

Boens, N., W. Qin, N. Basarić et al. Fluorescence lifetime standards for time and frequency domain fluorescence spectroscopy. *Analytical Chemistry* 79(5); 2007: 2137–49.

Boutet, J., L. Herve, M. Debourdeau et al. Bimodal ultrasound and fluorescence approach for prostate cancer diagnosis. *Journal of Biomedical Optics* 14(6); 2009: 064001.

Brancaleon, L., A. J. Durkin, J. H. Tu et al. In vivo fluorescence spectroscopy of nonmelanoma skin cancer. *Photochemistry and Photobiology* 73(2); 2001: 178–83.

Butte, P. V., A. N. Mamelak, M. Nuno et al. Fluorescence lifetime spectroscopy for guided therapy of brain tumors. *NeuroImage* 54(Suppl 1); 2011: S125–35.

Cantor, S. B., J.-M. Yamal, M. Guillaud et al. Accuracy of optical spectroscopy for the detection of cervical intraepithelial neoplasia: Testing a device as an adjunct to colposcopy. *International Journal of Cancer* 128(5); 2011: 1151–68.

Chandra, M., K. Vishwanath, G. D. Fichter et al. Quantitative molecular sensing in biological tissues: An approach to non-invasive optical characterization. *Optics Express* 14(13); 2006: 6157–71.

Chandra, M., J. Scheiman, D. Heidt et al. Probing pancreatic disease using tissue optical spectroscopy. *Journal of Biomedical Optics* 12(6); 2007a: 060501.

Chandra, M., R. H. Wilson, W.-L. Lo et al. Sensing metabolic activity in tissue engineered constructs. *Proceedings of SPIE* 6628, 2007b: 66280B.

Chandra, M., R. H. Wilson, J. Scheiman et al. Optical spectroscopy for clinical detection of pancreatic cancer. *Proceedings of SPIE* 7368, 2009: 73681G.

Chandra, M., J. Scheiman, D. Simeone et al. Spectral areas and ratios classifier algorithm for pancreatic tissue classification using optical spectroscopy. *Journal of Biomedical Optics* 15(1); 2010: 010514.

Chang, S. K., N. Marin, M. Follen, and R. Richards-Kortum. Model-based analysis of clinical fluorescence spectroscopy for *in vivo* detection of cervical intraepithelial dysplasia. *Journal of Biomedical Optics* 11(2); 2006: 024008.

Chang, C.-W., D. Sud, and M.-A. Mycek. Fluorescence lifetime imaging microscopy. In *Methods in Cell Biology*. Eds. Sluder, G. and D. Wolf. Vol. 81, Chapter 24: Elsevier Inc, 2007. Print.

Chang, C.-W., M. Wu, S. D. Merajver, and M.-A. Mycek. Improving Fret detection in living cells. *Proceedings of SPIE* 7370, 2009: 737007.

Chaudhari, A. J., S. Ahn, R. Levenson et al. Excitation spectroscopy in multispectral optical fluorescence tomography: Methodology, feasibility and computer simulation studies. *Physics in Medicine and Biology* 54(15); 2009: 4687.

Chen, L. C., W. R. Lloyd, S. Kuo et al. Label-free multiphoton fluorescence imaging monitors metabolism in living primary human cells used for tissue engineering. *Imaging, Manipulation, and Analysis of Biomolecules, Cells, and Tissues* X; 2012: 8225–18.

Chorvat, D. and A. Chorvatova. Multi-wavelength fluorescence lifetime spectroscopy: A new approach to the study of endogenous fluorescence in living cells and tissues. *Laser Physics Letters* 6(3); 2009: 175–93.

Clinicaltrials.gov. A service of the U.S. National Institutes of Health. Accessed. clinicaltrials.gov.

Crane, L. M. A., G. Themelis, H. J. G. Arts et al. Intraoperative near-infrared fluorescence imaging for sentinel Lymph node detection in Vulvar cancer: First clinical results. *Gynecologic Oncology* 120(2); 2011: 291–5.

Dabir, A. S., C. A. Trivedi, Y. Ryu, P. Pande, and J. A. Jo. Fully automated deconvolution method for on-line analysis of time-resolved fluorescence spectroscopy data based on an iterative Laguerre expansion technique. *Journal of Biomedical Optics* 14; 2009: 024030.

Darzynkiewicz, Z., X. Li, and J. Gong. Assays of cell viability: Discrimination of cells dying by apoptosis. *Methods in Cell Biology* 41; 1994: 15–38.

De Beule, P. A. A., C. Dunsby, N. P. Galletly et al. A hyperspectral fluorescence lifetime probe for skin cancer diagnosis. *Review of Scientific Instruments* 78(12); 2007: 123101–7.

de Leeuw, J., N. van der Beek, W. D. Neugebauer, P. Bjerring, and H. A. M. Neumann. Fluorescence detection and diagnosis of non-melanoma skin cancer at an early stage. *Lasers in Surgery and Medicine* 41(2); 2009: 96–103.

Delori, F. C., R. H. Webb, D. H. Sliney, and Institute American National Standards. Maximum permissible exposures for Ocular Safety (Ansi 2000), with emphasis on ophthalmic devices. *Journal of the Optical Society of America A, Optics, Image Science, and Vision* 24(5); 2007: 1250–65.

Digman, M. A., V. R. Caiolfa, M. Zamai, and E. Gratton. The Phasor approach to fluorescence lifetime imaging analysis. *Biophysical Journal* 94(2); 2008: L14–6.

Edell, E., S. Lam, H. Pass et al. Detection and localization of intraepithelial neoplasia and invasive carcinoma using fluorescence-reflectance bronchoscopy: An international, multicenter clinical trial. *Journal of Thoracic Oncology: Official Publication of the International Association for the Study of Lung Cancer* 4(1); 2009: 49–54.

Evers, D., B. Hendriks, G. Lucassen, and T. Ruers. Optical spectroscopy: Current advances and future applications in cancer diagnostics and therapy. *Future Oncology (London, England)* 8(3); 2012: 307–20.

Fang, C., D. Brokl, R. E. Brand, and Y. Liu. Depth-selective fiber-optic probe for characterization of superficial tissue at a constant physical depth. *Biomedical Optics Express* 2(4); 2011: 838–49.

Farah, C. S., L. McIntosh, A. Georgiou, and M. J. McCullough. Efficacy of tissue autofluorescence imaging (Velscope) in the visualization of oral mucosal lesions. *Head & Neck* 34(6); 2012: 856–62.

Fawzy, Y. S., and H. Zeng. Intrinsic fluorescence spectroscopy for endoscopic detection and localization of the endobronchial cancerous lesions. *Journal of Biomedical Optics* 13; 2008: 064022.

Fawzy, Y. S., M. Petek, M. Tercelj, and H. Zeng. *In vivo* assessment and evaluation of lung tissue morphologic and physiological changes from non-contact endoscopic reflectance spectroscopy for improving lung cancer detection. *Journal of Biomedical Optics* 11; 2006: 044003.

FDA Clears New Automated Probe Disinfector. 2011. Infection Control Today. Accessed April 20, 2012. http://www.infectioncontroltoday.com/news/2011/08/fda-clears-new-automated-probe-disinfector.aspx.

FDA, U.S. PMA Approvals. 2009. U.S. Food and Drug Administration. Accessed 2012. http://www.fda.gov/MedicalDevices/ProductsandMedicalProcedures/DeviceApprovalsandClearances/PMAApprovals/default.htm.

Ferrer-Roca, O. Telepathology and optical biopsy. *International Journal of Telemedicine and Applications* 2009: 740712.

Flusberg, B. A., E. D. Cocker, W. Piyawattanametha et al. Fiber-optic fluorescence imaging. *Nature Methods* 2(12); 2005: 941–50.

Frangioni, J. V. In vivo near-infrared fluorescence imaging. *Current Opinion in Chemical Biology* 7(5); 2003: 626–34.

Freeberg, J. A., D. M. Serachitopol, N. McKinnon et al. Fluorescence and reflectance device variability throughout the progression of a Phase Ii clinical trial to detect and screen for cervical neoplasia using a fiber optic probe. *Journal of Biomedical Optics* 12(3); 2007: 034015.

Fujiwara, M., T. Mizukami, A. Suzuki, and H. Fukamizu. Sentinel lymph node detection in skin cancer patients using real-time fluorescence navigation with Indocyanine green: Preliminary experience. *Journal of Plastic, Reconstructive & Aesthetic Surgery* 62(10); 2009: e373–8.

Gabrecht, T., A. Radu, M. Zellweger et al. Autofluorescence bronchoscopy: Clinical experience with an optimized system in head and neck cancer patients. *Medical Laser Application* 22(3); 2007: 185–92.

Gareau, D. S., P. R. Bargo, W. A. Horton, and S. L. Jacques. Confocal fluorescence spectroscopy of subcutaneous cartilage expressing green fluorescent protein versus cutaneous collagen autofluorescence. *Journal of Biomedical Optics* 9(2); 2004: 254–58.

Glanzmann, T., J.-P. Ballini, H. van den Bergh, and G. Wagnieres. Time-resolved spectrofluorometer for clinical tissue characterization during endoscopy. *Review of Scientific Instruments* 70(10); 1999: 11.

Gotoh, K., T. Yamada, O. Ishikawa et al. A novel image-guided surgery of hepatocellular carcinoma by indocyanine green fluorescence imaging navigation. *Journal of Surgical Oncology* 100(1); 2009: 75–9.

Graves, E. E., R. Weissleder, and V. Ntziachristos. Fluorescence molecular imaging of small animal tumor models. *Current Molecular Medicine* 4(4); 2004: 419–30.

Health, U.S. National Institutes of Fluorescence & reflectance imaging to detect oral neoplasia. Accessed May 31, 2012. http://clinicaltrials.gov/ct2/show/NCT00542373?term=fluorescence&rank=10.

Health, U.S. National Institutes of Fluorescence and reflectance spectroscopy during colposcopy in detecting cervical intraepithelial neoplasia and dysplasia in healthy participants with a history of normal pap smears. 2011. Accessed May 31, 2012. http://clinicaltrials.gov/ct2/show/NCT00084903?term=fluorescence&rank=11.

Health, U.S. National Institutes of Measurement of digital colposcopy for fluorescence spectroscopy of cervical intraepithelial neoplasia. 2009. Accessed May 31, 2012. http://clinicaltrials.gov/ct2/show/NCT00513123?term=fluorescence&rank=2.

Huang, B., M. Bates, and X. Zhuang. Super-resolution fluorescence microscopy. *Annual Review of Biochemistry* 78; 2009: 993–1016.

Huh, W. K., R. M. Cestero, F. A. Garcia et al. Optical detection of high-grade cervical intraepithelial neoplasia in vivo: Results of a 604-patient study. *American Journal of Obstetrics and Gynecology* 190(5); 2004: 1249–57.

Hundorfean, G., A. Agaimy, R. Atreya et al. Confocal laser endomicroscopy for characterization of Crohn's disease-associated duodenitis. *Endoscopy* 44(Suppl 2);UCTN 2012: E80.

Jawad, M. M., S. T. A. Qader, A. A. Zaidan et al. An overview of laser principle, laser-tissue interaction mechanisms and laser safety precautions for medical laser users. *International Journal of Pharmacology* 7(2); 2011: 149–60.

Kalchenko, V., Y. Kuznetsov, I. Meglinski, and A. Harmelin. Label free *in vivo* laser speckle imaging of blood and lymph vessels. *Journal of Biomedical Optics* 17(5); 2012: 050502.

Kara, M. A., R. S. DaCosta, C. J. Streutker et al. Characterization of tissue autofluorescence in Barrett's esophagus by confocal fluorescence microscopy. *Diseases of the Esophagus: Official Journal of the International Society for Diseases of the Esophagus/I S D E* 20(2); 2007: 141–50.

Kendrick, J. E., W. K. Huh, and R. D. Alvarez. Luma cervical imaging system. *Expert Review of Medical Devices* 4(2); 2007: 121–29.

Kirschner, R. L. The future of medical devices. *Pharmaceutical and Biomedical Project Management in a Changing Global Environment*. John Wiley & Sons, Inc., 2010. 345–57. Print.

Kobayashi, M., K. Shibuya, H. Hoshino, and T. Fujisawa. Spectroscopic analysis of the autofluorescence from human bronchus using an ultraviolet laser diode. *Journal of Biomedical Optics* 7(4); 2002: 603–8.

Konig, K., A. Ehlers, I. Riemann et al. Clinical two-photon microendoscopy. *Microscopy Research and Technique* 70(5); 2007: 398–402.

Konig, K., M. Weinigel, D. Hoppert et al. Multiphoton tissue imaging using high-Na microendoscopes and flexible scan heads for clinical studies and small animal research. *Journal of Biophotonics* 1(6); 2008: 506–13.

Kosaka, N., M. Ogawa, P. L. Choyke, and H. Kobayashi. Clinical implications of near-infrared fluorescence imaging in cancer. *Future Oncology (London, England)* 5(9); 2009: 1501–11.

Kuil, J., A. H. Velders, and F. W. B. van Leeuwen. Multimodal tumor-targeting peptides functionalized with both a radio- and a fluorescent label. *Bioconjugate Chemistry* 21(10); 2010: 1709–19.

Kumar, N., and N. Borth. Flow-cytometry and cell sorting: An efficient approach to investigate productivity and cell physiology in mammalian cell factories. *Methods (San Diego, Calif)* 56(3); 2012: 366–74.

Kumar, A. T. N., S. B. Raymond, A. K. Dunn, B. J. Bacskai, and D. A. Boas. A time domain fluorescence tomography system for small animal imaging. *IEEE Transactions on Medical Imaging* 27(8); 2008: 1152–63.

Kusunoki, Y., F. Imamura, H. Uda, M. Mano, and T. Horai. Early detection of lung cancer with laser-induced fluorescence endoscopy and spectrofluorometry. *Chest* 118(6); 2000: 1776–82.

Lakowicz, J.R. *Principles of Fluorescence Spectroscopy*. 3rd ed. New York: Springer, 2006. Print.

Lane, P., M. Follen, and C. MacAulay. Has fluorescence spectroscopy come of age? A case series of oral precancers and cancers using white light, fluorescent light at 405 nm, and reflected light at 545 nm using the Trimira Identafi 3000. *Gender Medicine* 9(1) (Suppl); 2012a: S25–35.

Lane, P., S. Lam, M. Follen, and C. MacAulay. Oral fluorescence imaging using 405-nm excitation, aiding the discrimination of cancers and precancers by identifying changes in collagen and elastic breakdown and neovascularization in the underlying stroma. *Gender Medicine* 9(1)(Suppl); 2012b: S78–S82.e8.

Lane, P. M., T. Gilhuly, P. Whitehead et al. Simple device for the direct visualization of oral-cavity tissue fluorescence. *Journal of Biomedical Optics* 11(2); 2006: 024006.

Leblond, F., S. C. Davis, P. A. Valdes, and B. W. Pogue. Pre-clinical whole-body fluorescence imaging: Review of instruments, methods and applications. *Journal of Photochemistry and Photobiology B* 98(1); 2010: 77–94.

Lee, B. T., A. Matsui, M. Hutteman et al. Intraoperative near-infrared fluorescence imaging in Perforator flap reconstruction: Current research and early clinical experience. *Journal of Reconstructive Microsurgery* 26(1); 2010: 59–65.

Lee, K. C. B., J. Siegel, S. E. D. Webb et al. Application of the stretched exponential function to fluorescence lifetime imaging. *Biophysical Journal* 81; 2001: 1265–74.

Lippincott-Schwartz, J. Emerging *in vivo* analyses of cell function using fluorescence imaging (*). *Annual Review of Biochemistry* 80; 2011: 327–32.

Liu, Q. Role of optical spectroscopy using endogenous contrasts in clinical cancer diagnosis. *World Journal of Clinical Oncology* 2(1); 2010: 14.

Liu, Q. Role of optical spectroscopy using endogenous contrasts in clinical cancer diagnosis. *World Journal of Clinical Oncology* 2(1); 2011: 50–63.

Lloyd, W., R. H. Wilson, C.-W. Chang, G. D. Gillispie, and M.-A. Mycek. Instrumentation to rapidly acquire fluorescence wavelength-time matrices of biological tissues. *Biomedical Optics Express* 1(2); 2010: 574–86.

Marcus, M. L., and B. R. Biersach. Regulatory requirements for medical equipment. *Instrumentation & Measurement Magazine, IEEE* 6(4); 2003: 23–29.

Marin, N. M., N. MacKinnon, C. MacAulay et al. Calibration standards for multicenter clinical trials of fluorescence spectroscopy for *in vivo* diagnosis. *Journal of Biomedical Optics* 11(1); 2006: 014010.

Martin, J. L., B. J. Norris, E. Murphy, and J. A. Crowe. Medical device development: The challenge for ergonomics. *Applied Ergonomics* 39(3); 2008: 271–83.

Mayinger, B., M. Jordan, T. Horbach, W. Hohenberger, and E. Hahn. Influence of collagen in endoscopic fluorescence spectroscopy for gastric cancer. *Gastrointestinal Endoscopy* 59(5); 2004a: 172.

Mayinger, B., M. Jordan, T. Horbach et al. Evaluation of *in vivo* endoscopic autofluorescence spectroscopy in gastric cancer. *Gastrointestinal Endoscopy* 59(2); 2004b: 191–98.

McGee, S., V. Mardirossian, A. Elackattu et al. Anatomy-based algorithms for detecting oral cancer using reflectance and fluorescence spectroscopy. *Annals of Otology, Rhinology, and Laryngology* 118 (11); 2009: 817–26.

McGinty, J., N. P. Galletly, C. Dunsby et al. Wide-field fluorescence lifetime imaging of cancer. *Biomedical Optics Express* 1(2); 2010: 627–40.

Medicine, Stanford. Cancer Overview. Accessed May, 31 2012. http://cancer.stanford.edu/information/cancerOverview.html.

Meier, J. D., H. Xie, Y. Sun et al. Time-resolved laser-induced fluorescence spectroscopy as a diagnostic instrument in head and neck carcinoma. *Otolaryngology— Head and Neck Surgery: Official Journal of American Academy of Otolaryngology-Head and Neck Surgery* 142(6); 2010: 838–44.

Miyashiro, I., N. Miyoshi, M. Hiratsuka et al. Detection of sentinel node in gastric cancer surgery by indocyanine green fluorescence imaging: Comparison with infrared imaging. *Annals of Surgical Oncology* 15(6); 2008: 1640–3.

Moghissi, K., K. Dixon, and M. R. Stringer. Current indications and future perspective of fluorescence bronchoscopy: A review study. *Photodiagnosis and Photodynamic Therapy* 5(4); 2008: 238–46.

Müller, M. G., I. Georgakoudi, Q. Zhang, J. Wu, and M. S. Feld. Intrinsic fluorescence spectroscopy in turbid media: Disentagling effects of scattering and absorption. *Applied Optics* 40 (25); 2001: 4633–46.

Muretta, J. M., A. Kyrychenko, A. S. Ladokhin et al. High-performance time-resolved fluorescence by direct waveform recording. *Review of Scientific Instruments* 81(10); 2010: 103101–103101-8.

Mycek, M.-A., and B. W. Pogue, eds. *Handbook of Biomedical Fluorescence*. New York, NY: Marcel Dekker, Inc., 2003. Print.

Mycek, M.-A., K. Schomacker, and N. Nishioka. Colonic polyp differentiation using time resolved autofluorescence spectroscopy. *Gastrointestinal Endoscopy* 48(4); 1998: 390–94.

Niemz, M. H. *Laser-Tissue Interactions: Fundamentals and Applications*. 3rd ed. Germany: Springer, 2007. Print.

Palmer, G.M. and N. Ramanujam. Monte Carlo-based inverse model for calculating tissue optical properties. Part I: Theory and validation on synthetic phantoms. *Applied Optics* 45; 2006: 1062–71.

Palmer, G. M. and N. Ramanujam. Monte Carlo-based model for the extraction of intrinsic fluorescence from Turbid media. *Journal of Biomedical Optics* 13(2); 2008: 024017–024017-9.

Park, S. Y., T. Collier, J. Aaron et al. Multispectral digital microscopy for *in vivo* monitoring of oral neoplasia in the hamster cheek pouch model of carcinogenesis. *Optics Express* 13(3); 2005: 749–62.

Pavlova, I., M. Williams, A. El-Naggar, R. Richards-Kortum, and A. Gillenwater. Understanding the biological basis of autofluorescence imaging for oral cancer detection: High-resolution fluorescence microscopy in viable tissue. *Clinical Cancer Research: An Official Journal of the American Association for Cancer Research* 14(8); 2008: 2396–404.

Pena, A. M., M. Strupler, T. Boulesteix, and M. C. Schanne-Klein. Spectroscopic analysis of keratin endogenous signal for skin multiphoton microscopy. *Optics Express* 13(16); 2005: 6268–74.

Poellinger, A., S. Burock, D. Grosenick et al. Breast cancer: Early- and late-fluorescence near-infrared imaging with indocyanine green—A preliminary study. *Radiology* 258(2); 2011: 409–16.

Pogue, B. W., S. L. Gibbs-Strauss, P. A. Valde et al. Review of neurosurgical fluorescence imaging methodologies. *IEEE Journal of Selected Topics in Quantum Electronics* 16(3); 2010: 493–505.

Poh, C. F., S. P. Ng, P. M. Williams et al. Direct fluorescence visualization of clinically occult high-risk oral premalignant disease using a simple hand-held device. *Head & Neck* 29(1); 2007: 71–6.

Poh, C. F., P. Lane, C. MacAulay, L. Zhang, and M. P. Rosin. The application of tissue autofluorescence in detection and management of oral cancer and premalignant lesions. In E. Rosenthal & K. R. Zinn (Eds.), *Optical Imaging of Cancer: Clinical Applications.* 1st ed. New York: Springer, pp. 101–18.

Pohl, H., T. Rösch, M. Vieth et al. Miniprobe confocal laser microscopy for the detection of invisible neoplasia in patients with Barrett's oesophagus. *Gut* 57(12); 2008: 1648–53.

Provenzano, P. P., K. W. Eliceiri, and P. J. Keely. Shining new light on 3d cell motility and the metastatic process. *Trends in Cell Biology* 19(11); 2009: 638–48.

Rahman, M., P. Chaturvedi, A. M. Gillenwater, and R. Richards-Kortum. Low-cost, multimodal, portable screening system for early detection of oral cancer. *Journal of Biomedical Optics* 13(3); 2008: 030502.

Rigel, D. S., J. Russak, and R. Friedman. The evolution of melanoma diagnosis: 25 years beyond the Abcds. *CA: A Cancer Journal for Clinicians* 60(5); 2010: 301–16.

Roblyer, D., R. Richards-Kortum, K. Sokolov et al. Multispectral optical imaging device for *in vivo* detection of oral neoplasia. *Journal of Biomedical Optics* 13(2); 2008: 024019.

Roy, R., S. Hohng, and T. Ha. A practical guide to single-molecule Fret. *Nature Methods* 5(6); 2008: 507–16.

Rutala, W. A. and D. J. Weber. Disinfection and sterilization in health care facilities: What clinicians need to know. *Clinical Infectious Diseases* 39(5); 2004: 702–09.

Rutala, W. A., M. F. Gergen, and D. J. Weber. Disinfection of a probe used in ultrasound-guided prostate biopsy. *Infection Control and Hospital Epidemiology* 28(8); 2007: 916–19.

Rutala, W. A., D. J. Weber, and Healthcare Infection Control Practices Advisory Committee (HICPAC). Guideline for disinfection and sterilization in healthcare facilities, 2008. Ed. Prevention, Centers for Disease Control and. www.cdc.gov, 2008. Print.

Scheer, M., J. Neugebauer, A. Derman et al. Autofluorescence imaging of potentially malignant mucosa lesions. *Oral Surgery, Oral Medicine, Oral Pathology, Oral Radiology, and Endodontics* 111(5); 2011: 568–77.

Services, US Department of Health and Human. Electro-optical sensors for the *in vivo* detection of cervical cancer and its precursors: Submission guidance for an Ide/Pma. Ed. FDA, Center for Devices and Radiological Health. www.fda.gov, 1997. Print.

Skala, M. C., K. M. Riching, D. K. Bird et al. *In vivo* multiphoton fluorescence lifetime imaging of protein-bound and free nicotinamide adenine dinucleotide in normal and precancerous epithelia. *Journal of Biomedical Optics* 12(2); 2007: 024014.

Thomas, S. S., S. Mohanty, J. L. Jayanthi et al. Clinical trial for detection of dental caries using laser-induced fluorescence ratio reference standard. *Journal of Biomedical Optics* 15(2); 2010: 027001.

Thong, P. S. P., M. Olivo, W. W. L. Chin et al. Clinical application of fluorescence endoscopic imaging using hypericin for the diagnosis of human oral cavity lesions. *British Journal of Cancer* 101(9); 2009: 1580–4.

Thong, P., M. Olivo, S. Tandjung et al. Review of confocal fluorescence endomicroscopy for cancer detection. *IEEE Journal of Selected Topics in Quantum Electronics* PP 99; 2011: 1–1.

Troyan, S. L., V. Kianzad, S. L. Gibbs-Strauss et al. The Flare((Tm)) intraoperative near-infrared fluorescence imaging system: A first-in-human clinical trial in breast cancer sentinel lymph node mapping. *Annals of Surgical Oncology* 16(10); 2009: 2943–52.

Truelove, E. L. Detecting oral cancer. *Journal of the American Dental Association* 141(6); 2010: 626–26.

Truelove, E. L., D. Dean, S. Maltby et al. Narrow band (light) imaging of oral mucosa in routine dental patients. Part I: Assessment of value in detection of mucosal changes. *General Dentistry* 59(4); 2011: 281–9; quiz 90-1, 319–20.

Urayama, P. K., W. Zhong, J. A. Beamish et al. A UV–visible fluorescence lifetime imaging microscope for laser-based biological sensing with picosecond resolution. *Applied Physics B: Lasers and Optics* 76(5); 2003: 483–96.

Utzinger, U. and R. R. Richards-Kortum. Fiber optic probes for biomedical optical spectroscopy. *Journal of Biomedical Optics* 8(1); 2003: 121–47.

van Dam, G. M., G. Themelis, L. M. A. Crane et al. Intraoperative tumor-specific fluorescence imaging in ovarian cancer by Folate Receptor-[Alpha] targeting: First in-human results. *Nature Medicine* 17 (10); 2011: 1315–19.

VELscope. Velscope. 2012. Accessed May 31, 2012. http://www.velscope.com/.

Vishwanath, K. and M.-A. Mycek. Do fluorescence decays remitted from tissues accurately reflect intrinsic fluorophore lifetimes? *Optics Letters* 29(3); 2004: 1512–14.

Vishwanath, K. and M.-A. Mycek. Time-resolved photon migration in bi-layered tissue models. *Optics Express* 13(19); 2005: 7466–82.

Vishwanath, K. and N. Ramanujam. Fluorescence spectroscopy in vivo. *Encyclopedia of Analytical Chemistry*. John Wiley & Sons, Ltd, 2011. Print.

Vo-Dinh, T., ed. *Biomedical Photonics Handbook*. Vol. 1. New York, NY: CRC Press, 2003. Print.

Volynskaya, Z., A. S. Haka, K. L. Bechtel et al. Diagnosing breast cancer using diffuse reflectance spectroscopy and intrinsic fluorescence spectroscopy. *Journal of Biomedical Optics* 13(2); 2008: 024012.

Wagnieres, G., W. Star, and B. Wilson. In vivo fluorescence spectroscopy and imaging for oncological applications. *Photochemistry and Photobiology* 68(5); 1998: 603–32.

Wagnieres, G. A., A. P. Studzinski, D. R. Braichotte et al. Clinical imaging fluorescence apparatus for the endoscopic photodetection of early cancers by use of Photofrin Ii. *Applied Optics* 36(22); 1997: 5608–20.

Weissleder, R. and M. J. Pittet. Imaging in the era of molecular oncology. *Nature* 452(71)87; 2008: 580–9.

Werner, S. G., H. E. Langer, S. Ohrndorf et al. Inflammation assessment in patients with arthritis using a novel *in vivo* fluorescence optical imaging technology. *Annals of the Rheumatic Diseases* 71(4); 2011: 504–10.

Wilson, R. H., M. Chandra, L.-C. Chen et al. Photon–tissue interaction model enables quantitative optical analysis of human pancreatic tissues. *Optics Express* 18(21); 2010: 21612–21.

Wilson, R. H., M. Chandra, J. Scheiman et al. Optical spectroscopy detects histological hallmarks of pancreatic cancer. *Optics Express* 17; 2009: 17502–16.

Wu, J., M. Feld, and R. Rava. Analytical model for extracting intrinsic fluorescence in Turbid media. *Applied Optics* 32(19); 1993: 3585–95.

Wu, Y. C., Y. X. Leng, J. F. Xi, and X. D. Li. Scanning all-fiber-optic endomicroscopy system for 3d nonlinear optical imaging of biological tissues. *Optics Express* 1(10); 2009: 7907–15.

Yasufuku, K. Early diagnosis of lung cancer. *Clinics in Chest Medicine* 31(1); 2010: 39–47, Table of Contents.

Zhang, C., T. Liu, Y. Su et al. A near-infrared fluorescent heptamethine indocyanine dye with preferential tumor accumulation for *in vivo* imaging. *Biomaterials* 31(25); 2010a: 6612–7.

Zhang, C., S. Wang, J. Xiao et al. Sentinel lymph node mapping by a near-infrared fluorescent heptamethine dye. *Biomaterials* 31(7); 2010b: 1911–7.

III

Biological Assays

Contributions of Technology to the Biological Revolution

Almost all disease states can be ultimately linked to some sort of dysfunction at the cellular or the molecular level. The ability to identify abnormalities in the cellular content of diseased tissues has expanded clinical medicine. New technologies to manipulate these processes are emerging every day. At the frontline of this technology is the variety of assays used in clinical laboratories to examine the basis of human disease.

The development of these tools began almost immediately following the identification of cells as the basic building blocks of living beings. Microscopic assessment of tissues and cells, combined with various staining techniques, has greatly enhanced disease diagnosis. Many of these techniques have been fine-tuned to the point of almost total automation. As the biological revolution and information age continue to mature, the ability of clinicians to obtain accurate diagnoses will continue to increase.

12

Biological Assays: Cellular Level

Clark T. Hung
Columbia University

Elena Alegre
Aguarón
Columbia University

Terri-Ann N. Kelly
Columbia University

Robert L. Mauck
University of Pennsylvania

12.1 Introduction

The goal of this chapter is to provide general background information targeted to the researcher who is considering the incorporation of living cells in their research. With the challenges associated with understanding fundamental mechanisms of cell behavior at the complex, *in vivo* level (most physiologic model), researchers have alternatively studied tissue explant cultures, cells cultured in three-dimensional (3D) cultures, and cells cultured in two-dimensional (2D) cultures (least physiologic model) (see Figure 12.1). Culturing cells *in vitro* facilitates the manipulation of the cell and affords the ability

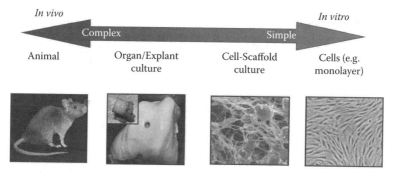

FIGURE 12.1   Schematic illustrating the continuum of *in vivo* to *in vitro* models, which progress from the most physiologic condition to the least physiologic experimental condition.

to prescribe the environment of the cell with specificity (e.g., chemical and physical). We begin with an overview of tissue culture techniques relevant to the design and practice of growing cells in culture. Issues of cell source, culture medium, cell phenotypic characterization, 2D and 3D cultures, and cryopreservation are specifically addressed. To characterize and monitor cell function and intracellular activities of cells growing in culture, researchers utilize a variety of powerful biological techniques, including microscopy, flow cytometry, and molecular biology assays. While many of the references and examples cited in this chapter are selected from the field of musculoskeletal research, which reflects the research focus of the authors of this chapter, the concepts described herein are generalizable to most cell and tissue types.

12.2 Cell Culture

12.2.1 Cell Sources

Cells for experimental investigations can be obtained from a variety of sources, including tissue-derived primary cell isolations, transformed cell lines, and clonal cell lines that are available from commercial cell banks (e.g., American Type Culture Collection and European Collection of Animal Cell Cultures). A cell line is derived from first-passaged primary cultures. The choice of cell source is often dictated by the nature of the scientific question or application that is pursued. Established cell lines have been well studied and offer less experimental variability with the benefits of ease of procurement and maintenance. However, cell lines are considered to be further removed (with respect to primary cells) from the physiologic situation.

12.2.1.1 Primary Cells

Primary cells constitute a cell population that are directly isolated from the tissue source (e.g., fragment or whole living tissue or organ) and is a term used to refer to the original cell culture prior to passage or subculture. These "normal" cells typically have a finite lifespan and exhibit contact inhibition, anchorage dependence, and changes in characteristics with increased age in culture. These cells can also generally be passaged for several generations depending on the cell type (e.g., 50 for normal human fibroblasts). *Transformed cells* represent cells derived from tumors or that are genetically altered (using carcinogenic agents or viruses). Transformed cells commonly exhibit infinite lifespan (are immortal or continuous) and may express numerous altered characteristics (Davis 1996).

12.2.1.2 Clonal Cell Lines

Clonal cell lines represent a population of cells that are derived from a single primary or transformed cell, resulting in a theoretically homogeneous phenotypic population. Over time, however, naturally occurring mutations can arise, causing inhomogeneity in the cell population, requiring recloning. Clonal selection permits a cell with specific characteristics to be isolated. In general, cloning of established cell lines is more successful than that of primary cells. While isolated primary cells behave most similarly to cells *in vivo*, since cell lines are maintained under artificial conditions *in vitro* (e.g., frozen or cultured), there exist several potential drawbacks associated with primary cells. Primary cultures rich in a particular cell type may be contaminated by extraneous cell types introduced during the isolation procedure, which gives rise to an inhomogeneous cell population. Progressive cell passaging may result in the propagation of the contaminant cells, permitting them to increase in proportion over time in culture. Although transformed cell lines are readily available, passaged routinely, and proliferate indefinitely, they may exhibit abnormal (e.g., nonphysiologic and loss of differentiated function) behavior as a result of their transformation.

Stem cells provide an alternative and continually expandable cell source. In particular, mesenchymal stem cells (MSCs) are pluripotent and have been isolated from bone marrow and a variety of other tissues, including perichondrium, periosteum, fat, muscle, and synovium (e.g., Caplan 1970; Dexter, Allen et al. 1977; Nevo, Robinson et al. 1998; Angela, Stering et al. 2001; Zuk, Zhu et al. 2001; Jankowski, Deasy et al. 2002; Pei, He et al. 2008; Wang, Dormer et al. 2010). Prompted by environmental cues [e.g., growth factors (Pittenger, Mackay et al. 1999; Majumdar, Wang et al. 2001; Erickson, Gimble et al. 2002) and physical stimulation (Elder, Goldstein et al. 2001; Altman, Horan et al. 2002)], these highly proliferative cells can differentiate toward a number of different cell lineages. Alternatively, differentiated cells may be transdifferentiated to other cell types, as in the case of dermal fibroblasts exhibiting a chondrocyte-like phenotype in micromass culture in the presence of lactic acid (Nicoll, Wedrychowska et al. 2001). Induced pluripotent stem (iPS) cells, as reviewed in Grskovic, Javaherian et al. (2011), have been derived from adult human dermal fibroblasts by transduction of four defined transcriptional factors: Oct3/4, Sox2, Klf4, and c-Myc (Takahashi, Tanabe et al. 2007) and sidestep the ethical issues related to embryonic stem cells. Subsequent studies have derived iPS cells from other cell types, including skin keratinocytes (Aasen, Raya et al. 2008).

12.2.1.3 Cell Passaging

Cell passaging refers to the process of subculturing cells (plating, trypsinization, and replating of cells) and is a method by which the cell number can be expanded in culture. Upon culturing the cells on an intermediary surface (e.g., tissue culture ware) and then releasing the cells through the disruption of cell–substrate attachments [using trypsin, mechanical cell scraping, or other agents such as versene or calcium-chelating molecules such as ethylene diaminetetraacetic acid (EDTA)], the cells are deemed passaged (i.e., cell passage 1). Each subsequent subculture increases the cell passage number. Trypsinization requires rinsing of cell monolayer with a serum-free medium (serum proteins competitively inhibit enzyme activity) such as phosphate-buffered saline (PBS) or Hank's balanced salt solution. Mild agitation may augment enzymatic activity. Cell lines are typically expanded (upon purchase) and then frozen for storage (at low passage). Cells may then be thawed and expanded at low passage for future use as needed (see Section 12.2.4). Cell passaging protocols can be used for cell expansion and to prime cells for subsequent 3D matrix production, as is performed for cartilage tissue engineering (Pei, Seidel et al. 2002; Ng, Lima et al. 2010; Sampat, O'Connell et al. 2011).

12.2.2 Techniques for Primary Cell Isolation

12.2.2.1 Enzymatic Cell Isolation Techniques

Enzymatic digestion techniques have been adopted for the isolation of a variety of cell types, including Schwann cells (Chafik, Bear et al. 2003), smooth muscle bladder cells (Haberstroh, Kaefer et al. 2002), dermal fibroblasts (Wilkins, Watson et al. 1994), endothelial cells (Levesque and Nerem 1985), and liver cells (Bhatia, Balis et al. 1999). In general, the tissue of interest is removed aseptically (e.g., using betadine, prepodyne, ethanol wash of the external surface, and a biological hood) from the source and the extraneous tissue dissected away. The tissue is typically minced so as to provide maximal surface area for subsequent enzymatic treatment. Enzymatic treatment, using an enzyme cocktail tailored to disrupt components of the particular tissue matrix with different levels of aggressiveness (trypsin, collagenase, hyaluronidase, protease, elastase, dispase), acts to dissociate the cells from their extracellular matrix (ECM). Agitation (e.g., orbital shaker, rocker plate, spinner flask) may also facilitate digestion of the tissue. Different enzymes may be applied simultaneously or sequentially.

Enzymatic digestion protocols need to be optimized for cell yield and cell viability. Tissue debris are separated from cells using a series of filtering steps, with a filter pore size dependent on the cell type [e.g., for chondrocytes typically ~20-μm filter (Hung, Henshaw et al. 2000) and for osteoblast-like cells a ~45-μm filter (Hung, Pollack et al. 1995)]. Centrifugation of cells ($1000 \times g$) after filtering permits the separation of the supernatant (enzyme solution) from the cells that have passed through the filter. The supernatant is aspirated and replaced with fresh serum-containing medium, and the cells resuspended via mild agitation to dislodge and disperse the cell pellet. Care must be taken to avoid overdigestion of tissue, which may result in decreased cell viability and altered function (e.g., from degradation of cell surface receptors). In some protocols, a serial digestion is performed, where sequential digestions are performed on the same harvested tissue, with differing cell populations released at varying time points. Extraction of osteoblast and osteoclast-like cells from neonatal rat calvaria (the skull cap) is an example of such a serial digestion (Luben, Wong et al. 1976; Brand and Hefley 1984; Wong 1990). After the isolation of cells, contaminant cells or mixed populations can be separated using a variety of techniques, including selective permeabilization (Modderman, Weidema et al. 1994), adherence/nonadherence to plating surfaces (Martin, Padera et al. 1998), and cell sorting (Chafik, Bear et al. 2003) (see Section 12.4). If distinct cell populations are spatially distributed (e.g., stratified tissues such as articular cartilage), the tissue can be dissected in appropriate sections and digested (Aydelotte and Kuettner 1988; Knight, Lee et al. 1998). Isolation techniques can also be designed to release units of cells and their associated matrix, such as chondrons that are found in articular cartilage (Poole 1997; Graff, Lazarowski et al. 2000; Knight, Ross et al. 2001). Mechanical isolation of cells is an alternative mechanism, but may be more susceptible to cell damage and lower yields.

12.2.2.2 Cell Migration Isolation Techniques

Migration of cells out of tissue explants offers an alternative method for isolating primary cells. Minced tissue is plated onto tissue culture-treated dishes and, over time, cells will crawl from the tissue and migrate onto the dish. Partial enzymatic digestion of the tissue as well as addition of chemokinetic/chemotactic factors in the medium may further help to expedite outgrowth. The migration technique avoids cell damage associated with the use of enzymes. However, if cells require subculture (such as for expansion), enzymatic exposure may be unavoidable. Cell isolation via migration has been used for a variety of cell types, including bone cells (Jonsson, Frost et al. 1999; Katzburg, Lieberherr et al. 1999; Voegele, Voegele-Kadletz et al. 2000), knee ligament fibroblasts (Nagineni, Amiel et al. 1992; Hung, Allen et al. 1997; Kobayashi, Healey et al. 2000) (see Figure 12.2), and skin epidermal cells (Wilkins, Watson et al. 1994). Although bone cells can be extracted from neonatal

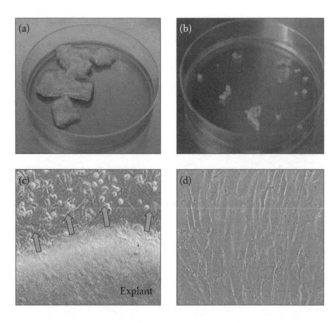

FIGURE 12.2 (See color insert.) Isolation of medial collateral ligament (MCL) knee fibroblasts from (a) tissue explant; (b) minced ligament tissue on polystyrene culture dish; (c) MCL fibroblasts migrating out-of-tissue explant; and (d) phase-contrast image of confluent MCL fibroblast monolayer.

calvarial tissue using enzymatic digestion, bone cells from mature bone, which is not amenable to enzymatic digestion due to its mineralized content, are also isolated using explant outgrowth cultures (Katzburg, Lieberherr et al. 1999).

12.2.2.3 Phenotypic Characterization

Once cells are isolated, they should be characterized for their expression of markers consistent with the desired cell type. Purity of the cell population may be assessed using molecular, biochemical, and histological analyses. The phenotypic expression of cells is, however, influenced by culturing conditions, including medium constituents, nature of plating substrate, 2D or 3D culture, and the physical environment. This characterization should also be performed periodically on passaged cells for the same reasons.

12.2.3 Culturing Techniques

12.2.3.1 Tissue Culture Media

Cells are cultured in an aqueous environment from which nutrients and wastes are exchanged via diffusion. A physiologic basal medium (e.g., for primary cells, Eagle's medium, Dulbecco's modified Eagle's medium, Alpha MEM, Ham's F-12 are used, whereas cell lines are typically grown in RPMI 1640, Iscove's modified Dulbecco's Medium, or DMEM/F12) is used to provide an aqueous environment for cells in culture, with cell-specific additives that may include amino acids, glucose, vitamins, and serum. Serum, typically used at 10% volume/volume, contains a cocktail of growth factors needed for cell proliferation and maintenance, albumin (the major protein component of serum), transferrin (the major iron transport protein in vertebrates), antiproteases, and attachment factors (e.g., fibronectin, laminin) (Davis 1996). The use of serum has potential disadvantages, including batch-to-batch variations (e.g., variable absolute and relative amounts of constituents) as well as the potential for contamination by bacteria, fungi and mycoplasm, and viruses. For these reasons, serum-free medium is often desirable (commercialization or basic science studies where batch-to-batch differences are not acceptable) (Dumont, Ionescu et al. 2000). These media are typically supplemented with factors such as growth regulators (e.g., insulin), transport proteins (e.g., transferrin), and trace elements (e.g., selenium) (Kisiday, Jin et al. 2002; Lima, Bian et al. 2007) (see Figure 12.3). In some cases, transient application of growth factors (e.g., for

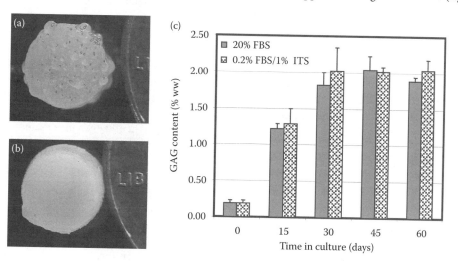

FIGURE 12.3 Chondrocyte-seeded agarose disks cultured in (a) DMEM supplemented with 20% FBS or (b) ITS (insulin, transferrin, and selenium) medium with 0.2% FBS after 45 days in culture. (c) Although there were differences in the overall appearance of the disks over culture time, chondrocyte synthesis of glycosaminoglycans was comparable for each medium used.

the first few weeks of culture) serves as a more potent stimulator of tissue development than continuous growth factor exposure (Lima, Bian et al. 2007; Byers, Mauck et al. 2008; Kalpakci, Kim et al. 2011; Ng, O'Conor et al. 2011). Cultureware (not presterilized commercially) are sterilized using alcohol (70% ethanol) rinse, ultraviolet irradiation (~20 min), ethylene oxide (caution should be used to ensure that the gas has completely dissipated before using), and growth medias are typically sterilized via sterile filtering (typically ~0.2-μm pore size). Certain liquids used in tissue culture (e.g., water) can also be autoclaved (with slow venting).

Acid and base maintenance of the medium is regulated using organic buffers such as HEPES (4-(2-hydroxyethyl)-1-piperazineethanesulfonic acid), TES (*N*,*N*,*N*′,*N*′-tetraethylsulfamide), and BES (*N*,*N*-bis(2-hydroxyethyl)-2-aminoethanesulfonic acid, *N*,*N*-bis(2-hydroxyethyl)taurine). Sodium bicarbonate buffer-supplemented medium in conjunction with a carbon dioxide incubator is a common buffering system for tissue culture. Phenol red is a convenient indicator of medium pH, where a more orange appearance indicates an acidic environment and a violet appearance indicates an alkaline environment. Physiologic osmolality (typically ~300 mOsmol/kg) must also be maintained, with some studies using bovine serum albumin (major protein component of serum) in place of serum in studies where growth factors (in serum) are undesirable. Antibiotics such as penicillin/streptomycin/neomycin as well as antifungal agents (e.g., amphotericin, nystatin) can be added. As a cautionary note, these agents may have unwanted effects on cells (such as creating pores in the cell membrane of cells; Marty and Finkelstein 1975); also as polycations, they may affect cell signaling pathways (Prentki, Deeney et al. 1986). Medium changes are typically performed daily or every other day depending on the metabolic activities of the cultures.

Tissue culture incubators add 5% carbon dioxide (95% air) and maintain a fully humidified environment at 37°C, thereby preventing medium evaporation and inadvertent changes to medium osmolality. Different oxygen tensions can also be used to create a hypoxic environment (Brighton, Schaffer et al. 1991; Hansen, Schunke et al. 2001; Bjork, Meier et al. 2012). Medium supplementation with cell-specific agents may be added to promote cell expansion, maintenance of cell phenotype function, or to optimize cell function synthetic activities (Wilkins, Watson et al. 1994; Einheber, Hannocks et al. 1995; Martin, Padera et al. 1998; Perka, Schultz et al. 2000). Metabolites (e.g., glucose, proline), cofactors (e.g., ascorbate), and mineral salts (e.g., calcium/magnesium that are important for enzyme function and cell adhesion) are often added as well (Wilkins, Watson et al. 1994). Media osmolarity may also be adjusted to modulate cell differentiation and activities, as has been done for cartilage tissue engineering, wherein chondrocytes normally see a hypertonic environment (Oswald, Ahmed et al. 2011; Oswald, Brown et al. 2011).

12.2.3.2 Determination of Cell Number and Viability

Once cells are suspended, serum-containing medium is added to quench the enzymatic activity followed by centrifugation. Upon removal of the supernatant (enzyme-culture medium), the cells can be resuspended in the tissue culture medium and counted. Counting of cells can be performed using the trypan blue exclusion assay with a light microscope and hemocytometer, yielding the percentage of viable cells and cell concentration (cells per mL). The molecular weight of trypan blue excludes it from crossing the membrane of healthy living cells, staining the interior of damaged or dead cells blue. Given the cell concentration for a cell suspension, cell plating (e.g., cells per unit area) or seeding densities (e.g., cells per unit volume) at desired levels can be achieved.

Assessment of cell viability in monolayer culture, explant tissues, and tissue constructs can be performed using commercially available live/dead assays and fluorescence microscopy (see Section 12.3), where calcein-AM (excitation 495 nm/emission 520 nm) stains living cells green and propidium iodide (PI) (excitation 530 nm/emission 615 nm) stains dead cells red. This assay is useful for a variety of applications such as the determination of nutrient limitations in 3D culture or tissue impact loading studies (Lucchinetti, Adams et al. 2002). For biochemical assays, there may also be a need to normalize cell activities by cell number rather than use absolute values. Changes or differences in the production of cell products, for example, may be due to variations in the cell number rather than differences in the inherent level of cell synthesis. Dyes that bind specifically to negatively charged DNA, such as

Hoechst 33258 dye, requires a cuvette or microplate fluorometer, or epifluorescence microscopy system equipped to quantify DNA. Using ultraviolet (365-nm wavelength) excitation, the bound Hoechst dye emits light at 465 nm. The fluorescence emission of Hoechst dye is 400-fold lower when bound to RNA than to DNA (Labarca and Paigen 1980). Cell number can be determined by using a conversion factor for DNA per cell (e.g., 7.7 pg of DNA/cell for chondrocytes; Kim, Sah et al. 1988). Such techniques are applicable to cell monolayers and tissues (Labarca and Paigen 1980; Kim, Sah et al. 1988). PicoGreen (excitation 480 nm/emission 520 nm) is another readily available dye for DNA quantification (Singer, Jones et al. 1997).

12.2.3.3 Two-Dimensional Cultures

Certain cells are anchorage-dependent and will grow and function only when adhered to physical surfaces. Cell attachment on tissue culture plastic, glass, or porous filter membranes commonly used for cultivating anchorage-dependent cells relies, in part, on medium constituents (such as fibronectin or laminin found in serum) that adsorb to the substrate surface and promote cell–substrate binding via integrins (Shyy and Chien 1997) (see Figure 12.2d). Surface treatments such as gas–plasma discharge (e.g., tissue culture plastic) or acid wash can also make the surface more wettable. Surface coating with biopolymers (poly-D-lysine, poly-L-lysine) or purified matrix components (e.g., fibronectin, laminin, or collagen I and IV) are also used to enhance cell adhesion, spreading, and proliferation on glass and plastic substrates. Biosubstrates such as the complex basement membrane extracellular matrix Matrigel can also be used (Davis 1996). Cell–cell interactions can be manipulated by cell seeding density (i.e., proximity of cells to one another). Preconfluent cultures are relatively sparse, while confluent cultures exhibit cell-to-cell abutment and limited cell division due to contact inhibition. Micropatterning techniques, which permit selective spatial adsorption of cell substrates at the microscale, have also been used to tailor design of cell shape, spatial organization of cells, and cell–cell interactions (Singhvi, Kumar et al. 1994; Bhatia, Yarmush et al. 1997; Chen, Mrksich et al. 1997; Thomas, McFarland et al. 1997; Bhatia, Balis et al. 1999; Takayama, McDonanld et al. 1999; Liu, Jastromb et al. 2001; Parker, Brock et al. 2002; Petersen, Spencer et al. 2002; Guo, Takai et al. 2006). Substrate stiffness can also influence cell differentiation (Engler, Sen et al. 2006).

12.2.3.4 Three-Dimensional Cultures

Behavior of cells in 2D and 3D may be different, with the latter representing, in most instances, a more physiologic environment (Benya and Shaffer 1982; Kolettas, Buluwela et al. 1995). Cells can be maintained in their physiologic 3D environment (preserving *in situ* cell–matrix and cell–cell interactions) using explant cultures. In explant cultures, whole or fragments of tissue are cultured and sustained by the growth medium in which they are submerged. Calvarial cultures (Linkhart and Mohan 1989), neuronal tissue slice cultures (Morrison III, Eberwine et al. 2000), cartilage explants (Sah, Kim et al. 1989), flexor tendons (Banes, Horesovsky et al. 1999), and bone (Takai, Mauck et al. 2004) have been maintained in such a manner. However, the complexities of the native tissue (e.g., inhomogeneous properties and variable cell populations) may not be suitable for certain studies. In such cases, there are several potential options for studying cells using a 3D culture system. Micromass cultures (adherent cells in high-density spot cultures) represent transition from 2D monolayer to 3D culture systems (Denker, Nicoll et al. 1995). Cells can be cultured as a cell suspension over a surface that discourages cell adhesion (e.g., tissue culture surface coated with agarose). Similarly, pellet cultures of centrifuged cells have been studied, permitting the study of 3D cell–cell interactions (Martin, Padera et al. 1998; Graff, Lazarowski et al. 2000).

Cells can also be suspended in biocompatible hydrogels (e.g., agarose; Benya and Shaffer 1982; Mauck, Soltz et al. 2000) (see Figure 12.3) or sodium alginate (Atala, Cima et al. 1993; Mauck, Soltz et al. 2000; Awad, Wickham et al. 2003), fibrin (Nixon, Fortier et al. 1999; Bjork, Meier et al. 2012), collagen (Kanda and Matsuda 1994; Andresen, Ledet et al. 1997; van Susante, Buma et al. 1998; Gillette, Jensen et al. 2008) (see Figure 12.4), silk (Chao, Yodmuang et al. 2010), polyethylene glycol (PEG) (Rice and Anseth 2007), and hyaluronan (Kim, Mauck et al. 2011; Erickson, Kestle et al. 2012), and

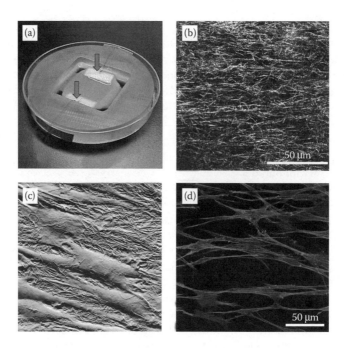

FIGURE 12.4 (**See color insert.**) (a) Polysulfone frame used to culture and mechanically stretch type I collagen gels seeded with cells. Velcro anchors permit mechanical loading via tension applied to sutures (arrows) to study contractile response of dermal fibroblasts to controlled loading conditions. (b) Confocal reflectance microscopy of collagen gel formed by dermal fibroblast alignment (488-nm laser epi-illumination, PMT using blue reflection filter, 60× 1.4 NA oil immersion objective at 1024 × 1024 pixels). (c) Contact mode atomic force microscopy (Bioscope, Digital Instruments), deflection image with 100 × 100 μm scan size, of dermal fibroblasts in 3D collagen gel matrix aligned as above. (d) Confocal microscopy image of rhodamine phalloidin labeling of actin cytoskeleton of dermal fibroblasts that were subjected to uniaxial tension (note parallel alignment with direction of loading). (Courtesy of Kevin D. Costa, Mt. Sinai School of Medicine.)

on commercially available microcarrier beads of various materials (e.g., collagen, dextran, polystyrene; Pollack, Meaney et al. 2000), as well as biocompatible scaffolds [e.g., poly glycolic acid (PGA), poly-L-lactide acid (PLLA), poly-ε-caprolactone (PCL)] that have different properties and confer varying cell–matrix interactions. An inherent advantage of hydrogel scaffold systems include their relative ease of cell seeding and achieving homogeneous cell distribution (Smetana 1993; Cushing and Anseth 2007). The latter will be reviewed in Chapter 15 and are designed to optimize the diffusion of nutrients and cell function in a physiologic 3D environment (Paige and Vacanti 1995). Cell expansion or tissue growth bioreactors can be used with cells alone or cells on carriers or seeded in scaffolds. Commercially available bioreactor systems for culturing cells include flow perfusion (e.g., hollow fiber) (Petersen, Fishbein et al. 2000) or permeation systems (Dunkelman, Zimber et al. 1995; Pazzano, Mercier et al. 2000), rotating-wall vessels (RWV) (Freed, Vunjak-Novakovic et al. 1993; Klement and Spooner 1993), spinner flasks, and mixing bioreactors (Vunjak-Novakovic, Freed et al. 1996; Carver and Heath 1999). These bioreactors are aimed at optimizing nutrient availability to cells and growing tissue constructs as well as providing a hydrodynamic environment conducive for cell/tissue growth (e.g., Gooch, Kwon et al. 2001). Other bioreactors are aimed at the application of specific mechanical stimuli for the study of cell physical regulation and long-term growth (Brown 2000; Mauck, Soltz et al. 2000; Takai, Mauck et al. 2004; Waldman, Spiteri et al. 2004; Tandon, Cannizzaro et al. 2009; Bian, Fong et al. 2010). Physical loading and growth factors may act synergistically (Mauck, Nicoll et al. 2003) and in some cases counteract one another (Lima, Bian et al. 2007; Syedain and Tranquillo 2011).

12.2.3.5 Cocultures

Cocultures of multiple cell types can also be maintained in 2D and 3D culture conditions, such as hepatocytes and 3T3 fibroblasts (Bhatia, Yarmush et al. 1997), neurons and Schwann cells (Einheber, Hannocks et al. 1995), smooth muscle cells and chondrocytes (Brown, Kim et al. 2000), smooth muscle cells and endothelial cells (Niklason, Gao et al. 1999), and endothelial cells and fibroblasts (Braddon, Karoyli et al. 2002), and different populations of chondrocytes (Kim, Sharma et al. 2003; Gan and Kandel 2007; Hwang, Varghese et al. 2008; Tan, Dong et al. 2011). In these systems, culture conditions may need to be optimized for maintaining multiple phenotypes simultaneously. Coculturing cells *in vitro* may reestablish physiologically relevant interactions or relationships between different types of cells that are necessary for optimal growth. Culture well inserts (such as the commercially available Transwell® system by Corning) are convenient for studying cell–cell interactions by providing physical separation of cell types while permitting paracrine signaling (Tan, Dong et al. 2011). In another form of coculture, feeder layers of fibroblasts are used to support growth and maintenance of embryonic stem cells (Tompkins, Hall et al. 2012).

12.2.4 Cryopreservation

Cryopreservation is important for long-term preservation and storage of living cells and tissues (Mazur 1984; Karlsson and Toner 1996; Cui, Dykhuizen et al. 2002). When freezing cells, the number of passages from the original must be minimized (i.e., early passage material available for new working stock). Cryopreservation involves dramatic temperature drops (e.g., from a physiologic +37°C to the boiling point of liquid nitrogen −196°C) accompanied by significant cell water efflux (as extracellular ice forms leading to hypertonic loading of cells) mediated by differential intracellular and extracellular electrolyte concentration during the freezing process (Karlsson and Toner 1996). The application of cryoprotectant agents has been used to minimize cell injury and to control ice formation during cell freezing.

12.2.4.1 Cryoprotectants

The most common cryoprotective agents are the cell-permeating dimethylsulfoxide (DMSO) and glycerol (5–10%, v/v, reagent grade or better) (Simione 1998). In general, cryoprotectants increase the osmolarity of the extracellular solution, thereby depressing the freezing point (to about −5°C) due to colligative effects and encouraging greater dehydration of the cells prior to intracellular freezing (thereby reducing intracellular ice formation). It is thought that permeating cryoprotectants reduce cell injury due to solution effects by reducing potentially harmful concentrations of electrolytes in the cell, stabilization of cell proteins or cell membranes, and increasing the viscosity of the extra- and intracellular solutions, thereby dramatically reducing the rates of ice nucleation and crystal growth (Karlsson and Toner 1996). In addition to these potential mechanisms above, DMSO and glycerol are thought to act by increasing the permeability of the membrane and thus the water flux across the cell membrane during the freezing and thawing process. The rate of water efflux is proportional to the magnitude of the driving force (i.e., the osmotic pressure difference across the membrane) and the permeability for the plasma membrane to water. Knowledge of the hydraulic permeability of the cell membrane is therefore useful in efforts to optimize protocols for cell freezing (Mazur 1965; McGann, Stevenson et al. 1988; Diller 1997; Martinez de Maranon, Gervais et al. 1997; Noiles, Thompson et al. 1997). Glycerol may be sterilized by autoclaving for 15 min at 121°C and 15 psig. DMSO must be sterilized by filtration using a 0.2-μm nylon syringe filter. Glycerol is less toxic than DMSO for most cells. DMSO is more penetrating and is usually preferred when using larger, more complex cells.

12.2.4.2 Freezing Process

Ice forms at different rates during the cooling process (Diller 1997). A uniform cooling rate of 1°C per minute from ambient temperature is effective for cryopreserving a wide variety of cells and can be performed effectively using commercially available freezing containers (e.g., Nalgene "Mr. Frosty"; Simione 1998). Despite the control applied to the cooling of the cells, most of the water present will freeze at approximately −2°C to −5°C. Cooling at too rapid rates is associated with intracellular ice formation, and membrane rupture due to osmotic fluxes. With too slow cooling, cell injury is thought to arise from exposure to highly concentrated intra- and extracellular solutions ("solution effects") or to mechanical interactions between cells and the extracellular ice (Lovelock 1957). Induction of gas bubble formation by intracellular ice, or osmotic effects due to the melting of intracellular ice during warming may also give rise to cell injury. Generally, the larger the cell, the more critical slow cooling becomes. In practice, cell suspensions are mixed in equal parts with cryoprotectant solution followed by equilibration for ~15 min at room temperature. Cells can be aliquoted into cryovials (sealed airtight) using 0.5–1 mL volume for commonly used vial sizes (1–2 mL), frozen at −20°C for several hours, and then transferred to −60°C to −80°C freezer for 48 h (acceptable for temporary short-term storage). Lastly, for long-term storage, vials should be transferred to liquid nitrogen (<−130°C) (Davis 1996).

12.2.4.3 Thawing Process

Cryoprotective chemicals can themselves be damaging to cells. For most cells, warming from the frozen state should occur as rapidly as possible until complete thawing is achieved. The contents of the vial should be transferred to fresh growth medium following thawing to minimize exposure to the cryoprotective agent. Furthermore, addition and removal of cryoprotectants before and after cryopreservation can cause damage to cells due to excessive osmotic forces. Accordingly, cryoprotectants are usually added and removed gradually, changing the concentration of the extracellular solution in a stepwise fashion. Thawing protocols using a gradual addition of medium to the frozen cells, dilution protocols, or initial resuspension in medium containing impermeant solutes of similar tonicity as the cryoprotectant solution may facilitate diffusion of the cryoprotectant out of the cell during the thawing process (Davis 1996). Generally, the greater the number of cells (typically 10^6–10^7 cells/mL) initially frozen down, the greater the recovery (Simione 1998). Ultimately, cell viability upon recovery will determine the efficacy of the cryopreservation protocol.

12.3 General Microscopy Principles

Various microscopy tools take advantage of interactions when light strikes an object, namely that incident light waves can be absorbed, reflected, refracted, polarized, diffracted, or cause resultant fluorescence. Several microscopy techniques that are commonly applied to the study of cells are presented in this section. Resolution of a microscope objective is defined as the smallest distance (R) that two distinct points on a specimen can be distinguished as such (Rayleigh criterion). Numerical aperture (NA) is a measure of the ability to gather light and resolve fine specimen detail at a fixed object distance. Resolution is given by the expression $R = 0.61\lambda/NA_{obj}$, where $NA = n \sin \alpha$ where α is the aperture angle, n = refractive index ($n_{air} = 0.95$, $n_{oil} = 1.35$–1.4), and λ is the wavelength of light (James and Tanke 1991). The limit of resolution is 100–250 nm, depending on the optics and wavelength of light applied (Weiss and Maile 1993). For bright-field applications, brightness varies as $(NA)^2/(magnification)^2$. The higher the NA of the total system, the better the resolution and the shallower the depth of field. Depth of field, related to the axial resolving power along the optical axis, is proportional to the inverse of $(NA)^2$ and is defined as the thickness of the optical section that can be brought into focus. Portions of the object out of this plane of focus (i.e., over and under the section) degrade the quality of the image. In practice, having a higher NA (horizontal resolving power) and magnification are accompanied by disadvantages of very shallow depth of field (more image degradation due to blurred background) and short working distance.

12.3.1 Light Microscopy

12.3.3.1 Brightfield

Bright-field microscopy, used with transmitted or reflected (epi-illuminated) light from a tungsten–halogen lamp source, is considered the most basic of the imaging modes, with illuminating rays entering the objective lens and "lighting up the background." The image results from illumination that falls on to the specimen emanating from within the aperture angle of the objective, while illumination from outside this angle provides a dark-ground image (Bradbury and Evennett 1996). Since bright-field microscopy provides low intrinsic contrast, other techniques have been developed for contrast enhancement. Contrast is the degree to which the object of interest is separated from its background in terms of color and/or brightness (Bradbury and Evennett 1996). Contrast results from interactions between the specimen with light. For opaque specimens (that absorb light), epi-illumination is used to illuminate the specimen from the same side of the objective. Transparent specimens (such as living cells and unstained specimens) are typically illuminated from the opposite side of the objective, or transmitted illumination. Using transmitted light, contrast of stained biological sections and histochemical reactions that yield colored end products arises from selective absorption of light in the visible spectrum (resulting in a change in color of the transmitted light). Contrast of transparent specimens can also be enhanced through the use of monochromatic light, in particular, green (550 nm, in the peak sensitivity range of the human visual perception system), that is well known to minimize chromatic and spherical aberrations in standard achromatic objectives (Bradbury and Evennett 1996).

Diffraction results when an object with regularly repeating features produces an orderly redistribution of light (diffraction pattern). It can also be used to enhance contrast of transparent microscopic specimens, in particular, accentuating borders (Bradbury and Evennett 1996). In *phase contrast*, the phase difference between the direct and diffracted rays (one-quarter wavelength out of phase), and their relative amplitudes, can be altered to produce conditions for interference and increased contrast (Bradbury and Evennett 1996). Phase contrast is ideal for contrast enhancement of nonabsorbing, thin objects that have no large refraction differences (or halos will be introduced) (see Figure 12.2d). *Differential interference contrast* (DIC) based on principles by Nomarski yields in-focus, high-contrast, shadowcast images of phase details in which the direction of shadowing is opposite for phase-advancing and phase-retarding details. Since DIC is an optical technique to generate contrast, the image may not reflect actual topography and features of the cell, thereby being more qualitative than quantitative. *Hoffman modulation contrast* is another contrast enhancement technique used to increase specimen visibility and contrast for unstained and living specimens, also creating a pseudo-3D effect. This technique is sensitive to gradients of optical path length and produces an asymmetrical, directional image with phase gradients in one direction always rendered bright and those in the opposite direction dark (Bradbury and Evennett 1996). Interpretation of modulation contrast should be made with the same caveat as DIC. Phase contrast, DIC, and modulation contrast are generally considered complementary techniques, with the latter more suitable for observation of relatively thick objects (James and Tanke 1991). In *dark-field microscopy*, samples are illuminated with a concentrated beam of light that is either directly transmitted or scattered when it hits an object. Directly transmitted light is not collected by the object; instead, the scattered light enters the lens producing an image where objects appear in high contrast against a dark background. This technique has been used to provide real-time visualization of individual microtubules, allowing direct examination of dynamic microtubule functions such as treadmilling and dynamic instability (Summers and Kirschner 1979; Horio and Hotani 1986; Hotani and Horio 1988; Hotani and Miyamoto 1990). More recently, variable bright–dark-field contrast, which utilizes alternating bright- and dark-field illumination, has been used to visualize complex 3D architecture with greater clarity than either technique can yield alone (Piper and Piper 2012). *Polarized light microscopy* is a contrast-enhancing technique that improves image quality of birefringent materials due to their anisotropic character.

12.3.2 Fluorescent Microscopy

12.3.2.1 Fluorophores

Fluorophores are molecules capable of absorbing and then reradiating secondary light (i.e., fluorescence). The latter continues as long as the excitation light is applied. When conjugated to antibodies, fluorophores bind specifically to targets and can be used for a variety of biological studies [e.g., identification of submicroscopic cellular components (see Figures 12.6 and 12.7), cell signaling, intracellular trafficking, gene reporter assays], as reviewed by Oksvold and coworkers (2002). The increased availability of applicable fluorophores has increased the popularity of live cell imaging (as reviewed in Wang, Shyy et al. 2008; Ishikawa-Ankerhold, Ankerhold et al. 2012). Here, fluorescently labeled proteins can also be introduced directly into living cells via transfection (using plasmid DNA), by infection (using viral vector delivery systems), by direct injection of fluorescently labeled proteins into the cells, or by passive and/or active cellular uptake mechanisms. These techniques provide valuable insights into the operation of internal cellular machinery in living cells and tissue such as cytoskeletal dynamics (Bulinski, Odde et al. 2001; Yarar, Waterman-Storer et al. 2005; Kelly, Katagiri et al. 2010), intracellular cell signaling (van Roessel and Brand 2002; Van Roessel, Hayward et al. 2002; Knight, Roberts et al. 2003), and gene expression (Chalfie, Tu et al. 1994; van Roessel and Brand 2002). Fluorescent probes can be used, for example, to monitor rapidly changing intracellular calcium levels in real time using the ratiometric dye fura-2AM (see Figure 12.5) (Grynkiewicz, Poenie et al. 1985; Tsien, Rink et al. 1985), as well as to monitor cytoskeletal microtubule reorganization in living cells using green fluorescent protein (GFP) (Chalfie, Tu et al. 1994) (see Figure 12.6). Upon absorbing the excitation light, electrons are raised to a higher vibrational energy state and rapidly (~billionth of a second) undergo a loss of vibrational energy, returning to the ground state with a simultaneous emission of fluorescent light. Owing to Stoke's shift, this emission wavelength is always of longer wavelength than the excitation wavelength (i.e., the emission spectrum is shifted to longer wavelengths than the excitation spectrum). Fluorophores have a peak excitation/fluorescence corresponding to a peak emission/fluorescence intensity, which can be applied in practice using a combination of appropriate excitation filters (which only permit specific excitation wavelengths to pass), dichromatic beamsplitters (or dichroic mirrors, designed to enhance or block transmission and reflectance at specific wavelengths), and barrier filters (designed to block or transmit fluorescence below and above a particular wavelength, respectively). The intensity of the fluorescence emission may be reduced (called fading) due to photobleaching or quenching. Irreversible decomposition of the fluorescent molecules due to light intensity in the presence of molecular oxygen is referred to as bleaching. Photobleaching gives rise to the technique of fluorescence recovery after photobleaching (FRAP). Quenching results from the transfer of energy to other so-called acceptor molecules (resonance energy transfer) in close proximity and serves as the basis for fluorescence resonance energy transfer (FRET).

12.3.2.2 Conventional Fluorescence Microscopy

In fluorescence microscopy with epifluorescence versus transmitted illumination, excitation light is focused by the objective on to the microscope field, after which the same lens system collects the fluorescence emission light (James and Tanke 1991). Since the brightness of the image in epifluorescence microscopy is proportional to $(NA_{obj})^4/(\text{magnification})^2$ (Herman 1996), an objective with the lowest overall magnification having acceptable resolution, with the highest light-gathering power yields the best results (Shotton 1993). Observation of thick specimens yields an image that is the sum of the sharp in-focus image and the blurred image of regions that are out of focus (see advantages of confocal microscopy). A xenon or mercury lamp that produces a high-intensity illumination (arc-discharge lamps have an average lifetime of 200 h) is usually used. Xenon lamps emit a spectrum of relatively constant intensity from ultraviolet to red, whereas mercury lamps have distinct peaks in the emission spectrum (360 nm—ultraviolet, 405 nm—violet, 546 nm—green, and 578 nm—yellow), making the latter generally more flexible with respect to useable fluorophores in fluorescence microscopy (James and Tanke 1991). The choice of *fluorophores* should be based on the compatibility of the excitation maximum with

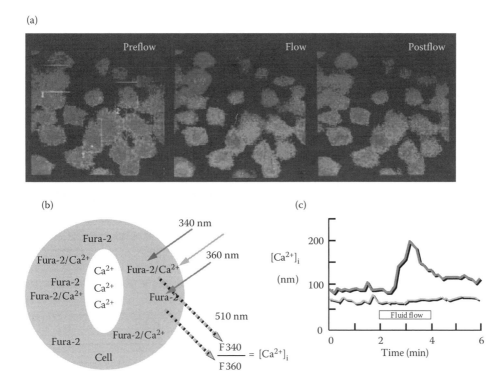

FIGURE 12.5 **(See color insert.)** Real-time intracellular calcium response, $[Ca^{2+}]_i$, of primary bone cells (derived from rat calvaria) subjected to fluid-induced shear stress (35 dynes/cm²), using the calcium indicator dye fura-2 and fluorescence microscopy, and a parallel-plate flow chamber (Allen et al. 2000). (a) Light intensity was recorded with a CCD camera and converted to pseudocolor (blue/green for low levels, orange/red high levels) for visualization and analysis. Olympus software was used to subtract background noise and single cells selected for analysis by boxing or tracing of the cell contour. (b) The cell-permeant acetoxymethyl ester form of the dye, Fura-2 AM, was loaded into living cells using DMSO and Pluronic®-127 in a buffered saline solution. Upon entering, the AM tail is cleaved by endogenous esterase activity, trapping fura-2 within the cytoplasm (Grynkiewicz et al. 1985; Tsien et al. 1985). (c) Fura-2 is a ratiometric dye, which has the advantage of minimizing artifacts of motion, uneven dye loading, and dye leakage. Rapidly alternating 340/360 nm excitation wavelengths are used to elicit 510-nm wavelength emissions from calcium-bound dye and the fura-2 isobestic point (calcium concentration insensitive), respectively (Hung et al. 1995). The ratio of the 510-nm emissions from the $R = 340$ nm/360 nm is compared to a calibration curve, providing a real-time measure to get $[Ca^{2+}]_i$.

light source (diode, lamp, or laser), resistance to photobleaching (when strong or prolonged excitation is needed), avoidance of overlapping emissions that can disrupt multistaining studies (see Figure 12.7a), and fluorophore size, which may impede staining penetration (Oksvold, Skarpen et al. 2002).

12.3.2.3 Scanning Confocal Fluorescence Microscopy

Scanning "confocal" microscopy derives from the fact that the illumination spot on the microscopic object is exactly imaged on the detector (James and Tanke 1991). The microscopic field is scanned by a small focused light spot that is directed across the microscopic object by computer-controlled mirrors. Lasers are used widely for confocal microcopy, serving as a high-intensity monochromatic light source (krypton, argon, neon, helium) (see Figures 12.4d, 12.6, and 12.7a. Compared to conventional fluorescence microscopy, confocal microscopy offers increased resolution, ~1.4-fold (White, Amos et al. 1987; Yuste, Lanni et al. 1999; Oksvold, Skarpen et al. 2002) and improved images for thicker specimens

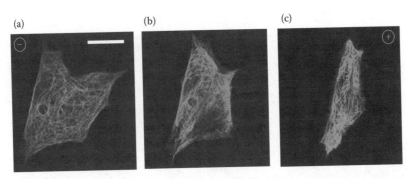

FIGURE 12.6 Confocal microscopy image of NIH3T3 fibroblasts (Faire et al. 1999) stably transfected with GFP-labeled microtubule associating protein that enables real-time visualization of changes to microtubule organization. Microtubule reorganization of a single cell in response to direct current electric field, at a field strength of 6 V/cm, at (a) time = 0 h, (b) time = 1.5 h, and (c) time = 3 h. The (+) and (−) indicate direction of field. Scale bar: 10 μm. (Courtesy of Grace Chao, National Taiwan University, and Clark Hung and Chloe Bulinski, Columbia University.)

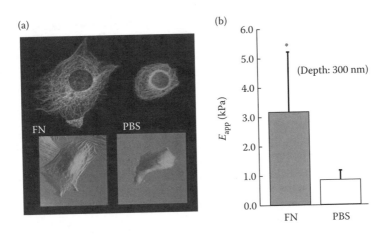

FIGURE 12.7 (**See color insert.**) MC3T3-E1 osteoblast-like cell line cultured on glass slides coated with fibronectin (FN) and control phosphate-buffered saline (PBS) (Takai et al. 2005). (a) Using confocal microscopy, double staining of actin and vinculin was performed to compare the cytoskeletal organization due to the different plating substrates. Cell topography images obtained from atomic force microscope (Digital Instruments) using the scanning mode and (b) complementary modulus data using the indentation mode were also performed (Costa and Yin 1999). A stronger development of cytoskeletal organization in the fibronectin-coated surface appears to correspond with increased spreading and stiffening of the cell.

(5–15 μm), which are a greater source of out-of-focus, extraneous background fluorescence. Confocal microscopy achieves this enhanced resolution by providing a controllable depth of field, elimination of out-of-focus information (from outside the plane of focus) using a pair of pinhole apertures, and the ability to acquire serial optical sections from thick specimens. From serial optical sections, 3D reconstruction using digital imaging tools of scans in the xz and xy-plane provides spatial localization of molecules of interest in the z-direction, which may not be discernible in the horizontal (xy) scan or conventional microscopy (e.g., localization at the surface or just below). The confocal has superior horizontal resolution, however, compared to vertical resolution. For applications requiring real-time acquisition under low-light conditions, conventional fluorescence microscopy system with a digital camera may yield better results. Low-light conditions necessitate a larger pinhole size that decreases the signal-to-noise ratio (permitting more background out-of-focus light) by reducing the resolution and increasing the thickness of optical sections. Application of fiber-optic monitoring coupled with confocal micros-copy (e.g., endoscope) has permitted imaging of gene expression *in vitro* and *in vivo* using GFP-labeled probes (Ilyin, Flynn et al. 2001; Al-Gubory and Houdebine 2006).

Although confocal microscopy offers improved resolution compared to conventional fluorescence microscopy, there is an associated loss of speed. With the increased availability of fluorescent biomo-lecular probes, live-cell imaging has gained increased popularity; however, single pinhole confocal microscopes are too slow (typically requiring 200 ms to 2 s to acquire an image) to accurately image these processes. As a result, a number of high-speed confocal microscopy techniques have emerged, including swept field (consisting of a linear array of pinholes), resonance scanning (containing a high-speed resonance scanner), spinning disk (consisting of a rotating circular disk with numerous pinholes), and slit-scanning (using a series of slit in lieu of the conventional pinholes), to address these issues (Conchello, Kim et al. 1994; Graf, Rietdorf et al. 2005; Wang, Babbey et al. 2005; Borlinghaus 2006; Murray, Appleton et al. 2007; Lu, Min et al. 2009). These modalities are designed to provide both high spatial and temporal resolution, thereby permitting real-time confocal imaging of dynamic biological processes such as cytoskeletal rearrangements during cell migration.

12.3.2.4 Multiphoton Confocal Microscopy

Multiphoton confocal microscopy involves the integration of confocal scanning microscopy with near-infrared long-wavelength multiphoton fluorescence excitation. In dual-photon microscopy, as an example, excitation arises from the simultaneous absorption of two photons by the fluorophore (Denk, Strickler et al. 1990; Yuste, Lanni et al. 1999; Majewska, Yiu et al. 2000). This reflects the fact that the radiation energy is linearly proportional to the inverse of the wavelength (Planck's law), meaning that radiation from a short wavelength has a higher quantum energy than long wavelengths (James and Tanke 1991). As an illustration, one 350-nm absorption (ultraviolet) is equivalent to two simultane-ously applied 700-nm absorptions (red). Pulsed lasers are used to increase the photon density necessary to achieve the latter with an average power that is only slightly higher than in confocal microscopy (Oksvold, Skarpen et al. 2002). This technique is ideal for applications with living cells and has several advantages over conventional confocal microscopy; it minimizes photobleaching and cell damage since two-photon excitation is restricted to only the focal point of the microscope. Additionally, sample pen-etration is increased (often 2–3-fold). The resolution in multiphoton confocal microscopy, however, may be decreased due to the longer excitation wavelengths used. Second harmonic generation (SHG) is a popular application of dual-photon microscopy, where the energies from the two photons are combined to generate a new photon at twice the energy when they encounter non-centrosymmetric structures such as collagen. This technique has been used to generate high-resolution images of collagen fiber ori-entation in tissues such as tendons (Theodossiou, Thrasivoulou et al. 2006; Houssen, Gusachenko et al. 2011), bone (Ambekar, Chittenden et al. 2012), cornea (Han, Zickler et al. 2004; Han, Giese et al. 2005), and cartilage (Mansfield, Yu et al. 2009; Werkmeister, de Isla et al. 2010; Mansfield and Winlove 2012).

12.3.2.5 Total Internal Reflection Fluorescence Microscopy

Total internal reflection fluorescence microscopy (TIRFM) is used to observe molecule–surface interac-tions (Axelrod, Thompson et al. 1982; Toomre and Manstein 2001). The technique is based on Snell's law, the principle that when light hits a less dense medium beyond a certain angle, it is reflected. Such interfaces exist at the culture dish/slide–cell interface or cell–culture medium interface. At a critical angle, the incident light is completely reflected. However, the reflection generates an electromagnetic field or evanescent wave (illuminating less than a few tenths of a micrometer) in the aqueous media that decays exponentially in magnitude with distance from the interface. For fluorescence applications, the exponential decay permits fluorophores beyond the surface (and out-of-focus plane) from being excited. This effect gives rise to a powerful methodology to visualize interface events with an increased signal-to-background ratio over traditional widefield techniques. Applications include study of ligand–receptor interactions, cytoskeletal and membrane dynamics, and cell/stratum contacts (Lanni, Waggoner et al. 1985; Truskey, Burmeister et al. 1992; Oksvold, Skarpen et al. 2002).

12.3.2.6 Fluorescence Recovery after Photobleaching

FRAP is a technique to monitor the mobility and dynamics of fluorescent proteins in cells and tissues (e.g., Axelrod, Koppel et al. 1976; Jain, Stock et al. 1990; Gribbon and Hardingham 1998; Reits and Neefjes 2001). Fluorescent-tagged proteins are irreversibly photobleached in a small region of the specimen using maximal laser intensity. The redistribution of nonbleached fluorescent molecules to the photobleached region is then visualized in the microscope to study molecular diffusion and binding.

12.3.2.7 Fluorescence Correlation Microscopy

Fluorescence correlation microscopy measures the emission fluctuations arising from diffusion of fluorescently labeled molecules in and out of a defined volume with a spatial resolution of fractions of femtoliters. These fluctuations reflect the average number of labeled molecules in the volume, as well as the characteristic diffusion time of each molecule across the defined volume (see review in Brock and Jovin 1998; Brock, Vamosi et al. 1999; Schille 2001).

12.3.2.8 Fluorescence Resonance Energy Transfer

FRET is a technique to detect direct protein–protein interactions and is made possible with fluorescent tags that have overlapping fluorescence spectra (Sato, Ozawa et al. 1999; Harpur, Wouters et al. 2001; Oksvold, Skarpen et al. 2002). When two overlapping fluorophores are located within 5 nm of one another, the excitation of the lower-wavelength fluorophore induces energy transfer to the neighboring higher-wavelength fluorophore (Clegg 1995). Alternatively, when energy is transferred to a nearby nonfluorescent molecule, an attenuation of the emission can be detected. Energy transfer also protects against fading, which can be exploited to detect protein interaction.

12.3.2.9 Fluorescence Lifetime Imaging Microscopy

Fluorescence lifetime imaging microscopy (FLIM) utilizes the decay rates of the fluorophore emissions to generate high spatial and temporal resolution images based on the fluorescence lifetimes of each fluorophore in the sample (Chang, Sud et al. 2007). These images are generated based on temporal changes in the intensity of each pixel as the sample decays, thereby allowing imaging of different material at a single excitation wavelength. Using this technique, we are able to resolve signals from fluorophores whose spectrum overlaps. FLIM is often combined with FRET since the lifetime of a fluorescent signal is dependent on fluorescent intensity and quenching. That is, resonant energy transfer between molecules reduces the emission lifetime of the donor fluorophores (Chen, Weeks et al. 2003; Wallrabe and Periasamy 2005).

12.3.3 Atomic Force Microscopy

Since its invention nearly two decades ago (Binnig, Quate et al. 1986), the atomic force microscope (AFM) has become the most widely used tool for studying mechanical properties of living cells. AFM involves tracking the deflection of a small (200–300-μm long) cantilever probe as its tip (~50-nm radius of curvature) scans, indents, or otherwise probes the sample. In particular, nanoindentation with AFM is well suited for cell mechanics applications due to its high sensitivity (sub-nanoNewton), high spatial resolution (submicron), and the ability to be used for real-time measurements in an aqueous cell culture environment in conjunction with an inverted microscope (see Figure 12.4b). In recent years, the AFM has been widely used to study the mechanical properties of cells and other soft samples. AFM has been used to monitor mechanical changes associated with platelet activation (Fritz, Radmacher et al. 1994; Walch, Ziegler et al. 2000), cell locomotion (Rotsch, Jacobson et al. 1999), myocyte contraction at the subcellular level (Domke and Radmacher 1998), and to examine the effects of chemical treatments that target specific cytoskeletal constituents (Wu, Kuhn et al. 1998; Rotsch and Radmacher 2000). Among these many studies, some findings have shown regional differences on the nucleus compared to the cytoplasm (Hoh and Schoenenberger 1994; A-Hassan, Heinz et al. 1998; Nagao and Dvorak 1999; Mathur, Truskey et al. 2000; Sato, Nagayama et al. 2000; Yamane, Shiga et al. 2000), the actin cytoskeleton appears to be largely responsible for cell stiffness (Henderson, Haydon et al. 1992; Haga, Beaudoin et al. 1998; Wu, Kuhn et al. 1998; Rotsch and Radmacher 2000; Sato, Nagayama et al. 2000), and in some cases, regions of altered perinuclear stiffness have been identified but could not be correlated with a specific cytoskeletal structure (A-Hassan, Heinz et al. 1998). Recent advances in AFM technology have yielded high-speed examination of dynamic biological processes such as myosin V walking along actin microfilaments (Ando, Uchihashi et al. 2007; Kodera, Yamamoto et al. 2010). In single-molecule force spectroscopy, AFM has also been used to study interactions between molecules and molecular assembly, where the AFM probe attaches to and pulls on individual molecules. This method has been especially useful for studying the interactions of proteins in cell membrane (see review in Noy 2011).

Only through computational methods can we accurately model the interaction between the AFM probe and the sample and ultimately extract a correct mathematical representation of the material properties of living cells. There are limitations, however, in the predominant method of analyzing AFM indentation data (the so-called Hertz theory; Love 1939; Weisenhorn, Maivald et al. 1992; Radmacher, Fritz et al. 1995) for studies of cells (Pourati, Maniotis et al. 1998; Mahaffy, Shih et al. 2000). Costa and Yin (1999) have published an analysis method for AFM indentation data, whereby accounting for the indenter geometry to compute an apparent elastic modulus as a function of indentation depth can reveal nonlinearity and heterogeneity of material properties from AFM indentation tests (Costa and Yin 1999). An important advantage of AFM over other cell mechanics techniques is the ability to combine high-resolution scanning with microindentation, which allows for direct correlation of local mechanical properties with underlying cytoskeletal structures (A-Hassan, Heinz et al. 1998; Rotsch and Radmacher 2000) (see Figure 12.7). In addition, the geometry of the probe tip can be modified relatively easily to allow for measurements over a range of length scales from submicron to tens of microns. This feature is especially useful for testing the applicability of continuum models for describing cellular mechanical properties.

12.3.4 Digital Microscopy

Digital imaging is the process of converting visual information into numeric form, which allows the study of cells and cellular events with quantitative temporal and spatial resolution (Herman 1996). Ultra-sensitive spatial cameras divide captured fluorescent images into an array of discrete picture elements (pixels), converting fluorescence intensity in each pixel to a number (e.g., 256 gray scale). These detectors permit quantitative analysis of variations in 2D intensity distributions in time (Herman 1996). The choice of electronic imaging detectors is a critical determinant in the range of light detection in fluorescence applications. Compared to traditional emulsion film, digital images can be acquired using a charge-coupled device (CCD) or video camera system. Another light-detecting device is the photomultiplier tube (PMT). PMTs are photoemissive devices in which the absorption of a photon results in electron emission. These devices offer high signal-to-noise ratio and are appropriate for weak signals and can be used for recording rapid events. During acquisition, frame averaging can further improve signal-to-noise ratio, boosting the detection of weakly fluorescent specimens.

Postprocessing of video or electronic image acquired through an optical microscope using various digital imaging techniques permits modification of the image that may enhance features as well as the information that is yielded. This postprocessing permits reversible, essentially noise-free modification of the image as an ordered matrix of integers rather than a series of analog variations in color and intensity (Oksvold, Skarpen et al. 2002). One such digital procedure is termed deconvolution (Conchello, Kim et al. 1994; Wang 1998), which can further increase the resolution of the confocal system up to approximately two-fold (Majewska, Yiu et al. 2000; Oksvold, Skarpen et al. 2002). In deconvolution, the point-spread function of the microscope optics is used to remove out-of-focus fluorescence when a series of images at variable depths is compared. In addition to improving image quality, digital tools can also be used to extract useful information about cell and tissue properties. For example, researchers have used cell nuclei (Schinagl, Ting et al. 1996), cell borders, as well as the material texture to obtain material properties of cells cultured in 3D scaffolds (Knight, van de Breevaart Bravenboer et al. 2002) and tissues (Guilak 1994; Guilak, Ratcliffe et al. 1995). Confocal reflection microscopy has been used to determine the fiber organization and mechanical properties of type I collagen 3D matrices (Roeder, Kokini et al. 2002) (see Figure 12.4c), whereas an optimized digital image correlation technique using both cell nuclei (epifluorescence microscopy) and material texture has been developed by Wang and coworkers for the determination of local strain fields and material properties of tissues and tissue-engineered constructs (Wang, Ateshian et al. 2002; Wang, Guo et al. 2002) (see Figure 12.8). Traction force microscopy has been used to quantify the forces that cells exerts upon their environment during cell migration and focal adhesion (Sabass, Gardel et al. 2008; Stricker, Sabass et al. 2010; Franck, Maskarinec et al. 2011; Koch, Rosoff et al. 2012). Fluorescent speckle microscopy (Waterman-Storer, Desai et al. 1998; Waterman-Storer 2002) and particle-tracking techniques (http://www.physics.emory.edu/faculty/weeks/idl) allow for the analysis

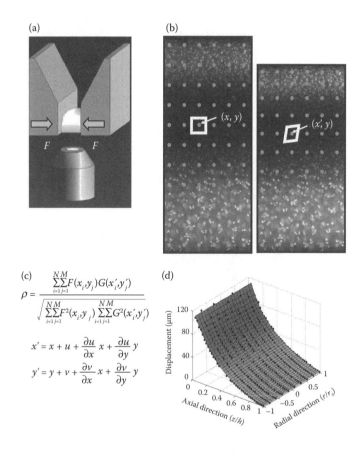

FIGURE 12.8 (a) Cell nuclei of chondrocytes in devitalized articular cartilage explant that have been labeled with Hoechst dye, which act as fiducial markers for the determination of deformation fields within the tissue under applied axial loading using impermeable platens (arrows). (b) Using epifluorescence microscopy (DAPI filter set) images obtained with a CCD camera, (c) optimized digital image correlation was used to track the displacement of nuclei within the tissue, smoothed, and then strain fields determined (Wang et al. 2002a,b). The program automatically tracks displacements with subpixel resolution (4× objective, 1 pixel = 1.66 μm). (d) Measured displacement fields in the axial (direction of loading) and radial (perpendicular to loading) directions.

and quantification of cytoskeletal dynamics within the living cells. These techniques have been used extensively to study the dynamic interplay between the extracellular environment and the intercellular machinery. For example, Kelly and colleagues (2010) have used fluorescent speckles of microtubule plus end binding protein to show that extracellular chondroitin-sulfate proteoglycan significantly alters microtubule polymerization rates and directionality when neuronal growth cones encounter CSPG-rich boundaries.

12.4 Flow Cytometry

12.4.1 Technical Background

Since its invention in the late 1960s, flow cytometry (FCM) has been the premier tool for single-cell analysis. FCM is complementary to fluorescence microscopy, being a powerful technique that offers the opportunity to measure multiple optical parameters, such as light scattering and fluorescence, of individual cells within heterogeneous populations at high throughput (rates of thousands of cells per second) (Shapiro 1995; Davey and Kell 1996). Flow cytometry presents some advantages: the analysis is fast, accurate (less than 5% instrumentation error), sensitive (detection as low as 100 cells per milliliter), and compatible with some staining methods providing great information at the single-cell level (Hammes and Egli 2010). Traditionally, single-cell analysis has focused on a few parameters in every experiment and nowadays advances in this technology (hardware, fluorochromes, and data analysis) permit us to quantify many parameters in every single cell from a representative population that led to the development of polychromatic flow cytometry.

The flow cytometer performs the analysis by passing the cells through a laser beam and capturing the light that emerges from each cell as it passes through (Shapiro 1995). The primary systems of flow cytometer consist of the fluidic system, where the laser and the sample intersect and the optics collect the resulting scatter and fluorescence, and take away the waste; the lasers, which are the light source for scatter and fluorescence; the optics, which gather and direct the light; the detectors, which receive the light; and the electronics and computer system, which convert the signals from the detectors into digital data and perform the necessary analyses (Carter and Meyer 1994; Shapiro 1995) (Figure 12.9a). FCM is the measurement of cell size, shape, and fluorescence as individual cells pass in single file within a hydrodynamically focused flow stream of sheath fluid or saline solution. The sample stream becomes compressed to roughly one cell in diameter. Most conventional flow cytometers will be detecting cells between 0.5 and 100 μm in diameter, although, using specialized systems, it is possible to detect particles outside this range such as viruses (Brussaard, Marie et al. 2000).

The measured parameters are generally of two different kinds: light scatter and fluorescence emission (Carter and Meyer 1994; Shapiro 1995). As a cell passes through the laser, it will refract or scatter light at all angles. Forward scatter (FSC), or low-angle light scatter, is the amount of light that is scattered in the forward direction as laser light strikes the cell and is proportional to the cell size. Light scattering at larger angles, for example, to the side, is caused by granularity and structural complexity inside the cell. This side-scattered light (SSC) is focused through a lens system and is collected by a separate detector, usually located 90° from the laser's path.

Fluorescence is a term used to describe the excitation of a fluorophore to a higher energy level followed by the return of that fluorophore to its ground state with the emission of light. Commonly, the most used fluorochromes include fluorescein isothiocyanate (FITC), rhodamine, phycoerythrin (PE), allophycocyanin (APC), and also PE- and APC-based tandem dyes and Alexa dyes (Lansdorp, Smith et al. 1991; Roederer, Kantor et al. 1996). Recently, the development of inorganic semiconductor nanocrystals, called quantum dots, and organic polymers (Service 2000) are available and led to 18-color cytometry (Chattopadhyay, Price et al. 2006).

If every single particle passing through the laser caused the instrument to collect data, most information would come from a very large number of small particles, such as platelets and debris. To prevent

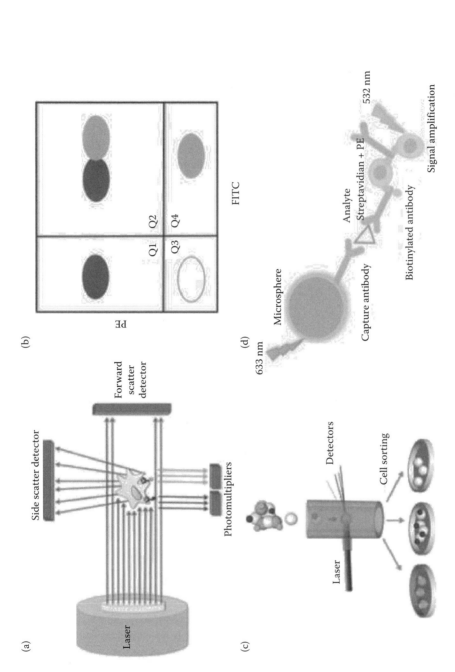

FIGURE 12.9 (a) Scheme of flow cytometry analysis. When the cells, suspended in the flow stream, pass through a laser beam, light scatter and fluorescence are collected. (b) Information obtained from a two-color (FITC/PE) dot plot. Cells showed in quadrant 1 (Q1) stain positive for the antigen detected with the PE-conjugated antibody. Cells showed in Q2 stain positive for both antigens. Cells appeared in Q3 would be negative cells and finally, those cells that are positive for the antigen detected with FITC-conjugated antibody are shown in Q4. (c) Cell sorting representation. FACS technology permits the individual isolation of cells from two or more populations. (d) Diagram for the detection of a molecule of interest by xMAP technology.

this, a threshold (or discriminator) is usually set on FSC such that a certain forward scatter pulse size must be exceeded for the instrument to collect the data.

As each cell passes through the laser beam of the cytometer, the detectors collect light intensity data for FSC, SSC, and each of the fluorescence channels. Data generated in an FCM system are collected in the flow cytometry standard format. Depending on the flow cytometer system, either pulse height or pulse area is used to quantify the size of the voltage pulse. Pulse width is used in certain special situations such as DNA analysis to distinguish single cells from double-cell events. Once the pulse has been converted into a numerical value, that value can be used to plot the intensity of the event.

The data can be displayed in 1D (histograms), 2D (dot, contour, or density plots) (Carter and Meyer 1994; Shapiro 1995), or 3D format. They can be generated using linear scaling or logarithmic scaling depending on the fluorescence intensities. Histograms display simple information; nevertheless, biological samples usually contain multiple cell populations, each with different characteristics. These populations can be better distinguished by looking at two parameters at once. Biexponential scale is another kind of scale available in digital flow cytometers that might assign fluorescence values that are below zero. This scale varies from logarithmic scaling at the upper end to linear scaling at the lower end. In a two-color experiment (FITC and PE) using a dot plot, four distinct populations emerge. Looking at the dot plot in terms of quadrants, cells with only bright orange fluorescence appear in the upper left quadrant. Cells with only green fluorescence appear in the lower right quadrant. Cells with both bright green and bright orange fluorescence appear in the upper right quadrant and finally, cells with both low green and low orange fluorescence appear in the lower left quadrant (Figure 12.9b). When multiple markers and multiple combinations are possible, data visualization in multiple dimensions becomes important (Roederer and Moody 2008).

Since it is possible to identify various populations within a cell sample based on their position in the FSC versus SSC plot, we can direct the analysis to consider only some of the events by drawing a region around. This is called gating and we can produce additional histograms and dot plots that help us to dissect subpopulations. The analysis based on a gated population exclude cells outside the gate. Since different fluorescent dyes have overlapping emission spectra, for proper interpretation of the data collected, they need to be compensated electronically and computationally. Most common programs used for flow cytometry data analysis include Cell Quest, FACSDiva, FlowJo, or WinMDI.

12.4.2 Applications

12.4.2.1 Immunophenotype

The study of the relation between phenotype and function across diverse cell types and tissues results is essential to understand cell biological actions and mechanisms of differentiation. Therefore, flow cytometry has shown a great number of applications in cell biology. The most common application is the characterization of the immunophenotype from different cell types. Specifically, the study of the immunophenotype is relevant in the case of stem cells such as MSCs due to the absence of single definitive marker for human MSCs (Pittenger and Martin 2004). Moreover, immunophenotyping has been widely used to characterize leukemias and other blood cancers (Huh and Ibrahim 2000).

12.4.2.2 Cell Growth and Division

12.4.2.2.1 Cell Cycle

Flow cytometry is also used to determine cell cycle and proliferation by examining the ploidy of a cell population. The amount of emitted fluorescence is proportional to the DNA amount. The most widely used dyes for cell-cycle analysis include PI or 4′,6-diamidino-2-phenylindole (DAPI), which required to be fixed or permeabilized and other live cell-cycle dyes such as Hoechst 33342, DRAQ5™, or CyTRAK Orange™, which does not require permeabilization. The acquisition should be in linear scale at lowest pressure and speed of injection, trying to get a good peak resolution. Analysis of ploidy could be applied in disease processes such as osteoarthritis (Kusuzaki, Sugimoto et al. 2001) as well as in the insurance of euploidy in tissue-engineered constructs that have been treated with high levels of growth factors (Kamil, Aminuddin et al. 2002).

12.4.2.2.2 Analysis of Division Rates

Another application involves the study of cell divisions using carboxyfluorescein succinimidyl ester (CFSE). CFSE is a fluorescent cell staining dye that is incorporated as carboxylfluorescein diacetate succinimidyl ester (CFDA-SE), which is a nonfluorescent. Once it enters the cytoplasm of cells, intracellular esterases remove the acetate groups to form CFSE, which can be retained by cells (Parish 1999) and permits the monitoring of cell proliferation due to its progressive halving within daughter cells following each cell division (Lyons and Parish 1994). Before the CFSE fluorescence is too low to be distinguished above the autofluorescence background, approximately 7–8 cell divisions can be identified. Furthermore, it could be combined with surface antigen staining (Nordon, Ginsberg et al. 1998; Portevin, Poupot et al. 2009; Oviedo-Orta, Perreau et al. 2010; Vaughan, Brackenbury et al. 2010) or study of viability (Tokalov, Henker et al. 2004).

12.4.2.3 Cellular Physiology and Metabolism

Flow cytometry could also serve for functional analysis such as reactive oxygen species (ROS) production. The most important cellular stress that occurs in cells in culture is the oxidative stress and ROS are directly associated with the energy metabolism of cells, indicator of cell state, and growth. Enzymatic activity, pH measurement, or Ca^{2+} measurement could also be assessed with this technique (O'Connor, Callaghan et al. 2001). For that purpose, it is necessary to use molecules whose fluorescence changes with redox state, enzymatic processing, pH, or ion concentration.

12.4.2.4 Apoptosis

Apoptosis could serve as a method to study cellular viability by flow cytometry. To assess whether cells entered into apoptosis, it is possible to study phosphatidylserine (PS) exposure (van Engeland, Nieland et al. 1998), DNA fragmentation, detection of activated caspases, or mitochondrial function. PS is detectable using annexin V-FITC (Koopman, Reutelingsperger et al. 1994). You can combine this with PI staining, being able to distinguish dead cells from live cells and early stage of apoptosis (negative for PI and positive for annexin V) from late stage of apoptosis (positive for PI and annexin V). If you study cell cycle in apoptotic cells, you could observe the characteristic subdiploid peak (cells with DNA content <2n). The most direct method is to use antibodies against the active form of caspases. Mitochondria are the key in apoptosis induced by cell stress due to the loss of mitochondrial membrane potential. There are fluorochromes that stay in the mitochondria if the mitochondrial membrane potential is preserved and vice versa, such as $DiOC_6(3)$ or other molecules that change their fluorescence depending on the aggregation state, such as JC-1.

12.4.2.5 Cell Sorting

Certain flow cytometers have the additional capability for allowing fluorescence-activated cell sorting (FACS) technology, where it is possible to separate cells individually from two or more cell populations according to defined flow cytometry parameters (Figure 12.9c). A cell is given a net charge by a short burst of electricity and then passes by charged plates as it falls, directing it into the correct sorting container (Carter and Meyer 1994). After the sorting process, these selected cells can be used for further analysis taking into account that cellular viability is variable depending on the cell type, time and sorting pressure, and collection media. The use of magnetic particles bound to antibodies to separate cells is known as magnetic activated cell sorting (MACS) and compared to FACS, which allows multiparametric sorting, you could only sort based on one parameter. However, MACS is fast and you can separate a huge amount of cells.

On the basis of FCM and ELISA technique emerged xMAP technology that uses 5.6-μm polystyrene microspheres, which are internally dyed with red and infrared fluorophores. This technology allows for the study of several analytes at the same time using a small amount of sample. The surface chemistry on the xMAP microspheres allows for simple chemical coupling of capture reagents such as antibodies, oligonucleotides, peptides, or receptors on the surface to bind the molecule of interest. Then, soluble proteins such as cytokines, nucleic acids (Dunbar 2006), or transcription factors could be detected. The excitation of one of the lasers of the instrument (633 nm, red), based on a flow cytometer, helps to identify the microspheres independently. Through the use of a second antibody against the molecule of interest and a second laser (532 nm, green), it is possible to detect whether the sample contains the analyte and at what concentration due to the interpolation off of the standard curve (Figure 12.9d). Thousands of microspheres are interrogated per second resulting in a multiplex analysis system capable of analyzing and reporting up to hundreds of different reactions in a single reaction vessel in just a few seconds per sample.

12.4.3 Future Directions

Flow cytometry is a powerful technique that has been used extensively in health bioscience applications for the understanding of immunology and stem-cell biology. Moreover, the technique is versatile and also has applications in other areas such as environmental microbiology, molecular biology, and genetics (Bergquist, Hardiman et al. 2009). However, for some aspects, it is necessary to have a higher level of multiparametric analysis of single cells that cannot be achieved with fluorescence technology due to the limitation of the number of spectrally resolvable fluorochromes. A new technology, called mass spectrometry, has emerged to offer single-cell analysis of at least 45 parameters in each cell, using stable isotopes of nonbiological metals such as samarium. This new technology has increasing understanding of cell expression and differentiation during hematopoiesis (Nolan 2011). In the near future, these cytometric technologies will remain important tools in cell biology.

12.5 Molecular Biology

12.5.1 Brief Background

The abundance, form, and function of the building blocks that comprise cellular architecture and activity originate in the genetic material contained in each cell nucleus. DNA, the double-stranded stable storehouse of cellular information, is transcribed in a regionally specific manner into temporary messages called heterogeneous nuclear RNA (hnRNA) (Purves, Orians et al. 1992b; Cambell 1993; Alberts, Bray et al. 1994a; Miesfeld 1999a). The production of these gene transcripts is controlled by a rich interplay of transcription factors and RNA polymerases interacting at the promoter region of a particular gene or set of genes (Purves, Orians et al. 1992a; Alberts, Bray et al. 1994a). The state and combination of these factors modulate the level to which transcription occurs. The transient message encoded in the hnRNA is processed into a final form, called messenger RNA (mRNA), which is transported from the nucleus to the cytoplasm (or into the lumen of the rough endoplasmic reticulum). Steady-state levels of mRNA are dictated by the rate of transcription, rate of degradation, and rate of transport of these molecules. In mammalian cells, the half-life of mRNA is on the order of hours to days, but can be modulated by alterations in cellular activity. Messenger RNA makes up 1–5% of the total RNA of a cell (Miesfeld 1999c) and is translated at ribosomes into proteins. Translation into protein involves the reading of the message (or open reading frame) in each mRNA via the transient coupling of transfer RNA molecules (tRNA) (Purves, Orians et al. 1992b; Cambell 1993; Alberts, Bray et al. 1994a). Transfer RNA molecules, each with a single amino acid in tow, arrive at the ribosome where the amino acid is covalently linked to its predecessor to sequentially form polypeptide structures or proteins.

When a cell undergoes an alteration in its physical and/or chemical environment, short-term (minutes to hours) transduction cascades signal these changes to the nucleus, wherein the numerous signals are summed up, resulting in long-term (hours to days) changes to the resting level of mRNA within the cell.

It is often of interest to understand the basal levels of gene expression, which defines the cellular phenotype, and how expression levels are modulated by alterations in the mechano-chemical environment. Most interesting is the control of gene expression at the promoter level, the amount and stability of mRNA produced, the sequence of these transcripts, and how changes in these parameters are affected by normal and disease processes. Using the ever-growing toolbox of molecular biology, one can assess the transcriptional state of isolated cells, cells in their native tissue environment, or cells in tissue-engineered composites.

12.5.2 Isolation Techniques

To assay the state and sequence of genetic material, it must first be extracted from a population of cells. The protocol for extraction is dependent on the type of genetic material to be extracted (DNA or RNA), as well as the size of the molecule and the milieu from which it is to be extracted. Extraction of DNA and RNA from single cells tends to be the most straightforward, with complications arising when extracting from tissues and other structures. In addition, care must be taken to ensure that contaminating DNA or RNA is not intermingled with extracted material and that the extraction protocol does not influence the hypothesis being tested.

12.5.2.1 DNA Isolation

The easiest method of extracting genomic DNA from a cluster of cells is by introducing a hypotonic media (i.e., distilled water), which causes the cell and nucleus to lyse, followed by boiling to free the genomic DNA (Innis, Gelfand et al. 1990a). Such preparations are crude, but are suitable for polymerase chain reaction (PCR) applications (see below). Where longer fragments of DNA are necessary, as in the case of making genomic libraries and cloning vectors, a number of techniques have been utilized that produce genomic fragments of varying sizes. To generate short stretches of DNA, which are suitable for most PCR applications, cells or tissue can be extracted with guanidine HCl with sodium acetate followed by ethanol precipitation (Sambrook, Fritsch et al. 1989a). If longer stretches of genomic DNA are required, cells may first be digested with proteinase K in a buffer containing sodium dodecyl sulfate (SDS, to solubilize proteins), EDTA (to chelate Mg, which inhibits DNases), and RNase (to remove contaminating RNA) (Sambrook, Fritsch et al. 1989a; Innis, Gelfand et al. 1990b). After digestion, DNA is extracted with phenol and precipitated in ethanol. Once dry, DNA can be resuspended in TE buffer (Tris–EDTA). This method generates genomic DNA strands ranging from 100 to 150 kb (Sambrook, Fritsch et al. 1989a). Finally, if even longer fragments are required, a similar procedure can be followed using a formamide-based denaturation buffer.

Another common requirement for DNA extraction involves the recovery of recombinant DNA vectors, such as plasmids, from host organisms. In this situation, it is critical that there be high yield of the fragment of interest with little contamination from the host organisms genomic DNA. The two most common methods for plasmid isolation are lithium chloride/Triton X-100 lysis and SDS-alkaline denaturation (Doyle 1996a). The lithium chloride/Triton X-100 method involves dissolving away part of the cell membrane, and then adding phenol:chloroform, which rapidly denatures proteins and causes cells to shrink, forcing the lower-molecular-weight plasmids out of the cell membrane while retaining the larger nuclear DNA within the cell. Alternatively, with SDS-alkaline denaturation, both plasmid and chromosomal DNA are denatured (at pH 11) followed by rapid neutralization with a high salt buffer in the presence of SDS. Under these conditions, genomic DNA hybridizes in an intrastrand manner, forming insoluble aggregates, while plasmid DNA renatures and remains in solution. The resulting "cleared" supernatant from either method can be further purified by ethanol precipitation and/or adsorption to silica resin columns (Doyle 1996a).

12.5.2.2 RNA Isolation

While the aforementioned techniques are reliable for DNA extraction, isolation of mRNA poses a more formidable challenge. By design, mRNA is a transient molecule, meant to convey its message in a spatially and temporally specific manner (Miesfeld 1999c). Breakdown of mRNA occurs naturally in solution due to hydrophilic attack of the 2′-hydroxyl group on the backbone. Additionally, endoribonucleases (RNases) that cleave the backbone are numerous (present in internal organelles as well as on skin) and extremely stable (Miesfeld 1999c). For this reason, RNase-free plastic ware and diethypyrocarbonate (DEPC)-treated glassware and water should be used at all times. Owing to the inherent instability of the molecule, most RNA extraction protocols begin with guanidine thiocyanate, a strong denaturant, and RNase inhibitor (Chomczynski and Sacchi 1987; Sambrook, Fritsch et al. 1989b; Doyle 1996b; Miesfeld 1999c). After complete dissolution of cells in 4 M guanidinium thiocyanate at pH 4, RNA will preferentially partition into the aqueous phase after mixing with phenol and chloroform (Chomczynski and Sacchi 1987). With centrifugation ($13,000 \times g$), proteins segregate to the lower organic phase, DNA molecules remain at the interface region, and RNA remains soluble and partitions into the upper aqueous phase. Soluble RNA can then be washed and precipitated in isopropanol, pelleted, and resuspended in RNase-free DEPC-treated water. The purity and amount of RNA can then be measured by the absorbance at 260 nm, and the ratio of absorbance at 260 and 280 nm; pure RNA has a ratio of 1.8–2.0 (Doyle 1996b). Alternatively, agarose-formaldehyde gels can be used to assess the amount and purity of RNA samples (Miesfeld 1999c). At this point, samples may be treated with RNase-free DNase to remove any contaminating genomic DNA, or further enriched for mRNA by a number of techniques (e.g., using oligo-dT affinity columns) (Re, Valhmu et al. 1995).

12.5.2.3 First Strand Synthesis

As RNA is quite unstable, even in DEPC-treated water, it is often useful to convert RNA to complementary DNA (cDNA), a much more stable molecule (Miesfeld 1999c). Complementary DNA is produced from RNA by a process called first strand synthesis. In this process, reverse transcriptase (RT), a DNA polymerase that uses RNA as a template, creates complementary strands of DNA based on the sequence of the RNA transcript (Miesfeld 1999c). The choice of RT is critical in determining the length of the cDNA products and their fidelity to the original transcript. In addition to an RNA template and deoxynucleotide phosphates (dNTPs), RT requires a primer from which to initiate transcription, which can be of three forms (Miesfeld 1999c). Oligo-dT priming makes use of the fact that mRNA is polyadenylated at the 3′ end of the molecule and therefore sequences of 12–18 dTs can be used to prime from the 3′ tail. While oligo-dT priming selects for messenger RNA, many transcripts have long stretches of noncoding sequences at the 3′ end and therefore cDNA products may lack open reading frames. Alternatively, random hexamers, as the name implies, use small random sequences to prime reverse transcription. This method generates a large cDNA library from a given RNA source, though nonmessenger RNA is transcribed as well. Finally, if the sequence of the gene of interest is well characterized, gene-specific primers (GSPs) can be designed to enrich the cDNA pool for a particular transcript.

12.5.2.4 Special Considerations

Extraction of mRNA from tissues can pose challenges not present during extraction from monolayer cultures. Extraction of mRNA from tissues with a low cellular density and high content of charged proteoglycans, such as articular cartilage, generally results in low yield. Furthermore, proteoglycans partition with mRNA and can interfere with RT and PCR reactions later (Re, Valhmu et al. 1995). To overcome these challenges, silica-based spin columns (Re, Valhmu et al. 1995; Martin, Jakob et al. 2001) have been reliably used to further purify the extracted mRNA. In the case of tissue-engineered equivalents, the 3D environment in which the cells were cultured can interfere with extraction techniques, especially in the case of hydrogels. In a recent paper (Hoemann, Sun et al. 2002), a multistep isolation technique involving both guanidinium hydrochloride and guanidinium isothiocyanate has been used to sequentially separate RNA, DNA, proteins, and other matrix molecules from both agarose and chitosan hydrogels. As is clear from this work, the extraction method used for different tissues and/or 3D culture environments should be optimized prior to the testing of experimental hypotheses.

12.5.3 Measurement Basics

The measurement of gene regulation in cultured cells is largely dependent on the hypothesis to be tested and the level of characterization of the gene of interest. When the gene sequence is known, and the objective is to evaluate the regulation of gene transcription in response to an applied stimulus, one can measure the amount and stability of mRNA levels (using the reverse transcriptase polymerase chain reaction, RT-PCR) or the regulation of promoter activity using reporter constructs. If the mechanism of interest is not well characterized, and the molecular mechanism is unknown, more complicated analytical steps must be taken. Methods such as differential display RT-PCR (DDRT-PCR) and subtractive hybridization allow for the identification of genes whose regulation changes without any knowledge of the gene *a priori*. An alternative to these methods involves the use of DNA microarrays, which allow many different genes to be monitored simultaneously, with the goal of finding genes whose regulation is clearly changed among a sea of other signals. In each of these situations, characterization must be done on multiple levels to ensure that the conclusions from any one technique are valid.

12.5.3.1 Promoter Studies

Functional maps of the promoter region of a particular gene can be constructed from multiple site-specific deletions in that region. In a cell line, this process can be quite arduous, and in an isolated tissue, impossible. For this reason, studies of promoter regulation are often performed using reporter constructs, such as plasmids. Once established, studies of this sort offer the advantage over the laborious (and expensive) procedures of mRNA extraction and quantitative PCR in their repeatability, and the ease with which changes to and deletions of the construct can be made (Miesfeld 1999b). To carry out studies of promoter architecture, the promoter region is isolated and cloned into an expression vector, such as a plasmid. In conjunction with the promoter is included a reporter gene. The reporter gene encodes an enzyme whose product is easily assayed (sensitive over several orders of magnitude), is not endogenous to the cell being tested, and does not interfere with the normal functioning of the cell (Miesfeld 1999b). One major assumption of this assay system is that the reporter mRNA is processed into protein in the same manner that the normal protein is processed (i.e., posttranscriptional control is not a major factor in gene regulation).

A number of enzymes exist that can effectively be used as reporter genes. Luciferase, the enzyme product of the luc gene in the firefly, catalyzes an ATP-dependent oxidation of its preferred substrate, luciferan (De Wet, Wood et al. 1987). This process generates a fluorescent emission that is in proportion to the amount of enzyme present and is detectable with a standard luminometer (Contag and Bachman 2002). β-Galactosidase is another common reporter gene and its presence can be assayed in cell extracts as well as identified by histochemical analysis to determine the location of expressed products (Miesfeld 1999). More recently, a reporter gene called GFP, isolated from the pacific-northwest jellyfish *Aequorea victoria*, has been developed for use in gene expression studies (Chalfie, Tu et al. 1994). Exposure of this protein and other color-shifted family members to UV light results in autofluorescence, making these proteins good *in vivo* markers of transfection (Schenk, Elliot et al. 1998). Another interesting use of GFP is that one can use the molecule to produce chimeric proteins that function normally and incorporate GFP, which allows one to monitor cellular dynamics in real time (Faire, Waterman-Storer et al. 1999) (as in Figure 12.6). Newer reporter molecules, called "fluorescent timers," are now commercially available that take advantage of time-dependent color shifts in fluorescent proteins to indicate not only the level of reporter expression, but also the lifetime of the molecule, allowing for a continuous monitoring of kinetic responses in gene expression in individual cells (Terskikh, Fradkov et al. 2000; van Roessel and Brand 2002).

After producing a plasmid containing the promoter of interest controlling a reporter gene, the construct must be amplified. Most plasmids for this purpose have origins of replication, as well as selectable markers. Competent cells are transformed with the plasmid and expanded under selective conditions (Miesfeld 1999b). After growing up a population of competent cells containing plasmid, stocks can be frozen down for future use or plasmid isolated and quantified directly (Doyle 1996a). Pure solutions of plasmids can then be transfected into primary cells or cell lines. Incorporation of plasmid is generally transient (lasting 2–3 days), and the plasmid becomes diluted through the process of cell division (Miesfeld 1999b). A subset of cells will randomly incorporate the plasmid into the genome, however, becoming stably transfected and can be used for numerous generations and long-term studies. In stably transfected lines, a drug resistance gene is often incorporated into the plasmid for selection purposes (Miesfeld 1999b). Transfection is mediated by a number of techniques, ranging from electroporation to viral infection. Most common nonviral methods use cationic carrier molecules (including calcium phosphate and DEAE dextran) that form an insoluble complex of DNA and carrier that is brought in close contact with the cell membrane and is then taken up by an unknown mechanism. Alternative strategies use positively charged liposomes that bring plasmid DNA in close contact with the negatively charged cell membrane and generally result in transfection at a higher efficiency (Felgner, Gadek et al. 1987). Finally, plasmid DNA can be incorporated into retroviral or adenoviral delivery systems that transfect with very high efficiencies. As with any technique, adequate controls must be undertaken to ensure that the measured response is not artifactual. For this reason, transfection studies are often carried out with two plasmids, with one reporter gene under the control of the sequence of interest while a second reporter is linked to a constitutively active promoter (as in Palmer, Chao et al. 2001).

Promoter studies have been carried out for the characterization of the regulatory elements of numerous genes. In our own laboratory, these techniques have been used to examine the functional composition of the aggrecan gene promoter region. By constructing a variety of promoter constructs with multiple deletions and rearrangements, Valhmu et al. examined the influence of the 3' and 5' untranslated regions (UTRs) on aggrecan gene expression (Valhmu, Palmer et al. 1998). Similar promoter constructs have been used to determine the regulatory regions of the promoter that respond to physical stimuli such as fluid-induced shear stress (Hung, Henshaw et al. 2000) and osmotic loading (Palmer, Chao et al. 2001; Hung, LeRoux et al. 2003) in articular chondrocytes (see Figure 12.10).

12.5.3.2 Polymerase Chain Reaction

Rather than measuring the rate of transcription, as is done with promoter studies, direct measures of mRNA levels can be undertaken with the PCR (Alberts, Bray et al. 1994b; Miesfeld 1999d). This reaction, first conceived by Kary Mullis in the 1980s (for which he won a Nobel Prize in 1993), is based on the amplification of specific regions of a DNA (or cDNA) template in a series of controlled reactions (Innis, Gelfand et al. 1990a). The PCR reaction mixture contains DNA (or cDNA), a pair of short sequence-specific primers, dNTPs, and a thermostable DNA polymerase (Saiki, Gelfand et al. 1988). The DNA polymerase (Taq) uses one strand of DNA as a template and extends from primers in the 5' to 3' direction. Repeated temperature changes, from 95°C to 55–65°C to 72°C, leads to the denaturing of DNA, annealing of primers, and extension of product. With each reaction, a doubling of product is accomplished, with up to a million-fold increase in the amount of target sequence in a short period of time. The initial phase of amplification is exponential (linear on a log plot) followed by a plateau phase that results from the stoichiometric limitations of reaction components and the decreased enzyme activity due to repeated exposure to high temperatures (Miesfeld 1999d). For each PCR reaction, there are a number of parameters to be optimized, including the choice of polymerase, the size and specificity of primers, the size of amplified product, and the annealing temperature. Conditions can be optimized for small (~100 bp) or large (up to 20 kb) products (see Invitrogen website). Different Taq molecules offer different processivities, extension rates, and error rates and should be chosen based on experimental needs. Furthermore, Taq can be kept separate from the reaction mixture until full denaturation has occurred, called "hot-start" PCR, to ensure that mispriming and amplification does not occur in early cycles (Miesfeld 1999d). Primers are designed to anneal to known complementary DNA sequences on the template and are themselves generally 18–24 nucleotides in length to ensure appropriate specificity and binding temperature. Primer concentration should be maintained at low levels and care should be taken to design intron-spanning primers where possible so as to differentiate mRNA from carryover

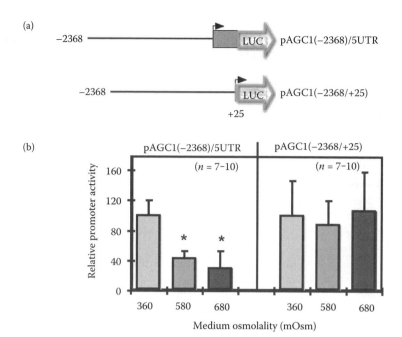

FIGURE 12.10 Promoter studies for aggrecan gene transcriptional regulation adapted from Palmer, Chao et al. (2001). (a) A luciferase reporter construct containing a 2.4-kb fragment of the promoter region with exon 1 (5′UTR) of the human aggrecan gene and a deletion construct in which exon 1 has been removed were transiently transfected in primary bovine chondrocytes. (b) Cultured chondrocytes exhibited decreased promoter activity in response to hypertonic loading (5 h with sucrose addition) that was dependent on exon 1 (which contains mechano-responsive elements).

genomic DNA. A number of web-based programs are available to aid in the design and selection of suitable primers and for calculations of annealing temperature. The product resulting from any new primer set should be characterized by restriction enzyme analysis or DNA sequencing to ensure amplification of the correct target.

12.5.3.3 Quantitative PCR

While PCR is adept at demonstrating the presence of specific gene products, it is also of interest to quantify the relative amounts of these products with a change in experimental conditions. Originally, this was done by Northern blots or RNase protection assays, both of which are labor-intensive procedures that require a large amount of RNA (Miesfeld 1999d). More recent techniques utilize the PCR reaction to quantify the resting level of transcript starting from relatively small amounts of total RNA. Meaningful results require an internal standard by which to normalize to varying levels of staring material (Souaze, Ntodue-Thome et al. 1996). This can be accomplished by normalizing to the starting total RNA amounts (quantified with spectroscopy) or by using endogenous control sequences that are constitutively expressed in all cells (such as GAPDH or actin). The drawback of such a method is that the use of different primers, the amplification of products of different lengths, and starting from drastically different copy numbers generally result in different amplification efficiencies (Miesfeld 1999d). In an alternative approach, both the sequence of interest and the normalizing sequence can be cloned into plasmids that are diluted to known concentrations and then amplified alongside experimental reactions to create a standard curve (Re, Valhmu et al. 1995). By making serial dilutions of sample mRNA, one can ensure that product intensities remain in the linear range of amplification (see Figure 12.11a) (Valhmu, Stazzone et al. 1998). Another method of normalizing is by using a complementary RNA sequence of

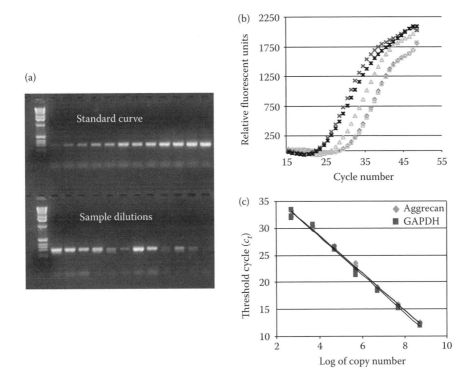

FIGURE 12.11 (a) Agarose gel electrophoresis of PCR products (aggrecan) with corresponding noncompetitive quantitative standard curve. (b) Increasing relative fluorescence intensity with increasing cycle number for real-time PCR for aggrecan of five experimental aggrecan samples (curves generated with the Biorad iCycler). (c) Copy number for these samples can be determined from an existing standard curve for aggrecan (and GAPDH also shown) relating the threshold cycle to the log of the initial copy number. (Courtesy of Wilmot Valhmu, University of Wisconsin.)

known concentration that differs from the native sequence by only a small insertion or deletion. Such vectors will generally have similar amplification efficiencies and can be used noncompetitively (in parallel reactions) or competitively (added to the same reaction mixture) (Becher-Andre and Hahlbrock 1989; Siebert and Larrick 1992). The competitive method has the benefit of using the same primer set and reaction mixture, and the amplified products can be run on the same gel. The StaRT PCR method (Willey, Crawford et al. 1998), which incorporates competitive transcripts for as many as 40 genes at known concentrations (both absolute and relative), has been proposed as a standardized quantitative competitive RT-PCR for the simultaneous analysis of multiple genes.

12.5.3.4 Real-Time PCR

While the standardization of traditional PCR techniques can minimize variations in quantitative RT-PCR results, significant errors may still occur between even experienced users within the same lab (Bustin 2002). Kinetic or "real-time" PCR offers better control and is significantly less variable and more specific than any conventional RT-PCR procedures because it does not rely on endpoint intensity measures of amplified product (Gibson, Heid et al. 1996; Bustin 2000). A number of fully automated machines are commercially available that can carry out many simultaneous reactions in a closed-tube environment, requiring no postamplification manipulation for the quantification of amplified product. This makes real-time PCR suited for high-throughput screening applications. Furthermore, the amplified sequence is generally shorter than 100 bp, and for this reason, reactions can be carried out more rapidly than with traditional PCR reactions.

The basic real-time PCR system is composed of a thermocycler combined with a system for monitoring fluorescent emission (Bustin 2000). The real-time PCR reaction is identical to the traditional reaction, with the inclusion of at least one further fluorescent probe. These probes can be DNA-binding dyes (such as SYBR Green), molecular beacons, hybridization probes, or hydrolysis probes (Bustin 2000). The simplest probe, the DNA-binding dye, fluoresces only when adhered to double-stranded DNA and therefore increases in amplified product lead directly to an increase in fluorescent intensity. This method offers no further specificity over traditional PCR methods, and fluorescence is dependent on product size, but the use of DNA dyes makes the quantification easier with obvious transitions from linear amplification to plateau levels of amplified products (see Figure 12.11a). Molecular beacons are specially designed probes, used in addition to standard primers, which provide a further level of specificity to the RT-PCR reaction. These probes form stem loop structures in free solution, but unwind and bind DNA when specific complementary sequences (located in the central region of the amplified product) are present. Each hairpin loop structure is designed with a fluorescent emitter and quencher molecule that are in close spatial association. This association effectively blocks emission unless a complementary sequence is present. Hybridization probes are similar, but are composed of two individual probe sequences designed to line up head to tail on the central region of the template. These probes contain two dyes with overlapping emission and absorption spectra, which interact in a FRET arrangement (see the previous section). When the sequences are lined up on a complementary sequence and excited, the emission of the first fluorophore excites the second, resulting in an emission at an even higher wavelength than that is detected by the device. Finally, hydrolysis probes are dual-labeled probes that quench one another and are designed to bind to the center of the amplified region. At each step of the amplification process, the 5′ exonuclease activity of the Taq DNA polymerase cleaves the probe, separating the two fluorophores from one another, allowing free emission (Bustin 2000). The level of probe binding DNA can be measured in real time, or at the terminus of each amplification cycle, and offers the advantage of using more than one set of primers and probes at the same time (so long as they have distinct spectral characteristics). Many commercial systems include a software package to aid in the design of probe sequences (e.g., http://www.premierbiosoft.com/molecular_beacons/taqman_molecular_beacons.html).

Each real-time PCR reaction is characterized by a Ct value (Bustin 2000), which is the fractional number of cycles at which the reporter fluorescent emission reaches a fixed threshold level in the exponential region of the amplification plot (see Figure 12.11a). This value can be correlated to the input mRNA amount, with larger starting amounts producing lower Ct values (Ct is inversely proportional to the log of the starting copy number) (Bustin 2000). This ensures that real-time PCR methods are not based on the measurement of the final amount of the amplified product and therefore do not have the same possibilities for error associated with standard PCR methods. In many systems, measurements are taken over the entire time course, and reactions can be halted before stoichiometric limitations of reaction components arise. As with standard PCR, relative measures compared to an internal control can be made (to actin, GAPDH, and/or ribosomal RNA), or standard curves can be constructed to determine the absolute copy number of a specific sequence (see Figure 12.11b). In most cases, the detection limit with real-time PCR is on the order of 50–500 copies of target mRNA (Bustin 2000).

While the initial monetary investment in reagents and instrumentation for carrying out real-time PCR may, at first, seem prohibitive, the ease of use, specificity, and reproducibility of results may actually lead to savings in the long run by overcoming the need for the time-consuming and numerous duplications involved in traditional RT-PCR analysis. For this reason, real-time PCR has been widely adopted, quickly becoming the gold standard in the field. Recent studies have used real-time PCR to quantify mRNA expression in chondrocytes and intervertebral disc cells in response to osmotic loading, both in monolayer cultures and in 3D hydrogels (Chen, Baer et al. 2002; Hung, LeRoux et al. 2003). It has additionally been employed to examine the changes in gene expression that occur with the onset of osteoarthritis (Martin, Jakob et al. 2001). Even more interesting, real-time PCR can be used reliably for single-cell analysis of mRNA in the cytoplasm of a single large cell (Liss 2002), or from individual tumor cells isolated with laser capture microdissection (LCM) (Emmert-Buck, Bonner et al. 1996).

12.5.3.5 Differential Display and Subtractive Hybridization

When less is known about the molecular mechanism involved in the response to an applied stimulus, other methodologies must be utilized. Two techniques can be used to identify gene transcripts present at different levels in different mRNA populations with no *a priori* knowledge of the sequence in question. Differential display reverse transcriptase PCR (DDRT-PCR) uses reverse transcription with oligo-dT primers directed to a subset of the RNA (Liang and Pardee 1992; Liang and Pardee 1998). In this method, oligo-dT primers, coupled with two additional dinucleotides (e.g., AGTTTT...), are used to generate cDNA in the presence of ^{32}P-dATP. The resulting radiolabeled products are then resolved on a gel, and differentially expressed bands excised. The purified cDNA is then reamplified with the same primer set, and the sequence determined by standard techniques. This reaction is carried out for all possible dinucleotide combinations to generate a full representation of the overall mRNA population. While the incidence of false positives can be high, DDRT-PCR offers one method for identifying genes that are differentially regulated and is a starting point for further characterization of unknown molecular mechanisms (Miesfeld 1999d). Another method with similar properties is subtractive suppression hybridization (SSH) (Diatchenko, Lau et al. 1996). This method combines cDNA normalization with selective PCR amplification to enrich differentially expressed gene transcripts. In this method, "tester" cDNA from the altered state and "driver" cDNA from control conditions are combined after a sequence of specific modifications (Miesfeld 1999d). After mRNA extraction and reverse transcription, two different pools of tester cDNA are digested to blunt ends, and then each population is annealed with a different primer adapter. Each pool is denatured separately and rapidly reassociated in the presence of excess driver cDNA. The two populations are then denatured a second time, combined, and again rapidly reassociated in the presence of excess driver cDNA. The resultant mixture of cDNA contains commonly expressed sequences that are mixed (without a full set of unique primer ends) while unique sequences remain paired to one another (with their unique primer ends). All sequences are then amplified with one primer each directed at each of the unique ends, and only those sequences that contain one of each end are amplified (Diatchenko, Lau et al. 1996). This procedure, while difficult in practice, can reduce the amplification of commonly expressed genes while enriching the amplification of rare transcripts up to 100–1000-fold (Miesfeld 1999d).

These methods have found numerous applications in identifying unknown molecular mechanisms. In a recent study, traditional RT-PCR methods were used to characterize changes in known genes in an experimental model of osteoarthritis, while DDRT-PCR was carried out to identify other candidate genes involved in the disease process (Bluteau, Gouttenoire et al. 2002). In another study, stem cells in micromass culture were converted to a chondrogenic phenotype in the presence of BMP-2, and the molecular mechanisms were monitored with subtractive hybridization (Izzo, Pucci et al. 2002). This study identified a number of previously unknown genes that were up- or down-regulated in conjunction with the phenotype conversion process. Additionally, SSH can be combined with DNA microarray technology (see below) to profile differential expressed genes with transformation (Sers, Tchernitsa et al. 2002).

12.5.3.6 DNA Microarrays

Given the observed changes in expression of so many genes in response to even the simplest chemical or mechanical stimulus, it is clear that a more global approach to expression profiling is warranted. Such a global approach will allow for a more complete picture to be taken at each experimental condition, enabling a better understanding of the underlying order of multiple gene regulation (Lander 1999). The recent development of DNA microarray technology has made this global analysis possible in an efficient and increasingly economical manner. DNA microarrays consist of many thousands of individual DNA (or cDNA) pieces affixed to a glass, silicon, or plastic substrate (Heller 2002). A variety of competing technologies allow for the placing up to one million test sequences on as little as 1–2 cm^2. These DNA sequences can be accurately fixed in place (either covalently or noncovalently), and the position and corresponding sequence indexed. The two most widely used chip fabrication technologies today involve either the physical delivery of known sequences to specific sites with inkjet or microjet deposition technology (Heller 2002) or the combination of photolithography and solid-phase oligonucleotide chemistry to build DNA sequences of up to 25 nucleotides up from the solid substrate (Gerhold, Rushmore et al. 1999). Target DNA (or cDNA) molecules from experimental conditions are fluorescently tagged and hybridized to the test sequences on the chip. The specificity of binding is controlled, and the position and the intensity of fluorescence are measured across the chip surface and related back to the location of specific sequences. Alternatively, mass spectroscopy can be used to read the chip.

Owing to the large amount of information contained in a single hybridized array, the construction of databases for the management of information is necessary. Mining and recognizing patterns within this data with bioinformatics may provide insights into the operation of genetic networks (Jain 2000). As with any new technology, however, care should be taken to ensure the correct design of studies (Graves 1999). One investigator likened the interpretation of DNA microarray results to "examining a battlefield and trying to infer the cause of war" (Gerhold, Rushmore et al. 1999). While the situation may not be so stark, care should be taken to ensure that studies are hypothesis-driven, with clear objectives, and an understanding of the variance in the system (Simon, Radmacher et al. 2002). Additionally, as there might be different amounts of cDNA in any one sample, data should be normalized to constitutively expressed genes (as with PCR methods).

The development of microarray technology has mirrored that of the computer microchip, with higher levels of data analysis possible in less surface area, with concurrent increases in sensitivity and selectivity. As with real-time PCR, cost was at first prohibitive, but has slowly decreased, and many research institutions now have established core facilities (Lander 1999). More than a dozen companies (Heller 2002) have some form of DNA microarray technology on the market, each with a niche application [e.g., high-density Gene Chips (www.affymetrix.com) and Atlas arrays (www.clontech.com) for expression profiling and genotyping, low-density microarrays from Orchid Biosciences, Inc. (www.orchid.com), for screening and diagnostics (Heller 2002)]. Chips can be used to sequence by hybridization (SBH), a procedure in which binding affinity is measured in the presence of single-nucleotide changes in sequence (Hychip; www.hyseq.com). For the analysis of gene expression profiles (Bucher 1999; Duggan, Bittner et al. 1999), chips are commercially available that incorporate expressed sequence tags (ESTs) representing individual tissues to whole organisms (e.g., all 6200 ESTs from *Saccharomyces cerevisiae*; Brown and Botstein 1999). The chips can be probed with experimental cDNA transcribed from mRNA that has been fluorescently tagged to allow for efficient visualization of hybridization. In such cases, the normal condition may be compared to pathologic conditions by using two separate fluorescent tags (one for each mRNA pool) and the relative expression levels viewed on the same chip (Brown and Botstein 1999). Such approaches have been successfully employed to examine the changes in gene expression in human cancers (Derisi, Penland et al. 1996; Hacia, Brody et al. 1996) and in cells responding to applied stimuli (e.g., hydrostatic pressure; Sironen, Karjalainen et al. 2002). Coupling this technology with unique single-cell acquisition devices (i.e., laser capture microscopy) allows for the characterization of subtle differences in individual cancer cells (Leethanakul, Patel et al. 2000). Microarrays are finding further use in the mapping of single-nucleotide polymorphisms (SNPs), in pharmacogenomic research (Debouck and Goodfellow 1999), and in forensic and genetic identification applications (Heller 2002).

12.6 Conclusions

For centuries, significant efforts have focused on gaining a better understanding of the *in vivo* situation by focusing studies on the basic building block of life, the cell (see Figure 12.12). With dynamic innovations in technology, it is becoming more permissible (and perhaps more routine) to extract greater volumes of data from smaller and smaller amounts of tissue and cells. Sophisticated culture techniques, spurred by interdisciplinary research efforts (e.g., biomaterials, bioengineering, and molecular and cell biology), have permitted investigators to modulate cell phenotype and expression levels in a manner similar to their native environment *in vivo*. Microscopy techniques, further being refined and optimized

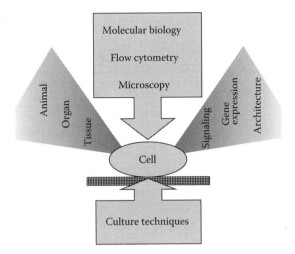

FIGURE 12.12 Paradigm for approaches to cellular-level biological assays. From whole-animal, organ, and tissue sources, tissue culture techniques form the foundation for the examination of cellular form and function. With microscopy, flow cytometry, and molecular biology techniques, new insights into cell signaling, gene expression pathways, and cell architecture can be gained.

with the incorporation of advances from the acquisition side and postprocessing side, enable noninvasive and dynamic interrogation of structure, protein interactions, ion fluxes, expression, and material characteristics of cells, providing new insights into intracellular and extracellular phenomena with continually growing sensitivity and specificity. Similarly, flow cytometry allows for identification and sorting of mixed cell populations, permitting the isolation of rare subsets of cells. Real-time PCR allows for transcripts from these isolated cells/tissues to be efficiently and reliably quantified. Finally, DNA microarrays allow for multiple genes to be monitored simultaneously. Just as technology and economies of scale spurred a revolution in the world of personal computing, so might DNA microarrays become commonplace in biological research, and may lead to a better understanding of the complexities of genetic regulation. Together, these tools represent a compendium of assays that permit researchers to examine questions regarding heterogeneity within populations (cells or molecules), instead of measuring attributes representing their average response. A new challenge that has emerged involves the interpretation of this plethora of information in the context of real physiologic situations, such as in the development and regulation of disease and pathology, as well as in efforts to engineer and grow replacement tissues.

Acknowledgments

This work was supported in part by the National Institutes of Health (NIAMS R01 AR46568 and AR060361, NIBIB 5P41EB002520, and R21 EB014382).

References

A-Hassan, E., W. F. Heinz et al. 1998. Relative microelastic mapping of living cells by atomic force microscopy. *Biophys. J.* **74**: 1564–1578.

Aasen, T., A. Raya et al. 2008. Efficient and rapid generation of induced pluripotent stem cells from human keratinocytes. *Nat. Biotechnol.* **26**(11): 1276–1284.

Al-Gubory, K. H. and L. M. Houdebine 2006. *In vivo* imaging of green fluorescent protein-expressing cells in transgenic animals using fibred confocal fluorescence microscopy. *Eur. J. Cell Biol.* **85**(8): 837–845.

Alberts, B., D. Bray et al. 1994a. Basic genetic mechanisms. *Molecular Biology of the Cell*. B. Alberts (ed). New York, Garland Publishing: pp. 223–290.

Alberts, B., D. Bray et al. 1994b. Recombinant DNA technology. *Molecular Biology of the Cell*. B. Alberts (ed). New York, Garland Publishing: pp. 291–334.

Allen, F. D., C. T. Hung et al. 2000. Mechano-chemical coupling in the flow-induced activation of intracellular calcium signaling in primary cultured bone cells. *J. Biomech.* **33**(12): 1585–1591.

Altman, G. H., R. L. Horan et al. 2002. Cell differentiation by mechanical stress. *FASEB J.* **16**: 270–272.

Ambekar, R., M. Chittenden et al. 2012. Quantitative second-harmonic generation microscopy for imaging porcine cortical bone: Comparison to SEM and its potential to investigate age-related changes. *Bone* **50**(3): 643–650.

Ando, T., T. Uchihashi et al. 2007. High-speed atomic force microscopy for observing dynamic biomolecular processes. *J. Mol. Recognit.* **20**(6): 448–458.

Andresen, J. L., T. Ledet et al. 1997. Keratocyte migration and peptide growth factors: The effect of PDGF, bFGF, EGF, IGF-I, aFGF and TGF-beta on human keratocyte migration in a collagen gel. *Curr. Eye Res.* **16**(6): 605–613.

Angela, F. E., A. I. Stering et al. 2001. Characterization of engraftable hematopoietic stem cells in murine long-term bone marrow cultures. *Exp. Hematol.* **29**: 643–652.

Atala, A., L. G. Cima et al. 1993. Injectable alginate seeded with chondrocytes as a potential treatment for vesicoureteral reflux. *J. Urol.* **150**: 745–747.

Awad, H. A., M. Q. Wickham et al. 2003. Chondrogenic differentiation of adipose-derived adult stem cells in agarose, alginate and gelatin scaffolds. *Biomaterials* **25**(16): 3211–3222.

Axelrod, D., D. E. Koppel et al. 1976. Mobility measurements by analysis of fluorescence photobleaching recovery kinetics. *Biophys. J.* **16**: 1055–1089.

Axelrod, D., N. L. Thompson et al. 1982. Total internal reflection fluorescence microscopy. *J. Microsc.* **129**: 19–28.

Aydelotte, M. and K. Kuettner 1988. Differences between sub-populations of cultured bovine articular chondrocytes. I. Morphology and cartilage matrix production. *Connect. Tissue Res.* **18**: 205–222.

Banes, A. J., G. Horesovsky et al. 1999. Mechanical load stimulates expression of novel genes *in vivo* and *in vitro* in avian flexor tendon cells. *Osteoarthritis Cartilage* **7**(1): 141–153.

Becher-Andre, M. and K. Hahlbrock 1989. Absolute mRNA quantitation using the polymerase chain reaction (PCR): A novel approach by a PCR aided transcript titration assay (PATTY). *Nucl. Acids Res.* **17**: 9437–9446.

Benya, P. D. and J. D. Shaffer 1982. Dedifferentiated chondrocytes reexpress the differentiated collagen phenotype when cultured in agarose gels. *Cell* **30**: 215–224.

Bergquist, P. L., E. M. Hardiman et al. 2009. Applications of FCM in environmental microbiology and technology. *Extremophiles* **13**(3): 389–401.

Bhatia, S. N., U. J. Balis et al. 1999. Effect of cell–cell interactions in preservation of cellular phenotype. Cocultivation of hepatocytes and nonparenchymal cells. *FASEB J.* **13**: 1883–1900.

Bhatia, S. N., M. L. Yarmush et al. 1997. Controlling cell interactions by micropatterning in co-cultures. Hepatocytes and 3T3 fibroblasts. *J. Biomed. Mater.* **34**: 189–199.

Bian, L., J. V. Fong et al. 2010. Dynamic mechanical loading enhances functional properties of tissue engineered cartilage using mature canine chondrocytes. *Tissue Eng. Part A* **16**(5): 1781–1790.

Binnig, G., C. F. Quate et al. 1986. Atomic force microscope. *Phys. Rev. Lett.* **56**: 930–933.

Bjork, J. W., L. A. Meier et al. 2012. Hypoxic culture and insulin yield improvements to fibrin-based engineered tissue. *Tissue Eng. Part A* **18**(7–8): 785–795.

Bluteau, G., J. Gouttenoire et al. 2002. Differential gene expression analysis in a rabbit model of osteoarthritis induced by anterior cruciate ligament (ACL) section. *Biorheology* **39**: 247–258.

Borlinghaus, R. T. 2006. MRT letter: High speed scanning has the potential to increase fluorescence yield and to reduce photobleaching. *Microsc. Res. Tech.* **69**(9): 689–692.

Bradbury, S. and P. Evennett 1996. *Contrast Techniques in Light Microscopy*. S. Bradbury and P. Evennett (eds). Oxford, BIOS Scientific Publishers Ltd.

Braddon, L. G., D. Karoyli et al. 2002. Maintenance of a functional endothelial cell monolayer on a fibroblast/polymer substrate under physiologically relevant shear stress conditions. *Tissue Eng.* **8**(4): 695–708.

Brand, J. S. and T. J. Hefley 1984. Collagenase and the isolation of cells from bone. *Cell Separation: Methods and Selected Applications*. T. G. Pretlow and T. P. Pretlow (eds). Orlando, Academic Press, Inc. **3**: pp. 265–283.

Brighton, C. T., J. L. Schaffer et al. 1991. Proliferation and macromolecular synthesis by rat calvarial bone cells grown in various oxygen tension. *J. Orthop. Res.* **9**: 847–854.

Brock, R. and T. M. Jovin 1998. Fluorescence correlation microscopy (FCM)-fluorescence correlation spectroscopy (FCS) taken into the cell. *Cell Mol. Biol.* **44**: 847–856.

Brock, R., G. Vamosi et al. 1999. Rapid characterization of green fluorescent protein fusion proteins on the molecular and cellular level by fluorescence correlation microscopy. *Proc. Natl. Acad. Sci. USA* **96**: 10123–10128.

Brown, T. 2000. Techniques for mechanical stimulation of cells *in vitro*: A review. *J. Biomech.* **33**: 3–14.

Brown, P. O. and D. Botstein 1999. Exploring the new world of the genome with DNA microarrays. *Nat. Genet.* **21**(Suppl): 33–37.

Brown, A. N., B. S. Kim et al. 2000. Combining chondrocytes and smooth muscle cells to engineer hybrid soft tissue constructs. *Tissue Eng.* **6**(4): 297–305.

Brussaard, C. P., D. Marie et al. 2000. Flow cytometric detection of viruses. *J. Virol. Methods* **85**(1–2): 175–182.

Bucher, P. 1999. Regulatory elements and expression profiles. *Curr. Opin. Struct. Biol.* **9**: 400–407.

Bulinski, J. C., D. J. Odde et al. 2001. Rapid dynamics of the microtubule binding of ensconsin *in vivo*. *J. Cell Sci.* **114**(Pt 21): 3885–3897.

Bustin, S. A. 2000. Absolute quantification of mRNA using real-time reverse transcription polymerase chain reaction assays. *J. Mol. Endocrinol.* **25**: 169–193.

Bustin, S. A. 2002. Quantification of mRNA using real-time reverse transcription PCR (RT-PCR): Trends and problems. *J. Mol. Endocrinol.* **29**: 23–39.

Byers, B. A., R. L. Mauck et al. 2008. Transient exposure to transforming growth factor beta 3 under serum-free conditions enhances the biomechanical and biochemical maturation of tissue-engineered cartilage. *Tissue Eng. Part A* **14**(11): 1821–1834.

Cambell, N. A. 1993. From gene to protein. *Biology*. Redwood City, CA, The Benjamin/Cummings Publishing Company, Inc.: pp. 316–343.

Caplan, A. I. 1970. Effects of the nicotinamide-sensitive teratogen 3-acetylpyridine on chick limb cells in culture. *Exp. Cell Res.* **62**: 341–355.

Carter, N. P. and E. W. Meyer 1994. Introduction to the principles of flow cytometry. *Flow Cytometry: A Practical Approach*. M. G. Ormerod (ed). Oxford, UK, Oxford University Press: pp. 1–25.

Carver, S. E. and C. A. Heath 1999. Influence of intermittent pressure, fluid flow, and mixing on the regenerative properties of articular chondrocytes. *Biotechnol. Bioeng.* **65**(3): 274–281.

Chafik, D., D. Bear et al. 2003. Optimization of Schwann cell adhesion in response to shear stress in an *in vitro* model for peripheral nerve tissue engineering. *Tissue Eng.* **9**(2): 233–241.

Chalfie, M., Y. Tu et al. 1994. Green fluorescent protein as a marker for gene expression. *Science* **263**: 802–805.

Chang, C. W., D. Sud et al. 2007. Fluorescence lifetime imaging microscopy. *Methods Cell Biol.* **81**: 495–524.

Chao, P. H., S. Yodmuang et al. 2010. Silk hydrogel for cartilage tissue engineering. *J. Biomed. Mater. Res. B Appl. Biomater.* **95**(1): 84–90.

Chattopadhyay, P. K., D. A. Price et al. 2006. Quantum dot semiconductor nanocrystals for immunophenotyping by polychromatic flow cytometry. *Nat. Med.* **12**(8): 972–977.

Chen, J., A. E. Baer et al. 2002. Matrix protein gene expression in intervertebral disc cells subjected to altered osmolarity. *Biochem. Biophys. Res. Comm.* **293**: 932–938.

Chen, C. S., M. Mrksich et al. 1997. Geometric control of cell life and death. *Science* **276**: 1425–1428.

Chen, D. T., E. R. Weeks et al. 2003. Rheological microscopy: Local mechanical properties from microrheology. *Phys. Rev. Lett.* **90**(10): 108301.

Chomczynski, P. and N. Sacchi 1987. Single-step method of RNA isolation by acid guanidinium thiocyanate–phenol–chloroform extraction. *Anal. Biochem.* **162**: 156–159.

Clegg, R. M. 1995. Fluorescence resonance energy transfer. *Curr. Opin. Biotechnol.* **6**: 103–110.

Conchello, J. A., J. J. Kim et al. 1994. Enhanced three-dimensional reconstruction from confocal scanning microscope images. II. Depth discrimination versus signal-to-noise ratio in partially confocal images. *Appl. Opt.* **33**(17): 3740–3750.

Contag, C. H. and M. H. Bachman 2002. Advances in *in vivo* bioluminescence imaging of gene expression. *Ann. Rev. Biomed. Eng.* **4**: 235–260.

Costa, K. D. and F. C. Yin 1999. Analysis of indentation: Implications for measuring mechanical properties with atomic force microscopy. *J. Biomech. Eng.* **121**(5): 462–471.

Cui, Z. F., R. C. Dykhuizen et al. 2002. Modeling of cryopreservation of engineered tissues with one-dimensional geometry. *Biotechnol. Prog.* **18**: 354–361.

Cushing, M. C. and K. S. Anseth 2007. Hydrogel cell cultures. *Science* **316**(25 May): 1133–1134.

Davey, H. and D. Kell 1996. FCM and cell sorting of heterogeneous microbial populations: The importance of single cell analyses. *Microbiol. Rev.* **60**: 641–696.

Davis, J. M., Ed. 1996. *Basic Cell Culture: A Practical Approach*. The Practical Approach Series. Oxford, Oxford University Press.

Debouck, C. and P. N. Goodfellow 1999. DNA microarrays in drug discovery and development. *Nat. Genet.* **21**(Suppl): 48–50.

Denk, W., J. H. Strickler et al. 1990. Two-photon laser scanning fluorescence microscopy. *Science* **648**: 73–76.

Denker, A. E., S. B. Nicoll et al. 1995. Formation of cartilage-like spheroids by micromass cultures of murine C3H10T1/2 cells upon treatment with transforming growth factor-beta 1. *Differentiation* **59**(1): 25–34.

Derisi, J., L. Penland et al. 1996. Use of a cDNA microarray to analyze gene expression patterns in human cancer. *Nat. Genet.* **14**: 457–460.

De Wet, J. R., K. V. Wood et al. 1987. Firefly luciferase gene: Structure and expression in mammalian cells. *Mol. Cell. Biol.* **7**: 725–737.

Dexter, T. M., T. D. Allen et al. 1977. Conditions controlling the proliferation of haemopoietic stem cells *in vitro*. *J. Cell Physiol.* **91**: 335.

Diatchenko, L., Y. Lau et al. 1996. Suppression subtractive hybridization: A method for generating differentially regulated or tissue-specific cDNA probes and libraries. *Proc. Natl. Acad. Sci. USA* **93**: 6025–6030.

Diller, K. R. 1997. Engineering-based contributions in cryobiology. *Cryobiology* **34**: 304–314.

Domke, J. and M. Radmacher 1998. Measuring the elastic properties of thin polymer films with the atomic force microscope. *Langmuir* **14**: 3320–3325.

Doyle, K (ed). 1996a. DNA purification. *Promega Protocols and Applications Guide*. pp. 75–90.

Doyle, K (ed). 1996b. RNA purification and analysis. *Promega Protocols and Applications Guide*. pp. 93–111.

Duggan, D. J., M. Bittner et al. 1999. Expression profiling using cDNA microarrays. *Nat. Genet.* **21**(Suppl): 10–14.

Dumont, J., M. Ionescu et al. 2000. Mature full-thickness articular cartilage explants attached to bone are physiologically stable over long-term culture in serum-free media. *Connect. Tissue Res.* **40**(4): 259–272.

Dunbar, S. A. 2006. Applications of luminex xMAP technology for rapid, high-throughput multiplexed nucleic acid detection. *Clin. Chim. Acta* **363**(1–2): 71–82.

Dunkelman, N. S., M. P. Zimber et al. 1995. Cartilage production by rabbit articular chondrocytes on polyglycolic acid scaffolds in a closed bioreactor system. *Biotech. Bioeng.* **46**: 299–305.

Einheber, S., M. J. Hannocks et al. 1995. Transforming growth factor-beta 1 regulates axon/Schwann cell interactions. *J. Cell Biol.* **129**(2): 443–458.

Elder, S. H., S. A. Goldstein et al. 2001. Chondrocyte differentiation is modulated by frequency and duration of cyclic compressive loading. *Ann. Biomed. Eng.* **29**(6): 476–482.

Emmert-Buck, M. R., R. F. Bonner et al. 1996. Laser capture microdissection. *Science* **274**: 998–1001.

Engler, A. J., S. Sen et al. 2006. Matrix elasticity directs stem cell lineage specification. *Cell* **126**(4): 677–689.

Erickson, G. R., J. M. Gimble et al. 2002. Chondrogenic potential of adipose tissue-derived stromal cells *in vitro* and *in vivo*. *Biochem. Biophys. Res. Commun.* **290**(2): 763–769.

Erickson, I. E., S. R. Kestle et al. 2012. High mesenchymal stem cell seeding densities in hyaluronic acid hydrogels produce engineered cartilage with native tissue properties. *Acta Biomater.* **8**(8): 3027–3034.

Faire, K., C. Waterman-Storer et al. 1999. E-MAP-115 (Ensconsin) associates dynamically with microtubules *in vivo* and is not a physiological modulator of microtubule dynamics. *J. Cell Sci.* **112**(4243–4255).

Felgner, P. L., T. R. Gadek et al. 1987. Lipofectin: A highly efficient, lipid-mediated DNA/transfection procedure. *Proc. Natl. Acad. Sci. USA* **84**: 7413–7417.

Franck, C., S. A. Maskarinec et al. 2011. Three-dimensional traction force microscopy: A new tool for quantifying cell-matrix interactions. *PLoS One* **6**(3): e17833.

Freed, L. E., G. Vunjak-Novakovic et al. 1993. Cultivation of cell–polymer cartilage implants in bioreactors. *J. Cell Biochem.* **51**: 257–264.

Fritz, M., M. Radmacher et al. 1994. Visualization and identification of intracellular structures by force modulation microscopy and drug induced degradation. *J. Vac. Sci. Technol. B* **12**: 1526–1529.

Gan, L. and R. A. Kandel 2007. *In vitro* cartilage tissue formation by co-culture of primary and passaged chondrocytes. *Tissue Eng.* **13**(4): 831–842.

Gerhold, D., T. Rushmore et al. 1999. DNA chips: Promising toys have become powerful tools. *Trends Biol. Sci.* **24**: 168–173.

Gibson, U. E., C. A. Heid et al. 1996. A novel method for real time quantitative RT-PCR. *Gen. Res.* **6**: 995–1001.

Gillette, B. M., J. A. Jensen et al. 2008. *In situ* collagen assembly for integrating microfabricated three-dimensional cell-seeded matrices. *Nat. Mater.* **7**(8): 636–640.

Gooch, K. J., J. H. Kwon et al. 2001. Effects of mixing intensity on tissue-engineered cartilage. *Biotechnol. Bioeng.* **72**(4): 402–407.

Graf, R., J. Rietdorf et al. 2005. Live cell spinning disk microscopy. *Adv. Biochem. Eng. Biotechnol.* **95**: 57–75.

Graff, R. D., E. R. Lazarowski et al. 2000. ATP release by mechanically loaded porcine chondrons in pellet culture. *Arthritis Rheum.* **43**(7): 1571–1579.

Graves, D. J. 1999. Powerful tools for genetic analysis come of age. *TIBTECH* **17**: 127–134.

Gribbon, P. and T. E. Hardingham 1998. Macromolecular diffusion of biological polymers measured by confocal fluorescence recovery after photobleaching. *Biophys. J.* **75**: 1032–1039.

Grskovic, M., A. Javaherian et al. 2011. Induced pluripotent stem cells—Opportunities for disease modelling and drug discovery. *Nat. Rev. Drug Discov.* **10**(12): 915–929.

Grynkiewicz, G., M. Poenie et al. 1985. A new generation of Ca^{2+} indicators with greatly improved fluorescence properties. *J. Biol. Chem.* **280**(6): 3440–3450.

Guilak, F. 1994. Volume and surface area measurement of viable chondrocytes *in situ* using geometric modeling of serial confocal sections. *J. Microsc.* **173**: 245–256.

Guilak, F., A. Ratcliffe et al. 1995. Chondrocyte deformation and local tissue strain in articular cartilage: A confocal microscopy study. *J. Orthop. Res.* **13**: 410–422.

Guo, X. E., E. Takai et al. 2006. Intracellular calcium waves in bone cell networks under single cell nanoindentation. *Mol. Cell Biomech.* **3**(3): 95–107.

Haberstroh, K. M., M. Kaefer et al. 2002. A novel *in-vitro* system for the simultaneous exposure of bladder smooth muscle cells to mechanical strain and sustained hydrostatic pressure. *J. Biomech. Eng.* **124**: 208–213.

Hacia, J., L. Brody et al. 1996. Detection of heterozygous mutations in BRCA1 using high density arrays and two colour fluorescence analysis. *Nat. Genet.* **14**: 441–447.

Haga, J. H., A. J. Beaudoin et al. 1998. Quantification of the passive mechanical properties of the resting platelet. *Ann. Biomed. Eng.* **26**(2): 268–277.

Hammes, F. and T. Egli 2010. Cytometric methods for measuring bacteria in water: Advantages, pitfalls and applications. *Anal. Bioanal. Chem.* **397**: 1083–1095.

Han, M., G. Giese et al. 2005. Second harmonic generation imaging of collagen fibrils in cornea and sclera. *Opt. Express* **13**(15): 5791–5797.

Han, M., L. Zickler et al. 2004. Second-harmonic imaging of cornea after intrastromal femtosecond laser ablation. *J. Biomed. Opt.* **9**(4): 760–766.

Hansen, U., M. Schunke et al. 2001. Combination of reduced oxygen tension and intermittent hydrostatic pressure: A useful tool in articular cartilage tissue engineering. *J. Biomech.* **34**(7): 941–949.

Harpur, A. G., F. S. Wouters et al. 2001. Imaging FRET between spectrally similar GFP molecules in single cells. *Nat. Biotechnol.* **19**: 167–169.

Heller, M. J. 2002. DNA microarray technology: Devices, systems, and applications. *Ann. Rev. Biomed. Eng.* **4**: 129–153.

Henderson, E., P. G. Haydon et al. 1992. Actin filament dynamics in living glial cells imaged by atomic force microscopy. *Science* **257**(5078): 1944–1946.

Herman, B. 1996. Fluorescence microscopy: State of the art. *Fluorescence Microscopy and Fluorescent Probes*. J. Slavik (ed). New York, Plenum Press.

Hoemann, C. D., J. Sun et al. 2002. A multivalent assay to detect glycosaminoglycan, protein, collagen, RNA, and DNA content of milligram samples of cartilage or hydrogel-based repair cartilage. *Anal. Biochem.* **300**: 1–10.

Hoh, J. H. and C.-A. Schoenenberger 1994. Surface morphology and mechanical properties of MDCK monolayers by atomic force microscopy. *J. Cell Sci.* **107**: 1105–1114.

Horio, T. and H. Hotani 1986. Visualization of the dynamic instability of individual microtubules by darkfield microscopy. *Nature* **321**(6070): 605–607.

Hotani, H. and T. Horio 1988. Dynamics of microtubules visualized by darkfield microscopy: Treadmilling and dynamic instability. *Cell Motil. Cytoskeleton* **10**(1–2): 229–236.

Hotani, H. and H. Miyamoto 1990. Dynamic features of microtubules as visualized by dark-field microscopy. *Adv. Biophys.* **26**: 135–156.

Houssen, Y. G., I. Gusachenko et al. 2011. Monitoring micrometer-scale collagen organization in rat-tail tendon upon mechanical strain using second harmonic microscopy. *J. Biomech.* **44**(11): 2047–2052.

Huh, Y. O. and S. Ibrahim 2000. Immunophenotypes in adult acute lymphocytic leukemia. Role of flow cytometry in diagnosis and monitoring of disease. *Hematol. Oncol. Clin. North Am.* **14**(6): 1251–1265.

Hung, C. T., F. D. Allen et al. 1997. Intracellular calcium response of ACL and MCL ligament fibroblasts to fluid-induced shear stress. *Cell. Signal.* **9**(8): 587–594.

Hung, C., D. Henshaw et al. 2000. Mitogen-activated protein kinase signaling in bovine articular chondrocytes in response to fluid flow does not require calcium mobilization. *J. Biomech.* **33**: 73–80.

Hung, C. T., M. A. LeRoux et al. 2003. Disparate aggrecan gene expression in chondrocytes subjected to hypotonic and hypertonic loading in 2D and 3D culture. *Biorheology* **40**: 61–72.

Hung, C. T., S. R. Pollack et al. 1995. Real-time calcium response of cultured bone cells to fluid flow. *Clin. Orthop.* **313**: 256–269.

Hwang, N. S., S. Varghese et al. 2008. Derivation of chondrogenically-committed cells from human embryonic cells for cartilage tissue regeneration. *PLoS ONE* **3**(6): e2498.

Ilyin, S. E., M. C. Flynn et al. 2001. Fiber-optic monitoring coupled with confocal microscopy for imaging gene expression *in vitro* and *in vivo*. *J. Neurosc. Meth.* **108**: 91–96.

Innis, M. A., D. H. Gelfand et al. (eds). 1990a. Amplification of genomic DNA. *PCR Protocols: A Guide to Methods and Applications.* New York, Academic Press, Inc.: pp. 13–20.

Innis, M. A., D. H. Gelfand et al. (eds). 1990b. Sample preparation from blood, cells, and other fluids. *PCR Protocols: A Guide to Methods and Applications.* New York, Academic Press, Inc.: pp. 146–152.

Ishikawa-Ankerhold, H. C., R. Ankerhold et al. 2012. Advanced fluorescence microscopy techniques—FRAP, FLIP, FLAP, FRET and FLIM. *Molecules* **17**(4): 4047–4132.

Izzo, M. W., B. Pucci et al. 2002. Gene expression profiling following BMP-2 induction of mesenchymal chondrogenesis *in vitro. Osteoarthritis Cartilage* **10**: 23–33.

Jain, K. K. 2000. Biotechnological applications of lab-chips and microarrays. *Trends Biotechnol.* **18**: 278–280.

Jain, R. K., R. J. Stock et al. 1990. Convection and diffusion measurements using fluorescence recovery after photobleaching and video image analysis: *In-vitro* calibration and assessment. *Microvasc. Res.* **39**: 77–93.

James, J. and H. J. Tanke 1991. *Biomedical Light Microscopy.* J. James and H. J. Tanke (eds). Dordrecht, Kluwer Academic Publishers.

Jankowski, R. J., B. M. Deasy et al. 2002. Muscle-derived stem cells. *Gene Ther.* **9**(10): 642–647.

Jonsson, K. B., A. Frost et al. 1999. Three isolation techniques for primary culture of human osteoblast-like cells: A comparison. *Acta Orthop. Scand.* **70**(4): 365–373.

Kalpakci, K. N., E. J. Kim et al. 2011. Assessment of growth factor treatment on fibrochondrocyte and chondrocyte co-cultures for TMJ fibrocartilage engineering. *Acta Biomater.* **7**(4): 1710–1718.

Kamil, S. H., B. S. Aminuddin et al. 2002. Tissue-engineered human auricular cartilage demonstrates euploidy by flow cytometry. *Tissue Eng.* **8**(1): 85–92.

Kanda, K. and T. Matsuda 1994. Mechanical stress-induced orientation and ultrastructural change of smooth muscle cells cultured in three-dimensional collagen lattices. *Cell Transplant.* **3**(6): 481–492.

Karlsson, J. O. M. and M. Toner 1996. Long-term storage of tissues by cryopreservation: Critical issues. *Biomaterials* **17**: 243–256.

Katzburg, S., M. Lieberherr et al. 1999. Isolation and hormonal responsiveness of primary cultures of human bone-derived cells: Gender and age differences. *Bone* **25**(6): 667–673.

Kelly, T. A., Y. Katagiri et al. 2010. Localized alteration of microtubule polymerization in response to guidance cues. *J. Neurosci. Res.* **88**(14): 3024–3033.

Kim, I. L., R. L. Mauck et al. 2011. Hydrogel design for cartilage tissue engineering: A case study with hyaluronic acid. *Biomaterials* **32**(34): 8771–8782.

Kim, Y. J., R. L. Sah et al. 1988. Fluorometric assay of DNA in cartilage explants using hoecsht 33258. *Anal. Biochem.* **174**: 168–176.

Kim, T. K., B. Sharma et al. 2003. Experimental model for cartilage tissue engineering to regenerate the zonal organization of articular cartilage. *Osteoarthritis Cartilage* **11**(9): 653–664.

Kisiday, J., M. Jin et al. 2002. Self-assembling peptide hydrogel fosters chondrocyte extracellular matrix production and cell division: Implications for cartilage tissue repair. *PNAS* **99**(15): 9996–10001.

Klement, B. J. and B. S. Spooner 1993. Utilization of microgravity bioreactors for differentiation of mammalian skeletal tissue. *J. Cell. Biochem.* **51**: 252–256.

Knight, M. M., D. A. Lee et al. 1998. The influence of elaborated pericellular matrix on the deformation of isolated articular chondrocytes cultured in agarose. *Biochem. Biophys. Acta* **1405**: 67–77.

Knight, M. M., S. R. Roberts et al. 2003. Live cell imaging using confocal microscopy induces intracellular calcium transients and cell death. *Am. J. Physiol. Cell Physiol.* **284**(4): C1083–C1089.

Knight, M. M., J. M. Ross et al. 2001. Chondrocyte deformation within mechanically and enzymatically extracted chondrons compressed in agarose. *Biochim. Biophys. Acta* **1526**(2): 141–146.

Knight, M. M., J. van de Breevaart Bravenboer et al. 2002. Cell and nucleus deformation in compressed chondrocyte-alginate constructs: Temporal changes and calculation of cell modulus. *Biochim. Biophys. Acta* **1570**: 1–8.

Kobayashi, K., R. M. Healey et al. 2000. Novel method for quantitative assessment of cell migration: A study on the motility of rabbit anterior cruciate (ACL) and medial collateral ligament (MCL) cells. *Tissue Eng.* **6**(1): 29–38.

Koch, D., W. J. Rosoff et al. 2012. Strength in the periphery: Growth cone biomechanics and substrate rigidity response in peripheral and central nervous system neurons. *Biophys. J.* **102**(3): 452–460.

Kodera, N., D. Yamamoto et al. 2010. Video imaging of walking myosin V by high-speed atomic force microscopy. *Nature* **468**(7320): 72–76.

Kolettas, E., L. Buluwela et al. 1995. Expression of cartilage-specific molecules is retained on long-term culture of human articular chondroctyes. *J. Cell Sci.* **108**: 1991–1999.

Koopman, G., C. P. Reutelingsperger et al. 1994. Annexin V for flow cytometric detection of phosphatidyl-serine expression on B cells undergoing apoptosis. *Blood* **84**(5): 1415–1420.

Kusuzaki, K., S. Sugimoto et al. 2001. DNA cytofluorometric analysis of chondrocytes in human articular cartilages under normal aging or arthritic conditions. *Osteoarthritis Cartilage* **9**(7): 664–670.

Labarca, C. and K. Paigen 1980. A simple, rapid, and sensitive DNA assay procedure. *Anal. Biochem.* **102**: 344–352.

Lander, E. S. 1999. Arrays of hope. *Nat. Genet.* **21**(Suppl): 3–4.

Lanni, F., A. S. Waggoner et al. 1985. Structural organization of interface 3T3 fibroblasts studied by total internal reflection fluorescence microscopy. *J. Cell. Biol.* **100**: 1091–1102.

Lansdorp, P. M., C. Smith et al. 1991. Single laser three color immunofluorescence staining procedures based on energy transfer between phycoerythrin and cyanine 5. *Cytometry* **12**(8): 723–730.

Leethanakul, C., V. Patel et al. 2000. Distinct pattern of expression of differentiation and growth-related genes in squamous cell carcinomas of the head and neck revealed by the use of laser capture micro-dissection and cDNA arrays. *Oncogene* **19**: 3220–3224.

Levesque, M. J. and R. M. Nerem 1985. Elongation and orientation of cultured endothelial cells in response to shear stress. *J. Biomech. Eng.* **107**: 341–347.

Liang, P. and A. B. Pardee 1992. Differential display of eukaryotic messenger RNA by means of the poly-merase chain reaction. *Science* **257**: 967–971.

Liang, P. and A. B. Pardee 1998. Differential display. A general protocol. *Mol. Biotechnol.* **10**(3): 261–267.

Lima, E. G., L. Bian et al. 2007. The beneficial effect of delayed compressive loading on tissue-engineered cartilage constructs cultured with TGF-B3. *Osteoarthritis Cartilage* **15**(9): 1025–1033.

Linkhart, T. and S. Mohan 1989. Parathyroid hormone stimulates release of insulin-like growth fac-tor (IGF-I) and IGF-II from neonatal mouse calvaria in organ culture. *Endocrinology* **125**(3): 1484–1491.

Liss, B. 2002. Improved quantitative real-time RT-PCR for expression profiling of individual cells. *Nucl. Acids. Res.* **30**(17): e89.

Liu, V. A., W. E. Jastromb et al. 2001. Engineering protein and cell adhesivity using PEO-terminated tri-block polymers. *J. Biomed. Mater. Res.* **60**: 126–134.

Love, A. E. H. 1939. Boussinesq's problem for a rigid cone. *Q. J. Math.* **10**: 161–175.

Lovelock, J. E. 1957. The denaturation of lipid–protein complexes as a cause of damage by freezing. *Proc. Royal Soc. London Ser. B* **147**: 427–433.

Lu, J., W. Min et al. 2009. Super-resolution laser scanning microscopy through spatiotemporal modula-tion. *Nano Lett.* **9**(11): 3883–3889.

Luben, R. A., G. L. Wong et al. 1976. Biochemical characterization with parathormone and calcitonin of isolated bone cells: Provisional identification of osteoclasts and osteoblasts. *Endocrinology* **99**: 526–534.

Lucchinetti, E., C. S. Adams et al. 2002. Cartilage viability after repetitive loading: A preliminary report. *Osteoarthritis Cartilage* **10**(1): 71–81.

Lyons, A. B. and C. R. Parish 1994. Determination of lymphocyte division by flow cytometry. *J. Immunol. Methods* **171**(1): 131–137.

Mahaffy, R. E., C. K. Shih et al. 2000. Scanning probe-based frequency-dependent microrheology of poly-mer gels and biological cells. *Phys. Rev. Lett.* **85**: 880–883.

Majewska, A., G. Yiu et al. 2000. A custom-made two-photon microscope and deconvolution system. *Pflugers Arch.-Eur. J. Physiol.* **441**: 398–408.

Majumdar, M. K., E. Wang et al. 2001. BMP-2 and BMP-9 promotes chondrogenic differentiation of human multipotential mesenchymal cells and overcomes the inhibitory effect of IL-1. *J. Cell Physiol.* **189**(3): 275–284.

Mansfield, J. C. and C. P. Winlove 2012. A multi-modal multiphoton investigation of microstructure in the deep zone and calcified cartilage. *J. Anat.* **220**(4): 405–416.

Mansfield, J., J. Yu et al. 2009. The elastin network: Its relationship with collagen and cells in articular cartilage as visualized by multiphoton microscopy. *J. Anat.* **215**(6): 682–691.

Martin, I., M. Jakob et al. 2001. Quantitative analysis of gene expression in human articular cartilage from normal and osteoarthritic joints. *Osteoarthritis Cartilage* **9**: 112–118.

Martin, I., R. F. Padera et al. 1998. *In vitro* differentiation of chick embryo bone marrow stromal cells into cartilaginous and bone-like tissues. *J. Orthop. Res.* **16**: 181–189.

Martinez de Maranon, I., P. Gervais et al. 1997. Determination of cell's water membrane permeability: Unexpected high osmotic permeability of *Saccharomyces cerevisiae*. *Biotechnol. Bioeng.* **56**: 62–70.

Marty, A. and A. Finkelstein 1975. Pores formed in lipid bilayer membranes by nystatin: Differences in its one-sided and two-sided action. *J. General Phys.* **65**: 515–526.

Mathur, A. B., G. A. Truskey et al. 2000. Atomic force and total internal reflection fluorescence microscopy for the study of force transmission in endothelial cells. *Biophys. J.* **78**(4): 1725–1735.

Mauck, R. L., S. B. Nicoll et al. 2003. Synergistic effects of growth factors and dynamic loading for cartilage tissue engineering. *Tissue Eng.* **9**(4): 597–611.

Mauck, R. L., M. A. Soltz et al. 2000. Functional tissue engineering of articular cartilage through dynamic loading of chondrocyte-seeded agarose gels. *J. Biomech. Eng.* **122**: 252–260.

Mazur, P. 1965. The role of cell membranes in the freezing of yeast and other single cells. *Ann. NY Acad. Sci.* **125**: 658–676.

Mazur, P. 1984. Freezing of living cells: Mechanisms and implications. *Am. J. Physiol.* **247**: 125–142.

McGann, L. E., M. Stevenson et al. 1988. Kinetics of osmotic water movement in chondrocytes isolated from articular cartilage and applications to cryopreservation. *J. Orthop. Res.* **6**: 109–115.

Miesfeld, R. L. 1999a. Biochemical basis of applied molecular genetics. *Appl. Mol. Genet.* R. L. Miesfeld (ed). New York, Wiley-Liss: pp. 3–29.

Miesfeld, R. L. 1999b. Expression of cloned genes in cultured cells. *Appl. Mol. Genet.* R. L. Miesfeld (ed). New York, Wiley-Liss: pp. 175–204.

Miesfeld, R. L. 1999c. Isolation and characterization of gene transcripts. *Appl. Mol. Genet.* R. L. Miesfeld (ed). New York, Wiley-Liss: pp. 115–120.

Miesfeld, R. L. 1999d. The polymerase chain reaction. *Appl. Mol. Genet.* R. L. Miesfeld (ed). New York, Wiley-Liss: pp. 143–172.

Modderman, W. E., A. F. Weidema et al. 1994. Permeabilization of cells of hemopoietic origin by extracellular ATP^{4-}: Elimination of osteoclasts, macrophages, and their precursors from isolated bone cell populations and fetal bone rudiments. *Calcif. Tissue Int.* **55**: 141–150.

Morrison III, B., J. H. Eberwine et al. 2000. Traumatic injury induces differential expression of cell death genes in organotypic brain slice cultures determined by complementary DNA array hybridization. *Neuroscience* **2000**: 131–139.

Murray, J. M., P. L. Appleton et al. 2007. Evaluating performance in three-dimensional fluorescence microscopy. *J. Microsc.* **228**(Pt 3): 390–405.

Nagao, E. and J. A. Dvorak 1999. Phase imaging by atomic force microscopy: Analysis of living homeothermic vertebrate cells. *Biophys. J.* **76**(6): 3289–3297.

Nagineni, C. N., D. Amiel et al. 1992. Characterization of the intrinsic properties of the anterior cruciate and medial collateral ligament cells: An *in vitro* cell culture study. *J. Orthop. Res.* **10**: 465–475.

Nevo, Z., D. Robinson et al. 1998. The manipulated mesenchymal stem cells in regenerated skeletal tissues. *Cell Transplant* **7**(1): 63–70.

Ng, K. W., E. G. Lima et al. 2010. Passaged adult chondrocytes can form engineered cartilage with functional mechanical properties: A canine model. *Tissue Eng. Part A* **16**(3): 1041–1051.

Ng, K. W., C. J. O'Conor et al. 2011. Transient supplementation of anabolic growth factors rapidly stimulates matrix synthesis in engineered cartilage. *Ann. Biomed. Eng.* **39**(10): 2491–2500.

Nicoll, S. B., A. Wedrychowska et al. 2001. Modulation of proteoglycan and collagen profiles in human dermal fibroblasts by high density micromass culture and treatment with lactic acid suggests change to a chondrogenic phenotype. *Connect. Tissue. Res.* **42**(1): 59–69.

Niklason, L. E., J. Gao et al. 1999. Functional arteries grown *in vitro*. *Science* **284**: 489–493.

Nixon, A. J., L. A. Fortier et al. 1999. Enhanced repair of extensive articular defects by insulin-like growth factor-I-laden fibrin composites. *J. Orthop. Res.* **17**(4): 475–487.

Noiles, E. E., K. A. Thompson et al. 1997. Water permeability, Lp, of the mouse sperm plasma membrane and its activation energy are strongly dependent on interaction of the plasma membrane with sperm cytoskeleton. *Cryobiology* **35**: 79–92.

Nolan, G. P. 2011. Flow cytometry in the post fluorescence era. *Best Pract. Res. Clin. Haematol.* **24**(4): 505–508.

Nordon, R. E., S. S. Ginsberg et al. 1998. High-resolution cell division tracking demonstrates the FLt3-ligand-dependence of human marrow CD34 + CD38− cell production *in vitro*. *Br. J. Haematol.* **98**(3): 528–539.

Noy, A. 2011. Force spectroscopy 101: How to design, perform, and analyze an AFM-based single molecule force spectroscopy experiment. *Curr. Opin. Chem. Biol.* **15**(5): 710–718.

O'Connor, J. E., R. C. Callaghan et al. 2001. The relevance of FCM for biochemical analysis. *IUBUB Life* **51**(4): 231–239.

Oksvold, M. P., E. Skarpen et al. 2002. Fluorescent histochemical techniques for analysis of intracellular signaling. *J. Histochem. Cytochem.* **50**(3): 289–303.

Oswald, E. S., H. S. Ahmed et al. 2011. Effects of hypertonic (NaCl) two-dimensional and three-dimensional culture conditions on the properties of cartilage tissue engineered from an expanded mature bovine chondrocyte source. *Tissue Eng. Part C Methods* **17**(11): 1041–1049.

Oswald, E. S., L. M. Brown et al. 2011. Label-free protein profiling of adipose-derived human stem cells under hyperosmotic treatment. *J. Proteome Res.* **10**(7): 3050–3059.

Oviedo-Orta, E., M. Perreau et al. 2010. Control of the proliferation of activated CD4+ T cells by connexins. *J. Leukoc. Biol.* **88**(1): 79–86.

Paige, K. T. and C. A. Vacanti 1995. Engineering new tissue: Formation of neo-cartilage. *Tissue Eng.* **1**: 97.

Palmer, G. D., P.-H. G. Chao et al. 2001. Time dependent aggrecan gene expression of articular chondrocytes in response to hyperosmotic loading. *Osteoarthritis Cartilage* **9**(8): 761–770.

Parish, C. R. 1999. Fluorescent dyes for lymphocyte migration and proliferation studies. *Immunology Cell Biol.* **77**(6): 499–508.

Parker, K. K., A. L. Brock et al. 2002. Directional control of lamellipodia extension by constraining cell shape and orienting cell tractional forces. *FASEB J.* **16**(10): 1195–1204.

Pazzano, D., K. A. Mercier et al. 2000. Comparison of chondrogensis in static and perfused bioreactor culture. *Biotechnol. Progr.* **16**(5): 893–896.

Pei, M., F. He et al. 2008. Engineering of functional cartilage tissue using stem cells from synovial lining: A preliminary study. *Clin. Orthop. Relat. Res.* **466**(8): 1880–1889.

Pei, M., J. Seidel et al. 2002. Growth factors for sequential cellular de- and re-differentiation in tissue engineering. *Biochem. Biophys. Res. Commun.* **294**(1): 149–154.

Perka, C., O. Schultz et al. 2000. The influence of transforming growth factor B1 on mesenchymal cell repair of full-thickness cartilage defects. *J. Biomed. Mater. Res.* **52**: 543–552.

Petersen, E. F., K. W. Fishbein et al. 2000. ^{31}P NMR spectroscopy of developing cartilage produced from chick chondrocytes in a hollow-fiber bioreactor. *Magn. Reson. Med.* **44**: 367–372.

Petersen, E. F., R. G. S. Spencer et al. 2002. Microengineering neocartilage scaffolds. *Biotechnol. Bioeng.* **78**: 802–805.

Piper, T. and J. Piper 2012. Variable bright-darkfield-contrast, a new illumination technique for improved visualizations of complex structured transparent specimens. *Microsc. Res. Tech.* **75**(4): 537–554.

Pittenger, M. F., A. M. Mackay et al. 1999. Multilineage potential of adult human mesenchymal stem cells. *Science* **284**: 143–147.

Pittenger, M. F. and B. J. Martin 2004. Mesenchymal stem cells and their potential as cardiac therapeutics. *Circ. Res.* **95**: 9–20.

Pollack, S. R., D. F. Meaney et al. 2000. Numerical model and experimental validation of microcarrier motion in a rotating bioreactor. *Tissue Eng.* **6**(5): 519–530.

Poole, C. A. 1997. Articular cartilage chondrons: Form, function and failure. *J. Anat.* **191**: 1–13.

Portevin, D., M. Poupot et al. 2009. Regulatory activity of azabisphosphonate-capped dendrimers on human CD4+ T cell proliferation enhances *ex-vivo* expansion of NK cells from PBMCs for immunotherapy. *J. Transl. Med.* **24**(7): 82.

Pourati, J., A. Maniotis et al. 1998. Is cytoskeletal tension a major determinant of cell deformability in adherent endothelial cells? *Am. J. Physiol.* **247**: C1283–C1289.

Prentki, M., J. T. Deeney et al. 1986. Neomycin: A specific drug to study the inositol-phospholipid signalling system? *FEBS* **197**(1–2): 285–288.

Purves, W. K., G. H. Orians et al. 1992a. Gene Expression in Eukaryotes. *Life: The Science of Biology*. Sunderland, MA, Sinauer Associates, Inc.: pp. 288–305.

Purves, W. K., G. H. Orians et al. 1992b. Nucleic Acids as the Genetic Material. *Life: The Science of Biology*. Sunderland, MA, Sinauer Associates, Inc.: pp. 236–266.

Radmacher, M., M. Fritz et al. 1995. Imaging soft samples with the atomic force microscope: Gelatin in water and propanol. *Biophys. J.* **69**: 264–270.

Re, P., W. B. Valhmu et al. 1995. Quantitative polymerase chain reaction assay for Aggrecan and Link protein gene expression in cartilage. *Anal. Biochem.* **225**: 356–360.

Reits, E. A. and J. J. Neefjes 2001. From fixed to FRAP: Measuring protein mobility and activity in living cells. *Nat. Cell Biol.* **3**: E145–E147.

Rice, M. A. and K. S. Anseth 2007. Controlling cartilaginous matrix evolution in hydrogels with degradation triggered by exogenous addition of an enzyme. *Tissue Eng.* **13**(4): 683–691.

Roeder, B. A., K. Kokini et al. 2002. Tensile mechanical properties of three-dimensional type I collagen extracellular matrices with varied microstructure. *J. Biomech. Eng.* **124**: 214–222.

Roederer, M., A. B. Kantor et al. 1996. Cy7PE and Cy7APC: Bright new probes for immunofluorescence. *Cytometry* **24**(3): 191–197.

Roederer, M. and M. A. Moody 2008. Polychromatic plots: Graphical display of multidimensional data. *Cytometry* **73**(9): 868–874.

Rotsch, C., K. Jacobson et al. 1999. Dimensional and mechanical dynamics of active and stable edges in motile fibroblasts investigated by using atomic force microscopy. *Proc. Natl. Acad. Sci. USA* **96**(3): 921–926.

Rotsch, C. and M. Radmacher 2000. Drug-induced changes of cytoskeletal structure and mechanics in fibroblasts: An atomic force microscopy study. *Biophys. J.* **78**(1): 520–535.

Sabass, B., M. L. Gardel et al. 2008. High resolution traction force microscopy based on experimental and computational advances. *Biophys. J.* **94**(1): 207–220.

Sah, R. L. Y., Y. J. Kim et al. 1989. Biosynthetic response of cartilage explants to dynamic compression. *J. Orthop. Res.* **7**: 619–636.

Saiki, R. K., D. H. Gelfand et al. 1988. Primer-directed enzymatic amplification of DNA with a thermostable DNA polymerase. *Science* **239**: 487–491.

Sambrook, J., E. J. Fritsch et al. 1989a. Analysis and cloning of eukaryotic genomic DNA. *Molecular Cloning: A Laboratory Manual*, Cold Springs Harbor Laboratory Press. **2**: pp. 9.4–9.59.

Sambrook, J., E. J. Fritsch et al. 1989b. Extraction, purification, and analysis of messenger RNA from eukaryotic cells. *Molecular Cloning: A Laboratory Manual*, Cold Springs Harbor Laboratory Press. **1**: pp. 7.3–7.87.

Sampat, S. R., G. O'Connell et al. 2011. Growth factor priming of synovium derived stem cells for cartilage tissue engineering. *Tissue Eng. Part A* **17**(17–18): 2259–2265.

Sato, M., K. Nagayama et al. 2000. Local mechanical properties measured by atomic force microscopy for cultured bovine endothelial cells exposed to shear stress. *J. Biomech.* 33(1): 127–135.

Sato, M., T. Ozawa et al. 1999. A fluorescent indicator for tyrosine phosphorylation-based insulin signaling pathways. *Anal. Chem.* 71: 3948–3954.

Schenk, P. M., A. R. Elliot et al. 1998. Assessment of transient gene expression in plant tissues using the green fluorescent protein as a reference. *Plant Mol. Biol. Rep.* 16: 313–322.

Schille, P. 2001. Fluorescence correlation spectroscopy and its potential for intracellular applications. *Cell. Biochem. Biophys.* 34(3): 383–408.

Schinagl, R. M., M. K. Ting et al. 1996. Video microscopy to quantitate the inhomogeneous equilibrium strain within articular cartilage during confined compression. *Ann. Biomed. Eng.* 24: 500–512.

Sers, C., O. I. Tchernitsa et al. 2002. Gene expression profiling in RAS oncogene-transformed cell lines and in solid tumors using subtractive suppression hybridization and cDNA arrays. *Adv. Enzyme Regul.* 42: 63–82.

Service, R. F. 2000. CHEMISTRY NOBEL: Getting a charge out of plastics. *Science* 290(5491): 425–427.

Shapiro, H. M. 1995. *Practical Flow Cytometry*. New York, Wiley-Liss.

Shotton, D. 1993. An introduction to the electronic acquisition of light microscope images. *Electronic Light Microscopy*. D. Shotton (ed). New York, John Wiley & Sons, Inc.: 355 pp.

Shyy, J. and S. Chien 1997. Role of integrins in cellular responses to mechanical stress and adhesion. *Curr. Opin. Cell Biol.* 9: 707–713.

Siebert, P. and J. Larrick 1992. Competitive PCR. *Nature* 359: 557–558.

Simione, M. S. 1998. *Cryopreservation Manual*. Rockville, MD, American Type Culture Collection (ATCC)-Nalge Nunc International Corp.

Simon, R., M. D. Radmacher et al. 2002. Design of studies using DNA microarrays. *Gen. Epidem.* 23: 21–36.

Singer, V. L., L. J. Jones et al. 1997. Characterization of PicoGreen reagent and the development of a fluorescence-based solution assay for double stranded DNA quantitation. *Anal. Biochem.* 249: 228–238.

Singhvi, R., A. Kumar et al. 1994. Engineering cell shape and function. *Science* 264: 696–698.

Sironen, R. K., H. M. Karjalainen et al. 2002. High pressure effects on cellular expression profile and mRNA stability. A cDNA array analysis. *Biorheology* 39: 111–117.

Smetana, K. 1993. Cell biology of hydrogels. *Biomaterials* 14(14): 1046–1050.

Souaze, F., A. Ntodue-Thome et al. 1996. Quantitative PCR: Limits and accuracy. *Biotechniques* 21: 280–285.

Stricker, J., B. Sabass et al. 2010. Optimization of traction force microscopy for micron-sized focal adhesions. *J. Phys. Condens. Matter* 22(19): 194104.

Summers, K. and M. W. Kirschner 1979. Characteristics of the polar assembly and disassembly of microtubules observed *in vitro* by darkfield light microscopy. *J. Cell Biol.* 83(1): 205–217.

Syedain, Z. H. and R. T. Tranquillo 2011. TGF-beta1 diminishes collagen production during long-term cyclic stretching of engineered connective tissue: Implication of decreased ERK signaling. *J. Biomech.* 44(5): 848–855.

Takahashi, K., K. Tanabe et al. 2007. Induction of pluripotent stem cells from adult human fibroblasts by defined factors. *Cell* 131(5): 861–872.

Takai, E., K. D. Costa et al. 2005. Osteoblast elastic modulus measured by atomic force microscopy is substrate dependent. *Ann. Biomed. Eng.* 33(7): 963–971.

Takai, E., R. L. Mauck et al. 2004. Osteocyte viability and regulation of osteoblast function in a 3D trabecular bone explant under dynamic hydrostatic pressure. *J. Bone Miner. Res.* 19(9): 1403–1410.

Takayama, S., J. C. McDonanld et al. 1999. Patterning cells and their environments using multiple laminar fluid flows in capillary networks. *Proc. Natl. Acad. Sci.* 96: 5545–5548.

Tan, A. R., E. Y. Dong et al. 2011. Coculture of engineered cartilage with primary chondrocytes induces expedited growth. *Clin. Orthop. Relat. Res.* 469(10): 2735–2743.

Tandon, N., C. Cannizzaro et al. 2009. Electrical stimulation systems for cardiac tissue engineering. *Nat. Protoc.* 4(2): 155–173.

Terskikh, A., A. Fradkov et al. 2000. Fluorescent timer: Protein that changes color with time. *Science* **290**: 1585–1588.

Theodossiou, T. A., C. Thrasivoulou et al. 2006. Second harmonic generation confocal microscopy of collagen type I from rat tendon cryosections. *Biophys. J.* **91**(12): 4665–4677.

Thomas, C. H., C. D. McFarland et al. 1997. The role of vitronectin in the attachment and spatial distribution of bone-derived cells on materials with patterned surface chemistry. *J. Biomed. Mater. Res.* **37**: 81–93.

Tokalov, S. V., Y. Henker et al. 2004. 8-Prenylnargingenin and derivatives in human cells. *Pharmacology* **71**(1): 46–56.

Tompkins, J. D., C. Hall et al. 2012. Epigenetic stability, adaptability, and reversibility in human embryonic stem cells. *Proc. Natl. Acad. Sci. USA*.

Toomre, D. and D. J. Manstein 2001. Lighting up the cell surface with evanescent wave microscopy. *Trends Cell Biol.* **11**: 298–303.

Truskey, G. A., J. S. Burmeister et al. 1992. Total internal reflection microscopy (TIRFM). *J. Cell Sci.* **103**: 491–499.

Tsien, R. Y., T. J. Rink et al. 1985. Measurement of cytosolic free Ca^{2+} in individual small cells using fluorescence microscopy with dual excitation wavelengths. *Cell Calcium* **6**: 145–157.

Valhmu, W. B., G. D. Palmer et al. 1998. Regulatory activities of the 5′- and 3′-untranslated regions and promoter of the human aggrecan gene. *J. Biol. Chem.* **273**(11): 6196–6202.

Valhmu, W. B., E. J. Stazzone et al. 1998. Load-controlled compression of articular cartilage induces a transient stimulation of aggrecan gene expression. *Arch. Biochem. Biophys.* **353**(1): 29–36.

van Engeland, M., L. J. Nieland et al. 1998. Annexin V-affinity assay: A review on an apoptosis detection system based on phosphatidylserine exposure. *Cytometry* **31**(1): 1–9.

van Roessel, P. and A. H. Brand 2002. Imaging into the future: Visualizing gene expression and protein interactions with fluorescent proteins. *Nat. Cell Biol.* **4**: E15–E20.

Van Roessel, P., N. M. Hayward et al. 2002. Two-color GFP imaging demonstrates cell-autonomy of GAL4-driven RNA interference in *Drosophila*. *Genesis* **34**(1–2): 170–173.

van Susante, J. L., P. Buma et al. 1998. Chondrocyte-seeded hydroxyapatite for repair of large articular cartilage defects. A pilot study in the goat. *Biomaterials* **19**(24): 2367–2374.

Vaughan, A. T., L. S. Brackenbury et al. 2010. Neisseria lactamica selectively induces mitogenic proliferation of the naive B cell pool via cell surface Ig. *J. Immunol.* **185**(6): 3652–3660.

Voegele, T. J., M. Voegele-Kadletz et al. 2000. The effect of different isolation techniques on human osteoblast-like cell growth. *Anticancer Res.* **20**(5B): 3575–3581.

Vunjak-Novakovic, G., L. E. Freed et al. 1996. Effects of mixing on the composition and morphology of tissue-engineered cartilage. *AIChE* **42**: 850–860.

Walch, M., U. Ziegler et al. 2000. Effect of streptolysin O on the microelasticity of human platelets analyzed by atomic force microscopy. *Ultramicroscopy* **82**(1–4): 259–267.

Waldman, S. D., C. G. Spiteri et al. 2004. Long-term intermittent compressive stimulation improves the composition and mechanical properties of tissue-engineered cartilage. *Tissue Eng.* **10**(9/10): 1323–1331.

Wallrabe, H. and A. Periasamy 2005. Imaging protein molecules using FRET and FLIM microscopy. *Curr. Opin. Biotechnol.* **16**(1): 19–27.

Wang, Y. L. 1998. Digital deconvolution of fluorescence images for biologists. *Methods Cell Biol.* **56**: 305–315.

Wang, C. C.-B., G. A. Ateshian et al. 2002a. An automated approach for direct measurement of strain distributions within articular cartilage under unconfined compression. *J. Biomech. Eng.* **124**: 557–567.

Wang, E., C. M. Babbey et al. 2005. Performance comparison between the high-speed Yokogawa spinning disc confocal system and single-point scanning confocal systems. *J. Microsc.* **218**(Pt 2): 148–159.

Wang, L., N. H. Dormer et al. 2010. Osteogenic differentiation of human umbilical cord mesenchymal stromal cells in polyglycolic acid scaffolds. *Tissue Eng. Part A* **16**(6): 1937–1948.

Wang, C. C.-B., X. E. Guo et al. 2002b. The functional environment of chondrocytes within cartilage subjected to compressive loading: Theoretical and experimental approach. *Biorheology* **39**(1–2): 39–45.

Wang, Y., J. Y. Shyy et al. 2008. Fluorescence proteins, live-cell imaging, and mechanobiology: Seeing is believing. *Annu. Rev. Biomed. Eng.* **10**: 1–38.

Waterman-Storer, C. 2002. Fluorescent speckle microscopy (FSM) of microtubules and actin in living cells. *Curr. Protoc. Cell Biol.* Chapter 4: Unit 4 10.

Waterman-Storer, C. M., A. Desai et al. 1998. Fluorescent speckle microscopy, a method to visualize the dynamics of protein assemblies in living cells. *Curr. Biol.* **8**(22): 1227–1230.

Weisenhorn, A. L., P. Maivald et al. 1992. Measuring adhesion, attraction, and repulsion between surfaces in liquids with an atomic force microscope. *Phys. Rev. B.* **45**: 11226–11232.

Weiss, D. G. and W. Maile 1993. Principles, practice, and applications of video-enhanced contrast microscopy. *Electronic Light Microscopy*. D. Shotton (ed). New York, Wiley-Liss, Inc.: pp. 105–140.

Werkmeister, E., N. de Isla et al. 2010. Collagenous extracellular matrix of cartilage submitted to mechanical forces studied by second harmonic generation microscopy. *Photochem. Photobiol.* **86**(2): 302–310.

White, J. G., W. B. Amos et al. 1987. An evaluation of confocal versus conventional imaging of biological structures by fluorescence light microscopy. *J. Cell Biol.* **105**: 41–48.

Wilkins, L. M., S. R. Watson et al. 1994. Development of bilayered living skin construct for clinical applications. *Biotech. Bioeng.* **43**: 747–756.

Willey, J. C., E. L. Crawford et al. 1998. Expression measurement of many genes simultaneously by quantitative PCR using standardized mixtures of competitive templates. *Am. J. Respir. Cell Mol. Biol.* **19**: 6–17.

Wong, G. 1990. Isolation and behavior of isolated bone-forming cells. *Volume 1: The Osteoblast and Osteocyte*. B. K. Hall (ed). Caldwell, Telford Press: pp. 494.

Wu, H. W., T. Kuhn et al. 1998. Mechanical properties of L929 cells measured by atomic force microscopy: Effects of anticytoskeletal drugs and membrane crosslinking. *Scanning* **20**(5): 389–397.

Yamane, Y., H. Shiga et al. 2000. Quantitative analyses of topography and elasticity of living and fixed astrocytes. *J. Electron. Microsc.* **49**(3): 463–471.

Yarar, D., C. M. Waterman-Storer et al. 2005. A dynamic actin cytoskeleton functions at multiple stages of clathrin-mediated endocytosis. *Mol. Biol. Cell* **16**(2): 964–975.

Yuste, R., F. Lanni et al., Eds. 1999. *Imaging: A Laboratory Manual*. Cold Spring Harbor, Cold Spring Harbor Press.

Zuk, P. A., M. Zhu et al. 2001. Multilineage cells from human adipose tissue: Implications for cell-based therapies. *Tissue Eng.* **7**(2): 211–228.

13

Histology and Staining

S.E. Greenwald
Queen Mary University of London

A.G. Brown
Queen Mary University of London

A. Roberts
Texas A&M University

F.J. Clubb, Jr.
Texas A&M University

13.1 Introduction

Histology is defined as "the science of organic tissue" although it is more commonly regarded as "that branch of biology or anatomy concerned with the minute structure of the tissues of plants or animals" (*Shorter Oxford Dictionary*, 1973). Although its origins have been ascribed to Aristotle, who distinguished between tissues and organs, histology as a modern science began with the invention of the compound microscope in about 1600 by Iansen and/or Gallileo, and developed with the evolution of microscopy. Nevertheless, many important observations using a single magnifying lens continued to be reported until the early eighteenth century, notably by van Leeuwenhoek (1791). Perhaps the first mention of tissue being composed of separate cells was by Robert Hooke (1665). For a detailed outline and chronology of the history of microscopy and histology, see, for instance, Kaiser (1985).

Although it is possible to visualize microscopic structures in living tissue, the amount of information obtainable is limited by the inability of visible light to penetrate most organisms beyond a depth of a millimeter or so. Consequently, two of the three main aims of practical histology are the preservation of dead tissue and the cutting of it into slices thin enough to be transparent. Under these conditions, it is often difficult to distinguish different parts of the specimen because they will generally have closely similar refractive indices (RIs). Therefore, the third aim is the staining of the specimen to make its components and structure distinguishable.

The acquisition from a living tissue or organ of a magnified image in which the composite materials can be identified and from which a pathologist can make a diagnosis or scientist can derive structural and functional information usually follows a well-defined sequence of steps, summarized in Figure 13.1. The aim of this chapter is to expand briefly on the major steps of the sequence.

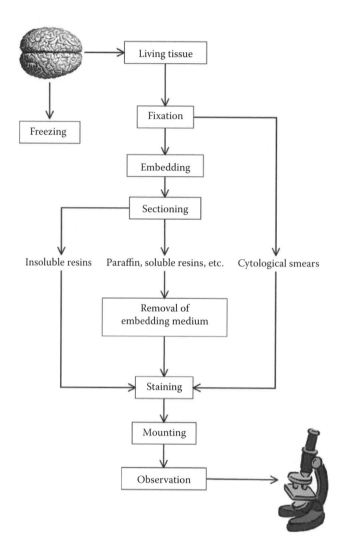

FIGURE 13.1 Summary of the basic steps in histological processing, from living tissue to a stained section suitable for LM. (Redrawn from Horobin, D.W. *Histochemistry: An Explanatory Outline of Histochemistry and Biophysical Staining*, 1982. Gustav Fischer Verlag, Stuttgart; Butterworth's, London. With the permission of the author and publishers.)

Ideally, the tissue would be examined in three dimensions. This obviously requires transparency which, for optical microscopy, limits specimen thickness to approximately 600 μm. To study opaque material or more deeply buried structures, the tissue must be sectioned and stained to produce the necessary contrast. Using tomographic software, digital images from contiguous or serial sections can be combined and a three-dimensional (3D) image reconstructed. In practice, however, 3D reconstruction is not commonly performed because processing the large number of sections required to produce a high-resolution image is extremely time-consuming. Furthermore, small but cumulative errors in aligning successive sections often result in distortion of the 3D geometry and render measurements of length and volume unreliable. During the last 20 years or so, advances in confocal microscopy (see below) have greatly reduced these problems, although structures more than 600 μm below the specimen surface cannot currently be visualized.

As an alternative approach used in cytological screening, where the pathologist is primarily interested in changes in the internal structure of cells, a small sample obtained, for instance, by a needle biopsy is smeared into a thin layer on a glass slide. After staining, these structures can be seen although information about the architecture of the tissue from which the cells are derived is lost.

13.2 Fixation

Unless it is to be examined very soon after death, nonliving tissue must be *fixed*. The aims of fixation and its effects are (Sanderson, 1994):

- To maintain geometry and dimension as close as possible to that of the living state.
- To prevent autolysis (destruction by enzymes released by dead or dying cells) and putrefaction (attack and digestion by microorganisms).
- To make tissue receptive to staining by changing its chemical composition. Unfixed tissue has little affinity for most histological stains. An additional benefit of the alteration in chemical structure is to produce differential changes in the RI of the tissue components, thus improving contrast when examined unstained.
- To harden tissue sufficiently to make it more resistant to damage caused by subsequent processing, without making it difficult to cut.

Fixatives can be divided into a number of groups (see below) on the basis of their chemical structure and mode of action (Baker, 1958b), the most important of which stabilize proteins, usually by promoting chemical cross-linking between their molecules forming a gel or a harder polymer surrounding softer components. In this way, the morphology of the tissue is stabilized, making it easier to cut clean sections. Fixation is normally achieved by immersion in an aqueous solution or, for small fragile specimens, by exposure to vapor. The time taken to fix a tissue depends on the rate at which the fixative diffuses, the temperature, and the nature of the tissue itself (Hopwood, 1990). The following empirical relationship between the distance penetrated by the fixative (*d*) and time (*t*) was derived by Medewar (1941) for noncoagulating fixatives and by Baker (1958b) for fixatives that coagulate protein (i.e., form cross-links):

$$d = kt^{0.5} \tag{13.1}$$

The value of k (at room temperature) for formaldehyde (Baker, 1958b) is 0.06 mm s$^{-0.5}$ from which it follows that it will take approximately 5 min for this fixative to penetrate 1 mm into a homogenous piece of soft tissue and nearly 8 h to penetrate 10 mm.

In *phase partition* fixation, the tissue is immersed in an aqueous solution of fixative in equilibrium with an organic solvent (Nettleton and McAuliffe, 1986) (e.g., 50% glutaraldehyde/water in heptane). It is suitable for delicate tissues and has been used to fix and study mucus on the surface of the trachea that would otherwise be dissolved by aqueous immersion (Sims et al., 1991).

To ensure rapid fixation of whole organs such as the lung, heart, liver, and so on, the fixative may be perfused through the vasculature (Rostgaard et al., 1993). For reliable morphometric analysis of blood vessels, the mean pressure should be close to *in vivo* values (Berry et al., 1993).

13.2.1 Types of Fixative

13.2.1.1 Aldehydes

This group includes formaldehyde, the most commonly used fixative for light microscopy (LM) and glutaraldehyde, favored for electron microscopy (EM). Formaldehyde is usually used as *formol saline,* a 4% solution of formaldehyde gas in water containing 0.9% sodium chloride. It reacts with protein by a condensation reaction, thereby forming cross-links frequently between lysine residues on the exterior of the protein chains (see, e.g., Baker (1958b) for a review of reaction mechanisms).

Fixation with glutaraldehyde often involving the formation of cross-links between pyridine residues (Hilger and Medan, 1987) greatly reduces the immunological activity of proteins, a process sometimes referred to as *denaturation* (Hardy et al., 1976), thus rendering them unsuitable for immunohistochemical staining (see below). On the other hand, it disturbs tissue morphology less severely than formaldehyde and is therefore preferred for investigations of cellular architecture. Acrolein, also an aldehyde, is often mixed with formaldehyde or glutaraldehyde. It has the advantage of rapid penetration and good preservation of morphology (Saito and Keino, 1976).

13.2.1.2 Oxidizing Agents

The most commonly used of this group, which includes potassium dichromate and potassium permanganate, is osmium tetroxide. The reactions of oxidizing agents with different types of tissue have been reviewed by Baker (1958b) and more recently by Kiernan (1990). With proteins, osmium tetroxide is thought to form cross-links. It is soluble in some lipids and is also reduced to the black dioxide by them, which then take on this color and become crumbly, making sections difficult to cut. In spite of these drawbacks and its slow rate of penetration (Baker, 1958a), osmium tetroxide "preserves the structure of the living cell better than any other primary or mixed fixative" (Baker, 1958b) and for this reason is used extensively both as a fixative and as a stain in EM. Dissolved in a nonaqueous fluorocarbon solvent (FC-72), osmium tetroxide has been used to demonstrate the ultrastructure of the glycocalyx, the fragile coating of vascular endothelial cells (Sims and Horne, 1994), and used alone in perfused brain tissue for both light and EM (Branchereau et al., 1995).

13.2.1.3 Alcohols and Acetone

Wine has been used as a preservative since antiquity and, according to Baker (1958b), was first used as a preservative for anatomical purposes in 1663 by Robert Boyle. Alcohols and ketones displace water from protein and dehydrate the tissue as a whole, causing it to harden and shrink. Their main advantage is rapid penetration and, when used at low temperature, preservation of enzymatic and immunological activity (Sato et al., 1986). Ethanol, in particular, is the fixative of choice when transporting specimens by air, as the airlines have little objection to their customers carrying large quantities of this material while remaining suspicious of other toxic chemicals.

13.2.1.4 Other Cross-Linking Agents

This group includes the carbodiimides introduced in 1971 (Kendall et al., 1971) and is seeing increasing use in light and EM owing to their speed of action and specificity (Tymianski et al., 1997), both as a mixture with glutaraldehyde (Willingham and Yamada, 1979) and alone (Panula et al., 1988). They have recently been shown to be suitable for denaturing heart valve tissue prior to implantation as they can easily be washed out and leave little or no toxic residues (Girardot and Girardot, 1996).

13.2.1.5 Heat

Heat alone will cause many proteins to coagulate. Microwaves usually in conjunction with fixatives allow rapid heating although the degree of fixation tends to be nonuniform with the center of the specimen less well-fixed (Leong and Duncis, 1986). For large specimens such as whole organs, preliminary fixation by immersion in 0.9% saline and irradiation to a temperature of 68–74°C are recommended. The process is completed by irradiating 2- to 3-mm pieces to a temperature of 50–68°C for about 2 min (Leong, 1994). In conjunction with formaldehyde or glutaraldehyde, fixation times as little as 10 s can be achieved, thus reducing the effects of autolysis, diffusion of cell contents, and retaining ultrastructural detail (Leong et al., 1985; Login and Dvorak, 1985). Among other applications in the histology laboratory, microwave irradiation has been used to preserve immunological activity (Leong et al., 1988), improve the quality of frozen sections, and enhance the staining of ultrathin sections for EM. Technical details are discussed at some length by Leong (1994) and general reviews may be found in Leong (1988, 1993).

13.2.1.6 Unknown Mechanism

Although fixatives have been in use for many decades, the mode of action of several standard fixatives is not well understood. These include mercuric chloride which penetrates tissue rapidly and coagulates proteins by reacting with many different amino acid residues (Hopwood, 1990), and picric acid which, in conjunction with other fixatives such as ethanol or formalin, also coagulates protein while leaving the tissue soft (Leong, 1994).

13.2.2 Postfixation

It has been found that staining intensity can be improved by following a period of primary fixation in formalin, for example, by a few hours of additional treatment with mercuric chloride (Leong, 1994). Sections are said to be easier to cut and to flatten more readily when postfixed (Hopwood, 1990). Staining of cell membranes in samples fixed in glutaraldehyde for EM is improved by postfixation with osmium tetroxide. Blocks destined for both light and electron microscopic investigation can be post-fixed with 10% formalin following standard glutaraldehyde treatment, which has the additional advantage of retaining intracellular structure (Tandler, 1990).

13.2.3 Fixation Artifacts

Tissue volume may change during fixation although the effects are small compared to shrinkage during embedding. They are discussed in the next section. Specimens that have been stored in formalin for extended periods under acidic conditions become suffused with fine brown crystals of *formalin pigment*, which is thought to be formed by the reaction of formalin with hematin derived from the hemoglobin of ruptured red blood cells (Baker, 1958b). Its formation is inhibited by the addition of 2% phenol to formalin. If it has formed, it may be removed by brief treatment with a saturated solution of picric acid in ethanol followed by washing in water (Drury and Wallington, 1980).

Although fixation normally stiffens the tissue as well as removing water, highly mobile inorganic ions as well as large molecules such as hemoglobin (Reale and Luciano, 1970) may diffuse through the specimen toward its periphery as it fixes, giving a spurious view of their localization and distribution. In metabolic studies of living tissue, the uptake of radioactively labeled amino acids or sugars may be studied postmortem by assessing the distribution of the label in the tissue. However, glutaraldehyde binds strongly to many of these substances and therefore their distribution in fixed tissue assessed microscopically by autoradiography (see below) may be altered. Finally, some tissue components, for example, lipids and mucopolysaccharides, do not react with commonly used fixatives and may be lost from the tissue during subsequent processing.

13.3 Tissue Processing

13.3.1 Dehydration and Clearing

Following fixation, tissue must be *processed* to render it suitable for embedding and subsequent sectioning. Paraffin wax, the most common embedding medium, is hydrophobic, whereas most fixatives are aqueous. Therefore, the material must first be dehydrated and then infiltrated with a solvent that is miscible with the embedding medium. Typically, the specimen is soaked initially in an aqueous solution of an alcohol and transferred at intervals through successively more concentrated solutions until no more water remains. The processing continues with the replacement of the alcohol by a transfer agent or transition medium (most commonly xylene or toluene) that is both miscible with alcohol and a solvent for paraffin wax, the embedding agent. Finally, xylene is replaced by molten wax resulting in complete infiltration of the tissue. Typically, the entire process under automatic control takes around 16 h (Gordon, 1990), although small specimens, for example, needle biopsies, may be processed in 3 h. Dehydration with ethanol and treatment with xylene usually cause hardening, which may be reduced by using *n*-butanol, and loss of lipids, which is minimized by rapid treatment.

13.3.2 Embedding

Ideally, the embedding medium should be of approximately equal density and resilience to the tissue to allow cutting of sections with minimal damage. The most commonly used medium is paraffin wax because it is easy to cut, cheap, and widely available. Sections thinner than 2 µm, which are needed to examine details of intracellular structure, require a tougher medium such as a plastic resin (see below), from which sections as thin as 0.2 µm can be cut. However, this requires harder and sharper knives made of tungsten carbide or glass (Bennett et al., 1976).

The tissue permeated with embedding medium is placed in a mold, oriented for subsequent sectioning (Arnolds, 1978; Hilger and Medan, 1987). The mold is filled with the medium and allowed to cool, usually under automatic control.

Other embedding media with lower melting points, such as polyethylene glycol or polyester waxes, can be used when excessive heat may reduce immunological activity of proteins that are to be stained by antibodies (see below). For hard and tough specimens or those that contain tissues with differing degrees of hardness, cellulose nitrate has traditionally been used. However, this material is chemically hazardous and is not suitable for sections less than 10-µm thick.

During the past 60 years, plastic resins have been developed for use as embedding media driven initially by the development of the electron microscope and the consequent need for a material which could produce sections no thicker than 80 nm and able to withstand temperatures to 200°C (Nunn, 1970a,b). Plastic resins are tougher than paraffin wax and therefore make it possible to cut thinner sections and harder tissues with less damage to the section. They also cause less shrinkage during processing than waxes. The three main types are briefly compared in Table 13.1.

TABLE 13.1 Summary of the Properties of Embedding Resins

Material	Advantages	Drawbacks
Epoxy (Spurr, 1969)	Tough, thin sections possible. Can be softened with plasticizers for easier sectioning. Shrinkage ≈ 3%	Curing temperature 60°C, may cause tissue damage. Hydrophobic, therefore, compatible with limited range of stains.
Acrylic (Murray, 1988; Litwin, 1985)	Low viscosity, easier processing. Monomer water miscible; polymer water permeable; dehydration not necessary, compatible with many stains.	Shrinkage ≈ 15%
Polyester	Low viscosity, easier processing. Some can be polymerized at low temperature with UV light, therefore, suitable for immunological staining (Altman, 1984). Controllable hardness to match tissue. More water compatible than epoxy, less than acrylic. Tolerant of electron beam; suitable for electron microscopy.	Small specimens only due to limited penetration of UV for curing. Non-UV-curable have curing temperature.

13.3.3 Tissue Shrinkage

Most fixatives cause tissue to shrink, largely owing to their dehydrating effect, although acidic fixatives including formalin lead, at least initially, to swelling driven by osmotic pressure (Baker, 1958a). However, as fixation and dehydration proceed, shrinkage inevitably occurs, different tissues being affected to different extents. The early literature (reviewed by Baker, 1958b) reports quite variable results on the effects of dehydration on different tissues. More recent reviews (see, e.g., Fox et al., 1985) and more modern studies agree that fixation and dehydration account for a reduction in volume of between 3% and 6% for whole organs (Iwadare et al., 1984) and up to 30% for individual cells (Ross, 1953). Removal of fixative by treatment with transition medium prior to embedding causes a further reduction of around 10% (with respect to the original volume). Finally, embedding in wax results in an overall reduction in volume of no less than 35%. Resin shrinks less than wax during the embedding process and, when sectioned and dried, may stretch. Overall, tissues in resin sections are nearer to their living dimensions than those embedded in wax, although factors such as temperature and section thickness affect the overall degree of volume change (Hanstede and Gerrits, 1983). In a study of arterial sections, Dobrin (1996) has shown that formalin fixation followed by dehydration and embedding in paraffin wax leads to a reduction in cross-sectional area of 19%. Fixation in McDowell's solution (recipe given in Dobrin, 1996) followed by embedding in glycol methacrylate (which avoids the need to further dehydrate the tissue) resulted in an overall increase in cross-sectional area as little as 4%. This procedure appears to be optimal at least for vascular histomorphometry. Clearly, when measuring absolute dimensions of cells or other tissue components, careful assessment of shrinkage is essential.

13.4 Cutting

13.4.1 Microtomes

The purpose of the microtome is to cut sections of known and uniform thickness through tissue, which is surrounded by and infused with embedding medium, in the form of a cuboidal or cylindrical block. At its most basic, the microtome consists of a chuck to hold the tissue block near to a blade that is passed over the specimen and that removes thin slices in the manner of a carpenter's plane. A later development allows the specimen to advance toward the blade, as each slice is removed, by rotating a lead screw attached to the chuck (Figure 13.2a). This type of rocking blade device was introduced in 1881 (Kaiser, 1985). In its modern form, the base sledge microtome (Figure 13.2b), the chuck-and-lead screw assembly, in a heavy casting moving on runners within a massive frame, is passed repeatedly under a fixed blade. Each pass of the specimen causes the lead screw to advance by a preset amount. The inertia of the rocking mechanism, which is usually driven by hand, minimizes chattering between the blade and the block, and consequent tearing of the specimen. The base sledge microtome is largely confined to the cutting of wax-embedded sections. As the sections are cut, they are pushed onto the upper surface of the blade where they collect as a delicate ribbon of crinkled sections. From time to time, these are carefully picked up and floated onto water held at some 10°C below the melting point of the wax. The softened wax is stretched by surface tension and the flattened section is transferred to a glass slide by passing the slide underneath the section, to which it adheres by surface tension, and lifting it out of the water (Figure 13.3). Thin resin sections suitable for EM may be flattened by exposure to xylene or chloroform vapor and picked up onto a fine metal grid rather than a glass slide (Nunn, 1970a,b). Slides with sections containing elastin such as blood vessels, skin, and lung tissue, which retain some resilience even after prolonged fixation and which tend to curl at the edges, are placed on a hot plate that softens the wax further and allows greater stretching. These processes require considerable manual dexterity and training. The tricks of the trade may be found in practical texts such as Bancroft and Stevens (1990) and Sanderson (1994).

In the rotary microtome, the chuck and embedded specimen block are moved against a blade in a reciprocating motion by a crank connected to a motor or manually driven flywheel. These widely used

FIGURE 13.2   (a) "Cambridge Rocker" microtome. This instrument, built in the late nineteenth century, was in routine use until the 1930s. The spring-loaded handle (A) is pulled toward the operator causing arm (B) to lift the specimen embedded in a wax block (E) above the edge of the blade (F). At the same time, a pawl engages in the toothed wheel causing the attached threaded rod to rotate lifting arm (D) and advancing the specimen toward the blade by a set amount which determines the section thickness. On releasing the handle, the specimen is drawn down over the blade, producing a thin slice which rides up onto the blade. (b) Base sledge microtome.

devices can cut sections varying in thickness from 0.5 to 25 μm in wax and resins and lend themselves well to automated section cutting (Vincent, 1991).

Sections once picked up onto the slides are placed in an incubator at just below the melting point of the wax for about 30 min. This dries the side and increases the adhesion between the section and the glass. Finally, they are soaked in xylene and, if they are to be stained with an aqueous dye, are rehydrated with aqueous ethanol and water. (If alcoholic dyes are to be used, the rehydration stage is omitted.)

Fixed or unfixed tissue can be hardened by freezing and cut in a *freezing microtome* designed or modified to maintain the specimen at a low temperature (−20°C to −40°C). This allows for the rapid assessment of biopsies obtained during a surgical procedure so that the surgeon's subsequent decisions may then be based on the pathologist's diagnosis. Frozen sections are also used for investigating tissues that are prone to rapid enzymatic degradation (e.g., liver, central nervous system) or that contain diffusible or soluble materials such as lipid.

In the vibrating blade microtome, or *vibratome,* a thin razor-like blade oscillating at mains frequency is advanced across the specimen glued to a stub, held in a chuck, and immersed in water, saline, or

FIGURE 13.3 Sections cut from a wax block floating on water and flattened by surface tension. A single slice has been lifted onto a glass slide.

fixative to prevent undue heating. The amplitude of the vibration and the speed at which the blade advances through the specimen can be adjusted to suit different types of tissue (Zelander and Kirkeby, 1978; Sallee and Russell, 1993). The major advantage of these devices is that they will cut cleanly through unfixed and unembedded tissue at room temperature. Their major drawback is that unprocessed specimens less than 20-μm thick cannot be cut because they tend to disintegrate.

13.4.2 Microtome Blades

The profile and sharpness of the blade is of critical importance in maintaining a clean cut and uniform section thickness. In paraffin sections under optimal conditions, the thickness of adjacent sections can vary by 10% in a standard 5-μm section and by as much as 50% in 1-μm sections. In tests on typical sections of nominal thickness of 5 μm, although the mean measured thickness was 4.8 μm (SD 0.14 μm), individual sections varied between 3 and 7 μm. In thinner sections (nominal thickness of 1 μm), the variation was as much as 50% (Merriam, 1957; Helander, 1983). Thickness variation in resin sections that is harder and more homogenous than wax is, however, considerably smaller (Helander, 1983). Nevertheless, thickness variation must be carefully assessed, especially in studies involving 3D reconstruction from serial sections (Bibb et al., 1993).

Microtome blades are typically made of high carbon steel and commonly have a plane wedge-shaped profile with a raked edge, although other profiles are used for particularly hard or soft materials (Ellis, 1994). Sharpening is a highly skilled procedure, details of which may be found in standard practical histology texts (Ellis, 1994; Sanderson, 1994). For harder materials such as teeth and bone, a cutting edge of sintered tungsten carbide is bonded to a steel body.

For very thin sections of hard material, glass knives made by fracturing a glass block under controlled conditions (Reid and Beesley, 1991) are suitable. They can be used in a base sledge, rotating blade, or specialized power-driven microtomes (e.g., the Ultramicrotome, Dupont Biomedical Products Division, Wilmington, Delaware) designed to cut sections from hard resin blocks as thin as 10 nm, suitable for EM. In this device, section thickness is controlled by advancing the block by a stepper motor under microprocessor control. Diamond knives may be used in place of glass with the advantage of increased sharpness and durability. The drawbacks are high cost and a cutting width limited to 4 mm.

13.4.3 Cutting Metal/Tissue Composites

When studying the pathology of the interaction between living tissue and metal implants such as joint prostheses or intravascular stents, it is frequently necessary to cut sections containing soft tissue in

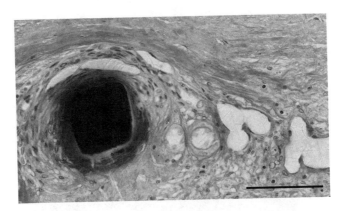

FIGURE 13.4 **(See color insert.)** An example of a plastic-embedded, microground section of a coronary artery section stained with hematoxylin and eosin/phyloxine B and containing a stent strut. Notice the clean interface between the plastic and the metal strut. Multinucleated giant cells surround the strut, while neurovascular buds, scattered macrophages/lymphocytes, and fibrous connective tissue can be seen adjacent to the strut. (Scale bar 100 μm.)

proximity to metals such as stainless steel or titanium (Figure 13.4). A standard technique developed to deal with these problems involves the following steps as shown in Figure 13.5 (Donath, 1985):

- Embed fixed specimen in resin to form a cylindrical block.
- Cut a disk (say, 5 mm in thickness) from the cylindrical block using a diamond-coated band saw.
- Glue disk to plastic base plate, typically with epoxy resin.
- Remove cutting marks by grinding one face of the disk flat using graded grinding paste and polish.
- Glue plastic slide to polished face of the disk.
- Cut the sandwich formed by the base plate, the disk, and the plastic slide with the band saw so that the distance between the plastic slide and the cut edge is slightly greater than the desired section thickness.
- Grind the cut face until the desired section thickness is achieved and polish.

This elaborate and time-consuming process minimizes the risk of crushing the soft tissue or tearing the metal out of the section while preserving its geometry and producing a section of known thickness and good optical quality. Sections as thin as 25 μm may be reliably produced.

To reduce the amount of time necessary to locate and section lesions and areas of interest when assessing the histological effects of implanted medical devices, the technique of microCT becomes invaluable. By using microCT to scan the tissue block after the embedding process, precise measurements can be taken to orient the plane of sectioning with areas of interest. Furthermore, by taking measurements of the entire block, targeted sections can be obtained without needing to section through the entire specimen in order to find the target site. This has the dual benefits of reducing both cost and processing time when employing plastic embedding and microgrinding techniques (Figure 13.6).

13.5 Staining

A biological stain has been defined as "a dye for making biological objects more clearly visible than they would be unstained" (Lillie, 1969). Initially, all dyes were of natural origin, obtained by extraction from plants such as crocus (saffron) (Leeuwenhoek, 1791) cited in Baker (1958b), the tree *Haematoxylon campechianum* (hematoxylin) (Waldeyer, 1863), and the cochineal beetle (carmine) (Goppert and Cohn, 1849), the latter being the first systematically to study dyed tissues with the microscope although, as mentioned above, von Leeuwenhoek certainly employed dyes. The early history of histological staining has been briefly reviewed by Lillie (1969) and Baker (1958b) and in more detail by, for instance, Lewis (1942).

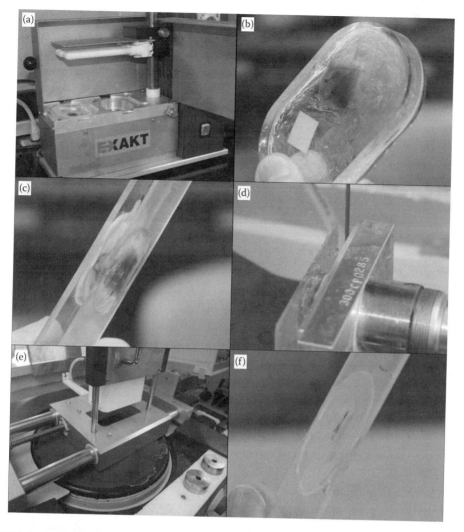

FIGURE 13.5 An illustration of the plastic-sectioning process also referred to as microgrinding or sawing/grinding. The tissue is infiltrated with resin and placed in a Teflon mold to harden in a light polymerizer (panel a). The block (panel b) is then removed from the mold and trimmed into a wafer. A "sandwich" of a supporting slide, the tissue wafer section, and the thin plastic slide are affixed to the face of the block using cyanoacrylate glue (panel c). A precision band saw equipped with a diamond blade is then used to cut the thin slide from the rest of the tissue block (panel d). The final slide is then ground down to the target thickness and polished on a wet rotary polisher (panel e) before it is stained (panel f).

The term *staining* is generally used to include any method of coloring tissue. This may be achieved by using a dye containing a chemical group (or groups) that binds with reactive sites in the tissue. This process is sometimes referred to as *dyeing* (Kiernan, 1990). Alternatively, the tissue may be allowed to absorb a solution of a coloring agent that remains in the tissue when the solvent evaporates in the manner of a coffee stain on a table cloth.

Baker (1958b) defines dyes as "aromatic, salt-like, crystalline solids that dissolve in aqueous solutions in the form of coloured ions which can attach themselves chemically to tissue components. When the attachment takes place they do not lose or change colour." This raises two questions: What makes ions colored? And how do they attach to tissues?

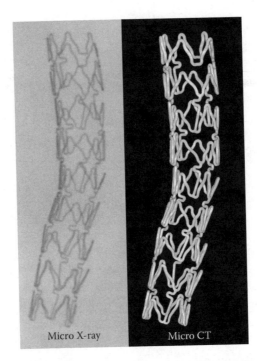

Micro X-ray   Micro CT

FIGURE 13.6 A comparison of the resolution of a microCT reconstruction and a micro x-ray radiograph of the same stented artery specimen. MicroCT is a powerful tool for the evaluation of medical devices *in situ*. From this image, an area of deflection of the stent and overexpansion of the struts in the middle of the stent can be observed and targeted for histology.

The majority of dyes used in histological staining are organic molecules containing conjugated double bonds consisting of electronic orbitals delocalized over several atoms (a chromogen) and polar residues which enable them to form ionic bonds with polar molecules in the tissue (auxochromes). The energy required to raise electrons in the delocalized orbitals to an excited state often corresponds to frequencies in the visible part of the EM spectrum, and most organic compounds with conjugated bonds (for instance, quinoids in which two of the hydrogen atoms on the benzene ring are replaced by an oxygen atom) are therefore colored.

There are many types of chemical bonds formed between the dye and the tissue. For instance, acidic dyes are commonly used to stain basic components in cellular cytoplasm and collagen in connective tissue, whereas basic dyes are more suitable for nucleic acids in the cell nucleus and other acidic moieties such as phospholipids or mucins. Neutral or amphoteric dyes such as hematoxylin require the presence of a *mordant*, usually a metal ion, which has the effect of making the amphoteric dye basic, thus strengthening the bond between the dye and the tissue. Mordants, in general, are able to bond chemically both with the tissue and the dye and are often used to ensure that the dye does not leach out during subsequent treatment such as using a second dye to stain another tissue component.

Perhaps the most frequently used coloring process in histology combines the dyes hematoxylin which, ideally, stains cell nuclei blue, and eosin, the "counterstain," which is taken up by the cytoplasm, rendering it red or pink (Figure 13.7).

Polychrome stains result when three or more dyes are applied sequentially to the same section to give a multicolored preparation. The classic trichrome technique described by Masson (1929) gives striking results, wherein the connective tissue protein collagen stains blue, while muscle cells are red (Figure 13.8), which also shows the results of the martius/scarlet/blue (MSB) stain. In Johansen's quadruple stain for plant tissue, for example, parasitic fungi stain green, cytoplasm stains orange, cellulose appears

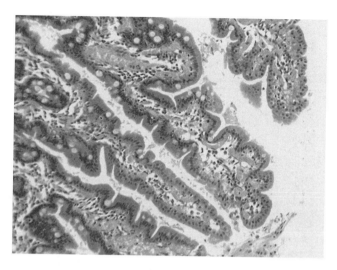

FIGURE 13.7 (**See color insert.**) Section through the lining of the small intestine stained with hematoxylin and eosin.

as a yellowish green, and lignin takes on a red color (Johansen, 1939). Polychrome stains have been reviewed by Culling et al. (1985).

The rubber-like protein elastin, which is unusually hydrophobic, is effectively stained by water-soluble dyes such as orcein or acid fuchsin which contain hydrophobic groups surrounded by molecular clusters of water stabilized by hydrogen bonding. A similar process takes place when the dye Congo red is used to demonstrate the presence of amyloid, a protein formed in the brain and other organs as a result of degenerative disease. When the hydrophobic groups in the dye and on the protein come together, the water clusters are destabilized and the entropy of the system is increased. This process has been termed hydrophobic bonding. This and other types of chemical bonding between the dye and the tissue are summarized in Horobin (1988, 1990).

In wax-embedded material, the wax is dissolved away after mounting the section on the slide and the tissue can then be exposed to a wide variety of aqueous or nonaqueous dye solutions. Resin-embedded material normally cannot be removed in this way, thus the variety of dyes is limited to those that are compatible with the resin used. Nevertheless, the number of resin-compatible dyes continues to increase and several polychrome varieties have been developed (see, e.g., Johansen, 1939; Scala et al., 1993). Acrylic resins, on the other hand, being water-soluble does not suffer from this incompatibility with aqueous dyes.

Other methods of coloring tissue include:

- The staining of lipid by nonpolar dyes such as Sudan Red dissolved in nonaqueous solvents (e.g., isopropanol). The degree of uptake depends on the relative solubility of the dye in solvent and tissue.
- Metallic impregnation used in EM to increase electron density.
- The Gram stain (see, e.g., Lillie, 1965) which exploits the difference between types of bacteria in the solubility of trapped dye molecules. Gram-positive bacteria retain the dye after treatment with iodine and appear blue, whereas Gram-negative organisms appear colorless and may then be visualized by treatment with a counterstain, usually neutral red, to the background. The presence or absence of this stain together with the shape of the bacterium provides the pathologist with a simple method of classifying microorganisms found in infectious diseases and infected wounds.

(a)

(b)

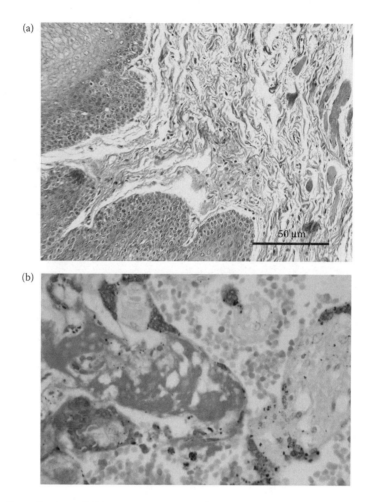

50 µm

FIGURE 13.8 (**See color insert.**) (a) Transverse section of the tongue treated with Masson's trichrome stain, showing muscle in red, connective tissue in blue/green, and cell nuclei in purple. (b) Section of placenta stained with MSB. With this stain, fibrin is red, other connective tissue is blue, and red blood cells are yellow.

- Vital staining in which a dye is taken up by living cells—for example, acridine orange stains intracellular DNA and connective tissue green, and intracellular RNA, red/orange. The dyes can be introduced via the vasculature, injected directly into muscle or, for freshly obtained biopsies, by immersion. Such stains are of particular value to the pathologist to demonstrate cell function in biopsies (Aschoff et al., 1982; Foskett and Grinstein, 1990).

13.5.1 Immunohistochemistry

A major problem with most staining techniques involving dyes is that staining intensity and even color are variable and interpretation depends on the knowledge and skill of the observer. The introduction of immunocytochemistry in the 1950s (Coons and Kaplan, 1950; Coons et al., 1955) largely overcame this, while at the same time introducing a different set of problems. In essence, immunocytochemistry relies on the highly specific reaction between an applied antibody and the tissue constituent to which it binds (antigen). To visualize the point of reaction, the antibody must carry a molecule (label) that enables a colored end product to be formed. Originally, these labels were enzymes, which were then demonstrated by standard histochemical staining techniques already in use for the visualization of enzymes in tissue.

Of those originally tried out, peroxidase and alkaline phosphatase have survived the test of time and are currently the most favored in routine use. In addition to enzymes, which remain the most commonly used labels for LM, fluorescent dyes, colloidal metals, or radioactive isotopes are used (Polak and Van Noorden, 1984).

Initially, a single-labeled primary antibody was used in what has come to be known as the direct technique. This method had the advantage that it was quick to carry out, and nonspecific reactions were minimized as only a single antibody was used. However, it suffered from the fact that each antigenic site had only one colored molecule attached to it giving little signal amplification. Through a number of intermediate methods, the avidin–biotin complex (ABC) techniques most favored today were developed. These rely on the fact that avidin has four binding sites available for reaction with biotin with which it has a high affinity. Biotin can also be coupled to an antibody as well as an enzyme label such as peroxidase.

The first step in carrying out the ABC technique is to apply an unlabeled primary antibody to the antigen of interest. The second step is to apply a secondary antibody labeled with biotin that reacts with the primary. For example, if the first antibody was raised in a rabbit (rabbit anti-antigen), the second antibody could be raised in biotinylated mouse anti-rabbit. The reason for using a secondary antibody rather than directly labeling the primary is that the general technique can then be applied to any antibody, no matter which animal it was raised in, simply by changing the secondary antibody species. The third step is to apply a preformed complex of avidin and peroxidase-labeled biotin. This is prepared such that not all of the biotin-binding sites on the avidin are occupied. These free sites are able to join to the biotin of the secondary antibody and also to other ABC molecules. In this way, the number of peroxidase molecules available for color formation is greatly increased providing the required signal amplification.

One of the major problems with the demonstration of antigenic sites is that they must be well preserved by the fixation process, but also be available for demonstration. This contradiction can be overcome by a number of poorly understood treatments grouped under the heading of antigen unmasking. Initially, enzymatic treatment was employed using enzymes such as trypsin and pronase. These are applied at the start of the technique under conditions that free the antigenic sites without destroying the structure of the remaining tissue. Enzymatic treatment has largely been replaced by heat-mediated antigen-retrieval techniques where the sections are heated in the presence of heavy metal salts or, more often, buffers such as citrate at pH 6.0. Heating is carried out using microwaves or by boiling in a pressure cooker or autoclave. This would seem to be a very harsh treatment but it would appear that antigens survive better using these treatments than when unmasked by enzymes.

Great care must be taken when employing unmasking techniques, as many antigens may be destroyed and larger proteins may be broken down to reveal molecules displaying antigenic sites leading to false-negative and false-positive results, respectively. It is, therefore, very important when introducing a new antibody into the laboratory that adequate tests are carried out using tissues known to contain the antigen for investigation and other tissues in which the antibody is lacking.

13.5.2 *In situ* Hybridization

A further development of the principles of immunocytochemistry has been the introduction of *in situ* hybridization (ISH) for the demonstration of DNA and RNA. This technique relies upon the hybridization of labeled single-stranded fragments of nucleic acid (probes) to complementary strands of nucleic acid located within the tissue sample. This has the distinct advantage over other methods of nucleic acid demonstration, in that it is the only technique that allows for their localization to specific cells within tissues. As with immunocytochemistry, the major problem with this technique is that the nucleic acids are masked in a complex matrix of other tissue elements which have been cross-linked by the process of fixation. In addition, DNA is already masked by being double-stranded, and one of these strands must be removed before hybridization can occur. As with immunocytochemistry, careful fixation preserves more of the nucleic acid of interest and gives better morphology, but also decreases the accessibility of the probe to the tissue. Similarly, mild protease treatment is employed; commonly proteinase K, and again the extent of this treatment must be determined in a series of trial runs.

Probe selection is an important step in carrying out successful hybridization, and a number of different probes are available to suit particular circumstances. Double-stranded DNA probes are available that contain both complementary strands that have been labeled. As there is no way of controlling which of the two strands will be removed from the tissue by the pretreatment, double-stranded probes have the advantage that either will do. However, they suffer from the fact that the two strands of the probe will re-anneal with themselves in solution, thereby reducing their sensitivity. Single-stranded DNA and RNA probes are available, both providing greater sensitivity. However, it is the introduction of oligonucleotide probes that has allowed ISH to mature into a technique available for use in routine laboratories. These are readily synthesized short lengths (normally 20–30 bases) of nucleotide that have the label incorporated during the production process. Their synthesis allows for the production of "designer" probes of known base sequence that can be used against specific regions on nucleic acid present in the tissue sample. They also have the advantage that their small size allows them to easily penetrate the spaces within the fixed tissue. However, because of their short length, the choice of base sequence must be carefully controlled to avoid mismatches occurring with other similar regions on nucleic acid of the target tissue.

An important factor that has made ISH such a useful technique is that the specificity of reaction can be very accurately controlled by varying the conditions under which hybridization is carried out. Thus, length and concentration of the probe, pH, temperature, and buffer composition are all factors that affect the sensitivity and specificity of the reaction. The variation of these factors determines the "stringency" under which the reaction occurs. Reactions carried out at high stringency conditions (high temperature, low salt, high formamide buffer concentrations) ensure that only reactions with high homology are stable. Under low-stringency conditions (low temperature, high salt, low formamide buffer concentrations), some degree of stable mismatching occurs giving nonspecific reactions. One might ask why high-stringency conditions are not employed all of the time. The answer is that, to a large extent, they are incompatible with the aims of keeping tissue sections on the slide and maintaining adequate tissue morphology because, for sufficient hybridization to occur, prolonged periods under harsh conditions are required. It is, therefore, left to the user to decide the level of mismatching that can be tolerated and in most cases moderate conditions over 4–6 h are used, with mismatches being removed by subsequent washing in buffer of high stringency.

Visualization of the label is carried out using the same techniques employed for immunocytochemistry. Currently, probes are mainly labeled with biotin or digoxigenin. The former can be demonstrated by an ABC technique, while the latter requires an antidigoxigenin antibody followed by ABC.

13.5.3 Autoradiography

Autoradiography, a method of locating small quantities of radioactivity in biological material, can trace its origins to the observation in 1896 by Henri Bequerel that uranium and its compounds are able to fog a nearby photographic plate (for an account of its early history, see Rogers, 1973). A radioactively labeled marker chosen for its ability to bind with the tissue or cell under investigation is administered to an experimental animal, a tissue culture or a cell culture system and is taken up by the target cells. The tissue is then fixed, embedded, and sectioned after which the section is placed in close contact with a specially prepared photographic emulsion, consisting of a suspension of silver bromide (AgBr) in gelatine bonded to the surface of the section. Radioactivity from the section is absorbed by the emulsion, producing free bromine ions which, in turn, yield free electrons. These are trapped by defects in the AgBr crystal lattice where they react with silver ions to produce atomic silver. This process is analogous to the exposure to light of a conventional photographic film. The free silver atoms act as nuclei for the formation of more atomic silver and during the development process, grains of silver are formed that are large enough to be detected microscopically. After development, the emulsion is treated with a fixative that dissolves the unreacted AgBr, rendering the emulsion transparent.

In principle, it is possible to measure the amount of radioactivity in the section by counting the silver grains although changes in pressure, temperature, or the effects of additional chemical reactions can all change their size as well as the number formed. To ensure the accuracy and repeatability of quantitative work, great care must be taken to standardize the experimental conditions. Ideally, an internal standard of known activity is processed in tandem with the test tissue or, where possible, actually incorporated with the tissue itself (Flitney, 1991).

When used in conjunction with conventional staining for light or EM, sites within a tissue or cell where a particular metabolic or synthetic process occurs can be related to microscopic structure and calibration is not normally necessary.

13.5.3.1 Isotopes Used

Of the isotopes in common use, which include [^{14}C], [^{35}S], and [^{125}I], tritium [^{3}H] is the most widely used for three reasons. First, hydrogen is found in most molecules of biological interest. Second, [^{3}H] is a β-emitter of low energy and therefore low penetrating power, giving sharper and higher resolution images. Third, it has a half-life of approximately 12 years, so the radioactive intensity does not change appreciably during the course of a typical experiment.

13.5.3.2 Emulsions

There are two general methods of applying emulsion to slides. *Stripping* (Doniach and Pelc, 1950) involves the following steps:

- The section, cut in the normal way, is mounted in the normal way on a slide precoated with gelatine and allowed to dry.
- A piece of emulsion is cut from the glass plate on which it is supplied and floated onto water, where it is left to absorb water for a few minutes.
- The slide and section are dipped into the water and withdrawn so that the emulsion is lifted out and covers the section.
- After drying, the emulsion is bonded to the slide and remains in close contact with the radioactive tissue.

Dipping, introduced in 1955 (Joftes and Warren, 1955), involves dipping the slide and the attached specimen into liquid emulsion, withdrawing it, allowing the excess liquid to drain off, and drying the preparation by evaporation. The advantages of dipping over stripping (Flitney, 1991) are:

- Closer contact between section and emulsion.
- Better control of AgBr crystal size and therefore resolution.
- Easier to make thin emulsion layers and hence better staining.
- Speed and ease of preparation lending itself to partial automation. The main drawback is the difficulty in producing an emulsion layer of uniform thickness, limiting the accuracy and repeatability of quantitative work.

For EM resin, sections are mounted on ultraclean microscope slides, pretreated with a very thin layer of emulsion (Salpeter and Bachmann, 1964) and stored during exposure. Owing to the thinness of the emulsion exposure times, approximately 10 times longer than comparable preparations for LM are necessary. The autoradiographs are developed and fixed on the slide and stained. Finally, they are transferred to electron microscope grids and coated with a 5-nm layer of carbon to minimize chemical reaction between the tissue and the emulsion.

13.5.3.3 Resolution

Resolution has been defined empirically as the distance in the plane of the section from the radioactive source at which the grain density is half its value directly above the source (Doniach and Pelc, 1950) or the radius of a circle centered on the source which contains half the grains produced by the source

(Bachmann and Salpeter, 1965) or, similarly, as the distance from a linear source of a strip parallel to the source which contains half the grains associated with the source (Salpeter et al., 1969). Factors affecting resolution include the size of the AgBr crystals and the thickness of the emulsion as well as the energy of the radioactive particles that determines the distance they penetrate into the emulsion. [³H], which emits β-particles with energies up to 18 keV, has, under optimal conditions, a resolution of 0.5 μm, whereas a higher energy emitter such as [¹⁴C] or [³⁵S] will have a resolution of 2–5 μm (Flitney, 1991).

13.5.3.4 Sensitivity

Sensitivity depends on emulsion thickness and *efficiency*, the fraction of β-particles captured by the emulsion and giving rise to detectable grains (typically 60–80% of those traveling toward the emulsion). As the thickness of the emulsion increases, the probability of a particle hitting an AgBr molecule, and thus ultimately producing a silver grain, increases. In general, increased sensitivity implies decreased resolution so, in practice, a compromise must be sought for individual experiments. Under ideal conditions, approximately 10^{-9} μCi can be detected. This is equivalent to around one disintegration of a [³H] atom per day. At this rate, however, exposure times of several weeks would be required, assuming that approximately 100 disintegrations are needed to produce a grain visible under the light microscope (Rogers, 1973).

13.5.3.5 Tissue Fixation

The aims of fixation, preservation of tissue structure, and no interference with subsequent staining are similar to those of conventional histology with the additional need of maintaining the resolution and sensitivity of the emulsion. Formalin and glutaraldehyde desensitize the emulsion, while fixatives that do not, such as methanol, tend to harden the tissue and disrupt its morphology (Flitney, 1991).

13.5.3.6 Staining

If staining is carried out before applying the photographic emulsion, care must be taken to compensate for loss of staining intensity during development and photographic fixation, whereas if the staining is performed after the autoradiography, stains must be chosen so as not to affect the stability of the silver grains. Suitable stains for each alternative are listed in Flitney (1991).

The techniques of immunohistochemistry and ISH may be combined with autoradiography to map the distribution of a particular antigen (Beckman et al., 1983), rates of cellular proliferation (Lacy et al., 1991), or the phase of DNA replication (Lockwood, 1980). However, as the range of available antibodies increases and their specificity improves, purely immunological techniques are finding favor over the combined approach, a tendency that is encouraged by the technical complexity of autoradiography as well as radiation safety issues.

13.6 Mounting

Once the stained tissue has been picked up onto a slide, it must be dehydrated and cleared in much the same way as the aqueous fixative was removed before embedding. Thus, water remaining from the staining process is removed by treatment with successively more concentrated solution of a water-miscible organic solvent such as ethanol or acetone, taking care to do this rapidly to avoid leaching out those stains that are soluble in these agents. Finally, ethanol or acetone is removed by treating with a transition agent such as xylene after which the section on the slide is ready for the application of the mounting agent, followed by a protective glass coverslip. Most resins used for embedding are only sparingly soluble in noncorrosive solvents in which case clearing is not attempted and the mounting agent is applied to the intact section.

The purpose of the mounting agent is to seal the space between the slide and the coverslip, keeping out air and moisture, and to "fine-tune" the optical properties of the entire preparation. With stained tissue, spherical aberration is minimized if the RI of the mounting agent is close to that of glass. For

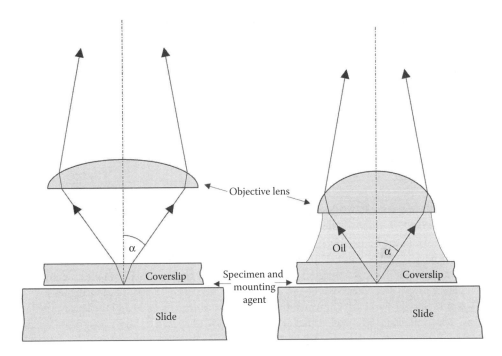

FIGURE 13.9 Ray diagrams for dry and oil immersion objectives showing how the oil (of RI similar to that of the glass coverslip) increases the "half-angle of acceptance" (α) and hence the resolving power of the lens.

unstained tissue, on the other hand, contrast is enhanced by choosing a mounting agent with an RI different to that of glass. The most common mounting agents for stained tissue are Canada Balsam and DPX, an artificial resin introduced in 1941 (Kirkpatrick and Lendrum, 1941) with RIs of 1.52, close to that of glass. More recently, methacrylate-mounting resins have been introduced that polymerize on exposure to light, do not undergo shrinkage, and do not cause fading of the stain with time (Silverman, 1986).

Aqueous mounting agents include Apathy's medium (RI 1.52), Farrant's medium (RI 1.42), and glycerol jelly (RI 1.4–1.47). Details of their composition and preparation are given in Sanderson (1994).

Figure 13.9 (left-hand side) shows that the presence of the coverslip between the section and the objective lens of the microscope can cause additional refraction at the glass/air interface. The resulting spherical aberration is only a significant problem for lenses with a numerical aperture greater than 0.5 (see below). These lenses are normally designed for use with coverslips of a standard thickness (0.17 mm). The effect of deviations from these standard conditions has been investigated by Rawlins (1992) and Pluta (1988). Objective lenses fitted with an adjustable *correction collar* in which the spacing between the component lenses is adjustable may also be used to compensate for variations in coverslip thickness or depth of the mounting medium (White, 1974).

13.7 Microscopy

13.7.1 Resolution

The most obvious function of microscopy is to produce a *magnified* image of a specimen. Of equal importance is the ability to resolve two closely spaced objects. Resolution or minimum resolved distance is "the least separation between two points at which they may be distinguished as separate" (Bradbury and Bracegirdle, 1998). The healthy human eye with a near point of 250 mm can resolve two objects 70 μm apart. To achieve higher resolution, a lens or lenses are required to reduce the near point. The resolution (*r*) of a lens that is restricted by interference between the light diffracted by two closely separated points on the object is expressed by

$$r = \frac{\lambda}{2NA}$$

(13.2)

where *l* is the wavelength of the illumination and *NA* the numerical aperture of the lens is given by the expression

$$NA = n\sin\alpha$$

(13.3)

where *n* is the RI of the medium between the object and the lens, and α is the half-angle of acceptance of the lens (Figure 13.9). Thus, resolving power may be increased by using shorter-wavelength illumination or by increasing the numerical aperture of the lens. Modern high-power objective lenses designed for use in air have acceptance angle approaching 90° and numerical apertures up to 0.95. In practice, the diffraction of light passing through the glass coverslip over the specimen into the air above reduces the acceptance angle. This effect can be reduced and the resolution increased by using an *immersion lens* (Figure 13.9, right-hand side) in which the space between the coverslip and the lens is filled with a medium such as oil whose RI matches that of the coverslip ($\approx$1.5).

13.7.2 Illumination

Most histological specimens viewed through the light microscope are illuminated by transmitted light, some of which is absorbed by dyes chosen to stain particular structures, which then appear as dark areas against a light background. When viewing living tissues or other materials that cannot be stained and that have an RI similar to that of the medium in which they are suspended or embedded, the structure is difficult to visualize (Figure 13.10a). To avoid this problem, *dark ground* or *dark field* illumination must be used. In this mode, the light source is prevented from entering the microscope objective lens directly by a suitably shaped mask. Thus, only light that has been reflected or refracted by parts of the specimen can pass through the objective and hence reach the eyepiece. In this way, the specimen appears as a light image against a dark background (Figure 13.10b). The main drawback of this technique is loss of fine detail and poor contrast.

Interference microscopy overcomes this problem by splitting the light beam into two different path lengths arranged so that the observed field in the absence of a specimen consists of closely spaced interference fringes. The presence of the specimen in the beam path will cause slight shifts in the phase of one or both beams, and the interference pattern is correspondingly changed.

Phase-contrast microscopes achieve a similar effect by passing an annular light beam through the specimen which is then shifted in phase by a similarly shaped *phase plate* in the objective lens system. Light that has been refracted by the specimen does not pass through the phase plate, and when combined with the phase-shifted beam produces an interference pattern which corresponds to the shape and structure of the specimen (Figure 13.10c).

In polarized LM, the birefringent properties of materials such as protein, bone, and lipid are exploited to enhance contrast with or without staining. A plain polarizing filter is positioned between the light source and the specimen, and a similar filter is placed between the objective and eyepiece lenses. One of the filters is rotated so that its plane of polarization is nearly at right angles to the other, giving a dark background in the field of view. The birefringent components of the specimen, having rotated the plane of polarization, will therefore appear lighter than the background (Figure 13.11). Both phase contrast and polarized light techniques allow visualization of unstained and or living tissue, although at the expense of high-contrast images.

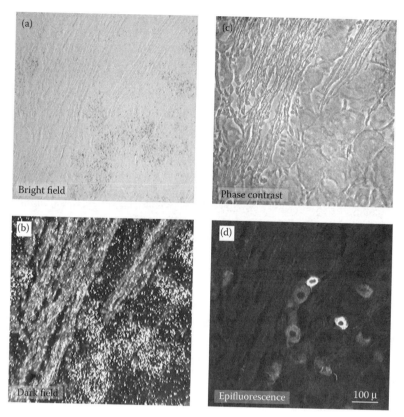

FIGURE 13.10   A brain section (10-μm thick cryostat section) is shown visualized with four different types of illumination. The section contains neuronal cell bodies and myelinated fiber tracts, and the neurons have been marked with two different specialized histochemical stains. One is an immunoflourescence marker which identifies a neuropeptide and the other is a radiolabeled oligonucleotide probe which identifies a receptor mRNA. The probe is revealed using the technique of autoradiography (see text). Under normal bright-field illumination (a), the tissue is difficult to see and the silver grains are just visible as faint dots. Phase-contrast illumination (b) reveals the tissue, and dark-field illumination (c) reveals the silver grains as bright white dots against a dark background (it also reveals the myelinated fiber tracts). Epifluorescence (d) illumination reveals the immunoflourescence marker. (Courtesy of Professor J.V. Priestley.)

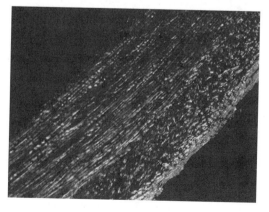

FIGURE 13.11   **(See color insert.)** Transverse section of a porcine carotid artery viewed under polarized light. The birefringent collagen fibers appear bright against a dark background.

13.7.3 Fluorescence Microscopy

Fluorescence is defined as the absorption of electromagnetic radiation of a specific wavelength (excitation) and its reemission at a longer wavelength. When a specimen containing naturally fluorescent material (e.g., vitamin A, chlorophyll, collagen), or one stained with a fluorescent compound having a specific affinity for a component of interest (i.e., a *fluorochrome*), is illuminated by visible or ultraviolet (UV) light of the appropriate wavelength, the fluorescence, which is generally much less intense than the exciting light, is normally masked by the background illumination. Common fluorochromes such as rhodamine or fluorescein are excited by UV light and emit in the visible part of the EM spectrum. The shorter wavelength of UV light improves the resolving power of the system (Equation 13.2). In early fluorescence microscopes, the specimen was illuminated by transmission in the conventional manner and the exciting light as well as background illumination was attenuated by suitable bandpass filters. In modern fluorescence microscopes, the specimen is normally visualized by incident light or *epi-illumination* (Figure 13.10d). In such instruments, the light is directed via a filter and a dichroic mirror (a mirror that reflects light of some wavelengths and transmits at others) through the objective and the fluorescence returns via a second filter through the mirror and into the eyepiece. By careful choice of the filter and the mirror, it is possible to visualize low-intensity fluorescence at a wavelength close to that of the exciting light.

In the technique of immunofluorescence originally reported in 1941 (Coons et al., 1941; Weller and Coons, 1954), a fluorescent stain is combined with an antibody having an affinity for an antigen in the specimen. The advantages of immunological specificity, fluorescent sensitivity, and short wavelength resolving power are thus combined.

13.7.4 Confocal Microscopy

A major drawback of optical microscopy, especially at high magnification, is the low depth of focus and the difficulty of visualizing objects below the surface of the section. For instance, at an overall magnification sufficient to examine a cell nucleus, say, 300¥, objects more than a few microns from the focal point will not be clearly resolved. In 1961, Minsky (1961) patented a device in which a pinhole is placed between the objective and the eyepiece, and the object is illuminated by a point source produced by a second pinhole in the light path. The pinholes allow light originating from object lying within or close to the focal plane to pass through, while blocking light from points remote from this plane. An image is then synthesized by moving the specimen in a raster pattern or by tandem scanning of the pinholes. The first successful use of such a system was reported by Egger and Petran (1967) who were able to examine unstained neural tissue; since then, neuroscientists have remained the most frequent and enthusiastic users of confocal microscopy (Fine et al., 1988). Commercial development was encouraged by the availability of compact and relatively low-cost lasers that provided a source of light sufficiently intense and compact to overcome the low image brightness inherent in any pinhole device. In modern instruments, which usually incorporate epi-illumination suitable for fluorescence techniques, the laser beam is scanned over the object (Carlsson et al., 1985) and the final image is captured by a high-definition video camera interfaced to a PC equipped with copious video memory. By moving the pinhole along the optical axis of the microscope, the position of the focal plane can be changed and a 3D tomographic image can be constructed. Integral software allows these images to be reconstructed, displayed from different viewpoints, and animated. With a combination of fluorescence induced by high-intensity excitation and high-sensitivity CCD cameras, objects as much as 600 μm below the surface of a specimen can be visualized.

More recently, it has been reported that by using a modified form of confocal fluorescence microscopy, structures smaller than the resolution limit imposed by Equation 13.3 may be resolved (Dyba and Hell, 2002), although this claim has been questioned (Stelzer, 2002). Nevertheless, there is little doubt that technical improvements will continue to extend the scope and utility of confocal microscopy for the foreseeable future.

13.7.5 Electron Microscopy

By the end of the nineteenth century, the resolving power of the light microscope was approaching its theoretical limit, that is, half the wavelength of blue light or approximately 0.2 μm (Equation 13.2). In 1931 (Kaiser, 1985), the introduction of the electron beam as a source of "illumination" led to the development of transmission electron microscopes (TEMs) with a resolution approaching 0.1 nm, with which it is possible to resolve the structure of intracellular organelles and even that of large molecules. In the TEM, electron-dense regions of the specimen scatter the electron beam, whereas electrons passing through the specimen are focused onto a fluorescent screen or photographic plate. The electron-dense regions thus appear dark against a light background. The degree of scatter depends on the energy of the electron beam, the atomic number of the scattering atoms, and the thickness of the specimen.

Scanning electron microscopy (SEM), developed commercially in the 1960s, has become an important tool for medical device evaluation owing to the pressing need to observe the surface morphology of the tissue-covering response to several types of medical implants. There are two types of SEM methodologies depending on the specimen chamber used: high-vacuum and low-vacuum chambers. Goodhew et al. (2001) have thoroughly reviewed SEM techniques, and a discussion of the differences in specimen preparation and SEM functionality can be found in Goldstein et al. (2003). In the SEM (Oatley, 1972), which has a lower resolution (5–10 nm) but a greater depth of focus, the specimen is scanned by the beam with a spot size of less than 10 nm using two pairs of electromagnetic deflection coils in a raster pattern and information is obtained primarily from scattered electrons and those produced by secondary emission. Unlike TEMs, SEM scopes utilize a detector that is located above the specimen rather than beneath it. Thus, it is possible to study surface details (from the outermost 1 μm) of a solid specimen whatever its thickness and to construct a pseudo-3D image of the surface. The difference between the size of the raster on the specimen and the display image on the screen is used to calculate the final magnification.

In addition to the scattered and secondary emission electrons, the primary beam gives rise to x-rays, the spectra of which are characteristic of the atoms from which they are emitted. By combining this information with the scattered electron signal, the spatial distribution of specific elements in the specimen may be mapped.

13.7.5.1 Specimen Preparation

Modern low-vacuum scanning electron microscopes (LVSEMs) require minimal specimen preparation compared to their high-vacuum SEM counterparts. Preparation of the specimen is similar to that required for LM and TEM. The specimen is fixed in 2% paraformaldehyde (w/v) and 2% glutaraldehyde (v/v). After fixation, the specimen is dehydrated with successively more concentrated solutions of ethyl alcohol in water. When absolute alcohol concentration is reached (i.e., 4% water), the specimen and device can be dried at room temperature and pressure or in a vacuum desiccator. The specimen is then affixed to an aluminum stub using double-sided copper or graphite tape after which it can be imaged.

13.7.5.2 Advantages of LVSEMs Over High-Vacuum (Conventional) SEMs

LVSEMs can image specimens in a chamber pumped to 18–40 Pa, while high-vacuum SEMs require pressures of 1×10^{-4} Pa. The ability to operate a scanning electron device at relatively high pressures is achieved by differential pumping so that the electron detector is maintained at a lower pressure than the specimen. Furthermore, the development of detectors capable of operating in a humid environment can, in some cases, avoid the need to dehydrate the specimens. Conversely, the extremely low pressure required by high-vacuum SEMs makes additional specimen processing such as sputter coating with gold or graphite and critical point drying essential to obtain good-quality images and to avoid the destruction of the tissue by the high vacuum. The advantage of LVSEMs is that after analysis (Figure 13.12), the specimen can be rehydrated and processed for light and/or TEM (Clubb et al., 2003). The coating and drying processes or standard SEM prevent the use of additional processing methodologies

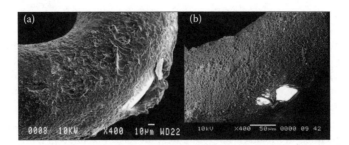

FIGURE 13.12 A comparison of high-vacuum SEM (a) and LVSEM (b) showing that LVSEM produces images of comparable quality to traditional high-vacuum techniques. However, the LVSEM specimen did not need the additional processing and coating procedures required by the conventional high-vacuum scanner on the left. It is therefore still in a condition for additional specimen evaluation; this includes LM and/or TEM.

such as plastic embedding and conventional histology because the gold or graphite coating is permanent and interferes with sectioning and embedding.

Owing to their expense, electron microscopes are not widely used in diagnostic histopathology, although they are increasingly being called upon to provide specialist diagnostic information.

References

Altman, L.G., Schneider, B.G., and Papermaster, D.S., Rapid embedding of tissues in Lowicryl K4M for immunoelectron microscopy, *J. Histochem. Cytochem.*, 32, 1217–1223, 1984.

Arnolds, W.J., Oriented embedding of small objects in agar-paraffin, with reference marks for serial section reconstruction, *Stain Technol.*, 53, 287–288, 1978.

Aschoff, A., Fritz, N., and Ilert, M., Axonal transport of fluorescent compounds in the brain and spinal cord of cat and rat, in *Axonal Transport in Physiology and Pathology*, Weiss, D.G. and Gorio, A., Eds., Springer, Berlin, 1982, p. 177.

Bachmann, L. and Salpeter, E.E., Autoradiography with the electron microscope. A quantitative investigation, *Lab. Invest.*, 14, 1041–1051, 1965.

Baker, J.R., Experiments on fixation, Unpublished observations, 1958a.

Baker, J.R., *Principles of Biological Microtechnique*, 1st ed., Methuen & Co. Ltd., London, 1958b.

Bancroft, J.D. and Stevens, A., *Theory and Practice of Histological Techniques*, Churchill Livingstone, Edinburgh, 1990.

Beckman, W.C.J., Stumpf, W.E., and Sar, M., Localisation of steroid hormones and their receptors. A comparison of autoradiographic and immunocytochemical techniques, in *Techniques in Imunocytochemistry*, Bullock, G.R. and Petrusz, P., Eds., Academic Press, London, 1983.

Bennett, H.S., Wyrick, A.D., Lee, S.W., and McNeil, J.H., Science and art in preparing tissues embedded in plastic for light microscopy, with special reference to glycol methacrylate, glass knives and simple stains, *Stain Technol.*, 51, 71–97, 1976.

Berry, C.L., Sosa-Melgarejo, J.A., and Greenwald, S.E., The relationship between wall tension, lamellar thickness and intercellular junctions in the fetal and adult aorta: Its relevance to the pathology of dissecting aneurysm, *J. Pathol.*, 169, 15–20, 1993.

Bibb, C.A., Pullinger, A.G., and Baldioceda, F., Serial variation in histological character of articular soft tissue in young human adult temporomandibular joint condyles, *Arch. Oral Biol.*, 38, 343–352, 1993.

Bradbury, S. and Bracegirdle, B., *Introduction to Light Microsocpy*, Royal Microscopical Society Handbooks, Vol. 42, BIOS Scientific Publishers, Oxford, 1998.

Branchereau, P., Van Bockstaele, E.J., Chan, J., and Pickel, V.M., Ultrastructural characterization of neurons recorded intracellularly *in vivo* and injected with lucifer yellow: Advantages of immunogold-silver vs. immunoperoxidase labeling, *Microsc. Res. Tech.*, 30, 427–436, 1995.

Carlsson, K., Danielsson, P.E., Lenz, R., Liljeborg, A., Majlof, L., and Aslund, N., Three-dimensional micros-copy using a confocal laser scanning microscope, *Opt. Lett.*, 10, 53–55, 1985.

Clubb, F., Jr., Coscio, M.R., Nichols, R., Rossi, A., Stagnoli, L., Sedlik, S., Koullick, E., Bergan, N., Fogt, E., Petersen, M., and McClay, C.B., Low vacuum scanning electron microscopy, a novel technique for integrated microscopic evaluation of implantable devices, *Microsc. Microanal.*, 9, 208–209, 2003.

Coons, A.H. and Kaplan, M.H., Localization of antigen in tissue cells, *J. Exp. Med.*, 91, 1–13, 1950.

Coons, A.H., Creech, H.J., and Jones, R.N., Immunological properties of an antibody containing fluores-cent groups, *Proc. Soc. Exp. Biol. Med.*, 47, 200, 1941.

Coons, A.H., Leduce, E.H., and Connolly, J.M., Studies on antibody production. I. A method for the his-tochemical demonstration of specific antibody and its application to a study of the hyperimmune rabbit, *J. Exp. Med.*, 102, 49–60, 1955.

Culling, C.F.A., Allison, R.T., and Barr, W.T., *Cellular Pathology Technique*, 4th ed., Butterworths, London, 1985.

Dobrin, P.B., Effect of histologic preparation on the cross-sectional area of arterial rings, *J. Surg. Res.*, 61, 413–415, 1996.

Donath, K., The diagnostic value of the new method for the study of undecalcified bones and teeth with attached soft tissue (Sage-Schliff (sawing and grinding) technique), *Pathol. Res. Pract.*, 179, 631–633, 1985.

Doniach, I. and Pelc, S.R., Autoradiographic technique, *Br. J. Radiogr.*, 23, 184–192, 1950.

Drury, R.A.B. and Wallington, E.A., *Carleton's Histological Technique*, 5th ed., Oxford University Press, Oxford, 1980.

Dyba, M. and Hell, S.W., Focal spots of size lambda/23 open up far-field fluorescence microscopy at 33 nm axial resolution, *Phys. Rev. Lett.*, 88, 163901-1–163901-4, 2002.

Egger, M.D. and Petran, M., New reflected-light microscope for viewing unstained brain and ganglion cells, *Science*, 157, 305–307, 1967.

Ellis, R.C., The microtome: Function and design, in *Laboratory Histopathology: A Complete Reference*, Woods, A.E. and Ellis, R.C., Eds., Churchill Livingstone, New York, 1994, pp. 4.4.1–4.4.23.

Fine, A., Amos, W.B., Durbin, R.M., and McNaughton, P.A., Confocal microscopy: Applications in neuro-biology, *Trends Neurosci.*, 11, 346–351, 1988.

Flitney, E., Autoradiography, in *Theory and Practice of Histological Techniques*, Bancroft, J.D. and Stevens, A., Eds., Churchill Livingstone, Edinburgh, 1991, pp. 645–665.

Foskett, J.K. and Grinstein, S., *Non Invasive Techniques in Molecular Biology*, Wiley-Liss, New York, 1990.

Fox, C.H., Johnson, F.B., Whiting, J., and Roller, P.P., Formaldehyde fixation, *J. Histochem. Cytochem.*, 33, 845–853, 1985.

Girardot, J.M. and Girardot, M.N., Amide cross-linking: An alternative to glutaraldehyde fixation, *J. Heart Valve Dis.*, 5, 518–525, 1996.

Goldstein, J.I., Newbury, D.E., Echlin, P., Joy, D.C., Romig, A.D., Jr., Lyman, C.E., Fiori, C., and Lifshin, E. *Scanning Electron Microscopy and X-ray Microanalysis*, Plenum Press, New York, 2003.

Goodhew, P.J., Humphreys, F.J., and Beanland, R., *Electron Microscopy and Microanalysis*, Taylor & Francis, New York, 2001.

Goppert, H.R. and Cohn, F., Uber die Rotation des Zellinhaltes von Nitella fiexilis, *Bot. Ztg.*, 7, 665–719, 1849.

Gordon, K.C., Tissue processing, in *The Theory and Practice of Histological Techniques*, Bancroft, J.D. and Stevens, A., Eds., Churchill Livingstone, Edinburgh, 1990, pp. 44–59.

Hanstede, J.G. and Gerrits, P.O., The effects of embedding in water-soluble plastics on the final dimensions of liver sections, *J. Microsc.*, 131, 79–86, 1983.

Hardy, P.M., Nicholls, A.C., and Rydon, H., The nature of the cross linking of proteins with glutaralde-hyde, *J. Chem. Soc.*, 1, 958–962, 1976.

Helander, K.G., Thickness variations within individual paraffin and glycol methacrylate sections, *J. Microsc.*, 132, 223–227, 1983.

Hilger, H.H. and Medan, D., A simple method for exact alignment of small paraffin embedded specimens to the cutting plane, *Stain Technol.*, 62, 282–283, 1987.

Hooke, R., *Micrographia* (facsimile edition), Royal Society London (1665), Dover, New York, 1961.

Hopwood, D., Fixation and fixatives, in *Theory and Practice of Histological Techniques*, Bancroft, J.D. and Stevens, A., Eds., Churchill Livingstone, Edinburgh, 1990, pp. 21–42.

Horobin, R.W., *Understanding Histochemistry: Evaluation and Design of Biological Stains*, Ellis Horwood, Chichester, 1988.

Horobin, R.W., An overview of the theory of staining, in *Theory and Practice of Histological Techniques*, Bancroft, J.D. and Stevens, A., Eds., Churchill Livingstone, Edinburgh, 1990, pp. 93–105.

Iwadare, T., Mori, H., Ishiguro, K., and Takeishi, M., Dimensional changes of tissues in the course of processing, *J. Microsc.*, 136, 323–327, 1984.

Joftes, D.L. and Warren, S., Simplified liquid emulsion radioautography, *J. Biol. Photogr. Assoc.*, 23, 145–151, 1955.

Johansen, D.A., A quadruple stain combination for plant tissues, *Stain Technol.*, 14, 125–128, 1939.

Kaiser, H.E., Functional comparative histology, *Gegenbaurs Morph Jahrb Lepzig*, 131, 815–862, 1985.

Kendall, P.A., Polak, J.M., and Pearse, A.G., Carbodiimide fixation for immunohistochemistry: Observations on the fixation of polypeptide hormones, *Experientia*, 27, 1104–1106, 1971.

Kiernan, J.A., *Histological and Histochemical Methods: Theory and Practice*, 2nd ed., Pergamon Press, Oxford, 1990.

Kirkpatrick, J. and Lendrum, A.C., Further observations on the use of synthetic resin as a substitute for Canada balsam: Precipitation of paraffin wax in medium and an improved plasticizer, *J. Pathol. Bacteriol.*, 53, 441–443, 1941.

Lacy, E.R., Kuwayama, H., Cowart, K.S., King, J.S., Deutz, A.H., and Sistrunk, S., A rapid, accurate, immunohistochemical method to label proliferating cells in the digestive tract. A comparison with tritiated thymidine, *Gastroenterology*, 100, 259–262, 1991.

Leeuwenhoek, A., *Epistolae Physiologicae Super Compluribus Naturae Arcanis*, Beman, Delft, 1791.

Leong, A.S., Microwave irradiation in histopathology, *Pathol. Annu.*, 23, 213–234, 1988.

Leong, A.S., Microwave techniques for diagnostic laboratories, *Scanning*, 15, 88–98, 1993.

Leong, A.S., Tissue and section preparation, in *Laboratory Histopathology: A Complete Reference*, Woods, A.E. and Ellis, R.C., Eds., Churchill Livingstone, New York, 1994, pp. 4.1.1–4.1.26.

Leong, A.S. and Duncis, C.G., A method of rapid fixation of large biopsy specimens using microwave irradiation, *Pathology*, 18, 222–225, 1986.

Leong, A.S., Daymon, M.E., and Milios, J., Microwave irradiation as a form of fixation for light and electron microscopy, *J. Pathol.*, 146, 313–321, 1985.

Leong, A.S., Milios, J., and Duncis, C.G., Antigen preservation in microwave-irradiated tissues: A comparison with formaldehyde fixation, *J. Pathol.*, 156, 275–282, 1988.

Lewis, F.T., The introduction of biological stains: Employment of Saffron by Vieussens and van Leeuwenhoek, *Anat. Rec.*, 83, 229–253, 1942.

Lillie, R.D., *Histopathologic Technic and Modern Histochemistry*, 2nd ed., McGraw-Hill, New York, 1965.

Lillie, R.D., *H.J. Conn's Biological Stains*, 8th ed., Williams & Wilkins, Baltimore, 1969.

Litwin, J.A., Light microscopic histochemistry on plastic sections, *Prog. Histochem. Cytochem.*, 16, 1–84, 1985.

Lockwood, A.H., Immunofluorescence radioautography. Simultaneous visualization of DNA replication and supramolecular antigens in individual cells, *Exp. Cell Res.*, 128, 383–394, 1980.

Login, G.R. and Dvorak, A.M., Microwave energy fixation for electron microscopy, *Am. J. Pathol.*, 120, 230–243, 1985.

Masson, P., Some histological methods. Trichrome stainings and their preliminary technique, *Bull. Int. Assoc. Med.*, 12, 75–90, 1929.

Medewar, P.B., The rate of penetration of fixatives, *J. R. Microsc. Soc.*, 61, 46–57, 1941.

Merriam, R.W., Determination of section thickness in quantitative microspectrophotometry, *Lab. Invest.*, 6, 28–43, 1957.

Minsky, M., Microscopy Apparatus, U.S. Patent No. 3013467, 1961.

Murray, G.I., Is wax on the wane? *J. Pathol.*, 156, 187–188, 1988.

Nettleton, G.S. and McAuliffe, W.G., A histological comparison of phase-partition fixation with fixation in aqueous solutions, *J. Histochem. Cytochem.*, 34, 795–800, 1986.

Nunn, R.E., Microtomy, staining and specialized techniques, Butterworths Laboratory aids, in *Electron Microscopy*, Baker, F.J., Ed., Butterworths, London, 1970a.

Nunn, R.E., Preparation of biological specimens, Butterworths Laboratory aids, in *Electron Microscopy*, Baker, F.J., Ed., Butterworths, London, 1970b.

Oatley, C.W., *The Scanning Electron Microscope*, Cambridge University Press, London, 1972.

Panula, P., Happola, O., Airaksinen, M.S., Auvinen, S., and Virkamaki, A., Carbodiimide as a tissue fixative in histamine immunohistochemistry and its application in developmental neurobiology, *J. Histochem. Cytochem.*, 36, 259–269, 1988.

Pluta, M., *Advanced Light Microscopy. Principles and Basic Properties*, Vol. 1, Elsevier, Amsterdam, 1988.

Polak, J.M. and Van Noorden, S., An introduction to immunocytochemistry: Current techniques and problems, in *Royal Microscopy Society Handbook*, Vol. 111, Oxford University Press, Oxford, 1984.

Rawlins, D.J., *Light Microscopy*, Bios Scientific, Oxford, 1992.

Reale, E. and Luciano, L., Fixation with aldehydes. Their usefulness for histological and histochemical studies in light and electron microscopy, *Histochemie*, 23, 144–170, 1970.

Reid, N. and Beesley, J.E., Sectioning and cryosectioning for electron microscopy, in *Methods in Electron Microscopy*, Glauert, A.M., Ed., Elsevier, Amsterdam, 1991.

Rogers, A.W., *Techniques of Autoradiography*, Elsevier, London, 1973.

Ross, K.F.A., Cell shrinkage caused by fixatives and paraffin wax embedding in ordinary cytological preparations, *Q. J. Microsc.*, 94, 125–139, 1953.

Rostgaard, J., Qvortrup, K., and Poulsen, S.S., Improvements in the technique of vascular perfusion fixation employing a fluorocarbon-containing perfusate and a peristaltic pump controlled by pressure feedback, *J. Microsc.*, 172, 137–151, 1993.

Saito, T. and Keino, H., Acrolein as a fixative for enzyme cytochemistry, *J. Histochem. Cytochem.*, 24, 1258–1269, 1976.

Sallee, C.J. and Russell, D.F., Embedding of neural tissue in agarose or glyoxyl agarose for vibratome sectioning, *Biotechnol. Histochem.*, 68, 360–368, 1993.

Salpeter, E.E. and Bachmann, L., Autoradiography with the electron microscope, *J. Cell. Biol.*, 22, 469–474, 1964.

Salpeter, M.M., Bachmann, L., and Salpeter, E.E., Resolution in electron microscope radioautography, *J. Cell. Biol.*, 41, 1–32, 1969.

Sanderson, J.B., *Biological Microtechnique*, Royal Microscopical Society, Microscopical Handbooks, Vol. 28, Bradbury, S., Ed., BIOS Scientific Publ., Oxford, 1994.

Sato, Y., Mukai, K., Watanabe, S., Goto, M., and Shimosato, Y., The AMeX method. A simplified technique of tissue processing and paraffin embedding with improved preservation of antigens for immunostaining, *Am. J. Pathol.*, 125, 431–435, 1986.

Scala, C., Preda, P., Cennacchi, G., Martinelli, G.N., Manara, G.C., and Pasquinelli, G., A new polychrome stain and simultaneous methods of histological, histochemical and immunohistochemical stainings performed on semithin sections of Bioacryl-embedded human tissues, *Histochem. J.*, 25, 670–677, 1993.

Shorter Oxford Dictionary, 3rd ed., Oxford University Press, Oxford, 1973.

Silverman, M., Light-polymerizing plastics as slide mounting media, *Stain Technol.*, 61, 135–137, 1986.

Sims, D.E. and Horne, M.M., Non-aqueous fixative preserves macromolecules on the endothelial cell surface: an *in situ* study, *Eur. J. Morphol.*, 32, 59–64, 1994.

Sims, D.E., Westfall, J.A., Kiorpes, A.L., and Horne, M.M., Preservation of tracheal mucus by nonaqueous fixative, *Biotechnol. Histochem.*, 66, 173–180, 1991.

Spurr, A.R., A low-viscosity epoxy resin embedding medium for electron microscopy, *J. Ultrastruct. Res.*, 26, 31–43, 1969.

Stelzer, E.H., Beyond the diffraction limit? *Nature*, 417, 806–807, 2002.

Tandler, B., Improved slides of semithin sections, *J. Electron. Microsc. Tech.*, 14, 285–286, 1990.

Tymianski, M., Bernstein, G.M., Abdel-Hamid, K.M., Sattler, R., Velumian, A., Carlen, P.L., Razavi, H., and Jones, O.T., A novel use for a carbodiimide compound for the fixation of fluorescent and nonfluorescent calcium indicators *in situ* following physiological experiments, *Cell Calcium*, 21, 175–183, 1997.

Vincent, J.F.V., Automating the microtome, *Microsc. Anal.*, 23, 19–21, 1991.

Waldeyer, R., Untersuchungen uber den Ursprung und den Verlauf des Axencylinders bei Wirbellosen und Wirbelthieren sowie uber dessen Endverhalten in der querdestreiften Muskelfaser, *Henle Pfeiffer Ztg Ration Med.*, 20, 193–256, 1863.

Weller, T.H. and Coons, A.H., Fluorescent antibody studies with agents of varicella and Herpes Zoster, *Proc. Soc. Exp. Biol. Med.*, 86, 789, 1954.

White, G.W., The correction of tube length to compensate for coverglass thickness variations, *Microscopy*, 32, 411–420, 1974.

Willingham, M.C. and Yamada, S.S., Development of a new primary fixative for electron microscopic immunocytochemical localization of intracellular antigens in cultured cells, *J. Histochem. Cytochem.*, 27, 947–960, 1979.

Zelander, T. and Kirkeby, S., Vibratome sections of difficult tissues, *Stain Technol.*, 53, 251–255, 1978.

(a)

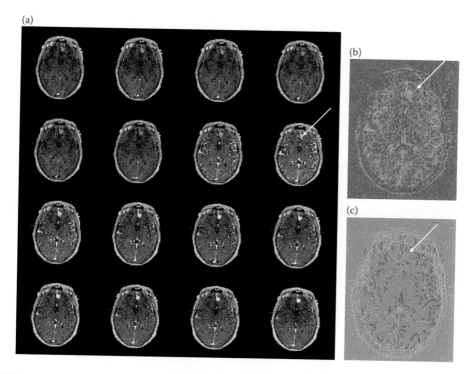

(b)

(c)

FIGURE 6.6 MR permeability assessment. (a) Sixteen of a series of dynamic T_1-weighted images acquired during passage of Gd-based contrast agent in a patient with an intracranial tumor (arrow). Kinetic modeling of the progressive positive enhancement yields estimates of (b) fractional blood volume and (c) microvascular permeability, both of which are elevated in this tumor.

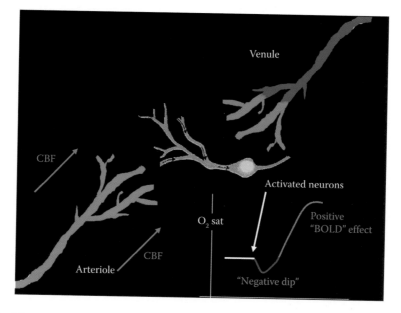

FIGURE 6.7 Schematic representation of the fMRI BOLD contrast mechanism. Neuronal activation leads to an increase in regional cerebral blood flow (CBF), without a concomitant increase in tissue oxygen consumption, leading to an increase in the oxygenation of the capillary bed and postcapillary venules. This increase in oxygenation can be visualized as an increase in signal on T_2^*-weighted images, as paramagnetic deoxyhemoglobin is displaced.

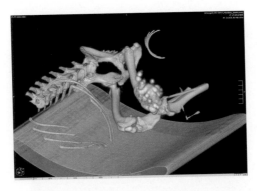

FIGURE 8.30 PET/CT volume rendering of a drug distribution. ^{90}Y is used to treat osteosarcoma in the distal femur of a dog. (Adapted from Zhou, J. et al. 2011. *J. Nucl. Med.*, 52(Supplement 1), 129.) (Texas A&M Institute for Preclinical Studies. With permission.)

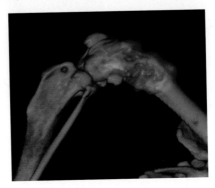

FIGURE 8.38 PET/CT of a canine osteosarcoma. Brightly colored areas indicate high-glucose SUV and correspond physically to the bone lysis and proliferation. (Texas A&M Institute for Preclinical Studies. With permission.)

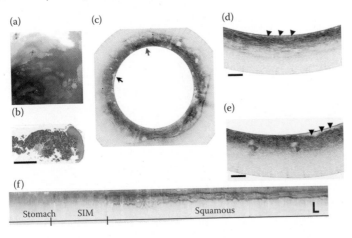

FIGURE 9.15 Barrett's esophagus with dysplasia (a) as viewed with videoendoscopy showing patchy mucosa consistent with SIM. (b) Histology image of biopsy taken from squamocolumnar junction diagnosed with intestinal metaplasia and low-grade dysplasia. (c) Cross-sectional OFDI image showing SIM without dysplasia [arrow, top; expanded in (d)] and with high-grade dysplasia [arrow, left; expanded in (e)]. (f) Longitudinal slice highlights the transition from gastric cardia, through a 9-mm segment of SIM and finally into squamous mucosa. Scale bars: 1 mm. (Reprinted from *Gastrointest Endosc*, **68**(4), Suter, M.J. et al., Comprehensive microscopy of the esophagus in human patients with optical frequency domain imaging, 745–753, Copyright 2008, with permission from Elsevier.)

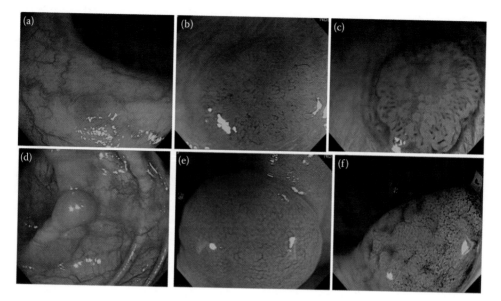

FIGURE 9.16 (a) Conventional white-light nonmagnifying video colonoscopic view of a colorectal IIa neoplastic polyp. (b) Magnifying (×100 magnification) NBI view of neoplastic polyp with remarkable superficial mesh-capillary pattern. (c) Magnifying chromoendoscopy of a IIa neoplastic polyp with indigo carmine showing IIIs and IIIL pit patterns. (d) Conventional view of a Is nonneoplastic polyp. (e) Magnifying NBI view of a Is nonneoplastic polyp without a superficial mesh-capillary pattern. (f) Magnifying chromoendoscopy of a Is nonneoplastic polyp with a type II pit pattern. (Reprinted with kind permission from Springer Science+Business Media: *Int J Colorectal Dis*, Comparative study of conventional colonoscopy, magnifying chromoendoscopy, and magnifying narrow-band imaging systems in the differential diagnosis of small colonic polyps between trainee and experienced endoscopist, **24**(12), 2009, 1413–1419, Chang, C.C. et al.)

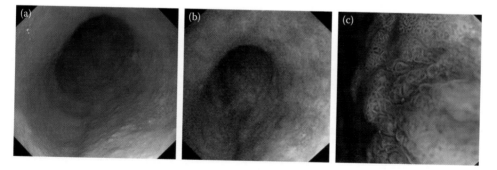

FIGURE 9.17 Images of Barrett's esophagus obtained using the endoscopic trimodality imaging system from a patient with multifocal high-grade dysplasia. (a) Only subtle irregularities are seen under conventional white-light endoscopy. (b) Multiple dark purple spots are visible on a green background with autofluorescence imaging. (c) Detailed imaging of a suspicious area using NBI reveals irregular mucosal and vascular patterns. All autofluorescence positive areas with irregular patterns on NBI contained high-grade dysplasia. (Reprinted from *Best Pract Res Cl Ga*, **22**(4), Curvers, W.L., R. Kiesslich, and J.J.G.H.M. Bergman, Novel imaging modalities in the detection of oesophageal neoplasia, 687–720, Copyright 2008, with permission from Elsevier.)

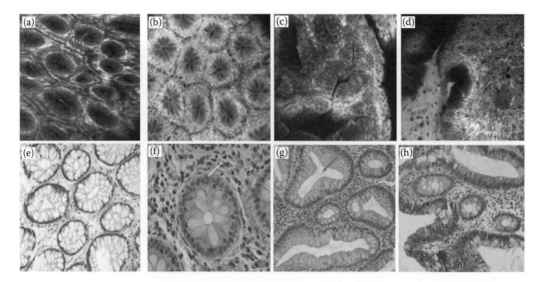

FIGURE 9.18 *In vivo* confocal endomicroscopy images of colon tissue stained with fluorescein. Normal crypt structure imaged (a) within the lamina propria readily identifies single capillaries labeled with fluorescein (red arrow) and (b) closer to the surface with visible mucin in goblet cells presenting as dark subcellular parts (blue arrow). (e) and (f) Corresponding histology. (c) Elongated crypts with corresponding star-shaped luminal openings (red arrow) were confirmed by histology (g) as hyperplastic tissue (red arrow). (d) Ulcerative colitis-associated neoplasia shows distorted crypt architecture with inflammation and dysplastic changes. (e) Histology revealed colitis-associated dysplasia with low-grade intraepithelial neoplasia. (Reprinted by permission from Macmillan Publishers Ltd., *Nat Clin Pract Oncol*, Kiesslich, R. et al., Technology insight: Confocal laser endoscopy for *in vivo* diagnosis of colorectal cancer, **4**(8): 480–490, 2007.)

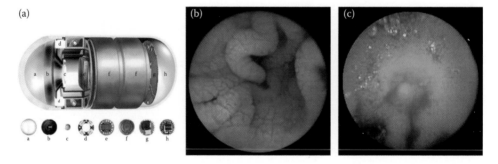

FIGURE 9.19 Wireless capsule endoscopy device and images. (a) The M2A capsule camera consisting of a disposable plastic capsule. The contents include an optical dome [a], a lens holder [b], a short focal-length lens [c], six white light-emitting diode illumination sources [d], a CMOS chip camera [e], two silver oxide batteries [f], a UHF band radio telemetry transmitter [g], and an antenna [h]. (Reprinted by permission from Macmillan Publishers Ltd., *Nat Rev Drug Discov*, Qureshi, W.A., Current and future applications of the capsule camera, **3**(5): 447–450, 2004.) Wireless capsule endoscopy is used in complicated celiac disease to identify mucosal abnormalities such as (b) mosaic pattern of the mucosa or (c) ulceration in the distal jejunum. (Reprinted from *Gastrointest Endosc*, **62**(1), Culliford, A. et al., The value of wireless capsule endoscopy in patients with complicated celiac disease, 55–61, Copyright 2005, with permission from Elsevier.)

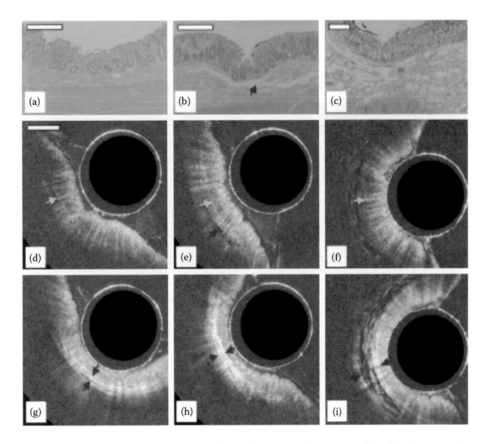

FIGURE 10.5 Sequences of endoscopic OCT (EOCT) images and corresponding histology sections, with increasing adipose tissue content in GI tract submucosa (bars = 0.5 mm). (a) Histologic section with lowest submucosal fat content. (b) Histologic section with medium submucosal fat content shown with the blue arrow. (c) Histologic section with highest submucosal fat content. Different probe pressure was applied while imaging and the effect was investigated. (d) EOCT image of (a) acquired while exerting minimal pressure with probe showing border between mucosa and submucosa (yellow arrow). (e) EOCT image of (b), acquired while exerting minimal pressure with probe, showing mucosa/submucosa (yellow arrow) and adipose band seen in (b) (blue arrow). (f) EOCT image of (c), acquired while exerting minimal pressure with probe, showing border between mucosa and submucosa (yellow arrow). (g) EOCT image of (a), acquired with high-probe pressure; thickness of submucosa indicated by red arrows. (h) EOCT image of (b), acquired with high-probe pressure; thickness of submucosa indicated by red arrows. (i) EOCT image of (c), acquired with high-probe pressure; thickness of submucosa is indicated by red arrows. In (d), (e), and (f), the tissue below the mucosa/submucosa line becomes darker and darker.

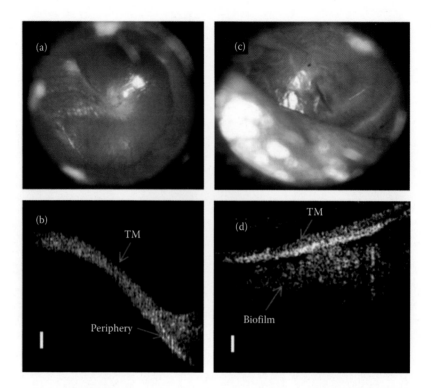

FIGURE 10.6 (a, b) Normal human ear. (a) Video otoscope image of the tympanic membrane (TM). (b) Cross-sectional OCT image of the TM. (c, d) Chronic middle-ear infection with a thick, highly scattering biofilm that is a major source of infection. (c) Video otoscopy image showing a less translucent TM. (d) Cross-sectional OCT image showing the lateral spatial extent of biofilm. Based on the scattering properties of the TM, its thickness was measured for diagnosis. (Scale bar in (b) and (d): 100 μm.) (Adapted from Nguyen, C.T. et al., *Proceedings of the National Academy of Sciences of the United States of America*, 2012. 109(24): 9529–9534. With permission.)

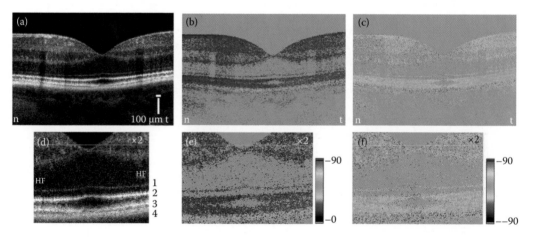

FIGURE 10.8 PS-OCT images of the fovea region of a healthy human subject. (a) Intensity image, (b) retardation image, (c) axis orientation image corresponding to horizontal axis; 2× magnified views are shown in (d)–(f). Numbers from 1 to 4 represent the posterior retinal layers. HF, Henle's fiber layer, n, nasal; t, temporal. (Adapted from Leitgeb, R., C.K. Hitzenberger, and A.F. Fercher, *Optics Express*, 2003. **11**(8): 889–894. With permission.)

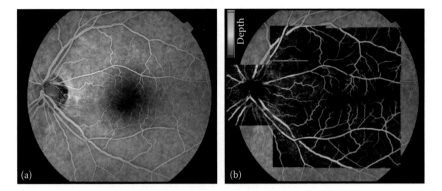

FIGURE 10.9 Large field-of-view stitched pvOCT imaging overlaid on a fundus FA. (a) Fluorescein angiography. (b) Depth color-coded imaging with the 10 volumes. Total image acquisition time for 10 volumetric images is approximately 35 s. (Adapted from Kim, D.Y. et al., *Biomedical Optics Express*, 2011. **2**(6): 1504–1513. With permission.)

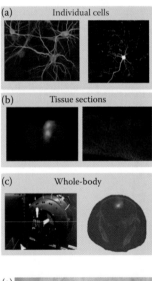

FIGURE 11.3 Fluorescence imaging was employed for *in vitro*, *ex vivo*, and *in vivo* preclinical studies. (a) Individual cells were imaged with two-photon microscopy that shows *in vitro* mouse hippocampal neuron and glial cells. Neurons and glial cells were transfected with green fluorescent protein (GFP) (green) and red fluorescent protein (red), respectively (courtesy of Dr. De Koninck). (b) Tissue sections, *ex vivo* sliced brain tissue sections, were imaged from a mouse that show glioma cells highlighted with fluorescence from GFP (green) and Protoporphyrin IX (PpIX, red). (c) Whole-body fluorescence tomography was employed to spatially map *in vivo* PpIX fluorescence from a brain tumor model in a mouse. (Reprinted from *Journal of Photochemistry and Photobiology B*, 98(1), F. Leblond et al., Preclinical whole-body fluorescence imaging: Review of instruments, methods and applications, 77–94, Copyright 2010, with permission from Elsevier, and Dr. Paul De Koninck, www.greenspine.ca.)

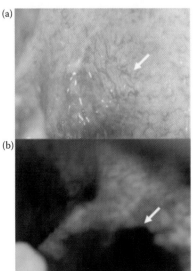

FIGURE 11.7 A pilot study was performed to detect oral malignancy with fluorescence imaging. (a) Under white-light imaging, there is little visible contrast between normal and cancerous tissue. (b) After excitation with 400–460-nm blue light, the endogenous fluorescence image contains a noticeable dark region (white arrow). After histological analysis, this spot was confirmed to be carcinoma *in situ*. (C.F. Poh et al., Direct fluorescence visualization of clinically occult high-risk oral premalignant disease using a simple hand-held device. *Head & Neck*. 2007. 29(1). Copyright Wiley-VCH Verlag GmbH & Co. KgaA. Reprinted with permission.)

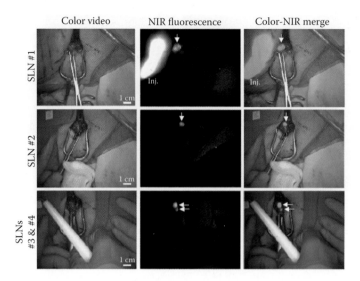

FIGURE 11.9 Images of breast cancer SLN mapping with NIR fluorescence imaging. White-light images (left column) are shown alongside NIR fluorescence (800 nm) images (middle column) after injection (Inj.) of a fluorescent dye. False-color (green) overlays of the NIR fluorescence are shown in the right column. Four SLNs (arrows) were identified for removal from the patient, whereas only three were detected with the current "gold standard," ^{99m}Tc-sulfur colloid lymphoscintigraphy. (Reprinted with permission from Springer Science+Business Media: The Flare((Tm)) Intraoperative near-infrared fluorescence imaging system: A first-in-human clinical trial in breast cancer sentinel lymph node mapping. *Annals of Surgical Oncology*, 16(10), 2009, S.L. Troyan et al.)

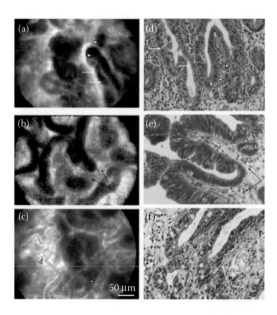

FIGURE 11.10 Barrett's esophagus with early-stage mucosal adenocarcinoma, imaged *in vivo* with confocal microscopic fluorescence endoscopy (a–c) and examined via histopathology using hematoxylin and eosin stains (d–f). White arrows (a, d) indicate the irregularly shaped epithelial lining with variable width, white triangles (a, d) show regions of higher cell density, black arrows (b, e) indicate gland fusion, and gray arrowheads (c, f) show blood vessels that are irregularly shaped and dilated. (Reprinted from H. Pohl et al. Miniprobe confocal laser microscopy for the detection of invisible neoplasia in Patients with Barrett's oesophagus. *Gut* 57(12); 2008. With permission from BMJ Publishing Group Ltd.)

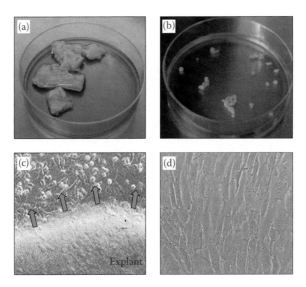

FIGURE 12.2 Isolation of medial collateral ligament (MCL) knee fibroblasts from (a) tissue explant; (b) minced ligament tissue on polystyrene culture dish; (c) MCL fibroblasts migrating out-of-tissue explant; and (d) phase-contrast image of confluent MCL fibroblast monolayer.

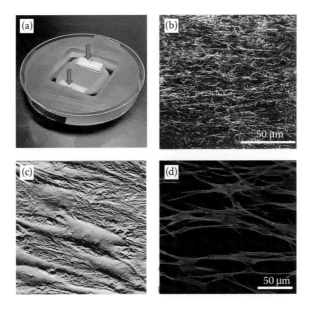

FIGURE 12.4 (a) Polysulfone frame used to culture and mechanically stretch type I collagen gels seeded with cells. Velcro anchors permit mechanical loading via tension applied to sutures (arrows) to study contractile response of dermal fibroblasts to controlled loading conditions. (b) Confocal reflectance microscopy of collagen gel formed by dermal fibroblast alignment (488-nm laser epi-illumination, PMT using blue reflection filter, 60× 1.4 NA oil immersion objective at 1024 × 1024 pixels). (c) Contact mode atomic force microscopy (Bioscope, Digital Instruments), deflection image with 100 × 100 μm scan size, of dermal fibroblasts in 3D collagen gel matrix aligned as above. (d) Confocal microscopy image of rhodamine phalloidin labeling of actin cytoskeleton of dermal fibroblasts that were subjected to uniaxial tension (note parallel alignment with direction of loading). (Courtesy of Kevin D. Costa, Mt. Sinai School of Medicine.)

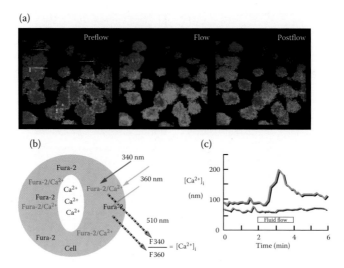

FIGURE 12.5 Real-time intracellular calcium response, $[Ca^{2+}]_i$, of primary bone cells (derived from rat calvaria) subjected to fluid-induced shear stress (35 dynes/cm²), using the calcium indicator dye fura-2 and fluorescence microscopy, and a parallel-plate flow chamber (Allen et al. 2000). (a) Light intensity was recorded with a CCD camera and converted to pseudocolor (blue/green for low levels, orange/red high levels) for visualization and analysis. Olympus software was used to subtract background noise and single cells selected for analysis by boxing or tracing of the cell contour. (b) The cell-permeant acetoxymethyl ester form of the dye, Fura-2 AM, was loaded into living cells using DMSO and Pluronic®-127 in a buffered saline solution. Upon entering, the AM tail is cleaved by endogenous esterase activity, trapping fura-2 within the cytoplasm (Grynkiewicz et al. 1985; Tsien et al. 1985). (c) Fura-2 is a ratiometric dye, which has the advantage of minimizing artifacts of motion, uneven dye loading, and dye leakage. Rapidly alternating 340/360 nm excitation wavelengths are used to elicit 510-nm wavelength emissions from calcium-bound dye and the fura-2 isobestic point (calcium concentration insensitive), respectively (Hung et al. 1995). The ratio of the 510-nm emissions from the $R = 340$ nm/360 nm is compared to a calibration curve, providing a real-time measure to get $[Ca^{2+}]_i$.

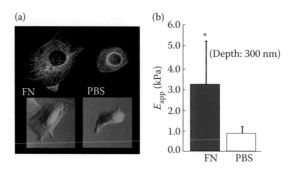

FIGURE 12.7 MC3T3-E1 osteoblast-like cell line cultured on glass slides coated with fibronectin (FN) and control phosphate-buffered saline (PBS) (Takai et al. 2005). (a) Using confocal microscopy, double staining of actin and vinculin was performed to compare the cytoskeletal organization due to the different plating substrates. Cell topography images obtained from atomic force microscope (Digital Instruments) using the scanning mode and (b) complementary modulus data using the indentation mode were also performed (Costa and Yin 1999). A stronger development of cytoskeletal organization in the fibronectin-coated surface appears to correspond with increased spreading and stiffening of the cell.

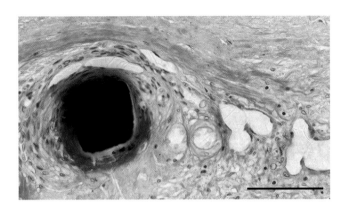

FIGURE 13.4 An example of a plastic-embedded, microground section of a coronary artery section stained with hematoxylin and eosin/phyloxine B and containing a stent strut. Notice the clean interface between the plastic and the metal strut. Multinucleated giant cells surround the strut, while neurovascular buds, scattered macrophages/lymphocytes, and fibrous connective tissue can be seen adjacent to the strut. (Scale bar 100 μm.)

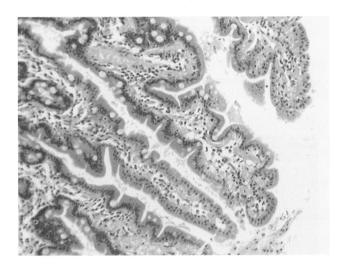

FIGURE 13.7 Section through the lining of the small intestine stained with hematoxylin and eosin.

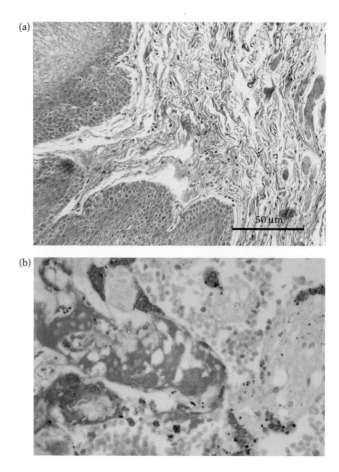

FIGURE 13.8 (a) Transverse section of the tongue treated with Masson's trichrome stain, showing muscle in red, connective tissue in blue/green, and cell nuclei in purple. (b) Section of placenta stained with MSB. With this stain, fibrin is red, other connective tissue is blue, and red blood cells are yellow.

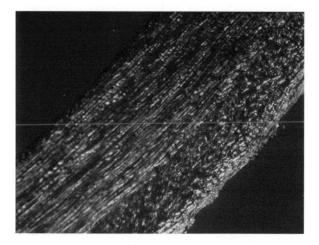

FIGURE 13.11 Transverse section of a porcine carotid artery viewed under polarized light. The birefringent collagen fibers appear bright against a dark background.

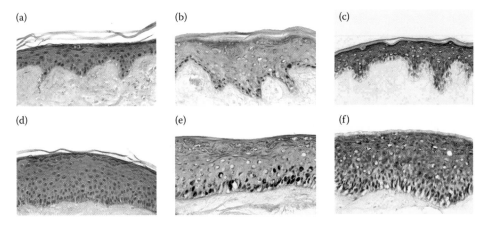

FIGURE 17.3 KGF induces hyperproliferation and delays differentiation in a skin-equivalent model system. Sections of skin equivalents unmodified (a, b, c) and genetically modified to express KGF (d, e, f) were cultured for 7 days at the air/liquid interface. Cryosections were stained with hematoxylin and eosin (H&E; a, d), a nuclear proliferation antigen (Ki67; brown; b, e), and the differentiation marker keratin 10 (K10; brown; c, f). Ki67 and K10 sections were counterstained with hematoxylin (blue) (magnification 40×). Note the increased thickness and basal cell density of the KGF-modified tissues. Also note that in unmodified tissues, proliferation is confined to the basal layer, while in KGF-modified tissues, proliferating cells extend up to three to four suprabasal layers. The same suprabasal layers that contain proliferating cells are also negative for the differentiation marker K10, suggesting that KGF alters the spatial control of proliferation and differentiation in the genetically modified tissues. (From Andreadis, S.T. et al. *FASEBJ.*, 15, 898, 2001.)

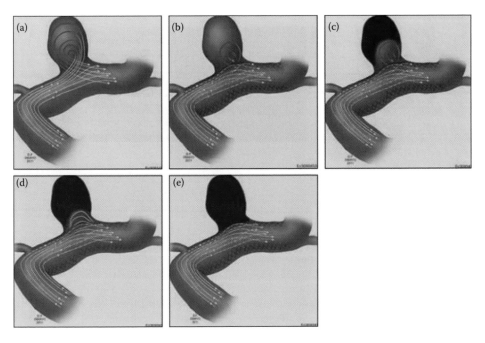

FIGURE 20.10 Prior to the insertion of the flow diverter, whirling flow can be seen in the aneurysm (a). After deploying the flow diverter, there is a gradual decrease in the amount of whirling flow inside the aneurysmal sac (b) followed by thrombus formation (c, d) before eventual resolution of the flow (e). (From Lanzino, G. 2011. *Endovascular Today* 10(9), 35–36.)

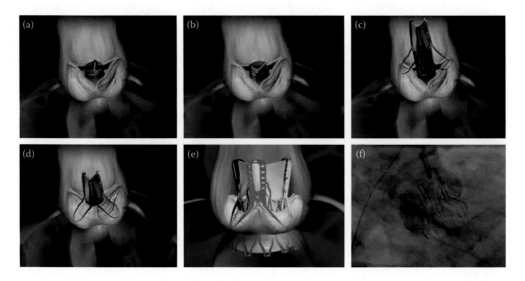

FIGURE 20.14 Anatomical positioning, Medtronic Engager system shown. The posts are rotated until they are aligned with the native leaflets (a,b). Once aligned, the support arms are released directly above the sinuses (c). The system is pulled back until the support arms engage (d). The system is then fully deployed by removing the sheath (e) as seen *in situ* (f). (From Falk, V., Walther, T., et al. 2011. *Eur Heart J* 32(7), 878–887.)

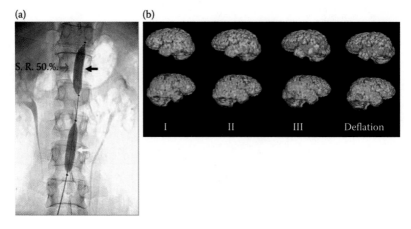

FIGURE 21.8 Increase in global cerebral perfusion after treatment with the NeuroFlo™ catheter. Original text: The NeuroFlo catheter. (a) Fluoroscopic images demonstrate the suprarenal (*black arrow*) and infrarenal (*white arrow*) balloons. (b) Positron-emission tomography demonstrates progressive increase in flow during and after balloon inflation in a patient with a right MCA stroke. Note the global increase in perfusion in both nonischemic (upper panel) and ischemic (lower panel) hemispheres. (Courtesy of Professor Wolf-Dieter Heiss, Max Planck Institute for Neurologic Research, Cologne, Germany.) (From Nogueira, R. G., L. H. Schwamm, et al. 2009. Endovascular approaches to acute stroke, part 1: Drugs, devices, and data. *American Journal of Neuroradiology* **30**(4): 649–661, reprinted with the permission of the American Society of Neuroradiology.)

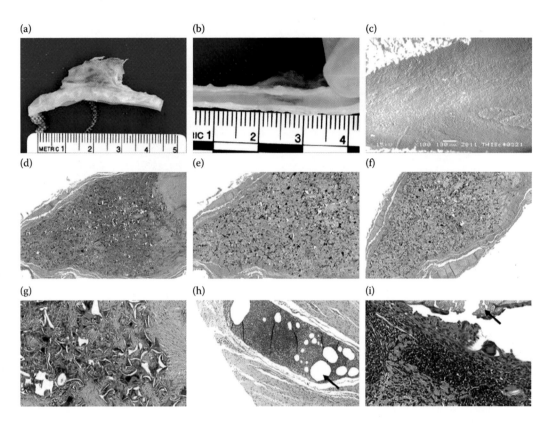

FIGURE 21.11 Preliminary pathology of SMP foams implanted for 90 days within a porcine animal. Original text: SEM and pathology results of implanted foams: (a) Gross picture of dissected aneurysm and parent vessel, (b) gross picture of healing that took place between aneurysm and parent artery intersection (en face), (c) SEM of endothelial cell morphology across the ostium (100× magnification), (d) hematoxylin and eosin (H&E) cross-section of bisected artery and aneurysm sac (4× magnification), (e) trichrome cross-section of bisected aneurysm sac (4× magnification), (f) PTAH cross-section of bisected artery and aneurysm sac (3.5× magnification), (g) H&E cross-section of bisected artery and aneurysm sac (20× magnification), (h) H&E detail of the FDA-approved suture material, polypropylene, indicated by the black arrow (10× magnification), and (i) H&E detail of the FDA-approved suture material, silk, indicated by the black arrow (40× magnification). (Reprinted with kind permission from Springer Science+Business Media: Rodriguez, J. N., Y. J. Yu et al. 2012. Opacification of shape memory polymer foam designed for treatment of intracranial aneurysms. *Annals of Biomedical Engineering* **40**(4): 883–897.)

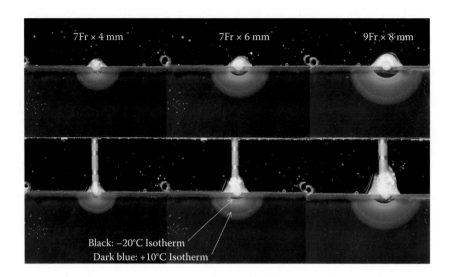

FIGURES 22.10 Freezing capability *in vitro* for different tip size and orientation (lateral vs. vertical orientation).

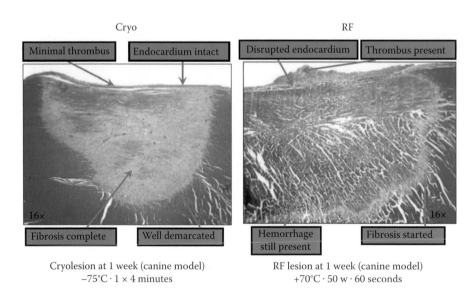

FIGURE 22.14 Cryo versus RF lesions. (From Haqqani, H.M., Mond, H.G. *Pacing Clin Electrophysiol* 2009 32(10):1336–1353. With permission.)

14

Radioimmunoassay: Technical Background

Eiji Kawasaki
Nagasaki University
Hospital

14.1 Introduction

The technique of radioimmunoassay (RIA), first developed in 1960 by Berson and Yalow for the measurement of insulin, has expanded to include the detection of other biological agents [1–8]. RIAs are based on the ability of an unlabeled antigen (Ag) to inhibit the binding of a labeled antigen (Ag*) by antibody (Ab):

$$Ag\text{--}Ab \underset{}{\overset{Ag}{\rightleftharpoons}} \ Ab \ \underset{}{\overset{Ag^*}{\rightleftharpoons}} \ Ag^* \cdot Ab$$

The process may be viewed as a simple competition in which Ag reduces the amount of free Ab, by decreasing the availability of Ab to Ag*. When the assay is performed, Ag* and Ab are incubated together in the presence and absence of samples containing Ag. After equilibration, free Ag* and Ag* · Ab are separated. The commonly used separation procedures include solid-phase absorption, precipitation of Ag*Ab complexes with either a second Ab or a salt, and chromatoelectrophoresis. Ag* · Ab (or free Ag*) is then determined by comparing the diminished Ag* binding of the sample to that of a standard curve obtained by adding graded, known amounts of Ag to Ag* and Ab. A new standard curve is determined in each assay to allow for variation in Ag binding from assay to assay. Radioimmunological methods combine the extreme sensitivity of detection of isotopically labeled compounds with the high specificity of immunological reactions. Thus, on use of radioisotopes, the detection limit is improved up to 10^7-fold over physicochemical analytical methods. Recently, in some fields of internal medicine, especially in autoimmune disease, RIA used to measure autoantibodies associated with disease prediction, diagnosis, and progression has been improved and simplified using a small amount of serum [9,10].

14.2 Principle

RIA is a general method by which the concentration of virtually any substance can be determined. The principle on which it is based is summarized in the competing reactions shown in Figure 14.1. The concentration of Ag in the unknown sample is obtained by comparing its inhibitory effect on the binding of Ag* to a limited amount of Ab with the inhibitory effect of known standards. A typical RIA is performed by the simultaneous preparation of standard and unknown mixtures in test tubes. To these tubes are added a fixed amount of Ag* and a fixed amount of antiserum. After an appropriate reaction time, the antibody-bound (B) and -free (F) fractions of the Ag* are separated by one of the many different techniques. The B/F ratios in the standards are plotted as a function of the concentration of Ag ("standard curve") and the concentration of Ag in the unknown sample is determined by comparing the observed B/F ratio with the standard curve (Figure 14.2). The radioactive isotopes most frequently used for labeling are ^{3}H, ^{14}C, ^{35}S, ^{57}Co, ^{75}Se, ^{125}I, and ^{131}I (Table 14.1). Of these, ^{125}I offers useful characteristics for labeling and is very widely used.

The RIA principle is not only limited to immune systems, but can also be extended to systems in which in place of the specific Ab, there is a specific reactor (i.e., a binding substance) that might be, for instance, a specific binding protein in plasma [11], an autoantibody [12], an enzyme [13], or a tissue receptor site [14]. For example, concentration of Ab in the unknown sample can be obtained by measuring the binding to an appropriate Ag*. Recently, for *in vitro* assays, the quantity of samples and the number of items to be measured are rapidly increasing. To meet this trend, the equipment for RIA has been semiautomated or automated.

14.3 RIA Techniques

The essential requirements for RIA include suitable reactants (Ag* and specific Ab) and some techniques for separating the Ab bound from free Ag*, since under the usual conditions of assay, the Ag–Ab complexes do not spontaneously precipitate.

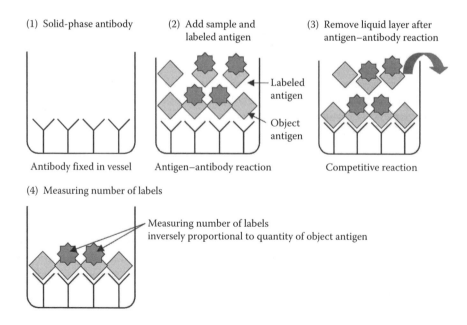

FIGURE 14.1 Principles of RIA.

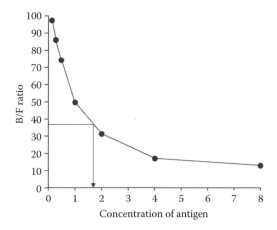

FIGURE 14.2 Standard curve for the assay of Ag. Concentration of Ag in the unknown sample is determined by comparing the observed B/F ratio as shown.

TABLE 14.1 Radioactive Isotopes Used for Labeling in Radioimmunoassay

Radioisotope	Half-Life	Energy	Detection Method
^{3}H	12.3 years	β	Liquid scintillation
^{14}C	5730 years	β	Liquid scintillation
^{35}S	87.4 days	β	Liquid scintillation
^{57}Co	270 days	γ	Scintillation crystal
^{75}Se	120.4 days	γ	Scintillation crystal
^{125}I	60 days	γ	Scintillation crystal
^{131}I	8 days	β, γ	Scintillation crystal

14.3.1 Radioactive Marker

14.3.1.1 Radiolabeled Ag

The first requirement for an RIA is the preparation of a highly purified Ag that can be radiolabeled or "tagged" without producing any loss of immunoreactivity. Since most polypeptide hormones contain at least one tyrosine residue, they can be labeled with a radioisotope of iodine (e.g., ^{125}I or ^{131}I). The radioiodine usually substitutes onto a tyrosine residue. The radioisotopes of iodine have the advantage of higher specific activities than can be found in ^{3}H or ^{14}C. Since the isotopic abundance of ^{125}I is close to 100%, and the isotopic abundance of ^{131}I is not more than 15–30% at the time of receipt into the laboratory [15], the shorter half-life of ^{131}I confers no advantage, and ^{125}I has been the radioiodine isotope of choice. The specific activity of a ^{125}I-labeled hormone may be increased by increasing the number of radioiodine substitutions. However, it has been shown that the more highly iodinated molecules have diminished immunoreactivity as well as increased susceptibility to damage (radiodegradation) [16,17]. The latter appears to have arisen from radiation self-damage within the molecule. Isotopes ^{3}H and ^{14}C can be used for labeling; however, because they emit extremely low-energy β-rays, a liquid scintillation counter is used to make measurements with these two isotopes.

Recent advances of molecular biology techniques allow for developing the cell-free protein synthesizing and labeling system [18]. With *in vitro* transcription/translation system using reticulocyte lysate, wheat germ extract, or *Escherichia coli* extract, one can directly prepare Ag* from the plasmid DNA containing Ag cDNA using a radioactive amino acid (e.g., ^{35}S-methionine, ^{3}H-leucine).

Ag* often needs to be purified for separating from the free isotope. There are various techniques available for purification. Adsorption column chromatography on powdered cellulose is a rapid assay [6,19]. For more extensive purification, one must resort to a separation involving dialysis, gel filtration (using a molecular sieve), or ion-exchange chromatography [20–22]. Inorganic iodine resin has also been used to absorb the unreacted [131]I [23]. After purification, one should determine the absolute quality of Ag required in a particular assay for high sensitivity. It is important that this quantity must be kept at a minimum. Therefore, it is desirable to produce a high specific activity of the radiolabeled Ag. If the Ag* is to be stored for considerable time, it is usually kept at 2–4°C (e.g., steroids) or it may be quickly frozen for storage at −20°C (e.g., polypeptides). After storage, the Ag must be checked for changes in immunoreactivity and radioactivity before use in an assay. The actual conditions for storing the Ag* depend on the particular Ag.

14.3.1.2 Radiolabeled Ab

Wide and coworkers [24] and Mikes and Hales [25,26] have pointed out the possible advantage of radio-iodinating the Ab instead of the Ag. The larger molecular weight of immunoglobulin and the presence of multiple tyrosines on the molecule permit the introduction of multiple radioiodine molecules without detrimental effects on Ab activity. Iodination of Ab rather than Ag may be especially advantageous if the Ag is easily damaged during iodination or lacks readily iodinatable tyrosines. The major drawback of RIA using radioiodinated Ab is likely to come in the strong propensity of iodinated antibodies (Abs) to adhere nonspecifically to glassware and insoluble resin. Thus, the selection of the immunoadsorbent is likely to be critical if this method is to be successful. Nonspecific binding of the iodinated Ab to the resin can be diminished by preparing fragment antigen-binding (Fab) fragments of the labeled Ab [27], but this will reduce the functional avidity of the Ab for the resin and may adversely affect assay sensitivity. Another problem is the requirement for substantial amounts of Ag to prepare the resin, which includes the use of this approach for antigens (Ags) that are in short supply.

14.3.2 Specific Ab

The second prerequisite for an RIA is the production of a suitable antiserum. The Abs are a group of serum proteins that are also referred to as γ-globulins or immunoglobulins. Most of these immunoglobulins belong to the IgG class, while the other classes are termed IgA, IgM, IgD, and IgE. Since these immunoglobulins possess not only Ab reaction sites, but also antigenic determinant sites, the immunoglobulins themselves can serve as Ags when injected into a "foreign" animal. The Ag* must be highly purified to avoid interaction of labeled contaminants with nonspecific Ab. The Ag-binding sites appear to reside on the H and L chains of the IgG molecule. While the titer, or concentration of the Ab is important, the main criterion for establishing a suitable antiserum is the energy of interaction between the Ag and Ab, or the specificity and affinity for the Ag being assayed. For most clinical chemists who are interested in performing RIA procedures, it would be more advantageous to procure the specific antiserum from a laboratory or commercial supplier that has the facilities required for the generation and evaluation of Abs.

The important limiting factors for the development of highly sensitive RIAs are Ab affinity, avidity, specificity, and cross-reactivity. Ab affinity is the strength of the reaction between a single antigenic determinant and a single combining site on the Ab. It is the sum of the attractive and repulsive forces operating between the antigenic determinant and the combining site of the Ab. Affinity is the equilibrium constant that describes the Ag–Ab reaction as illustrated below. Most Abs have a high affinity for their Ags.

$$K_{eq} = \frac{[Ag - Ab]}{[Ag] \times [Ab]}$$

Avidity is a measure of the overall strength of binding of an Ag with many antigenic determinants and multivalent Abs. Avidity is influenced by both the valence of the Ab and the valence of the Ag.

Avidity is more than the sum of the individual affinities. Specificity refers to the ability of an individual Ab-combining site to react with only one antigenic determinant or the ability of a population of Ab molecules to react with only one Ag. In general, there is a high degree of specificity in Ag–Ab reactions. Abs can distinguish differences in (1) the primary structure of an Ag, (2) isomeric forms of an Ag, and (3) secondary and tertiary structures of an Ag. Cross-reactivity refers to the ability of an individual Ab-combining site to react with more than one antigenic determinant or the ability of a population of Ab molecules to react with more than one Ag. Cross-reactions arise because the cross-reacting Ag shares an epitope in common with the immunizing Ag or because it has an epitope that is structurally similar to the one on the immunizing Ag (multispecificity).

The general method of inducing Ab formation is to inject into a number of animals the pure Ag mixed with "Freund's adjuvant" [28,29]. Freund's adjuvant is a mixture of mineral oil, waxes, and killed bacilli that enhances and prolongs the antigenic response. Small peptides (molecular weight 1,000–5,000) or nonpeptidal substances that are not themselves antigenic may be rendered so by coupling to a large protein. A variety of methods may be employed to bind the small molecules to immunogenic carriers [30–32]. The antisera can be stored for long periods of time under proper conditions. Repeated freezing and thawing should be avoided and all antisera should be stored and properly diluted. Reports vary as to the best temperature for antiserum storage, some researchers prefer –80°C, others –20°C, or –15°C [33]. Once thawed, the sera is best maintained at 4°C.

14.3.3 Separation Methods

The third requirement for an RIA is a suitable method for complete and rapid separation of the bound Ag from the free Ag. In addition, a separation procedure that permits further association and dissociation of the reactants will seriously impair the effectiveness of the assay. Regardless of the method of separation chosen, it must be reproducible, simple to perform, and economically feasible. Table 14.2 lists

TABLE 14.2 Methods and Materials for the Separation of Antibody-Bound and -Free Antigen

Separating Action	Method/Material
Precipitation of the Ag–Ab complex	Ammonium sulfate
	Sodium hydrogen sulfate
	Zirconyl phosphate
	Ethanol
	2-Propanol
	Dioxane
	Polyethylene glycol (PEG)
Partition of the components due to their different mobility and molecular size	Chromatography
	Electrophoresis
	Gel filtration
	Ultracentrifugation
Adsorption of free Ag to solid-phase materials	Charcoal (also bound to dextran or albumin)
	Cellulose
	Sephadex (cross-linked dextrans)
	Sepharose (beaded agarose)
	Silicates
	Ion-exchange resins
	Polymerized antibodies
Binding of the antibody to a solid phase	Antibody chemically bound to polymer carrier-coated tubes or beads
Immunological complex formation with a second antibody	Double-antibody technique
	Second antibody chemically bound, for example, to activated
	Cellulose

a variety of techniques that have been used for the separation of Ab-bound and -free Ag*. The use of so many methods is a tribute to the imagination and versatility of investigators in the field, and is also due to the recognition that no single method has proved completely satisfactory for all Ags. The following are some examples of the separation method:

1. *Precipitation with ammonium sulfate:*
   Ammonium sulfate (33–50% final concentration) will precipitate immunoglobulins but not many Ags. Thus, this can be used to separate the immune complexes from free Ag. This has been called the Farr technique.

2. *Anti-immunoglobulin Ab:*
   The addition of a second Ab directed against the first Ab can result in the precipitation of the immune complexes and thus the separation of the complexes from free Ag. This has been called the double antibody method (Figure 14.3). This method is widely applicable and gives highly satisfactory results provided careful attention is given to the possible pitfalls. The diluted antiserum containing the first Ab must contain sufficient γ-globulin to give a macroscopic precipitate with an excess of a second Ab (usually 2–5 μg/mL). To minimize the unnecessary expenditure of the second Ab and to avoid nonspecific binding of Ag* to the bulky precipitate, the quantity of the precipitate is adjusted so as to be just enough to give good precipitation of radioactivity. In the event that the first Ab has a low titer, the expenditure of the second Ab will be prohibitive and other methods must be used.

3. *Immobilization of the Ab:*
   The Ab can be immobilized onto the surface of a plastic bead or can be coated onto the surface of a plastic plate and thus the immune complexes can be easily separated from the other components by simply washing the beads or the plate [34]. This is the most common method used today and is referred to as solid-phase RIA (Figure 14.3). A number of methods have been used for attaching Ab to the solid phase. Formation of a stable resin–protein complex may require an initial chemical reaction to activate the resin followed by exposure to the Ab. In conjunctions involving Ab, ideally γ-globulin fractions rather than whole serum are used in the second step to minimize binding of extraneous proteins to the resin. Activation of the resin is commonly accomplished with bromoacetyl bromide or cyanogen bromide [35]. Alternatively, stable azide or aromatic amine residues can be introduced into a resin and the activation can be accomplished just prior to the introduction of the protein. Resins containing primary aliphatic amino or carboxyl groups combine with Ab in the presence of carboxyl-activating reagents. Protein A or protein G, which binds the Fc portion of immunoglobulin, can be used instead of the Ab. The commonly used resins include agarose, Sephadex, and cellulose.

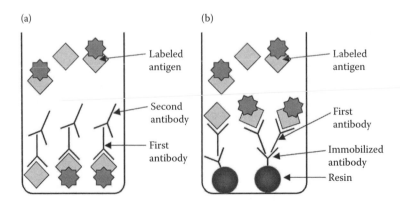

FIGURE 14.3 Separation of Ab-bound and -free Ag*. (a) Double antibody method and (b) solid phase method.

Solid-phase systems have the advantage of simplicity and rapidity. However, specific problems may arise, such as difficulties in dispersion of the resin. The establishment of equilibrium, although rapid in some systems, may require continuous mixing over a period of many hours in others.

14.4 Validation of an RIA Procedure

The RIA differs from traditional bioassay in that it is an immunochemical procedure, which is not affected by biological variability of the test system. The measurement depends only upon the interaction of chemical reagents in accordance with the low mass action. However, nonspecific factors do interfere in chemical reactions, and cross-reacting prohormones, molecular fragments, and related hormonal Ags can alter the specificity of the immune reaction.

To ensure reliability, the results obtained, and thus to guarantee the quality of an RIA, it is necessary to know a number of characteristic data. These include the specificity and sensitivity, and above all the accuracy, reproducibility, and precision of an assay. The specificity of an RIA is essentially determined by the Ab. It can be impaired by cross-reactions with similar substances and fragments thereof, and also by the separating step. Serum factors, pH value, additives such as albumin, buffer, heparin, or preservatives can also lower the specificity. The sensitivity depends on the specificity of the Ab and the specific radioactivity of the tracer. The sensitivity can be regarded as the special case of the accuracy at zero concentration. The sensitivity of assays in which the immunological reaction is irreversible or measurement is performed at nonequilibrium can occasionally be raised by delayed addition of the radioactively labeled substance [36].

The accuracy attainable in a determination depends on three errors of various magnitudes. Apart from experimental errors arising in several pipetting steps and interference of the reaction, errors also occur in measuring the radioactivity and on evaluation of the various samples. Owing to the different nature of these errors, it is clear that the accuracy cannot be the same in each portion of the curve. However, multiple determinations and long counting time can do much to reduce random errors and thus lead to a more accurate result.

Reproducibility is another criterion of quality, including intra- and interassay reproducibility. Duplicate or multiple determinations on samples of an assay and the resulting deviations from the average value yield the intra-assay reproducibility. It normally lies below 5%. The interassay reproducibility is the coefficient of variation obtained on determination of the same sample in several different assay runs using different reagents. Its value lies below 10% for accurate and reproducible RIA [37].

14.5 Applications

In the years since the development of RIA, the application of the technique has brought profound changes to medicine and biology. Figure 14.4 shows the schematic methods for radioligand-binding assay currently for detecting autoantibodies against recombinant Ag in sera [38]. For the measurement of autoantibodies, the cDNA of autoantigens are required to prepare the radiolabeled autoantigens. Such cDNAs are usually obtained from appropriate cDNA library using plaque hybridization technique or appropriate mRNA using reverse-transcriptase–PCR (polymerase chain reaction) method. The cloned cDNAs are then subcloned into the plasmid vectors suitable for *in vitro* transcription/translation system. Radiolabeled autoantigens are prepared using *in vitro* transcription/translation system with reticulocyte lysates and a radioactive amino acid (e.g., ^{35}S-methionine, ^{3}H-leucine). For detecting autoantibodies to recombinant autoantigens, we are using a 96-well plate filtration technology and a microplate direct β-counter (Figure 14.5) [10,38]. After determining the incorporation rate of the radioisotope by trichloroacetic acid (TCA) precipitation method, *in vitro*-translated radiolabeled protein (20,000 cpm of TCA precipitable protein) is incubated with patients' serum (5 μL for duplicate) at a 1:25 dilution overnight at 4°C, and the resulting immunocomplexes are precipitated with 25 μL of protein A-Sepharose in the 96-well plate. After the washing step using

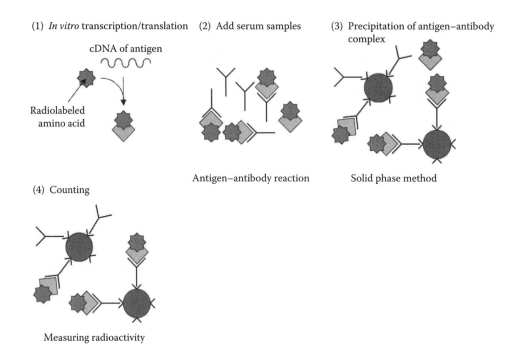

FIGURE 14.4 Schematic methods for radioligand binding assay.

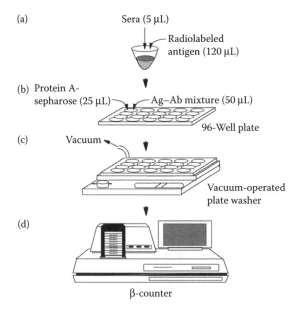

FIGURE 14.5 High-throughput RIA for autoantibodies to recombinant autoantigens. (From Kawasaki, E. and Eisenbarth, G. S., *Front. Biosci.,* 5, E181, 2000. With permission.)

a vacuum-operated 96-well plate washer, radioactivity is determined directly with a 96-well plate β-counter. The adaptation of the assay to a 96-well plate and semiautomated 96-well counting allows a single person to analyze more than 40,000 samples per year. Table 14.3 lists *in vitro* transcribed and translated recombinant autoantigens that we have successfully utilized for disease diagnosis and prediction using radioligand binding assay.

TABLE 14.3 Autoantibody Radioassays Based on the *In Vitro* Transcription/Translation of Autoantigens

Worked	Did Not Work
GAD65	Proinsulin
ICA512/IA-2	ICA69
Phogrin/IA-2beta	H^+, K^+-ATPase
Carboxypeptidase H	
21-hydroxylase	
CYP2D6	
Transglutaminase	
Zinc transporter-8	

Source: Modified from Kawasaki, E. and Eisenbarth, G. S., *Front. Biosci.,* 5, E181, 2000. With permission.

A similar assay format can be used to measure several autoantibodies at the same time using recombinant autoantigens labeled with different radioisotopes. We have developed a combined IA-2, GAD65 autoantibody radioassay utilizing [^{3}H]-labeled GAD65 and [^{35}S]-labeled IA-2 that allows simultaneous detection and discrimination of both autoantibody specificities [39]. *In vitro*-translated [^{3}H]-GAD65 and [^{35}S]-IA-2 were mixed and incubated with serum in a tube and the radioactivity was counted in a 96-well plate with channel windows set for each radionucleotide after protein-A Sepharose precipitation. The combined assay gave essentially identical results to those obtained in the single radioassays. The simplicity of this assay with dual determination of the two autoantibodies utilizing 5 µL of sera, 96-well membrane separation of autoantibody-bound labeled autoantigen, and 96-well β-counting facilitates the rapid screening of thousands of samples.

References

1. Berson, S. A. et al., Insulin-I^{131} metabolism in human subjects: Demonstration of insulin binding globulin in the circulation of insulin treated subjects, *J. Clin. Invest.*, 35, 170, 1956.
2. Berson, S. A. and Yalow, R. S., Kinetics of reaction between insulin and insulin-binding antibody, *J. Clin. Invest.*, 36, 873 (abstract), 1957.
3. Berson, S. A. and Yalow, R. S., Isotopic tracers in the study of diabetes, *Adv. Biol. Med. Phys.*, 6, 349, 1958.
4. Berson, S. A. and Yalow, R. S., Recent studies on insulin-binding antibodies, *Ann. NY. Acad. Sci.*, 82, 338, 1959.
5. Yalow, R. S. and Berson, S. A., Assay of plasma insulin in human subjects by immunological methods, *Nature*, 184, 1648–1649, 1959.
6. Yalow, R. S. and Berson, S. A., Immunoassay of endogenous plasma insulin in man, *J. Clin. Invest.*, 39, 1157, 1960.
7. Yalow, R. S. and Berson, S. A., *Introduction and General Considerations*, Philadelphia, J.B. Lippincott Co., 1971, 1.
8. Yalow, R. S. and Berson, S. A., *Fundamental Principles of Radioimmunoassay Techniques in Measurement of Hormones*, Amsterdam, Excepta Medica Foundation, 1971, 16.
9. Kawasaki, E. and Eisenbarth, G. S., High-throughput radioassays for autoantibodies to recombinant autoantigens, *Front. Biosci.*, 5, E181, 2000.
10. Maniatis, A. K. et al., Rapid assays for detection of anti-islet autoantibodies: Implications for organ donor screening, *J. Autoimmun.*, 16, 71, 2001.
11. Murphy, B. E. P., Engelberg, W. and Pattee, C. J., Simple method for the determination of plasma corticoids, *J. Clin. Endocr.*, 23, 293, 1963.
12. Carr, R. I., Wold, R. T., and Farr, R. S., Antibodies to bovine gamma globulin (BGG) and the occurrence of a BGG-like substance in systemic lupus erythematosus sera, *J. Allergy Clin. Immunol.*, 50, 18, 1972.
13. Rothenberg, S. P., A radio-enzymatic assay for folic acid, *Nature*, 206, 1154, 1965.

14. Lefkowitz, R. J. et al., ACTH receptors in the adrenal: Specific binding of ACTH-125I and its relation to adenyl cyclase, *Proc. Natl. Acad. Sci. USA,* 65, 745, 1970.
15. Yalow, R. S. and Berson, S. A., Labelling of proteins—Problems and practices, *Trans. NY. Acad. Sci.,* 28, 1033, 1966.
16. Berson, S. A. and Yalow, R. S., *Recent Advances in Immunoassay of Peptide Hormones in Plasma,* Amsterdam, Excepta Medica Foundation, 1969, 50.
17. Berson, S. A. and Yalow, R. S., Iodoinsulin used to determine specific activity of iodine-131, *Science,* 152, 205, 1966.
18. Jackson, R. J. and Hunt, T., Preparation and use of nuclease-treated rabbit reticulocyte lysates for the translation of eukaryotic messenger RNA, *Meth. Enzymol.,* 96, 50, 1983.
19. Berson, S. A. and Yalow, R. S., Preparation and purification of human insulin-I131; binding to human insulin-binding antibodies, *J. Clin. Invest.,* 40, 1803, 1961.
20. Desranleau, R., Gilardeau, C., and Chretien, M., Radioimmunoassay of ovine beta-lipotropic hormone, *Endocrinology,* 91, 1004, 1972.
21. Catt, K. J. and Cain, M. C., Measurement of angiotensin II in blood, *Lancet,* 2, 1005, 1967.
22. Midgley, A. R., Jr., Radioimmunoassay: A method for human chorionic gonadotropin and human luteinizing hormone, *Endocrinology,* 79, 10, 1966.
23. Saxena, B. B. et al., Radioimmunoassay of human follicle stimulating and luteinizing hormones in plasma, *J. Clin. Endocrinol. Metab.,* 28, 519, 1968.
24. Wide, L., Bennich, H., and Johansson, S. G., Diagnosis of allergy by an *in-vitro* test for allergen antibodies, *Lancet,* 2, 1105, 1967.
25. Miles, L. E. and Hales, C. N., Labelled antibodies and immunological assay systems, *Nature,* 219, 186, 1968.
26. Miles, L. E. and Hales, C. N., The preparation and properties of purified 125-I-labelled antibodies to insulin, *Biochem. J.,* 108, 611, 1968.
27. Hubacek, J., Kubicek, R., and Vojacek, K., Immunoassay of human luteinizing hormone using univalent radioactive antibodies, *J. Endocrinol.,* 51, 91, 1971.
28. Freund, J., The effect of paraffin oil and Mycobacteria on antibody formation and sensitization, *Am. J. Clin. Pathol.,* 21, 645, 1951.
29. Freund, J., Some aspects of active immunization, *Annu. Rev. Microbiol.,* 1, 291, 1947.
30. Erlanger, B. F. and Beiser, S., Antibodies specific for ribonucleosides and ribonucleotides and their reaction with DNA, *Proc. Natl. Acad. Sci. USA,* 52, 68, 1964.
31. Talamo, R. C., Haber, E., and Austen, K. F., Antibody to bradykinin: Effect of carrier and method of coupling on specificity and affinity, *J. Immunol.,* 101, 333, 1968.
32. Richards, F. M. and Knowles, J. R., Glutaraldehyde as a protein cross-linkage reagent, *J. Mol. Biol.,* 37, 231, 1968.
33. Thoeneycroft, I. H. et al., *Preparation and Purification of Antibodies to Steroids,* New York, Appleton-Century-Crofts, 1970, 63.
34. Wide, L., Radioimmunoassays employing immunosorbents, *Acta. Endocrinol. Suppl. (Copenh),* 142, 207, 1969.
35. Robbins, J. B., Haimovich, J., and Sela, M., Purification of antibodies with immunoadsorbents prepared using bromoacetyl cellulose, *Immunochemistry,* 4, 11, 1967.
36. Rodbard, D. et al., Mathematical analysis of kinetics of radioligand assays: Improved sensitivity obtained by delayed addition of labeled ligand, *J. Clin. Endocrinol. Metab.,* 33, 343, 1971.
37. Rodbard, D., Statistical quality control and routine data processing for radioimmunoassays and immunoradiometric assays, *Clin. Chem.,* 20, 1255, 1974.
38. Sera, Y. et al., Autoantibodies to multiple islet autoantigens in patients with abrupt onset type 1 diabetes and diabetes diagnosed with urinary glucose screening, *J. Autoimmun.,* 13, 257, 1999.
39. Kawasaki, E. et al., Evaluation of islet cell antigen (ICA) 512/IA-2 autoantibody radioassays using overlapping ICA512/IA-2 constructs, *J. Clin. Endocrinol. Metab.,* 82, 375, 1997.

IV

Biomaterials and Tissue Engineering

Role of Technology in Biomolecular Revolution

The information revealed by revolutionary undertakings, such as the human genome project, has provided unprecedented insights into the inner workings of the most basic structures of living tissues. With much of the molecular architecture defined, attention has turned to manipulating these structures for therapeutic purposes.

Techniques have been developed to transplant genetic material into a variety of living tissues. The reliability of animal models for human disease conditions has been enhanced by genetic manipulation.

Disease treatment strategies can be more directly developed for maximum effectiveness. While there are significant ethical issues involved, the potential to grow entire organs for implantation is exciting. The future of many areas of disease treatment will be affected by developments in the field of tissue engineering. As clinicians and patients alike realize the potential of such therapies, technologies will be developed that provide effective disease treatment.

15

Clinically Applied Biomaterials: Soft Tissue

Shelby Buffington
Texas A&M University

Mary Beth Browning
Texas A&M University

Elizabeth
Cosgriff-Hernández
Texas A&M University

15.1 Introduction

Soft tissues are defined as tissues that connect, support, or surround other structures and organs such as tendons, ligaments, skin, muscles, and blood vessels. Two of the primary components of soft tissues are elastin and collagen. Elastin provides elasticity to the tissues and collagen provides form and strength [1]. These tissues are incompressible, are generally anisotropic, and often demonstrate viscoelastic properties such as stress relaxation, creep, and hysteresis. Annually, several million Americans are afflicted with soft tissue loss due to trauma or disease. Treatment options include reconstruction with tissue grafts and prosthetics [2]. The first step to adequately replacing a tissue is to determine the physiological loadings and native mechanical properties [1]. The physical properties of the biomaterial can then be designed to match the native tissue by selecting appropriate material properties. Fabrication and processing techniques strongly influence graft properties and can be used to further tune properties to match the target tissue [3,4].

For a material to replace the function of a native tissue, it must be mechanically similar to the original tissue. The simplest representation of these properties is the elastic modulus. Representative values of the moduli of select soft tissues and common natural materials used in reconstruction are listed in Table 15.1. Metals and ceramics are typically poor choices to replace soft tissue due to unsuitably high moduli. Additionally, it is important to match the viscoelastic properties of natural tissue. Polymeric biomaterials have moduli and viscoelastic properties that can be tuned to match soft tissues. There are two classifications of clinically useful polymeric biomaterials: synthetic (man-made) and natural polymers. Synthetic polymers have the advantage of being mass-produced with physical properties that can be tuned to a specific application. However, synthetic polymers are only capable of interacting with the

TABLE 15.1 Mechanical Properties of Select Soft Tissues and Common Natural Materials Used in Reconstruction

Polymer	Modulus	Tensile Strength (MPa)	Ultimate Elongation (%)	Reference
Collagen	1.8–46 MPa	0.9–7.4	24–68	[71]
Crosslinked collagen	40–80 MPa	47–72	12–16	[71]
ACL	345 MPa	36	15	[72]
Tendon	1.5 GPa	150	12	[73]
Spider silk	11–13 GPa	875–972	17–18	[74]

native tissues of the body via nonspecific interactions, which limits tissue integration. An alternative to synthetic polymers is natural polymers that are derived from proteins commonly found in the body (collagen, fibrin, or hyaluronic acid) or outside the body (chitosan, alginate). Natural polymers derived from proteinous sources have the advantage of being similar in composition to the tissues that they are replacing. Therefore, these materials can interact with host cells via native interactions to facilitate integration with the surrounding tissue. Fabrication of these natural polymers is restricted to processes that do not denature the protein and reduce the bioactivity of the material. These materials also suffer from batch variability and there is rising concern of animal-derived materials eliciting undesired immune responses [5–7].

This chapter provides a brief overview of different polymeric biomaterials used in soft tissue applications. It begins with a discussion on polymer classifications and properties, then provides a general overview of fabrication techniques, and concludes with a selection of clinically applied biomaterials for soft tissue applications.

15.2 Polymeric Biomaterials

15.2.1 Synthetic Polymers

There are numerous classification schemes for polymers. Carothers originally classified polymers into condensation and addition polymers based on the compositional difference between the monomer and corresponding polymer [8]. Addition polymers are characterized by the formation of the polymer from monomers without the loss of a small molecule. The major addition polymers are formed via the polymerization of vinyl monomers that contain the carbon–carbon double bond. In contrast, condensation polymers are formed from multifunctional monomers with the elimination of a small molecule such as water. As a result, the structure of the repeating unit of the polymer chain differs from the monomeric structure. Condensation polymers have also been defined as polymers with repeating units joined together by functional groups such as esters, urethanes, and amides [9]. Polymer nomenclature can be based on source (e.g., polyethylene) or polymer structure (e.g., polyurethane). For the more complex polymer structures, the International Union of Pure and Applied Chemistry (IUPAC) structure-based nomenclature system is utilized, which is based on the selection of a preferred constitutional repeating unit following a specific set of rules for naming organic polymers [8]. For example, polystyrene is the commonly used source-based name and poly(1-phenylethylene) is the IUPAC name.

Polymer properties are strongly dependent on polymer variables, including structure, architecture, molecular weight, crystallinity, and glass transition temperature (T_g). The chemical structure can be related to key properties such as degradation profile and solubility as well as indicate the intermolecular forces present between chains. Polymers can be classified as linear, branched, or crosslinked depending on the chain architecture. Unlike other low-molecular-weight compounds, most polymers exhibit characteristics of both crystalline and amorphous solids. As such, the two major types of transition temperatures are the crystalline melting temperature (T_m) and the glass transition temperature (T_g). The T_g

TABLE 15.2 List of Common Polymeric Biomaterials Used in Soft Tissue Applications

Polymer Name	Current Areas of Research and Applications
Nondegradable Polymeric Biomaterials	
Polypropylene	ACL replacement, sutures, hernia repair meshes [7,75,76]
Polyacrylamide	Facial contouring [77]
Polytetrafluoroethylene	Vascular grafts, ACL replacement [7]
Polydimethylsiloxane	Contact lenses [62]
Polyurethane	Mitral valve replacement, vascular graft, shape memory and drug-eluting stents, wound dressing [78–80]
Polyethylene terephthalate	Vascular grafts, ACL replacement [7,75]
Polyacrylate	Surgical adhesive, drug delivery [60]
Biodegradable Polymeric Biomaterials	
Polydioxanone	Sutures, bone fixation, wound clips [13]
Polycaprolactone	Contraceptives, surgical staples, drug delivery, drug-eluting stents [5,81]
Polyhydroxybutyrate	Drug delivery, orthopedic applications, stents, sutures [5]
Poly(glycolic acid), poly(lactic acid), and copolymers	Drug delivery, drug-eluting stents, tissue-engineered vascular grafts, sutures [45,60,81,82]

is the temperature at which the amorphous domains of the polymer transition from a hard and relatively brittle state into a rubber-like state. The key determinant for the utility of a polymer in a given application is its mechanical behavior or deformation characteristics under stress. A wide range of mechanical properties can be achieved by synthesizing polymers with different combinations of crystallinity, crosslinking, T_g, and T_m [9]. A list of common polymeric biomaterials used in soft tissue applications is provided in Table 15.2.

15.2.2 Biodegradable Polymers

Polymers can be used in either permanent or temporary devices. Permanent implants are present throughout the lifetime of the patient and replace the function of the tissue. These implants can suffer from long-term complications such as implant-associated infection and stress shielding, which can cause complications or necessitate revision surgery. Temporary devices are designed to replace function but only last for the duration of soft tissue healing and therefore eliminate long-term complications. During the period of healing, the implant is slowly resorbed through a process of polymer degradation and clearance of degradation products. Degradation refers to a chemical process resulting in the cleavage of covalent bonds. Chain scission results in reduced molecular weight and solubilization of polymer fragments leading to polymer erosion. Erosion is defined as the physical changes in size, shape, or mass of a device [5]. Biodegradable polymers used clinically such as polyesters and polyanhydrides typically degrade via hydrolytic degradation. The main modulators of degradation rate are the reactivity of the hydrolytically labile bonds, polymer hydrophobicity, and crystallinity. The two primary modes of hydrolytic degradation are bulk degradation and surface erosion. Bulk degradation occurs when the rate of water ingress is greater than the rate of hydrolysis, which results in degradation occurring throughout the polymer system. This causes a rapid loss in mass and mechanical properties, which can lead to premature device failure. Surface erosion occurs when the rate of water ingress into the material is slower than the rate of hydrolysis. Some applications require that the polymer exhibit hydrolytically labile bonds as well as sufficient hydrophobicity to limit water diffusion. Surface erosion results in a gradual reduction of thickness while the polymer maintains bulk mechanical properties. This profile is advantageous in systems that replace mechanical function and drug delivery.

15.2.3 Natural Polymers

Natural polymers are biomacromolecules with complex hierarchical structures that are correlated with their specific function. For example, proteins have primary (amino acid sequence), secondary (regular substructures), tertiary (3D structure), and quaternary (protein complexes) structures. The three primary classes of biomacromolecules are polysaccharides such as chitin and cellulose, polynucleotides such as DNA and RNA, and polypeptides or proteins such as collagen, elastin, and silk. The higher-order structures of these molecules are stabilized by relatively weak intermolecular bonding forces that are readily denatured by many of the high temperatures and harsh solvents used in synthetic polymer processing. Although limited in processing, these natural polymers have the advantage of innate cues that facilitate tissue integration. Biomaterials made from extracellular matrix have been used in skin grafts and ligament repair for decades [10]. Three of the most commonly used natural polymers are collagen, fibrin, and silk (Table 15.3).

15.2.3.1 Collagen

Collagen is the most abundant protein in animals, making up to 25–35% of the whole-body protein content. It is the major fibrous protein in the extracellular matrix and the main component of connective tissue [11]. Owing to its abundance, it was one of the first proteins isolated from animal sources and utilized clinically. Collagen has the distinctive feature of a regular arrangement of amino acids, Gly–Pro–X or Gly–X–Hyp. Each of the three collagen subunits form left-handed helices that are twisted together into a right-handed triple helix. These supercoils crosslink together to form collagen myofibrils, which then assemble into collagen fibrils that finally aggregate into the collagen fibers and networks of tissues [12]. Collagen has a wide variety of applications including cosmetic surgery and wound dressings. Heat denaturation of collagen separates the three tropocollagen strands into globular domains. This process is used to generate gelatin, which is used commonly in pharmaceutical and cosmetic industries.

15.2.3.2 Fibrin

Fibrin is a derivative of fibrinogen, a glycoprotein that is a key factor in the coagulation cascade and acts as a provisional matrix for cell infiltration in wound healing. Fibrinogen is a hexameric molecule containing three sets of dimers (α, β, and γ) that are linked with disulfide bonds [13]. Thrombin converts fibrinogen to fibrin, which aggregates into a fibrin mesh that makes up the core of the clot. Coagulation factor XIIIa further stabilizes the clot by crosslinking the fibrin. After clot formation and hemostasis are achieved, fibrin serves as a provisional matrix for leukocytes, endothelial cells, and fibroblasts to begin the remodeling phase of wound healing. The clotting capability and wound healing functions of fibrinogen are commonly utilized in medical applications such as hemostats and tissue sealants. Fibrin glue is a two-component system of fibrinogen and thrombin that is injected through a two-barrel syringe with a mixing head and used as a tissue sealant [13].

15.2.3.3 Silk

Silk is produced by a variety of creatures in nature; however, the silk from silkworms and spiders is most commonly used in clinical applications. Silkworms produce fibers 10–25 μm in diameter composed of

TABLE 15.3 List of Natural Polymers with Current and Future Clinical Applications

Polymer Name	Current Areas of Research and Applications
Silk	Ligament grafts, sutures [83]
Collagen	Facial contouring, ligament grafts [7]
Fibrin	Surgical glue, hemostat [6,84]

two proteins: a light chain and a heavy chain. These proteins are present in equal amounts and linked by a single disulfide bond. Silk is then coated with a hydrophobic protein that is removed during purification [14]. Silk is one of the most mechanically robust natural polymers, which makes it ideal for load-bearing applications such as ligament grafts. Silk fibers undergo proteolytic degradation, typically losing tensile properties in 1 year and degrading completely in 2 years.

15.3 Fabrication and Processing

The properties of a polymeric device depend on both its chemical structure and its 3D architecture. Processing techniques can be utilized to tune material properties for a specific application. A variety of fabrication processes can be used to make polymers in the form of foams, fibers, or solid constructs. This section describes a selection of common fabrication techniques used for soft tissue applications.

15.3.1 Thermal Processing

During thermal processing, high temperatures are used to generate a polymer melt that is then either extruded or cast into a mold. The polymer must be stable at high temperatures, which precludes this processing technique for temperature-sensitive components such as proteins. In extrusion, a polymer melt is heated in a chamber and then forced through a hole while being cooled with jets of air. This produces polymer rods or tubes and is used to produce GORE-TEX® vascular grafts [15]. Another option is injection molding where the melted polymer is forced through a nozzle into a mold at high pressures. Once the mold is filled, the polymer melt is cooled and solidified into the desired shape. Injection molding is fast and allows for the production of a wide variety of 3D architectures.

15.3.2 Fibers

Fibrous medical devices can be divided into four categories based on the organization of the fibers: knitted, woven, braided, and nonwoven. Biomedical textiles have been used in a wide variety of soft tissue applications, including vascular grafts, ligament/tendon grafts, surgical meshes, and hernia repair [15]. Fibers can be either generated with heat in melt spinning or spun from solvent in wet spinning and electrospinning. In melt spinning, the polymer is heated above its melting temperature and then extruded through a small hole in a spinneret. The fibers are then drawn uniaxially, cooled, and wound onto spools [16]. In wet or dry spinning, fibers are produced from extrusion of a polymer solution from a spinneret. The solvent is removed either via evaporation in dry spinning or via solvent exchange in wet spinning by submerging the spinneret in a chemical bath that causes the polymer to precipitate. Electrospinning utilizes an applied electric charge to draw micro- or nanofibers from a polymer solution followed by evaporation of the solvent to form solid fibers [17]. A variety of fiber bonding setups can be used to generate different device geometries with the most popular being sheets such as those used for wound dressings and tubes used for vascular grafts.

15.3.3 Porous Foams

Porous monoliths are used in a variety of medical devices from nerve guidance channels to drug-delivery vehicles. The most common techniques to produce porous devices include gas foaming, particulate leaching, thermal-induced phase separation, and solid free-form fabrication. Gas foaming is based on the growth of gas bubbles throughout the polymer melt through either a physical or a chemical blowing process. Physical blowing processes utilize high pressures to generate a gas-saturated polymer phase and upon depressurization, pore nucleation, growth, and coalescence occur. Alternatively, chemical blowing processes utilize a blowing agent that is added to the polymer and decomposes during thermal processing leading to the generation of gas such as carbon dioxide [18]. One limitation of gas blowing is the common formation of a skin or nonporous film layer, which can limit tissue integration into the device. Another method to produce highly porous foams is particulate leaching. In this process, a porogen, most commonly salt, wax, or sugar, is mixed into a polymer solution, and the solvent is removed leaving a polymer matrix embedded with the porogen. During the second stage, the porogen is removed using a secondary solvent that selectively dissolves the porogen and not the polymer [19]. The pore size and foam properties can be controlled through careful selection of the porogen size, concentration, and geometry. Thermally induced phase separation, or TIPS, uses the thermodynamic demixing of a polymer–solvent solution to induce phase separation. The polymer solution usually divides into a polymer-rich phase and a polymer-poor phase by cooling the solution below the bimodal solubility curve [20]. The polymer-poor phase is then removed through typical solvent-removal techniques leaving a variety of foam architectures from discrete, spherical pores to microchannels. TIPS offers the ability to control material pore size by varying processing conditions such as selected temperatures and rate of thermal changes [21]. More complex architectures can be achieved using new techniques in solid free-form fabrication. These strategies control the microarchitecture by integrating computer-aided design with a variety of polymer processing techniques such as selective laser sintering, inkjet printing, and photolithography [22].

15.3.4 Hydrogels

Hydrogels are crosslinked polymeric networks capable of absorbing large amounts of water—from 20% to 1000% of their dry weight [23–24]. Owing to their highly tunable properties, established biocompatibility, and high permeability, hydrogels formed from both natural and synthetic polymers have been of great interest to biomedical scientists for many years [23–28]. The biomedical applications of hydrogels range from wound dressings and contact lenses to drug-delivery vehicles and tissue-engineered scaffolds [29–35]. Many hydrogels can be crosslinked under mild conditions, allowing *in situ* polymerization in direct contact with tissues and cells [36–39]. This enables facile implementation of cell-laden scaffolds with complex geometries [24,40]. Furthermore, hydrogels have highly tunable mechanical properties that can be matched to those of soft tissues [24,28,41].

15.4 Sterilization

Before a device can be implanted into the body, it must be sterilized to decrease the risk of infection. Owing to the thermal sensitivity of most polymers, steam or dry heat sterilization are not typically utilized. The most common sterilization techniques of polymeric devices are ethylene oxide sterilization and radiation sterilization. In ethylene oxide sterilization, pathogen death is triggered by permanent chemical alterations in their nucleic acids. Ethylene oxide treatment is advantageous due to its rapid efficiency, penetration, and broad spectrum action; however, thorough degassing is needed to remove toxic residues prior to implantation. Sterilization with gamma radiation results in pathogen elimination through photon-induced changes at the molecular level that cause pathogen death or render the organism incapable of reproduction. It is rapid, effective, and compatible with many materials but can result in chain scission or crosslinking depending on the dose and chemical structure of the polymer [15]. The sterility assurance level is used to test the sterility of the device. This is done by culturing implants in nutrient media after sterilization several times and quantifying bacterial growth.

15.5 Applications

15.5.1 Temporary Support Devices

Temporary support devices, such as sutures, bone fixation devices, and surgical adhesives, provide artificial support to the natural tissue bed when it has been weakened by disease, injury, or surgery. Biostable support devices must be removed once healing has occurred, whereas resorbable support devices degrade as the tissue heals, providing a gradual transfer of the full mechanical load from the device to the tissue. Sutures, one of the most widely used temporary support devices, are designed to promote wound closure and healing. Several synthetic polymers are currently employed as biodegradable suture materials (e.g., polyglycolic acid); however, the first sutures were fashioned from natural polymers. For example, silk has been employed as a suture material for centuries due to its impressive mechanical properties, natural biodegradability, and biocompatibility [42]. Similarly, degradable collagen sutures have been in use for several decades, and fibrin glue is currently used as a topical hemostat and tissue sealant in place of sutures [6,43]. When designing temporary support devices that are resorbable, the rate of polymer degradation must be tuned to match that of natural tissue healing [5]. If the material degrades too quickly and fails before the tissue is ready to take on the full load, then further tissue damage could result. Conversely, if the polymer degradation rate is too slow, then this could in turn cause stress shielding and hinder tissue healing. Thus, much of current research is focused on developing temporary support devices with controllable degradation rates to prevent these types of failures.

15.5.2 Vascular Grafts

Vascular grafts are used to replace or bypass occluded vessels and as arteriovenous grafts for hemodialysis access. Autologous vessels from the patients' own body are the current gold standard for vascular bypass grafts, but these are unavailable in up to 20% of patients [44]. In these cases, two main synthetic vascular graft options are clinically available. GORE-TEX grafts are solid tubular films made of expanded poly(tetrafluoroethylene) (ePTFE). The highly compact structure of ePTFE provides a highly crystalline polymer. This results in scaffolds with high stiffness and thermal stability. GORE-TEX vascular grafts are currently used in 75% of medium-diameter vessel replacement surgeries [3]. The second synthetic option that is currently available is the Dacron® vascular graft made of poly(ethylene terephthalate) (PET). PET is also a very crystalline material with a high modulus, and woven Dacron grafts are used in 95% of large-diameter and 20% of medium-diameter vessel replacements [3]. Both PET and ePTFE have high failure rates in small-diameter (<4 mm) applications due to restenosis that is caused by thrombogenicity, calcium deposition, lack of endothelialization, or infection. Thus, there exists a large clinical need for alternative vascular graft materials for the cases in which autologous vessels are unavailable and synthetic grafts fail [45].

Allografts and xenografts were initially pursued as potential vascular graft materials; however, these options presented concerns of disease transfer and possible immune rejection [7]. To address this need, researchers are turning toward tissue engineering as a potential solution. A clinical trial was carried out in Japan in 2003 on a tissue-engineered vascular graft composed of a 50:50 copolymer of lactide and ε-caprolactone-implanted in children. In preparation for the surgery, a piece of saphenous vein was harvested from the patients, and cells from it were cultured for 2 months. Before implantation, the cells were seeded onto the synthetic scaffold and cultured for 1 week. In this study, alternative grafts were fabricated from patients' bone marrow cells that were seeded directly onto the scaffold in the surgical suite. This provided a vascular graft that could be implanted on the day of the cell harvest. No complications from thrombosis or restenosis were observed, and no mortalities occurred [45]. Other groups have demonstrated clinical success in the creation of tissue-engineered vascular grafts based on patient cells and still others are working toward synthetic polymer-based grafts that improve on current options by providing biomechanical properties that better match those of native vasculature and appropriate cell–material interactions to promote long-term success.

15.5.3 Skin Grafts

There is a large need for a permanent physiologic replacement for skin, and extensive research has been conducted to develop improved skin grafts. Synthetic polymer-based skin grafts can provide biologically inert scaffolds that control fluid loss, reduce bacterial entry, and adhere to the wound throughout healing, but natural polymer skin grafts have also been utilized with clinical success. A bovine collagen-based skin graft was one of the first to undergo human clinical trials in 1981. This graft promoted the growth of a new dermal layer, but a second surgery was required for the addition of an epidermis. While this was promising, the graft only proved to be successful at repairing the barrier between the patients and their environment and was not capable of replacing the secondary functions of the tissue [46].

Today, several skin grafts are available on the medical market. One example is AlloDerm®, which is an acellular dermal graft manufactured by LifeCell. The graft is composed of cadaveric dermis that serves as a scaffold for the growth of recipient tissue. Alloderm is also used in a variety of alternative applications, including abdominal wall reconstruction. Integra™ provides another acellular dermal regeneration template that became available in 1996 [46]. This scaffold can also be used in adhesion prevention during surgery wherein a thin polymeric film or mesh is placed between adhesion-prone tissues to prevent the formation of scar tissue. The graft material is designed to degrade away once the healing process is complete and adhesion prevention is no longer needed [46]. Current skin graft research is focused on creating grafts that reduce scarring and that can be used over larger areas of skin.

15.5.4 Ligament Replacements

Ligaments and tendons serve as stabilizers for joints and bones. They are incapable of natural regeneration following injury due to their avascular structure, and it is difficult to successfully replace them because of the high physiological loads to which they are subjected. The anterior cruciate ligament (ACL), which serves as the primary stabilizer of the knee, is particularly susceptible to ruptures or tears that can cause pain and discomfort, joint instability, and eventually degenerative joint disease [47–49]. The current gold standard for these procedures is autologous tendons, but harvesting these grafts presents significant donor site pain, and there is limited autologous graft material. In efforts to improve the performance of autologous grafts, ligament augmentation devices have been developed. For example, the Kennedy Ligament-Augmentation Device is a ribbon of woven polypropylene that fits over an autograft to provide additional mechanical support and stability. However, these devices experience several complications, including a delay in graft maturation due to stress shielding and limited long-term success [50]. GORE-TEX produces a synthetic ACL graft made from woven ePTFE that has been implanted into over 18,000 patients worldwide. However, material fatigue and lack of neotissue growth has resulted in high failure rates of these grafts [51,52]. Dacron also developed an ACL graft aiming to reduce the stress shielding seen with previous synthetic grafts. Four strands of PET were woven tightly together and encased in a sleeve of loosely woven PET to encourage tissue growth. Unfortunately, the Dacron graft was subject to rupture and elongation, possibly due to permeation of synovial fluid and connective tissue [53,54].

15.5.5 Articular Cartilage Replacements

Articular cartilage is the strong lubricating cartilage between moving joints such as the knee or elbow. Osteoarthritis disrupts the structure and function of this highly organized tissue in approximately 27 million Americans. Owing to its lack of vascularization, cartilage lacks the native ability to completely self-repair, and cartilage injuries are one of the primary causes of joint replacement surgery later in life. To remedy this, attempts have been made to fill cartilage defects with a variety of natural and synthetic scaffolds [55]. Owing to the role of cartilage in lubricating joints, hydrogels have been extensively studied as potential cartilage replacements. SaluCartilage, produced by Salumedica, is a poly(vinyl alcohol) (PVA) hydrogel plug available in a variety of sizes. Clinical trials have shown reduced pain in patients with no evidence of graft loosening after 2 years. Similarly, Carticept Medical™ developed a PVA hydrogel as a cartilage replacement therapy for full-thickness chondral defects. These gels have been tailored to provide similar wear, strength, and coefficient of friction values to native articular cartilage. Both of these products are available on the European market, with efforts to obtain FDA approval for use in the

United States [56]. Additionally, tissue engineering has emerged as a potential strategy for improved cartilage repair.

15.5.6 Nerve Guidance Channels

Large defects, or gaps, in peripheral nerves are incapable of regeneration and require graft-facilitated repair. The current gold standard for nerve replacement is autografts harvested from another part of the patient's body. This results in a loss of donor site function, and these grafts can fail due to a mismatch in dimensions and modality. Thus, there is a clinical need for a synthetic nerve guidance channel. A critical first step in the sequence of nerve repair is the proper formation of an aligned fibrin matrix between the nerve stumps [57]. To achieve this, a saline-filled polymer channel has been used to bridge the gap between damaged nerves. This technique has successfully stimulated nerve repair in humans [57]. SaluMedica was one of the first companies to obtain market approval for a synthetic nerve cuff in 2001. These scaffolds are made of poly(vinyl alcohol) hydrogel tubes up to 6.35-cm long that have undergone a repeated freeze/thaw cycle to provide grafts with similar strength, durability, and flexibility to native nerve tissues [58]. Currently, AxoGen® produces a peripheral nerve allograft made from decellularized extracellular matrix processed from human peripheral nerve tissue. This allograft combines the ease and availability of the off-the-shelf products with the natural nerve structure of an autograft. During the healing process, the device promotes and supports remodeling and growth of natural tissue [59]. While these products show promise, the majority of progress made in nerve grafts has been limited to peripheral nerves, and there are no viable options available for use in the central nervous system. Additionally, large gaps have proven difficult to bridge, and a great deal of current research is focused on developing nerve grafts for large defects.

15.5.7 Drug Delivery

Improved drug-delivery systems can help to increase efficacy and reduce toxicity of drugs, and they could serve to enhance convenience and patient compliance. There are two primary types of drug release that can be utilized to maintain a drug's therapeutic index, or its effective concentration range: temporal and distribution control. In temporal release, the drug is delivered at designated times during treatment. This approach works best with drugs that are quickly metabolized so that they remain within their therapeutic index. Distribution control involves the release of the drug to the precise site of activity in the body. This is useful when a drug's therapeutic index is very narrow and can help to reduce the chance of unwanted side effects [60].

Hydrogels, in particular, serve as ideal drug-delivery vehicles, as they provide several mechanisms by which to control release, including swelling, degradation, and stimuli response (e.g., pH, light, temperature) [61]. Among the many hydrogel-based products on the drug-delivery market are SQZ Gel™, an oral medication that is composed of chitosan and poly(ethylene glycol). SQZ Gel provides pH-sensitive controlled release of diltiazem hydrochloride to treat hypertension. To initiate and continue cervical ripening during labor induction, Cervidil®, a poly(ethylene oxide) and polyurethane-based hydrogel, is available to provide a controlled *in vivo* release of dinoprostone. Moraxen is a hydrogel suppository available in the EU that provides slow release of morphine sulfate to relieve end-stage cancer pain [61]. Hydrogel drug-delivery systems can be utilized in an extensive range of applications to maximize drug use within the therapeutic range to provide increased convenience and improved treatment for patients.

15.5.8 Contact Lenses

Contact lenses have been under development for over 100 years. Significant advances and improvements have been made, but a completely ideal solution remains to be developed. A contact lens must maintain a stable, continuous tear film for clear vision, resist the deposition of tear film components, sustain natural hydration, maintain permeability to oxygen and ions, and be nonirritating and comfortable [62]. The first glass contact lens was introduced in the 1800s, and the first commercial contact lenses fabricated from PMMA were introduced in 1936. It was later determined that the cornea requires oxygen to maintain its clarity, structure, and function. In light of this, researchers turned their focus toward improving the oxygen permeability of contact lenses. This led to the development of siloxane hydrogels made from PDMS [62]. Since then, commercially available contacts such as Focus Night & Day Lenses® and PureVision® have oxygen-permeable lenses, and serve as an improvement over past systems. However, these materials have caused new problems, such as epithelial arcuate lesions and mucin balls [63]. Current research aims to develop contact lens materials with a lower modulus to improve fit and comfort for the user [62].

15.6 Summary and Future Directions

The field of clinically applied biomaterials in soft tissue applications is mainly focused on polymeric devices due to the similarity of their viscoelastic properties to native tissue. These properties can be further tuned using different fabrication techniques to match a biomaterial to a specific application. The selection of biomaterials discussed above provides examples of current clinical applications in soft tissues. The ever-increasing demands for soft tissue repair and reconstruction has engaged researchers to design materials with more complex properties and responses. Two of the most promising areas of research to address these needs are stimuli-responsive, or smart materials, and tissue engineering.

15.6.1 Smart Materials

Intelligent or smart materials exhibit a significant change in one or more properties in response to an external stimuli such as stress, temperature, moisture, pH, and electric or magnetic fields [64]. Examples of such materials include smart hydrogels and shape memory polymers. Smart hydrogels exhibit different swelling behavior based on their environmental parameters such as temperature and pH. These materials are currently being researched for targeted drug-delivery applications, specifically for cancer and diabetes treatment. Antigen-responsive hydrogels are another example of gels that swell in the presence of an excess of a specific antigen allowing for the release of a specific drug or protein [65]. Shape memory polymers have the ability to return from a deformed, temporary shape to a permanent geometry when triggered by an external stimulus. The most common example is temperature-sensitive polymer systems that exhibit a collapsed state above or below the polymer's transition temperature. Another example is materials that are liquids at room temperature but harden in response to an external stimulus such as an increase in temperature or light. The material can then be injected into the patient in a conformed state, but will harden into a permanent device [65]. Preclinical studies have begun on a shape memory polymer vascular stent and heart valve replacements. Thrombectomy devices and devices to treat aneurysms are also currently in production [66]. Shape memory sutures that knot themselves in response to an increase in temperature have already been deployed in the clinic [65].

15.6.2 Tissue Engineering

Tissue engineering seeks to harness the body's natural ability to heal through the use of a resorbable polymer matrix. This eliminates the need for permanent implanted devices and thus reducing long-term complications and restoring the full function of the regenerated tissue. The traditional tissue engineering paradigm combines isolated cells with appropriate bioactive agents in a biomaterial scaffold. The biomaterial scaffold sustains functionality during regeneration and serves as a template for the necessary cellular interactions. Scaffold design is naturally guided by the physiological properties of the target tissue such as the elasticity of skin or the high modulus and strength of bone. The scaffold is seeded with cells prior to implantation or recruits cells from adjacent tissue after implantation. As such, critical design criteria also include proper cellular cues to guide regeneration and pore interconnectivity to allow for proper nutrient/waste transport [67]. Many tissue-engineered skin grafts that utilize a range of natural and synthetic biomaterials and cell sources are currently available on the market [68]. Additionally, clinical trials for tissue-engineered vascular grafts have been carried out with promising results, and a collagen-based cartilage tissue engineering scaffold is clinically available in Europe [69–70]. Preclinical trials of a wide variety of tissue-engineered grafts have shown promise, and this approach could provide improved treatment options in the future.

References

1. Humphrey, J.D., Review paper: Continuum biomechanics of soft biological tissues. *Proceedings of the Royal Society*, 2003. 459(2029): 3–46.
2. Baier, J., *Hyaluronic Acid Hydrogel Biomaterials for Soft Tissue Engineering Applications, in Chemical Engineering*. 2003, Austin, TX: The University of Texas at Austin.
3. Puskas, J.E. and Y. Chen, Biomedical application of commercial polymers and novel polyisobutylene-based thermoplastic elastomers for soft tissue replacement. *Biomacromolecules*, 2004. 5(4): 1141–1154.
4. Buttafoco, L. et al., Electrospinning of collagen and elastin for tissue engineering applications. *Biomaterials*, 2005. 27(5): 724–734.
5. Ratner, B. et al., *Biomaterials Science: An Introduction to Materials in Medicine*. 2nd ed. 2004, San Diego, CA: Elsevier Inc.
6. Patel, S., E.C. Rodriguez-Merchan, and F.S. Haddad, The use of fibrin glue in surgery of the knee. *The Journal of Bone and Joint Surgery*, 2010. 92(10): 1325–1331.
7. Lee, C.H., A. Singla, and Y. Lee, Biomedical applications of collagen. *International Journal of Pharmaceutics*, 2001. 221(1–2): 1–22.
8. Odian, G., *Principles of Polymerization*. 4th ed. 2004, Hoboken, NJ: John Wiley & Sons, Inc. 832.
9. Stevens, M.P., *Polymer Chemistry: An Introduction*. 1999, New York, NY: Oxford University Press.
10. Cornwell, K.G., A. Landsman, and K.S. James, Extracellular matrix biomaterials for soft tissue repair. *Clinics in Podiatric Medicine and Surgery*, 2009. 26(4): 507–523.
11. Shoulders, M.D. and R.T. Raines, Collagen structure and stability. *Annual Reviews in Biochemistry*, 2009. 78: 929–958.
12. Khor, E., Methods for the treatment of collagenous tissues for bioprostheses. *Biomaterials*, 1997. 18(2): 95–105.
13. Standeven, K.F., R.A.S. Ariens, and P.J. Grant, The molecular physiology and pathology of fibrin structure/function. *Blood Reviews*, 2005. 19(5): 275–288.
14. Vepari, C. and D.L. Kaplan, Silk as a biomaterial. *Progress in Polymer Science*, 2007. 32(8–9): 991–1007.
15. Temenoff, J.S. and A.G. Mikos, *Biomaterials: The Intersection of Biology and Materials Science*. 2008, Upper Saddle River, NJ: Pearson Prentice Hall Bioengineering.
16. Matabola, K. et al., Single polymer composites: A review. *Journal of Materials Science*, 2009. 44(23): 6213–6222.
17. Dalton, P.D. et al., Patterned melt electrospun substrates for tissue engineering. *Biomedical Materials*, 2008. 3(3): 1–11.
18. Ji, C. et al., Fabrication of poly-DL-lactide/polyethylene glycol scaffolds using the gas foaming technique. *Acta Biomaterialia*, 2012. 8(2): 570–578.
19. Johnson, T. et al., Fabrication of highly porous tissue-engineering scaffolds using selective spherical porogens. *Bio-Medical Materials and Engineering*, 2010. 20(2): 107–118.
20. Nam, Y.S. and T.G. Park, Porous biodegradable polymeric scaffolds prepared by thermally induced phase separation. *Journal of Biomedical Materials Research*, 1999. 47(1): 8–17.

21. Guan, J. et al., Preparation and characterization of highly porous, biodegradable polyurethane scaffolds for soft tissue applications. *Biomaterials*, 2005. **26**(18): 3961–3971.

22. Sirringhaus, H., Inkjet printing of functional materials. *MRS Bulletin*, 2003. **28**(11): 802.

23. Hubbell, J.A., Bioactive biomaterials. *Current Opinion in Biotechnology*, 1999. **10**: 123–129.

24. Nguyen, K.T. and J.L. West, Photopolymerizable hydrogels for tissue engineering applications. *Biomaterials*, 2002. **23**: 4307–4314.

25. Griffith, L.G. and G. Naughton, Tissue engineering—Current challenges and expanding opportunities. *Science*, 2002. **295**: 1009–1014.

26. Lee, K.Y. and D.J. Mooney, Hydrogels for tissue engineering. *Chemical Review*, 2001. **101**(7): 1869–1880.

27. Hoffman, A.S., Hydrogels for biomedical applications. *Advanced Drug Delivery Reviews*, 2002. **43**: 3–12.

28. Rakovsky, A. et al., Poly(ethylene glycol)-based hydrogels as cartilage substitutes: Synthesis and mechanical characteristics. *Journal of Applied Polymer Science*, 2008. **112**: 390–401.

29. Deible, C.R. et al., Molecular barriers to biomaterial thrombosis by modification of surface proteins with polyethylene glycol. *Biomaterials*, 1998. **19**(20): 1885–1893.

30. Greenwald, R.B. et al., Effective drug delivery by PEGylated drug conjugates. *Advanced Drug Delivery Reviews*, 2003. **55**(2): 217–250.

31. Queen, D. et al., Burn wound dressings—A review. *Burns*, 1987. **13**(3): 218–228.

32. Burdick, J.A. and K.S. Anseth, Photoencapsulation of osteoblasts in injectable RGD-modified PEG hydrogels for bone tissue engineering. *Biomaterials*, 2002. **23**: 4315–4323.

33. Mann, B. et al., Smooth muscle cell growth in photopolymerized hydrogels with cell adhesive and proteolytically degradable domains: Synthetic ECM analogs for tissue engineering. *Biomaterials*, 2001. **22**: 3045–3051.

34. Bryant, S.J. and K.S. Anseth, Controlling the spatial distribution of ECM components in degradable PEG hydrogels for tissue engineering cartilage. *Journal of Biomedical Materials Research Part A*, 2003. **64A**(1): 70–79.

35. Stosich, M.S. and J.J. Mao, Adipose tissue engineering from human adult stem cells: Clinical implications in plastic and reconstructive surgery. *Plastic and Reconstructive Surgery*, 2007. **119**(1): 71–83.

36. Bryant, S.J. and K.S. Anseth, Hydrogel properties influence ECM production by chondrocytes photoencapsulated in poly(ethylene glycol) hydrogels. *Journal of Biomedical Materials Research*, 2002. **59**: 63–72.

37. Namba, R.M. et al., Development of porous PEG hydrogels that enable efficient, uniform cell-seeding and permit early neural process extension. *Acta Biomaterialia*, 2009. **5**: 1884–1897.

38. Shoichet, M.S. et al., Stability of hydrogels used in cell encapsulation: An *in vitro* comparison of alginate and agarose. *Biotechnology and Bioengineering*, 1996. **50**: 374–381.

39. Weber, L.M. et al., PEG-based hydrogels as an *in vitro* encapsulation platform for testing controlled beta-cell microenvironments. *Acta Biomaterialia*, 2006. **2**: 1–8.

40. Qui, Y. and K. Park, Environment-sensitive hydrogels for drug delivery. *Advanced Drug Delivery Reviews*, 2001. **53**: 321–339.

41. Bryant, S.J., G.D. Nicodemus, and I. Villanueva, Designing 3D photopolymer hydrogels to regulate biomechanical cues and tissue growth for cartilage tissue engineering. *Pharmaceutical Research*, 2008. **25**(10): 2379–2386.

42. Altman, G.H. et al., Silk-based biomaterials. *Biomaterials*, 2003. **24**(3): 401–416.

43. Postlethwait, R.W., Polyglycolic acid surgical suture. *Archives of Surgery*, 1970. **101**(4): 489–494.

44. Darling, R.C. and R.R. Linton, Durability of femoropopliteal reconstructions: Endarterectomy versus vein bypass grafts. *The American Journal of Surgery*, 1972. **123**(4): 472–479.

45. Matsumura, G. et al., Successful application of tissue engineered vascular autografts: Clinical experience. *Biomaterials*, 2003. **24**(13): 2303–2308.

46. Wood, B.C., C.N. Kirman, and J.A. Molnar, Skin grafts. *Medscape Reference: Drugs, Diseases, and Procedures*, Web MD, 4 January 2011. Web. 2 February 2012.

47. Vunjak-Novakovic, G. et al., Tissue engineering of ligaments. *Annual Reviews in Biochemistry*, 2004. **6**: 131–156.

48. Pennisi, E., Tending tender tendons. *Science*, 2002. **295**(5557): 1011.

49. Albright, J.C., J.E. Carpenter, B.K. Graf, and J.C. Richmond, Knee and leg: Soft-tissue trauma. *Orthopaedic Knowledge Update*, 1999. **6**: 533–559.

50. Kumar, K. and N. Maffulli, The ligament augmentation device: An historical perspective. *Arthroscopy: The Journal of Arthroscopic & Related Surgery*, 1999. **15**(4): 422–432.

51. Indelicato, P.A., M.S. Pascale, and M.O. Huegel, Early experience with the GORE-TEX polytetra-fluoroethylene anterior cruciate ligament prosthesis. *The American Journal of Sports Medicine*, 1989. **17**(1): 55–62.

52. Thomson, L.A. et al., Biocompatibility of particles of GORE-TEX® cruciate ligament prosthesis: An investigation both *in vitro* and *in vivo*. *Biomaterials*, 1991. **12**(8): 781–785.

53. Lopez-Vasquez, E., J.A. Juan, E. Vila, and J. Debon, Reconstruction of the anterior cruciate ligament with a Dacron prosthesis. *Journal of Bone and Joint Surgery*, 1991: 294–300.

54. Richmond, J.C., C.J. Manseau, and R. Patz, Anterior cruciate reconstruction using a Dacron ligament prosthesis. A long-term study. *The American Journal of Sports Medicine*, 1992. **20**(1): 24–28.

55. Downes, S. et al., The regeneration of articular cartilage using a new polymer system. *Journal of Materials Science: Materials in Medicine*, 1994. **5**(2): 88–95.

56. McNickle, A.G., M.T. Provencher, and B.J. Cole, Overview of existing cartilage repair technology. *Sports Medicine and Arthroscopy Review*, 2008. **16**(4): 196–201.

57. Clements, I.P. et al., Thin-film enhanced nerve guidance channels for peripheral nerve repair. *Biomaterials*, 2009. **30**(23–24): 3834–3846.

58. Bell, J.H.A. and J.W. Haycock, Next generation nerve guides: Materials, fabrication, growth factors, and cell delivery. *Tissue Engineering Part B: Reviews*, 2012. **18**(2): 116–128.

59. Inc, A. *Avance Nerve Graft: The Natural Connection.* 2010 [cited April 19, 2012]; Available from: http://www.axogeninc.com/nerveGraft.html.

60. Uhrich, K.E. et al., Polymeric systems for controlled drug release. *Chemical Reviews*, 1999. **99**(11): 3181–3198.

61. Gupta, P., K. Vermani, and S. Garg, Hydrogels: From controlled release to pH-responsive drug delivery. *Drug Discovery Today*, 2002. **7**(10): 569–579.

62. Nicolson, P.C. and J. Vogt, Soft contact lens polymers: An evolution. *Biomaterials*, 2001. **22**(24): 3273–3283.

63. Morgan, P.B. and N. Efron, Comparative clinical performance of two silicone hydrogel contact lenses for continuous wear. *Clinical and Experimental Optometry*, 2002. **85**(3): 183–192.

64. Roy, I. and M.N. Gupta, Smart polymeric materials: Emerging biochemical applications. *Chemistry & Biology*, 2003. **10**(12): 1161–1171.

65. Langer, R. and D.A. Tirrell, Designing materials for biology and medicine. *Nature*, 2004. **428**(6982): 487–492.

66. Sokolowski, W. et al., Medical applications of shape memory polymers. *Biomedical Materials*, 2007. **2**(1): S23.

67. Nicodemus, G.D. and S.J. Bryant, Cell encapsulation in biodegradable hydrogels for tissue engineering applications. *Tissue Engineering, Part B*, 2008. **14**(2): 149–165.

68. Groeber, F. et al., Skin tissue engineering—*In vivo* and *in vitro* applications. *Advanced Drug Delivery Reviews*, 2011. **63**(4–5): 352–366.

69. Peck, M. et al., The evolution of vascular tissue engineering and current state of the art. *Cells, Tissues, Organs*, 2012. **195**(1–2): 144–158.

70. Stoddart, M.J., Cells and biomaterials in cartilage tissue engineering. *Regenerative Medicine*, 2009. **4**(1): 81–98.

71. Pins, G.D. et al., Self-assembly of collagen fibers. Influence of fibrillar alignment and decorin on mechanical properties. *Biophysical Journal*, 1997. **73**(4): 2164–2172.

72. Butler, D.L., M.D. Kay, and D.C. Stouffer, Comparison of material properties in fascicle-bone units from human patellar tendon and knee ligaments. *Journal of Biomechanics*, 1986. **19**(6): 425–432.

73. Matthews, J.A. et al., Electrospinning of collagen nanofibers. *Biomacromolecules*, 2002. **3**(2): 232–238.

74. Cunniff, P.M. et al., Mechanical and thermal properties of dragline silk from the spider Nephila clavipes. *Polymers for Advanced Technologies*, 1994. **5**(8): 401–410.

75. Guidoin, M.-F. et al., Analysis of retrieved polymer fiber based replacements for the ACL. *Biomaterials*, 2000. **21**(23): 2461–2474.

76. Korenkov, M. et al., Randomized clinical trial of suture repair, polypropylene mesh or autodermal hernioplasty for incisional hernia. *British Journal of Surgery*, 2002. **89**(1): 50–56.

77. von Buelow, S., D. von Heimburg, and N. Pallua, Efficacy and safety of polyacrylamide hydrogel for facial soft-tissue augmentation. *Plastic and Reconstructive Surgery*, 2005. **116**(4): 1137–1146.

78. Ota, K., et al. Clinical application of modified polyurethane graft to blood access. *Artificial Organs*, 1991. **15**(6): 449–453.

79. Wache, H.M. et al., Development of a polymer stent with shape memory effect as a drug delivery system. *Journal of Materials Science: Materials in Medicine*, 2003. **14**(2): 109–112.

80. Kricheldorf, H.R. et al., Polylactides—Synthesis, characterization and medical application. *Macromolecular Symposia*, 1996. **103**(1): 85–102.

81. Waksman, R., Update on bioabsorbable stents: From bench to clinical. *Journal of Interventional Cardiology*, 2006. **19**(5): 414–421.

82. Tsuji, T. et al., Biodegradable stents as a platform to drug loading. *Acute Cardiac Care*, 2003. **5**(1): 13–16.

83. Fan, H. et al., *In vivo* study of anterior cruciate ligament regeneration using mesenchymal stem cells and silk scaffold. *Biomaterials*, 2008. **29**(23): 3324–3337.

84. Krishnan, L.K. et al., Comparative evaluation of absorbable hemostats: Advantages of fibrin-based sheets. *Biomaterials*, 2004. **25**(24): 5557–5563.

16

Active Materials

Kathryn E. Smith
MedShape, Inc

Carl P. Frick
University of Wyoming

David L. Safranski
MedShape, Inc

Christopher M. Yakacki
University of Colorado Denver

Ken Gall
Georgia Institute of Technology

16.1 Introduction

Active materials represent a class of materials that are able to undergo a physical or chemical change upon exposure to a stimulus. They can be activated by temperature, force, electric current, water, and/or chemical moieties. In their "active" form, these materials can change shape, apply force, degrade, or interact with biological tissues. Given these unique properties, there is a growing interest in the use of these "smart" materials in biomedical applications as these materials can provide additional novel functions to an implant or device. For example, a shape-memory material can alter its shape, allowing for an implant to be inserted minimally invasively in a compact shape and then activated into a larger form upon implantation to serve its physiological function.

Although there are many modes of activity that have been demonstrated, only a few have translated to clinical application. As shown in Figure 16.1, those active materials in clinical use (or "on the cusp") can be classified as being

1. Mechanically active
2. Bioactive
3. Electroactive

Active materials consist of metals, polymers, and ceramics, but certain material types have superior functional properties over others. For instance, bioactive materials are predominantly composed of ceramics, although metals and polymers can exhibit bioactivity after undergoing certain surface treatments. Mechanically active materials primarily relate to either shape-memory polymers (SMPs) or alloys. Electroactive materials are composed of either ionic or electronic electroactive polymers (EAPs) and are the least developed for biomedical applications compared with the other active materials. It is

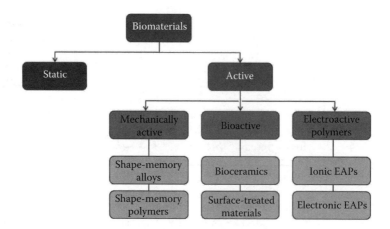

FIGURE 16.1 Schematic of various classes of active materials.

important to note that there are no EAPs currently in clinical use, but attempts at commercializing devices with these materials are ongoing and worth mentioning for the purposes of this chapter.

While active materials offer many advantages as components of medical devices, they also bring with them a new set of challenges pertaining to processing and packaging. This chapter will provide a broad overview of the various types of active materials that have reached clinical application. For each type, the specific activation mechanism will be discussed in relation to the materials' chemistry, structure, and processing. Next, some current clinical applications for each will be highlighted along with the current limitations that need to be addressed in future research and development.

16.2 Mechanically Active Materials: Shape-Memory Alloys

A shape-memory material is one that can recover inelastic strain in response to an outside stimulus. The shape-memory effect has been observed in all three classes of engineering materials (i.e., ceramics, metals, and polymers). A qualitative comparison is shown in Table 16.1.[1] Ceramics that can be considered shape-memory mostly fall into the category of piezoelectric. Such crystalline materials can become electrically polarized when subjected to a mechanical stress and then conversely change shape when under an applied electric field.[2] However, ultimately, the recoverable strain levels for even the most ideal piezoelectric ceramics struggle to get above 1%, and the necessary applied electric field makes them largely inappropriate for biomedical devices.

Conversely, several types of biomedical devices fabricated from shape-memory alloys and polymers have been suggested.[3-6] The inherent material behavior between these two classes of shape-memory materials are in stark contrast from one another, which from a design perspective is beneficial as it allows for the selection of materials with an extremely broad range of relevant material properties. In general, SMPs exhibit recoverable strains on the order of tens to hundreds of percent strain,[7-9] are comparatively inexpensive, are easy to manufacture, are compatible with magnetic resonance imaging, and

TABLE 16.1 Comparison of Classes of Shape-Memory Materials

Shape-Memory Material	Relative Activation Timescale	Recoverable Strain Levels (%)	Relative Stress Levels	Relative Cost
Ceramics	Fast	0.1–1	Large	High
Alloys	Moderate	2–12	Large	High
Polymers	Slow	30–400	Small	Low

Source: Adapted from Juhasz JA, Best SM. *J Mater Sci* 2012;47:610–624; Frayssinet P. In: Poitout DG, ed. *Biomechanics and Biomaterials in Orthopedics.* London: Springer; 2004:101.

can be degradable and/or used for drug delivery. As shape-memory behavior in polymers is a function of material processing and not an intrinsic property,[10] many types of polymer systems have the opportunity to be made into an SMP. However, their poor thermal conductivity inherent to their polymer nature makes shape recovery time relatively slow (e.g., several minutes), low radio-opacity makes imaging of small volumes difficult via x-ray, and they are often not considered for actuator or high-strength purposes due to the low stress during recovery.

Unlike their polymer counterpart, shape-memory alloys activate quickly (e.g., several seconds),[11] exhibit the largest work output per volume of any known actuator,[12] and several devices made from shape-memory alloys have already received the US Food and Drug Administration (FDA) clearance and are available in the marketplace. Perhaps most importantly, unlike shape-memory ceramics and polymers, alloys have the capability of spontaneous strain recovery upon removal of applied load[13]; however, the amount of strain recovery achievable is far less compared with SMPs. While the limited strain recovery and high cost of manufacturing are of some concern, the widespread use of shape-memory alloys has primarily been hindered by a limited understanding of the complex microstructural mechanisms that dictate the shape-memory response.

16.2.1 Shape-Memory Alloy Basics

Shape-memory alloys are extremely unique relative to other metals in that they have the ability to recover relatively large amounts of applied deformation. This deformation recovery manifests itself in one of two ways: the shape-memory effect and pseudoelasticity.[13] The shape-memory effect refers to the materials' ability to store the applied deformation for a period of time, and subsequently recover the deformation upon exposure to a stimulus, typically heat. Pseudoelasticity refers to spontaneous deformation recovery upon mechanical unloading. Relative to other shape-memory materials, alloys are well established in both medical and other engineering applications. They are currently utilized in a wide variety of biomedical devices ranging from vascular to orthopedic to orthodontic applications.[4,14–16]

The shape-memory behavior in metal alloys is inherently due to a reversible thermoelastic martensitic solid-state phase transformation. To illustrate this concept, consider an oversimplified example of a shape-memory alloy at relatively "high" temperature where the metal is in its parent phase (often termed austenite), typically consisting of a cubic crystal structure. As the temperature is decreased, this cubic structure is no longer energetically favored, and the atomic structure transitions to a low-symmetry martensitic phase. Figure 16.2 schematically illustrates the austenite-to-martensite phase transformation upon cooling. The transformation initiates at the martensite start temperature (M_s) and is nearly 100% transformed at the martensite finish temperature (M_f). Analogously, upon reheating, the phase transformation is entirely

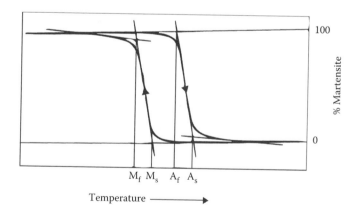

FIGURE 16.2 Martensite fraction as a function of temperature.

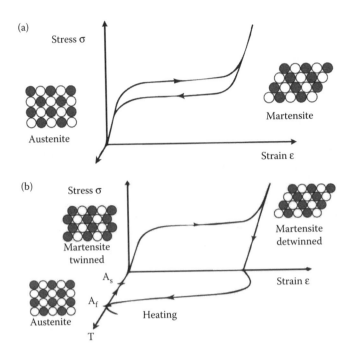

FIGURE 16.3 Schematic illustrations of (a) pseudoelastic and (b) shape-memory stress-induced phase transformations. (Adapted from Michael Kaack, Elastic behavior of NiTi shape memory alloys. Dissertation, Ruhr-Universität Bochum, 2002.)

reversible, as shown in Figure 16.2. It is important to note that the austenite start and finish temperatures (A_s and A_f) are significantly higher than M_s and M_f. This hysteresis is due to the internal energy of the incumbent crystal structure, which must be overcome for the phase transformation to propagate.

In addition to decreasing temperature, it is also possible to create martensite by applying an external load. Inducing the martensitic phase transformation via deformation is inherently different from doing so by thermal means. The martensite variants will orientate preferentially in the direction of the applied stress. This allows the alloy to rearrange its atomic lattice structure to accommodate relatively large amounts of inelastic deformation. Stress–strain behavior will exhibit initial elastic loading, followed by a stress plateau associated with nucleation and propagation of the oriented martensite phase. Depending on the temperature at which deformation was performed, one of two outcomes is possible, qualitatively illustrated in Figure 16.3: (1) if martensite is energetically unfavorable (i.e., temperatures above A_f), then the material will spontaneously recover the applied deformation exhibiting pseudoelastic behavior; (2) if martensite is energetically stable (i.e., a testing temperature below A_s), the oriented martensite remains stable upon unloading and the deformation is stored. Upon heating above the austenite transformation temperatures, the crystal structure switches back to austenite, and the deformation is recovered, exhibiting the shape-memory effect. It is important to note that even if the stress is applied at temperatures below M_f, shape-memory behavior will be observed. The twinned martensite will become oriented in the direction of the applied forces, allowing for the shape recovery upon heating.

16.2.2 Nickel–Titanium Shape-Memory Alloys

Over the last 50 years, more than 30 shape-memory alloys have been identified, and the list continues to grow.[17] Unfortunately, the majority of these alloys exhibit relatively low amounts of recoverability, exhibit transformation temperatures too high for practical use, or consist of rare/precious metals, limiting their

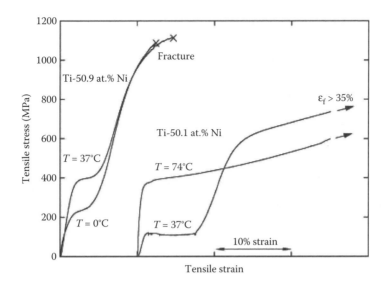

FIGURE 16.4 Strain-to-failure behavior of wire formed from Ti-50.1 at.% Ni and Ti-50.9 at.% Ni. Each material was cold worked above 30% area reduction, then heat-treated at 350°C for 1.5 h. (Taken from Gall K et al. *J Biomed Mater Res A* 2005;73:339–348.)

practical use in commercial applications. Only a few alloys have been developed into novel devices that have become commercially viable. These successful shape-memory metals stem from either copper-based alloys or nickel–titanium-based alloys. Of these two groups, nickel–titanium (NiTi) is more heavily utilized due to its favorable mechanical and electrochemical properties. Specifically, NiTi (often termed nitinol) can recover much higher strains than other shape-memory alloys, observed to be as much as 12%.[18] NiTi has outstanding strength,[19,20] corrosion resistance,[21,22] fatigue behavior,[23,24] and biocompatibility.[25,26] However, specific material properties of NiTi shape-memory alloys are extremely sensitive to chemical composition, prior thermal heat treatment, and deformation processing history. For example, strain-to-failure plots for two conventional compositions of NiTi wire are shown in Figure 16.4.[20] The compositions vary by less than one atomic percent (50.1 vs. 50.9 at.% Ni); however, A_f for the 50.9 at.% occurs at a temperature 25°C lower than the other composition. Carefully observing the stress–strain behavior of the 50.9 at.% Ni tested at 37°C illustrates pseudoelastic behavior with a martensite transformation stress of approximately 400 MPa. Deformation beyond the martensite transformation range results in elastic deformation, followed by plastic deformation and fracture at approximately 1100 MPa, at a strain of approximately 12%. In contrast, the 50.1 at.% Ni wire demonstrates shape-memory behavior at body temperature, a martensite transformation stress less than 200 MPa, a martensitic yield stress of approximately 650 MPa, and a fracture strain greater than 35%. The stark contrast in isothermal mechanical behavior is due to the residual dislocation density induced by cold working and the formation of Ti_3Ni_4 precipitates in the nickel-rich material. However, a thorough discussion of the microstructural mechanisms that dictate the mechanical behavior of NiTi alloys is beyond the scope of this work; for a comprehensive review of NiTi metallurgy, the reader is referenced to Ref. [27].

16.2.3 Clinical Applications of NiTi

NiTi is used as the primary component in a wide variety of biomedical devices, based largely on the martensitic phase transformation. Devices include self-expanding stents, superelastic tubing, and orthopedic porous scaffolds, just to name a few. Below, several specific examples are discussed, with an emphasis on highlighting a unique aspect of mechanical behavior of NiTi.

16.2.3.1 Pseudoelastic Deployment

From a practical perspective, the large strain recovery associated with pseudoelasticity gives NiTi an enormous advantage over traditional metal biomedical devices. In comparison to stainless steel, cobalt–chromium, or titanium alloys, which exhibit approximately 0.5–1% elastic strain, typical recovery values for polycrystalline NiTi, properly treated, is 6–12%. Even in comparison to other shape-memory materials, which require some stimulus to recover deformation, pseudoelasticity is a relatively unique attribute. As modern medicine strives toward less invasive procedures, instruments and devices that can pass through relatively small openings and then self-expand into a desired shape become extremely useful.[15]

An example to illustrate the advantages of pseudoelasticity in biomedical devices is the StarClose vascular closure system developed by Abbott Vascular.[28–32] Vascular closure devices are frequently used to close the small hole in the artery after a cardiovascular procedure or endovascular surgery requiring a catheterization. Traditionally, the standard technique to achieve hemostasis is accomplished by manual compression, which involves 10–20 min of sustained pressure over the puncture site followed by bed rest for a period of 3–6 h. Relative to compression, a wide range of vascular closure devices ranging from collagen/thrombin-based products to staples/clips have demonstrated reduced time to hemostasis and facilitated ambulation.[33] StarClose utilizes a NiTi clip, which employs pseudoelasticity to extraluminally pull the arteriotomy hole shut. The device is deployed using a "through-the-sheath" method, and is designed in a star-like configuration to exhibit a circumferential, extravascular closure over several seconds. Figure 16.5 illustrates the deployment of the StarClose device. A study involving 596 patients showed that the median time to achieve homeostasis was decreased from 19.60 min to 20s using StarClose relative to traditional compression.[29]

16.2.3.2 Shape-Memory Deployment

In addition to pseudoelasticity, it is also possible to utilize the shape-memory effect for biomedical NiTi device deployment. An example of this is a nitinol vena cava filter, such as the TrapEase (Cordis) or the

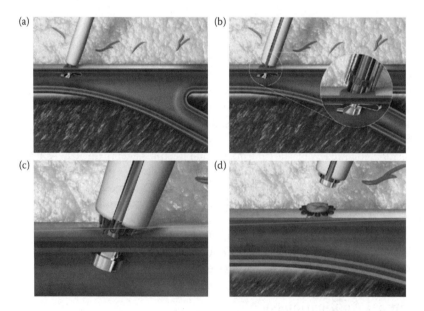

FIGURE 16.5 Deployment of StarClose: (a) The StarClose device is inserted into the sheath. The vessel locator button is depressed. The device slides out until resistance is felt. (b) The advancement of the thumb advancer completes the splitting of the sheath. The device is raised to an angle of slightly less than 90°. (c) The clip is deployed. (d) The device is retracted. (Taken from Hon L-Q et al. *Curr Probl Diagn Radiol* 38:33–43. Copyright 2009, with permission from Elsevier.)

Simon Filter (Bard). Inferior vena cava filters are often implanted for the prevention of a pulmonary embolism as part of the treatment of venous thromboembolic disease.[34] A pulmonary embolism can occur when one or more emboli break away from a blood clot in a vein and are carried to the lung, which may lead to heart failure and death. Pulmonary embolisms are a significant cause of morbidity and mortality, accounting for more than 140,000 deaths per year in the United States alone.[35] Traditional treatment involves anticoagulant drugs to reduce blood clotting; however, for a select number of patients for whom anticoagulation has failed or patients at high risk, a filter can be placed in the vena cava designed to catch relatively large embolism before they reach the lungs.[36,37] The emboli are held in the filter where they will gradually dissolve over time. Filters made from NiTi are ideally suited for this application, as they can be removed after a period of time,[38] and they can be deployed through low-profile insertion systems such as the antecubital vein (in the elbow) where deployment is more easily controlled.[15] The NiTi filters are collapsed in a saline-chilled catheter while in the martensitic state, and when positioned in the deployment site, the flow of chilled saline is stopped. The filter is naturally warmed by the surrounding blood and expands safely to its previously defined shape, typically 3–8 times the catheter diameter.[14]

16.2.3.3 Constant Applied Load

The stress plateau associated with the martensitic phase transformation shown in Figure 16.3 is in and of itself an explicit advantage of pseudoelastic NiTi alloys. Once the plateau has been reached, NiTi can experience a dramatic increase in deformation without an accompanying increase in stress. As a consequence, strain localization (i.e., kinking) is prevented by creating a somewhat uniform strain, not possible with conventional metals.[15] Several devices utilize this concept, such as guide wires for angioplasty.[14] Perhaps the most classic example utilizing the stress plateau in NiTi are orthodontic archwires, first reported in 1971.[39] Conventional metals used for archwires, such as stainless steel, function by applying a significant force to the patient's teeth to correct their position. As the teeth are corrected, the applied forces quickly relax, requiring periodic readjustment by an orthodontist. Wires manufactured from NiTi could provide continuous forces over a much larger range of displacement relative to conventional wire, which is considered ideal for orthodontic treatment.[40] It is estimated that NiTi is used for over 30% of archwires.[14]

Joint fusion is another emerging application that takes advantage of NiTi's stress plateau. In particular, MedShape, Inc. has recently cleared an intramedullary nail that contains an internal pseudoelastic NiTi element. Ankle arthrodesis (fusion) of the tibio-talo-calcaneal joints is a salvage procedure designed to provide a stable and pain-free alternative to amputation for patients with severe ankle arthropathy. Currently, the two surgical options are an external fixation frame or an intramedullary nail to provide compression across the joints. During fusion, bone resorption occurs at the joint, which necessitates sustained compression to maintain bone–bone contact for proper fusion. External fixation frames can adapt to this change; however, complications arise from pin tract infections and patient noncompliance. Intramedullary nails do not have the previous complications, but cannot sustain their compressive load in response to bone resorption, resulting in nonunion. In contrast, the internal NiTi element of this new nail applies a constant load across a large strain due to its pseudoelastic recovery plateau, allowing it to apply sustained compression on the joint for up to 6 mm of bone resorption.[41]

16.2.3.4 Shape-Memory Actuation

In addition to pseudoelasticity and the shape-memory effect, it is also possible to constrain stress-induced martensite and utilize NiTi as an actuator upon heating.[42,43] From a materials perspective, utilizing NiTi as an actuator yields a larger force per unit volume than any other known actuator.[12] NiTi is capable of being resistively heated to initiate the shape-memory effect, and using small material volumes allows for relatively fast cooling such that a low bias force can re-deform the device.[11,44] Utilizing NiTi in this manner has been proposed for various prosthetic and orthotic applications.[45] Traditionally, prosthetic devices are static structures with limited adaptability. State-of-the-art prosthetics may offer a motorized actuation; however, these systems are relatively noisy, complex, and heavy. Recently, shape-memory NiTi wire bundles have been investigated for use as actuation for finger joints, with an emphasis

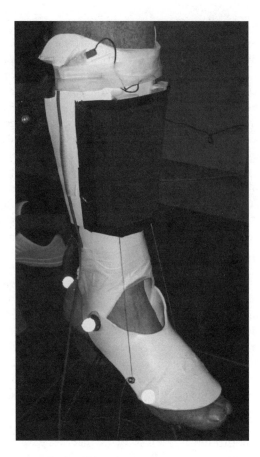

FIGURE 16.6 Prototype of a shape-memory-activated device promotion ankle dorsiflexion. (Taken from Pittaccio S et al. *J Mater Eng Perform* 2009;18:824–830.)

on improving actuation speed by forced cooling using a conductive heat sinking technique.[46–48] NiTi has also been proposed for actuation use in a device that is geared toward rehabilitation of the ankle joint shown in Figure 16.6.[49] Ankle rehabilitation is a common practice for several neuromuscular diseases, including the roughly 5 million people permanently disabled by stroke events.[49]

16.2.4 NiTi Considerations for Clinical Use

Despite the relative success of shape-memory alloy biomedical devices, the full potential of NiTi has yet to be realized. The primary limitations of NiTi are primarily related to its microstructure. The underlying and interwoven mechanisms that dictate the martensitic phase transformation heavily influence stress–strain behavior, and are extremely difficult for novice users to appropriately manipulate (e.g., Figure 16.4). Also, concerns associated with its biocompatibility stem from the relatively large nickel content, but with appropriate surface treatment, researchers have shown that nickel leaching is less than stainless steel leaching. Below is a more thorough discussion of a handful of common pitfalls associated with the use of NiTi.

16.2.4.1 Fatigue Behavior

For conventional metals, repeated mechanical loading over many cycles leads to crack nucleation, growth, and catastrophic fracture, a process known as metal fatigue. For NiTi alloys, fatigue is

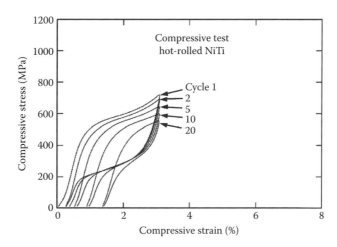

FIGURE 16.7 Low-cycle stress–strain behavior of pseudoelastic NiTi hot-rolled bar. (Taken from Frick CP et al. *Metall Mater Trans A-Phys Metall Mater Sci* 2004; 35A: 2013–2025.)

complicated by the martensitic phase transformation. Despite numerous investigations, much about fatigue behavior in NiTi remains ambiguous. Considering basic macroscopic low-cycle fatigue of pseudoelastic NiTi, several universal observations have been made that include a decrease in the martensitic phase transformation stress, an increase in the phase transformation slope, a decrease in the hysteresis, and an accumulation of residual strain.[24,50–54] In general, low-cycle fatigue of NiTi is considered to be superior to conventional metals because the martensitic phase transformation allows for relatively large recoverable strains, which would otherwise lead to fracture. However, the martensitic phase transformation can significantly degrade with only a handful of repeated loading cycles. Figure 16.7 illustrates a pseudoelastic NiTi rod cyclically loaded to only 3% strain for 20 cycles at room temperature; the residual 1.7% strain is associated with permanent deformation that cannot be recovered upon heating.[50] This finding is a critical concern for any NiTi device that intends to utilize the shape-memory effect or pseudoelasticity for more than one cycle. Specific low-cycle fatigue behavior of NiTi is heavily dependent on a wide variety of microstructural mechanisms associated with composition, heat treatment, loading direction, mean stress, and control mode. High-cycle fatigue behavior and life of NiTi is also different from conventional metals and must be accounted for in-device design.[24,55–57]

16.2.4.2 Biocompatibility

NiTi has gained widespread clinical acceptance as a biomaterial; however, concerns still exist about the high nickel content. Nickel is widely believed to trigger an allergic response in a small fraction of patients.[58] Numerous *in vitro* and *in vivo* studies have investigated the biocompatibility of NiTi.[59–61] In general, results demonstrate that the TiO_2 layer naturally formed on the NiTi surface[62] serves as a protective barrier[26] and that NiTi demonstrates corrosive traits on par with Ti alloys when the surface is treated appropriately. Nickel ion content was measured to be consistent with that found in drinking water,[26,63] and generally considered safe. Nonetheless, numerous researchers have investigated non-nickel shape-memory alloys for potential biomedical use[64–67]; however, the mechanical properties of NiTi remain unmatched. For example, interesting results have been demonstrated by studies that have replaced nickel with another biocompatible or bioactive element. Structural and mechanical investigation of NiTiPt shape-memory alloy wires demonstrate a much smaller stress hysteresis and smaller temperature dependence of the transformation stress relative to NiTi.[68] The natural antibacterial function of copper and silver has served as a motivation for mixing with NiTi to form NiTiCu or NiTiAg.[69–71] Cumulatively, results have shown good mechanical behavior, which is promising.

16.3 Mechanically Active Materials: SMPs

16.3.1 Potential for SMPs in Biomedical Applications

Polymeric biomaterials have become increasingly popular for use in biomedical applications over their metallic counterparts for a myriad of reasons, including their relatively low cost, ease of manufacturability, mechanical and biological tailor-ability, biodegradability, and radio-transparency. Polymeric biomaterials have found their way in almost every facet of biomedical devices from biodegradable sutures, polymeric heart valves, spinal fusion cages, and orthopedic interference screws. In each of these examples, there is a mechanical component to the device: sutures must close and secure wounds and lacerations, polymeric heart valves must restore hemodynamic function, spinal cages must provide separation and stability between two vertebrae, and orthopedic screws must provide fixation. Furthermore, each device has its own degree of inherent difficulties regarding implantation: sutures must be tied with the appropriate amount of tension, heart valves must be compacted to a smaller size for transcather delivery, spinal cages must navigate through a dense area of vital nerves and organs, and interference screws must be inserted without twisting or damaging the fixated graft. Lastly, all of these devices are required to work for an extended period of time within a patient whose anatomy, lifestyle, and health can change over time.

SMPs offer enormous potential for biomedical device advancement by addressing the challenges faced in many clinical applications. SMPs can be tailored to provide specific levels of mechanical support or action. SMPs promote minimally invasive surgery and overcome implantation challenges by allowing for large, bulky devices to be significantly compacted, inserted through a small incision, and return to their original size once in place. Lastly, SMP devices offer the ability to adapt to biological changes such as bone resorption or even pediatric growth to eliminate the need for surgical revisions and replacements. The purpose of this section is to introduce the reader to SMPs, their requirements for use, and highlight several current and potential clinical applications.

16.3.2 Shape-Memory Effect in Polymers

SMPs are defined by their ability to undergo the shape-memory effect. The shape-memory effect in polymers is not an intrinsic material property but rather the result of proper conditioning of a polymer sample with a specific structure.[8,72] In general, a polymer sample can demonstrate the shape-memory effect in the following manner (Figure 16.8)[73]:

1. Heat the sample to the vicinity of or above a transition temperature (T_{trans}).
2. Deform to and hold at a desired temporary shape.
3. Cool well below T_{trans}.
4. Store indefinitely at a temperature below T_{trans}.
5. Reheat to the vicinity of or above T_{trans}.

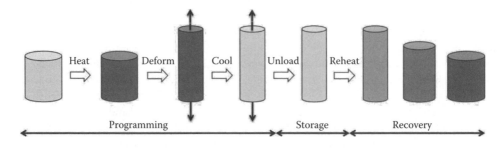

FIGURE 16.8 Illustration of the shape-memory effect in polymers following the three stages: programming, storage, and recovery.

These steps can be divided into three stages: programming (1–3), storage (4), and recovery (5). The first stage of shape-memory programming involves heating a polymer sample near or above a thermal transition. Because the temperature can be associated with a glass (T_g) or melting (T_m) transition, T_{trans} is often used to cover both scenarios. Samples are heated to T_{trans} to increase chain mobility, ease of processing, and strain-to-failure. Cooling below T_{trans} essentially "freezes" the polymer sample in the desired temporary shape. In amorphous SMPs, a sudden volumetric contraction associated with a reduction in free-volume occurs as the polymer is cooled below T_g. In semicrystalline SMPs, a portion of the polymer structure will crystallize when cooled below T_m. In either event, the polymer chains become kinetically restricted from large-scale rearrangement, and thus the polymer is "frozen" into a temporary configuration.

Once programmed, SMPs are stored below T_{trans} for an indefinite amount of time. The storage temperature should be well below T_{trans} to prevent partial or premature shape recovery. For example, for an amorphous SMP network, viscoelastic recovery will occur if the sample is stored at a temperature near the onset of the glass transition. For clinical applications of SMPs, proper storage conditions should be maintained over the shelf life of the device. Sterilization must also be considered, as it is possible that the high temperatures associated with many sterilization procedures could trigger recovery.

Recovery is the final stage in the shape-memory effect and occurs when the stored SMP is exposed to a stimulus that ultimately increases chain mobility. Heat is used as the primary stimulus to unrestrict the polymer chains via an increase in free volume or melting of crystalline domains; however, solvents can also be used as a plasticizer to increase chain mobility and trigger the shape-memory effect. Regardless of the activation method, once the chains are released, they seek to return to their equilibrium configuration through entropy elasticity.[74] If there are no constraints on the SMP, the sample will viscoelastically return to its original, permanent shape, often referred to as unconstrained or free-strain recovery. If constraint is applied to the SMP, the sample will generate a stress in the direction toward its original, permanent shape, often referred to as constrained or fixed-strain recovery. In practical applications, the recovery conditions of an SMP are never fully unconstrained or constrained, but rather partially constrained.[10]

There has been some debate as to what exactly constitutes an SMP. Some researchers consider only those polymers that induce shape recovery through a first-order transition, such as a melting transition to be an SMP, while shape-recovery via a second-order transition, such as a glass transition, is said to be a by-product of viscoelasticity. However, both cases exploit a reversible thermal transition to alter polymer chain mobility to restrict and allow shape recovery driven by entropy elasticity. Another consideration is that the majority of all polymers can demonstrate at least a small amount of shape recovery if programmed correctly, although it would be impractical to consider the majority of polymers as SMPs since the same argument could be used for metals, which all recover some amount of applied strain. For these reasons, researchers generally are willing to consider both T_g- and T_m-based polymers as official SMPs. Here, an SMP is a polymer that has a distinct thermal transition and the ability to *efficiently* store and recover a significant amount of strain. SMPs can be classified by their structure, which dictates the characteristics of their thermal transitions. Figure 16.9 illustrates the four classifications of SMPs with respect to their thermal transitions. It should be noted that each transition is distinct in nature and experiences a drop in modulus of 2–3 orders of magnitude. Every type of SMP contains some degree of chemical or physical crosslinking. Physical crosslinks are created by the separated crystalline phases in Type III and IV SMPs. Crosslinking is a necessary requirement in SMPs as they serve to remember the low-energy equilibrium state (i.e., original and permanent shape) throughout the shape-memory effect.

16.3.3 Shape-Memory Considerations for Clinical Use

16.3.3.1 Programming

Programming is the first stage of the shape-memory effect. One should consider the conditions at which the temporary deformations are programmed into the polymer, as they can play a significant role in the storage and recovery characteristics of the polymer. For SMPs utilizing a melting transition as the

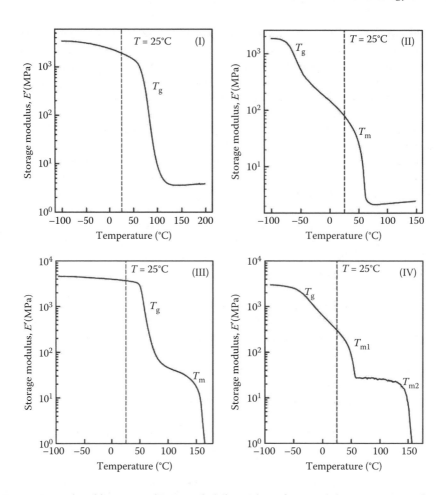

FIGURE 16.9 Examples of four types of SMPs with different shape-fixing and shape-recovery mechanisms as a function of dynamic mechanical behavior. SMPs are classified as (I) chemically crosslinked amorphous thermosets, (II) chemically crosslinked semicrystalline rubbers, (III) physically crosslinked thermoplastics, and (IV) physically crosslinked block copolymers. (Liu C, Qin H, Mather PT. Review of progress in shape-memory polymers. *J Mater Chem* 2007;17:1543–1558. Reproduced by permission of The Royal Society of Chemistry.)

thermal transition, such as type IV physically crosslinked block copolymers, programming should be performed above the melting temperature of the soft segment (T_{m1}). If the soft segment is not melted and reformed into the crystalline state, the achievable strains and shape fixity of the polymer are severely limited. Therefore, the melting of the soft segment serves to maximize the achievable stored strains and increase strain fixity. It should be noted that the programming temperature should not exceed that of the melting temperature of the hard segment (T_{m2}), as this would create a new equilibrium configuration of the SMP and essentially create a new permanent shape.

In amorphous networks, the programming temperature has a greater influence on the shape-memory behavior. Traditionally, it is recommended that amorphous SMP networks be programmed at temperatures above T_g. Deformation in the rubbery state increases chain mobility and ease of processing. The strain-to-failure of networks has been shown to increase as the polymer heats from the glassy state into the glass transition; however, as a network continues to be heated into the rubbery regime, the strain-to-failure will decrease. This effect has been linked to the "failure envelop" that exists for elastomeric networks.[75] As a consequence, there exists an optimal deformation temperature, which maximizes the programmable strain in SMP networks. This optimal temperature has been shown to exist within the glass transition.[76,77]

Many amorphous networks can still show significant achievable strains when programmed below T_g. Low-temperature deformation ($T_d < T_g$) requires greater levels of force and shows slightly lower levels of strain fixity due to elastic recoil. The major advantage of low-temperature deformation is that the activation temperature can then be reduced. Studies have shown that low-temperature deformation ($T_d < T_g$) results in an SMP with an activation temperature ~50°C lower when compared to the same SMP that has undergone higher-temperature deformation ($T_d \geq T_g$).[78] This effect is attributed to the low-temperature deformation storing more internal energy within the network, which in turn requires a lower activation temperature to make recovery thermodynamically and kinetically favorable. A recent study has shown that this effect can be exploited to program SMP networks with multiple shape-memory effects. For example, if a sample is deformed at T_{d1} and then further deformed at T_{d2}, where $T_{d1} > T_{d2}$, the sample will show two distinct shape-memory effects when reheated.[79] Lastly, low-temperature deformation will often result in a stress overshoot when recovered in a constrained environment. Figure 16.10 compares a sample deformed at $T_d < T_g$ compared to a sample deformed at $T_d > T_g$. The amount of stress generated by the low-T_d sample reaches a maximum before returning the stress level of the high-T_d sample. This effect has been attributed to lower activation temperature for the SMP coupled with thermal expansion.

16.3.3.2 Sterilization and Storage

After an SMP has been programmed, it must be sterilized and stored before it can be used clinically. Many polymeric biomedical devices have shelf lives up to 2 years, so it is important to understand the effects of these two processes on shape-memory behavior. An appropriate sterilization method should be selected that does not erase the programming step or alter the polymer chemistry or recovery behavior. Furthermore, appropriate storage conditions must be met to avoid premature activation of the shape-memory effect.

There are a wide variety of methods that have been approved to sterilize polymer-based biomedical devices. While all sterilization methods should be examined to avoid any adverse influence on the structure and properties of any implant material, extra attention must be given to SMPs. As previously noted, the majority of SMPs are thermally activated. For use in clinical applications, the activation temperature of SMPs are usually within the range of $T_{body} \leq T_{activation} \leq T_{body} + 30°C$ so as to avoid any denaturation of tissues surrounding the device. Unfortunately, most sterilization techniques operate at temperatures above 50°C, well above the activation temperature to trigger the shape-memory effect. As a result, SMP devices typically cannot be sterilized without some sort of constraint to prevent premature activation from occurring. If constraints are used, additional safety precautions must be taken,

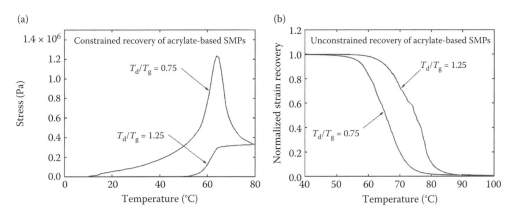

FIGURE 16.10 The effect of deformation temperature (T_d) on (a) unconstrained and (b) constrained recovery. Deformation conditions are given as a ratio of T_d to T_g. (Gall K et al. Thermomechanics of the shape memory effect in polymers for biomedical applications. *J Biomed Mater Res A* 2005;73:339–348. Copyright Wiley-VCH Verlag GmbH & Co. KGaA. Reproduced with permission.)

including analyzing the fixtures that are in contact with the device to ensure no contamination. For example, ethylene oxide and low-temperature plasma (LTP) sterilization methods require the gas or plasma to come into contact with the surface of the device. Any external constraints on the device must not prevent the method from sterilizing the surface properly.

Also the selected sterilization method must not affect the polymer chemistry or structure. Gamma and e-beam irradiation techniques are energy-based methods of sterilization; however, these methods have been shown to induce crosslinking in many polymer systems, an effect that can alter the shape-memory properties. Irradiation-induced crosslinking not only raises the required activation temperature of SMPs, but also forms new permanent crosslinks in the temporary programmed shape, which can impede shape recovery. Other methods, such as LTP, have been shown to alter the surface chemistry of certain polymer systems. For example, a study of acrylate-based SMPs revealed that a cytotoxic response occurs after the networks are sterilized using LTP.[80]

For SMPs that are programmed around a glass transition, long-term storage may result in a phenomenon known as physical aging or densification. Physical aging is a process of structural relaxation associated with rapidly cooling a polymer sample through its glass transition.[81] The glass transition is defined by a sudden increase in free volume when heated. During cooling, the free volume collapses. If cooling takes place rapidly, the polymer does not have significant time to achieve an equilibrium configuration, and a finite amount of free volume (known as nonequilibrium volume) will remain. Over time, the polymer chains will slowly relax toward its equilibrium configuration and decrease the amount of nonequilibrium volume present. Hence, this process is often referred to as densification. Physical aging has been shown to increase both the activation temperature (by 5–9°C) and initial rate of recovery (by up to nine times) in amorphous SMP networks stored over a 1-year period (Figure 16.11).[82] Therefore, one should consider the effects of long-term storage and physical aging on SMPs with amorphous domains, as the performance of the device may be altered as a function of time.

16.3.3.3 Activation

After careful programming to a desired shape, appropriate sterilization treatment, and possible months in storage, the activation of the shape-memory effect is the final stage in the shape-memory cycle. This stage is associated with the implantation or implementation of an SMP medical device. Multiple

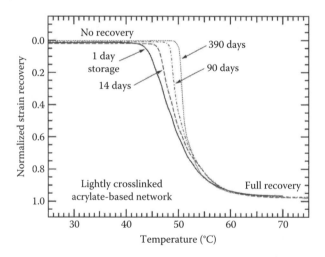

FIGURE 16.11 The effect of long-term storage on the unconstrained recovery behavior of chemically crosslinked amorphous thermosets. (Ortega AM et al. Effect of crosslinking and long-term storage on the shape-memory behavior of (meth)acrylate-based shape-memory polymers. *Soft Matter* 2012;8:3381–3392. Reproduced by permission of The Royal Society of Chemistry.)

techniques have been proposed to activate the shape-memory effect, including the use of body temperature, laser light, alternating magnetic fields, and even water absorption.

Heating has been the primary chosen method of activation of the shape-memory effect in polymers. SMPs have the potential to be activated at body temperature, offering an elegant solution as the body is naturally regulated at $T{\sim}36°C$. Storage conditions for SMP medical devices should be $\leq 22°C$ to prevent premature activation. The implantation and implementation conditions should also be considered if this approach is desired. For example, an SMP suture for wound closure would likely not reach T_{body} naturally due to a lower temperature at the surface of the skin. In this example, a heating gun or hot pack that could apply localized heat to the device might be a suitable direct heating method. In orthopedic applications, such as rotator cuff or anterior cruciate ligament (ACL) repairs, room-temperature saline is constantly pumped through the joint, consequently lowering the effective temperature at the implantation site and possibly delaying activation until the surgery is completed. For cardiovascular applications, navigating through the tortuous blood vessel pathways to reach the implantation site may take a significant amount of time, thus creating the potential that the device could prematurely activate while inside the delivery instrumentation.

Laser light is an alternative method of direct heating of SMP devices.[83,84] This method is advantageous for applications where it is desired that the device remain stable at body temperature and only activate once in place, such as cardiovascular applications. Under these conditions, an SMP must be programmed such that $T_{trans} > T_{body}$. In this method, a laser is passed through an SMP device using total internal reflectance. It should be noted that this application is limited to devices with low masses and/or slender geometries, such as wires and coils. Direct heat activation can also be achieved using electricity.[85] With this approach, thin wires are incorporated throughout the device and an electrical current is passed through the wires generating heat within the device. Carbon nanofiber SMP composites may also be used to generate heat with electrical current; however, one needs to consider how to prevent the current from transferring from the device to the patient if an electrically conductive composite is used.[86]

Indirect methods of heating have also been proposed to activate SMP devices.[87–89] Magnetic nanoparticles, such as magnetite (Fe_3O_4), can be dispersed throughout a polymer matrix. After implantation, the device is then subjected to inductive heating via an alternating magnetic field.[90,91] As the magnetic particles try to align themselves with the direction of the magnetic field, a hysteresis associated with the particles changing direction is created that results in heat generation (Figure 16.12). This method is particularly advantageous for devices that need to be activated long after being implanted. For example,

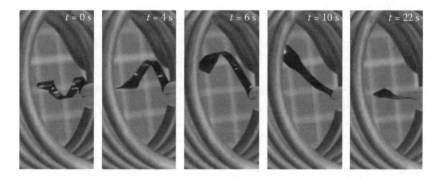

FIGURE 16.12 Series of photographs showing the activation of the shape-memory effect via inductive heating. The sample recovers from a temporary, corkscrew shape to its original flat shape. The polyurethane-based sample contained 10 wt.% magnetite particles and was exposed to a magnetic field of $f = 258$ kHz and $H = 30$ kA/m. (Reproduced with permission from Mohr R et al. Initiation of shape-memory effect by inductive heating of magnetic nanoparticles in thermoplastic polymers. *Proceedings of the National Academy of Sciences of the United States of America* 103:3540–3545. Copyright 2006 National Academy of Sciences, USA.)

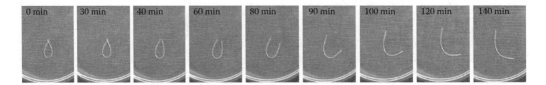

FIGURE 16.13 Series of photographs showing water-driven shape recovery in a polyurethane SMP. The sample has an initial dry T_g of 35°C, which decreased to approximately 0°C with the uptake of water. Consequently, the decrease in T_g and disruption of hydrogen bonding with water update activates the shape-memory effect. (Reprinted with permission from Huang WM et al. Water-driven programmable polyurethane shape memory polymer: Demonstration and mechanism. *Applied Physics Letters* 86:114105. Copyright 2005, American Institute of Physics.)

using indirect heating, an SMP medical device could be reactivated to adjust to a change in patient anatomy without the need for a surgical operation. If considering this activation method, it is important to consider that the equipment requirements to create an alternating magnetic field can be substantial.

While most methods of activation involve heating the SMP to a temperature above T_{trans}, an inverse method of activation would be to lower $T_{trans} < T_{body}$. Solvents, such as water, can be used as a plasticizer that effectively lower the T_g of amorphous networks when absorbed and consequently activating shape-memory recovery (Figure 16.13).

The amount of time required for an SMP medical device to recover can be a critical factor. For devices with a high T_g or requiring immediate shape recovery, mechanical force can be used to help drive shape recovery. Deformation processing, such as extrusion and rolling, have been shown to create anisotropy in the alignment of the polymer chains. Shape-memory programming is essentially a deformation processing technique that induces large amounts of chain alignment. As a result, there is an advantageous mechanical anisotropy in the direction of returning a programmed SMP toward its permanent shape. It has been shown that programmed SMPs require significantly less force to be deformed toward their permanent shape as compared to identical, nonprogrammed samples experiencing the same deformations. This process of mechanically driven shape recovery returns the polymer toward it equilibrium configuration and serves to relieve stored strains.[92]

16.3.4 Clinical Applications of SMPs

16.3.4.1 Soft Tissue Fixation

In orthopedics, SMPs offer the potential to create novel fixation and anchoring devices to overcome the limitations of traditional devices. For example, screw-based devices cannot adapt to any loosening experienced at the insertion site due to either cyclic loading or osteolysis. In the case of interference screws, which are often used for ACL reconstruction, the device commonly twists and lacerates the ACL graft during insertion. An early embodiment of an SMP interference device can be seen in Figure 16.14. In the figure, an SMP plug was programmed to be extended longitudinally and compressed radially. In this configuration, the plug could easily be placed within a bone tunnel (represented by a glass tube) along with a soft tissue graft. As the SMP interference plug is heated to body temperature, the shape recovery attempts to return the plug to its original geometry. As a result, the plug expands radially and exerts a normal force on the tendon, thus creating an interference fit. In this approach, the insertion of the device does not involve a twisting motion and allows the position of the graft to be verified before interference fixation occurs. Furthermore, the device does not contain threads, which have been shown to lacerate the graft upon insertion.

An interference fixation device based upon SMP technology has been developed and recently cleared through the FDA. The ExoShape™ interference device by MedShape, Inc. is an example of utilizing the shape-memory effect in high-strength polymers for orthopedic fixation (Figure 16.15). The device is manufactured from PEEK, a semicrystalline thermoplastic material. The device is comprised of two

FIGURE 16.14 Series of photographs showing a prototype shape-memory interference device. The device was programmed to a long, slender geometry and placed inside a glass tube along with a bovine tendon. As the SMP device warmed to body temperature, it expanded radially to create an interference fit to fixate the tendon. (Some photographs from Yakacki CM et al. Strong, tailored, biocompatible shape-memory polymer networks. *Adv Funct Mater* 2008;18:2428–2435. Copyright Wiley-VCH Verlag GmbH & Co. KGaA. Reproduced with permission.)

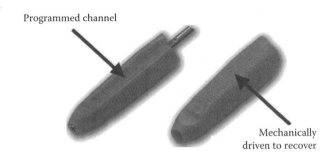

FIGURE 16.15 Picture of the MedShape ExoShape™ shape-memory interference device. The device is made from a shape-memory PEEK sheath and programmed with two channels (top channel shown) to accommodate a soft tissue in a bone tunnel. The device is mechanically driven to its original, permanent shape by inserting a PEEK cone (not shown) inside the sheath. The sheath is held in place during insertion by a metal bar, which is removed once shape recovery has occurred.

components: an outer sheath and inner cone. The outer sheath is programmed with two channels running along its body to accommodate the soft tissue graft during insertion. The high T_g of PEEK (~150°C) allows for the device to be sterilized and stored without any concerns of premature activation. Rather than relying on thermal mechanisms, shape recovery is driven mechanically by inserting an inner cone into the sheath. As the inner cone is inserted, the outer sheath expands radially and creates an interference fit. The sheath is held in place during recovery by a metallic bar, which is then removed after recovery has occurred.

16.3.4.2 Cardiovascular Stents

Stents are expandable scaffolds designed to prevent vasospasms and restenosis of a vessel after balloon angioplasty. Caused by coronary artery disease (atherosclerosis), stenosis is defined as the narrowing of a blood vessel, which causes plaque build-up within the arterial wall and constriction of blood flow. Stenting is designed to eliminate elastic recoil and adverse remodeling caused by angioplasty alone.[93] Polymers have been used with stenting primarily as a coating for metallic stents. These coatings have been designed to elute therapeutics, such as sirolimus and paclitaxel, to help reduce hyperplasia caused by smooth muscle cell proliferation. A pure polymer stent has been proposed to increase biocompatibility, biodegradability, drug loading, compliance matching, and ease of fabrication. Other benefits include the ability for molecular surface engineering as well as applying the shape-memory effect. A shape-memory stent can be programmed to deploy more gently than balloon expandable stents and pseudoelastic stents, while also providing improved strain recovery capacity, suggesting the use of smaller delivery instrumentation. Furthermore, shape-memory offers the potential for continued expansion, which could be critical in pediatric stenting.

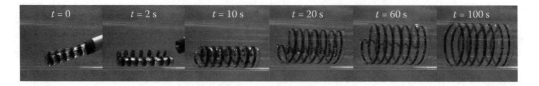

FIGURE 16.16   Series of photographs showing an SMP stent with $T_g = 52°C$ being deployed from a 6F catheter into a 22-mm glass tube at body temperature. (Reprinted from *Biomaterials*, 28, Yakacki CM et al. Unconstrained recovery characterization of shape-memory polymer networks for cardiovascular applications, 2255–2263, 2007, with permission from Elsevier.)

Wache et al. were the first to report on the development of an SMP stent as a means to enhance drug delivery.[94] In their approach, a thermoplastic polyurethane stent was injection molded and programmed for shape memory via an extrusion process. The SMP stents were determined to be stable when stored at or below 8°C and could be activated at body temperature. Yakacki et al. investigated the use of amorphous acrylate-based networks for SMP stenting.[10] In their approach, programming was performed by rolling the stent into a compact geometry and placing the device within a catheter (Figure 16.16). These stents had T_g values between 52°C and 55°C, allowing the stents to be relatively stable when stored at room temperature while still having the ability to activate at body temperature. Alternatives to designing stents that activate at body temperature have also been explored. Baer et al. created a stent using a thermoplastic polyurethane (DiAPLEX MM5520, Mitsubishi Heavy Industries).[95] The material had a T_g of 55°C, but a very broad transition causing activation to not occur until 40–45°C. As a result, activation was achieved by applying laser light, in which an 810 nm diode laser was used to photothermally activate the stent.

Interestingly, the first study and first clinical trial of an SMP stent occurred somewhat as an unexpected artifact during a clinical trial of a poly(L-lactic acid) (PLLA) stent. During a study of the Igaki-Tamai stents in 2000, the authors documented the stent's ability for self-expansion.[96] The deployment time was recorded as a function of temperature, with the stent deploying in 0.2, 13, and 1200 s at 70°C, 50°C, and 37°C, respectively. However, because the full programming and storage conditions of the PLLA stent were not reported, it is difficult to separate shape recovery from viscoelasticity. Most important though, the authors demonstrated the feasibility and efficacy of a pure polymer stent manufactured from a shape-memory material. This device was successfully implanted into 15 patients and monitored over 6 months, showing no major cardiac events.

16.3.4.3 Anuerysm Treatment

Intracranial (cerebral) aneurysms are localized, "balloon-like" dilations of an arterial wall due to thinning and weakening of the vessel. If ruptured, bleeding into the subarachnoid space can occur, leading to hemorrhagic stroke, brain damage, and even death. Anuerysm rupture occurs in 1 out of 10,000 people in the United States annually.[97,98] Endovascular treatment of anuerysms was introduced in 1995 using the Guglielmi coil technique. This repair method places metallic coils into the anuersym to induce a clotting response and seal off the aneurysm from the artery. Ideally, endothelium regrowth will then occur at the "neck" or entry point of the anuerysm.[99] Although over 200,000 patients have been treated with this technique worldwide, there are some limitations to using coils, including low packing volume of the coils, migration of loose coils, coil compaction, subsequent formation of new side aneursyms, and possible puncturing of the anuerysm during placement.[100] Furthermore, coil-based treatment is not preferred with anuerysms having neck sizes greater than 4 mm.

SMPs offer novel solutions to the problems inherent to traditional Guglielmi repair. In particular, SMP foams are now being developed for treatment for intracranial anuersysm. SMP foams are ideal for applications requiring both extremely high volume compression and recovery ratios. An SMP foam could overcome the limitations of metallic coils by allowing for the placement of a single compacted

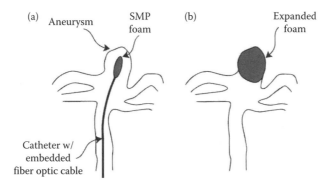

FIGURE 16.17 Illustration of how SMP foams could be used to treat an aneurysm. A foam is compacted and placed within an anuerysm in (a), and activated to expand and fill the anuerysm using laser light through a fiber optic cable in (b). (With kind permission from Springer Science+Business Media: *Ann Biomed Eng*, Vascular dynamics of a shape memory polymer foam aneurysm treatment technique, 35, 2007, 1870–1884, Ortega J et al.) (Original caption: Treatment of an intracranial aneurysm with SMP foam. (a) Catheter delivery of a compressed piece of SMP foam to the aneurysm. (b) Fully expanded SMP foam within the post-treatment aneurysm.)

device that would expand gently and safely to fill the entire anuerysm cavity. It would also help one to decrease surgery time, create a porous matrix for cell invasion and neointima formation, and better seal off the aneurysm neck for endotheliazation. An illustration of how an SMP foam could treat an anuerysm can be seen in Figure 16.17.

The first reported results for SMP foam treatment came from a canine model investigating carotid aneurysms.[101] The polyurethane-based SMP foam demonstrated improved angiographic scores after 3 weeks, while histological results showed thick neointima on the surface of the foam, sealing most of the aneurysm neck. The study suggested that higher scores might have been reached if the foams were fully deployed at body temperature. The T_g of the foams were 60°C, but the exact programming conditions of the foams were not reported, beyond a statement that an 8 mm long and 4 mm in diameter foam was compacted down into a 6F catheter.

Recent advances in SMP foams for anuerysm treatment have investigated a myriad of factors to achieve optimal recovery conditions. Research from Maitland et al. have investigated the use of lasers to fully activate shape memory.[100–103] The addition of tungsten has also been investigated as a means to increase the radiopacity for imaging during placement and follow up. The T_g of the foams can also be varied from 60°C to 45°C to better aid in shape recovery. A recent report suggests that a high T_g polyurethane foam may be able to achieve complete activation via water absorption. In this study, a polyurethane foam was synthesized from hexamethylene diisocyanate, *N,N,N′,N′*-tetrakis(2-hydroxypropyl) ethylenediamine, and triethanolamine.[100] The foams showed a decrease in T_g from 60°C to 5°C within 10 h when exposed to 100% humidity levels at 37°C. This rapid decrease in T_g due to water absorption could be used to fully activate the shape-memory effect after the foam has been implanted in the body.

16.4 Bioactive Materials

16.4.1 Fundamentals of Bone Formation with Materials

Long-term viability of an implant relies on its ability to form a stable interface with the host tissue. Implants that do not form a chemical or biological bond are subjected to continuous micromotion and development of a fibrous capsule layer that can compromise implant function and potentially damage surrounding tissue. Bioactive materials have gained much attention in the biomedical community because they have the ability to address these issues facing more inert materials. A bioactive material is defined as any material that is able to elicit or modulate a biological response, by producing a chemical or physical stimulus.[104] The ultimate goal is to facilitate fast healing and, in some instances, regeneration of the injured tissue due to the implantation. Many bioactive materials also degrade over time in response to the physiological environment, promoting tissue infiltration into the space previously occupied by the implant material. Ideally, bioactive materials should have a controlled release of their stimuli that occurs in tandem with a sequence of cellular changes at the injury site. Much of the work with these materials has been focused on tuning their chemical and structural properties to achieve the desired biological response at the appropriate time.

Bioactive materials have found the most use in orthopedic applications due to the fact that many materials demonstrate an intrinsic ability to stimulate bone growth. Bone formation and implant integration result from a series of reactions that occur in parallel on the material surface and in the surrounding physiological environment. A simple illustration of these processes is shown in Figure 16.18. For bioactive materials that undergo a chemical reaction, the primary steps include surface dissolution, apatite precipitation, and ion exchange.[105] When exposed to physiological fluid, ions are released from the bulk material and replaced with hydrogen ions. The increased ion concentration creates a saturated environment that contributes to hydroxycarbonate apatite (HCA) formation on the material surface. The released ions also act as inducing agents to recruit immature cells to the injury site and promote their differentiation into adult osteoblasts that can produce new bone tissue. Under *in vitro* conditions, increasing calcium and phosphate ion concentrations to cell culture medium caused a dose-dependent effect on osteoblast viability and differentiation[106–108] by inducing several intracellular signaling pathways via receptors on the cell membrane.[108] This cell signaling pathway is distinct from the signaling pathway initiated through biological agents, such as bone morphogenic protein (BMP). Simultaneously, proteins and cells are adhering and interacting with the material surface in a manner that further contributes to surface dissolution and apatite precipitation. In addition to chemical stimuli, materials can also exhibit bioactivity by having certain physical surface properties. Grain size, crystallinity, porosity, and roughness have all been shown to be influential in promoting bone formation.[109]

Bone-forming bioactive materials can be classified as being osteoinductive and/or osteoconductive. Osteoinduction refers to the recruitment of immature pluripotent cells and the subsequent stimulation of these cells by an inducing agent (i.e., released ions) that promotes their maturation into adult bone-forming osteoblasts. Osteoconduction refers to the ability of adult cells to lay down new bone matrix.

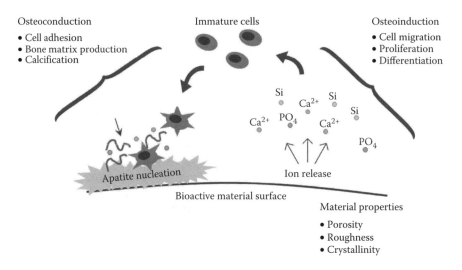

FIGURE 16.18 Illustration of how bioactive materials promote bone formation through osteoinduction and osteoconduction.

These cells originate from either previously differentiated preosteoblasts or mesenchymal stem cells and are able to adhere to a material surface. Thus, in this context, osteoconduction depends on osteoinduction to have previously occurred to some extent.[110] With regard to implant materials, osteoconduction also refers to the ability to lay down bone directly on the material surface without the creation of a fibrous tissue layer. It was previously thought that osteoinduction could only be achieved by having certain biological molecules, such as BMP, present. Injecting BMPs is a common clinical therapy used to promote fast bone growth, but this approach also has its limitations. Determining the correct dosage and localizing the drug to the injury site are just a few of the current challenges.[109] Thus, materials with intrinsic osteoinductivity serve as an advantageous alternative to biologic therapies to promote bone regeneration.

Bioactive materials, such as bioceramics, have a long history of clinical application, beginning with the use of porcelain in dental crowns in the late eighteenth century. However, their osteoinductive potential was not truly realized until the early 1960s, when it was discovered that Pyrex glass tubes, when implanted subcutaneously, could calcify and eventually promote bone formation.[111] Almost a decade later, the subcutaneous calcification of 2-hydroxyethyl methacrylate (2HEMA) sponges in a porcine model was reported.[112] To date, 2HEMA is the only known synthetic polymer to exhibit any inherent bioactivity, though the exact mechanism is still unknown.[109] Bioactivity can be achieved across all material classes either through their inherent chemistry or through certain processing conditions. The remainder of this section will provide examples of bioactive materials that have found success in clinical applications.

16.4.2 Bioactive Glass

First developed in 1969, bioactive glass (i.e., Bioglass) consists of a silicate network with sodium, calcium oxide, and P_2O_5. It exhibits both osteoconductive and osteoinductive properties as well as an ability to stimulate soft tissue growth.[113,114] The mechanism related to its bioactivity has been thoroughly discussed in the literature.[113–117] Briefly, bioglass promotes bone growth by undergoing a surface dissolution under physiological conditions that results in a high surface concentration of silica and the formation of an HCA layer.[117] In an *in vivo* model, this bond has been shown to have equivalent strength to natural bone. The silicon, in the presence of calcium, also acts as a nucleation site for apatite and has been associated with osteogenesis and at the beginning stages of calcification.[115] It is also thought that collagen fibers from soft tissue can also bond with the silica and HCA-rich surface.[118,119]

16.4.2.1 Chemistry and Processing

Bioglass is processed using a combination of melting and sol–gel techniques. Highly pure quartz or silica sand, reactive-grade sodium and/or potassium carbonates, and calcium carbonates are melted at temperatures between 1200°C and 1240°C, annealed to reduce thermal stresses, and then cooled to 400–500°C.[119] The material is then processed through a sol–gel method, where a sol phase is prepared and then transformed into a solid phase (i.e., the gel). The inorganic materials are polymerized by the reaction of metal alkoxides. The resulting silicon alkoxide is then hydrolyzed into a silica gel network with calcium salts added to serve as apatite initiators.[119] Sol–gel processing is a relatively newer technique that allows for better tailoring of composition and structure as well as the ability to produce the material in a variety of forms, including particulates, fibers, foams, and coatings.[116] It is important that a stable concentration of silicon and phosphorus be incorporated, as too much of either can inhibit calcification.[115] 4S5S Bioglass (45 wt.% SiO_2, 24.5 wt.% Na_2O, 24.5 wt.% CaO, and 6 wt.% P_2O_5) is considered the most bioactive glass and is believed to best promote stem cell differentiation and gene expression resulting in enhanced bone growth.[120]

16.4.2.2 Clinical Applications of Bioactive Glass

Bioglass is primarily used as a surface to promote bone integration, though it is currently being explored as a scaffold for soft tissue engineering applications.[121] It was first cleared in 1985 as a hearing loss device used to replace bones of the middle ear. These implants have demonstrated long-term stability thought to be the result of the strong interfacial bond between the glass, bone, and tympanic membrane.[119,122] 4S5S Bioglass has also been used in dental applications to aid in preserving the jawbone of patients after tooth extraction. Termed an endosseous ridge maintenance implant (ERMI), the material is injection molded into a conical shape that fits in the space previously occupied by the tooth root. Because of Bioglass' ability to bond to both bone and soft tissue, ERMI implants remain very stable and have better survivability compared with other bioactive materials.[113,123] Bioglass has seen the most success when processed into particulate form that can be injected to promote fast bone formation. In dental applications, the fast bone formation prevents epithelial tissues from migrating down the tooth, a common problem with these procedures.[113] The particulates can also be applied in putty or injectable form to serve numerous other clinical procedures, including bone grafting, spine fusion, and revisions on joint replacements.

Despite its unique osteoinductive properties, Bioglass is very brittle and exhibits low fracture toughness. Efforts to improve the mechanical properties have included adding low contents of alkali oxides to the material to form a bioactive glass-ceramic.[124] The final product is a polycrystalline material with a very fine microstructure. The most notable glass-ceramic (GC) material is apatite–wollastonite (A–W). AW-GC has two crystalline phases composed of 38 wt.% hydroxyfluorapatite ($Ca_{10}(PO_4)_6(O,F_2)$) and 34 wt.% wollastonite ($CaO \cdot SiO_2$) that provides the material with a bending strength and fracture toughness that is four times higher compared with traditional Bioglass.[119] Though its bioactivity is less than Bioglass, AW-GC can still promote bone formation and bonding and has been successfully used in bulk, granular, and porous forms, as vertebral prostheses, bone grafts, and coatings for joint replacement implants.[125]

16.4.3 Calcium Phosphates

Calcium phosphates (CaPs) are synthetic bioceramics that have a structure that closely mimics the inorganic part of calcified tissues, including bone and teeth. Besides displaying both osteoconductivity and osteoinductivity, CaPs can also resorb in a physiological environment within a time frame that aligns with the bone healing process.[113] Their bioactivity is related to the release of Ca^{2+}, PO_4^{3-}, and HPO_4^{2-} ions from the bulk material, creating a supersaturated extracellular environment that results in the precipitation of carbonated apatites and their subsequent incorporation with other inorganic and organic matrix components.[108] Though their bioactivity is less than bioglass, CaPs have higher mechanical properties enabling their use in more load-bearing environments.

16.4.3.1 Chemistry and Processing

Table 16.2 shows the most common CaPs in clinical use. The calcium-to-phosphate (Ca/P) ratio dictates the level of solubility and bioreactivity of the ceramic. For instance, decreasing the Ca/P ratio increases solubility. CaPs with a ratio less than 1.0 are considered not suitable for the body.[119] In acidic environments (pH less than 6.5), CaPs that are relatively stable at physiological pH become more soluble and reactive. Despite the numerous compositions that can be produced, hydroxyapatite (HA) and tricalcium

TABLE 16.2 Calcium Phosphates That Can Be Used in Medical Devices

Name	Abbreviation	Formula	Ca/P Ratio	Solubility
Dicalcium phosphate dehydrate	DCPD	$CaHPO_4 \cdot 2H_2O$	1.0	1.26×10^{-7}
Octacalcium phosphate	OCP	$Ca_8H_2(PO_4)_6 \cdot 5H_2O$	1.33	2.51×10^{-97}
Tricalcium phosphate	TCP	$Ca_3(PO_4)_2$	1.5	3.16×10^{-26}
				1.25×10^{-29}
Hydroxyapatite	HA	$Ca_{10}(PO_4)_6(OH)_2$	1.67	2.35×10^{-59}
Tetracalcium phosphate	TTCP	$Ca_4(PO_4)_2O$	2.00	1×10^{-38}–1×10^{-44}
Native bone			1.5–1.67	

Source: Adapted from Juhasz JA, Best SM. *J Mater Sci* 2012;47:610–624.; Frayssinet P. In: Poitout DG, ed. *Biomechanics and Biomaterials in Orthopedics.* London: Springer; 2004:101.

phosphate (TCP) are the two primary CaPs that are currently used in clinical applications. HA naturally occurs in bone and teeth and is the most stable calcium phosphate under ambient conditions, as indicated by its high Ca/P ratio (1.67). However, in acidic environments (pH < 6.5), its solubility has been shown to increase. TCP, also known as whitlokite, is often used as a substitute for HA when an application requires faster resorption. Depending on the processing conditions, TCP can be produced in two phases: α-TCP and β-TCP. β-TCP is formed from HA by undergoing specific thermal treatments with higher sintering temperatures resulting in more TCP precipitation. β-TCP can be transformed into α-TCP by heating to 1125°C and then quickly cooling back to room temperature. β-TCP has a more close-packed molecular orientation and lower solubility compared with α-TCP and is more commonly used clinically, primarily in biodegradable macroporous ceramics. On the other hand, owing to its high reactivity, α-TCP is mostly used as a fine powder in preparing bone cements.[126] Oftentimes, to achieve specific biodegradability and bioactivity, a biphasic CaP system (BCP) containing both TCP and HA at various ratios is produced to take advantage of the contrasting solubilities of the two individual CaPs.[127]

Depending on the application, CaPs can be manufactured into a variety of physical forms via aqueous precipitation and sol–gel processing.[119] Aqueous precipitation involves initiating a reaction between a calcium salt and alkaline phosphate or between a calcium hydroxide/calcium carbonate and phosphoric acid. Once the precipitates are formed, they are usually dry pressed and sintered to fuse the powder into a dense, crystallized material. The material then undergoes sol–gel processing similar to Bioglass to produce powders, solids, and thin films. Like the processing method for Bioglass, sol–gel methods fabricate CaP materials with a homogenous molecular composition and controllable Ca/P ratio. Ion contaminates can alter the crystal structure and behavior of the formed CaP and thus is highly undesirable.[128] It is important to note that processing can greatly influence the physical and chemical properties of CaPs. These changes in physicochemical properties have a direct effect on the material's biological performance. For instance, sintering temperature can affect the crystal size and porosity, which in turn directly influences degradability and bone growth.

16.4.3.2 Clinical Applications of CaPs

Given the tailorable bioactivity and ability to be easily manufactured into multiple forms, CaPs have found use in a broad range of orthopedic applications. Since the 1980s, bioceramics have been applied as coatings on metal implants used in joint replacement and dental repair to promote implant–bone integration. These coatings are particularly important at locations where a strong interface with bone is required to prevent implant loosening. Femoral stems for hip joints as well as the femoral and tibial components of knee joints are just some of the implants manufactured with CaP coatings. CaP coatings provide better fixation to load-bearing implants compared to using poly(methyl methacrylate) (PMMA) cement, the alternative approach. This stronger bonding is due to the material's osteoconductive properties that promote bone deposition directly on the implant surface.[119] Coatings are usually applied via plasma spraying, where heated CaP particles are projected at a high speed onto the implant, or by solution deposition, in which the coating nucleates across the implant.[129] The processing conditions alter the physical properties of the coating and thus greatly influence its performance. Specifically, it is desired that the coating have low porosity, high cohesive strength, high crystallinity, and good chemical/phase stability.[129,130] These properties also dictate the dissolution rate of the coating, allowing for the degradation to be tailored to the rate of bone formation. Thickness is also another important property to control as 50–75 µm coatings demonstrate better strength than thicker coatings.[129] Although CaP coatings are widely used, it is still unclear as to whether CaP coatings add any clinical benefit to the total joint implants. Some studies suggest that HA coatings help reduce tibial implant micromotion, an early sign of implant loosening, by providing a "biological seal."[131,132] However, no difference in bone formation or clinical outcome was observed for femoral stems for total hip arthroplasty coated with HA[133,134] suggesting these coatings only offer some benefit under certain load-bearing environments.

CaPs have also found utility when mixed with a degradable polymer matrix to form a polymer–ceramic composite. These composite materials are used in implants where fast bone healing is desired

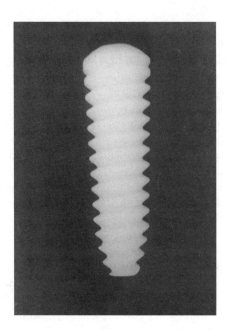

FIGURE 16.19   Image of biocomposite interference screw made out of TCP and PLLA.

such as soft tissue fixation applications. One example is an interference screw used in ACL reconstruction (Figure 16.19). The screw is inserted alongside the ACL graft (either soft tissue or bone–tendon–bone) in a bone tunnel and serves to fixate the graft against the bone. The biocomposite device consists of a degradable polymer matrix such as PLLA dispersed with either HA or TCP. The added HA or TCP is thought to enhance bone growth allowing for faster tissue healing and patient recovery.[136] Several biocomposite fixation devices are commercially available (Table 16.3) for use in applications such as rotator cuff repair, tendon transfer, and ACL reconstruction. The majority of these implants have TCP incorporated due to its higher solubility compared with HA. While biocomposite fixation devices have been used for almost a decade, there are few clinical studies that report their ability to completely degrade and support bone regeneration.[135,136] On the other hand, numerous complications have been reported with biocomposite fixation devices. A search through the FDA Manufacturer and User Facility Device Experience (MAUDE) database for reports filed on complications pertaining to ACL fixation devices reveals that 67% of device-related complications involve a biocomposite fixation device (Figure 16.20).[137]

TABLE 16.3   Fixation Devices Commercially Available That Contain HA and/or TCP

Product	Device	Material	Company
Biocomposite Screw	Interference screw	CaP	Arthrex
Osteotite	Bone screw	HA	Orthofix
Osteoraptor	Suture anchor	HA	Smith and Nephew
Biosure HA	Interference screw	HA	Smith and Nephew
ComposiTCP 30	Interference screw	β-TCP	Biomet
Matryx	Interference screw	β-TCP	Conmed
Bilok	Interference screw	TCP	Arthrocare
Milagro BR	Interference screw	TCP	Depuy
Bio-IntraFix	Soft tissue fastener	TCP	Depuy
Biocryl	Interference screw	TCP	Smith and Nephew
Lupine BR	Suture anchor	TCP	Smith and Nephew

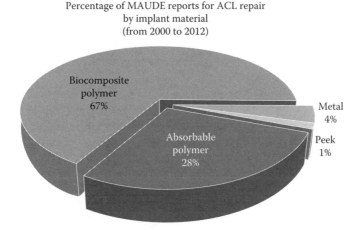

Percentage of MAUDE reports for ACL repair
by implant material
(from 2000 to 2012)

Biocomposite
polymer
67%

Metal
4%

Absorbable
polymer
28%

Peek
1%

FIGURE 16.20 Diagram illustrating the percentage of ACL device-related complications reported in the FDA MAUDE database by implant material. Absorbable polymers represent implant materials made purely of a degradable polymer with no added CaP. A total of 448 reports were found. PEEK, poly (ether ether ketone).

Almost 50% of these reports relate to implant fracture, either intra- or postoperative, and is suggested to be due, in part, to the inherent brittleness of CaP–polymer composites. Furthermore, almost 30% of reports mention device-related adverse tissue reactions, including inflammation and osteolysis.[135] These findings suggest that more evidence is needed to fully determine the utility of CaP-based biocomposite devices as a viable fixation method.

CaPs are also used frequently as bone graft substitutes and/or cements to promote bone regeneration or fusion at an injury site. Most synthetic grafts are composed of biphasic CaP (HA/TCP) as this allows for easy tailorability of degradation and bioactivity by changing the relative concentrations of HA and TCP. CaP cements consist of a solid dispersed active phase (filler) and a hardening liquid (binder) and are often injectable to allow for their use in minimally invasive procedures. Setting time and injectability are two key characteristics to consider and will vary depending on the specific application. The setting time is dependent upon the acidity of the dispersed phases and relative chemical activity with the binder while injectability is dictated by the relative amounts of the liquid and solid phases.[138] CaP bone substitutes offer several advantages over autografts or allografts because they require a less invasive surgery, are available in large quantities, and do not risk host rejection or the transmission of diseases (as is sometimes the case with allografts). Table 16.4 shows a range of bone substitutes that are commercially available. Despite the numerous compositions and forms that can be achieved, CaP bone substitutes still have room for improvement. Obtaining the appropriate viscosity to allow for quick injection is a current limitation for many of the injectable systems.[139] A large variability in properties also exists among the different commercial products, which is not always taken into account by clinicians when selecting a graft, but can greatly affect the biological properties. For instance, some cements have low porosity that greatly reduces resorption and bone ingrowth, limiting their use to applications that do not require fast integration.[139] Furthermore, given their extreme brittleness, CaP substitutes can only be used in low load-bearing applications.[120]

16.4.4 Alternative Approaches to Gaining Bioactivity

While some ceramics display intrinsic bioactivity, metals and polymers (except for 2HEMA) only become bioactive after undergoing certain surface treatments. Titanium is a prime example of how bioactivity can be created in a proclaimed "bioinert" material and is shown in Figure 16.21. Titanium, like most other metals, is usually covered by a thin, passive oxide layer. When exposed to an alkali

TABLE 16.4   Examples of Commercially Available Bone Graft Substitutes

Product	Form	Material	Company
Calcibon	Granules	CaP	Biomet
Beta-bsm	Injectable paste	CaP	DJO Global
Hydroset	Injectable paste	CaP	Stryker
CopiOs Bone Void Filler	Sponge or paste	CaP/Collagen	Zimmer
Pro-Stim	Injectable	$CaPO_4$/$CaSO_4$	Wright Medical
Endobon	Granules	HA	Biomet
NanoXIM-HAp200	Spray powder	HA	Fluidinova
Kyphon ActivOs 10	Cement	HA	Medtronic
OsteoMax	Granules, injection putty, block	HA/TCP/$CaSO_4$	Orthofix
Mimix	Paste	HA/α-TCP	Biomet
OpteMx	Granules, blocks	HA/TCP	Exactech
BoneSave	Granules	HA/TCP	Stryker
Mozaik	Granules	β-TCP	Integra
JAX TCP		β-TCP	Smith and Nephew
Conduit TCP	Granules	TCP	Depuy
Pro-Dense	Injectable	TCP/$CaSO_4$	Wright Medical

Note: $CaSO_4$, calcium sulfate.

solution, this layer dissolves while the titanium material is simultaneously being hydrated. The hydrated Ti species react with alkali ions to form an alkali titanate hydrogel layer. While this layer is unstable upon formation, it can be stabilized by heating above 600°C. When exposed to physiological fluids, the alkali ions from the layer lead to saturation of ions in the environment while the surface layer hydrates, inducing apatite formation.[140] *In vivo* studies have demonstrated that heat and alkali-treated titanium have the ability bond with bone, suggesting their use as an alternative fixation method to cementing.[141]

Besides undergoing a chemical reaction, materials can also become bioactive by changing their surface structure across multiple size scales. For example, nanometer-sized texture on bioactive glass is considered to be a critical factor in improving solubility by increasing the surface area-to-volume ratio to allow for more rapid ion dissolution. The texture also offers more initiation sites for apatite nucleation allowing for faster bone deposition.[142] Adding surface textures also has utility with other classes of materials. When titanium is sand-blasted and acid-etched, a surface roughness with both micro- and nano-scale features is created that is correlated to better *in vivo* bone formation around the implant.[143] Based on these findings, both dental and orthopedic implants are now being fabricated with specific roughness patterns in an attempt to ensure better fixation and longevity in the patient. Many materials also exhibit osteoinductive behavior when fabricated to have certain macro and/or micropores.[144] In particular, porous titanium has been shown to promote osteogenesis and bone formation *in vivo*; however, the exact mechanism related to the added porosity is still unclear.[144,145]

16.5 Electroactive Materials

EAPs are an emerging class of active polymers that use an electrical stimulus to facilitate a shape change. EAPs offer several benefits in comparison to other mechanical actuators, such as higher strains, tunable compliance, low density, scalability, easy processing, and low cost. Linear and bending actuations are the most frequently studied methods and depend upon the device's geometry. Linear actuators are typically composed from neat films or fibers; however, bending actuators use a bilayer composite of an EAP and a passive layer.[146] In general, there are two types of EAPs: ionic and electronic. While many forms of EAP are being studied, this section will focus on the types of EAP that are currently being developed for clinical use. More extensive reviews of EAP can be found in Refs. 147 and 148.

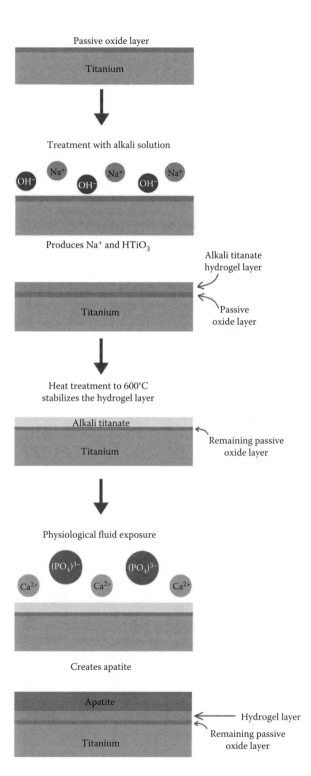

FIGURE 16.21 Schematic of the chemical reaction on the surface of Ti after alkali and heat treatments. The resultant alkali titanate layer induces apatite formation upon exposure to physiological fluid. (From Kokubo T. *Thermochim Acta* 1996;280:479–490.)

16.5.1 Ionic EAPs

Ionic EAPs depend upon the transport of ions and solvent for activation. They require a low voltage (1–5 V) for activation. They can achieve high strains, but exhibit slow response times and low actuation forces. They can be used under aqueous conditions, which is particularly beneficial in biomedical applications. The two main types of ionic EAPs are conjugated polymers (CPs) and ionic polymer–metal composites (IPMCs).

16.5.1.1 Conjugated Polymers

CPs are inherently conducting polymers that change properties when a voltage is applied. They can undergo both linear and bending actuation and display changes in conductivity, color, hydrophilicity, permeability, modulus, and volume upon being actuated. Compared to other EAPs, CPs display relatively large strains, low stresses, and operate at low voltages (<1 V).[149] Two common CPs are polypyrrole and polyaniline (Figure 16.22), but CP systems can also be composed of an electrolyte sandwiched between two CP electrodes.

CP actuation takes place via a reversible electrochemical redox reaction. The reversible redox reaction leads to subtraction or addition of charge from the polymer, which then leads to ion transport out of or into the polymer to balance the charge. Actuation via a volume change then occurs as a result of the ion transport and subsequent solvent transport (Figure 16.23). Most often, a molecular dopant, such as a simple anion, balances the charge on the polymer backbone, but if the anion is trapped by the polymer structure, then a smaller cation from the electrolyte medium will be incorporated into the polymer backbone.[149] The rate of actuation is dependent upon the rate of ion transport. For example, the choice of dopant can greatly affect the actuator performance, where smaller dopants will allow for faster actuation since they can diffuse faster, but larger dopants enable larger strains.[150]

Within the last decade, major improvements have been made in the mechanical performance of CP actuators. Actuation strains can now reach near 40% at rates up to 15%/s while the maximum stresses equal to 100 MPa can be obtained.[149–152] CPs offer several advantages as clinically applied materials because they have shown long-term biocompatibility during implantation, demonstrated improved tissue growth, and can be made biodegradable.[153] Another benefit is their actuation control, where the actuator can be in its original, final, or varying intermediate positions. Unfortunately, certain trade-offs in properties and performance have prevented any one material from exhibiting all the desired performance criteria. For example, CP actuators that achieve the highest stresses exhibit the lowest actuation strains. For more information, a database of actuator performance is tracked by the University of British Columbia.[154]

CP-based devices are currently under development for a number of clinical applications. Prototypes of prosthetic muscles, fingers, pumps, and muscle fibers based on polypyrrole are being investigated.[155] A steerable catheter is being developed using polyaniline fibers that are embedded in the walls of a standard catheter.[156] This device would allow for enhanced visualization during minimally invasive cardiac procedures and reduce surgical time. A CP-based expanding microtube is under development to aid in the reconnection of small blood vessels and to hold open the ear drum.[153] The CP tube would be reduced to a small diameter and inserted into each side of the dissected blood vessel. Upon activation, the tube would expand and open the blood vessels, which would allow for faster surgical repair. A CP-based microvalve to treat urinary incontinence is also being investigated, where the CP actuator would open the valve to allow urine passage.[153] A steerable cochlear implant is being developed to facilitate easy

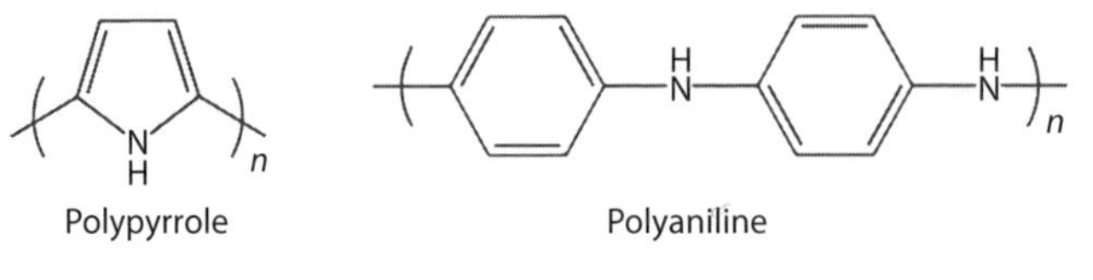

Polypyrrole   Polyaniline

FIGURE 16.22 Chemical structures of polypyrrole and polyaniline.

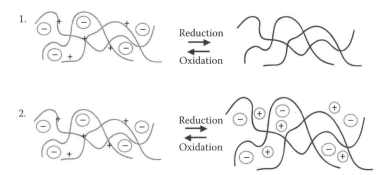

FIGURE 16.23 (1) Oxidized CP has positive charges located on polymer backbone and is swollen with dopant anions. When polymer is reduced, mobile anions leave polymer causing polymer to shrink. (2) Oxidized CPs have positive charges located on polymer backbone and is swollen with dopant anions that are trapped. When the polymer is reduced, mobile cations from electrolyte cause further swelling and volumetric expansion.

delivery of electrodes into the inner ear, which will enhance the implant's restorative hearing function.[157] While none of these applications are directly replacing muscle, they are being applied to multiple biomedical fields.

16.5.1.2 Ionic Polymer–Metal Composites

IPMCs are another type of ionic EAPs. They are formed from an ionic polymer gel coated with flexible metal electrodes. IPMCs can be actuated with low voltages (1–3 V) and work well in wet environments.[158] Bending is the most common actuation method and occurs in the following manner. An ionic polymer is coated on both sides by a metal, often platinum or gold. The composite is solvated and then counter cations are added to balance the charge due to the anionic nature of the ionic polymer. When current is applied, mobile cations and water bound to the cations diffuse toward the cathode, causing the polymer to swell and the IPMC to bend toward the anode (Figure 16.24). A slow relaxation toward the cathode will subsequently occur due to water diffusion. IPMC actuators cannot achieve the level of mechanical performance of CP actuators, with reported strain, strain rates, and sustained stresses only reaching 3%, 3%/s, and 15 MPa, respectively.[159] These low strains limit the applications of IMPCs.

Most ionic polymers are perfluorinated ionic polymers or copolymers of polytetrafluoroethylene and perfluorinated vinyl ether sulfonate. The chemical structures of two of the most common ionic

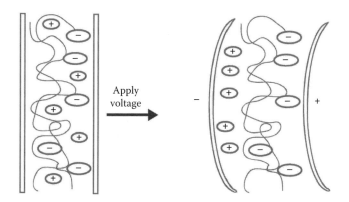

FIGURE 16.24 Ionic polymer with covalently bonded anionic has charged balanced by mobile counter cations. As voltage is applied to electrodes, mobile counter cations are attracted to cathode, inducing a bending toward anode.

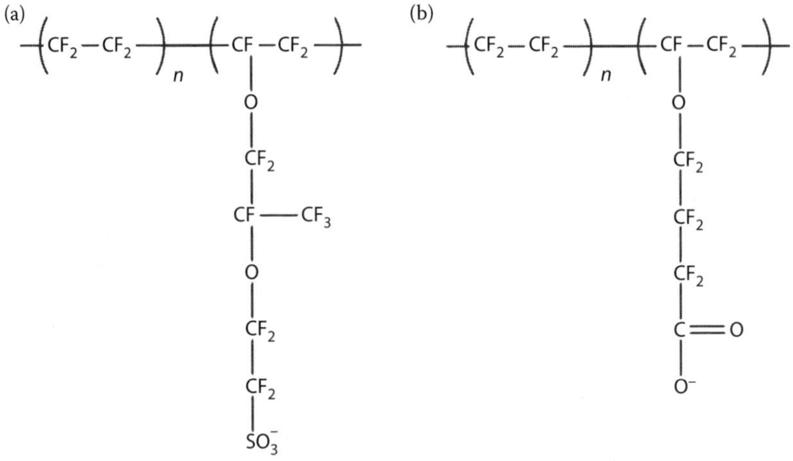

FIGURE 16.25 Chemical structure of Nafion (a) and Flemion (b).

polymers used for IPMC, Nafion and Flemion, are found in Figure 16.25. They can be solvent cast or molded using standard polymer processing techniques from raw thermoplastic pellets. The mechanical performance is determined by a number of factors, including polymer properties, water uptake, cation size and chemistry, and metal plating process. The length of the side chain and end functionality can be altered to change the properties like the ion exchange density, which results in Flemion having higher conductivity than Nafion.[158] For Nafion, the water uptake decreases as the hydrophobicity of the cation increases, thus decreasing actuation. The ionic conductivity depends upon the cation size and chemistry where the conductivity decreases as the cation size increases and alkali cations have higher conductivity than alkyl ammonium cations.[160]

The properties of the metal electrodes on the outer surface of the ionic polymer are important in determining the actuation performance. In particular, the metal electrodes should have sufficient adhesion to the polymer, large conductivity, and high interfacial area.[158] Platinum and gold are the most commonly used electrode materials because of their wide potential window, corrosion resistance, and stiffness. Considered an expensive process, standard plating occurs as such[161]:

1. The ionic polymer is soaked in a metal salt solution where the metal diffuses to each anionic site.
2. The soaking solution is changed to a chemical reducing agent that cannot penetrate the polymer structure, so the metal is reduced at the polymer's surface.
3. The desired counter cation is exchanged into the polymer.

Newer deposition methods, including physically loading the metal particles and then electroplating, have lowered the cost while maintaining performance.[162]

An advantage of IPMCs is their low power requirements. However, the voltage cannot exceed 1.23 V; otherwise, electrolysis will occur causing reduced counter cation migration. Another disadvantage is that the manufacturing materials costs are relatively high, where platinum, gold, and Nafion are approximately priced at $120, $100, and $7 per gram, respectively. Nevertheless, IPMCs are being considered for use as a steerable catheter and heart compression implant. Several studies have looked at different designs for prototype steerable catheters where either the catheter has an active IPMC internal guide wire that bends the catheter around it or the catheter is made of a tubular IPMC that can bend in multiple directions.[163–165] To address congestive heart failure, prototype IPMC devices are being developed to mechanically assist weak hearts to continue pumping by exerting mechanical forces on the outside of the heart. These prototype devices, either in a four-finger configuration or a band, would compress the heart via electrical stimulation to promote proper blood flow.[166]

16.5.2 Electronic EAPs

Electronic EAPs depend upon electric fields for activation. Unlike ionic EAPs, they require high voltage (>150 V/μm) for activation, which necessitates a local power source either external or implanted along with the device. They have fast response times and varying mechanical properties, but must be used in air; thus, they must be packaged and protected from the aqueous environment, making them less desirable for biomedical use.[167] The two main types of electronic EAP are dielectric elastomers and piezoelectric polymers.

16.5.2.1 Dielectric Elastomers

While ionic EAPs rely on the movement of ions for shape change, electronic EAPs move directly in response to an applied electric field. Dielectric elastomers can be used to create an EAP actuator by coating the top and bottom surfaces of the dielectric elastomer with compliant electrodes, essentially making a compliant capacitor. When a voltage is applied to the electrodes, the attractive Coulombic forces apply an electromechanical pressure (Maxwell stress) to the elastomer, causing the actuator to decrease in thickness and increase in lateral dimensions (Figure 16.26). Equation 16.1 describes how the effective pressure, P, exerted on the dielectric elastomer, relates to the dielectric constant of the polymer, ε_r, the permittivity of free space, ε_o, the applied electric field, E, the applied voltage, V, and the polymer thickness, t[168]:

$$P = \varepsilon_r \varepsilon_o (V/t)^2 = \varepsilon_r \varepsilon_o E^2 \tag{16.1}$$

Silicones and acrylic elastomers are the most common dielectric elastomers while the compliant electrodes are typically made of graphite spray, carbon grease, or graphite powder.[167,169]

The actuator's strain is improved by prestraining the elastomer either biaxially or uniaxially.[170] This serves to decrease the thickness, which lowers the voltage necessary for actuation and allows for a higher maximum electric field. Another way to induce large strains is to increase the voltage; however, high electric fields can result in electrical breakdown and damage to the dielectric elastomer. The reported performance for acrylic and silicone elastomers can be found in Table 16.5.[168] In general, dielectric elastomers achieve higher strains compared to other EAP actuators, but lower pressures.[171] Despite their enhanced mechanical performance, dielectric EAPs have disadvantages, particularly the required high voltages needed for operation (300 V to 5 kV). Although high voltage is not a problem for biomedical

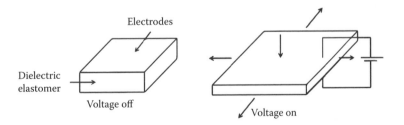

FIGURE 16.26 Actuation mechanism of dielectric elastomer actuator.

TABLE 16.5 Performance Characteristics of Acrylic and Silicone Dielectric Elastomers

Property	Acrylic	Silicone
Maximum strain (%)	380	120
Maximum pressure (MPa)	7.2	3.0
Maximum electric field (MV/m)	440	350

devices, current leakage into neighboring tissue needs to be addressed by proper packaging that could add bulk and deter use in minimally invasive surgeries. Unlike some of the ionic EAPs, dielectric EAPs are expected to be very cost efficient due to the low raw material costs of silicone and acrylics.

With regard to their clinical application, EAPs are principally being explored as prostheses to replace injured muscle. Ideally, these materials should exhibit a mechanical performance within the range of the native tissue. Specifically, it is desired that EAPs exhibit the following properties[171]:

- Linear actuation
- Strains in excess of 40%
- Response time in milliseconds
- Maximum stresses near 0.3 MPa
- Operate in an aqueous environment

In addition, dielectric elastomers must have low modulus, high dielectric constant, and high electrical breakdown. A dielectric elastomer actuator in the form of a sheet is being developed as an artificial diaphragm muscle to potentially treat patients suffering from spinal cord injuries and nerve damage that lose diaphragm control. The actuator is implanted under tension so that it is in the contracted muscle state. A voltage is then applied to expand the sheet, which causes an upward motion and maximum exhalation. These devices have been successfully tested in small and large animal models, but further developments with regard to packaging and volume actuation still need to be addressed.[172,173] Another possible application is for facial muscle repair, especially in defective eyelids that are unable to fully open or close. An implantable eyelid sling has been developed and tested in cadaveric specimens, and demonstrated complete eyelid closure.[174] However, further development remains on reliable water-tight packaging, local power supply, and long-term performance.

16.5.2.2 Piezoelectric Polymers

Piezoelectric EAPs have recently emerged as candidate materials for clinical applications. The piezoelectric effect is a linear electromechanical interaction, where a noncentrosymmetric crystalline material or semicrystalline polymer exhibits a buildup of electrical charge due to an applied stress. The reverse piezoelectric effect is when a material undergoes a deformation strain due to an applied electric field. The most common piezoelectric EAP that display meaningful amounts of strain are ferroelectric poly(vinylidene fluoride) (PVDF), its copolymers, such as poly(vinylidene fluoride-trifluoroethylene) (PVDF-TrFE), and odd-numbered nylons.[175] A ferroelectric polymer has the ability to spontaneously switch polarization, which can be reoriented with an applied electric field. Ferroelectric polymers show enhanced piezoelectric properties, where the piezoelectric coefficient is 10 times that of nonferroelectric piezoelectric materials.[176] PVDF does not initially display piezoelectric properties because the semicrystalline domains are nonferroelectric α phase (nonpolar), but upon mechanical loading, the semicrystalline domains are converted to the ferroelectric β phase (polar).[175] The randomly oriented ferroelectric β phase semicrystalline domains are then poled in a high electric field to create a net dipole moment and the material becomes piezoelectric. Poling occurs by plating the polymer with electrodes and applying an electric field or by a corona discharge from a needle.[175]

PVDF and its copolymers have the advantages of high stresses (45 MPa), rapid cycle speed, and can be easily processed and shaped. The associated high electric fields (200 MV/m) for activation and limited actuation strains (7%) are its main disadvantages[159,167] and are currently limiting the ability of PVDF systems to be applied in biomedical devices.

References

1. Gall K, Kreiner P, Turner D, Hulse M. Shape-memory polymers for microelectromechanical systems. *J Microelectromech Syst* 2004;13:472–483.
2. Heywant W, Lubitz K, Wersing W. *Piezoelectricity*. Berlin: Springer; 2008.

3. Duerig TW, Pelton AR, Stockel D. The utility of superelasticity in medicine. *Bio-Med Mater Eng* 1996;6:255–266.

4. Pelton AR, Stockel D, Duerig TW. Medical uses of nitinol. *Materials Science Forum* 2000; 327–328: 63–70.

5. Mano JF. Stimuli-responsive polymeric systems for biomedical applications. *Adv Eng Mater* 2008;10:515–527.

6. Yakacki CM, Gall K. Shape-memory polymers for biomedical applications. In: Andreas Lendlein, ed.; *Shape-Memory Polymers* 2010:147–175.

7. Lendlein A, Kelch S. Shape-memory polymers. *Angew Chem Int Ed* 2002;41:2034–2057.

8. Liu C, Qin H, Mather PT. Review of progress in shape-memory polymers. *J Mater Chem* 2007;17:1543–1558.

9. Mather PT, Luo XF, Rousseau IA. Shape memory polymer research. *Annu Rev Mater Res* 2009;39:445–471.

10. Yakacki CM, Shandas R, Lanning C, Rech B, Eckstein A, Gall K. Unconstrained recovery characterization of shape-memory polymer networks for cardiovascular applications. *Biomaterials* 2007;28:2255–2263.

11. Shin DD, Mohanchandra KR, Carman GP. Development of hydraulic linear actuator using thin film SMA. *Sens Actuators A Phys* 2005;119:151–156.

12. Krulevitch P, Lee AP, Ramsey PB, Trevino JC, Hamilton J, Northrup MA. Thin film shape memory alloy microactuators. *J Microelectromech Syst* 1996;5:270–282.

13. Otsuka K, Wayman CM. *Shape Memory Materials.* Cambridge, United Kingdom: Cambridge University Press; 1998.

14. Duerig T, Pelton A, Stockel D. An overview of nitinol medical applications. *Mater Sci Eng A* 1999;273–275:149–160.

15. Morgan NB. Medical shape memory alloy applications—the market and its products. *Mater Sci Eng A Struct Mater Prop Microstruct Process* 2004;378:16–23.

16. Elahinia MH, Hashemi M, Tabesh M, Bhaduri SB. Manufacturing and processing of NiTi implants: A review. *Prog Mater Sci* 2012;57:911–946.

17. Shaw JA, Churchill CB, Iadicola MA. Tips and tricks for characterizing shape memory alloy wire: Part 1—Differential scanning calorimetry and basic phenomena. *Exp Tech* 2008;32:55–62.

18. Shaw JA, Kyriakides S. Thermomechanical aspects of niti. *J Mech Phys Solids* 1995;43:1243–1281.

19. Sehitoglu H, Karaman I, Anderson R et al. Compressive response of NiTi single crystals. *Acta Mater* 2000;48:3311–3326.

20. Gall K, Tyber J, Brice V, Frick CP, Maier HJ, Morgan N. Tensile deformation of NiTi wires. *J Biomed Mater Res Part A* 2005;75A:810–823.

21. Rondelli G. Corrosion resistance tests on NiTi shape memory alloy. *Biomaterials* 1996;17:2003–2008.

22. Figueira N, Silva TM, Carmezim MJ, Fernandes JCS. Corrosion behaviour of NiTi alloy. *Electrochim Acta* 2009;54:921–926.

23. Eggeler G, Hornbogen E, Yawny A, Heckmann A, Wagner M. Structural and functional fatigue of NiTi shape memory alloys. *Mater Sci Eng A Struct Mater Prop Microstruct Process* 2004; 378:24–33.

24. Gall K, Tyber J, Wilkesanders G, Robertson SW, Ritchie RO, Maier HJ. Effect of microstructure on the fatigue of hot-rolled and cold-drawn NiTi shape memory alloys. *Mater Sci Eng A Struct Mater Prop Microstruct Process* 2008;486:389–403.

25. Shabalovskaya SA. On the nature of the biocompatibility and on medical applications of NiTi shape memory and superelastic alloys. *Bio-Med Mater Eng* 1996;6:267–289.

26. Es-Souni M, Fischer-Brandies H. Assessing the biocompatibility of NiTi shape memory alloys used for medical applications. *Anal Bioanal Chem* 2005;381:557–567.

27. Otsuka K, Ren X. Physical metallurgy of Ti-Ni-based shape memory alloys. *Prog Mater Sci* 2005;50:511–678.

28. Hermiller J, Simonton C, Hinohara T et al. Clinical experience with a circumferential clip-based vascular closure device in diagnostic catheterization. *J Invasive Cardiol* 2005;17:504–510.

29. Hermiller JB, Simonton C, Hinohara T et al. The StarClose (R) vascular closure system: Interventional results from the CLIP study. *Catheter Cardiovasc Interv* 2006;68:677–683.

30. Gray BH, Miller R, Langan EM, Joels CS, Yasin Y, Kalbaugh CA. The utility of the StarClose arterial closure device in patients with peripheral arterial disease. *Ann Vasc Surg* 2009;23:341–344.

31. McTaggart RA, Raghavan D, Haas RA, Jayaraman MV. StarClose vascular closure device: Safety and efficacy of deployment and reaccess in a neurointerventional radiology service. *Am J Neuroradiol* 2010;31:1148–1150.

32. Spiliopoulos S, Katsanos K, Karnabatidis D, Diamantopoulos A, Nikolaos C, Siablis D. Safety and efficacy of the StarClose vascular closure device in more than 1000 consecutive peripheral angio-plasty procedures. *J Endovascular Ther* 2011;18:435–443.

33. Hon L-Q, Ganeshan A, Thomas SM, Warakaulle D, Jagdish J, Uberoi R. Vascular closure devices: A comparative overview. *Curr Probl Diagn Radiol* 2009;38:33–43.

34. Rousseau H, Perreault P, Otal P et al. The 6-F nitinol TrapEase inferior vena cava filter: Results of a prospective multicenter trial. *J Vasc Interv Radiol* 2001;12:299–304.

35. Stavropoulos SW, Clark T, Jacobs D et al. Placement of a vena cava filter with an antecubital approach. *Acad Radiol* 2002;9:478–481.

36. Chung J, Owen RJT. Using inferior vena cava filters to prevent pulmonary embolism. *Can Fam Phys* 2008;54:49–55.

37. Kalva SP, Chlapoutaki C, Wicky S, Greenfield AJ, Waltman AC, Athanasoulis CA. Suprarenal inferior vena cava filters: A 20-year single-center experience. *J Vasc Interv Radiol* 2008;19:1041–1047.

38. Van Ha TG, Chien AS, Funaki BS et al. Use of retrievable compared to permanent inferior vena cava filters: A single-institution experience. *Cardiovasc Interv Radiol* 2008;31:308–315.

39. Andreasen GF, Hilleman TB. Evaluation of 55 cobalt substituted nitinol wire for use in orthodontics. *J Am Dent Assoc* 1971;82:1373–1375.

40. Kusy RP. A review of contemporary archwires: Their properties and characteristics. *Angle Orthod* 1997;67:197–207.

41. Yakacki CM, Gall K, Dirschl DR, Pacaccio DJ. Pseudoelastic intramedullary nailing for tibio-talo-calcaneal arthrodesis. *Expert Rev Med Devices* 2011;8:159–166.

42. Huang W. On the selection of shape memory alloys for actuators. *Mater Des* 2002;23:11–19.

43. Van Humbeeck J. Non-medical applications of shape memory alloys. *Mater Sci Eng A Struct Mater Prop Microstruct Process* 1999;273:134–148.

44. Nespoli A, Besseghini S, Pittaccio S, Villa E, Viscuso S. The high potential of shape memory alloys in developing miniature mechanical devices: A review on shape memory alloy mini-actuators. *Sens Actuators A Phys* 2010;158:149–160.

45. Henderson E, Buis A. Nitinol for prosthetic and orthotic applications. *J Mater Eng Perform* 2011;20:663–665.

46. O'Toole KT, McGrath MM. Mechanical design and theoretical analysis of a four fingered prosthetic hand incorporating embedded SMA bundle actuators. In: Ardil C, ed. *Proceedings of World Academy of Science, Engineering and Technology*, Vol 25; 2007:142–149.

47. O'Toole KT, McGrath MM, Coyle E. Analysis and evaluation of the dynamic performance of SMA actuators for prosthetic hand design. *J Mater Eng Perform* 2009;18:781–786.

48. O'Toole KT, McGrath MM, Hatchett DW. Transient characterisation and analysis of shape memory alloy wire bundles for the actuation of finger joints in prosthesis design. *Mechanika* 2007:65–69.

49. Pittaccio S, Viscuso S, Rossini M, et al. SHADE: A shape-memory-activated device promoting ankle dorsiflexion. *J Mater Eng Perform* 2009;18:824–830.

50. Frick CP, Ortega AM, Tyber J, Gall K, Maier HJ. Multiscale structure and properties of cast and deformation processed polycrystalline NiTi shape-memory alloys. *Metall Mater Trans A-Phys Metall Mater Sci* 2004;35A:2013–2025.

51. Gall K, Maier H. Cyclic deformation mechanisms in precipitated NiTi shape memory alloys. *Acta Mater* 2002;50:4643–4657.

52. Hurley J, Ortega AM, Lechniak J, Gall K, Maier HJ. Structural evolution during the cycling of NiTi shape memory alloys. *Z Metallk* 2003;94:547–552.

53. Melton KN, Mercier O. Fatigue of niti thermoelastic martensites. *Acta Metall* 1979;27:137–144.

54. Miyazaki S, Imai T, Igo Y, Otsuka K. Effect of cyclic deformation on the pseudoelasticity characteristics of ti-ni alloys. *Metall Trans A Phys Metall Mater Sci* 1986;17:115–120.

55. McKelvey AL, Ritchie RO. Fatigue-crack growth behavior in the superelastic and shape-memory alloy Nitinol. *Metall Mater Trans A-Phys Metall Mater Sci* 2001;32:731–743.

56. Robertson SW, Ritchie RO. *In vitro* fatigue-crack growth and fracture toughness behavior of thin-walled superelastic Nitinol tube for endovascular stents: A basis for defining the effect of crack-like defects. *Biomaterials* 2007;28:700–709.

57. Soboyejo W, Milne I, Ritchie RO, Karihaloo B. Fatigue of biomaterials/biomedical systems. In: Ian Milne, ed.; *Comprehensive Structural Integrity*. Oxford: Pergamon; 2003:443–465.

58. Al-Waheidi EM. Allergic reaction to nickel orthodontic wires: A case report. *Quintessence International (Berlin, Germany: 1985)* 1995;26:385–387.

59. Shabalovskaya SA. Surface, corrosion and biocompatibility aspects of Nitinol as an implant material. *Bio-Med Mater Eng* 2002;12:69–109.

60. Assad M, Yahia LH, Rivard CH, Lemieux N. *In vitro* biocompatibility assessment of a nickel-titanium alloy using electron microscopy *in situ* end-labeling (EM-ISEL). *J Biomed Mater Res* 1998;41:154–161.

61. Castleman LS, Motzkin SM, Alicandri FP, Bonawit VL, Johnson AA. Biocompatibility of nitinol alloy as an implant material. *J Biomed Mater Res* 1976;10:695–731.

62. Chan CM, Trigwell S, Duerig T. Oxidation of an NiTi alloy. *Surf Interface Anal* 1990;15:349–354.

63. Ryhanen J, Niemi E, Serlo W et al. Biocompatibility of nickel-titanium shape memory metal and its corrosion behavior in human cell cultures. *J Biomed Mater Res* 1997;35:451–457.

64. Biesiekierski A, Wang J, Gepreel MAH, Wen C. A new look at biomedical Ti-based shape memory alloys. *Acta Biomater* 2012;8:1661–1669.

65. Kim HY, Satoru H, Kim JI, Hosoda H, Miyazaki S. Mechanical properties and shape memory behavior of Ti-Nb alloys. *Mater Trans* 2004;45:2443–2448.

66. Fukui Y, Inamura T, Hosoda H, Wakashima K, Miyazaki S. Mechanical properties of a Ti-Nb-Al shape memory alloy. *Mater Trans* 2004;45:1077–1082.

67. Hosoda H, Fukui Y, Inamura T, Wakashima K, Miyazaki S, Inoue K. Mechanical properties of Ti-base shape memory alloys. In: Chandra T, Torralba JM, Sakai T, eds. *Thermec'2003*, Pts 1–5. Zurich-Uetikon: Trans Tech Publications Ltd; 2003:3121–3125.

68. Lin B, Gall K, Maier HJ, Waldron R. Structure and thermomechanical behavior of NiTiPt shape memory alloy wires. *Acta Biomater* 2009;5:257–267.

69. Zheng YF, Zhang BB, Wang BL et al. Introduction of antibacterial function into biomedical TiNi shape memory alloy by the addition of element Ag. *Acta Biomater* 2011;7:2758–2767.

70. Gil FJ, Planell JA. Effect of copper addition on the superelastic behavior of Ni-Ti shape memory alloys for orthodontic applications. *J Biomed Mater Res* 1999;48:682–688.

71. Grossmann C, Frenzel J, Sampath V, Depka T, Eggeler G. Elementary transformation and deformation processes and the cyclic stability of NiTi and NiTiCu shape memory spring actuators. *Metall Mater Trans A-Phys Metall Mater Sci* 2009;40A:2530–2544.

72. Behl M, Lendlein A. Shape-memory polymers. *Mater Today* 2007;10:20–28.

73. Gall K, Yakacki CM, Liu Y, Shandas R, Willett N, Anseth KS. Thermomechanics of the shape memory effect in polymers for biomedical applications. *J Biomed Mater Res A* 2005;73:339–348.

74. Liu C, Mather PT. Thermomechanical characterization of a tailored series of shape memory polymers. *J Appl Med Polym* 2002;6:47–52.

75. Smith TL. Ultimate tensile properties of elastomers. I. Characterization by a time and temperature independent failure envelope. *J Polym Sci A Gen Papers* 1963;1:3597–3615.

76. Yakacki CM, Willis S, Luders C, Gall K. Deformation limits in shape-memory polymers. *Adv Eng Mater* 2008;10:112–119.

77. Smith KE, Temenoff JS, Gall K. On the toughness of photopolymerizable (Meth)acrylate networks for biomedical applications. *J Appl Polym Sci* 2009;114:2711–2722.

78. Xie T, Page KA, Eastman SA. Strain-based temperature memory effect for nafion and its molecular origins. *Adv Funct Mater* 2011;21:2057–2066.

79. Xie T. Tunable polymer multi-shape memory effect. *Nature* 2010;464:267–270.

80. Yakacki CM, Lyons MB, Rech B, Gall K, Shandas R. Cytotoxicity and thermomechanical behavior of biomedical shape-memory polymer networks post-sterilization. *Biomed Mater* 2008;3:015010.

81. Mijovic J, Nicolais L, D'Amore A, Kenny JM. Principal features of structural relaxation in glassy polymers. A review. *Polym Eng Sci* 1994;34:381–389.

82. Ortega AM, Yakacki CM, Dixon SA, Likos R, Greenberg AR, Gall K. Effect of crosslinking and long-term storage on the shape-memory behavior of (meth)acrylate-based shape-memory polymers. *Soft Matter* 2012;8:3381–3392.

83. Maitland DJ, Metzger MF, Schumann D, Lee A, Wilson TS. Photothermal properties of shape memory polymer micro-actuators for treating stroke. *Lasers Surg Med* 2002;30:1–11.

84. Small Wt, Buckley PR, Wilson TS, Loge JM, Maitland KD, Maitland DJ. Fabrication and characterization of cylindrical light diffusers comprised of shape memory polymer. *J Biomed Opt* 2008; 13:024018.

85. Small Wt, Wilson TS, Buckley PR et al. Prototype fabrication and preliminary *in vitro* testing of a shape memory endovascular thrombectomy device. *IEEE Trans Biomed Eng* 2007;54:1657–1666.

86. Leng JS, Lv HB, Liu YJ, Du SY. Synergic effect of carbon black and short carbon fiber on shape memory polymer actuation by electricity. *J Appl Phys* 2008;104:104917.

87. Mohr R, Kratz K, Weigel T, Lucka-Gabor M, Moneke M, Lendlein A. Initiation of shape-memory effect by inductive heating of magnetic nanoparticles in thermoplastic polymers. *Proc Natl Acad Sci USA* 2006;103:3540–3545.

88. Weigel T, Mohr R, Lendlein A. Investigation of parameters to achieve temperatures required to initiate the shape-memory effect of magnetic nanocomposites by inductive heating. *Smart Mater Struct* 2009;18:9.

89. Buckley PR, McKinley GH, Wilson TS et al. Inductively heated shape memory polymer for the magnetic actuation of medical devices. *IEEE Trans Biomed Eng* 2006;53:2075–2083.

90. Razzaq MY, Anhalt M, Frormann L, Weidenfeller B. Thermal, electrical and magnetic studies of magnetite filled polyurethane shape memory polymers. *Mater Sci Eng A* 2007;444:227–235.

91. Yakacki CM, Satarkar NS, Gall K, Likos R, Hilt JZ. Shape-memory polymer networks with Fe_3O_4 nanoparticles for remote activation. *J Appl Polym Sci* 2009;112:3166–3176.

92. Yakacki CM, Nguyen TD, Likos R, Lamell R, Guigou D, Gall K. Impact of shape-memory programming on mechanically-driven recovery in polymers. *Polymer* 2011;52:4947–4954.

93. Sigwart U, Puel J, Mirkovitch V, Joffre F, Kappenberger L. Intravascular stents to prevent occlusion and restenosis after trans-luminal angioplasty. *New Engl J Med* 1987;316:701–706.

94. Wache HM, Tartakowska DJ, Hentrich A, Wagner MH. Development of a polymer stent with shape memory effect as a drug delivery system. *J Mater Sci Mater Med* 2003;14:109–112.

95. Baer GM, Small Wt, Wilson TS et al. Fabrication and *in vitro* deployment of a laser-activated shape memory polymer vascular stent. *Biomed Eng Online* 2007;6:43.

96. Tamai H, Igaki K, Kyo E et al. Initial and 6-month results of biodegradable poly-L-lactic acid coronary stents in humans. *Circulation* 2000;102:399–404.

97. Rinkel GJE, Djibuti M, Algra A, van Gijn J. Prevalence and risk of rupture of intracranial aneurysms: A systematic review. *Stroke* 1998;29:251–256.

98. Currie S, Mankad K, Goddard A. Endovascular treatment of intracranial aneurysms: Review of current practice. *Postgrad Med J* 2011;87:41–50.

99. Koebbe CJ, Veznedaroglu E, Jabbour P, Rosenwasser RH. Endovascular management of intracranial aneurysms: Current experience and future advances. *Neurosurgery* 2006;59:S93–S102; discussion S103–S113.

100. Rodriguez JN, Yu YJ, Miller MW et al. Opacification of shape memory polymer foam designed for treatment of intracranial aneurysms. *Ann Biomed Eng* 2012;40:883–897.

101. Metcalfe A, Desfaits AC, Salazkin I, Yahia L, Sokolowski WM, Raymond J. Cold hibernated elastic memory foams for endovascular interventions. *Biomaterials* 2003;24:491–497.

102. Maitland DJ, Small W, Ortega JM et al. Prototype laser-activated shape memory polymer foam device for embolic treatment of aneurysms. *J Biomed Opt* 2007;12:030504.

103. Small W, Buckley PR, Wilson TS et al. Shape memory polymer stent with expandable foam: A new concept for endovascular embolization of fusiform aneurysms. *IEEE Trans Biomed Eng* 2007;54: 1157–1160.

104. Williams DF. *The Williams Dictionary of Biomaterials*. Liverpool, UK: Liverpool University Press; 1999.

105. Ducheyne P, Qiu Q. Bioactive ceramics: The effect of surface reactivity on bone formation and bone cell function. *Biomaterials* 1999;20:2287–2303.

106. Meleti Z, Shapiro IM, Adams CS. Inorganic phosphate induces apoptosis of osteoblastlike cells in culture. *Bone* 2000;27:359–366.

107. Dvorak MM. Physiological changes in extracellular calcium concentration directly control osteoblast function in the absence of calciotropic hormones. *Proc Natl Acad Sci USA* 2004;101:5140–5145.

108. Habibovic P, Barralet JE. Bioinorganics and biomaterials: Bone repair. *Acta Biomater* 2011;7:3013–3026.

109. Barradas AM, Yuan H, van Blitterswijk CA, Habibovic P. Osteoinductive biomaterials: Current knowledge of properties, experimental models and biological mechanisms. *Eur Cell Mater* 2011;21:407–429; discussion 429.

110. Albrektsson T, Johansson C. Osteoinduction, osteoconduction, and osseointegration. *Eur Spine J* 2001;10.

111. Selye H, Lemire Y, Bajusz E. Induction of bone, cartilage, and hemopoietic tissue by subcutaneously implanted tissue diaphragms. *Roux' Arch Entwicklungsmech* 1960;151:572–585.

112. Winter GD, Simpson BJ. Heterotopic bone formed in a synthetic sponge in the skin of young pigs. *Nature* 1969;223:88–90.

113. Hench LL. Bioceramics. *J Am Ceram Soc* 1998;81:1705–1728.

114. Hench LL, Polak JM. Third-generation biomedical materials. *Science* 2002;295:1014–1017.

115. Hench LL. Bioceramics—from concept to clinic. *J Am Ceram Soc* 1991;74:1487–1510.

116. Hench LL. The story of Bioglass (R). *J Mater Sci-Mater Med* 2006;17:967–978.

117. Hench LL, Splinter RJ, Allen WC, Greenlee TK. Bonding mechanisms at the interface of ceramic prosthetic materials. *J Biomed Mater Res* 1971;2:117.

118. Hench LL, Wilson J. Surface-active biomaterials. *Science* 1984;226:630–636.

119. Juhasz JA, Best SM. Bioactive ceramics: Processing, structures and properties. *J Mater Sci* 2012;47:610–624.

120. Chevalier J, L. G. Ceramics for medical applications: A picture for the next 20 years. *J Eur Ceram Soc* 2009;29:1245–1255.

121. Verrier S, Blaker JJ, Maquet V, Hench LL, Boccaccini AR. PDLLA/Bioglass composites for soft-tissue and hard-tissue engineering: An *in vitro* cell biology assessment. *Biomaterials* 2004;25: 3013–3021.

122. Merwin GE, Atkins JS, Wilson J, Hench LL. Comparison of ossicular replacement materials in a mouse ear model. *Otolaryngol Head Neck Surg* 1982;90:461–469.

123. Stanley HR, Hall MB, Clark AE, King CJ, 3rd, Hench LL, Berte JJ. Using 45S5 bioglass cones as endosseous ridge maintenance implants to prevent alveolar ridge resorption: A 5-year evaluation. *Int J Oral Maxillofac Implants* 1997;12:95–105.

124. Kokubo T, Ito S, Shigematsu M, Sakka S, Yamamuro T. Fatigue and lifetime of bioactive glass ceramic a-w containing apatite and wollastonite. *J Mater Sci* 1987;22:4067–4070.

125. Ido K, Asada Y, Sakamoto T, Hayashi R, Kuriyama S. Radiographic evaluation of bioactive glass-ceramic grafts in postero-lateral lumbar fusion. *Spinal Cord* 2000;38:315–318.

126. Carrodeguas RG, De Aza S. Alpha-tricalcium phosphate: Synthesis, properties and biomedical applications. *Acta Biomater* 2011;7:3536–3546.

127. Daculsi G, Laboux O, Malard O, Weiss P. Current state of the art of biphasic calcium phosphate bioceramics. *J Mater Sci-Mater Med* 2003;14:195–200.

128. Nelson DGA, Barry JC, Shields CP, Glena R, Featherstone JDB. Crystal morphology, composition, dissolution behavior of carbonated apatites prepared at controlled pH and temperature. *J Coll Int Sci* 1989;130:467–479.

129. Dumbleton J, Manley MT. Hydroxyapatite-coated prostheses in total hip and knee arthroplasty. *J Bone Joint Surg Am* 2004;86-A:2526–2540.

130. Tsui YC, Doyle C, Clyne TW. Plasma sprayed hydroxyapatite coatings on titanium substrates. Part 1: Mechanical properties and residual stress levels. *Biomaterials* 1998;19:2015–2029.

131. Bercovy M, Beldame J, Lefebvre B, Duron A. A prospective clinical and radiological study comparing hydroxyapatite-coated with cemented tibial components in total knee replacement. *J Bone Joint Surg-Br Vol* 2012;94B:497–503.

132. Voigt JD, Mosier M. Hydroxyapatite (HA) coating appears to be of benefit for implant durability of tibial components in primary total knee arthroplasty. *Acta Orthop* 2011;82:448–459.

133. Steens W, Schneeberger AG, Skripitz R, Fennema P, Goetze C. Bone remodeling in proximal HA-coated versus uncoated cementless SL-Plus(A (R)) femoral components: A 5-year follow-up study. *Arch Orthop Trauma Surg* 2010;130:921–926.

134. Gandhi R, Davey JR, Mahomed NN. Hydroxyapatite coated femoral stems in primary total hip arthroplasty a meta-analysis. *J Arthroplast* 2009;24:38–42.

135. Konan S, Haddad FS. A clinical review of bioabsorbable interference screws and their adverse effects in anterior cruciate ligament reconstruction surgery. *Knee* 2009;16:6–13.

136. Macarini L, Milillo P, Mocci A, Vinci R, Ettorre GC. Poly-L-lactic acid–hydroxyapatite (PLLA-HA) bioabsorbable interference screws for tibial graft fixation in anterior cruciate ligament (ACL) reconstruction surgery: MR evaluation of osteointegration and degradation features. *Radiol Med* 2008;113:1185–1197.

137. Khan I, Smith N, Jones E, Finch DS, Cameron RE. Analysis and evaluation of a biomedical polycarbone urethane tested in an *in vitro* study and an ovine arthroplasty model. Part I: Materials selection and evaluation. *Biomaterials* 2005;26:621–631.

138. Barinov SM, Komlev VS. Calcium phosphate bone cements. *Inorg Mater* 2011;47:1470–1485.

139. Low KL, Tan SH, Zein SH, Roether JA, Mourino V, Boccaccini AR. Calcium phosphate-based composites as injectable bone substitute materials. *J Biomed Mater Res B Appl Biomater* 2010;94:273–286.

140. Kokubo T. Formation of biologically active bone-like apatite on metals and polymers by a biomimetic process. *Thermochim Acta* 1996;280:479–490.

141. Nishiguchi S, Nakamura T, Kobayashi M, Kim HM, Miyaji F, Kokubo T. The effect of heat treatment on bone-bonding ability of alkali-treated titanium. *Biomaterials* 1999;20:491–500.

142. Pereira MM, Hench LL. Mechanisms of hydroxyapatite formation on porous gel-silica substrates. *J Sol-Gel Sci Technol* 1996;7:59–68.

143. Schwartz Z, Raz P, Zhao G et al. Effect of micrometer-scale roughness of the surface of Ti6Al4V pedicle screws *in vitro* and in vivo. *J Bone Joint Surg Am* 2008;90:2485–2498.

144. Fujibayashi S, Neo M, Kim HM, Kokubo T, Nakamura T. Osteoinduction of porous bioactive titanium metal. *Biomaterials* 2004;25:443–450.

145. Fukuda A, Takemoto M, Saito T et al. Osteoinduction of porous Ti implants with a channel structure fabricated by selective laser melting. *Acta Biomater* 2011;7:2327–2336.

146. Pei Q, Inganas O. Conjugated polymers and the bending cantilever method: Electrical muscles and smart devices. *Adv Mater* 1992;4:277–278.

147. Carpi F, Smela E, eds. *Biomedical Applications of Electroactive Polymer Actuators*. UK: John Wiley & Sons; 2009.

148. Bar-Cohen Y, ed. *Electroactive Polymer (EAP) Actuators as Artificial Muscles: Reality, Potential, and Challenges*. 2 ed. USA: SPIE Press; 2004.

149. Spinks GM, Alici G, McGovern S, Xi B, Wallace GG. Conjugated polymer actuators: Fundamentals. In: Carpi F, Smela E, eds. *Biomedical Applications of Electroactive Polymers Actuators*. UK: John Wiley & Sons; 2009:195–228.

150. Hara S, Zama T, Takashima W, Kaneto K. Free-standing gel-like polypyrrole actuators doped with bis(perfluoroalkylsulfonyl)imide exhibiting extremely large strain. *Smart Mater Struct* 2005;14:1501–1510.

151. Ding J, Liu L, Spinks GM, Zhou D, Wallace GG. High performance conducting polymer actuators utilizing a tublar geometry and helical wire interconnects. *Synth Met* 2003;138:391–398.

152. Spinks GM, Mottaghitalab V, Bahrami-Samani M, Whitten PG, Wallace GG. Carbon nanotube reinforced polyaniline fibers for high strength artificial muscles. *Adv Mater* 2006;18:637–640.

153. Smela E. Conjugated polymer actuators for biomedical applications. *Adv Mater* 2003;15:481–494.

154. Actuator Selection Tool. (Accessed at http://www.actuatorweb.org.)

155. (Accessed at http://www.eamex.co.jp/index_e.html.)

156. Mazzoldi A, De Rossi D. Conductive polymer based structures for a steerable catheter. *Proc SPIE* 2000;3987:273–280.

157. Shoa T, Madden JD, Munce NR, Yang VXD. Steerable catheters. In: Carpi F, Smela E, eds. *Biomedical Applications of Electroactive Polymer Actuators*. UK: John Wiley & Sons; 2009:229–248.

158. Asaka K, Oguro K. IPMC Actuators: Fundamentals. In: Carpi F, Smela E, eds. *Biomedical Applications of Electroactive Polymer Actuators*. UK: John Wiley & Sons; 2009:103–119.

159. Madden JD, Vandesteeg NA, Anquetil PA et al. Artificial muscle technology: Physical principles and naval prospects. *IEEE J Ocean Eng* 2004;29:706–728.

160. Asaka K, Fujiwara N, Oguro K, Onishi K, Sewa S. State of water and ionic conductivity of solid polymer electrolyte membranes in relation to polymer actuators. *J Electroanal Chem* 2001;505:24–32.

161. Nemat-Nasser S, Thomas CW. Ionomeric polymer-metal composites. In: Bar-Cohen Y, ed. *Electroactive Polymer (EAP) Actuators as Artificial Muscles: Reality, Potential, and Challenges*. Bellingham, Washington: SPIE; 2004:171–230.

162. Shahinpoor M, Kim KJ. Novel physically loaded and interlocked electrode developed for ionic polymer-metal composites. In: Bar-Cohen Y, ed. *Smart Structures and Materials 2001: Electroactive Polymer Actuators and Devices*. Bellingham, Washington: SPIE; 2001:174–181.

163. Guo S, Fukuda T, Arai F, Oguro K, Negoro M, Nakamura T. Micro active guide wire catheter using ICPF actuator. *IEEE International Conference on Intelligent Robotics and Systems* 1995;2:172–177.

164. Guo S, Fukuda T, Kosuge K, Arai F, Oguro K, Negoro M. Micro catheter system with active guide wire-structure, experimental results and characteristic evaluation of active guide wire catheter using ICPF actuator. *5th International Symposium on Micro machine and Human Science Proceedings* 1994:191–197.

165. Oguro K, Fujiwara N, Asaka K, Onishi K, Sewa S. Polymer electrolyte actuator with gold electrodes. *Proceedings of the SPIE 6th Annual International Symposium on Smart Structures and Materials*. Newport Beach, CA. 1999:64–71.

166. Shahinpoor M. Implantable heart-assist and compression devices employing an active network of electrically-controllable ionic polymer-metal nanocomposites. In: Carpi F, Smela E, eds. *Biomedical Applications of Electroactive Polymer Actuators*. UK: John Wiley & Sons; 2009:137–159.

167. Samatham R, Kim KJ, Dogruer D et al. Active polymers: An overview. In: Kim KJ, Tadokoro S, eds. *Electroactive Polymers for Robotic Applications: Artificial Muscles and Sensors*. London: Springer; 2007.

168. Kornbluh R, Heydt R, Pelrine R. Dielectric elastomer actuators: Fundamentals. In: Carpi F, Smela E, eds. *Biomedical Applications of Electroactive Polymer Actuators*. UK: John Wiley & Sons; 2009:387–393.

169. Carpi F, Chiarelli P, Mazzoldi A, De Rossi D. Electromechanical characterisation of dielectric elastomer planar actuators: Comparative evaluation of different electrode materials and different counterloads. *Sens Actuators A* 2003;107:85–95.

170. Pelrine R, Kornbluh R, Pei Q, Joseph J. High-speed electrically actuated elastomers with strain greater than 100%. *Science* 2000;287:836–839.

171. Kornbluh R, Pelrine R, Pei Q et al. Application of dielectric elastomer EAP actuators. In: Bar-Cohen Y, ed. *Electroactive Polymer (EAP) Actuators as Artificial Muscles: Reality, Potential, and Challenges*. Bellingham, Washington: SPIE; 2004:529–581.

172. Bashkin JS, Heim J, Prahlad H, Kornbluh R, Pelrine R, Elefteriades J. Medical device applications of dielectric elastomer-based artificial muscle actuators. In: *Medical Device Materials IV: Proceedings of the 2007 Materials and Processes for Medical Devices Conference*; 2007; CA: ASM; 2007.

173. Bashkin JS, Kornbluh R, Prahlad H, Wong-Foy A. Biomedical applications of dielectric elastomer actuators. In: Carpi F, Smela E, eds. *Biomedical Applications of Electroactive Polymer Actuators*. UK: John Wiley & Sons; 2009:395–410.

174. Tollefson TT, Senders CW. Restoration of eyelid closure in facial paralysis using artificial muscle: Preliminary cadaveric analysis. *Laryngoscope* 2007;117:1907–1911.

175. Zhang Q, Huang C, Xia F, Su J. Electric EAP. In: Bar-Cohen Y, ed. *Electroactive Polymer (EAP) Actuators as Artificial Muscles: Reality, Potential, and Challenges*. Bellingham, Washington: SPIE; 2004:95–148.

176. Li Z, Cheng Z. Piezoelectric and electrostrictive polymer actuators: Fundamentals. In: Carpi F, Smela E, eds. *Biomedical Applications and Electroactive Polymer Actuators*. UK: John Wiley & Sons; 2009:319–334.

177. Huang WM, Yang B, An L, Li C, Chan YS. Water-driven programmable polyurethane shape memory polymer: Demonstration and mechanism. *Appl Phys Lett* 2005;86:114105.

178. Yakacki CM, Shandas R, Safranski D, Ortega AM, Sassaman K, Gall K. Strong, tailored, biocompatible shape-memory polymer networks. *Adv Funct Mater* 2008;18:2428–2435.

179. Ortega J, Maitland D, Wilson T, Tsai W, Savas O, Saloner D. Vascular dynamics of a shape memory polymer foam aneurysm treatment technique. *Ann Biomed Eng* 2007;35:1870–1884.

180. Frayssinet P. In: Poitout DG, ed. *Biomechanics and Biomaterials in Orthopedics*. London: Springer; 2004:101.

17

Gene Therapy for Tissue Engineering

Pedro Lei
State University of New York at Buffalo

Stelios T. Andreadis
State University of New York at Buffalo

17.1 Introduction

Tissue engineering applies the principles and methods of engineering and life sciences toward the development of tissue substitutes to restore, maintain, or improve tissue function.[1-3] The field of tissue engineering is motivated by the tremendous need for transplantation of human tissue. Engineered tissues can also be used as realistic biological models to obtain fundamental understanding of the structure–function relationships under normal and disease conditions and as toxicological models to facilitate drug development and testing.

To engineer tissues in the laboratory, cells must grow on three-dimensional scaffolds, which provide the right geometric configuration, mechanical support, and bioactive signals that promote tissue growth and differentiation. The cells may come from the patient (autologous), another individual (allogeneic), or a different species (xenogeneic). Cell sourcing may be overcome by the use of adult or embryonic stem cells that have the capacity for self-renewal and can differentiate into multiple cell types, thus providing an unlimited supply of cells for tissue and cellular therapies. Application of stem cells in tissue engineering requires control of their differentiation into specific cell types, which in turn depends on the fundamental understanding of the factors that affect stem cell self-renewal and lineage commitment.[4,5]

Alternatively, implantation of biomaterials at the site of tissue injury may be used to stimulate the surrounding cells to regenerate the severed tissue. Examples include induction of nerve regeneration, dermal wound healing, and neovascularization (i.e., formation of new blood vessels) at the injured sites. The main challenge in this approach is to recreate the conditions that induce tissue regeneration instead of repair. Since embryonic tissues heal without scarring, understanding the molecular differences between fetal and adult wound healing and cell–cell interactions during embryological development could lead to significant advances in the engineering of a regenerative environment for successful healing.

The challenge in engineering living tissues lies in creating the right macroscopic (tissue) architecture and function starting from microscopic components (cells and molecules). Part of the challenge may be addressed by appropriate design of bioactive scaffolds that provide the appropriate molecular signals and mechanical environment to guide cellular function.

Alternatively, cellular function may be directed by molecular engineering at the most fundamental level, the genome. Gene delivery can be applied in tissue engineering to impart new functions or enhance existing cellular activities in tissue substitutes.[6] This is achieved by genetic modification of cells that will be part of the implant or gene transfer to the site of injury to facilitate *in situ* tissue regeneration. Cells can be genetically engineered to express a variety of molecules, including growth factors that induce cell growth/differentiation or cytokines that prevent an immunologic reaction to the implant. Therefore, gene delivery has the potential to improve the quality of tissue substitutes by altering the genetic basis of the cells that make up the tissues.

17.2 Gene Therapeutics

Gene therapy is the transfer of genes into cells to achieve a therapeutic effect. Expression of the transferred gene(s) can result in the synthesis of therapeutic proteins and potentially the correction of biochemical defects. Although gene therapy was initially developed for the treatment of genetic diseases, the majority of ongoing clinical trials involve various gene transfer technologies for the treatment of a wide variety of cancers and infectious diseases (Table 17.1).

In general, two classes of methods are employed to deliver genes into target cells. They are broadly classified as viral and nonviral.[7,8] Viral methods, which use recombinant viruses as gene delivery vehicles, are used in the majority of clinical trials (Table 17.2). The genome of recombinant viruses has been modified by deletion of some or all viral genes and replacement with foreign therapeutic or marker genes. The major types of recombinant viruses that are currently used in gene therapy include retrovirus, lentivirus (HIV-based), adenovirus, and adeno-associated virus (AAV).

Nonviral methods include delivery of DNA using physical and chemical means. Physical methods such as electroporation and particle acceleration (gene gun) facilitate entry into target cells, but compromise cell viability and therefore may not be appropriate for use in tissue engineering. On the other hand, delivery of DNA complexed with lipids or polymers has met with some success *in vitro* and is employed in a significant fraction of current clinical trials (Table 17.2). Moreover, natural and synthetic

TABLE 17.1 Current Gene Therapy Clinical and Disease Targets

Disease	Clinical Trials	
	Number	%
Cancer	1186	64.4
Monogenic	161	8.7
Infectious	147	8.0
Vascular	155	8.4
Gene marking	50	2.7
Inflammatory	13	0.7
Neurological	36	2.0
Ocular	28	1.5
Healthy volunteers	42	2.3
Others	25	1.4
Total	1843	100

Source: Information compiled from the website of the *Journal of Gene Medicine* (http://www.wiley.com/legacy/wileychi/genmed/clinical/).

TABLE 17.2   Current Gene Therapy Clinical and Disease Targets

Disease	RV	LV	AV	AAV	Naked DNA	LP	Other
Cancer	209	15	323	19	198	89	333
Monogenic	52	22	18	38	5	15	11
Infectious	44	12	19	2	14	1	55
Vascular	3		55	3	83	5	6
Gene marking	48		1		1		
Inflammatory	5			3	1		4
Neurological	2	2	1	14	13		4
Ocular	1	3	2	13	4		5
Healthy volunteers			3		13		26
Others	3	1	6		9	1	5
Total	367	55	428	92	340	111	450
Percent of all trials	19.9	3.0	23.2	5.0	18.4	6.0	24.4

Source: Information compiled from the website of the *Journal of Gene Medicine* (http://www.wiley.com/legacy/wileychi/genmed/clinical/).

Note: AAV, adeno-associated virus; AV, adenovirus; LP, lipofection; LV, lentivirus; RV, retrovirus.

TABLE 17.3   Characteristics of the Most Common Gene Transfer Vehicles

Properties	RV	LV	AV	AAV	Plasmid DNA
Titer (infectious particles/mL)	10^5–10^7	10^5–10^7	10^{10}–10^{13}	10^{12}	10^7
Integrates into host genome?	Yes	Yes	No	Yes	No
Persistence of gene expression	Years	Years	Months	Years	Weeks
Stability	No	No	Yes	Yes	Yes
Maximum transgene size (kb)	7–8	7–8	36	4–5	Unlimited
Immunogenicity	No	No	Yes	Yes	No
Gene transfer to nondividing cells	No	Yes	Yes	Yes	Yes
Potential for gene transfer to stem cells	Yes	Yes	No	Yes	No

Note: RV, retrovirus; LV, lentivirus; AV, adenovirus; AAV, adeno-associated virus.

polymeric materials have been used as scaffolds for DNA delivery *in vivo* to promote tissue regeneration. Here, we review the most commonly used viral and nonviral gene transfer vehicles, their use in tissue engineering, and the major limitations that need to be overcome to realize the potential of gene therapy in regenerative medicine. A comparison of the main characteristics of viral and nonviral technologies is given in Table 17.3.

17.2.1 Viral Gene Delivery

17.2.1.1 Recombinant Retroviruses

Retroviruses are enveloped particles with a diameter of approximately 100 nm. The lipid bilayer contains the envelope glycoproteins that confer the host range of the virus by interacting with receptors on the surface of target cells.[9] The bilayer surrounds a nucleocapsid that contains two single-stranded RNA molecules and enzymes necessary for virus replication. The genome of retroviruses has three genes: *gag*, which encodes the major capsid protein; *pol*, which encodes the enzymes reverse transcriptase and integrase that participate in early events in viral replication, as well as a protease that is used in the processing of viral proteins; and *env*, which encodes the envelope glycoprotein. In addition, the viral genome contains regulatory sequences that are necessary for transcription [long terminal repeats (LTRs)] and an *cis*-acting sequence (Ψ) that mediates packaging of the RNA into the viral capsid.

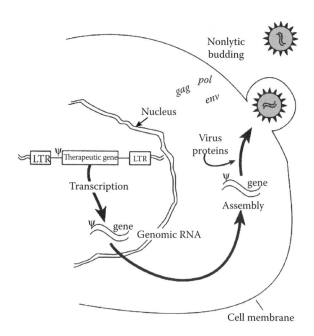

FIGURE 17.1   Schematic of retrovirus production from a packaging cell. (From Andreadis, S.T. et al., *Biotechnol. Prog.*, 15, 1, 1999. With permission.)

Recombinant retroviruses are produced by a two-part system: a retroviral vector and a packaging cell line (Figure 17.1).[10] The recombinant vector is derived from a provirus, usually the Moloney murine leukemia virus (MMuLV), in which a therapeutic or marker gene has replaced the *env, pol,* and *gag* sequences. The latter are provided by the packaging cells to enable the formation of the viral particle. The recombinant vector is transfected into the packaging cells, where it is transcribed and the resulting RNA is recognized by viral proteins and incorporated into the viral capsid. The capsid containing the RNA genome and all the enzymatic activities of the virus buds from the cell surface, acquiring its plasma membrane and envelope glycoproteins. Recombinant retroviruses are able to transfer genes to target cells, but they cannot replicate because they lack the genes that encode for the structural proteins.

Gene transfer is a multistep process that begins with the diffusion of the virus particle to the cell surface, binding of the virus to specific cell surface receptors, and internalization into the cell cytoplasm (Figure 17.2). Following entry into the cell, the RNA genome is reverse transcribed into a double-stranded DNA molecule, by the combined action of reverse transcriptase and RNase H. The newly synthesized viral DNA is transported to the nucleus where it is covalently joined to the genomic DNA of the host cell to form an integrated provirus.[11] Once integrated, the provirus is stable and is inherited by daughter cells as any other autosomal gene. The newly integrated gene serves as a template for synthesis of mRNA and protein that the target cells may not normally express.

Recombinant retroviruses are ideal vehicles for gene delivery, as they have several advantages over other gene transfer technologies: (1) they have a broad host range, (2) the transferred genes are stably integrated into the chromosomes of the host, resulting in permanent modification of the transduced cells, and (3) the transferred gene is transmitted without rearrangements. The disadvantages of recombinant retroviruses include (1) limited size of the genes that they can accommodate (7–8 kb), (2) low transduction efficiencies, and (3) requirement for cell division for integration into the genome of the target cell. One advancement in overcoming these limitations is the development of recombinant lentivirus that enables introduction of genes into nondividing cells.[12–16]

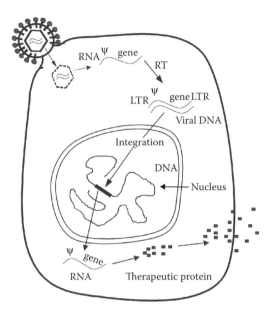

FIGURE 17.2 Schematic of the steps involved in retroviral gene transfer to a target cell.

17.2.1.2 Recombinant Lentivirus

Lentiviruses belong to a family of complex pathogenic retroviruses that include members such as human immunodeficiency virus (HIV-1). Lentiviruses have a complex genome, which in addition to the structural genes *gag, pol*, and *env* contain other regulatory (*rev, tat*) and accessory genes (*vif, vpr, vpu, nef*). The regulatory genes are essential for virus replication. *Tat* encodes for a protein, which binds to the viral LTR and promotes RNA synthesis. *Rev* encodes for a protein, which binds a region of the viral RNA known as rev-responsive element (RRE) and promotes transport of the RNA from the nucleus to the cytoplasm.[17] The accessory genes are essential for pathogenesis but not for virus replication.[18]

The first-generation recombinant lentivirus is produced by cotransfection of three plasmids.[14] The first plasmid contains all viral genes except for the gene that encodes for the envelope glycoprotein, *env*, which is provided by a separate plasmid. The *env* may be substituted with the envelope glycoprotein of amphotropic retrovirus or the envelope glycoprotein of vesicular stomatitis virus (VSV-G), which confers higher stability and broader tropism.[15] The third plasmid encodes for the therapeutic protein and contains sequences necessary for packaging, reverse transcription, and integration into the genome of the target cells.

The first plasmid has been modified to reduce the probability of recombination and increase the safety of lentiviral vectors. These modifications have led to the development of the second-generation recombinant lentivirus in which all accessory genes were deleted to eliminate the virulence of any replication-competent virus that may be present in the viral preparations.[19] Further gains in safety were afforded by the third-generation constructs, in which the sequences encoding for *tat* and *rev* were also deleted and *rev* was provided by a fourth plasmid under the control of the rous sarcoma virus (RSV) promoter. The elimination of six genes that are essential for replication and pathogenesis as well as a deletion at the U3 region of the 3′LTR (i.e., self-inactivating vector)[20] significantly reduced the risk associated with the use of recombinant lentivirus as a tool for gene transfer and for treatment of disease.

The main advantage of recombinant lentivirus is its ability to transfer genes to nondividing, quiescent target cells, including neurons and hematopoietic stem cells. Similar to murine retrovirus, lentivirus integrates stably into the genome of the target cells. But unlike murine retrovirus, there is no evidence for lentivirus promoter shut-off *in vivo*.[18] However, the efficiency of lentiviral gene transfer remains low and its efficacy has not yet been evaluated in clinical trials due to safety concerns.

17.2.1.3 Recombinant Adenoviruses

Adenoviruses are icosahedral, nonenveloped viral particles with a diameter of 70–100 nm. Their genetic material is double-stranded DNA that is 36 kb pairs long. Two short inverted terminal repeats (ITRs) at each end of the DNA are necessary for viral replication. The genome consists of five early (E1–E5) and five late genes (L1–L5) named based on their expression before or after DNA replication. The early genes encode for proteins that activate transcription of the late genes, which in turn encode for structural proteins that make up the viral capsid.[21]

The first generation of recombinant adenovirus is derived from mutant adenovirus in which the E1 gene is deleted. The E1⁻ mutants can be grown in 293 cells that express the E1 gene that provides the lost function. The resulting adenovirus can infect multiple cell types and express the transgene but it cannot replicate in a cell that does not contain the E1 gene. In addition to E1, the E3 region of the adenoviral genome is often deleted to allow packaging of approximately 8 kb of foreign DNA without affecting the viral titer.[22]

The major advantages of adenoviral vectors are that they can infect nondividing cells and can be grown to very high titers (10^{10}–10^{12} CFU/mL). This results in very efficient gene transfer in tissues with terminally differentiated cells, such as the brain or lung. However, these vectors can only transfer small genes (less than 7 kb) and their DNA does not integrate, resulting in transient gene expression. The major disadvantage that hinders clinical application of adenoviruses is their immunogenicity. Unfortunately, the immune response is raised not only against the virus but also against the therapeutic protein, possibly due to marked elevation of inflammatory cytokines such as IL-6.[23,24] These toxic effects may limit the use of adenovirus *in vivo*, especially when efficient treatment requires repeated administration of the virus.[25–27]

17.2.1.4 Recombinant Adeno-Associated Viruses

AAVs are nonenveloped, single-stranded DNA viruses that do not cause disease in humans or animals. Their DNA is short, containing less than 5 kb pairs, and encodes for two proteins: *Rep*, which is necessary for virus replication and *cap*, a structural protein, which forms the icosahedral capsid that houses the genome. The genome is flanked by ITRs, which are necessary for replication, packaging, and integration of AAV.

Production of AAV requires the presence of a helper virus for replication.[28] The helper function may be provided by other viruses, including adenovirus, herpes, and vaccinia virus. Production involves cotransfection of 293 or HeLa cells with two plasmids. One plasmid contains the therapeutic gene flanked by the viral ITRs, and the other contains the *rep* and *cap* genes but lacks the packaging sequences. Subsequently, the transfected cells are infected with wild-type adenovirus that provides the helper functions. Two days later, the cells are lysed, the wild-type adenovirus is heat inactivated, and AAV viral particles are purified with density gradient centrifugation or column chromatography, yielding highly concentrated viral preparations (10^{12}–10^{13} AAV particles from ~10^9 cells[28]).

The major drawback of AAV is the potential of contamination of the final product by wild-type adenovirus. Other disadvantages of this system include the short foreign DNA (4–5 kb) that can be incorporated in these vectors, and the imprecise integration that often results in tandem repeats of the transferred gene. However, these vectors have certain characteristics that make them particularly attractive, namely, stable integration in a specific location in the cellular genome (chromosome 19) and the ability to transfer genes to nondividing cells.

17.2.2 Nonviral Gene Delivery

The second class of gene delivery methods is nonviral. These methods include the use of naked DNA or DNA complexed with cations, cationic polymers, liposomes, or combinations of lipids and polymers to enhance the efficiency of gene transfer.[29] Physical methods such as electroporation and particle acceleration are also employed to facilitate DNA transport into cells and tissues.[30] Moreover, the use of natural and synthetic biomaterials for delivery of DNA *in vivo* has met with some success. This approach, which pertains to *in situ* tissue regeneration, is reviewed in detail below (see Section 17.3.2).

Cationic polymers that are employed as DNA carriers include peptides,[31–34] poly(ethyleneimine),[35–39] poly-L-lysine (PLL),[40] modified PLL with histidine[41–44] or imidazole,[45] and dendrimers.[46–48] The positively charged polymers interact with the negatively charged DNA and form a complex. The DNA–polymer complex has a net positive charge facilitating the interaction of DNA with the target cells. Additional modification such as conjugation of cationic polymers with ligands endows the complex with targeting specificity. Several ligands have been used to achieve targeting, including asialoglycoprotein,[40,49] transferrin,[50,51] antibodies,[52,53] fibroblast growth factor (FGF),[54] and epidermal growth factor.[55]

Cationic lipids are also used as DNA carriers.[56,57] Cationic lipids (e.g., DOTAP) contain a hydrophobic group that ensures their assembly into bilayer vesicles and a charged group (e.g., amine) that mediates the electrostatic interactions with the negatively charged DNA. Addition of neutral lipids (e.g., DOPE) has been shown to increase the transfection efficiency possibly by facilitating endosomal escape.[58] Liposome-mediated gene transfer, better known as lipofection, has enjoyed wide use resulting in commercial production of gene delivery kits and initiation of several clinical trials (Table 17.2).

The nonviral gene delivery systems are prepared by mixing the solution of plasmid DNA with the solution of polymers, lipids, or both, resulting in complexes of colloidal size (50–1000 nm) and positive charge.[29] Gene transfer is initiated by the incubation of these complexes with the target cells.

The main advantages of nonviral technologies include (1) unlimited size of DNA that can be transferred, (2) lack of immune response (unless proteins are inserted into the lipids to convey specificity), and (3) safety due to their inability to replicate or recombine to produce infectious particles.[59] The main disadvantages include (1) very low efficiencies of gene transfer, especially *in vivo*, (2) short-lived transgene expression, and (3) cytotoxicity.[35,60,61] These shortcomings may hinder the use of nonviral technologies in gene therapy of stem cells and tissue engineering applications that require sustained gene expression.

17.3 Genetic Modification for Tissue Engineering

Gene transfer can be used in tissue engineering in two settings. First, genes can be introduced into cells that are subsequently used to prepare three-dimensional tissue substitutes, such as skin, cartilage, and bone. Alternatively, genes can be delivered *in vivo* to enhance *in situ* tissue regeneration. The first approach is useful when tissues must be replaced to restore or improve function. Interestingly, genetically modified tissues may also be used as biological models for studying the role of effector molecules on organ/tissue development. The second approach can be used when the defects are relatively small or engineered tissues are not yet available (e.g., peripheral nerve injury).

17.3.1 Genetically Engineered Tissues

In this setting, cells are genetically modified using one of the viral or nonviral methods described above and then used to engineer tissue substitutes. Genetic modification of tissues may be used to enhance tissue function or improve existing cellular activities for treatment of genetic diseases. Interestingly, genetically modified tissues may also be used as ectopic sites for protein production and delivery into the systemic circulation. In this case, *permanent* genetic modification is required to provide long-lasting effects. Consequently, the most suitable vectors are the recombinant viruses that can mediate permanent gene transfer such as retrovirus, lentivirus, and AAV.

In other cases, genetic modification of the engineered tissue may be used to increase the rate of graft survival by inducing ingrowth of new vessels (angiogenesis) or suppressing immune rejection. These applications may require *transient* gene expression until the transplant integrates with the surrounding tissue or until wound healing is complete. Therefore, adenoviruses or nonviral technologies of gene delivery may be more appropriate.

17.3.1.1 Genetically Engineered Skin

The epidermis is an attractive target for gene therapy because it is easily accessible and shows great potential as an ectopic site for protein delivery *in vivo*. The cells that are primarily used to recreate skin substitutes are epidermal keratinocytes and/or dermal fibroblasts, which can be genetically modified with viral or nonviral vectors.[6,62] Genetically modified cells are then used to prepare three-dimensional skin equivalents, which, when transplanted into animals, act as *in vivo* "bioreactors" that produce and deliver the desired therapeutic proteins either locally or systemically. Local delivery of proteins may be used for the treatment of genetic diseases of the skin or wound healing of burns or injuries, while systemic delivery may be used for correction of systemic diseases such as hemophilia or diabetes.

Retroviral gene transfer to epidermal cells has been used for the correction of genetic diseases of the skin such as epidermolysis bullosa and lamellar ichthyosis.[63–66] The same technology has been used to genetically modify keratinocytes to express wound-healing factors such as PDGF-A,[67] IGF-I,[68] and keratinocyte growth factor (KGF)[69] to examine their potential in enhancing wound healing. Gene transfer of angiogenic factors using retroviral[70,71] and nonviral approaches[72] has also been used to increase the vascularization of transplanted skin substitutes *in vivo*.

Genetically modified tissues may also be used as "bioreactors" to deliver proteins to the systemic circulation. The main property that makes the skin an ideal target for gene delivery is the ease of accessibility for transplantation or tissue removal if adverse effects occur. Some of the first studies have used recombinant retroviruses to genetically modify human keratinocytes with genes encoding for human growth hormone (hGH),[73] apoE,[74] and clotting factor.[75] In all these cases, genetically modified cells expressed the transgenes and secreted functional proteins to the systemic circulation of grafted animals. Since then, other investigators have used recombinant retroviruses to introduce genes into human keratinocytes for the correction of metabolic[76,77] and systemic diseases.[78,79]

Genetically engineered tissues can also be employed as models for studying tissue development and physiology. We prepared skin equivalents that were genetically modified to express KGF,[69] a protein that plays an important role in tissue morphogenesis and wound healing.[80,81] The modified tissues showed dramatic changes in the three-dimensional organization of the epidermis, including hyperthickening and flattening of the corrugations of the dermoepidermal junction (rete ridges) (Figure 17.3a and d). KGF increased proliferation of basal cells, induced proliferation in the suprabasal cell compartment that is normally quiescent (Figure 17.3b and e), and delayed differentiation (Figure 17.3c and f). This study demonstrated that the expression of a single growth factor is able to mediate many of the events associated with epidermal growth and differentiation.[69] It also demonstrated the usefulness of "transgenic" engineered tissues in studying the effects of a single gene in tissue development. Although transgenic animal models have been successfully employed for many years, "transgenic" engineered tissues may provide an alternative approach for better-controlled and quantitative studies. Furthermore, deletion of some genes may be lethal for animal embryos but it may still be possible to study gene knockouts using "transgenic" tissue equivalents.

17.3.1.2 Genetically Engineered Cartilage

Articular cartilage, the tissue that forms the surface of the joints has limited regenerative capacity. As a result, tissue-engineering approaches may be very useful in restoring loss of cartilage due to trauma, osteoarthritis, or inflammatory diseases such as rheumatoid arthritis. Genetic modification may be used to reduce inflammation or increase the regenerative capacity of engineered cartilage ultimately restoring function and reducing pain.

Interleukin-1 (IL-1) is a key mediator of degradation of the extracellular matrix of cartilage associated with osteoarthritis. In one study, chondrocytes that were genetically modified with adenovirus to express IL-1 receptor antagonist (IL-1RA) were able to protect osteoarthritic cartilage by blocking IL-1.[82] In another study, synovial cells that were transduced with a retrovirus encoding IL-1RA were also able to protect the joint from IL-1-mediated degradation.[83]

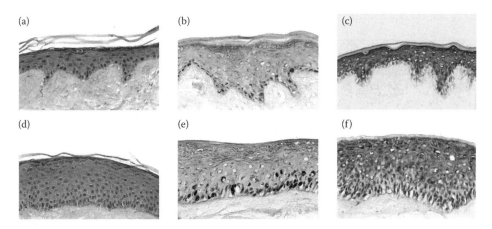

FIGURE 17.3 (See color insert.) KGF induces hyperproliferation and delays differentiation in a skin-equivalent model system. Sections of skin equivalents unmodified (a, b, c) and genetically modified to express KGF (d, e, f) were cultured for 7 days at the air/liquid interface. Cryosections were stained with hematoxylin and eosin (H&E; a, d), a nuclear proliferation antigen (Ki67; brown; b, e), and the differentiation marker keratin 10 (K10; brown; c, f). Ki67 and K10 sections were counterstained with hematoxylin (blue) (magnification 40×). Note the increased thickness and basal cell density of the KGF-modified tissues. Also note that in unmodified tissues, proliferation is confined to the basal layer, while in KGF-modified tissues, proliferating cells extend up to three to four suprabasal layers. The same suprabasal layers that contain proliferating cells are also negative for the differentiation marker K10, suggesting that KGF alters the spatial control of proliferation and differentiation in the genetically modified tissues. (From Andreadis, S.T. et al. *FASEBJ.*, 15, 898, 2001.)

Other studies used nonviral technologies to transfer genes that promote the chondrocytic phenotype, such as TGF-β1 or parathyroid hormone-related protein (PTHrP). When the modified chondrocytes were seeded into polylactic acid (PLA) scaffolds and placed into rabbit femoral defects, they expressed the transgenes for at least 2 weeks.[84] Similarly, cartilage cells transfected with a plasmid encoding for IGF-1 synthesized proteoglycan and type II collagen, two extracellular matrix proteins indicative of mature chondrocytes. IGF-I overexpressing chondrocytes also exhibited higher levels of proliferation and proteoglycan synthesis following transplantation *in vivo*.[85] These studies suggest that gene transfer may enhance the function of engineered cartilage and ultimately reduce inflammation and promote tissue regeneration.

17.3.1.3 Genetically Engineered Bone

Every year, more than 800,000 bone grafting procedures are performed in the United States to reconstruct or replace bone affected by trauma, pathological degeneration, or congenital deformity. Although many synthetic and natural materials, for example, ceramics, polymers, and collagen, are used to replace bone, these materials have markedly different mechanical properties than the human bone, often resulting in implant failure.[86] Tissue engineering may provide an alternative treatment by combining biodegradable materials with cells that can differentiate along the osteogenic pathway and provide the lost function.

Human bone marrow-derived mesenchymal stem cells represent a pluripotent population of cells that serve as precursors for osteoprogenitor cells.[87] The potential of mesenchymal stem cells as a source in tissue engineering is highly augmented by the lack of immunogenicity and their high-growth potential. Gene therapy can serve as a means to guide differentiation of these cells into bone or cartilage by the expression of signaling molecules such as bone morphogenetic proteins (BMPs). A number of studies have demonstrated that delivery of genes encoding for BMPs promotes bone formation *in vitro* and *in vivo*.

Human mesenchymal stem cells were effectively transduced with recombinant retrovirus encoding for green fluorescence protein[88] as well as functional proteins such as human factor VIII.[89] Other studies used recombinant adenovirus to transfer genes encoding for bone morphogenetic protein (BMP-2) to mesenchymal stem cells. The genetically modified cells differentiated into bone and cartilage cells and corrected a bone defect after implantation *in vivo*. In contrast, control cells expressing β-galactosidase differentiated into fibrous tissue. Most importantly, the expression of BMP-2 by mesenchymal stem cells derived from an aged osteoporotic patient formed the bone ectopically in subcutaneous sites similar to cells derived from young healthy patients.[90] In another study, genetically modified bone marrow osteoprogenitor cells expressing BMP-2 were seeded onto polylactic-*co*-glycolic acid (PLGA) scaffolds to engineer the bone tissue. Although engineered cells secreted BMP-2 only for 8 days, they differentiated into osteoblasts *in vitro* and produced the bone tissue after implantation *in vivo*.[91]

Periosteal cells, which contain osteoprogenitor and chondroprogenitor cells, were also transduced with a retrovirus encoding for bone morphogenetic protein 7 (BMP-7). When the modified cells were delivered into rabbit cranial defects using polyglycolic acid (PGA) as a scaffold, they promoted bone healing in 12 weeks.[92] Likewise, W-20 stromal cells (a marrow-derived cell line) were transduced with a retrovirus encoding for the BMP-2 gene and transplanted into the pocket of the quadriceps muscles of severe combined immune-deficient (SCID) mice using polymer–ceramic scaffolds (poly[lactide-*co*-glycolide]-hydroxyapatite). In contrast to unmodified control cells, genetically modified cells induced ectopic bone formation around the implant.[93]

It is evident from the above studies that gene transfer may be used to guide differentiation of osteoprogenitor and mesenchymal stem cells along the osteogenic pathway and may even restore the osteogenic potential of cells derived from osteoporotic patients. Delivery of genes encoding for angiogenic factors (e.g., FGF, vascular endothelial growth factor [VEGF]) may further enhance the potential of engineered bone by providing the necessary blood supply to maintain viability and support the function of the newly formed tissue, especially for treatment of large size defects.

17.3.1.4 Genetically Engineered Vessels

One the most common and severe forms of heart disease is atherosclerosis, the narrowing of the vessels.[94] The current treatment involves coronary artery bypass graft surgery using a patient's mammary artery or a sephanous vein. Tissue-engineered vessels may provide an alternative source to the currently employed native vessels in bypass surgery. To accomplish this goal, blood vessel substitutes must meet certain performance criteria, including nonthrombogenicity, vasoactivity, and mechanical properties that match those of native vessels.[94]

Gene therapy can be used to improve the properties of engineered vessels by enhancing secretion of nonthrombogenic compounds, expression of extracellular matrix molecules (e.g., elastin) that provide strength and elasticity, or by engineering immune acceptance into allogeneic cells.[94] In one study, endothelial cells were genetically modified with a recombinant retrovirus to express an antithrombotic protein, tissue plasminogen activator (TPA). The genetically modified cells were seeded onto 4 mm Dacron grafts and were either exposed to a nonpulsatile flow system *in vitro* or transplanted as femoral and carotid interposition grafts *in vivo*. In both cases, genetically modified cells showed decreased adherence, suggesting that gene therapy strategies for decreasing the thrombosis of engineered vessels may require expression of molecules that lack proteolytic activity.[95]

Viral and nonviral vehicles have also been used to transfer genes into large and small vessels *in vivo*. A number of studies showed that gene transfer of NO synthase (eNOS) increased NO bioavailability and improved vascular reactivity of surgically removed vessels from several animal models, including hypertensive rats and hypercholesterolimic rabbits.[96–99] However, gene transfer is restricted to the endothelial cell layer unless the vessel wall is damaged.[100]

To improve the efficiency of gene transfer to the underlying smooth muscle cells, standard angioplasty was used to increase the pressure in the lumen of the vessel, resulting in significant enhancement of plasmid DNA and oligonucleotide delivery.[101] In addition to pressure, application of electric fields has been used to facilitate gene transfer by creating transient pores in cell membranes to allow macromolecules into the cell cytoplasm. Electroporation increased DNA delivery to large[102] and small vessels,[103] with peak gene transfer achieved at a field strength of 200 V/cm. In particular, in small vessels, gene expression was observed in all cell layers, including endothelial, smooth muscle, and adventitial cells.[103] However, gene expression was short lived, declining to background levels by 5 days after treatment.

17.3.1.4.1 Gene Transfer to Modulate Angiogenesis

Gene transfer to small vessels has been used to inhibit angiogenesis as a means to hinder tumor growth or increase vascularization to treat ischemia and promote wound healing. For example, gene transfer of VEGF and FGF receptors inhibited angiogenesis in pancreatic tumors.[104] On the other hand, adenoviral gene transfer of angiogenic factors such as FGF and VEGF increased angiogenesis in several tissues, including skeletal muscle,[105] ischemic rabbit hindlimbs,[106] and adipose tissue.[107]

One of the most important challenges facing tissue engineering is vascularization of engineered tissues after implantation *in vivo*. Since oxygen diffuses only a few hundred microns before it is consumed,[108] new vessels must form quickly to supply oxygen and nutrients to transplanted tissues. One way of promoting new vessel formation is to genetically modify the engineered tissue to express angiogenic factors. Indeed some studies demonstrated that overexpression of VEGF by genetically modified skin substitutes promoted early vascularization following transplantation onto athymic mice.[70-72] Similarly, meniscal cells were genetically modified to express hepatocyte growth factor (HGF) using recombinant adenovirus. When the genetically modified cells were seeded onto PLGA scaffolds and transplanted into athymic mice, angiogenesis increased significantly.[109] These studies suggest that gene therapy can be used to promote vascularization of engineered tissues ultimately improving the clinical outcome of tissue replacement therapy.

17.3.2 *In Situ* Gene Delivery for Tissue Regeneration

In this setting, biomaterials are directly applied to the site of injury to promote and guide tissue regeneration. The biomaterials may be decorated with bioactive molecules such as adhesion peptides or growth factors to promote cellular responses. Enzymatic recognition sites can be engineered into these molecules to facilitate control release.[110] Nonetheless, it has been difficult to maintain full bioactivity of the proteins released from controlled delivery systems mainly due to protein instability. Gene delivery may overcome this problem, as infiltrating cells may uptake the genes and produce the therapeutic protein(s) continuously. However, gene delivery by injection may result in loss of the genes from the site of administration and from subsequent degradation by nucleases.[111,112] Alternatively, bioactive materials may be employed as gene delivery vehicles and at the same time serve as scaffolds to promote tissue regeneration.[113] Biomaterials that have been used to deliver plasmid DNA include gelatin nanospheres,[114] collagen,[115,116] poly(ethylene-*co*-vinyl acetate) (EVAc),[117] and poly(lactide-*co*-glycolide),[118] fibrin,[119,120] and chitosan.[121]

Efficient gene delivery has been demonstrated using natural and synthetic biomaterials. Plasmid DNA encoding for human parathyroid hormone was delivered to a canine bone defect in a collagen sponge yielding significant bone regeneration in a dose- and time-dependent manner.[115] Similarly, collagen-embedded DNA encoding for platelet-derived growth factor (PDGF-A or PDGF-B) increased granulation tissue, epithelialization, and wound closure in an ischemic rabbit ear model.[122] In another study, gelatin/alginate nanospheres conjugated with human transferring-enhanced gene transfer to mouse tibialis muscle as compared to naked DNA.[114]

Synthetic biomaterials have also been used as DNA carriers. Synthetic sutures coated with DNA were used to deliver the alkaline phosphatase gene to rat skeletal muscle and canine myocardium.[123] Poly(lactide-*co*-glycolide) matrices were also used to deliver the PDGF gene into skin wounds resulting in significantly increased vascularization and granulation tissue formation up to 4 weeks postwounding.[118]

Similarly, PLGA nanoparticles were shown to encapsulate DNA efficiently and exhibited sustained release over a period of 4 weeks.[124] These examples are encouraging as they demonstrate the feasibility of *in vivo* gene delivery using bioactive materials. Further advances in biomaterial and vector design are required to achieve controlled release for prolonged periods of time, protection of DNA from nuclease degradation, and targeting of the plasmid to specific cell types of the wound. Strategies that employ liposomes or DNA-condensing agents may be integrated into this approach to release the DNA from the endosomes or target it to the nucleus.[125]

Bioactive matrix may also be used to deliver recombinant viruses to the site of injury. Encapsulation of viral particles in biomaterials may increase viral stability, reduce immunogenicity, and achieve targeted gene transfer only to the cells that infiltrate the wound bed. Indeed, encapsulation in gelatin/alginate microspheres protected adenovirus particles from degradation and the release kinetics could be controlled by changes in the microsphere formulation.[126] Delivery of a PDGF-BB encoding adenovirus increased granulation tissue formation and neovascularization of full-thickness wounds. Conjugation of adenoviral particles with fibroblast growth factor (FGF2) further increased the potency of the preparation by targeting cellular uptake through the FGF receptor.[127,128] Additionally, encapsulation of adenovirus in PLGA matrices reduced immunogenicity and decreased inactivation by neutralizing antibodies, thus facilitating repeating virus administrations that may be required for a therapeutic effect.[129–131]

Although temporary genetic modification (adenovirus) may be required in most cases to ensure no transgene expression after the healing is complete, permanent genetic modification (retrovirus, AAV) may be advantageous in the treatment of chronic wounds such as diabetic ulcers.[132] In the environment of the wound, retroviral gene transfer may be facilitated by the natural propensity of the wound-infiltrating cells to divide. More studies are required to establish matrix-assisted retroviral delivery as an efficient strategy to enhance tissue regeneration especially in chronic wounds.

17.4 Challenges Facing Gene Delivery

17.4.1 Low Efficiency of Gene Transfer

The major challenges of gene transfer are to deliver therapeutic levels of genes to specific target cells and to control the level and the length of time of transgene expression. While each gene transfer technology has its own idiosyncrasies, they all have to undergo a series of physicochemical steps to deliver their cargo successfully. These steps include transport to the target cells, binding to the cell surface, internalization into the cell cytoplasm, entry into the nucleus, and for some viral vectors (such as retrovirus, lentivirus, and AAV), integration of the transgene into the genome of the target cells. Each one of these steps can potentially limit the efficiency of gene transfer. Here, we will briefly discuss the physicochemical factors that govern gene transfer. For a more detailed review on the subject, see Ref. 133.

17.4.1.1 Transport to the Target Cells

The first step in gene transfer is the transport of viral or nonviral vectors to the cell surface. The size of viral particles and many complexes of DNA with cationic lipids and polymers are in the order of 100 nm, yielding diffusion coefficients in the order of 10^{-8} cm^2/s.[134,135] As a result, transport to the cell surface is limited by Brownian motion.[135,136] The slow diffusion in conjunction with the short half-life (5–7 h[135–137]) limits the distance that retroviral particles can diffuse before they lose biological activity. Mathematical modeling and experiments showed that only particles in close proximity to the target cells (within 500 μm) contribute to gene transfer. Increasing the depth of retroviral supernatant beyond 500 μm did not increase the efficiency of gene transfer further.[135] Therefore, methods to overcome the diffusion limitation and increase the probability of virus–cell encounter may increase the efficiency of gene transfer. Methods such as centrifugation of viral particles onto target cells[138] and convective flow of the virus past a monolayer of target cells have shown considerable promise for gene delivery to some cell types.[139–141]

In addition, fibronectin (FN) was used to immobilize retroviral particles and increase their surface concentration facilitating virus–cell interactions.[142,143] Using mathematical modeling and experiments, we optimized the time, temperature, and virus adsorption to FN yielding 10-fold enhancement of the gene transfer efficiency.[144] We have also identified the mechanism of retrovirus binding to FN. Our results indicate that retrovirus binds FN through virus-associated heparan sulfate (HS) and that gene transfer with immobilized retrovirus does not depend on polycations, for example, polybrene (PB).[145] A better understanding of these interactions may provide insight into virus–cell interactions and may lead to a more rational design of transduction protocols.

17.4.1.2 Binding to the Cell Surface

Several physicochemical forces, including electrostatic, van der Waals, osmotic, and steric, mediate the initial interaction of viral and nonviral vectors with the cell surface.[133] Electrostatic forces play an important role since polycations such as PB, protamine sulfate (PS), PLL, and polyethyleneimine (PEI) are necessary for viral and nonviral gene transfer. These positively charged polymers screen the electrostatic repulsion between the negatively charged virus or DNA and the target cells ultimately promoting binding and gene transfer. On the other hand, negatively charged proteoglycans (e.g., chondroitin sulfate) were found to inhibit retroviral gene transfer.[146]

Following an initial nonspecific interaction, viruses have evolved to bind to receptor(s) that are ubiquitously expressed on the cell surface. For instance, amphotropic retrovirus binds to a phosphate symporter,[147,148] AAV binds integrin $\alpha v\beta 5$,[149] and adenovirus binds several integrins, including $\alpha v\beta 5$, $\alpha v\beta 3$, and $\alpha v\beta 1$.[150–153] On the other hand, plasmid DNA and liposomes bind nonspecifically unless engineered to possess a ligand, for example, transferrin that mediates binding to a specific receptor.[50,51]

Molecular engineering has been used to increase gene transfer and to target viruses and nonviral vectors to specific cell types. Retroviruses pseudotyped with gibbon–ape leukemia virus envelope protein,[154] T-cell leukemia virus envelope protein,[155] and parts of the erythropoietin molecule (Epo)[156] have shown altered host range. Interestingly, retrovirus pseudotyped with the G glycoprotein of VSV-G[157] showed resistance to shear when concentrated by ultracentrifugation, yielding preparations of more than 10^9 infectious particles per milliliter. In another approach, an antibody to melanoma-associated antigen was fused to the amphotropic murine leukemia virus envelope and was used to target recombinant retroviruses to melanoma cells.[158] Similarly, a bispecific antibody targeted AAV to megakaryocytes.[159] Ultimately, a combination of physical methods and molecular engineering may be necessary to increase adsorption to the cell surface, promote internalization, and achieve targeting specificity.

17.4.1.3 Intracellular Trafficking and Transgene Expression

Once in the cell, viral and nonviral vectors undergo a series of steps that eventually lead to gene expression. In general, there are two pathways of entry into the cell cytoplasm: fusion with the plasma membrane and endocytosis. Amphotropic retrovirus follows the first pathway, while adenovirus, AAV, and plasmid DNA follow the second route.

Retrovirus internalization into the cell is followed by uncoating, reverse transcription, nuclear entry, and integration of the newly synthesized DNA into the genome of the target cell. Entry into the nucleus depends on the position of the cell in the cell cycle,[160,161] leading to complicating dynamics that have been described by a mathematical model that accounts for the intracellular retrovirus trafficking and the kinetics of the cell cycle.[162] This model identified intracellular decay as the rate-limiting step and is very useful in the design of experiments to determine the intracellular half-life of recombinant retrovirus.[162,163] These studies suggested that strategies to induce cells into the cell cycle or to increase the nuclear entry would increase the efficiency of gene transfer. Indeed, lentiviruses, which contain nuclear localization signals (NLS) in several of their proteins (e.g., matrix, integrase, and vpr proteins), exhibit efficient entry into the nucleus.[164] As a result, they can transduce slowly dividing hematopoietic stem cells[165,166] and even nondividing neurons.[14,15]

Mathematical models have been proposed to describe the process of binding, internalization, and intracellular trafficking for several other viruses. These models were used to determine the mode of binding—monovalent versus multivalent—and extract key physical parameters such as receptor number and affinity.[167,168] Others have also included postbinding steps such as endocytosis, lysosomal routing, and nuclear accumulation to identify rate-limiting steps and predict kinetic parameters of the infection process.[169,170] This approach may prove useful in understanding the overall dynamics of the gene transfer process and identifying key targets for intervention to overcome the bottlenecks and increase the efficiency of gene transfer.

Adenoviruses, AAVs, and nonviral vectors internalize through endocytosis. Entry of adenovirus and AAV is facilitated by further interactions with several integrins on the cell surface[149–153,171] and requires activation of intracellular mediators such as phosphatidylinositol-3-kinases (PI3K).[171] Escape from the endosome, which is induced by the low endosomal pH, is a major resistance to successful gene transfer. Indeed, fusogenic lipids such as DOPE,[172] polycations, for example, PEI,[35,173] and fusogenic peptides,[174,175] enhance endosomal escape and transgene expression.

Entry into the nucleus is a major resistance to gene transfer, not only for retrovirus but also for plasmid DNA. Sequences that contain NLS were shown to promote entry of the DNA into the nucleus and transgene expression.[32,176,177] Others used DNA sequences (e.g., portions of the SV40 enhancer) that bind transcription factors, which harbor NLS and facilitate the entry of the plasmid DNA into the nucleus.[178] Sequences with binding sites for cell-specific transcription factors were also used to afford target cell specificity. In one study, plasmids containing portions of the smooth muscle gamma actin promoter entered the nucleus and increased gene expression in smooth muscle cells, but not in fibroblasts or CV1 cells.[179]

These studies suggest that understanding the rate-limiting steps of gene transfer may facilitate the development of strategies that improve the transduction efficiency and promote targeted and controlled delivery. One of the most important targets of gene therapy is the population of cells with the highest potential for tissue regeneration, namely, the stem cells.

17.4.2 Gene Therapy to Stem Cells

Continuously renewing tissues such as skin, blood, and bone contain cells with different growth potential and at different stages of differentiation: slowly dividing stem cells that continue to proliferate for the lifetime of the tissue, progenitor cells that divide fast but are limited to a finite number of cell divisions before their progeny must commit to differentiate, and cells that are committed to differentiation along a certain lineage, which will eventually reach full maturity and die. Although certain gene transfer technologies, such as retroviruses, allow for permanent genetic modification of target cells, differentiation and eventually loss of the transduced cells from the engineered tissue may result in temporary transgene expression (Figure 17.4). Therefore, to achieve stable long-term gene expression, it is critical that stem cells are transduced with high efficiency. Additionally, some studies suggest that stem cells do not express molecules that are recognized by the immune system and therefore appear to be immunoprivileged.[180] The potential of stem cells to establish universal donor cells make them ideal targets for both gene therapy and tissue engineering.

Retrovirus is the most appropriate vector for gene delivery into stem cells due to its ability to integrate into the genome of the target cells and thus be a part of the genome of daughter cells for all future generations. However, retroviral gene transfer depends on the cell cycle[160,181] and the intracellular half-life of retroviruses results in low efficiency of gene transfer to the slowly dividing stem cells.[161–163] Although lentiviruses also integrate and can transduce even nondividing stem cells, difficulties with safety make their use problematic. Safety may also be a problem with preparations of AAV that may contain infectious adenovirus. Therefore, efforts to increase the efficiency of retroviral gene transfer to stem cells are needed to provide a clinically acceptable means of stem cell gene therapy.

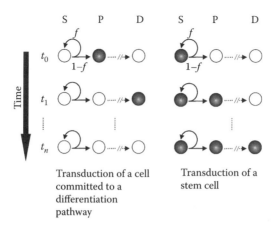

FIGURE 17.4 Retroviral gene transfer to stem cells is required in long-term transgene expression. Stem cells (S) give rise to other stem cells with probability, f, and to progenitor cells (P) with probability $1-f$. In turn, progenitor cells give rise to cells that are committed to terminal differentiation (D). Retroviral gene transfer may result in permanent genetic modification as the transgene integrates into the genome of transduced cells (filled circles) and is inherited by daughter cells as any other autosomal gene. However, differentiated cells have a finite life span resulting in loss of the transgene(s) from the lineage and temporary genetic modification. In contrast, when stem cells are transduced, the pool of genetically modified stem cells is maintained through self-renewal. The modified stem cells give rise to modified progenitor and differentiated cells sustaining long-term transgene expression. Filled and open circles represent genetically modified and nonmodified cells, respectively.

Gene transfer to stem cells can be achieved by either isolation or expansion of stem cells that are subsequently genetically modified or by targeting stem cells in cultures containing stem and differentiated cells. The first approach requires knowledge of specific stem cell markers (e.g., hematopoietic stem cells may be enriched 100- to 1000-fold by selecting cells that express CD34 but not CD38 and major histocompatibility class II [HLA-DR][182]) that can be used to obtain pure stem cells from tissues or heterogeneous cultures. Even more important is maintenance and expansion of the genetically modified stem cells to large numbers necessary for clinical applications. This is a major challenge facing the field of tissue engineering since for most tissues, including the skin and the hematopoietic system, it is still not clear how to maintain and expand the stem cell pool *in vitro*.[4] Despite limited knowledge in stem cell maintenance and expansion, the use of combinations of growth factors and cytokines increased gene transfer by promoting proliferation of the slowly growing stem cells.[182] Interestingly, the presence of stroma improved the efficiency of retroviral gene transfer into hematopoietic progenitor and stem cells as compared to the use of growth factors alone.[183–185]

Another promising way to transduce stem cells is by the use of extracellular matrix molecules to maintain the stem cell phenotype and at the same time colocalize stem cells and retroviral particles. Recombinant retroviruses have been shown to bind to the heparin-binding domain of FN[142,143] and promote gene transfer to T-lymphocytes[186,187] and hematopoietic stem cells.[188–191] This method has also shown promising results in two clinical trials.[192,193]

Previously, our results showed that retroviral gene transfer to epidermal keratinocytes correlated with the levels of $\alpha 5$ and $\beta 1$ integrin expression and that the efficiency of gene transfer was significantly higher on FN.[194] These findings are potentially important for gene therapy since integrins have been associated with epidermal stem cell phenotype.[195–197] Based on these observations, we hypothesized that retroviral transduction on FN may increase the efficiency of gene transfer to epidermal stem cells. Our long-term growth assay and *in vitro* bioengineered skin substitutes provide definitive evidence to support this hypothesis.[198] Preliminary *in vivo* transplantation of genetically modified skin equivalents in our laboratory also showed efficient gene transfer to epidermal stem cells (data not shown).

17.4.3 Physiologically Controlled Gene Therapy

The majority of gene transfer vehicles provide constitutive (always on) gene expression that may be advantageous for treatment of genetic diseases. However, tissue-engineering applications may require physiologically regulated gene expression by a subset of cells in the regenerated tissue. For example, a promising approach for the treatment of intimal hyperplasia involves suppression of smooth muscle cell proliferation.[199] Cytostatic genes such as p21 or p27 can suppress proliferation of smooth muscle as well as the neighboring endothelium. This nonspecific action may lead to adverse effects, especially in damaged vessels, where endothelial cells must proliferate to repopulate the luminal surface. Spatial and temporal control of gene expression requires the design of a new class of gene delivery vehicles that may be regulated by host mechanisms or by administration of secondary agents.

Tissue-specific transcriptional elements, such as promoters/enhancers, can be used to restrict transgene expression in specific cells. Young and Dean[199] have proposed the promoters of smooth muscle myosin heavy chain,[200] smooth muscle alpha actin,[201] gamma actin,[202] and calponin[203] as potential candidates for this approach. On the other hand, endothelial-specific promoters for genes such as von Willebrand factor,[204] flk-1/KDR,[205] or endothelin-1[206] maybe used to restrict gene expression in the lumen of the vessels. Others have inserted the liver-specific promoters such as human alpha 1-antitrypsin (hAAT), and murine albumin (mAIb) into retroviral vectors. *In vivo* gene delivery resulted in highly upregulated and long-term transgene expression as compared to the expression driven by a constitutive Pol-II promoter.[207] Similarly, hybrid promoters have been used for gene transfer into epidermal keratinocytes. One such promoter, containing the cytomegalovirus (CMV) immediate early enhancer/promoter and regulatory elements of the human keratin 5 (hK5), was inserted into the backbone of a retroviral vector encoding for factor IX. This hybrid promoter increased the levels of factor IX in the plasma of mice by two- to threefold as compared to a CMV promoter alone.[208]

Synthetic promoters that can be regulated by the administration of exogenous agents show much promise for controlled gene expression. For example, a tetracycline-dependent promoter has been constructed by two regulatory elements: the tetracycline resistance *tet* operon embedded in a CMV promoter and a hybrid activator protein (tTA). Using this system, it was shown that gene expression could be tightly regulated by the addition of tetracycline in a dose-dependent way.[209–212] Such promoters have been incorporated into retroviral[213] and adenoviral vectors[214,215] to achieve tetracycline-dependent regulated transgene expression *in vitro* and *in vivo*. These genetic switches could control the temporal and spatial expression of the transgene quantitatively and reversibly,[211] ultimately providing physiological control of transgene expression.

17.5 Summary

In summary, gene therapy is currently used in the field of tissue engineering as a means to enhance the clinical performance of tissue substitutes. Delivery of genes can be used to guide differentiation along a specific lineage, improve the properties of the engineered tissue, reduce or eliminate immune reaction to transplanted tissues, treat genetic diseases, promote tissue regeneration, or transform engineered tissues into "bioreactors" for systemic delivery of therapeutic proteins *in vivo*. As a result, genetically modified tissues can be thought of as the next generation of tissue substitutes.

Viral and nonviral vehicles are currently used in gene therapy. Although several studies have demonstrated the usefulness of these vectors, many challenges facing gene therapy must be overcome to achieve clinically relevant results. Understanding the rate-limiting steps of the gene transfer process will certainly facilitate the production of improved vectors and engineering of more efficient physicochemical strategies to deliver genes.

Finally, genetically engineered tissue substitutes will have a tremendous impact as biological models of tissue physiology and disease states. They can serve as realistic models to understand the molecular and cellular mechanisms of biological processes such as differentiation,[69] three-dimensional migration,[216] and tumor metastasis. They may also be employed as toxicological models for testing of molecular and gene therapeutics. In combination with the new technologies of genomics, proteomics, and bioinformatics, they have an enormous potential to impact biology, medicine, and bioengineering by integrating global molecular information with three-dimensional cellular organization and tissue architecture.

References

1. Langer, R. and Vacanti, J.P., Tissue engineering, *Science*, 260, 920, 1993.
2. Berthiaume, F. and Yarmush, M.L., Tissue engineering, In *Biomedical Engineering Handbook*, Bronzino, G.J., ed., CRC Press, New York, 1995.
3. Griffith, L.G. and Naughton, G., Tissue engineering—Current challenges and expanding opportunities, *Science*, 295, 1009, 2002.
4. Zandstra, P.W. and Nagy, A., Stem cell bioengineering, *Annu. Rev. Biomed. Eng.*, 3, 275, 2001.
5. Watt, F.M. and Hogan, B.L., Out of Eden: Stem cells and their niches, *Science*, 287, 1427, 2000.
6. Morgan, J.R. and Yarmush, M.L., Gene therapy in tissue engineering, In *Frontiers in Tissue Engineering*, Patrick, C.W. J., Mikos, A.G., and McIntire, L.V., eds., Pergamon Press, New York, 1998, 278.
7. Anderson, W.F., Human gene therapy, *Science*, 256, 808, 1992.
8. Mulligan, R.C., The basic science of gene therapy, *Science*, 260, 926, 1993.
9. Hunter, E., Viral entry and receptors, In *Retroviruses*, Coffin, J.M., Hughes, S.H., and Varmus, H.E., eds., Cold Spring Harbor Laboratory Press, Cold Spring Harbor, New York, 1997, 71.
10. Andreadis, S.T., Roth, C.M., Le Doux, J.M., Morgan, J.R., and Yarmush, M.L., Large-scale processing of recombinant retroviruses for gene therapy, *Biotechnol. Prog.*, 15, 1, 1999.
11. Goff, S.P., Intracellular trafficking of retroviral genomes during the early phase of infection: Viral exploitation of cellular pathways, *J. Gene Med.*, 3, 517, 2001.
12. Kafri, T, Blomer, U., Peterson, D.A., Gage, F.H., and Verma, I.M., Sustained expression of genes delivered directly into liver and muscle by lentiviral vectors, *Nat. Genet.*, 17, 314, 1997.
13. Miyoshi, H., Takahashi, M., Gage, F.H., and Verma, I.M., Stable and efficient gene transfer into the retina using an HIV-based lentiviral vector, *Proc. Natl. Acad. Sci. USA*, 94, 10319, 1997.
14. Naldini, L., Blomer, U., Gage, F.H., Trono, D., and Verma, I.M., Efficient transfer, integration, and sustained long-term expression of the transgene in adult rat brains injected with a lentiviral vector, *Proc. Natl. Acad. Sci. USA*, 93, 11382, 1996.
15. Naldini, L., Blomer, U., Gallay, P., Ory, D., Mulligan, R., Gage, F.H., Verma, I.M., and Trono, D., *In vivo* gene delivery and stable transduction of nondividing cells by a lentiviral vector, *Science*, 272, 263, 1996.
16. Reiser, J., Harmison, G., Kluepfel-Stahl, S., Brady, R.O., Karlsson, S., and Schubert, M., Transduction of nondividing cells using pseudotyped defective high-titer HIV type 1 particles, *Proc. Natl. Acad. Sci. USA*, 93, 15266, 1996.
17. Buchschacher, G.L., Jr. and Wong-Staal, F., Development of lentiviral vectors for gene therapy for human diseases, *Blood*, 95, 2499, 2000.
18. Vigna, E. and Naldini, L., Lentiviral vectors: Excellent tools for experimental gene transfer and promising candidates for gene therapy, *J. Gene Med.*, 2, 308, 2000.
19. Zufferey, R., Nagy, D., Mandel, R.J., Naldini, L., and Trono, D., Multiply attenuated lentiviral vector achieves efficient gene delivery *in vivo*, *Nat. Biotechnol.*, 15, 871, 1997.
20. Zufferey, R., Dull, T., Mandel, R.J., Bukovsky, A., Quiroz, D., Naldini, L., and Trono, D., Self-inactivating lentivirus vector for safe and efficient in vivo gene delivery, *J Virol.*, 72, 9873, 1998.
21. Wilson, J.M., Adenovirus-mediated gene transfer to liver, *Adv. Drug Deliv. Rev.*, 46, 205, 2001.

22. Mizuguchi, H., Kay, M.A., and Hayakawa, T., Approaches for generating recombinant adenovirus vectors, *Adv. Drug Deliv. Rev.*, 52, 165, 2001.

23. Lozier, J.N., Metzger, M.E., Donahue, R.E., and Morgan, R.A., Adenovirus-mediated expression of human coagulation factor IX in the rhesus macaque is associated with dose-limiting toxicity, *Blood*, 94, 3968, 1999.

24. Lozier, J.N., Csako, G., Mondoro, T.H., Krizek, D.M., Metzger, M.E., Costello, R., Vostal, J.G., Rick, M.E., Donahue, R.E., and Morgan, R.A., Toxicity of a first-generation adenoviral vector in rhesus macaques, *Hum. Gene Ther.*, 13, 113, 2002.

25. Yang, Y., Su, Q., and Wilson, J.M., Role of viral antigens in destructive cellular immune responses to adenovirus vector-transduced cells in mouse lungs, *J. Virol.*, 70, 7209, 1996.

26. Kaplan, J.M., St. George, J.A., Pennington, S.E., Keyes, L.D., Johnson, R.P., Wadsworth, S.C., and Smith, A.E., Humoral and cellular immune responses of nonhuman primates to long-term repeated lung exposure to Ad2/CFTR-2, *Gene Ther.*, 3, 117, 1996.

27. Chirmule, N., Raper, S.E., Burkly, L., Thomas, D., Tazelaar, J., Hughes, J.V., and Wilson, J.M., Readministration of adenovirus vector in nonhuman primate lungs by blockade of CD40-CD40 ligand interactions, *J. Virol.*, 74, 3345, 2000.

28. Ponnazhagan, S., Curiel, D.T., Shaw, D.R., Alvarez, R.D., and Siegal, G.P., Adeno-associated virus for cancer gene therapy, *Cancer Res.*, 61, 6313, 2001.

29. Brown, M.D., Schatzlein, A.G., and Uchegbu, I.F., Gene delivery with synthetic (non viral) carriers, *Int. J. Pharm.*, 229, 1, 2001.

30. Hodgson, C.P., The vector void in gene therapy, *Biotechnology*, 13, 222, 1995.

31. Mahat, R.I., Monera, O.D., Smith, L.C., and Rolland, A., Peptide-based gene delivery, *Curr. Opin. Mol. Ther.*, 1, 226, 1999.

32. Subramanian, A., Ranganathan, P., and Diamond, S.L., Nuclear targeting peptide scaffolds for lipofection of nondividing mammalian cells, *Nat. Biotechnol.*, 17, 873, 1999.

33. Vaysse, L., Guillaume, C., Burgelin, I., Gorry, P., Ferec, C., and Arveiler, B., Proteolipidic vectors for gene transfer to the lung, *Biochem. Biophys. Res. Commun.*, 290, 1489, 2002.

34. Vaysse, L., Burgelin, I., Merlio, J.P., and Arveiler, B., Improved transfection using epithelial cell line-selected ligands and fusogenic peptides, *Biochim. Biophys. Acta*, 1475, 369, 2000.

35. Boussif, O., Lezoualc'h, F., Zanta, M.A., Mergny, M.D., Scherman, D., Demeneix, B., and Behr, J.P., A versatile vector for gene and oligonucleotide transfer into cells in culture and *in vivo*: Polyethylenimine, *Proc. Natl. Acad. Sci. USA*, 92, 7297, 1995.

36. Ferrari, S., Pettenazzo, A., Garbati, N., Zacchello, E., Behr, J.P., and Scarpa, M., Polyethylenimine shows properties of interest for cystic fibrosis gene therapy, *Biochim. Biophys. Acta*, 1447, 219, 1999.

37. Godbey, W.T., Wu, K.K., Hirasaki, G.J., and Mikos, A.G., Improved packing of poly(ethylenimine)/DNA complexes increases transfection efficiency, *Gene Ther.*, 6, 1380, 1999.

38. Godbey, W.T., Wu, K.K., and Mikos, A.G., Tracking the intracellular path of poly(ethylenimine)/DNA complexes for gene delivery, *Proc. Natl. Acad. Sci. USA*, 96, 5177, 1999.

39. Erbacher, P., Bettinger, T., Belguise-Valladier, P., Zou, S., Coll, J.L., Behr, J.P., and Remy, J.S., Transfection and physical properties of various saccharide, poly (ethylene glycol), and antibody-derivatized polyethylenimines (PEI), *J. Gene Med.*, 1, 210, 1999.

40. Wu, G.Y. and Wu, C.H., Receptor-mediated *in vitro* gene transformation by a soluble DNA carrier system, *J. Biol. Chem.*, 262, 4429, 1987.

41. Midoux, P. and Monsigny, M., Efficient gene transfer by histidylated polylysine/pDNA complexes, *Bioconjug. Chem.*, 10, 406, 1999.

42. Chen, Q.R., Zhang, L., Stass, S.A., and Mixson, A.J., Co-polymer of histidine and lysine markedly enhances transfection efficiency of liposomes, *Gene Ther.*, 7, 1698, 2000.

43. Benns, J.M., Choi, J.S., Mahato, R.I., Park, J.S., and Kim, S.W., pH-sensitive cationic polymer gene delivery vehicle: N-Ac-poly(L-histidine)-graft-poly(L-lysine) comb shaped polymer, *Bioconjug. Chem.*, 11, 637, 2000.

44. Chen, Q.R., Zhang, L., Stass, S.A., and Mixson, A.J., Branched co-polymers of histidine and lysine are efficient carriers of plasmids, *Nucleic Acids Res.*, 29, 1334, 2001.

45. Putnam, D., Gentry, C.A., Pack, D.W., and Langer, R., Polymer-based gene delivery with low cytotoxicity by a unique balance of side-chain termini, *Proc. Natl. Acad. Sci. USA*, 98, 1200, 2001.

46. Kukowska-Latallo, J.F., Bielinska, A.U., Johnson, J., Spindler, R., Tomalia, D.A., and Baker, J.R., Jr., Efficient transfer of genetic material into mammalian cells using starburst polyamidoamine dendrimers, *Proc. Natl. Acad. Sci. USA*, 93, 4897, 1996.

47. Bielinska, A.U., Chen, C., Johnson, J., and Baker, J.R., Jr., DNA complexing with polyamidoamine dendrimers: Implications for transfection, *Bioconjug. Chem.*, 10, 843, 1999.

48. Hudde, T., Rayner, S.A., Comer, R.M., Weber, M., Isaacs, J.D., Waldmann, H., Larkin, D.F., and George, A.J., Activated polyamidoamine dendrimers, a nonviral vector for gene transfer to the corneal endothelium, *Gene Ther.*, 6, 939, 1999.

49. Wu, G.Y., Wilson, J.M., Shalaby, F., Grossman, M., Shafritz, D.A., and Wu, C.H., Receptor-mediated gene delivery *in vivo*. Partial correction of genetic analbuminemia in Nagase rats, *J. Biol. Chem.*, 266, 14338, 1991.

50. Zenke, M., Steinlein, P., Wagner, E., Cotten, M., Beug, H., and Birnstiel, M.L., Receptor-mediated endocytosis of transferrin-polycation conjugates: An efficient way to introduce DNA into hematopoietic cells, *Proc. Natl. Acad. Sci. USA*, 87, 3655, 1990.

51. Wagner, E., Zenke, M., Cotten, M., Beug, H., and Birnstiel, M.L., Transferrin-polycation conjugates as carriers for DNA uptake into cells, *Proc. Natl. Acad. Sci. USA*, 87, 3410, 1990.

52. Shimizu, N., Chen, J., Gamou, S., and Takayanagi, A., Immunogene approach toward cancer therapy using erythrocyte growth factor receptor-mediated gene delivery, *Cancer Gene Ther.*, 3, 113, 1996.

53. Chen, J., Gamou, S., Takayanagi, A., Ohtake, Y., Ohtsubo, M., and Shimizu, N., Targeted *in vivo* delivery of therapeutic gene into experimental squamous cell carcinomas using anti-epidermal growth factor receptor antibody: Immunogene approach, *Hum. Gene Ther.*, 9, 2673, 1998.

54. Sosnowski, B.A., Gonzalez, A.M., Chandler, L.A., Buechler, Y.J., Pierce, G.F., and Baird, A., Targeting DNA to cells with basic fibroblast growth factor (FGF2), *J. Biol. Chem.*, 271, 33647, 1996.

55. Schaffer, D.V. and Lauffenburger, D.A., Optimization of cell surface binding enhances efficiency and specificity of molecular conjugate gene delivery, *J. Biol. Chem.*, 273, 28004, 1998.

56. Felgner, P.L. and Ringold, G.M., Cationic liposome-mediated transfection, *Nature*, 337, 387, 1989.

57. Felgner, P.L., Gadek, T.R., Holm, M., Roman, R., Chan, H.W., Wenz, M., Northrop, J.P., Ringold, G.M., and Danielsen, M., Lipofection: A highly efficient, lipid-mediated DNA-transfection procedure, *Proc. Natl. Acad. Sci. USA*, 84, 7413, 1987.

58. Farhood, H., Serbina, N., and Huang, L., The role of dioleoyl phosphatidylethanolamine in cationic liposome mediated gene transfer, *Biochim. Biophys. Acta*, 1235, 289, 1995.

59. Crystal, R.G., Transfer of genes to humans: Early lessons and obstacles to success, *Science*, 270, 404, 1995.

60. Filion, M.C. and Phillips, N.C., Toxicity and immunomodulatory activity of liposomal vectors formulated with cationic lipids toward immune effector cells, *Biochim. Biophys. Acta*, 1329, 345, 1997.

61. Godbey, W.T., Wu, K.K., and Mikos, A.G., Poly(ethylenimine)-mediated gene delivery affects endothelial cell function and viability, *Biomaterials*, 22, 471, 2001.

62. De Luca, M. and Pellegrini, G., The importance of epidermal stem cells in keratinocyte-mediated gene therapy (editorial), *Gene Ther.*, 4, 381, 1997.

63. Choate, K.A., Medalie, D.A., Morgan, J.R., and Khavari, P.A., Corrective gene transfer in the human skin disorder lamellar ichthyosis, *Nat. Med.*, 2, 1263, 1996.

64. Freiberg, R.A., Choate, K.A., Deng, H., Alperin, E.S., Shapiro, L.J., and Khavari, P.A., A model of corrective gene transfer in X-linked ichthyosis, *Hum. Mol. Genet.*, 6, 927, 1997.

65. Page, S.M. and Brownlee, G.G., An *ex vivo* keratinocyte model for gene therapy of hemophilia B, *J. Invest. Dermatol.*, 109, 139, 1997.

66. Dellambra, E., Vailly, J., Pellegrini, G., Bondanza, S., Golisano, O., Macchia, C., Zambruno, G., Meneguzzi, G., and De Luca, M., Corrective transduction of human epidermal stem cells in laminin-5-dependent junctional epidermolysis bullosa, *Hum. Gene Ther.*, 9, 1359, 1998.

67. Eming, S.A., Lee, J., Snow, R.G., Tompkins, R.G., Yarmush, M.L., and Morgan, J.R., Genetically modified human epidermis overexpressing PDGF-A directs the development of a cellular and vascular connective tissue stroma when transplanted to athymic mice—Implications for the use of genetically modified keratinocytes to modulate dermal regeneration, *J. Invest. Dermatol.*, 105, 756, 1995.

68. Eming, S.A., Snow, R.G., Yarmush, M.L., and Morgan, J.R., Targeted expression of insulin-like growth factor to human keratinocytes: Modification of the autocrine control of keratinocyte proliferation, *J. Invest. Dermatol.*, 107, 113, 1996.

69. Andreadis, S.T., Hamoen, K.E., Yarmush, M.L., and Morgan, J.R., Keratinocyte growth factor induces hyperproliferation and delays differentiation in a skin equivalent model system, *FASEB J.*, 15, 898, 2001.

70. Supp, D.M., Supp, A.P., Bell, S.M., and Boyce, S.T., Enhanced vascularization of cultured skin substitutes genetically modified to overexpress vascular endothelial growth factor, *J. Invest. Dermatol.*, 114, 5, 2000.

71. Supp, D.M. and Boyce, S.T., Overexpression of vascular endothelial growth factor accelerates early vascularization and improves healing of genetically modified cultured skin substitutes, *J. Burn. Care Rehabil.*, 23, 10, 2002.

72. Rio, M.D., Larcher, F., Meana, A., Segovia, J., Alvarez, A., and Jorcano, J., Nonviral transfer of genes to pig primary keratinocytes. Induction of angiogenesis by composite grafts of modified keratinocytes overexpressing VEGF driven by a keratin promoter, *Gene Ther.*, 6, 1734, 1999.

73. Morgan, J.R., Barrandon, Y., Green, H., and Mulligan, R.C., Expression of an exogenous growth hormone gene by transplantable human epidermal cells, *Science*, 237, 1476, 1987.

74. Fenjves, E.S., Smith, J., Zaradic, S., and Taichman, L.B., Systemic delivery of secreted protein by grafts of epidermal keratinocytes: Prospects for keratinocyte gene therapy, *Hum. Gene Ther.*, 5, 1241, 1994.

75. Gerrard, A.J., Hudson, D.L., Brownlee, G.G., and Watt, F.M., Towards gene therapy for haemophilia B using primary human keratinocytes, *Nat. Genet.*, 3, 180, 1993.

76. Fenjves, E.S., Schwartz, P.M., Blaese, R.M., and Taichman, L.B., Keratinocyte gene therapy for adenosine deaminase deficiency: A model approach for inherited metabolic disorders, *Hum. Gene Ther.*, 8, 911, 1997.

77. Sullivan, D.M., Jensen, T.G., Taichman, L.B., and Csaky, K.G., Ornithine-delta-aminotransferase expression and ornithine metabolism in cultured epidermal keratinocytes: Toward metabolic sink therapy for gyrate atrophy, *Gene Ther.*, 4, 1036, 1997.

78. Fenjves, E.S., Yao, S.N., Kurachi, K., and Taichman, L.B., Loss of expression of a retrovirus-transduced gene in human keratinocytes, *J. Invest. Dermatol.*, 106, 576, 1996.

79. Meng, X., Sawamura, D., Tamai, K., Hanada, K., Ishida, H., and Hashimoto, I., Keratinocyte gene therapy for systemic diseases. Circulating interleukin 10 released from gene-transferred keratinocytes inhibits contact hypersensitivity at distant areas of the skin, *J. Clin. Invest.*, 101, 1462, 1998.

80. Werner, S., Smola, H., Liao, X., Longaker, M.T., Krieg, T., Hofschneider, P.H., and Williams, L.T., The function of KGF in morphogenesis of epithelium and reepithelialization of wounds, *Science*, 266, 819, 1994.

81. Finch, P.W., Cunha, G.R., Rubin, J.S., Wong, J., and Ron, D., Pattern of keratinocyte growth factor and keratinocyte growth factor receptor expression during mouse fetal development suggests a role in mediating morphogenetic mesenchymal-epithelial interactions, *Dev. Dyn.*, 203, 223, 1995.

82. Baragi, V.M., Renkiewicz, R.R., Jordan, H., Bonadio, J., Hartman, J.W., and Roessler, B.J., Transplantation of transduced chondrocytes protects articular cartilage from interleukin 1-induced extracellular matrix degradation, *J. Clin. Invest.*, 96, 2454, 1995.

83. Bandara, G., Mueller, G.M., Galea-Lauri, J., Tindal, M.H., Georgescu, H.I., Suchanek, M.K., Hung, G.L., Glorioso, J.C., Robbins, P.D., and Evans, C.H., Intraarticular expression of biologically active interleukin 1-receptor-antagonist protein by *ex vivo* gene transfer, *Proc. Natl. Acad. Sci. USA*, 90, 10764, 1993.

84. Goomer, R.S., Deftos, L.J., Terkeltaub, R., Maris, T., Lee, M.C., Harwood, F.L., and Amiel, D., High-efficiency nonviral transfection of primary chondrocytes and perichondrial cells for *ex vivo* gene therapy to repair articular cartilage defects, *Osteoarthritis Cartilage*, 9, 248, 2001.

85. Madry, H., Zurakowski, D., and Trippel, S.B., Overexpression of human insulin-like growth factor-I promotes new tissue formation in an *ex vivo* model of articular chondrocyte transplantation, *Gene Ther.*, 8, 1443, 2001.

86. Laurencin, C.T., Ambrosio, A.M., Borden, M.D., and Cooper, J.A., Jr., Tissue engineering: Orthopedic applications, *Annu. Rev. Biomed. Eng.*, 1, 19, 1999.

87. Rose, F.R. and Oreffo, R.O., Bone tissue engineering: Hope vs hype, *Biochem. Biophys. Res. Commun.*, 292, 1, 2002.

88. Marx, J.C., Allay, J.A., Persons, D.A., Nooner, S.A., Hargrove, P.W., Kelly, P.F., Vanin, E.F., and Horwitz, E.M., High-efficiency transduction and long-term gene expression with a murine stem cell retroviral vector encoding the green fluorescent protein in human marrow stromal cells, *Hum. Gene Ther.*, 10, 1163, 1999.

89. Chuah, M.K., Brems, H., Vanslembrouck, V., Collen, D., and Vandendriessche, T., Bone marrow stromal cells as targets for gene therapy of hemophilia A, *Hum. Gene Ther.*, 9, 353, 1998.

90. Turgeman, G., Pittman, D.D., Muller, R., Kurkalli, B.G., Zhou, S., Pelled, G., Peyser, A., Zilberman, Y., Moutsatsos, I.K., and Gazit, D., Engineered human mesenchymal stem cells: A novel platform for skeletal cell mediated gene therapy, *J. Gene Med.*, 3, 240, 2001.

91. Partridge, K., Yang, X., Clarke, N.M., Okubo, Y, Bessho, K., Sebald, W., Howdle, S.M., Shakesheff, K.M., and Oreffo, R.O., Adenoviral BMP-2 gene transfer in mesenchymal stem cells: *In vitro* and *in vivo* bone formation on biodegradable polymer scaffolds, *Biochem. Biophys. Res. Commun.*, 292, 144, 2002.

92. Breitbart, A.S., Grande, D.A., Mason, J.M., Barcia, M., James, T., and Grant, R.T., Gene-enhanced tissue engineering: Applications for bone healing using cultured periosteal cells transduced retrovirally with the BMP-7 gene, *Ann. Plast. Surg.*, 42, 488, 1999.

93. Laurencin, C.T., Attawia, M.A., Lu, L.Q., Borden, M.D., Lu, H.H., Gorum, W.J., and Lieberman, J.R., Poly(lactide-*co*-glycolide)/hydroxyapatite delivery of BMP-2-producing cells: A regional gene therapy approach to bone regeneration, *Biomaterials*, 22, 1271, 2001.

94. Nerem, R.M. and Seliktar, D., Vascular tissue engineering, *Annu. Rev. Biomed. Eng.*, 3, 225, 2001.

95. Dunn, P.F., Newman, K.D., Jones, M., Yamada, I., Shayani, V., Virmani, R., and Dichek, D.A., Seeding of vascular grafts with genetically modified endothelial cells. Secretion of recombinant TPA results in decreased seeded cell retention *in vitro* and *in vivo*, *Circulation*, 93, 1439, 1996.

96. Kullo, I.J., Mozes, G., Schwartz, R.S., Gloviczki, P., Crotty, T.B., Barber, D.A., Katusic, Z.S., and O'Brien, T., Adventitial gene transfer of recombinant endothelial nitric oxide synthase to rabbit carotid arteries alters vascular reactivity, *Circulation*, 96, 2254, 1997.

97. Mozes, G., Kullo, I.J., Mohacsi, T.G., Cable, D.G., Spector, D.J., Crotty, T.B., Gloviczki, P., Katusic, Z.S., and O'Brien, T., *Ex vivo* gene transfer of endothelial nitric oxide synthase to atherosclerotic rabbit aortic rings improves relaxations to acetylcholine, *Atherosclerosis*, 141, 265, 1998.

98. Alexander, M.Y., Brosnan, M.J., Hamilton, C.A., Fennell, J.P., Beattie, E.C., Jardine, E., Heistad, D.D., and Dominiczak, A.F., Gene transfer of endothelial nitric oxide synthase but not Cu/Zn superoxide dismutase restores nitric oxide availability in the SHRSP, *Cardiovasc. Res.*, 47, 609, 2000.

99. Sato, J., Mohacsi, T., Noel, A., Jost, C., Gloviczki, P., Mozes, G., Katusic, Z.S., O'Brien, T., and Mayhan, W.G., *In vivo* gene transfer of endothelial nitric oxide synthase to carotid arteries from hypercholesterolemic rabbits enhances endothelium-dependent relaxations, *Stroke*, 31, 968, 2000.

100. Armeanu, S., Pelisek, J., Krausz, E., Fuchs, A., Groth, D., Curth, R., Keil, O., Quilici, J., Rolland, P.H., Reszka, R., and Nikol, S., Optimization of nonviral gene transfer of vascular smooth muscle cells *in vitro* and *in vivo*, *Mol. Ther.*, 1, 366, 2000.

101. von der Leyen, H.E., Braun-Dullaeus, R., Mann, M.J., Zhang, L., Niebauer, J., and Dzau, V.J., A pressure-mediated nonviral method for efficient arterial gene and oligonucleotide transfer, *Hum. Gene Ther.*, 10, 2355, 1999.
102. Matsumoto, T., Komori, K., Shoji, T., Kuma, S., Kume, M., Yamaoka, T., Mori, E., Furuyama, T., Yonemitsu, Y., and Sugimachi, K., Successful and optimized *in vivo* gene transfer to rabbit carotid artery mediated by electronic pulse, *Gene Ther.*, 8, 1174, 2001.
103. Martin, J.B., Young, J.L., Benoit, J.N., and Dean, D.A., Gene transfer to intact mesenteric arteries by electroporation, *J. Vasc. Res.*, 37, 372, 2000.
104. Compagni, A., Wilgenbus, P., Impagnatiello, M.A., Cotten, M., and Christofori, G., Fibroblast growth factors are required for efficient tumor angiogenesis, *Cancer Res.*, 60, 7163, 2000.
105. Gowdak, L.H., Poliakova, L., Wang, X., Kovesdi, I., Fishbein, K.W., Zacheo, A., Palumbo, R., Straino, S., Emanueli, C., Marrocco-Trischitta, M., Lakatta, E.G., Anversa, P., Spencer, R.G., Talan, M., and Capogrossi, M.C., Adenovirus-mediated VEGF(121) gene transfer stimulates angiogenesis in normoperfused skeletal muscle and preserves tissue perfusion after induction of ischemia, *Circulation*, 102, 565, 2000.
106. Gowdak, L.H., Poliakova, L., Li, Z., Grove, R., Lakatta, E.G., and Talan, M., Induction of angiogenesis by cationic lipid-mediated VEGF165 gene transfer in the rabbit ischemic hindlimb model, *J. Vasc. Surg.*, 32, 343, 2000.
107. Magovern, C.J., Mack, C.A., Zhang, J., Rosengart, T.K., Isom, O.W., and Crystal, R.G., Regional angiogenesis induced in nonischemic tissue by an adenoviral vector expressing vascular endothelial growth factor, *Hum. Gene Ther.*, 8, 215, 1997.
108. Avgoustiniatos, E.S. and Colton, C.K., Effect of external oxygen mass transfer resistances on viability of immunoisolated tissue, *Ann. N.Y. Acad. Sci.*, 831, 145, 1997.
109. Hidaka, C., Ibarra, C., Hannafin, J.A., Torzilli, P.A., Quitoriano, M., Jen, S.S., Warren, R.F., and Crystal, R.G., Formation of vascularized meniscal tissue by combining gene therapy with tissue engineering, *Tissue Eng.*, 8, 93, 2002.
110. Hubbell, J.A., Bioactive biomaterials, *Curr. Opin. Biotechnol.*, 10, 123, 1999.
111. Choate, K.A. and Khavari, P.A., Direct cutaneous gene delivery in a human genetic skin disease, *Hum. Gene Ther.*, 8, 1659, 1997.
112. Levy, M.Y., Barron, L.G., Meyer, K.B., and Szoka, F.C., Jr., Characterization of plasmid DNA transfer into mouse skeletal muscle: Evaluation of uptake mechanism, expression and secretion of gene products into blood, *Gene Ther.*, 3, 201, 1996.
113. Bonadio, J., Tissue engineering via local gene delivery: Update and future prospects for enhancing the technology, *Adv. Drug Deliv. Rev.*, 44, 185, 2000.
114. Truong-Le, V.L., August, J.T., and Leong, K.W., Controlled gene delivery by DNA-gelatin nanospheres, *Hum. Gene Ther.*, 9, 1709, 1998.
115. Bonadio, J., Smiley, E., Patil, P., and Goldstein, S., Localized, direct plasmid gene delivery *in vivo*: Prolonged therapy results in reproducible tissue regeneration, *Nat. Med.*, 5, 753, 1999.
116. Fang, J., Zhu, Y.Y., Smiley, E., Bonadio, J., Rouleau, J.P., Goldstein, S.A., McCauley, L.K., Davidson, B.L., and Roessler, B.J., Stimulation of new bone formation by direct transfer of osteogenic plasmid genes, *Proc. Natl. Acad. Sci. USA*, 93, 5753, 1996.
117. Luo, D., Woodrow-Mumford, K., Belcheva, N., and Saltzman, W.M., Controlled DNA delivery systems, *Pharm. Res.*, 16, 1300, 1999.
118. Shea, L.D., Smiley, E., Bonadio, J., and Mooney, D.J., DNA delivery from polymer matrices for tissue engineering, *Nat. Biotechnol.*, 17, 551, 1999.
119. Michlits, W., Mittermayr, R., Schafer, R., Redl, H., and Aharinejad, S., Fibrin-embedded administration of VEGF plasmid enhances skin flap survival, *Wound Repair Regen.*, 15, 360, 2007.
120. Lei, P., Padmashali, R.M., and Andreadis, S.T., Cell controlled and spatially arrayed gene delivery from fibrin hydrogels, *Biomaterials*, 30, 3790, 2009.

121. Hagiwara, K., Kishimoto S., Ishihara, M., Koyama, Y., Mazda, O., and Sato, T., *In vivo* gene transfer using pDNA/chitosan/chondroitin sulfate ternary complexes: Influence of chondroitin sulfate on the stability of freeze-dried complexes and transgene *in vivo*, *J. Gene Med.*, 2013. doi: 1002/jgm.2694 [Epub ahead of print].

122. Tyrone, J.W., Mogford, J.E., Chandler, L.A., Ma, C., Xia, Y., Pierce, G.F., and Mustoe, T.A., Collagen-embedded platelet-derived growth factor DNA plasmid promotes wound healing in a dermal ulcer model, *J. Surg. Res.*, 93, 230, 2000.

123. Labhasetwar, V., Bonadio, J., Goldstein, S., Chen, W., and Levy, R.J., A DNA controlled-release coating for gene transfer: Transfection in skeletal and cardiac muscle, *J. Pharm. Sci.*, 87,1347, 1998.

124. Cohen, H., Levy, R.J., Gao, J., Fishbein, I., Kousaev, V., Sosnowski, S., Slomkowski, S., and Golomb, G., Sustained delivery and expression of DNA encapsulated in polymeric nanoparticles, *Gene Ther.*, 7, 1896, 2000.

125. Madsen, S. and Mooney, D.J., Delivering DNA with polymer matrices: Applications in tissue engineering and gene therapy, *Pharm. Sci. Technol. Today*, 3, 381, 2000.

126. Kalyanasundaram, S., Feinstein, S., Nicholson, J.P., Leong, K.W., and Garver, R.I., Jr., Coacervate microspheres as carriers of recombinant adenoviruses, *Cancer Gene Ther.*, 6, 107, 1999.

127. Chandler, L.A., Gu, D.L., Ma, C., Gonzalez, A.M., Doukas, J., Nguyen, T., Pierce, G.F., and Phillips, M.L., Matrix-enabled gene transfer for cutaneous wound repair, *Wound Repair Regen.*, 8, 473, 2000.

128. Chandler, L.A., Doukas, J., Gonzalez, A.M., Hoganson, D.K., Gu, D.L., Ma, C., Nesbit, M., Crombleholme, T.M., Herlyn, M., Sosnowski, B.A., and Pierce, G.F., FGF2-targeted adenovirus encoding platelet-derived growth factor-B enhances *de novo* tissue formation, *Mol. Ther.*, 2, 153, 2000.

129. Beer, S.J., Matthews, C.B., Stein, C.S., Ross, B.D., Hilfinger, J.M., and Davidson, B.L., Poly (lacticglycolic) acid copolymer encapsulation of recombinant adenovirus reduces immunogenicity *in vivo*, *Gene Ther.*, 5, 740, 1998.

130. Chillon, M., Lee, J.H., Fasbender, A., and Welsh, M.J., Adenovirus complexed with polyethylene glycol and cationic lipid is shielded from neutralizing antibodies *in vitro*, *Gene Ther.*, 5, 995, 1998.

131. Matthews, C., Jenkins, G., Hilfinger, J., and Davidson, B., Poly-L-lysine improves gene transfer with adenovirus formulated in PLGA microspheres, *Gene Ther.*, 6, 1558, 1999.

132. Falanga, V., Wound healing and chronic wounds, *J. Cutan. Med. Surg.*, 3, S1, 1998.

133. Palsson, B.O. and Andreadis, S., The physico-chemical factors that govern retrovirus-mediated gene transfer, *Exp. Hematol.*, 25, 94, 1997.

134. Salmeen, I., Rimai, L., Luftig, R.B., Liebes, L., Retzer, E., Rich, M., and McCormick, J.J., Hydrodynamic diameters of murine mammary, rous sarcoma and feline leukemia RNA tumor viruses: Studies by laser beat frequency light-scattering spectroscopy and electron microscopy, *J. Virol.*, 17, 584, 1976.

135. Andreadis, S., Lavery, T., Davis, H.E., Le Doux, J.M., Yarmush, M.L., and Morgan, J.R., Toward a more accurate quantitation of the activity of recombinant retroviruses: Alternatives to titer and multiplicity of infection, *J. Virol.*, 74, 3431, 2000.

136. Chuck, A.C., Clarke, M.F., and Palsson, B.O., Retroviral infection is limited by Brownian motion, *Hum. Gene Ther.*, 7, 1527, 1996.

137. LeDoux, J.M., Davis, H.E., Yarmush, M.L., and Morgan, J.R., Kinetics of retrovirus production and decay, *Biotech. Bioeng.*, 63, 654, 1999.

138. Bunnell, B.A., Muul, L.M., Donahue, R.E., Blaese, R.M., and Morgan, R.A., High-efficiency retroviral-mediated gene transfer into human and nonhuman primate peripheral blood lymphocytes, *Proc. Natl. Acad. Sci. USA*, 92, 7739, 1995.

139. Chuck, A.S. and Palsson, B.O., Consistent and high rates of gene transfer can be obtained using flow-through transduction over a wide range of retroviral titers, *Hum. Gene Ther.*, 7, 743, 1996.

140. Chuck, A.S. and Palsson, B.O., Membrane adsorption characteristics determine the kinetics of flow-through transductions, *Biotech. Bioeng.*, 51, 260, 1996.

141. Bertolini, F., Battaglia, M., Corsini, C., Lazzari, L., Soligo, D., Zibera, C., and Thalmeier, K., Engineered stromal layers and continuous flow culture enhance multidrug resistance gene transfer in hematopoietic progenitors, *Cancer Res.*, 56, 2566, 1996.

142. Hanenberg, H., Xiao, X.L., Dilloo, D., Hashino, K., Kato, I., and Williams, D.A., Colocalization of retrovirus and target cells on specific fibronectin fragments increases genetic transduction of mammalian cells, *Nat. Med.*, 2, 876, 1996.

143. Hanenberg, H., Hashino, K., Konishi, H., Hock, R.A., Kato, I., and Williams, D.A., Optimization of fibronectin-assisted retroviral gene transfer into human CD34+ hematopoietic cells, Hum. Gene Ther., 8, 2193, 1997.

144. Bajaj, B., Lei, P., and Andreadis, S.T., High efficiencies of gene transfer with immobilized recombinant retrovirus: Kinetics and optimization, *Biotechnol. Prog.*, 17, 587, 2001.

145. Lei, P., Bajaj, B., and Andreadis, S.T., Retrovirus-associated heparan sulfate mediates immobilization and gene transfer on recombinant fibronectin, *J. Virol.*, 76(17), 8722, 2002.

146. LeDoux, J.M., Morgan, J.R., and Yarmush, M.L., Proteoglycans secreted by packaging cell lines inhibit retrovirus infection, *J. Virol.*, 70, 6468, 1996.

147. Kavanaugh, M.P., Miller, D.G., Zhang, W., Law, W., Kozak, S.L., Kabat, D., and Miller, A.D., Cell-surface receptors for gibbon ape leukemia virus and amphotropic murine retrovirus are inducible sodium-dependent phosphate symporters, *Proc. Natl. Acad. Sci. USA*, 91, 7071, 1994.

148. Kozak, S.L., Siess, D.C., Kavanaugh, M.P., Miller, A.D., and Kabat, D., The envelope glycoprotein of an amphotropic murine retrovirus binds specifically to the cellular receptor/phosphate transporter of susceptible species, *J. Virol.*, 69, 3433, 1995.

149. Summerford, C., Bartlett, J.S., and Samulski, R.J., Alpha V beta 5 integrin: A co-receptor for adeno-associated virus type 2 infection., *Nat. Med.*, 5, 78, 1999.

150. Wickham, T.J., Mathias, P., Cheresh, D.A., and Nemerow, G.R., Integrins alpha v beta 3 and alpha v beta 5 promote adenovirus internalization but not virus attachment, *Cell*, 73, 309, 1993.

151. Wickham, T.J., Filardo, E.J., Cheresh, D.A., and Nemerow, G.R., Integrin alpha v beta 5 selectively promotes adenovirus mediated cell membrane permeabilization, *J. Cell Biol.*, 127, 257, 1994.

152. Davison, E., Diaz, R.M., Hart, I.R., Santis, G., and Marshall, J.F., Integrin alpha 5 beta 1-mediated adenovirus infection is enhanced by the integrin-activating antibody TS2/16, *J. Virol.*, 71, 6204, 1997.

153. Li, E., Brown, S.L., Stupack, D.G., Puente, X.S., Cheresh, D.A., and Nemerow, G.R., Integrin alpha v beta 1 is an adenovirus coreceptor, *J. Virol.*, 75, 5405, 2001.

154. Miller, A.D., Garcia, J.V., von Suhr, N., Lynch, C.M., Wilson, C., and Eiden, M.V., Construction and properties of retrovirus packaging cells based on gibbon ape leukemia virus, *J. Virol.*, 65, 2220, 1991.

155. Wilson, C., Reitz, M.S., Okayama, H., and Eiden, M.V., Formation of infectious hybrid virions with gibbon ape leukemia virus and human T-cell leukemia virus retroviral envelope glycoproteins and the *gag* and *pol* proteins of Moloney murine leukemia virus, *J. Virol.*, 63, 2374, 1989.

156. Kasahara, N., Dozy, A.M., and Kan, Y.W., Tissue-specific targeting of retroviral vectors through ligand-receptor interactions, *Science*, 266, 1373, 1994.

157. Burns, J.C., Friedmann, T., Driever, W., Burrascano, M., and Yee, J.K., Vesicular stomatitis virus G glycoprotein pseudotyped retroviral vectors: Concentration to very high titer and efficient gene transfer into mammalian and non-mammalian cells, *Proc. Natl. Acad. Sci. USA*, 90, 8033, 1993.

158. Martin, F., Neil, S., Kupsch, J., Maurice, M., Cosset, F., and Collins, M., Retrovirus targeting by tropism restriction to melanoma cells, *J. Virol.*, 73, 6923, 1999.

159. Bartlett, J.S., Kleinschmidt, J., Boucher, R.C., and Samulski, R.J., Targeted adeno-associated virus vector transduction of nonpermissive cells mediated by a bispecific F(ab'gamma)2 antibody, *Nat. Biotechnol.*, 17, 181, 1999.

160. Andreadis, S., Fuller, A.O., and Palsson, B.O., Cell cycle dependence of retroviral transduction: An issue of overlapping time scales, *Biotechnol. Bioeng.*, 58, 272, 1998.

161. Roe, T., Reynolds, T.C., Yu, G., and Brown, P.O., Integration of murine leukemia virus DNA depends on mitosis, *EMBO J.*, 12, 2099, 1993.

162. Andreadis, S. and Palsson, B.O., Kinetics of retrovirus mediated gene transfer: The importance of the intracellular half-life of retroviruses, *J. Theor. Biol.*, 182, 1, 1996.

163. Andreadis, S., Brott, D.A., Fuller, A.O., and Palsson, B.O., Moloney murine leukemia virus-derived retroviral vectors decay intracellularly with a half-life in the range of 5.5 to 7.5 h, *J. Virol.*, 71, 7541, 1997.

164. Lewis, P.F. and Emerman, M., Passage through mitosis is required for oncoretroviruses but not for the human immunodeficiency virus, *J. Virol.*, 68, 510, 1994.

165. Case, S.S., Price, M.A., Jordan, C.T., Yu, X.J., Wang, L., Bauer, G., Haas, D.L. et al., Stable transduction of quiescent CD34(+) CD38(−) human hematopoietic cells by HIV-1-based lentiviral vectors, *Proc. Natl. Acad. Sci. USA*, 96, 2988, 1999.

166. Guenechea, G., Gan, O.I., Inamitsu, T., Dorrell, C., Pereira, D.S., Kelly, M., Naldini, L., and Dick, J.E., Transduction of human CD34(+) CD38(−) bone marrow and cord blood-derived SCID-repopulating cells with third-generation lentiviral vectors, *Mol. Ther.*, 1, 566, 2000.

167. Wickham, T.J., Granados, R.R., Wood, H.A., Hammer, D.A., and Shuler, M.L., General analysis of receptor-mediated viral attachment to cell surfaces, *Biophys. J.*, 58, 1501, 1990.

168. Wickham, T.J., Shuler, M.L., Hammer, D.A., Granados, R.R., and Wood, H.A., Equilibrium and kinetic analysis of *Autographa californica* nuclear polyhedrosis virus attachment to different insect cell lines, *J. Gen. Virol.*, 73, 3185, 1992.

169. Dee, K.U., Hammer, D.A., and Shuler, M.L., A model of the binding, entry, uncoating, and RNA synthesis of *Semliki* forest virus in baby hamster kidney (BHK-21) cells, *Biotechnol. Bioeng.*, 46, 485, 1995.

170. Dee, K.U. and Shuler, M.L., A mathematical model of the trafficking of acid-dependent enveloped viruses: Application to the binding, uptake, and nuclear accumulation of baculovirus, *Biotechnol. Bioeng.*, 54, 1997.

171. Sanlioglu, S., Benson, P.K., Yang, J., Atkinson, E.M., Reynolds, T., and Engelhardt, J.F., Endocytosis and nuclear trafficking of adeno-associated virus type 2 are controlled by racl and phosphatidylinositol-3 kinase activation, *J. Virol.*, 74, 9184, 2000.

172. Xu, Y. and Szoka, F.C., Jr., Mechanism of DNA release from cationic liposome/DNA complexes used in cell transfection, *Biochemistry*, 35, 5616, 1996.

173. Godbey, W.T., Wu, K.K., and Mikos, A.G., Poly(ethylenimine) and its role in gene delivery, *J. Control. Release*, 60, 149, 1999.

174. Plank, C., Oberhauser, B., Mechtler, K., Koch, C., and Wagner, E., The influence of endosome-disruptive peptides on gene transfer using synthetic virus-like gene transfer systems, *J. Biol. Chem.*, 269, 12918, 1994.

175. Nishikawa, M., Yamauchi, M., Morimoto, K., Ishida, E., Takakura, Y., and Hashida, M., Hepatocyte-targeted *in vivo* gene expression by intravenous injection of plasmid DNA complexed with synthetic multi-functional gene delivery system, *Gene Ther.*, 7, 548, 2000.

176. Zanta, M.A., Belguise-Valladier, P., and Behr, J.P., Gene delivery: A single nuclear localization signal peptide is sufficient to carry DNA to the cell nucleus, *Proc. Natl. Acad. Sci. USA*, 96, 91, 1999.

177. Branden, L.J., Mohamed, A.J., and Smith, C.I., A peptide nucleic acid-nuclear localization signal fusion that mediates nuclear transport of DNA, *Nat. Biotechnol.*, 17, 784, 1999.

178. Dean, D.A., Dean, B.S., Muller, S., and Smith, L.C., Sequence requirements for plasmid nuclear import, *Exp. Cell Res.*, 253, 713, 1999.

179. Vacik, J., Dean, B.S., Zimmer, W.E., and Dean, D.A., Cell-specific nuclear import of plasmid DNA, *Gene Ther.*, 6, 1006, 1999.

180. Caplan, A.I., and Bruder, S.P., Mesenchymal stem cells: Building blocks for molecular medicine in the 21st century, *Trends Mol. Med.*, 7, 259, 2001.

181. Miller, D.G., Adam, M.A., and Miller, A.D., Gene transfer by retrovirus vectors occurs only in cells that are actively replicating at the time of infection, *Mol. Cell. Biol.*, 10, 4239, 1990.

182. Lutzko, C., Dube, I.D., and Stewart, A.K., Recent progress in gene transfer into hematopoietic stem cells, *Crit. Rev. Oncol. Hematol.*, 30, 143, 1999.

183. Moore, K.A., Deisseroth, A.B., Reading, C.L., Williams, D.E., and Belmont, J.W., Stromal support enhances cell-free retroviral vector transduction of human bone marrow long-term culture-initiating cells, *Blood*, 79, 1393, 1992.
184. Bienzle, D., Abrams-Ogg, A.C., Kruth, S.A., Ackland-Snow, J., Carter, R.F., Dick, J.E., Jacobs, R.M., Kamel-Reid, S., and Dube, I.D., Gene transfer into hematopoietic stem cells: Long-term maintenance of *in vitro* activated progenitors without marrow ablation, *Proc. Natl. Acad. Sci. USA*, 91, 350, 1994.
185. Xu, L.C., Kluepfel-Stahl, S., Blanco, M., Schiffmann, R., Dunbar, C., and Karlsson, S., Growth factors and stromal support generate very efficient retroviral transduction of peripheral blood CD34+ cells from Gaucher patients, *Blood*, 86, 141, 1995.
186. Pollok, K.E., Hanenberg, H., Noblitt, T.W., Schroeder, W.L., Kato, I., Emanuel, D., and Williams, D.A., High-efficiency gene transfer into normal and adenosine deaminase-deficient T lymphocytes is mediated by transduction on recombinant fibronectin fragments, *J. Virol.*, 72, 4882, 1998.
187. Dardalhon, V., Noraz, N., Pollok, K., Rebouissou, C., Boyer, M., Bakker, A.Q., Spits, H., and Taylor, N., Green fluorescent protein as a selectable marker of fibronectin-facilitated retroviral gene transfer in primary human T lymphocytes, *Hum. Gene Ther.*, 10, 5, 1999.
188. Moritz, T., Dutt, P., Xiao, X., Carstanjen, D., Vik, T., Hanenberg, H., and Williams, D.A., Fibronectin improves transduction of reconstituting hematopoietic stem cells by retroviral vectors: Evidence of direct viral binding to chymotryptic carboxy-terminal fragments, *Blood*, 88, 855, 1996.
189. Conneally, E., Eaves, C.J., and Humphries, R.K., Efficient retroviral-mediated gene transfer to human cord blood stem cells with *in vivo* repopulating potential, *Blood*, 91, 3487, 1998.
190. Dao, M.A., Hashino, K., Kato, I., and Nolta, J.A., Adhesion to fibronectin maintains regenerative capacity during *ex vivo* culture and transduction of human hematopoietic stem and progenitor cells, *Blood*, 92, 4612, 1998.
191. Kiem, H.P., Andrews, R.G., Morris, J., Peterson, L., Heyward, S., Allen, J.M., Rasko, J.E., Potter, J., and Miller, A.D., Improved gene transfer into baboon marrow repopulating cells using recombinant human fibronectin fragment CH-296 in combination with interleukin-6, stem cell factor, FLT-3 ligand, and megakaryocyte growth and development factor, *Blood*, 92, 1878, 1998.
192. Cavazzana-Calvo, M., Hacein-Bey, S., de Saint Basile, G., Gross, F., Yvon, E., Nusbaum, P., Selz, F., Hue, C., Certain, S., Casanova, J.L., Bousso, P., Deist, F.L., and Fischer, A., Gene therapy of human severe combined immunodeficiency (SCID)-X1 disease, *Science*, 288, 669, 2000.
193. Abonour, R., Williams, D.A., Einhorn, L., Hall, K.M., Chen, J., Coffman, J., Traycoff, C.M., Bank, A., Kato, I., Ward, M., Williams, S.D., Hromas, R., Robertson, M.J., Smith, F.O., Woo, D., Mills, B., Srour, E.F., and Cornetta, K., Efficient retrovirus-mediated transfer of the multidrug resistance 1 gene into autologous human long-term repopulating hematopoietic stem cells, *Nat. Med.*, 6, 652, 2000.
194. Bajaj, B., Behshad, S., and Andreadis, S.T., Retroviral gene transfer to human epidermal keratinocytes correlates with integrin expression and is significantly enhanced on fibronectin, *Hum. Gene Ther.*, 13(15), 1821, 2002.
195. Jones, P.H. and Watt, F.M., Separation of human epidermal stem cells from transit amplifying cells on the basis of differences in integrin function and expression, *Cell*, 73, 713, 1993.
196. Jones, P.H., Harper, S., and Watt, F.M., Stem cell patterning and fate in human epidermis, *Cell*, 80, 83, 1995.
197. Zhu, A.J., Haase, I., and Watt, F.M., Signaling via beta 1 integrins and mitogen-activated protein kinase determines human epidermal stem cell fate *in vitro*, *Proc. Natl. Acad. Sci. USA*, 96, 6728, 1999.
198. Bajaj, B., Lei, P., and Andreadis, S.T., Efficient gene transfer to human epidermal keratinocytes on fibronectin: *In vitro* evidence for transduction of epidermal stem cells, *Mol. Ther.*, 11, 969, 2005.
199. Young, J.L. and Dean, D.A., Nonviral gene transfer strategies for the vasculature, *Microcirculation*, 9, 35, 2002.
200. Manabe, I. and Owens, G.K., CArG elements control smooth muscle subtype-specific expression of smooth muscle myosin *in vivo*, *J. Clin. Invest.*, 107, 823, 2001.

201. Owens, G.K., Loeb, A., Gordon, D., and Thompson, M.M., Expression of smooth muscle-specific alpha-isoactin in cultured vascular smooth muscle cells: Relationship between growth and cyto-differentiation. *J. Cell Biol.*, 102, 343, 1986.

202. Kovacs, A.M. and Zimmer, W.E., Molecular cloning and expression of the chicken smooth muscle gamma-actin mRNA, *Cell Motil. Cytoskeleton*, 24, 67, 1993.

203. Samaha, F.F., Ip, H.S., Morrisey, E.E., Seltzer, J., Tang, Z., Solway, J., and Parmacek, M.S., Developmental pattern of expression and genomic organization of the calponin-hl gene. A contractile smooth muscle cell marker, *J. Biol. Chem.*, 271, 395, 1996.

204. Ferreira, V., Assouline, Z., Schwachtgen, J.L., Bahnak, B.R., Meyer, D., and Kerbiriou-Nabias, D., The role of the 5′-flanking region in the cell-specific transcription of the human von Willebrand factor gene, *Biochem. J.*, 293, 641, 1993.

205. Patterson, C., Perrella, M.A., Hsieh, CM., Yoshizumi, M., Lee, M.E., and Haber, E., Cloning and functional analysis of the promoter for KDR/flk-1, a receptor for vascular endothelial growth factor, *J. Biol. Chem.*, 270, 23111, 1995.

206. Lee, M.E., Bloch, K.D., Clifford, J.A., and Quertermous, T., Functional analysis of the endothelin-1 gene promoter. Evidence for an endothelial cell-specific cis-acting sequence, *J. Biol. Chem.*, 265, 10446, 1990.

207. Hafenrichter, D.G., Wu, X., Rettinger, S.D., Kennedy, S.C., Flye, M.W., and Ponder, K.P., Quantitative evaluation of liver-specific promoters from retroviral vectors after *in vivo* transduction of hepatocytes, *Blood*, 84, 3394, 1994.

208. Page, S.M. and Brownlee, G.G., Differentiation-specific enhancer activity in transduced keratinocytes: A model for epidermal gene therapy, *Gene Ther.*, 5, 394, 1998.

209. Gossen, M. and Bujard, H., Tight control of gene expression in mammalian cells by tetracycline-responsive promoters, *Proc. Natl. Acad. Sci. USA*, 89, 5547, 1992.

210. Furth, P.A., St Onge, L., Boger, H., Gruss, P., Gossen, M., Kistner, A., Bujard, H., and Hennighausen, L., Temporal control of gene expression in transgenic mice by a tetracycline-responsive promoter, *Proc. Natl. Acad. Sci. USA*, 91, 9302, 1994.

211. Kistner, A., Gossen, M., Zimmermann, F., Jerecic, J., Ullmer, C., Lubbert, H., and Bujard, H., Doxycycline-mediated quantitative and tissue-specific control of gene expression in transgenic mice, *Proc. Natl. Acad. Sci. USA*, 93, 10933, 1996.

212. Baron, U., Gossen, M., and Bujard, H., Tetracycline-controlled transcription in eukaryotes: Novel transactivators with graded transactivation potential, *Nucleic Acids Res.*, 25, 2723, 1997.

213. Paulus, W., Baur, I., Boyce, F.M., Breakefield, X.O., and Reeves, S.A., Self-contained, tetracycline-regulated retroviral vector system for gene delivery to mammalian cells, *J. Virol.*, 70, 62, 1996.

214. Harding, T.C., Geddes, B.J., Noel, J.D., Murphy, D., and Uney, J.B., Tetracycline-regulated transgene expression in hippocampal neurons following transfection with adenoviral vectors, *J. Neurochem.*, 69, 2620, 1997.

215. Harding, T.C., Geddes, B.J., Murphy, D., Knight, D., and Uney, J.B., Switching transgene expression in the brain using an adenoviral tetracycline-regulatable system, *Nat. Biotechnol.*, 16, 553, 1998.

216. Geer, D.J., Swartz, D.D., and Andreadis, S.T., Fibrin promotes migration in a three-dimensional *in vitro* model of wound regeneration, *Tissue Eng.*, 8(5), 787, 2002.

18

Bioactive Scaffold Design for Articular Cartilage Engineering

Eric M. Darling
Brown University

Kyriacos A. Athanasiou
University of California, Davis

18.1 Introduction

Tissue engineering is increasingly poised to provide more viable alternatives to surgical procedures for the repair of orthopedic defects, such as focal articular cartilage lesions. Cartilage appears to be a relatively simple tissue, yet it is not amenable to regeneration with the techniques available to date. It is alymphatic and aneural, has no contact with blood, and exhibits low cellularity. The main constituents of cartilage are proteoglycans and type II collagen. Many researchers are currently pursuing the goal of creating functional cartilage, both *in vitro* and *in vivo*. By creating a tissue *in vitro*, the time required for healing is decreased, while the probability of a successful recovery is increased. Scaffolds that are designed to be used *in vivo* have the ability to reduce healing time even more dramatically because surgery can be done arthroscopically. However, current injectable materials do not fulfill all the structural and biological conditions needed to function in a mechanically loaded environment.

The physiological loading environment of cartilage in a diarthrodial joint is quite complicated, but one can consider it as consisting primarily of direct compression. As stated in a review by Darling and Athanasiou,[1] loads in the human knee range from 5 to 15 MPa. An implanted scaffold or tissue-engineered construct has to be able to withstand these forces until the new tissue is able to support itself. For *in vitro* tissue engineering approaches, the scaffold has to be able to withstand the mechanical environment in which it is cultured, whether it is hydrostatic pressure, direct compression, or just a static environment. In addition to the mechanical constraints, the scaffold has to degrade in such a way that it transfers the load gradually onto the newly formed tissue. This is especially important for *in vivo* applications. The degradation properties as well as the biocompatibility of the degradation products are an important concern when choosing an implant material.

Cartilage implants or tissue-engineered constructs experience a relatively low immune response in the knee compared to the rest of the body. Since articular cartilage is avascular, there is no simple way for lymphocytes, antigens, and complement proteins to attack the foreign construct. Mild inflammatory responses do occur, though, and a more biocompatible scaffold is always preferred. Biocompatibility, as well as the material properties of the construct, can be controlled based on what components are selected for the scaffold.

In addition to the material's mechanical properties and biocompatibility, a scaffold should elicit a beneficial response from the actual cells. Cell–substrate interactions play an important role in the attachment, viability, and biological response of the cells once they are seeded onto the scaffold. By using a modified scaffold, cells can be guided onto a desired developmental track.

Growth factors (GFs) and other bioactive agents are increasingly investigated for their ability to stimulate articular cartilage formation *in vitro* and *in vivo*. The addition of these molecules can significantly improve the tissue growth, but their incorporation into the scaffold can be a difficult process. Additionally, the activity of a GF is usually on the order of days, which restricts its use *in vivo*.

It is the belief of the authors that the ideal scaffold for tissue engineering articular cartilage is obtained by combining bioactive molecules, such as adhesion proteins, peptides, and GFs, with a mechanically sound scaffold material that has the proper viscoelastic properties, strength, and degradation characteristics. This review will highlight the materials that are most commonly used today for tissue engineering cartilage. Table 18.1 lists the representative researchers and their corresponding scaffolds, including

TABLE 18.1 Listing of Representative Authors and Their Scaffold Approaches

First Author, Year	Scaffold Material	Protein/Peptide	Growth Factor
Murphy, 2001[2] (*in vitro*)	Alginate		
Mauck, 2000[6] (*in vitro*)	Agarose		
Hunter, 2002[7] (*in vitro*)	Collagen I		
Lee, 2000[8] (*in vitro*)	Collagen II		
Sechriest, 2000[12] (*in vitro*)	Chitosan–GAG		
Silverman, 1999[14] (*in vivo*)	Fibrin glue		
Solchaga, 2000[64] (*in vivo*)	Hyaluronan		
Freed, 1998[29] (*in vitro*)	PGA		
Grande, 1997[28] (*in vitro*)	PLA		
Athanasiou, 1998[20] (review)	PLGA		
Honda, 2000[36] (*in vivo*)	PCL		
Bryant, 2002[49] (*in vitro*)	PEG/PEO		
Suggs, 1999[43] (*in vivo*)	PPF-PEG		
Ameer, 2002[51] (*in vitro*)	Fibrin glue and PLGA		
Marijnissen, 2002[50] (*in vivo*)	Alginate and PLGA		
Bhati, 2001[63] (*in vitro*)	Trimethylene carbonate–PGA	Fibronectin	
Solchaga, 2002[64] (*in vivo*)	Hyaluronan	Fibronectin	
Thomas, 1997[65] (*in vitro*)	Modified quartz (monolayer)	Vitronectin	
Makihira, 1999[66] (*in vitro*)	Plastic (monolayer)	Cartilage matrix protein	
Alsberg, 2001[79] (*in vivo*)	Alginate	RGD (covalent)	
Jo, 2000[69] (*in vitro*)	PPF-PEG	GRGD (covalent)	
Quirk, 200l (*in vitro*)	PLA	RGD (bound to PLL and adsorbed to PLA)	
Carlisle, 2000 (*in vitro*)	PCL and PLLA	RGD (pulsed plasma deposition)	
Eid, 2001[34] (*in vitro*)	PLGA	RGD (adsorption)	
Elisseeff, 2000[47] (*in vitro*)	PEO		TGF-β (PLGA encapsulated)
Mann, 2001[83] (*in vitro*)	PEG		TGF-β (tethered)
Athanasiou, 1997[24] (*in vivo*)	PLGA		TGF-β (physically entrapped)
Toolan, 1996 (*in vitro*)	Collagen I		FGF (soluble)

any modifications made to the materials using bioactive molecules. A few novel polymers will also be discussed based on their promise as scaffold materials. Surface modification and the incorporation of GFs will be discussed as an integral part of the scaffold design.

18.2 Scaffold Materials

The base scaffold material is the main component when designing a tissue-engineering implant. It should fulfill three main requirements: have an interconnected network that allows for efficient diffusion of nutrients and room for tissue growth, be biocompatible and bioresorbable, with degradation characteristics that match the rate of new tissue formation, and allow for the attachment, proliferation, and differentiation of chondrocytes that are seeded onto its surface. The last requirement can be fulfilled by using bioactive molecules attached to an inert scaffold, but this option will be discussed later in this chapter.

Another major concern when making a scaffold is whether it will be used *in vitro* or *in vivo*. If the scaffold is implanted immediately, it must possess the proper mechanical characteristics for the loading environment to retain the implant's shape and partially shield the chondrocytes from the surrounding forces. Injectable scaffold materials have the additional problem of needing to form a mechanically stable implant after entering the body without killing the seeded cells. Typically, injectable materials are photocrosslinked to obtain a high enough stiffness to function under physiological loads, but this process often involves chemicals that are toxic to the cells. Research is ongoing to find a successful approach to this problem.

A cell-seeded scaffold that is cultured *in vitro*, however, does not need extensive structural integrity. Once the tissue forms, the engineered construct should have its own mechanical characteristics that do not depend on the partially degraded scaffold. With this approach, the scaffold exists only to support the cells for a period of weeks while the new tissue forms. By the time the construct is implanted, the new matrix should be able to function sufficiently under the native loading environment.

Scaffold materials can be sorted into four main categories: natural polymers, synthetic polymers, hydrogels, and composites. There is overlap between the categories, since many natural polymers are hydrogels, and some of those have been used in combination with others to form composite scaffolds. Natural polymers are found in living organisms and can be extracted with relative simplicity. Synthetic polymers are created using chemical processes that are sometimes more complicated, but these polymers are often more conducive to having their material properties changed than natural polymers. Hydrogels can be synthesized, but can also occur naturally. They are primarily composed of fluid that significantly swells the polymer network to form a biphasic construct. Composite scaffolds combine two or more materials into one scaffold to take advantage of the special characteristics intrinsic to each substance.

New biomaterials are synthesized constantly, but extensive chemical and physical characterization has to be conducted before a material can be used with living cells. The materials summarized in this section have been well characterized and used in experiments involving living cells. Not all have been used with chondrocytes, but their biocompatibility has been proved using other cell types. However, different cells often prefer different materials, and what might work for endothelial cells might not work for chondrocytes. This section summarizes materials that have been used for many tissue engineering applications; so the functional cell type is indicated along with any results.

18.2.1 Natural Polymers

The preference for using natural materials for a scaffold is the belief that many biological materials elicit little or no immune response. Among the natural materials used in cartilage engineering are alginate, agarose, chitosan, fibrin glue, type I and II collagens, and hyaluronic acid-based materials. Each material has its specific advantages, but the results vary widely depending on the additional culturing conditions.

Alginate is a polysaccharide extracted from algae and is used to encapsulate cells within the scaffold rather than having them attached to the surface. Encapsulation maintains a chondrocyte's rounded morphology, allowing the redifferentiation of a cell that has been cultured in the monolayer.[2] Besides the ability to encapsulate cells, the main advantage to using alginate is its biocompatibility, as reviewed by Hutmacher.[3] For successful use *in vivo*, alginate has to be carefully purified using filtration, precipitation, or extraction. Alginate grafts have major disadvantages as well. They do not degrade rapidly *in vivo*, which can cause problems as the new tissue starts to grow. Long-term implants are not possible because the scaffold loses its functionality within a year.[3] Unlike synthetic polymers, its degradation characteristics cannot be tailored to fit a set timeline.

Agarose is also a polysaccharide but is derived from seaweed and exhibits a temperature-sensitive solubility in water that is used to encapsulate cells.[3] Like alginate, it is biocompatible and provides a three-dimensional environment that helps maintain the chondrocyte's phenotype during culture. Agarose also has degradation properties similar to alginate, thus not allowing control of the scaffold's lifetime in culture. It does not degrade rapidly enough for many *in vitro* experiments. However, several studies have used agarose as a scaffold material when stimulating chondrocytes with direct compression.[4,5] Agarose transmits the applied mechanical forces to the chondrocytes, which stimulates the cells to produce more extracellular matrix proteins than static controls.[6]

One of the most common natural scaffold materials is collagen, specifically type I collagen. Collagen is an extracellular matrix protein that is the major component in connective tissues. It has been studied intensely because of its abundant presence in native tissues. Of course, it has to be purified like most other natural materials to make it less antigenic before it is seeded with chondrocytes. Studies using type I collagen scaffolds in conjunction with direct compression[7] or cross-linked proteoglycans[8,9] showed better results than controls. However, type I collagen scaffolds alone resulted in the dedifferentiation of seeded chondrocytes.[3] Conversely, the use of type II collagen scaffolds helps retain the cells' phenotypes,[10] but fabricating type II collagen scaffolds is a much more difficult and expensive process.

Chitin and chitosan, which are often derived from crab shells, are semicrystalline polymers that have a high degree of biocompatibility *in vivo*, as reviewed by Hutmacher.[3] The molecular structure of chitosan is similar to many glycosaminoglycans (GAGs), which may give it the ability to have interactions with GFs and adhesion proteins.[3] The degradation characteristics of chitosan are controlled by the degree of deacetylation within the polymer. Unlike alginate and agarose, scaffolds fabricated from chitosan can degrade rapidly *in vivo* depending on the polymer's deacetylation.[3] Researchers can vary the degree of deacetylation to produce a scaffold that degrades over a period of months instead of years so that the new tissue can fill the scaffold's space. The porosity can also be controlled, which affects the scaffold's overall strength and elasticity.[11] Studies based on chitosan scaffolds have been conducted with promising results. Sechriest and associates[12] cross-linked chondroitin sulfate with chitosan to form a scaffold that promotes the chondrocytic phenotype when bovine articular chondrocytes were seeded onto it. Endothelial and smooth muscle cell attachment and growth were seen on a dextran sulfate–chitosan composite, heparin–chitosan composite, and chitosan material alone, but a GAG–chitosan composite inhibited attachment and growth.[13] These results indicate that the inclusion of specific proteoglycans in chitosan scaffolds can dramatically change the overall characteristics of the scaffold. As with other biomaterials, chondrocyte response on chitosan has not been well characterized. However, the response of other cell types shows that it could be a beneficial material for articular cartilage engineering.

Fibrin glue has also been used as a carrier for cells. It is often used in conjunction with other scaffold materials, which will be discussed later in Section 18.2.4. Fibrin glue is made by mixing fibrinogen with thrombin and allowing it to solidify. This material is advantageous because it is completely biodegradable and can be injected. However, injectable carriers such as fibrin glue have little mechanical strength, which is a problem when used in articular cartilage engineering. Researchers have studied chondrocytes in fibrin glue[14] alone and with alginate[15,16] or collagen.[17] Biochemical results did not show a dramatic difference from other scaffold materials.

Hyaluronan (HA) or hyaluronic acid is a polysaccharide that has been used for cartilage engineering applications. HA is a natural material that lubricates articulating joints. It can be cross-linked to form a scaffold that can then be seeded with chondrocytes. One of the main advantages to using HA is that it is also injectable. As with fibrin glue, it can be used in irregularly shaped defects, and implantation is minimally invasive. However, the potential of using HA as a solid, porous scaffold has also been investigated. Grigolo and associates[18] used a HA derivative as a prefabricated scaffold *in vivo* and found histological results to be better than controls. Another *in vivo* experiment by Solchaga and associates[19] found that cross-linked HA sponges produced better histological results and integration with the host tissue than benzylated HA, which was, in turn, better than untreated defects.

18.2.2 Synthetic Polymers

Synthetic scaffold materials are man-made polymers that can be fabricated in a laboratory. Unlike natural polymers, synthetics have the advantage of being flexible in varying their physical and chemical properties. The mechanical and degradation characteristics of a polymer can be altered depending on the chemical composition of the macromolecule. This allows researchers the ability to design a scaffold that degrades over a set period, while still retaining a portion of its strength.

The most prevalent synthetic polymers for scaffolding are polyglycolides, polylactides, and their copolymers.[20–25] These polymers can be fabricated as porous scaffolds or nonwoven meshes and felts. Polyglycolic acid (PGA), probably the most common synthetic polymer used in cartilage engineering, is an alpha polyester that degrades by hydrolytic scission. Total degradation, defined as the time to complete mass loss, usually occurs in 4–12 months, which is short in comparison to other polyesters used as implants.[26] Its degradation products are naturally reabsorbed into the body, making it useful as a biocompatible material for many medical applications. PGA has a tensile strength of 57 MPa and a modulus of 6.5 GPa,[26] but these properties do not translate directly to the scaffold stiffness because the material is formed into a mesh or felt instead of a solid structure. PGA has been used extensively as a suture material when copolymerized with polylactic acid (PLA) and also as a tissue engineering scaffold material. It can be fabricated as a porous scaffold through a salt-leaching process that controls the porosity and interconnectivity of the pores. More commonly, PGA is used as a mesh or felt for cartilage engineering purposes. Since PGA is often extruded into thin polymer strands (~13 µm in diameter),[27] it must be molded into nonwoven mesh discs to be used as a scaffold. This provides a highly porous environment in which cells can be seeded. However, the structure of the scaffold prevents its immediate use in loading–bearing environments. The tissue would have to fill much of the void space to give the construct sufficient mechanical integrity. In articular cartilage engineering studies, Grande and associates[28] found that PGA promoted proteoglycan synthesis to a greater extent than PGA/PLA copolymers or collagen matrices. Freed and associates[27,29,30] have also used PGA scaffolds extensively in their research, obtaining good extracellular matrix production along with predictable polymer degradation.

PLA is another alpha polyester used extensively in the medical field and, as with PGA, has been approved by the food and drug administration (FDA) for implantation in humans. In general, PLA degrades slower than PGA. The total degradation time for PLA ranges from 12 months to over 2 years.[31] Alone, PLA can have a tensile strength between 11.4 and 72 GPa and a modulus ranging from 0.6 to 4 GPa, depending on which isomer is used in the polymer.[26] PLA exists in two stereoisometric forms, which gives rise to four different types of PLA: poly(D-lactide), poly(L-lactide) (PLLA), poly(D, L-lactide), and poly(mesolactide).[32] Poly(D,L-lactide) is an amorphous polymer that is used primarily for drug delivery. However, the D and L monomers form semicrystalline polymers, the materials used primarily in cartilage engineering. Like PGA, PLA is used primarily as a nonwoven mesh for tissue engineering applications, although many researchers have investigated porous PLA scaffolds for their application in orthopedics. When observing cell attachment, Ishaug-Riley and associates[33] found that fewer chondrocytes attached to PLLA than to PGA initially, but both surfaces allowed extensive proliferation of the cells, giving similar total cell numbers at the confluence.

Polylactic-*co*-glycolic acid (PLGA) is a copolymer that is composed of varying ratios of PGA and PLA. The material properties of the copolymer can be controlled depending on the ratio of each monomer present in the macromolecule. For example, a 75/25 ratio of PLA/PGA has a degradation time of

4–5 months, while a 50/50 ratio of the monomers has a total degradation time of only 1–2 months.[31] The modulus of elasticity of these copolymers stays relatively constant at 2 GPa with no major variation due to the monomer ratio.[31] Again, this value is the tensile modulus measured for the solid material, not a scaffold, which would be considerably less stiff. The biocompatibility of PLGA has been proven in numerous clinical trials using it as an implant in large- and small-animal models.[24,34] Like both PGA and PLA, the copolymer can be fabricated as a nonwoven mesh or felt when used as a cartilage-engineering scaffold, as well as a highly porous, solid scaffold. The interconnectivity of the pores, pore size, and void fraction can all be adjusted during the fabrication process to give the construct the proper architecture to function successfully as a tissue-engineering scaffold.[35]

Another synthetic polymer that has been used in cartilage engineering is polycaprolactone (PCL). Its degradation and strength characteristics fit well for orthopedic applications, and it also has good biocompatibility. PCL has a tensile strength between 19 and 27 MPa and a modulus of 0.34 GPa[26] and is most commonly formed into a porous scaffold through a salt-leaching process. Like other polyesters, PCL degrades through hydrolytic scission, but it can take as long as 24 months to totally degrade the material.[31] Because of its resistance to rapid hydrolysis, PCL is often copolymerized with other materials to produce the desired degradation characteristics.[33] Honda and associates[36] used poly (L-lactide-epsilon-caprolactone) as a biodegradable sponge that was implanted into nude mice, and after 4 weeks, histology data showed the formation of cartilage-like structures in the construct.

A newer trend in synthetic polymers is to control and restrict the attachment of cells and proteins on the scaffold. Polyethylene glycol (PEG), also known as polyethylene oxide (PEO), is a polymer that resists the adsorbance of proteins and cells onto its structure because it is very hydrophilic. By using PEG in a copolymer, researchers can take advantage of this property and control the cell attachment characteristics of the scaffold. The primary reason for using PEG is to enhance the biocompatibility of the copolymer.[37] With increased hydrophilicity, antibodies and other proteins find it difficult to attach to the construct, thereby lessening any immune response. PEG alone has a compressive modulus ranging from 200 to 500 kPa, with the higher modulus corresponding to a higher molecular weight.[38] PEG is versatile in its ability to be copolymerized. Work has been done with PEG–PLA constructs,[39] PEG–poly(propylene fumarate) (PPF),[37,40–44] PEG dimethacrylate,[45] and lactide-based PEG networks.[46] These copolymers often have better degradation characteristics than PEG alone, while still retaining good biocompatibility. PEG–PPF samples have been found to lose 40–60% of their mass in 1 week,[41] whereas PEG does not degrade at all over that duration.[44] For articular cartilage engineering, significant degradation of the scaffold is necessary to provide room for new tissue formation. Oftentimes, copolymerization is the only viable solution when working with specific polymers.

18.2.3 Hydrogels

Hydrogels are a subclass of natural and synthetic polymers that are characterized by being composed mainly of fluid and having a network of polymer chains that allow the macroscopic construct to swell significantly when placed in a polar, liquid solution. Alginate, agarose, and fibrin glue are all natural hydrogels, and many PEG-based polymers are classified as synthetic hydrogels. As stated before, the major characteristic of this subclass of polymers is their ability to swell significantly in water. Hydrogels can have up to 99% water by volume while still retaining their shape. This highly hydrated composition is reminiscent of native articular cartilage, both having a large liquid phase that permeates a solid phase, composed of either polymer or extracellular matrix.

Hydrogels are used in cartilage engineering to encapsulate cells and GFs in a polymer network (Figure 18.1).[47] This process effectively immobilizes the cells and encourages differentiation in chondrocytes by forcing them to retain a rounded shape. GFs that have been encapsulated are present throughout the scaffold, removing any complications that might be present due to concentration gradients in the media. By encapsulating the cells in a polymer matrix, researchers aim to replicate the mechanotransduction present in native cartilage. Meshes and felts do not transmit external loads to the cells in the same way

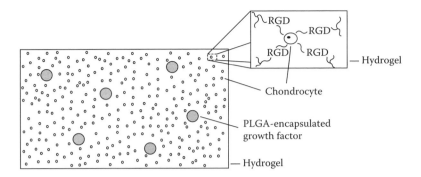

FIGURE 18.1 This is one possible approach to bioactive scaffold design. The base polymer is a hydrogel with RGD covalently attached. PLGA-encapsulated growth factors are dispersed throughout the scaffold. Chondrocytes retain their spherical shape within the hydrogel and are stimulated by the peptides and growth factors.

as hydrogels because stress exists only along the fibers if nothing is present in the void spaces. However, hydrogels exert a controlled compressive force on the encapsulated cells that is similar to physiological conditions.[4]

Experiments focusing on the effect of direct compression on chondrocytes most often use agarose or a similar hydrogel.[6,48] The reason for this is twofold. First, dynamic compression is usually conducted at frequencies of approximately 1 Hz.[1] At this speed, the scaffold material must have sufficient elasticity and recovery from compressive deformation to stay in contact with the loading platen. Hydrogels can easily be fabricated to have this property, whereas a fibrous mesh cannot. Second, as discussed before, the mechanotransduction through a gel may be more similar to native cartilage than other scaffold architectures. Since the overall goal is cartilage regeneration, researchers want to replicate the native loading environment as closely as possible.

Hydrogels are being tested by many researchers for their use in both *in vitro* and *in vivo* cartilage engineering applications.[15,37,41,42,47,49,50] In addition to their suitability for mechanical stimulation studies, hydrogels are ideal as injectable scaffolds because their mass is composed primarily of water.[43] However, major problems still exist when encapsulating cells in a three-dimensional hydrogel. Almost all cross-linking agents are cytotoxic at the levels needed to produce a sufficiently stiff synthetic scaffold. In addition, none of the natural hydrogels have sufficient stiffness to function immediately *in vivo*. The encapsulation process has to be mild enough to have high viability while still producing the necessary mechanical properties. For *in vitro* experiments in general, studies suggest that fibrous, mesh scaffolds support chondrogenesis to a greater degree than hydrogels, indicating that hydrogels might not be the best scaffold material for articular cartilage engineering *in vitro*.[1]

18.2.4 Composite Scaffolds

Composite scaffolds consist of two or more of the previously discussed materials used together to produce a better scaffold. For example, Ameer and associates[51] encapsulated chondrocytes in fibrin glue, which was subsequently used to fill the void volumes of PLGA meshes. This approach produced 2.6 times more GAG (wet weight composition) after 4 weeks than PLGA alone. The authors hypothesized that the fibrin network more effectively retains the GAG within the construct by resisting diffusion out of the construct. However, there was no difference between the amount of collagen in the composite and the PLGA scaffolds.

Marijnissen and associates[52] studied two different composite scaffolds. Cells seeded in alginate were combined with either PLGA or demineralized bone matrix (DBM) and then implanted in nude mice for 8 weeks. Results showed that the PLGA–alginate composite produced type II collagen while the

DBM–alginate composite did not. This shows that the type of material used can have an impact on the type of products secreted by the cells. Scaffold materials should be carefully selected based on the tissue that is to be grown. For example, articular cartilage constructs should contain little or no type I collagen, while meniscal cartilage should mostly have type I collagen. Therefore, to achieve articular cartilage regeneration, materials that stimulate type I collagen and not type II collagen should be avoided.

Fibers, either of the same or different bulk material, have also been used in composite scaffolds to improve structural support. Slivka and associates[53] reinforced PLGA with PGA fibers. The compressive modulus and yield strength of the material was improved by up to 20% using this method. Carbon fibers have also been used with satisfactory results for filling defects *in vivo*. Kus and associates[54] implanted carbon fiber scaffolds in human knee defects and had a 71% success rate, while Brittberg and associates[55] used carbon fiber to fill human knee defects and had an 83% success rate. Both studies based the effectiveness of the procedure on qualitative measures of pain at the defect site after several years.

18.3 Surface Modification

The surface of a scaffold can be modified in several ways to make it more or less attractive to cells. Protein coating, peptide attachment, and micropatterning have been used in various applications to control the attachment of cells on the surface. All these procedures are based on the integrin–receptor properties that exist between cells and extracellular matrix proteins. Chondrocytes express specific integrins, many of which are present on other types of cells.[56] These integrins attach to a corresponding protein, effectively adhering the cell to whatever the protein is adsorbed onto. Studies have shown that $\alpha_1\beta_1$, $\alpha_2\beta_1$, $\alpha_5\beta_1$, $\alpha_V\beta_5$, $\alpha_V\beta_3$, and $\alpha_3\beta_1$ are all integrins that exist on chondrocytes, the latter two being more predominant in the superficial zone of articular cartilage than in the deep zone.[57,58] Modifying biomaterials to take advantage of these integrins allows researchers the ability to control cellular attachment to a scaffold.

18.3.1 Protein Coating

Protein coating is a common method for changing the attachment characteristics of a surface. By using a scaffold material that is hydrophobic, proteins can be readily adsorbed to the surface. The type of coating can affect what cell types attach, along with changing the construct's biocompatibility. The most common adhesion-promoting proteins that are present in the body are collagen, thrombospondin, osteopontin, bone sialoprotein, fibronectin, vitronectin, fibrinogen, von Willebrand factor, laminin, entactin, and tenascin.[56,59] A protein is chosen for surface coating depending on the target cell's integrin receptors. Chondrocytes, based on the integrins present, can attach to collagen, laminin, fibronectin, vitronectin, fibrinogen, bone sialoprotein, thrombospondin, and von Willebrand factor.[60] These extracellular matrix molecules convey mechanical and chemical stimuli in the native tissue. Tissue-engineered constructs should be able to replicate this function, using either supplemental or naturally secreted proteins. Collagen and fibronectin are the most common proteins used for cartilage applications. The reasoning is that if a scaffold is coated with a specific protein, then only cells that express the corresponding integrin can attach to the surface.

Collagen has been discussed previously as a scaffold material. It is used primarily as the main component of the scaffold and not as a coating. However, monolayer studies have shown that chondrocytes readily attach to collagen surfaces as in the native tissue.[61] When type II collagen fragments are introduced to a chondrocyte-on-collagen culture, cell attachment is inhibited, indicating the binding of cell integrins to the fragments rather than the surface.[62] Collagen is much more effective as a matrix material than as a coating; so the number of studies on this application is limited.

Extracellular matrix proteins promote the haptotactic and chemotactic motility of chondrocytes, which can be used to increase the migration of cells into a scaffold.[58] Bhati and associates[63] found that cell ingrowth and attachment was increased when a polymer scaffold was coated with fibronectin. The best results came when a coated scaffold was seeded and cultured in a spinner flask, but even under static conditions, cell attachment was greater for the coated scaffolds. Solchaga and associates[64] used an HA scaffold coated with fibronectin as an implant in rabbits and observed total tissue ingrowth into the plug.

Vitronectin has also been studied as a scaffold coating, although focus has been on the attachment of bone-derived cells rather than chondrocytes. Thomas and associates[65] found that vitronectin, not fibronectin, was the protein that controlled osteoblast attachment and spreading. The media without fibronectin had no effect on cell attachment, but media without vitronectin greatly reduced attachment and spreading. Whether this effect holds for chondrocytes has yet to be determined.

A molecule that has not been studied extensively but holds promise as a coating material is cartilage matrix protein (CMP). As reviewed by Makihira and associates,[66] CMP is a noncollagenous protein that is expressed almost exclusively in cartilage. It binds to aggrecan and type II collagen and has been found to involve the $\alpha_1\beta_1$ integrin during chondrocyte adhesion. Makihira and associates[66] studied the adhesion and spreading of chondrocytes on a CMP-coated dish. CMP enhanced both attachment and spreading on the surface, and there was an even more dramatic increase when type II collagen was added to the media. Makihira and associates propose that CMP would be more appropriate in cartilage applications than other proteins because it is specific to chondrocytes.

Protein adsorbance can be varied depending on the type of base material for the scaffold. Unless the protein is cross-linked to the surface, which would likely reduce its effectiveness, it will eventually disassociate from the scaffold. The more hydrophobic the material, the stronger the attachment of the protein; so most base scaffold materials for coating applications are very hydrophobic. For two-dimensional surfaces, the protein adsorbance can be controlled using self-assembled monolayers. The surface chemistry can be varied to produce hydrophobic and hydrophilic regions, which will change the protein adsorbance to the surface.[67] For most three-dimensional applications, the only way to modulate surface adsorbance is to choose an appropriate base material for the scaffold.

For some applications, only two dimensions are needed. As far as cartilage engineering is concerned, a functional scaffold has to be three-dimensional. However, many preliminary experiments are carried out in the monolayer. If the concept of protein coating is taken further, then cell patterns can be made on a material's surface. Areas of the material can be coated while others are left blank. If the underlying material does not promote attachment of cells by itself, then the cells will only attach to the areas that have the protein. This concept can be implemented using full proteins, but more often, short peptide sequences are used instead.

18.3.2 Peptide Attachment

Peptides are short amino acid sequences derived from large, adhesion proteins. The reason cells attach to a protein-coated surface is because of amino acid recognition sequences located in the protein structure. Researchers have determined which sequences are recognized by cells and use that knowledge to create scaffolds incorporating peptide sequences instead of full proteins. The major advantage of this technique is that a peptide can be permanently attached to a material without seriously affecting its function. Proteins on the other hand can lift off from a surface in an unpredictable manner. In general, the attachment and surface properties of a biomaterial can be strictly controlled using peptides instead of full proteins.

A peptide is most often bound to a material using a covalent bond. This securely attaches the sequence to the surface, assuring localized effects due to the peptide. One problem that can occur is a reduction in peptide activity. By attaching the peptide to a linker chain, it can be moved away from the surface of the material and allowed more flexibility in its function.[56] The biomaterials chosen as base scaffolds often have properties that inhibit the adherence of cells and proteins. This forces the attachment characteristics of the material to be totally dependent on the bound peptides. However, the steric hindrance of the base material requires peptides to be far enough away from the surface for binding between cellular integrins and peptides to occur.

Another design parameter is the density of peptides on the material. Increasing the concentration of peptides will increase cellular attachment, but it will also decrease cell motility on the scaffold.[56] A balance between promoting attachment and allowing cell motility has to be found for specific cell types and applications. Migration of cells into a material will not occur if the peptide concentrations are too high. However, a migration response to extracellular matrix proteins can be mediated by clustering integrins on the material.[58] This approach can be useful if an implant wants to encourage the ingrowth of native cells.

The most common peptide sequence used for cell attachment is Arg–Gly–Asp (RGD), which corresponds to a portion of the fibronectin protein.[56] It is present in many species of both plants and animals, which hints at its longevity in biological life.[68] RGD is useful in tissue engineering because it has an almost generic attachment with many integrins. Of the 20 known integrins, 8–12 of them recognize the RGD sequence in their ligands.[59] In chondrocytes, RGD binds strongly to $\alpha_5\beta_1$, $\alpha_v\beta_5$, and $\alpha_v\beta_3$ and weakly to $\alpha_3\beta_1$. This corresponds with both the superficial and deep-zone chondrocytes present in cartilage.

Many peptides used for tissue engineering incorporate the RGD sequence but are more than three amino acids long. Researchers have used peptide sequences such as **GRGD**,[69] **GRGDSP**,[70] and CGGNGEPRGDTYRAY[71] to utilize the benefits of the RGD sequence. A peptide is chosen based on what protein it is modeled after (the last sequence is from bone sialoprotein) and how it will be attached to the scaffold. For example, a short peptide sequence such as GRGD can be used on the end of hydrophilic spacers when modifying a hydrophobic bulk material.[69]

Various attachment peptides have been studied extensively for use in tissue engineering and drug-delivery applications. Some of the recognition sequences that have known activities are: KGD ($\alpha_{IIb}\beta_3$), PECAM ($\alpha_v\beta_3$), KQAGDV ($\alpha_{IIb}\beta_3$), LDV ($\alpha_4\beta_1$ $\alpha_4\beta_7$) YGYYGDALR and FYFDLR ($\alpha_2\beta_1$), and RLD/KRLDGS ($\alpha_v\beta_3$ and $\alpha_M\beta_2$).[59,68,72] Other peptide sequences have been studied, but specific integrin relationships were not found. For chondrocytes, the functional sequences of those listed (not including RGD) are PECAM, YGYYGDALR, FYFDLR, and RLD/KRLDGS with weak binding occurring with KQAGDV and LDV.[59]

The attachment of peptides to a biomaterial can be controlled to form desired patterns on the surface. This process is known as micropatterning and has been used for attachment studies in the monolayer.[73–76] The peptides are bound to the surface in such a way as to restrict either a single cell's shape or a community of cells' pattern. Areas that have the peptide will allow integrin binding, while blank areas will not allow cell attachment. If the peptides are cell specific, then different regions will promote attachment for different cells.

Micropatterning is often used to control the shape of a cell by restricting the degree of spreading. A cell's shape governs whether it will grow, die, or differentiate.[76] When a cell spreads more, there is an increase in cell survival and proliferation, but when a cell cannot spread and balls up on a surface, it enters apoptosis and dies. A patterned surface that supports neither growth nor apoptosis will induce the cells to differentiate.[75] A cell at this stage will express proteins and produce extracellular matrix in greater amounts than in either other state. Physically, the spreading is controlled by placing spacer ligands apart at different distances. The cell will spread further if the ligands are spaced apart than if they are grouped in one small clump. If the spacers are too far apart, then the cells will not spread because the focal adhesions cannot be made.

Proliferation is controlled in the same manner. If the peptide concentration is too high, then the cells do not move apart because they are bound too tightly to the surface, leaving little room for cells produced through division. If the peptide concentration is too low, then the cells still cannot migrate because there is no place to form focal adhesions, resulting in overcrowding like before. A balance has to be found to allow sufficient migration of the cells over the entire surface.

Although the majority of peptide studies have been conducted on two-dimensional surfaces, there is evidence suggesting that peptides are also beneficial in three-dimensional scaffolds.[77,78] Alginate that contained the RGD peptide promoted cell adhesion and spreading as well as increased differentiation compared to controls.[79] However, Mann and associates[80,81] found that PEG hydrogels modified with adhesion peptides decreased proliferation throughout the construct and decreased extracellular matrix secretions as well. However, this result could be caused by using too high a concentration of peptides. Future research can investigate this hypothesis. Unlike hydrogels, porous scaffolds have surfaces that can be modified similarly to two-dimensional materials. Research on the benefits of peptides in these situations is limited, but it appears that it would be just as beneficial as in two dimensions.

18.4 Growth Factor Inclusion

Another way that a scaffold can be modified is to include GFs. Proteins such as TGF-β, IGF, PDGF, HGF, and FGF all produce an effect on the proliferation and differentiation of chondrocytes. These proteins will likely be required to engineer functional articular cartilage. GFs stimulate cells to secrete extracellular matrix in greater amounts than without them. For example, Bonnassar and associates[82] used IGF-I to increase protein and proteoglycan synthesis in cartilage explants 90% and 120%, respectively. Other experiments confirm the dramatic effect of GFs on extracellular matrix synthesis in explants and cell-seeded constructs.

The results due to GFs make their use a certainty in future cartilage engineering studies. Problems arise, however, when attempting to include GFs in a cartilage scaffold. In *in vitro* experiments, GFs can be added to the media to stimulate the cells, but under *in vivo* conditions, GFs must somehow be incorporated into the scaffold for extended effects. The *in vitro* experiments might also benefit from the inclusion of GFs in the scaffold because stimulants would be dispersed evenly throughout the construct, whereas before, diffusion might keep a significant amount of the GFs outside the scaffold.

There are three main methods of GF inclusion: encapsulation, covalent bonding, and in solution. Encapsulation requires the use of a hydrogel either as the entire scaffold or as a GF carrier. This means that the whole construct can be made of the same hydrogel, such as alginate or PEG, or only one component of a composite scaffold is a hydrogel, such as using alginate gel in a PGA felt. Nonhydrogel scaffolds by themselves usually have too high porosity to effectively retain GFs in the construct, but Athanasiou and associates[24] used physical entrapment to load TGF-β into PLGA scaffolds (65–75% porous) and obtained positive results. The most common method of including a GF in a hydrogel is to load it into a microparticle and then encapsulate the microparticle in the hydrogel.[47] The protein is released into the hydrogel in a predictable fashion, thereby stimulating the chondrocytes seeded in the scaffold. Microparticle encapsulation can also be used in fibrous scaffolds as a form of GF delivery.

Covalent bonding is another way to include a GF in a scaffold. Immobilizing proteins often decreases their effectiveness because access to the active regions of the molecule is restricted, but Mann and associates[83] covalently bonded TGF-β1 to PEG and found no decrease in its ability to stimulate the cells. This approach appears to be an effective method for physical inclusion of the GF in the scaffold if the material is not a hydrogel. The difficulty lies in attaching GFs to various base materials while not losing their activity and ability to affect matrix synthesis.

Including GFs in the media solution is the most common and simplest method of stimulation for *in vitro* experiments. This approach obviously would not work for an *in vivo* study. The GF is added in controlled amounts to the media and replaced along with the media. The activity of the GF does not decrease because it is continuously replaced, which is not the case in the other two methods. However, a larger amount of GF is used because of its continuous replenishment, making this approach expensive for use in large, fluid-dominated bioreactors.

Including GFs in the articular cartilage engineering process is undoubtedly a must, but it still remains to be seen which GFs are best for chondrogenesis. Certain GFs might stimulate proliferation, while others promote synthesis of the various extracellular molecules. Blunk and associates[84] found that IGF-I increased GAG fractions, TGF-β1 increased total collagen, and IL-4 minimized the GAG-depleted region on an engineered constructs surface. It is possible that several GFs will be necessary to encourage seeded chondrocytes into producing native levels of extracellular matrix. Whether they are included in the solution, encapsulated in a gel, or covalently bonded to a scaffold, GFs have to be a part of a bioactive scaffold design.

18.5 Conclusion

The creation of a bioactive scaffold will increase the chances of producing a functional articular cartilage construct *in vitro* or *in vivo*. By controlling the attachment of cells and stimulating their differentiation, scaffolds can promote the formation of the correct type of tissue in the correct location. Initially, the materials have to be chosen based on the characteristics that are desired. Degradation time, strength, and elasticity are the physical properties that can be controlled by using various base materials for the scaffold. To a limited degree, attachment and differentiation can also be affected by the scaffold material. However, the surface properties are more effective when the biomaterial is modified with proteins or peptides. Biocompatibility and cell attachment are the two major benefits to altering the surface of the scaffold. Some results also indicate an effect on proliferation and differentiation. To get the most out of the seeded cells, GFs should be included in the scaffold design. They can be added as a supplement in the media, but incorporating them into the actual scaffold will assure an interaction between the GFs and the cells. The combinations and permutations available between the different parameters described in this chapter leave plenty of room for future research. By using the proper combination of scaffold components, researchers will move closer to the goal of successfully engineering a functional articular cartilage construct.

References

1. Darling, E.M. and Athanasiou, K.A., Articular cartilage bioreactors and bioprocesses, *Tissue Eng.*, 9, 9, 2003.
2. Murphy, C.L. and Sambanis, A., Effect of oxygen tension and alginate encapsulation on restoration of the differentiated phenotype of passaged chondrocytes, *Tissue Eng.*, 7, 791, 2001.
3. Hutmacher, D.W., Scaffold design and fabrication technologies for engineering tissues—State of the art and future perspectives, *J. Biomater. Sci. Polym. Ed.*, 12, 107, 2001.
4. Saris, D.B., Mukherjee, N., Berglund, L.J., Schultz, E.M., An, K.N., and O'Driscoll, S.W., Dynamic pressure transmission through agarose gels, *Tissue Eng.*, 6, 531, 2000.
5. Elder, S.H., Goldstein, S.A., Kimura, J.H., Soslowsky, L.J., and Spengler, D.M., Chondrocyte differentiation is modulated by frequency and duration of cyclic compressive loading, *Ann. Biomed. Eng.*, 29, 476, 2001.
6. Mauck, R.L., Soltz, M.A., Wang, C.C., Wong, D.D., Chao, P.H., Valhmu, W.B., Hung, C.T., and Ateshian, G.A., Functional tissue engineering of articular cartilage through dynamic loading of chondrocyte-seeded agarose gels, *J. Biomech. Eng.*, 122, 252, 2000.
7. Hunter, C.J., Imler, S.M., Malaviya, P., Nerem, R.M., and Levenston, M.E., Mechanical compression alters gene expression and extracellular matrix synthesis by chondrocytes cultured in collagen I gels, *Biomaterials*, 23, 1249, 2002.
8. Lee, C.R., Breinan, H.A., Nehrer, S., and Spector, M., Articular cartilage chondrocytes in type I and type II collagen-GAG matrices exhibit contractile behavior *in vitro*, *Tissue Eng.*, 6, 555, 2000.
9. van Susante, J.L.C., Pieper, J., Buma, P., van Kuppevelt, T.H., van Beuningen, H., van Der Kraan, P.M., Veerkamp, J.H., van den Berg, W.B., and Veth, R.P.H., Linkage of chondroitin-sulfate to type I collagen scaffolds stimulates the bioactivity of seeded chondrocytes *in vitro*, *Biomaterials*, 22, 2359, 2001.
10. Nehrer, S., Breinan, H.A., Ramappa, A., Shortkroff, S., Young, G., Minas, T., Sledge, C.B., Yannas, I.V., and Spector, M., Canine chondrocytes seeded in type I and type II collagen implants investigated *in vitro*, *J. Biomed. Mater. Res.*, 38, 95, 1997.

11. Madihally, S.V. and Matthew, H.W., Porous chitosan scaffolds for tissue engineering, *Biomaterials*, 20, 1133, 1999.

12. Sechriest, V.F., Miao, Y.J., Niyibizi, C., Westerhausen-Larson, A., Matthew, H.W, Evans, C.H., Fu, F.H., and Suh, J.K., GAG-augmented polysaccharide hydrogel: A novel biocompatible and biodegradable material to support chondrogenesis, *J. Biomed. Mater. Res.*, 49, 534, 2000.

13. Chupa, J.M., Foster, A.M., Sumner, S.R., Madihally, S.V., and Matthew, H.W., Vascular cell responses to polysaccharide materials: *In vitro* and *in vivo* evaluations, *Biomaterials*, 21, 2315, 2000.

14. Silverman, R.P., Passaretti, D., Huang, W., Randolph, M.A., and Yaremchuk, M.J., Injectable tissue-engineered cartilage using a fibrin glue polymer, *Plast. Reconstr. Surg.*, 103, 1809, 1999.

15. Almqvist, K.F., Wang, L., Wang, J., Baeten, D., Cornelissen, M., Verdonk, R., Veys, E.M., and Verbruggen, G., Culture of chondrocytes in alginate surrounded by fibrin gel: Characteristics of the cells over a period of eight weeks, *Ann. Rheum. Dis.*, 60, 781, 2001.

16. Perka, C., Arnold, U., Spitzer, R.S., and Lindenhayn, K., The use of fibrin beads for tissue engineering and subsequential transplantation, *Tissue Eng.*, 7, 359, 2001.

17. Perka, C., Schultz, O., Lindenhayn, K., Spitzer, R.S., Muschik, M., Sittinger, M., and Burmester, G.R., Joint cartilage repair with transplantation of embryonic chondrocytes embedded in collagen-fibrin matrices, *Clin. Exp. Rheumatol.*, 18, 13, 2000.

18. Grigolo, B., Roseti, L., Fiorini, M., Fini, M., Giavaresi, G., Aldini, N.N., Giardino, R., and Facchini, A., Transplantation of chondrocytes seeded on a hyaluronan derivative (hyaff-11) into cartilage defects in rabbits, *Biomaterials*, 22, 2417, 2001.

19. Solchaga, L.A., Yoo, J.U., Lundberg, M., Dennis, J.E., Huibregtse, B.A., Goldberg, V.M., and Caplan, A.I., Hyaluronan-based polymers in the treatment of osteochondral defects, *J. Orthop. Res.*, 18, 773, 2000.

20. Athanasiou, K.A., Agrawal, C.M., Barber, F.A., and Burkhart, S.S., Orthopaedic applications for PLA-PGA biodegradable polymers, *Arthroscopy*, 14, 726, 1998.

21. Athanasiou, K.A., Niederauer, G.G., Agrawal, C.M., and Landsman, A.S., Applications of biodegradable lactides and glycolides in podiatry, *Clin. Pediatr. Med. Surg.*, 12, 475, 1995.

22. Athanasiou, K.A., Singhal, A.R., Agrawal, C.M., and Boyan, B.D., *In vitro* degradation and release characteristics of biodegradable implants containing trypsin inhibitor, *Clin. Orthop.*, (315), 272, 1995.

23. Athanasiou, K.A., Niederauer, G.G., and Agrawal, C.M., Sterilization, toxicity, biocompatibility and clinical applications of polylactic acid/polyglycolic acid copolymers, *Biomaterials*, 17, 93, 1996.

24. Athanasiou, K.A., Korvick, D., and Schenck, R.C., Biodegradable implants for the treatment of osteochondral defects in a goat model, *Tissue Eng.*, 3, 363, 1997.

25. Vunjak-Novakovic, G. and Freed, L.E., Culture of organized cell communities, *Adv. Drug Deliv. Rev.*, 33, 15, 1998.

26. Daniels, A.U., Chang, M.K., and Andriano, K.P., Mechanical properties of biodegradable polymers and composites proposed for internal fixation of bone, *J. Appl. Biomater.*, 1, 57, 1990.

27. Freed, L.E., Grande, D.A., Lingbin, Z., Emmanual, J., Marquis, J.C., and Langer, R., Joint resurfacing using allograft chondrocytes and synthetic biodegradable polymer scaffolds, *J. Biomed. Mater. Res.*, 28, 891, 1994.

28. Grande, D.A., Halberstadt, C., Naughton, G., Schwartz, R., and Manji, R., Evaluation of matrix scaffolds for tissue engineering of articular cartilage grafts, *J. Biomed. Mater. Res.*, 34, 211, 1997.

29. Freed, L.E., Hollander, A.P., Martin, I., Barry, J.R., Langer, R., and Vunjak-Novakovic, G., Chondrogenesis in a cell–polymer–bioreactor system, *Exp. Cell Res.*, 240, 58, 1998.

30. Freed, L.E., Vunjak-Novakovic, G., and Langer, R., Cultivation of cell-polymer cartilage implants in bioreactors, *J. Cell Biochem.*, 51, 257, 1993.

31. Middleton, J.C. and Tipton, A.J., Synthetic biodegradable polymers as orthopedic devices, *Biomaterials*, 21, 2335, 2000.

32. Hutmacher, D.W., Goh, J. C., and Teoh, S.H., An introduction to biodegradable materials for tissue engineering applications, *Ann. Acad. Med. Singapore*, 30, 183, 2001.

33. Ishaug-Riley, S.L., Okun, L.E., Prado, G., Applegate, M.A., and Ratcliffe, A., Human articular chondrocyte adhesion and proliferation on synthetic biodegradable polymer films, *Biomaterials*, 20, 2245, 1999.

34. Eid, K., Chen, E., Griffith, L., and Glowacki, J., Effect of RGD coating on osteocompatibility of PLGA-polymer disks in a rat tibial wound, *J. Biomed. Mater. Res.*, 57, 224, 2001.

35. Zeltinger, J., Sherwood, J.K., Graham, D.A., Muller, R., and Griffith, L.G., Effect pore size and void fraction on cellular adhesion, proliferaion, and matrix desposition, *Tissue Eng.*, 7, 557, 2001.

36. Honda, M., Yada, T., Ueda, M., and Kimata, K., Cartilage formation by cultured chondrocytes in a new scaffold made of poly (L-lactide-epsilon-caprolactone) sponge, *J. Oral Maxillofac. Surg.*, 58, 767, 2000.

37. Suggs, L.J., Shive, M.S., Garcia, C.A., Anderson, J.M., and Mikos, A.G., *In vitro* cytotoxicity and *in vivo* biocompatibility of poly (propylene fumarate-co-ethylene glycol) hydrogels, *J. Biomed. Mater. Res.*, 46, 22, 1999.

38. Zimmermann, J., Bittner, K., Stark, B., and Mulhaupt, R., Novel hydrogels as supports for *in vitro* cell growth: Poly(ethylene glycol)- and gelatine-based (meth)acrylamidopeptide macromonomers, *Biomaterials*, 23, 2127, 2002.

39. Anseth, K.S., Metters, A.T., Bryant, S.J., Martens, P.J., Elisseff, J.H., and Bowman, C.N., *In situ* forming degradable networks and their application in tissue engineering and drug delivery, *J. Control Release*, 78, 199, 2002.

40. Fisher, J.P., Vehof, J.W., Dean D., van der Waerden, J.P., Holland, T.A., Mikos, A.G., and Jansen, J.A., Soft and hard tissue response to photocrosslinked poly(propylene fumarate) scaffolds in a rabbit model, *J. Biomed. Mater. Res.*, 59, 547, 2002.

41. Suggs, L.J., Krishnan, R.S., Garcia, C.A., Peter, S.J., Anderson, J.M., and Mikos, A.G., *In vitro* and *in vivo* degradation of poly(propylene fumarate-*co*-ethylene glycol) hydrogels, *J. Biomed. Mater. Res.*, 42, 312, 1998.

42. Suggs, L.J., Kao, E.Y., Palombo, L.L., Krishnan, R.S., Widmer, M.S., and Mikos, A.G., Preparation and characterization of poly (propylene fumarate-*co*-ethylene glycol) hydrogels, *J. Biomater. Sci. Polym. Ed.*, 9, 653, 1998.

43. Suggs, L.J. and Mikos, A.G. Development of poly(propylene fumarate-*co*-ethylene glycol) as an injectable carrier for endothelial cells, *Cell Transplant.*, 8, 345, 1999.

44. Temenoff, J.S., Anthanasiou, K.A., Lebaron, R.G., and Mikos, A.G., Effect of poly(ethylene glycol) molecular weight on tensile and swelling properties of oligo(poly(ethylene glycol) fumarate) hydrogels for cartilage tissue engineering, *J. Biomed. Mater. Res.*, 59, 429, 2002.

45. Bryant, S.J. and Anseth, K.S., The effects of scaffold thickness on tissue engineered cartilage in photocrosslinked poly(ethylene oxide) hydrogels, *Biomaterials*, 22, 619, 2001.

46. Han, D.K., Park, K.D., Hubbell, J.A., and Kim, Y.H., Surface characteristics and biocompatibility of lactide-based poly(ethylene glycol) scaffolds for tissue engineering, *J. Biomater. Sci. Polym. Ed.*, 9, 667, 1998.

47. Elisseeff, J., McIntosh, W., Anseth K., Riley, S., Ragan, P., and Langer, R., Photoencapsulation of chondrocytes in poly (ethylene oxide)-based semi-interpenetrating networks, *J. Biomed. Mater. Res.*, 51, 164, 2000.

48. Buschmann, M.D., Gluzband, Y.A., Grodzinsky, A.J., and Hunziker, E.B., Mechanical compression modulates matrix biosynthesis in chondrocyte/agarose culture, *J. Cell Sci.*, 108, 1497, 1995.

49. Bryant, S.J. and Anseth, K.S., Hydrogel properties influence ECM production by chondrocytes photoencapsulated in poly (ethylene glycol) hydrogels, *J. Biomed. Mater. Res.*, 59, 63, 2002.

50. Marijnissed, W.J., van Osch, G.J., Aigner, J., van der Veen, S.W., Hollander, A.P., Verwoerd-Verhoef, H.L., Woven scaffold for cartilage tissue engineering, *Biomaterials*, 23, 1511, 2002.

51. Ameer, G.A., Mahmood, T.A., and Langer, R., A biodegradable composite scaffold for cell transplantation, *J. Orthop. Res.*, 20, 16, 2002.

52. Marijnissen, W.J., van Osch, G.J., Aigner, J., Verwoerd-Verhoef, H.L., and Verhaar, J.A., Tissue-engineered cartilage using serially passaged articular chondrocytes. Chondrocytes in alginate, combined *in vivo* with a synthetic (E210) or biologic biodegradable carrier (DBM), *Biomaterials*, 21, 571, 2000.

53. Slivka, M.A., Leatherbury, N.C., Kieswetter, K., and Niederauer, G.G., Porous, resorbable, fiber-reinforced scaffolds tailored for articular cartilage repair, *Tissue Eng.*, 7, 767, 2001.

54. Kus, W.M., Gorecki, A., Strzelczyk, P., and Swiader, P., Carbon fiber scaffolds in the surgical treatment of cartilage lesions, *Ann. Transplant.*, 4, 101, 1999.

55. Brittberg, M., Faxen, E., and Peterson, L., Carbon fiber scaffolds in the treatment of early knee osteoarthritis. A prospective 4-year followup of 37 patients, *Clin. Orthop.*, (307), 155, 1994.

56. LeBaron, R.G. and Athanasiou, K.A., Extracellular matrix cell adhesion peptides: Functional applications in orthopedic materials, *Tissue Eng.*, 6, 85, 2000.

57. Woods, V.L., Jr., Schreck, P.J., Gesink, D.S., Pacheco, H.O., Amiel, D., Akeson, W.H., and Lotz, M., Integrin expression by human articular chondrocytes, *Arthritis Rheum.*, 37, 537, 1994.

58. Shimizu, M., Minakuchi, K., Kaji, S., and Koga, J., Chondrocyte migration to fibronectin, type I collagen, and type II collagen, *Cell Struct. Funct.*, 22, 309, 1997.

59. Ruoslahti, E., RGD and other recognition sequences for integrins, *Annu. Rev. Cell Dev. Biol.*, 12, 697, 1996.

60. Hubbell, J.A., Matrix effects, In *Principles of Tissue Engineering*, Lanza, R., Langer, R., and Chick, W., eds., R.G. Landes Company, Austin, 1997.

61. Lee, V., Cao, L., Zhang, Y., Kiani, C., Adams, M.E., and Yang, B.B., The roles of matrix molecules in mediating chondrocyte aggregation, attachment, and spreading, *J. Cell Biochem.*, 79, 322, 2000.

62. Jennings, L., Wu, L., King, K.B., Hammerle, H., Cs-Szabo, G., and Mollenhauer, J., The effects of collagen fragments on the extracellular matrix metabolism of bovine and human chondrocytes, *Connect. Tissue Res.*, 42, 71, 2001.

63. Bhati, R.S., Mukherjee, D.P., McCarthy, K.J., Rogers, S.H., Smith, D.E, and Shalaby, S.W., The growth of chondrocytes into a fibronectin-coated biodegradable scaffold, *J. Biomed. Mater. Res.*, 56, 74, 2001.

64. Solchaga, L.A., Gao, J., Dennis, J.E., Awadallah, A., Lundberg, M., Caplan, A.I., and Goldberg, V.M., Treatment of osteochondral defects with autologous bone marrow in a hyaluronan-based delivery vehicle, *Tissue Eng.*, 8, 333, 2002.

65. Thomas, C.H., McFarland, C.D., Jenkins, M.L., Rezania, A., Steele, J.G., and Healy, K.E., The role of vitronectin in the attachment and spatial distribution of bone-derived cells on materials with patterned surface chemistry, *J. Biomed. Mater. Res.*, 37, 81, 1997.

66. Makihira, S., Yan, W., Ohno, S., Kawamoto, T., Fujimoto, K., Okimura, A., Yoshida, E., Noshiro, M., Hamada, T., and Kato, Y., Enhancement of cell adhesion and spreading by a cartilage-specific noncollagenous protein, cartilage matrix protein (CMP/Matrilin-1), via integrin alpha 1 beta 1, *J. Biol. Chem.*, 274, 11417, 1999.

67. Prime, K.L. and Whitesides, G.M., Self-assembled organic monolayers: Model systems for studying adsorption of proteins at surfaces, *Science*, 252, 1164, 1991.

68. Koivunen, E., Wang, B., Dickinson, C.D., and Ruoslahti, E., Peptides in cell adhesion research, *Methods Enzymol.*, 245, 346, 1994.

69. Jo, S., Engel, P.S., and Mikos, A.G., Synthesis of poly(ethylene glycol)-tethered poly(propylene fumarate) and its modification with GRGD peptide, *Polymer*, 41, 7595, 2000.

70. Pierschbacher, M.D. and Ruoslahti, E., Variants of the cell recognition site of fibronectin that retain attachment-promoting activity, *Proc. Natl. Acad. Sci. USA*, 81, 5985, 1984.

71. Rezania, A., Thomas, C.H., Branger, A.B., Waters, C.M., and Healy, K.E., The detachment strength and morphology of bone cells contacting materials modified with a peptide sequence found within bone sialoprotein, *J. Biomed. Mater. Res.*, 37, 9, 1997.

72. Pasqualini, R., Koivunen, E., and Ruoslahti, E., Peptides in cell adhesion: Powerful tools for the study of integrin–ligand interactions, *Braz. J. Med. Biol. Res.*, 29, 1151, 1996.

73. Britland, S., Clark, P., Connolly, P., and Moores, G., Micropatterned substratum adhesiveness: A model for morphogenetic cues controlling cell behavior, *Exp. Cell Res.*, 198, 124, 1992.

74. Lorn, B., Healy, K.E., and Hockberger, P.E., A versatile technique for patterning biomolecules onto glass coverslips, *J. Neurosci. Methods*, 50, 385, 1993.

75. Dike, L.E., Chen, C.S., Mrksich, M., Tien, J., Whitesides, G.M., and Ingber, D.E., Geometric control of switching between growth, apoptosis, and differentiation during angiogenesis using micropatterned substrates, *in vitro, Cell Dev. Biol. Anim.*, 35, 441, 1999.

76. Chen, C.S., Mrksich, M., Huang, S., Whitesides, G.M., and Ingber, D.E., Geometric control of cell life and death, *Science*, 276, 1425, 1997.

77. Mann, B.K., Gobin, A.S., Tsai, A.T., Schmedlen, R.H., and West, J.L., Smooth muscle cell growth in photopolymerized hydrogels with cell adhesive and proteolytically degradable domains: Synthetic ECM analogs for tissue engineering, *Biomaterials*, 22, 3045, 2001.

78. Gobin, A.S. and West, J.L., Cell migration through defined, synthetic ECM analogs, *FASEB J.*, 16, 751, 2002.

79. Alsberg, E., Anderson, K.W., Albeiruti, A., Franceschi, R.T., and Mooney, D.J., Cell-interactive alginate hydrogels for bone tissue engineering, *J. Dent. Res.*, 80, 2025, 2001.

80. Mann, B.K. and West, J. L. Cell adhesion peptides alter smooth muscle cell adhesion, proliferation, migration, and matrix protein synthesis on modified surfaces and in polymer scaffolds, *J. Biomed. Mater. Res.*, 60, 86, 2002.

81. Mann, B.K., Tsai, A.T., Scott-Burden, T., and West, J.L., Modification of surfaces with cell adhesion peptides alters extracellular matrix deposition, *Biomaterials*, 20, 2281, 1999.

82. Bonassar, L.J., Grodzinsky, A.J., Frank, E.H., Davila, S.G., Bhaktav, N.R., and Trippel, S.B., The effect of dynamic compression on the response of articular cartilage to insulin-like growth factor-I, *J. Orthop. Res.*, 19, 11, 2001.

83. Mann, B.K., Schmedlen, R.H., and West, J.L., Tethered-TGF-beta increases extracellular matrix production of vascular smooth muscle cells, *Biomaterials*, 22, 439, 2001.

84. Blunk, T., Sieminski, A.L., Gooch, K.J., Courter, D.L., Hollander, A.P., Nahir, A.M., Langer, R., Vunjak-Novakovic, G., and Freed, L.E., Differential effects of growth factors on tissue-engineered cartilage, *Tissue Eng.*, 8, 73, 2002.

V

Interventional Disease Treatment

Contributions of Technology to Direct Disease Treatment

The role of technology in healthcare (Make consistent throughout) certainly goes far beyond the initial diagnosis of disease. It is only basic human nature to desire to fix a problem once it has been identified. Exciting new technologies for the direct treatment of disease have thus been readily accepted.

There is perhaps no direct healthcare interface with a higher need for reliable technology development than the surgical suite. The act of entering the human body for direct manipulation cannot be undertaken without tremendous care. A wide variety of technologies have made surgical procedures more reliable, less traumatic, and shorter in duration. The savings in mortality and morbidity are added to the direct savings from less-expensive procedures.

The nature of any surgical procedure requires constant monitoring of the patient during and after the procedure. While reliable monitoring technologies have allowed the surgeon to pay more attention to the procedure itself, human observation is still required. This has resulted in an interesting hybrid of automated devices that provide constant feedback and controllability of anesthetics.

Advances in materials, know-how have led to the development of fantastic technologies that allow for the minimally invasive treatment of a variety of diseases. The ability to design miniature devices that are deployed via small catheters has revolutionized patient care. Procedures that once took hours of highly invasive surgery and required weeks of recovery are now being replaced with nearly outpatient-delivered procedures. The pharmaceutical industry has been quick in adapting many of the technologies presented in other sections of this book in order to develop new medicines. New chemical pathways to attack disease are being identified and are utilized to produce astonishing results.

The continued development of treatment technologies will provide tremendous benefits for patients in terms of improved safety, shorter hospital stay, improved post-procedure care, and faster recovery; while at the same time, healthcare providers will enjoy the benefits of increased efficiency, reliability, and speed.

19

Anesthesia Monitoring Devices

Spencer S. Kee
University of Texas

Elizabeth Rebello
University of Texas

Jagtar Singh Heir
University of Texas

Ronaldo V.
Purugganan
University of Texas

19.1 Introduction

Standards for Basic Anesthetic Monitoring have been published by the American Society of Anesthesiologists and are effective since July 1, 2011 [1 2011 #74]. The first standard is the presence of qualified anesthesia personnel. The next standard is continuous monitoring of the patient's oxygen, ventilation, circulation, and temperature. The layout of this chapter will look at each monitoring mode before considering additional ancillary devices. Both standards are required for the practice of safe anesthesia. Appropriate monitoring of the patient allows the anesthesia personnel to anticipate and correct the patient's deviation from normality with judicious intervention.

19.2 Oxygenation

Measurement of the oxygen concentration in the inspired gases and in the patient's blood will be dealt with in this section. The foundation of measuring oxygen in a gaseous sample depends on measuring a property of oxygen that distinguishes itself from the other gases present.

19.2.1 Oxygen Measurement in Gaseous Form

19.2.1.1 Raman Spectrometer

When a photon collides with a molecule, there is usually zero energy transfer and the photon is re-emitted with the same wavelength and frequency. However, about 1 in 10 million (10^7) collisions results in some energy absorption by the molecule and the photon is reemitted with a lower frequency and longer wavelength (lower energy). The lower frequency emission is known as Raman scattering. The change in the frequency is related to the intermolecular bonding and thus identifies the different components present in the gas mixture. Mapping each lower frequency and its intensity that is Raman scattered will plot the component gases present and their relative concentrations. This analyzer requires a laser to provide a high concentration of photons to collide with the gas molecules so that the resulting lower frequencies can be detected [2]. The response time of the analyzer is fast enough to give breath by breath detail; however, cost and maintenance relegate this analyzer to laboratory work rather than routine clinical practice.

19.2.1.2 Paramagnetic Oxygen Analyzer

Paramagnetism is a property where a substance is rendered attractive to an applied magnetic field but when the field is discontinued, the property disappears. Diamagnetism is a property where a substance is rendered repellent to an applied magnetic field and the property is lost as soon as the field is discontinued. Oxygen is paramagnetic, whereas nitrogen is diamagnetic; the paramagnetic oxygen analyzer takes advantage of these two properties. A nitrogen-filled dumbbell is suspended to lie inside a magnetic field. The suspending wire has a mounted mirror so that any movement displacing the dumbbell sphere away from the magnetic field is exaggerated. When no oxygen is present in the sample, the system is calibrated to zero and the sphere lies inside the magnetic field. When oxygen is introduced into the sample, the oxygen molecules displace the nitrogen-filled spheres from their zero balanced position. This is exaggerated by the mounted mirror. An electrical coil is positioned and calibrated so that, when activated, the dumbbell is restored to the zero position. The current required to restore the dumbbell to the zero position is a function of the oxygen partial pressure. Thus, by measuring the restoring feedback current, the partial pressure of oxygen is known. This is very accurate and has a fast rate of response, so this device can measure inspired and exhaled concentrations breath by breath [3].

19.2.1.3 Galvanic Oxygen Analyzer (Fuel Cell)

Fuel cells are chemical batteries that generate a current proportional to the oxygen available to react with the components in the electrodes; thus, the higher the oxygen concentration, the higher the current and the faster the chemicals in the electrode are consumed. An oxygen permeable membrane separates the electrolyte gel from the gas sample. As oxygen diffuses into the electrolyte, it reacts with the electrode components and generates a current that is proportional to the amount of oxygen available. It has a slow response time so it should be positioned at the inhalational limb of the anesthesia circuit. There is no warm-up time for this analyzer but it does require calibration before use [4].

19.2.1.4 Polarographic Oxygen Analyzer (Clark Electrode)

This device also generates a current that is directly proportional to the oxygen present. However, the response time is much faster because an external power source is used to energize this reaction. Both galvanic and polarographic analyzers have electrodes, electrolyte solutions, and oxygen permeable membranes but the external power source differentiates the polarographic cell from the galvanic cell. In this instance, oxygen atoms accept electrons from the cathode and transport the electron to the anode in a hydroxide solution. The external power is used to polarize the electrodes. There is a warm-up time for this device for the electrodes to stabilize, and calibration is essential before use.

19.2.2 Oxygen Measurement in Blood

19.2.2.1 Partial Pressure of Oxygen in Solution

The pressure of oxygen dissolved in blood can be measured directly through the Clark electrode where Teflon acts as an oxygen permeable membrane. The electrolyte solution is heated to body temperature (as this affects the solubility of blood) and a stirrer maintains oxygen equilibration with the blood sample [5]. The cathode is platinum and the anode is silver.

19.2.2.2 Oximetry

This is the basis of the pulse oximeter. Different species of hemoglobin have different light absorption spectra, particularly hemoglobin and oxyhemoglobin. By selecting different wavelengths to transilluminate the tissue, the total amount of hemoglobin and the relative amount of oxyhemoglobin can be derived. Modern pulse oximeters use two or more light-emitting diodes with one photo diode acting as a sensing transducer. A variable fluctuation is essential in determining the absorption of the wavelength by the hemoglobin moiety, as all tissues will absorb some of each wavelength. Therefore, the pulsatile portion of the absorbed light will correspond to the arterial hemoglobin portion of the blood, as the venous and capillary hemoglobin will be relatively constant. The pulsatile portion of the absorption spectrum will be seen in all frequencies that are absorbed by hemoglobin. Pulse oximeters will not work in patients who are on cardiopulmonary bypass because the blood flow through the tissues is constant rather than intermittent like a pulse. It is this pulsating element of the absorption spectrum that allows for determination of the amount of total hemoglobin that is present in the tissue. Once an estimate of the total hemoglobin is made, the saturation of that hemoglobin can be derived from ratio adjustments. The ratio of the fluctuating parts of both wavelengths is applied to an internal algorithm that computes the arteriolar oximetry. If we define the following conditions, we will be able to derive an equation for that ratio (R).

$$AC660 = \text{fluctuating absorption at 660-nm wavelength}$$

$$DC660 = \text{background absorption at 660-nm wavelength}$$

$$AC940 = \text{fluctuating absorption at 940-nm wavelength}$$

$$DC940 = \text{background absorption at 940-nm wavelength}$$

$$R = (AC660/DC660)/(AC940/DC940)$$

The value R is then applied to a "look-up table" that has pulse oximeter saturation values derived from healthy human volunteers. For example if $R = 1$, then the saturation will be 85%. The accuracy of the system is 1–2%. The software inside the oximeter corrects errors by increasing the light emitted when a poor quality signal is received at the photo diode receiver, longer averaging periods to reduce motion artifacts, correcting for the ambient light received at the photo diode, and ensuring a closed fit between tissue and photo diode. However, there are a few caveats. Errors in pulse oximetry occur when there is more than 5% of other species present, such as carboxyhemoglobin, methemoglobin, and sickle-cell hemoglobin. The machine will "read" carboxyhemoglobin as oxyhemoglobin and will thus over-read the true saturation [6]. Methemoglobinemia will falsely lower the perceived saturation readout, and if the true saturation should drop, the presence of methemoglobinemia will "mask" the desaturation so the readout remains unchanged [7]. The presence of injected dyes will cause artifacts, as the dyes will absorb some of the wavelengths employed. Methylene blue and, to a lesser extent, indocyanine green cause a perceived drop in saturation readout when there has been no change [8]. The pulse oximeter also becomes inaccurate in low cardiac output (CO) states such as shock, heart failure, and hypothermia [9]. When the probe is positioned incorrectly, the light from the light source to the photo diode may go through the tissues at a grazing incidence. This incidental light will effectively decrease the "signal-to-noise" ratio so that real changes in saturation are masked. This effect is known as the "penumbra" effect and can be addressed by correct repositioning of the probe [10]. Pulse oximeters are also subject to interference with electrocautery devices and mechanical factors such as blood pressure cuffs and motion artifacts [11]. Newer generations of pulse oximeters use adaptive digital filtering such as pioneered by Masimo Corporation (Irvine, CA), which reduces motion artifact and adjusts for venous blood pulsation [12]. Furthermore, additional wavelengths used by newer oximeters will also calculate the circulating hemoglobin concentration and account for the methemoglobin and carboxyhemoglobin values in order to correct the true hemoglobin saturation.

19.3 Ventilation

There are two parts in monitoring ventilation: monitoring what the ventilator delivers to the patient along with what comes from the patient. These general tasks break down into monitoring ventilation volumes, pressures and rates, as well as exhaled carbon dioxide and anesthesia agents.

19.3.1 Ventilation Parameters

The majority of modern ventilators have built-in alarms that are triggered when readings deviate from default parameters. These parameters are derived from pressure, volume, and time. Digital signal processors use these readings to derive further ventilation patterns such as pressure–flow loops, pressure–volume curves, flow–volume curves, respiratory compliance, static circuit compliance, and leak tests for breathing circuit integrity, disconnection alarms, sustained pressure alarms, and so forth. Pressure is measured by pressure transducers.

19.3.1.1 Volume

This can be measured in a number of ways. Volume monitors are usually based on the principle of a rotameter, a vane/turbine rotating in a flow of gas. They tend to underestimate the volume recorded at very low tidal volumes and overestimate large tidal volumes, a result of inertia in their moving parts. An optical counter will determine how many times the vanes are turned and thus the patient's tidal volume can be calculated.

Flow monitors also measure volume by integrating the measured flow with time. The following are flow monitors. Fleisch pneumotachographs consist of a bundle of capillary tubes placed parallel with the airflow. The flow within the capillary tubes is laminar and thus the pressure drop measured across the capillary tubes is linearly related to gas flow [13].

Screen-type pneumotachographs are much lighter and less bulky than the Fleisch type and measure a pressure drop across the screen mesh. The pressure drop is nonlinear as the flow across the mesh is turbulent (flow across orifices). The characteristics of the mesh are very sensitive to deposition of moisture or debris across the mesh, and the mesh becomes less reliable when moisture and debris accumulate.

Vortex-type pneumotachographs are usually incorporated into ventilators. An ultrasonic beam is disrupted by vortices that are generated when laminar flow is broken up by struts placed across the tube. The formation of vortices is related to the velocity and thus to the flow.

Hot-wire anemometers pass a current along a conductor to maintain the conductor at constant temperature. When gases flow across the wire, heat is lost and more current is needed to restore the set temperature [14]. The increase in current can be measured and related to the flow. However, the presence of mucus and droplets affect the heating characteristics of the wire and render the device less accurate.

Ultrasonic flow meters are mounted coaxially and measure the change in speed in a wave that is pulsed parallel to the direction of flow [15].

19.3.2 Capnography

Capnography provides a trace of the carbon dioxide content within each respiratory cycle. Carbon dioxide can be detected by mass spectrometry or Raman analysis, and both of these methods have been discussed earlier (see oxygen analyzer). However, the majority of capnographs rely on infrared absorption. Carbon dioxide, like nitrous oxide and water vapor, absorbs infrared radiation. Selection of a specific wavelength (4.3 mm) will lessen the collateral absorption effect. Nitrous oxide will absorb infrared at 4.5 mm; thus, the use of highly specific filters reduces the collateral absorption. However, nitrous oxide will affect the infrared absorption of carbon dioxide by a phenomenon called collision broadening where the spectral absorption peaks of a gas (carbon dioxide) are broadened by the presence/collision of another gas (nitrous oxide) [16]. This can be corrected for if the concentration of nitrous oxide is known. For example, if nitrous oxide is present at 70%, then the perceived carbon dioxide readout must be corrected by a factor of 0.90, and similarly if nitrous oxide is present at 50%, then the carbon dioxide readout needs a correction factor of 0.94 to derive the true carbon dioxide concentration [17]. Of the infrared absorption analyzers, there are two types of capnographs: sidestream and mainstream. All capnographs must be calibrated periodically to prevent equipment drift.

19.3.2.1 Sidestream Capnographs

These capnographs continuously aspirate a sample of the respiratory gases and analyze the carbon dioxide content of that sample. Setting the wrong aspiration rate, leaks within the sampling system, and water vapor condensation, causing blockages and dust particles, are all sources of error. Water traps are designed to prevent the entry of liquids into the measuring chamber as liquids have a high infrared absorbance and will greatly interfere with the results. After the gases have been analyzed, the waste gases may be returned to the anesthesia gas circuit on the exhalation side before the soda-lime canister.

19.3.2.2 Mainstream Capnographs

These capnographs analyze the whole of the respiratory gases at one point in the anesthesia circuit. The measuring chamber (cuvette) is placed as close as possible to the endotracheal tube. The chamber is heated to 40°C to prevent liquid condensation of the water vapor present in exhaled gases. The chamber is heavy and needs to be supported away from the patient's face as it may cause burns or cause pressure ulceration. Particulate matter such as mucus or saliva from the airway will cause errors by absorbing infrared if they enter the cuvette. However, the response time of this type of analyzer is much faster than with the sidestream capnographs, mass spectrometers, or Raman analyzers. There are also no issues about sampling rate flow time, blockages in the aspirating tubing, or leaks or malfunction of the aspirating pump, which are all difficulties present in the other types of capnographs.

19.4 Circulation

Heart rate and blood pressure measurements are two indices of circulation. Heart rate may be measured by electrocardiogram (EKG) or pulse oximetry. Blood pressure measurement may be obtained indirectly using cuffs which are placed around an extremity, indirect continuous techniques which may produce variable results, and directly by transducing the arterial pressure from an invasive intra-arterial catheter [18].

19.4.1 The Electrocardiogram

The EKG is a noninvasive measure of the electrical activity of the heart. Heart rate is calculated by identifying the R wave in the QRS complex and measuring the R–R interval between adjacent R waves to derive the heart rate. Due to beat-to-beat variability in the heart rate, an average of the heart rate over the last 5–15 s is used. Telemetry devices have a moving average filter which places more weight on recent changes in rate. The averaging interval varies depending on new changes in the rate measured. The EKG usually has 3–12 leads with standard location of lead placement. Lead II best demonstrates the *p* wave and is used to monitor the rhythm of the heart. Lead V5 displays ST segment depression and elevation which may play a role in ischemia. Multiple factors contribute to errors in heart rate calculation and include: improper electrode placement, poor electrical conduction across the skin due to dehydrated conducting gel, poor cable integrity to conduct the evoked potential to the monitor, noncardiac muscle activity that also generates electrical potentials, electrocautery instruments that generate radio frequency waves, 60 Hz interference from fluid warmers, inappropriate signal gain, and inappropriate R wave signal threshold.

19.4.2 Blood Pressure

Blood pressure may be monitored by different methods: using noninvasive cuff, noninvasive continuous devices, and invasive blood pressure monitors.

19.4.2.1 Noninvasive Blood Pressure Measurement

When using a noninvasive cuff, the correct size of the cuff and the position of the pneumatic bladder are important in obtaining accurate readings. The bladder width for indirect blood pressure monitoring should be approximately 40% of the circumference of the extremity and the bladder length should be sufficient to encircle at least 80% of the extremity [19]. Small cuff size, loosely applied cuffs, and an extremity below heart level result in a falsely elevated measurement. Falsely low estimates occur with a cuff too large, an extremity above heart level, or after quick deflations. The cuff should be applied to have a snug fit to the limb and the bladder must be deflated prior to measurement. Complications from the cuff include bruising, nerve palsies, limb edema, thrombophlebitis, and compartment syndrome [20,21].

Methods utilizing the blood pressure cuff may be manual or automated. Manual cuff blood pressure measurement depends on palpation or auscultation of Korotkoff sounds which estimates systolic and diastolic blood pressures. Korotkoff sounds are created from turbulent flow within an artery from the inflation and deflation of the blood pressure cuff [22]. Systolic blood pressure is designated by the first Korotkoff sound. Diastolic blood pressure is described by the disappearance of the sounds. Due to the subjective variability of the appearance and disappearance of Korotkoff sounds, measurement of systolic and diastolic blood pressure using manual techniques may be prone to errors. In the perioperative environment, automated cuffs are the standard for blood pressure monitoring. Automated machines depend on the principle of oscillometry which measures mean blood pressure by detecting the point of maximal fluctuations in cuff pressure produced while deflating a blood pressure cuff. Cuff pressure is inflated above the systolic pressure and slowly deflated in an incremental stepwise manner. At each step of deflation, the pressure is stabilized and small fluctuations in the occluding pressure are measured. These small fluctuations are the result of a partial column of blood passing under the occluding cuff. Their oscillations vary with the occluding cuff pressures. Peak oscillations correlate closely with the mean arterial pressure. The systolic and diastolic pressures are extrapolated from the peak oscillations. Automated blood pressure monitors may under-interpret the systolic pressure and over-interpret the diastolic blood pressure readings.

19.4.2.2 Noninvasive Continuous Blood Pressure Measurement

New methods of continuous noninvasive blood pressure monitoring have been developed. These devices provide a continuous blood pressure estimate and a display of the arterial blood pressure trace. Although these techniques provide noninvasive continuous blood pressure monitoring, the accuracy of such measurements can be inconsistent. [18,23–25].

19.4.2.3 Invasive Blood Pressure Measurement

An invasive blood pressure monitor consists of a catheter and attached tubing which measures the pressure wave within that artery. The radial artery is the most common artery accessed; however, the femoral, brachial, and dorsalis pedis arteries may be used.

The invasive blood pressure line gives continuous beat-to-beat data and allows for the sampling of arterial blood. The rigid high-pressure fluid-filled tubing transmits the force of the pulse wave to a pressure transducer that converts the displacement of a silicon crystal into voltage changes [22]. The waveform can display information about the regularity of the pulse, variations of the blood pressure with changes in respiration, and the status of the circulating volume. Disposable transducers utilize a simple calibration method to zero the transducer to atmospheric pressure. The natural frequency of the system and its damping coefficient are important determinants to whether the system will overshoot or undershoot the true systolic blood pressure. One way to measure this at the patient interface is to perform a fast flush test. After several fast flushes, the flush artifact is examined. The distance between two adjacent peaks or troughs should be constant and represents the natural frequency of the system. It is calculated by the paper speed (mm/s) divided by the horizontal distance between two flush artifacts (mm) to give a frequency (f/s = f Hz). The higher the frequency value, the less likely the whole system will reverberate. Systems with a natural frequency of less than 7 Hz are unsatisfactory. Examining the flush artifact and measuring the vertical heights of two successive waves will derive the damping coefficient. The ratio of these two successive vertical heights will render an amplitude ratio. This ratio should be between 0.05 and 0.10 as this will equate with a damping coefficient of 0.6–0.7, which would be ideal for a pressure transducer system.

Marked false increases and decreases in blood pressure may be seen with kinking of the catheter or tubing, and incorrect placement or calibration of the transducer which should be at the level of the right atrium. Other factors which contribute to erroneous measurements include: improper selection of arterial cannula, partial obstruction from the inflated blood pressure cuff to the artery, positioning the patient on the operating table so that the artery to the limb is partially obstructed, incorrect compliant pressure tubing connecting arterial cannula to transducer, air bubbles within the cannula or pressure tubing, and overdamping or underdamping of the pressure wave.

Invasive arterial monitors are more expensive than the noninvasive cuff. A skilled provider is necessary to perform the procedure. Monitoring pressure once the arterial line is placed also requires training and vigilance. Complications from an intra-arterial line can cause severe morbidity and include arterial thrombosis, embolism, vasospasm, local and systemic infection, iatrogenic blood loss, hematoma, nerve and vessel injury, ischemia, and necrosis [26–31].

19.5 Temperature Monitoring

Monitoring body temperature is a standard of anesthetic care. Anesthesiologists should strive to maintain a normal core temperature in patients undergoing anesthesia. Patients receiving anesthesia should have temperature monitored when clinically significant changes in body temperature occur [32 effective July 1, 2011 #74]. Temperature should be continually observed because monitoring changes in temperature in anesthetized patients will permit for the detection of hypothermia or hyperthermia. Heat loss and malignant hyperthermia are two conditions at opposite ends of the temperature spectrum that require intervention.

Two types of body temperatures are recorded: peripheral and core temperatures. Central core temperatures can be measured by using probes which may be placed into the nasopharynx, esophagus, rectum, bladder, ear canal, trachea, or pulmonary artery [33]. Peripheral temperature may be obtained via measurement from the skin including but not limited to the axilla, forehead, abdomen, nose, groin, and limbs.

Thermoregulation is the body's ability to maintain its temperature within certain limits even when the environmental temperature may fluctuate. Core temperature is maintained by heat conserving or dissipating mechanisms. General and regional anesthesia inhibit the body's ability to maintain thermoregulation [34,35]. Thermoregulatory responses are based on a physiologically weighted average reflecting changes in the mean body temperature. Mean body temperature is estimated by the following equation [22]:

$$\text{Mean temperature} = 0.85 \, T \, \text{core} + 0.5 \, T \, \text{skin}$$

Temperature may be measured by various technologies including thermistors, thermocouples, infrared readers for tympanic membrane, liquid crystal thermometry for skin, infrared thermometry of the skin over the superficial temporal artery, and thermometers. Thermistors and thermocouples serve as temperature transducers. Thermistors are semiconductors that respond to temperature changes by changing their electrical resistance. Their resistance is inversely proportional with increasing temperature. Thermocouples are two unlike metals joined in a circuit. The current measured is directly proportional to the temperature difference between the two metal junctions. Infrared tympanic membrane readers may be less reliable in terms of accuracy because the measurement is dependent on an unobstructed external auditory canal. In addition, temperature strips on the skin may identify peripheral vasoconstriction but they do not correlate with core temperature measurements [36]. Infrared thermometry of the skin over the superficial temporal artery does not accurately measure core temperature [37]. The most accurate means of measuring core temperature are invasive and include esophageal probes, probes in urinary Foley catheters, and pulmonary artery probes.

Heat loss may result from many different sources. Surgical exposure and the cool temperature in the operating room are two environmental factors. Both general and spinal anesthesia cause vasodilation, which promotes heat loss [34,35]. General anesthesia also reduces metabolic rate which further contributes to heat loss. Core temperature may fall by 0.5–1.5°C within the first hour of general anesthesia. Heat may also be lost from evaporation of sweat or other fluids. For every gram of water evaporated, 0.58 kcal of heat is lost [22]. Heat losses also result from radiation, conduction, and convection.

Changes in mean temperature are most reliably measured by core temperature sites. During most surgeries, temperature differences between these sites are small. However, when anesthetized patients are being cooled and later rewarmed, changes in rectal temperature lag behind other probes. Rewarming a patient is best achieved by measuring temperature at several locations.

19.6 CO Monitors

Rapid advancements in the medical and surgical care of the critically ill have led to an effort to improve existing technologies that monitor patient physiological parameters. These technologies aid the clinician by providing information on the current physiological state of the patient and gauging the effects of therapeutic interventions, which may, in turn, influence care rendered. CO is a commonly determined variable in critical care scenarios and in specific patients undergoing high-risk surgery. CO may be considered a global indicator for cardiac function and subsequently, oxygen delivery. In fact, many therapeutic interventions ranging from drug administration to volume management are aimed at optimizing CO and other associated hemodynamic variables. In this section, we will discuss different modalities for measuring CO, from traditional invasive to current "minimally" (or less) invasive devices.

19.6.1 Thermodilution: Pulmonary Artery Catheter

Perhaps no technique has been more studied than thermodilution cardiac output (TCO), obtained via the injection of a known quantity of fluid into a pulmonary artery catheter (PAC) and measuring the change in temperature as the blood traverses the heart. In essence, this is a modification of the Stewart–Hamilton equation [38,39]:

$$Q = \frac{I}{\displaystyle\int_0^\infty C_I \, dt}$$

where Q is the cardiac output, I the amount of indicator, and $\int_0^\infty C_I \, dt$ the integral of indicator concentration with respect to time.

However, it was Fegler who is the first to adapt the indicator dilution principle previously described by Stewart and Hamilton, by utilizing thermal indicators to obtain measurements of CO of anesthetized animals in 1954 [40]. TCO has essentially become the clinical gold standard for measurement of CO against which all other monitors measuring CO are compared. Although temperature is continuously measured at the tip of PAC in the TCO technique, there is also a second thermistor at the injector port;

hence, the net change in temperature is measured as a known quantity bolus of injectate passes through the PAC. Computers then utilize the modified Stewart–Hamilton equation:

$$Q = \frac{(T_B - T_I) * K}{\int_0^\infty \Delta T_B(t)\mathrm{d}t}$$

where Q is the cardiac output (l/min), T_B the blood temperature, T_I the injectate temperature, K the computation constant, and $\int_0^\infty \Delta T_B(t)\mathrm{d}t$ the integral of temperature change with respect to time. Computer software does the laborious calculations thereby accounting for multiple factors such as volume of the injectate, heat of blood and the injectate, catheter size and composition, specific gravity of both the blood and injectate, as well as the area under the blood temperature curve to give measurements of CO [41,42]. To obtain a TCO measurement, typically a quantity of cold fluid (2.5, 5, or 10 mL) is injected as a bolus into the proximal CVP lumen of the PAC, and the net change in the pulmonary artery blood temperature is recorded by the thermistor at the catheter tip. The blood temperature curve is displayed in real time so that unwanted artifacts can be recognized, and the measurement can be ignored. Typically this injection is repeated rapidly for a total of three times, so that an average can be obtained to get the most accurate result. Fluid injected in the PAC is preferred to be below body temperature to get the most reliable readings, as the degree of temperature change is inversely proportional to CO; that is, temperature change is significant if CO is low and vice versa.

Although the TCO technique is commonly used, there are multiple pitfalls associated with its use, which can be broadly categorized into those related with the operator, the patient's pathology or the injected indicator. Stated most accurately, the TCO technique measures the change in temperature of blood flow from the right ventricular outflow to the pulmonary artery outflow. Therefore, when a patient has either a significant right-to-left or left-to-right shunt, inherently the outputs of the right ventricle and the left ventricle cannot be equal, thus leading to inaccurate estimates of CO. Furthermore, the TCO technique relies on complete mixing of the injectate (hence rapid bolus is desired) before the blood is delivered to distal tip PAC thermistor. This brings into question the accuracy of readings obtained via TCO for patients with significant tricuspid regurgitation because thermodilution curves will have longer decay times as there will be recirculation of the thermal signal between the right atrium and the right ventricle [43]. Furthermore, the PAC is an expensive device to use and in certain clinical conditions (i.e., patients with cardiac masses, dysrhythmias), the use of PAC can be potentially disadvantageous.

There are other pitfalls that are potentially associated with the use of the TCO technique. For instance, if one is mindful of the fact that CO is inversely proportional to the area under the thermodilution curve, and other volumes are inadvertently used, then the operator can potentially over or underestimate the CO. Hence if the preprogrammed computation constant (K) was meant for a 10 mL injectate, but rather half that volume was instead injected, the resulting area would be smaller and the CO would be falsely doubled. Furthermore, to maximize precision of CO measurements, ideally both the volume as well as the rate of injection of the injectate, should be kept constant. Likewise, there is controversy as to which phase of respiratory cycle the TCO measurements should be obtained as CO can vary significantly throughout the respiratory cycle. This is particularly true when a patient is mechanically ventilated [44,45]. Hence, it is desired that the bolus be injected through the same phase of the respiratory cycle, which is clinically difficult to accomplish on a reliable basis. Moreover, this is not a true estimate of CO, which is optimally obtained by multiple injections in all phases of the respiratory cycle to obtain a CO that is more reflective of the true CO. To counter some of these problems, the continuous cardiac output (CCO) technique utilizes embedded thermal filaments, which emit heat from within the PAC to essentially accomplish the same objective.

Several different thermal techniques are currently used to generate the heat in commercially available PACs. For example, pulsed thermodilution uses a coiled right ventricular filament, which is randomly heated. In contrast, yet another technique employs heat applied to the thermistor at the distal tip of PAC while the right ventricle outflow cools the tip [46,47]. Hence, depending on the CCO PAC, different proprietary algorithms analyze the thermal signal measured by the thermistor at the tip of the catheter to arrive at an average CO value. The value given is updated every 30–60 sec representing the average CO over the previous 5–15 min; thus, current CCO PACs display a continual, frequently updated CO numerical values, which are not truly a real-time monitor, despite claims to the contrary [48]. In fact, investigators have shown average *in vivo* time delays in critically ill patients to be 9.3, 10.5, and 11.8 min for a 50%, 75%, and 90% sudden change in CO, respectively [49]. Therefore, some clinicians argue that these CCO PACs should more correctly be called "continual" monitors, as they are less than optimal for detecting and measuring sudden CO changes.

CCO PACs have rapidly gained wider clinical use because they are easier to use as no injectate is needed, thus reducing mechanical work, operator variability, chance of infection, and obviate the concern relating to which phase of the respiratory cycle the injectate was bolused. In summation, although CCO PACs provide a more reproducible and precise measurement of CO in comparison to the TCO, they have an inherent lag since the derived values come from the previous 5–15 min. This compels the provider to be vigilant of other monitors such as the EKG or direct arterial monitoring [50].

19.6.2 Esophageal Aortic Doppler Ultrasound

Use of Doppler technology to estimate blood flow has increased in the past few decades, whether it is via esophageal aortic Doppler ultrasound, transesophageal, or transthoracic echocardiograms. The esophageal Doppler (ED) takes advantage of the anatomical close proximity of the esophagus to the aorta, thus giving a stroke volume (SV), which is estimated from a measurement of stroke distance and aortic cross-sectional area. As stated in earlier chapters, Dopplers measure specifically velocity, not flow, therefore certain assumptions are made when estimating CO when using an ED. It is assumed that the aorta is a uniform cylinder and flow through it is estimated by multiplying the cross-sectional area (CSA) by velocity (V_f). As blood flow in the aorta is dynamic and pulsatile, the velocity will change over time, therefore V_f can be defined as the area under the curve of velocity–time graph. The area is then calculated as the integral of the velocity curve with respect to time (dV/dt ($Vf \, dt$) from the start of aortic blood flow T_0 to the end of flow T_1) [51]. This distance traveled by the blood during systole (measured in cm) is then multiplied by the CSA, which is derived from nomograms or direct measurement depending on the ED device.

In contrast to the commercially available transesophageal echocardiogram probes (TEE, *discussed in the next section*) these probes are much smaller (CardioQ/Medicina TECO (Deltex Medical Ltd, Chichester, UK); HemoSonic100 (Arrow, Reading, Pennsylvania, USA), hence easier to insert in the esophagus, thereby potentially reducing the incidence of injury (Figures 19.1 and 19.2).

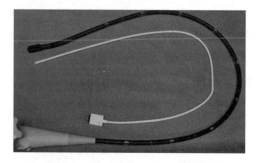

FIGURE 19.1   Side by side comparison of transesophageal echocardiogram (dark) vs esophageal Doppler probe (white).

FIGURE 19.2 Detail of the distal tip of each probe: note the further bulbous dilation of the transesophageal echocardiogram probe compared to the slim angled tip of the Doppler probe (white).

These probes are typically the size of a nasogastric tube, with a Doppler transducer which is either a 4 MHz continuous or 5 MHz pulsed wave, depending on the device type. This probe is usually advanced to mid-thoracic level with the transducer probe being parallel to the descending aorta. However, despite being smaller and leaner than the conventional TEE probes, these probes can still be cumbersome to use and proper probe positioning can have quite a steep learning curve. Also these probes are most often used in patients who are intubated, as insertion and the need for ongoing manipulation in an awake patient can be challenging. Furthermore, the Doppler signal can intermittently be of poor quality due to signal detection problems hence reducing the accuracy of the derived CO. Moreover, the use of nomograms to determine aortic cross-sectional area can be unreliable due to patient pathophysiological conditions such as calcification of the aortic valve leaflets and annulus, or ascending aortic aneurysms involving the actual aortic valve. To address this, some commercial companies have introduced other probes (HemoSonic100), which allow for a more accurate measurement of aortic diameter to counter the limitation of the conventional aortic ED probes. These new probes utilize M-mode echocardiography, which better demarcates the aortic walls in comparison to the conventional ED probes.

Despite improvements in technology leading to better ultrasonic signals and thus more accurately derived COs, studies have shown inconsistent results in past few years; when ED was compared with other methods of obtaining CO, some studies demonstrated good results while others failed to show good correlation [52–57]. The main advantages of this type of probe continue to be that it is potentially less traumatic on the patient due to its smaller size, and it does not require mechanical tracing of cross-sectional area at the site of quantification as is needed with the conventional TEE. Furthermore, if the probe is aligned properly (typically the Doppler beam is desired to be within 20° of axial flow), this probe has the additional advantage over the TEE Doppler probe of giving continuous CO values.

Additionally, the ED has the ability to measure FT_c which is a value that represents the time from the beginning of the aortic waveform upstroke to its return to baseline. Specifically FT_c is the systolic flow time corrected for a HR of 60 bpm and is used as an index of estimating preload. However, some investigators have advocated using SV as the preferred variable to assess fluid status as opposed to FT_c, since value interpretation may be complicated by its inverse relationship to systemic vascular resistance (SVR) [58]. For instance, in certain conditions such as heart failure, hypothermia, or excessive use of vasopressors, SVR may be increased, while FT_c is reduced, which might prompt the provider to give fluid without an improvement in CO. Similarly, in certain pathophysiological conditions (pulmonary embolus, pericardial tamponade, or mitral stenosis) where there is an impediment in LV filling, giving additional fluid might actually push the patient on the Starling curve so that the risk of pulmonary edema increases [59]. Finally, one can estimate SVR on certain models if the clinician manually enters mean arterial pressure and CVP; thus, the clinician can potentially determine if the FT_c is falsely decreased or increased. Whether the clinician chooses to use SV or FT_c, the ED offers the advantage of potentially providing the clinician with a monitor to determine whether fluid challenges actually optimize preload or not.

Although the ED is a monitoring device that can aid in optimization of goal-directed therapy, it has several limitations. For instance, ED measures only descending aortic flow, hence accounting for only about 70% of CO, while the remaining 30% goes to the brachiocephalic and coronary vessels which has to be accounted for with a correction factor (K-factor). Albeit this may be a valid assumption in healthy patients; with certain changes in metabolic activity or hemodynamic status such as those in the setting of aortic coarctation, aortic cross clamp application, or pregnancy, this 70:30 ratio may not hold true [51,60]. In summation, although ED is a tool that is used in monitoring goal-directed therapy however, its application may be limited in clinical scenarios with low CO. Therefore, further clinical trials are warranted to evaluate its reliability and accuracy in those patients requiring vasopressor support or those who have dynamic changes in SVR.

19.6.3 Transesophageal Echocardiography

Echocardiography utilizes the Doppler effect to measure changes in the frequency of an ultrasonic signal, which is emitted from the probe into various parts of the body and reflected; strictly speaking, echocardiography does not truly measure flow nor pressure but rather estimates the velocity of the blood. Assuming blood flow in large vessels (left ventricular outflow tract, normal cardiac valves) is constant, laminar, and has a flat flow velocity, meaning cells in the center of the vessel are flowing at the same speed as cells in the peripheral sides, then flow can be estimated via echocardiography [61]. In physiological terms, this is the volume of blood moved during each cardiac cycle or SV, which is an important index of cardiac performance. Therefore, applying the continuity equation, which states blood flow is constant throughout the heart in the absence of valve dysfunction or shunting, the echocardiographer can estimate SV. Hence if blood moves a distance s over time t, then velocity can be obtained by dividing s/t. Additionally, flow (Q) can be defined as the product of V and cross-sectional area (CSA), if we assume the shape of the vessel to be a rigid uniform cylinder [62].

1. $V = s/t$
2. $Q = CSA \times V$

Hence, this calculated volume of the cylinder conceptually represents SV and knowing that CO is the product of SV and heart rate (HR), the echocardiographer can obtain CO.

3. $CO = SV \times HR$

Therefore, the echocardiographer can first calculate SV by obtaining the instantaneous velocities during systole, which are then traced from the spectral display and the ultrasound machine software package then automatically calculates the time–velocity integral (TVI) (Figure 19.3).

4. $SV = CSA \times TVI$

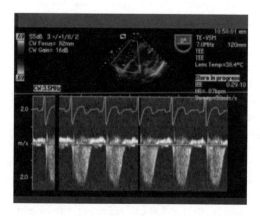

FIGURE 19.3 Echocardiogram display showing view through the heart (top), the EKG (middle), and velocities (bottom).

FIGURE 19.4 Cylindrical representation to display parameters, r = radius (cm), h = height (cm) to derive volume (cm³).

Conventionally the cross-sectional area shape at the left ventricular outflow tract is assumed to be a circle, therefore area is defined as πr^2, most commonly measured in the mid-esophageal long axis view (Figure 19.4).

However, other shapes such as an ellipse and equilateral triangle have been used at other cardiac structures such as mitral valve annulus and the aortic valve [62]. Therefore, the formula used to calculate the area at those cardiac structures would have to be adjusted. Most importantly, in order to obtain optimal measurements, the Doppler signal must be oriented parallel to the outflow tract. In fact, the results are conventionally considered unreliable if the angle of incidence of the beam upon the object is greater than 20°.

Alternatively, another less commonly used method for measuring CO is utilizing the Simpson formula, which is especially helpful in irregularly shaped ventricles [63]. This formula models the left ventricle as a series of stacked decreasing cylindrical discs capped at the apex by an elliptical disc. Conveniently, most echocardiographic software does the calculations needed to measure the volume of disks, thus avoiding laborious manual calculations [64]. Since the operator is able to trace the irregular border of the left ventricle, and manipulate the TEE probe to avoid apex–base foreshortening, this gives the Simpson method a major advantage as no assumptions are made about the shape of the left ventricle geometry.

Although the TEE can provide the clinician with a measurement of CO as explained earlier, this can be quite time-consuming. Furthermore, a prerequisite for obtaining this measurement or other indices is familiarity with the use of TEE, associated software, as well as a significant degree of echocardiographer skill and knowledge. Therefore, the TEE is used more commonly to assess global function (ischemia, valve dysfunction, volume, other functional or structural abnormalities, etc.) rather than measure a single isolated CO. If the desired value is a CCO value, this is not feasible with commonly used software; rather, the TEE offers a validated and alternative technique to confirm other methods of determining CO.

19.6.4 Pulse Contour Analysis (PCA-CO)

The determination of CO by pulse contour analysis (PCA) is based on theories fielded by Otto Frank in 1899. Frank proposed that circulatory system physiology was analogous to a "Windkessel" or air chamber model, where a pulsatile pump (the heart) delivers flow via tubes (blood vessels) through chambers containing air (compressible air representing vessel compliance). Simply put, the Windkessel model describes the flow of blood, driven by the heart, through elastic compartments. Through further analysis, Frank also noted that changes in circulatory pressure had a mathematical relationship to SV [65]. This concept was first to introduce the possibility of using pressure as a surrogate for SV when calculating CO via the traditional equation CO = HR × SV, where HR = heart rate and SV = stroke volume. In fact, Erlanger and Hooker advanced Frank's concepts in 1904 by proposing a correlation not only between volume and change in pressure, but also between CO and arterial pulse pressure wave contour [66]. These concepts, combined with modern computer processing technology, allowed the development of the algorithms currently used in PCA–CO monitoring devices. Three most commonly used systems: PiCCOplus (Pulsion Medical Systems, Munich, Germany), LiDCOplus/rapid (LiDCO, Cambridge, UK), and Vigileo/Flotrac/EV1000 (Edwards Lifesciences, Irvine, CA, USA) will be discussed.

19.6.4.1 PiCCOplus (Pulsion Medical Systems, Munich, Germany)

The PiCCOplus system incorporates a transpulmonary thermodilution calibrated PCA algorithm for CCO determination. It was the first of the PCA–CO devices available for patient care, and its algorithm is based on a modified Wesseling algorithm:

$$\text{Stroke volume} = x \int_{\text{systole}} [P(t)/\text{SVR} + C_{(p)}dP/dt]dt$$

where P is the pressure, t the time, SVR the systemic vascular resistance, $C_{(p)}$ the compliance (corrected for arterial pressure), and CF the calibration factor (derived by transpulmonary thermodilution). This modification of the Wesseling algorithm addresses the issue of the human vasculature's nonlinear compliance [65].

Essentially, the Wesseling algorithm is based on the premise that SV determines the contour of the arterial pressure waveform; therefore, integration of the area under a pulse pressure time curve from end systole to end diastole, yields SV [67]. SV may then be used to approximate CO via the traditional CO formula, as described. The PiCCO system uses a proprietary thermistor equipped arterial catheter for pressure detection and subsequent approximation of SV. This arterial catheter is ideally placed in the femoral artery, but longer catheters may be used when arterial access is limited to the upper extremities. Calibration of the PiCCO algorithm is accomplished by saline transpulmonary thermodilution with calculation of the patient's CO, using a modified Stewart–Hamilton principle. Central venous catheterization is required for this calibration regimen. Validation studies comparing PiCCO to the gold standard, pulmonary artery catheter, showed a close correlation between the two monitoring modalities [68,69]. However, in hemodynamically unstable patients, recalibration may be required more frequently to preserve the system's accuracy.

19.6.4.2 LiDCOplus/rapid (LiDCO, Cambridge, UK)

The LidCOplus system uses a "pulse power" (PulseCO) algorithm with lithium (Li)-based transpulmonary dilution calibration to provide CCO monitoring. Although usually classified under PCA, LidCO's PulseCO algorithm is more specifically a *pulse power* analysis algorithm, based on the law of conservation of mass. In this algorithm, the SV entering the arterial vasculature during systole minus the blood flow out, results in a change of power in the arterial system [70,71]. Combined with correction for arterial compliance and a calibration process, this concept may be further developed to assume linearity between net power and net flow. This is in contrast to the arterial waveform morphology analysis performed in PCA–CO. The PulseCO algorithm incorporates these principles and approximates SV through autocorrelation. Autocorrelation is a signal-processing technique that determines that portion of the entire beat where the SV exerts its influence on the arterial vasculature (change in power). This relationship between the power function of a standardized volume waveform and the arterial pressure waveform is represented by the equation:

$$\text{Volume} = \text{CF} \times 250 \times e^{-k \times P}$$

where CF is the calibration factor, 250 the aortic/arterial vasculature "nominal" saturation (nVmax) in mL, P the pressure (mmHg), and k the exponential function representing compliance. Li calibration optimizes accuracy of the system by determining a patient's actual V_{max}. This value allows for the appropriate adjustment of nV_{max} to better individualize the algorithm [72]. Once volume is determined, CCO may be calculated (CO = SV × HR), as described earlier.

The LiDCOplus requires venous (central or peripheral) access for transpulmonary dilution and an arterial line for pressure detection and Li sampling. Limitations of the technology include patients on Li therapy, certain muscle relaxants (quaternary ammonium ion), and women in their first trimester of pregnancy. Validation studies for the LiDCO system have been mixed; some studies showing acceptable correlation with the pulmonary artery catheter while others showed inadequate correlation [73–75].

The LiDCOrapid system is a modification of the LiDCOplus optimized for rapid assessment. Although it incorporates the PulseCO algorithm, it does not employ a Li calibration feature. Instead, a nomogram based on patient demographics approximates the V_{max} for SV determination. Although not as accurate as the LiDCOplus (for absolute values), it serves as an expedient way to track CO trends in scenarios (e.g., operating room) where a full LiDCOplus setup may be impractical. The LiDCOrapid requires an arterial line for its PulseCO algorithm but does not require central venous catheterization.

19.6.4.3 FloTrac/Vigileo/EV1000 (Edwards Lifesciences, Irvine, CA, USA)

The FloTrac/Vigileo system is a PCA–CO device employing a specialized "FloTrac" arterial pressure transducer to detect and process pressure waveforms for its "arterial pressure-based cardiac output" (APCO) method of calculating CCO. APCO is calculated via the following formula:

$$APCO = PR \times (\sigma_{AP} * \chi),$$

where χ is a multivariate parameter that scales pulse pressure proportionally to the effects of vascular tone (resistance and compliance). χ is determined using measured hemodynamics (HR, MAP, σ_{AP}), demographics (age, gender, height, weight) base estimates of large vessel compliance, and by an analysis of arterial waveform morphology (skewness, kurtosis). σ_{AP} is the standard deviation of arterial pulse pressure (mm Hg) and is proportional to pulse pressure. σ_{AP} is determined using a proprietary algorithm where the arterial waveform is analyzed over a 20 s time interval at a frequency of 100 times/s. This provides 2000 data points from which σ_{AP} is determined. The conversion of pressure (σ_{AP}) to SV is achieved by multiplying σ_{AP} and χ. Lastly, PR represents the pulse rate of *perfused* beats. The APCO equation may be considered analogous to the traditional equation, CO = HR × SV, where APCO ≅ CO [76–78].

FloTrac/Vigileo does not require manual calibration or central venous access for basic functions (e.g., continuous CO, SV), and its ease of use makes it amenable for rapid assessment scenarios. The latest generation system, designated the EV1000, has algorithm updates to better maintain measurement fidelity during situations of extreme changes in vascular tone (which was a limitation in the earlier versions). Also, arrhythmia filters have been enhanced to filter out as much as 5–7 irregular beats every 20 s to help maintain a robust signal. Validation studies have been satisfactory; however, some studies have shown limited correlation in hemodynamically unstable patients [78,79]. Further studies with the latest generation algorithm are warranted.

PCA devices are currently the most commonly used minimally invasive CO monitors in the critical care and perioperative settings. Advantages include less invasiveness, CCO, and ease of setup (LiDCOrapid and Vigileo/FloTrac). Limitations include the need for a robust continuous arterial pressure signal as well as errors due to arrhythmias, rapid body temperature changes, cardiac shunts, and cardiac valvular abnormalities (i.e., aortic insufficiency or regurgitation). Furthermore, mixed validation results, especially of the LiDCO and Vigileo/Flotrac system, suggest that these devices may be more appropriately used as "trend" monitors, as opposed to absolute value CO monitors.

19.6.5 Applied Fick Principle

19.6.5.1 NICO System (Philips Respironics, Murraysville, PA, USA)

The NICO system uses an applied Fick principle algorithm to derive CCO values. The Fick equation assumes that the total uptake or release of a substance (e.g., oxygen consumption) from an organ (e.g., the heart, lungs, etc.) is the product of the blood flow (CO) to the organ and the arteriovenous concentration difference of an indicator (e.g., O_2, CO_2, etc.):

$$VO_2 = CO (Ca_{O_2} - Cv_{O_2})$$

where VO_2 is oxygen consumption, CO is cardiac output, Ca_{O_2} is the arterial content of oxygen, Cv_{O_2} is the mixed venous oxygen content, and $Ca_{O_2} - Cv_{O_2}$ is the arteriovenous concentration difference of oxygen (O_2). The Fick principle may use a variety of indicators, thus substituting carbon dioxide (CO_2) for O_2 (for ease of detection from patients' exhaled breaths), the equation may be modified as follows:

$$CO = VCO_2 / \left(Cv_{CO_2} - Ca_{CO_2} \right)$$

where VCO_2 is carbon dioxide elimination, Ca_{CO_2} the arterial content of CO_2, Cv_{CO_2} the venous content of CO_2, and $Cv_{CO_2} - Ca_{CO_2}$ the arteriovenous CO_2 concentration difference. Subtracting inspired from expired CO_2 concentration approximates VCO_2, and Ca_{CO_2} may be estimated from end-tidal CO_2, under the condition that there are no diffusion abnormalities. However, Cv_{CO_2} determination poses a more difficult challenge. Therefore, the NICO system employs a differential CO_2 Fick partial rebreathing method that allows the elimination of Cv_{CO_2} as a variable. The NICO gas sampling system consists of a rebreathing loop, a rebreathing valve, a mainstream infrared CO_2 sensor, and a fixed orifice differential pressure pneumotachometer. Partial rebreathing is triggered every 3 min via the rebreathing loop and valve. The Fick CO equations for periods of "non" rebreathing and rebreathing may be combined; and because of CO_2's slow time constant for diffusion and the body's large stores of CO_2, the equation may be modified to nullify Cv_{CO_2}. As a result:

$$CO = \Delta VCO_2 / \Delta Ca_{CO_2}$$

where ΔVCO_2 is the change in CO_2 elimination and ΔCa_{CO_2} is the change in alveolar blood CO_2 from "non" rebreathing to rebreathing states. The NICO algorithm calculates CO from these measured variables, and also takes into account the shunted portion of blood (via Nunn's isoshunt plots) that circulates through the cardiopulmonary circulation [80].

Limitations of the NICO system include the need for patient intubation/mechanical ventilation, the necessity that patients be able to tolerate periods of rebreathing, the exclusion of pulmonary disease states that would increase dead space, and the need for $PaCO_2$ to be greater than 30 mmHg and the CO_2–Hgb dissociation curve to be linear for accurate CO determination. Validation studies have been mixed, with the underestimation of CO (as compared to pulmonary artery catheterization thermodilution) being the most common drawback [81–83].

19.6.6 Thoracic Electrical Bioimpedance and Bioreactance

Thoracic electrical bioimpedance and bioreactance offer a method of CCO monitoring that is totally noninvasive.

19.6.6.1 Thoracic Electrical Bioimpedance (BioZ*, CardioDynamics, San Diego, USA)

Thoracic electrical bioimpedance (TEB) is based on the principle that applied electrical current may be used to detect impedance variations in the thorax caused by pulsatile flow-induced blood volume variations (and hence, CO). The applied high-frequency, low-amplitude electrical current is directed through electrodes positioned in the neck and thorax, and is sensed from outer to inner electrodes. The electrical resistance, or impedance, is determined from this interaction between thorax and applied current and is inversely proportional to the volume of fluid in the thoracic cavity [84]. Since the majority of this fluid volume may be attributed to aortic blood flow (SV), SV may be calculated using impedance, the resistivity of blood, ventricular ejection time, distance between the electrodes, and the maximum negative slope of the bioimpedence signal [81]. This calculated SV may then be used to determine CO via the traditional formula (CO = HR × SV, as described earlier).

Unfortunately, first-generation TEB has had many limitations that precluded its use as a routine CO monitor. Abnormal tissue edema, respiratory variations affecting thoracic fluid volume, arrhythmias, induced changes in cardiac contractility, electrocautery, and incorrect electrode placement hamper TEB's ability to determine accurate CO values [51]. Newer generation devices such as the BioZ* by CardioDynamics employ revised algorithms that feature improved signal processing/filtering, EKG triggering/arrhythmia sensing, and respiratory filtering to improve signal robustness and data accuracy. Validation studies have been mixed and have emphasized the limitations outlined earlier [85–87].

19.6.6.2 Thoracic Bioreactance (NICOM, Cheetah Medical, Inc., Indianapolis, IN, USA)

Thoracic bioreactance was developed as an improvement to existing bioimpedance technology. In contrast to TEB, phase shifts of delivered oscillating currents are detected and correlated to volume changes (in the thorax) via changes in electrical capacitive and inductive properties, or bioreactance. The bioreactance method preserves more signal fidelity and is less prone to external interference than TEB and initial validation studies are promising [88–91]. Limitations include hypotensive states, in which algorithm assumptions may fall short and lead to decreased accuracy [88].

19.6.7 Conclusion

Despite the wide variety of cardiac monitoring modalities available, the ideal monitor—quick, accurate, precise, cost-efficient, continuous, easy (to use and interpret), noninvasive, and safe for the patient—is yet to be developed. However, the range of existing technology provides the clinician a choice of the best method of hemodynamic monitoring for any given scenario. However, it must be emphasized that data gained from CO monitoring must be combined with other hemodynamic parameters (i.e., systemic vascular resistance, SV (or pulse pressure) variation, mixed-venous oxygen saturation, etc.) to provide an accurate assessment of a patient's hemodynamic condition.

References

1. Anesthesiologists, A.S.o., *Standards for Basic Anesthetic Monitoring*, 1986, amended 2010, effective July 1, 2011: http://www.asahq.org/For-Members/Standards-Guidelines-and-Statements.aspx.
2. Van Wagenen, R.A. et al., Dedicated monitoring of anesthetic and respiratory gases by Raman scattering. *J Clin Monit*, 1986. **2**(4): 215–22.
3. Hill, R.W., Determination of oxygen consumption by use of the paramagnetic oxygen analyzer. *J Appl Physiol*, 1972. **33**(2): 261–3.
4. Torda, T.A. and G.C. Grant, Test of a fuel cell oxygen analyzer. *Br J Anaesth*, 1972. **44**(10): 1108–12.
5. Severinghaus, J.W. and P.B. Astrup, History of blood gas analysis. IV. Leland Clark's oxygen electrode. *J Clin Monit*, 1986. **2**(2): 125–39.
6. Barker, S.J. and K.K. Tremper, The effect of carbon monoxide inhalation on pulse oximetry and transcutaneous PO_2. *Anesthesiology*, 1987. **66**(5): 677–9.
7. Barker, S.J., K.K. Tremper, and J. Hyatt, Effects of methemoglobinemia on pulse oximetry and mixed venous oximetry. *Anesthesiology*, 1989. **70**(1): 112–7.
8. Scheller, M.S., R.J. Unger, and M.J. Kelner, Effects of intravenously administered dyes on pulse oximetry readings. *Anesthesiology*, 1986. **65**(5): 550–2.
9. Lawson, D. et al., Blood flow limits and pulse oximeter signal detection. *Anesthesiology*, 1987. **67**(4): 599–603.
10. Barker, S.J. et al., The effect of sensor malpositioning on pulse oximeter accuracy during hypoxemia. *Anesthesiology*, 1993. **79**(2): 248–54.
11. Striebel, H.W. and F.J. Kretz, The functional principle, reliability and limitations of pulse oximetry. *Anaesthesist*, 1989. **38**(12): 649–57.
12. Goldman, J. et al., Masimo signal extraction pulse oximetry. *J Clin Monit Comput*, 2000. **16**(7): 475–483.

13. von der Hardt, H. and C.H. Zywietz, Reliability in pneumotachographic measurements. *Respiration*, 1976. **33**(6): 416–24.

14. Lundsgaard, J.S., J. Grønlund, and N. Einer-Jensen, Evaluation of a constant-temperature hot-wire anemometer for respiratory-gas-flow measurements. *Med Biol Eng Comput*, 1979. **17**(2): 211–5.

15. Blumenfeld, W., S.Z. Turney, and R.J. Denman, A coaxial ultrasonic pneumotachometer. *Med Biol Eng*, 1975. **13**(6): 855–60.

16. Raemer, D.B. and I. Calalang, Accuracy of end-tidal carbon dioxide tension analyzers. *J Clin Monit*, 1991. **7**(2): 195–208.

17. Kennell, E.M., R.W. Andrews, and H. Wollman, Correction factors for nitrous oxide in the infrared analysis of carbon dioxide. *Anesthesiology*, 1973. **39**(4): 441–3.

18. Findlay, J. et al., Vasotrac (R) arterial blood pressure and direct arterial blood pressure monitoring during liver transplantation. *Anesth Analg*, 2006. **102**(3): 690–693.

19. Pickering, T. et al., Recommendations for blood pressure measurement in humans and experimental animals—Part 1: Blood pressure measurement in humans—A statement for professionals from the Subcommittee of Professional and Public Education of the American Heart Association Council on High Blood Pressure Research. *Circulation*, 2005. **111**(5): 697–716.

20. Sy, W.P., Ulnar nerve palsy possibly related to use of automatically cycled blood pressure cuff. *Anesth Analg*, 1981. **60**(9): 687–8.

21. Celoria, G., J.A. Dawson, and D. Teres, Compartment syndrome in a patient monitored with an automated blood pressure cuff. *J Clin Monit*, 1987. **3**(2): 139–41.

22. Barash, P.G. et al., *Clinical Anesthesia*. 2009, Philadelphia, PA: Lippincott Williams & Wilkins.

23. Smith, N.T., K.H. Wesseling, and B. de Wit, Evaluation of two prototype devices producing noninvasive, pulsatile, calibrated blood pressure measurement from a finger. *J Clin Monit*, 1985. **1**(1): 17–29.

24. Janelle, G.M. and N. Gravenstein, An accuracy evaluation of the T-Line Tensymeter (continuous noninvasive blood pressure management device) versus conventional invasive radial artery monitoring in surgical patients. *Anesth Analg*, 2006. **102**(2): 484–90.

25. Hirschl, M.M. et al., Accuracy and reliability of noninvasive continuous finger blood pressure measurement in critically ill patients. *Crit Care Med*, 1996. **24**(10): 1684–9.

26. Weiss, B.M. and R.I. Gattiker, Complications during and following radial artery cannulation: A prospective study. *Intensive Care Med*, 1986. **12**(6): 424–8.

27. Davis, F.M. and J.M. Stewart, Radial artery cannulation. A prospective study in patients undergoing cardiothoracic surgery. *Br J Anaesth*, 1980. **52**(1): 41–7.

28. Jones, R.M. et al., The effect of method of radial artery cannulation on postcannulation blood flow and thrombus formation. *Anesthesiology*, 1981. **55**(1): 76–8.

29. Silver, M.J. et al., Evaluation of a new blood-conserving arterial line system for patients in intensive care units. *Crit Care Med*, 1993. **21**(4): 507–11.

30. Downs, J.B. et al., Hazards of radial-artery catheterization. *Anesthesiology*, 1973. **38**(3): 283–6.

31. Band, J.D. and D.G. Maki, Infections caused by arterial catheters used for hemodynamic monitoring. *Am J Med*, 1979. **67**(5): 735–41.

32. Clutton-Brock, J. Skin temperature as a clinical aid during anaesthesia. *Proc R Soc Med*, 1947. 40(10): 607–608.

33. Yamakage, M. et al., The utility of tracheal temperature monitoring. *Anesth Analg*, 1993. **76**(4): 795–9.

34. Sessler, D.I., Central thermoregulatory inhibition by general anesthesia. *Anesthesiology*, 1991. **75**(4): 557–9.

35. Ozaki, M. et al., Thermoregulatory thresholds during epidural and spinal anesthesia. *Anesthesiology*, 1994. **81**(2): 282–8.

36. Vaughan, M.S., R.C. Cork, and R.W. Vaughan, Inaccuracy of liquid crystal thermometry to identify core temperature trends in postoperative adults. *Anesth Analg*, 1982. **61**(3): 284–7.

37. Mangat, J. et al., A comparison of technologies used for estimation of body temperature. *Physiol Meas*, 2010. **31**(9): 1105–18.

38. Hamilton, W., J. Moore, and J. Kinsman, Studies on the circulation. IV. Further analysis of the injection method, and changes in hemodynamics under physiologic and pathological conditions. *Am J Physiol*, 1932. **99**: 534.

39. Stewart, G.N., Researches on the circulation time and on the influences which affect it. *J Physiol*, 1897. **22**(3): 159–83.

40. Fegler, G., Measurement of cardiac output in anaesthetized animals by a thermodilution method. *Q J Exp Physiol CMS*, 1954. **39**(3): 153–64.

41. Weisel, R.D., R.L. Berger, and H.B. Hechtman, Current concepts measurement of cardiac output by thermodilution. *New Engl J Med*, 1975. **292**(13): 682–4.

42. Wong, M., A. Skulsky, and E. Moon, Loss of indicator in the thermodilution technique. *Catheter Cardio Diag*, 1978. **4**(1): 103–9.

43. Cigarroa, R.G. et al., Underestimation of cardiac output by thermodilution in patients with tricuspid regurgitation. *Am J Med*, 1989. **86**(4): 417–20.

44. Okamoto, K. et al., Effects of intermittent positive-pressure ventilation on cardiac output measurements by thermodilution. *Crit Care Med*, 1986. **14**(11): 977–80.

45. Snyder, J. and D. Powner, Effects of mechanical ventilationon the measurement of cardiac output by thermodilution. *Crit Care Med*, 1982. **10**: 677.

46. Yelderman, M.L. et al., Continuous thermodilution cardiac output measurement in intensive care unit patients. *J Cardiothor Vasc An*, 1992. **6**(3): 270–4.

47. Mihm, F.G. et al., A multicenter evaluation of a new continuous cardiac output pulmonary artery catheter system. *Crit Care Med*, 1998. **26**(8): 1346–50.

48. Nelson, L.D., The new pulmonary artery catheters: Continuous venous oximetry, right ventricular ejection fraction, and continuous cardiac output. *New Horiz*, 1997. **5**(3): 251–8.

49. Leibowitz, A.B. and J.M. Oropello, The pulmonary artery catheter in anesthesia practice in 2007: An historical overview with emphasis on the past 6 years. *Sem Cardiothor Vasc Anesth*, 2007. **11**(3): 162–76.

50. Lazor, M.A. et al., Evaluation of the accuracy and response time of STAT-mode continuous cardiac output. *J Cardiothor Vasc Anesth*, 1997. **11**(4): 432–6.

51. Funk, D.J., E.W. Moretti, and T.J. Gan, Minimally invasive cardiac output monitoring in the perioperative setting. *Anesth Analg*, 2009. **108**(3): 887–97.

52. Kim, K. et al., Comparison of cardiac outputs of major burn patients undergoing extensive early escharectomy: Esophageal Doppler monitor versus thermodilution pulmonary artery catheter. *J Trauma*, 2004. **57**(5): 1013–7.

53. Hullett, B. et al., A comparison of CardioQ and thermodilution cardiac output during off-pump coronary artery surgery. *J Cardiothor Vasc Anesth*, 2003. **17**(6): 728–32.

54. Sharma, J. et al., Hemodynamic measurements after cardiac surgery: Transesophageal Doppler versus pulmonary artery catheter. *J Cardiothor Vasc Anesth*, 2005. **19**(6): 746–50.

55. Leather, H.A. and P.F. Wouters, Oesophageal Doppler monitoring overestimates cardiac output during lumbar epidural anaesthesia. *Br J Anaesth*, 2001. **86**(6): 794–7.

56. Sinclair, S., S. James, and M. Singer, Intraoperative intravascular volume optimisation and length of hospital stay after repair of proximal femoral fracture: Randomised controlled trial. *BMJ*, 1997. **315**(7113): 909–12.

57. Gan, T.J. et al., Goal-directed intraoperative fluid administration reduces length of hospital stay after major surgery. *Anesthesiology*, 2002. **97**(4): 820–6.

58. Bundgaard-Nielsen, M. et al., Flow-related techniques for preoperative goal-directed fluid optimization. *Brit J Anaesth*, 2007. **98**(1): 38–44.

59. Singer, M., The FTc is not an accurate marker of left ventricular preload. *Intens Care Med*, 2006. **32**(7): 1089; author reply 1091.

60. Laupland, K.B. and C.J. Bands, Utility of esophageal Doppler as a minimally invasive hemodynamic monitor: A review. *Can J Anaesth*, 2002. **49**(4): 393–401.

61. Perrino, A.C., Jr., S.N. Harris, and M.A. Luther, Intraoperative determination of cardiac output using multiplane transesophageal echocardiography: A comparison to thermodilution. *Anesthesiology*, 1998. **89**(2): 350–7.

62. Sidebotham, D., A. Merry, and M. Legget, *Practical Perioperative Transoesophageal Echocardiography*. 2003, Edinburgh: Butterworth-Heinemann.

63. Ryan, T. et al., The agreement between ventricular volumes and ejection fraction by transesophageal echocardiography or a combined radionuclear and thermodilution technique in patients after coronary artery surgery. *J Cardiothor Vasc Anesth*, 1996. **10**(3): 323–8.

64. Mathew, J.M.D. and C. Ayoub, *Clinical Manual and Review of Transesophageal Echocardiography*. 2005, New York: McGraw-Hill Professional.

65. Cecconi, M., A. Rhodes, and G.D. Rocca, From arterial pressure to cardiac output. *Intens Care Med*, 2008(Section XV): 591–601.

66. Erlanger, J. and D. Hooker, An experimental study of blood-pressure and of pulse-pressure in man. *Johns Hopkins Hosp Rep*, 1904. **12**: 145–378.

67. Wesseling, K.H. et al., Computation of aortic flow from pressure in humans using a nonlinear, three-element model. *J Appl Physiol*, 1993. **74**(5): 2566–73.

68. Goedje, O. et al., Continuous cardiac output by femoral arterial thermodilution calibrated pulse contour analysis: Comparison with pulmonary arterial thermodilution. *Crit Care Med*, 1999. **27**(11): 2407–12.

69. Button, D. et al., Clinical evaluation of the FloTrac/Vigileo system and two established continuous cardiac output monitoring devices in patients undergoing cardiac surgery. *Br J Anaesth*, 2007. **99**(3): 329–36.

70. Rhodes, A. and R. Sunderland, Arterial pulse pressure analysis: The LiDCOplus system, in *Functional Hemodynamic Monitoring* M.R. Pinsky and D. Payen, Editors. 2004, Berlin: Springer-Verlag, pp. 183–192.

71. Pearse, R.M., K. Ikram, and J. Barry, Equipment review: An appraisal of the LiDCO plus method of measuring cardiac output. *Crit Care*, 2004. **8**(3): 190–5.

72. Jonas, M., J. Fennell, and C. Brudney, Haemodynamic optimisation of the surgical patient revisited. *Anaesth Int*, 2008. **2**(1): 41.

73. Cecconi, M. et al., A prospective study to evaluate the accuracy of pulse power analysis to monitor cardiac output in critically Ill patients. *BMC Anesthesiol*, 2008. **8**: 3.

74. Hamilton, T.T., L.M. Huber, and M.E. Jessen, PulseCO: A less-invasive method to monitor cardiac output from arterial pressure after cardiac surgery. *Ann Thorac Surg*, 2002. **74**(4): S1408–12.

75. McCoy, J.V. et al., Continuous cardiac index monitoring: A prospective observational study of agreement between a pulmonary artery catheter and a calibrated minimally invasive technique. *Resuscitation*, 2009. **80**(8): 893–7.

76. Hamm, J.B. et al., Assessment of a cardiac output device using arterial pulse waveform analysis, Vigileo, in cardiac surgery compared to pulmonary arterial thermodilution. *Anaesth Intensive Care*, 2010. **38**(2): 295–301.

77. Frazier, J. and F. Hatib. *Getting ml/beat from mmHg" Arterial Pressure-based Cardiac Output The Edwards FloTrac Algorithm*. [pdf] 2008 [cited 2012 May 19]; Available from: http://www.edwards.com/eu/products/mininvasive/Pages/flotracalgorithm.aspx?Flotrac3G=1.

78. Vasdev, S. et al., Arterial pressure waveform derived cardiac output FloTrac/Vigileo system (third generation software): Comparison of two monitoring sites with the thermodilution cardiac output. *J Clin Monit Comput* , 2012. **26**(2): 115–20.

79. Monnet, X. et al., Third-generation FloTrac/Vigileo does not reliably track changes in cardiac output induced by norepinephrine in critically ill patients. *Br J Anaesth*, 2012. **108**(4): 615–22.

80. Lee, A.J., J.H. Cohn, and J.S. Ranasinghe, Cardiac output assessed by invasive and minimally invasive techniques. *Anesthesiol Res Pract*, 2011. **2011**: 475151.

81. Ng, J.M. et al., Evaluation of partial carbon dioxide rebreathing cardiac output measurement during thoracic surgery. *J Cardiothor Vasc Anesth*, 2007. **21**(5): 655–8.

82. Odenstedt, H., O. Stenqvist, and S. Lundin, Clinical evaluation of a partial CO_2 rebreathing technique for cardiac output monitoring in critically ill patients. *Acta Anaesth Scand,* 2002. **46**(2): 152–9.

83. Nilsson, L.B., N. Eldrup, and P.G. Berthelsen, Lack of agreement between thermodilution and carbon dioxide-rebreathing cardiac output. *Acta Anaesth Scand*, 2001. **45**(6): 680–5.

84. Sathyaprabha, T.N. et al., Noninvasive cardiac output measurement by transthoracic electrical bioimpedence: Influence of age and gender. *J Clin Monitor Comput*, 2008. **22**(6): 401–8.

85. Appel, P.L. et al., Comparison of measurements of cardiac output by bioimpedance and thermodilution in severely ill surgical patients. *Crit Care Med*, 1986. **14**(11): 933–5.

86. Zacek, P. et al., Thoracic electrical bioimpedance versus thermodilution in patients post open-heart surgery. *Acta Med*, 1999. **42**(1): 19–23.

87. Clancy, T.V. et al., Cardiac output measurement in critical care patients: Thoracic electrical bioimpedance versus thermodilution. *J Trauma,* 1991. **31**(8): 1116–20; discussion 1120–1.

88. Keren, H., D. Burkhoff, and P. Squara, Evaluation of a noninvasive continuous cardiac output monitoring system based on thoracic bioreactance. *Am J Physiol-Heart C*, 2007. **293**(1): H583–9.

89. Squara, P. et al., Comparison of monitoring performance of bioreactance vs. pulse contour during lung recruitment maneuvers. *Crit Care*, 2009. **13**(4): R125.

90. Raval, N.Y. et al., Multicenter evaluation of noninvasive cardiac output measurement by bioreactance technique. *J ClinMonitor Comput*, 2008. **22**(2): 113–9.

91. Marque, S. et al., Comparison between Flotrac-Vigileo and Bioreactance, a totally noninvasive method for cardiac output monitoring. *CritCare*, 2009. **13**(3): R73.

20

Minimally Invasive Cardiovascular Technologies

Michael R. Moreno
Texas A&M University

Christie Bergerson
Texas A&M University

Danika Hayman
Imperial College London

James E. Moore, Jr.
Imperial College London

20.1 Introduction

Cardiovascular disease remains the most common cause of death in western countries despite considerable advances in treatment strategies. In the twentieth century, technologies such as the heart/lung bypass machine were developed to enable surgeons to operate on virtually any part of the system, including the heart itself. While these procedures have saved countless lives, they suffer from relatively high mortality and morbidity rates due to their invasiveness. Coronary bypass procedures often take several hours in the surgical suite, followed by several days to weeks of recovery in intensive care wards. The cost to the patient and/or their insurance carrier for these procedures exceeds $56,000 (Chapman and Liu, 2011). Many of the most severely diseased patients are excluded from surgery because of the danger associated with the procedure itself.

An effort began in the 1960s to move toward less-invasive procedures to treat cardiovascular disease states. Advances in materials science, imaging, and pharmaceuticals have led to a wide variety of devices to treat pathologies such as occlusive atherosclerotic plaques, heart valve deficiencies, aneurysms, deep venous thrombosis, and most recently, congestive heart failure. These procedures are less traumatic to the patient, require fewer hospital resources, and shorten recovery times. In most vascular applications, a small entry point is established in the femoral artery, located near the groin. Catheters are inserted and guided to the disease location. The patient may be under general or local anesthesia. This is advantageous in neural procedures, in which the patient can be asked certain questions or asked to perform certain tasks to monitor neural function. Once the procedure is completed, the entry point is closed, and the patient is sent to recovery. In many cases, the patient goes home within days to resume normal activities. The cost of such procedures varies from approximately one-third less than the equivalent surgical procedure to as little as a 7% savings (Iribarne, 2011). The cost of coronary artery bypass graft (CABG) procedures using robotic assistance is actually comparable to the cost of the traditional surgical approach. While the postoperative costs are significantly lower, the cost of the actual surgery is significantly higher (Poston et al., 2008). The move toward less-invasive procedures continues to occur in parallel with an overall trend of increasing numbers of total cardiovascular procedures. From 1997 to 2007, the total number of cardiovascular operations and procedures rose 27%. In 2009, there were approximately 163,149 procedures involving a CABG in the United States. In 2007, an estimated 622,000 percutaneous coronary interventions (PCIs) (previously referred to as percutaneous transluminal coronary angioplasty (PTCA) procedures) were performed (Heart Disease and Stroke Statistics, 2011).

The purpose of this section is to describe some of the technologies that have emerged for minimally invasive treatment of cardiovascular pathologies. It is important to recognize that new devices are being developed constantly. However, a sufficient ground level of these devices has emerged that demonstrates the feasibility of the minimally invasive approach. These exciting technologies are increasing in application because both clinicians and patients recognize their potential benefits.

20.2 Angioplasty

Percutaneous transluminal angioplasty (PTA) involves the deployment of a small balloon inside a blocked vessel using a catheter. The PTA procedure begins by introducing a delivery sheath into a superficial artery, such as the femoral artery. A guidewire is then pushed up to the disease site using a guiding catheter. Guidewires are typically less than 0.4 mm in diameter, and serve as the delivery vehicle for the remainder of the devices to be delivered to the diseased vessel. The guiding catheter is withdrawn so that the PTA balloon catheter, stent delivery catheter, or other device can be guided to the disease site. The balloon is then inflated to widen the blockage (Figure 20.1). PTA is the most common therapeutical intervention worldwide. It was developed in response to the need for a less invasive, less traumatic treatment for patients who could not tolerate open surgical bypass. It is hoped that the vessel remains open when the balloon is removed, although this is not always the case. If the vessel is not permanently deformed by the balloon inflation, it will simply return to its previous blocked state (elastic recoil). Even if the artery remains open initially, it may occlude weeks or months later due to tissue proliferation. This process, termed *restenosis*, occurs in approximately 40% of patients (Fleisch and Meier, 1999). The history of PTA can be traced back to the pioneering work of Charles Dotter in the 1960s.

It is believed that angioplasty enlarges the lumen through permanent deformation of the plaque and healthy artery. Postmortem inspection of arteries treated with PTA reveals fracture of the plaque near its thinnest margin (Zarins et al., 1982). Balloon inflation can also provoke damage to the healthy artery structure, including disruption of the internal and/or external elastic laminae. The risk of restenosis

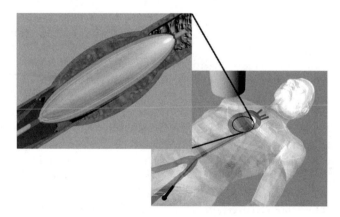

FIGURE 20.1 Illustration of a PTA procedure. Entry into the systemic arterial circulation is gained via the femoral artery in the groin. A small diameter catheter is threaded to the blockage site with the aid of fluoroscopy, and the balloon is inflated to spread open the artery lumen.

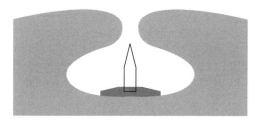

FIGURE 20.2 Angioplasty cutting balloon detail. Radially protruding blades are placed on the exterior surface of the balloon. These blades are contained within the folds of the undeployed balloon as shown. Once the balloon is in place, the blades push into the plaque, promoting fracture to facilitate expansion of the lumen.

increases with greater arterial injury because of the increase in hyperplasia, a natural response to injury (Farb et al., 2002). Realizing that plaque fracture is key to permanent lumen expansion, cutting balloon catheters have been developed that include radially protruding blades that cut into the plaque as the balloon is inflated (Figure 20.2).

Important design considerations for guidewires include the ability to deliver them to the vessel reliably and quickly while minimizing the trauma to the vessel wall. The extent of potential injury is primarily a function of the stiffness of the guidewire, though coating the guidewire with a low friction material such as Teflon can reduce potential trauma. The guidewire should move precisely when pushed without buckling but at the same time be able to negotiate tortuous vessels with ease. It is often desirable to have a guidewire that is relatively stiff along most of its length, but which has a more flexible end to avoid vessel trauma and dissection. Core-to-tip guidewires are often designed using a step tapered metallic core and a 3-cm-long flat, soft shapeable atraumatic tip that is welded or soldered to the tapered distal core. When the design is not core-to-tip, the intermediate or tapered section is connected to the shaft using a glued hypo-tube and a 3-cm-long flat, soft shapeable atraumatic tip welded or soldered to the tapered distal core. The 3-cm tip is often covered with a radio-opaque coil for improved visibility in fluoroscopy, as well as bending flexibility. Materials used for guidewire core designs include but are not limited to SS304V (vacuum re-melt for improved purity and homogeneity stainless steel), Cr–Co alloys (i.e., MP35N, Elgiloy, L605, etc.), and superelastic Ni–Ti (Nitinol). Other than Nitinol, these materials need to be stress relieved and hardened by precipitation or an age-hardening heat-treatment process for improved kink resistance.

Once a suitable guidewire has been delivered to the diseased vessel, the PTA balloon catheter is placed concentrically over the guidewire and pushed directly to the blockage. These balloons come in a range of sizes, from small coronary sizes (2–4 mm) to large diameter balloons for peripheral vessels (15–20 mm). Balloons are typically designed to expand to a specified diameter at a deployment pressure of approximately 10 atm, with very little further expansion at higher pressures. This allows the operator to predetermine the final expanded diameter. A variety of complex balloon shapes can be fabricated from polyethylene to completely fill various body cavities. Other balloon materials are low- and high-density polyethylene, polyvinyl chloride (PVC), and nylon. However, when lower precision balloons are desired, elastomeric balloons are used that conform to the lumen shape as they expand. There are numerous balloon shapes, for example, conical balloon, spherical balloon, dog bone balloon, conical/square long balloon, stepped balloon, offset balloon, and so on, which could be fabricated/formed in different diameters and lengths. A specific shaped balloon may be used depending upon the anatomical site to be treated, the requirements of the treatment process, or both. Platinum or gold marker bands are attached at each end of the balloon for distinctive tip radiopacity. This allows the operator to position the balloon precisely prior to inflation.

Important design considerations for balloon catheters extend beyond their ability to expand the lumen of the diseased vessel. The "trackability" of a catheter refers to its ability to track through tortuous vessels, or how easily the catheter follows over the guidewire through a tortuous vessel. Obviously this is a function of catheter sliding friction force (frictional resistance between the guidewire, catheter, and vessel wall), constructing elements/materials, and geometrical configurations (shaft diameter and length, column strength, lateral flexibility). Small diameter catheters have greater flexibility, but lack column strength. This makes it necessary to use stronger shaft materials, distal tapering and/or reinforcing with braids, coils, or stiffening wires. The crossability of a catheter refers to the ability of its distal end to traverse the lesion to be treated. Thus, the goals are to design low-profile catheters with small diameter distal tips, and balloons with high inflation ratios (inflated to deflated diameters). These strategies are augmented with polymeric coatings to reduce friction, and minimizing the amount of adhesive used for balloon–catheter attachment. The adhesion spots create relatively stiff sections that impede crossability. Finally, gradual transitions in catheter stiffness and matching properties of catheters to guidewires help to optimize crossability.

In many cases, clinicians determine that some of the atherosclerotic plaque must be removed to increase the long-term chances for an open lumen. This process is referred to as "debulking." There are a variety of technologies developed for this purpose, but the degree of success has been limited. Despite considerable development work in the 1980s and 1990s, laser angioplasty has never enjoyed widespread clinical use. This procedure uses high laser energy to vaporize plaque by breaking molecular bonds. The guidewire is positioned across the target coronary lesion, and the laser catheter is activated and advanced slowly through the lesion. In a laser balloon angioplasty device used in coronary angioplasty, the laser is mounted inside the balloon and transmits light through the balloon walls. Mechanical atherectomy catheters aim to physically destroy and remove, rather than displace, the plaque material. The design of this type of device requires pulverization and vacuum aspiration of material to prevent thromboembolic events. In a subset of these devices, directional coronary atherectomy, a catheter/cutter is pushed against arterial plaque. A related application of such technologies is the removal of thrombus from bypass and AV fistula grafts (thrombectomy).

An emerging technology for the treatment of atherosclerotic plaques is the drug eluting balloon (DEB). DEBs are used in conjunction with traditional PTA procedures to slow the development of restenosis and are usually deployed after a traditional PTA balloon has been used to enlarge the lumen (Scheller et al., 2006). Multiple DEB designs are available (De Labriolle et al., 2009). One type of DEB is similar in design to traditional PTA balloons; however, the balloon surface is coated with a drug, usually paclitaxel. The drug is adhered to the balloon surface either embedded in a hydrophilic matrix, a rapidly bioabsorbable matrix, or contained in micropores (Scheller et al., 2006; De Labriolle et al., 2009). The DEB is folded to limit the amount of drug that is lost as the balloon is directed to the site of the lesion. Upon inflation, usually 60 s in duration, the majority of the drug is released and approximately 10% is retained in the artery wall. In the case of the bioabsorbable matrix the drug and matrix are both transferred to the artery lumen where the matrix degrades and the drug is absorbed. Another type of DEB is the double balloon catheter (Herdeg et al., 2007). Double balloon catheters have been previously used to treat other diseases including cancer. This catheter consists of two balloons designed to span the length of the lesion. Once the balloons are inflated the drug is released into the space between the balloons to transfer to the artery wall. Although DEBs represent a short-term exposure to the antirestenosis drug both animal and clinical studies have shown that restenosis rates are significantly reduced by the use of DEBs (Marzullo et al., 2011). Additionally, the rates of reendothelialization are unaffected which suggests that DEBs may shortly be a popular tool for the treatment of atherosclerosis (Speck et al., 2006). Indeed this technology offers many advantages over more traditional technologies: (1) uniform drug delivery across the lesion surface, (2) high, short-term dose delivered immediately after artery injury but absent during long-term healing and reendothelialization, (3) can be deployed in smaller vessels, and (4) can be used in conjunction with stents (Waksman and Pakala, 2009).

20.3 Stents

First described in the modern era by Dotter (1969), stents were designed to improve upon the limited success of balloon angioplasty for the treatment of occlusive vascular disease. A stent is a tubular scaffold that is expanded and deployed inside the artery in order to prop open the lumen. Dotter attempted placement of plastic stents in canine popliteal arteries that thrombosed after 24 h. Subsequent animal studies employing both stainless-steel coil stents and nitinol coils were completed successfully (Dotter et al., 1983). The first human use of stents occurred in 1986 (Sigwart et al., 1987).

The variety of stent designs conceived or developed is staggering. A search of the U.S. Patent and Trademark Office database in late 2002 revealed more than 4500 patents including the term "stent." By 2012, that number had grown to 15,890, but not all of these patents were for vascular devices. Most vascular stents can be classified as either balloon-expandable or self-expanding. Balloon-expandable stents are mounted over an angioplasty balloon, delivered to the diseased vessel site, and expanded through inflation of the balloon. Balloon-expandable stents are often made of 316 stainless steel or other noncorrosive, stiff alloy metals (e.g., tantalum, or Cr–Co super alloys Elgiloy, MP35N, and L605). The stent is permanently or plastically deformed to the desired diameter, and the balloon is removed (Figure 20.3). Self-expanding stents are produced at the desired final diameter, and spring-loaded into a much smaller catheter (Figure 20.4). Once the stent is delivered to the diseased vessel, the catheter sheath is withdrawn, and the stent expands elastically outward into the artery. The need for a high degree of elastic deformation requires that self-expanding stents be made of more elastic materials such as Nitinol, a nickel–titanium alloy with shape memory capability (Duerig et al., 2000). Nitinol can be strained approximately 8% and remain elastic if processed correctly, compared to less than 0.5% for stainless steel. Its shape memory properties have also been exploited in the design of stents. Polymer stents (bioresorbable scaffolds) are also under development (Tsuii et al., 2001). The ability of a polymer stent to degrade safely once the artery has remodeled is desirable, but much development work remains to be done for this process to be dependable.

The key parameters in stent design are numerous, and in some cases conflicting (Table 20.1). The first important issue is that the material be biocompatible, since it will be exposed directly to the blood stream. To be deployed in small narrowings, a high expansion ratio (deployed diameter/crimped diameter) is required. The crimped diameter of the stent is often referred to as its profile, which should be as low as possible. The expansion of the stent from the crimped to the deployed state must result in as little axial foreshortening as possible or it will be difficult to predict the final position of the stent. Deploying the stent often requires navigating tortuous vessels or entering branches with an acute angle. Thus, the stent must have good longitudinal flexibility. Propping open a previously blocked passage while remaining in place following deployment requires a high degree of radial strength and accurate positioning of the stent requires that it be sufficiently radiopaque. These two considerations generally require a minimum amount of material in the stent that may compromise the low profile requirement. Thus, there must be a trade-off in the design process. Other considerations include the ability to retrieve the stent in the case of mal deployment, and minimal interference with CT and MRI imaging. The delivery system

FIGURE 20.3 Illustration of a balloon expanded stent deployed inside a stenotic artery. The atherosclerotic plaque material is displaced and held outward by the more rigid stent structure.

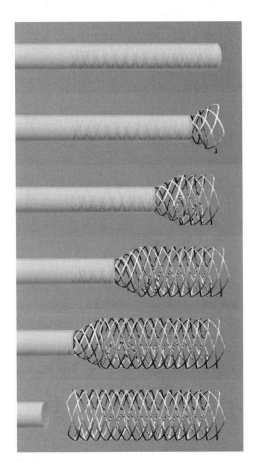

FIGURE 20.4 Illustration of a self-expanding stent being deployed. The sheath is withdrawn, releasing the spring-loaded stent to its full diameter.

must work well with the particular stent design, and must hold the stent in place in a way that does not injure or snag on the artery wall as it is guided to the disease site.

Perhaps the most important design characteristic for stents is that they do not provoke the development of a new blockage either acutely through thrombosis or well after deployment, due to hyperplasia. Unfortunately, there is little direct information on how stent design affects these processes. The reaction of the body to a stent implanted in an artery is a multistage process (Edelman and Rogers, 1998). First, the exposure of the subendothelium and the stent material to the blood stream may initiate thrombus formation (Figure 20.5). This process, which takes place within minutes/hours, includes an aggregation of platelets, fibrin, and erythrocytes. The degree of platelet adhesion depends not only on the surface characteristics of the stent, but also on the strut configuration. Areas of flow stagnation, which depend heavily on strut design, influence the degree of platelet adhesion (Robaina et al., 2003). *In vitro* studies have shown that areas subjected to intermittent flow stagnation and low wall shear stress are conducive to greater platelet accumulation. The second stage of the reaction to the stent is inflammation. The peak of this process occurs approximately 4–14 days following stent implantation. Deposits of surface adherent and tissue infiltrating monocytes can be seen around stent struts, demonstrating the degree to which the struts are injuring the wall. These monocytes release cytokines, mitogens, and tissue growth factors that further increase neointimal formation. The third stage is the proliferation of vascular smooth muscle cells in the media and neointima. This process depends on the stent material, as well as the stress placed on the artery wall by the stent. The final stage of arterial adaptation is remodeling. One can think

TABLE 20.1   Design Considerations for Vascular Stents

Characteristic	Comment
Outward radial force	Necessary to prop open artery. Excessive radial force can injure artery wall and promote hyperplasia over time
Material	Must be biocompatible and noncorrosive. Typical: SS316, Co/Cr alloys, nitinol, polymers
Radiopacity	Depends on material properties and total mass present. More radiopaque markers or coatings of different material may be added to stent
MRI artifacts	Depends on material properties
Profile	Diameter of catheter containing crimped stent. Must be as low as possible
Expansion ratio	Ratio of deployed to crimped diameter. Higher is better
Expansion mode	Typically balloon or self-expanding
Foreshortening	Ratio of deployed length to crimped length. Should be 1 for accurate placement
Longitudinal flexibility	Necessary to navigate tortuous arteries. Deployment in curved arteries should not injure the wall at the ends of stent
Surface treatment	Highly polished and passivated to reduce thrombus and inflammation
Surface coating	Polymers with embedded anti-inflammatory drugs show promise
Manufacturing method	Typically laser-machined from tube stock. Also knitted mesh and coils. Thermal treatments may be necessary
Fatigue resistance	Must be able to withstand at least 10 years of cyclic fatigue

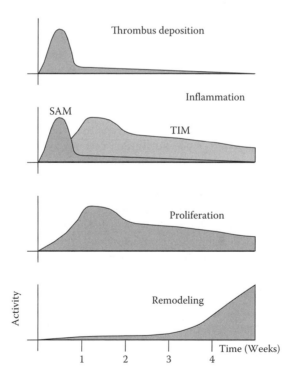

FIGURE 20.5   Response of the artery wall to the presence of a stent. The process begins with thrombus deposition, followed quickly by inflammation. Surface adherent monocytes (SAMs) respond to the injury presented by the stent, then enter the wall as tissue infiltrating monocytes (TIMs). Cellular proliferation (mainly smooth muscle) provides additional tissue to shore up stress concentrations in the artery wall. Finally, the remodeling stage takes place in which the artery attempts to redefine a new homeostatic state. (Adapted from Edelman, E.R. and Rogers, C. 1998. *Am J Cardiol* 81, 4E–6E.)

of this phase as the artery's attempt to reach a new homeostatic state in the presence of the persistent injury caused by the stent. The final thickness of the neointima depends heavily on the degree to which the stent injures the artery wall, as indicated by the disruption of the internal and external elastic laminae (Rogers and Edelman, 1995).

Acute thrombosis was a common cause of failure for stenting procedures in the early 1990s, but advances in stenting technology and aggressive antiplatelet drug regimens have reduced these failures to nearly negligible levels. The deployment of a stent with or without balloon inflation substantially denudes the endothelial layer of the artery, which normally provides antithrombotic protection. The complete reestablishment of the endothelial layer takes weeks to months, leaving a considerable time gap for clinically significant thrombosis to form. The potential for thrombus is enhanced by the fact that the stent itself is a foreign material that provokes platelet attachment and buildup. Electrochemical surface treatments have reduced the risk associated with the stent itself. However, the use of systemic antiplatelet drugs has played a more dominant role in reducing acute stent thrombosis. Administration of heparin (an indirect thrombin inhibitor) during stenting procedures was common in the early days of stenting. This has largely been displaced by Bivalirudin, a direct thrombin inhibitor derived from the saliva of leeches. This drug has a highly predictable behavior in a wide spectrum of patients, and a half-life of approximately 25 min (Warkentin et al., 2008). Platelet glycoprotein inhibitors such as Abciximab are also used during procedures. Abciximab is a glycoprotein IIb/IIIa receptor antagonist that blocks platelet interactions. Postprocedure, standard dual antiplatelet therapy is generally required. This involves administration of aspirin and a thienopyridine; Clopidogrel being the most commonly used. This class of drugs inhibits the P2Y12 receptor to prevent platelet activation. Ticlopidine can be effective when used in conjunction with aspirin, but suffers from a number of undesirable side effects. Clopidogrel, like Ticlopidine, inhibits platelet activation, but has fewer side effects. However, genetic variations in liver enzymes and drug–drug interactions with proton-pump inhibitors can decrease the effectiveness of Clopidogrel in some patients. The combination of aspirin and Clopidogrel (whose dosages can be optimized for a particular patient) far outperforms the use of either alone, and has become the standard of care. Cessation of dual antiplatelet therapy is the single most important predictor of stent thrombosis (Abualsaud and Eisenberg, 2010).

In the months following stent implantation, the growth of tissue into the stented region can cause a new blockage, a process called in-stent restenosis. This process is responsible for the clinical failure of 20–30% of bare metal stenting procedures within 6 months following implantation. Among the factors that influence the risk for restenosis is stent design. In a clinical study of more than 4500 coronary stent implantations (Kastrati et al., 2001), it was shown that restenosis rates for different stent designs vary from 20% to 50% (Table 20.2). The exact relationship between stent design and the development of intimal hyperplasia is not well understood. It is generally thought that the percent area covered by the stent and the mechanical mismatch between the stent and artery wall are important factors. The characteristics of the original atherosclerotic plaque also play an important role.

The problem of restenosis is being addressed by coating stents with polymers in which drugs have been embedded. The polymers provide a stable matrix into which drugs are uniformly distributed and released over a specific period of time, typically days to weeks (De Luca et al., 2012). Although most stent-coating technology is closely guarded, several coating techniques have been described. The most common polymer coating that is often applied to the surface of the stent struts uses nonerodable and nonbiodegradable methacrylate and/or ethylene base copolymers. The other biodegradable and nonbiodegradable polymers often employed include nylon, polyurethane, silicone, polyethylene terphthalate, polyglycolic acid/polylactic acid (PLGA), polycaprolactone, polyhydroxybutyrate valerate, poly(*n*-butyl methacrylate), polyorthoester, poly(ethylene-*co*-vinyl acetate), and polyethyleneoxide/polybuty-leneterephthalate (Uhrich et al., 1999). One important factor in the choice of polymers is the adherence of the polymer to the stent. With deployment, mechanical stress/strain may have a tendency to delaminate the coating or cause uneven drug distribution (Hwang et al., 2001). Other factors involve uniform distribution of the drug for expected efficacy.

TABLE 20.2 Restenosis Rates of Different Stent Designs Implanted in Human Coronary Arteries[a]

Stent	Restenosis Rate (%)	Design (Strut Thickness)
Guidant Multi-Link	20	
Jomed Jostent	25.8	
J&J Palmaz-Schatz	29.0	
PURA-A	30.9	
Inflow Steel	37.3	
NIR	37.8	
Inflow gold	50.3	

Source: Adapted from Kastrati, A. et al. 2001. *Am J Cardiol* 877, 34–39.

[a] All stents included in this study are laser machined from 316 L stainless steel, and balloon expanded. The Inflow gold stent is coated with gold to improve radiopacity.

Four drugs are currently used in the coatings of commercially available drug eluting stents. Paclitaxel (Taxol), previously used as an anticancer agent that reduces cellular proliferation, migration, and signal transduction, has demonstrated *in vitro* and *in vivo* success. Early clinical trial data showed that stents coated with this drug have demonstrated reduced restenosis rates (Grube and Büllesfeld, 2002). Sirolimus (Rapamycin) is a natural microlide immunosuppressant that inhibits vascular smooth muscle cell proliferation through blocking the cellular transition from the Gl to the S phase (Gallo et al., 1999). This proliferation is prevented when Sirolimus binds to its cytosolic receptor, FKBP12, and blocks the activation of T cells, the lymphocytes responsible for cell-mediated immunity. Through this mechanism, Sirolimus also prevents smooth muscle cell migration, thus interrupting the cascade of cellular activity leading to neointimal hyperplasia causing restenosis (Sousa et al., 2001). Early clinical studies of stents coated with Sirolimus showed near zero restenosis rates (Degertekin et al., 2002). This drug has also been systemically applied to patients for the prophylactic prevention of renal transplant rejection (Poon et al., 2002). Everolimus, a modified form of Sirolimus, also inhibits vascular smooth muscle cell proliferation by preventing cells from exiting the G1 phase (Schuler et al., 1997; Lavigne et al., 2012). Clinical trials carried out to 6 months have shown that Everlimus eluting stents effectively reduce neo-intimal proliferation compared to bare metal stents (Tsuchiya et al., 2006). Another Rapamycin-derived drug, Zotarolimus, has also been investigated as an antiproliferative treatment for arterial smooth muscle cells. Zotarolimus decreases smooth muscle cell proliferation through a mechanism similar to that of Sirolimus; however, the shorter half-life of Zoterolimus suggests that it will produce less immunosuppression than Sirolimus (Chen et al., 2007).

Wide adoption of both Paclitaxel-coated and Sirolimus-coated stents have made it possible to compare the efficacy of drug eluting stents to their bare metal stent precursors. Early assessments of risk and patient health (<6 months to 1 year after implantation) indicate that drug eluting stents outperform their bare metal stent counterparts. After the first year this trend reverses itself and an increasing risk of myocardial infarction, stent thrombosis, or target vessel revascularization is observed (Lagerqvist et al., 2007; Kalesan et al., 2012). There is speculation that this late increase in negative outcomes may be due to delayed healing caused by the polymer coatings or insufficient duration of Clopidrogel therapy post-stent placement (Grube et al., 2004; Lagerqvist et al., 2007). Development of biodegradable coatings, as well as investigation of additional drug options, seek to solve this issue.

An emerging technology in the treatment of atherosclerotic plaques is the bioresorbable scaffold (biodegradable stent). Although nonbioresorbable stents (described above) have many advantages over PTCA alone, restenosis and neointimal hyperplasia remain a problem. One theory is that the occurrence of stenosis years after stent implantation may be attributable to an inflammation reaction to permanently implanted stent struts (Ormiston and Serruys, 2009). Although stents prevent elastic recoil and tack down lumen dissections, positive remodeling of the artery is complete after approximately 6 months, making the continued presence of the stent unnecessary. Bioresorbable scaffolds would also be advantageous in situations where a permanent stent would eventually interfere with artery function (i.e., pediatric procedures) or additional interventions (i.e., target artery revascularization). Bioresorbable polymers such as polylactide (D- and L-), polyglycolide, poly ε-caprolactone, and their copolymers as well as bioresorbable metals such as magnesium alloys and iron alloys are being investigated as possible materials for bioresorbable scaffolds. These materials are chosen based on their mechanical properties, biocompatibility, and degradation rates. Ideally the scaffolds should be constructed from a material that can withstand high radial stresses and cyclic strains, undergo large amounts of deformation to allow the stent to expand upon delivery to the site of the stenosis, and exhibit low-yield strains to prevent elastic recoil (Onuma et al., 2011). Additionally, the material and its degradation products should cause little or no damage to the blood vessel and degrade within a suitable amount of time (>6 months). Bioresorbable scaffolds can also serve as vehicles for drug delivery. Drugs, such as the ones described above, can be embedded in the scaffold matrix and released upon resorption (Serruys et al., 2009). Small clinical trials, such as the ABSORB trial, indicate that bioresorbable scaffolds are a promising alternative to the current stent technology (Serruys et al., 2009; Onuma et al., 2011). The ABSORB trial demonstrated the patency of poly-L-lactide scaffolds in human patients. Two years after implantation, the scaffolds had been completely resolved with no evidence of restenosis; however, there was evidence of stent recoil. A second bioresorbable scaffold was designed which changed the strut spacing and geometry in an attempt to prevent recoil. The ABSORB cohort B trial demonstrated that the redesigned scaffold successfully prevented recoil (Gomez-Lara et al., 2010). Widespread clinical use over time will reveal whether these exciting new devices really solve the problems of restenosis and neointimal hyperplasia.

20.4 Aneurysm Treatment

The technologies developed for the treatment of PTA and stenting are applicable to other areas of cardiovascular disease as well, including aneurysms. In 1991, Juan Parodi treated a patient with an

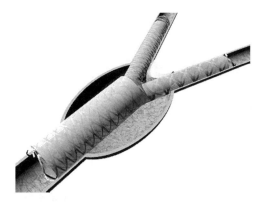

FIGURE 20.6 Illustration of an endograft deployed to seal off an AAA. The graft material clots off, sealing the aneurysm from being subjected to arterial pressure, and reducing the risk of rupture.

abdominal aortic aneurysm (AAA) with a balloon expandable stent to which a vascular graft had been attached (Parodi et al., 1991). This stent graft, or endograft, was deployed via a percutaneous catheter approach. The idea was to seal the aneurysm from arterial pressure and reduce the risk of its rupture (Figure 20.6). AAAs, the 13th leading cause of death in the United States, manifest as a local ballooning of the aortic wall that forms 1–2 cm distal to the renal arteries. The aneurysm often extends into the iliac arteries. The large radius and reduced wall thickness in the aneurysm make it prone to rupture, often a fatal event.

Since Parodi's initial clinical experience in 1990, there have been numerous endografts developed for AAA treatment. There are eight designs currently in use and are available in a range of sizes. The important design considerations include the ability to remain sealed against healthy portions of the artery wall proximal and distal to the aneurysm, ability to clot off the porous graft material, construction using radiopaque materials and minimal delivery profile. It is also important to recognize that aneurysm geometry is highly individual. The device must be prepared to treat a wide variety of vessel diameters, lengths, and degrees of tortuosity. Although Parodi's initial design was a single tube, the extension of many aneurysms into the iliac arteries required adaptation to bifurcated designs. A typical approach is to deploy the main body and one leg of the endograft via one femoral artery, then deploy the other leg of the endograft via the contralateral femoral artery. It is also possible to extend a single tapered graft into one iliac artery, and then implant a femoral cross-over graft surgically to supply the contralateral side. The contralateral side must then be sealed using an iliac occluder plug. Regardless of the particular design, the ability to make a reliable seal between the endograft components is crucial. Both balloon expanded and self-expanding stents have been used to make endografts. In some designs, the stents extend the whole length of the endograft, while in others short stents are included only at the proximal and distal attachment regions (Figure 20.7). Helical staples which secure the body of the unsupported endograft to the wall of the aneurysmal sac are used with the Aptus system (Melas et al., 2012). The proximal attachment site is particularly worrisome because the aneurysm may extend proximally to the level of the renal arteries. The length of "healthy neck" in the aorta between the renal arteries and the beginning of the aneurysm is an important criterion in determining patient eligibility for this treatment. This has led endograft designers to devise ways of securing endografts above the renal arteries without blocking their outflow. The graft material is typically PTFE or Dacron, and is attached to the graft material with sutures or other means.

Many of the same philosophies can be used when treating thoracic abdominal aneurysms. Two years after Parodi's landmark procedure, Dake et al. (1994) pioneered the use of similar technology for the descending thoracic aorta, but its development has lagged behind that of its renowned cousin due to lower incidence rates. TEVAR devices, or thoracic endovascular aortic repair devices, are used for the

(a)   (b)

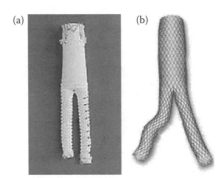

FIGURE 20.7 Illustrations of two endografts used to treat abdominal aortic aneurysms. The Guidant Ancure device (a) is an "unsupported" design in which stents are placed at the proximal and distal ends, but not the middle portion. The Medtronic Aneurx device (b) is a fully supported device in which the stents run the entire length.

treatment of aneurysms in the ascending, arch, descending, and thoracoabdominal aorta. Endographs are traditionally made of nitinol self-expanding stents with a Dacron or woven polyester graft attached to the struts, much like the AAA endograft but without a bifurcation. Complications with treating these regions of the aorta include bird beaking, where the graft fails to conform to the bend of the aortic arch and instead sticks out into the lumen, and aortic branch management to avoid blocking side branches of the aortic arch (Figure 20.8). Branch management techniques available today are adjunctive, where a smaller graft is sewn into the main endograft at the site of the branch. Current innovative efforts are focused around this problem (Lanzino, 2011).

The failures of endografting have been classified into four categories (Table 20.3). Type I failures are leaks that occur at the proximal or distal attachment sites. Type II failures occur when the aneurysm sac receives pressurization from communicating circulation, typically from arteries that connect to the inferior mesenteric artery. Type III failures are seen when leaks occur at endograft component junctions, or there is some other breach in the structural integrity of the middle portion of the endograft. Type IV failures include sac pressurization due to continued graft material porosity. The occurrence of these different failures depends on endograft design and patient-specific conditions (van Marrewijk et al., 2002). Type I and III failures can be treated with the deployment of additional endograft components. In many cases, sleeves of bare or graft-coated stents are inserted over the trouble area. There is considerable controversy concerning the treatment of Type II endoleaks. Coil embolization can be performed at the time of endograft implantation or later, but access to these often small communicating arteries is difficult. Many clinicians prefer to monitor these cases for changes in aneurysm morphology

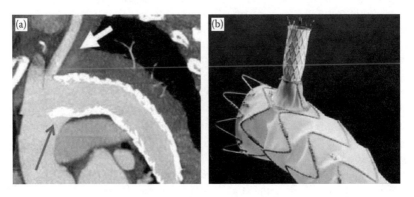

FIGURE 20.8 (a) An example of bird beaking as seen in the aortic arch. (Criado, F.J. 2011. *Endovascular Today* 10(11), 34–47.) (b) Medtronic's branched stent graft, designed to allow flow through the vessels branching off of the aorta.

TABLE 20.3 Failure Classifications for AAA Endografts

Endoleak Type	Description
I	Attachment site leak
IA	Proximal end of graft
IB	Distal end of graft
IC	Iliac occluder (plug)
II	Branch leaks from communicating arteries
IIA	One branch inlet only
IIB	Complex, flow-through, or multiple branches
III	Graft defect
IIIA	Junctional leak or modular disconnection
IIIB	Fabric disruption (hole)
IV	IV Graft porosity

rather than actively treat the endoleaks. As a last resort, the patient can undergo conversion to surgical treatment, in which an artificial graft is sewn in place of the diseased aorta. Endograft design is evolving to minimize the risk of failure, but progress is impeded by lack of knowledge concerning the truly physiological mechanical conditions to which endografts are subjected. In addition to tortuous geometries and cycling pressures, which are included in FDA required device testing, endografts are subjected to complex hemodynamic forces (Liffman et al., 2001). Changes in vessel geometry that occur during normal physiological movement such as hip flexion may be important as well and these changes are beginning to be understood to the point that they can be accounted for during fatigue testing.

Minimally invasive devices have also been developed to treat aneurysms forming in the cerebral circulation. Ruptures of cerebral aneurysms eventually kill approximately half of those afflicted. It is desirable in this case to form a thrombus inside the aneurysm so that the structural weakness is sealed off and eventually covered with neointima. This can be accomplished with the catheter-based deployment of a thin, metallic wire. Once a sufficient length of the "coil" has been deployed into the aneurysm, an electric charge is applied to the end of the catheter, breaking it off from the catheter. Platinum is often used for the wire material (Uda et al., 1998; Hope et al., 1999). For these devices, being able to advance, retract, and reposition the coils prior to detaching them is important to the success of the procedure. This has been a focal point for innovation over the past 10 years and is now standard in these devices. Platinum coiling has not enjoyed the success it was initially anticipated to have. The coils have a low packing density (Gaba, 2006), meaning they only fill up 20–30% of the aneurismal sac when properly deployed and the platinum is bioinert, meaning it does not attract fibrous material necessary to stabilize the clot that forms after the coils are introduced (Kallmes, 2002). Matrix bioactive coils were developed to add a bioactive component to the traditional platinum coil. They consist of traditional platinum coil material surrounded by PGLA, a known bioresorbable polymer that induces tissue formation. Accelerated thrombus formation and healing does occur after treatment with these coils, but overall only a marginal improvement in patient outcomes has been seen (Piotin, 2012). Hydrogel coils were developed to increase the packing density and have appreciated success in clinical trials. The gel-like hydrophilic polymer that swells in blood surrounding the platinum wire conforms to the inner geometry of the aneurismal sac, even at low pressures. It is believed that a correlation between the packing density and the success of the treatment exists. Hydrogels are attractive because of their ability to expand 5–10× their original diameter (Figure 20.9), which means lower profile introductory catheters and the ability to treat larger aneurysms with less coil material. Hydrocoils routinely see 85% occlusion of the aneurysm with shorter coil lengths than bare platinum coils (Gaba et al., 2006) but the hydrogel itself can only be worked with for a short amount of time before it is set and cannot be repositioned (Kallmes and Fujiwara, 2002). Other solutions such as thermosetting shape memory polymers around wires are being investigated as a way to get increased expansion ratios and ultimately lower profile devices.

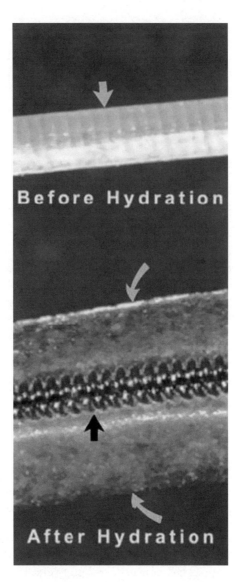

FIGURE 20.9 A platinum coil wrapped with a bioactive hydrogel. (Top) The coil in its inactive state. The indentations from the windings of the underlying platinum coil can be seen in the outer edges of the hydrogel (straight white arrow). (Bottom) After activation the hydrogel expands and becomes translucent. The smooth outer edges are denoted by the curved white arrows and the platinum coil is denoted by the black arrow. (Adapted from Kallmes, D.F. and Fujiwara, N.H. 2002. *AJNR Am J Neuroradiol* 23(9): 1580–1588.)

The placement of a stent or a flow diverter across the neck of the aneurysm is also a treatment option (Figure 20.10). In this case, the blood flow in the aneurysm is slowed by the stent mesh or flow diverter, causing thrombus to form (Lieber and Gounis, 2002). The device also aids in forming neointima. The use of a flow diverter in conjunction with metal coils is a common method of treating aneurysms where the coils are more likely to migrate, such as those with wide necks. Early clinical trials have shown that the use of flow diverter not only causes the blood inside the aneurismal sac to clot, but it leads to shrinkage and eventually complete retraction of the aneurysm (Lanzino, 2011).

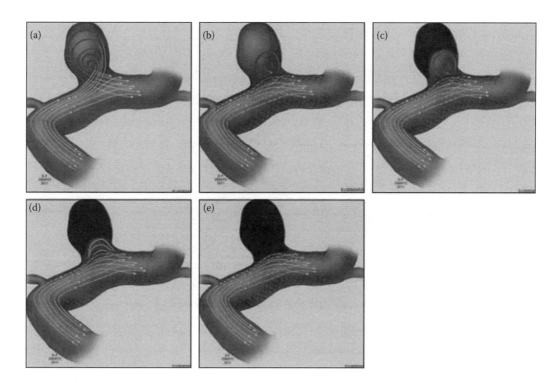

FIGURE 20.10 **(See color insert.)** Prior to the insertion of the flow diverter, whirling flow can be seen in the aneurysm (a). After deploying the flow diverter, there is a gradual decrease in the amount of whirling flow inside the aneurysmal sac (b) followed by thrombus formation (c, d) before eventual resolution of the flow (e). (From Lanzino, G. 2011. *Endovascular Today* 10(9), 35–36.)

20.5 Embolic Filters

There are a variety of clinical situations in which embolic material shedding from a plaque, thrombosis, or intervention site need to be trapped and prevented from reaching distal tissues. If emboli from a deep-vein thrombosis reach the heart or lungs, the results can be deadly. Deep-vein thrombosis can result from extended periods in which the legs are kept stationary, such as postoperative situations. Emboli can also result from clinical interventions to treat occlusive atherosclerotic plaques. Balloon angioplasty and stenting provoke plaque fracture, thus the possibility that pieces will break off and lodge downstream in a smaller artery. This is a particular concern for carotid stenting, where emboli can cause serious brain ischemia (Sievert and Rabe, 2002). Emboli may also result during the treatment of stenoses in saphenous vein coronary bypass grafts (Morales and Heuser, 2002).

Emboli from deep-vein thrombosis are responsible for an estimated 200,000 deaths per year in the United States (Zwaan et al., 1995). Systemic delivery of heparin and thrombolysis agents has been successful in treating some emboli; however, the effectiveness of systemic drug treatments is limited by the decreased blood flow to lesion sites and increased risk of bleeding complications (Lin et al., 2010). With the use of catheter-based treatment options a variety of alternatives to systemic drug treatments are available. Vena cava filters deployed via a catheter approach have been in use since the 1960s to prevent emboli from reaching the pulmonary circulation. The most common current design is the Greenfield filter (Figure 20.11). It is thought that the filter captures larger emboli in its middle section, which are then broken down by the body's natural thrombolytic process (Greenfield and Proctor, 2000). Competing designs have emerged that feature different capture basket designs and deployment schemes. Problems

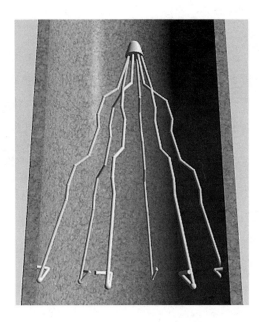

FIGURE 20.11   Illustration of the Greenfield vena cava filter. This device is deployed percutaneously to prevent emboli from deep-vein thrombi from reaching the heart and pulmonary circulation.

with permanent filters include vena cava thrombus rates of up to 19% (Ferris et al., 1993). More recent designs have emerged that can be withdrawn, dissolved, or converted to simple stents once the danger of emboli is thought to have subsided (Stecker et al., 2001). Testing of all filter designs *in vitro* with artificial emboli have reported trapping rates ranging from 22% to 98%, depending on filter design and flow conditions (Lorch et al., 1998; Finol et al., 2008). Although retrievable vena cava filters have been approved for both temporary and permanent placement *in vivo*, recent clinical data have prompted the FDA to release guidelines urging for the removal of these retrievable devices once they are no longer needed (FDA, 2012). A 2011 review of the clinical trials carried out to date suggest that the incidence of complications associated with the use of retrievable filters increase with the length of time the filter is left in the body (Angel et al., 2011). The most commonly observed complications included artery perforation, filter migration, and filter fracture; 90% of these complications were observed after 30 days (Angel et al., 2011). In addition to filters, other catheter-based treatment methods have been developed to directly remove the emboli. Thrombectomies, similar to the debulking procedures described for the removal of atherosclerotic lesions, apply a mechanical stimulus, rheological stimulus, or ultrasound pulse directly to the emboli to break up and dislodge it (Lee et al., 2004; Lin et al., 2010). In some cases, a thrombolytic agent is also delivered through the catheter to work in conjunction with the thrombectomy device; this method, pharmacomechanical thrombectomy, has the advantages of allowing the use of lower drug concentrations which can be delivered directly to the lesion site (Hilleman and Razavi, 2008; Parikh et al., 2008; Lin et al., 2010). These methods are accompanied by a gentle aspiration to remove the debris, and in some cases, the lesion site is isolated from the vascular network by occlusion balloons for the duration of the procedure (Lin et al., 2010). These devices have shown promising clinical results and have improved patient recovery.

Another class of temporary embolic protection devices has emerged to augment endovascular treatments, the so-called distal protection devices. Because they remain in place only as long as the intervention itself, designs for distal protection devices ranges from total occlusion balloons to basket-like porous membranes (Figure 20.12). *In vitro* studies of distal protection device effectiveness have indicated that these filters effectively capture a large percentage of the debris dislodged by angioplasty and

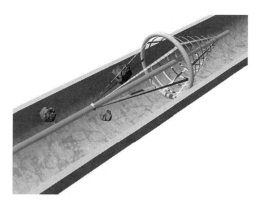

FIGURE 20.12 Illustration of a distal protection device deployed in an artery. Emboli from a procedure being performed proximal to the device are captured, then removed when the device is recaptured into the catheter.

stenting procedures; however, none of the devices evaluated captured all the debris present (Müller-Hülsbeck et al., 2003; Finol et al., 2008; Siewiorek et al., 2009). Additionally, although neurological events, such as strokes, are associated with the presence of debris, the absence of debris is difficult to predict indicating that distal protection devices should be used in all cases (Sprouse et al., 2005). However, these devices can also cause damage to the vessel wall (Müller-Hülsbeck et al., 2005). One clinical trial found that distal protection devices were associated with an increased risk of thrombosis and target vessel revascularization at the 15 month follow-up (Kaltoft et al., 2010). Although this may be due, at least in part, to a mismatch between the artery and device diameters it still indicates a need for caution when employing these devices (West et al., 2010). Another embolic protection device is the proximal protection device. Like the distal total occlusion balloons these devices inflate to prevent blood flow through the lesion sight during the angioplasty or stenting procedure. In both cases, debris is removed through gentle aspiration (Mauri et al., 2006). However, the proximal occlusion balloons are placed proximal to the atherosclerotic lesion which eliminates the need for unnecessary equipment to cross the lesion site before the debris collection device is in place (Haeck et al., 2009). Clinical trials for this type of device have found evidence of improved microvascular flow; however, no difference in major cardiac and cerebral events, infarct size, or left ventricular function was observed (Haeck et al., 2009, 2010).

20.6 Cardiac Ablation Catheters

Cather-based ablation therapies represent a highly effective, minimally invasive treatment for cardiac arrhythmias (Lustgarden et al., 1999; Bella et al., 2001; Keane, 2002). Pharmacological therapy, a traditional first line of defense, has limited success in preventing future arrhythmic episodes which is problematic since continued arrhythmia can cause pathological structural changes in the arteries and the heart (Wijffels et al., 1995; Pappone et al., 2011; Thomas and Sanders, 2012). Additionally, techniques ranging from the radiofrequency-based to laser-enhanced approaches have simplified surgical treatment and avoided several of the drawbacks of surgery (Boersma et al., 2012). Catheter-based ablation therapy is quickly becoming the standard treatment for arrhythmias due to its high rate of success and its effectiveness in wide populations (Leitch and Barlow, 2012; Melo et al., 2012; Thomas and Sanders, 2012).

Ablation technology destroys small areas of the myocardium, which exhibit irregular electrical activity, and produces scar tissue in its place. This eliminates the arrhythmogenic areas and allows the surrounding tissue to regain its normal excitation profile. Radio-frequency-based catheter ablation is currently the most widely used type of catheter ablation (Challa and Mansour, 2011). This technique uses radio-frequency energy to generate heat (>50°C) at the tissue site (Swanson et al., 2011). Radio-frequency-based catheter ablation has a relatively safe record, producing precisely defined lesions.

However, the technique still has some significant disadvantages, primary among which is the limited tissue depth which can be treated with this type of energy. Recent research suggest that radio-frequency ablation with concomitant saline infusion is one way to increase the lesion size (Mittleman et al., 1995; Watanabe et al., 2002; Sapp et al., 2006). Additionally, the energy required to produce these lesions can cause endocardial disruption. This is often due to the primary requirement of continuous contact between the electrode and the endocardium during ablation.

Alternative ablation techniques include both cryogenic and thermal ablation (microwave-, infrared-, and laser-based) approaches. Catheter-based cryoablation uses freezing, rather than heating to destroy the abnormal areas of the myocardium. Recently, cryoablation has been used specifically in the percutaneous transvascular mapping and ablation of arrhythmias (Klein et al., 2008; Namdar et al., 2012). Some of the potential advantages of cryoablation with regard to radiofrequency ablation include a stable adhesion of the cryothermal catheter to the endocardium throughout the process, reduced incidence of endocardial thrombus formation at the site of cryoablation, and freedom from electrical interference by intracardiac echo imaging during ablation. The latter effect is often seen during radiofrequency ablation. Another advantage of cryoablation catheters is that they can be used to reversibly interrupt electrical function by cooling small regions of the heart to 0°C. This allows the physician to cool selected regions and monitor cardiac function for improvement before irreversibly damaging the tissue (Braunwald and Bonow, 2012). The second method, microwave ablation, uses electromagnetic radiation rather than radiofrequencies. Because of this, microwave does not require a contact dependency, unlike radiofrequency- or cryo-based ablation (Maessen et al., 2002; Williams et al., 2002). Recently, microwave ablation has been used as an intraoperative tool in the surgical maze procedure and evaluated as an alternative for radiofrequency epicardial ablation in minimally invasive access surgery (Knaut et al., 2001; Benussi et al., 2002). However, while a review of the clinical trials to date indicates that there is some evidence that microwave ablation is comparable to radio-frequency ablation there is no compelling evidence that it confers significant advantages over radio-frequency ablation (Macdonald et al., 2012). Finally, in order to apply a highly focused beam of energy, laser ablation is being evaluated. Early studies in the use of laser ablation met with significant difficulties. Lesions created through the use of high-energy pulsed lasing (Nd-YAG, i.e., infrared wavelength) carried a risk of crater formation (Lee et al., 1985). This risk was addressed by the use of a higher frequency laser, argon (630 nm) vs. Nd-YAG (1064 nm), which reduced the creation of craters (Sakena et al., 1989). These results were encouraging for the potential application of laser ablation. This application of the continuous lower energy diode laser facilitates the ability to heat and create controlled and precisely located lesions uniformly. To this end, the ability to focus a narrow beam will enhance refined lesions through more versatile devices.

Ablation balloons are one novel technology that has recently been developed. Both cryo and laser ablation balloons have been developed for the treatment of atrial fibrillation. The stated clinical goal in these cases is to isolate the pulmonary vein which is known to be a primary location for arrythmogenic sites (Haïssaguerre et al., 1998). The use of ablation balloons allows for quick isolation of the arrythmogenic sites. A cryoballoon is delivered to the site via catheter and inflated with nitrous oxide for 240–300 s. Clinical trials of the cryoballoon have demonstrated successful treatment of atrial fibrillation; however, direct comparison with radio-frequency catheter ablation and increased clinical use are necessary before the advantages or disadvantages of this technique can be fully known (Neumann et al., 2008; Challa and Mansour, 2011). Laser ablation balloons treat the arrythmogenic sites with laser energy that is directed independently of the balloon (Metzner et al., 2011). Although clinical trials to date are promising, longer term follow ups and more data are needed to determine the true safety and efficacy of this technology (Challa and Mansour, 2011; Metzner et al., 2011).

Radio-frequency ablation has quickly become the clinical gold standard in the treatment of cardiac arrhythmia. However, further testing and development of the alternative technologies and energy sources described above may change this fact, as the risks associated with radio-frequency ablation (i.e., endocardial disruption) may be reduced with these alternative technologies.

20.7 Heart Valves

The main pathology associated with the mitral valve in the developed world has transitioned from stenosis to regurgitation (Bergsland et al., 2011). Traditionally, this problem is treated with a full sternotomy to replace the valve, but minimally invasive options are gaining favor with surgeons. In a minimally invasive valve replacement, an 18-26F catheter is inserted through the apex or through the femoral vein and into the left ventricle. Inside the catheter is the crimped replacement valve attached to a stent. The catheter is positioned in the diseased valve and a balloon inflates the stent which props the existing leaflets back and out of the way. The use of robotics to assist with the surgery, coupled with intra-aortic balloon occlusion, allow this procedure to truly be minimally invasive (Falk et al., 1999; Reichenspurner et al., 1999).

The safety of the minimally invasive approach for mitral valve replacement is comparable to the open procedure commonly used, but it also appreciates decreased pain perception, atrial fibrillation, sternal infection, and length of hospital stay. There is an increased risk of neurological complications, groin infection, phrenic nerve palsy, and a slightly higher risk of aortic dissection associated with minimally invasive procedures (Falk et al., 2011). Phrenic nerve palsy is thought to have come from forceful pulling from the pericardial retraction sutures which are sometimes used to increase exposure of the left atrium (Falk et al., 2011). This is not a common complaint when robotic assistance is used for these types of surgeries. A conglomeration of studies spanning 12 years and including 1282 patients found that the only significant risk factor for a stroke was the use of retrograde perfusion, a method for delivering oxygenated blood via the veins, in high-risk patients with diseased aortas. Neurological complications including stroke occurred in 2.3% of the patients. Incision location does not appear to be a risk factor for neurological complications, as previously suspected (Grossi et al., 2012). Minimally invasive procedures for aortic valve replacement have been shown to be just as safe as the traditional full sternotomy approach, with the main advantages being shorter hospital stays and requiring less blood product. Though these results are encouraging, the present generation of minimally invasive replacement valves is used mainly for high-risk patients and not for the general populous.

Alfieri mitral valve repair uses the edge-to-edge technique, meaning the surgeon clips a prolapsing leaflet to another nonprolapsing leaflet, creating a double-orificed valve. The MitraClip is widely used for this procedure and is the best known of this type of device (Figure 20.13). While the MitraClip leaves the natural valve in place, several entirely artificial products exist. Bovine leaflets with nitinol stent supports as are seen in the Engager and the Lotus Valve are the most popular combination, though other combinations are being tested. The Acurate has porcine leaflets with nitinol supports that extend all the way to the aortic root, and the direct flow valve has bovine leaflets and Dacron supports. Innovative efforts are being put toward positioning the valves during minimally invasive surgery. The direct flow valve has independently inflatable leaflets and supports, while the Engager has three positioning arms to use the patient's anatomy to assist in alignment.

FIGURE 20.13 The MitraClip System and steerable catheter from Abbott Vascular.

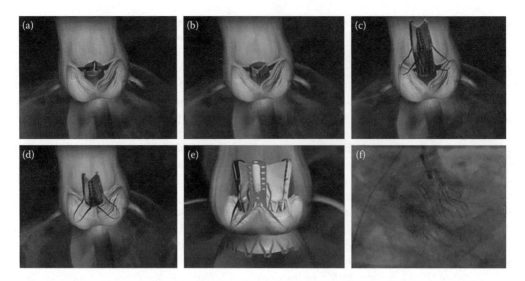

FIGURE 20.14 (**See color insert.**) Anatomical positioning, Medtronic Engager system shown. The posts are rotated until they are aligned with the native leaflets (a,b). Once aligned, the support arms are released directly above the sinuses (c). The system is pulled back until the support arms engage (d). The system is then fully deployed by removing the sheath (e) as seen *in situ* (f). (From Falk, V., Walther, T., et al. 2011. *Eur Heart J* 32(7), 878–887.)

All of these devices have completed or begun clinical trials. European clinical trials for MitraClip involving 51 patients with moderate-to-severe regurgitation had promising results. Most patients saw improvement within 3 h of device deployment. Healthy remodeling of the left ventricle was seen at the 6- and 12-month time points. There was an 18% mortality rate associated with the procedure (Falk et al., 2011). The Engager is the newest redesign of the Embracer; a valve manufactured by Medtronic whose design led to injury in 10% of its patients during clinical trials (Armstrong and Rogers, 2011). The Engager is an artificial aortic valve (Figure 20.14). A European clinical trial involved 30 octogenarians and the mean survival rate at 6 months was 56.7%, but many of the mortalities were not device related. Accurate placement was confirmed in all but one patient, though there was a high incidence rate of aortic dissection, presumably due to the delivery method. The delivery system has been redesigned for upcoming clinical trials (Brosky, 2008). Successful implantation of Sadra Medical's Lotus Valve has been performed in a four-patient preliminary feasibility study. The Lotus is an artificial aortic valve that features a stent construct that shortens as it is deployed to decrease cell size and increase radial strength (Armstrong and Rogers, 2011). A one-year follow up on the first patient to have this valve implanted combined with the success of the procedure in the four newest patients is promising (Medical Device Daily, 2010). The Acurate aortic valve saw successful implantation in 95% of its patients during the first clinical trial involving 40 octogenarians. The 30-day mortality rate was 12.5% with three of those deaths unrelated to the surgery (Bijuklic et al., 2011). The direct flow valve just finished the EVEREST II randomized clinical trial in the United States which showed it to be just as effective as traditional surgery in patients without atrial fibrillation (Figure 20.15) (Treede et al., 2010). Twenty-two patients underwent implantation with the direct flow aortic valve in a European trial. This trial saw a 12.9% mortality rate at 6 months (Herrmann et al., 2012). A 2-year follow up shows that the positioning, shape, and performance remain satisfactory (Maisano et al., 2011).

Several devices for treating regurgitation from different perspectives exist. Coapsys has a minimally invasive device for changing the geometry of the left ventricle in an effort to stop mitral regurgitation. Two pads are placed on either side of the ventricle, with the posterior pad underneath the mitral valve and the anterior pad directly across the ventricle and shifted distally. A transventricular string is tightened between the two pads which realigns the papillary muscles for good closure between the leaflets,

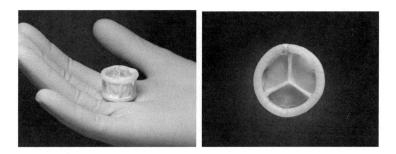

FIGURE 20.15 The direct flow valve percutaneous aortic valve features bovine leaflets attached to an inflatable polyester cuff (direct flow).

thus decreasing or eliminating regurgitation (Bizzarri et al., 2010). Techniques for the minimally invasive implantation of artificial chordae have just entered clinical trials. This is useful for regurgitation caused by rupture or elongation of the natural chordae. Novel methods for implanting these new chords without the use of a cardiopulmonary bypass machine, or via beating heart bypass, exist but they have not yet begun clinical trials (Treede et al., 2010; Bergsland et al., 2011). Procedures still using CPB are the standard. Some allow the lengths of the chordae to be adjusted after the patient is off bypass, thus minimizing bypass and cross-clamping time (Treede et al., 2010).

20.8 Cardiac Support and Assist

Recent advancements in cardiac support and assist technology include the CardiacSTAR device recently developed by CorInnova Inc. This device is deployed within the pericardial sac and is delivered via a 1½″ diameter tube using minimally invasive procedures. It is capable of simultaneously providing both adjustable passive support and synchronous active assist. Assist is provided via direct cardiac compression, thus the device is nonblood contacting. The device has two distinct components, an inner passive chamber that is filled with saline via a subcutaneous injection port to provide

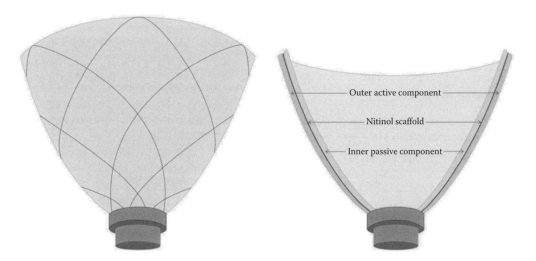

FIGURE 20.16 Structure of the Criscione device. (Left) The orientation and geometry of the nitinol scaffold. (Right) Cross-section of the device showing the inner passive saline component (bottom label), the outer active air component (top label), and the scaffold between the two. (From Moreno, M.R. et al. 2011b. *ASME J Med Dev* 5, 041008-1–041008-9.)

passive support and constraint (Figure 20.16). Saline can be added or removed post implant, thus the support and constraint is adjustable. There is also an outer chamber that is connected to a pneumatic driveline. Air is shuttled to and from the outer chamber to provide direct cardiac compression. The magnitude of assist can be adjusted by adjusting the driving pressure. A nitinol scaffold between the inner and outer chambers provides structural support and assists with diastolic recoil. ECG sensors are incorporated into the device and are used to trigger the active component. Another important feature is that the device is held in place by an intrinsic pneumatic attachment and does not require suturing or other external fixation. Animal safety studies have demonstrated proof of concept and the nonobligatory feature of the device. In an esmolol failure model using sheep, the device successfully restored full cardiac output and stroke work. Another unique and important feature of the device is the chamber geometry, which has been designed such that the device conforms to the end-diastolic and end-systolic shapes of the heart. This results in a reduction of device-induced aberrant strain patterns in the myocardium. It is hypothesized that restoring and/or maintaining healthy kinematics and strain patterns may be conducive to cardiac remodeling that is restorative or rehabilitative in nature (Moreno et al., 2011a,b).

References

Abualsaud, A.O. and Eisenberg, M.J. 2010. Perioperative management of patients with drug eluting stents. *JACC Cardiovasc Interv* 3, 131–42.

Angel, L.F., Tapson, V., Galgon, R.E., Restrepo, M.I., and Kaufman, J. 2011. Systematic review of the use of retrievable inferior vena cava filters. *J Vasc Interv Radiol* 22, 1522–1530.e3.

Armstrong, E.J. and Rogers, J.H. 2011. Next-generation transcatheter aortic valves. *Cardiac Interv Today* September/October, 66–69.

Bella, P.D., De Ponti, R., Uriarte, J.A.S., Tondo, C., Klersy, C., Carbucicchio, C., Storti, C., Riva, S., and Longobardi, M. 2001. Catheter ablation and antiarrhymic drugs for haemodynamically tolerated postinfarction ventricular tachycardia. *Eur Heart J* 23, 414–424.

Benussi, S., Nascimbene, S., Agricola, E., Calori, G., Calvi, S., Caldarola, A., Oppizzi, M., Casati, V., Pappone, C., and Alfieri, O. 2002. Surgical ablation of atrial fibrillation using the epicardial radiofrequency approach: Mid-term results and risk analysis. *Ann Thorac Surg* 74, 1050–1057.

Bergsland, J., Mujanovic, E. et al. 2011. Minimally invasive repair of the mitral valve: Technological and clinical developments. *Minim Invasive Ther Allied Technol* 20(2), 72–77.

Bijuklic, K., Tuebler, T. et al. 2011. Midterm stability and hemodynamic performance of a transfemorally implantable nonmetallic, retrievable, and repositionable aortic valve in patients with severe aortic stenosis. Up to 2-year follow-up of the direct-flow medical valve: A pilot study. *Circ Cardiovasc Interv* 4(6), 595–601.

Bizzarri, F., Tudisco, A. et al. 2010. Different ways to repair the mitral valve with artificial chordae: A systematic review. *J Cardiothorac Surg* 5, 22.

Boersma, L.V.A., Castella, M., van Boven, W., Berruezo, A., Yilmaz, A., Nadal, M., Sandoval, E., Calvo, N., Brugada, J., Kelder, J. et al. 2012. Atrial fibrillation catheter ablation versus surgical ablation treatment (FAST): A 2-center randomized clinical trial. *Circulation* 125, 23–30.

Braunwald, E., and Bonow, R.O. 2012. *Braunwald's Heart Disease: A Textbook of Cardiovascular Medicine* (Philadelphia: Saunders).

Brosky, J. 2008. Report from Europe: French agency backing IVD Tracker theranostics project. *Med Device Daily* 8.

Challa, P.K. and Mansour, M. 2011. Atrial fibrillation: Update on ablation strategies and technology. *Curr Cardiol Rep* 13, 394–398.

Chapman, R.H. and Liu, L.Z. 2011. Determining initial and follow-up costs of cardiovascular events in a US managed healthcare population. *BMC Cardiovasc Disord* 11(11), 1–10.

Chen, Y.-W., Smith, M.L., Sheets, M., Ballaron, S., Trevillyan, J.M., Burke, S.E., Rosenberg, T., Henry, C., Wagner, R., Bauch, J. et al. 2007. Zotarolimus, a novel sirolimus analogue with potent anti-proliferative activity on coronary smooth muscle cells and reduced potential for systemic immunosuppression. *J Cardiovasc Pharmacol* 49, 228–235.

Criado, F.J. 2011. The TEVAR landscape in 2012. *Endovasc Today* 10(11), 34–47.

Dake M.D., Miller D.C., Semba C.P. et al. 1994. Transluminal placement of endovascular stent-grafts for the treatment of descending thoracic aortic aneurysms. *N Engl J Med* 331, 1729–1734.

Degertekin, M., Regar, E., Tanabe, K., Lee, C.H., and Serruys, P.W. 2002. Sirolimus eluting stent in the treatment of atherosclerosis coronary artery disease. *Minerva Cardioangiol* 50(5), 405–418.

De Labriolle, A., Pakala, R., Bonello, L., Lemesle, G., Scheinowitz, M., and Waksman, R. 2009. Paclitaxel-eluting balloon: From bench to bed. *Catheter Cardiovasc Interv* 73, 643–652.

De Luca, G., Dirksen, M.T., Spaulding, C., Kelbaek, H., Schalij, M., Thuesen, L., van der Hoeven, B., Vink, M.A., Kaiser, C., Musto, C., Chechi, T., Spaziani, G., Diaz de la Llera, L.S., Pasceri, V., Di Lorenzo, E., Violini, R., Cortese, G., Suryapranata, H., and Stone G.W. 2012. Drug-eluting vs bare-metal stents in primary angioplasty: A pooled patient-level meta-analysis of randomized trials. *Arch Intern Med* 172, 611–21; discussion 621–2.

Dotter, C.T. 1969. Transluminally-placed coilspring endarterial tube grafts: Long term patency in canine popliteal artery. *Invest Radiol* 4, 329–332.

Dotter, C.T., Buschman, R.W., McKinney, M.K., and Rösch, J.1983. Transluminal expandable nitinol coil stent grafting: Preliminary report. *Radiology* 147, 259–260.

Duerig, T.W., Tolomeo, D.E., and Wholey, M. 2000. An overview of superelastic stent design. *Min Invas Ther Allied Technol* 9(3/4), 235–246.

Edelman, E.R. and Rogers, C. 1998. Pathobiologic responses to stenting. *Am J Cardiol* 81, 4E–6E.

Falk, V., Autschbach, R. et al. 1999. Computer-enhanced mitral valve surgery: Toward a total endoscopic procedure. *Semin Thorac Cardiovasc Surg* 11(3), 244–249.

Falk, V., Cheng, D.C. et al. 2011. Minimally invasive versus open mitral valve surgery: A consensus statement of the international society of minimally invasive coronary surgery (ISMICS) 2010. *Innovations (Phila)* 6(2), 66–76.

Falk, V., Walther, T. et al. 2011. Transapical aortic valve implantation with a self-expanding anatomically oriented valve. *Eur Heart J* 32(7), 878–887.

Farb, A., Weber, D.K., Kolodgie, F.D., Burke, A.P., and Virmani, R. 2002. Morphological predictors of restenosis after coronary stenting in humans. *Circulation* 105(25), 2974–2980.

FDA. 2012. *Removing Retrievable Inferio Vena Cava Filters: Initial Communication* (U.S. Food and Drug Administration).

Ferris, E.J., McCowan, T.C., Carver, D.K., and MacFarland, D.J. 1993. Percutaneous inferior vena cava filters: follow-up of seven designs in 320 patients. *Radiology* 188, 851–856.

Fleisch, M. and Meier, B. 1999. Management and outcome of stents in 1998: Long-term outcome. *Cardiol Rev* 7, 215–218.

Finol, E.A., Siewiorek, G.M., Scotti, C.M., Wholey, M.H., and Wholey, M.H. 2008. Wall apposition assessment and performance comparison of distal protection filters. *J Endovasc Therapy* 15, 177–185.

Gaba, R.C. 2006. Embolization of intracranial aneurysms with hydrogel-coated coils versus inert platinum coils: Effects on packing density, coil length and quantity, procedure performance, cost, length of hospital stay, and durability of therapy. *Stroke* 37(6), 1443–450.

Gallo, R., Padurean, A., and Jayaramon, T. 1999. Inhibition of intimal thickening after balloon angioplasty in porcine coronary arteries by targeting regulators of the cell cycle. *Circulation* 99, 2164–2170.

Gomez-Lara, J., Brugaletta, S., Diletti, R., Garg, S., Onuma, Y., Gogas, B.D., van Geuns, R.J., Dorange, C., Veldhof, S., Rapoza, R. et al. 2010. A comparative assessment by optical coherence tomography of the performance of the first and second generation of the everolimus-eluting bioresorbable vascular scaffolds. *Eur Heart J* 32, 294–304.

Greenfield, L.J. and Proctor, M.C. 2000. The percutaneous greenfield filter: Outcomes and practice patterns. *J Vasc Surg* 32(5), 888–893.

Grossi, E.A., Loulmet, D.F. et al. 2012. Evolution of operative techniques and perfusion strategies for minimally invasive mitral valve repair. *J Thorac Cardiovasc Surg* 143(4 Suppl), S68–70.

Grube, E. and Büllesfeld, L. 2002. Initial experience with Paclitaxel-coated stents. *J Interven Cardiol* 15, 471–476.

Grube, E., Sonoda, S., Ikeno, F., Honda, Y., Kar, S., Chan, C., Gerckens, U., Lansky, A.J., and Fitzgerald, P.J. 2004. Six- and twelve-month results from first human experience using everolimus-eluting stents with bioabsorbable polymer. *Circulation* 109, 2168–2171.

Haeck, J.D., Kuijt, W.J., Koch, K.T., Bilodeau, L., Henriques, J.P., Rohling, W.J., Baan, J., Jr, Vis, M.M., Nijveldt, R., van Geloven, N. et al. 2010. Infarct size and left ventricular function in the PRoximal Embolic Protection in Acute myocardial infarction and Resolution of ST-segment Elevation (PREPARE) trial: Ancillary cardiovascular magnetic resonance study. *Heart* 96, 190–195.

Haeck, J.D.E., Koch, K.T., Bilodeau, L., Van der Schaaf, R.J., Henriques, J.P.S., Vis, M.M., Baan, J., Jr, Van der Wal, A.C., Piek, J.J., Tijssen, J.G.P. et al. 2009. Randomized comparison of primary percutaneous coronary intervention with combined proximal embolic protection and thrombus aspiration versus primary percutaneous coronary intervention alone in ST-segment elevation myocardial infarction: The PREPARE (PRoximal Embolic Protection in Acute myocardial infarction and Resolution of ST-Elevation) study. *JACC Cardiovasc Interv* 2, 934–943.

Haïssaguerre, M., Jaïs, P., Shah, D.C., Takahashi, A., Hocini, M., Quiniou, G., Garrigue, S., Le Mouroux, A., Le Métayer, P., and Clémenty, J. 1998. Spontaneous initiation of atrial fibrillation by ectopic beats originating in the pulmonary veins. *N Engl J Med* 339, 659–666.

Herdeg, C., Göhring-Frischholz, K., Geisler, T., May, A., Haase, K.K., and Gawaz, M. 2007. GENIE catheter for liquid local drug delivery. *EuroIntervention* 3, 286–288.

Herrmann, H.C., Gertz, Z.M. et al. 2012. Effects of atrial fibrillation on treatment of mitral regurgitation in the EVEREST II (Endovascular Valve Edge-to-Edge Repair Study) randomized trial. *J Am Coll Cardiol* 59(14), 1312–1319.

Hilleman, D.E., and Razavi, M.K. 2008. Clinical and economic evaluation of the Trellis-8 infusion catheter for deep vein thrombosis. *J Vasc Interv Radiol* 19, 377–383.

Hope, J.K., Byrne, J.V. et al. 1999. Factors influencing successful angiographic occlusion of aneurysms treated by coil embolization. *AJNR Am J Neuroradiol* 20(3): 391–399.

Hwang, C., Wu, D., and Edelman, E. 2001. Physiological transport forces govern drug distribution for stent-based delivery. *Circulation* 104, 600–605.

Iribarne, L., Easterwood, R., Chan, E.Y.H., Yang, J., Soni, L., Russo, M.J., Smith, C.R., and Argenziano, M. 2011. The golden age of minimally invasive cardiothoracic surgery: Current and future perspectives. *Future Cardiology* 7(3), 333–46.

Kalesan, B., Pilgrim, T., Heinimann, K., Raber, L., Stefanini, G.G., Valgimigli, M., da Costa, B.R., Mach, F., Luscher, T.F., Meier, B. et al. 2012. Comparison of drug-eluting stents with bare metal stents in patients with ST-segment elevation myocardial infarction. *Eur Heart J* 33, 977–987.

Kallmes, D.F. 2002. New expandable hydrogel-platinum coil hybrid device for aneurysm embolization. *Am J Neuroradiol* 23 , 1580–588.

Kallmes, D.F. and Fujiwara, N.H. 2002. New expandable hydrogel-platinum coil hybrid device for aneurysm embolization. *AJNR Am J Neuroradiol* 23(9): 1580–1588.

Kaltoft, A., Kelbaek, H., Kløvgaard, L., Terkelsen, C.J., Clemmensen, P., Helqvist, S., Lassen, J.F., and Thuesen, L. 2010. Increased rate of stent thrombosis and target lesion revascularization after filter protection in primary percutaneous coronary intervention for ST-segment elevation myocardial infarction: 15-month follow-up of the DEDICATION (Drug Elution and Distal Protection in ST Elevation Myocardial Infarction) trial. *J Am Coll Cardiol* 55, 867–871.

Kastrati, A., Mehilli, J., Dirsschinger, J., et al. 2001. Restenosis after coronary placement of various stent types. *Am J Cardiol* 877, 34–39.

Keane, D. 2002. New catheter ablation techniques for the treatment of cardiac arrhythmias. *Cardiac ElectroPhy Rev* 6, 341–348.

Klein, G., Oswald, H. et al. 2008. Efficacy of pulmonary vein isolation by cryoballoon ablation in patients with paroxysmal atrial fibrillation. *Heart Rhythm* 5(6), 802–806.

Knaut, M., Tugtekin, M., Spitzer, S.G., Karolyi, L., Boehme, H., and Schueler, S. 2001. Curative treatment of chronic atrial fibrillation in patients with simultaneous cardiosurgical diseases with intraoperative microwave ablation. *JACC* 109A(abstr).

Lagerqvist, B., James, S.K., Stenestrand, U., Lindbäck, J., Nilsson, T., and Wallentin, L. 2007. Long-term outcomes with drug-eluting stents versus bare-metal stents in Sweden. *New Engl J Med* 356, 1009–1019.

Lanzino, G. 2011. Flow diversion for intracranial aneurysms. *Endovascular Today* 10.9 , 35–36.

Lavigne, M.C., Grimsby, J.L., and Eppihimer, M.J. 2012. Antirestenotic mechanisms of everolimus on human coronary artery smooth muscle cells: Inhibition of human coronary artery smooth muscle cell proliferation, but not migration. *J Cardiovasc Pharmacol* 59, 165–174.

Lee, B.I., Gottdiener, J.S., Fletcher, R.D., Rodriguez, E.R., and Ferrans, V.J. 1985. Transcatheter ablation: Comparison between laser photoablation and electrode shock ablation in the dog. *Circulation* 71, 579–586.

Lee, M.S., Singh, V., Wilentz, J.R., and Makkar, R.R. 2004. AngioJet thrombectomy. *J Invasive Cardiol* 16, 587–591.

Leitch, J. and Barlow, M. 2012. Radiofrequency ablation for pre-excitation syndromes and AV nodal re-entrant tachycardia. *Heart, Lung Circ* 21(6–7), 376–385.

Lieber, B.B. and Gounis, M.J. 2002. The physics of endoluminal stenting in the treatment of cerebrovascular aneurysms. *Neurol Res* 24(Suppl. 1), S33–S42.

Liffman, K., Lawrence-Brown, M., Semmens, J.B., Bui, A., Rudman, M., and Hartley, D. 2001. Analytical modeling and numerical simulation of forces in an endoluminal graft. *J Endovasc Ther* 8, 358–371.

Lin, P.H., Ochoa, L.N., and Duffy, P. 2010. Catheter-directed thrombectomy and thrombolysis for symptomatic lower-extremity deep vein thrombosis: Review of current interventional treatment strategies. *Perspect Vasc Surg Endovasc Ther* 22, 152–163.

Lorch, H., Zwaan, M., Kulke, C., and Weiss, H.D. 1998. In vitro studies of temporary vena cava filters, *Cardiovasc Interv Radiol* 21, 146–150.

Lustgarden, D., Keane, D., and Ruskin, J. 1999. Cryothermal catheter ablation: Mechanism of tissue injury and clinical results. *Prog Cardiovasc Dis* 41, 481–498.

Macdonald, D.R.W., Maruthappu, M., and Nagendran, M. 2012. How effective is microwave ablation for atrial fibrillation during concomitant cardiac surgery? *Interact Cardiovasc Thoracic Surg* 15(1), 122–127.

Maessen, J.G., Nijis, J.F.M.A., Smeets, J.L.R.M., Vainer, J., and Mochtar, B. 2002. Beating-heart treatment of atrial fibrillation with microwave ablation. *Ann Thorac Surg* 74, S1307–S1311.

Maisano, F., Cioni, M. et al. 2011. Beating-heart implantation of adjustable length mitral valve chordae: Acute and chronic experience in an animal model. *Eur J Cardiothorac Surg* 40(4), 840–847.

Marzullo, R., Aprile, A., Biondi-Zoccai, G., Politi, L., Leuzzi, C., and Sangiorgi, G. 2011. Drug-eluting balloon technology. *Cardiac Interv Today* 5, 40–49.

Mauri, L., Rogers, C., and Baim, D.S. 2006. Devices for distal protection during percutaneous coronary revascularization. *Circulation* 113, 2651–2656.

Mayer, D., Pfammatter, T., Rancic, Z., Hechelhammer, L., Wilhelm, M., Veith, F.J., and Lachat, M. 2009. 10 Years of emergency endovascular aneurysm repair for ruptured abdominal aortoiliac aneurysms: Lessons learned. *Ann Surg* 249(3), 510–15.

Medical Device Daily. 2010. Report from Europe: Dupont, DSM form new JV for advanced surgical biomaterials. *Med Device Daily* 8.

Melas, N., Perdikides, T. et al. 2012. Helical EndoStaples enhance endograft fixation in an experimental model using human cadaveric aortas. *J Vasc Surg* 55(6): 1726–1733.

Melo, S.L. de, Scanavacca, M.I., Pisani, C., Darrieux, F., Hachul, D., Hardy, C., Camargo, P.R., Atik, E., and Sosa, E.A. 2012. Radiofrequency ablation of childhood arrhythmia: Observational registry in 125 children. *Arquivos Brasil Cardiol* 98(6), 514–518.

Metzner, A., Schmidt, B., Fuernkranz, A., Wissner, E., Tilz, R.R., Chun, K.R.J., Neven, K., Konstantinidou, M., Rillig, A., Yoshiga, Y. et al. 2011. One-year clinical outcome after pulmonary vein isolation using the novel endoscopic ablation system in patients with paroxysmal atrial fibrillation. *Heart Rhythm* 8, 988–993.

Mittleman, R.S., Huang, S.K., de Guzman, W.T., Cuénoud, H., Wagshal, A.B., and Pires, L.A. 1995. Use of the saline infusion electrode catheter for improved energy delivery and increased lesion size in radiofrequency catheter ablation. *Pacing Clin Electrophysiol* 18, 1022–1027.

Morales, P.A. and Heuser, R.R. 2002. Embolic protection devices. *J Interven Cardiol* 15, 485–490.

Moreno, M.R., Biswas, S, Harrison, L.D., Miller, M.W., Fossum, T.W., Nelson, D.A., and Criscione, J.C. 2011a. Development of a non-blood contacting cardiac assist and support device: An *in vivo* proof of concept study. *ASME J Med Dev* 5, 041007-1–041007-9.

Moreno, M.R., Biswas, S, Harrison, L.D., Pernelle, G., Miller, M.W., Fossum, T.W., Nelson, D.A., and Criscione, J.C. 2011b. Assessment of minimally invasive device that provides simultaneous adjustable cardiac support and active synchronous assist in an acute heart failure model. *ASME J Med Dev* 5, 041008-1–041008-9.

Müller-Hülsbeck, S., Jahnke, T., Liess, C., Glass, C., Grimm, J., and Heller, M. 2003. Comparison of various cerebral protection devices used for carotid artery stent placement: An *in vitro* experiment. *J Vasc Interv Radiol* 14, 613–620.

Müller-Hülsbeck, S., Stolzmann, P., Liess, C., Hedderich, J., Paulsen, F., Jahnke, T., and Heller, M. 2005. Vessel wall damage caused by cerebral protection devices: Ex vivo evaluation in porcine carotid arteries. *Radiology* 235, 454–460.

Namdar, M., Chierchia, G.B. et al. 2012. Isolating the pulmonary veins as first-line therapy in patients with lone paroxysmal atrial fibrillation using the cryoballoon. *Europace* 14(2), 197–203.

Neumann, T., Vogt, J., Schumacher, B., Dorszewski, A., Kuniss, M., Neuser, H., Kurzidim, K., Berkowitsch, A., Koller, M., Heintze, J. et al. 2008. Circumferential pulmonary vein isolation with the cryoballoon technique results from a prospective 3-center study. *J Am Coll Cardiol* 52, 273–278.

Onuma, Y., Serruys, P.W., Gomez, J., de Bruyne, B., Dudek, D., Thuesen, L., Smits, P., Chevalier, B., McClean, D., Koolen, J. et al. 2011. Comparison of *in vivo* acute stent recoil between the bioresorbable everolimus-eluting coronary scaffolds (revision 1.0 and 1.1) and the metallic everolimus-eluting stent. *Catheter Cardiovasc Interv* 78, 3–12.

Ormiston, J.A., and Serruys, P.W.S. 2009. Bioabsorbable coronary stents. *Circ: Cardiovasc Interv* 2, 255–260.

Pappone, C., Vicedomini, G., Augello, G., Manguso, F., Saviano, M., Baldi, M., Petretta, A., Giannelli, L., Calovic, Z., Guluta, V. et al. 2011. Radiofrequency catheter ablation and antiarrhythmic drug therapy: A prospective, randomized, 4-year follow-up trial: The APAF study. *Circ Arrhythm Electrophysiol* 4, 808–814.

Parikh, S., Motarjeme, A., McNamara, T., Raabe, R., Hagspiel, K., Benenati, J.F., Sterling, K., and Comerota, A. 2008. Ultrasound-accelerated thrombolysis for the treatment of deep vein thrombosis: Initial clinical experience. *J Vasc Interv Radiol* 19, 521–528.

Parodi, J.C., Palmaz, J.C. et al. 1991. Transfemoral intraluminal graft implantation for abdominal aortic aneurysms. *Ann Vasc Surg* 5(6): 491–499.

Piotin, M. 2012. Intracranial aneurysm coiling with PGLA-coated coils versus bare platinum coils: Long-term anatomic follow-up. *Interv Neuroradiol* 54, 345–48.

Poon, M., Badimon, J.J., and Fuster, V. 2002. Overcoming restenosis with sirolimus: From alphabet soup to clinical reality. *Lancet* 359, 619–622.

Poston, R.S., Tran, R., Collins, M., Reynolds, M., Connerney, I., Reicher, B., Zimrin, D., Griffith, B.P.and Bartlett S.T. 2008. Comparison of economic and patient outcomes with minimally invasive versus traditional off-pump coronary artery bypass grafting techniques. *Trans. Meeting Am Surg Assoc* 126, 281–89.

Reichenspurner, H., Boehm, D. et al. 1999. Minimally invasive mitral valve surgery using three dimensional video and robotic assistance. *Semin Thorac Cardiovasc Surg* 11(3), 235–243.

Robaina, S., Jayachandran, B., He, Y., Frank, A.O., Moreno, M.R., Schoephoerster, R., and Moore, J.E. Jr. 2003. Platelet adhesion to simulated stented surfaces under physiologic flow conditions. *J Endovascular Therapy* 10(5), 978–986.

Roger, V.L. 2011. Heart Disease and Stroke Statistics—2011 Update: A Report from the American Heart Association Statistics Committee and Stroke Statistics Subcommittee. *Circulation* 123, E18–E209.

Rogers, C. and Edelman E.R. 1995. Endovascular stent design dictates experimental restenosis and thrombosis. *Circulation* 91(12), 2995–3001.

Sakena, S., Gielchinsky, J.S., and Tullo, N.G. 1989. Argon laser ablation of malignant ventricular tachycardia associated with coronary artery disease. *Am J Cardiol* 64, 1298–1304.

Sapp, J.L., Cooper, J.M., Zei, P., and Stevenson, W.G. 2006. Large radiofrequency ablation lesions can be created with a retractable infusion-needle catheter. *J Cardiovasc Electrophysiol* 17, 657–661.

Scheller, B., Hehrlein, C., Bocksch, W., Rutsch, W., Haghi, D., Dietz, U., Böhm, M., and Speck, U. 2006. Treatment of coronary in-stent restenosis with a paclitaxel-coated balloon catheter. *N Engl J Med* 355, 2113–2124.

Schuler, W., Sedrani, R., Cottens, S., Häberlin, B., Schulz, M., Schuurman, H.J., Zenke, G., Zerwes, H.G., and Schreier, M.H. 1997. SDZ RAD, a new rapamycin derivative: Pharmacological properties *in vitro* and in vivo. *Transplantation* 64, 36–42.

Serruys, P.W., Ormiston, J.A., Onuma, Y., Regar, E., Gonzalo, N., Garcia-Garcia, H.M., Nieman, K., Bruining, N., Dorange, C., Miquel-Hébert, K. et al. 2009. A bioabsorbable everolimus-eluting coronary stent system (ABSORB): 2-year outcomes and results from multiple imaging methods. *The Lancet* 373, 897–910.

Sievert, H. and Rabe, K. 2002. Role of distal protection during carotid stenting. *J Interven Cardiol* 15, 499–504.

Siewiorek, G.M., Wholey, M.H., and Finol, E.A. 2009. *In vitro* performance assessment of distal protection filters: Pulsatile flow conditions. *J Endovasc Ther* 16, 735–743.

Sigwart, U., Puel, J., Mirkovitch, V., Joffre, F., and Kappenberger, L. 1987. Intravascular stents to prevent occlusion and restenosis after transluminal angioplasty. *N Engl J Med* 316, 701–706.

Sousa, J.E., Costa, M.A., Abizaid, A.C. et al. 2001. Sustained suppression of neointimal proliferation by Sirolimus-eluting stents. One-year angiographic and intravascular ultrasound follow up. *Circulation*, 104, 2007–2011.

Speck, U., Scheller, B., Abramjuk, C., Breitwieser, C., Dobberstein, J., Boehm, M., and Hamm, B. 2006. Neointima inhibition: Comparison of effectiveness of non-stent-based local drug delivery and a drug-eluting stent in porcine coronary arteries. *Radiology* 240, 411–418.

Sprouse, L.R., Peeters, P., and Bosiers, M. 2005. The capture of visible debris by distal cerebral protection filters during carotid artery stenting: Is it predictable? *J Vasc Surg* 41, 950–955.

Stecker, M.S., Barnhart, W.H., and Lang, E.V. 2001. Evaluation of a spiral nitinol temporary inferior vena caval filter. *Acad Radiol* 8(6), 484–493.

Swanson, D.K., Smith, W.J., Ibrahim, T., and Wechsler, A.S. 2011. Tissue temperature feedback control of power: The key to successful ablation. *Innovations (Phila)* 6, 276–282.

Thomas, S.P., and Sanders, P. 2012. Catheter ablation for atrial fibrillation. *Heart, Lung Circ* 21(6–7), 395–401.

Treede, H., Tubler, T. et al. 2010. Six-month results of a repositionable and retrievable pericardial valve for transcatheter aortic valve replacement: The direct flow medical aortic valve. *J Thorac Cardiovasc Surg* 140(4), 897–903.

Tsuchiya, Y., Lansky, A.J., Costa, R.A., Mehran, R., Pietras, C., Shimada, Y., Sonoda, S., Cristea, E., Negoita, M., Dangas, G.D. et al. 2006. Effect of Everolimus-eluting stents in different vessel sizes (from the Pooled FUTURE I and II Trials). *Am J Cardiol* 98, 464–469.

Uda, K., Goto, K. et al. 1998. Embolization of cerebral aneurysms using Guglielmi detachable coils— problems and treatment plans in the acute stage after subarachnoid hemorrhage and long-term efficiency. *Neurol Med Chir (Tokyo)* 38(3): 143–152.

Uhrich, K., Cannizzaro, S.M., Langer, R.S., and Shakesheff, K.M. 1999. Polymeric systems for controlled drug release. *Chem Rev* 99, 3181–3198.

van Marrewijk, C., Buth, J. et al. 2002. Significance of endoleaks after endovascular repair of abdominal aortic aneurysms: The EUROSTAR experience. *J Vasc Surg* 35(3): 461–473.

Waksman, R., and Pakala, R. 2009. Drug-eluting balloon: The comeback kid? *Circ Cardiovasc Interv* 2, 352–358.

Warkentin, T.E., Greinacher, A., and A. Koster. 2008. Bivalirudin. *Thromb Haemost* 99, 830–839.

Watanabe, I., Masaki, R., Min, N., Oshikawa, N., Okubo, K., Sugimura, H., Kojima, T., Saito, S., Ozawa, Y., and Kanmatsuse, K. 2002. Cooled-tip ablation results in increased radiofrequency power delivery and lesion size in the canine heart: Importance of catheter-tip temperature monitoring for prevention of popping and impedance rise. *J Interv Card Electrophysiol* 6, 9–16.

West, N.E.J., Hoole, S.P., and McCormick, L. 2010. Are all distal protection devices created equal? *J Am Coll Cardiol* 56, 745–746; author reply 746.

Wijffels, M.C., Kirchhof, C.J., Dorland, R., and Allessie, M.A. 1995. Atrial fibrillation begets atrial fibrillation. A study in awake chronically instrumented goats. *Circulation* 92, 1954–1968.

Williams, M.R., Knaut, M., Berube, D., and Oz, M.C. 2002. Application of microwave energy in cardiac tissue ablation: From in vitro analyses to clinical use. *Ann Thorac Surg* 74, 1500–1505.

Zarins, C.K., Lu, C.T., Gewertz, B.L., Lyon, R.T., Rush, D.S., and Glagov, S. 1982. Arterial disruption and remodeling following balloon dilatation. *Surgery* 92(6), 1086–1095.

Zwaan, M., Kagel, C., Marienhoff, N., Weiss, H.D., Grimm, W., Eberhard, I., and Schweider, G. 1995. Erste Erfahrungen mit temporaren vena cava filtern. *Fortschr Rontgenstr* 163, 74–79.

21

Neuroendovascular Medical Devices

Andrea D.
Muschenborn*
Texas A&M University

Jennifer N.
Rodriguez*
Texas A&M University

Jonathan Hartman
*Kaiser Permanente
Sacramento Medical Center*

Duncan J. Maitland
Texas A&M University

21.1 Introduction

This chapter will focus on the major points of the development process and current devices used in the prevention and treatment of hemorrhagic stroke due to intracranial saccular aneurysms (ISAs) and arteriovenous malformations (AVMs) as well as treatment of ischemic stroke due to neurovascular occlusions. Also, a brief overview of common manufacturing and bench-top testing techniques will be presented.

The neurovasculature is located within the brain and consists of arteries, capillaries, and veins. In general, the neurovasculature functions in the same way as the other vascular systems; however, there are some important differences. One major difference is that cerebral arteries lack the external elastic lamina and have a much thicker adventitia and thinner media when compared to muscular arteries such as the aorta (Humphrey 2002; Kumar et al. 2007). These structural changes within the neurovasculature play a role in the pathogenesis of the vasculature diseases of the brain that are treated with the devices in which we will focus on throughout the rest of this chapter. The fast progression of these devices in the past 50 years has been enabled by parallel advancements in fields such as material science, imaging, and computer engineering. Many diseases that were traditionally treated surgically with high risk can now be treated endovascularly with advantages in safety and recovery time.

---

* These authors contributed equally to this work.

21.1.1 Diseases of the Neurovasculature

21.1.1.1 Intracranial Aneurysms

Aneurysms are abnormal dilatations of a portion of the vascular wall (Figure 21.1). They can occur at any age, but most commonly occur in the adult female population (Deshaies and Boulos 2011). There are two types of intracranial aneurysms: fusiform aneurysms and ISAs. The first are dilatations of a straight segment of artery without an obvious neck and often occur in the basilar artery. The second are spherical in shape, have a neck, and commonly arise at the apex of a bifurcation located within the Circle of Willis (Figure 21.1) (Humphrey 2002; Kumar et al. 2007). ISAs are further classified by their diameter and size of neck. Small, large, and giant ISAs have diameters 4–10, 11–25, and >25 mm, respectively. Necks smaller than 4 mm are classified as small-necked, and necks greater than 4 mm as wide-necked ISAs (Murayama et al. 2003b). Fusiform aneurysms usually remain asymptomatic until they press on adjacent tissue. In contrast, ISAs often remain asymptomatic until they rupture resulting in spontaneous subarachnoid hemorrhage (SAH), also known as a hemorrhagic stroke. Of these patients, approximately 35–50% die (Humphrey 2002; Szikora et al. 2006). Of those who do survive, 50% regain functional independence, while 30% are permanently disabled, and 20% require institutional care (Carod-Artal and Egido 2009).

There is little agreement on the pathogenesis of ISAs; however, major contributing factors are distinct structural and geometrical features of the cerebral vasculature, inherited genetic risks, environmental stimuli, and hemodynamic interactions. ISAs have structural irregularities at the apex of their dome, which render them prone to a focal weakening of the wall upon a mechanical insult. The walls of small ISAs are primarily composed of a thin layer of slightly thickened endothelial cells, type I collagen and deposits of glasslike fibers known as hyaline, making them less compliant (Stehbens 1975; Kumar et al. 2007).

One of the geometric factors that make them more prone to rupture involves the radius of curvature of ISAs. After the radius of curvature of the vessel changes from infinite to finite, ISAs enlarge. However, it can be shown that a generalized spherical geometry does not enlarge or rupture as a result of limit point or dynamic instabilities (Humphrey and Delange 2004). Therefore, this leads to the possibility that the final and often fatal state of these lesions is triggered when hemodynamically induced wall stresses exceed the strength of the wall, resulting in rupture.

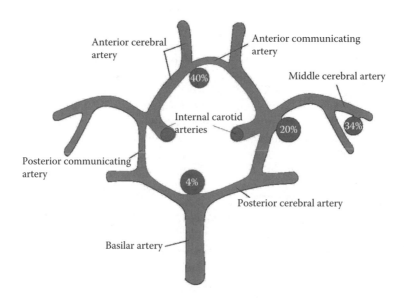

FIGURE 21.1   Location and site of aneurysm formation in the Circle of Willis.

Taking into account the amount of stress a device exerts on the aneurysm wall while filling is crucial for design engineers. For example, Marks et al. (1996) found that the packing of coils during placement can result in a significant distortion of the original 3D shape of the coils, which can lead to increased forces on the wall of the aneurysm.

21.1.1.2 Arteriovenous Malformations

Cerebral AVMs are vascular lesions that consist of abnormal direct connections between the arterial and venous systems (Figure 21.2) (Deshaies and Boulos 2011). The majority of AVMs remain asymptomatic during life. Those that are symptomatic typically present in younger patients (20–40 years). Symptoms include hemorrhage (>50%), seizures (20–25%), headaches (15%), and focal neurological deficits (5%) (Fiorella et al. 2006; Deshaies and Boulos 2011). AVMs are congenital lesions, suggested to originate within the first 3 months of gestation (Mullan et al. 1996). Although genetic factors remain largely unexplored, some AVMs are associated with congenital or hereditary syndromes such as Rendu–Osler–Weber and Klippel–Trenaunay syndromes (Deshaies and Boulos 2011). The primary goal of treatment is to prevent hemorrhage by complete eradication of the lesion.

21.1.1.3 Neurovascular Occlusions

Ischemic stroke is caused by occlusions of the neurovasculature due to embolisms, hypoperfusion, and focal occlusions. Embolisms can be formed due to atherosclerotic plaque rupture and local platelet aggregation as well as due to cardiac anomalies such as atrial fibrillation, valvular disease, and heart failure (Deshaies and Boulos 2011). Hypoperfusion is attributed to severe stenosis of cerebral arteries due to large-artery atherosclerosis. Focal occlusions mainly arise due to atherosclerosis of small vessels and vasospasm and result in what is known as lacunar stroke. Hypertension, diabetes, hyperlipidemia, and family history are major risk factors associated with neurovascular occlusions (Deshaies and Boulos 2011).

21.1.2 Treatment History of Neurovascular Malformations

The first described presence of cerebral malformations dates back to the early eighteenth century (Morgagni and Tissot 1779); however, treatment of vascular lesions of the head and neck did not begin until approximately two centuries later (Prestigiacomo 2006). The twentieth century witnessed an array of different attempts to treat cerebrovascular malformations, some more successful than others.

Some of the early devices developed to treat vascular malformations involved the idea of slowly occluding the patent vessel with different clips coated with thrombogenic materials. These devices included Matas's band from aluminum strips, double fascia band. Another version of these types of

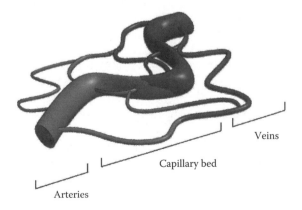

Veins

Capillary bed

Arteries

FIGURE 21.2 Schematic showing the early stage of an AVM. The arterial system is directly connected to the venous system, rather than via a capillary bed like in the healthy vasculature.

devices included the Neff Clamp developed in 1904, which consisted of using absorbable catgut around the clamp's rubber bands for a slow but progressive occlusion of arteries (Neff 1911; Prestigiacomo 2006). Similar devices later developed included the Dott, Crutchfield, Selverstone, and Kindt clamps and were used up until the 1970s (Lehecka 2009).

Beyond implantable devices, an implant that was derived from human tissue was also used to occlude vascular malformations. In 1930, Brooks implanted harvested muscle to embolize a carotid-cavernous fistula. This procedure laid the foundation for the later development and progression of surgery that was more focused on treating malformations via intravascular methods (Debrun et al. 1975).

Different particulates and materials were explored as embolizing agents to thrombose vascular malformations. In the 1960s, methyl methacrylate spheres (Luessenhop and Spence 1960), metal pellets (Doppman et al. 1968), horse hair (Gallagher 1963), Gelfoam (Ishimori et al. 1967) and Bucrylate® (Ethicon, New Jersey) (Zanetti and Sherman 1972) were injected into the carotid artery (Luessenhop and Spence 1960) and spinal cord AVMs (Doppman et al. 1971).

Techniques other than filling methods were also used in the 1960s to treat vascular malformations. The concept of using magnets for endovascular navigation (Rosomoff 1966; Alksne et al. 1967; Yodh et al. 1968) was investigated with the intention of using direct current for thrombosis was explored (Werner et al. 1941; Mullan et al. 1969). Rothenberg et al., proposed using balloons for the treatment of aneurysms (Prestigiacomo 2006). Charles Dotter hypothesized that catheters had the potential not only to diagnose arterial illness, but also to use them to treat atherosclerosis for angioplasty (Dotter and Judkins 1989). In the 1970s, these balloons underwent different modifications including the "calibrated leak" balloon (Kerber 1976) and the caged "korsett balloon catheter" (Porstmann 1973).

Surgical methods to treat vascular abnormalities were developed in parallel to the intravascular methods. In 1937, Walter Dandy placed a silver clip at the base of an aneurysm. This effort by Dandy, marked a new era in the treatment of aneurysms and was followed by many modifications of clips that would be used in the years to come (Dandy 1938; Richling 2006). Surgical clipping became the standard for treating vascular malformations prior to proven endovascular treatment methods.

However, it should be mentioned that modern endovascular treatments for neurovascular diseases include interventions performed with the use of imaging modalities and catheters. True catheter-based approaches did not begin until the 1960s, when catheter–cerebral angiography was brought to America, nearly 30 years after being developed in Europe (Moniz 1933; Kerber 2006; Prestigiacomo 2006). This decade was a revolutionary period of great technological advancements in endovascular neurosurgery.

21.1.3 Treatment History of Neurovascular Occlusions

The first surgical embolectomies of intracranial arteries date back to 1956; however, it was not until 1996 that the U.S. Food and Drug Administration (FDA) approved the drug recombinant tissue plasminogen activator (rtPA) (Gobin et al. 2004). To date, rtPA is the only FDA-approved drug to treat ischemic stroke, nonetheless it is only administered to 2–5% of the patients because of the narrow treatment time window and other exclusion criteria (Brown et al. 2005). Percutaneous mechanical embolectomy devices are an alternative treatment option and are often the only treatment option for patients. One of their greatest advantages is the rapid recanalization time. Different types of mechanical embolectomy devices have been developed. Examples include snare devices (e.g., Amplatz Goose Neck Snare, ev3, California), laser-based systems (e.g., Endovascular Photo-Acoustic Recanalization System, EndoVasix Inc., California), ultrasound-based devices (e.g., MicroLysUs Catheter, EKOS Corp., Washington), retrieval devices (e.g., Merci Retriever®, Stryker, California), aspiration devices (e.g., AngioJet, Possis Medical Inc., Minnesota) and some use a combination of two or more types (e.g., Penumbra System®, Penumbra Inc., California) (Stead et al. 2008). However, some of these devices are used off label and others have not received FDA approval yet.

In 2001, the FDA gave clearance to market the Concentric Retriever (Concentric Medical, California) based on substantial equivalence to the Amplatz Goose Neck Microsnare (ev3, California) and the

Attractor Endovascular Snare (Target Therapeutics, California) for removal of emboli or foreign bodies in distal peripheral vessels. Three years later, the Merci Retriever® got clearance to market based on substantial equivalence to the Concentric Retriever (Becker and Brott 2005). This was the first device cleared by the FDA intended to restore blood in the cerebral vasculature for patients experiencing ischemic stroke ineligible to receive rtPA. Subsequently, the Penumbra System® (Penumbra Inc., California) was approved in 2008 based on substantial equivalence to the Concentric Balloon Guide Catheter (Concentric Medical, California). The most recent device approved in the United States to treat ischemic stroke is the Solitaire™ FR stent (ev3, California).

21.1.4 Advancements in Other Fields That Have Contributed toward New Device Development

Parallel advancements in imaging and material science have paved the way for more developed neuroendovascular techniques and devices for medical intervention. Many diseases of the neurovasculature that were traditionally treated surgically, with high risk, are now treated with much lower complications with the aid of these new treatments. These advancements have given rise to the development of less invasive microsurgical devices which make it possible for interventional radiologists to treat sensitive areas of the neurovasculature with advantages in safety and recovery time (Deshaies and Boulos 2011; Wong 2011). These advancements have been very beneficial to the patients, as treatments have evolved from potentially high-risk surgeries, with long recovery times, to outpatient treatments that are safer and have less cost. This increase in safety is important due to the sensitive areas of the brain (i.e., base of the brain) that are commonly treated with these devices.

21.1.5 Materials Designed for Devices to Treat the Neurovasculature

Materials used for the fabrication of these interventional devices are chosen such that their blood contacting surface is biocompatible for their intended use; resistant to blood clotting or initiate blood clotting. Blood clotting is a complex process and beyond the scope of this chapter, but it should be stated that "Blood coagulation is generally considered in terms of contributions from features such as platelet activation, intrinsic, extrinsic, and common pathways, and the control systems participating in thrombus inhibition and fibrinolysis" (Courtney et al. 1994). Design of a neurovascular device should take into account the ailment and the desired outcome of the treatment. In the instance of arterial malformations, it is desirable for the implanted material to promote clotting at the site of deployment to isolate and stabilize the lesion from the rest of the vasculature. Materials selected for these devices would most likely have at least a surface chemistry that is relatively hydrophilic and polar [i.e., polyethylene glycol (PEG) coating, or "PEGylation" of a material to reduce protein adhesion to the devices]. In addition to surface modifications of blood contacting devices, anticoagulants such as heparin are commonly used. Another method of decreasing the initiation of clotting involves the addition of antifouling molecules like poly(ethylene oxide) to the device. A main determining factor for the necessary surface chemistry and rigorous attempts at controlling clotting of a neurovasculature device is determined by the amount of time in which the device will be within the vasculature.

21.1.5.1 Biocompatibility of Materials Used

Beyond evaluating the initial clotting or lack thereof while in the body, the process of developing a neuroendovascular device must consider lifetime, or length of time the device will remain in the body, and the environment in which the device will encounter *in vivo*. Devices that will remain in the body for an extended period of time must be biocompatible to the host; it must not have a negative impact on the patient while in use. Sometimes the optimal biocompatibility includes a controlled inflammatory response. For example, in the case of embolic devices, the desired response includes a short-lived inflammatory reaction, followed by an encapsulation of the device, thereby isolating it from the rest of the vasculature. In this instance, the host initially considers the material inflammatory for a short period of time, that is, less than 1 week (Kumar et al. 2007), but then resolves this state of active inflammation by walling the material off from itself, or encapsulating the material and ceasing the inflammation process, normally taking weeks to months to complete this process.

For this example, the device caused a chronic inflammation process marked by the presence of mononuclear cells (macrophages, lymphocytes, and plasma cells), with the subsequent destruction of the surrounding tissues, and ending with resolution involving angiogenesis of new blood vessels and fibrosis of the tissue (Kumar et al. 2007). This new tissue is granulation tissue, which can appear as early as the first day of implant, and includes fibroblasts and the promotion of fibroblast and endothelial cell proliferation, angiogenesis of new blood vessels, and a loosely connected extracellular matrix (ECM) (Kumar et al. 2007). The granulation tissue eventually undergoes the process of fibrosis. Fibrosis of the tissue involves a decrease in the amount of fibroblasts (the cells responsible for deposition of collagen and generation of the extracellular matrix) and new blood vessels (angiogenesis) (Kumar et al. 2007). The remaining fibroblasts take on the role of depositing more extracellular matrix and deposition of collagen (Kumar et al. 2007). The tissue is now transitioning from active granulation tissue to a scar (Kumar et al. 2007). The scar no longer needs the same support from cellular components and the tissue becomes devascularized (Kumar et al. 2007). This scar is composed of collagen, elastic tissue, and spindle-shaped fibroblasts (Kumar et al. 2007).

If the device will remain in the body, it is desirable for the material to promote endothelialization over blood contacting surfaces. Examples of devices that may promote this type of healing include some bare metal stents and polymeric aneurysm filling materials (Szikora et al. 2006; Rodriguez et al. 2012). As mentioned before, healing may involve only encapsulation of the device, and not involve full healing of the surrounding tissue. If prior to encapsulation the device begins to degrade, it is important to characterize the degradation products and their toxicity to the patient. Implantable devices need to be evaluated for their bulk stability and leachability of toxic substances; this requires more extensive chemical evaluation beyond simply evaluating their surface chemistry. A full list of appropriate biocompatibility tests that should be used to evaluate medical devices are outlined in the International Standards Organization (ISO) 10993 series (1–20) of standards.

21.1.5.2 Device Visualization and Monitoring

The development of neuroendovascular devices requires that the materials chosen for blood contacting applications and implants must not only be biocompatible, but at least a portion of the device must also be radio-opaque while in the body. Fluoroscopy, a projection imaging modality based on x-ray technology, is most commonly used by interventional radiologists to position devices within the brain. Due to the location of the sometimes inaccessible and sensitive regions of the brain that are treated, it is essential that the devices be visible via fluoroscopic imaging methods. To be radio-opaque via fluoroscopy, the materials used must be able to attenuate x-rays. Attenuation of x-rays requires that the material be thick and dense enough to scatter x-rays, characterized by the mass absorption coefficient of an element, which is the sum of the absorption of a cross-section and depth of a material with which an x-ray interacts (Figure 21.3). Radio-opacity is usually accomplished by having a high-z element incorporated somewhere into the device, and the visible portion of the device is the high-z element itself. This form of contrast agent is known as a direct contrast agent (Brown and Semelka 2010). A high-z element is one that is located within the fifth row or lower on the periodic table. These elements, usually metals and include the noble or precious metals [gold (Au), silver (Ag), platinum (Pt), iridium (Ir), palladium (Pd), rhodium (Rh), ruthenium (Ru), and osmium (Os)], but not always, and can also include salts or metals of elements such as iodine, tantalum (Hampikian et al. 2006), tungsten (Rodriguez et al. 2012), barium, and bismuth. These are some of the commonly used materials for making devices or solutions radio-opaque in the body, mainly due to their proven biocompatibility or resistance to chemical corrosion.

Metal devices that are made of platinum, stainless steel, or nitinol are inherently radio-opaque, due to the density and the thickness of the device being able to attenuate x-rays. However, not all devices

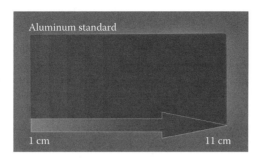

FIGURE 21.3 Fluoroscopic image of an aluminum standard demonstrating the increase in radio-opacity with increasing thickness of material.

used for neurovascular interventions are made of metal. It may be advantageous to develop a medical device designed and fabricated out of polymeric materials due to their mechanical tunabilty and biocompatibility. Unfortunately, polymeric materials are usually not able to attenuate a sufficient number of x-rays to be visible inside of the body, usually due to size (e.g., not thick enough to be more dense than the surrounding tissue, especially within the skull). Due to these limitations, devices are either doped with high-*z* element fillers or are fashioned with metal bands on the distal and/or proximal ends of the devices. These metal bands, or commonly called radio-opaque markers, are usually made of platinum.

Previously proposed fillers for primarily nonmetallic materials, include tungsten (Rodriguez et al. 2012), bismuth (Mohn 2012) and tantalum (Hampikian et al. 2006). These materials have been shown to be not only radio-opaque (Hampikian et al. 2006; Rodriguez et al. 2012) when added to polymeric materials, but preliminary results have shown favorable biocompatibility as well (Rodriguez et al. 2012). In volume percentages of 4%, tungsten has been shown to be an adequate filler material to attenuate enough x-rays to be visible *in vivo* (Rodriguez et al. 2012).

Neurovascular interventional procedures performed using fluoroscopy require the use of contrast agent to be able to visualize the vasculature and flow. A commonly used contrast agent involves a sodium iodide solution injected into the vasculature while cine, or successive fluoroscopic images, are acquired.

Other imaging techniques used to diagnose arterial malformations and blockages include computed tomography (CT) and magnetic resonance imaging (MRI). CT imaging, similar to fluoroscopy, is based on x-ray technology and obtains images based on the linear attenuation coefficient of each material, with dense tissues such as bone, scattering much of the x-rays. The greater the level of scattering, the greater the amount of contrast that is generated in comparison to the background; dense tissues show up white and materials of lower densities exhibit a much lower contrast in comparison to the background, usually seen as dark gray. These images appear opposite to fluoroscopy, where the higher the attenuation of the material, the darker it appears within the image. Another difference between fluoroscopy and CT is that CT is a 3D imaging modality. Although, it should be noted that three-dimensional fluoroscopy is available, and has the ability to generate images very similar to CT. Due to the naturally lower contrast in less dense tissues, CT can be used to diagnose vascular malformations, but in comparison to MRI it has less differentiation of soft tissues and often lower resolution. The main benefits of using CT over MRI to diagnose patients with malformations or blockages of the vascular system are: the greatly decreased imaging time, and relatively less expensive.

MRI involves the use of a main magnetic field applied to a patient to initially align the proton spins of the nuclei of atoms within a tissue; this molecule is usually hydrogen, abundant in water and fat of tissues. After the protons of either molecule within the tissue have been aligned (either parallel or antiparallel) with the main magnetic field, radio frequency (RF) pulses and gradients are applied to alter the spins of the tissue by increasing the energy state of the protons. These perturbations of the spins are temporary, as these processes were quantum absorptions of energy, and by Gibbs free energy the protons prefer to be in the lower energy state and eventually realign with the main magnetic field. This temporary state happens when the protons are excited to a higher energy, and then drop to a lower energy as they release energy that is proportional to the main magnetic field. As these processes are occurring, a coil adjacent to the tissue receives the energy released during the relaxation of the atoms back to their normal low-energy state, this energy is the free inductions decay (FID), or the MRI signal. The RF pulse causes the protons to be excited and gradients are used to slightly alter the frequency of the proton based on location, thereby allowing the position of the released energy to be determined within the MRI signal. Use of the Fourier transform allows these data to be converted from the time to frequency domain, which enables the reconstruction of a visible image.

Image formation comes from the detection of a signal given off by the tissues at particular time points. Two of the most common times to acquire images are T1 and T2 relaxation times. T1 relaxation time, or spin–lattice relaxation time, is the time required for the z component, or longitudinal component, of the net magnetization of the tissue to decay to 63% of its original value following an RF pulse (Brown and Semelka 2010). This time is when a signal is released by the tissue and recorded to form an image, where water appears dark and fat appears bright. Contrast agents for T1 decay include gadolinium-based molecules. An alternative imaging method includes T2 imaging. T2 relaxation time, also known as spin–spin and transverse relaxation time, is the time required for the transverse component of the net magnetization of the tissues to decay to 37% of its initial value, which is irreversible (Brown and Semelka 2010). This time is when a signal is released by the tissue and recorded to form an image, where water appears light and fat appears dark. T1 imaging is good at providing definition of anatomy, while T2 imaging is good for visualizing pathologies within the brain. This difference in imaging techniques can be overcome with the use of a gadolinium contrast agent for T1 imaging.

Although MRI does distinguish soft tissues quite well in the absence of a contrast agent, with the use of T1 imaging, sometimes it may be difficult to differentiate pathological from normal tissues by this technique. In this case, an MRI contrast agent would be used. Unlike fluoroscopy or CT, the contrast agent is not what is visible in the image, however, what is visible is the contrast agent's influence on the proximal hydrogen atoms within water molecules affecting the relaxation times (Brown and Semelka 2010).

From this information, it is reasonable to assume that images from MRI will be different than images obtained by the previously mentioned x-ray-based technologies (fluoroscopy and CT). In fact, MRI has the ability to generate images with high resolution in the soft tissues, making them very valuable to doctors that want to visualize abnormalities within the brain. Therefore, it is reasonable to understand why it would be beneficial for a neurointerventionalist to have access to such images.

Although the inherent contrast via MRI could pose to be superior in the soft tissues of the brain, the use of MRI for positioning of neurovascular devices is not widely used due to the limited compatibility of most interventional devices and length of imaging times. In fact, devices that are to be used with an MRI should be tested for compatibility with the large magnetic fields. With most clinical MRIs being either 1.5 or 3 T, it is important that medical device developers investigate the compatibility of their prospective device with a magnetic resonant environment using the following ASTM standards: ASTM F2052, ASTM F2119, ASTM F2182, and ASTM F2213, which evaluate magnetically induced displacement force, MR image artifacts from passive implants, RF-induced heating, and magnetically induced torque on medical devices in the magnetic resonance environment.

Owing to the value of MRI in evaluating pathologies of the brain, these testing standards have previously been used to evaluate aneurysm clips and coils, and interventional instruments such as catheters and guidewires. Based on results from these tests it can be concluded that the device may be compatible with MRI, and may be implanted via MRI imaging, but it is more likely that MRI would be used to diagnose pathology or to track the lifetime of the endovascular implant. This can be advantageous due to the increased resolution and the lack of use of ionizing energy, as excessive exposure to ionizing radiation can be harmful to the patient.

Other MRI imaging techniques involving the visualization of flow are used to diagnose and evaluate ischemic and potentially hemorrhagic presenting patients by diffusion-weighted imaging (DWI) (Moseley et al. 1990) and magnetic resonance angiography (MRA). DWI is a diffusion imaging technique, used after ischemic stroke, due to its ability to be sensitive to the changes in flow of the lesion (Moseley et al. 1990). MRA imaging is used for diagnosing narrowing of arteries and aneurysms, commonly with the use of a gadolinium contrast agent or with the use of a fluid visualization imaging technique called time-of-flight flow sequence which images blood that has moved into the imaging plane (Haacke 1999). Therefore, development of neurovascular devices should include efforts to not only evaluate devices for MRI compatibility, but also design them to be compatible.

21.2 Current Neuroendovascular Devices

21.2.1 Devices to Prevent/Treat Hemorrhagic Stroke: Filling Devices

21.2.1.1 Coils

Metal coils were designed to treat aneurysms with the primary goal of embolization, that is, occlusion of the aneurysm with initially thrombus and eventually fibrous tissue. Embolization of the malformation would cause exclusion from the normal circulation to reduce the risk of rupture and subsequent hemorrhage (Figure 21.4). Although, coils were used to occlude aortic aneurysms more than a century ago (Prestigiacomo 2006), it was not until the mid-twentieth century that coils were used as a means to deliver current to electro thrombose aneurysms (Werner et al. 1941; Mullan et al. 1969), and the first reported introduction of platinum coils into a cerebral aneurysm did not happen until 1988 by Hilal et al. (1988). Soon thereafter, Guido Guglielmi began developing techniques to improve coil delivery. In collaboration with Target Therapeutics, the first generation of electrolytically detachable coils, termed Guglielmi Detachable Coils (GDCs) after its inventor, was developed in 1990 (Prestigiacomo 2006). Electrolytically detachable coils use electrical current to melt the connection between the delivery wire and the platinum coils, leaving the implanted coils inside the aneurysm to allow for embolization.

Before the invention of endovascular coils, the standard treatment approach was surgical clipping of the aneurysm neck (Dandy 1938). In the 1960s, when balloons were proposed as embolic aneurysm fillers, the focus of aneurysm treatment shifted from surgical means to embolization by filling methods (Debrun et al. 1975; Lusis 2000).

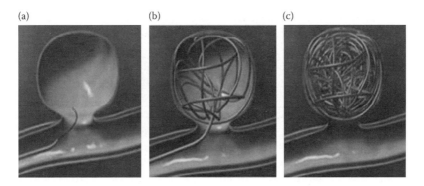

(a) (b) (c)

FIGURE 21.4 GDCs used to treat an aneurysm. A microcatheter is used to deliver the coils (a), they are detached using an electrolytic process (b), and more coils are delivered in an attempt to fill the aneurysm (c). Original text: Detachable platinum coils being used to treat an aneurysm: (a) a microcatheter is used to deliver the platinum coils to the aneurysm; (b) the coils are detached using an electrolytic process; (c) more coils are added to fill the aneurysm and allow blood to coagulate, forming a permanent seal. (From Cowley, A. and B. Woodward 2011. A healthy future: Platinum in medical applications. *Platinum Metals Review* **55**(2): 98–107, reprinted with the permission of Johnson and Matthey Plc.)

Helical platinum coils are now standard for endovascular treatment of aneurysms and other malformations (Deshaies and Boulos 2011). Platinum was used for several reasons; it is a soft metal that can be easily bent, has been shown to recover its shape, and is biocompatible when implanted (Murayama et al. 1999b).

In addition to the previously mentioned reasons, platinum was a desirable material for the initial application because of its sensitivity to low electrical current, making it easily detachable from steel guidewires. However, common problems with using bare platinum coils include incomplete filling of aneurysms and coil compaction. Complete occlusion of wide-necked aneurysms is achieved in only 15–35% of the cases (Fernandez Zubillaga et al. 1994; Viñuela et al. 1997). Over time, bare platinum coils tend to compact resulting in recanalization (i.e., recirculation of blood flow to the aneurysm dome posttreatment) and need for treatment repeat (Guglielmi et al. 1992; Murayama et al. 1999b). In an 11-year-long study, in which recanalization was defined as more than 10% increase in contrast filling of the treated aneurysm with bare platinum coils, the recanalization rate of small aneurysms with small necks, small aneurysms with wide necks, large aneurysms, and giant aneurysms was 5.1%, 20%, 35.3%, and 59.1%, respectively (Murayama et al. 2003a).

Recently, advances have been made to the original helical coil design. For example, nonhelical, complexly shaped coils have been introduced, including the Complex Fill™ (Cordis, Florida) and the GDC®360° (Stryker, California). These nonhelical systems improve on the low packing density achieved by traditional GDCs. By increasing packing density, these newer devices lower the rate of recanalization of the aneurysm dome (Kallmes and Fujiwara 2002; Piotin et al. 2003).

There is evidence that platinum coils are not sufficiently bioactive to produce a rapid inflammatory response (Szikora et al. 2006), necessary for the initiation of clotting throughout the aneurysm dome. This initial inflammatory response is desired to quickly cut off an aneurysm from the parent vessel circulation. One way to increase the bioactivity of the coils is to coat them in biodegradable polymer to promote an inflammatory response. The Matrix system introduced in 2003, by Murayama et al., was the first attempt at increasing the inflammatory response with polymer coating of coils (Murayama et al. 2003b). A further benefit to the polymer coating is the potential for a reduced risk of perforation of the aneurysm wall during placement.

Hydrogels are a form of smart materials that have the ability to change their shape when a stimuli is applied. They consist of polymer networks that are composed of at least two components: a three-dimensional nonsoluble polymer network that forms the structure, and a water-soluble intertwined polymer material that stays within the structure via ionic interactions.

Another potential technique to overcome recanalization of treated aneurysm is the use of hydrogel-coated platinum coils. Hydrogels are a form of smart materials that have the ability to change their shape when a stimulus is applied. They consist of polymer networks that are formed from at least two components: a 3D nonsoluble polymer network that constitutes the structure, and a water-soluble intertwined polymer material that stays within the structure via ionic interactions. Hydrogels are composed of polymeric chains that are cross-linked by physical and chemical bonding, tethering, and entanglements. Physical cross-links are hydrogen bonding and Van der Waals forces. Chemical bonding includes covalent or ionic bonds. These are known as physical gels and chemical gels, respectively. Tethering is the branching of chains off a polymer backbone or substrate, which form entanglements when they interact. These networks are considered highly water insoluble. Hydrogels are stabilized by ionic interactions, and consist of a 3D network of hydrophilic polymer chains that are cross-linked together and saturated with a liquid. This liquid is trapped within the 3D network of polymers through surface tension. In the specific case of hydrogels, water is the liquid that is suspended between the polymer networks. Due to their high content of water, these gels are very flexible and can swell to volumes much greater than that of the polymer network alone, making these materials ideal candidates for tissue engineering and other biomedical engineering applications. These gels are sensitive to stimuli, such as pH, temperature, ionic strength, solvent composition, pressure, and electrical potentials. These polymers are also used in drug-delivery systems. Micro Vention Inc. (California) uses a hybrid device of hydrogel on platinum coils, called Intelligel™. These coils are delivered endovascularly in a similar manner to traditional GDCs. However, unlike GDCs, the hydrogel portion of these coils expands while in contact with blood, tripling the coil diameter after approximately 20 min (Cloft 2006; White et al. 2008). With the aid of the hydrogel coating, these coils are able to more densely pack the aneurysm dome.

In addition to detachable coils, there are injectable coils [e.g., Berenstein Liquid Coil® (Stryker, California)] that are used in occlusions of arteriovenous fistulae within the nidus of AVMs (Fiorella et al. 2006).

21.2.1.2 Embolic Liquids

Liquids used for the purpose of embolization have a history of over a century. The first attempt to use a liquid as an embolic material dates back to 1904, when Dawbarn used liquid paraffin to embolize a malignant tumor in the external iliac artery (Berenstein et al. 2006). Sano et al. (1975) used polymerizing silicone to occlude arteriovenous shunts Zannetti and Sherman (1972) used Bucrylate (Ethicon, New Jersey) to embolize AVMs and aneurysms. Bucrylate, is a 4-carbon cyanoacrylate, initially conceived as a suture substitute. A similar material called TruFill® (Cordis, Florida), was developed; used to embolize malformations (Kerber 2006). In 1990, Taki et al., reported the use of ethylene vinyl alcohol copolymer (EVAL), as a liquid embolizing agent for aneurysm therapy (Taki et al. 1990). EVAL underwent some modifications and became what is now Onyx® (ev3, California) (Richling 2006). Onyx is a dimethyl sulfoxide (DMSO) solvent-based system ethylene vinyl alcohol (EVOH) copolymer (Fiorella et al. 2006). A few years later, Yakes et al., reported positive outcomes of using ethanol (ETOH) in the embolization of AVMs (Yakes et al. 1996). To date, only TruFill and Onyx are FDA approved as embolic liquids within the United States.

Cyanoacrylates are liquid adhesive polymer agents that allow for deep penetration and permanent embolization with durable occlusion of malformations. They are introduced as liquid monomers through a small, flexible catheter, and subsequently polymerize upon contact with an anion-containing solution, such as the hydroxyl groups in blood (Fiorella et al. 2006). In contrast with cyanoacrylates, EVOH is a nonadhesive polymeric agent that reduces the risk of the catheter adhering to the injected polymer.

ETOH occludes malformations by dehydrating the endothelium, causing the formation of fractures within the vessel wall that extend to the level of the internal elastic lamina which result in acute thrombosis (Yakes et al. 1996). A drawback to using this embolizing method is that it causes brain edema, requiring high doses of steroids before and after the procedure. A high level of experience is required to perform ETOH embolization. Therefore, due to comfort and widespread experience with cyanoacrylates, clinicians are generally reluctant to use ETOH (Fiorella et al. 2006). Both cyanoacrylates and EVOH, are radiolucent, not dense enough to impede x-rays in fluoroscopic imaging. Cyanoacrylates are usually mixed with methodized oil and EVOH with tantalum for radio-opacity.

21.2.1.3 Embolic Particles

Different particulates have been explored as embolizing agents. Lussenhop and Spence introduced handmade methyl methacrylate spheres into the carotid artery (Luessenhop and Spence 1960). Doppman embolized spinal cord AVMs by using pellets of stainless steel (Doppman et al. 1971). Currently in the market there are embolization particles made with polyvinyl alcohol (PVA) (Cook Medical, Indiana) and "embospheres," such as the Embozene® microspheres (CeloNova Biosciences, Texas). The latter consist of a hydrogel core with the company's proprietary polymer Polyzene®-F on the exterior.

21.2.2 Combination Devices: Treat Hemorrhagic and Ischemic Stroke

21.2.2.1 Stents

A single method or device is not always used in the treatment of aneurysms. Stents can be placed to aid in aneurysm treatment at various stages of aneurysm filling (coil or liquid treatment), or alone. Stents provide important stabilization to the parent vessel, while promoting fibrous growth around the neck of the aneurysm. This is particularly true in wide-necked aneurysms (an aneurysm with a neck opening diameter greater than the dome diameter) (Wajnberg et al. 2009), because without a stent to stabilize the filler, it may migrate out of the aneurysm. Stents are also used to treat other diseases such as extracranial arterial occlusive disease, intracranial arterial stenosis, and spinal AVMs (Deshaies and Boulos 2011). The invention of the self-expanding nitinol stent in 2002 has proved to be a major advancement in neurovascular treatment. These stents can be placed endovascularly in smaller and more tortuous vessels than previously possible (Henkes et al. 2002). Commonly used nitinol stents include the Neuroform™ (Stryker, California), Enterprise™ (Cordis, Florida), and the Wingspan™ (Stryker, California) (Zaidat 2006). Recently, self-expanding stents were used as mechanical embolectomy devices. For example, the Solitaire™. FR stent (ev3, California), was approved by the FDA in March of 2012, for the treatment of ischemic stroke (Figure 21.5).

Stents are also used in stent-assisted angioplasty. This variation of percutaneous transluminal angioplasty procedure is performed to treat intracranial stenosis which is responsible for 8–10% of ischemic strokes (Qureshi et al. 2003). Both balloon-expandable and/or self-expandable stents are used. Recently,

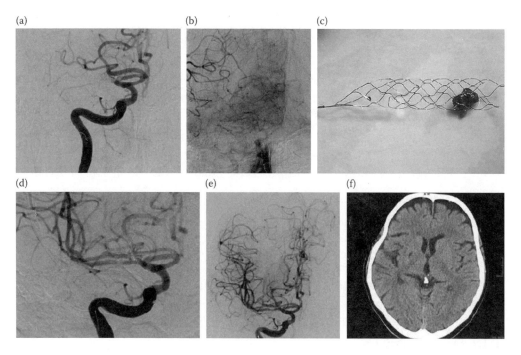

FIGURE 21.5 Treatment of an embolic occlusion of the right middle cerebral artery (MCA) using a Solitaire™ stent. Original text: Embolic occlusion of the right middle cerebral artery (MCA), treated by mechanical thrombectomy using a Solitaire stent. (a) Digital subtraction angiography (DSA) with injection of the right internal carotid artery (ICA) shows embolic occlusion of the right M1 segment. (b) In the late phase of this DSA run with injection of the right ICA, the leptomeningeal collaterals between the right anterior and MCA were visualized. (c) Solitaire stent loaded with thrombus, which was removed from the right M1 segment. (d, e) DSA with injection of the right ICA after Solitaire thrombectomy confirms the recanalization of the short M1 and both M2 segments (d, magnified view) without vasospasm or distal emboli (e, whole head run). There might be a minor remnant of the previously occluding thrombus in the inferior trunk to the MCA/M2 (d). (f) CT 22 months after the Solitaire thrombectomy of the right MCA shows a small postischemic scar of the right basal ganglia. (From Perez, M. A., E. Miloslavski, et al. 2011. Intracranial thrombectomy using the solitaire stent: A historical vignette. *Journal of Neurointerventional Surgery*, reprinted with the permission of BMJ Publishing Group.)

the FDA approved the Neurolink® (Guidant, Indiana), a drug eluting, self-expanding stent, indicated for patients who experience a recurrence in intracranial stroke. Despite the potential benefits of the use of endovascular stenting, a common major drawback among the available stents, especially with respect to self-expandable stents, is the difficulty or lack of repositioning the stent after deployment (Biondi et al. 2007). Also, microemboli get released during stenting and can cause new lesions in the brain. Distal protection devices such as occlusive balloons, filter devices, or flow-reversal devices are used to retain particles that can cause harm (Qureshi 2004).

21.2.2.2 Flow Diverters

Like stents, flow diverters treat aneurysms from outside the aneurysm dome, within the lumen of the parent vessel. Flow diverters do exactly what their name implies: divert flow away from an aneurysm dome, reduce the risk of rupture while providing an environment that allows for the stagnation of blood within the aneurysm sac. Stagnation of blood flow promotes the formation of clot, which like previously mentioned devices, isolates the aneurysm from the parent vessel (Gobran Riad et al. 2007; Augsburger et al. 2009; Sadasivan et al. 2009). Flow diverters first appeared in 2006; initially they were simply modified stents. Further study has shown that there is a range of porosity values, or mesh cell fineness, that provide the best opportunity for stagnation and clot formation. Available flow diverters include the Pipeline™ Embolization Device (ev3, California) and the SILK Flow diverter (Balt Extrusion, France) (Wong 2011).

21.2.3 Devices to Treat Ischemic Stroke

21.2.3.1 Mechanical Embolectomy Devices

Acute ischemic strokes occur when loss of blood flow to an area of the brain due to an occluded blood vessel. The only noninvasive FDA-approved treatment technique is the use of recombinant tissue plasminogen activator (rtPA) administered intravenously. Unfortunately, only approximately 2% of patients are eligible to receive rtPA due to its long list of exclusion factors. Included in this list is the requirement of administering the treatment immediately within three hours of stroke onset (Brown et al. 2005). In the 1990s, a more direct solution was developed. Endovascular techniques such as mechanical thrombectomy and clot disruption were introduced. With these techniques, the treatment time window is greater than the three hours with rtPA, allowing for clinicians to use it to treat more cases (Nesbit et al. 2004).

Mechanical embolectomy is defined as the physical removal of a thrombus using a catheter. Such devices can be further classified as snare, clot retrieval, or thromboaspiration devices. In addition to the Solitaire™ stent (ev3, California), the only other devices FDA approved for the treatment of stroke are the Merci Retriever® system (Stryker, California) and the Penumbra System® (Penumbra Inc., California). The first uses a nitinol corkscrew shaped tip on the end of a catheter to grasp the thrombus and pull it out. The second uses clot disruption and aspiration techniques (Figure 21.6). Other devices that use different clot disrupting techniques such as ultrasound, photoacoustic energy, and saline solution jets are being investigated (Qureshi 2004).

21.2.3.2 Balloon Angioplasty

This technique, used mostly for the treatment of intracranial arterial stenosis, involves a balloon mounted on the tip of the microcatheter, which is then inserted through the femoral artery and navigated into the lesion site in the brain. Once the microcatheter is advanced, the balloon is inflated to widen the arterial walls and increase blood circulation. While the widening of the artery is important, the goal is the prevention of stroke by increasing blood flow to ischemic areas. Balloon angioplasty has also been reported successful in the treatment of acute stroke (Figure 21.7) (Mori 1999; Song 2002; Deshaies and Boulos 2011).

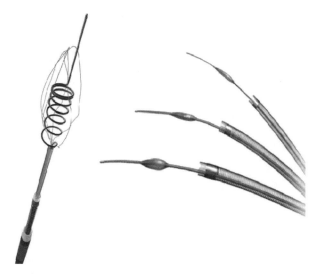

FIGURE 21.6   Merci Retriever® (Stryker, California) (left) and variety of Penumbra System® devices (Penumbra Inc., California) (right). (Reprinted with the permission of Stryker and Penumbra Inc.)

In 2005, the FDA approved the NeuroFlo™ catheter (CoAxia, Minnesota) to treat cerebral ischemia in patients with symptomatic vasospasm. The device is a dual balloon catheter designed to partially occlude the aorta above and below the origin of the renal arteries. This device has been reported to increase global cerebral perfusion within minutes of balloon inflation with no significant increase in mean arterial pressure (Figure 21.8). These results indicated this treatment as a potential candidate to treat acute ischemic stroke (Nogueira et al. 2009; Deshaies and Boulos 2011).

21.2.4 Required Equipment for Endovascular Treatment

21.2.4.1 Catheter Lab (Imaging Suite)

A catheter lab or Cath-lab is the fluoroscopic imaging suite in the hospital where neurointerventional procedures are performed by a radiologist. The fluoroscopic imaging suite is required for positioning devices; it has an angiographic table set up for imaging based on x-ray technology. To image the vasculature, a radio-opaque dye or contrast needs to be injected, as blood is not a good material to impede x-rays.

21.2.4.2 Introducers

An introducer is a small tube that provides access to the intravascular space via an incision in the skin. It is composed of a stopcock, a hub, a dilator, and a sheath shaft. The stopcock is connected to a flush line and can be used to inject contrast or draw blood. Most hubs have a built-in hemostatic valve that prevents blood loss while devices are introduced. The dilator is a removable inner cannula that is coaxially inserted into the sheath, while it is advanced through the skin. The dilator makes it easier to insert the introducer and helps prevent "snow plowing," that is, burying the tip of the introducer into the vessel wall. The sheath shaft is a hollow tube between the tip of the introducer and the hub, through which devices are inserted into the target vessel. Introducers are sized based on their inner diameter (Deshaies and Boulos 2011).

21.2.4.3 Catheters

Catheters have a tip, a hub, and a shaft. They are inserted via an introducer, and are sized by their outer diameter. There are three different types of catheters: diagnostic catheters, guide catheters, and microcatheters (Deshaies and Boulos 2011).

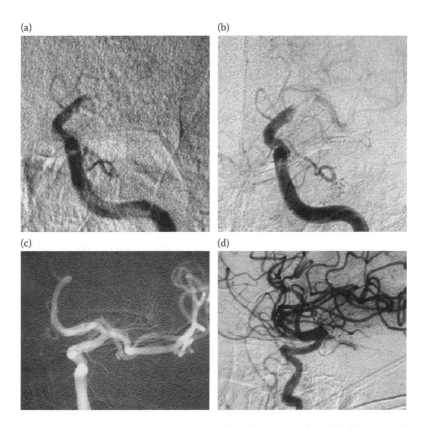

FIGURE 21.7 Case of an embolic occlusion of the supraclinoid ICA treated with balloon angioplasty. Original text: Case 1. (a) Left ICA angiogram, frontal projection, obtained after the intravenous t-PA infusion confirms the persistent supraclinoid ICA occlusion. (b) Left ICA angiogram, frontal projection, obtained after intravenous and intra-arterial thrombolysis shows minimal improvement with a continued supraclinoid left ICA and left M1 segment MCA clot burden. Note the course of the microcatheter through the clot. (c and d) Control left ICA angiograms, frontal (c) and lateral (d) projections, obtained after balloon angioplasty of the left supraclinoid ICA and left M1 segment MCA show complete recanalization of the carotid T. (From Song. 2002. Intracranial balloon angioplasty of acute terminal internal carotid artery occlusions. *American Journal of Neuroradiology* **23**: 1308–1312, reprinted with the permission of the American Society of Neuroradiology.)

Although any catheter could be used for a diagnostic angiogram, diagnostic catheters are usually softer to avoid arterial dissection or embolus dislodgement. They are available in recursive (e.g., angled glide) and nonrecursive (e.g., Simmons 1) shapes (Figure 21.9) (Deshaies and Boulos 2011).

Guide catheters have a braided stainless-steel core of decreasing stiffness that runs along their entire length. This makes them stiffer while in a tortuous path to allow it to maintain its configuration when smaller catheters and/or devices are passed through them. They decrease in stiffness from the proximal to distal end of the segmented catheter. This design helps in the navigation through tortuous vessels, while reducing the risk of vessel damage. Due to their large sizes and stiffness, they are rarely manipulated forward without a guidewire.

In addition to traditional guide catheters, there are balloon-guide catheters which have a compliant balloon mounted at the tip. The only FDA-approved balloon guide catheter is the Merci® Balloon Guide Catheter (Stryker, California) used to arrest blood flow in the internal carotid artery (ICA). This balloon catheter is commonly used during embolus removal (Deshaies and Boulos 2011).

Microcatheters are designed for intracranial use; their size must be less than or equal to 3 French (~1 mm, outer diameter) in order to access these vessels. In addition to being small, they are characterized

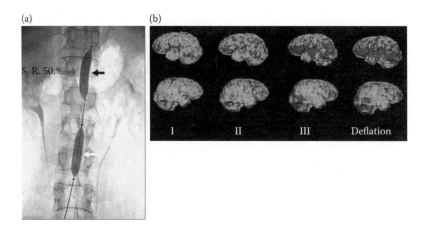

FIGURE 21.8   (**See color insert.**) Increase in global cerebral perfusion after treatment with the NeuroFlo™ catheter. Original text: The NeuroFlo catheter. (a) Fluoroscopic images demonstrate the suprarenal (*black arrow*) and infrarenal (*white arrow*) balloons. (b) Positron-emission tomography demonstrates progressive increase in flow during and after balloon inflation in a patient with a right MCA stroke. Note the global increase in perfusion in both nonischemic (upper panel) and ischemic (lower panel) hemispheres. (Courtesy of Professor Wolf-Dieter Heiss, Max Planck Institute for Neurologic Research, Cologne, Germany.) (From Nogueira, R. G., L. H. Schwamm, et al. 2009. Endovascular approaches to acute stroke, part 1: Drugs, devices, and data. *American Journal of Neuroradiology* **30**(4): 649–661, reprinted with the permission of the American Society of Neuroradiology.)

by their tapered shaft, hydrophilic coating, braided metal core shaft, distal markers, and preshaped or custom tips (shapeable with steam) (Deshaies and Boulos 2011). An alternative version of a microcatheter is a flow-guided catheter, which lacks the braided metal core.

21.2.4.4 Guidewires

Guidewires are typically used in neuroendovascular procedures as a means of navigating through the vasculature. They have a metal core surrounded by a wire coil giving it stiffness at the proximal end and more flexible at the distal tip. Microwires are a subset of guidewires that have diameters between 0.007″

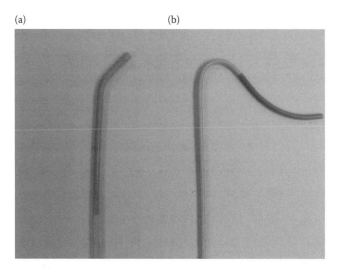

FIGURE 21.9   Angled glide catheter (a) and Simmons 1 catheter (b).

and 0.021″. Guidewires are used to navigate catheters, balloons, and other devices into distal precerebral and intracranial vessels. These wires have soft tips that either come preshaped or can be shaped on the table. Both types of wires are steered using a torque device, that is, a small plastic device that grips onto the wire, and can be easily held and manipulated by the interventional neuroradiologist (Deshaies and Boulos 2011).

21.2.4.5 Balloons

There are both compliant and noncompliant balloons used in neuroendovascular procedures. Compliant balloons are low-pressure balloons designed to conform to the vessel lumen, typically used for temporary vessel occlusion or balloon-assisted aneurysm coiling. Noncompliant balloons, on the other hand, are designed to attain a particular size and shape at a given pressure with the intent to force the vessel lumen to conform to their shape, and they are typically used in balloon angioplasty. Rapid exchange or monorail balloons are mounted inside a short catheter attached to a pusher hypotube (typically a stainless steel or nitinol tube). It is used to exchange devices over shorter microwires, allowing for faster and safer exchanges (Deshaies and Boulos 2011).

21.3 Premarket Hemorrhagic Stroke Devices

21.3.1 Shape Memory Polymer Embolic Foams

Shape memory polymers (SMPs) are smart materials that become malleable when heated above their transition temperature. If deformed and then cooled below their transition temperature, they are able to maintain a stable secondary shape. Upon reheating above transition temperature, they recover their primary shape (Lendlein and Kelch 2002; Gall et al. 2004; Ortega 2007; Maitland 2009; Singhal et al. 2010; Small et al. 2010; Rodriguez et al. 2012). These characteristics make SMPs an ideal candidate material for endovascular devices due to their ability to be crimped to small sizes and then actuated to recover their primary shape *in vivo*. These materials can also be foamed and fabricated to fill any desired volume making them ideal to fill arterial malformations such as aneurysms (Figure 21.10) (Maitland 2011). These materials are low density, compliant and open-celled (Singhal et al. 2012). Their maximum circumferential stress during deployment is less than 1/10 of the minimum wall-breaking stress of an aneurysm (Hwang et al. 2011). Additionally, they can fully pack the aneurysm and they can reduce the recurrence of aneurysm formation due to their physical properties (Maitland 2011). In a preliminary aneurysm animal model study, these devices were doped with a radio-opaque filler to be seen using fluoroscopy, filled the full volume of the aneurysm sac, were stable *in vivo* (did not migrate into the parent vessel) and were completely endothelialized at the base of the aneurysm at 90 days (Figure 21.11) (Rodriguez et al. 2012).

21.3.2 SMP Stents

Commercially available neurovascular stents are made out of stainless steel, cobalt, and shape memory alloys (SMAs), typically nitinol. Complications arise due to the high compliance mismatch between the arterial wall and metal stents (Baer et al. 2007). For example, a stainless-steel stent has about 3 times the radial stiffness of a similarly designed self-expandable nitinol stent. Although self-expandable stents have lower radial stiffness, they are usually oversized to avoid acute recoil, which refers to the narrowing that occurs in a stent after being deployed as a result of the compressive force exerted by the elastic vessel wall. This oversizing oftentimes leads to complications as well (Deshaies and Boulos 2011). SMPs offer the geometric changes necessary for catheter delivery and have lower moduli than metals, which would reduce the compliance mismatch between the stent and arterial wall. In addition, navigation of a stent with a lower modulus through the tortuous vasculature would be easier (Baer et al. 2007). Baer et al. fabricated laser-etched SMP polyurethane stents and measured full collapse pressures

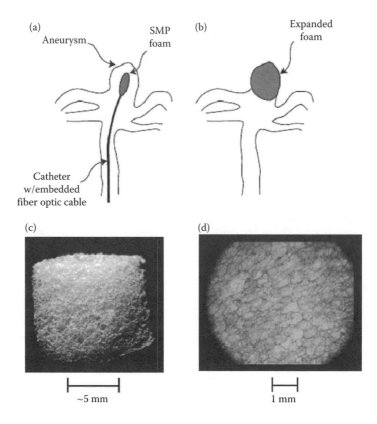

FIGURE 21.10 Concept of treatment of an intracranial aneurysm with SMP foam. Original text: Treatment of an intracranial aneurysm with SMP foam. (a) Catheter delivery of a compressed piece of SMP foam to the aneurysm. (b) Fully expanded SMP foam within the posttreatment aneurysm. (c) A sample of SMP foam in a primary, expanded shape. (d) Close-up of the SMP foam. (Reprinted with kind permission from Springer Science+Business Media: Ortega, J., D. Maitland, et al. 2007. Vascular dynamics of a shape memory polymer foam aneurysm treatment technique. *Annals of Biomedical Engineering* **35**(11): 1870–1884.)

of 10.5 psi. Although this value is higher than the 4.7 psi collapsing pressure exerted by vasospastic arteries, it is on the lower end compared to other metal stents. This would make them fall under the "lower pressure" stents category (Baer et al. 2007). However, the benefit of working with polymers is that their mechanical properties can be tuned to meet specific design criteria. Unlike nitinol, whose mechanical properties are restricted by a maximum deformation of 8%, SMPs can be deformed to much higher degrees and can reach up to 1100% elongation (Lendlein and Kelch 2002). Furthermore, SMPs have the potential for two-way and even triple shape-memory effect ideal for stent repositioning, whereas SMAs do not (Lendlein and Kelch 2002; Ware et al. 2012). Figure 21.12 is an example of an SMP stent.

21.4 Device Development

21.4.1 Fabrication

Fabrication of medical devices is generally preceded by prototyping. A prototype is usually a working model that lays the basis for later stages of a device. It can be made via conventional machining equipment or via rapid prototyping (RP) equipment. RP is a computer run mechanical machining process which provides the means of fabricating a "physical" model from raw materials guided by

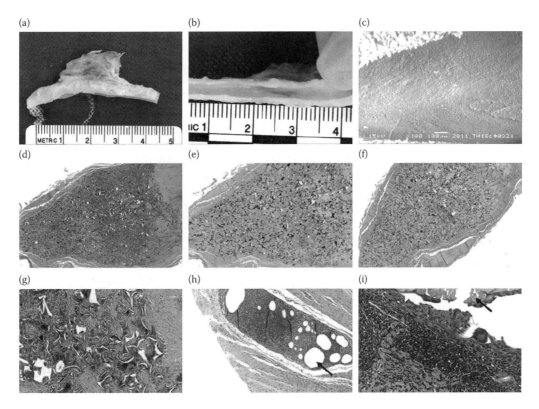

FIGURE 21.11 **(See color insert.)** Preliminary pathology of SMP foams implanted for 90 days within a porcine animal. Original text: SEM and pathology results of implanted foams: (a) Gross picture of dissected aneurysm and parent vessel, (b) gross picture of healing that took place between aneurysm and parent artery intersection (en face), (c) SEM of endothelial cell morphology across the ostium (100× magnification), (d) hematoxylin and eosin (H&E) cross-section of bisected artery and aneurysm sac (4× magnification), (e) trichrome cross-section of bisected aneurysm sac (4× magnification), (f) PTAH cross-section of bisected artery and aneurysm sac (3.5× magnification), (g) H&E cross-section of bisected artery and aneurysm sac (20× magnification), (h) H&E detail of the FDA-approved suture material, polypropylene, indicated by the black arrow (10× magnification), and (i) H&E detail of the FDA-approved suture material, silk, indicated by the black arrow (40× magnification). (Reprinted with kind permission from Springer Science+Business Media: Rodriguez, J. N., Y. J. Yu et al. 2012. Opacification of shape memory polymer foam designed for treatment of intracranial aneurysms. *Annals of Biomedical Engineering* **40**(4): 883–897.)

a three-dimensional computer-aided design (CAD) file in either an additive or subtractive manner (Otto and Wood 2000). Additive RP adds material in layers until the final shape is produced (e.g., Stratasys' 3D printer, Minnesota). Subtractive RP is a form of computer numeric-control (CNC) machining, whereby material is removed using cutting tools until the final shape is produced (e.g., Roland's SRP™, Japan).

Examples of common fabrication techniques used in endovascular medical devices are injection molding, laser machining, microelectron discharge machining (EDM), and micro-CNC machining. Injection molding consists of heating the molding material (plastic or metal) to a liquid state and injecting it into a mold. A good example of the use of this technique in making endovascular devices is in the fabrication of catheters. The catheter shaft is placed in the mold of the hub, and the hub is injection molded around it, eliminating the use of adhesives.

Laser machining is the most common fabrication method for stents and stent-like devices (Martinez and Chaikof 2011). The cutting modality can be either direct write, for example, the laser is focused

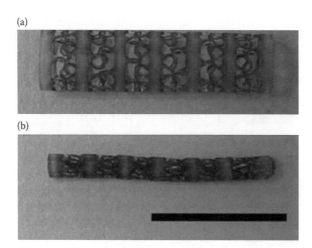

FIGURE 21.12 SMP stent fabricated by laser micromachining. Expanded (a) and collapsed (b) forms. Scale bar = 1 cm. Original text: As previously shown [32], SMP stent fabricated by laser micromachining. The struts are designed to provide flexibility in both the (a) expanded and (b) collapsed forms. Scale bar = 1 cm. The outer diameter of the expanded stent is 4.4 mm with a wall thickness of 200 μm. (From Maitland, E. A. 2009. Design and realization of biomedical devices based on shape memory polymers, reprinted with the permission of Cambridge University Press.)

directly on the material being cut, or masked projection, for example, the laser is passed through a mask before hitting the substrate. Both modalities can be done via sublimation, melting, or oxygen reaction. Due to the low cutting speed of sublimation, melting and oxygen reaction remain the industry standards, even though sublimation results in less impurity adherence.

The most commonly used lasers for metals and polymeric materials are the excimer laser (which stands for excited dimer laser), the neodymium: yttrium–aluminum–garnet (Nd: YAG) laser and the femtosecond laser (Martinez and Chaikof 2011). Carbon dioxide (CO_2) lasers have also been used to machine stents. However, they are limited by their beam width size and the preponderance to induce chemical changes in the substrate (Lootz et al. 2001).

Micro-EDM and micro-CNC machining are mostly used for the machined components of devices when more expensive lasers are either not available or not appropriate for the material. For example, a micro-CNC machine can be used to shape embolic foam devices based on patient-specific aneurysm geometries.

21.4.2 Bench-Top Testing

Bench-top testing is commonly used to evaluate a medical device *in vitro*. These bench-top studies are conducted under conditions similar to the *in vivo* environment as a means to show proof of concept of a device prior to animal studies, or the premarket approval process. Medical device regulatory agencies do not go into great detail about the tests that medical devices should undergo, but do provide guidelines which need to be met prior to using the devices in a clinical setting. As it is written in section 820.30 of the Code of Federal Registry (CFR) volume 8, "... each manufacturer shall establish its own procedure for validating device design. It shall include testing of production units under actual or simulated use conditions" (CFR 21 vol 8 section 820.30 Design controls, Code of Federal Register). For neuroendovascular devices, bench-top testing usually involves the need of phantom vessel models with which device properties can be characterized.

21.4.2.1 Model Fabrication

For neurovascular devices it is necessary to have a realistic model of the vasculature to be used in flow studies. Even though *in vivo* conditions are difficult to be reproduced experimentally, the ability to simulate a stable endovascular environment and variable flow conditions are key in the optimization of a device. Furthermore, such models can also be used to get feedback from clinicians prior to scheduling animal trials.

Anatomic vascular models can be ideal or realistic. Ideal models are models that are generated from an idealized geometry, either for ease of fabrication or for an ideal condition to test devices, which are scaled to their desired size. Anatomical models are generated from 3D data sets, obtained from segmented or isolated 3D models, from clinical imaging of patients. These models are a more realistic representation of size, scale, and tortuosity of the vasculature. However, due to variability in vasculature from person to person, ideal models may be more representative of the general population and therefore, both types of models are valid for testing and developing devices. From the patient's 3D data set, a 3D surface is segmented using imaging software. This surface is edited to remove small branches from the parent vessels and to ease fabrication with in a CAD software package. Within the same CAD software package, a model is saved out in a format such as STL. This 3D model only describes the surface geometry of the object without any representation of color or texture or other CAD attributes. This file is then used in a 3D printer or stereolithography machine to print the positive mold of the vasculature.

The lost wax technique is a common method to make exact copies of the vasculature based off of a positive model generated from a 3D CAD model. With this technique, two types of silicone models can be made—rigid and soft (Figures 21.13 and 21.14). Rigid vascular models can be made by casting liquid silicone around a wax model of a section of the vasculature. After the silicone has been cured, the wax is melted and drained out of the silicone block. Alternatively, brushing a few layers of liquid silicone onto

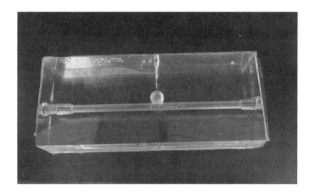

FIGURE 21.13   Rigid silicone aneurysm model. (Courtesy of Wonjun Hwang, 2012.)

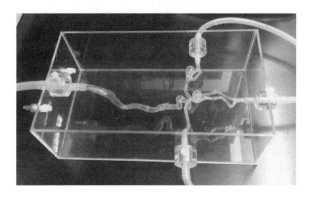

FIGURE 21.14   Soft silicone aneurysm model (Elastrat, Switzerland).

a similar wax model, and then melting and draining out the wax can create soft models. The resulting models are optically clear and can be used for multiple steps in the device development and testing process.

Sugiu et al., used these methods to create rigid and soft vascular aneurysm models to train practitioners on how to use both electrically (GDC) and mechanically (detach coil system (DCS)) detachable platinum coils, as well as for surgical clipping of aneurysms. These models can be used as phantoms with all available medical imaging modalities, as well as for flow experiments using particle image velocimetry (PIV) (Sugiu et al. 2003).

Other materials are also used to make anatomical models of the human vasculature. For example, Asakura et al., made polyvinyl alcohol (PVA) models to test a clot-retrieving wire. One of their observations was that thrombi did not attach to the PVA surfaces as they would *in vivo,* probably due to the greater smoothness of the PVA model compared to the vasculature (Asakura et al. 2007). Another example of alternative models involves glass blowing to create optically clear glass models of the vasculature.

21.5 Conclusions

Medical devices to treat diseases of the neurovasculature have progressed over the past 50 years, mainly due to developments in other fields, including imaging and materials. Treatments for these diseases are somewhat based on the decisions of the treating physician, as they have license to use devices off-label (in a manner that they are not prescribed for). New devices are on the horizon and have been spurred by advancements in all fields involved, probably more so due to new biomaterials that offer better healing responses and a better match of the native mechanics of the surrounding tissues that are treated. Knowledge of all aspects of the diseases treated, including characterization of the physical and chemical environment of these lesions will help further advance this field of medicine.

Acknowledgments

The authors would like to acknowledge David Szafron for his help on research of current devices, Tony Boyle and John Horn for their contributions toward the device development section, Jason Szafron and Jordan Conway for their editorial contributions, and Wonjun Hwang for the picture of his aneurysm phantom for Figure 21.13. This work was supported by the National Institutes of Health/National Institute of Biomedical Imaging and Bioengineering Grant R01EB000462.

References

Alksne, J. F., A. G. Fingerhut et al. 1967. Magnetic probe for the stereotactic thrombosis of intracranial aneurysms. *Journal of Neurology, Neurosurgery & Psychiatry* **30**(2): 159–162.

Asakura, F., H. Yilmaz et al. 2007. Preclinical testing of a new clot-retrieving wire device using polyvinyl alcohol hydrogel vascular models. *Neuroradiology* **49**(3): 243–251.

Augsburger, L., M. Farhat et al. 2009. Effect of flow diverter porosity on intraaneurysmal blood flow. *Urban & Vogel.* **19**: 204–214.

Baer, G. M., W. Small et al. 2007. Fabrication and *in vitro* deployment of a laser-activated shape memory polymer vascular stent. *BioMedical Engineering OnLine* **6**(1): 43.

Becker, K. J. and T. G. Brott 2005. Approval of the MERCI clot retriever a critical view. *Stroke* **36**(2): 400–403.

Berenstein, A., J. K. Song et al. 2006. Personal accounts of the evolution of endovascular neurosurgery. *Neurosurgery* **59**(5): S3.

Biondi, A., V. Janardhan et al. 2007. Neuroform stent-assisted coil embolization of wide-neck intracranial aneurysms: Strategies in stent deployment and midterm follow-up. *Neurosurgery* **61**(3): 460–469 410.1227/1201.NEU.0000290890.0000262201.A0000290899.

Brown, D. L., W. G. Barsan et al. 2005. Survey of emergency physicians about recombinant tissue plasminogen activator for acute ischemic stroke. *Annals of Emergency Medicine* **46**(1): 56–60.

Brown, M. A. and R. C. Semelka. 2010. *MRI: Basic Principles and Applications*. Hoboken, New Jersey, Wiley.

Carod-Artal, F. J. and J. A. Egido 2009. Quality of life after stroke: The importance of a good recovery. *Cerebrovascular Diseases* **27**(1): 204–214.

Cloft, H. J. 2006. HydroCoil for Endovascular Aneurysm Occlusion (HEAL) study: Periprocedural results. *American Journal of Neuroradiology* **27**(2): 289–292.

Courtney, J. M., N. M. K. Lamba et al. 1994. Biomaterials for blood-contacting applications. *Biomaterials* **15**(10): 737–744.

Cowley, A. and B. Woodward 2011. A healthy future: Platinum in medical applications. *Platinum Metals Review* **55**(2): 98–107.

Dandy, W. E. 1938. Intracranial aneurysm of the internal carotid artery: Cured by operation. *Annals of Surgery* **107**(5): 654.

Debrun, G., P. Lacour et al. 1975. Inflatable and released balloon technique experimentation in dog—Application in man. *Neuroradiology* **9**(5): 267–271.

Deshaies, E. and A. Boulos. 2011. *Handbook of Neuroendovascular Surgery*, New York, Thieme Medical Pub.

Doppman, J. L., G. Di Chiro et al. 1968. Obliteration of spinal-cord arteriovenous malformation by percutaneous embolisation. *Lancet* **1**(7540): 477.

Doppman, J. L., G. Di Chiro et al. 1971. Percutaneous embolization of spinal cord arteriovenous malformations. *Journal of Neurosurgery* **34**(1): 48–55.

Dotter, C. and M. Judkins 1989. Transluminal treatment of arteriosclerotic obstruction. Description of a new technic and a preliminary report of its application. 1964. *Radiology* **172**(3): 904–920.

Fiorella, D., F. C. Albuquerque et al. 2006. The role of neuroendovascular therapy for the treatment of brain arteriovenous malformations. *Neurosurgery* **59**(5): S3.

Gall, K., P. Kreiner et al. 2004. Shape-memory polymers for microelectromechanical systems. *Journal of Microelectromechanical Systems* **13**(3): 472–483.

Gallagher, J. P. 1963. Obliteration of intracranial aneurysms by pilojection. *The Journal of the American Medical Association* **183**(4): 231–236.

Gobin, Y. P., S. Starkman et al. 2004. MERCI 1 a phase 1 study of mechanical embolus removal in cerebral ischemia. *Stroke* **35**(12): 2848–2854.

Gobran Riad, H., K. Wakhloo Ajay et al. 2007. Blood flow diverters for the treatment of intracranial aneurysms. Pennslyvania/United States, Florida/United States, Virginia/United States, A61F2/06; A61F2/06. **11/225,285**.

Guglielmi, G., F. Viñuela et al. 1992. Endovascular treatment of posterior circulation aneurysms by electrothrombosis using electrically detachable coils. *Journal of Neurosurgery* **77**(4): 515–524.

Haacke, E. M. B., F. Robert, M. Thompson, and R. Venkatesan. 1999. *Magnetic Resonance Imaging: Physical Principles and Sequence Design*. New York, John Wiley & Sons.

Hampikian, J. M., B. C. Heaton et al. 2006. Mechanical and radiographic properties of a shape memory polymer composite for intracranial aneurysm coils. *Proceedings of the First TMS Symposium on Biological Materials Science* **26**(8): 1373–1379.

Henkes, H., A. Bose et al. 2002. Endovascular coil occlusion of intracranial aneurysms assisted by a novel self-expandable nitinol microstent (Neuroform). *Interventional Neuroradiology* **8**(2): 107–120.

Hilal, S., A. Khandji et al. 1988. Synthetic fiber-coated platinum coils successfully used for the endovascular treatment of arteriovenous malformations, aneurysms and direct arteriovenous fistulas of the CNS. *American Journal of Neuroradiology* **9**: 1030.

Humphrey, J. D. 2002. *Cardiovascular Solid Mechanics: Cells, Tissues, and Organs*, New York, Springer.

Humphrey, J. D. and S. L. Delange. 2004. *An Introduction to Biomechanics: Solids and Fluids, Analysis and Design*, New York, Springer-Verlag.

Hwang, W., B. L. Volk et al. 2011. Estimation of aneurysm wall stresses created by treatment with a shape memory polymer foam device. *Biomechanics and Modeling in Mechanobiology* **11**(5): 1–15.

Ishimori, S., M. Hattori et al. 1967. Treatment of carotid-cavernous fistula by gelfoam embolization. *Journal of Neurosurgery* **27**(4): 315.

Kallmes, D. F. and N. H. Fujiwara. 2002. New expandable hydrogel-platinum coil hybrid device for aneurysm embolization. *American Journal of Neuroradiology* **23**(9): 1580–1588.

Kerber, C. 1976. Balloon catheter with a calibrated leak. *Radiology* **120**(3): 547–550.

Kerber, C. 2006. History of endovascular neurosurgery: A personal view. *Neurosurgery* **59**(5): S3.

Kumar, V., A. Abbas et al. 2007. *Robbins Basic Pathology*. Philadelphia, PA, Saunders Elsevier.

Lehecka, M. 2009. *Distal Anterior Cerebral Artery Aneurysms*. PhD dissertation, University of Helsinki.

Lendlein, A. and S. Kelch 2002. Shape memory polymers. *Angewandte Chemie International Edition* **41**(12): 2034–2057.

Lootz, D., D. Behrend et al. 2001. Laser cutting: Influence on morphological and physicochemical properties of polyhydroxybutyrate. *Biomaterials* **22**(18): 2447–2452.

Luessenhop, A. J. and W. T. Spence. 1960. Artificial embolization of cerebral arteries. Report of use in a case of arteriovenous malformation. *The Journal of the American Medical Association* **172**(11): 1153–1155.

Lusis, A. 2000. Atherosclerosis. *Nature* **407**(6801): 233–241.

Maitland, D. J., W. Small IV, P. Singhal, W. Hwang, J. N. Rodriguez, F. Clubb, and T. S. Wilson. 2009. In *Proceedings of the Material Research Society Symposium*, San Fransisco, 2009. Edited by A. Lendlein and P. Shastri, 1190: NN06-01.

Maitland, E. A. 2009. Design and realization of biomedical devices based on shape memory polymers.

Marks, M. P., C. Tsai et al. 1996. In vitro evaluation of coils for endovascular therapy. *American Journal of Neuroradiology* **17**(1): 29–34.

Martinez, A. W. and E. L. Chaikof. 2011. Microfabrication and nanotechnology in stent design. *Wiley Interdisciplinary Reviews: Nanomedicine and Nanobiotechnology* **3**(3): 256–268.

Mohn, D., M. Zehnder, T. Imfeld, and W. J. Stark 2012. Radio-opaque nanosized bioactive glass for potential root canal application: Evaluation ofradiopacity, bioactivity and alkaline capacity. *International Endontic Journal* **43**: 210–217.

Moniz, E. 1933. Cerebral angiography. *The Lancet* **222**(5751): 1144–1147.

Morgagni, G. and S. A. D. Tissot. 1779. *De Sedibus et Causis Morborum per Anatomen Indagatis*.

Mori. 1999. Balloon angioplasty for embolic total occlusion of the middle cerebral artery and ipsilateral carotid stenting in an acute stroke stage. *American Journal of Neuroradiology* **20**: 1462–1464.

Moseley, M. E., Y. Cohen et al. 1990. Early detection of regional cerebral ischemia in cats: Comparison of diffusion- and T2-weighted MRI and spectroscopy. *Magnetic Resonance in Medicine* **14**(2): 330–346.

Mullan, S., S. Mojtahedi et al. 1996. Embryological basis of some aspects of cerebral vascular fistulas and malformations. *Journal of Neurosurgery* **85**(1): 1–8.

Mullan, S., C. Reyes et al. 1969. Stereotactic copper electric thrombosis of intracranial aneurysms. *Progress in Neurological Surgery* **3**: 193–211.

Murayama, Y., Y. L. Nien et al. 2003a. Guglielmi detachable coil embolization of cerebral aneurysms: 11 years' experience. *Journal of Neurosurgery* **98**(5): 959–966.

Murayama, Y., S. Tateshima et al. 2003b. Matrix and bioabsorbable polymeric coils accelerate healing of intracranial aneurysms: Long-term experimental study. *Stroke* **34**(8): 2031–2037.

Murayama, Y., F. Viñuela et al. 1999. Embolization of incidental cerebral aneurysms by using the Guglielmi detachable coil system. *Journal of Neurosurgery* **90**(2): 207–214.

Neff, J. M. 1911. A method for gradual automatic occlusion of the larger blood-vessels at one operation. *Journal of the American Medical Association* **57**(9): 700.

Nesbit, G. M., G. Luh et al. 2004. New and future endovascular treatment strategies for acute ischemic stroke. *Journal of Vascular and Interventional Radiology* **15**(1, Part 2): S103–S110.

Nogueira, R. G., L. H. Schwamm et al. 2009. Endovascular approaches to acute stroke, part 1: Drugs, devices, and data. *American Journal of Neuroradiology* **30**(4): 649–661.

Ortega, J., D. Maitland et al. 2007. Vascular dynamics of a shape memory polymer foam aneurysm treatment technique. *Annals of Biomedical Engineering* **35**(11): 1870–1884.

Ortega, J. M., D. J. Maitland, T. S. Wilson, W. Tsai, O. Savas, and D. Saloner. 2007. A shape memory polymer dialysis needle adapter for the reduction of hemodynamic stress within arteriovenous grafts. *IEEE Transactions on Biomedical Engineering* **54**(9): 1722–1724.

Otto, K. N. and K. L. Wood 2000. *Product Design*, Prentice-Hall, Englewood Cliffs, NJ.

Perez, M. A., E. Miloslavski et al. 2011. Intracranial thrombectomy using the solitaire stent: A historical vignette. *Journal of Neurointerventional Surgery* 1–4.

Piotin, M., A. Iijima et al. 2003. Increasing the packing of small aneurysms with complex-shaped coils: An *in vitro* study. *American Journal of Neuroradiology* **24**(7): 1446–1448.

Porstmann, W. 1973. A new corset balloon catheter for dotter's transluminal recanilization with special reference to obliterations of the pelvic arteries. *Radiologia Diagnostica* **14**(2): 239.

Prestigiacomo, C. J. 2006. Historical perspectives: The Microsurgical and endovascular treatment of aneurysms. *Neurosurgery* **59**(5): S3.

Qureshi, A. I. 2004. Endovascular treatment of cerebrovascular diseases and intracranial neoplasms. *The Lancet* **363**(9411): 804–813.

Qureshi, A. I., M. F. K. Suri et al. 2003. Stroke-free survival and its determinants in patients with symptomatic vertebrobasilar stenosis: A multicenter study. *Neurosurgery* **52**(5): 1033.

Richling, B. 2006. History of endovascular surgery: Personal accounts of the evolution. *Neurosurgery* **59**(5): S3.

Rodriguez, J. N., Y. J. Yu et al. 2012. Opacification of shape memory polymer foam designed for treatment of intracranial aneurysms. *Annals of Biomedical Engineering* **40**(4): 883–897.

Rosomoff, H. 1966. Stereomagnetic occlusion of intracranial aneurysm: Principle and application. *Trans Am Neurol Assoc* **91**: 330–331.

Sadasivan, C., L. Cesar et al. 2009. Treatment of rabbit elastase-induced aneurysm models by flow diverters: Development of quantifiable indexes of device performance using digital subtraction angiography. *IEEE Transactions on Medical Imaging* **28**(7): 1117–1125.

Sano, K., M. Jimbo et al. 1975. Artificial embolization of inoperable angioma with polymerizing substance. Pia, H. W., Gleave, J. R. W., Grote, E., Zierski, J., eds, *Cerebral Angiomas*. Springer, Berlin, pp. 222–229.

Singhal, P., J. N. Rodriguez et al. 2012. Ultra low density and highly crosslinked biocompatible shape memory polyurethane foams. *Journal of Polymer Science Part B: Polymer Physics*. **50**(10): 724–737.

Singhal, P., T. S. Wilson et al. 2010. *Controlling the Physical Properties of Random Network Based Shape Memory Polymer Foams*, San Fransisco, CA, Cambridge University Press.

Small, I. V. W., P. Singhal et al. 2010. Biomedical applications of thermally activated shape memory polymers. *Journal of Materials Chemistry* **20**(17): 3356.

Song. 2002. Intracranial balloon angioplasty of acute terminal internal carotid artery occlusions. *American Journal of Neuroradiology* **23**: 1308–1312.

Stead, L. G., R. M. Gilmore et al. 2008. Percutaneous clot removal devices in acute ischemic stroke: A systematic review and meta-analysis. *Archives of Neurology* **65**(8): 1024.

Stehbens, W. 1975. Flow in glass models of arterial bifurcations and berry aneurysms at low Reynolds numbers. *Experimental Physiology* **60**(3): 181–192.

Sugiu, K., J. B. Martin et al. 2003. Artificial cerebral aneurysm model for medical testing, training, and research. *Neurologia Medico-Chirurgica* **43**(2): 69–73.

Szikora, I., P. Seifert et al. 2006. Histopathologic evaluation of aneurysms reated with Guglielmi detachable coils or matrix detachable microcoils. *American Journal of Neuroradiology* **27**(Journal Article): 283–288.

Taki, W., Y. Yonekawa et al. 1990. A new liquid material for embolization of arteriovenous malformations. *American Journal of Neuroradiology* **11**(1): 163–168.

Viñuela, F. G., Duckwiler, M. Mawad, G. Guglielmi. 1997. Detachable coil embolization of acute intracranial aneurysm: Perioperative anatomical and clinical outcome in 403 patients. *Journal of Neurosurgery* **86**(3): 475–482.

Wajnberg, E., J. M. de Souza et al. 2009. Single-center experience with the neuroform stent for endovascular treatment of wide-necked intracranial aneurysms. *Surgical Neurology* **72**(6): 612–619.

Ware, T., K. Hearon et al. 2012. Triple-shape memory polymers based on self-complementary hydrogen bonding. *Macromolecules* **45**(2): 1062–1069.

Werner, S. C., A. H. Blakemore et al. 1941. Aneurysm of the internal carotid artery within the skull. *Journal of the American Medical Association* **116**(7): 578.

White, P. M., S. C. Lewis et al. 2008. HydroCoil Endovascular Aneurysm Occlusion and Packing Study (HELPS Trial): Procedural safety and operator-assessed efficacy results. *American Journal of Neuroradiology* **29**(2): 217–223.

Wong. 2011. Evolution of intracranial aneurysm treatment: From hunterian litigation to the flow diverter. **15**(1): 16–17, 18, 19, 20.

Yakes, W. F., P. Rossi et al. 1996. Arteriovenous malformation management. *Cardiovascular and Interventional Radiology* **19**(2): 65–71.

Yodh, S. B., N. T. Pierce et al. 1968. A new magnet system for 'intravascular navigation'. *Medical and Biological Engineering and Computing* **6**(2): 143–147.

Zaidat, S. O. 2006 New "Wingspan" Stent used in Arteries. http://www.froedtert.com/HealthResources/ReadingRoom/EveryDay/May-July2006Issue/NewStentUsedinBrainArteries.htm, 9 April 2011.

Zanetti, P. H. and F. E. Sherman. 1972. Experimental evaluation of a tissue adhesive as an agent for the treatment of aneurysms and arteriovenous anomalies. *Journal of Neurosurgery* **36**(1): 72–79.

Zubillaga, A. F., G. Guglielmi, F. Vi, and G. R. Duckwiler. 1994. Endovascular occlusion of intracranial aneurysms with electrically detachable coils: Correlation of aneurysm neck size and treatment results. *American Journal of Neuroradiology* 15(5): 815–820.

22

Electro-Physiological Solutions for Cardiac Disease

Lawrence J.
Mulligan
Medtronic, Inc.

Jean-Pierre Lalonde
Medtronic, Inc.

Paul A. Iaizzo
University of Minnesota

Michael D. Eggen
Medtronic, Inc.

22.1 Cardiac Pacing and Defibrillation

22.1.1 Indications for Cardiac Pacing and Defibrillation

The heart functions as a physiological pump to circulate blood through the pulmonary and systemic circulations. The relative rate of blood flow circulated by the heart either through the lungs or through the rest of the body is characterized by the cardiac output (CO, L/min), which is defined as a product of the heart rate (HR, beats per minute or bpm) and the stroke volume (SV, L), or $CO = HR \times SV$. An insufficient CO during physical activity causes tissue hypoxia, a condition where oxygen delivery is less than the metabolic demand, resulting in symptoms such as premature muscle fatigue, dizziness, and shortness of breath. There are many physiological and pathological conditions that may result in an insufficient cardiac output. For example, some cardiac arrhythmias, such as sinus bradycardia (i.e., HRs < 60 bpm) and heart block, will result in an inappropriate cardiac output owing to a slower than required HR [1]. For these bradyarrhythmias, an implantable cardiac pacing system is almost always used as a therapy to restore relative HRs to more physiological levels, thereby also restoring appropriate cardiac outputs. As such, the guidelines for implantation of cardiac pacemakers and defibrillators have been established by a joint committee formed by the American College of Cardiology (ACC), the American Heart Association (AHA), and the Heart Rhythm Society [2]. A summary of the indications for the implantation of cardiac pacemakers and defibrillators are shown in Tables 22.1 and 22.2.

As such, the three most common indications for the implantation of pacing and defibrillation systems are bradyarrhythmias, tachyarrhythmias, and cardiac resynchronization therapy. As discussed, bradyarrhythmias are conditions in which the heart beats at a lower rate insufficient to provide an appropriate CO, and can be treated by an implantable pacing system. In general, tachyarrhythmias, such as ventricular tachycardia and ventricular fibrillation, compromise hemodynamic function by either causing the heart to beat at a higher than normal rate (i.e., ventricular tachycardia results in impaired ventricular filling and therefore reduced SV), or by causing the heart to beat in an unsynchronized, chaotic manner

TABLE 22.1   ACC/AHA/HRS Indications for Cardiac Pacing

- Sinus node dysfunction
- Acquired atrioventricular block
- Chronic bifascicular and trifascicular block
- Atrioventricular block associated with acute myocardial infarction
- Hypersensitive carotid sinus and neurocardiogenic syncope
- Prevention and termination of atrial and ventricular tachyarrhythmias
- Atrial fibrillation
- Congenital heart disease
- Hypertrophic obstructive cardiomyopathy
- Long-QT syndrome
- Idiopathic dilated cardiomyopathy
- Cardiac transplantation
- Cardiac sarcoidosis
- Cardiac resynchronization therapy

Source: Adapted from Epstein, A.E. et al. *Circulation* 2008;117(21):e350–e408; Finn, B., ed. *Exposing Electronics*. 2000, Boca Raton, FL: Taylor & Francis, Inc.

TABLE 22.2   ACC/AHA/HRS Indications for Implantable Cardioverter-Defibrillator Therapy

- Prevention of cardiac arrest and sustained ventricular tachycardia
- Coronary artery disease
- Nonischemic dilated cardiomyopathy
- Hypertrophic cardiomyopathy
- Arrhythmogenic right ventricular dysplasia/cardiomyopathy
- Genetic arrhythmia syndromes
- Syncope with inducible sustained ventricular tachycardia
- Long-QT syndrome
- Noncompaction of the left ventricle.
- Primary electrical disease (idiopathic ventricular fibrillation, short-QT syndrome, Brugada syndrome, and catecholaminergic polymorphic ventricular tachycardia)
- Idiopathic ventricular tachycardias
- Advanced heart failure and cardiac transplantation

Source: Adapted from Epstein, A.E. et al. *Circulation* 2008;117(21):e350–e408; Finn, B., ed. *Exposing Electronics*. 2000, Boca Raton, FL: Taylor & Francis, Inc.

(ventricular fibrillation), which results in no net ejection of blood from the ventricles. Unlike bradyarrhythmias, which only require an implantable pulse generator (IPG or pacemaker) to locally stimulate the heart to restore relative HRs, tachyarrhythmias generally require an implantable cardioverter defibrillator (ICD or defibrillator) to deliver a high-energy shock so as to defibrillate the entire heart. As such, for the category of heart failure patients needing cardiac resynchronization therapy (CRT) to restore cardiac output impaired by dyssynchronous ventricular activation, a CRT system can be implanted that synchronizes the ventricular activation and can restore appropriate hemodynamic function. Specific details regarding these three types of systems (IPG, ICD, and CRT) are described later.

22.2 Implantable Pacing and Defibrillation Systems

22.2.1 Overview

The first portable battery powered pacemaker was developed by Earl Bakken and used in postsurgical pediatric patients by C. Walton Lillehei in 1958 at the University of Minnesota [3]. Bakken invented the portable pacemaker at the request of Lillehei whose pediatric patients undergoing cardiac surgery to repair congenital defects would develop temporary complete heart block, and thus were dependent on pacemakers driven by AC power (which was undesirable because of potential power failures). Since then pacing technology has evolved significantly, so as to manage many aspects of cardiac rhythm.

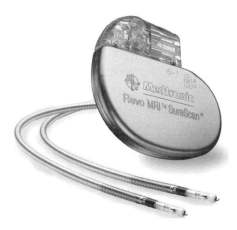

FIGURE 22.1 Dual-chamber pacemaker shown with two active fixation pacing leads.

As mentioned earlier, implantable pacing and defibrillation systems are typically used for the treatment of bradyarrhythmias, tachyarrhythmias, and for CRT. The implantable systems include either a pulse generator (IPG or pacemaker), an implantable cardioverter defibrillator (ICD or defibrillator), or a CRT device and associated pacing and/or defibrillation leads. The IPGs, ICDs, and CRT devices are commonly implanted subcutaneously in the left pectoral region, but can also be implanted in the right pectoral region dependent on physician preference or anatomical restraints (such as limited venous access in the left pectoral region). As such, most leads are implanted endocardially, using either a cephalic cutdown technique or a subclavian puncture approach. Once access is gained through either the cutdown or puncture technique, the leads are typically delivered using stylets, catheters, and/or guidewires. The target implant sites for the pacing leads are commonly the right atrium, right ventricle, and/or coronary sinus (for transvenous left ventricular pacing during CRT).

A typical IPG contains a computer, lithium iodide battery, and an integral pulse generator housed in a hermetically sealed titanium case (Figure 22.1). For connection to the pacing leads, the IPG contains a header block (usually made of polyurethane) and electrically insulated feedthroughs which provide an interface between the pacing leads and the internal circuitry. The leads are inserted into cavities within the header block and are secured by setscrews to maintain electrical connectivity during the lifetime of the device. As such, the connections are standardized across the pacing industry such that leads and devices can be interchanged across manufacturers, where the IS-1 connector standard is used for pacing lead connections. IPGs designed to pace either the atrium only (for sinus node dysfunction, etc.) or the ventricle only (for atrioventricular block, etc.) are designated as single chamber pacemakers and contain a single IS-1 connection port. IPGs with two IS-1 connection ports having the ability to synchronously pace the right atrium and the right ventricle are designated as dual-chamber devices.

A typical ICD (Figure 22.2) includes the same components as an IPG, as ICDs not only have the same pacing capabilities, but also include a set of larger capacitors that are charged so as to deliver a high voltage shock to defibrillate the heart. Due to the larger capacitors and batteries needed for defibrillation, the geometric sizes of the ICDs are larger than IPGs. The connections for high voltage leads are also standardized, where DF-1 connectors and the newer DF-4 connectors are used. A typical ICD will contain the necessary connections for a single defibrillation lead placed in the right ventricle and a pacing lead placed in the right atrium.

22.2.2 CRT Devices

For CRT devices (discussed in more detail later), there is an extra connection for an LV lead such that the right atrium, right ventricle, and left ventricle can be paced to synchronize the chambers of the heart.

FIGURE 22.2   Implantable cardioverter defibrillator (ICD).

These CRT-P devices are designed specifically for patients who are candidates for CRT, but who are not indicated for an implantable defibrillator. Patients who are candidates for CRT and an implantable defibrillator can receive a CRT-D device which has both CRT and defibrillation capabilities.

As such, the longevity of these devices is dependent on several factors including: the programmed pacing output, number of leads paced, percent pacing, pacing rate, lead impedance, battery size, and internal current drain [4]. Device longevity for modern pacemakers is usually 8–10 years.

22.2.3 Pacing and Defibrillation Leads

Pacing leads provide an electrical connection between the pulse generator and heart through an insulated wire and are designed to conduct low voltage pacing pulses (<10 V) in order to locally stimulate the myocardium to cause contraction. Additionally, leads are required to provide low, stable, acute, and chronic pacing thresholds, and to provide sensing that detects cardiac activity. The stimulation threshold or pacing threshold is the lowest electrical output that will consistently depolarize (capture) the myocardium, when that stimulus occurs outside the heart's refractory period [5]. A low pacing threshold is desired in order to increase battery longevity, and is indicative of a good electrode–tissue contact. As such, for an active fixation lead, a pacing threshold at implant <1.5 V can be considered adequate [6]. For sensing, the EGM amplitude should be >1.5–2.0 mV in the atrium and at least 5–6 mV in the ventricle at the time of lead implantation [7]. Moreover, the fixation mechanism for pacing leads can either be active or passive. For active fixation leads (Figure 22.1), a helix is typically screwed into the myocardium which secures the lead in place, and also serves as the cathode for electrical stimulation. For bipolar pacing and sensing, the anode (ring electrode) is usually spaced ~1–2 cm proximal from the tip electrode. It should be noted that when connected to a pacemaker, the pacemaker shield can commonly serve as the anode if programmed in unipolar mode. Figure 22.3 demonstrates the implantation of an active fixation lead within the right ventricle of an *ex vivo* swine heart model [8]. The helix in most active fixation leads can be retracted in order to prevent any unnecessary tissue damage from navigating the lead through the venous anatomy and across the cardiac structures (tricuspid valve, etc.). As such, leads designed in this manner are said to have extendable–retractable helices. Active fixation leads without extendable–retractable helices are termed fixed screw leads, and are typically delivered via a catheter to the right atrium or right ventricle.

Passive fixation leads are designed with polymer tines which engage in narrow cavities within the trabeculation of the heart in order to secure the electrode in place. As opposed to active fixation leads,

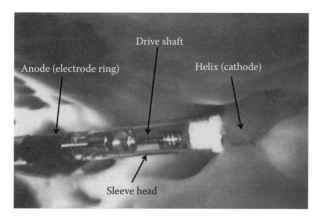

FIGURE 22.3 Distal end of an active fixation pacing lead positioned in the right ventricle of an isolated swine heart (Medtronic model 5076). The helix engagement with the myocardium is clearly seen.

these tined leads do not require active deployment by the clinician, and are thus termed "passive" fixation leads. An example of a passive fixation lead is shown in Figure 22.4, and implantation of such a lead in the atrium of an isolated human heart model is shown in Figure 22.5 [9]. For passive fixation leads, the cathode is usually positioned in the center of the tines to enable stable contact with the myocardium. Note, since passive fixation leads are dependent upon engaging with trabeculation, they cannot be placed just anywhere in the heart, but only where there is a presence of relatively normal trabeculation.

Transvenous endocardial defibrillation leads (ICD leads) are designed to deliver high-voltage defibrillation shocks to terminate lethal arrhythmias such as ventricular fibrillation and sustained ventricular tachycardia, and also function to provide low-voltage bradycardia therapy (pacing and sensing intracardiac signals) as previously discussed [10]. Similar to pacing leads, defibrillation leads can be active or passive fixation, and also contain pacing and sensing electrodes. Uniquely, defibrillation leads contain shock coils, which function as electrodes to deliver high-voltage shocks (of upto ~35 J) in order to defibrillate the heart. Shown in Figure 22.6 is a schematic of an active fixation defibrillation lead, and its associated placement within the right ventricle is shown in Figure 22.7. As such, ICD leads are typically available in both single- and dual-coil designs, where single-coil ICD leads only contain a single shock coil in the RV position (RV coil), and dual-coil designs contain shock coils in both the RV and the SVC positions, and allow for more electrode configuration options for defibrillation. For intracardiac EGM

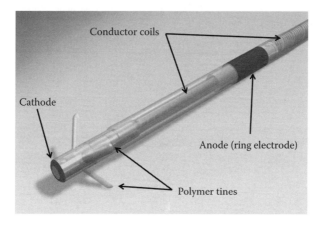

FIGURE 22.4 Passive fixation pacing lead (Medtronic model 4074).

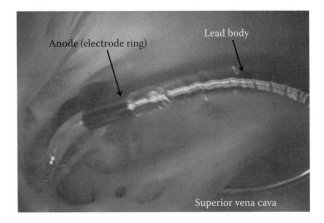

FIGURE 22.5 Passive fixation pacing lead positioned in the right atrium of an isolated human heart. The tine engagement between the pectinate muscles within the atrial appendage is clearly seen.

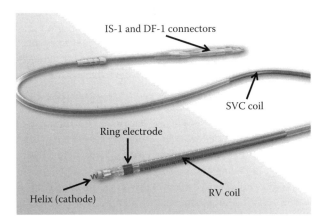

FIGURE 22.6 Dual coil ICD lead (Medtronic model 6947).

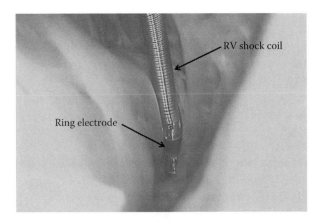

FIGURE 22.7 Active fixation ICD lead positioned in the right ventricular apex of an isolated swine heart (Medtronic model 6947).

sensing, the typical configurations are integrated bipolar and bipolar. For integrated bipolar sensing, the EGM is sensed between the tip electrode and RV distal shock coil, and for true bipolar sensing, the EGM is sensed between the tip electrode and the ring electrode similar to a conventional pacing lead.

After implantation, defibrillation conversion testing is conducted in order to determine if the system (ICD lead and device) is capable of sensing and terminating ventricular fibrillation (VF). The most common method of defibrillation conversion testing requires at least two VF inductions with subsequent conversion success using shock energies of at least 10 J below the maximum shock energy output of the ICD [11]. In a similar fashion to conventional bradycardia leads, the pacing threshold is also measured at the tip electrode with similar acceptability criteria.

22.3 Cardiac Resynchronization Therapy

In the mid-1990s, a new solution for treating patients with heart failure (HF) and a wide QRS was investigated. Patients with NYHA class III/IV and left bundle branch block (QRS >120 ms) who were in refractory HF underwent an acute electro-physiological pacing procedure. Right ventricular and atrial leads were connected with a lead placed in the left heart coronary venous system. Patients were then paced to "synchronize" the left and right ventricles. The outcome of this investigation demonstrated that CO increased and pulmonary capillary wedge pressures (PCWP) decreased with pacing on and that this change was reproducible [12]. This improvement in cardiac function provided the foundation for the development of cardiac resynchronization therapy (CRT). The results from the MIRACLE trial provided the necessary evidence for the FDA to approve this new revolutionary therapy [13]. The impact of this therapy led the FDA to classify the therapy as class 1. This classification allows CRT to be used without exception in patients who meet the appropriate guidelines. The functional benefit of this therapy is shown in Table 22.3.

Over the life of CRT, as with all medical therapies, a subset of patients appears to show no benefit or response from the treatment. However, the definitions of benefit range from the patient feeling better, a very subjective measure, to their being alive at 5 years, a very impressive outcome as 50% of patients with HF are dead at 5 years. However, with an implanted cardiac device, compared with medication, the titration of the therapy is all or none. The issue of response has led many investigations focus on methods to improve the response rate. The response rate in the MIRACLE trial of 67% was a composite rate, as was seen in the major HF pharmaceutical trials.

Deriving benefit from CRT is dependent on many variables, both electrical and mechanical. There is general agreement regarding what is needed for a response rate greater than 67% but significant disagreement exists in how to describe these variables. It has been shown that patients presenting with QRS < 150 ms, with ischemic HF and/or scar, little mechanical dyssynchrony, poor options for lead placement and heterogenous electrical – mechanical activation would have a lower likelihood of response compared with patients presenting with a QRS > 150 ms, absence of ischemic HF or scar, significant mechanical dyssynchrony, several options for lead placement, and homogenous electrical–mechanical activation (see Table 22.4).

The issue of CRT response became visible as early as 2002 [14]. In an attempt to improve the response rate, echocardiography was employed both as a preimplant tool and at the follow-up. Several hundred

TABLE 22.3 Functional Benefits of CRT

⇑ 6-min walking distance
⇑ Health-related quality-of-life score
⇑ Peak oxygen consumption
⇓ Hospitalizations for decompensated heart failure
⇓ New York Heart Classification

Note: ⇑ increased; ⇓ decreased.

TABLE 22.4 Impact of Implant Variables on CRT Response

Response Rate (%)	QRS (ms)	Etiology	Presence of Mechanical Dyssynchrony	LV Lead Position	Electro-Mechanical Activation Pattern
<67	<150	Ischemic and/or scar	Little	Less than desired	Heterogeneous
>67	>150	Dilated with/without scar	Large	Good to great	Homogeneous

research efforts led to unique single-center solutions. However, the development of echocardiographic parameters for use worldwide was not accomplished. In an elegant set of papers by Sengupta et al. [15,16], several observations are presented that provide insights into the possible reasons why echocardiography failed or was unable to be part of the solution. It has become more obvious that solutions may need to be patient-focused versus population-focused.

The response rate to CRT has remained stable over the 11 years that the therapy has existed. Over this time, there has been a significant improvement in the LV lead delivery systems as well as new LV leads, leading to faster implant times, lower dislodgment rates, and lower rates of phrenic nerve stimulation but the impact on the response rate has been small yet remaining similar to the major HF pharmaceutical trials.

Multielectrode LV leads and navigation tools are a current focus of the medical device industry [17, 18]. The ability of these tools to impact the response issue is currently under investigation.

Importantly, the success of these new options has to be considered in a different manner than the original intent of the MIRACLE trial. Current efforts are trying to build on the success rate of 67% compared to proving a therapy works. As a result, the impact of new tools is intended to add the 67% success rate and as such requires greater focus on the individual patient versus on a population.

22.4 Cryoablation

Since 1970, cryoablation has been used as a means to control arrhythmias during an open-heart surgery: primarily for the treatment of supraventricular tachyarrhythmias [19]. Subsequently from 1995, successful cryoablative techniques have evolved toward endovascular applications, due to rapid developments and applications in the catheterization methodologies. More specifically, the use of cryocatheters offers a much less invasive approach, as well as maintaining the advantages of cryoadhesion and cryomapping capabilities, which will be discussed later. It is important to note, however, that these methods will never totally replace the advantages of cryosurgery: as in many cases of open-heart surgery, it is still required (valve replacement, pacemaker, etc.) and/or remains the best therapeutic option.

22.4.1 Device Description

22.4.1.1 Surgical Cryoablation Probe

A typical surgical probe consists of an insulated tube followed by a cooling segment. Both are distinctly malleable and can be shaped manually to accommodate anatomical idiosyncrasies. Such probes are generally used epicardially to create linear lesions, as well as to electrically isolate the pulmonary veins (PVs). For this latter purpose, an adaptable clamp is provided to pinch the vein and efficiently produce contiguous transmural lesions around the PV [20]. The argon gas used within such systems is capable of providing a cooling down the temperatures as low as −144°C: that is, through the Joule–Thomson evolution as the refrigerant passes across a capillary tube from a high to a low pressure. Because this process has a relatively low-cooling power capability, temperature may vary in the case of beating-heart surgery, depending on the contact surface of the probe with the tissue, tissue thickness, and local blood flow circulation underneath, hence offsetting the technique's usefulness in endovascular applications (Figure 22.8).

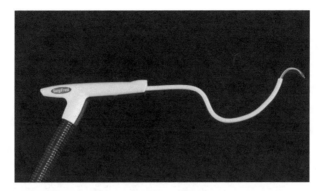

FIGURE 22.8 Surgical probe (Cardioblate® CryoFlex™ shown in example).

22.4.1.2 Focal Cryoablation Catheters

These steerable devices have been designed for perinodal ablation via cardiac assessment through the vasculature where precision and safety are paramount. The cryomapping capabilities of these catheters and the possibilities to chill tissues to −30°C to temporarily stop electrical conduction allow for ablation-site confirmations prior to creating permanent blocks [21,22]. Combined with the cryoadhesion effects/benefit, where the cooling segment adheres to the tissue, this catheter approach becomes the perfect tool to use for critical ablation sites [23]. While the coldest achievable temperature is in the range of −80°C, compared to −144°C of the surgical probe, this has been shown to be sufficient to create transmural lesions. More specifically, as the heat load on the pericardial surface is low (i.e., where no blood is circulating), the ice front propagates transmurally toward the epicardial surface [24,25] (Figure 22.9).

Note that with a longer cooling segment it becomes easier to produce linear lesions; through successive applications. As an example, the 9Fr device shown in Figure 22.9 is generally not only used to create the so-called "flutter lines," the "isthmus lines," and the "roof lines," but can also be used to ablate the pulmonary vein ostia. Shown in Figure 22.10 is an *in vitro* comparison of the relative impact of the tip size and orientation on the resultant cooling area: a thermal paper model immersed in a 37°C water bath was employed, where the blue surface indicates temperatures below 10°C.

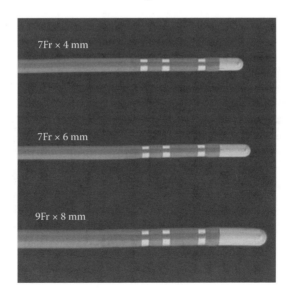

FIGURE 22.9 Typical focal ablation devices.

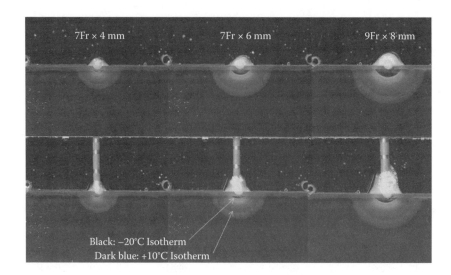

FIGURES 22.10 (**See color insert.**) Freezing capability *in vitro* for different tip size and orientation (lateral vs. vertical orientation).

Shown in Figure 22.11 are video images of the Freezor® MAX taken before cryoablation (left) and during cryoablation (right), within a physiological buffer perfused swine heart (using Visible Heart methodologies). Interestingly, in these studies we could observe how the ice forms around the tip of the catheter and that it remained fully fixed (frozen onto the myocardium) even when one tried rigorously to dislodge the catheter.

22.4.1.3 Cryoballoon Device

The most recent and perhaps the most sophisticated design of the cryo tools is the cryoballoon device that has been developed to treat atrial fibrillation by ablating as well as possible a total pulmonary vein ostium. The construction of this balloon catheter is an over-the-wire design where a central lumen can be used to insert tools that facilitate the passage through the septal wall and to guide the catheter toward the pulmonary vein. Once aligned with the vein, a concentric dual balloon arrangement located at the distal end of the catheter is inflated and pushed toward the ostium to occlude the vein to be treated. To assess the quality of occlusion, a contrast agent can be injected through the central lumen: thus, the accumulation of that contrast distal to the balloon can be visualized with fluoroscopy and its persistence is generally a strong indication of a good occlusion (see Figure 22.12). Typically, at this point, the cryoablation is initiated and sustained for 4 min. In general, a second freeze of the same vein is recommended, but wiring the vein differently to ensure transmural contiguous lesions all around the total ostium.

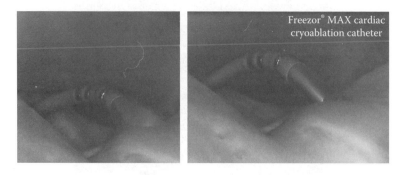

FIGURE 22.11 Picture of the Freezor MAX, *ex vivo* swine heart, before and during cryoablation. (Pictures provided by the Visible Heart Lab, University of Minnesota. With permission.)

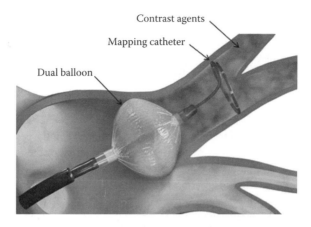

FIGURE 22.12 Arctic Front® balloon device schematic occluding the vein.

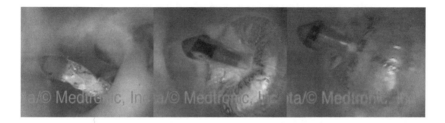

FIGURE 22.13 Images of an Arctic Front device before inflation, during inflation, and during ablation in a swine heart *ex vivo* model. (Pictures provided by the Visible Heart Lab, University of Minnesota. With permission.)

Shown in Figure 22.13 are several examples of the cryoablation applications within a reanimated swine heart: that is, one can appreciate the cryoballoon being placed, inflated, and then turning on the cryo. Starting from the picture on the left, the balloon was positioned within a pulmonary vein ostium and was thus ready to be inflated. In this image, opposite to the vein one can observe the atrial septum from which the catheter passes through and a tight curve is commonly required to position the catheter within a given vein. The middle image shows the balloon inflated, fully obstructing the vein. Finally on the right, one can observe the balloon catheter device in cryoablation mode, where ice formation (of the clear perfusate) can be observed in the periphery of the balloon in contact with the vein ostium.

22.4.1.4 Fundamental Cryo Mechanisms Leading to Cellular Destruction

It is important to note that if one is to use the cryo mechanism for cellular destruction most effectively, it becomes essential to fully understand the process by which cryo affects various types of tissues. Although the details are still being debated, the following discussion summarizes some of the basic understandings of the underlying cryomechanism leading to cellular destruction.

22.4.1.5 Extracellular Ice Crystallization

The phenomenon of "extracellular ice crystallization" occurs at a slow cooling rate. As a given tissue freezes at a low cooling rate, ice crystals typically grow around the cells and the solution surrounding the cells becomes hypertonic (salt concentration >0.9%). To equilibrate within this new environment, the cells must release water, and thus shrink. As the concentration keeps increasing, equilibrium is maintained by the solute passing into the cells. It is then considered that cell destruction occurs upon thawing, where the extra concentrations of solute in the cells force a large intake of water that damages the cell structures [26].

22.4.1.6 Intracellular Ice Crystallization

In contrast when high cooling rates occur, there is ice formation both on the inside and outside of the cells before osmotic equilibrium can be achieved, thus causing the rupture of cell membrane. This phenomenon is known to be more lethal than extracellular ice formation [26].

22.4.1.6.1 Implications

The tissues within the close proximity of a cryodevice are generally exposed to both colder temperatures and faster cooling rates. Therefore, intracellular ice can be considered as the primary cause of death for the cells that are near the cryointerface (catheter tip or balloon). This in turn may explain, why there is importance placed on designing cryo devices that can achieve lower temperature and faster cooling rate: that is, to maximize intracellular ice formations. In contrast, cells distal from the cryointerface are subject to both "warmer" temperatures and slower cooling rates. Therefore, any cell death there occurs through the water absorption during the thawing phase. It should be noted that this latter situation may not be lethal, especially when the cells are further away from the ice formation: that is, it has been noted that only a certain percentage of the cells will go through apoptosis (programmed cell death). Therefore when indicated, a second freeze–thaw cycle may maximize cell destruction more effectively than a single cycle, for the cells away from the cryodevice interface.

22.4.1.7 Cryo versus RF Lesions: Histological Effect on the Connective Tissue Matrix

Figure 22.14 shows an example of typical histological sections of myocardial tissues, following either the application of cryo or RF ablative therapies. More specifically in this study, using a canine model in which the animals were sacrificed at 7 days, the ablation dose presented for cryo is 4 min at −75°C, while the dose for RF is 1 min at 70°C and 50 W. It should be noted that these are typical applications of thermal therapies for such cardiac ablative clinical procedures. As one can observe, the cryoablation lesions are well circumscribed and form discrete lesions with sharp borders, dense areas of fibrotic tissue, and contraction band necrosis. Yet, it is important to note that the surrounding myocardium following cryotherapy is generally intact with minimal thrombus formation. In contrast, following the RF therapy, the lesions are characterized by intralesional hemorrhage and ragged edges that are less clearly

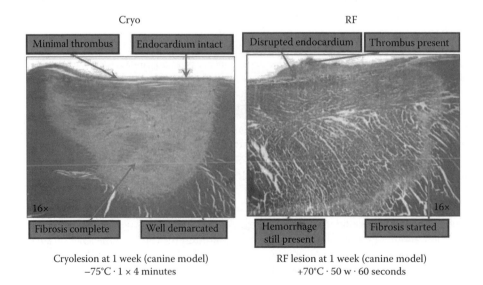

Cryolesion at 1 week (canine model)
−75°C · 1 × 4 minutes

RF lesion at 1 week (canine model)
+70°C · 50 w · 60 seconds

FIGURE 22.14 (**See color insert.**) Cryo versus RF lesions. (From Haqqani, H.M., Mond, H.G. *Pacing Clin Electrophysiol* 2009 32(10):1336–1353. With permission.)

demarcated from the underlying normal myocardium. Furthermore, note that the replacement fibrosis was confined to the outer margin of the RF lesions, suggesting a slower post ablation healing response to RF energy [27].

References

1. Hesselson, A.B. Basic ECG refresher, *Simplified Interpretation of Pacemaker ECGs*, 2008. Oxford, UK: Blackwell Publishing Ltd.
2. Epstein, A.E. et al. ACC/AHA/HRS 2008 Guidelines for Device-Based Therapy of Cardiac Rhythm Abnormalities: A Report of the American College of Cardiology/American Heart Association Task Force on Practice Guidelines (Writing Committee to Revise the ACC/AHA/NASPE 2002 Guideline Update for Implantation of Cardiac Pacemakers and Antiarrhythmia Devices). *Circulation* 2008;117(21):e350–e408.
3. Finn, B., ed. *Exposing Electronics*. 2000, Boca Raton, FL: Taylor & Francis, Inc.
4. Scherer, M, Ezziddin, K, Klesius, A, Skupin, M, Helms, S, Moritz, A, Olbrich, H.G. Extension of generator longevity by use of high impedance ventricular leads, *Pacing Clin Electrophysiol* 200124(2):206–11.
5. Furman, S., Escher, D.J.W., *Principles and Techniques of Cardiac Pacing*, New York, NY: Harper & Row, 1970, p. 53.
6. Saxonhouse, S.J, Conti, J.B, Curtis, A.B. Current of injury predicts adequate active lead fixation in permanent pacemaker/defibrillation leads, *J Am Coll Cardiol* 2005 45(3):412–7.
7. Ellenbogen, Kay, Lau, Wilkoff. *Clinical Cardiac Pacing, Defibrillation, and Resynchronization Therapy*, 3rd edition., Philadelphia, PA: W.B. Saunders, 2007, p. 79.
8. Chinchoy, E, Soule, C.L, Houlton, A.J, Gallagher, W.J, Hjelle, M.A, Laske, T.G, Morissette, J, Iaizzo, P.A. Isolated four-chamber working swine heart model, *Ann Thorac Surg* 2000 70(5):1607–14.
9. Hill, A.J, Laske, T.G, Coles, J.A Jr, Sigg, D.C, Skadsberg ND, Vincent SA, Soule CL, Gallagher WJ, Iaizzo PA. *In vitro* studies of human hearts, *Ann Thorac Surg* 2005 79(1):168–77.
10. Haqqani, H.M, Mond, H.G. The implantable cardioverter-defibrillator lead: Principles, progress, and promises, *Pacing Clin Electrophysiol* 2009 32(10):1336–53.
11. Day, J.D, Olshansky, B, Moore, S, Brown, S, Stolen, K.Q, Lerew, D.R; INTRINSIC RV study investigators. High defibrillation energy requirements are encountered rarely with modern dual-chamber implantable cardioverter–defibrillator systems, *Europace* 2008 10(3):347–50.
12. Leclercq, C et al. Acute hemodynamic effects of biventricular DDD pacing in patients with end-stage heart failure, *J Am Coll Cardiol* 1998 32(7):1825–31.
13. Abraham, W.T et al. Cardiac resynchronization in chronic heart failure. MIRACLE study group. Multicenter InSync randomized clinical evaluation, *N Engl J Med* 2002 13;346(24):1845–53.
14. Reuter, S. et al. Comparison of characteristics in responders versus nonresponders with biventricular pacing for drug-resistant congestive heart failure, *Am J Cardiol* 2002 1;89(3):346–50.
15. Sengupta, P.P. et al. Twist mechanics of the left ventricle: Principles and application, *JACC Cardiovasc Imaging* 2008 1(3):366–76.
16. Sengupta, P.P. Following the flow in chambers, *Heart Fail Clin* 2008 4(3):325–32.
17. Ma, Y.L. An integrated platform for image-guided cardiac resynchronization therapy, *Phys Med Biol* 2012 21;57(10):2953–68
18. Shetty, A.K. The acute hemodynamic response to LV pacing with in individual branches of the coronary sinus using a quadripolar lead, *Pacing Clin Electrophysiol* 2012 35(2):196–203.
19. Slama Ra et al: Surgical creation of an auriculoventrical block and implantation of a stimulator in 2 cases of irreducible rhythm disorders, *Arch Mal Coeur Vaiss* 1967 60:406–422.
20. Doll et al. Intraoperative cryoablation with a new argon probe, *Ann Thorac Surg* 2004;77:1460–1462.
21. Wong, T. et al. Clinical usefulness of cryomapping for ablation of tachycardias involving perinodal tissue, *J Inter Card Electro Physiol* 2004 10(2):153–158.

22. Allan, C. Skanes, MD et al. Cryothermal ablation of the slow pathway for the elimination of atrio-ventricular nodal reentrant tachycardia, *Circulation* 2000 102:2856–2860.

23. Mats Jensen-Urstad et al. High success rate with cryomapping and cryoablation of atrioventricular nodal reentrytachycardia, *PACE* 2006 29:487.

24. Richard, B. Schuessler et al. Animal studies of epicardial atrial ablation, *Heart Rhythm* 2009 6(12, S1): S31–S45.

25. Thomas, A. Pilcher, M.D et al. Contrasting effects of convective flow on catheter ablation lesion size: Cryo versus radiofrequency energy, *PACE* 2008 31:300–307.

26. Orpwood, R.D. Biophysical and engineering aspects of cryosurgery, *Phys Med Biol* 1981 26:555.

27. Khairy, P. et al. Lower incidence of thrombus formation with cryoenergy versus radiofrequency catheter ablation, *Circulation* 2003 107:2045–2050.

23

Stereotactic Procedures

Michael D. Weil
Sirius Medicine, LLC

23.1 Introduction

Stereotaxy employs minimally invasive techniques to precisely locate a target within the body. This requires rendering the internal target and organs relative to an external reference in three-dimensional space. A system of coordinates links the defined exterior and the viewed internal anatomy. The imaging can be performed with diagnostic x-rays, magnetic resonance imaging, or ultrasound. The external reference can be a frame attached to the patient, or marked points on the patient's surface anatomy.

Stereotactic systems can be used to diagnose or treat. Diagnostic procedures can be performed on small lesions with greater precision than conventional biopsy. Diagnostic accuracy is thereby improved. Therapy using stereotaxy can be done with surgery or radiation. The methods permit approach to inaccessible regions of the body with minimal disruption of the intervening tissues. As a result, there is less potential for morbidity and less discomfort for the patient.

Stereotactic procedures are employed in a wide range of applications, from basic scientific research to common clinical problems.[1] The research includes neurological monitoring[2,3] and gene therapy. The clinical spectrum of usage ranges from cancer diagnosis and treatment to the evaluation and treatment of movement disorders.

23.2 Methods

It is necessary to measure and define the three-dimensional volume around a target (Figure 23.1). An external frame is attached to the patient or, alternatively, prominent surface anatomy is marked ("frameless"). The target and body are next imaged along with the external reference points. The important structures in the space are outlined. The distances of the anatomy relative to the reference frame can then be measured (Figure 23.2). An orthogonal coordinate system (x-, y-, and z-axis) is delineated according to these measurements. The points of interest, outside and inside the patient, are then known by their position in the three-dimensional stereotactic space and given coordinates relative to the reference. Though the target is unseen by the clinician, it can be readily located and

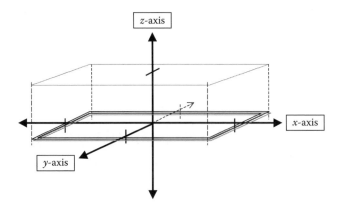

FIGURE 23.1   Stereotactic space is defined relative to a fixed, external reference plane. The coordinate system, x-, y-, and z-axis, is measured in millimeters from the reference plane to define a stereotactic volume (dotted lines). The volumes of the target and normal surrounding structures within the region are defined and localized to these coordinates by planning software.

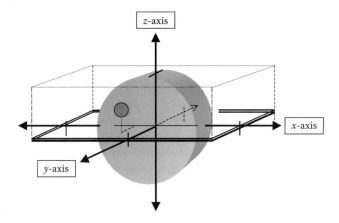

FIGURE 23.2   The stereotactic volume characterizes the anatomy (cylinder) and the target (dark circle). Distances from the reference markers are established by obtaining an image of the lesion within the frame or other orientation points. The coordinates of the lesion at depth can then be targeted from an access point on the surface.

approached by starting at a known point on the patient's surface and proceeding to the coordinates of the target (Figure 23.3).

23.2.1 Frame or Frameless

The methods require an image of the internal structures to be correlated to an accessible external reference. The orientation can be done within a frame to define the borders and coordinates of stereotactic space. Various commercial frames are on the market and most commonly used for neurosurgical procedures. The frame is secured to the patient, often by screws to the bony anatomy, which is then secured to the immobilization mechanism. The images of the patient within the frame can next be scanned into various software.[4] The program can then be used to locate the anatomy based on the distance from the frame. Sophisticated planning systems can then design trajectories of approach for surgery or radiation. The same techniques can be performed by marking the surface anatomy with points (fiducials) that will be visible to the clinician and on the image. Thus, stereotaxy can be formed without the frame as well.

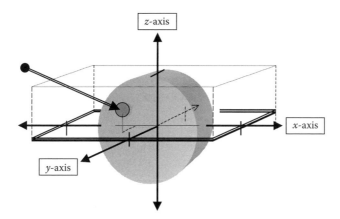

FIGURE 23.3 Trajectory of an instrument or beam is designed through an accessible point on the patient's surface to the coordinates of the target. The coordinates of the surface, interior anatomy, and the target are relative to the frame, or skin markers. Planning software navigates the path (double-lined arrow) and can guide the procedure.

The surface is delineated with skin markers,[5] or natural anatomical projections. Various frameless systems are commercially available and are very precise.[6]

Whether the intent of the procedure is biopsy or ablation, the operator must have a mechanism to guide the instrument to the target. In the case of radiosurgery,[7] it is reasonably uncomplicated to aim the beam at the coordinates of the target because the stereotactic frame is attached to the treatment table. The patient lying on the treatment table is aligned with lasers in the room via the frame's coordinate system so that the isocenter of the beam intersects the location of the target. On the other hand, when the lesion is approached with an instrument in the surgeon's hand, a navigation arrangement is needed to know where the needle or scalpel is relative to the target. The so-called neuronavigation is accomplished with room sensors that track the instruments.[8] The registration process of the fiducials is critical to the accuracy of these neuronavigation systems. The accuracy of the guidance system is similar irrespective of whether a frameless or frame-based reference system is employed.[9]

23.3 Clinical Applications

23.3.1 Movement Disorders

Tremor suppression can be accomplished with stimulation, which is as effective as coagulation but with less adverse effects, and the possibility of bilateral operations. Thalamotomy may remain an option in selected patients.[10] By using stereotaxy with MRI and electrophysiology, it is possible to record[11] and stimulate single-neuron activity deep within the brain. The electrodes placed at suitable locations can be stimulated with high frequency to treat tremor of different origins.[12] The targets include the ventral intermediate (VIM) region of the thalamus, the subthalamic nucleus (STN), and the medial pallidum (Gpi). Improvement of tremor, rigidity, and akinesia translates into less medication and its associated dyskinesia. In Parkinson's disease,[13] the STN has become the preferred site of stimulation in carefully selected patients, often with the use of ventriculography to improve the accuracy of the stereotactic coordinates.

Stereotactic intracerebral EEG stimulations and recordings can be used to study epilepsy along with other modalities.[14] By mapping somatosensory-evoked potentials via stereotactically placed intracortical electrodes, it is possible to study complex projections.[15] The use of standard stereotactic atlases can be expanded with software adapted to compare individuals to averaged anatomy.[16]

Stereotactic surgical techniques can safely reach and ablate deep structures in the brain such as the thalamus. Ventral intermediate (VIM) thalamotomies have been used to treat medically intractable essential tremor,[17] with a 60% success rate and low morbidity. Tremor in multiple sclerosis has been reported to respond to stereotactic surgery of the thalamus, zona incerta, and subthalamic nuclei.[18] Stereotactic surgery has been performed with reasonable outcomes in some Parkinson's patients as well. Ventrolateral (VL) thalamotomies and posteroventral (PV) pallidotomies can improve dyskinesias, though without a significant decrease in the levodopa dose.[19] Following stereotactic pallidotomy, patients with Parkinson's disease have been noted to have improvement in perception.[20]

Involuntary, abnormal head movements, and pain characterize cervical dystonia. The mechanisms are poorly understood but complex forms of the condition have been reported to respond to stereotactic surgery with bilateral pallidotomy or globus pallidus deep brain stimulation.[21] Unresponsive Tourette's syndrome has been treated with stereotactic surgical zona incerta (ZI) and VL thalamotomy.[22] There is a significant risk of complications with bilateral surgery.

23.3.2 Mass Lesions

Orientation in an operative field in the brain ordinarily requires adequate exposure, so the surgeon can identify familiar structures. Image-guided and computer-assisted navigation systems can display accurate spatial information directly from patient images.[23] These data permit delineation of normal anatomy and pathological lesions with much less manipulation. Therefore, the technique readily lends itself to brain biopsy[24] and drainage of abscesses.[25] Alternatively, the technique can be used with endoscopy. Neuronavigation has been successfully incorporated into surgery for brain and spine masses with improvement in the postoperative course.[8] Computer-assisted guidance has been reported to be of use in transsphenoidal parasellar surgery.[27]

Radiosurgery offers great promise and many advantages over conventional techniques.[28] However, to date there have been no randomized clinical trials to confirm the early hopeful data.[29] Radiosurgical beams, that is, radiation that is highly conformal to the target, can be delivered to a selected point in a stereotactic volume by several methods. Specially outfitted linear accelerators and the gamma knife (multiple cobalt sources) deliver photons. While several specialty centers can treat with protons, there is a huge disadvantage in cost of the protons. Modern accelerators rigged with micro-, multileaf collimation can deliver conformal radiation that is comparable to protons.[30] In rare situations, when a lesion is too intimately involved with a critical structure, protons may give a better dose distribution. Intensity modulation along with these dynamic micro-, multileaf collimators can significantly decrease the impact of x-ray beams overshooting the target with unintended consequences.[31]

Delivery of fractionated radiation with stereotactic setups, that is, fractionated stereotactic radiotherapy, improves reproducibility of multiple treatments,[32] and may be appropriate in selected circumstances.[33,34]

Radiosurgery is playing an increasingly important role in the management of brain metastases when the primary disease is controlled.[35] The role of whole brain radiotherapy with its significant morbidity is challenged by radiosurgery.[36,37]

23.3.3 Vascular Malformations

Endovascular embolization, stereotactic radiosurgery, and microsurgery can be used alone or in combination to effectively treat arteriovenous malformations (AVMs) of the brain.[38] Results with stereotactic radiosurgery for the treatment of benign intracranial mass lesions such as AVMs[39] have been very good in terms of toxicity and local control. Only 5% of the patients have been reported to suffer adverse symptoms, with a 71% obliteration rate at 2 years. AVMs can reappear after total occlusion with radiosurgery on occasion, particularly in pediatric patients.[40] After being declared cured, 10 of 48 patients had clinical symptoms from AVMs. Radiosurgical procedures for AVMs can be repeated with acceptable risk.[41] Control decreases for bigger lesions. In the case of large AVMs (>3-cm diameter), there was a greater complication rate and significantly less chance of obliteration.[42] The majority of long-term complications after radiosurgery for AVMs[43] were associated with radiation injury to the brain. Minimal sequelae after radiosurgery and no history of bleeding resulted in significantly less symptoms. Other vascular pathology, such as cerebral cavernous malformations, are presently treated with microsurgical techniques rather than radiosurgery.[44]

23.3.4 Palliation

Posteromedial thalamotomy with stereotactic navigation under MR guidance has been used to successfully relieve intractable pain.[45] Radiosurgery is more effective for idiopathic trigeminal neuralgia[46] than for tumor-related facial pain. Radiosurgery produced control in ≥70% of patients treated for trigeminal neuralgia.[47]

Insertion of Ommaya reservoirs for intrathecal chemotherapy in difficult technical scenarios is facilitated by stereotactic techniques.[48]

23.3.5 Breast

Mammography and biopsy of suspicious lesions can be readily performed with stereotactic methods,[49] where the breast and lesion can be viewed three-dimensionally. Surgical excisional biopsy of nonpalpable suspicious breast lesions are more invasive and likely more costly than stereotactic large-core needle biopsy.[50] Stereotaxy can be employed for breast biopsy in place of wire localization for nonpalpable abnormalities on mammogram.[51] The procedure is accurate[52] with less surgical trauma. Stereotactic biopsy avoided a surgical procedure in nearly half of the patients in this study. Amorphous calcifications on mammogram remain a diagnostic dilemma. Additionally, there is still risk of false-negative exams,[53] which might be lessened with improved procedures.[54–56] It is expected to be the biopsy technique of choice in the near future for lesions discovered by mammogram.

23.4 Complications

The use of stereotactic navigation is not totally without risk. Seizures can be induced by brain procedures.[24] Bleeding can occur at a biopsy site requiring craniotomy.[57] Tumor seeding along a stereotactic biopsy tract, though rare, has been reported.[58,59] In addition, there have been reports of surgical instruments transferring the infectious agents of variant Creuztfeldt–Jakob disease, though blood-borne agents portend the greatest risk of clinically significant contamination.[60]

It is less difficult to radiate after surgery than vice versa. In a survey of a small group of acoustic neuroma patients retreated with microsurgery after radiosurgery, the results were not good.[61] This is likely the result of the technical hurdles from scarring following high dose delivered to a small region.

23.5 New Directions

It may be possible in the near future to perform real-time tissue identification during stereotactic brain biopsies and functional neurosurgery.[62] Cannulas containing microsensors can discriminate cancer from benign tissue via optical scattering spectroscopy *in vivo*. Stereotactic biopsy of parasitic brain lesions, for example, neurocysticercosis, has been considered the standard of care. However, serodiagnosis might become a reasonable alternative as access to the biopsy techniques is not available in many areas.[63]

Stereotactic methods are commonly employed for brain research, such as electrode placement,[64] or creating precise lesions.[65] Models for tissue transplant into the brain with stereotactic guidance have also been reported,[66] along with early clinical data.[67] Characterization of neuroanatomy is possible by delivering neurotropic viruses into specific regions of the brain with stereotactic guidance.[68] Direct injection of gene therapy into gliomas may be performed routinely with stereotaxy as the efficiency of the vectors is improved.[69]

The impact of precision treatment is often difficult to document initially. Mapping of complex three-dimensional dose distributions of stereotactic radiation treatment can be achieved with polymer gel dosimetry,[70,71] or with radiochromic film.[72]

23.6 Discussion

Stereotactic neurosurgery has been practiced for the past half-century.[73] Since its inception, the field has expanded to epilepsy, movement disorders, radiosurgery, and frameless techniques, and applied to other sites such as spine and breast. The treatment of Parkinson's disease, for instance, has progressed from open surgical techniques in the 1950s to recording and stimulation with deeply implanted electrodes.[74] These advances were possible with the accuracy of locating the lesions under stereotactic guidance.

The methodology of stereotaxis and image guidance drastically reduce the potential for harming the patient with invasive procedures in critical areas. However, improved therapeutic modalities will drastically enhance the impact of these less-invasive delivery systems. Progress in viral vectors for gene therapy[75] or combinations of treatments could offer the next opportunity for growth of the field of stereotaxy and image-guided navigation.

References

1. Ohye, C., The idea of stereotaxy toward minimally invasive neurosurgery, *Stereotact. Funct. Neurosurg.*, 74(3–4), 185–193, 2000.
2. Stefani, A. et al., Subdyskinetic apomorphine responses in globus pallidus and subthalamus of parkinsonian patients: Lack of clear evidence for the 'indirect pathway,' *Clin. Neurophysiol.*, 113(1), 91–100, 2002.
3. Hua, S.E. et al., Microelectrode studies of normal organization and plasticity of human somatosensory thalamus, *J. Clin. Neurophysiol.*, 17(6), 559–574, 2000.
4. Slavin, K.V., Anderson, G.J., and Burchiel, K.J., Comparison of three techniques for calculation of target coordinates in functional stereotactic procedures, *Stereotact. Fund. Neurosurg.*, 72(2–4), 192–195, 1999.
5. Wolfsberger, S. et al., Anatomical landmarks for image registration in frameless stereotactic neuronavigation, *Neurosurg. Rev.*, 25(1–2), 68–72, 2002.
6. Benardete, E.A., Leonard, M.A., and Weiner, H.L., Comparison of frameless stereotactic systems: Accuracy, precision, and applications, *Neurosurgery*, 49(6), 1409–1415; discussion 1415–1416, 2001.
7. Chang, S.D. and Adler, J.R., Jr., Current status and optimal use of radiosurgery, *Oncology (Huntingt).*, 15(2), 209–216; discussion 219–221, 2001.
8. Haberland, N. et al., Neuronavigation in surgery of intracranial and spinal tumors, *J. Cancer Res. Clin. Oncol.*, 126(9), 529–541, 2000.
9. Steinmeier, R. et al., Factors influencing the application accuracy of neuronavigation systems, *Stereotact Funct. Neurosurg.*, 75(4), 188–202, 2000.
10. Speelman, J.D. et al., Stereotactic neurosurgery for tremor, *Mov. Disord.*, 17(Suppl. 3), S84–S88, 2002.
11. Benazzouz, A. et al., Intraoperative microrecordings of the subthalamic nucleus in Parkinson's disease, *Mov. Disord.*, 17(Suppl. 3), S145–S149, 2002.
12. Benabid, A.L. et al., Deep brain stimulation of the corpus luysi (subthalamic nucleus) and other targets in Parkinson's disease. Extension to new indications such as dystonia and epilepsy, *J. Neurol.*, 248(Suppl. 3), III37–III47, 2001.
13. Benabid, A.L. et al., Deep brain stimulation of the subthalamic nucleus for Parkinson's disease: Methodologic aspects and clinical criteria, *Neurology*, 55(12), S40–S44, 2000.
14. Baciu, M. et al., Functional MRI assessment of the hemispheric predominance for language in epileptic patients using a simple rhyme detection task, *Epileptic Disord.*, 3(3), 117–124, 2001.

15. Barba, C. et al., Stereotactic recordings of median nerve somatosensory-evoked potentials in the human pre-supplementary motor area, *Eur. J. Neurosci.*, 13(2), 347–356, 2001.

16. Berks, G., Pohl, G., and Keyserlingk, D.G., Three-dimensional-VIEWER: An atlas-based system for individual and statistical investigations of the human brain, *Methods Inf. Med.*, 40(3), 170–177, 2001.

17. Akbostanci, M.C., Slavin, K.V., and Burchiel, K.J., Stereotactic ventral intermedial thalamotomy for the treatment of essential tremor: Results of a series of 37 patients, *Stereotact. Funct. Neurosurg.*, 72(2–4), 174–177, 1999.

18. Alusi, S.H. et al., Stereotactic lesional surgery for the treatment of tremor in multiple sclerosis: A prospective case-controlled study, *Brain*, 124(Pt. 8), 1576–1589, 2001.

19. Aguiar, P.M. et al., Motor performance after posteroventral pallidotomy and VIM-thalamotomy in Parkinson's disease: A 1-year follow-up study, *Arq. Neuropsiquiatr.*, 58(3B), 830–835, 2000.

20. Barrett, A.M. et al., Seeing trees but not the forest: Limited perception of large configurations in PD, *Neurology*, 56(6), 724–729, 2001.

21. Adler, C.H. and Kumar, R., Pharmacological and surgical options for the treatment of cervical dystonia, *Neurology*, 55(12), S9–S14, 2000.

22. Babel, T.B., Warnke, P.C., and Ostertag, C.B., Immediate and long term outcome after infrathalamic and thalamic lesioning for intractable Tourette's syndrome, *J. Neurol. Neurosurg. Psychiatry*, 70(5), 666–671, 2001.

23. Suess, O. et al., Intracranial image-guided neurosurgery: Experience with a new electromagnetic navigation system, *Acta Neurochir. (Wien)*, 143(9), 927–934, 2001.

24. Yu, X. et al., Stereotactic biopsy for intracranial space-occupying lesions: Clinical analysis of 550 cases, *Stereotact. Funct. Neurosurg.*, 75(2–3), 103–108, 2000.

25. Strowitzki, M., Schwerdtfeger, K., and Steudel, W.I., Ultrasound-guided aspiration of brain abscesses through a single burr hole, *Minim. Invasive Neurosurg.*, 44(3), 135–140, 2001.

26. Gumprecht, H., Trost, H.A., and Lumenta, C.B., Neuroendoscopy combined with frameless neuronavigation, *Br. J. Neurosurg.*, 14(2), 129–131, 2000.

27. Kacker, A. et al., Transphenoidal surgery utilizing computer-assisted stereotactic guidance, *Rhinology*, 39(4), 207–210, 2001.

28. Weil, M.D., Advances in stereotactic radiosurgery for brain neoplasms, *Curr. Neurol. Neurosci. Rep.*, 1(3), 233–237, 2001.

29. Haines, S.J., Moving targets and ghosts of the past: Outcome measurement in brain tumour therapy, *J. Clin. Neurosci.*, 9(2), 109–112, 2002.

30. Baumert, B.G. et al., A comparison of dose distributions of proton and photon beams in stereotactic conformal radiotherapy of brain lesions, *Int. J. Radiat. Oncol. Biol. Phys.*, 49(5), 1439–1449, 2001.

31. Benedict, S.H. et al., Intensity-modulated stereotactic radiosurgery using dynamic micro-multileaf collimation, *Int. J. Radiat. Oncol. Biol. Phys.*, 50(3), 751–758, 2001.

32. Albeit, H. et al., Patient position reproducibility in fractionated stereotactically guided conformal radiotherapy using the BrainLab mask system, *Strahlenther Onkol.*, 177(5), 264–268, 2001.

33. Andrews, D.W. et al., Stereotactic radiosurgery and fractionated stereotactic radiotherapy for the treatment of acoustic schwannomas: Comparative observations of 125 patients treated at one institution, *Int. J. Radiat. Oncol. Biol. Phys.*, 50(5), 1265–1278, 2001.

34. Aoyama, H. et al., Treatment outcome of single or hypofractionated single-isocentric stereotactic irradiation (STI) using a linear accelerator for intracranial arteriovenous malformation, *Radiother. Oncol.*, 59(3), 323–328, 2001.

35. Arnold, S.M. and Patchell, R.A., Diagnosis and management of brain metastases, *Hematol Oncol. Clin. North Am.*, 15(6), 1085–1107, vii, 2001.

36. Weil, M.D., Stereotactic radiosurgery for brain tumors, *Hematol. Oncol. Clin. North Am.*, 15(6), 1017–1026, 2001.

37. Chen, J.C. et al., Stereotactic radiosurgery in the treatment of metastatic disease to the brain, *Neurosurgery*, 47(2), 268–279; discussion 279–281, 2000.

38. Fleetwood, I.G. and Steinberg, G.K., Arteriovenous malformations, *Lancet*, 359(9309), 863–873, 2002.

39. Flickinger, J.C., Kondziolka, D., and Lunsford, L.D., Radiosurgery of benign lesions, *Semin. Radiat. Oncol.*, 5(3), 220–224, 1995.

40. Lindqvist, M. et al., Angiographic long-term follow-up data for arteriovenous malformations previously proven to be obliterated after gamma knife radiosurgery, *Neurosurgery*, 46(4), 803–808; discussion 809–810, 2000.

41. Maesawa, S. et al., Repeated radiosurgery for incompletely obliterated arteriovenous malformations, *J. Neurosurg.*, 92(6), 961–970, 2000.

42. Miyawaki, L. et al., Five year results of LINAC radiosurgery for arteriovenous malformations: Outcome for large AVMS, *Int. J. Radiat. Oncol. Biol. Phys.*, 44(5), 1089–1106, 1999.

43. Flickinger, J.C. et al., A multi-institutional analysis of complication outcomes after arteriovenous malformation radiosurgery, *Int. J. Radiat. Oncol. Biol. Phys.*, 44(1), 67–74, 1999.

44. Bertalanffy, H. et al., Cerebral cavernomas in the adult. Review of the literature and analysis of 72 surgically treated patients, *Neurosurg. Rev.*, 25(1–2), 1–53; discussion 54–55, 2002.

45. Balas, I.I. et al. [In Process Citation], *Rev. Neurol.*, 31(6), 531–533, 2000.

46. Chang, J.W. et al., The effects of stereotactic radiosurgery on secondary facial pain, *Stereotact. Funct. Neurosurg.*, 72(Suppl. 1), 29–37, 1999.

47. Pollock. B.E. et al., The Mayo Clinic gamma knife experience: Indications and initial results [see comments], *Mayo Clin. Proc.*, 74(1), 5–13, 1999.

48. Al-Anazi, A. and Bernstein, M., Modified stereotactic insertion of the Ommaya reservoir. Technical note, *J. Neurosurg.*, 92(6), 1050–1052, 2000.

49. Bagnall, M.J. et al., Predicting invasion in mammographically detected microcalcification, *Clin, Radiol.*, 56(10), 828–832, 2001.

50. Buijs-van der Woude, T. et al., Cost comparison between stereotactic large-core-needle biopsy versus surgical excision biopsy in The Netherlands, *Eur. J. Cancer*, 37(14), 1736–1745, 2001.

51. Carr, J.J. et al., Stereotactic localization of breast lesions: How it works and methods to improve accuracy, *Radiographics*, 21(2), 463–473, 2001.

52. Berg, W.A. et al., Biopsy of amorphous breast calcifications: Pathologic outcome and yield at stereotactic biopsy, *Radiology*, 221(2), 495–503, 2001.

53. Adler, D.D. et al., Follow-up of benign results of stereotactic core breast biopsy, *Acad. Radiol.*, 7(4), 248–253, 2000.

54. Bergaz, F. et al., Clip placement facilitating the approach to breast lesions, *Eur. Radiol.*, 12(2), 471–474, 2002.

55. Atallah, N. et al., Stereotaxic excisional biopsy of non-palpable breast lesions by the ABBI (Advanced Breast Biopsy Instrumentation) technique. Advantages. Disadvantages. Indications. Apropos of 67 cases, *J. Med. Liban*, 48(2), 70–76, 2000.

56. Ancona, A., Caiffa, L., and Fazio, V., Digital stereotactic breast microbiopsy with the mammotome: Study of 122 cases, *Radiol Med. (Torino)*, 101(5), 341–347, 2001.

57. Daszkiewicz, P., [In Process Citation]. *Neurol. Neurochir. Pol.*, 35(5), 899–905, 2001.

58. Aichholzer, M. et al., Epidural metastasis of a glioblastoma after stereotactic biopsy: Case report, *Minim. Invasive Neurosurg.*, 44(3), 175–177, 2001.

59. Steinmetz, M.P. et al., Metastatic seeding of the stereotactic biopsy tract in glioblastoma multiforme: Case report and review of the literature, *J. Neurooncol.*, 55(3), 167–171, 2001.

60. Brown, P., The risk of blood-borne Creutzfeldt–Jakob disease, *Dev. Biol. Stand.*, 102, 53–59, 2000.

61. Battista, R.A. and Wiet, R.J., Stereotactic radiosurgery for acoustic neuromas: A survey of the American Neurotology Society, *Am. J. Otol.*, 21(3), 371–381, 2000.

62. Andrews, R. et al., Multimodality stereotactic brain tissue identification: The NASA smart probe project, *Stereotact. Funct. Neurosurg.*, 73(1–4), 1–8, 1999.

63. Bedi, S., Prasad, A., and Anand, K.S., Neurocysticercal serodiagnosis—Updated, *J. Indian Med. Assoc.*, 99(2), 96, 98–99, 2001.

64. Akaike, K. et al., Regional accumulation of ^{14}C-zonisamide in rat brain during kainic acid-induced limbic seizures, *Can. J. Neurol. Set.*, 28(4), 341–345, 2001.

65. Bechmann, I. et al., Reactive astrocytes upregulate Fas (CD95) and Fas ligand (CD95 L) expression but do not undergo programmed cell death during the course of anterograde degeneration, *Glia*, 32(1), 25–41, 2000.

66. Barami, K. et al., Transplantation of human fetal brain cells into ischemic lesions of adult gerbil hippocampus, *J. Neurosurg.*, 95(2), 308–315, 2001.

67. Mauser, R.A. et al., Bilateral human fetal striatal transplantation in Huntington's disease, *Neurology*, 58(5), 687–695, 2002.

68. Janson, C.G. and During, M.J., Viral vectors as part of an integrated functional genomics program, *Genomics*, 78(1–2), 3–6, 2001.

69. Alavi, J.B. and Eck, S.L., Gene therapy for high grade gliomas, *Expert Opin. Biol. Ther.*, 1(2), 239–252, 2001.

70. Berg, A., Ertl, A., and Moser, E., High-resolution polymer gel dosimetry by parameter selective MR-microimaging on a whole body scanner at 3T, *Med. Phys.*, 28(5), 833–843, 2001.

71. Audet, C. et al., CT gel dosimetry technique: Comparison of a planned and measured three-dimensional stereotactic dose volume, *J. Appl. Clin. Med. Phys.*, 3(2), 110–118, 2002.

72. Bazioglou, M. and Kalef-Ezra, J., Dosimetry with radiochromic films: A document scanner technique, neutron response, applications, *Appl. Radiat. Isot.*, 55(3), 339–345, 2001.

73. Gildenberg, P.L., History of the American Society for Stereotactic and Functional Neurosurgery, *Stereotact. Funct. Neurosurg.*, 72(2–4), 77–81, 1999.

74. Gillingham, J., Forty-five years of stereotactic surgery for Parkinson's disease: A review, *Stereotact. Funct. Neurosurg.*, 74(3–4), 95–98, 2000.

75. Lee, E.J., Thimmapaya, B., and Jameson, J.L., Stereotactic injection of adenoviral vectors that target gene expression to specific pituitary cell types: Implications for gene therapy, *Neurosurgery*, 46(6), 1461–1468; discussion 1468–1469, 2000.

24

Extracorporeal Blood-Filtering Technologies

Jeong Chul Kim
Ospedale San Bortolo

*International Renal
Research Institute*

Manish Kaushik
Ospedale San Bortolo

*International Renal
Research Institute*

Claudio Ronco
Ospedale San Bortolo

*International Renal
Research Institute*

24.1 Introduction

To remove toxins from the body, various blood-purification techniques have been used in clinical practice. The general approach of blood purification is to use a semipermeable membrane that selectively removes the solute by diffusion, convection, and, optionally, adsorption. Blood-purification techniques can be broadly divided into two categories: intracorporeal therapies and extracorporeal therapies. In peritoneal dialysis, an example of an intracorporeal therapy, the native peritoneal membrane, which covers most of the intra-abdominal organs and forms the lining of the peritoneal cavity, functions as a semipermeable membrane. In contrast, during extracorporeal therapy, polymer membranes and sorbent cartridges are used to remove toxins and excess body water by concentration-driven, pressure-driven, electrochemical force, and specific antibody-driven operations. Currently, various combinations of hollow-fiber membrane and sorbents support or replace internal organ functions. This chapter focuses on basic principles and clinical applications of extracorporeal blood-purification techniques to provide insights into contemporary and emerging blood-filtering techniques.

24.2 Basic Setup of Extracorporeal Blood-Filtering Techniques

Extracorporeal blood purification requires four main elements: vascular access, blood-filtering unit, dialysis/replacement fluid, and machine for circulation and monitoring (Figure 24.1).

24.2.1 Vascular Access

Vascular access (VA) is the prerequisite and Achilles' heel of extracorporeal blood-purification therapy. VA gives access to the internal milieu of the patient via the bloodstream, permitting blood purification. The patency of VA is of paramount importance for chronic kidney disease (CKD) patient on thrice-weekly long-term hemodialysis therapy.

Depending on the indication of their use, VA may be temporary (acute or midterm) or permanent (chronic) [1]. Acute VA is generally indicated for starting immediate dialysis in an emergency context when a permanent VA has not been created previously, in patients with acute kidney injury who require renal replacement therapy (RRT), in patients with intoxication amenable to removal by extracorporeal blood purification, or for any other indication requiring extracorporeal blood purification. The usual option for acute VA is a central venous catheter (CVC), also called VA catheter. The temporary VA catheters generally used for acute indications are made of semirigid polymer material. Catheter rigidity permits a nontunneled direct percutaneous introduction of the catheter into a large central vein, under local anesthesia. Several types of acute catheters are presently available in the market. Catheters are usually specified by polymer structure, manufacturing design, presence or absence of side holes, single or double lumen, and external extension tubing part.

Bridging VA is considered a temporary solution between acute and permanent VA. This midterm VA is indicated to solve the blood access problem and to ensure the continuity of RRT in a VA failure situation. There are two options available: venovenous angioaccess or arteriovenous (AV) access. Venovenous angioaccess includes two main options: one is the tunneled catheter and the other is the subcutaneous port catheter device. Tunneled catheters differ from nontunneled ones in that they have subcutaneous tunneling and an anchoring system during catheter insertion. They are made of soft biocompatible polymer and have a large-bore lumen to permit a high flow rate, are dual lumen (arterial and venous lumens), are tunneled in the subcutaneous tissue, and are tightly anchored in the subcutaneous tissue, either with a cuff or a U-shaped suture, and their distal tips are usually located in the superior vena cava or in the right atrium. Based on the chemotherapy port concept, venous port catheter devices have also been developed [2,3]. They are a port valve system made of titanium containing one passageway,

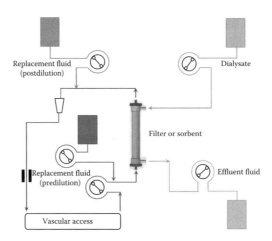

FIGURE 24.1   Main elements of extracorporeal blood-purification therapy.

subcutaneously implanted and connected to one or two large-bore silicon catheters. As an AV midterm VA, a synthetic polytetrafluoroethylene (PTFE) or polyurethane (PU) AV graft has been used when an autologous AV fistula has failed or cannot be created, or used easily.

As a long-term VA, several types of AV fistula have been described. AV fistula is a surgically created passageway between an artery and a vein. During hemodialysis, the volume of blood is too great for the veins to handle, so a vein must be enlarged. An artery and a vein, usually in the arm above or below the elbow, are sewn together, to create a fistula, and arterial pressure eventually enlarges the vein. The enlarged vein can accommodate a large needle. Both the artery and the vein dilate and elongate in response to the greater blood flow and shear stress, but the vein dilates more and becomes "arterialized" and repeated access by needles is feasible. Though AV graft and in special circumstances, tunneled VA catheters, may also be considered as a permanent VA, AF fistula at this moment is recognized as the most reliable long-term VA [4].

24.2.2 Blood-Filtering Unit

24.2.2.1 Parallel-Plate Dialyzers

In parallel-plate dialyzers, several layers of flat sheet membranes are stacked, supported by thin plates. There were initially some advantages such as lower thrombogenicity and ease of sterilization by ethylene oxide. However, usage of parallel-plate dialyzers has recently declined because hollow-fiber dialyzers are small, efficient, and suitable for reuse.

24.2.2.2 Hollow-Fiber Dialyzers

A contemporary hollow-fiber dialyzer consists of a housing containing a single membrane fiber bundle (Figure 24.2). The bundle is embedded at both ends in polyurethane, which also fixes the bundle within the casing. The end surfaces are covered by end caps, which contain the blood inlet and outlet ports. The two dialysis fluid ports are positioned on the housing. The structural design of both the hollow-fiber hemodialyzer and the dialysis membrane material are of considerable importance. Hollow-fiber membranes are manufactured in wet, dry–wet, or dry-spinning process. Table 24.1 shows a family tree of dialysis membranes. Until the late 1960s, only membranes manufactured from regenerated cellulose were available. In the early 1970s, the interest in removing middle molecules resulted in the search for improved materials and the clinical introduction of new modalities of treatment such as hemofiltration. These

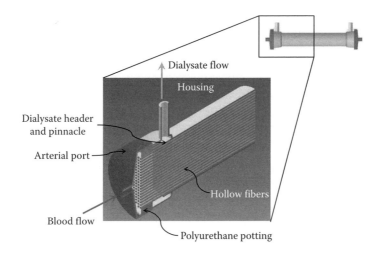

FIGURE 24.2 Structure of hollow-fiber dialyzer and design factors.

TABLE 24.1   Family of Hollow-Fiber Membrane for Hemodialysis

Cellulosic membranes	Substituted/modified cellulose	Hemopheane®
		SMC®
		PEG
		Excebrane®
	Acetylated cellulose	Cellulose acetate
		Cellulose diacetate
		Cellulose triacetate
	Regenerated cellulose	Cuprophan®
		Bioflux®
		Cuprammonium rayon
		SCE
		G-O-P DIAFIL®
Synthetic membranes	Hydrophilized copolymers	PPC Gambrane®
		PMMA
		PAN AN69
		PAN DX
		SPAN
	Hydrophilic/hydrophobic copolymers (blended)	PA
		PS
		Helixone®
		APS®
		Diapes®
		Arylane®
		PEPA®
		Arylane
		Polyamix™
	Hydrophilic copolymers	EVAL

modalities used synthetic membranes prepared from engineered thermoplastics such as polysulfones (PS), polyamides, and polyacrylonitrile (PAN) polymers.

The main performance characteristics of each dialyzer are determined by the size and design of the fiber bundle included. There are some designs that facilitate efficient flow distributions and mass transport by modification in housing structure [5] and fiber structure [6]. The composition of the potting compound has changed over the years. The main motivation was to minimize risks associated with toxic substances, which may evolve after sterilization of the polyurethane resin (PUR). In particular, irradiation with β- or γ-beams may lead to the by-product of carcinogens.

24.2.2.3 Sorbent Cartridges

Since the early 1960s, sorbents have been used in an attempt to increase the efficiency of dialysis, or replace it, in the management of uremia. Additionally, hemoperfusion, a process that allows direct contact of blood from the patient with sorbents has been used to treat drug and chemical intoxication as well as fulminant hepatic encephalopathy. Sorbents used in hemoperfusion devices are activated carbons, ion exchange resins, or nonion macroporous resins. Sorbents can be divided into two large categories (Table 24.2): (i) those that have hydrophobic properties and therefore adsorb the molecules dissolved in the solution in contact with the sorbent and (ii) those that eliminate solutes by chemical affinity. Charcoal and nonionic macroporous resin are within the first category. Charcoal is produced both from biological substances such as coconut shells or peach pits and from nonbiological substances such as petroleum. The charcoal is activated by controlled oxidation in air–carbon dioxide, or steam. Adsorption into charcoal occurs through its pores, and its efficiency depends on the total number of pores and their radius. Activated charcoals are available in many forms, beginning with uncoated granular carbons; most of the available devices now use coated activated charcoal either as granular

TABLE 24.2 Types of Sorbents and Their Use in Clinical Practice

Sorbent	Type	Applications
Nonselective	Charcoal	Poisoning
	Uncharged resins	Chronic renal failure
Selective	Hydrophobic resins	Rabdomyolysis
	Powdered sorbent	Hepatic failure
	Microsphere-based detoxification system	HIV
	Polymyxin-B	Sepsis
	Polyethyleneimine	Drug overdose and poisoning
		LDL apheresis

charcoal coated with cellulose nitrate (collodion) polymer and albumin or as heparinized copolymer. Other devices containing charcoals are prepared with extruded charcoal coated with cellulose acetate or with methacrylic hydrogel. Others use spherical charcoals derived from petroleum, coated with poly-hema solutions, or derived from pyrrolized inorganic resins. The nonionic resins consist of macroporous cross-linked polystyrene amberlite series; for example, XAD-2 and XAD-4 were available in the United States and Europe for the treatment of poisoning. Recently, a modified divinylbenzene resin coated for hemocompatibility has been used [7]. The nonionic macroporous resins are very similar to charcoal and microsphere agglomerates, which adsorb the toxins they eliminate in their surface. The sorbents that eliminate substances by chemical affinity are fundamentally ion exchange resins, which exchanges one ion for another of the same electrical charge.

24.2.3 Dialysis/Replacement Fluid

In extracorporeal blood-purification therapy, dialysis solution, the so-called, dialysate is used to enhance the removal of low-molecular-weight toxins. Because the goal of dialysis is to restore the composition of the body's fluid environment toward normal, dialysate compositions are set to approximate normal values in the body. Moreover, dialysate composition is a factor strongly affecting cardiovascular stability during treatment. In hemodialysis, 500–800 mL/min of dialysate flow is used generally and depending on the treatment modality, ultrapure dialysate/replacement fluid may be additionally required for infusion or replacement. Water represents more than 95% of dialysate, and a dialysis patient is exposed to 300–400 L of water a week, in contrast to a normal person ingesting 14 L per week. To avoid acute or chronic side effects, quality standards for dialysis water and dialysis fluids have been proposed. AAMI and European Pharmacopoeia standards have been widely used or referenced [8,9]. A standard water treatment system is made of a pretreatment section, including softeners, granular activated carbon, and microfilters, followed by a final treatment section. To polish the pretreated water a deionizer or reverse osmosis (RO) may be used. Usually two RO modules in series are preferred to preserve microbiological quality of the treated water. Recently, with increasing use of ultrapure dialysate, defined as sterile, nonpyrogenic fluid obtained by cold filtration, ultrafiltration of final dialysate is done by installing a polysulfone or polyamide ultrafilter on the dialysate line of the dialysis machine. It is also of importance to assure proper water quality by regular maintenance and a quality control program.

24.2.4 Machine for Extracorporeal Therapy

A machine for extracorporeal blood purification is in charge of blood circulation and monitoring of the treatment system. It is equipped with pumps for blood and dialysate supply, biosensors, and safety system. Because of the paramount importance of safety, international authorities and committees have developed regulations and standards. The basic regulation within the European Union (EU) is Council Directive 93/42/EEC, commonly known as "Medical Device Directive (MDD)." This directive defines essential requirements now used as the basis for the registration (CE marking) of medical devices within the EU. These essential requirements demand that any risk that may be associated with medical devices must be weighed against the benefit for the patient, and must be compatible with a high level of protection of health and safety. Recently, bio-feedback control systems for blood volume, body temperature, and ultrafiltration have been adopted to improve the tolerance of treatment [10–12].

24.3 Transport Principles in Blood-Filtering Technologies

24.3.1 Diffusion

Diffusion is defined as the migration of molecules by random motion from a region of higher concentration to a region of lower concentration. The rate of diffusion per unit area (J/A) is proportional to the concentration gradient (ΔC), which is the driving force.

$$J/A = -K_0 \, (\Delta C)$$

where K_0 is a constant, the overall mass transfer coefficient that characterizes the resistances of layers limiting diffusion of solute across the dialyzer membrane. In hemodialysis, it is useful to use the mass transfer area coefficient (K_0A) to compare diffusion performance of whole dialyzers. K_0A is a property of both the solute and the dialysis membrane with units of mL/min.

24.3.2 Convection

Convective transport is defined as solute movement that results from bulk movement of solvent, usually in response to differences in hydrostatic pressure. Hydraulic pressure causes molecules to move from a region of high pressure to a region of low pressure. For solutes that are much smaller than the membrane pores, the movement of solvent carries the solute with it at the same rate. For larger solutes, however, movement of solutes may be relatively restricted. The magnitude of this restriction can be expressed as the sieving coefficient (SC).

$$SC = \frac{C_r}{C_d}$$

where C_r is the mean concentration mass receiving stream (i.e., dialysate compartment) and C_d is the mean concentration mass donating stream (i.e., blood compartment). The fluid transport in hollow-fiber membrane (J_f) is defined as ultrafiltration and can be described as

$$J_f = K_{UF} \cdot \mathrm{TMP}$$

where K_{UF} is the ultrafiltration coefficient that represents the hydraulic permeability of the membrane and $\mathrm{TMP} = P_B - P_D - \pi$ (transmembrane pressure), where P_B and P_D are hydrostatic pressures of blood and dialysate, respectively, and π is the oncotic pressure.

24.3.3 Adsorption

Dialysis membrane may interact with solutes, causing them to adhere or, when present in high concentrations such as serum albumin, to coat the membrane, reducing membrane permeability to other solutes. The albumin coating effect occurs immediately after exposure to blood or serum and accounts in part for the lower K_0A [13]. However, because of high blood concentration, adherence of albumin to the membrane does not reduce its overall concentration. On the other hand, solutes present in much lower concentrations may be substantially removed by adsorption to the membrane. For example, the adherence of beta-2 microglobulin and endotoxin to polysulfone membrane during standard dialysis significantly enhances their clearances [14,15]. Sorbents, by taking advantages of large surface area, can be additionally used to adsorb the molecules dissolved in the solution in contact with the sorbent and by the following steps [16]: (a) external (interphase) mass transfer of the solute from the bulk fluid by convection through a thin film or boundary layer, to the outer surface of the sorbent; (b) internal (interphase) mass transfer of the solute by pore diffusion from the outer surface of the sorbent to the inner surface of the internal porous structure; (c) surface diffusion along the porous surface; and (d) adsorption of the solute onto the porous surface. Physical adsorption may occur thanks to van der Waals forces and chemical adsorption occurs due to chemical affinity.

24.4 Factors Influencing Solute Transport

24.4.1 Blood Flow

In low-flux hemodialysis, solutes are mostly removed by diffusion. For some easily dialyzed solutes, removal by diffusion is a near-linear function of blood flow. For other solutes, the removal is primarily membrane-dependent, and independent of blood flow (Figure 24.3). When membrane permeability is high, solutes are removed quickly, in the proximal part of the hollow fiber. Increase in blood flow enhances solute removal by extending the gradient further. Blood flow distribution in hollow-fiber bundle could also influence solute transport in a dialyzer. Blood flow distribution is determined by arterial port design of the dialyzer and blood flow rate [17].

24.4.2 Hematocrit and Blood Water Content

The nonaqueous fraction of whole blood is not dialyzable, so, as the patient's hematocrit increases, solute clearance falls, due to the decrease in flow of "blood water" through the dialyzer. An expression can be derived for blood water flow:

$$Q_{BW} = Q_B[0.72\gamma(hct) + 0.93(1 - hct)]$$

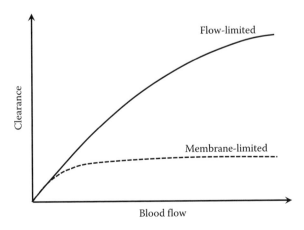

FIGURE 24.3 Relationship between blood flow and solute clearance in hollow-fiber dialyzer.

where Q_B is whole blood flow into the dialyzer, Q_{BW} is blood water flow, *hct* is the fractional red cell volume of whole blood, 0.93 is the water fraction of plasma, and 0.72 is the effective water fraction of the red cell. To completely account for the hematocrit effect, the fraction of red cell volume available for dialysis (γ) must be added to the formula. For urea, γ has been measured at 1.11 due to a reversely bound pool of urea within the red cell [18]. For creatinine, γ is approximately 0.50, and for phosphorous, γ is essentially zero.

24.4.3 Dialysate Flow

The flow of dialysate enhances solute removal by maintaining the concentration gradient across the dialysis membrane. When blood and dialysate flow are countercurrent, solute removal is maximal and the relationship between dialysate flow and clearance is similar to that of blood flow as mentioned earlier. However, in contrast to blood side, stagnation and channeling of dialysate flow lead to nonuniform dialysate flow distribution. Fiber structure, fiber packing density, and flow baffle design influence dialysate flow distribution in a dialyzer [6,19].

24.4.4 Membrane Properties

Dialysis membranes manufactured in the past were composed of a thin uniform meshwork of modified cellulose fibers with an effective pore size that depended on the density of fibers. Modern synthetic membranes are often asymmetric; that is, they are constructed from a highly porous, relatively thick polymer that offers little resistance to diffusion to the overlying thin layer that constitutes the diffusion barrier. For small dialyzable solutes, the rate of diffusion across a porous membrane depends on the number and geometry of pores as well as the charge and surface area of membrane [20,21]. During extracorporeal blood purification using hollow-fiber membrane, blood-membrane surface interection, swelling of the membrane and membrane surface potential affect the effective surface area for solute transport.

24.4.5 Solute Properties and Protein Binding

The transport of solute across dialysis membranes depends on the interaction of the solute with the membrane, so properties of both must be considered. The most important property of solute is molecular size, but water solubility, charge, and molecular shape also influence transport. Serum albumin serves as a transport protein for hydrophilic solutes such as fatty acids that are poorly soluble in aqueous solutions. Just as binding to serum proteins may affect removal of drugs by dialysis, binding of uremic toxins to albumin can inhibit the therapeutic effectiveness of dialysis, causing even small toxins to behave like larger poorly dialyzable compounds [22].

24.4.6 Convection

In addition to solute permeability, membrane porosity can be measured in terms of water permeability. The latter is usually expressed as an ultrafiltration coefficient (K_{UF}), the filtration flow rate per unit of applied hydraulic pressure:

$$K_{UF} = Q_f / \Delta P$$

Removal of solute by filtration (convection) during dialysis slightly augments removal by diffusion. In reality, during extracorporeal therapy, both diffusion and convection occur simultaneously along the same membrane. The decrease in solute concentration due to diffusion means the capacity for solute removal by convection falls along the dialyzer length. Conversely, the fall in blood flow rate due to filtration along the dialyzer length interferes with diffusion. A simple model proposed for this phenomenon is [23]

$$K_D = K_{D0} + Q_f T$$

where T, termed the transmittance, represents the mL/min increase in clearance for each mL/min of filtration. During filtration, on the other hand, concentration polarization and cake layer formation on the surface of dialysis membrane impair the effectiveness of filtration. High blood flow rate may serve to reduce the thickness of the secondary layer by increasing shear forces at the membrane surface.

24.5 Clinical Applications

24.5.1 Blood Filtering in Kidney Diseases

There are various extracorporeal renal replacement techniques using hollow-fiber membranes (Figure 24.4).

24.5.1.1 Low-Flux Hemodialysis

Conventionally, hemodialysis membranes are classified according to water permeability. The ultrafiltration coefficient K_{UF} is a general parameter that represents water permeability of dialyzers. Low-flux dialyzers have ultrafiltration coefficient less than 8–10 mL/h/mmHg [24]. Low-flux hemodialysis focuses on removal of low-molecular-weight solutes by diffusion. During low-flux hemodialysis, fluid transport across dialysis membrane does not influence hematocrit level in hollow fibers and the probability of backfiltration is low.

24.5.1.2 High-Flux Hemodialysis

High-flux membranes are more porous due to larger pore size, increased surface area, and higher pore number. Generally, dialyzers that have an ultrafiltration coefficient of more than 20 mL/h/mmHg are

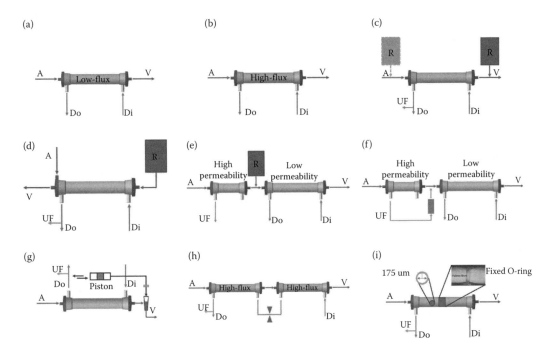

FIGURE 24.4 Various extracorporeal RRT using hollow-fiber dialyzers. (a) Low-flux hemodialysis, (b) high-flux hemodialysis, (c) hemodiafiltration, (d) middilution hemodiafiltration, (e) paired filtration dialysis, (f) hemodiafiltration with reinfusion, (g) push/pull hemodiafiltration, (h) double high-flux dialysis, and (i) enhanced internal filtration dialysis.

called high-flux dialyzers, which allow more fluid transport across dialysis membrane by convection. During high-flux hemodialysis, direct filtration, fluid transport from blood compartment to dialysate compartment occurs in the proximal part of hemodialyzer while backfiltration occurs in the distal part. Volumetric ultrafiltration control contributes to increase in dialysate pressure. Both low- and middle-molecular-weight solutes are removed by diffusion and convection during high-flux hemodialysis.

24.5.1.3 Hemodiafiltration

Owing to advances in dialysis membrane, ultrafiltration control system and online production of large amounts of ultrapure dialysate, hemodiafiltration (HDF) is becoming popular to enhance middle-molecular-weight solute removal [25]. HDF allows high filtration volume (5–20 L/session) necessitating infusion or replacement. Dialysis membranes that have an ultrafiltration coefficient of more than 50 mL/h/mmHg are adopted in this dialysis modality.

24.5.1.4 Middilution Hemodiafiltration

To reduce spontaneous BF in a hollow-fiber module, middilution hemodiafiltration (MD-HDF) uses a specially designed filter with a blood header cap that contains fibers arranged in outer annular and inner core region [26]. Blood enters the side port of the blood header and flows down to the outer annular region of fibers and mixes with the replacement fluid at the other end, where blood flow is countercurrent to dialysate. The diluted blood then flows back along the central core of fibers to exit at the central port of the blood header, where blood flow is cocurrent to dialysate.

24.5.1.5 Paired Filtration Dialysis

Paired filtration dialysis (PFD) involves high-permeable membrane and low-permeable membrane to separate diffusion and convection [27]. The first filter allows hemofiltration via a high-permeable membrane followed by the infusion of sterile replacement fluid before blood flow through the second filter to undergo dialysis across a low-permeable membrane.

24.5.1.6 Hemofiltration with Reinfusion

HFR is modified from PFD to avoid exogenous reinfusion. The ultrafiltrate from the hemofilter is passed through a sorbent cartridge to retain uremic solutes and is reinfused as a replacement fluid. Endogenous reinfusion avoids losses of nutrients, amino acids, hormones, and vitamins during dialysis [28].

24.5.1.7 Push/Pull Hemodiafiltration

To enhance DF and BF, a double-cylinder piston pump is attached to the dialysis outlet pathway at one end and the venous chamber at the other [29]. As this pump pulls out the dialysate from the dialysis pathway, it creates a negative TMP and effects the filtration in the hemodiafilter. It simultaneously lowers the air–fluid level in the venous air chamber. As the pump pushes the dialysate back into the dialysate outlet pathway, it creates a positive TMP and affects BF in the hemodiafilter. It simultaneously elevates the air–fluid level in the air chamber. Since the time taken for the extracorporeally circulating blood to pass the hemodiafilter is approximately 40 s, the blood is concentrated and diluted many times before it leaves the hemodiafilter (approximately 25 times) and body fluid replacement volume will exceed 120 L during a 4-h session. The synchronized lowering and elevation of air–fluid level in the venous chamber with the filtration and BF, respectively, is to circumvent the variation in the blood flow returned to the patient's body.

24.5.1.8 Double High-Flux Dialysis

To shorten treatment time, two high-flux dialyzers are placed in series. A flow restrictor in the dialysate pathway between the two dialyzers allows a low dialysate pressure in the proximal dialyzer favoring DF. The high dialysate pressure, generated upstream to the flow restrictor, in the distal dialyzer favors BF, which thus acts as the replacement fluid. The high dialysate compartment pressure of the distal dialyzer favors BF, which thus acts as the replacement fluid [30].

24.5.1.9 Enhanced Internal Filtration Dialysis

By various modifications in filter or fiber geometry, higher convection transport is achieved during enhanced internal filtration dialysis (EIFD). The dialyzer may be modified by the introduction of an O-ring to enhance the dialysate pressure differentials between the proximal and distal parts of the dialyzer [31]. A lower dialysate pressure in the proximal part favors filtration, while a higher dialysate pressure in the distal part of the dialyzer favors BF. Minor alterations in the fiber diameter can accentuate the pressure drop along the fiber and hence favor DF in the proximal part and BF in the distal part [32]. Higher fiber density ratio and housing structure with a wholly surrounding baffle and a slope facilitate the uniform diffusion of the dialysate [33].

24.5.2 Blood Filtering in Acute Liver Failure

Acute liver failure (ALF) or acute chronic liver failure is associated with a high mortality of 28%. The prognosis of ALF has been substantially improved by orthotopic liver transplantation (OLT). However, the scarcity of organs and the lifelong need for immunosuppression and its attending adverse effects precludes OLT as an option in all cases. Moreover, 40% of the episodes of ALF may recover, with medical therapy and the regenerative ability of the liver, thereby making an irreversible treatment like OLT redundant.

The main metabolic functions of the normal liver are detoxification, biotransformation, excretion, and synthesis. Therefore, ALF is characterized by the accumulation of several toxic substances, including bilirubin, ammonia, glutamine, glutamate, aromatic amino acids, free fatty acids, lactate, phenols, mercaptans, endogenous benzodiazepines, and proinflammatory cytokines. These toxins are responsible for the clinical manifestations of cerebral edema, hepatic encephalopathy, jaundice, pruritis, sepsis, and so on. On the other hand, decreased synthesis of coagulation factors and proteins can cause coagulopathy, ascites, immune disorders, and so on. Cerebral edema leading to raised intracranial pressure and brain herniation is the principal cause of mortality in patients with ALF and is attributed to be secondary to hyperammonemia. Ammonia produced by the urease-producing bacteria in the colon is normally broken down into urea by the liver. In cases of ALF, there is inadequate conversion of ammonia to urea and there is extracellular and intracellular hyperammonemia. Intracellular ammonia is converted to glutamine, in the brain astrocytes, leading to increased intracellular osmolality and brain edema. Similarly, the conversion of extracellular ammonia to glutamate eventually triggers nitric oxide synthase release and cerebral vasodilatation. It has been observed that hyperammonemia precedes brain herniation by 1–2 days, allowing an opportunity for extracorporeal detoxification [34].

Extracorporeal liver assist may be offered to patients as a support to the recovery of injured liver with medical therapy or as a bridge to a more definitive treatment like OLT. Based on the previous discussion, liver assist should essentially involve two components—first, detoxification of blood and second, the replacement of the more important metabolic functions of the liver. Consequently, liver-assist devices can be classified into two broad categories:

1. *Artificial liver (AL):* Involves the detoxification of blood by circulating it extracorporeally against physical/chemical gradient involving albumin or through sorbents.
2. *Bioartificial liver (BAL):* In addition to the detoxification of blood, it involves the circulation of plasma through a hepatocyte-housing bioreactor to replace the metabolic functions of the liver. These are cellular-based techniques for supporting liver function and are in various phases of clinical evaluation. For further information, the reader should consult textbooks on liver diseases.

24.5.2.1 Artificial Liver

The toxins involved in the pathogenesis of liver failure represent both water-soluble (ammonia, uremic toxins) and albumin- or protein-bound toxins (bilirubin, bile acids, aromatic amino acids, free fatty acids, benzodiazepines) [35]. Thus, unlike in renal failure where the toxins are mainly water soluble, modifications in dialysis or other additional techniques to enhance the clearance of protein-bound toxins have to be implemented in patients with ALF. Second, the available AL devices incorporate multiple removal techniques in conjunction, either in series or in parallel, for effective detoxification of blood. The commonly applied techniques of toxin removal include [36]:

1. *Plasmapheresis or plasma exchange*: Separation of plasma by the centrifugation method or via membrane filtration facilitates a nonselective removal of all noncellular components of blood.
2. *Plasma fractionation*: Filtration of plasma through very high-permeability membranes (molecular weight cut-off of >70–100 kDa); removes high-molecular-weight substances.
3. *Hemofiltration*: Filtration through medium- to high-permeability membranes (molecular weight cut-off between 15 and 70 kDa).
4. *Hemodialysis*: Removal of water-soluble, low-molecular-weight substances by diffusion across a low-permeability membrane.
5. *Albumin dialysis or aided transfer*: Albumin in the dialysate or bound to the filter membrane enhances the clearance of albumin-bound toxins.
6. *Adsorption*: Circulating blood over a sorbent material like charcoal, neutral resin, or anion-exchange resin removes protein-bound toxins.

The two commercially available and more extensively used AL devices are the molecular adsorbent recirculating system (MARS; Gambro Lundia, Lund, Sweden) and the Prometheus system (Fresenius Medical Care, Bad Homburg, Germany).

24.5.2.2 Molecular Adsorption Recirculating System

MARS essentially is a modification of a normal dialysis circuit, with the interposition of a closed albumin dialysate circuit between the blood (flowing in dialyzer 1) and the bicarbonate-based dialysate (flowing in dialyzer 2) (Figure 24.5a). The MARS platform is a single pump platform that circulates albumin dialysate in a closed loop. The MARS monitor is mounted onto a hemodialysis or CRRT machine that maintains the blood and bicarbonate dialysate flows. Blood exits the patient via a central vein vascular access and flows through a high-flux albumin-coated polysulfone hemodialyzer (dialyzer 1; membrane thickness 100 nm; pore size 50 kDa; surface area 2.1 m^2). The dialysate side of dialyzer 1 is bathed by the albumin dialysate that is continuously pumped by the MARS monitor. Thus, in dialyzer 1, there is an effective exchange of protein-bound toxins in blood with the albumin in the dialysate and dialysis of water-soluble toxins across the membrane. The toxin-rich or "spent" albumin dialysate leaves dialyzer 1 and then flows across the "blood" compartment of a low-flux polysulfone hollow-fiber dialyzer (dialyzer 2). The dialysate compartment of dialyzer 2 is bathed by the bicarbonate dialysate pumped by the HD or CRRT machine. The water-soluble uremic toxins are dialyzed across the gradient established by the bicarbonate dialysate, which is discarded. The removal of water-soluble toxins in dialyzer 2 constitutes the first step in the regeneration of the "spent" albumin dialysate. The next step in the regeneration of albumin dialysate is by adsorption and involves its flow through an activated charcoal column followed in series by its flow across an anion exchange resin. This regenerated albumin dialysate is now available for the next cycle of flow through the dialysate compartment of dialyzer 1. As is evident, the albumin dialysate flows in a "closed" circuit and its capacity to remove protein-bound toxins in the blood is dependent upon the regenerating ability of the activated charcoal column and anion exchange resin. The typical intermittent MARS treatment session is implemented for 6–8 h with the blood flow and albumin dialysate flow rates set between 150 and 200 mL/min and the bicarbonate buffered dialysate flow rate set between 300 and 500 mL/min. Heparin is used to maintain circuit anticoagulation with a targeted

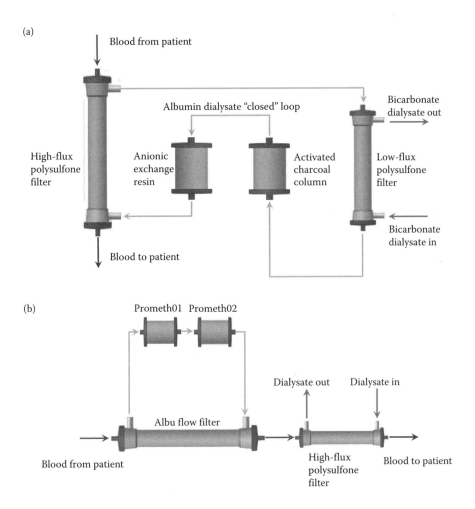

FIGURE 24.5 Schematic diagrams of (a) molecular adsorbent recirculating system (MARS) and (b) Prometheus system.

activated clotting time between 160 and 190 s. In select cases, it is possible to run the circuits heparin free [37]. MARS appears to be a safe procedure though it carries the potential of adverse effects including inherent risks of catheterization for vascular access, mild thrombocytopenia and disseminated intravascular coagulation, and hypoglycemia. MARS also eliminates both water- and protein-bound drugs and may alter their levels in the blood, thereby compromising their clinical efficacy.

24.5.2.3 Fractionated Plasma Separation and Adsorption (Prometheus System)

The Prometheus system differs from the aforementioned MARS in applying albumin-permeable membrane to separate albumin-bound toxins and thereby obviating the need for exogenous albumin (Figure 24.5b). Blood exits the patient via a central vein vascular access and flows through a filter with a high-molecular-weight cut-off polysulfone membrane (AlbuFlow; 250 kDa cut-off; albumin sieving coefficient 0.6). This allows albumin and all albumin-bound substances to be separated from the blood across the membrane and into the dialysate compartment. This separated plasma is subsequently purified by the removal of albumin-bound toxins by flowing it through two adsorbent columns connected in series. The first column, Prometh01, is a neutral resin column and the second column, Prometh02, is an anion exchange resin column. Following this, the "endogenous" albumin is returned to the patient along with the blood. The blood exiting the AlbuFlow flows through a regular high-flux polysulfone dialyzer where the water-soluble toxins are removed by dialysis against bicarbonate buffered dialysate. The Prometheus circuit is integrated on a modified hemodialysis unit (4008H). The blood flow is set at around 200 mL/min, the dialysate flow is at 300–500 mL/min and the flow of separated plasma in the secondary circuit at 300 mL/min). The circuit is anticoagulated with heparin or citrate.

In comparison to MARS, studies have demonstrated that the Prometheus system has better clearance rates and reduction ratios for both water-soluble and albumin-bound substances, except for bile acids [38]. In another study comparing the two procedures, mean arterial pressure and peripheral resistance was improved only in patients treated with MARS [39]. Other than a transient and reversible decrease in mean arterial pressure and white blood cell count [40], treatment with Prometheus was safe.

24.5.2.4 Other Artificial Liver Devices

Besides the two commonly used systems discussed earlier, other techniques are also being evaluated. Single-pass albumin dialysis (SPAD) is performed on a regular dialysis setup using a high-flux albumin-impermeable membrane. The removal of protein-bound substances is facilitated by the addition of albumin to the dialysate side [41,42]. Selective plasma filtration (SEPET) involves the selective removal of albumin-bound and water-soluble substances using a medium pore (cut-off 100 kDa) filter. The "spent" plasma is discarded and replaced by an electrolyte solution, 5% albumin, and fresh frozen plasma [43]. Slow plasma exchange plus high flow dialysate continuous HDF using a filter with PMMA membrane is used in Japan for the management of ALF and has shown promising results [44].

24.5.3 Blood Filtering in Respiratory Diseases

Mechanical ventilation in patients with respiratory failure may be associated with ventilator-induced lung injury. Lung protective ventilator strategies are met with a worrisome consequence of hypercapnia and hypercapnic acidosis. Technological development over the years, in particular, the introduction of microfiber nonmicroporous poly-4-methyl-1-pentelene membrane, has allowed low blood flow extracorporeal carbon dioxide removal (ECCO2R). Unlike extracorporeal membrane oxygenation (ECMO), which depending on how it is implemented has the capability to totally replace cardiac and respiratory function, ECCO2R is used as an adjunct with mechanical ventilation to allow for ultralung protective ventilation in severe respiratory failure. In select cases of hypercapnic respiratory failure, ECCO2R may even successfully obviate the need for mechanical ventilation.

Low blood flow ECCO2R may be established via an arteriovenous or a venovenous route. Arteriovenous CO_2 removal is represented by Novalung (GmbH, Hechingen, Germany), a single-use, high-molecular-weight heparin-coated, ultracompact gas exchange system, consisting of a very low resistance but highly efficient PMP membrane lung (pressure drop of 10 mmHg at blood flow of 2 L/min) [45]. The surface area available for gas exchange is 1.3 m² and O_2 flows can be regulated at 1–15 L/min, based on the patient's $PaCO_2$. Blood is accessed via the femoral artery and returned via the femoral vein (15–21 Fr catheters). Extracorporeal blood flow generated by the arteriovenous pressure gradient of the patient causes a significant amount of AV shunting, between 1 and 2.5 L/min (up to one-quarter of cardiac output). This is continuously monitored by an ultrasound sensor. Therefore, for optimal functioning, a MAP of >70 mmHg (with or without vasopressor support) and a cardiac index >3 L/min/m² is required [46]. The CO_2 removal capacity is 80–200 mL/min (50% of body's CO_2 production) and is proportional to the $PaCO_2$, sweep gas flow, and blood flow [45]. The contraindications for AVCO2R are cardiac failure, shock, heparin-induced thrombocytopenia, and severe peripheral vascular disease.

The Decap/Decapsmart ECCO2R device, operating in venovenous mode, is a modified renal replacement circuit incorporating a neonatal polypropylene membrane lung (0.3 m²), coupled in series with a polysulfone hemofilter (1.35 m²). The blood flow into the membrane is aided by a nonocclusive roller pump (maximum 450 mL/min), whereby CO_2 is eliminated by diffusion against a concentration gradient, created by sweep gas flow of 6–8 L/min of O_2. After exiting the membrane, blood is ultrafiltered by a hemofilter, connected in series, before returning to the body. The ultrafiltrate is recirculated back into the membrane prior to membrane inflow, by another roller pump (0–150 mL/min). The process of recirculation not only enhances the removal of dissolved CO_2 in the plasma water but also dilutes the blood premembrane, which helps one to reduce the anticoagulation need. The hemofilter also contributes to increased resistance within the membrane, thereby reducing chances of bubble formation [47]. Prolung (Estor, Italy) membrane lung is similar to DECAP, but does not have the hemofilter connected in series.

As the majority of CO_2 is transported in blood as bicarbonate, dialytic clearance of CO_2 in the form of bicarbonate is an appropriate option. Earlier experiments have shown that this strategy was capable of removing 26–38 mL of CO_2/100 mL blood flow [48] and the newer hydrophilic polysulfone membranes also remove CO_2 at rates of up to 15% of the CO_2 metabolic production rate [49]. However, respiratory dialysis is limited by the options available to correct the ensuing metabolic acidosis secondary to bicarbonate removal. Sodium hydroxide, trishydroxy-methylaminomethane (THAM), and other organic anions have been tried as bicarbonate replacement, but their use has been hindered by metabolic acidosis, hyperchloremia, hemolysis, and fluid gain [50].

24.5.4 Blood Filtering in Sepsis

The pathophysiology of sepsis is generally accepted as involving proinflammatory and anti-inflammatory mediators. Most of these immune mediators of sepsis are water soluble and belong to the medium-molecular-weight category of substances (5–50 kDa) and thus are amenable to extracorporeal removal by techniques of diffusion, convection, or adsorption. However, given the high generation rates of these mediators, conventional RRT doses may not be able to influence adequate removal. Consequently, various modalities have been attempted to increase the removal of these inflammatory mediators.

1. *Continuous high-flux hemodialysis (CHFD)*: This method uses a highly permeable dialyzer, with blood and dialysate flowing in countercurrent direction. The ultrafiltrate volume is controlled by a pump and thus filtration in the proximal part of the dialyzer is balanced by an equal volume of backfiltration of the ultrafiltrate in the distal part of the dialyzer. This obviates the need for replacement fluid while providing increased convective clearance of middle molecules without compromising urea clearance (Ref. 1, p. 884).
2. *High-volume hemofiltration (HVHF)*: HVHF attempts blood purification by the convective removal of mediators using high ultrafiltration volumes of 50–100 mL/kg/h [51–53]. The application of HVHF is usually done at centers with extensive experience in CRRT. It requires a good vascular access to accommodate high blood flows, bicarbonate buffered replacement fluid to be administered as pre- and postdilution (33–67%) to achieve a best compromise between loss of treatment efficacy and optimization of blood flow, and the use of synthetic large surface area biocompatible high-flux (Kuf 30–40 mL/mm Hg/h) membrane dialyzer. Also, appropriate heating of the replacement fluid to maintain body temperature control is important. To circumvent some of the operational logistics and effects of HVHF, pulse HVHF was proposed as an alternative, wherein pulses of HVHF at an ultrafiltration rate of 85 mL/kg/h is applied for intervals of 6–8 h. This pulse of HVHF is preceded and followed by the implementation of conventional CRRT at doses of 35 mL/kg/h [54,55].
3. *High cut-off hemodialysis or hemofiltration*: This modality involves the performance of hemodialysis or hemofiltration using conventional equipment and doses, but using high cut-off (HCO) membranes. These membranes are highly porous and allow for the removal of substances of molecular weights between 15 and 60 kDa with greater efficiency. This modality has demonstrated better cytokine removal characteristics [56], but expectedly has been associated with increased albumin losses [57].
4. *Hemoadsorption*: This technique utilizes the capability of sorbents to attract substances of varying molecular weights, including substances exceeding molecular weight cut-off of high-flux membranes, and binding with them through hydrophobic interactions, electrostatic attractions,

hydrogen bonds, or van der Waals forces. Modification of pore structure and size can impart selectivity to the adsorbent resins, while an outer biocompatible layer reduces bioincompatibility.

Endotoxemia has been associated with the severity of sepsis secondary to Gram-negative bacteria. Polymyxin B is an antibiotic that is able to compromise the bacterial outer membrane of Gram-negative bacteria and bind lipopolysaccharide, thereby neutralizing its toxic effects. Perfusion of blood through a cartridge in which polymyxin B is immobilized to polystyrene fibers has been used in the management of patients with Gram-negative sepsis. This technique has demonstrated improvement in the hemodynamic parameters and a reduction in vasopressor support; however, no survival advantage could be convincingly demonstrated [58].

5. *Plasma therapy and coupled plasma filtration and adsorption (CPFA)*: Plasma therapy involves plasmapheresis and plasma exchange. In plasmapheresis, the plasma is separated by a centrifugal pump or filtered by a plasma filter, following which it is reprocessed, to facilitate the removal of components of interest, by passing it through adsorbent columns. The reprocessed plasma is then returned to the patient, thus eliminating the need for the replacement of plasma. In contrast, in plasma exchange, the separated or filtered plasma is discarded and is substituted in the body with the replacement fluid (e.g., 5% albumin) or fresh frozen plasma. The process of CPFA, whereby the filtered plasma is passed through a nonspecific adsorbent column placed in series downstream of the plasma filter, helps enhance the nonselective removal of circulating soluble mediators potentially involved in the pathogenesis of sepsis and could improve hemodynamic stability over CRRT [59,60]. Using a monoclonal or polyclonal antibody-coated resin column, coupled plasma filtration immunoadsorption could improve the removal of specific mediators.

24.6 Emerging Technologies for Blood Purification

24.6.1 Hemodialysis Technologies

Researchers have tried to develop a wearable artificial kidney (WAK) with the dream of improving dialysis patients' quality of life. Recently, Gura et al. developed a belt-type wearable kidney, the WAK [61]. Patients were connected to the WAK via their usual vascular access. The WAK regenerates dialysate using REDY® sorbent cartridge with only 375-mL dialysate volume. Here, 0.6-m^2 high-flux hemodialyzer was used and the total system weight was 2.3 kg. This device was powered by two standard 9 V batteries and applied for 4–8 h to eight patients with mean blood and dialysate flow rates of 58.6 and 47.1 mL/min, respectively. However, this device showed some limitations such as clotting of vascular access and needle dislodgment associated with the device's mobility. Inaccurate electrolytes, acid–base control, and ammonium ion accumulation in the sorbent cartridge are also typical problems of REDY cartridge. Further modification has been made by adopting a double-channel pulsatile pump to enhance convective transport. The ammonium accumulation problem was resolved by increasing the pH of the dialysate [62]. Further modifications of the WAK are ongoing with the integration of electrical control and safety system.

24.6.2 Peritoneal Dialysis

In contrast to hemodialysis, because vascular access is not required in peritoneal dialysis (PD), it might be appropriate for a wearable device. Ronco et al. proposed a concept of wearable PD device (ViWAK: Vicenza Wearable Artificial Kidney) as a possible alternative to automated peritoneal dialysis (APD) or continuous ambulatory peritoneal dialysis (CAPD). Based on sorbent cartridge technology, double-lumen PD catheter, and handheld computer for remote control, it performs a continuous-flow PD [63]. However, the system requires the addition of an injection system for glucose and bicarbonate and sorbent for small-molecule removal. Lee and Roberts proposed an automated wearable artificial kidney (AWAK) using modified SORB® cartridge for dialysate regeneration. This device regenerates a PD solution of

4 L/h, which is 8–12 times the current rate. The cycler weighs 0.55 lb and there are two types of sorbent cartridges according to the usage (1.65 lb for 7-h treatment with a flow rate of 2 L/h and 3.75 lb for 12-h treatment with a flow rate of 4 L/h). However, proteinaceous components and the accumulation of fibrin in the spent peritoneal dialysate requires a minifilter in series with sorbent and the packing density of the sorbent should be optimized considering adsorption capacity and flow resistance of the recycling circuit [64]. Currently, the manufacturer is preparing for clinical trials of AWAK on pediatric patients.

24.6.3 Hemofiltration

Gura et al. modified the WAK as a lightweight wearable continuous ambulatory ultrafiltration device consisting of hemofilter, pulsatile pump, and two micropumps to control anticoagulant and ultrafiltration volume [65]. It was applied to six patients with fluid overload for 6 h using a central venous dual-lumen catheter. Blood flow averaged at 116 mL/min, the ultrafiltration rate ranged from 120 to 288 mL/h with about 150 mmol of sodium removed. Recently, Ronco suggested a vest-type wearable device for ultrafiltration, WAKMAN: an easy-to-wear system compatible with daily activities and life allowing a fully ambulatory treatment [66]. It is equipped with an integrated pump-hemofilter unit, which is extremely compact and lightweight, a unit for the control/monitoring system utilizing wireless communication technology, and a rechargeable battery placed in the back. The maximum flow rate of blood pump and UF pump is 50 mL/min and 300 mL/h, respectively, and disposable blood pressure sensor (−50–300 mmHg) and blood leak detector were integrated in the pump-hemofilter unit. This unit can be easily connected to the control unit by means of a single connector. The jacket also includes two waste bags ($0.75 \times 2 = 1.5$ L) that can be easily replaced in case of long-lasting treatments. The wireless remote control system with 2.4" color touch screen and loudspeaker allows the user to monitor the treatment parameters, start or stop pumps, and manage alarm conditions. However, at this moment, this system is a prototype and requires more modification and evaluation in clinical trials in the future. These devices could have a major impact on the quality of life of fluid-overloaded patients with heart failure.

In current clinical practice, the application of dialysis equipment to pediatric patients weighing below 10 kg is of great concern about side effects of extracorporeal therapy. The cardio-renal, pediatric dialysis emergency machine (CARPEDIEM) project was designed to create the basis for the conception of an RRT equipment specifically dedicated to newborns and small infants with a weight range of 2.5–9.9 kg and with a body surface area from 0.15 to 0.5 m² [67]. The miniaturization effort has currently resulted in the development of three extracorporeal circuits with new polysulfone hemofilters with surface areas of 0.075, 0.147, and 0.245 m² and priming volumes of 27.2, 33.5, and 41.5 mL. The blood pump flow rate ranges from 5 to 50 mL/min. The maximum total achievable UF and dialysis/hemofiltration rates range from 5 mL/min with the largest hemofilter, down to 2.5 mL/min for the smallest one. This technical setup is considered to be able to meet the target of small solute clearance of 2 L/h/1.73 m² and/or 25–35 mL/kg/h in patients weighing <10 kg [68]. The CARPEDIEM project is aimed to achieve five short- and medium-term goals: (i) identify optimal prescriptions and technical requirements for neonatal CRRT; (ii) design a dedicated equipment; (iii) manufacture such a machine and make possible the large-scale production of its disposable material; (iv) validate its use in clinical practice; and (v) develop a multicenter trial to define ideal prescription and application of neonatal CRRT. Currently, CARPEDIEM is CE-marked and ready for clinical trial.

24.6.4 Fractionated Plasma Separation and Adsorption

As a project of the Framework Programme (FP7) for research and technological development, a Dutch company is developing ICT-enabled wearable artificial kidney (iNephron) and personal renal care system with a target dimension of $10 \times 6 \times 4$ cm³ and weight less than 2 kg. Based on fractionated plasma separation and adsorption (FPSA), a high-flux filter separates blood from plasma, which is purified by nanostructured sorbents and then returned to the blood; so no dialysate fluid is needed.

24.6.5 Intracorporeal Approaches

As an intracorporeal approach, slow and continuous intracorporeal plasmapheresis (SCIP) was developed for direct intravenous hemofilter filtration, with a special hollow-fiber morphology composed of four layers providing several performance advantages over conventional hollow fibers and showed clinical feasibility for 72 h with an ultrafiltration volume of 3 L in animal experiments [69]. However, it needs to combine with an extracorporeal hemofilter to support congestive heart failure patients.

The concept of implantable artificial kidney (IAK) introduced by Fissel et al. is another paradigm toward a continuously functioning artificial kidney [70]. It supposes the application of MEMS (microelectromechanical system) technology, silicon nanoporous membranes with a highly monodisperse pore size distribution to overcome the limitations of conventional polymer membranes. Humes suggested the concept of renal assist device (RAD) to replace not only filtrative but also metabolic and endocrinologic functions of the kidney [71,72]. It is composed of hemofilter and hollow-fiber cartridges containing human tubular cells derived from donor organs unsuitable for human transplantation and shows promising results in clinical trials in acute kidney failure patients. By integrating with RAD technology to incorporate cellular function, with connection to the iliac vessel and the bladder, IAK is under development.

24.7 Summary

Blood-filtering technologies have supported functions of vital organs by various combinations of hollow-fiber technology and sorbent technology. Diffusion, convection, and adsorption are the main mechanisms of toxin removal. Vascular access is the most complicated issue in clinical practice, which should be resolved in the near future to improve the patient's quality of life with wearable medical devices. Although the development of new polymer membranes and sorbents improves treatment adequacy and biocompatibility, mimicking metabolic and endocrinologic functions of the kidney remains to be resolved. Dialysate quality monitoring and control is another important issue for extracorporeal RRT. Biofeedback control system takes account of physiological signals from the patient to adjust treatment parameters. However, these extracorporeal blood-purification techniques support rather than replace internal organ functions since they cannot serve such a wide range of physiologic needs. In conclusion, for sustainable development of blood-purification technology, efforts of biomedical engineers to integrate new technologies are essential.

References

1. Canaud B, Desmeules S: Vascular access for hemodialysis; in Horl WH, Koch KM, Lindsay RM, Ronco C, Winchester JF (eds): *Replacement of Renal Function by Dialysis*. Dordrecht, Kluwer Academic Publishers, 2004, pp. 203–230.
2. Beathard GA, Posen GA: Initial clinical results with the lifesite hemodialysis access system. *Kidney Int* 2000;58:2221–2227.
3. Canaud B, My H, Morena M, Lamy-Lacavalerie B, Leray-Moragues H, Bosc JY, Flavier JL, Chomel PY, Polaschegg HD, Prosl FR, Megerman J: Dialock: A new vascular access device for extracorporeal renal replacement therapy. Preliminary clinical results. *Nephrol Dial Transplant* 1999;14:692–698.
4. Astor BC, Eustace JA, Powe NR, Klag MJ, Fink NE, Coresh J: Type of vascular access and survival among incident hemodialysis patients: The choices for healthy outcomes in caring for ESRD (choice) study. *J Am Soc Nephrol* 2005;16:1449–1455.
5. Ronco C, Bowry SK, Brendolan A, Crepaldi C, Soffiati G, Fortunato A, Bordoni V, Granziero A, Torsello G, La Greca G: Hemodialyzer: From macro-design to membrane nanostructure; the case of the fx-class of hemodialyzers. *Kidney Int Suppl* 2002:126–142.

6. Ronco C, Brendolan A, Crepaldi C, Rodighiero M, Scabardi M: Blood and dialysate flow distributions in hollow-fiber hemodializers analyzed by computerized helical scanning technique. *J Am Soc Nephrol* 2002;13 Suppl 1:S53–61.

7. Ronco C, Brendolan A, Winchester JF, Golds E, Clemmer J, Polaschegg HD, Muller TE, La Greca G, Levin NW: First clinical experience with an adjunctive hemoperfusion device designed specifically to remove beta(2)-microglobulin in hemodialysis. *Blood Purif* 2001;19:260–263.

8. ANSI/AAMI. Water treatment equipment for hemodialysis applications (rd62:2011). AAMI, Arlington, VA, 2001.

9. Iacovazzi M, Oreste N, Sardelli P, Barrettara B, Grasso S: Extracorporeal carbon dioxide removal for additional pulmonary resection after pneumonectomy. *Minerva Anestesiol* 2012;78:381–384.

10. Santoro A, Mancini E: Blood volume monitoring systems and biofeedback. *Contrib Nephrol* 2002: 233–244.

11. van der Sande FM, Kooman JP, Leunissen KM: Blood temperature monitor: A novel tool in the management of dialysis-induced hypotension. *Contrib Nephrol* 2002:245–253.

12. Pedrini LA, De Cristofaro V, Pagliari B, Filippini M, Ruggiero P: Optimization of convection on hemodiafiltration by transmembrane pressure monitoring and biofeedback. *Contrib Nephrol* 2002:254–259.

13. Langsdorf LJ, Krankel LG, Zydney AL: Effect of blood-membrane interactions on solute clearance during hemodialysis. *ASAIO J* 1993;39:M767–772.

14. Clark WR, Macias WL, Molitoris BA, Wang NH: Membrane adsorption of beta 2-microglobulin: Equilibrium and kinetic characterization. *Kidney Int* 1994;46:1140–1146.

15. Bender H, Pflazel A, Saunders N, Czermak P, Catapano G, Vienken J: Membranes for endotoxin removal from dialysate: Considerations on feasibility of commercial ceramic membranes. *Artif Organs* 2000;24:826–829.

16. Ronco C, Bordoni V, Levin NW: Adsorbents: From basic structure to clinical application. *Contrib Nephrol* 2002:158–164.

17. Kim JC, Kim JH, Sung J, Kim HC, Kang E, Lee SH, Kim JK, Min BG, Ronco C: Effects of arterial port design on blood flow distribution in hemodializers. *Blood Purif* 2009;28:260–267.

18. Grossmann DF, Kopp KF, Frey J: Transport of urea by erythrocytes during haemodialysis. *Proc Eur Dial Transplant Assoc* 1968;4:250–253.

19. Poh CK, Hardy PA, Liao Z, Huang Z, Clark WR, Gao D: Effect of flow baffles on the dialysate flow distribution of hollow-fiber hemodializers: A nonintrusive experimental study using MRI. *J Biomech Eng* 2003;125:481–489.

20. Leypoldt JK, Cheung AK: Characterization of molecular transport in artificial kidneys. *Artif Organs* 1996;20:381–389.

21. Morti SM, Zydney AL: Protein-membrane interactions during hemodialysis: Effects on solute transport. *ASAIO J* 1998;44:319–326.

22. Gulyassy PF, Depner TA: Impaired binding of drugs and endogenous ligands in renal diseases. *Am J Kidney Dis* 1983;2:578–601.

23. Depner TA, Garred L: Solute transport mechanisms in dialysis; in Horl WH, Koch KM, Lindsay RM, Ronco C, Winchester JF (eds): *Replacement of Renal Function by Dialysis*. Dordrecht, Kluwer Academic Publishers, 2004, pp 73–93.

24. Clark WR, Hamburger RJ, Lysaght MJ: Effect of membrane composition and structure on solute removal and biocompatibility in hemodialysis. *Kidney Int* 1999;56:2005–2015.

25. Ronco C: Evolution of hemodiafiltration. *Contrib Nephrol* 2007;158:9–19.

26. Krieter DH, Collins G, Summerton J, Spence E, Moragues HL, Canaud B: Mid-dilution on-line haemodiafiltration in a standard dialyser configuration. *Nephrol Dial Transplant* 2005;20:155–160.

27. Ghezzi PM, Frigato G, Fantini GF, Dutto A, Meinero S, Cento G, Marazzi F, D'Andria V, Grivet V: Theoretical model and first clinical results of the paired filtration-dialysis (pfd). *Life Support Syst* 1983;1 Suppl 1:271–274.

28. Meloni C, Ghezzi PM, Cipriani S, Petroni S, Tozzo C, Tatangelo P, Rossini B, Rossi V, Cecilia A, Casciani CU: Hemodiafiltration with post-dilution reinfusion of the regenerated ultrafiltrate: A new on-line technique. *Clin Nephrol* 2005;63:106–112.

29. Shinzato T, Maeda K: Push/pull hemodiafiltration. *Contrib Nephrol* 2007;158:169–176.

30. von Albertini B: Double high-flux hemodiafiltration. *Contrib Nephrol* 2007;158:161–168.

31. Ronco C, Orlandini G, Brendolan A, Lupi A, La Greca G: Enhancement of convective transport by internal filtration in a modified experimental hemodialyzer: Technical note. *Kidney Int* 1998;54:979–985.

32. Dellanna F, Wuepper A, Baldamus CA: Internal filtration—Advantage in haemodialysis? *Nephrol Dial Transplant* 1996;11 Suppl 2:83–86.

33. Tomo T, Matsuyama M, Nakata T, Kadota J, Toma S, Koga N, Fukui H, Arizono K, Takamiya T, Matsuyama K, Ueyama S, Shiohira Y, Uezu Y, Higa A: Effect of high fiber density ratio polysulfone dialyzer on protein removal. *Blood Purif* 2008;26:347–353.

34. Kramer L, Kodras K: Detoxification as a treatment goal in hepatic failure. *Liver Int* 2011;31 Suppl 3:1–4.

35. Krisper P, Stadlbauer V, Stauber RE: Clearing of toxic substances: Are there differences between the available liver support devices? *Liver Int* 2011;31 Suppl 3:5–8.

36. Carpentier B, Gautier A, Legallais C: Artificial and bioartificial liver devices: Present and future. *Gut* 2009;58:1690–1702.

37. Tan HK, Yang WS, Choong HL, Wong KS: Albumin dialysis without anticoagulation in high-risk patients: An observational study. *Artif Organs* 2012;36:E83–88.

38. Krisper P, Haditsch B, Stauber R, Jung A, Stadlbauer V, Trauner M, Holzer H, Schneditz D: *In vivo* quantification of liver dialysis: Comparison of albumin dialysis and fractionated plasma separation. *J Hepatol* 2005;43:451–457.

39. Laleman W, Wilmer A, Evenepoel P, Elst IV, Zeegers M, Zaman Z, Verslype C, Fevery J, Nevens F: Effect of the molecular adsorbent recirculating system and prometheus devices on systemic haemodynamics and vasoactive agents in patients with acute-on-chronic alcoholic liver failure. *Crit Care* 2006;10:R108.

40. Rifai K, Ernst T, Kretschmer U, Bahr MJ, Schneider A, Hafer C, Haller H, Manns MP, Fliser D: Prometheus—A new extracorporeal system for the treatment of liver failure. *J Hepatol* 2003;39:984–990.

41. Peszynski P, Klammt S, Peters E, Mitzner S, Stange J, Schmidt R: Albumin dialysis: Single pass vs. Recirculation (mars). *Liver* 2002;22 Suppl 2:40–42.

42. Sauer IM, Goetz M, Steffen I, Walter G, Kehr DC, Schwartlander R, Hwang YJ, Pascher A, Gerlach JC, Neuhaus P: *In vitro* comparison of the molecular adsorbent recirculation system (mars) and single-pass albumin dialysis (spad). *Hepatology* 2004;39:1408–1414.

43. Rozga J, Umehara Y, Trofimenko A, Sadahiro T, Demetriou AA: A novel plasma filtration therapy for hepatic failure: Preclinical studies. *Ther Apher Dial* 2006;10:138–144.

44. Nakae H, Yonekawa C, Wada H, Asanuma Y, Sato T, Tanaka H: Effectiveness of combining plasma exchange and continuous hemodiafiltration (combined modality therapy in a parallel circuit) in the treatment of patients with acute hepatic failure. *Ther Apher* 2001;5:471–475.

45. Muller T, Lubnow M, Philipp A, Bein T, Jeron A, Luchner A, Rupprecht L, Reng M, Langgartner J, Wrede CE, Zimmermann M, Birnbaum D, Schmid C, Riegger GA, Pfeifer M: Extracorporeal pumpless interventional lung assist in clinical practice: Determinants of efficacy. *Eur Respir J* 2009;33:551–558.

46. Florchinger B, Philipp A, Klose A, Hilker M, Kobuch R, Rupprecht L, Keyser A, Puhler T, Hirt S, Wiebe K, Muller T, Langgartner J, Lehle K, Schmid C: Pumpless extracorporeal lung assist: A 10-year institutional experience. *Ann Thorac Surg* 2008;86:410–417; discussion 417.

47. Terragni PP, Del Sorbo L, Mascia L, Urbino R, Martin EL, Birocco A, Faggiano C, Quintel M, Gattinoni L, Ranieri VM: Tidal volume lower than 6 ml/kg enhances lung protection: Role of extracorporeal carbon dioxide removal. *Anesthesiology* 2009;111:826–835.

48. Mancini P, 2nd, Whittlesey GC, Song JY, Salley SO, Klein MD: CO_2 removal for ventilatory support: A comparison of dialysis with and without carbonic anhydrase to a hollow fiber lung. *ASAIO Trans* 1990;36:M675–678.

49. Czermak P, Razcuhn B, Walz M, Catapano G: Feasibility of continuous CO_2 removal with hydrophilic membranes at low blood flow rates. *Int J Artif Organs* 2005;28:264–269.

50. Cressoni M, Zanella A, Epp M, Corti I, Patroniti N, Kolobow T, Pesenti A: Decreasing pulmonary ventilation through bicarbonate ultrafiltration: An experimental study. *Crit Care Med* 2009;37:2612–2618.

51. Honore PM, Jamez J, Wauthier M, Lee PA, Dugernier T, Pirenne B, Hanique G, Matson JR: Prospective evaluation of short-term, high-volume isovolemic hemofiltration on the hemodynamic course and outcome in patients with intractable circulatory failure resulting from septic shock. *Crit Care Med* 2000;28:3581–3587.

52. Joannes-Boyau O, Rapaport S, Bazin R, Fleureau C, Janvier G: Impact of high volume hemofiltration on hemodynamic disturbance and outcome during septic shock. *ASAIO J* 2004;50:102–109.

53. Cornejo R, Downey P, Castro R, Romero C, Regueira T, Vega J, Castillo L, Andresen M, Dougnac A, Bugedo G, Hernandez G: High-volume hemofiltration as salvage therapy in severe hyperdynamic septic shock. *Intensive Care Med* 2006;32:713–722.

54. Brendolan A, D'Intini V, Ricci Z, Bonello M, Ratanarat R, Salvatori G, Bordoni V, De Cal M, Andrikos E, Ronco C: Pulse high volume hemofiltration. *Int J Artif Organs* 2004;27:398–403.

55. Ratanarat R, Brendolan A, Piccinni P, Dan M, Salvatori G, Ricci Z, Ronco C: Pulse high-volume hemofiltration for treatment of severe sepsis: Effects on hemodynamics and survival. *Crit Care* 2005;9:R294–302.

56. Uchino S, Bellomo R, Goldsmith D, Davenport P, Cole L, Baldwin I, Panagiotopoulos S, Tipping P: Super high flux hemofiltration: A new technique for cytokine removal. *Intensive Care Med* 2002;28:651–655.

57. Haase M, Bellomo R, Baldwin I, Haase-Fielitz A, Fealy N, Davenport P, Morgera S, Goehl H, Storr M, Boyce N, Neumayer HH: Hemodialysis membrane with a high-molecular-weight cutoff and cytokine levels in sepsis complicated by acute renal failure: A phase 1 randomized trial. *Am J Kidney Dis* 2007;50:296–304.

58. Cruz DN, Antonelli M, Fumagalli R, Foltran F, Brienza N, Donati A, Malcangi V, Petrini F, Volta G, Bobbio Pallavicini FM, Rottoli F, Giunta F, Ronco C: Early use of polymyxin b hemoperfusion in abdominal septic shock: The euphas randomized controlled trial. *J Am Med Sci* 2009;301:2445–2452.

59. Bellomo R, Tetta C, Ronco C: Coupled plasma filtration adsorption. *Intensive Care Med* 2003;29:1222–1228.

60. Formica M, Olivieri C, Livigni S, Cesano G, Vallero A, Maio M, Tetta C: Hemodynamic response to coupled plasmafiltration-adsorption in human septic shock. *Intensive Care Med* 2003;29:703–708.

61. Davenport A, Gura V, Ronco C, Beizai M, Ezon C, Rambod E: A wearable haemodialysis device for patients with end-stage renal failure: A pilot study. *Lancet* 2007;370:2005–2010.

62. Gura V, Macy AS, Beizai M, Ezon C, Golper TA: Technical breakthroughs in the wearable artificial kidney (WAK). *Clin J Am Soc Nephrol* 2009;4:1441–1448.

63. Ronco C, Fecondini L: The vicenza wearable artificial kidney for peritoneal dialysis (viwak pd). *Blood Purif* 2007;25:383–388.

64. Lee DB, Roberts M: A peritoneal-based automated wearable artificial kidney. *Clin Exp Nephrol* 2008;12:171–180.

65. Gura V, Ronco C, Nalesso F, Brendolan A, Beizai M, Ezon C, Davenport A, Rambod E: A wearable hemofilter for continuous ambulatory ultrafiltration. *Kidney Int* 2008;73:497–502.

66. Ronco C: The wakman project for ambulatory treatment of heart failure: World conference on portable-wearable and miniaturized systems for dialysis and ultrafiltration. Vicenza, 2010.

67. Ricci Z, Ronco C: Technical advances in renal replacement therapy. *Semin Dial* 2011;24:138–141.

68. Ronco C, Garzotto F, Ricci Z: Ca.R.Pe.Di.E.M. (cardio-renal pediatric dialysis emergency machine): Evolution of continuous renal replacement therapies in infants. A personal journey. *Pediatr Nephrol* 2012;27:1203–1211.

69. Handley Jr HH, Gorsuch R, Levin NW, Ronco C: Intravenous catheter for intracorporeal plasma filtration. *Blood Purif* 2002;20:61–69.

70. Fissell WH, Roy S: The implantable artificial kidney. *Semin Dial* 2009;22:665–670.

71. Humes HD, MacKay SM, Funke AJ, Buffington DA: Tissue engineering of a bioartificial renal tubule assist device: *In vitro* transport and metabolic characteristics. *Kidney Int* 1999;55:2502–2514.

72. Humes HD, Weitzel WF, Bartlett RH, Swaniker FC, Paganini EP, Luderer JR, Sobota J: Initial clinical results of the bioartificial kidney containing human cells in ICU patients with acute renal failure. *Kidney Int* 2004;66:1578–1588.

VI

Monitoring

IV

25

Point-of-Care Monitoring Devices in Critical–Emergency–Disaster Care

Nam Tran
*University of California,
Davis*

Gerald J. Kost
*University of California,
Davis*

25.1 Introduction

25.1.1 Definition of Point-of-Care Testing

Point-of-care testing (POCT) is defined as medical testing at or near the site of patient care [1]. This definition encompasses devices used for near-patient and bedside testing. POCT devices can be used in a variety of settings, including clinics, intensive care units (ICUs), alternative care facilities, emergency departments, and in the field. *The goal of POCT is to accelerate therapeutic turnaround times, the time from test order to initiation of therapy, and to improve patient outcomes through evidence-based treatment decisions* [1].

POCT plays an important role in medical monitoring for critical, emergency, and disaster care. Medical monitoring provides continuous (e.g., electrocardiogram, pulse oximetry) or serial (e.g., blood glucose testing) assessment of a disease or condition over the course of time. The serial and continuous nature of medical monitoring during life-threatening conditions, especially in acute and resource-limited settings (e.g., in the field or at the site of disaster response) requires rapid analytical turnaround times only attainable by POCT. This chapter summarizes the underlying analytical principles, capabilities, and applications of point-of-care (POC) monitoring devices for critical care, emergency medicine, and disaster applications.

25.1.2 POC Monitoring Device Formats and Test Clusters

25.1.2.1 POCT Format

POC device formats include handheld, portable, transportable, benchtop, and *in vivo/ex vivo* instruments (Figure 25.1) [1]. Handheld devices, by definition, fit in the palm of the hand. These handheld tests may be reusable and require battery power for testing. Some handheld devices are disposable. Disposable devices incorporate low-cost components to facilitate single-use features. Portable devices are larger than handheld devices—up to the size of a laptop computer. These portable devices often serve as a reader for test cards and cassettes. Power is provided through an alternating current (AC) adaptor and/or internal batteries. Transportable devices require a cart to provide mobility. In contrast, benchtop are immobile and must be situated on a bench for POC applications. An external power source is necessary for both testing formats. Lastly, *in vivo* and ex vivo systems provide POC capabilities within and outside of the patient, respectively. Electrical power for *in vivo/ex vivo* systems comes from the attached monitor via an external power source.

25.1.2.2 POC Test Clusters

POC devices can test a variety of analytes, including cardiac biomarkers, electrolytes, hematology, hemostasis, metabolites, and pathogen detection [1–3]. Some analytes are combined into "test clusters" [1] to provide information for a specific disease or condition. For example, cardiac troponin I/T, myoglobin, and creatine kinase myoglobin isoforms (CK-MB) are often clustered together for diagnosing acute myocardial infarction (AMI). Table 25.1 summarizes test clusters found on commercially available POC critical, emergency, and disaster care monitoring devices.

25.2 POCT Analytical Principles

POC monitoring devices incorporate traditional biosensor technology to detect and measure analytes from a variety of clinical sample types such as whole blood, plasma, capillary blood, saliva, and urine

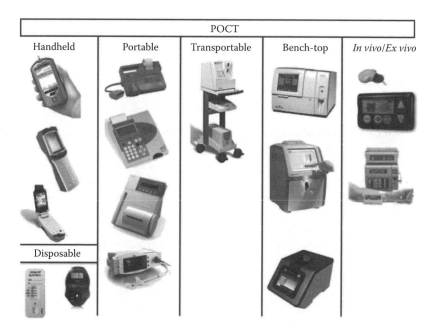

FIGURE 25.1 POCT formats. Visual depiction of commercially available POC tests stratified into the five main types of instrument formats. Handheld devices include disposable tests.

TABLE 25.1 POC Monitoring Devices for Critical, Emergency, and Disaster Care

Manufacturer	Instrument (Format)	Analyte(s)	Methodology(s)	Sample Type(s)
Abaxis www.abaxis.com	Piccolo Blood Analyzer (benchtop)	Albumin, ALP, ALT, amylase, AST, creatinine, Ca^{2+}, Cl^-, K^+, Mg^{2+}, HDL, cholesterol, CK, glucose, total protein, bilirubin, phosphorous, total CO_2, GGT, uric acid, triglycerides	Amperometry Potentiometry	Whole blood, plasma
Abbott www.abbott.com	i-STAT (handheld)	ACT, pH, pCO_2, pO_2, Na^+, K^+, Cl^-, Ca^{2+}, creatinine, PT/INR, CK-MB, BNP, cTnI	Amperometry Potentiometry Immunoassay	Whole blood, plasma
	Freestyle Navigator (handheld)	Continuous glucose	Amperometry	Subcutaneous fluids
	Precision PCx (handheld)	Glucose	Amperometry	Whole blood, capillary
Alere www.alere.com	ePOC (handheld)	pH, pCO_2, pO_2, Na^+, K^+, Ca^{2+}, Glu, Hct	Amperometry Potentiometry	Whole blood
	Triage Meter (portable)	BNP, cTnI, CK-MB, D-dimer, myoglobin, NGAL	Immunoassay	Whole blood, plasma
Bayer www.bayer.com	A1cNow + (handheld, disposable)	HbA1c	Photometry	Whole blood, capillary
Echo Therapeutics www.echotx.com	tCGM (handheld)	Continuous glucose	Amperometry	Transcutaneous
ITC www.itcmed.com	IRMA TruPoint (portable)	pH, pCO_2, pO_2, Na^+, K^+, Cl^-, Ca^{2+}, glucose, BUN, Hct, creatinine	Amperometry Potentiometry	Whole blood, plasma
Masimo www.masimo.com	Radical-7 (portable)	SpO_2, SpHb, SpMet, SpOC, SpCO, pulse rate	Photometry	Noninvasive
	Rad-87 (portable)	SpO_2, SpHb, SpMet, SpOC, SpCO, pulse rate	Photometry	Noninvasive
	Rad-57 (handheld)	SpO_2, SpHb, SpMet, SpOC, SpCO, pulse rate	Photometry	Noninvasive
	Pronto-7 (handheld)	SpO_2, SpHb, pulse rate	Photometry	Noninvasive
	Pronto (handheld)	SpO_2, SpHb, pulse rate	Photometry	Noninvasive
Nonin www.nonin.com	LifeSense (portable)	Continuous $EtCO_2$	Photometry	Noninvasive
	Onyx II (handheld)	SpO_2, pulse rate	Photometry	Noninvasive
	PalmSAT 2500 (portable)	SpO_2, pulse rate	Photometry	Noninvasive
Nova Biomedical www.novabiomedical. com	Critical Care Xpress (benchtop)	Na^+, K^+, Ca^{2+}, Cl^-, Mg^{2+}, pH, pO_2, pCO_2, HCO_3^-, BUN, creatinine, lactate, glucose, Hb, Hct, total bilirubin, co-oximetry	Amperometry Potentiometry Photometry	Whole blood, plasma

continued

TABLE 25.1 (**continued**) POC Monitoring Devices for Critical, Emergency, and Disaster Care

Manufacturer	Instrument (Format)	Analyte(s)	Methodology(s)	Sample Type(s)
	pHO$_x$ Ultra (transportable)	Na$^+$, K$^+$, Ca^{2+}, Cl$^-$, Mg^{2+}, pH, pO$_2$, pCO$_2$, HCO$_3^-$, BUN, creatinine, lactate, glucose, Hb, Hct, total bilirubin, co-oximetry	Amperometry Potentiometry Photometry	Whole blood, plasma
	StatSensor Creatinine (handheld)	Creatinine Hematocrit (autocorrection)	Amperometry	Whole blood, capillary
	StatStrip Glucose (handheld)	Glucose Hematocrit (autocorrection)	Amperometry	Whole blood, capillary
	StatStrip Glu/Ket (handheld)	Glucose Beta-hydroxybutyrate Hematocrit (autocorrection)	Amperometry	Whole blood, capillary
	StatStrip Lactate (handheld)	Lactate Hematocrit (autocorrection)	Amperometry	Whole blood
Response Biomedical www.responsebio.com	RAMP Reader (portable)	cTnI, CK-MB, myoglobin	Immunoassay	Whole blood
Roche Diagnostics www.roche.com	Accu-Chek Inform II (handheld)	Glucose	Amperometry	Whole blood, capillary
	Cardiac Reader (portable)	cTnT, CK-MB, D-dimer, NT-proBNP	Immunoassay	Whole blood
	Cobas h232 (handheld)	cTnT, CK-MB, myoglobin D-dimer, NT-proBNP	Immunoassay	Whole blood

Abbreviations: ALP, alkaline phosphatase; ALT, alanine aminotransferase; aPTT, activated partial thromboplastin time; AST, aspartate aminotransferase; BNP, B-type natriuretic peptide; BUN, blood urea nitrogen; CK, creatine kinase; CK-MB, creatine kinase-myoglobin; cTnI, cardiac troponin I; cTnT, cardiac troponin T; GGT, gamma glutamyl transferase; GOT, glutamic oxaloacetic transaminase; GPT, glutamic pyruvic transaminase; Hb, hemoglobin; HbA1c, hemoglobinA1c; Hct, hematocrit; HDL, high-density lipoprotein; INR, international normalized ratio; Lac, lactate; NGAL, neutrophil gelatinase-associated lipocalin; NT-proBNP, N-terminal pro-BNP; pCO$_2$, partial pressure of carbon dioxide; PDAO, protamine dose assay; pO$_2$, partial pressure of oxygen; PT, prothrombin time; SpCO, pulse oximeter carboxyhemoglobin; SpHb, pulse oximeter hemoglobin; SpMet, pulse oximetry, methemoglobin; SpO$_2$, pulse oximeter oxygen saturation; SpOC, pulse oximetry oxygen content.

[2,3]. These POC biosensors can be classified as amperometric, photometric, and potentiometric. The analytical principles for these three methods are described below.

25.2.1 Amperometric Biosensors

Amperometry is the measurement of a steady-state current in an electrolytic cell. The current is generated by the reduction or oxidation (redox) of the target analyte species [2,3]. An applied potential may or may not be used. Redox occurs through a chemical reaction, resulting in the generation of electrons. The movement of charge (current) from this reaction is stoichiometrically related and proportional to the concentration of the target analyte. Amperometry is used for detecting a variety of analytes, most notably, creatinine, lactate, and glucose. Analyte-specific enzymes are used to catalyze the redox reaction.

25.2.1.1 Creatinine Testing

Creatinine, for example, requires a series of enzymes (i.e., creatinine amidohydrolase, creatinine amidinohydrolase, and sarcosine oxidase) to catalyze the conversion of creatinine and water to generate peroxide, which disassociates to diatomic oxygen, hydrogen cations, and free electrons (Equations 25.1 through 25.4) [4]. The biosensor electrode quantifies the creatinine levels by measuring the generated current.

$$\text{Creatinine} + H_2O \xrightarrow{\text{Creatinine amidohydrolase}} \text{Creatinine} \qquad (25.1)$$

$$\text{Creatinine} + H_2O \xrightarrow{\text{Creatinine amidinohydrolase}} \text{Sarcosine} + \text{Urea} \qquad (25.2)$$

$$\text{Sarcosine} + O_2 + H_2O \xrightarrow{\text{Sarcosine oxidase}} \text{Formaldehyde} + \text{Glycine} + H_2O_2 \qquad (25.3)$$

$$H_2O_2 \rightarrow O_2 + 2H^+ + 2e^- \qquad (25.4)$$

25.2.1.2 Glucose and Lactate Testing

Glucose and lactate testing occurs through a similar redox reaction [2,5,6]. For glucose testing, glucose oxidase (GO) or glucose dehydrogenase (GD) are used to catalyze the redox reaction (Figure 25.2). Both enzymes require carrier molecules [e.g., pyrroloquinolone quinone (PQQ)] or reaction intermediates to shuttle electrons from the enzymatic layer to the electrode for analysis. Figure 25.3 illustrates

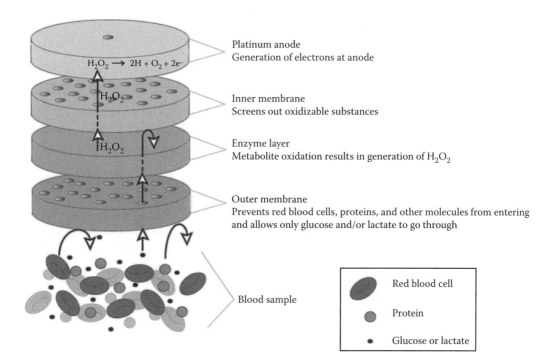

FIGURE 25.2 Glucose biosensor schematic. The figure illustrates a typical glucose oxidase (GO)-based glucose biosensor. The outer sensor membrane serves to filter out unwanted compounds (e.g., red blood cells, proteins) and allowing glucose to diffuse to the GO enzyme layer. The enzyme catalyzes the reduction–oxidation reaction to generate hydrogen peroxide (H_2O_2), which passes through the inner membrane and quickly disassociates to H^+, O_2, and two electrons. The anode measures the electrons. Lactate testing follows a similar scheme, where the enzyme layer uses lactate oxidase.

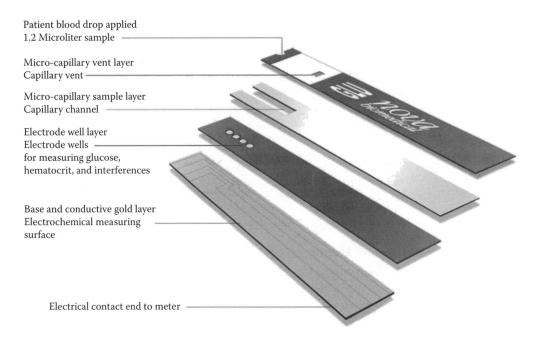

Patient blood drop applied
1.2 Microliter sample

Micro-capillary vent layer
Capillary vent

Micro-capillary sample layer
Capillary channel

Electrode well layer
Electrode wells
for measuring glucose,
hematocrit, and interferences

Base and conductive gold layer
Electrochemical measuring
surface

Electrical contact end to meter

FIGURE 25.3 Hematocrit correcting glucose biosensor. The figure illustrates a modern POC glucose biosensor. Blood is applied to the tip and drawn via capillary action over the electrodes. In this design, four electrodes are used—one for measuring glucose, two for interference compounds, and one reference electrode.

a commercially available POC glucose biosensor. For lactate testing, lactate oxidase (Equations 25.5 and 25.6) is used to catalyze the redox reaction. Lactate dehydrogenase can also be used. The lactate biosensor layout is similar to the glucose biosensor and is illustrated in Figure 25.2 [1–4,7].

$$\text{Lactate} + O_2 \xrightarrow{\text{Lactate oxidase}} \text{Pyruvic acid} + H_2O_2 \qquad (25.5)$$

$$H_2O_2 \rightarrow O_2 + 2H^+ + 2e^- \qquad (25.6)$$

25.2.2 Photometric Biosensors

Photometry relies on optical reflectance, detection, or absorbance to measure an analyte species or biological process [2,8]. For reflectance-based optical systems, the difference in wavelength between the initial incident light and the reflected light is measured for a specific analyte. The change in light is the result of a chemical reaction or biological process. Regardless of the mechanism, a photodetector is used to determine the wavelength to measure the sample. Optical detection methods rely on a detector to detect the presence of light when in the presence of a specific analyte or biological process (e.g., coagulation). Absorbance-based systems follow the Beer–Lambert law where absorbance at a specific wavelength (A) is proportional to the concentration (c) of analyte, for a given path length (l) and extinction coefficient (e) (Equation 25.7).

$$A = elc \qquad (25.7)$$

25.2.2.1 Glucose Monitoring

Reflectance photometry is used in some POC blood glucose meters [7]. Glucose reacts with a dye found on the test strip. As before, an enzyme catalyzes the redox reaction. As the reaction proceeds, the color changes relative to the concentration of glucose found in the sample. The color intensity is measured by photometry and the difference in the wavelength between the initial incident light and reflected light is used to calculate glucose concentration.

25.2.2.2 Pulse Oximetry

Pulse oximetry enables noninvasive oxygen saturation monitoring at the point of care. Oxygen saturation determination by pulse oximetry relies on the Beer–Lambert law and the difference in absorbance of red and infrared (IR) light for deoxyhemoglobin and oxyhemoglobin (Figure 25.4a) [1,2]. Oxyhemoglobin absorbs more IR light compared to red. Conversely, deoxyhemoglobin absorbs more

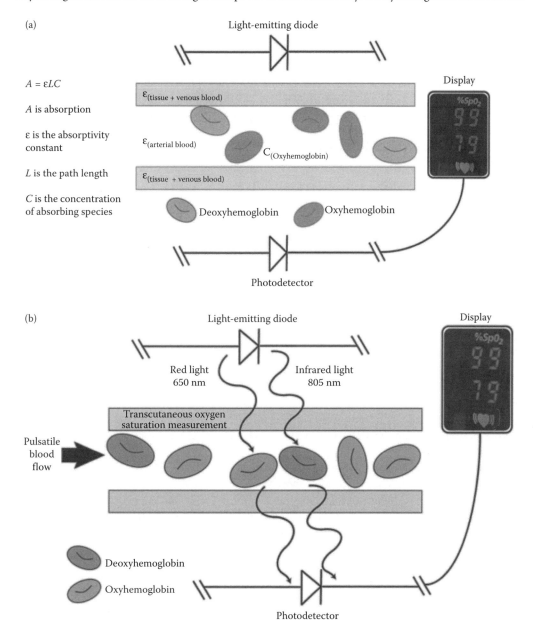

FIGURE 25.4 (a) Illustration of the relationship of the Beer–Lambert law with pulse oximetry. (b) Operating principle of pulse oximetry. Red (650 nm) and infrared light (805 nm) are emitted by a light-emitting diode (LED). Red light is reflected by oxyhemoglobin. IR is reflected by deoxyhemoglobin. The red and IR light detected by the photodetector to generate a oxygen saturation value. Additionally, pulsatile blood flow alters the red/IR ratio over time and is used to calculate pulse rate.

red light compared to IR. The red/IR ratio measured by the pulse oximetry is compared to a blood gas-derived standard curve to derive an oxygen saturation value (Figure 25.4b). Recent pulse oximetry devices (Radical-7, Masimo, Irvine, CA) incorporate additional wavelengths (up to 12) of light to improve accuracy as well as measure other analytes, including carboxyhemoglobin and hemoglobin using the same analytical principles [9,10]. The "pulse" component of pulse oximetry measures the pulsatile blood flow by measuring the increasing and decreasing amounts of absorbance that corresponds to blood volume. Absorbance is treated as a wave function where the resultant wave is used to calculate the changes in blood volume through the skin [2,3].

25.2.3 Potentiometric Biosensors

Potentiometry is used extensively in clinical chemistry for measuring electrolytes [2,3,5]. These potentiometric measurements require the development of an electrical potential difference between a transducer and a reference electrode. Electrodes are connected electrically through the sample in a potentiometric circuit and follow the Nernst equation (Equation 25.8):

$$E_0 = \frac{RT}{nF} \ln K \tag{25.8}$$

where E_0 is the standard potential of the cell, R is the gas law constant, T is temperature, K is the equilibrium constant, F is Faraday's constant, and n is the number of equivalents of electricity.

The reference electrode is responsible for providing a constant half-cell potential while the indicator electrode generates a variable potential, depending on the activity or concentration of an analyte. The electrical potential difference is measured using a high-input impedance voltmeter, thus potentiometry measures ion activity.

25.2.3.1 Electrolyte Testing and Ion Selective Electrodes

H^+, Na^+, K^+, Ca^{2+}, Cl^-, Mg^{2+}, and Li^+ are measured by potentiometry via ion-selective electrodes (ISE) [2,3,5,6]. Ion-selective membranes allow specific electrolytes to be measured in clinical samples. These membranes may be composed of glass or polyvinyl chloride (PVC). For example, electrodes constructed of H^+-sensitive glass facilitate the measurement of sample pH. This selectivity toward H^+ is the result of alkali metal ions on the glass surface generating a potential difference when reacted with hydrogen cations. Within the glass body are two electrodes: (a) the measuring electrode and (b) the reference electrode.

Alternately, PVC electrodes are often used for measuring Na^+, K^+, Ca^{2+}, Cl^-, Mg^{2+}, and Li^{2+}. PVC is less rugged than glass but has proved to be more versatile since PVC is easily adapted to many roles. Traditionally, these PVC membranes are composed of a 1–3 weight percent neutral ionophore or ion exchanger. These ionophores or ion exchangers enable ion selectivity. The selected ions then interact with the electrodes to generate an electrical potential that is measured.

pO_2 biosensors utilize a Clark potentiometric electrode. The Clark electrode consists of a silver (Ag) anode and platinum (Pt) cathode and is covered by an oxygen-permeable membrane [5]. The electrodes sit inside a silver chloride electrolyte solution. The reaction at the anode with the electrolyte generates electrons for the reaction at the cathode (Equations 25.9 and 25.10):

$$\text{Ag anode:} \quad 4\,Ag + 4Cl^- \rightarrow 4\,AgCl + 4e^- \tag{25.9}$$

$$\text{Pt cathode:} \quad O_2 + 4H^+ \rightarrow 4H_2O \tag{25.10}$$

pCO_2 monitoring is an adaptation of potentiometry [5]. These sensors are composed of a CO_2-permeable membrane containing a buffer solution and a pH electrode. The measurement of pCO_2 or total CO_2 depends on the activity of H^+ derived from the reaction between CO_2 and H_2O, and the dissolution of H_2CO_3.

25.2.4 Immunosensors

The detection of cardiac injury markers relies on immunosensors. Immunosensors monitor antigen–antibody interactions, where either an antigen or antibody immobilized on a surface allows for the detection and quantification of the analyte of interest [11,12]. These immunosensors ultimately monitor specific interactions between antigen and antibody by detecting a change in a physical property. Immunosensors can detect analytes directly or indirectly. In direct mode, the immunosensor detects real-time physical changes during immune complex formation. Additionally, direct mode can be performed without antibody labeling, whereas the indirect mode requires signal-generating labels attached to the analytes to provide more sensitive and versatile detection when incorporated into the immune complex. Various labels are used in indirect immunosensors. These include enzymes (e.g., alkaline phosphatase), fluorescent dyes, and fluorescence resonance energy transfer (FRET).

25.3 POC Monitoring Applications

25.3.1 Acute Myocardial Infarction

Heart disease remains the leading cause of death in the United States [13]. Each year, ~1.5 million cases of AMI occur in the United States, where the yearly incidence rate is approximately 600 cases per 100,000 people [14]. Missed diagnosis of AMI is associated with adverse clinical outcomes and expense from malpractice suits compared to any other condition [15]. Rapid identification of patients at high risk for AMI by rapid cardiac biomarker testing improves outcome [16]. In clinical practice, ruling out AMI from in the ED requires electrocardiography and serial cardiac biomarker testing; thus, POC may be advantageous in this clinical setting.

AMI also occurs frequently following disasters [17,18]. Leor et al. [19] reported a modest yet significant increase (35%, from 149 to 201, $P = 0.01$) in the number of emergency department admissions for AMI immediately after the 1994 California Northridge Earthquake. Interestingly, hospitals located ≤15 miles from the earthquake epicenter had significantly ($P = 0.01$) higher AMI admissions than those located ≥15 miles. Similar observations were reported following the 1995 Great Hanshin-Awaji earthquake in Japan [20]. In the end, delays in recognizing AMI adversely affect outcomes in these challenging settings [21]. Disaster response teams are poised to identify AMI early and to quickly initiate appropriate and lifesaving interventions. Highly sensitive quantitative cardiac biomarker assays and serial testing to rapidly rule out of AMI in these austere settings [21].

25.3.2 Renal Function

Acute kidney injury (AKI) is the rapid loss of renal function. The incidence of AKI has increased from 60 to 500 events per 100,000 patients over the last decade and occurs most frequently in critical and emergency care settings [22,23]. Based on the risk, injury, failure, loss, and end-stage (RIFLE) criteria for assessing AKI severity [24], investigators have reported a mortality of 18.9, 36.1, and 46.5% in risk, injury, and failure classes, respectively [25,26]. Crush injuries in disasters place patients at risk for AKI. The development of AKI is the second most common cause of death after large earthquakes and other natural disasters [24].

Chronic kidney disease (CKD) is the slow loss of renal function over time due to a variety of diseases or conditions, including autoimmune disorders, birth defects, certain toxic chemicals, glomerulonephritis, injury/trauma, kidney stones, infections, and vascular disease. The prevalence of CKD in adults in the United States was estimated to be 11% (19.2 million): 3.3% (5.9 million) had stage 1, 3% (5.3 million) had stage 2, 4.3% (7.6 million) had stage 3, 0.2% (400,000) had stage 4, and 0.2% (300,000) had stage 5 [27]. The prevalence of CKD increased from 10% in 1988–1994 to 13.1% in 1999–2004, making it likely that patients CKD will appear in critical, emergency, and disaster care settings. Later studies show the prevalence of chronic renal failure increasing to 104% between the years 1990 and 2001 [28]. As CKD progresses, dialysis becomes necessary for management. Mortality rates associated with hemodialysis are poor with the life expectancy of patients entering into hemodialysis being markedly shortened. In 2003, more than 69,000 dialysis patients died due to end-stage renal disease (ESRD).

Monitoring of serum creatinine and urine output are common methods for assessing renal function, determining AKI severity in the critically ill, and managing CKD [29]. POC whole-blood creatinine monitoring has recently become feasible due to the development of novel four-well amperometric biosensor that automatically corrects for interfering substances to provide accurate measurements (Table 25.1). Neutrophil gelatinase-associated lipocalin (NGAL) is another method for determining renal function. This biomarker is released during renal ischemia and has been shown to be predictive of renal failure in ICU patients. When coupled to B-type natriuretic peptide (BNP), NGAL measurements can be used to evaluate the dysfunction of the cardio-renal axis in both AKI and CKD. Both BNP and NGAL can be detected by immunosensors. Commercial BNP/NGAL POC tests are shown in Table 25.1.

The clinical applications of renal function testing are not limited to only critical and emergency care settings. Novel POC devices such as creatinine and NGAL/BNP enable patient side monitoring in challenging disaster conditions. AKI in disasters can also occur in disasters. Crush injuries in disasters have been reported as recently as the 2011 ChristChurch Earthquake in New Zealand [30], and can progress quickly to severe AKI. The management of crush injuries requires rapid renal function testing to determine risk for AKI and provide life-saving interventions to preserve or restore renal function and mitigate adverse sequlae (e.g., acid–base disturbances).

25.3.3 Diabetes and Glycemic Control

Bedside and home glucose monitoring represents the standard of care for managing diabetics and tight glycemic control (TGC) in ICU patients. Glucose monitoring helps diabetics adjust therapy and diet. Landmark studies conducted in 1978 and 1993 have verified the clinical impact of POC glucose monitoring in diabetic patients [31,32]. These studies codified the use of POC glucose monitoring as a tool for preventing hypoglycemia and adjusting medications, dietary regimens, and physical activity to achieve glycemic goals [33–35].

Alternately, in the ICU, bedside glucose monitoring is used to guide intensive insulin therapy (IIT) to maintain TGC. TGC has been shown by the seminal Leuven [36], to improve outcomes in surgical ICU patients. TGC protocols require frequent (e.g., hourly) or continuous monitoring to adjust insulin infusion rates. Subsequent studies suggest that POC glucose monitoring accuracy may impact glycemic control [37,38]. Inaccurate glucose measurements could lead to increased risk for hypoglycemic events, glycemic variability, and insulin rates [38].

25.3.4 Sepsis

Sepsis is a systemic response to infection. Severe sepsis kills one in four globally and carries a mortality ranging from 28% to 50% [39]. If treated inadequately, sepsis progresses rapidly to include shock and death. Early goal-directed therapy improves outcomes in severe sepsis and septic shock by providing hemodynamic support and anti-infective therapy [40]. Acute fluid resuscitation serves as the primary method for restoring tissue perfusion during severe sepsis and septic shock. Central venous oxygen saturation ($ScvO_2$) measurements have been used to guide fluid resuscitation. More recently, studies show serial lactate measurements guiding acute resuscitation to have comparable outcomes as patients being resuscitated based on $ScvO_2$. In the study by Jones et al. [41] 300 patients presenting with severe sepsis or evidence of hypoperfusion, or septic shock, were enrolled and randomized to receive either $ScvO_2$- versus lactate-directed resuscitation. The study observed no difference in administered treatments, mortality, and adverse events between the $ScvO_2$ and serial lactate groups.

POC lactate monitoring serves as a convenient method for guiding goal-directed resuscitation in severe sepsis and septic shock. POC lactate devices show good correlation when compared to whole-blood analyzers (Radiometer ABL 725, Radiometer Medical A/S, Abbott i-STAT, Nova Lactate Plus) [42]. More recently, POC lactate testing devices incorporate a novel gold biosensor design that autocorrects for interfering substances such as hematocrit, oxygen tension, and oxidizing substances to improve accuracy and precision—making lactate monitoring for sepsis both feasible and potentially efficacious at the point of care.

25.4 Interfering Substances

Interfering substances lead to discrepant and erroneous values [3,8]. These confounding factors affect preanalytical, analytical, and postanalytical steps involved with POCT. Examples of confounding factors include user error, hematocrit, oxidizing substances, sample temperature, and oxygen tension.

25.4.1 Glucose Monitoring

Confounding factors affecting glucose monitoring has been well documented and is illustrated in Figure 25.5. These confounding factors have been attributed to inadequate glycemic control and adverse outcomes in critically ill populations [37,38].

25.4.1.1 Hematocrit

Hematocrit is the proportion of red blood cells relative to plasma. Tang et al. [43] found increases in hematocrit decreased glucose meter readings, and conversely, decreases in hematocrit were found to increase glucose values. Hematocrit interference is due to is diffusion, plasma water content, and mechanical impendence. High hematocrit can decrease the amount of glucose that diffuses into the

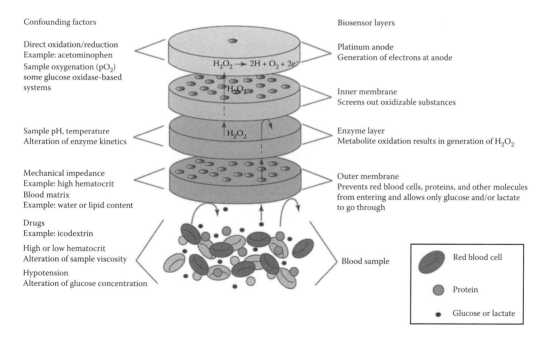

FIGURE 25.5 Glucose biosensor confounding factors. Confounding factors effecting glucose biosensors. These confounding factors include oxidizing and reducing substances, oxygen tension, pH, temperature, hematocrit, water and lipid content, medications, and hypotension.

biosensor. High hematocrit may also mechanically impede the biosensor and reduce glucose readings. Low hematocrit samples facilitate increased diffusion of glucose toward the biosensor, thereby leading to higher readings on the glucose biosensors.

Some manufacturers address the hematocrit problem by prelysing red blood cells before testing (HemoCue) [8] or by compensating for hematocrit effects through software-based autocorrection [44]. Newer systems (Nova StatStrip Glucose, Nova Biomedical, Waltham, MA) now measure hematocrit and automatically correct the glucose measurements to yield results comparable to plasma glucose laboratory methods [44].

25.4.1.2 pO_2 Effects

Critically ill patients may receive oxygen therapy; therefore, blood pO_2 levels may exceed normal levels. Early GO-based systems showed inaccuracies at extreme pO_2 levels [45]. The pO_2 effect was the result of the hydrogen peroxide (H_2O_2) intermediate during the GO redox reaction. Since H_2O_2 spontaneously disassociates into hydrogen cations, diatomic oxygen, and electrons, exogenous oxygen will drive the reaction equilibrium toward the reactants (H_2O_2) based on the law of mass action. As pO_2 increased (>100 mm Hg), the dissociation reaction would become less favorable, thus causing lower glucose readings. Recent GO-based systems (e.g., StatStrip, NOVA Biomedical, Waltham, MA) use a modified GO enzyme to alleviate pO_2 effects and minimize this confounding factor [44].

25.4.1.3 Hypotension and Hypoperfusion

Both GD- and GO-based systems have been shown to significantly differ ($P < 0.001$) when testing capillary whole-blood samples from hypotensive patients in comparison to a reference analyzer [46]. In contrast, venous samples tested on glucose meters did not show statistically significant differences when compared to venous samples tested on laboratory analyzers. These results suggest hypotension and hypoperfusion effecting glucose meter accuracy regardless of analytical process.

25.4.1.4 Sample Source

Capillary blood is commonly used for POC glucose monitoring. Studies show variations in accuracy when comparing capillary versus venous blood samples [47]. More recently, greater accuracy is observed with arterial samples versus capillary blood; however, older glucose meter models exhibit similar performance between the two sample types. Alternately, venous glucose results exhibit improved accuracy compared to capillary blood for GD-based glucose meters.

25.4.1.5 Specimen Matrix

Lipid, proteins, and cellular components may impede photometric systems by increasing the turbidity of the sample. These specimen matrix effects may also alter the viscosity and hence reduce the glucose concentration. A recent study by Lyons et al. [48] suggests plasma water and perhaps red cell water content may also affect glucose meter accuracy. Glucose is highly solubilized in the water phase of blood; thus, changes in blood water content during critical illness will alter the glucose distribution. A correction factor of 1.11 accounts for the plasma water effect to yield a "plasma glucose-like" result from whole-blood glucose measurements [49].

25.4.1.6 Drugs

Oxidizing and reducing drugs may interfere with glucose sensors. Tang et al. [50] showed high doses of ascorbic acid to yield lower glucose values on GO-based systems. Acetaminophen increased glucose values in the tested GD-based systems and on some GO-based systems. High doses of dopamine increased glucose values in GD-based systems while mannitol increased glucose values in GO-based systems. Icodextrin, an osmotic agent used for peritoneal dialysis, has also been shown to generate falsely high results due to the drug being metabolized to maltose [51]. Maltose is indistinguishable from glucose on GD/PQQ-based systems and causing adverse outcomes, including death.

25.4.1.7 pH and Temperature

Since most glucose meters convert glucose into a detectable signal via enzyme-catalyzed reactions, pH and temperature may impact enzyme kinetics and lead to erroneous results. The study by Kilpatrick et al. [52] reported significant deviations in glucose measurements for samples having a pH <6.95 and >7.85. Sample and patient temperature may also adversely affect glucose test results. Oberg et al. [53] showed cold temperatures producing discrepant results; however, the effects of fever are currently unknown instruments.

25.4.1.8 Other Factors

User error and environmental conditions may contribute to inaccuracies. Operator error still exists despite increasingly user-friendly instruments and testing protocols [3]. Environmental conditions have also been shown to significantly affect POC instruments. In a study by Kost et al. [54], conditions identified in posttsunami Thailand, and posthurricane Katrina New Orleans to be unacceptable for routine POC glucose meter use due to the high humidity and temperatures. The effects of abnormal temperature and humidity were quantified in a 2007 study showing that exposure to extreme conditions lead to erroneous quality control results and impact patient care.

25.4.2 Pulse Oximetry

Noninvasive devices are also affected by confounding factors. Pulse oximeters, for example, are susceptible to photointerference and hypoperfusion. These interferences have resulted in pulse oximeter failure rates of 5–50%, and potentially reduce clinical sensitivity and specificity from 0.92 to 0.74 and 0.90 to 0.84, respectively [55,56]. Photointerference occurs in the presence of ambient light, sensor motion, surgical diathermy, and carboxyhemoglobin or methylene blue. Hypoperfusion from cardiovascular disease, hypoxia, sepsis, or respiratory diseases can also lead to aberrant oxygen saturation results.

25.5 Future Trends

In the future, POC monitoring devices will offer additional mobile technology features (e.g., informatics, connectivity, artificial intelligence) including those found on smartdevices (e.g., iPhone, iPad, Android devices, etc.). Recent devices have implemented Bluetooth™ connectivity for data transfer in complex disaster settings [57]. Through the process of miniaturization, detection methods normally found in the laboratory will be translated to the point of care and facilitate highly accurate, sample-processing-free patient monitoring. For example, nuclear magnetic resonance (NMR) measures a physical property of absorbing and releasing electromagnetic radiation by specific atomic nuclei (e.g., protons) when subjected to pulsed radiofrequency while exposed to a magnetic field. The spin–spin relaxation time, or T2-time, serves as a specific measurement of proton localization within a given sample. Since protons are abundant in the water phase of whole blood and other biological specimens, water molecules serve as proximity sensors for multiplex measurements [58,59]. This T2-time magnetic resonance (T2MR) technology has been applied to hemostasis monitoring (T2 Biosystems, Cambridge, MA) and enables the measurement of clotting time, clot strength, fibrinolysis, and platelet function at the bedside—thus, effectively merging four independent clinical tests (prothrombin time, activated partial thromboplastin time, thromboelastography, and platelet function assay) into one point of care device [60]. The T2MR method can also be applied to other diseases, including glucose testing for diabetes, and pathogen detection during sepsis.

25.6 Summary

The continued development of new POC monitoring devices will increase the number of possible applications in critical, emergency, and disaster care settings. Therefore, the following conclusions can be drawn.

- *Clinical applications*: POC monitoring devices have numerous applications in mobile, emergency, critical, and disaster care environments. Both invasive and noninvasive devices provide many benefits; however, limitations should be considered. POCT will allow monitoring devices to become increasingly mobile. Applications at the patient bedside, in rural areas, at disaster locations, and in mobile medicine are the benefits generated by POCT. POC glucose testing can improve outcomes in diabetics and critically ill patients. The fusion of immunoassays and biosensors has enabled the concept of immunosensors. Immunosensors allow the detection of proteins such as markers for cardiac and renal injury. Photometry provides unique detection capabilities and enables noninvasive monitoring of oxygen saturation, co-oximetry, hemoglobin levels, and pulse rate.
- *Future biosensor technologies*: Advances in connectivity and the continued miniaturization of instruments will provide faster, cheaper, intelligent, and less-invasive disposable multifunction biosensors for analytes.

Acknowledgments

This work was supported by the University of California, Davis, Point-of-Care Testing Center for Teaching and Research, and the National Institute of Biomedical Imaging and Bioengineering Point-of-Care Technologies grant (U54 EB007959; G.J.K., principal investigator, National Institutes of Health). This content is solely the responsibility of the authors and does not necessarily represent the official view of the National Institute of Biomedical Imaging and Bioengineering or the National Institutes of Health.

References

1. Kost GJ. Goals, guidelines and principles for point-of-care testing. In: Kost GJ, ed. *Principles and Practice of Point-of-Care Testing*. Philadelphia: Lippincott Williams and Wilkins, 2002: pp. 3–12.
2. Tran NK, Promptmas C, Kost GJ, Biosensors, miniaturization, and noninvasive techniques, In *Clinical Diagnostic Technology: The Total Testing Process, Preanalytical, Analytical, and Post-Analytical Phases*, eds. KW Cook, C Lehmann, L Schoeff, R Williams, AACC, Washington, DC, 145–184, vol. 3 of 3, Chapter 7, 2006.
3. Kost GJ, Tran NK, Louie RF. 2008. Point-of-care testing: Principles, practice, and critical-emergency-disaster medicine. *Encyclopedia of Analytical Chemistry*. Published Online: 29 September 2008.
4. Kost GJ, Vu HT, Inn M et al. Multicenter study of whole-blood creatinine, total carbon dioxide content, and chemistry profiling for laboratory and point-of-care testing in critical care in the United States. *Crit Care Med* 2000;28:2378–2389.
5. D'Orazio P. Biosensors in clinical chemistry. *Clin Chim Acta* 2003;334:41–69.
6. D'Orazio P. Electrochemical sensors: A review of techniques and applications in point-of-care testing. *Point Care* 2004;3:49–59.
7. Kost GJ, Nguyen TH, Tang Z. Whole-blood glucose and lactate. Trilayer biosensors, drug interference, metabolism, and practice guidelines. *Arch Pathol Lab Med* 2000;124:1128–34.
8. Tang Z, Louie RF, Kost GJ. Principles and performance of point-of-care testing instruments. In: *Principles and Practice of Point-of-Care Testing*. Kost GJ, ed. Lippincott Williams and Wilkins, Philadelphia, 2002: pp. 67–92.
9. Kost GJ, Tran NK. Continuous noninvasive hemoglobin monitoring: The standard of care and future impact. *Crit Care Med* 2011;39:2369–2371.
10. Masimo website: www.masimo.com, accessed on July 23, 2012.
11. Luppa PB, Sokoll LJ, Chan DW. Immunosensors—Principles and applications to clinical chemistry. *Clin Chim Acta* 2001;314:1–26.
12. Morgan CL, Newman DJ, Price CP. Immunosensors: Technology and opportunities in laboratory medicine. *Clin Chem* 1996;42:193–209.

13. Center for Disease Control and Prevention Data Statistics. http://www.cdc.gov/DataStatistics/archive/heartdisease.html, Accessed on July 17, 2012.
14. Zafari AM, Yang EH. Myocardial Infarction. Medcape website: http://emedicine.medscape.com/article/155919-overview#a0156, Accessed on July 17, 2012.
15. Schull MJ, Vermeulen MJ, Stukel TA. The risk of missed diagnosis of acute myocardial infarction associated with emergency department volume. *Ann Emerg Med* 2006;48:647–655.
16. Storrow AB, Apple FS, Wu AB, et al. National Academy of Clinical Biochemistry laboratory medicine practice guidelines: Point of care testing, oversight and administration of cardiac biomarkers for acute coronary syndrome. http://www.aacc.org/SiteCollectionDocuments/NACB/LMPG/acute_heart/ACS_PDF_chapter5.pdf, Accessed on July 17, 2012.
17. Miller AC, Arquilla B. Chronic diseases and natural hazards: Impact of disasters on diabetic, renal, and cardiac patients. *Prehospital Disast Med* 2008;23:185–194.
18. Gold LS, Kane LB, Sotodehnia N et al. Disaster events and the risk of sudden cardiac death: A Washington State investigation. *Prehospital Disast Med* 2007;4:313–317.
19. Leor J, Kloner RA. The Northridge earthquake as a trigger for acute myocardial infarction. *Am J Cardiol* 1996;77:1230–1232.
20. Ogawa K, Tsuji I, Shiono K et al. Increased acute myocardial infarction mortality following the 1995 Great Hanshin-Awaji earthquake in Japan. *Int J Epidemiol* 2000;29:449–455.
21. Hamm CW, Bassand J, Agewall S et al. ESC guidelines for the management of acute coronary syndromes in patients presenting without persistent ST-segment elevation. *Eur Heart J* 2011;32:2999–3054.
22. Hsu CY, McCulloch CE, Fan D et al. Community-based incidence of acute renal failure. *Kidney Int* 2007;72:208–212.
23. Waikar SS, Curhan GC, Wald R et al. Declining mortality in patients with acute renal failure, 1988 to 2002. *J Am Soc Nephrol* 2006;17:1143–1150.
24. Bellomo R, Ronco C, Kellum JA et al. Acute renal failure—definition, outcome measures, animal models, fluid therapy and information technology needs: The Second International Consensus Conference of the Acute Dialysis Quality Initiative (ADQI) Group. *Crit Care* 2004;8:R204–R212.
25. Abosaif NY, Tolba YA, Heap M et al. The outcome of acute renal failure in the intensive care unit according to RIFLE: Model application, sensitivity and predictability. *Am J Kidney Dis* 2005;46:1038–1048.
26. Ahlstrom A, Kuitunen A, Peltonen S et al. Comparison of 2 acute renal failure severity scores to general scoring systems in the critically ill. *Am J Kidney Dis* 2006;48:262–268.
27. Center for Disease Control and Prevention Mortality Morbidity Weekly Report 2004; 53: 918–920. http://www.cdc.gov/mmwr/preview/mmwrhtml/mm5339a3.htm, Accessed on July 17, 2012.
28. Center for Disease Control and Prevention Mortality Morbidity Weekly Report 2007; 56: 161–165. http://www.cdc.gov/mmwr/preview/mmwrhtml/mm5608a2.htm, Accessed on July 17, 2012.
29. Hawkins R. New biomarkers for acute kidney injury and the cardio-renal syndrome. *Korean J Lab Med* 2011;31:72–80.
30. Wall Street Journal website: http://blogs.wsj.com/dispatch/2011/02/21/updates-fromthe-new-zealand-earthquake/. Accessed on March 14, 2012.
31. Danowksi TS, Sunder JH. Jet injection of insulin during self-monitoring of blood glucose. *Diabetes Care* 1978;1:27–33.
32. Reichard P, Nilsson B-Y, Rosenqvist U. The effect of long-term intensified insulin treatment on the development of microvascular complications of diabetes mellitus. *N Engl J Med* 1993;329:304–309.
33. American Diabetes Association. Standards of medical care in diabetes—2008. *Diabetes Care* 2008;31(Suppl):S12–S54.
34. AACE Diabetes Mellitus Clinical Practice Guidelines Task Force. American Association of Clinical Endocrinologists medical guidelines for clinical practice for the management of diabetes mellitus. *Endocr Pract* 2007;13(Suppl 1):4–68.
35. Medscape website: http://www.medscape.org/viewarticle/581962, Accessed on July 19, 2012.

36. Van den Berghe G, Wouters P, Weekers F et al. Intensive insulin therapy in critically ill patients. *N Engl J Med* 2001;8:1359–1367.

37. Klonoff DC. Regulatory controversies surround blood glucose monitoring. *J Diabetes Sci Technol* 2010;4:231–235.

38. Godwin Z, Bockhold J, Bomze L, Tran NK. Hematocrit effects leads to inadequate glycemic control and insulin dosing in adult burn patients. 2013 [submitted].

39. Angus DC, Wax RS. Epidemiology of sepsis: An update. *Crit Care Med* 2001;29:S109–S116.

40. Rivers E, Nguyen B, Havstad S et al. Early goal-directed therapy in the treatment of severe sepsis and septic shock. *N Engl J Med* 2001;345:1368–1377.

41. Jones AE, Shapiro NI, Trzeciak S et al. Lactate clearance vs. central venous oxygen saturation as goals of early sepsis therapy: A randomized clinical trial. *JAMA* 2010;303:739–746.

42. Karon BS, Scott R, Burritt MF, Santrach PJ. Comparison of lactate values between point-of-care and central laboratory analyzers. *Am J Clin Pathol.* 2007;128:168–171.

43. Tang Z, Lee JH, Louie RF, Kost GJ. Effects of different hematocrit levels on glucose measurements with handheld meters for point-of-care testing. *Arch Pathol Lab Med* 2000;124:1135–1140.

44. Rao LV. Jakubiak F. Sidwell JS. Winkelman JW. Snyder ML. Accuracy evaluation of a new glucometer with automated hematocrit measurement and correction. *Clin Chim Acta* 2005;356:178–183.

45. Tang Z, Louie RF, Lee JH, Lee DM, Miller EE, Kost GJ. Oxygen effects on glucose meter measurements with glucose dehydrogenase- and oxidase-based test strips for point-of-care testing. *Crit Care Med* 2001;29:1062–1070.

46. Sylvain HF, Pokorny ME, English SM, et al. Accuracy of fingerstick glucose values in shock patients. *Am J Crit Care* 1995;4:44.

47. Kuwa K, Makayama T, Hoshimo T, Tominaga M. Relationships of glucose concentrations in capillary whole blood, venous whole blood and venous plasma. *Clin Chim Acta* 2001;307:187.

48. Lyon ME, Lyon AW. Patient acuity exacerbates discrepancy between whole blood and plasma methods through error in molality to molarity conversion: "Mind the gap!" *Clin Biochem* 2011;44:412–417.

49. D'Orazio P, Burnett RW, Fogh-Anderson N et al. Approved IFCC recommendation on reporting results for blood glucose. *Clin Chem* 2005;51:1573–1576.

50. Tang Z, Du Z, Louie RF, Kost GJ. Effects of drugs on glucose measurements with handheld glucose meters and a portable glucose analyzer. *Am J Clin Pathol* 2000;113:75–86.

51. Perera NJ, Stewart PM, Williams PF. The danger of using inappropriate point-of-care glucose meters in patients on icodextrin dialysis. *Diabetes Med* 2011;28:1272–1276.

52. Kilpatrick ES, Rumley AG, Smith EA. Variations in sample pH and pO_2 affect ExacTech meter glucose measurements. *Diabetes Med* 1994;11:506.

53. Oberg D, Ostenson CG. Performance of glucose dehydrogenase- and glucose oxidase-based blood glucose meters at high altitude and low temperatures (letter). *Diabetes Care* 2005;28:1261.

54. Kost GJ, Tran NK, Tuntideelert M, Kulrattanamaneeporn S, Peungposop N. Katrina, the tsunami, and point-of-care testing: Optimizing rapid response diagnosis in disasters, *Am J Clin Pathol* 2006;126:513.

55. Lee WW, Mayberry K, Crapo R, Jenson RL. The accuracy of pulse oximetry in the emergency department. *Am J Emerg Med* 2000;18:427–431.

56. Trivedi NS, Ghouri AF, Lai E, Shah NK, Barker SJ. Pulse oximeter performance during desaturation and resaturation: A comparison of seven models. *J Clin Anesth* 1997;9:184–188.

57. Kost GJ, Sakaguchi A, Curtis C, Tran NK, Katip P, Louie RF. Enhancing crisis standards of care using innovative point-of-care testing. *Am J Disaster Med* 2011;6:351–368.

58. Lowery TJ, Palazzolo R, Wong SM et al. Single-coil multisample, proton relaxation method for magnetic relaxation switching assays. *Anal Chem* 2008;80:118–123.

59. Haun J, Yoon TJ, Lee H, Weissleder R. Magnetic nanoparticle biosensors. *Wiley Interdiscip Rev Nanomed Nanobiotechnol* 2010;2:291–304.

60. T2 Biosystems website. www.t2bio.com, Accessed on July 23, 2012.

VII

Recovery

26

Neural Prostheses for Movement Restoration

Dejan B. Popović
University of Belgrade

26.1 Introduction

A neural prosthesis (NP) is an assistive system that provides/augments a function that was lost/diminished because of the injury or disease of the nervous system (Popović and Sinkjær, 2000; Peckham 2005). External electrical stimulation is the most frequently used instrument within an NP. More precisely, functional electrical stimulation (FES) delivers bursts of pulses (electrical charge). This charge targets neural systems and elicits activity that is followed by muscle contractions leading to a function. FES could be understood as a pacemaker of movement and a bypass of the missing communications between the central and peripheral sensory–motor systems (Popović et al., 2009).

Figure 26.1 shows the principle of FES-based NP. After an injury or disease (e.g., stroke, spinal cord injury) of the central nervous system (CNS), parts of the sensory–motor systems will remain intact (above the lesion); yet, others will be affected, causing paralysis. The paralysis can be spastic (muscles are innervated but not volitionally controlled) or flaccid (muscles are denervated). Innervated muscles are the best candidates for effective NP application. In parallel, many sensory pathways preserve their connections to the CNS, but their activity does not reach the appropriate centers within the CNS, and exteroception and proprioception are compromised. An NP can also restore some elements of the sensory systems.

26.2 Skeletal Muscles: Actuators

The motor system of the human comprises three interrelated anatomical systems: the skeletal system, the muscle system, which supplies the power to move the skeleton, and the nervous system, which

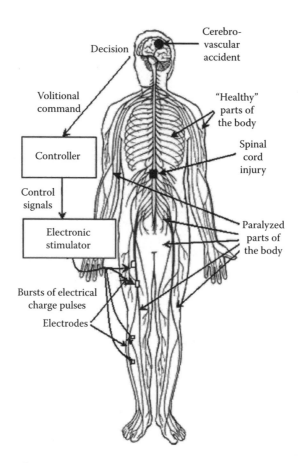

FIGURE 26.1   Schematic of a motor NP. The components of the NP are controller, electronic stimulator, sensory feedback (not shown), and electrodes. NP acts as a bypass for the diminished or missing transmission of neural signals to and from the CNS.

directs and regulates the activity of the muscles. Humans use about 350 pairs of skeletal muscles of many shapes and sizes that are situated across more than 300 joints, being attached to bones via tendons. A shortening of the muscle brings the muscle ends closer to each other and changes the angle at the joint across which it acts.

26.2.1 Activation of Skeletal Muscles

The muscle activation is linked to the activity of neural cells. A single electrical pulse produces a single action potential. Since each action potential is followed by an absolute refractory period, action potentials cannot summate, but remain separate and discrete. The neurons do not conduct impulses at rates as high as the absolute refractory periods would suggest. A fiber with spikes lasting about 0.4 ms might be expected to conduct impulses at a frequency of 2500 per second, but its upper limit will be closer to 1000 per second. Conduction frequencies rarely approximate their possible maxima. Motoneurons usually conduct at frequencies of 20–40, rarely as high as 50, impulses per second, although, at the start of a maximum contraction, rates greater than 100 Hz have been recorded. Upper-limit frequencies for sensory neurons normally lie between 100 and 200 impulses per second although auditory neurons may conduct between 800 and 1000 impulses per second. Information is conveyed by the presence or absence of an action potential, as well as by the frequency of action potentials. Velocity of conduction depends on the diameter of the fiber. It can be said that the conduction velocity is proportional to the diameter of the axon, and it is in the range of 50 m/s. The largest motor and sensory nerve fibers (diameters about 20 μm) have conduction velocities up to 120 m/s. In small unmyelinated fibers, the velocities are from 0.7 to 2 m/s. Large fibers have lower stimulus thresholds compared with small fibers.

26.2.2 Contractile Force

The force developed by a contracting muscle is influenced by several factors: characteristics of the stimulus, the length of the muscle both at the time of stimulation and during the contraction, and the speed at which the muscle is changing its length.

When a single pulse is applied directly to a motoneuron, the corresponding muscle fiber will respond in an all-or-none fashion. Increasing the intensity of the pulse will not increase the magnitude of the response. The all-or-none response of the muscle fiber is determined by the all-or-none character of its excitation and not by any all-or-none limitations inherent in the contractile mechanism itself. When a single adequate pulse is applied to a whole muscle, the muscle will respond with a quick contraction, followed immediately by relaxation. Such a response is called a twitch. Its magnitude will vary with the number of muscle fibers, which respond to the stimulus, and this will vary directly with the intensity of the pulse up to a finite maximal intensity.

The twitch is an indication of force development by the muscle. After a short latent period, the force becomes evident and rises to a peak, it then declines over a slightly longer time course to zero. A single electrical pulse must have a certain minimal intensity to be effective. The minimal effective intensity is called the threshold or minimal stimulus. Subthreshold and subminimal refer to a stimulus of inadequate intensity. As the intensity of the single pulse is increased above the minimum, then the contractile force in the muscle increases progressively as a result of the activation of more and more muscle fibers. Once the intensity is at the level to excite all fibers, the force remains at maximum. The weakest stimulus intensity that evokes maximal contraction is called the maximal stimulus.

A weak but adequate pulse with a rapid rate of rise from zero to its preset intensity will evoke a stronger contraction than a pulse of the same intensity with a slower rise. A minimal rate is required even for an intense stimulus. If intensity rises too gradually, there will be no response at all; the stimulus is then ineffectual. For any stimulus of adequate intensity, the more abruptly it is applied, the greater will be the response it evokes within the limits of the muscle's capacity. The greater the intensity, the less rapidly it needs to produce a given level of response.

The relationship of intensity and duration of single-current pulses in the production of a barely perceptible contraction is presented in the intensity–duration curve (Figure 26.2, left panel). For all points (I–T) right and above the intensity–duration curve, the contraction will occur; yet, there will be no contraction if the I–T point is left and below the I–T curve.

The frequency of stimulation (number of pulses per second) determines the pattern and magnitude of a contraction. If the pulses are delivered within an interval that places successive stimuli during the relaxation phase of the preceding response, then the contraction approaches a tremor. With a period between pulses short enough to allow for restimulation during the contraction phase, the force becomes smooth and is called fused contraction. If stimulation is continued, the contraction peak is maintained at this level. Such a response is known as tetanus or tetanic contraction. The force developed in response to repetitive pulses is greater than that evoked by a single pulse of the same magnitude (Figure 26.2, right panel). When stimulation ceases, contraction terminates, and the fiber relaxes, the force subsides quickly to zero.

The initial length of a muscle, that is, its length at the time of stimulation, influences the magnitude of its contractile response to a given stimulus. The relationship of force to muscle length may be

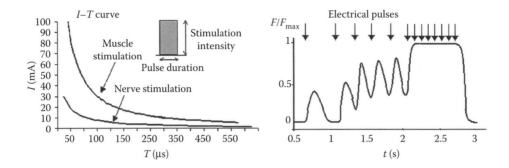

FIGURE 26.2 Left panel: The intensity–pulse duration (*I–T*) curve for direct stimulation of the muscle and stimulation of the muscle via its neural supply. The minimal current required for stimulation is termed the *reobasis*, and the duration that is required to elicit activation when the stimulation pulse has the intensity twice greater than the *reobasis* is the *chronaxia*. Right panel: The effect of the repetition rate (frequency) of stimulation on the force developed in isometric conditions. The first applied stimulus is a single pulse resulting in a twitch; four pulses per second results with increasing, yet pulsating force, and finally 20 pulses per second results with non-pulsatile (fused) muscle contraction. If the repetition rate is further increased, the force will reach the maximum level (tetanus).

presented graphically in the form of a force–length curve, in which force in an isolated muscle is plotted against a series of muscle lengths from less than to greater than the resting length (Figure 26.3, left panel). Both the passive elastic force (curve 1) exerted by the elastic components in the passively stretched muscle and the total force (curve 2) exerted by the actively contracting muscle are typically included.

In concentric contraction, the relationship is evidenced by the decrease in velocity as the load is increased. Shortening velocity is maximal with zero loading and reflects the intrinsic shortening speed of the contractile material. In eccentric contraction, values for shortening velocity become negative, and the muscle's ability to sustain the force increases with increased speed of lengthening, but not to the extent that might be expected from extrapolation of the shortening curve (Figure 26.3, right panel).

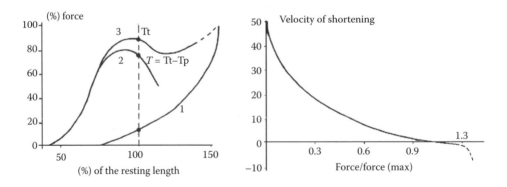

FIGURE 26.3 Left panel: Force versus length curve for an isolated muscle. (1) Passive elastic tension, (2) total force, and (3) force obtained by subtracting of passive force from the total force. These curves suggest that the muscle force depends on the joint angle. Maximum force can be generated at muscle lengths that are shorter than the relaxed length of the muscle. Right panel: Relationship between the normalized muscle force and velocity of shortening of the muscle. Negative velocity relates to eccentric contraction, while the positive velocity relates to active contraction. The faster the shortening, the smaller is the force, and vice versa.

26.2.3 Muscle Function

Movement results from the interaction of muscular force, gravity, and other external forces, which impinge on bones. The muscles rarely act singly; rather, groups of muscles interact in many ways so that the desired movement is accomplished. Whenever a muscle causes movement by shortening, it is functioning as a mover or an agonist. If the observed muscle makes the major contribution to the movement, that muscle is named as the prime mover. The muscles whose actions are opposite to that of a prime mover are called antagonists.

Synergistic action has been defined as a cooperative action of two or more muscles in the production of a desired movement. A synergist is a muscle that cooperates with the prime mover so as to enhance the movement. Two muscles acting together to produce a movement, which neither could produce alone, may be classed as conjoint synergists. The *sine qua non* of an effective coordinate movement involves greater stabilization of the more proximal joints so that the distal segments move effectively. The greater the amount of force to be exerted by the open end of the arm or leg, the greater is the amount of stabilizing force that is needed at the proximal joints.

When a joint is voluntarily fixed rather than stabilized, there is, in addition to immobilization, a rigidity or stiffness resulting from the strong isometric contraction of all muscles crossing that joint. These muscles will forcefully resist all external efforts to move that joint. Economy of movement involves the use of minimal stabilizing synergy (minimum effort).

When a muscle crosses more than one joint, it creates torque at each of the joints crossed whenever it generates force. The torque it exerts at any given instant depends on two factors: the instantaneous length of the moment arm at each joint and the corresponding amount of force that the muscle is exerting.

26.3 Functional Electrical Stimulation

The literature dealing with electrical stimulation frequently uses terms other than FES, such as functional neuromuscular stimulation (FNS), aiming to precisely describe the structures that are activated by electrical stimulation.

FES can be delivered using monopolar or bipolar configuration. In the bipolar configuration, two electrodes are positioned in vicinity of the structure that should be stimulated. In the monopolar configuration, active electrodes (cathodes) are positioned in the vicinity of the structure to be stimulated, while a single common electrode (anode) is positioned distant to the stimulated structure, but along the neural pathway to the CNS.

FES delivers trains of the electrical charge pulses, mimicking to an extent the natural flow of excitation signals generated by the CNS in nonimpaired structures. FES effects can be modeled with a simple electrical circuit: generator, electrodes, and tissue. The tissue is an ionic conductor with an impedance of about 10–100 Ω, and electrodes are capacitive conductors whose electrical properties depend on many variables, but their impedance is from 500 Ω to 5 kΩ, and they induce a phase shift of about 10–30°. The generator is the source of electrical charge and it can work as a current- or voltage-regulated device.

The amplitude and duration of stimulus pulses, output impedance of the generator, and impedance of electrodes determine the electrical charge that will be delivered to the neuromuscular structure. The electrical charge delivered to the stimulated structure depends on the impedance of electrodes to the tissue interface when the stimulators that have a low-output impedance, or the so-called constant-voltage or voltage-regulated devices, are applied. This is the reason to use the current-regulated electronic stimulator so that the consequences of typical impedance changes can be ignored. Figure 26.4 (left panel) shows the patterns of the voltage and current for the current-regulated (a) and voltage-regulated (b) stimulators. Since the electrode–skin interface has electrocapacitive properties, the voltage regulated the stimulation results with uncontrolled electrical current; thus, although the voltage may be substantial, the actual charge delivered to the tissue may be very small. This may result in pain, yet no or very weak

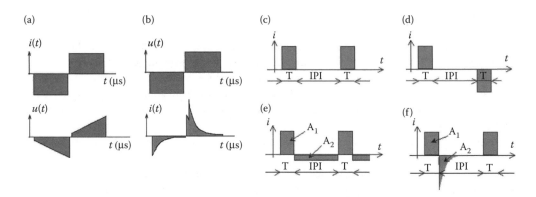

FIGURE 26.4 Left panel: The current- and voltage-controlled stimulation pulses: (a) current-regulated stimulation and (b) voltage-regulated stimulation. The top row shows the output from the stimulator and the bottom row shows the current and voltage applied to the tissue. Right panel: Modalities of stimulation: (c) monophasic, (d) biphasic symmetrical, (e) biphasic balanced, and (f) biphasic compensated pulses. The area A2 (electrical charge) has to be at least 80% of the size of the area A1 for safe stimulation (minimal tissue damage and electrode corrosion). T, pulse duration; IPI, interpulse interval.

muscle contraction. In contrast, current-regulated stimulators precisely control the charge delivered to the tissue.

The stimulus waveform selected for the excitation process must take into consideration the desired physiological effect (action potential generation), potential damage to the tissue, and potential degradation of the electrode (Mortimer, 1981; Scheiner et al., 1990; McCreery et al., 1995). The waveform selected is generally rectangular. A nonrectangular pulse could be utilized, but the rise time must be sufficiently fast so that the nerve membrane does not accommodate and fails to open its channels. The stimulus waveform may be unidirectional (monophasic) or bidirectional (biphasic) as shown in Figure 26.4.

Biphasic stimulus is recommended for several reasons. Surface stimulation is more comfortable with biphasic than monophasic stimulation. For implanted electrodes, the potential for damage to the tissue will be lessened with the biphasic stimulus. Tissue damage is significantly related to the pH change at the electrode–tissue interface (Scheiner et al., 1990). Although reactions at the electrodes are not completely reversed with the biphasic stimulation, this stimulus allows significantly greater charge injection before tissue damage is encountered (Mortimer, 1981). The shape of the secondary pulse is important. One would like the electrode reactions to be totally reversible, suggesting a rapid current reversal in the secondary phase. Balanced charge stimuli are generally used in which equal charge is delivered in each half-cycle.

The surface area of the electrode is important in defining the safety of the stimulation. Large surface areas will diffuse the current and may not affect the excitation desired. A small surface area may result in a high charge density and current density. Although, absolute safe levels of stimulation have not been established for all electrode types and all stimulus waveforms, the values presented by Grill and Mortimer (2000) provide information about the estimated safe levels for neural stimulation. With intramuscular electrodes, safe stimulation is demonstrated at a charge density of $2.0\ \mu C/mm^2$ in the cathodic phase using balanced biphasic monopolar stimulation with a current-regulated stimulus with a capacitively coupled recharge phase (Grill and Mortimer, 2000).

Either amplitude modulation (AM) or pulse width (duration) modulation (PWM) governs the recruitment. A comparison of the method of recruitment modulation to be selected should include consideration of recruitment linearity, charge injected (for safety reasons), and ease of implementation and control of the circuitry that generates the stimulus pulses. This comparison has been made for intramuscular electrodes. The results were that, in consideration of recruitment, the differences between

AM and PWM were small. PWM utilized a slightly lower charge density than AM to evoke a response of equal magnitude, a result that has been theoretically confirmed. Since timing circuits (i.e., regulating pulse width) can be easily constructed and controlled with a resolution of 1 μs or less, many designers of the stimulator use this technique.

As shown in the amplitude–duration (*I–T*) curve (Figure 26.2), relatively short-stimulus rectangular pulses result in the muscle nerve being excited. Much larger charge is required to stimulate the muscle directly. Therefore, FES utilizes short pulses, generally less than 200 μs, resulting in the activation of the neurons.

The threshold, as said earlier, for the excitation of peripheral nerve fibers is proportional to the diameter of the fiber. Since the nerve is composed of a mixture of afferent and efferent fibers with a spectrum of fiber diameters, short pulses of constant amplitude will excite large afferent and efferent fibers. Longer pulses may also excite smaller fibers, including afferents normally carrying information of noxious stimuli, and therefore may be painful to the subject. For this reason and to minimize the electrical charge injection, short pulse duration is preferred.

In a physiological contraction, the recruitment order is fixed; slow, fatigue-resistant motor units are active at a lower voluntary effort than larger, fast, fatigable units. In an electrically induced recruitment, the recruitment order is not known *a priori*, but depends upon the variables of position and geometry as well as the fiber size. An inverse order of electrically induced recruitment is typical when applying FES; the largest fibers are being easily excited, compared with small fibers. The recruitment of nerve fibers with increasing stimulus pulse amplitude or duration is nonlinear as shown in Figure 26.5. For this reason, a linear increase of muscle output force cannot be achieved by a linear change in the input (Crago et al., 1974; Popović et al., 1991).

In most human upper extremity muscles, the fusion occurs at less than 20 pulses per second (pps). Increasing the stimulus frequency above the fusion frequency to the level of tetanus results in a further increase in force (Figure 26.6). Up to 40% or 50% of the maximum muscle force may be regulated by temporal summation from the fusion to the tetanus. Temporal summation leads to temporal modulation being inversely proportional with the frequency (*f*) of stimulation. To grade the force, the muscle temporal modulation of the interpulse interval (IPI = 1/*f*), or a combination of recruitment

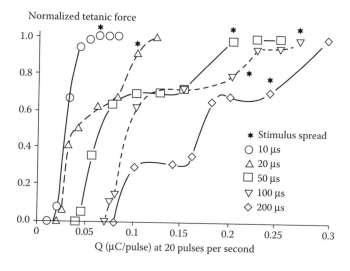

FIGURE 26.5 An example of the recruitment curves (relation between the stimulation strength and the force generated). The data are from the recordings of the force developed in *Medial Gastrocnemius m* of a cat. Shorter pulses required less charge to elicit a force, but the recruitment range was steep. Longer pulses were less effective and show nonlinear output.

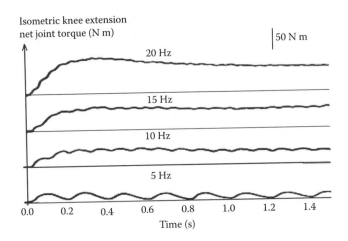

FIGURE 26.6 The isometric knee extension joint torque in a paraplegic subject (spinal cord injury at the level T8) elicited by stimulating *Quadriceps m.* with surface electrodes. The stimulation parameters were pulse duration $T = 300$ μs and current intensity $I = 100$ mA.

and temporal modulation can be selected. The fusion frequency in the muscles of lower extremities is somewhat higher ($\approx$25 pps) compared to the muscles in upper extremities.

26.4 Instrumentation for FES

A functional diagram of the FES system (Figure 26.7) shows the main components of an NP: (1) electrodes, (2) stimulator, (3) sensors, (4) command interface, and (4) controller.

26.4.1 Electrodes

Surface electrodes are placed on the skin surface over the area where the stimulus is to be delivered (Navarro et al., 2005). The criteria for surface electrodes are: even distribution of current, flexibility to maintain good skin contact, ease of application and removal, and suitable mounting for days without irritation of the skin. A surface electrode integrates three elements: the conductor, the interfacial layer, and the adhesive. Usage of the conductive polymer and conductive adhesives proved to be effective for clinical and home usage. Surface electrodes for most applications have a rather large surface area of 5 cm² or more (Figure 26.8, left). More recently, a new concept of surface electrodes has been introduced (Popović-Bijelić et al., 2005; Kuhn et al., 2010). The multipad (matrix) electrodes (Figure 26.8, right panel) allow distributed and asynchronous activation of different motoneurons, thereby allowing

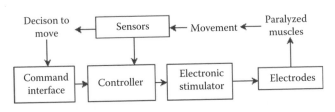

FIGURE 26.7 Components of an FES system: command interface, controller, electronic stimulator, electrodes, and sensors. The decision to move is at the voluntary level of the user. The arrow from sensors to the decision to move indicates biofeedback.

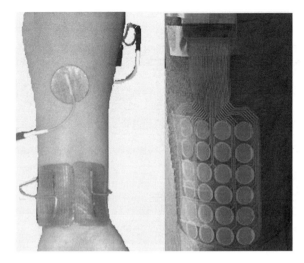

FIGURE 26.8 Surface electrodes made as a multilayer structure. The bottom (skin side) layer is a polymer with strong adhesive properties that sticks to the skin, the middle layer is a conductive layer (metal) connected with the lead, and the top (external) layer is the isolating material. The left panel shows the conventional surface electrode and the right panel shows the multipad surface electrode on the forearm.

selective and less fatigable activation of paralyzed muscles (Popović and Popović, 2009; Malešević et al., 2010).

The stimulus parameters required for the activation by using surface electrodes depend on the stimulus waveform, the surface area, electrode materials, placement, skin impedance, and other factors. Typically, for the rectangular pulse waveform frequently used, threshold values are 30 mA or greater for a pulse width of 100–300 μs. Stimulus pulses shorter than 50 μs cause stronger unpleasant sensation (pain); thus, they are not used. The frequency of stimulation depends on the application but it is typically between 16 and 50 pps.

The primary limitation encountered with surface electrodes is that small muscles generally cannot be selectively activated, and deep muscles cannot be activated without first exciting muscles that are more superficial. Furthermore, fine gradation of force can be difficult because the relative movement between the electrode and the muscle will alter the stimulation–force relationship. The physical movement of the electrode can cause such movement, from length changes in the muscle induced by the contraction process, or from internal change in the nerve–electrode geometry during isometric contraction. Pain is also a limiting factor in applying surface electrodes in subjects with preserved sensations.

Subcutaneous electrodes can be divided into those in which the electrode is secured to the muscle exciting the motoneurons, and those, which are contacting the nerve that contains the motoneurons. The advantages of subcutaneous electrodes over surface electrodes are better selectivity, repeatable excitation, and permanent positioning. The sensation to the users is much more comfortable since the electrodes are placed away from the pain receptors, and the current amplitude is much lower. The potential disadvantage of implanted electrode is the damage that can result from improper design and implantation (e.g., irreversible deleterious effects to the neural tissue, physical failure of an electrode requiring an invasive revision procedure) (Mortimer, 1981; Scheiner et al., 1990).

Subcutaneous electrodes that are secured to the muscle are of two types: the intramuscular electrode (Bowman and Erickson, 1985; Handa et al., 1989a; Smith et al., 1994), which can be injected using a hypodermic needle either nonsurgically through the closed skin or through an open incision, and the epimysial electrode (Grandjean and Mortimer, 1986), which is fixed to the muscle surface and must be placed surgically.

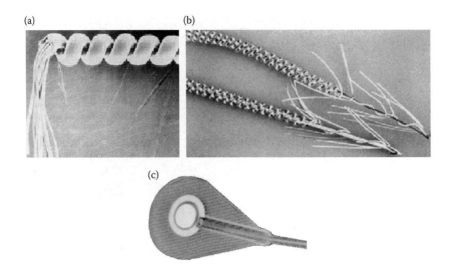

FIGURE 26.9   (a and b) Two versions of intramuscular electrodes. The left panel shows an electrode made of multistranded stainless steel with Teflon insulation. The electrode is coiled, and the tip of the electrode is bared from the insulation. The right panel shows the Peterson-type electrodes with the core (surgical thread) for minimizing the breakage of the electrodes. The multistranded Teflon-insulated stainless-steel wire is rapped around the core. The tip of the electrode shows the wires when the insulation was taken off. (c) An example of the epimysial electrode. The electrode is made out of platinum (disk with the diameter of 0.5 cm) that is welded to the stainless-steel wire. The stainless-steel wire is Teflon insulated and pulled through a silastic tube filled with silastic. The electrode is surgically positioned at the surface of the muscle and sutured to the fascia by using the Dacron-reinforced silastic.

The intramuscular electrode is a helical coil fabricated from a multiple-strand wire (Figure 26.9, left panels). Such a configuration provides a structure, which is able to sustain multiple flexions without fracture. Generally, nonnoble alloys are employed (e.g., type 316 L stainless steel) and the wire insulation is Teflon. A hook formed at the end of the coil anchors the electrode in place.

Intramuscular electrodes generally elicit a maximal muscular contraction with an amplitude of about 20 mA and pulse duration of about 200 µs. The charge is on the order of 10% of the stimulus charge required by surface electrodes. The impedance of the intramuscular electrode is typically 300 Ω, but the entire load impedance of the tissue and surface anode may be as high as 1.5 kΩ.

The epimysial electrode (Grandjean and Mortimer, 1986) is a disk-shaped metal with a reinforced polymer for shielding the surface away from the muscle and for suturing to the muscle (Figure 26.9, right panel). The electrode is surgically placed on the muscle near the motor point. The conductive surface of the disk is 3 mm in diameter. In contrast to the intramuscular electrode, because the placement is surgical and the small size is not so essential, the lead may have a more mechanical redundancy than the intramuscular lead. The impedance and physiological characteristics of this electrode are similar to the intramuscular electrodes.

Nerve electrodes have the potential for producing the most desired physiological response (Navarro et al., 2005). The electrode must be designed with an appreciation for the sensitivity of the nerve to mechanical trauma, manifest by swelling, its longitudinal mobility during muscle movement, and the necessity of maintaining a constant orientation between the nerve fibers and the electrode.

The nerve electrodes are characterized by their placement relative to the nerve: encircling or intraneural. Cuff electrodes (Figure 26.10) come in a variety of configurations; they all have a longitudinal opening to allow installation on the nerve without damaging it. The cuff is formed of a polymer (usually silicone rubber or polyamide). The metal electrodes in the cuff were originally circumferential around the nerve. The encircling cuff electrodes typically require one-tenth the charge required with muscle

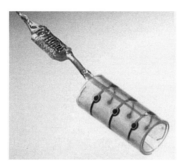

FIGURE 26.10 The self-wrapping multipolar electrode for selective recording/stimulation of the whole nerve. (Courtesy of Thomas Stieglitz, Fraunhofer Institute, St Ingbert, Germany, 2000.)

(intramuscular or epimysial) electrodes, with maximal responses elicited with stimulus amplitudes on the order of 2 mA and pulse durations of 300 µs.

Multipolar cuff electrodes are made to allow selective stimulation, that is, possible steered stimulation of various fascicles within the same nerve (Qi et al., 1999; Navarro et al., 2005).

Intraneural electrodes utilize a conductor that invades the epineurium. Many studies have been done with intraneural electrodes. There is sufficient evidence that maximal contraction is elicited at stimulation levels an order of magnitude lower than with nerve cuff electrodes (amplitude 200 µA, pulse duration 300 µs).

Branner et al. (2001) presented the Utah slanted electrode array (USEA) that was redesigned to be implanted into peripheral nerves (Figure 26.11). Its goal was to provide such an interface that could be useful in rehabilitation. The stimulation capabilities of the USEA were evaluated in acute experiments in cat sciatic nerve. The recruitment properties and the selectivity of stimulation were examined by determining the target muscles excited by stimulation via each of the 100 electrodes in the array and using force transducers to record the force produced in these muscles. It was shown in the results that groups of up to 15 electrodes were inserted into individual fascicles. Stimulation slightly above the threshold was selective to one muscle group for most individual electrodes. At higher currents, coactivation of agonist but not antagonist muscles was observed in some instances.

To maximize the usage of the preserved neural systems after the injury of CNS, it could be effective to stimulate higher CNS structures, rather than to directly activate the last-order motoneurons. There is evidence in humans that the neural circuitry of the spinal cord is capable of generating complex behaviors with coordinated muscle activity (Mushahwar and Horch, 1997, 2000). Two methods are equally

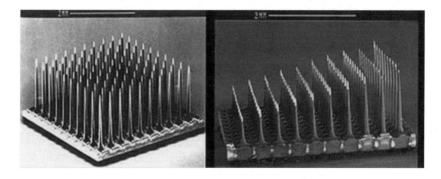

FIGURE 26.11 The Utah array electrode (left panel) and the Utah slant array (right panel) for intraneural stimulation/recording. The electrode is made of silicone with the tips covered with platinum black.

appealing: intraspinal microstimulation enables direct activation of spinal neurons, and afferent input makes synaptic contacts on spinal neurons; thus, they could also be used for the stimulation of the CNS.

26.4.2 Stimulators

An electronic stimulator for FES application has to be a self-contained device with a low power consumption, small, light, and must have the simplest possible user interface (Donaldson, 1986; Smith et al., 1987; Buckett et al., 1988; Minzly et al., 1993; Ilić et al., 1994; Bijak et al., 1999). The stimulator should be programmable, and the programming should be done wirelessly, although the use of wires is acceptable. The stimulator needs a setup mode, the mode of programming when communicating with the host computer. Once the programming is finished, the stimulator should be turned to the autonomous mode.

An electronic stimulator must have the following elements: power source or communication link to a remote unit that delivers energy for operation, DC/DC converter, output stage that generates current-regulated or voltage-regulated pulses, and the controller that defines the shape, intensity, and frequency of pulses. The DC/DC is a system that generates the needed high voltage for stimulation. In some cases, the required voltage must be as high as 300 V; yet in others, it is within 10 V. The output stages are parts that secure that the pulse applied will be effective, yet not harmful.

Implantable stimulators for FES may be separated into single- or multichannel devices. Single-channel implants, which have been fabricated, are all radio frequency (RF)-powered and RF-controlled devices. They use relatively few discrete components and have a receiving antenna, which is integrated into the circuitry. The packaging materials are epoxy or glass–ceramic. The most common single-channel configuration is one in which lead wires are used to place the electrode away from the site of the receiver unit.

Alternatively, the electrodes may be an integral part of the packaging of the circuitry, allowing the entire device to be placed adjacent to the nerve such as it is in the Ljubljana-designed implantable foot-drop system (Strojnik et al., 1987). Two alternative schemes have been considered for multimuscle excitation. It is possible, in principle, to use several one-channel units that are controlled from one controller (Strojnik et al., 1993; Cameron et al., 1997), or to use a single implantable stimulator that will connect with multiple electrodes (Kobetič et al,, 1999; Holle et al., 1984; Peckham, 1988; Rushton, 1990; Scott et al., 1996; Strojnik et al., 1990; Smith et al., 1996; Davis et al., 1997).

The wireless single-channel stimulator has been developed for extensive use in restoring motor functions, and up to now, animal experiments show great promise (Cameron et al., 1997). The original BION (an injectable microstimulator for functional electrical stimulation), single-channel wireless stimulator was sealed with glass beads and uses an anodized tantalum and surface-activated iridium electrodes to minimize tissue damage (Figure 26.12). The current version uses ceramics and titanium packaging and possibly a rechargeable battery.

The diameter of the unit is about 2 mm, and the length of the whole device is less than 20 mm. The BION is powered by inductive coupling from an external coil at 2 MHz. The difficulty of using implantable units without batteries is the low efficiency of energy transmission. To transmit energy, the emitting and receiving antenna must be close and aligned; this being very difficult if a stimulator is injected into a deep muscle.

The use of long leads should be eliminated after the technique of selective nerve stimulation, including potential stimulation of the spinal cord directly, or spinal roots is perfected. The experiments conducted at Aalborg University with the two-channel fully implanted RF-driven stimulators (Haugland et al., 1999) integrated in the cuff electrode demonstrates that this is a viable technique.

An appealing method to use implantable electrodes has been presented by Gan and Prochazka (2010). The new method eliminates the implanted stimulator and wires passing through the skin. Stimulus pulse trains are passed between two surface electrodes placed on the skin. An insulated lead with conductive terminals at each end is implanted inside the body. One terminal is located under the cathode surface electrode and the other is attached to a nerve targeted for stimulation. A fraction (10–15%) of the current flowing between the surface electrodes is routed through the implanted lead. The nerve is stimulated when the amount of routed current is sufficient. In the experiments, stimulation of the nerve innervating the ankle

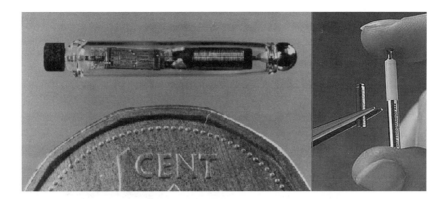

FIGURE 26.12 BION—the microstimulator that integrates electrodes. The electronic circuitry is hermetically sealed with glass beads in the glass capillary tube (left panel). The BION is packaged in ceramics and titanium (right panel).

flexors produced graded force over the full physiological range at amplitudes below threshold for evoking muscle contractions under the surface electrodes. The implants remained stable for over 8 months.

26.4.3 Sensors

Artificial sensors for NP applications should provide both the system and the user information regarding the conditions of the NP. In some cases, it is not obvious that the user does need instant information (e.g., if automatic execution follows the desired trajectory); yet, if anything unexpected is happening, the sensory warning may prevent catastrophic consequences. Sensors are needed in FES systems for the command interface (e.g., activating the NP, changing the mode of operation, etc.).

The sensor system to be used should provide information of various kinds, such as the contact force or pressure over the area of contact (grasping, standing, and walking), the position of the joints (prehension, reaching, standing, and walking), and perhaps the activity of the muscle. The dynamic range, resolution, and frequency response of sensors must be determined upon the application. For example, force sensors for walking and standing must withstand several times body weight under dynamic loading and joint position sensors must allow unrestrained movement over the entire range of motion of the joint.

The constraints imposed on the sensors for FES systems are significant; they must be cosmetically acceptable and easy to mount, they should be self-contained, have low power consumption, and must provide adequate information. In most available FES systems, sensors are placed externally. The sensor positioned at the surface of the body is not a suitable solution for many situations (e.g., an external-force sensor on the digits of the hand requires donning and needs a cable to communicate with the control box, and it should work in variable temperature conditions and hazardous environment). The alternative is to use implanted sensors. They have to meet the same performance specifications while functioning in a more hostile environment. These sensors should communicate with the remote control box and must be powered via RF link.

The use of miniature sensors that use micromachining microelectromechanical systems (MEMS) has been recently introduced for estimating joint angles and angular velocities.

26.4.4 Muscle and Nerve Activity (EMG and ENG) as the Sensory Information

Signals that can be used for NP (Figure 26.13) are the recordings from a muscle, that is, an electromyogram (EMG) and an electroneurogram (ENG) (Sinkjær et al., 2003). The signals are recorded by using electrodes that are positioned over a muscle being able to generate EMG (Frigo et al., 2000). The simplest

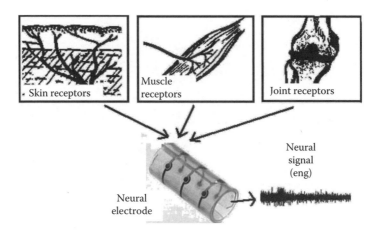

FIGURE 26.13 The natural sensors for neural prostheses. Most of the receptors (skin, muscle, and joint) generate signals although they are not connected to the appropriate centers in the CNS. These signals can be recorded with the adequate electrodes.

processing that can produce a control signal is the amplification, rectification, and bin integration (Saxena et al., 1995). The obtained integrated signal can be compared with the preset value (threshold), and the results can be applied as a switch within a state machine (Saxena et al., 1995). A comparator with hysteresis works good enough eliminating the on–off bouncing because of the nonvoluntary variation of the processed EMG signal.

A laboratory version of the myoelectric control of a grasping system has been developed by the ParaCare group at the ETH, Zurich, Switzerland (Popović et al., 2001). The system controls grasping based on (1) analog EMG control that uses recordings from the deltoid muscle of the ipsilateral arm and (2) discrete EMG control (i.e., digitally processed EMG based on the coded series of integrated, rectified, and compared with a threshold signal). It is possible to stimulate the same muscle that is used for control (Vossius et al., 1987; Thorsen et al., 1999). Hart et al. (1998) used the EMG signals from three surface electrodes (bipolar recordings) placed over the wrist extensor muscles to control the grasping NP. The proportional signals were determined based upon the amplitude of the EMG recordings.

One of the common problems in the recording of the activity from muscles is the stimulation artifact (Knaflitz and Merletti, 1988). As described above, the way around this problem is to apply the circuitry that will blank the response for a period that is longer than the stimulation artifact. However, the stimulation contaminates the natural signals even at the muscle that is remote from the stimulation. The blanking circuitry is sufficient to eliminate the effects that could prevent recording and signal processing; yet, additional filtering can improve the canceling of the noise imposed by stimulation (Thorsen, 1999).

ENG obtained from a cuff electrode (Stein et al., 1975; Hoffer and Loeb, 1980; Hoffer, 1988; Hoffer et al., 1995) is a viable signal for NP (Haugland and Sinkjær, 1995; Upshaw and Sinkjær, 1998; Haugland et al., 1999; Inmann et al., 2001). In the current implants at Aalborg University, Denmark, a telemetric amplifier system that amplifies the ENG recorded from a tripolar nerve cuff electrode and transmits it from the implanted amplifier through the skin to an external control unit was demonstrated to work good enough in both walking and grasping NP. The output is raw ENG, bandpass filtered from 800 to 8000 Hz. The implant is powered by RF induction and will operate for coil-to-coil separations up to 30 mm. Input-referred noise is not more than 1 μV (r.m.s.). The device is small enough to be implanted in the extremities of human adults. The method used for artifact suppression in nerve cuff electrode recordings is intimately related to the method for the extraction of the envelope of the signal as described above for the EMG (Haugland and Hoffer, 1994a, b) and is essentially based on rectification

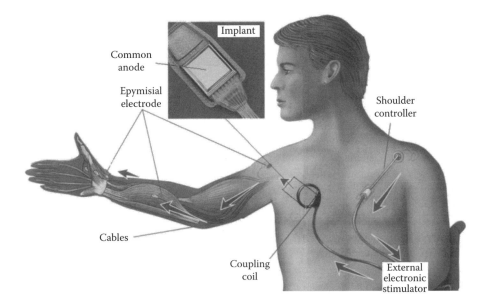

FIGURE 26.14 Drawing of the Freehand system with the nerve electrode for the detection of sliding (feedback signal). The external, microcontroller-based unit generates the control signals that are sent via RF link using the inductive coupling (antenna). The external unit also sends the energy to the implanted unit using the same RF link.

followed by the integration of the signal during noise-free periods (bin integration). Initially, the signal from the amplifier is high-pass filtered to remove the remaining EMG and noise contamination of the nerve signal, which is still present even after passing the filter in the amplifier. The signal is then rectified and bin integrated (Figure 26.14).

An early example of using EMG and ENG for NP in animals demonstrated the feasibility of this approach (Popović et al., 1993; Nikolić et al., 1994), which provided amplification, filtering, and bin integration; thus, real-time use of natural sensors for control was facilitated.

The control signals could also be derived from recordings from the afferent pathways (Yoshida and Horch, 1996). Riso et al. (2000) used a rabbit ankle as a model for the human ankle. They characterized the responses evoked in a pair of complementary mixed nerves (the tibial and peroneal components of the sciatic nerve) that carry muscle afferents from the main ankle extensor and ankle flexor muscles. Simultaneous recordings were obtained using tripolar cuffs installed around the two nerves.

26.4.5 Cortical Recordings (Computer–Brain Interface) as the Command Signal

Recently, the possible use of recordings from higher cortical structures is attracting attention from many researchers. One possible method is to use the focal recordings using the EEG instrumentation (McFarland et al., 1998; Wolpaw et al., 1998; Vaughan et al., 1998). The other, rather invasive technique is much more attractive because it will provide direct access to the structures that are directly delegated for sensory–motor functions. A multielectrode system with 100 needles made out of silicone has been tested for both recording and stimulating brain and peripheral neural structures.

Norman with colleagues tested the system in animal experiments aiming to provide a possible tool for visual prostheses (Normann et al., 1999; Rousche and Normann, 1999). The electrode is only $2.5 \times 2.5 \times 2$ mm³ and provides 100 independent stimulation or recording points. An overview of the current trends can be found in recent review papers by Kennedy et al. (2011) and Lebedev and Nicolelis (2011).

26.4.6 Controller

A controller for NP is a system that has to be programmable to provide the following functions: (1) use data from analog and analog sensors in real time for sensory-driven feed-forward or closed-loop control, (2) use switches for finite state control, and (3) generate signals that define the shape, amplitude, duration, and repetition rate of impulses to be delivered. Today, many stimulation patterns are needed that require sophisticated programming. A controller must produce sufficient number of output signals that accommodates the stimulator number of channels. Today, most microcontrollers can do the simple real-time control; yet, most are not capable of executing complex model-based control algorithms. Digital signal processors can be used, yet their power consumption at this stage is still questionable for portable, self-contained systems, especially implantable systems.

The controller should communicate with the host computer. This communication today can be easily accomplished by infrared communication, RF transmission, or using universal serial bus (USB)- and serial wire-based communication link. The host computer needs an interface that allows the user to efficiently program the stimulation protocol in the sense of the number of channels to be used, the number and type of sensors and switches that will be applied, timing sequences, and special events. The host computer interface must allow the user to potentially control the pulse amplitude and duration, repetition rate, and rise and fall time of the strength of stimulation.

26.5 Neural Prostheses for Restoring Upper and Lower Extremity Functions

During the last 50 years, many NP for the restitution of upper and lower extremity functions have been developed; yet, only few reached the production of even prototype series for clinical studies. The main reason is that the application of motor NP is extremely demanding, and that in many cases, the pathology does not allow sufficient improvement in function to make an NP effective enough. Many neural prostheses are used for therapeutic purposes. The advancements of technology, better understanding of sensory–motor systems, and greatly improved computer power that facilitates the implementation of much more complex control schema raise great hopes for the future.

26.5.1 Neural Prostheses for Reaching and Grasping

NPs for upper limbs are being developed to establish and augment the independence for the user. The target population for many years is tetraplegics with diminished, yet preserved shoulder and elbow functions, lacking the wrist control and grasping ability (Triolo et al., 1996).

Controlling the movement of the whole arm also received attention (Lemay and Grill, 1997; Crago et al., 1998; Grill and Peckham, 1998; Popović and Popović, 1998a, b, 2000, 2001; Smith et al., 1996). The feasibility of restoring the upper extremity function to individuals with high-cervical spinal cord injury (C4 or higher) using FES and/or reconstructive surgery has been evaluated. Externally controlling movement to these individuals is challenging and not very successful for two primary reasons: (1) the number of retained voluntary functions is so low that there is a minimal opportunity to substitute for lost functions or even to use these motions to control external devices and (2) individuals with C3 or C4 level spinal cord injuries exhibit extensive denervation of the shoulder and elbow muscles, which limits the possibility of using FES to restore movement.

The first grasping system used to provide prehension and release (Long and Masciarelli, 1963) had a splint with a spring for closure and electrical stimulation of the thumb extensor for release. Vodovnik and collaborators (Reberšek and Vodovnik, 1973) suggested the use of a simple two-channel stimulation system and a position transducer. The follow-up of the initial FES system use was systematically continued in Japan (Handa et al., 1989b). Subjects have been implanted with up to 30 percutaneous intramuscular electrodes per extremity. These electrodes were connected to a 30-channel device that delivers a preprogrammed sequence of stimulation pulses. The preprogrammed sequence was prepared based on the activity of muscles recorded in able-bodied subjects.

A different method has been developed at Ben Gurion University, Israel (Nathan, 1989). Twelve bipolar stimulation channels and a splint have been used to control the elbow, wrist, and hand functions (Nathan and Ohry, 1990). Prochazka et al. (1997) suggested the use of wrist position to control the stimulation of muscles to enhance the tenodesis grasping, and designed a device, the so-called Bionic glove. A sensor is used to detect the wrist movement, and trigger the opening and closing of the hand. The user volitionally triggers the opening of the hand by flexing the wrist for an angle that is bigger than a preselected threshold value. The closing is initiated by extending the wrist to an angle that is bigger than a preselected threshold. A dead zone (hysteresis) allows movement of the wrist once the "open" or "close" stimulation pattern is activated. In the clinical evaluation (Popović et al., 1999), it was indicated that the stimulation is beneficial to tetraplegic subjects (therapy effects and orthotic assistance), but that the overall acceptance rate for long-time use is at about 30% of potential users. One of the conclusions was that the control, donning, and doffing should be improved, as well as its cosmetic appearance.

The Belgrade grasping/reaching system (BGS) is a four-channel system (Ilić et al., 1994) allowing two modalities of grasping: side and palm grasps by generating opposition and control of the elbow movement. A preprogrammed sequence is triggered using a switch interface. The BGS also comprises a reaching controller. The initial version of the system used the so-called scaling law; the new version uses synergistic control based on mapping of the angular velocities at the shoulder and elbow joints (Popović and Popović, 1998a, b, 2001). The advanced version of the system has been developed and is known by the name UNAFET 4 and UNAFET8 (UNA Sistemi, 2012). The system that provides excellent operation is now available at the market as the Bioness H200 (Ness H200, 2012).

The most advanced, and by all standards the most promising grasping NP follows from more than 20 years of dedicated work by Peckham and coworkers from Case Western Reserve University (CWRU), Cleveland, Ohio (Peckham et al., 1980a, b). A fully implantable stimulation device is approved for human use under the name Freehand system, distributed by the NeuroControl Corp., Cleveland, Ohio. The system has an external unit (control box, power source) and an implantable part (remotely powered and controlled stimulator, leads with epimysial electrodes, and a sensory system) (Smith et al., 1987).

The objective of this system is to provide grasp and release for individuals with C5 and C6 level spinal cord injuries. Coordinated electrical stimulation of the paralyzed forearm and hand muscles is used to provide lateral (key pinch) and palmar grasp patterns. The subjects obtain proportional control of grasp opening and closing by voluntary movement of either the shoulder or the wrist. An external transducer is mounted on the chest to measure the shoulder motion, or on the dorsum of the wrist to measure the wrist flexion/extension. The novel version of the wrist sensor can be implanted (Kilgore et al., 1997; Hart et al., 1998). The control signal is sent to an external control unit, which converts the signal into the appropriate stimulation signals for each electrode. These signals are sent across an inductive link to an implanted stimulator receiver, which generates the stimulus to the appropriate electrode. Seven epimysial electrodes, sewn onto the muscle surface through surgical exposure, are used for muscle excitation. Sensory feedback regarding the control state is provided through an eighth implanted electrode placed in an area of normal sensation.

The electrode leads are tunneled subcutaneously to the implanted stimulator located in the pectoral region. The surgical procedures to enhance both voluntary and stimulated hand functions are often performed in conjunction with the stimulator implantation. More than 200 tetraplegics have received the Freehand NP at more than a dozen sites around the world. The subjects have demonstrated the ability to grasp and release objects and to perform activities of daily living more independently when using the NP. The subjects utilize the device at home on a regular basis.

The initial implementation of the advanced stimulator/telemeter provides 10 channels of stimulation and one implanted joint angle transducer. The stimulator/telemeter unit has been implanted in several individuals. Clinical tests of the implanted joint angle transducer began in 1997. The transducer has been implanted in the wrist, allowing the individual to control grasp opening and closing through voluntary movement of the wrist.

The Freehand system utilizes the individually tuned cocontraction map to provide palmar and lateral grasp. The user voluntarily selects between these two grasps and controls the prehension proportionally. The additional feature of the system is the ability to "lock" the stimulation pattern at any of the preprogrammed combinations. Visual feedback and experience gained through the usage help the subject to perform daily activities (Wijman et al., 1990). The joystick mounted on the contralateral shoulder or some other position at the body controls the preprogrammed sequence of stimulation. The palmar grasp starts from the extended fingers and thumb (one end position of the joystick), followed by the movement of the thumb to opposition and flexing of the fingers (other terminal position of the joystick). The lateral grasp starts from the full extension of the fingers and the thumb, followed by the flexion of the fingers, and adduction of the thumb. The system is applicable if the following muscles can be stimulated: *Extensor Pollicis Longus m*, *Flexor Pollicis Longus m*, *Adductor Pollicis m*, *Opponens m*, *Flexor Digitorum Profundus m.* and *Superficialis m*, and *Extensor Digitorum Communis m.* An important feature of the grasping system is related to daily fitting of the joystick (zeroing the neutral position), and going to hold mode from the movement mode. The hold mode is the regime where the muscle nerves are stimulated at the level at which the same force is maintained, and the user selects the level.

26.5.2 Neural Prostheses for Restoring Standing and Walking

NP was first used for walking assistance in Ljubljana, Slovenia (Gračanin et al., 1967, 1969; Vodovnik et al., 1967, 1981; Kralj et al., 1980; Bajd et al., 1982, 1983, 1989; Kralj and Bajd, 1989). The available surface NP systems use various numbers of stimulation channels.

The simplest NP to assist walking, from a technical point, is a single-channel system. This system is only suitable for stroke subjects and a limited group of incomplete spinal cord-injured subjects. These individuals can perform limited ambulating with the assistance of the upper extremities without an FES system, although this ambulating may be modified and/or impaired. The FES in these humans is used to activate a single muscle group. The first demonstrated application of this technique was in stroke subjects (Gračanin et al., 1969), even though the original patent came from Liberson et al. (1961). The stimulation is applied to ankle dorsiflexors so that the "foot-drop" can be eliminated.

The Odstock-dropped foot stimulator (ODFS) (Odstock, 2012) is a technology that follows the work of Liberson et al. (1961). It is a single-channel stimulator providing electrical stimulation to the *Tibialis Anterior m.* or the common peroneal nerve (Granat et al., 1996; Maležič et al., 1984; Stein et al., 1993; Sweeney and Lyons, 1999; Wieler et al., 1999). The stimulation, timed to the gait cycle by using a switch placed in the shoe, causes ankle dorsiflexion with some eversion and/or could elicit a flexor withdrawal reflex (Burridge et al., 1997a, b). The reflex comprises ankle dorsiflexion, hip, and knee flexion, with external rotation of the hip. Adjusting the electrode position and stimulation amplitude may vary the components of the movement. The ODFS stimulator gives an asymmetrical biphasic output of maximum amplitude 80 mA, with a 300 μs pulse duration, and an IPI of 40 ms. Stimulation is applied by means of skin-surface electrodes placed, typically, over the common peroneal nerve as it passes over the head of the fibula bone and over the motor point of *Tibialis Anterior m.* The output of the stimulator is normally triggered after the heel rises from the ground on the affected side and continues until the heel strike occurs. Stimulation can also be triggered by the heel strike from the contralateral leg, which is a useful function if the heel contact is unreliable on the affected side. It may also be possible to stimulate for a fixed time, again useful if the heel contact was inconsistent. The device with advanced technology has been introduced by Bioness (Ness L300, 2012).

Stein and colleagues (Dai et al., 1996) have designed an appealing self-contained system. This system was translated into a commercial device known as Walkaide (Walkaide, 2012). The system integrates a single-channel stimulator and a tilt sensor; thus, it eliminates a foot switch, which was not fully reliable, because it easily generated false triggering and malfunctioning. The dual-channel stimulator is used to correct eversion/inversion of the foot in addition to the dorsiflexion (Hansen, 2001). This approach was translated into a commercial product named Actigait (2012).

A multichannel system with a minimum of four channels of FES is required for ambulating of a subject with a complete motor lesion of lower extremities and preserved balance and upper body motor control (Kralj and Bajd, 1989). Appropriate bilateral stimulation of the quadriceps muscles locks the knees during standing (Bajd et al., 1982). Stimulating the common peroneal nerve while switching off the quadriceps stimulates a flexion of the leg (swing phase of the gait). Sufficient arm strength must be available to provide balance in parallel bars (clinical application), and with a rolling walker or crutches (daily use of FES). These systems evolved into a commercial product called Parastep-1R (Graupe and Kohn, 1997; Jacobs et al., 1997; Klose et al., 1997; Parastep, 2012).

Multichannel percutaneous systems for gait restoration with many channels were suggested (Marsolais and Kobetič, 1983, 1987; Kobetič and Marsolais, 1994; Kobetič et al., 1997). The main advantage of these systems is the plausibility to activate many different muscle groups. A preprogrammed stimulation sequence is used for timing and intensity variation of up to 16 channels. The sequence is cloned from the EMG pattern recorded in able-bodies subjects. A 16-channel stimulator has been applied over as much as 64 percutaneous electrodes. Some electrodes are stimulating at very low-intensity muscles that are still voluntarily controlled in the abdominal region to provide proprioceptive feedback to the user about the operation of the system. The experience of the Cleveland research team suggested that 48 channels are required for a human with a complete spinal cord injury at the mid-thoracic level to achieve a reasonable walking pattern. Initially, the method used excluded external bracing, and claimed that FES per se will be adequate, but this method has been changed toward combining the stimulation with some orthotics. Fine-wire intramuscular electrodes were positioned by using subdermal needles, and were tunneled toward a point where an external connector from several electrodes is attached (externally). These intramuscular electrodes serve as the cathodes, positioned close to the motor point of selected muscles. Knee extensors (*Rectus Femoris m, Vastus Medialis m, Vastus Lateralis m*, and *Vastus Intermedius m.*), hip flexors (*Sartorius m, Tensor Fasciae Latae m, Gracilis m*, and *Iliopsoas m.*), hip extensors (*Semimembranosus m, Gluteus Maximus m.*), hip abductors (*Gluteus Medius m.*), ankle dorsiflexors (*Tibialis Anterior m, Peroneus Longus m.*), ankle plantar flexors (*Gastrocnemius Lateralis m, Gastrocnemius Medialis m, Plantaris m*, and *Soleus m.*), and paraspinal muscles are selected for activation. A surface electrode has been used as a common anode. Interleaved pulses are delivered with a multichannel, battery-operated, portable stimulator. The hand controller allows for the selection of the walking modality. The application was investigated in complete spinal cord lesions and in stroke subjects. The same strategy and selection criteria for implantation were used for both stroke and spinal cord injury (SCI) subjects. This system allowed selected humans to walk up to 1.1 m/s, and walk distances of almost 1000 m. The preprogrammed stimulation sequence included the following walking modes: standing up, sitting down, quiet standing, walking, walking stairs or curb, and walking backwards. Recent research somewhat changed the initial emphasis of research with implanted systems toward therapeutic applications (Dutta et al., 2009, Mushahwar et al., 2007).

26.5.3 Hybrid Assistive Systems

To minimize problems due to muscle fatigue and to increase safety, a combination of a mechanical orthosis and FES was suggested. The resulting orthosis is called hybrid assistive system or hybrid orthosis (Popović et al., 1989; Tomović et al., 1995). The support, stability of the joints, and constraint to unwanted motion of the joints are provided by the mechanical component of the orthosis, while FES provides propulsion. Schwirtlich and Popović (1984) suggested a hybrid orthosis, which consisted of the self fitting modular orthosis (SFMO) and surface electrodes FES to provide the knee extension and swing of the leg during walking (Schwirtlich and Popović, 1984).

Andrews and Bajd (1984) suggested two variants of hybrid orthoses. One consisted of a combination of a pair of simple plastic splints used to maintain the knee extension and a two-channel stimulator per leg. One channel stimulated gastrosoleus muscles, and the other provided flexor withdrawal response. Since the knee was held in extension, only dorsiflexion of the foot and flexion of the hip were obtained. The other hybrid system comprised the knee–ankle–foot orthosis (KAFO) and two-channel stimulation per leg. The quadriceps and common peroneal nerve were stimulated on each side. The quadriceps stimulation caused knee extension, and the peroneal nerve stimulation produced synergistic flexor response. The mechanical brace incorporated knee joint locks, which were remotely controlled by a solenoid actuator or a Bowden cable. Andrews also described a short-leg orthosis in combination with FES. Andrews and associates (1988) suggested a different way of using the foot reaction orthoses (FRO) in conjunction with FES. This consisted of a rigid ankle foot orthosis, a multichannel stimulator with surface electrodes, body-mounted sensors, a "rule-based" controller, and an electrocutaneous display for supplementary feedback.

Solomonow and associates (1989) suggested a hybrid assistive system that combines FES and the reciprocating gait orthosis. The reciprocating gait orthosis was jointly developed by the Louisiana State University Medical Center and Durr-Fillauer Medical Inc. (Douglas et al., 1983). The orthosis includes trunk support, a pelvic assembly, and bilateral KAFO. A rigid pelvic assembly consisting of a pelvic band bears hip joints. KAFO has posterior offset of the knee joints with drop locks on the lateral sides. The essential feature of the system is the reciprocating mechanism connecting the hip joints. This coupling provides hip joint stability during standing by preventing simultaneous flexion of both hips; yet, it allows flexion of one hip and simultaneous extension of the other when a step is taken. The four-channel FES of the hybrid system is used for standing up, and swinging of the legs. Surface stimulation of *Rectus Femoris* and hamstrings was applied. The current research in the domain of robotics presents many more devices that can be combined with the FES. However, all of them are in the pioneering stage of development, and many more aspects need to be studied before they actually can be translated into the clinic and home.

References

Actigait, Neurodan A/S, Denmark, http://www.neurodan.com/actigait.asp (accessed May 15, 2012).

Andrews, B.J. and T. Bajd. 1984. Hybrid orthosis for paraplegics. In: *Advances in External Control of Human Extremities VIII, (Suppl)*, ed. D.B. Popović, ETAN, Belgrade, pp. 55–57.

Andrews, B.J., Baxendale, R.H., Barnett, R., Philips, G.F., Yamazaki, T., Paul, J.P., and P.A. Freeman. 1988. Hybrid FES orthosis incorporating closed loop control and sensory feedback. *J Biomed Eng* 10:189–95.

Bajd, T., Kralj, A., and R. Turk. 1982. Standing-up of a healthy subject and paraplegic patients. *J Biomech* 15:1–10.

Bajd, T., Kralj, A., Turk, R., Benko, H., and J. Šega. 1983. The use of a four channel electrical stimulator as an ambulatory aid for paraplegic patients. *Phys Ther* 63:1116–20.

Bajd, T., Kralj, A., Turk, R., Benko, H., and J. Šega. 1989. Use of functional electrical stimulation in the rehabilitation of patients with incomplete spinal cord injuries. *J Biomed Eng* 11:96–102.

Bijak, M., Sauerman, S., Schmutterer, C., Lanmueller, H., Unger, E., and W. Mayr. 1999. A modular PC-based system for easy setup of complex stimulation patterns. In: *Proceedings of the Fourth International Conference IFESS*, Sendai, pp. 24–28.

Bowman, R.B. and R.C. Erickson. 1985. Acute and chronic implantation of coiled wire intraneural electrodes during cyclical electrical stimulation. *Ann Biomed Eng* 13:75–93.

Branner, A., Stein, R.B., and R.A. Normann. 2001. Selective stimulation of cat sciatic nerve using an array of varying length microelectrodes. *J Neurophysiol* 85:1585–94.

Buckett, J.R., Peckham, P.H., Thrope, G.B., Braswell, S.D., and M.W. Keith. 1988. A flexible, portable system for neuromuscular stimulation in the paralyzed upper extremity. *IEEE Trans Biomed Eng* 35:897–904.

Burridge, J.H., Taylor, P.N., Hagan, S., and I.D. Swain. 1997a. Experience of clinical use of the Odstock dropped foot stimulator. *Artif Organs* 21:254–60.

Burridge J.H, Taylor P.N, Hagan S., Wood D.E., and Swain I.D. 1997b. The effects of common peroneal stimulation on the effort and speed of walking: A randomized controlled trial with chronic hemiplegic patients. *Clin Rehabil* 11:201–10.

Cameron, T., Loeb, G.E., Peck, R.A., Schulman, J.H., Strojnik, P., and P.R. Troyk. 1997. Micromodular implants to provide electrical stimulation of paralyzed muscles and limbs. *IEEE Trans Biomed Eng* 44:781–90.

Crago, P.E., Memberg, W.D., Usey, M.K., Keith, M.W., Kirsch, R.F., Chapman, G.J., et al. 1998. An elbow extension neuroprosthesis for individuals with tetraplegia. *IEEE Trans Rehabil Eng* 6:1–6.

Crago, P.E., Peckham, P.H., Mortimer, J.T., and J. Van der Meulen. 1974. The choice of pulse duration for chronic electrical stimulation via surface, nerve and intramuscular electrodes. *Ann Biomed Eng* 2:252–64.

Dai, R., Stein, R.B., Andrews, B.J., James, K.B., and M. Wieler. 1996. Application of tilt sensors in functional electrical stimulation. *IEEE Trans Rehabil Eng* 4:63–72.

Davis, R., Houdayer, T., Andrews, B.J., Emmons, S., and J.P. Patrick. 1997. Paraplegia: Prolonged closed-loop standing with implanted nucleus FES-22 stimulator and Andrews' foot-ankle orthosis. *Stereotact Funct Neurosurg* 69:281–87.

Donaldson, N. 1986. A 24-output implantable stimulator for FES. In: *Proceedings of the Second Vienna International Workshop Functional Electrostimulation*, Vienna, pp. 197–200.

Douglas, R., Larson, P.F., D' Ambrosia, R., and R.E. McCall. 1983. The LSU reciprocating gait orthosis. *Orthopedics* 6:34–9.

Dutta, A., Kobetič, R., and R.J. Triolo. 2009. Gait initiation with electromyographically triggered electrical stimulation in people with partial paralysis. *J Biomech Eng* 131:8–9 . http://dx.doi.org/10.1115/1.3086356.

Frigo, C., Ferrarin, M., Frasson, W., Pavan, E., and R. Thorsen. 2000. EMG signals detection and processing for on-line control of functional electrical stimulation. *J Electromyogr Kinesiol* 10:351–60.

Gan, L.S. and A. Prochazka. 2010. Properties of the stimulus router system: A novel neural prosthesis. *IEEE Trans Biomed Eng* 57:450–59.

Gračanin, F., Kralj, A., and S. Reberšek. 1969. Advanced version of the Ljubljana functional electronic peroneal brace with walking rate controlled tetanization. In: *Advances in External Control of Human Extremities II*, ETAN, Belgrade, pp. 487–500.

Gračanin, F., Prevec, T., and J. Trontelj. 1967. Evaluation of use of functional electronic peroneal brace in hemiparetic patients. In: *Advances in External Control of Human Extremities I*, ETAN, Belgrade, pp. 198–210.

Granatd M.H., Maxwell, D.J., Ferguson, A.C.B., Lees, K.R., and J.C. Barbenel. 1996. Peroneal stimulator: Evaluation for the correction of spastic drop foot in hemiplegia. *Arch Phys Med Rehabil* 77:19–24.

Grandjean, P.A., and J.T. Mortimer. 1986. Recruitment properties of monopolar and bipolar epimysial electrodes. *Ann Biomed Eng* 14:53–66.

Graupe, D. and K.H. Kohn. 1997. Transcutaneous functional neuromuscular stimulation of certain traumatic complete thoracic paraplegics for independent short-distance ambulation. *Neurol Res* 19:323–33.

Grill, W.M., and J.T. Mortimer. 2000. Neural and connective tissue response to long-term implantation of multiple contact nerve cuff electrodes. *J Biomed Mater Res* 50:215–26.

Handa, Y., Hoshimiya, H., Iguchi, Y., and T. Oda. 1989a. Development of percutaneous intramuscular electrode for multichannel FES system. *IEEE Trans Biomed Eng* 36:705–10.

Handa, Y., Ohkubo, K., and N. Hoshimiya. 1989b. A portable multi-channel functional electrical stimulation (FES) system for restoration of motor function of the paralyzed extremities. *Automedica* 11:221–32.

Hansen, M. 2001. *Machine Learning Techniques for Control of Functional Electrical Stimulation Using Natural Sensors*. PhD Thesis, Center for Sensory-Motor Interaction, Aalborg University, Denmark.

Hart, R.L., Kilgore, K.L., and P.H. Peckham. 1998. A comparison between control methods for implanted FES hand-grasp systems. *IEEE Trans Rehab Eng* 6:208–18.

Haugland, M., and J.A. Hoffer. 1994a. Slip information provided by nerve cuff signals: Application in closed-loop control of functional electrical stimulation. *IEEE Trans Rehab Eng* 2:29–36.

Haugland, M. and J.A. Hoffer. 1994b. Artifact-free sensory nerve signals obtained from cuff electrodes during functional electrical stimulation of nearby muscles. *IEEE Trans Rehab Eng* 2:37–9.

Haugland, M., Lickel, A., Haase, J., and T. Sinkjær. 1999. Control of FES thumb force using slip information obtained from the cutaneous electroneurogram in quadriplegic man. *IEEE Trans Rehab Eng* 7:215–27.

Haugland, M., and T. Sinkjær. 1995. Cutaneous whole nerve recordings used for correction of footdrop in hemiplegic man. *IEEE Trans Rehab Eng* 3:307–17.

Hoffer, J.A. 1988. Closed loop, implant sensor, functional electrical stimulation system for partial restoration of motor functions. United States Patent [19], Patent Number 4,750,499.

Hoffer, J.A., Stein, R.B., Haugland, K., Sinkjær, T., Durfee, W.K., Schwartz, A.B., Loeb, G.E., and C. Kantor. 1995. Neural signals for command control and feedback in functional electrical stimulation. *J Rehabil* 33:145–57.

Holle, J., Frey, M., Gruber, H., Kern, H., Stohr, H., and H. Thoma. 1984. Functional electrostimulation of paraplegics: Experimental investigations and first clinical experience with an implantable stimulation device. *Orthopaedics* 7:1145–60.

Ilić, M., Vasiljević, D., and D.B. Popović. 1994. A programmable electronic stimulator for FES systems. *IEEE Trans Rehabil Eng* 2:234–39.

Inmann, A. 2001. *Natural Sensory Feedback for FES Controlled Hand Grasp.* PhD Thesis, Center for Sensory-Motor Interaction, Aalborg University, Denmark.

Jacobs, P.L., Nash, M.S., Klose, K.J., Guest, R.S., Needham-Shropshire, B.M., and B.A. Green. 1997. Evaluation of a training program for persons with SCI paraplegia using the Parastep 1 ambulation system: Part 2. Effects on physiological responses to peak arm ergometry. *Arch Phys Med Rehabil* 78:794–98.

Kennedy, P., Andreasen, D., Bartels, J., Ehirim, P., Mao, H., Velliste, M., Wichmann, T., and J. Wright. 2011. Making the lifetime connection between brain and machine for restoring and enhancing function. *Progress in Brain Research* 194:1–26.

Kilgore, K.L., Peckham, P.H., Keith, M.W., Thrope, G.B., Wuolle, K.S., Bryden, A.M., and R.T. Hart. 1997. An implanted upper-extremity neuroprosthesis. Follow-up of five patients. *J Bone Joint Surg Am* 79:533–41.

Klose, K.J., Jacobs, P.L., Broton, J.G., Guest, R.S., Needham-Shropshire, B.M., Lebwohl, N., et al. 1997. Evaluation of a training program for persons with SCI paraplegia using the Parastep 1 ambulation system: Part 1. Ambulation performance and anthropometric measures. *Arch Phys Med Rehabil* 78:789–93.

Knaflitz, M. and R. Merletti. 1988. Suppression of stimulation artifacts from myoelectric-evoked potential recordings. *IEEE Trans Biom Eng* 35:758–63.

Kobetič, R. and E.B. Marsolais. 1994. Synthesis of paraplegic gait with multichannel functional electrical stimulation. *IEEE Trans Rehab Eng* 2:66–79.

Kobetič, R., Triolo, R.J., and E.B. Marsolais. 1997. Muscle selection and walking performance of multi-channel FES systems for ambulation in paraplegia. *IEEE Trans Rehab Eng* 5:23–9.

Kobetič, R., Triolo, R., Uhlir, J.P., Bieri, C., Wibowo, M., Polando, G., Marsolais, E.B., Davis, J.A. Jr., Ferguson, K.A., and M. Sharma. 1999. Implanted functional electrical stimulation system for mobility in paraplegia: A follow-up case report. *IEEE Trans Rehab Eng* 7:390–98.

Kralj, A. and T. Bajd. 1989. *Functional Electrical Stimulation: Standing and Walking after Spinal Cord Injury.* CRC Press, Boca Raton, Florida.

Kralj, A., Bajd, T., and R. Turk. 1980. Electrical stimulation providing functional use of paraplegic patient muscles. *Med Prog Technol* 7:3–9.

Kuhn, A., Keller, T., Lawrence, M., and M. Morari. 2010. The influence of electrode size on selectivity and comfort in transcutaneous electrical stimulation of the forearm. *IEEE Trans Neural Syst Rehabil Eng* 18:255–62.

Lebedev, A.M. and M.A.L. Nicolelis. 2011. Toward a whole-body neuroprosthetic. *Progress in Brain Research* 194:47–60.

Lemay, M.A. and W.M. Grill. 1999. Spinal force fields in the cat spinal cord (abstract). *Soc Neurosci* 25:1396.

Liberson, W.F., Holmquest, H.J., Scott, D., and A. Dow. 1961. Functional electrotherapy: Stimulation of the peroneal nerve synchronized with the swing phase of the gait in hemiplegic patients. *Arch Phys Med Rehab* 42:101–05.

Long II, C. and C.V. Masciarelli. 1963. An electrophysiologic splint for the hand. *Arch Phys Med Rehab* 44:499–503.

Malešević, N., Popović, L., Schwirtlich, L., and D.B. Popović. 2010. Distributed low-frequency electrical stimulation delays muscle fatigue compared to conventional stimulation. *Muscle Nerve* 42:556–62.

Maležič, M., Stanič, U., Kljajić, M., Aćimović, R., Krajnik, J., Gros, N., and M. Stopar. 1984. Multichannel electrical stimulation of gait in motor disabled patients. *Orthopedics* 7:1187–95.

Marsolais, E.B. and R. Kobetič. 1983. Functional walking in paralyzed patients by means of electrical stimulation. *Clin Orthop* 175:30–6.

Marsolais, E.B. and R. Kobetič. 1987. Functional electrical stimulation for walking in paraplegics. *J Bone Jt Surg* 69:728–33.

McCreery, D.B., Agnew, W.F., Yuen, T.G., and L.A. Bullara. 1995. Relationship between stimulus amplitude, stimulus frequency and neural damage during electrical stimulation of sciatic nerve of cat. *Med Biol Eng Comput* 33:426–29.

McFarland, D.J., McCane, L.M., and J.R. Wolpaw. 1998. EEG-based communication and control: Short-term role of feedback. *IEEE Trans Rehabil Eng* 6:7–11.

Minzly, J., Mizrahi, J., Isakov, E., Susak, Z., and M. Verbeke. 1993. A computer controlled portable stimulator for paraplegic patients. *J Biomed Eng* 15:333–38.

Mortimer, T. 1981. Motor prosthesis. In: *Handbook of Physiology*. Ed. V.B. Brooks (Sect 1, vol 11, part 1(5)) Am Physiol Soc, Bethesda, pp. 155–87.

Mushahwar, V.K. and K.W. Horch. 1997. Proposed specifications for a lumbar spinal cord electrode array for control of lower extremities in paraplegia. *IEEE Trans Rehabil Eng* 5:237–43.

Mushahwar, V.K. and K.W. Horch. 2000. Muscle recruitment through electrical stimulation of the lumbosacral spinal cord. *IEEE Trans Rehab Eng* 8:22–8.

Mushahwar, V.K., Jacobs, P.K., Normann, R.A., Triolo, R.J., and N. Kleitman. 2007. New functional electrical stimulation approaches to standing and walking, *J Neural Eng* 4:18, doi:10.1088/1741-2560/4/3/S05.

Nathan, R.H. 1989. An FNS-based system for generating upper limb function in the C4 quadriplegic. *Med Biol Eng Comp* 27:549–56.

Nathan, R.H. and A. Ohry. 1990. Upper limb functions regained in quadriplegia: A hybrid computerized neuromuscular stimulation system. *Arch Phys Med Rehabil* 71:415–21.

Navarro, X., Krueger, T.B., Lago, N., Micera, S., Stieglitz, T., and P. Dario. 2005. A critical review of interfaces with the peripheral nervous system for the control of neuroprostheses and hybrid bionis systems. *J Peripher Nerv Syst*, 10:229–58.

Ness L300, Bioness, Inc., http://www.bionessinc.com (accessed May 2012).

Ness H200, Bioness, Inc., http://www.bionessinc.com (accessed May 2012).

Nikolić, Z.M., Popović, D.B., Stein, R.B., and Z. Kenwell. 1994. Instrumentation for ENG and EMG recordings in FES systems. *IEEE Trans Biomed Eng* 41:703–06.

Normann, R.A., Maynard, E.M., Rousche, P.J., and D.J. Warren. 1999. A neural interface for a cortical vision prosthesis. *Vision Res* 39:2577–87.

Odstock dropped foot stimulator, NDS Medical, http://www.odfs.com (accessed May 2012).

Parastep, Sigmedics Inc. http://www.sigmedics.com (accessed May 2012).

Peckham, P.H. 1988. Functional electrical stimulation. In *Encyclopedia of Medial Devices and Instrumentation*. Ed. J. Webster, John Wiley, New York, pp. 1341–58.

Peckham, P.H., 2005. Functional electrical stimulation for neuromuscular applications. *Annu Rev Biomed Eng* 7:327–60.

Peckham, P.H., Mortimer, J.T., and E.B. Marsolais. 1980a. Controlled prehension and release in the C5 quadriplegic elicited by functional electrical stimulation of the paralyzed forearm muscles. *Annals Biomed Eng* 8:369–88.

Peckham, P.H., Marsolais, E.B., and J.T. Mortimer. 1980b. Restoration of the key grip and release in the C6 quadriplegic through functional electrical stimulation. *J Hand Surg* 5:464–69.

Popović, D.B., Gordon, T., Rafuse, V., and A. Prochazka. 1991. Properties of implanted electrodes for functional electrical stimulation. *Ann Biomed Eng* 19:303–16.

Popović, M.R., Keller, T., Pappas, I., Dietz, V., and M. Morari. 2001. ETHZ-ParaCare grasping and walking neuroprostheses. *IEEE Med Biol Eng Mag* 20(1):82–93.

Popović, D.B. and M.B. Popović. 1998a. Tuning of a nonanalytical hierarchical control system for reaching with FES. *IEEE Trans Biomed Eng* 45:203–12.

Popović, D.B. and M.B. Popović. 1998b. Belgrade grasping system. *J Electron* (Banja Luka, Bosnia) 2: 21–8.

Popović, D.B. and M.B. Popović. 2000. Nonanalytical control for assisting reaching in humans with disability. In *Control of Posture and Movement: Neuro-Musculo-Skeletal Interaction and Organization Principles*, eds. J.M. Winters and P.E. Crago. Springer-Verlag, London, UK, pp. 535–48.

Popović, D.B. and M.B. Popović. 2001. Control for an elbow neuroprosthesis: Cloning biological synergies. *IEEE Med Biol Eng Mag* 20:74–81.

Popović, D.B. and M.B. Popović. 2009. Automatic determination of the optimal shape of a surface electrode: Selective stimulation. *J Neurosci Methods* 178(1):174–81.

Popović, D.B. and T. Sinkjær. 2000. *Control of Movement for the Physically Disabled: Control for Rehabilitation Technology*. Springer, London.

Popović, D.B., Sinkjær, T., and M.B. Popović. 2009. Electrical stimulation as a means for achieving recovery of function in stroke patients, *J Neuro Rehabil* 25:45–58.

Popović, D.B., Stein, R.B., Jovanović, K.L., Rongching, D., Kostov, A., and W.W. Armstrong. 1993. Sensory nerve recording for closed-loop control to restore motor functions. *IEEE Trans Biomed Eng* 40:1024–31.

Popović, D.B., Stojanović, A., Pjanović, A., Radosavljević, S., Popović, M.B., Jović, S., and D. Vulović. 1999. Clinical evaluation of the Bionic glove. *Arch Phys Med Rehabil* 80:299–304.

Popović, D.B., Tomović, R., and L. Schwirtlich. 1989. Hybrid assistive system: Neuroprosthesis for motion. *IEEE Trans Biomed Eng* 36:729–38.

Popović-Bijelić, A., Bijelić, G., Jorgovanović, N., Bojanić, D., Popović, D.B., and M.B. Popović. 2005. Multi-field surface electrode for selective electrical stimulation. *Art Organs* 29:448–52.

Prochazka, A., Gauthier, M., Wieler, M., and Z. Kenwell. 1997. The Bionic glove: An electrical stimulator garment that provides controlled grasp and hand opening in quadriplegia. *Arch Phys Med Rehabil* 78:608–14.

Reberšek, S. and L. Vodovnik. 1973. Proportionally controlled functional electrical stimulation of hand. *Arch Phys Med Rehabil* 54:378–82.

Riso, R.R., Fahard, K., Jensen, W., and T. Sinkjær. 2000. Nerve cuff recordings of muscle afferent activity from tibial and peroneal nerves in rabbit during passive ankle motion. *IEEE Trans Rehab Eng* 8:121–32.

Rousche, P.J. and R.A. Normann. 1999. Chronic intracortical microstimulation (ICMS) of cat sensory cortex using the Utah intracortical electrode array. *IEEE Trans Rehabil Eng* 7:56–68.

Rushton, D.N. 1990. Choice of nerves roots for multichannel leg controller implant. In: *Advances in External Control of Human Extremities X*, ed. D.B. Popović, Nauka, Belgrade, pp. 99–108.

Saxena, S., Nikolić, S., and D.B. Popović. 1995. An EMG controlled FES system for grasping in tetraplegics. *J Rehabil Res Dev* 32:17–23.

Scheiner, A., Mortimer, J.T., and U. Roessmann. 1990. Imbalanced biphasic electrical stimulation: Muscle tissue damage. *Ann Biomed Eng* 18:407–25.

Schwirtlich, L., and D.B. Popović. 1984. Hybrid orthoses for deficient locomotion. In *Advances in External Control of Human Extremities VIII*, ed. D.B. Popović, ETAN, Belgrade, pp. 23–32.

Scott, T.R.D., Peckham, P.H., and K.L. Kilgore. 1996. Tri-state myoelectric control of bilateral upper extremity neuroprosthesis for tetraplegic individuals. *IEEE Trans Rehab Eng* 4:251–63.

Sinkjær, T., Haugland, M., Inmann, A., Hansen, M., and K.D. Nielsen. 2003. Biopotentials as command and feedback signals in functional electrical stimulation systems. *Med Eng Phys* 25:29–40.

Smith, B.T., Betz, R.R., Mulcahey, M.J., and R.J. Triolo. 1994. Reliability of percutaneous intramuscular electrodes for upper extremity functional neuromuscular stimulation in adolescents with C5 tetraplegia. *Arch Phys Med Rehabil* 75:939–45.

Smith, B.T., Mulcahey, B.J., and R.P. Betz. 1996. Development of an upper extremity FES system for individuals with C4 tetraplegia. *IEEE Trans Rehab Eng* 4:264–70.

Smith, B.T., Peckham, P.H., Keith, M.W., and D.D. Roscoe. 1987. An externally powered, multichannel, implantable stimulator for versatile control of paralyzed muscle. *IEEE Trans Biomed Eng* 34:499–508.

Solomonow, M., Baratta, R., Hirokawa, S., Rightor, N., Walker, W., Beaudette, P., Shoji, H., and R. D'Ambrosia. 1989. The RGO generation II: Muscle stimulation powered orthosis as a practical walking system for thoracic paraplegics. *Orthopedics* 12:1309–15.

Stein, R.B., Bélanger, M., Wheeler, G., Wieler, M., Popović, D.B., Prochazka, A., and L. Davis. 1993. Electrical systems for improving locomotion after incomplete spinal cord injury: An assessment. *Arch Phys Med Rehabil* 74:954–59.

Stein, R.B., Charles, D., Davis, L., Jhamandas, J., Mannard, A., and T.R. Nichols. 1975. Principles underlying new methods for chronic neural recording. *Canad J Neurol Sci* 2:235–44.

Strojnik, P., Aćimović-Janežič, R., Vavken, E., Simić, V., and U. Stanič. 1987. Treatment of drop foot using an implantable peroneal underknee stimulator. *Scand J Rehabil Med* 19:37–43.

Strojnik, P., Schulman, J., Loeb, G., and P. Troyk. 1993. Multichannel FES system with distributed microstimulators. In: *Proceedings of the Annual International Conference of IEEE EMBS*, Atlanta, GA, USA, pp. 1352–53.

Strojnik, P., Whitmoyer, D., and J. Schulman. 1990. An implantable stimulator for all seasons. In: D.B. Popović, ed. *Advances in External Control of Human Extremities X*, Nauka, pp. 335–344.

Sweeney, P.C. and G.M. Lyons. 1999. Fuzzy gait event detection in a finite state controlled FES drop foot correction system. *J Bone Joint Surg* (BR) 81:91–93.

Thorsen, R. 1999. An artefact suppressing fast-recovery myoelectric amplifier. *IEEE Trans Biomed Eng* 46:764–66.

Thorsen, R., Ferrarin, M., Spadone, R., and C. Frigo. 1999. Functional control of the hand in tetraplegics based on residual synergistic EMG activity. *Artif Organs* 23:470–73.

Tomović, R., Popović, D.B., and R.B. Stein. 1995. *Nonanalytical Methods for Motor Control.* World Scientific Publication, Singapore.

Triolo, R., Nathan, R., Handa, Y., Keith, M., Betz, R.R., Carroll, S., and C. Kantor. 1996. Challenges to clinical deployment of upper limb neuroprostheses. *J Rehabil Res Dev* 33:111–22.

UNAsystems, Belgrade, http://www.unasistemi.com (accessed May 15, 2012).

Upshaw, B. and T. Sinkjær. 1998. Digital signal processing algorithms for the detection of afferent nerve activity recorded from cuff electrodes. *IEEE Trans Rehab Eng* 6:172–81.

Vaughan, T.M., Miner, L.A., McFarland, D.J., and J.R. Wolpaw. 1998. EEG-based communication: Analysis of concurrent EMG activity. *Electroencephalogr Clin Neurophysiol* 107:428–33.

Vodovnik, L., Bajd, T., Kralj, A., Gračanin, F., and P. Strojnik. 1981. Functional electrical stimulation for control of locomotor systems. *CRC Crit Rev Bioeng* 6:63–131.

Vodovnik, L., Crochetiere, W.J., and J.B. Reswick. 1967. Control of a skeletal joint by electrical stimulation of antagonists. *Med Biol Eng* 5:97–109.

Vossius, G., Mueschen, U., and H.J. Hollander. 1987. Multichannel stimulation of the lower extremities with surface electrodes. In *Advances in External Control of Human Extremities IX*, Ed. D.B. Popović, ETAN, Belgrade, pp. 193–203.

Walkaide, Innovative Neurotronics, Inc., http://www.walkaide.com (accessed May 15, 2012).

Wieler, M., Stein, R.B., Ladouceur, M., Whittaker, M., Smith, A.W., Naaman, S., et al. 1999. Multicenter evaluation of electrical stimulation systems for walking. *Arch Phys Med Rehabil* 80:495–500.

Wijman, A.C., Stroh, K.C., Van Doren, C.L., Thrope, G.B., Peckham, P.H., and M.W. Keith. 1990. Functional evaluation of quadriplegic patients using a hand neuroprosthesis. *Arch Phys Med Rehabil* 32:1053–57.

Wolpaw, J.R., Ramoser, H., McFarland, D.J., and G. Pfurtscheller. 1998. EEG-based communication: Improved accuracy by response verification. *IEEE Trans Rehabil Eng* 6:326–33.

Yoshida, K. and K.W. Horch. 1996. Closed-loop control of ankle position using muscle afferent feedback with functional neuromuscular stimulation. *IEEE Trans Biomed Eng* 43:167–76.

27

Pharmaceutical Technical Background on Drug-Delivery Methods

Robert S. Litman
*Nova Southeastern
University*
Ohio State University

27.1 Introduction

In the evolution of drug development and manufacturing, drug-delivery systems have risen to the forefront in the latest of pharmaceutical advances. There are many new pharmacological entities discovered each year, each with its own unique mechanism of action. Each drug will demonstrate its own pharmacokinetic profile. This profile may be changed by altering the drug-delivery system to the target site. For

647

a drug to demonstrate its pharmacological activity, it must be absorbed, transported to the appropriate tissue or target organ, penetrate into the responding subcellular structure, and elicit a response or change an ongoing process. The drug may be simultaneously or sequentially distributed to a variety of tissues, bound or stored, further metabolized to active or inactive products, and eventually excreted from the body. A delivery system that may have an effect upon the absorption, distribution, metabolism, or excretion of a pharmaceutical entity may then affect the potency, half-life, potential for drug interactions, and side-effect profile of that specific entity. Drug-delivery systems are developed to enhance the desired pharmacological effect at specific target sites, while reducing the probability of drug interactions and unwanted side effects.

Other reasons for the development of new drug-delivery systems include the masking of unpleasant tastes, inability of a patient to swallow a specific dosage form, protecting components from atmospheric degradation, controlling the site of drug release, prolonging or delaying the absorption of the drug moiety, improving the physical appearance of the drug, and changing the physical surface characteristics of the active ingredients. The following text will present a number of pharmaceutical drug-delivery systems used in the treatment of a variety of disease states.

27.2 Central Nervous System: Drug Delivery

27.2.1 Challenges to Delivery: Blood–Brain Barrier

The central nervous system (CNS) consists of the brain and the spinal cord. Drug delivery to the brain is challenging because of the blood–brain barrier (BBB). The BBB is present in the brain of all vertebrates and is a system that protects the brain from substances in the blood. Because of the presence of the BBB, over 98% of new drugs discovered for the CNS do not penetrate the brain following systemic administration. The BBB is composed of

1. The continuous endothelium of the capillary wall
2. A relatively thick basal laminal surrounding the external face of the capillary
3. The bulbous feet of the astrocytes that cling to the capillaries

The capillary endothelial cells are almost seamlessly joined all around by tight junctions, making them the least permeable capillaries in the entire body. This relative impermeability of the brain capillaries constitutes most of the BBB.[1] In addition, once having traversed this barrier of the capillary endothelial cell, the drug must then penetrate the glial cells that envelop the capillary structure. Cerebral endothelial cells also express ATP-dependent transmembrane glycoproteins involved in active transport of substances to outside the cell.[2] There are several theories as to what factors affect permeability into the brain. Factors such as lipophilicity, molecular size, polarity, and hydrogen bonding have been studied as methods to predict the penetration capacity of a drug into the BBB.[3]

The cerebral spinal fluid (CSF) is a plasma-like fluid that fills the cavities of the CNS and surrounds the CNS externally, protecting the brain and the spinal cord. Passage of chemical substances into the CSF is controlled by the blood–CSF barrier. This barrier is created by the ependymal cells of the choroid plexus.[4]

The choroid plexus (which is located in the 3rd and 4th ventricles of the brain) has the ability to secrete substances out through an active transport system. Thus, attempts at accessing the brain through the CSF may be unsuccessful due to the protective nature of the choroid plexus. Also, it cannot be inferred that a given drug crosses the BBB just on the basis of its distribution into the CSF.[5]

Access to the CNS can be gained through direct or indirect methods. For drugs with the ability to penetrate the BBB, possible routes of administration are described below as indirect routes of administration. Direct routes of administration are attempts to bypass the BBB and gain access to the CNS tissue.

Indirect routes of administration include

- Intravenous (i.v.), intra-arterial
- Intraperitoneal
- Digestive tract
- Lung
- Skin
- Nasal
- Intramuscular (i.m.)
- Subcutaneous (s.c.)
- Sublingual
- Buccal
- Rectal

27.2.2 Indirect Routes of Administration

As previously described, access to the CNS can be achieved through indirect routes of administration; in addition to having to overcome the BBB, such routes of administration are subject to other bodily methods that may decrease or negate the amount of drug that penetrates the brain. For example, the drugs may be subject to being metabolized by the liver, excreted by the kidney, or acted upon by enzymes in the intestine or lung, all of which would result in a decrease in the amount of circulating drug available to the brain. All these routes of administration access the CNS through systemic absorption; in other words, access is gained to the bloodstream, which then attempts to cross the BBB.

Examples of drugs that cross the BBB include clonidine and propranolol, both antihypertensives. Propranolol can be administered via the oral route or the i.v. route. Extensive first-pass metabolism through the liver makes the oral dose necessitate a much higher than the injectable dose. Clonidine is available for administration orally, as a transdermal patch and even for epidural use for intractable pain. Other well-known drugs able to cross the BBB include the opiate analgesics such as morphine, selective serotonin reuptake inhibitors such as Prozac®, and benzodiazepines such as Valium. These drugs are available in multiple dosage forms ranging from injectable to oral to rectal gels. Fentanyl, a potent analgesic, is available in transdermal patches and buccal formulations and generally used to treat cancer pain.

Still, the BBB and the blood–CSF barriers remain the largest challenge in developing drugs to effectively treat CNS disorders. Therefore, numerous ways to circumvent these barriers such as direct delivery to the CNS or attempted interruption of the BBB system are being researched.

27.2.3 Direct Routes of Administration

27.2.3.1 Nasal Drug Delivery

The nasal route of administration bypasses the BBB. Common drugs of addiction such as cocaine or amphetamine derivative may rapidly enter the brain by the nasal route. Nasal drug intake appears to be a fast and effective route of administration, suitable for drugs that must act rapidly and are taken in small amounts. Examples include antimigraine drugs such as Imitrex® nasal spray and analgesics such as Stadol® nasal spray. Unfortunately, frequent use of this route of administration may lead to complications such as mucosal damage that can lead to infections. Also, some patients may lose the ability to smell.

27.2.3.2 Epidural Drug Delivery

Drugs administered into the epidural space to reach the spinal cord must traverse the dura mater and the arachnoid mater, and then enter the CSF to reach the spinal cord gray and white matter. This occurs through simple diffusion. Epidural infusion and anesthesia is a common tool in the United States for pain relief from contractions during labor. Pharmacological means to prevent the redistribution of drug to the systemic circulation or to prolong the drug effect have been used. One such example is the addition of epinephrine to local anesthetics and epidural opioids. The addition of epinephrine has been shown to improve the quality and prolong the duration of epidural anesthesia and analgesia. Other pharmaceutical modifiers of redistribution include preparations that provide slow-release "depot" formulations, encapsulating drugs in liposomes, embedding drugs in biodegradable polymers, or using drugs that are themselves nearly insoluble in aqueous solutions.[6]

27.2.3.3 Intrathecal Drug Delivery

This method involves direct injection into the CSF via a spinal needle or catheter. Drug injection into the CSF results in mixing of the drug product and the CSF, which does not occur with epidurally administered drugs. Drugs enter the CSF as a solution, instead of as individual molecules in the case of epidural administration. This causes the density of the solution and the patient's position to be the most important factors with regard to where the drug initially gets distributed along the spinal cord. The more dense the solution, the more the tendency to move down the spinal column until complete mixture with the CSF makes the solution isobaric. Hydrophobic drugs have poor permeability into the CSF. One way to improve the permeability is to increase their aqueous solubility.

Research into compounds such as cyclodextrins shows that combining drugs, for example, sufentanil with cyclodextrin, increases its permeability into the meninges more than twofold *in vitro*. However, it is also possible that when used *in vivo* the redistribution of drugs into unintended sites might occur. Rigorous toxicology trials in animals are still necessary in this area.[6,7]

27.2.3.4 Intracerebroventricular Drug Delivery

Invasive brain drug-delivery systems have been the most widely used for circumventing the BBB drug-delivery problem. This invasive strategy requires either a craniotomy by a neurosurgeon or access to the carotid artery by an interventional radiologist. The neurosurgical-based systems include intracerebroventricular (ICV) infusion of drugs or intracerebral implants of biodegradable polymers.[8] Unfortunately, drug penetration following ICV injection is minimal.[5] Also, because of the one-way flow of CSF in the brain following ICV injection, the distribution of drug to both sides of the brain following ICV injection would require the placement of catheters in both lateral ventricles.[5] ICV has been used primarily in three treatment areas: chemotherapy for brain tumors, treatment of infections of the CNS,[9] and delivery of analgesics in the setting of intractable pain.[10] In the case of chemotherapy, direct injection into the CNS also avoids significant systemic toxicity by limiting chemotherapy exposure to the CNS.

27.2.3.5 CNS and Drug Targeting

Drug targeting describes a process for attempting site-specific delivery of drugs. For drugs needing access to the CNS for their action to be exerted, the need for drug targeting and avoidance of the BBB is an obvious one.

Drug targeting has been classified into three types[11]:

1. Delivery to a discrete organ or tissue
2. Targeting to a specific cell type (e.g., tumor cells vs. normal cells)
3. Delivery to a specific intracellular compartment in the target cells (e.g., lysosomes)

Targeting can be achieved through different methodologies: biological agents that are selective to a particular site in the body, preparation of prodrug that becomes active once it reaches the target site, and use of a biologically inert macromolecular carrier system that directs a drug to a specific site in the body. For example, 3.85% carmustine (Gliadel®)-impregnated polymers consisting of CPP:SA have shown improved survival in patients with high-grade recurrent gliomas (brain tumor).[8] Carmustine itself crosses the BBB; however, the polymers allow higher tissue levels to be achieved with minimal systemic side effects. The polymer wafers are placed after the surgical removal of the brain tumor itself in the cavity left behind once the tumor is removed. Drug targeting is a promising field of research to aid in drug delivery to the CNS.

27.3 Cardiovascular System: Drug Delivery

Cardiovascular disease has become an important cause of morbidity and mortality as the population ages. The new millennium holds even greater promise as genetic engineering produces new and even more effective drugs and devices to prevent and treat patients with cardiovascular disease.[12] In this section, the latest cardiovascular drug-delivery methods will be discussed.

27.3.1 Chronotherapeutics

Chronotherapeutic medications deliver medications in concentrations that vary according to changes in physiological need.[13,14] In hypertension, chronotherapeutic medications deliver the drug in highest concentrations during the morning period, when blood pressure is at its greatest, and in lesser concentrations at nighttime, when blood pressure is at its lowest.[13,14] Verapamil, a calcium channel blocker, has been marketed in two formulations that use novel delivery systems to provide chronotherapy: Verelan PM™ (Schwarz Pharma Inc, Milwaukee, WI) and Covera-HS™ (Pharmacia Corp, Peapack, NJ).[13] Verelan PM uses chronotherapeutic oral drug absorption system (CODAS) technology, and Covera-HS uses the controlled-onset, extended-release (COER-24) delivery system.[13] The CODAS delivery system incorporates a 4- to 5-h delay in drug delivery followed by an extended drug release, with a peak concentration occurring approximately 11 h after administration, which is designed for bedtime dosing.[13] Trough concentrations occur approximately 4 h after dosing.[13] Each capsule contains numerous pellets that consist of an inert core surrounded by active drug and rate-controlling membranes that combine water-soluble and water-insoluble polymers.[13] As the pellets lie in the gastrointestinal (GI) tract, water washes over the pellets, slowly dissolving the water-soluble polymer and allowing the drug to diffuse through pores in the coating.[13] The water-insoluble polymer continues to provide a barrier that allows the drug to be dosed every 24 h.[13] For the Covera-HS™, the outermost component of the (COER-24) delivery system is a semipermeable membrane that regulates the absorption of water into the tablet.[13] Water is absorbed from the GI tract at a fixed rate until the second layer, or delay coat, is reached.[13] The second layer then absorbs water and temporarily prevents the passage of water into the inner core of the tablet.[13] This process delays drug release for approximately 4–5 h while the patient is sleeping, when blood pressure is lower.[13] When sufficient moisture has been absorbed, a third layer expands by osmosis, pushing verapamil out of the tablet at a constant rate that adequately controls patient's blood pressure during the morning hours.[13] Continued absorption of water and ongoing osmotic expansion of the third layer provide for extended release of drug and once-daily dosing.[13]

In conclusion, chronotherapeutic verapamil formulations provide effective 24-h control of blood pressure.[13,14] The delay in drug release avoids problems with excessive blood pressure lowering during sleep and provides blood pressure control in the late morning and early afternoon hours when blood pressure is at its highest.[13,14]

27.3.2 Grafts

Since 1952, when the first vascular prostheses were constructed out of the fabric Vinyon N, many researchers have focused upon the production of an ideal synthetic vascular graft.[15] Restenosis complicates the outcome of most of the interventional cardiovascular procedures for relieving coronary artery obstruction.[16] The pathogenesis of restenosis is incompletely understood, but seems to be due to a number of factors that include acute fibrin thrombus, platelet binding, smooth muscle proliferation, and inflammation.[16] Thus, the search for the ideal graft that will maintain a long patency is still on. NO-containing cross-linked polyethylenimine microspheres that release NO with a half-life of

51 h have been applied to vascular grafts to prevent thrombosis and restenosis. The incorporation of the NO to the polymeric matrices has shown powerful antiplatelet activity in cardiovascular grafts.[17] Alternatively, liposomal drug-delivery systems bearing arginine–glycine–aspartic acid (RGD) peptides on the surface could emulate the function of fibrinogen in binding GPIIb–IIIa on activated platelets, and therefore represent a means to target liposome-encapsulated anticoagulant or antiplatelet effects to discrete regions of the cardiovascular system.[18] The RGD peptide utilized has demonstrated to inhibit fibrinogen binding to platelets, which is one of the factors for restenosis.[18] Owing to their relative chemical simplicity, peptides are versatile ligands for use in liposomal drug delivery to molecular targets within the cardiovascular system.[18] Poly(ethylene oxide) (PEO)-grafted phospholipids were shown to dramatically increase liposome survival in the circulation by avoiding rapid reticuloendothelial system uptake.[18] Recently, investigators inserted silastic tubing into the peritoneal cavity of rats and rabbits.[15] The resulting inflammatory reaction to the silastic covered the tubes with layers of myofibroblasts, collagen matrix, and a monolayer of mesothelial cells. By withdrawing the tubing from the laminar, multicellular remains of the inflammatory response and then everting this biological tube, these investigators discovered synthetic arteries with architecture similar to that of native blood vessels.[15]

The mesothelial cells mimicked endothelial cells, the myofibroblasts acted as smooth muscle cells lying in a collagen bed, and the entire structure was surrounded by a collagenous adventitia.[15] The vessels also developed structures similar to high-volume myofilaments that are able to respond to pharmacological agonists.[15] In conclusion, the presumably higher patency rates and longer half-lives of tissue-engineered vascular prostheses and the use of liposomes as potentially advantageous targeted drug carriers for such intravascular applications will be the future for the treatment of restenosis.[15,18]

27.4 Orthopedic: Drug Delivery

27.4.1 Metabolic Bone Diseases

The skeletal system is affected by a host of disorders, including osteoporosis, osteomalacia, renal osteodystrophy, Paget's disease, osteomyelitis, and numerous others. Osteoporosis, a disorder frequently encountered as bone loss associated with aging in postmenopausal women, is now recognized as a major health issue in the United States. Osteoporosis is defined as a universal, gradual reduction in bone mass to a point where the skeleton is compromised.[19] The majority of current treatments for osteoporosis are limited to antiresorptive therapy that slow bone turnover and loss, rather than building new bone mass. Current therapeutic prophylactic and treatment alternatives for osteoporosis include hormone replacement therapy, calcitonin, bisphosphonates, and selective estrogen receptor modulators.

27.4.1.1 Indirect Routes of Administration

Indirect routes of administration include nasal, transdermal, i.v., oral, and s.c.

27.4.1.2 Nasal

Calcitonin is a polypeptide hormone composed of 32 amino acids secreted by the parafollicular cells of the thyroid gland. Although the actions of calcitonin on bone are still not completely understood, it inhibits bone resorption by decreasing the number of osteoclasts and their resorptive activities and limiting osteocytic osteolysis.[20] A synthetic nasal calcitonin formulation is available (Miacalcin®). Calcitonin is also available for s.c. injection.

27.4.1.3 Transdermal

Hormone replacement therapy (i.e., estrogen or estrogen–progestin combinations) has been shown to be beneficial in postmenopausal women at risk for osteoporosis. Women exposed to estrogen therapy for 7–10 years have a 50% reduction in the incidence of osteoporotic fractures.[21] Estrogen therapy for osteoporosis is available as oral tablets and transdermal patches. The patch itself is usually composed of three layers: a backing layer, an adhesive layer, and a protective release layer.

Transdermal patches generally have one of two different designs of the drug compartment: (1) form-fill and seal design and (2) monolith design. In form-fill and seal design, the drug is contained as a liquid or a semisolid reservoir in shallow pouches within the backing layer. The monolith design is further subdivided into the peripheral adhesive laminate structure and solid-state laminate structure; in the peripheral adhesive design, the active delivery area is generally much less than the patch size, and in the solid-state laminate structure, the active delivery area is identical to the size of the patch. Monolith patches are uniform in composition throughout. They can be cut to smaller sizes without compromising their basic delivery function.[21] There are numerous transdermal estrogen patches available ranging from 25 to 100 mcg of delivery per day. Generally, the patches are changed every 72 h; now several weekly patches have been introduced into the market.

27.4.1.4 Intravenous

Bisphosphonates are a class of drugs indicated to treat Paget's disease, hypercalcemia of malignancy, and osteoporosis. They can be given orally and intravenously to humans. Clinical studies of the feasibility of bisphosphonate transdermal delivery and direct delivery to bone via prodrug are being conducted. The i.v. route, though available, is seldom used for the treatment of osteoporosis. This is because large amounts and rapid injection of i.v. bisphosphonates can result in kidney failure.

27.4.1.5 Oral

Bisphosphonates are the mainstay of osteoporosis treatment in the oral setting. The mechanism of action of the bisphosphonates is based on their affinity to bone mineral hydroxyapatite. The bisphosphonates bind strongly to the calcium phosphate crystals and inhibit their growth, aggregation, and dissolution. The biological effects of the bisphosphonates in calcium-related disorders are due to their incorporation in bone, enabling direct interaction with osteoclasts and/or osteoblasts.[20] The oral route, as can be imagined, is the most preferred route for chronic drug therapy. The major disadvantage of the clinically utilized bisphosphonates is their poor bioavailability (less than 1%) due to their hydrophilic nature and their side effects of GI irritation; in addition, food can further suppress absorption, as much as four- to fivefold. Therefore, these drugs are taken on empty stomach. Attempts at circumventing the bioavailability issue have included absorption enhancement through the use of EDTA, development of prodrugs that are lipophilic, and prodrugs that would use carrier-mediated transport systems.[22] EDTA can improve the absorption of these compounds by directly enhancing intestinal permeability. Unfortunately, EDTA damages the mucosal integrity and cannot be used in humans. The other approaches are still in the initial stages of research, none yet available for marketed use in humans.

27.4.1.6 Bone Infections

The term "osteomyelitis" describes any infection involving bone. Osteomyelitis represents a difficult infection to treat for various reasons. First, it has a tendency to be chronic and recurrent, and second, there is a need to deliver high concentrations of antimicrobial agents in the blood to achieve adequate levels in bone.[23] Traditional therapy includes intravenously administered antimicrobials (to avoid issues with bioavailability). Since the classic organisms that infect bone are of a Gram-positive nature, the most frequently used antimicrobials are cephalosporins, extended-spectrum penicillins, and vancomycin for resistant species. Three major considerations are critical in managing these infections in bone:

1. Spectrum of activity of agent chosen
2. Ability of antimicrobial to penetrate and reach the site of infection
3. Duration of therapy

27.4.1.7 Direct Routes of Administration

Other treatment options used for treatment of osteomyelitis include the use of ceramic composites as implantable systems. The treatment of osteomyelitis as previously discussed is a complicated process involving surgical removal of dead bone tissue and prolonged systemic antibiotics. Hydroxyapatite cement systems have been developed to deliver drug to the skeletal tissue at therapeutic concentrations without causing systemic toxicity.[24] These hydroxyapatite cement formulations are loaded with antibiotics such as cephalexin and then placed directly at the site of infection or fracture.

Tricalcium phosphate and amino acid antibiotic composite ceramics and poly(methyl methacrylate) (PMMA) antibiotic-impregnated beads have also been used to treat osteomyelitis with success.[24,25]

27.4.1.8 Drug Targeting

Targeting of bisphosphonate release from bone has been the subject of recent study. This drug-delivery system is based on the concept of a site-specific bisphosphonate prodrug. The system, which is used only in animal trials thus far, is called the osteotropic drug-delivery system. This approach is based on the chemical adsorption of the prodrug to the mineral component, hydroxyapatite.[26] Also, the subject of recent study was the use of the osteotropic drug-delivery system to deliver diclofenac (a nonsteroidal anti-inflammatory drug). This study in rats showed that once the prodrug complex was injected into the animals, it was predominantly distributed to the skeleton. This study showed hope for this approach for highly potent and nontoxic therapy of diclofenac with less frequent medication administration.[27] Studies of other nonsteroidal anti-inflammatory drugs are also being conducted.[28] Future studies will elucidate if these will be of value in humans.

27.5 Muscular System: Drug Delivery

Muscles of the skeletal system in our body have adapted to contract for us to carry out daily functions of motion. Contraction of these muscle fibers is achieved when these cells are stimulated by nerve impulses. Acetylcholine is the neurotransmitter released presynaptically from axon terminals at the neuromuscular junction causing an electrical activation of the skeletal fibers. The interaction of acetylcholine with receptor proteins causes a change in membrane structure that results in the opening of sodium and potassium channels, leading to depolarization.[29]

This perfect mechanism of motion is sometimes affected by autoimmune diseases. Myasthenia gravis is a disease characterized by episodic muscle weakness caused by loss or dysfunction of acetylcholine receptors. Current treatments for myasthenia gravis are limited to relieving the symptoms or immunosuppressing the pathogenesis.[30] Anticholinesterase muscle stimulants such as neostigmine and pyridostigmine have been formulated to inhibit the destruction of acetylcholine by cholinesterase, therefore allowing constant stimulation of postsynaptic cells, leading to muscle contraction.[30]

Access to the skeletal muscle can be gained through indirect (oral, i.v.) or direct (i.m.) routes. As can be imagined, oral route is the preferred route for chronic drug therapy; however, the major disadvantage of these agents are the poor bioavailability (<10%). Poor bioavailability requires frequent administration of large doses, leading to many GI adverse events.[31]

Parenteral routes are also available. These routes of administration are most desirable for treatment when patients have difficulty swallowing or are undergoing a myasthenic crisis.[31] No clinical trials have been done concluding which route of administration is preferred. If the i.v. route is chosen over the i.m., then medication should be infused slowly to be able to look for cholinergic reactions (i.e., bradycardia).[32]

Neostigmine and pyridostigmine also have other therapeutic indications. Both agents can be used for the reversal of nondepolarizing muscle relaxants.[31] Nondepolarizing muscle relaxants such as mivacurium, vecuronium, and pancuronium are used in adjunct to general anesthesia to facilitate endotracheal intubation or to provide skeletal muscle relaxation during surgery or mechanical ventilation.[31] These agents act by antagonizing acetylcholine by competitively binding to cholinergic site on motor endplate, leading to inhibition of skeletal muscle movement.[29] All nondepolarizing neuromuscular blockers are only available in injection form, i.v. use is recommended due to tissue irritation caused by i.m. administration.[33] Duration of action is the only factor that sets a difference between these agents. Currently available short-acting agents include mivacurium and cisatricurium, intermediate acting agents are atracurium, rocuronium, and vancuronium, and long-acting agents include pancuronium and pipecuronium. Generally, short-acting neuromuscular blockers are preferred since limited complications related to prolonged or excessive blockade are avoided.[31] Numerous reports have described the use of neuromuscular blocking agents to facilitate mechanical ventilation; however, none of these reports have compared neuromuscular blockers to placebo.[33]

Currently, there are other pharmacological agents available that affect the skeletal muscle system. These agents are classified into centrally (i.e., baclofen, cyclobenzaprine) or direct-acting (i.e., dantrolene) muscle relaxants.[31] Baclofen is a widely used centrally acting agent for the management of spasticity associated with multiple sclerosis or spinal cord lesion. Baclofen functions by inhibiting transmission of reflexes at the spinal cord by hyperpolarization.[34] Baclofen is also available in direct (intrathecal) or indirect (oral) routes of administration; the intrathecal administration involves a direct injection into the CSF. This route of administration is indicated for the management of severe spasticity of spinal cord origin for patients who are unresponsive to oral baclofen therapy or who experience intolerable CNS side effects at effective doses.[34] When used intrathecally, Baclofen is given as single bolus test dose or for chronic use, only in an implantable pump approved by the Food and Drug Administration (FDA) specifically for baclofen administration.[31]

Oral route is preferred for the rest of the patients because it is convenient and also absorption is rapid from the GI tract. Dantrolene is another muscle relaxant used for the management of spinal cord spasticity, but, as opposed to baclofen, dantrolene acts directly on skeletal muscle. This agent works by interfering with the release of calcium ions from the sarcoplasmic reticulum.[35] Dantrolene is administered orally to manage this condition. A major disadvantage of this route of administration is the slow and incomplete absorption from the GI tract. Injections are also available, but this route of administration is reserved for the treatment of malignant hyperthermia.[35]

27.6 Sensory: Drug Delivery

Advances in biopharmaceutical technology have led to sophisticated drug-delivery devices that allow drugs to be delivered through the skin and mucous membranes of the mouth. Transdermal drug delivery requires that drug molecules have biphasic solubility.[36] The transdermal drug needs lipid solubility to pass through the first layer of skin, the stratum corneum, and aqueous solubility to move through the dermis.[36] The drug must contain high potency, low molecular weight, and insignificant cutaneous metabolism, and the skin must be able to tolerate long-term contact with the drug.[36] Most transdermal drug systems have a rate-controlling membrane that can be a disadvantage because of the slow systemic absorption of the drug.[36] One method used to increase the absorption rate of drugs through the skin is that of iontophoresis.[36] Iontophoresis is defined as "the introduction of ions of soluble salts into the skin or mucosal surfaces of the body by mean of an electric current."[36] Once one activates the current, electrons flow through the skin beneath the electrode being attracted by the oppositely charged electrode on the other side of the skin.[36] The use of ultrasound (sonophoresis), defined as sound of frequency greater than 20 kHz, has also been considered to improve the delivery of transdermal medications.[37] In addition to the elevation of skin temperature, sonophoresis is also reported to induce an increase in pore size and in the formation of small gaseous pockets within cells (cavitation), which is thought to be the predominant mechanism by which low-frequency ultrasound promotes skin penetration enhancement and probably accounts for the enhanced transport of polar molecules.[37] Yet another method being considered is electroporation, which uses high-voltage short-duration pulses to open up new pathways through the stratum corneum, which is thought to create localized regions of membrane permeabilization by producing aqueous pathways in lipid membrane bilayers.[37,38] In this section, different medications and drug-delivery devices will be discussed that deal with the sensation of pain.

27.6.1 Ultrasound

In this particular study, ultrasound was applied continuously at a frequency of 1 MHz, and at an intensity of 2 W/cm² using an ultrasound generator to deliver radiolabeled lidocaine, a local anesthetic, through a piece of stratum corneum.[39] Stratum corneum permeability was enhanced by a factor of about 9 due to ultrasound application, which led to an enhanced diffusion coefficient of most molecules, by a factor ranging from 2.6 to 15 depending on the molecule.[39] Ultrasound is thought to disrupt lipid bilayers, thus allowing a higher rate of solute diffusion.[39] On the other hand, ultrasound in some cases enhanced partition coefficient of some solutes by up to 60% and at the same time decreased the partition coefficient of some drugs by 30%.[39] In any case, the enhanced diffusion coefficients outweigh the decreases in partition coefficients.[39] In another study, ultrasound was used to enhance the permeability of fentanyl, a transdermal opioid agonist; in this experiment, ultrasound was applied using a frequency of 20 kHz with the diameter of the ultrasound probe being 1.3 cm², pulsed for 1 h or continuously for 10 min.[40] "When ultrasound was used in pulsed mode, the diffusion flux of fentanyl was 35-fold greater than controls; however, diffusion flux calculated 7 h after the end of ultrasound exposure was not significantly different from controls."[40] Microscopic study of the skin after ultrasound revealed no damage to the stratum corneum.[40]

The administration of ultrasound to a transdermal patch might allow self-regulation of pain by the patient. Further studies will be needed to conclude the efficacy and safety of ultrasound in the delivery of transdermal medications.

27.6.2 Iontophoresis/Electroporation

In this study, the transdermal delivery of buprenorphine, a synthetic opiate analgesic, was assisted using iontophoresis and/or electroporation.[38] A current of 0.5 mA/cm² was applied for 4 h and sampling was continued for 24 h.[38] The amount delivered under anode was much higher than that delivered under cathode due to the fact that buprenorphine has a positive charge at a pH of 4.[38] Electroporation alone was unable to enhance transport of buprenorphine across the skin.[38] On the other hand, electroporation and iontophoresis combined to produce a delivery, which was over six times higher than that achieved by electroporation alone and about twice that achieved by iontophoresis alone.[38] In conclusion, the combination of iontophoresis and electroporation may be used in the future to control drug release and to increase drug permeation across the skin.

27.7 Digestive System: Drug Delivery

In general, the digestive system is made up of the stomach, small intestine, and large intestine. The fasting pH of the stomach is about 2–6, while in the presence of food, the stomach pH is about 1.5–2; thus, basic drugs are solubilized rapidly in the presence of stomach acid.[41] Stomach emptying is influenced by the food content and osmolality. Food often slows down the gastric emptying time, usually allowing an oral medication anywhere from 3 to 6 h to empty out of the stomach.[41] The duodenum, the upper portion of the small intestine, is the optimum site for drug absorption.[41] This is because of the unique anatomy of the duodenum, which possesses microvilli that provide a large surface area for drugs to passively diffuse through.[41] The ileum, which is the terminal part of the small intestine, also plays a role in the absorption of hydrophobic drugs.[41] On the other hand, the large intestine lacks the microvilli of the small intestine, so it is very limited in drug absorption.[41] In this section, new delivery methods to the digestive tract will be discussed.

27.7.1 GI Stents

Expandable metal stents have been approved by the FDA for the treatment of GI obstruction due to cancer and could possibly be used in other benign diseases.[42] In the past, plastic stents have been used since they are relatively inexpensive compared to metal stents.[43] In this study by Knyrim et al., where two groups of patients either received plastic or metal stents for esophageal obstruction, complications of device placement and functioning were significantly more frequent in the plastic-prosthesis group than in the metal stent group.[43] In addition, metal stents are easier to place and require less dilation than plastic stents, and are less expensive after cost analysis.[43] The major problem with metal stents is tumor ingrowth, but this can be treated by laser ablation, which can be done in an outpatient setting.[43] These gastric stents are made up of different metal alloys and are put in by gastroenterologists.[43] The stents collapse to 3 mm in diameter at placement but can then expand up to 16 mm after being positioned in the stricture.[43] These metal stents are used for esophageal carcinoma, in which all other treatment options have failed to produce any relief of dysphagia.[42] Dysphagia is usually relieved in up to 90% of patients who have metal stents placed and in all the patients in the Knyrim et al. trial.[42,43] Esophageal expandable metal stents are also used to treat tracheoesophageal fistulas due to cancer and may increase the survival of the patient, and that is why it is considered the primary treatment option.[42] Stents may also be used in the upper GI tract and for cancerous large-bowel obstruction.[42] In the future, biodegradable stents could be used for the treatment of benign diseases. Stents that release chemotherapeutic agents or radiation to facilitate tumor regression, may also be developed in the near future.[42]

27.7.2 Colonic Drug Delivery

There are many different designs of colonic delivery systems, and targeting has been achieved in several ways: coating drugs with pH-sensitive polymer, coating drugs with bacterial degradable polymers, using prodrugs, and delivering drugs through bacterial degradable matrixes.[44] The colon-targeted delivery capsule was recently developed at Tanabe Co. Ltd., and was designed by making three different layers along with an organic acid that is used as a pH adjusting agent along with the medication.[44]

The three different layers are an enteric-coated layer, a hydrophilic layer, and an acid-soluble layer.[44] Using this system, the drug does not release until at least 5 h without regard to fed or fasted patients.[44] Electrostatic interaction between polyanions and polycation led to the formation of polyelectrolyte complexes (PECs), which can provide a greater barrier to drug release in the upper gastrointestinal tract (GI tract) than either material alone.[45] Thus, pectin, a polyanion, and chitosan, a polycation, can be used together to better improve drug delivery to the colon.[45] In this study, the optimum ratio of PEC was 10:1 weight ratio of pectin to chitosan.[45] Another study found that the ratio of pectin to chitosan to hydroxypropyl methylcellulose (HPMC) of 3:1:1 also had the potential of colonic selective delivery.[46] The delivery to the colon would ultimately occur when the bacterial enzymes commence to break down the pectin and the medication is released.[45,46] In another study using 5-aminosalicylic acid (5-ASA), which is used to treat ulcerative colitis, it was found that using an 80:20 pectin–HPMC coating mixture provided an intermediate erosion pattern for the colonic delivery of 5-ASA tablets.[47]

This is promising to patients who have ulcerative colitis, whose current treatment options include rectally applied foams, suppositories, and enemas.[47]

27.8 Pulmonary: Drug Delivery

Inhalation is one of the oldest modes of drug delivery dating back to the earliest days of medical history. Medications were added to boiling water for patients to inhale.[22] Many advances have come from the renewed interest in this form of drug delivery.

Direct administration using inhalation as a method of drug delivery to the respiratory tract has become well established in the treatment of lung disease. This route has several advantages. Medication is delivered directly to the tracheobronchial tree, allowing for rapid and predictable onset of action; the first-pass effect is circumvented; degradation within the GI tract is avoided; much lower dosages than by the oral route can be administered with equivalent therapeutic efficacy, minimizing the potential for undesired side effects; and it can be used as an alternative route to avoid potential drug interactions when two or more medications are used concurrently.

For many years, theophylline was the gold standard for the treatment of asthma. It is now known that asthma is an inflammatory process best treated on a chronic basis with corticosteroids. Nonetheless, theophylline continues to be used to treat asthma. It is available as an injectable to be administered via i.v. infusion, as controlled-release tablets, and liquid suspensions and rectal suppositories.

Bronchodilators and corticosteroids are the mainstay of treatment for asthma and chronic obstructive pulmonary disease (COPD). Administration via inhalation reduces systemic exposure of these compounds and unwanted systemic side effects. Bronchodilators exert their action by relaxing airway smooth muscle. Numerous compounds are available on the market with varying degrees of duration of action. Some bronchodilators are available for injectable use such as terbutaline, isoproterenol, and epinephrine. Others are available as oral solutions or tablets such as albuterol. Unfortunately, systemic use of these drugs to treat respiratory conditions results in unwanted systemic side effects. These side effects include conditions such as tachycardia; therefore, direct administration into the lungs is desirable.

Corticosteroids affect the inflammation caused by airway diseases. Corticosteroids are available for injectable use as i.v. or i.m. depot injections, as oral solutions and tablets. Again, systemic use of corticosteroids can lead to many undesirable side effects, including osteoporosis, hyperglycemia, and electrolyte imbalances.

27.8.1 Indirect Routes of Administration

Indirect routes of administration for agents available to treat the respiratory system include i.v., i.m., oral, and rectal. A discussion of direct methods of administration follows.

27.8.2 Direct Routes of Administration

Direct routes of administration through the mouth and into the lungs can be categorized in the following manner:

1. Nebulizers (ultrasonic or jet)
2. Metered dose inhalers (MDIs)
3. Dry powder inhalers
4. Administration through chest tube into the pleural cavity

The respiratory tract consists of multiple generations of branching airways (pharynx, larynx, trachea, bronchi, bronchioles, and alveoli) that progressively decrease in diameter but increase in number and total surface area. The large surface area of bronchioles and alveoli facilitates the rapid absorption of inhaled drugs.[48]

27.8.2.1 Jet Nebulizers

Jet nebulizers use compressed gas (air or oxygen) from a compressed gas cylinder, hospital air-line, or electrical compressor to convert a liquid into a spray. The aerosol leaving the nebulizer is diluted by atmospheric air and inhaled through a face mask or mouthpiece. The ability of an aerosol to penetrate the respiratory tract is directly related to its efficacy. The most important property to possess that governs penetration and deposition in the respiratory tract is particle size. Particles must be less than 5 μm and preferably less than 2 μm for alveolar deposition.[49] For the most part, drugs are in aqueous solution form when available for nebulization. There are multidose preparations available, but most nebulizer formulations are packaged as sterile, isotonic preservative-free unit doses. Examples include albuterol, *n*-acetylcysteine, and cromolyn sodium, all drugs used in the treatment of asthma.

27.8.2.2 Ultrasonic Nebulizers

In ultrasonic nebulizers, the energy required to atomize a liquid comes from a piezoelectric crystal, usually a man-made ceramic material, vibrating at high frequency.[48]

Commercially available ultrasonic nebulizers produce aerosol droplets that are often significantly larger than those produced by jet nebulizers. The absence of droplets with size less than 2 μm suggests that such nebulizers may be inappropriate for applications requiring that the drug penetrates into the most peripheral lung regions.[50]

As previously mentioned, most solutions for nebulization are aqueous; however, drugs poorly soluble in water can be formulated as suspensions. For example, Pulmicort Repulses® consists of a suspension of the corticosteroid budesonide; in general, ultrasonic nebulizers are less efficient and more variable in delivering suspensions than jet nebulizers.[50] Nebulizers are established devices for the delivery of therapy to the lungs. They have advantages over other systems in the elderly and pediatric population. This advantage stems from the fact that the drug may be inhaled during normal tidal breathing through a mouthpiece or face mask. Other delivery methods require coordination of inhalation and activation of the device, which would be unsuitable for the very elderly or for the very young.

27.8.2.3 Metered Dose Inhalers

The MDI is currently the most widely used inhalation delivery device. This is due to its portability, durability, long shelf-life, cost effectiveness, and relative ease of use. Unlike nebulizers, MDIs require metering and dispensing in coordination with the patient's inspiratory cycle. Therefore, successful lung delivery depends on a patient's ability to operate the inhalation device properly. Current improvements include the use of spacer devices. One of the biggest challenges with MDIs is to reduce the amount of drug that is deposited in the oropharyngeal area instead of the lungs. Spacer devices have the ability to reduce the speed of the emitted aerosol cloud and reduce the deposition of drug in the throat by as much as 45%.[51]

Drugs available for delivery through these devices include β-agonists for smooth muscle relaxation, corticosteroids, and mast cell stabilizers. MDIs available on the market containing these products are numerous. As previously mentioned, all these drugs treat asthma and COPD.

27.8.2.4 Dry Powder Inhalers

Prior to 1987, aerosolized MDIs were delivered via systems that relied on chlorofluorocarbon (CFC) propellant systems. There was a subsequent ban on all nonmedical uses of these CFC products that could deplete the ozone layer. Pharmaceutical manufacturers were encouraged to investigate other propellant systems; in addition to researching new propellants following the ban on CFC, pharmaceutical companies began to develop inhalable drugs in new forms such as dry powders.[52]

There are several different designs for dry powder inhalation devices (DPIDs). One such design is the Spinhaler; in this device, a gelatin capsule filled with drug and excipient is mounted in a rotor upon which are several small fan blades. The capsule is pierced by two small needles by sliding the outer casing of the inhaler relative to the inner casing. When the patient inhales, the capsule rotates rapidly and empties its content.[48] Some devices such as the Diskhaler® make it possible for the patient to know how many doses remain. This is an advantage over the MDI, which does not have this capacity. As with MDI, these devices rely on the patient's inspiratory effort; in frail, elderly patients or small children, this can be an issue with adequate drug delivery. The drug formulations for these devices exist as capsules or disks. The majority of the drug formulations again consist of β-agonists, corticosteroids, and mast cell stabilizers. A novel formulation being investigated with these devices is the use of vaccines by inhalation. It is postulated that there is potential for enhanced biological efficacy since pulmonary delivery may produce mucosal immunity superior to that produced by parenteral administration of vaccines.

Studies have shown the safety and efficacy of measles vaccine delivered as a liquid aerosol from a nebulizer. A powder formulation of measles vaccine has been formulated for aerosol delivery in feasibility study. Challenges remain such as the development of appropriate delivery technology and reduction in the hygroscopic nature of the formulation.[53]

27.8.2.5 Chest Tube Administration

Malignant pleural effusions are a common complication in advanced malignancy. Metastatic lung and breast cancer account for 75% of the cases.[54] In this procedure, a small bore thoracostomy tube is placed under local anesthesia. Then, a sclerosing agent is instilled through the tube. Agents that have been used with success include bleomycin, doxycycline, and talc poudrage. The sclerosing agent instillation, if successful, will stop the accumulation of the fluid in the lung.

27.9 Ear, Nose, and Throat: Drug Delivery

27.9.1 Ear

Diseases of the ear, most commonly infections, are very prevalent in children. Extensive use of antibiotics for this indication has led to marked antimicrobial resistance. Recently, the American Academy of Otolaryngology convened to set consensus on the treatment of common ear ailments. These included chronic suppurative otitis media, otitis externa, and tympanostomy tube otorrhea. The consensus was that in the absence of systemic infection or serious underlying disease, topical antibiotics alone should constitute the first-line treatment.[55]

27.9.1.1 Indirect Routes of Administration

Indirect routes of administration include oral, i.v., and i.m. antibiotics. Using these routes, the antibiotic reaches the systemic blood circulation, which then enters the middle and inner ear.

27.9.1.2 Direct Routes of Administration

Direct administration includes the use of antibiotic and anti-inflammatory ear drops such as Cipro® HC Otic Suspension or Otobiotic Otic (which contains a mixture of antibiotic and anti-inflammatory). Solutions such as Otocain contain topical analgesics (benzocaine) to alleviate ear pain.

Novel direct delivery devices being studied in animals include a biodegradable support matrix incorporating a therapeutically releasable amount of antibiotic. This device is then inserted into the middle ear and is capable of drug delivery for 3 months. Progression into human studies may show potential for this device to be used as a source of inner ear drug delivery.[56]

Intratympanic therapy for Meniere's disease has also been studied. Gentamicin solution of 0.5 mL has been injected transtympanically using a tuberculin syringe and a 27-gauge long needle. Patients treated in the study had good response to treatment with over 50% having complete control of their vertigo.[57]

27.9.2 Nose

Drugs have been administered nasally for both topical and systemic action. Common ailments that affect the nose and are treated with topical therapy include allergic rhinitis, congestion, and sinusitis. Acute sinusitis is a condition manifesting inflammation and infection, usually of the frontal and maxillary sinuses. Goals of therapy are to improve drainage in the blocked sinuses and resolve the infection. Steam inhalation can cause vasoconstriction and help with drainage as can topical vasoconstrictors such as phenylephrine spray.

27.9.2.1 Indirect Routes of Administration

Indirect routes of administration include oral and i.v. use of antibiotics.

27.9.2.2 Direct Routes of Administration

Direct routes of administration include the use of nasal sprays and, for more accurate dosing, mechanical pumps and pressurized aerosol systems. Allergic rhinitis is a prevalent condition that frequently appears in patients who also have asthma. It is best treated by nasal corticosteroids. The use of these medications directly to the nasal mucosa avoids exposing the body to systemic levels of corticosteroids and their potential side effects.

One of the simplest and oldest methods of nasal drug delivery is the use of a device to administer solutions via dropper. This system, while being cost-effective and easy to manufacture, has the disadvantage of not accurately measuring drugs because of its inability to control the exact volume delivered. Squeezed nasal bottles are mainly used as a delivery device for decongestants, such as Afrin® spray. They function by pressing the bottle and pushing the air inside the bottle through a simple jet, which atomizes a certain volume of fluid. Metered-dose pump sprays (MDPSs) allow for the application of a defined dose with a high dosing accuracy. These systems consist of the container, the pump with the valve, and the actuator.[58]

Powder dosage forms are under study and include inhaled insulin. Although dry powders have advantages over liquid formulations, they are infrequently used in nasal drug delivery. Nasal gels have also been studied as a means of prolonging drug contact with the nasal mucosa. One product is marketed as vitamin B-12 (Nascobal® gel).

27.9.3 Throat

Infection of the throat is a common ailment as is cough. Many throat infections are self-limiting and require treatment only with analgesics. Streptococcal throat infections, however, do require antimicrobial treatment because of their potential to damage the heart valves. Treatment for these infections consists of systemic oral antibiotics; the drug of choice is a simple penicillin. Cough is another common ailment that is treated with different medications depending on the cause of the cough. For cough from the common cold and allergic rhinitis, oral antihistamines and decongestants along with ipratropium nasal spray can be used. If the cough is due to COPD, a 2-week trial of oral corticosteroids can be utilized.[59]

Other over-the-counter treatment modalities include liquid sprays that contain topical anesthetics for throat pain, such as Orasept® throat spray, or lozenges such as Cepacol®.

A novel treatment in throat disorders involves the use of botulinum toxin to treat spasmodic dysphonias. This condition is a disorder that results in the patient having either a strained or strangled voice or a breathy, whispery voice. The toxin is injected directly into the posterior cricoarytenoid muscle, paralyzing the muscle that is causing the spasm and resulting in relief of the condition.[60]

27.10 Lymphatic System: Drug Delivery

The human body is exposed to a wide variety of diseases of benign and malignant origin. These diseases affect the lymphatic system at the early stages of the process. Cancers, as well as many infections (viral, bacterial, or fungal), spread by lymphatic dissemination.[61] The high prevalence of lymph node involvement in these diseases is not surprising because the primary function of the lymphatic tissue is to provide the body's immune response.

Appropriate diagnosis and treatment of diseases affecting the lymph nodes depend on the availability of drugs that are retained by the lymph nodes. Effective accumulation of drugs in the lymph nodes can be achieved by intralymphatic or interstitial administration.[62] Since there is a high variability of lymphatic networks and drainage routes, systemic administration would always be preferred and currently represents a focus in lymphotropic drug design.[63]

The interstitial space constitutes a significant barrier to the diffusion of macromolecules and particulates and may limit the rate and extent of drainage from the site of administration. Macromolecules dissolved in the interstitial fluid are readily drained by lymphatics; therefore, the limiting factor to macromolecular transfer from blood capillary to lymph is capillary permeability. The rate of macromolecule and particle extravasation depends on the size and structure. Macromolecules may be transferred in liquid phase, while some proteins are transported by transcytosis-associated receptors. The nature and the size of lymphatic vessels vary along the route.[62]

Access routes to the lymph nodes can be accomplished through lymphatic vessels or blood vessels. Drug delivery through the lymphatic vessels is highly efficient; interstitial macromolecules are cleared from the injection site almost exclusively by lymphatics, but do not return to blood capillaries. Direct drug delivery through blood vessels to intranodal tissue is not very effective unless homing receptors are utilized.[61] Different local administration routes have been utilized to deliver diagnostic or therapeutic drugs to the lymph nodes.

For intralymphatic administration, a peripheral lymphatic vessel has to be cannulated by surgical cut-down. This method results in high concentration of drug in the lymph nodes, but it is only limited to draining lymph nodes, leading to an uneven drug distribution. This method has been used for x-ray lymphography and CT. Agents used include iodized oils.[62]

Interstitial administration does not require cannulation of a lymphatic vessel, and therefore is easier to perform. Agents can be injected at any accessible anatomical site, where they penetrate from the interstitial space into small lymphatic vessels through intercellular gaps of lymphatic endothelium. The agents are then transported through the network of lymphatic vessels to peripheral lymph nodes. The disadvantage of the interstitial route is that it provides low, unreliable drug delivery to mediastinal and abdominal lymph nodes, which are involved in the majority of carcinomas.[64]

Oral delivery of lipophilic drugs to the lymph nodes is associated with the formation and transport of lipoproteins, which are formed after the adsorption of lipid digestion products. Lymphatic transport of polar drugs after intestinal absorption is lower because of their preferential absorption by blood.

Intra-arterial injections of drugs carrying particulates that are too large to pass through capillary vessels result in high local tissue concentration of the released drugs, which can then diffuse to local lymph nodes. This method is used to chemoembolize tumors rather than to treat metastatic diseases.[62]

In addition to locally administered drugs, there are systemic agents that can be injected intravenously and can accumulate in the lymph nodes by different mechanisms. Low-molecular-weight compounds such as gallium citrate accumulates in normal lymphatic tissue and lymphomas; however, because of its nonspecific biodistribution, it is not used very often except for imaging of lymphomas, tumor recurrence, or sarcoid imaging. Metalloporphyrins have also been used for selective lymphatic imaging and MRI of human colon carcinoma.[65]

Radiolabeled lymphocytes are another source for clinical use in detecting sites of inflammation and lymphoreticular malignancies. Usually, lymphocytes are incubated with an indium oxide complex and the resulting labeling efficiency is 50–60%.[63]

Homing receptors of lymphocytes are responsible for cell accumulation in lymphatic tissue and inflammations (where homing molecules are also expressed). Because the homing process is highly selective, it is possible that the future availability of vector molecules with specificity to lymphocyte adhesion glycoproteins will allow efficient drug delivery.

Colloidal iron oxide particles have been studied as diagnostic MR contrast agents. It has been found that after i.v. administration, some dextran-coated colloids accumulate in lymph nodes in much higher concentrations than any other particulate.[65]

Lymph nodes are an easy target for drug delivery through intralymphatic or interstitial administration with local concentrations achievable. These administration routes are rarely used because of the unreliable and highly variable delivery to different lymph node groups. Because of these factors, system carriers are being developed for lymph node delivery. The two classes of agents are (1) those that have long circulation times and are able to extravasate into the interstitium and then are cleared by lymphatics (i.e., dextran-covered particles of dextran-based graft copolymers) and (2) those targeting lymph node-specific lymphocyte homing receptors or antigens.[61]

27.11 Reproductive System: Drug Delivery

The female reproductive tract is divided into external and internal genitalia. Parts of the female anatomy include the pelvis, bladder, urethra, and vagina.[66] For reproduction, the human endometrium must receive hormonal signals that prepare endometrium for implantation.[66] When conception does not occur, these signals initiate mechanisms that lead to menses and controlled regeneration of this tissue, and the cycle repeats itself again.[66] In contrast, the male reproductive system is composed of the scrotal sac, testes, genital ducts, accessory glands, and penis.[67] The scrotal sac performs an important role in maintaining the testes at a temperature about 2°C below the temperature of the internal organs so that spermatogenesis can occur.[67] In the following section, new reproductive system drug-delivery methods will be discussed.

27.11.1 Contraceptive Implants

Norplant™ (The Population Council) and Jadelle are the only subdermal implants currently available in the United States, even though there are several more that are currently being used in other countries.[68]

Norplant consists of six capsules that release the progestin, levonorgestrel, for at least 5 years; after this time, there is still 69% of the original steroid load left in the silastic capsules to act as a safety margin for women who do not remove the implants after the recommended 5 years.[68–70] The system consists of six silastic capsules, 34-mm long and 2.4 mm in diameter.[69] Jadelle™ (The Population Council) also uses levonorgestrel just like Norplant.[70] The only difference between the products is that Jadelle uses two silastic rods instead of six capsules, thus making insertion and removal easier for the Jadelle implant.[70]

Each Jadelle implant is 4.4-cm long and is composed of a polydimethylsiloxane elastomer covered by silicone rubber tubing.[68] Jadelle is approved by the FDA for 3 years of use, but has been shown to be effective for up to 5 years.[68] The next implant that is currently used in other countries is the Implanon™ (NV Organon, Oss, The Netherlands).[70] Implanon is a single-rod implant that is 4 cm long and 2 mm in diameter.[68] It contains 68 mg of 3-keto-desogestrel (etonogestrel) in an ethylene vinyl acetate (EVA) polymer core surrounded by an EVA membrane with a contraceptive dose maintained for 3 years.[68,69] Its contraceptive efficacy is excellent since there has not been even one report of a single pregnancy in over 5000 woman-years of experience with this product.[70] Yet another implant used in other countries is the Uniplant™. Uniplant is a single silastic 3.5-cm-long, 2.4-mm-diameter capsule containing 55 mg nomegestrol acetate developed as a single 1-year contraceptive implant.[68–70] An alternative approach to resolving the difficulties of implant removal that is considered a minor surgery and usually takes about 20 min is to eliminate the need for removal altogether.[68–70] Capronor™ is a 40-mm rod containing levonorgestrel in an E-caprolactone polymer.[68] The polymer releases levonorgestrel about 10 times faster than silastic and thereby one implant can achieve adequate serum concentrations, instead of normally two implants.[68]

The implant is biodegradable and it appears to remain for about 1 year.[68] Another form of biodegradable implants is pellets.[68] These pellets are expected to dissolve within 2 years of application, but are impossible to take out after several months of implantation.[68] In conclusion, some of the advantages of progestin implants are long unattended use, efficacy, no compliance issues, and lower levels of progestin when compared to oral contraceptives.[70] Disadvantages are that a minor surgery is required for insertion and removal and there is a high cost associated with the method if early removal is performed.[70]

27.11.2 Contraceptive Patch

Ortho Evra/Evra™ (Janssen Pharmaceutica, NV Belgium) is the only available female contraceptive transdermal patch in the market. The matrix patch that is 20 cm^2 is thin and consists of three layers: an outer protective layer of polyester; a medicated, adhesive middle layer; and a clear, polyester release liner that is removed prior to patch application.[71,72] The patch is designed to deliver 150 μg of norelgestromin and 20 µg of ethinyl estradiol daily to the systemic circulation.[71–73] The patch can be applied to the buttocks, upper outer arm, lower abdomen, or upper torso (excluding the breast).[71] Because the patch is replaced weekly on the same day of the week for three consecutive weeks (followed by 1 week patch-free) and the next patch cycle begins on the same day, it makes patient compliance easier when compared with oral contraceptives.[73]

27.11.3 Male Contraceptive

Androgen therapy is predominantly used for replacement in primary hypogonadism but may be used for male contraception.[74] Desogestrel DSG (300 µg daily) and transdermal testosterone (T0) (5 mg daily) patches were given to male patients for 24 weeks.[75] Using this regimen, 71% of the patients were azoospermia (no sperm) by the end of week 12.[75] This regimen was not as effective as using IM T enanthate.[75] The lower efficacy of the transdermal T is likely to be due to failure of the transdermal T system in maintaining circulating T levels consistently in the required range.[75] In conclusion, the use of desogestrel and IM T enanthate might lead to male contraception in the near future.

27.12 Emerging Technologies

27.12.1 Nanotechnology/Nanomedicine

Nanotechnology or more specifically nanomedicine encompasses the science of diagnosing, treating, and preventing disease and injury, relieving pain and suffering, and preserving and improving health, using microscopic tools and the molecular knowledge of the human organism.

According to the FDA, nanomedicine is the understanding and control of matter at dimensions between 1 and 100 nm, where unique phenomena enable novel applications (most pharmaceutical applications of nanotechnology nanoparticles averaging 1000 nm).

Current nanotechnology research includes therapeutic and diagnostic applications that will revolutionize the medical and more specifically the pharmaceutical industries in the near future.

Nanoencapsulation of medicines in a small polymer or lipid matrix will facilitate the passage through the GI lining, reaching the bloodstream where the therapeutic entity will be released. Nanoparticles may facilitate the oral administration of drugs that are currently only administered by injection.

Nanoparticles, many as small as a virus, may penetrate diseased cells and enable more effective diagnostic techniques and more efficient therapeutic treatment options. In summary, nanomedicine will improve a patient's quality of care by facilitating early detection and more effective treatments, with far less drug-related side effects.

Current research in nanomedicine is quite varied encompassing realms that include liposomes, nanocrystals, magnestic nanoparticles, gold nanoparticles, quantum dots, microspheres, carbon nanotubes, polymeric nanopartcles, and a variety of other nanoapplications. Most of these products are still in preclinical trials, while others have already found themselves on the market, providing many clinical therapeutic applications,

27.12.2 Lipid-Based Nanoparticles

Lipid-based nanoparticles include liposomes. Liposomes are composed of one or more concentric lipid bilayers encapsulating an inner aqueous core. The outer layer is composed primarily of a hydrophobic, phospholipid layer that protects the encapsulated therapeutic entity from harmful environment until reaching the target tissue, where the therapeutic entity is released. The outer surface of the liposome can be altered and functionalized to enhance the drug-delivery system.

The liposomal carrier can increase the circulation time and the polymer coating may provide a shielding quality to the drug, by avoiding the immune system. The liposomal encapsulation may limit the deleterious side effects of the payload drug by avoiding healthy tissues and targeting the desired tissues. There are currently eight FDA-approved products utilizing lipid-based nanoparticles.

27.12.3 Polymeric Nanoparticles

Polymeric nanoparticles provide the solution to some of the most persistent challenges in drug delivery, drug solubility, drug stability, drug half-life, and reduction of toxicity to nontarget tissues.

Polymeric nanoparticles are defined by their structure and polymer composition, with the therapeutic agent conjugated to the surface, or centered in the core of the nanoparticle.

Polymer-based nanoparticles have the ability to alter the release profile of the drug, control the targeting of the therapeutic agent through active or passive targeting, while minimizing its degradation within the bloodstream. The majority of FDA-approved agents utilize PEGylation (polyethylene glycol), in which there is a covalent attachment of the polymer PEG to the therapeutic drug or drug carrier. By controlling the structure of the PEG molecules, the molecular weight and the number of chains as well as the attachment chemistry PEGylation can protect the active therapeutic agent from being seen and acted upon by the immune system, and can reduce renal filtration and elimination, altering the biodistribution of the therapeutic agent by increasing the circulation half-life (this in turn will enable a reduction in total dose and a reduction of frequency of administration).

This has great importance for the delivery of naturally unstable and short half-life agents such as proteins, enzymes, hormones, and antibodies.

27.12.4 Polymeric Micelles

Much akin to the spheroidal structure of liposomes, micelles are aggregates of surfactants or polymers dispersed in an aqueous solution but do not have an internal aqueous phase, like that of liposomes.

Polymeric micelles have the same advantages as other polymeric and liposomal formulations, which is the protection of the therapeutic entity from degradation and an increase in circulation half-life.

In general, polymeric micelles are made with two different polymers, a hydrophilic shell that is responsible for colloidal stability and protects the therapeutic ingredients, and a hydrophobic core polymer that protects the ingredients either physically or chemically or both.

27.12.5 Nanoparticles/Microparticles

Microencapsulation has been widely used in the pharmaceutical industry for many years.

Microspheres and microcapsules (collectively referred to as microparticles) have been used in many FDA-approved medications to enhance drug delivery. These drug-delivery systems contain a variety of polymers, including both biodegradable and nonbiodegradable. Microparticles are used in drug formulations to control the release of the drug, to protect sensitive therapeutics from degradation, and to allow for surface functionality of the microparticle for targeting and delivery. The degradation rate and drug release rate of microparticles are controlled by the material choice, porosity, surface properties, and the size of the microparticles. Although they are larger than typical nanoparticles, microparticles are usually found on a scale of microns rather than nanometers, but have similar properties and functions as those particles that fit within the exact definition of nanotechnology.

27.12.6 Other Nanomedicines

There are emerging technologies utilizing protein-based nanoparticles. Similar to polymeric nanoparticles, natural proteins may also be used to improve the pharmacokinetic profile and reduce the toxicity of current drug formulations. These formulations may be stabilized utilizing a patient's own serum albumin, the protein faction of the blood.

Nanoemulsions utilized in drug-delivery systems are another field of research in the realm of nanomedicine. Nanoemulsions are stabilized nano-sized oil droplets emulsified in water. The oil droplets can range in size from 10 to 500 nm in diameter and may act as carriers for water-insoluble therapeutic entities and drug compounds.

The future applications of nanomedicine in drug-delivery systems are limitless. There are literally thousands of potential products based on nanotechnology in the preclinical developmental stages.

With continued innovations, the number and type of products utilizing nanotechnology will continue to grow by leaps and bounds.

References

1. Marieb, E., The central nervous system, in *Human Anatomy & Physiology*, 4th ed., Fox, D., Ed., Benjamin/Cummings Science Publishing, Menlo Park, CA, 1997, pp. 405–455.
2. Minn, A et al., Drug metabolism in the brain: Benefits and risks, in *The Blood-Brain Barrier and Drug Delivery to the CNS*, Begley, D., Bradbury, M., and Kreuter, J., Eds., Marcel Dekker, New York, 2000, pp. 145–170.
3. Bradbury, M., History and physiology of the blood-brain barrier in relation to delivery of drugs to the brain, in *The Blood-Brain Barrier and Drug Delivery to the CNS*, Begley, D., Bradbury, M., and Kreuter, J., Eds., Marcel Dekker, New York, 2000, pp. 1–8.
4. Madaras-Kelly, K et al., Central nervous system infections, in *Pharmacotherapy: A Pathophysiologic Approach*, 3rd ed., Dipiro, J et al., Eds., Appleton-Lange, Norwalk, CT, 1997, pp. 1971–1993.
5. Partridge, W.M., Invasive brain drug delivery, in *Brain Drug Targeting: The Future of Brain Drug Development*, Cambridge University Press, New York, 2001, pp. 13–35.
6. Bernards, C.M., Epidural and intrathecal drug movement, in *Spinal Drug Delivery*, Yaksh, T., Ed., Elsevier, New York, 1999, pp. 239–252.
7. Wallace, M.S., Human spinal drug delivery: Methods and technology, in *Spinal Drug Delivery*, Yaksh, T., Ed., Elsevier, New York, 1999, pp. 345–370.
8. Haroun, R.I. and Brem, H., Local drug delivery, *Curr. Opin. Oncol.*, 12, 187, 2000.
9. Scheld, W.M., Drug delivery to the central nervous system: General principles and relevance to therapy for infections of the central nervous system, *Rev. Infect. Dis.*, 11S, 1669, 1989.
10. Harbaugh, R.E, Saunders, R.L., and Reeder, R.F., Use of implantable pumps for central nervous system drug infusions to treat neurological disease, *Neurosurgery*, 23, 693, 1988.
11. Kumar, M. and Banker, G., Biological processes and events involved in drug targeting, in *Modern Pharmaceutics*, Vol. 72, Banker, G. and Rhodes, C., Eds., Marcel Dekker, New York, 1996, pp. 613–625.
12. Zaret, B.L., Berliner, R.W., Moser, M., et al., Cardiovascular drugs, in *Yale University School of Medicine Heartbook*, Goetz, D.M., Ed., Hearst Books, New York, 1992, chap. 23.
13. Smith, D.H.G., Pharmacology of cardiovascular chronotherapeutic agents, *Am. J. Hypertension*, 14(9), Suppl. 1, S296, 2001.
14. Smolensky, M.H. and Portaluppi, F., Chronopharmacology and chronotherapy of cardiovascular medications: Relevance to prevention and treatment of coronary heart disease, *Am. Heart J.*, 137(4), S14, 1999.
15. Sahil, A.P. and Edelman, E.R., Endothelial cell delivery for cardiovascular therapy, *Adv. Drug Delivery Rev.*, 42(1–2), 139, 2000.
16. Levy, R.J., Labhasetwar, V., Song, C., Lerner, E., et al., Polymeric drug delivery systems for treatment of cardiovascular calcification, arrhythmias, and restenosis, *J. Controlled Release*, 36, 137, 1995.
17. Ignarro, L.J., Napoli, C., and Loscalzo, J., Nitric oxide donors and cardiovascular agents modulating the bioactivity of nitric oxide: An overview, *Circ. Res.*, 90(1), 21, 2002.
18. Lestini, B.J., Sagnella, S.M., Xu, Z., Shive, M.S., et al., Surface modification of liposomes for selective cell targeting in cardiovascular drug delivery, *J. Controlled Release*, 78, 1–3, 235, 2002.

19. O' Connell, M.B. and Bauwens, S.F., Osteoporosis and osteomalacia, in *Pharmacotherapy: A Pathophysiologic Approach*, 3rd ed., Dipiro, J et al., Eds., Appleton-Lange, Norwalk, CT, 1997, pp. 1689–1716.

20. Patton, J.S., Pulmonary delivery of drugs for bone disorders, *Adv. Drug Deliv. Rev.*, 42, 239, 2000.

21. Ramachandran, C. and Fleisher, D., Transdermal delivery of drugs for the treatment of bone diseases, *Adv. Drug Deliv. Rev.*, 42, 197, 2000.

22. Newman, S.P. and Busse, W.W., Evolution of dry powder inhaler design, formulation, and performance, *Respir. Med.*, 96, 293, 2002.

23. Norden, C., Gillespie, W., and Nade, S., Principles of management, in *Infections in Bones and Joints*, Blackwell Scientific, Cambridge, 1994, pp. 115–118.

24. Dash, A. and Cudworth, G., Therapeutic applications of implantable drug delivery systems, *J. Pharmacol. Toxicol. Methods*, 40, 1, 1998.

25. Alonge, T.O. and Fashina, A.N., Ceftriaxone-PMMA beads—A slow release preparation? *Int. J. Clin. Practice*, 54, 353, 2000.

26. Fujisaki, J et al., Osteotropic drug delivery system (ODDS) based on bisphosphonic prodrug. IV. Effects of osteotropic estradiol on bone mineral density and uterine weight in ovariectomized rats, *J. Drug Targeting*, 4, 129, 1997.

27. Hirabayashi, H et al., Bone-specific delivery and sustained release of diclofenac, a non-steroidal anti-inflammatory drug, via bisphosphonic prodrug based on the osteotropic drug delivery system, *J. Controlled Release*, 70, 183, 2001.

28. Otsuka, M. and Nakahigashi, Y., A novel skeletal drug delivery system using self-setting calcium phosphate cement. VII. Effect of biological factors on indomethacin release from cement loaded on bovine bone, *J. Pharm. Sci.*, 83, 1569, 1994.

29. Goodman Gilman, A., Goodman, L.S., and Gilman, A., Drugs acting at synaptic and neuroeffector junctional site, in *Goodman and Gilman's The Pharmacological Basis of Therapeutics,* Macmillan, New York, 1980, chap. 2.

30. Fauci, A.S et al., Myasthenia gravis and other diseases of the neuromuscular junction, in *Harrison's Principles of Internal Medicine,* Dan L. Lango et al., Eds., McGraw-Hill, New York, 1998, chap. 382.

31. Kastrup, E.K et al., Central nervous system agents, in *Facts and Comparisons,* Wolters Kluwer, St. Louis, 2002, chap. 7.

32. Briassoulis, G., Continuous neostigmine infusion in post-thymectomy juvenile myasthenic crisis, *J. Child Neurol.*, 15, 747, 2002.

33. Murray, M.J et al., Clinical practice guidelines for sustained neuromuscular blockade in the adult critically ill patients, *Crit. Care Med.*, 30, 1, 2002.

34. Nielsen, J.F., Baclofen increases the soleus stretch reflex threshold in the early swing phase during walking in spastic multiple sclerosis, *Multiple Sclerosis*, 6, 105, 2000.

35. Borasio, G.D., Palliative care in amyotrophic lateral sclerosis, *Neurol Clin.*, 19, 829, 2001.

36. Ashburn, M.A. and Rice, L.J., *The Management of Pain*, 1st ed., Churchill Livingstone, New York, 1998, chap. 11.

37. Naik, A., Yogeshvar, N.K., and Guy, R.H., Transdermal drug delivery: Overcoming the skin's barrier function, *Pharm. Sci. Technol. Today*, 3(9), 318, 2000.

38. Bose, S., Ravis, W.R., Lin, Y., Zhang, L., Hofmann, G.A., et al., Electrically-assisted transdermal delivery of buprenorphine, *J. Controlled Release*, 73(2–3), 197, 2001.

39. Mitragotri, S., Effect of therapeutic ultrasound on partition and diffusion coefficients in human stratum corneum, *J. Controlled Release*, 71(1), 23, 2001.

40. Boucaud, A., Machet, L., Arbeille, B., and Machet, M.C., In vitro study of low-frequency ultrasound-enhanced transdermal transport of fentanyl and caffeine across human and hairless rat skin, *Int. J. Pharm.*, 228(1–2), 69, 2001.

41. Shargel, L. and Yu, B.C., *Applied Biopharmaceutics and Pharmacokinetics*, 3rd ed., Appleton-Lange, Norwalk, CT, 1993, chap. 7.

42. Baron, T.H., Expandable metal stents for the treatment of cancerous obstruction of the gastrointestinal tract, *N. Engl. J. Med.*, 344(22), 1681, 2001.

43. Knyrim, K., Wagner, H.J., Bethge, N., Keymling, M., and Vakil, N.A., Controlled trial of an expansile metal stent for palliation of esophageal obstruction due to inoperable cancer, *N. Engl. J. Med.*, 329, 1302, 1993.

44. Ishibashi, T., Pitcairn, G.R., Yoshino, H., Mizobe, M., et al., Scintigraphic evaluation of a new capsule-type colon specific drug delivery system in healthy volunteers, *J. Pharm. Sci.*, 87(5), 531, 1998.

45. Macleod, G.S., Collett, J.H., and Fell, J.T., The potential use of mixed films of pectin, chitosan, and HPMC for bimodal drug release, *J. Controlled Release*, 58, 303, 1999.

46. Macleod, G.S., Collett, J.H., Fell, J.T., Sharma, H.L., et al., *Int. J. Pharm.*, 187(2), 251, 1999.

47. Turkoglu, M. and Ugurlu, T., *In* vitro evaluation of pectin-HPMC compression coated 5-aminosalicylic acid tablets for colonic delivery, *Eur. J. Pharm. Biopharm.*, 53(1), 65, 2002.

48. Timsina, M.P et al., Drug delivery to the respiratory tract using dry powder inhalers, *Int. J. Pharm.*, 101, 1, 1994.

49. McCallion, O et al., Jet nebulisers for pulmonary drug delivery, *Int. J. Pharm.*, 130, 1, 1996.

50. Taylor, M. and McCallion, O., Ultrasonic nebulisers for pulmonary drug delivery, *Int. J. Pharm.*, 153, 93, 1997.

51. Keller, M., Innovations and perspectives of metered dose inhalers in pulmonary drug delivery, *Int. J. Pharm.*, 186, 81, 1999.

52. Anderson, P., Delivery options and devices for aerosolized therapeutics, *CHEST*, 120, 89s, 2001.

53. LiCalsi, C et al., A powder formulation of measles vaccine for aerosol delivery, *Vaccine*, 19, 2629, 2001.

54. Diacon, A.H., Prospective randomized comparison of thorascopic talc poudrage under local anesthesia versus bleomycin instillation for pleurodesis in malignant pleural effusions, *Am. J. Respir. Crit. Care Med.,* 162, 1445, 2000.

55. Hannley, M.T., Denney, J.C., and Holzer, S.S., Use of ototopical antibiotics in treating 3 common ear diseases, *Otolaryngology*, 122, 934, 2000.

56. Goycoolea, M.V., Extended middle ear drug delivery. A new concept; a new device, *Acta Otolaryngol.*, 493, 119, 1992.

57. Quaranta, A., Intratympanic therapy for Meniere's disease: Effect of administration of low concentration of gentamicin, *Acta Otolaryngol.*, 121, 387, 2001.

58. Kublik, H. and Vidgren, M.T., Nasal delivery systems and their effect on deposition and absorption, *Adv. Drug Deliv. Rev.*, 29, 157, 1998.

59. Irwin, R.S. and Madison, J.M., Primary care: The diagnosis and treatment of cough, *N. Engl. J. Med.*, 343, 1715, 2000.

60. Neuenschwander, M.C. and Prtibitkin, E.A., Botulinum toxin in otolaryngology: A review of its actions and opportunities for use, *ENT-Ear Nose Throat J.*, 79, 799, 2000.

61. Swartz, M.A., The physiology of the lymphatic system, *Adv. Drug Deliv. Rev.*, 50, 3, 2001.

62. Porter, C.J., Transport and absorption of drugs via the lymphatic system, *Adv. Drug Deliv. Rev.*, 50, 1, 2001.

63. Porter, C.J., Lymphatic transport of proteins after subcutaneous injection: Implications of animal model selection, *Adv. Drug Deliv. Rev.*, 50, 157, 2001.

64. Porter, C.J., Intestinal lymphatic drug transport: An update, *Adv. Drug Deliv. Rev.*, 50, 61, 2001.

65. Swart, P.J., Homing of negatively charged albumins to the lymphatic system: General implications for drug targeting to peripheral tissues and reservoirs, *Biochem. Pharmacol.*, 58, 1425, 1999.

66. Bernhisel, M.A., Braly, P.S., Branch, W.D., and Bristow, R.E., *Danforth's Obstetrics and Gynecology*, 8th ed., Lippincott, Williams & Wilkins, Philadelphia, 1999, chap. 3.

67. Pizzorno, J.E., *Textbook of Natural Medicine*, 2nd ed., Churchill Livingstone, New York, 1999, 1378, 74.

68. Kovalevsky, G. and Barnhart, K., Norplant and other implantable contraceptives, *Clin. Obstet. Gynecol.*, 44(1), 92, 75, 2001.

69. Jordan, A., Toxicology of progestogens of implantable contraceptives for women, *Contraception*, 65(1), 3, 2002.

70. Croxatto, H.B., Progestin implants, *Steriods*, 65, 681, 2000.

71. Audet, M.C., Moreau, M., Koltun, W.D., Waldbaum, A.S., et al., Evaluation of contraceptive efficacy and cycle control of a transdermal contraceptive patch vs an oral contraceptive: A randomized controlled trial, *J. Am. Med. Assoc.*, 285(18), 2347, 2001.

72. Abhrams, L.S., Skee, D.M., Natarajan, J., Wong, F.A., et al., Pharmacokinetics of norelgestromin and ethinyl estradiol delivered by a contraceptive patch (Ortho Evra/Evra™) under conditions of heat, humidity, and exercise, *J. Clin. Pharmacol.*, 41, 1301, 2001.

73. Dittrich, R., Parker, L., Rosen, J.B., Shangold, G., et al., Transdermal contraception: Evaluation of three transdermal norelgestromin/ethinyl estradiol doses in a randomized, multicenter, dose-response study, *Am. J. Obstet. Gynecol.*, 186(1), 15, 2002.

74. Anderson, R.A., Martin, C.W., Kung, A.W., Everington, D., et al., 7Alpha-methyl-19-nortestosterone maintains sexual behavior and mood in hypogonadal men, *J. Clin. Endocrinol. Metab.*, 84(10), 3556, 1999.

75. Hair, W.M., Kitteridge, K., O'Connor, D.B., and Wu, F.C., A novel male contraceptive pill-patch combination: Oral desogestrel and transdermal testosterone in the suppression of spermatogenesis in normal men, *J. Clin. Endocrinol. Metab.*, 86(11), 5201, 2001.

VIII

Alternative and Emerging Techniques

The incorporation of technology in so many areas of patient care has inspired many researchers who strive to develop radically new ideas. New treatment strategies, as well as new applications of existing technologies, have emerged from seemingly limitless imagination. The degree to which these therapies become accepted as commonplace depends on many factors. However, proof of efficacy remains the dominating determinant.

The therapies outlined in these chapters are chosen from a wide range of ideas as case studies in innovation. It would be beyond the scope of this volume to attempt to summarize the domain of alternative and emerging medical technologies. Nevertheless, one can see the promise that technology brings to health care in the applications described.

28

Hyperbaric Oxygen Therapy

Brent DeGeorge
*University of Virginia
Health System*

Thomas J. Gampper
*University of Virginia
Health System*

28.1 Introduction

Hyperbaric oxygen (HBO) therapy is the delivery of 100% oxygen at pressures greater than atmospheric pressure, which by convention is considered to be 760 mmHg or 1 atm at sea level. The therapeutic effects are dependent upon the increase in both pressure and oxygen delivered to tissues.

Although the primary oxygen-delivery mechanism in humans is hemoglobin, which binds, carries, and delivers up to four molecules of oxygen per hemoglobin molecule, a certain amount of oxygen is dissolved freely in plasma. Therefore, the content of oxygen in the arterial blood (CaO_2) is dictated by the following relationship: oxygen content = oxygen carried by hemoglobin + oxygen dissolved in plasma. More specifically, arterial oxygen content can be calculated by the following equation: CaO_2 = [1.34 (hemoglobin)(% oxygen saturation)] + [0.003 (PO_2)]. The amount of dissolved oxygen is reflected by the PO_2, which is proportional to the percent of inspired oxygen (FiO_2), the solubility of oxygen in plasma, and the pressure gradient between the atmosphere (P_{atm}) and water vapor pressure (P_{H_2O}) with the following relationship: PO_2 = FiO_2 (P_{atm} − P_{H_2O}). Consequently, PO_2, and therefore the dissolved oxygen, will increase in proportion to the increase in atmospheric pressure. Hemoglobin becomes fully saturated at a PO_2 of approximately 100 mmHg. Under ambient conditions, hemoglobin can carry roughly 20.1 mL of oxygen per 100 mL of blood, whereas the plasma dissolved oxygen concentrations of 0.3 mL of oxygen per 100 mL of blood. When the inspired oxygen concentration is raised to 100% and the ambient pressure is raised to 3 atm, the amount of dissolved oxygen rises to approximately 6 mL of oxygen per 100 mL of blood [1]. This equals the volume of oxygen extracted by the human tissues at rest, which means HBO conditions can theoretically support aerobic metabolism without utilizing the hemoglobin transport mechanism [1,2].

The increased partial pressure of oxygen in the plasma has a myriad of effects on cellular physiology. Under conditions of hyperoxygenation, oxygen delivery to tissues becomes markedly more efficient. Oxygen dissolved in the plasma can reach the relatively ischemic tissues inaccessible to red blood cells. Furthermore, the deformability of red blood cells is augmented by hyperbaric oxygenation allowing for more efficient oxygen delivery in regions of threatened microcirculation. Normoxic tissues respond to hyperbaric conditions with contraction of vascular smooth muscle and consequent vasoconstriction. This phenomenon does not occur in ischemic areas resulting in a relative increase in perfusion pressure in these areas.

The augmented oxygen delivery under hyperbaric conditions protects tissues by ameliorating reperfusion injury with decreased generation of reactive oxygen species, increased tissue synthesis of free radical scavengers, decreased lipid peroxidation of cellular membranes, and decreased activation of neutrophil inflammatory cascades. HBO treatment has a distinct antimicrobial effect through the improvement in the efficacy of neutrophils for bacterial killing, and the hyperoxic effect of HBO improves the efficacy of antibiotics, including aminoglycosides and antimetabolites. HBO therapy directly affects the gene expression of a variety of signaling molecules implicated in cytoprotection against oxidative stress, angiogenesis, collagen synthesis, and epithelialization. The supra-atmospheric pressure exerts mechanical effects during HBO treatment to decrease the size of gas bubbles in the circulation. This is described by the pressure–volume relationship in Boyle's law; pressure = 1/vol, and accounts for the efficacy of HBO in the setting of gas embolism and gas gangrene.

Hyperbaric therapy was first documented in 1664, when a British physician, Dr. Henshaw, attempted to treat chronic illnesses, such as arthritis, with hypobaric pressure, and acute illnesses, such as fever, with hyperbaric pressures [3]. The discovery of oxygen waited another 100 years until described by Joseph Priestly in 1775 [4]. By the 1800s, hyperbaric chambers were in fashion with Europe's wealthy class, despite the scarcity of scientific evidence to support actual medical benefit. It was not until 1917 that Drager described the application of HBO therapy for decompression sickness and it was first clinically applied by Behnke in the 1930s [5]. Then, in 1960, the Dutch cardiac surgeon Boerema published the landmark article, *Life without Blood*, which described an experimental group of pigs who underwent serial replacement of their blood with plasma expanders. In a powerful demonstration of the potential of HBO, these pigs were able to lead active lives without any hemoglobin while maintained at 3 atm of pressurized oxygen [6]. Following this, research interest and clinical applications expanded in a dramatic fashion. In response to this growth, the Undersea Medical Society was founded in 1967 with the mission to evaluate and support only those applications based on scientific methodology. This group continues today as the Undersea and Hyperbaric Medical Society (UHMS) to actively support research and education and publishes a triennial committee review of approved HBO indications [7,8]. Table 28.1 contains a list of indications from these sources.

The absolute contraindications to HBO treatment include untreated pneumothorax, the use of certain medications and chemotherapeutic agents, and prematurity [9,10]. Doxorubicin in combination with HBO has been shown to be cardiotoxic in animal models and must be withheld 7 days prior to treatment. Platinum-containing chemotherapeutic agents, such as cisplatin and carboplatin, have been shown to negatively affect wound healing. Disulfiram or antabuse inhibits the production of superoxide dismutase, resulting in increased toxic reactive oxygen species generation and therefore is contraindicated with HBO. Bleomycin has detrimental effects on pulmonary function and must be withheld for 3 months before HBO therapy. The relative contraindications to HBO include any history of the following: pulmonary lesions on chest x-ray or computed tomography (CT) imaging, spontaneous pneumothorax, prior thoracic surgery, chronic obstructive pulmonary disease, congestive heart failure, recent prior ear or eye surgery, or optic neuritis. Also, the following medications are relatively contraindicated with HBO treatment, including narcotics, insulin, steroids, nitroprusside, nicotine replacement, and hydrocarbon-based ointments and gels.

TABLE 28.1 Current Indications for HBO Therapy

Air or gas embolism

Carbon monoxide poisoning complicated by cyanide poisoning

Clostridial myositis and myonecrosis

Crush injury compartment syndrome

Decompression sickness

Arterial insufficiencies

Central retinal artery occlusion

Severe anemia

Intracranial abscess

Necrotizing soft-tissue infections

Refractory osteomyelitis

Delayed radiation injury with soft tissue and bony necrosis

Compromised grafts and flaps

Acute thermal burns

Idiopathic sensorineural hearing loss

Indications under Investigation

Stroke

Traumatic brain injury

Cerebral palsy

Migraine headache

Acute myocardial infarction

Chronic Lyme disease

Invasive fungal infections

HBO is delivered in either a mono- or multiplace chamber. The monoplace chamber is usually constructed of a clear acrylic shell that accommodates a single person (Figure 28.1). Transparent walls provide the patient with a more reassuring and less "closed-in" feeling, while allowing the clinical staff a clear view of the patient. During monoplace therapy, the patient is positioned either supine or slightly reclined, and the chamber is filled with 100% oxygen; therefore, face masks or oxygen tents are not

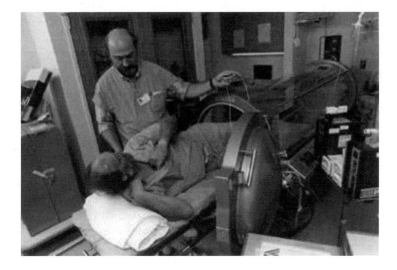

FIGURE 28.1 Representative image of the monoplace. HBO therapy chamber at the University of Virginia Medical Center.

required. Pass-through ports in the chamber wall allow the clinical staff the ability to provide electronic monitoring, IV pumps, and other medical equipment. However, the clinical staff must remain outside the chamber during treatment; consequently, monoplace chambers are unsuitable for patients requiring continuous nursing or provider assistance. The operator remains outside to control the chamber pressure and observe the patient for any adverse effects of pressurization and hyperoxygenation. The maximum pressure obtainable in a monoplace unit is 3 atm because of the physical properties of the chamber. Multiplace units have the advantage of accommodating two or more people at a time, including medical staff that can attend to patients and perform therapeutic maneuvers. Patients receive 100% oxygen either through a face mask or a head tent, while staff members within the chamber are only subjected to pressurized air (21% oxygen). Multiplace chambers are especially useful in treating decompression sickness and air emboli because they can exceed 6 atm of pressure. The disadvantages of these larger chambers include the high initial cost, large space requirements, and the need to closely monitor the staff time in the pressurized environment to avoid decompression sickness.

28.2 Indications and Outcomes for HBO Treatment

28.2.1 Air or Gas Embolism and Decompression Sickness

Air embolism can result from traumatic cardiothoracic or vascular injuries, as well as complicate central venous catheter placement or exchange, hemodialysis, cardiothoracic surgery, and mechanical ventilation. Injury occurs when environmental air is entrained into the vascular system through an injured vessel. It can also occur during uncontrolled scuba-diving ascents or mountain-climbing ascents above 5500 m (altitude decompression sickness) [11]. Damage results from the decrease in solubility that a dissolved gas experiences as the atmospheric pressure and hence its partial pressure decreases. When the solubility of the gas in the plasma decreases, nitrogen (as in decompression sickness of divers) or oxygen reforms gas bubbles within the circulatory system and tissues faster than the ventilatory system can disperse the excess molecules. Treatment with HBO acts mechanically through the Boyle's law; pressure–volume relationship, which reduces bubble size, corrects hypoxia, and provides a controlled return to ambient pressure with effective ventilation. The symptoms of air embolism may vary from rash to weakness, paralysis, altered mental status, and seizures, with death possible in severe cases. A high level of suspicion based upon the clinical setting is critical for the early diagnosis and treatment of air embolism. Treatment begins with 100% oxygen delivered at 2.5–3 atm for 2–4 h. The best outcomes are achieved when therapy is initiated within 6 h of the symptom onset [12]. Further treatments are given until clinical symptoms resolve or fail to show continued improvement.

28.2.2 Carbon Monoxide and Cyanide Poisoning

Carbon monoxide (CO) and cyanide poisoning occur as the products of combustion and fumes that are inhaled from an accident or suicide attempt. Injury results from the extreme affinity with which CO binds to hemoglobin. Since it does so at a rate 240 times greater than oxygen, CO effectively displaces oxygen and prevents its transportation. In addition, CO shifts the hemoglobin–oxygen dissociation curve to the left, which further decreases the release of oxygen to the tissues. Myoglobin oxygen receptors are affected in a manner similar to those for hemoglobin, which completes the overall effect of profound hypoxia [13–15]. The degree of CO poisoning may manifest clinically as mild, moderate, or severe. With mild CO exposure, symptoms can mimic those of influenza, with headaches, dizziness, nausea, and vomiting; moderate cases cause confusion, blurred vision, weakness, tachypnea/tachycardia, or ataxia; and severe exposure results in chest pain, palpitations, electrocardiographic ischemia, dysrhythmias, hypotension, syncope, pulmonary edema, disorientation, seizures, obtundation, coma, and finally death. Neurological presentations may be acute, recurrent after a lucid interval of 1–40 days, or delayed (from which only 50% of patients fully recover) [16]. Laboratory studies may assist in the diagnosis of

CO poisoning. Normal values for blood carboxyhemoglobin (COHb) range from 0% to 10%, with the upper range seen in cigarette smokers and city dwellers. Since CO moves to the intracellular environment, however, blood levels of COHb after hours of exposure will not accurately reflect the degree of CO intoxication. The half-life of hemoglobin-bound CO averages 5 h under standard conditions (21% oxygen at 1 atm). This may be shortened to 1.5 h by increasing the percentage of inspired oxygen to 100% via a nonrebreathing mask and is further reduced by HBO therapy to 23 min [17]. HBO treatment in CO poisoning is significantly more effective at higher pressures (2.8–3.0 atm compared to 2.0 atm) and when started shortly after exposure [18–20]. In one study, HBO reduced mortality from 29.1% to 13.5% when initiated within 6 h of CO exposure. The incidence of delayed neurological presentations was also similarly reduced from 36% to 0.7% when HBO was instituted within 6 h of exposure [21].

In contrast to CO, cyanide produces tissue hypoxia by blocking intracellular oxidative phosphorylation. This is manifested clinically as respiratory depression, metabolic acidosis with increased lactic acid levels, and convulsions. Primary therapy consists of thiosulfate (12.5 mg intravenous for adults and 1.65 mg/kg for children), sodium nitrate (300 mg for adults and 0.33 mg/kg for children), 100% oxygen by mechanical ventilation, and anticonvulsants. In cases of inhalational cyanide poisoning refractory to treatment, HBO therapy should be initiated because of the potential for undiagnosed concomitant CO poisoning. In addition, studies have shown that mice survive lethal doses of cyanide when treated with HBO at 2.8 atm [22,23].

28.2.3 Clostridial Myositis/Myonecrosis and Necrotizing Soft-Tissue Infections

Because of their rapidly progressive nature and high mortality rates (up to 30–50%), soft-tissue infections with gas-forming bacteria (e.g., clostridial myositis/myonecrosis) or involvement of the fascial planes necrotizing fasciitis are true surgical emergencies [24,25]. Although intravenous antibiotics and surgical debridement of necrotic tissue is the first line of therapy, HBO has proven beneficial in reducing mortality from 42% to 7% in matched cases of necrotizing fasciitis of the perineum (Fournier's gangrene) [25]. In other studies of clostridial soft-tissue infections, the rate of amputation was decreased from 50% to 24% when HBO was added to surgical debridement and antibiotic therapy [26]. A 15-year retrospective review of clostridial infection reported similar outcome benefits with the addition of HBO therapy [27]. The improvement seen with HBO is achieved via several mechanisms. Areas of localized hypoxia are predisposed to infection because oxygen is needed for neutrophil production of the free radicals necessary to kill bacteria [28,29]. HBO, therefore, reverses local hypoxia and promotes neutrophil-mediated phagocytosis [30]. Some anaerobic bacteria, such as *Clostridium perfringens*, also lack the antioxidant enzymes dismutase and catalase that degrade superoxide and hydrogen peroxide. This makes them susceptible to the free radicals produced by neutrophils in the setting of HBO therapy [31]. HBO also inhibits production of the *Clostridium* α-toxin that is responsible for tissue liquefaction, septic shock, hemolytic anemia, disseminated intravascular coagulation, renal failure, and cardiotoxicity [32,33]. In addition, HBO may be bacteriostatic for *Escherichia coli* and *Pseudomonas* [34–36]. The recommended hyperbaric treatments begin at 2.5–3 atm, twice daily for 90 min, and continue based on clinical response [37].

28.2.4 Intracranial Abscess

Although the literature is scant and limited to case reports, the same mechanisms (enhanced host defenses, direct bactericidal/bacteriostatic effects) that make HBO therapy beneficial in myonecrosis apply to the treatment of brain abscesses as well. The infectious organisms in these reported cases include *Clostridium* (for which HBO was ineffective) and fungal infections with mucormycosis [38–42]. The hyperbaric protocol consists of twice daily treatments at 2.5 atm for 90 min in advanced cases, with the total number of treatments ranging between 5 and 20 [43,44].

28.2.5 Refractory Osteomyelitis

The standard treatment of osteomyelitis includes surgical debridement of devitalized bone coupled with antibiotics. Intraoperative bone biopsies are obtained for culture to determine the causative organism(s) and select the appropriate antibiotic regimen based on bacterial sensitivities. In some cases, however, debridement and antibiotics are unsuccessful in eradicating the infection because of the localized tissue hypoxia. Animal models in infected rabbit tibiae show a decrease in oxygen tension when compared to normal controls (23 and 45 mmHg, respectively). As noted above, decreased oxygen levels impair the bactericidal action of neutrophils. This condition is reversed with the increased oxygen levels achieved during HBO treatments. Multiple human studies have confirmed the efficacy of HBO in refractory osteomyelitis when combined with secondary surgical debridement and antibiotics, with reports of 70–89% disease-free rates [45–49]. One study failed to confirm that these results were nonrandomized, however, and its authors felt that the conservative initial debridement may have contributed to the treatment failures [50]. The current recommended treatment protocol is 2.0 atm for 90 min twice daily for inpatients, daily for outpatients, or if the infection is under control, and 2.5 atm if *Pseudomonas* or *E. coli* is the causative organism [45].

28.2.6 Acute Traumatic Ischemia, Crush Injury, and Reperfusion Injury

Interruption of blood flow deprives the dependent tissues of oxygen and nutrients. This results in multiple sequelae that must be recognized to prevent further morbidity. The vascular compromise may be apparent, as in the case of open extremity fractures with loss of pulses, or occult, as with electrical injuries, crush injuries, or compartment syndrome. Compartment syndrome exists when the pressure within an anatomic compartment exceeds the capillary perfusion pressure of 20 mmHg. At this point, nutritive blood flow to tissues ceases.

The recognition of vascular compromise is classically described in terms of the 5 Ps: Pain out of proportion to the clinical situation (the earliest clinical symptom, usually within the first 30 min), paresthesias in the distribution of nerves within the affected compartment, paralysis/weakness of involved muscles, poikilothermia (coolness when compared to other areas), and pulselessness (a very late finding). In addition, compartment syndrome may have the additional "P" of pressure, presenting as tense, shiny skin. The subsequent time course of hypoxic tissues involves functional muscle weakness at 2–4 h, decreased capillary integrity and soft-tissue edema within 3 h, and irreversible muscle damage with the onset of myoglobinuria in 4–6 h. Restoration of perfusion is critical to stop this progression, but with the return of blood flow comes a further increase in tissue edema. This occurs because the damaged capillaries allow the fluid to leak into the interstitium. Treatment of compartment syndrome includes fasciotomy, which is the release of the fascia enveloping the affected tissues. Fasciotomy can be performed prophylactically at the time of vascular repair in anticipation of complications or definitively to treat an existing compartment syndrome.

HBO is used in the setting of ischemia to limit edema, support hypoxic tissues, and limit the extent of reperfusion injury. Edema is limited because hyperoxia causes a reflexive vasoconstriction that slows further extravasation [51]. Tissue oxygenation is supported by diffusion from the hyperoxygenated plasma into the soft tissues, improving neutrophil phagocytic activity, and fibroblast proliferation (that is inhibited when oxygen levels are less than 30 mmHg) [52]. In a postischemic animal model, adenosine triphosphatase (ATPase), phosphocreatine kinase, and lactate levels from the HBO-treated group were similar to levels in uninjured controls. Despite the restoration of blood flow, the non-HBO-treated postischemic group continued to have high lactate levels indicative of the ongoing anaerobic metabolism [53]. Other studies have shown that oxygen-free radicals interact with cell membrane lipid layers to produce lipid peroxidation, a reperfusion event that is decreased by HBO [54–56]. HBO's greatest protective effect, however, is the inhibition of neutrophil adherence to postcapillary venules [54,57].

A special category of traumatic ischemia includes amputation of the ear or nose, for which successful salvage has been reported utilizing HBO [58, 59]. In cases where the small caliber of vessels in the amputated tissue precludes microvascular repair, the reattached tissue behaves as a composite graft. Animal studies transferring a composite graft of skin, subcutaneous tissue, and cartilage to the opposite ear demonstrated postoperative HBO therapy to significantly improve graft survival [60]. In experimental studies, grafts require approximately 72 h for neovascularization to occur [61]. Because of their low oxygen levels, ischemic composite grafts are at risk for decreased ATP, increased osmolarity and edema, and further impairment of imbibition and inosculation [62]. HBO stimulates macrophages to secrete angiogenic factors that then increase fibroblastic activity, forming a new fibrous matrix, and stimulating inosculation [63]. In fact, capillaries have been shown to grow three times farther into pedicled guinea-pig flaps treated with HBO [64]. Evidence also shows that fibroblasts behave as facultative anaerobes and intermittent episodes of hypoxia (as occur between HBO treatments) have a stimulatory effect on collagen synthesis [65]. The HBO protocol is 2.5 atm for 90 min three times daily in the first 24 h, then twice daily for 2–4 days if treating compartment syndrome postfasciotomy, or 5–7 days for crush injury or composite graft of amputated tissues.

28.2.7 Compromised Skin Grafts and Flaps

Human clinical studies have shown a beneficial effect of HBO on skin grafts and flaps, especially when compromised or when placed on poorly vascularized tissue beds (90% salvage rate) [66]. A prospective, randomized study of 48 consecutive patients receiving skin grafts demonstrated an 84% skin graft survival rate with HBO therapy compared to 62% graft survival without HBO [67]. The mechanisms responsible for flap and graft salvage are those described for acute traumatic injury and reperfusion injury. In flaps with evidence of ischemia, measured oxygen tensions are commonly less than 15 mmHg [68]. This is harmful because fibroblast synthesis, collagen production, and angiogenesis require oxygen tensions of 30–40 mmHg [63]. The commonly used protocols include 2–2.5 atm for 90 min twice daily tapered to 120 min daily once clinical improvement is sustained.

28.2.8 Tumor Sensitization to Radiotherapy

At present, in the United States, 1.5 million new cancers are diagnosed annually, and approximately 50% of these cases will be managed with radiotherapy. Solid tumors contain regions of relative tissue hypoxia within their substance, and these regions have been shown to be resistant to radiotherapy. Intensive clinical and preclinical efforts have focused on overcoming the radioresistance of these regions, including the administration of agents and conditions that sensitize the tumors to radiotherapy [69]. To this end, a number of centers worldwide have developed the technology to simultaneously deliver HBO therapy and radiation therapy. The dose of HBO per treatment session in these studies was standardized at 3 atm and 100% oxygen between 30 and 40 min; however, the fractionation of external beam radiotherapy varied widely between two sessions separated by 3 weeks versus 40 sessions separated by 8 weeks and with the dose varying between 2600 and 7000 rad [70]. The overall effectiveness of these studies is controversial; however, several small randomized clinical trials demonstrate the reduction of 5-year mortality for cancers of the head and neck, uterine/cervical, and urinary bladder. Furthermore, locoregional recurrence has been shown to be reduced in the head and neck and cervical cancer [70]. Given the findings of improved overall mortality and improved locoregional control, more investigation will be required to identify the optimum combination of radiation dose and fractionation with HBO regimen.

28.2.9 Complications of Radiation Therapy and Prevention of Osteoradionecrosis

An additional issue associated with tumor radiation therapy is the detrimental effects on adjacent tissues. There are progressive changes in the irradiated arterioles and capillaries, beginning with thickening of the vessel walls, followed by endarteritis and vasculitis, and ending with vessel obliteration.

The soft tissues become fibrotic and occasionally painful. The time course for these changes is 6 months to 5 years after the ionizing radiation treatments [71]. This creates a hypoxic and hypocellular zone that exhibits poor wound healing following even minor trauma or surgery. HBO therapy addresses this by creating a wound oxygen gradient sufficient to stimulate neoangiogenesis and reverses the effects of radiation [64,72].

Bone is particularly sensitive to irradiation because its high density results in absorption of a greater proportion of the delivered dosage. Osteoporosis and osteonecrosis are not uncommon late findings months to years after treatment. In the head-and-neck cancer, for example, the mandible is often included in the radiation field. Even procedures as minor as tooth extraction within this zone can precipitously lead to osteoradionecrosis (ORN), osteomyelitis, and pathological fractures. These are particularly difficult to treat and prove refractory to the conventional therapy of debridement and antibiotics in over half of the cases [73]. HBO therapy has been successfully applied to these wounds with reported healing rates of up to 93% [74–77]. Prophylactic HBO has also been shown to be more effective than penicillin in preventing ORN of the mandible in previously irradiated patients, with incidences of 5.4% and 30%, respectively [77,78]. The accepted protocol for ORN prophylaxis in an irradiated mandible is 20 presurgical treatments at 2.4 atm for 90 min followed by another 10 postsurgical treatments.

There are also reports of improvement in soft-tissue radiation injury. Women undergoing breast-conservation therapy for breast cancer (lumpectomy + radiation) may have late sequelae common to radiation-induced soft-tissue changes. Patients treated with HBO at 2.4 atm for 90 min daily for an average of 25 sessions showed a significant reduction in postirradiation pain, erythema, and edema when compared to untreated controls ($p < 0.001$) [79]. Radiation prostatitis is likewise responsive to HBO therapy, typically applied as an extended treatment course consisting of 90 min daily treatments at 2.4 atm for 4–8 weeks [80–82].

28.2.10 Thermal Burns

The beneficial effects of HBO in thermal injury include edema reduction (improved capillary integrity and hyperoxic vasoconstriction), support of marginal tissue in the zone of stasis, and an increased rate of neovascularization. When compared to control patients, burn patients receiving hyperbaric treatments demonstrated a 30% decrease in mortality, a 50% increase in wound-healing rate (in total body surface area burns of less than 50%, the mean healing time was 19.7 days vs. 43.8 days in the untreated group), and required 35% less fluid during the initial resuscitative period [83]. In another series, HBO reduced the average hospital length of stay from 73 to 43 days and surgical procedures from 4.2 to 2.4, with an overall cost reduction of 34% [84]. The recommended treatment parameters are 2.0 atm for 90 min three times in the first 24-h postburn, and then twice daily for 5–45 treatments depending upon patient response and the extent of burn.

28.2.11 Enhanced Healing in Selected Problem Wounds

In addition to wounds in an irradiated field, HBO therapy may benefit other chronic nonhealing wounds [85]. Diabetic patients with nonhealing lower extremity wounds have benefited from HBO as a part of a comprehensive diabetic wound care program, reducing major amputation rates from 40.5% to 23.5% in one study and similarly in others [86–89]. A later randomized controlled study by the same investigators found similar results with the HBO-treated group demonstrating an increase in transcutaneous oxygen measurements (TCOMs) on the dorsum of the foot [89]. This would seem to indicate that collateral neovascularization of the foot was stimulated by HBO therapy. TCOM studies are often obtained before initiating therapy to predict the possible benefit from HBO intervention [90,91]. One method of testing includes exposure to supplemental normobaric oxygen via a nonrebreathing mask for 10 min. When the resulting transcutaneous oxygen values increase by at least 100 mmHg over the baseline, the patient is likely to benefit from HBO therapy [92]. The recommended protocol is 2.0 atm for 90 min twice daily if the wound is gangrenous or otherwise daily for 14 days. When improvement is evident, HBO is continued Monday through Friday, with a possible 7-day halt in treatments to monitor whether healing progresses without HBO therapy.

28.2.12 Acute Blood Loss Anemia

The acute loss of a significant portion of the blood volume leads to circulatory collapse. Immediate blood transfusion is not always possible due to issues such as blood-type incompatibility or religious beliefs. In these situations, HBO therapy may be used in the interim until blood-type cross-matching is complete or erythropoiesis is adequate. As previously noted, there is sufficient oxygen dissolved in the plasma under hyperbaric conditions (3 atm and 100% oxygen) to sustain life in the absence of red blood cells [1,2,6]. Multiple reports describe the successful use of HBO in cases of organ dysfunction secondary to acute blood loss [93,94]. Since wide variations exist in the tolerance to anemia, treatment is based upon patient symptoms rather than an arbitrary hemoglobin/hematocrit laboratory value.

28.2.13 Fracture Healing and Nonunion

Impaired fracture healing and nonunion is defined as absent evidence of bony healing of 6 months duration following fracture, and the presence of relative tissue ischemia, infection, large bony gaps, and mechanically unstable injuries all contribute to poor bony healing. In animal models, HBO has been shown to improve bone synthesis and the resorption of dead or abnormal bone. Several small clinical studies have demonstrated a potential therapeutic benefit of hyperbaric oxygen therapy (HBOT) in the setting of established nonunion or in the setting of severe orthopedic injury with unstable posttraumatic injuries. Typical treatment regimens involve a dose of HBOT at 100% oxygen between 1.5 and 3.0 atm for periods between 60 and 120 min once or more daily for a variable time course with the resultant improved histological and radiological evidence of fracture union.

28.2.14 Neurological Injury and Dysfunction

Neurological impairments in the form of traumatic injury and acquired vascular insufficiency have been the focus of multiple research preclinical studies and clinical trials recently. Cerebral vascular accidents or strokes, defined as an abrupt neurological impairment resulting from either ischemic impairment in blood supply to the brain or hemorrhage resulting from focal bleeding in the brain, are the second leading cause of death worldwide [95]. Sustained local tissue hypoxia within the brain will result in loss of neuronal membrane integrity, the accumulation of oxygen-free radicals and metabolites, and ultimately neuronal cell death.

HBO therapy offers the potential to improve local tissue oxygenation and ameliorate cerebral edema in the setting of acute neurological injury resulting in neuroprotection of the ischemic penumbra, or area of the brain at risk of ischemia [96]. Several studies have demonstrated an improvement in global cerebral functioning as assessed through validated functional scales; however, no effect on mortality poststroke has been observed [97,98]. Typically, HBO dose consisting of 100% oxygen between 1.5 and 3.0 atm for periods of 60 or 120 min, once or twice daily, until clinical effects are observed [99].

In the United States, traumatic brain injury accounts for 50,000 deaths annually with motor vehicle accidents, falls, and violence representing the most common etiologies. Following traumatic brain injury, the brain undergoes a primary injury through direct damage to neuronal tissue, followed by secondary injury due to local tissue malperfusion, free radical generation, and cerebral edema. This secondary injury is the therapeutic target for hyperbaric therapy in the setting of traumatic brain injury. The dose of HBO typically ranged from 100% oxygen at 1.5–2.5 atm for 40–60 min from once to 10 times daily, until clinical effects are observed. Treatment with HBO has been shown to reduce the relative risk of death in patients following significant closed head injury with a number needed to treat of seven [100,101]. HBOT has also been shown in small clinical studies to reduce intracranial pressure, improve Glascow Coma Scale score, and decrease unfavorable neurological outcomes [102]. Additional studies will be required to identify the optimum dose, interval, and number of treatments to provide the best neurological outcome.

Migraine headache is also an evolving area of application of HBO therapy. The current therapeutic options fall into two broad categories: acute or abortive therapy and chronic or preventative therapy. HBO therapy has been demonstrated to have the potential to relieve acute migraine headache pain, with standard treatment involving 100% oxygen at 2.5 atm for 30 min in a single treatment. However, HBOT demonstrated no efficacy for decreasing the frequency of migraine headaches or migraine-associated symptoms such as nausea and vomiting [103].

28.2.15 Acute Coronary Syndrome

Worldwide, cardiovascular disease is the leading cause of death, accounting for more than 12% of all deaths [104]. Acute coronary syndrome is defined as ischemic chest pain with or without the presence of myocardial infarction evidence by leakage of cardiac enzymes and electrocardiogram changes. The physiology of acute coronary syndrome involves myocardial ischemia leading to infarction and death of myocardial tissue. The potential of HBO to augment the oxygen tension in regions of myocardial malperfusion has led investigators to attempt to administer HBOT in the setting of acute coronary syndrome in animal and human models since the 1960s [105]. In several small clinical trials, the administration of HBOT resulted in a significant decrease in mortality, troponin I elevation, and improved left ventricular contractile function. Furthermore, several studies demonstrate relative reductions in the second myocardial infarction and postinfarction arrhythmias. The administration of HBOT was performed from within 12 h of diagnosis to up to 10 days postdiagnosis, and the dose of HBOT in these studies involved the administration of 100% oxygen between 2 and 3 atm ranging from 30 to 120 min with a variable total number of treatments from a single session to 16 treatments over 2 days [106]. Large randomized controlled clinical trials will be required to identify how HBOT can be possibly combined with cardiac catheterization and the existing standard of care for the management of acute coronary syndrome [107].

28.2.16 Complications of HBO Therapy

Complications of HBO therapy fall into two major categories: barotrauma and oxygen toxicity. Barotrauma of the ear is the most common complication from treatment, with published reports placing the incidence anywhere from 5% to 82% [108]. This occurs when the eustachian tube dysfunction does not allow adequate pressure equalization under hyperbaric conditions. Barotrauma may also affect any air-filled space, including the facial sinuses, where "sinus squeeze" can cause severe pain. Pulmonary barotrauma, however, is the most concerning pressure-related complication of HBO treatments. Potentially, life-threatening pneumothorax or pneumomediastinum may occur. This is more likely in the setting of preexisting pulmonary diseases such as emphysema and chronic obstructive pulmonary disease.

Oxygen toxicity is dependent upon the partial pressure of inspired oxygen and the duration of exposure. At 3 atm, toxicity occurs after 3 h, but at 4 atm, only 10–40 min of exposure are necessary before unwanted side effects occur [109]. The pathophysiology is most likely related to oxygen-free radical mechanisms [110], although the effects are reversible and clinical protocols avoid these levels. The four most important manifestations of oxygen toxicity are: lowered seizure threshold, diffuse pulmonary alveolar injury, myopia, and increased rate of cataract formation. A history of seizure disorder predisposes patients to oxygen-induced seizures with HBO treatments. This necessitates the maintenance of antiepileptic medications within therapeutic levels during the course of treatment [111]. Other factors that lower the seizure threshold can also increase the risk of oxygen-induced seizures, including fever, nicotine, aspirin, and alcohol. At the typical treatment pressure of 2.4 atm, the incidence of

HBO-induced seizures is approximately 1/10,000 [112]. There are no reports of pulmonary toxicity in patients with healthy lungs who receive HBO at 1.5–2 atm for periods of less than 1 h [113].

Under normobaric conditions, more than 24 h of treatment with 100% oxygen are needed for pulmonary toxicity to occur. This phenomenon usually manifests first with a reversible reduction in vital capacity from an oxygen-dependent inhibition of surfactant production or from a depletion of natural antioxidants. Myopia, commonly called near-sightedness, is frequently reported during HBO therapy and averages a change of three diopters. This side effect, however, is transitory and reverts to the patient's baseline vision within 6 months of completing therapy [43,114]. Since there are few antioxidants in the lens, oxidation of lens proteins may result in the formation of aggregates. An increased rate of cataract formation can also occur, especially in patients with preexisting cataracts [114].

28.3 Summary

HBO is an efficacious therapeutic and prophylactic modality for a number of disease processes. Although historically used as a panacea for well-being and health improvement, critical scientific investigation has yielded specific uses that decrease both patient morbidity and mortality. As laboratory and human clinical trials evaluate further applications, HBO therapy will continue to expand and more patients may hope to benefit from this unique manipulation of the environment. At present, more than 6600 peer-reviewed publications on the potential therapeutic efficacy of HBO have been produced, including nearly 500 publications in the past year. As the body of scientific evidence supporting HBO therapy continues to evolve, high-quality randomized-controlled clinical trials will be required to elevate this therapeutic modality from an adjunctive to a primary role in the treatment of the disease.

Acknowledgments

Former graduates of the University of Virginia-Department of Plastic Surgery Training Program, Dr. Tonja Weed and Dr. Timothy Bill, integrally participated in the development of previous versions of this chapter.

References

1. C. J. Lambertsen et al., Oxygen toxicity: effects in man of oxygen inhalation at 1 and 3.5 atmospheres upon blood gas transport, cerebral circulation and cerebral metabolism. *Journal of Applied Physiology* **5**, 471, March, 1953.
2. S. S. Kety, C. F. Schmidt, The effects of altered arterial tensions of carbon dioxide and oxygen on cerebral blood flow and cerebral oxygen consumption of normal young men. *The Journal of Clinical Investigation* **27**, 484, July, 1948.
3. N. Henshaw, *Aero-Chalinos, or, a Register for the Air in Five Chapters: For the Better Preservation of Health, and Cure of Diseases, After a New Method.* (C. U. Library, London, 1664).
4. A. Faulconer, T. E. Keys, *Foundations of Anaesthesiology* (Springfield, IL, 1965).
5. A. R. Behnke, The effect of oxygen on man at pressures from 1–4 atmospheres. *American Journal of Physiology* **110**, 563, 1935.
6. I. Boerema et al., Life without blood. *Nederlands Tijdschrift Voor Geneeskunde* [*Netherlands Journal of Medicine*] **104**, 949, May 7, 1960.
7. S. L. Thom, *Hyperbaric Oxygen Therapy Indications.* (Undersea and Hyperbaric Medical Society, Durham, NC, pp. 12, 2008).
8. H. Takahashi, S. Kobayashi, New indications for hyperbaric oxygen therapy and its complication. *Advances in Oto-Rhino-Laryngology* **54**, 1, 1998.
9. J. Kot, D. Mathieu, Controversial issues in hyperbaric oxygen therapy: A European Committee for Hyperbaric Medicine Workshop. *Diving and Hyperbaric Medicine: The Journal of the South Pacific Underwater Medicine Society* **41**, 101, June, 2011.

10. A. S. Toklu, S. Korpinar, M. Erelel, G. Uzun, S. Yildiz, Are pulmonary bleb and bullae a contraindication for hyperbaric oxygen treatment? *Respiratory Medicine* **102**, 1145, August, 2008.

11. I. J. Blumen, M. K. Abernethy, M. J. Dunne, Flight physiology. Clinical considerations. *Critical Care Clinics* **8**, 597, July, 1992.

12. R. E. Moon, D. F. Gorman, Treatment of the decompression disorders. B. Benett and D. H. Elliott eds., *The Physiology and Medicine of Diving* (W.B. Saunders, Philadelphia, PA, 4th ed., 1993).

13. K. C. Chen, J. J. McGrath, Response of the isolated heart to carbon monoxide and nitrogen anoxia. *Toxicology and Applied Pharmacology* **81**, 363, December, 1985.

14. M. D. Ginsberg, R. E. Myers, B. F. McDonagh, Experimental carbon monoxide encephalopathy in the primate. II. Clinical aspects, neuropathology, and physiologic correlation. *Archives of Neurology* **30**, 209, March, 1974.

15. B. A. Wittenberg, J. B. Wittenberg, P. R. Caldwell, Role of myoglobin in the oxygen supply to red skeletal muscle. *The Journal of Biological Chemistry* **250**, 9038, December 10, 1975.

16. R. Bartlett, Carbon monoxide poisoning. L. D. Winchester and J. F. Haddad eds., *Clinical Management of Poisoning and Drug Overdose* (W. B. Saunders, Philadelphia, PA, 2nd ed., 1998).

17. R. A. M. Myers, Carbon monoxide poisoning. L. D. Winchester and J. F. Haddad eds., *Clinical Management of Poisoning and Drug Overdose* (W. B. Saunders, Philadelphia, PA, 2nd ed., 1998).

18. H. Nikkanen, A. Skolnik, Diagnosis and management of carbon monoxide poisoning in the emergency department. *Emergency Medicine Practice* **13**, 1, February, 2011.

19. J. C. Raphael et al., Trial of normobaric and hyperbaric oxygen for acute carbon monoxide intoxication. *Lancet* **2**, 414, August 19, 1989.

20. K. A. Vishwanathan, S. A. Rizvi, R. Pahuja, T. K. Pandey, Subacute sequelae of carbon monoxide poisoning. *The Journal of the Association of Physicians of India* **37**, 349, May, 1989.

21. M. Goulon et al., Carbon monoxide poisoning and acute anoxia due to inhalation of coal gas and hydrocarbons: 302 cases, 273 treated by hyperbaric oxygen at 2 atm. *Annales de Medecine Interne* [*Annals of Internal Medicine*] **120**, 335, May, 1969.

22. W. G. Skene, Effect of hyperbaric oxygen in cyanide poisoning. I. Brown and B. Cox eds., *Proceedings of the 3rd International Congress on Hyperbaric Oxygen* (National Academy of Science, Washington, DC, 1966).

23. J. L. Way et al., Effect of oxygen on cyanide intoxication. IV. Hyperbaric oxygen. *Toxicology and Applied Pharmacology* **22**, 415, July, 1972.

24. R. J. Balcerak, J. M. Sisto, R. C. Bosack, Cervicofacial necrotizing fasciitis: Report of three cases and literature review. *Journal of Oral and Maxillofacial Surgery: Official Journal of the American Association of Oral and Maxillofacial Surgeons* **46**, 450, June, 1988.

25. R. S. Hollabaugh, Jr., R. R. Dmochowski, W. L. Hickerson, C. E. Cox, Fournier's gangrene: Therapeutic impact of hyperbaric oxygen. *Plastic and Reconstructive Surgery* **101**, 94, January, 1998.

26. C. R. Hitchcock, F. J. Demello, J. J. Haglin, Gangrene infection: New approaches to an old disease. *The Surgical Clinics of North America* **55**, 1403, December, 1975.

27. G. B. Hart, R. C. Lamb, M. B. Strauss, Gas gangrene. *The Journal of Trauma* **23**, 991, November, 1983.

28. T. K. Hunt, H. Hopf, Z. Hussain, Physiology of wound healing. *Advances in Skin & Wound Care* **13**, 6, May–June, 2000.

29. D. R. Knighton, B. Halliday, T. K. Hunt, Oxygen as an antibiotic. A comparison of the effects of inspired oxygen concentration and antibiotic administration on *in vivo* bacterial clearance. *Archives of Surgery* **121**, 191, February, 1986.

30. J. T. Mader, G. L. Brown, J. C. Guckian, C. H. Wells, J. A. Reinarz, A mechanism for the amelioration by hyperbaric oxygen of experimental staphylococcal osteomyelitis in rabbits. *The Journal of Infectious Diseases* **142**, 915, December, 1980.

31. G. B. Hill, S. Osterhout, Experimental effects of hyperbaric oxygen on selected clostridial species. II. *In vitro* studies in mice. *The Journal of Infectious Diseases* **125**, 26, January, 1972.

32. D. Kaye, Effect of hyperbaric oxygen on Clostridia *in vitro* and *in vivo*. *Proceedings of the Society for Experimental Biology and Medicine* **124**, 360, February, 1967.

33. U. Van, Inhibition of toxin production in *Clostridium perfringens in vitro* by hyperbaric oxygen. *Antonie van Leeuwenhoek* **31**, 181, 1965.

34. D. E. Boehm, K. Vincent, O. R. Brown, Oxygen and toxicity inhibition of amino acid biosynthesis. *Nature* **262**, 418July 29, 1976.

35. O. R. Brown, Reversible inhibition of respiration of *Escherichia coli* by hyperoxia. *Microbios* **5**, 7, January, 1972.

36. M. K. Park, K. H. Muhvich, R. A. Myers, L. Marzella, Hyperoxia prolongs the aminoglycoside-induced postantibiotic effect in *Pseudomonas aeruginosa*. *Antimicrobial Agents and Chemotherapy* **35**, 691, April, 1991.

37. R. L. Scher, Hyperbaric oxygen therapy for necrotizing cervical infections. N. Yanagha and T. Nakashima eds., *Hyperbaric Oxygen Therapy in Otorhinolaryngology* (S. Karger AG, Basel, Switzerland, 1998).

38. V. K. Anand, G. Alemar, J. A. Griswold, Jr., Intracranial complications of mucormycosis: An experimental model and clinical review. *The Laryngoscope* **102**, 656, June, 1992.

39. L. Couch, F. Theilen, J. T. Mader, Rhinocerebral mucormycosis with cerebral extension successfully treated with adjunctive hyperbaric oxygen therapy. *Archives of Otolaryngology—Head and Neck Surgery* **114**, 791, July, 1988.

40. B. J. Ferguson, T. G. Mitchell, R. Moon, E. M. Camporesi, J. Farmer, Adjunctive hyperbaric oxygen for treatment of rhinocerebral mucormycosis. *Reviews of Infectious Diseases* **10**, 551, May–June, 1988.

41. A. J. Keogh, Clostridial brain abscess and hyperbaric oxygen. *Postgraduate Medical Journal* **49**, 64, January, 1973.

42. J. C. Price, D. L. Stevens, Hyperbaric oxygen in the treatment of rhinocerebral mucormycosis. *The Laryngoscope* **90**, 737, May, 1980.

43. F. K. Butler, Jr., C. Hagan, H. Murphy-Lavoie, Hyperbaric oxygen therapy and the eye. *Undersea & Hyperbaric Medicine: Journal of the Undersea and Hyperbaric Medical Society, Inc,* **35**, 333, September–October, 2008.

44. R. W. Pelton, E. A. Peterson, B. C. Patel, K. Davis, Successful treatment of rhino-orbital mucormycosis without exenteration: The use of multiple treatment modalities. *Ophthalmic Plastic and Reconstructive Surgery* **17**, 62, January, 2001.

45. H. Andel, M. Felfernig, D. Andel, W. Blaicher, W. Schramm, Hyperbaric oxygen therapy in osteomyelitis. *Anaesthesia* **53** (Suppl 2), 68, May, 1998.

46. M. Cimsit, G. Uzun, S. Yildiz, Hyperbaric oxygen therapy as an anti-infective agent. *Expert Review of Anti-Infective Therapy* **7**, 1015, October, 2009.

47. J. C. Davis, J. D. Heckman, J. C. DeLee, F. J. Buckwold, Chronic non-hematogenous osteomyelitis treated with adjuvant hyperbaric oxygen. *The Journal of Bone and Joint Surgery. American Volume* **68**, 1210, October, 1986.

48. C. G. Kaide, S. Khandelwal, Hyperbaric oxygen: Applications in infectious disease. *Emergency Medicine Clinics of North America* **26**, 571, May, 2008.

49. C. Wang, S. Schwaitzberg, E. Berliner, D. A. Zarin, J. Lau, Hyperbaric oxygen for treating wounds: A systematic review of the literature. *Archives of Surgery* **138**, 272, March, 2003.

50. J. L. Esterhai, Jr. et al., Adjunctive hyperbaric oxygen therapy in the treatment of chronic refractory osteomyelitis. *The Journal of Trauma* **27**, 763, July, 1987.

51. M. B. Strauss et al., Reduction of skeletal muscle necrosis using intermittent hyperbaric oxygen in a model compartment syndrome. *The Journal of Bone and Joint Surgery. American Volume* **65**, 656, June, 1983.

52. T. K. Hunt, B. Zederfeldt, T. K. Goldstick, Oxygen and healing. *American Journal of Surgery* **118**, 521, October, 1969.

53. G. Nylander, D. Lewis, H. Nordstrom, J. Larsson, Reduction of postischemic edema with hyperbaric oxygen. *Plastic and Reconstructive Surgery* **76**, 596, October, 1985.

54. S. R. Thom, Functional inhibition of leukocyte B2 integrins by hyperbaric oxygen in carbon monoxide-mediated brain injury in rats. *Toxicology and Applied Pharmacology* **123**, 248, December, 1993.

55. S. R. Thom, M. E. Elbuken, Oxygen-dependent antagonism of lipid peroxidation. *Free Radical Biology & Medicine* **10**, 413, 1991.

56. W. A. Zamboni et al., Morphologic analysis of the microcirculation during reperfusion of ischemic skeletal muscle and the effect of hyperbaric oxygen. *Plastic and Reconstructive Surgery* **91**, 1110, May, 1993.

57. W. A. Zamboni et al., The effect of acute hyperbaric oxygen therapy on axial pattern skin flap survival when administered during and after total ischemia. *Journal of Reconstructive Microsurgery* **5**, 343, October, 1989.

58. T. J. Bill, M. A. Hoard, T. J. Gampper, Applications of hyperbaric oxygen in otolaryngology head and neck surgery: Facial cutaneous flaps. *Otolaryngologic Clinics of North America* **34**, 753, August, 2001.

59. L. S. Nichter, D. T. Morwood, G. S. Williams, R. J. Spence, Expanding the limits of composite grafting: A case report of successful nose replantation assisted by hyperbaric oxygen therapy. *Plastic and Reconstructive Surgery* **87**, 337, February, 1991.

60. F. Zhang et al., Effect of hyperbaric oxygen on survival of the composite ear graft in rats. *Annals of Plastic Surgery* **41**, 530, November, 1998.

61. L. C. Mc, Composite ear grafts and their blood supply. *British Journal of Plastic Surgery* **7**, 274, October, 1954.

62. G. Nylander, H. Nordstrom, D. Lewis, J. Larsson, Metabolic effects of hyperbaric oxygen in postischemic muscle. *Plastic and Reconstructive Surgery* **79**, 91, January, 1987.

63. D. R. Knighton et al., Oxygen tension regulates the expression of angiogenesis factor by macrophages. *Science* **221**, 1283, September 23, 1983.

64. Y. J. Tai, B. C. Birely, M. J. Im, J. E. Hoopes, P. N. Manson, The use of hyperbaric oxygen for preservation of free flaps. *Annals of Plastic Surgery* **28**, 284, March, 1992.

65. D. R. Knighton, I. A. Silver, T. K. Hunt, Regulation of wound-healing angiogenesis-effect of oxygen gradients and inspired oxygen concentration. *Surgery* **90**, 262, August, 1981.

66. H. I. Friedman, M. Fitzmaurice, J. F. Lefaivre, T. Vecchiolla, D. Clarke, An evidence-based appraisal of the use of hyperbaric oxygen on flaps and grafts. *Plastic and Reconstructive Surgery* **117**, 175S, June, 2006.

67. D. J. Perrins, Influence of hyperbaric oxygen on the survival of split skin grafts. *Lancet* **1**, 868, April 22, 1967.

68. P. J. Sheffield, Measuring tissue oxygen tension: A review. *Undersea & Hyperbaric Medicine: Journal of the Undersea and Hyperbaric Medical Society, Inc,* **25**, 179, Fall, 1998.

69. G. Delaney, S. Jacob, C. Featherstone, M. Barton, The role of radiotherapy in cancer treatment: Estimating optimal utilization from a review of evidence-based clinical guidelines. *Cancer* **104**, 1129, September 15, 2005.

70. M. Bennett, J. Feldmeier, R. Smee, C. Milross, Hyperbaric oxygenation for tumour sensitisation to radiotherapy. *Cochrane Database System Review,* **2005**, 4, CD005007, 2005.

71. J. J. Feldmeier et al., Hyperbaric oxygen in the treatment of delayed radiation injuries of the extremities. *Undersea & Hyperbaric Medicine : Journal of the Undersea and Hyperbaric Medical Society, Inc,* **27**, 15, Spring, 2000.

72. S. Bayati, R. C. Russell, A. C. Roth, Stimulation of angiogenesis to improve the viability of prefabricated flaps. *Plastic and Reconstructive Surgery* **101**, 1290, April, 1998.

73. P. M. Tibbles, J. S. Edelsberg, Hyperbaric-oxygen therapy. *The New England Journal of Medicine* **334**, 1642, June 20, 1996.

74. S. M. Bentzen, J. Overgaard, Patient-to-patient variability in the expression of radiation-induced normal tissue injury. *Seminars in Radiation Oncology* **4**, 68, April, 1994.

75. G. B. Hart, E. G. Mainous, The treatment of radiation necrosis with hyperbaric oxygen (OHP). *Cancer* **37**, 2580, June, 1976.

76. E. G. Mainous, P. J. Boyne, Hyperbaric oxygen in total rehabilitation of patients with mandibular osteoradionecrosis. *International Journal of Oral Surgery* **3**, 279, 1974.

77. R. E. Marx, J. R. Ames, The use of hyperbaric oxygen therapy in bony reconstruction of the irradiated and tissue-deficient patient. *Journal of Oral and Maxillofacial Surgery: Official Journal of the American Association of Oral and Maxillofacial Surgeons* **40**, 412, July, 1982.

78. R. E. Marx, R. P. Johnson, S. N. Kline, Prevention of osteoradionecrosis: A randomized prospective clinical trial of hyperbaric oxygen versus penicillin. *Journal of American Dental Association* **111**, 49, July, 1985.

79. U. M. Carl, J. J. Feldmeier, G. Schmitt, K. A. Hartmann, Hyperbaric oxygen therapy for late sequelae in women receiving radiation after breast-conserving surgery. *International Journal of Radiation Oncology, Biology, Physics* **49**, 1029, March 15, 2001.

80. J. Bem, S. Bem, A. Singh, Use of hyperbaric oxygen chamber in the management of radiation-related complications of the anorectal region: Report of two cases and review of the literature. *Diseases of the Colon and Rectum* **43**, 1435, October, 2000.

81. M. A. Dall'Era, N. B. Hampson, R. A. Hsi, B. Madsen, J. M. Corman, Hyperbaric oxygen therapy for radiation induced proctopathy in men treated for prostate cancer. *The Journal of Urology* **176**, 87, July, 2006.

82. C. Oliai et al., Hyperbaric oxygen therapy for radiation-induced cystitis and proctitis. *International Journal of Radiation Oncology, Biology, Physics*, March 20, 2012.

83. G. B. Hart et al., Treatment of burns with hyperbaric oxygen. *Surgery, Gynecology & Obstetrics* **139**, 693, November, 1974.

84. P. Cianci et al., Adjunctive hyperbaric oxygen in the treatment of thermal burns. An economic analysis. *The Journal of Burn Care & Rehabilitation* **11**, 140, March–April, 1990.

85. W. A. Zamboni, H. P. Wong, L. L. Stephenson, M. A. Pfeifer, Evaluation of hyperbaric oxygen for diabetic wounds: A prospective study. *Undersea & Hyperbaric Medicine: Journal of the Undersea and Hyperbaric Medical Society, Inc,* **24**, 175, September, 1997.

86. G. Baroni et al., Hyperbaric oxygen in diabetic gangrene treatment. *Diabetes Care* **10**, 81, January–February, 1987.

87. E. Faglia, G. C. Baroni, F. Favales, G. Ballerio, Treatment of diabetic gangrene by hyperbaric oxygen therapy. *Journees Annuelles De Diabetologie de l'Hotel-Dieu*, 209, 1987.

88. E. Faglia et al., Change in major amputation rate in a center dedicated to diabetic foot care during the 1980s: Prognostic determinants for major amputation. *Journal of Diabetes and Its Complications* **12**, 96, March–April, 1998.

89. E. Faglia et al., Adjunctive systemic hyperbaric oxygen therapy in treatment of severe prevalently ischemic diabetic foot ulcer. A randomized study. *Diabetes Care* **19**, 1338, December, 1996.

90. T. R. Harward, J. Volny, F. Golbranson, E. F. Bernstein, A. Fronek, Oxygen inhalation—Induced transcutaneous PO_2 changes as a predictor of amputation level. *Journal of Vascular Surgery: Official Publication, the Society for Vascular Surgery and International Society for Cardiovascular Surgery, North American Chapter* **2**, 220, January, 1985.

91. R. E. Pecoraro, J. H. Ahroni, E. J. Boyko, V. L. Stensel, Chronology and determinants of tissue repair in diabetic lower-extremity ulcers. *Diabetes* **40**, 1305, October, 1991.

92. R. E. Grolman, D. K. Wilkerson, J. Taylor, P. Allinson, M. A. Zatina, Transcutaneous oxygen measurements predict a beneficial response to hyperbaric oxygen therapy in patients with nonhealing wounds and critical limb ischemia. *The American Surgeon* **67**, 1072, November, 2001.

93. J. E. Greensmith, Hyperbaric oxygen reverses organ dysfunction in severe anemia. *Anesthesiology* **93**, 1149, October, 2000.

94. G. B. Hart, Exceptional blood loss anemia. Treatment with hyperbaric oxygen. *JAMA: The Journal of the American Medical Association* **228**, 1028, May 20, 1974.

95. B. Ovbiagele, M. N. Nguyen-Huynh, Stroke epidemiology: Advancing our understanding of disease mechanism and therapy. *Neurotherapeutics: The Journal of the American Society for Experimental Neurotherapeutics* **8**, 319, July, 2011.

96. J. G. McCormick, T. T. Houle, H. A. Saltzman, R. C. Whaley, R. C. Roy, Treatment of acute stroke with hyperbaric oxygen: Time window for efficacy. *Undersea & Hyperbaric Medicine: Journal of the Undersea and Hyperbaric Medical Society, Inc,* **38**, 321, September–October, 2011.

97. S. Y. Chen et al., Improvement of clinical outcome and cerebral perfusion in a patient of atherosclerotic cerebral infarction after repetitive hyperbaric oxygen treatment—A case report and literature review. *Undersea & Hyperbaric Medicine: Journal of the Undersea and Hyperbaric Medical Society, Inc,* **38**, 375, September–October, 2011.

98. D. Michalski, W. Hartig, D. Schneider, C. Hobohm, Use of normobaric and hyperbaric oxygen in acute focal cerebral ischemia—A preclinical and clinical review. *Acta Neurologica Scandinavica* **123**, 85, February, 2011.

99. M. H. Bennett, J. Wasiak, A. Schnabel, P. Kranke, C. French, Hyperbaric oxygen therapy for acute ischaemic stroke. *Cochrane Database System Review,* **2005**, 3, CD004954, 2005.

100. L. Huang, A. Obenaus, Hyperbaric oxygen therapy for traumatic brain injury. *Medical Gas Research* **1**, 21, 2011.

101. T. Sahni, M. Jain, R. Prasad, S. K. Sogani, V. P. Singh, Use of hyperbaric oxygen in traumatic brain injury: Retrospective analysis of data of 20 patients treated at a tertiary care centre. *British Journal of Neurosurgery* **26**, 202, April, 2012.

102. M. H. Bennett, B. Trytko, B. Jonker, Hyperbaric oxygen therapy for the adjunctive treatment of traumatic brain injury. *Cochrane Database System Review,* **2004**, CD004609, 2004.

103. M. H. Bennett, C. French, A. Schnabel, J. Wasiak, P. Kranke, Normobaric and hyperbaric oxygen therapy for migraine and cluster headache. *Cochrane Database System Review,* **2008**, CD005219, 2008.

104. From the Centers for Disease Control and Prevention. Mortality from coronary heart disease and acute myocardial infarction—United States, 1998. *JAMA: The Journal of the American Medical Association* **285**, 1287, March 14, 2001.

105. R. E. Whalen, H. A. Saltzman, Hyperbaric oxygenation in the treatment of acute myocardial infarction. *Progress in Cardiovascular Diseases* **10**, 575, May, 1968.

106. M. H. Bennett, J. P. Lehm, N. Jepson, Hyperbaric oxygen therapy for acute coronary syndrome. *Cochrane Database System Review,* CD004818, 2011.

107. M. Sharifi et al., Usefulness of hyperbaric oxygen therapy to inhibit restenosis after percutaneous coronary intervention for acute myocardial infarction or unstable angina pectoris. *The American Journal of Cardiology* **93**, 1533, June 15, 2004.

108. R. R. Marsh, C. J. Lambertsen, D. M. Schwartz, J. M. Clark, R. F. Wetmore, Auditory and vestibular function in hyperbaric oxygen. *Otolaryngology—Head and Neck Surgery: Official Journal of American Academy of Otolaryngology—Head and Neck Surgery* **93**, 390, June, 1985.

109. D. Torbati, A. Torbati, Blood glucose as a predictive measure for central nervous system oxygen toxicity in conscious rats. *Undersea Biomedical Research* **13**, 147, June, 1986.

110. L. Frank, J. Summerville, D. Massaro, Protection from oxygen toxicity with endotoxin. Role of the endogenous antioxidant enzymes of the lung. *The Journal of Clinical Investigation* **65**, 1104, May, 1980.

111. L. H. Fenton, M. B. Robinson, Repeated exposure to hyperbaric oxygen sensitizes rats to oxygen-induced seizures. *Brain Research* **632**, 143, December 31, 1993.

112. C. Plafki, P. Peters, M. Almeling, W. Welslau, R. Busch, Complications and side effects of hyperbaric oxygen therapy. *Aviation, Space, and Environmental Medicine* **71**, 119, February, 2000.

113. J. M. Clark, C. J. Lambertsen, Rate of development of pulmonary O_2 toxicity in man during O_2 breathing at 2.0 atm. *Journal of Applied Physiology* **30**, 739, May, 1971.

114. G. Margolis, I. W. Brown, Jr., Hyperbaric oxygenation: The eye as a limiting factor. *Science* **151**, 466, January 28, 1966.

29

Image-Guided Thermal Therapy

Robert Staruch
Sunnybrook Research Institute

Kullervo Hynynen
Sunnybrook Research Institute

Rajiv Chopra
University of Texas Southwestern Medical Center

29.1 Introduction

Thermal therapy refers to the application of heat to treat disease. While heat has been used to destroy tumors since ancient times, research in the past few decades has deepened our understanding of the biological basis for its therapeutic effects. This knowledge has provided clear clinical goals for using heat both to destroy unwanted tissue and to enhance or potentiate the effects of other therapies. During this time, a diverse array of technologies for heat generation have been developed and refined to provide precise control of energy deposition deep within the body via interstitial, intracavitary, and completely noninvasive devices. However, until recently, reliance on interstitial temperature measurement probes continued to make thermal therapy an invasive technique when targeting nonsuperficial tissues. The introduction of noninvasive temperature-measurement techniques such as magnetic resonance (MR) thermometry has enabled entirely noninvasive thermal therapy, opening the possibility of using quantitative measurements of temperature distribution in a feedback loop to automatically control three-dimensional heating patterns.

This chapter provides an overview of the field of image-guided thermal therapy, focusing on image-controlled thermal therapies utilizing ultrasound for tissue heating and magnetic resonance imaging (MRI) for real-time temperature control. The first section covers the physical principles of thermal therapy, including the relevant biological and physiological responses of tissue to both mild hyperthermia and high-temperature thermal therapy. The second section discusses the role for image guidance in thermal therapy, focusing on MRI temperature monitoring. The third section describes existing methods of delivering heat to a targeted region of tissue, with emphasis on noninvasive energy deposition using ultrasound, and the recent development of real-time temperature control of ultrasound thermal therapy.

29.2 Thermal Therapy: Biological Rationale and Bioeffects

By increasing tissue temperature above 40°C, heat can be used to elicit a variety of cellular and physiological effects that can either work in concert with other therapies or be used to induce a direct cytotoxic effect. Two broad categories exist to classify these different therapeutic regimes. Mild hyperthermia refers to tissue heating at temperatures of 40–45°C used primarily to enhance and localize the effects of other therapies. Modest temperature elevations maintained for minutes to hours can be used to increase blood flow, induce chemo- or radio-sensitization, increase vessel permeability, or trigger the action of temperature-sensitive drugs. High-temperature thermal therapy, sometimes called thermal ablation, is used to achieve irreversible tissue destruction by applying temperatures greater than 50°C over timescales of seconds to minutes. With recent improvements in the ability to monitor and control treatment, there has been increased clinical interest in both regimes as precisely controlled minimally or noninvasive approaches for the treatment of localized tumors.

29.2.1 Mild Hyperthermia

29.2.1.1 Cytotoxic Effects

The mechanism of cell killing by heat occurs through the inactivation of proteins caused by heat-induced conformational changes [1,2]. In the mild hyperthermia regime, protein unfolding temporarily renders cells vulnerable to other stresses such as ionizing radiation and chemotherapy [3]. Cells respond to such damage by increasing production of heat-shock proteins, which act as molecular chaperones that bind to and protect denatured proteins and guide their correct folding [2]. However, prolonged heating can overwhelm this response leading to the formation of aggregates of denatured cellular proteins, which can block essential cellular processes and trigger apoptotic cell death [4]. The cytotoxic effects of hyperthermia are enhanced in tumors, whose disorganized vasculature is less effective at removing heat. In poorly vascularized regions, the resulting hypoxia and acidity of the tumor microenvironment makes cancer cells more susceptible to heat-induced cell death [5]. However, the cellular response to mild hyperthermia depends strongly on both the duration of heating and the exposure temperature. At 43–45°C, heating must be maintained for more than an hour to achieve a high degree of cell kill. Temperatures less than 43°C are generally ineffective at killing cells, but are useful in the enhancement of other therapies, as discussed below.

29.2.1.2 Radiosensitization

Studies on the sensitivity of cells to heat and radiation have shown that the two modalities work synergistically to produce an enhanced therapeutic effect compared to either modality alone [6]. At temperatures of 41–44°C, hyperthermia enhances radiotherapy through tumor reoxygenation [7–10], as well as the inhibition of enzymes involved in DNA damage repair [11]. However, it has been shown that cells surviving heating to temperatures >43°C acquire resistance to heat killing for the 12–24-h period after the heating is completed [12]. This thermotolerance is thought to be mediated by heat-shock proteins [13], which act as molecular chaperones that bind to denatured proteins to prevent their aggregation with neighboring proteins, which would result in irreversible damage. Thus, optimal effects occur when hyperthermia and radiation are delivered at the same time, and clinical techniques for simultaneous heat and radiation treatment are being developed [14,15]. Recently, veterinary [16] and prospective clinical trials with heat applied before radiation have demonstrated that increases in cumulative heat exposure, even at mild exposure temperatures of 39–41°C result in improved radiotherapy outcomes [17–19].

29.2.1.3 Chemosensitization

Hyperthermia enhances the cytotoxicity of anticancer drugs through a variety of mechanisms [20–22]. As in radiotherapy, mild heating has been shown to inhibit DNA repair of chemotherapy-induced damage [2]. Heat-induced inhibition of the transmembrane drug efflux proteins responsible for multiple-drug resistance allows drugs to remain within the cell at effective concentrations [3,20]. For alkylating and adduct-forming agents, the increase in kinetic energy provided by hyperthermia accelerates reaction rates related to drug action [21]. However, these effects vary widely between chemotherapeutic agents and are sensitive to heating duration, temperature, and drug dose, as well as the relative timing of hyperthermia and drug administration. For example, infusion of cisplatin or bleomycin during 30 min of heating to 43°C shows thermal enhancement ratios of 2–4 over drug or heat alone [23,24]. However, for certain anticancer drugs including doxorubicin and actinomycin D, if hyperthermia is administered before the drug, activated heat-shock proteins protect targeted cellular proteins, thereby decreasing therapeutic effect [22]. For several agents, including doxorubicin, hyperthermia sensitization exhibits a temperature-threshold effect, with strong thermal enhancement for thermal exposures of 30 min at 43°C versus 41°C [20,24]. Doxorubicin's thermal chemosensitization is especially pronounced for local doxorubicin concentrations higher than 10 mg/kg [20], which would be lethal if delivered systemically but could possibly be administered using localized drug release [25]. In recent clinical trials, hyperthermia has improved treatment outcomes when combined with liposomal doxorubicin and paclitaxel for patients with locally advanced breast cancer [26], and with combination etoposide, ifosfamide, doxorubicin chemotherapy in the treatment of soft-tissue sarcoma [27].

29.2.1.4 Blood Flow

In normal tissue, blood vessels respond to localized heating by the relaxation of vascular smooth muscle cells, causing vasodilation and increased flow of normothermic blood to the heated region [28]. This tissue-dependent thermoregulatory response can cause blood flow to reach 2–20 times its resting rate [29], increasing with temperature up to a threshold at which vascular damage occurs and blood flow is rapidly decreased [30]. Angiogenic tumor vessels lack smooth muscle and innervation, reducing their ability to autoregulate. As a result, heat-induced increases in tumor blood flow are smaller in magnitude and have a lower threshold for vessel collapse. In a rabbit ear window preparation, blood flow in normal tissue increased with heating up to 45.7°C for 60 min, while tumor vessels experienced shutdown following 30 min at 43°C [30]. In several rodent tumors, heating to 40–42°C causes immediate increases in tumor blood flow and oxygenation that persist for up to 24 h, with acute vessel damage occurring at 43°C or higher held for 30 min or more [28,31]. Human tumor cells and tumor vasculature demonstrate higher resilience to thermal damage when compared to rodent tumors [28,32]. In human tumors, the ability of blood vessels to increase perfusion to cool tissue increases with temperature up to 42–43°C [33,34], and even transient heat-induced decreases in blood flow occur only at temperatures greater than 44°C [33]. During ultrasound hyperthermia, tumor blood flow either stays constant or increases at least transiently during 30–60 min of heating to 42.5°C, with increases starting approximately 10 min after the onset of heating [35]. Clinically, increased blood flow in response to heat makes tumors more sensitive to radiation by increasing their oxygenation [36] and it can also improve drug delivery to tumors.

29.2.1.5 Vessel Permeability

Angiogenic tumor vasculature is known to be ill formed and chaotic, with regions of higher permeability than that of normal tissue, heterogeneously distributed along tumor microvessels [37]. The "leakiness" of tumor vessels is exploited by nanoparticle drug carriers such as liposomes, designed to preferentially leak out of tumor vessels and to achieve localized accumulation in the tumor [38]. In animal tumors, hyperthermia has been shown to enhance nanoparticle extravasation through changes in pore size [39,40], where endothelial cells suffer a transient disaggregation of microtubular proteins that acts to change the morphology of the cytoskeleton causing the cell to "shrink." This increases the occurrence and size of pores in the vessel wall, enabling and increasing extravasation of normally impermeable particles as large as 400 nm [39,41]. Exposures as short as 10 min can enhance permeability for several hours [42–45], which corresponds to the time required for heat-shock protein-mediated reassembly of microtubular proteins [40]. In rodent tumor studies, 30–60 min at 42°C has been shown to increase extravasation up to 50 times [46], with both the maximum particle size and rate of extravasation increasing with thermal dose up to the threshold for vascular collapse [40]. By increasing the amount of nanoparticle extravasation from tumor vessels, mild hyperthermia can selectively increase drug delivery to tumor cells.

29.2.1.6 Immunomodulation

Finally, hyperthermia is a potent modulator of both the innate and adaptive immune systems [47]. Fever range heating to 39–40°C affects the development and function of antigen-presenting cells, T cells, and natural killer cells [48]. Temperatures of 40–43°C induce a heat-shock response that increases tumor detection through expression of membrane proteins recognized by antigen-presenting cells. With heat-induced cell death, dying cells release tumor-specific antigens bound to heat-shock proteins that can be trafficked to the lymph nodes, inducing antitumor immunity through a T-cell response.

29.2.2 High-Temperature Thermal Therapy

The goal of high-temperature thermal therapy is to destroy tissue by elevating temperatures within a targeted volume above the threshold required to produce thermal coagulation. Thermal coagulation is a process where cellular and tissue structural proteins undergo rapid denaturation, resulting in irreversible structural and chemical changes within the intra- and extracellular environment of tissue. These effects occur at temperatures above approximately 50°C for a range of tissue types [49–51], with a cell survival of 0.01% after 5 s of heating at 57°C [52]. Although cell death occurs rapidly above the coagulation temperature threshold, there are a range of distinct structural changes, physiological effects, and time-dependent biological responses that occur between 50°C and 100°C. In this regime, mechanisms for heat-induced tissue destruction include ablation, vaporization, coagulation, enzyme denaturation, and the appearance of a red zone of damaged but viable tissue at the periphery of the thermal lesion. If one considers the region of thermal damage produced by a focal deposition of energy with temperature decreasing radially from the center, a number of distinct regions can be seen, as reviewed by Thomsen [53].

29.2.2.1 Ablation and Carbonization

These effects occur rapidly in situations where the temperature elevation generated in tissue is in excess of 100°C. While the term thermal ablation is often used to refer broadly to high-temperature thermal therapies, in this context ablation is defined as the physical removal of tissue components and can be seen as the presence of craters, holes, and defects within a heated region. Carbonization results in the formation of a thin layer of black residue on the surface of thermal lesion resulting from reduction of tissue components to carbon. These biological changes significantly alter tissue optical, acoustic, and physical properties, potentially hampering the delivery of energy for heating and should thus be avoided.

29.2.2.2 Water Vaporization

As temperatures approach and exceed 100°C in a heated volume, water vapor is formed, and steam collects in small pockets forming vacuoles in tissue. This process of boiling results in dessication of tissue, with the presence of small vacuoles throughout. With the use of high-intensity focused ultrasound at high-pressure amplitudes, shock waves formed by nonlinear wave propagation at the focus can enhance heating to cause tissue boiling in a matter of milliseconds [54].

29.2.2.3 Structural Protein Thermal Coagulation

Structural proteins (collagen, elastin) are found in the extracellular matrix and cytoskeleton of all cells. Upon thermal coagulation, a number of distinct changes occur corresponding to the denaturation of these structural proteins. The semicrystalline network of collagen fibers in tissue is disrupted during coagulation, reduced to an amorphous field of denatured protein [53]. The region of thermal coagulation in most tissues appears visually as lighter in color, and more opaque. Temperatures above 60°C are usually required to observe coagulation in tissue, and the coagulated volume is often referred to as the thermal lesion.

29.2.2.4 Vital Enzyme Denaturation

The denaturation process undergone by functional proteins within cells is similar to that of the structural proteins, but occurs at lower temperatures. Denatured proteins form aggregates that block essential cellular processes and lead to cell death [2]. These lethal effects occur beyond the visible boundary of the thermal lesion and can be detected with vital staining techniques.

29.2.2.5 Red Thermal Damage Zone

A red zone appears in tissue peripheral to the coagulation boundary in thermal lesions within 30 s after heating. This site of initial inflammation is characterized by damage to the vasculature, and a number of host physiological responses to heat such as hemorrhage, increased blood flow, and/or hemostasis. Evaluation of the full extent of thermal damage can be made 1–5 days after heating, when all irreversibly damaged cells undergo cell death. The type of necrosis undergone by cells depends on the initial zone they were located in (structural coagulation, red zone, etc.) and includes effects such as coagulative necrosis, ischemic necrosis, and apoptosis. The extent of thermal necrosis seems to be related to the outer boundary of the red thermal damage zone in tissue [53].

29.2.3 Dosimetry

Accurate delivery of mild hyperthermia or high-temperature thermal therapy is predicted on quantitative thermal dosimetry. It is critical to understand the cytotoxic effect of temperature on tissue in a quantitative way so that thermal therapies can be associated with a predictable and repeatable therapeutic response. Two predictors of thermal damage are commonly used: thermal dose based on Arrhenius rate process models and critical temperature thresholds. Both descriptions require knowledge of the temperature distribution over time in a target volume. This has been a long-standing limitation of dosimetry in thermal therapy due to the technological limitations of thermometry. Until recently, dosimetry was based on invasive point measurements of temperature from a few thermocouples implanted in the target region. The development of noninvasive thermometry has provided the ability to make spatially resolved temperature measurements, which can be used to monitor and control thermal treatments, as well as to make accurate predictions of the occurrence and extent of thermal damage.

29.2.3.1 Thermal Dose

In thermal therapy, both temperature and time are important variables determining cell death. A wide range of studies have identified a logarithmic relationship between temperature and time for an equivalent cytotoxic effect. The thermal dose model is based on empirical observations in cells from temperatures of 43°C to 57°C that for an increase in temperature by 1°C, the time required to achieve a particular level of cell death could be reduced by approximately one-half [5]. This response to heating can be modeled as a change of state reaction following an Arrhenius relationship [55]. To compare heating regimens of different temperatures and durations and predict their therapeutic effect, the model calculates a thermal isoeffective dose that normalizes hyperthermia treatments to a number of minutes required at a reference temperature, arbitrarily chosen as 43°C, to achieve an equal biological effect. The cumulative equivalent minutes at 43°C (CEM43) of a given heat exposure is calculated as [56]

$$\text{CEM43} = \sum_{t=0}^{t=final} R^{(43-T)} \Delta t, R = \begin{cases} 0.25, & T < 43°C \\ 0.50, & T \geq 43°C \end{cases} \tag{29.1}$$

where R is a constant equal to 0.25 below 43°C and 0.5 above 43°C, T is the average temperature during time Δt, and CEM43 is the number of minutes at 43°C required to achieve the equivalent cell kill. This formulation allows many different treatment regimens to be standardized into a single unit and compared in terms of cell death.

Thermal dose thresholds for various degrees of thermal damage have been measured in many tissues [57]. The onset of thermal damage occurs in muscle at 30 CEM43 [51], while complete necrosis occurs in various tissues at 50–240 equivalent minutes [58]. A threshold of 240 CEM43 is commonly used as a conservative treatment goal to ensure thermal coagulation [50,51]. While thermal dose is based on empirical observations of heat-induced cell kill, it has also proved useful in describing sublethal biological effects, including blood flow increases and shutdown [7,9], as well as tumor radio- and chemosensitization [5,59].

29.2.3.2 Critical Temperature

At temperatures greater than 54°C, the 240 CEM43 threshold is reached in a few seconds, and cell death occurs rapidly by the thermal coagulation of structural proteins [53] rather than internal cellular processes, as discussed in Sections 2.2.3 and 2.2.4. For high-temperature thermal therapies of this type, the temporal resolution of noninvasive temperature measurement techniques such as MR thermometry is insufficient to provide a meaningful estimate of thermal dose. Furthermore, normally distributed noise in temperature measurements results in systemic overestimation of thermal dose. A number of studies have concluded that the selection of a threshold temperature is sufficient to predict thermal damage, and temperatures between 52°C and 62°C have been used as practical treatment goals for thermal coagulation therapy [49,51,55,60–62].

29.2.4 Heat Conduction and Thermal Modeling

Heating patterns in tissue during thermal therapy are influenced by heat conduction, the removal of heat by blood flow, and the deposition of energy from heating applicators. Heat conduction transports heat to regions of lower temperature. It depends on temperature gradients within the heated region and the thermal conductivity of the material. Blood flow tends to remove heat from tissue at a rate that is dependent on the density, size, and flow rate of the vessels, as well as the temperature of the tissue. The elevation of temperatures during heating is determined by the rate of energy deposition from a heating applicator. These effects have been modeled by Pennes in the bio-heat transfer equation [63]:

$$k\nabla^2 T(r,t) + Q(r,t) - w_b c_b [T(r,t) - T_b(r,t)] = \rho_t c_t \frac{\partial T(r,t)}{\partial t} \tag{29.2}$$

where k is the thermal conductivity of tissue (W/m°C), $T(r,t)$ the temperature, T_b the temperature of blood, $Q(r,t)$ the power deposited by an energy source (W/m^3), w_b the blood perfusion (kg/m^3·s), c_b the specific heat capacity of blood (J/kg°C), c_t the specific heat of tissue, and ρ_t the density of tissue (kg/m^3). The first term on the left-hand side describes the process of heat conduction, and energy deposition is described by the second term. The third term represents an approximation to the interaction between tissue and blood perfusion, assuming the absence of large blood vessels. More sophisticated models based on the spatial arrangement of blood vessels have been proposed [64]. Tissue thermal properties for water, blood, muscle, and bone are provided in Table 29.1.

TABLE 29.1   Physical Parameters Used to Model the Effects of Thermal Conduction in Tissue

Material Property	Water	Blood	Muscle	Bone
Density ρ (kg m^{-3})	1000	1030	1060	1700
Specific heat capacity of tissue c_t (J kg^{-1}°C^{-1})	4180	3620	3720	1600
Thermal conductivity k (W m^{-1}°C^{-1})	0.615	0.5	0.537	0.6
Blood perfusion w_b (kg m^{-3} s^{-1})	0.0	—	0.5–5.0	0.0

Source: Duck FA. *Physical Properties of Tissue: A Comprehensive Reference Book.* London: Academic Press; 1990.

The influence of heat loss due to blood flow has a major impact on the outcome of thermal therapy [65]. For temperature elevations of a few degrees, the vascular network can regulate temperature by increasing the amount of blood flowing through the heated volume. This increase in blood flow is tissue-specific and can be spatially variable and temperature-dependent, resulting in the need to adjust the spatial power deposition from heating devices to maintain a uniformly heated tissue volume [66]. This effect has been a major issue for hyperthermia and places strong emphasis on the ability to measure dynamic temperature distributions during heating for online power adjustment. During high-temperature thermal therapy, increased blood flow in targeted tissues can result in the preservation of perivascular cells which can be a problem for cancer treatment; it can also transport heat downstream to cause thermal damage outside the target zone. If sufficient heating is achieved, vessels in coagulated regions are destroyed and flow ceases. This vascular shutdown can result in faster rates of heating due to inadequate removal of energy and must be taken into account to avoid tissue vaporization or ablation. Coagulation of blood vessels can also affect the supply of oxygen and nutrients to downstream tissues outside of the target region. In some thermal coagulation therapies, a slight build-up of energy occurs in untargeted normal tissue. To avoid thermal damage in these regions, a delay is required to allow normal tissue to return to resting body temperature through the cooling effects of conduction and blood flow.

Modeling heat transport in tissues is important for thermal therapy treatment planning and control [67]. Knowledge of the effect of blood flow helps predict the power distribution required to achieve a desired temperature distribution without causing unintended thermal damage in neighboring healthy tissue.

29.2.5 Summary

The cellular response to heat is both time- and temperature-dependent and varies between different tissues and tumor types. High-temperature thermal therapy uses temperatures greater than 55°C to cause thermal coagulation, where extracellular protein denaturation results in homogenization of tissue structure and cell death in a matter of seconds. Mild hyperthermia at temperatures of 40–45°C is typically used in combination with other therapies, to localize drug delivery or render cells more vulnerable to ionizing radiation and chemotherapy [1,10,68]. The dependence of mild heating on the variable cooling effects of blood flow, and the potential for damage to normal tissue during high-temperature thermal therapy, makes it important to monitor temperature elevations in tissue during the application of thermal therapy.

29.3 Image Guidance for Thermal Therapy

Thermal therapy is typically delivered using a minimally or noninvasive approach. As such, visualization of targeted regions and surrounding structures is incomplete. In image-guided interventions, imaging techniques are used to guide the procedure in five important ways: planning, targeting, monitoring, controlling, and assessing treatment response [69]. The following subsections describe the various roles for image guidance in thermal therapy, with special reference to noninvasive energy deposition using focused ultrasound guided by MRI.

29.3.1 Treatment Planning and Targeting

Essential to any image-guided procedure is the ability to identify a target region and register the energy deposition relative to the surrounding anatomy. For energy delivery from interstitial devices inserted into the body, this capability is met by visualizing the therapeutic device, which can be achieved using a wide range of medical imaging technology, including ultrasound, MRI, CT, and x-ray fluoroscopy. When MRI is used, device materials must be carefully selected for patient safety and to avoid image artifacts. Some types of surgical-grade stainless steel produce artifacts 5–10 times larger than the actual size of the device, rendering accurate image guidance and localization impossible [70]. With materials having a magnetic susceptibility close to that of water, artifacts are avoided and interventional devices are easily distinguished from surrounding tissue in MR images, appearing as a signal void due to the absence of water. However, the list of available materials is limited to ceramics, plastics, and a few metals, making the design of MRI-compatible equipment and hardware a significant challenge [70].

For noninvasive heating devices, targeting is best achieved by visualizing the effects of the energy deposition itself, typically by monitoring temperature changes during a low-power heating test. The need for direct visualization of energy deposition arises from the potential for deflection of the heating focus due to spatially heterogeneous energy absorption within tissue. For example, when ultrasound is directed into the body from externally focused transducers, complex tissue interfaces affect the position of the ultrasound focus, and unintended hot spots occur if there is highly absorbing bone or highly reflecting air-filled lung or bowels in the beam path. While the ability to focus acoustic energy into tissue has been known since the 1940s [71], the inability to know the actual position of the focus limited the clinical utility of focused ultrasound before the development of image guidance techniques. MR is particularly well-suited for guiding noninvasive heating techniques, as it is the only clinically accepted method to noninvasively quantify temperature changes deep within the tissue. As discussed in the following section, MR thermometry can be used to identify the location of the ultrasound focus during low-power exposures [72], as well as to monitor temperatures in the target and normal tissues during treatment [73]. In the case of focused ultrasound, MR can also be used to directly image the tissue displacements caused by absorbed ultrasound power in a technique known as magnetic resonance acoustic radiation force imaging (MR-ARFI) [74]. This enables localization of the site of energy deposition without the need for heating, a capability which is important in sensitive tissues such as the brain.

29.3.2 Treatment Monitoring and Control: MR Thermometry

The rate of energy deposition in thermal coagulation therapy is high, and at temperatures sufficient for thermal coagulation, tissue destruction is indiscriminate and happens within seconds. Therefore, a critical component of thermal therapy is the monitoring of energy delivery during heating to observe the growth of the thermal lesion and to minimize damage to normal tissue. This can be accomplished through measurement of the temperature distribution in the target volume during heat delivery.

The requirements of thermometry for thermal coagulation therapy are related to the rate at which heat is applied and spatial variation of temperatures. However, general guidelines for the clinical requirements of thermometry techniques are a temperature resolution of at least 1°C, a spatial resolution of less than 1 cm, and a temporal resolution of a few seconds.

Traditionally, temperature measurements during hyperthermia are made by implanted thermocouples or fiber-optic probes [19,75–78]. This form of thermometry has high temporal resolution and temperature sensitivity at each of the implanted probes, but provides only a coarse sampling of the spatially varying temperature distribution. Furthermore, implanted probes interact with energy deposition to create temperature measurement artifacts and distort the energy absorption pattern in tissue [79]. The development of MR thermometry has provided a noninvasive method to make spatially resolved temperature measurements and has begun to replace implanted probes for guidance of hyperthermia.

Noninvasive thermometric techniques have been proposed using various imaging modalities including x-ray computed tomography [80,81], microwave radiation [82], ultrasound [83], and MRI [84]. Of these techniques, MR thermometry has seen the furthest development and is the only one in clinical use [85,86]. Several tissue properties that can be measured using MRI vary with temperature and can be used for thermometry, including spin–lattice relaxation time T1 [84], water diffusion [87], equilibrium magnetization [88], and the water proton resonance frequency shift [89] which can be measured using MR spectroscopy [90] or phase mapping [91]. For thermal therapy, the phase-difference proton resonance frequency shift technique has several advantages over the other techniques and is by far the most commonly used. The PRF varies linearly with temperature and is insensitive to tissue type and thermal damage [92]. Using this technique at 1.5 T, temperature information can be obtained with millimeter spatial resolution and approximately ±1°C accuracy in an imaging time of approximately 1 s, thus meeting the requirements for thermal therapy. However, as discussed below, the phase difference technique measures relative temperature changes and requires an external reference to compute estimates of absolute tissue temperature. In situations where absolute temperature measurements are essential, or where direct temperature measurement using MRI is difficult due to motion or weak temperature-related signal changes, temperature-sensitive exogenous contrast agents can be used to indicate the spatial extent of tissues in which therapeutic temperatures have been achieved [86]. Several lipid-based nanoparticle contrast agents have been developed that release MRI contrast agents when their local temperature is raised above the melting temperature of the lipid [93–96], which can be set to a therapeutic temperature of interest.

29.3.2.1 PRF Technique

The physical basis for the proton resonance frequency shift is the subtle change in local magnetic field around water protons caused by the decreasing number of hydrogen bonds per water molecule with increasing temperature due to increased mobility of water molecules [89]. In MRI, the resonance frequency of a water proton is proportional to the magnetic field it experiences. As temperature increases and the kinetic energy of water molecules overcomes the strength of the hydrogen bonds between them, water protons dissociate from the oxygen atoms of their neighbors and experience a greater shielding by their own electrons, thus experiencing a smaller local magnetic field and hence Larmor frequency ω_0. During the echo time between MR signal excitation and measurement, the small frequency difference results in an accumulated phase offset proportional to the change in temperature. Images of these phase offsets can be acquired from gradient echo or echo-planar MR image acquisitions. However, phase variations due to heating are much smaller than the background phase variations that arise from the static magnetic field distortions caused by the susceptibility of objects placed in the bore. Phase differences arising from relative temperature changes can be isolated from static phase variations by subtracting phase images acquired before and during heating [91]:

$$\Delta T = \frac{\phi_2 - \phi_1}{\gamma \cdot TE \cdot \alpha \cdot B_o} \tag{29.3}$$

This relationship between temperature change (ΔT) and phase difference ($\phi_2 - \phi_1$) depends only on the main magnetic field strength (B_o, often 1.5 or 3.0 T), the prescribed echo time (TE, typically 5–30 ms) the gyromagnetic constant (($\gamma/2\pi$) = 42.58 MHz/T), and the proton resonance frequency shift coefficient (α = 0.01 ppm/°C). The PRF shift coefficient is constant for a wide range of temperatures (0–100°C) and is insensitive to tissue type for water-based tissues, even after thermal coagulation [92]. Phase images can be rapidly acquired (typically 0.1–5 s) in multiple slices through the use of gradient echo-pulse sequences, maintaining sufficient signal-to-noise ratio (SNR) for typical temperature measurement uncertainty of less than 1°C at 1.5 T.

The biggest challenge with the phase subtraction technique is the time-varying change in magnetic field due to breathing and motion. Object motion in the image plane causes misregistration of the image subtraction, while motion anywhere near the image plane can alter the susceptibility distribution to produce artifactual phase changes that mask the temperature signal. This poses major difficulties for temperature measurement near moving organs such as the heart and the liver, or in the vicinity of the inflating and deflating lungs. However, there are some very promising solutions. The self-referenced method [97] estimates the background phase from the current image eliminating the need for a reference image subtraction and thus is not affected by sudden displacements between images. For periodic displacements, the multi-baseline method collects an "atlas" of rapidly acquired background phase images sampling the entire range of possible displacements, and the correct reference for each subsequent image is selected during treatment [98].

The phase difference technique is also sensitive to slow variations in the main magnetic field; this magnetic field "drift" can be corrected by subtracting the phase measured in image regions where the temperature is known to be constant [92,99]; however, scanner-specific spatial variation in field drift can make accurate correction a technical challenge [100,101]. Another important limitation is that this mode of temperature-dependence applies only to water protons and works well only in tissues with high water content; in tissues containing a mix of water and lipid, the weak temperature dependence of susceptibility confounds the PRF measurement and some form of lipid suppression should be used [102,103]. Finally, because the technique measures a relative temperature change, it requires accurate temperature information when the background images are acquired. Despite these limitations, phase difference MR thermometry provides the ability to measure dynamic temperature distributions noninvasively during thermal therapy, enabling accurate dosimetry and damage predictions. This has been used successfully in humans to guide thermal therapy in many tissues, including the uterus [104], brain [105–107], breast [108,109], rectum [110], and prostate gland [111].

An example of temperature measurements using the PRF is shown in Figure 29.1. In this experiment, the temperature distribution transverse to an externally focused ultrasound transducer was measured during heating, depicting the shape of the heating pattern.

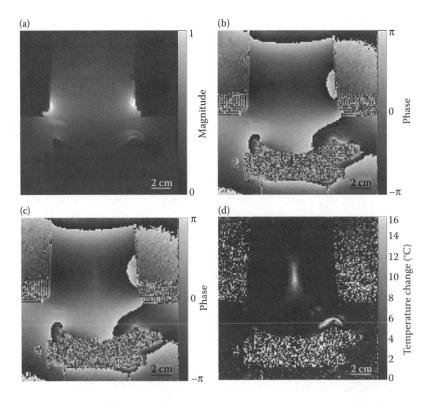

FIGURE 29.1 Temperature measurements made with MRI using the proton resonance frequency shift phase-difference technique. (a) Magnitude reconstruction of gradient echo acquisition depicts transverse slice through an externally focused ultrasound transducer situated in a bath of degassed, deionized water, as it heats a focal region in a gel phantom above. (b, c) Phase image reconstructions acquired before and after application of ultrasound energy. (d) Subtraction of phase images in (b) and (c), scaled to relative temperature by Equation 29.3 to identify in this case a 16°C temperature rise caused by focused ultrasound heating.

29.3.3 Assessment of Treatment Outcome

Immediately after thermal therapy, imaging can be used to detect the effects of thermal coagulation. Areas identified to be undertreated can be reheated to ensure complete coagulation of the targeted volume. While temperature images acquired during heating can be used to calculate a cumulative thermal dose map for predicting the extent of thermal damage, their summation requires adequate temporal sampling and accurate registration to a common reference position. A direct imaging method capable of distinguishing the thermal lesion from viable tissue is desirable. Fortunately, a number of imaging modalities are capable of indicating the extent and severity of heat-induced damage during thermal therapy.

One imaging approach to measure tissue damage after thermal coagulation is to delineate zones of decreased perfusion caused by changes in the vasculature after thermal therapy. The dynamics of signal enhancement by injected imaging contrast agents is altered in regions whose blood supply has been destroyed, an effect that has been observed with multiple imaging modalities [112] including MRI [113], CT [114], and ultrasound [115]. With dynamic contrast-enhanced MRI using gadolinium-based contrast agents, thermal lesions appear immediately after ablation as dark, nonperfused regions on T1-weighted images [51]. In breast cancers resected after thermal ablation, Gianfelice and coworkers identified a strong correlation between DCE-MRI parameters and histological extent of thermal damage [116,117]. In the treatment of uterine fibroids with focused ultrasound, nonperfused tissue identified using contrast-enhanced T1-weighted imaging corresponds to regions where a lethal thermal dose is measured with MR thermometry [104], with the exception of cases where large vessels are destroyed causing ischemia in downstream unheated tissue. NPV has also been associated with overall fibroid shrinkage and symptom relief [118].

Another approach is to directly identify coagulation-induced changes in tissue properties by changes in signal intensity. Both ultrasound and MRI have been investigated extensively in this area to understand the various changes in tissue parameters with heating [119,120]. Most MR parameters, including T1, T2, and the apparent diffusion coefficient, change upon coagulation [120,121]. The magnitude and temperature dependence of these changes varies with tissue type, but the changes are large enough for standard imaging sequences to detect differences between normal and coagulated tissue. In particular, T2-weighted images depict thermal lesions in muscle as regions of low-signal intensity surrounded by a ring whose intensity increases over the course of 1–2 h after heating [51,122]. This peripheral enhancement has been shown to correlate with histologically identified edema and inflammation [123,124]. Recently, the extent of treated volumes identified using diffusion-weighted imaging have been demonstrated to correlate well with contrast-enhanced T1-weighted imaging [125,126], and is thought to provide information on treatment-related changes in intra- and extracellular water motion related to coagulation of structural proteins [126].

An interesting area of recent research has been the development of thermally mediated image-guided drug delivery [38,127]. In this approach, temperature-sensitive macromolecular constructs are loaded with both chemotherapeutic agents as well as an MR contrast agent [128–131]. Upon heating, both the drug and the tracer are released, providing a change in imaging contrast in the vicinity of the released drug. Imageable drug carrier development has been supported by the use of MR imaging techniques capable of both temperature measurement and detection of gadolinium contrast enhancement [132]. This combination, though still in preclinical development, would enable the clinician to noninvasively assess and direct drug accumulation and release in an approach known as "drug dose painting" [133].

29.3.4 Summary

Image guidance is essential for the safe and accurate delivery of thermal therapy. Heating devices are typically positioned without direct visualization of either the target or the applicator, and energy deposition can be significantly altered by spatial variations in tissue properties. These uncertainties are exacerbated for noninvasive energy delivery, necessitating visualization of target regions, guided localization of energy delivery, immediate monitoring of local effects, and noninvasive evaluation of thermal damage and treatment outcome. Several imaging modalities (MRI, ultrasound, CT) can satisfy one or more of these requirements. However, MRI is especially well-suited for guiding thermal therapy. It is nonionizing, and has excellent anatomic resolution using a variety of contrast mechanisms that are sensitive both to malignancy and tissue changes caused by thermal damage, enabling accurate treatment planning and verification. It also has the capability to noninvasively quantify temperature changes deep within the tissue. The temporal and three-dimensional spatial temperature information provided by MR thermometry allows us to noninvasively estimate the delivered time–temperature history, and is the most important feature of MRI for guiding thermal therapy.

29.4 Delivery Strategies for Image-Guided Thermal Therapy

While treatment monitoring and assessment are important components of thermal therapy, treatment success depends on the ability to control the spatial energy deposition in tissue. Tissue heating is typically accomplished using one of three basic routes. Interstitial heating involves the delivery of energy from devices inserted directly into or adjacent to the tumor of interest. Intracavitary devices can be inserted into body cavities such as the esophagus, vagina, urethra, and rectum, offering a less invasive approach. Finally, external devices can transmit energy through the body to generate heating noninvasively at a target location. Within each category of therapeutic device, a number of energy sources have been investigated, including laser, microwave (MW), radiofrequency (RF), and ultrasound.

These techniques are discussed below, with an emphasis on technology for noninvasive energy delivery. For noninvasive thermal therapy, image guidance is a necessity, not only providing thermal dosimetry and evaluation of thermal damage, but enabling the accurate targeting of energy deposition to the correct region in tissue. Within this category, high-intensity focused ultrasound is highlighted for its unique capability to localize heating to precise volumes of tissue, which allows ultrasound to be used both for noninvasive hyperthermia and high-temperature thermal therapy.

29.4.1 Interstitial Devices for Thermal Ablation

Interstitial high-temperature thermal therapy has been applied to treat a variety of neoplasms [134–137]. The most extensive clinical experience is in the use of RF electrodes, which are commonly used to coagulate primary and secondary malignancies of the liver [138–141], as well as tumors in the bone [142–144], lung [145,146], and kidney [147,148]. Interstitial RF heating is typically performed at 460–480 kHz. Between 0.3 and 30 MHz, the primary mechanism of absorption of electromagnetic energy in tissue is resistive loss near the electrode, with heating by thermal conduction at the lesion periphery [149]. Below about 100 kHz, undesirable stimulation of nerve and muscle fibers occurs, and above 30 MHz resistive heating is no longer the main mechanism of action [149]. Interstitial RF applicators have the ability to generate thermal lesions with diameters between 3 and 4 cm *in vivo* through the use of multi-tined electrodes [150], but require a ground return for current, either implanted in tissue or attached to the skin surface [149]. Also, tissue dessication at temperatures approaching the vaporization threshold can decrease tissue conductivity in the region surrounding the applicator, which in turn adversely affects the penetration of RF energy into tissue. Direct saline injections to increase tissue conductivity, and active device cooling to prevent charring have been investigated to overcome these effects [137].

While RF electrodes are the most pervasive class of devices for interstitial thermal ablation, microwave, laser, and ultrasound devices are also in clinical use [137]. Interstitial microwave antennae typically operate at frequencies of 915 or 2450 MHz, where the predominant mode of propagation for electromagnetic waves is radiative and tissue absorption occurs primarily due to dielectric losses [134,137,151]. With a larger active heating region than RF electrodes, interstitial microwave antennae are capable of producing larger lesions with reduced sensitivity to tissue-specific thermal conductivity and the presence of large blood vessels [152]. Laser fibers provide ablation zones of 1–2 cm in diameter with inherently CT and MR-compatible devices, and have found use in the treatment of small tumors [153–155], most notably in the brain [107,156,157]. Interstitial ultrasound applicators have been developed for a number of applications as reviewed by Lafon [158], including rectal hyperthermia [18], ablation of targets in the brain [159], spine [160], liver [161,162], esophagus [163], heart [164–166], and the cervix [167], along with a variety of devices designed for prostate hyperthermia and ablation [158,168]. Interstitial ultrasound applicators offer control over the three-dimensional shape of the heating pattern through the use of multielement piezoceramic transducer configurations for control along the axis of the device [169], with sectored [170] or planar elements [171] providing directional control, multi-frequency transducers allowing control over treatment depth [172], and rotational control enabling rapid conformal heating [173,174].

For interstitial thermal therapy in soft tissues, device insertion is typically guided by ultrasound imaging, with assessment of thermal coagulation extent using contrast-enhanced CT or MRI [69]. MR thermometry during RF [175,176] and MW [177] ablation have been developed, and MR-guided RF ablation has been demonstrated clinically [178]. MRI has long been used to guide laser thermal therapy [179,180], and quantitative temperature maps have recently been used clinically as feedback for automatically controlling laser thermal therapy in the brain [107,181,182]. This form of closed-loop treatment has also been used to achieve MRI-controlled conformal ultrasound heating of the prostate using transurethral intracavitary ultrasound applicators [111,183].

29.4.2 Noninvasive Electromagnetic Hyperthermia

The absorbed power density (SAR, units W/kg) for an electromagnetic wave in a dissipative medium is given by

$$\text{SAR} = \frac{1}{2}\frac{\sigma}{\rho}|E|^2 \tag{29.4}$$

where σ is the electrical conductivity (S/m), ρ is the density of tissue (kg/m^3), and E is the electric field. At frequencies below 30 MHz, used by interstitial RF electrodes, SAR drops rapidly with distance from the heating applicator, and the primary mechanism for temperature elevation in a target volume is heat conduction. The short wavelength at microwave frequencies (10, 4.5, and 1.7 cm in muscle at commonly used frequencies of 433, 915, and 2450 MHz, respectively) enables using phased applicators to focus energy [184]. However, at these frequencies, microwaves have limited depth of penetration (2–3 cm) and suffer strong reflections at soft-tissue interfaces. This makes microwave applicators particularly well-suited for hyperthermia of superficial tumors [17,26,185], and conformal microwave arrays have recently been developed to treat large-area superficial disease such as diffuse breast cancer recurrence at the chest wall [186].

Regional heating of deep-seated tumors is achieved noninvasively using externally applied electromagnetic hyperthermia with antenna arrays that operate between the RF and MW regimes, at frequencies of 100–150 MHz [184]. Clinical devices of this type have been important in the clinical combination of mild hyperthermia with radiation and chemotherapy, achieving temperatures of 40–42°C in various deep-seated tissues, including soft-tissue sarcoma [27], pelvic tumors [187,188], and tumors of the head and neck [77]. However, heterogeneous energy propagation, weak focusing abilities in this frequency range, and the sensitivity of mild hyperthermia applications to temperature variations are serious limitations [189], and make it desirable to guide electromagnetic hyperthermia with real-time temperature imaging as enabled by MR thermometry. Preservation of µW power MR signals (at 63.87 MHz for 1.5 T) without affecting the delivery of kiloWatt power EM hyperthermia (at 100–150 MHz) requires careful suppression of all spectral contaminants in the hyperthermia antenna amplifier system (especially at the MR resonance frequency), along with elimination of the antenna's operating frequency from the MR receive pathway [190]. Simultaneous RF hyperthermia and MR imaging at 1.5 T has been demonstrated using a large regional heating device at 100 MHz [190,191] and an extremity applicator at 140 MHz [192]. Clinically, these hybrid systems have enabled the use of real-time PRF thermometry to monitor temperatures achieved during hyperthermia of soft-tissue sarcomas [193] and recurrent rectal carcinoma [110]. Feedback control algorithms have been proposed [194] and demonstrated in gel phantoms [195].

29.4.3 Focused Ultrasound

29.4.3.1 Physical Principles

Ultrasound is a mechanical wave that propagates within a medium inducing particle vibrations at frequencies between 20 kHz and 20 MHz, which are capable of penetrating several centimeters in tissue and eliciting a variety of biological effects. Low-intensity pulsed exposures of a few milliseconds in length produce pressure elevations that can be used with ultrasound contrast agents to enhance blood vessel permeability without increasing temperature. Higher intensity pulsed exposures can be used to cause localized shock waves and cavitation that can break kidney or gall stones for lithotripsy [196] and liquefy tissues in a technique referred to as histotripsy [197]. Continuous-wave exposures of 1–10 s can rapidly increase temperatures above the coagulation threshold for high-temperature thermal therapy. Longer continuous-wave exposures of several minutes at lower power can be used to achieve mild heating in the hyperthermia regime.

As ultrasound travels through tissue it is attenuated through absorption, with the energy of the induced particle vibrations being converted to heat. Ultrasound-induced heating is dependent on the absorption characteristics of tissue and transducer geometry. The absorbed power (W/mm³) in tissue from an ultrasound beam is given by

$$\text{SAR} = \frac{\alpha p_o^2}{\rho c} \tag{29.5}$$

where α is the absorption coefficient in tissue (Np/m), ρ is the density of tissue (kg/m³), c is the speed of sound in tissue (m/s), and p_o is the pressure amplitude at a given point (N/m²). This expression is valid for a continuous sound field propagating in a medium with zero shear viscosity [198]. The absorption of ultrasound is primarily due to the macromolecular components of tissue and increases with protein content, with a small contribution due to higher-order structural organization [199]. Ultrasound absorption increases with frequency, and can be described by the relationship,

$$\alpha = \alpha_o f^n \tag{29.6}$$

where α_o is the absorption of ultrasound at 1 MHz, f is the ultrasound frequency in MHz, and n is the exponent of the frequency dependence, ranging from 1 to 1.3 for most soft tissues [200].

29.4.3.2 Treatment Delivery

In focused ultrasound therapy, a transducer is positioned external to the body, coupled to tissue using a water bath or other coupling medium. The ultrasound energy from the transducer is focused to a point several centimeters into the body, accomplished with a geometrically focused or phased array transducer. Frequencies between 1 and 2 MHz are used to enable sufficient depth of penetration while maintaining a tight focal volume less than 1 cm³ [201]. For short exposures where temperature rise is mostly independent of perfusion and conduction effects [202], increased energy deposition at the focus enables the heating of millimeter-sized targets several centimeters beneath the skin, with negligible temperature increase between the transducer and the focus.

Focused ultrasound surgery uses a series of short, high-intensity ultrasound exposures to achieve thermal coagulation of a tumor without surgical incision. An exposure time of approximately 10 s achieves thermal coagulation of a 10 mm³ focal volume for a spherically focused ultrasound transducer [202]. Following coagulation of one volume, the transducer focus is translated to a new location and the process is repeated until complete coagulation of the tumor is achieved. Between sonications, a delay is applied to allow the tissue between the transducer and the focus to return to body temperature. This waiting time is necessary to avoid thermal buildup in normal tissue due to thermal conduction from the focus and due to energy deposition outside the focus, but significantly increases treatment time. Complete coagulation of tumors greater than 3 cm in diameter using this approach requires hundreds of sonications and impractical treatment times greater than 10 h [203]. This issue has been mitigated through the use of optimized treatment planning and phased array transducers with larger focal volumes [204,205] but remains a serious limitation of point-by-point thermal ablation using focused ultrasound.

One method proposed to heat regions larger than the size of the ultrasound focus for rapid tumor ablation or spatially uniform hyperthermia is to move the focal point along a circular trajectory, taking advantage of the thermal conductivity of tissue to heat the target interior [206–208] (Figure 29.2). The ultrasound focus can be scanned during continuous sonication to heat large tumor volumes by either mechanically repositioning a single element transducer, or by electronically steering the focus generated

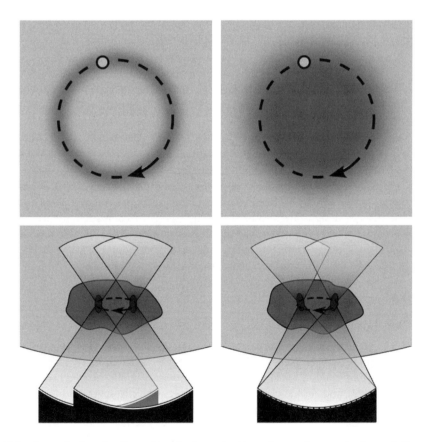

FIGURE 29.2 Strategies for using thermal conduction to heat large tissue volumes using focused ultrasound. Top: Moving a small focal point along the periphery of a targeted region allows thermal conductivity of tissue to heat the target interior. Bottom: With externally focused ultrasound transducers, the focus can translate during continuous sonication by either mechanically repositioning a single element transducer, or by electronically steering the focus generated by the hundreds of small, independently driven transducer elements in a phased array.

by the hundreds of small, independently driven transducer elements in a phased array [209]. However, with heating taking place over more than a few seconds, temperatures achieved in tissue depend not only on the ultrasound frequency and applied power, but also on thermal conduction in tissue, as well as spatially and temporally varying blood flow and ultrasound absorption. When the heating duration is long, perfusion significantly reduces the achieved temperature, and is known to vary with duration of applied heating [35,210]. The resulting difficulty in predicting temperature elevation can be overcome by automatically regulating the applied energy using feedback control from temperature measurements made during heating [208,211–213]. While feedback control based on thermocouple temperature measurements has been used clinically [75], noninvasive thermometry using MRI promises to enable widespread application of controlled heating.

The integration of MR imaging into focused ultrasound therapy [214] combines the precision of ultrasound power deposition with MR imaging for treatment guidance, monitoring, and evaluation [73]. In this approach, the location of the ultrasound beam is visualized by depositing a small amount of power resulting in a moderate temperature increase of 3–5°C [72]. This change in temperature, localized to the focal volume, can be measured with MR thermometry to reveal the beam location, which is often difficult to predict as complex tissue interfaces can affect the position of the ultrasound focus, and undesired hot spots can occur if there is bone or air in the beam path. In MR-guided focused ultrasound surgery [214], knowledge of the size of a thermal lesion created in a single exposure allows prescription of a number of target locations such that successive thermal lesions overlap slightly to create a continuous region of thermal damage [215]. During heating, MR thermometry is used to monitor the evolution of the temperature generation in tissue, and to calculate the delivered thermal dose in order to predict the extent of thermal lesion created in a single deposition [122]. After the treatment is complete, the thermal lesion is imaged with MRI to compare with the tumor volume and assess the outcome of the therapy [113]. In volumetric ablation and mild hyperthermia applications, MR images acquired during treatment can be used to automatically modulate power deposition for control of the spatial temperature distribution, as discussed in Section 29.4.2.4.

The use of ultrasound heating technology in the MR imaging environment is a technical challenge first overcome by Hynynen et al. [216,217]. Strong magnetic fields require ultrasound transducers built using MRI-compatible materials, with remaining ferromagnetic objects like impedance matching networks kept at a safe distance away from the magnet bore. For transducer positioning, traditional electromagnetic motors are commonly replaced by piezoceramic actuators and optical encoders [218], but hydraulic and lead-screw designs are also used. Recently, an MRI-compatible robotic arm has been developed for accurate positioning of focused ultrasound transducers [219]. MR scanner rooms are electrically shielded, and all signals driving the transducer and positioning system must be filtered and connected into the room's Faraday cage through a grounded RF penetration panel. Finally, customized imaging coils are often used to improve image SNR in the target volume for accurate thermometry while providing an acoustic window for sonication into the body.

29.4.3.3 Clinical Applications

A number of US-guided and MRI-guided focused ultrasound systems are either clinically approved or under investigational use for several applications in oncology as reviewed by Tempany [220], as well as in the ablation of uterine fibroids [221], and in a growing number of applications in neurology [222], cardiovascular disease [223], and cosmetic surgery [224,225]. Historically, several other systems achieved clinical success within the institution at which they were developed [75,226,227], but did not gain widespread adoption.

To date, the biggest impact has been in the MRI-guided ablation of uterine fibroids using phased array transducers [221]. There is now roughly 10 years of experience with the Insightec Exablate focused ultrasound system, demonstrating in over 7000 patients that noninvasive point-by-point thermal coagulation of these benign but painful masses is safe and effective [220]. Following thermal coagulation, fibroid tissue is resorbed and the mass shrinks, providing pain relief. When compared to the 1 month of recovery for traditional treatment by hysterectomy, and 1 week of recovery for uterine artery emboliza-tion, the 1 day of rest following focused ultrasound therapy is an attractive option for the nearly one-third of all women who develop fibroids. However, as in all focused ultrasound techniques, particular care must be taken to ensure a viable acoustic window from outside the skin into the fibroid devoid of ultrasound-reflecting air-filled bowels or ultrasound-absorbing scar tissue or bone [228,229]. The exit path of the beam must also be considered, as bone in the beam path will rapidly absorb and reflect energy causing heating of surrounding soft tissue. Early on, the FDA placed restrictions on the volume of tissue that was allowed to be targeted in a uterine fibroid treatment [230]. Ablating large volumes in a point-by-point approach is also a practical challenge as the patient must remain relatively still in the MRI for many hours. However, long-term clinical results indicate that outcomes are improved when larger fractions of the fibroid volume are coagulated or necrose as a result of thermal damage to feed-ing vessels [118,230]. To rapidly ablate large fibroid volumes, the Philips Sonalleve MRI-guided focused ultrasound system uses rapid electronic beam steering in a volumetric rather than point-by-point soni-cation approach, where regions as large as 20 mm × 20 mm × 50 mm can be coagulated before applying a delay for normal tissue cooling [124,231].

The earliest human trial of MRI-guided focused ultrasound surgery was in the ablation of benign masses in the breast [108]. This study was followed by a series of clinical trials for the treatment of local-ized breast cancer [116,232]. Progress was delayed by difficulties in MR thermometry in breast tissue containing water and fat affected by changes in magnetic susceptibility caused by the motion of air in the nearby lungs. However, impressive results from Japan [109,233] have renewed interest in the use of focused ultrasound as a noninvasive lumpectomy in carefully selected patients [234], and have led to advances in ultrasound devices [235,236] and MR thermometry [237–240] that may extend treatment to more women.

While not approved in the United States, there is extensive international experience in ultrasound-guided ablation of the prostate using transrectal phased arrays [241,242]. In this approach, ultrasound is used to identify target locations and to detect tissue destruction following heating to ablative tem-peratures [243]. As this technique also uses a point-by-point ablation strategy, it requires several hours of patient cooperation. While it does not require expensive magnet time, it lacks accurate thermom-etry and dosimetry and instead relies upon biochemical and biopsy endpoints, leaving some uncer-tainty regarding assessment of treatment outcome [244]. Recently, an MRI-controlled transurethral approach using a directional planar transducer array was investigated in a series of patients under a treat-and-resect protocol [183]. As in preliminary canine studies [174,245], automatic adjustment of the device rotation rate and ultrasound frequency provided uniform tissue coagulation within a precisely defined region exposed to a threshold temperature of 55°C. In those studies, damage boundaries pre-dicted by MR imaging and thermal dosimetry corresponded with histologically determined necrosis, demonstrating the value of MR guidance as an indicator of treatment success [245]. By applying energy continuously from within the prostatic urethra, there is no need for treatment delays to allow cooling of intervening tissue as in the transrectal approach, thereby reducing treatment time from 2 to 3 h to approximately 30 min. Transrectal ultrasound treatments conducted in the MRI are also under inves-tigation to overcome the limitation of poor image guidance associated with existing clinical systems [246–248].

Another recent success for focused ultrasound surgery has been in pain palliation for patients with radiation-resistant bone metastases [249,250]. While not intended to eradicate the tumor, focused ultra-sound is used to ablate nerves at the bone surface to provide durable pain relief. Clinical results in patients with painful bone tumors originating from late-stage breast, prostate, and lung cancer sug-gest that MRI-guided focused ultrasound appears to provide a noninvasive, repeatable technique for improving quality of life [249]. Compared with radiation therapy, the pain relief offered with thermal ablation is immediate and long lasting, which is important for patients with late-stage advanced disease. Similar responses have been observed with other minimally invasive thermal techniques such as RF ablation [144] and cryotherapy [251].

Finally, the development of large hemispherical phased arrays with many hundreds of elements has enabled focusing of ultrasound through the intact skull [252]. By spreading the energy over a large surface, and using frequencies lower than 1 MHz, heating of the scalp and brain surface due to the high ultrasound absorption of the skull can be reduced to tolerable levels [253,254]. Estimates of the skull geometry and density from pretreatment CT or MR imaging allows calculation of appropriate phase delays for all transducer elements by simulating the effect of the skull on an imaginary point source radiating outwards from the desired focal spot [255,256]. With commercial hemispherical arrays of 512–1024 elements operating at 0.2–1.0 MHz, a focal spot less than 10 mm^3 can be noninvasively produced within the brain [257,258]. These clinical devices are being used for thermal coagulation of brain tumors [106] as well as deep brain structures responsible for specific pathologies such as chronic pain [105] and essential tremor [259].

Current challenges include the development of clinically viable techniques for delivering ultrasound energy to moving targets and tissues lacking a direct acoustic window, such as tumors of the liver, kidney, and pancreas. Clinically, externally focused ultrasound ablation in the liver and kidney have been performed under ultrasound guidance using controlled apnea [260], single lung ventilation [261], or surgical removal of the ribs [262] to overcome respiratory motion and bone heating in fully anesthetized patients. These series demonstrated that HIFU (high-intensity focused ultrasound) is feasible and has similar efficacy to other ablation techniques, but online temperature monitoring may reduce the number of patients suffering minor skin burns, and provide improved intra-procedural treatment verification [263]. Preliminary MR-guided liver HIFU treatments have been performed in carefully selected patients using fast single-slice MR thermometry and controlled apnea [264,265], but clinical outcomes are yet to be reported.

The clinical results have been encouraging; however, several limitations of focused ultrasound have limited its application to specialized clinical situations. Noninvasive thermal therapy using focused ultrasound requires the presence of a suitable acoustic path through soft tissue, free of air or bone. The acoustic reflection coefficients of air and bone are high, blocking penetration of ultrasound energy to a target volume. In the case of bone, reflection and absorption can cause excessive heating of the bone–tissue interface [266,267]. This makes focused ultrasound heating difficult to perform in areas where there may be obstructing bone or air, such as the abdomen, prostate, heart, or lungs. In the case of the brain, adjustments to the phase of the ultrasound signal can be made to compensate for the beam disruption due to the skull bone [254]. For noninvasive heating of abdominal organs, transcostal sonication can result in various degrees of thermal damage to the ribs and other tissues along the beam path [268]. Several groups are investigating sonication schemes with phased array transducers that selectively deactivate array elements, to focus energy between the ribs to targets in the liver or pancreas [269–272]. For abdominal organs that move during patient respiration, motion of the target volume relative to the ultrasound transducer is also a complicating factor. Translation of the target volume relative to the external ultrasound transducer must be accounted for during heating to maintain targeting accuracy and adequate heating. MR thermometry using the proton resonant frequency shift technique is compromised by motion; however, combinations of multi-baseline and self-referenced thermometry techniques appear to be well-suited to mitigate this issue [98,273]. The ability to move the ultrasound focus dynamically during treatment to follow the target is another technical challenge currently under investigation [274–276], as discussed in the next section.

29.4.3.4 MRI-Controlled Ultrasound Thermal Therapy

Figure 29.3 illustrates the differences between MRI-guided and MRI-controlled ultrasound thermal therapy. In a typical MR-guided treatment, anatomical MR images are used in treatment planning to prescribe a target region and temperature or thermal dose. Energy is deposited with a predefined power, spatial deposition pattern, and duration, and the resulting temperature elevation is measured with MRI, verifying treatment outcome by thermal dose or other indicators of thermal damage. However, even when energy can be selectively delivered to targets deep within tissue, spatial variations in energy absorption and blood flow make it technically challenging to maintain temperature uniformity in a

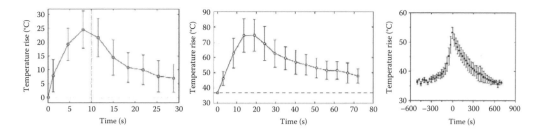

FIGURE 29.3 Comparison of temperature vs. time for MRI-guided vs. MRI-controlled approaches to high-temperature ultrasound thermal therapy. Left: Temperature elevation as a function of time at the hottest voxel of 63 sonications delivered to a breast fibroadenoma using MRI-guided externally focused ultrasound. The points are the mean temperature elevations, and the error bars the standard deviation. In this situation and the following one, treatment parameters were selected by the operator prior to each sonication and temperature elevations were monitored using MR thermometry [KH 2001]. Middle: Temperature measured at the focus for 23 sonications in one uterine fibroid treatment using MRI-guided externally focused ultrasound [Tempany2003]. Right: Temperature vs. time along all the target boundary points across eight patients treated using MRI-controlled transurethral ultrasound thermal therapy. With online temperature control, energy deposition parameters are adjusted automatically during treatment based on rapidly acquired MR temperature measurements, achieving desired temperatures with greater accuracy and reproducibility [RC2012].

heated target over time. Clinically, temperature elevations achieved with focused ultrasound for a given applied power have been shown to vary by 30–40% both between different patients and between individual sonications for one patient due to beam aberration at tissue interfaces and variations in energy absorption [104]. Temporally, temperature fluctuations with a standard deviation of 1°C result in a 25% increase in the estimate of thermal dose [277], making it important to minimize both temperature measurement uncertainty [60,277,278] and actual temperature fluctuations [211]. Spatial temperature variations are especially difficult to avoid in the presence of large blood vessels [65], and even more so when scanning a small heating focus, making spatial control of applied energy a necessity [211,213]. These variations are perhaps less influential in thermal ablation as long as a certain threshold is reached. However, for mild hyperthermia applications such as thermally triggered drug release, tissue-specific heating is necessary to achieve localized drug delivery and avoid systemic drug release. Furthermore, precise control of temperature elevation is required for optimal drug release without prematurely preventing drug supply by damaging tumor vessels.

In an MRI-controlled approach, temperature measurements are made rapidly and used as input to a feedback control algorithm that automatically updates the location, duration, and power of energy deposition during treatment in order to track the prescribed target temperatures, ensuring that a desired spatial and temporal heating pattern is achieved and maintained.

The temporal, spatial, and temperature resolution requirements for thermometry used in closed-loop feedback control depend on the type of treatment, and tradeoffs can be made. In point-by-point thermal ablation, large temperature increases in small focal regions over short durations require high spatial and temporal resolution, but the requirement for temperature resolution is relaxed, as the main goal is to exceed a thermal dose or temperature threshold. For volumetric ablation where a large region is rapidly heated to high temperatures all at once, temporal resolution is important, but the spatial resolution requirement is relaxed. Clinically, a simple on–off control approach has been used to achieve uniform coagulation of uterine fibroids using externally focused ultrasound [124,279]. By electronically steering a focal point along concentric circles and moving outwards from one circle to the next once the boundary temperature reaches 57°C, this simple approach makes efficient use of inward thermal conduction to achieve precise, reproducible lesions up to 16 mm in diameter [124]. A similar threshold-based approach has been used for transurethral ultrasound ablation of the prostate, where a planar ultrasound transducer heats the prostate from within the urethra and is slowly rotated

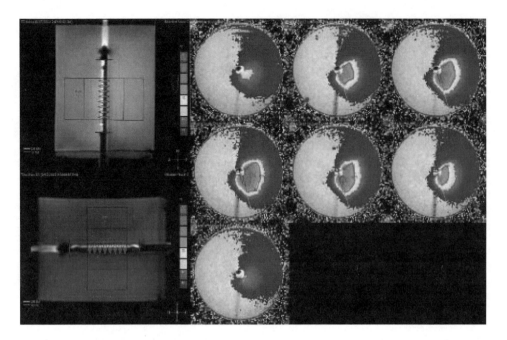

FIGURE 29.4 MRI-controlled ultrasound heating in a gel phantom using a multi-element transurethral applicator developed for minimally invasive thermal therapy of prostate cancer. Left: Coronal (top) and sagittal (bottom) T2-weighted treatment planning images displaying overlays used to identify applicator position and thermometry image slices relative to the target volume. Right: Maximum temperature distribution achieved during MRI-controlled ultrasound heating of a user-defined three-dimensional sector in a gel phantom, acquired in 5 mm thick transverse planes for each element of the device. The 52°C isotherm (dark gray area to the right of center) matches well with the targeted contours defined independently for each slice. The contoured regions are heated simultaneously, with applied power to each element and overall device rotation rate determined automatically by temperatures measured during the treatment using MRI.

as the target boundary reaches a threshold temperature (Figure 29.4) [183]. However, for mild heating to enhance chemotherapy, smaller temperature elevations are maintained over a long duration, with conduction effects spreading the heat over time. In this case, the requirements for temporal and spatial resolution are relaxed, but precise temperature resolution is required, as fluctuations of 1–2°C can cause dramatic changes in biological effect.

For mild hyperthermia under MRI feedback control, several control algorithms are under preclinical development [274,280–282]. In one approach, the energy to be deposited at a series of points covering a targeted three-dimensional tissue volume is prescribed using a customized proportional–integral–derivative feedback control algorithm that accounts specifically for the spread of heat by thermal conduction [274,283]. The proportional term corrects for instantaneous error between the measured and the target temperatures. The derivative term acts to increase power when the target temperature is changing, and the integral term uses the sum of past temperature errors to gradually adjust the applied power. An additional term adds energy to replace heat dissipated by thermal conduction, and the prescribed power is adjusted for heating efficiency. Every time a temperature image is acquired, an energy deposition is prescribed and delivered by electronically repositioning the focus of a phased-array transducer through a series of short sonications at discrete locations, all in the time between MRI temperature measurements. This technique is also being investigated for controlled heating of moving targets in the abdomen such as tumors of the liver and kidney [284], and eventually pancreas. For periodic respiratory motion, motion-corrected MR thermometry can be achieved using rapid image acquisition and a multiple-baseline technique [98]. Target tracking is achieved by modeling the periodic motion to predict the target's future location, and adjusting the spatial deposition pattern by rapidly switching the location of the transducer focus. Advanced delivery approaches may eventually allow for noninvasive ultrasound ablation or hyperthermia in abdominal organs.

MRI-controlled ultrasound hyperthermia has been applied in preclinical studies of localized drug release from temperature-sensitive, drug-carrying nanoparticles called liposomes [129,280,285]. Liposomes remain intact in the body, but at temperatures greater than 41°C rapidly release their chemotherapeutic payload while passing through vessels of the heated tumor, allowing free drug to leave the vasculature and accumulate in the targeted tissue [38]. Figure 29.5 depicts the preclinical hyperthermia setup for a rabbit using an MRI-compatible positioning system to mechanically scan the 1 mm focus of an ultrasound transducer for heating a 10 mm diameter target region. A proportional–integral controller uses MR temperature images acquired every 5 s to adjust the applied power, maintaining a target temperature of 43°C for 20 min, with temperatures in surrounding unheated regions remaining below 39°C. Thermosensitive liposomal doxorubicin was infused intravenously during MRI-controlled heating, and drug concentrations measured in excised tissue samples demonstrated a 16-fold increase in doxorubicin deposited in heated muscle versus unheated muscle in the contralateral thigh.

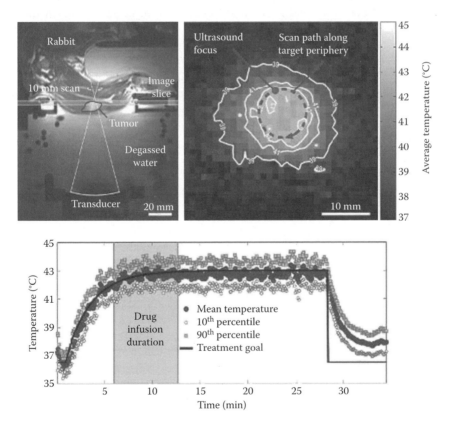

FIGURE 29.5 Preclinical experiment demonstrating MRI-controlled focused ultrasound hyperthermia for thermally mediated drug delivery in rabbit tumors. Top left: Axial T2-weighted MR image of tumor-bearing rabbit positioned for sonication with a mechanically positioned focused ultrasound transducer. Overlay shows coronal image plane in which continuously acquired MR temperature measurements were used to automatically control treatment. Top right: Time-averaged MRI temperature measurements in coronal image plane through tumor, demonstrating heating pattern achieved using online control of energy deposition. Desired temperature was 43°C in the displayed 10 mm diameter targeted region. Bottom: Target region temperature versus time measured using MR thermometry. Duration of thermosensitive liposome infusion is shown.

This work provided evidence that MRI-controlled focused ultrasound hyperthermia can be used with temperature-sensitive drug carriers to achieve image-guided localized drug release [280].

29.5 Conclusions

The field of thermal therapy is evolving rapidly as a deeper understanding of the thermal biology of the tissue is obtained, and technology for improved delivery and monitoring of treatments is developed. Recently, there has been increased investigation into noninvasive heating with focused ultrasound energy, both for thermal coagulation therapy and for localized drug delivery.

Imaging technology plays an important role in all aspects of thermal therapy. The localization of energy deposition to the correct location within a target volume can be achieved with imaging, as well as the capability to monitor the temperature distribution across the heated region. Thermal damage predictions can be made based on the measured temperature distributions, and images of the region of thermal damage can be obtained after the treatment is complete.

Tissue heating can be achieved with interstitial, intracavitary, or noninvasive devices delivering RF, microwave, laser, or ultrasound energy. Each strategy has various advantages and limitations; however, there is a gradual evolution toward heating technologies that provide the maximum amount of control over the shape of the heating pattern. Focused ultrasound provides a means of noninvasively heating targets deep within the body to either high temperatures for direct thermal damage or mild temperatures to enhance the effects of radiation or chemotherapy. MR image guidance enables accurate target definition and treatment assessment, with MR thermometry allowing localization of heating and temperature monitoring during treatment. However, for a given ultrasound power and duration, temperature elevations vary considerably due to spatial and temporal variations in tissue absorption properties and blood perfusion. Feedback control of ultrasound heating parameters based on MR temperature measurements increases the reproducibility of thermal exposures, for more predictable lesions or to maintain optimal conditions for enhanced drug delivery. MRI-guided focused ultrasound is gaining clinical acceptance for a number of indications, and early clinical results suggest the feasibility of MRI-controlled therapies for targets such as uterine fibroids and prostate cancer. Ongoing research aims to enable the accurate treatment of tumors in moving organs and provide precise control of heat-triggered drug delivery.

References

1. Lepock JR. How do cells respond to their thermal environment? *International Joural of Hyperthermia* 2005;21(8):681–687.
2. Roti Roti JL. Cellular responses to hyperthermia (40–46 degrees C): Cell killing and molecular events. *International Journal of Hyperthermia* 2008;24(1):3–15.
3. Kampinga HH. Cell biological effects of hyperthermia alone or combined with radiation or drugs: A short introduction to newcomers in the field. *International Journal of Hyperthermia* 2006;22(3):191–196.
4. Lepock JR. Role of nuclear protein denaturation and aggregation in thermal radiosensitization. *International Journal of Hyperthermia* 2004;20(2):115–130.
5. Dewey WC. Arrhenius relationships from the molecule and cell to the clinic. *International Journal of Hyperthermia* 1994;10(4):457–483.
6. Horsman MR. Angiogenesis and vascular targeting: Relevance for hyperthermia. *International Joural of Hyperthermia* 2008;24(1):57–65.
7. Brizel DM, Scully SP, Harrelson JM, Layfield LJ, Dodge RK, Charles HC et al. Radiation therapy and hyperthermia improve the oxygenation of human soft tissue sarcomas. *Cancer Research* 1996;56(23):5347–5350.
8. Song CW, Park H, Griffin RJ. Improvement of tumor oxygenation by mild hyperthermia. *Radiation Research* 2001;155(4):515–528.

9. Jones EL, Prosnitz LR, Dewhirst MW, Marcom PK, Hardenbergh PH, Marks LB et al. Thermo-chemoradiotherapy improves oxygenation in locally advanced breast cancer. *Clinical Cancer Research* 2004;10(13):4287–4293.

10. Vujaskovic Z, Song CW. Physiological mechanisms underlying heat-induced radiosensitization. *International Journal of Hyperthermia* 2004;20(2):163–174.

11. Kampinga HH, Dikomey E. Hyperthermic radiosensitization: Mode of action and clinical relevance. *International Journal of Radiation Bioliogy* 2001;77(4):399–408.

12. Dewey WC. Interaction of heat with radiation and chemotherapy. *Cancer Research* 1984;44(10 Suppl):4714s–4720s.

13. Calderwood SK, Ciocca DR. Heat shock proteins: Stress proteins with Janus-like properties in cancer. *International Journal of Hyperthermia* 2008;24(1):31–39.

14. Straube W, Klein E, Moros E, Low D, Myerson R. Dosimetry and techniques for simultaneous hyperthermia and external beam radiation therapy. *International Journal of Hyperthermia* 2001;17(1):48–62.

15. Moros EG, Penagaricano J, NovaK P, Straube WL, Myerson RJ. Present and future technology for simultaneous superficial thermoradiotherapy of breast cancer. *International Journal of Hyperthermia* 2010;26(7):699–709.

16. Thrall DE, LaRue SM, Yu DH, Samulski T, Sanders L, Case B et al. Thermal dose is related to duration of local control in canine sarcomas treated with thermoradiotherapy. *Clinical Cancer Research* 2005;11(14):5206–5214.

17. Jones EL, Oleson JR, Prosnitz LR, Samulski TV, Vujaskovic Z, Yu DH et al. Randomized trial of hyperthermia and radiation for superficial tumors. *Journal of Clinical Oncology* 2005;23(13):3079–3085.

18. Hurwitz MD, Hansen JL, Prokopios-Davos S, Manola J, Wang Q, Bornstein BA et al. Hyperthermia combined with radiation for the treatment of locally advanced prostate cancer long-term results from Dana–Farber Cancer Institute Study 94–153. *Cancer* 2011;117(3):510–516.

19. Franckena M, Stalpers LJA, Koper PCM, Wiggenraad RGJ, Hoogenraad WJ, Duk JDPv et al. Long-term improvement in treatment outcome after radiotherapy and hyperthermia in locoregionally advanced cervix cancer: An update of the Dutch Deep Hyperthermia Trial. *International Journal of Radiation Oncology Biology Physics* 2008;70(4):1176–1182.

20. Hahn GM, Braun J, Har-Kedar I. Thermochemotherapy: Synergism between hyperthermia (42–43 degrees) and adriamycin (of bleomycin) in mammalian cell inactivation. *Proceedings of The National Academy of Sciences, USA* 1975;72(3):937–940.

21. Hahn GM. *Hyperthermia and Cancer.* New York: Plenum Press; 1982.

22. Issels RD. Hyperthermia adds to chemotherapy. *European Journal of Cancer* 2008;44(17):2546–2554.

23. Hahn GM. Potential for therapy of drugs and hyperthermia. *Cancer Research* 1979;39(6 Pt 2): 2264–2268.

24. Marmor JB. Interactions of hyperthermia and chemotherapy in animals. *Cancer Research* 1979; 39(6 Pt 2):2269–2276.

25. Mayer LD, Shabbits JA. The role for liposomal drug delivery in molecular and pharmacological strategies to overcome multidrug resistance. *Cancer Metastasis Review* 2001;20(1–2):87–93.

26. Vujaskovic Z, Kim DW, Jones E, Lan L, Mccall L, Dewhirst MW et al. A phase I/II study of neo-adjuvant liposomal doxorubicin, paclitaxel, and hyperthermia in locally advanced breast cancer. *International Journal of Hyperthermia* 2010;26(5):514–521.

27. Issels RD, Lindner LH, Verweij J, Wust P, Reichardt P, Schem B et al. Neo-adjuvant chemotherapy alone or with regional hyperthermia for localised high-risk soft-tissue sarcoma: A randomised phase 3 multicentre study. *Lancet Oncology* 2010;11(6):561–570.

28. Pence DM, Song CW. Effects of heat on blood flow. In: Anghileri LJ, Robert J, editors. *Hyperthermia in Cancer Treatment.* 1st ed. Boca Raton, FL: CRC Press, Inc; 1986. pp. 1–16.

29. Jain RK, Wardhartley K. Tumor blood-flow—Characterization, modifications, and role in hyperthermia. *IEEE Transactions on Sonics and Ultrasonics* 1984;31(5):504–526.

30. Dudar TE, Jain RK. Differential response of normal and tumor microcirculation to hyperthermia. *Cancer Research* 1984;44(2):605–612.

31. Song CW. Effect of local hyperthermia on blood-flow and microenvironment—A review. *Cancer Research* 1984;44(10):4721–4730.

32. Hahn GM, Ning SC, Elizaga M, Kapp DS, Anderson RL. A comparison of thermal responses of human and rodent cells. *International Journal of Radiation Bioliogy* 1989;56(5):817–825.

33. Acker JC, Dewhirst MW, Honore GM, Samulski TV, Tucker JA, Oleson JR. Blood perfusion measurements in human tumors—Evaluation of Laser Doppler Methods. *International Journal of Hyperthermia* 1990;6(2):287–304.

34. Waterman FM, Tupchong L, Nerlinger RE, Matthews J. Blood flow in human tumors during local hyperthermia. *International Journal of Radiation, Oncology Biology and Physics* 1991;20(6):1255–1262.

35. Anhalt DP, Hynynen K, Roemer RB. Patterns of changes of tumour temperatures during clinical hyperthermia: Implications for treatment planning, evaluation and control. *International Journal of Hyperthermia* 1995;11(3):425–36.

36. Vujaskovic Z, Poulson JM, Gaskin AA, Thrall DE, Page RL, Charles HC et al. Temperature-dependent changes in physiologic parameters of spontaneous canine soft tissue sarcomas after combined radiotherapy and hyperthermia treatment. *International Journal of Radiation Oncology Biology and Physics* 2000;46(1):179–185.

37. Yuan F. Transvascular drug delivery in solid tumors. *Seminar on Radiation Oncology* 1998;8(3):164–175.

38. Grüll H, Langereis S. Hyperthermia-triggered drug delivery from temperature-sensitive liposomes using MRI-guided high intensity focused ultrasound. *Journal of Controlled Release* 2012; 161(2):317–327.

39. Kong G, Braun RD, Dewhirst MW. Hyperthermia enables tumor-specific nanoparticle delivery: Effect of particle size. *Cancer Research* 2000;60(16):4440–4445.

40. Kong G, Braun RD, Dewhirst MW. Characterization of the effect of hyperthermia on nanoparticle extravasation from tumor vasculature. *Cancer Research* 2001;61(7):3027–3032.

41. Fujiwara K, Watanabe T. Effects of hyperthermia, radiotherapy and thermoradiotherapy on tumor microvascular permeability. *Acta Pathologica Japan* 1990;40(2):79–84.

42. Huang SK, Stauffer PR, Hong K, Guo JW, Phillips TL, Huang A et al. Liposomes and hyperthermia in mice: Increased tumor uptake and therapeutic efficacy of doxorubicin in sterically stabilized liposomes. *Cancer Research* 1994;54(8):2186–2191.

43. Ning S, Macleod K, Abra RM, Huang AH, Hahn GM. Hyperthermia induces doxorubicin release from long-circulating liposomes and enhances their anti-tumor efficacy. *International Journal of Radiation, Oncology, Biology, and Physics* 1994;29(4):827–834.

44. Gaber MH, Wu NZ, Hong K, Huang SK, Dewhirst MW, Papahadjopoulos D. Thermosensitive liposomes: Extravasation and release of contents in tumor microvascular networks. *International Journal of Radiation Oncology Biology and Physics* 1996;36(5):1177–1187.

45. Wu NZ, Braun RD, Gaber MH, Lin GM, Ong ET, Shan S et al. Simultaneous measurement of liposome extravasation and content release in tumors. *Microcirculation* 1997;4(1):83–101.

46. Kong G, Dewhirst MW. Hyperthermia and liposomes. *International Journal of Hyperthermia* 1999;15(5):345–370.

47. Zhang H, Mehta K, Cohen P, Guha C. Hyperthermia on immune regulation: A temperature's story. *Cancer Letters* 2008;271(2):191–204.

48. Calderwood SK, Theriault JR, Gong J. How is the immune response affected by hyperthermia and heat shock proteins? *International Journal of Hyperthermia* 2005;21(8):713–716.

49. Robinson TC, Lele PP. Analysis of lesion development in brain and in plastics by high-intensity focused ultrasound at low-megahertz frequencies. *Journal of the Acoustic Society of America* 1972;51(4):1333–1351.

50. Damianou C, Hynynen K. The effect of various physical parameters on the size and shape of necrosed tissue volume during ultrasound surgery. *Journal of the Acoustic Society of America* 1994;95(3):1641–1649.

51. McDannold NJ, King RL, Jolesz FA, Hynynen KH. Usefulness of MR imaging-derived thermometry and dosimetry in determining the threshold for tissue damage induced by thermal surgery in rabbits. *Radiology* 2000;216(2):517–523.

52. Borrelli MJ, Thompson LL, Cain CA, Dewey WC. Time-temperature analysis of cell killing of bhk cells heated at temperatures in the range of 43.5-Degrees-C to 57.0-Degrees-C. *International Journal of Radiation Oncology Biology Physics* 1990;19(2):389–399.

53. Thomsen S. Mapping of thermal injury in biologic tissues using quantitative pathologic techniques. *Thermal Treatment of Tissue with Image Guidance, Proceedings of* 1999;3594:82–95.

54. Canney MS, Khokhlova VA, Bessonova OV, Bailey MR, Crum LA. Shock-induced heating and millisecond boiling in gels and tissue due to high intensity focused ultrasound. *Ultrasound Medicine and Biology* 2010;36(2):250–267.

55. Yung JP, Shetty A, Elliott A, Weinberg JS, McNichols RJ, Gowda A et al. Quantitative comparison of thermal dose models in normal canine brain. *Medical Physics* 2010;37(10):5313–5321.

56. Sapareto SA, Dewey WC. Thermal dose determination in cancer therapy. *International Journal of Radiation, Oncology, Biology and Physics* 1984;10(6):787–800.

57. Yarmolenko PS, Moon EJ, Landon C, Manzoor A, Hochman DW, Viglianti BL et al. Thresholds for thermal damage to normal tissues: An update. *International Journal of Hyperthermia* 2011; 27(4):320–343.

58. Dewhirst MW, Viglianti BL, Lora-Michiels M, Hanson M, Hoopes PJ. Basic principles of thermal dosimetry and thermal thresholds for tissue damage from hyperthermia. *International Journal of Hyperthermia* 2003;19(3):267–294.

59. Dewhirst MW, Vujaskovic Z, Jones E, Thrall D. Re-setting the biologic rationale for thermal therapy. *International Journal of Hyperthermia* 2005;21(8):779–790.

60. Graham SJ, Chen L, Leitch M, Peters RD, Bronskill MJ, Foster FS et al. Quantifying tissue damage due to focused ultrasound heating observed by MRI. *Magnetic Resonance in Medicine* 1999;41(2):321–328.

61. Peters RD, Chan E, Trachtenberg J, Jothy S, Kapusta L, Kucharczyk W et al. Magnetic resonance thermometry for predicting thermal damage: An application of interstitial laser coagulation in an in vivo canine prostate model. *Magnetic Resonance in Medicine* 2000;44(6):873–883.

62. Chopra R, Luginbuhl C, Weymouth AJ, Foster FS, Bronskill MJ. Interstitial ultrasound heating applicator for MR-guided thermal therapy. *Physics in Medicine and Biology* 2001;46(12):3133–3145.

63. Pennes HH. Analysis of tissue and arterial blood temperatures in the resting human forearm. *Journal of Applied Physiology* 1948;1(2):93–122.

64. Kotte ANTJ, van Leeuwen GMJ, Lagendijk JJW. Modelling the thermal impact of a discrete vessel tree. *Physics in Medicine and Biology* 1999;44(1):57–74.

65. Kolios MC, Worthington AE, Holdsworth DW, Sherar MD, Hunt JW. An investigation of the flow dependence of temperature gradients near large vessels during steady state and transient tissue heating. *Physics in Medicine and Biology* 1999;44(6):1479–1497.

66. Akyurekli D, Gerig LH, Raaphorst GP. Changes in muscle blood flow distribution during hyperthermia. *International Journal of Hyperthermia* 1997;13(5):481–496.

67. Lagendijk JJ. Hyperthermia treatment planning. *Physics in Medicine and Biology* 2000;45(5):R61–R76.

68. Urano M, Kuroda M, Nishimura Y. For the clinical application of thermochemotherapy given at mild temperatures. *International Journal of Hyperthermia* 1999;15(2):79–107.

69. Goldberg SN, Grassi CJ, Cardella JF, Charboneau JW, Dodd GD, Dupuy DE et al. Image-guided tumor ablation: Standardization of terminology and reporting criteria. *Radiology* 2005;235(3): 728–739.

70. Schenck JF. The role of magnetic susceptibility in magnetic resonance imaging: MRI magnetic compatibility of the first and second kinds. *Medical Physics* 1996;23(6):815–850.

71. Lynn JG, Zwemer RL, Chick AJ, Miller AE. A new method for the generation and use of focused ultrasound in experimental biology. *Journal of General Physiology* 1942;26(2):179–93.

72. Chung AH, Hynynen K, Colucci V, Oshio K, Cline HE, Jolesz FA. Optimization of spoiled gradient-echo phase imaging for in vivo localization of a focused ultrasound beam. *Magnetic Resonance in Medicine* 1996;36(5):745–752.

73. Cline HE, Hynynen K, Hardy CJ, Watkins RD, Schenck JF, Jolesz FA. MR temperature mapping of focused ultrasound surgery. *Magnetic Resonance in Medicine: Official Journal of the Society of Magnetic Resonance in Medicine/Society of Magnetic Resonance in Medicine* 1994;31(6):628–636.

74. McDannold N, Maier SE. Magnetic resonance acoustic radiation force imaging. *Medical Physics* 2008;35(8):3748–3758.

75. Hynynen K, Shimm D, Anhalt D, Stea B, Sykes H, Cassady JR et al. Temperature distributions during clinical scanned, focused ultrasound hyperthermia treatments. *International Journal of Hyperthermia* 1990;6(5):891–908.

76. Juang T, Stauffer PR, Neuman DG, Schlorff JL. Multilayer conformal applicator for microwave heating and brachytherapy treatment of superficial tissue disease. *International Journal of Hyperthermia* 2006;22(7):527–544.

77. Paulides MM, Bakker JF, Linthorst M, van der Zee J, Rijnen Z, Neufeld E et al. The clinical feasibility of deep hyperthermia treatment in the head and neck: New challenges for positioning and temperature measurement. *Physics in Medicine and Biology* 2010;55(9):2465–2480.

78. de Bruijne M, Van der zee J, Ameziane A, Van Rhoon GC. Quality control of superficial hyperthermia by treatment evaluation. *International Journal of Hyperthermia* 2011;27(3):199–213.

79. Hynynen K, Edwards DK. Temperature measurements during ultrasound hyperthermia. *Medical Physics* 1989;16(4):618–626.

80. Fallone BG, Moran PR, Podgorsak EB. Non-invasive thermometry with a Clinical x-ray CT Scanner. *Medical Physics* 1982;9(5):715–721.

81. Bruners P, Pandeya GD, Levit E, Roesch E, Penzkofer T, Isfort P et al. CT-based temperature monitoring during hepatic RF ablation: Feasibility in an animal model. *International Journal of Hyperthermia* 2012;28(1):55–61.

82. Leroy Y, Bocquet B, Mamouni A. Non-invasive microwave radiometry thermometry. *Physiology Measurements* 1998;19(2):127–148.

83. Arthur RM, Straube WL, Trobaugh JW, Moros EG. Non-invasive estimation of hyperthermia temperatures with ultrasound. *International Journal of Hyperthermia* 2005;21(6):589–600.

84. Parker DL, Smith V, Sheldon P, Crooks LE, Fussell L. Temperature distribution measurements in two-dimensional NMR imaging. *Medical Physics* 1983;10(3):321–325.

85. Denis de Senneville B, Quesson B, Moonen CT. Magnetic resonance temperature imaging. *International Journal of Hyperthermia* 2005;21(6):515–531.

86. Rieke V, Butts Pauly K. MR thermometry. *Journal of Magnetic Resonance Imaging* 2008;27(2):376–390.

87. Le Bihan D, Delannoy J, Levin RL. Temperature mapping with MR imaging of molecular diffusion: Application to hyperthermia. *Radiology* 1989;171(3):853–857.

88. Chen J, Daniel BL, Pauly KB. Investigation of proton density for measuring tissue temperature. *Journal of Magnetic Resonance Imaging* 2006;23(3):430–437.

89. Hindman JC. Proton resonance shift of water in gas and liquid states. *Journal of Chemical Physics* 1966;44(12):4582–4592.

90. Dickinson R, Hall A, Hind A, Young I. Measurement of changes in tissue temperature using MR imaging. *Journal of Computer Assisted Tomography* 1986;10(3):468–472.

91. Ishihara Y, Calderon A, Watanabe H, Okamoto K, Suzuki Y, Kuroda K et al. A precise and fast temperature mapping using water proton chemical shift. *Magnetic Resonance in Medicine* 1995;34(6):814–823.

92. Peters RD, Hinks RS, Henkelman RM. Ex vivo tissue-type independence in proton-resonance frequency shift MR thermometry. *Magnetic Resonance Medicine* 1998;40(3):454–459.

93. McDannold N, Fossheim SL, Rasmussen H, Martin H, Vykhodtseva N, Hynynen K. Heat-activated liposomal MR contrast agent: Initial in vivo results in rabbit liver and kidney. *Radiology* 2004;230(3):743–752.

94. Salomir R, Palussiere J, Fossheim SL, Rogstad A, Wiggen UN, Grenier N et al. Local delivery of magnetic resonance (MR) contrast agent in kidney using thermosensitive liposomes and MR imaging-guided local hyperthermia: A feasibility study in vivo. *Journal of Magnetic Resonance Imaging* 2005;22(4):534–540.

95. Lindner L, Reinl H, Schlemmer M, Stahl R, Peller M. Paramagnetic thermosensitive liposomes for MR-thermometry. *International Journal of Hyperthermia* 2005;21(6):575–588.

96. Shuhendler AJ, Staruch R, Oakden W, Gordijo CR, Rauth AM, Stanisz GJ et al. Thermally-triggered 'off-on-off' response of gadolinium-hydrogel-lipid hybrid nanoparticles defines a customizable temperature window for non-invasive magnetic resonance imaging thermometry. *Journal of Controlled Release* 2012;157(3):478–484.

97. Rieke V, Vigen KK, Sommer G, Daniel BL, Pauly JM, Butts K. Referenceless PRF shift thermometry. *Magnetic Resonance in Medicine* 2004;51(6):1223–1231.

98. Roujol S, Ries M, Quesson B, Moonen C, de Senneville BD. Real-time MR-thermometry and dosimetry for interventional guidance on abdominal organs. *Magnetic Resonance in Medicine: Official Journal of the Society of Magnetic Resonance in Medicine/Society of Magnetic Resonance in Medicine* 2010;63(4):1080–1087.

99. Depoorter J, Dewagter C, Dedeene Y, Thomsen C, Stahlberg F, Achten E. Noninvasive MRT thermometry with the proton-resonance frequency (PRF) method—In-vivo results in human muscle. *Magnetic Resonance in Medicine: Official Journal of the Society of Magnetic Resonance in Medicine/Society of Magnetic Resonance in Medicine* 1995;33(1):74–81.

100. El-Sharkawy AM, Schar M, Bottomley PA, Atalar E. Monitoring and correcting spatio-temporal variations of the MR scanner's static magnetic field. *MAGMA* 2006;19(5):223–236.

101. Wyatt C, Soher B, Maccarini P, Charles HC, Stauffer P, Macfall J. Hyperthermia MRI temperature measurement: Evaluation of measurement stabilisation strategies for extremity and breast tumours. *International Journal of Hyperthermia* 2009;25(6):422–433.

102. Kuroda K, Oshio K, Chung AH, Hynynen K, Jolesz FA. Temperature mapping using the water proton chemical shift: A chemical shift selective phase mapping method. *Magnetic Resonance in Medicine: Official Journal of the Society of Magnetic Resonance in Medicine/Society of Magnetic Resonance in Medicine* 1997;38(5):845–851.

103. de Zwart JA, Vimeux FC, Delalande C, Canioni P, Moonen CT. Fast lipid-suppressed MR temperature mapping with echo-shifted gradient-echo imaging and spectral–spatial excitation. *Magnetic Resonance in Medicine* 1999;42(1):53–59.

104. McDannold N, Tempany CM, Fennessy FM, So MJ, Rybicki FJ, Stewart EA et al. Uterine leiomyomas: MR imaging-based thermometry and thermal dosimetry during focused ultrasound thermal ablation. *Radiology* 2006;240(1):263–272.

105. Martin E, Jeanmonod D, Morel A, Zadicario E, Werner B. High-intensity focused ultrasound for noninvasive functional neurosurgery. *Annals of Neurology* 2009;66(6):858–861.

106. McDannold N, Clement GT, Black P, Jolesz F, Hynynen K. Transcranial magnetic resonance imaging-guided focused ultrasound surgery of brain tumors: Initial findings in 3 patients. *Neurosurgery* 2010;66(2):323–332.

107. Carpentier A, McNichols RJ, Stafford RJ, Guichard J, Reizine D, Delaloge S et al. Laser thermal therapy: Real-time MRI-guided and computer-controlled procedures for metastatic brain tumors. *Lasers in Surgery and Medicine* 2011;43(10):943–950.

108. Hynynen K, Pomeroy O, Smith DN, Huber PE, McDannold NJ, Kettenbach J et al. MR imaging-guided focused ultrasound surgery of fibroadenomas in the breast: A feasibility study. *Radiology* 2001;219(1):176–185.

109. Furusawa H, Namba K, Thomsen S, Akiyama F, Bendet A, Tanaka C et al. Magnetic resonance-guided focused ultrasound surgery of breast cancer: Reliability and effectiveness. *Journal of the American College of Surgeons* 2006;203(1):54–63.

110. Gellermann J, Wlodarczyk W, Hildebrandt B, Ganter H, Nicolau A, Rau B et al. Noninvasive magnetic resonance thermography of recurrent rectal carcinoma in a 1.5 Tesla hybrid system. *Cancer Research* 2005;65(13):5872–5880.

111. Chopra R, Colquhoun A, Burtnyk M, N'djin WA, Kobelevskiy I, Boyes A et al. MR imaging-controlled transurethral ultrasound therapy for conformal treatment of prostate tissue: Initial feasibility in humans. *Radiology* 2012;265(1):303–313.

112. Meloni MF, Goldberg SN, Livraghi T, Calliada F, Ricci P, Rossi M et al. Hepatocellular carcinoma treated with radiofrequency ablation: Comparison of pulse inversion contrast-enhanced harmonic sonography, contrast-enhanced power Doppler sonography, and helical CT. *American Journal of Roentgenology* 2001;177(2):375–380.

113. Vykhodtseva N, Sorrentino V, Jolesz FA, Bronson RT, Hynynen K. MRI detection of the thermal effects of focused ultrasound on the brain. *Ultrasound Medicine and Biology* 2000;26(5):871–880.

114. Purdie TG, Lee TY, Iizuka M, Sherar MD. Dynamic contrast enhanced CT measurement of blood flow during interstitial laser photocoagulation: Comparison with an Arrhenius damage model. *Physics in Medicine and Biology* 2000;45(5):1115–1126.

115. Solbiati L, Ierace T, Tonolini M, Cova L. Guidance and monitoring of radiofrequency liver tumor ablation with contrast-enhanced ultrasound. *European Journal of Radiology* 2004;51:S19–S23.

116. Gianfelice D, Khiat A, Amara M, Belblidia A, Boulanger Y. MR imaging-guided focused ultrasound surgery of breast cancer: Correlation of dynamic contrast-enhanced MRI with histopathologic findings. *Breast Cancer Research and Treatment* 2003;82(2):93–101.

117. Khiat A, Gianfelice D, Amara M, Boulanger Y. Influence of post-treatment delay on the evaluation of the response to focused ultrasound surgery of breast cancer by dynamic contrast enhanced MRI. *British Journal of Radiology* 2006;79(940):308–314.

118. Lenard ZM, McDannold NJ, Fennessy FM, Stewart EA, Jolesz FA, Hynynen K et al. Uterine leiomyomas: MR imaging-guided focused ultrasound surgery—Imaging predictors of success. *Radiology* 2008;249(1):187–194.

119. Damianou CA, Sanghvi NT, Fry FJ, Maass Moreno R. Dependence of ultrasonic attenuation and absorption in dog soft tissues on temperature and thermal dose. *Journal of the Acoustic Society of America* 1997;102(1):628–634.

120. Graham SJ, Stanisz GJ, Kecojevic A, Bronskill MJ, Henkelman RM. Analysis of changes in MR properties of tissues after heat treatment. *Magnetic Resonance in Medicine* 1999;42(6):1061–1071.

121. Taylor BA, Elliott AM, Hwang KP, Hazle JD, Stafford RJ. Correlation between the temperature dependence of intrinsic MR parameters and thermal dose measured by a rapid chemical shift imaging technique. *NMR Biomedicine* 2011;24(10):1414–1421.

122. Chung AH, Jolesz FA, Hynynen K. Thermal dosimetry of a focused ultrasound beam in vivo by magnetic resonance imaging. *Medical Physics* 1999;26(9):2017–2026.

123. Goldberg SN, Gazelle GS, Compton CC, Mueller PR, Tanabe KK. Treatment of intrahepatic malignancy with radiofrequency ablation—Radiologic-pathologic correlation. *Cancer* 2000;88(11):2452–2463.

124. Kohler MO, Mougenot C, Quesson B, Enholm J, Le Bail B, Laurent C et al. Volumetric HIFU ablation under 3D guidance of rapid MRI thermometry. *Medical Physics* 2009;36(8):3521–3535.

125. Pilatou MC, Stewart EA, Maier SE, Fennessy FM, Hynynen K, Tempany CMC et al. MRI-based thermal dosimetry and diffusion-weighted imaging of MRI-guided focused ultrasound thermal ablation of uterine fibroids. *Journal of Magnetic Resonance and Imaging* 2009;29(2):404–411.

126. Jacobs MA, Gultekin DH, Kim HS. Comparison between diffusion-weighted imaging, T-2-weighted, and postcontrast T-1-weighted imaging after MR-guided, high intensity, focused ultrasound treatment of uterine leiomyomata: Preliminary results. *Medical Physics* 2010;37(9):4768–4776.

127. Tashjian JA, Dewhirst MW, Needham D, Viglianti BL. Rationale for and measurement of liposomal drug delivery with hyperthermia using non-invasive imaging techniques. *International Journal of Hyperthermia* 2008;24(1):79–90.

128. Viglianti BL, Abraham SA, Michelich CR, Yarmolenko PS, MacFall JR, Bally MB et al. in vivo monitoring of tissue pharmacokinetics of liposome/drug using MRI: Illustration of targeted delivery. *Magnetic Resonance in Medicine* 2004;51(6):1153–1162.

129. de Smet M, Heijman E, Langereis S, Hijnen NM, Grull H. Magnetic resonance imaging of high intensity focused ultrasound mediated drug delivery from temperature-sensitive liposomes: An in vivo proof-of-concept study. *Journal of Control Release* 2011;150(1):102–110.

130. Negussie AH, Yarmolenko PS, Partanen A, Ranjan A, Jacobs G, Woods D et al. Formulation and characterisation of magnetic resonance imageable thermally sensitive liposomes for use with magnetic resonance-guided high intensity focused ultrasound. *Interntional Journal of Hyperthermia* 2011;27(2):140–155.

131. Tagami T, Foltz WD, Ernsting MJ, Lee CM, Tannock IF, May JP et al. MRI monitoring of intratumoral drug delivery and prediction of the therapeutic effect with a multifunctional thermosensitive liposome. *Biomaterials* 2011;32(27):6570–6578.

132. Hey S, de Smet M, Stehning C, Grull H, Keupp J, Moonen CTW et al. Simultaneous T1 measurements and proton resonance frequency shift based thermometry using variable flip angles. *Magnetic Resonance in Medicine* 2012;67(2):457–463.

133. Ponce AM, Viglianti BL, Yu D, Yarmolenko PS, Michelich CR, Woo J et al. Magnetic resonance imaging of temperature-sensitive liposome release: Drug dose painting and antitumor effects. *Journal of National Cancer Institute* 2007;99(1):53–63.

134. Dodd GD, Soulen MC, Kane RA, Livraghi T, Lees WR, Yamashita Y et al. Minimally invasive treatment of malignant hepatic tumors: At the threshold of a major breakthrough. *Radiographics* 2000;20(1):9–27.

135. Dupuy DE, Goldberg SN. Image-guided radiofrequency tumor ablation: Challenges and opportunities—Part II. *Journal of Vascular and Interventional Radiology* 2001;12(10):1135–1148.

136. Haemmerich D, Laeseke PF. Thermal tumour ablation: Devices, clinical applications and future directions. *International Journal of Hyperthermia* 2005;21(8):755–760.

137. Ahmed M, Brace CL, Lee FT, Goldberg SN. Principles of and advances in percutaneous ablation. *Radiology* 2011;258(2):351–369.

138. Livraghi T, Goldberg SN, Lazzaroni S, Meloni F, Solbiati L, Gazelle GS. Small hepatocellular carcinoma: Treatment with radio-frequency ablation versus ethanol injection. *Radiology* 1999;210(3):655–661.

139. Livraghi T, Goldberg SN, Lazzaroni S, Meloni F, Ierace T, Solbiati L et al. Hepatocellular carcinoma: Radio-frequency ablation of medium and large lesions. *Radiology* 2000;214(3):761–768.

140. Tateishi R, Shiina S, Teratani T, Obi S, Sato S, Koike Y et al. Percutaneous radiofrequency ablation for hepatocellular carcinoma—An analysis of 1000 cases. *Cancer* 2005;103(6):1201–1209.

141. Lau WY, Lai ECH. The current role of radiofrequency ablation in the management of hepatocellular carcinoma: A systematic Review. *Annals of Surgery* 2009;249(1):20–25.

142. Dupuy DE, Safran H, Mayo-Smith WW, Goldberg SN. Radiofrequency ablation of painful osseous metastatic disease. *Radiology* 1998;209P:1108.

143. Callstrom MR, Charboneau JW, Goetz MP, Rubin J, Wong GY, Sloan JA et al. Painful metastases involving bone: Feasibility of percutaneous CT- and US-guided radio-frequency ablation. *Radiology* 2002;224(1):87–97.

144. Dupuy DE, Liu DW, Hartfeil D, Hanna L, Blume JD, Ahrar K et al. Percutaneous radiofrequency ablation of painful osseous metastases a multicenter American College of Radiology Imaging Network Trial. *Cancer* 2010;116(4):989–997.

145. Dupuy DE, Zagoria RJ, Akerley W, Mayo-Smith WW, Kavanagh PV, Safran H. Percutaneous radiofrequency ablation of malignancies in the lung. *American Journal of Roentgenology* 2000;174(1):57–59.

146. Simon CJ, Dupuy DE, DiPetrillo TA, Safran HP, Grieco CA, Ng T et al. Pulmonary radiofrequency ablation: Long-term safety and efficacy in 153 patients. *Radiology* 2007;243(1):268–275.

147. McGovern F, Wood B, Goldberg S, Mueller P. Radio frequency ablation of renal cell carcinoma via image guided needle electrodes. *Journal of Urology* 1999;161(2):599–600.

148. Gervais D, McGovern F, Arellano R, McDougal W, Mueller P. Radiofrequency ablation of renal cell carcinoma: Part I, indications, results, and role in patient management over a 6-year period and ablation of 100 tumors. *American Journal of Roentgenology* 2005;185(1):64–71.

149. Goldberg SN, Dupuy DE. Image-guided radiofrequency tumor ablation: Challenges and opportunities—Part I. *Journal of Vascular and Interventional Radiology* 2001;12(9):1021–1032.

150. Haemmerich D, Lee FT. Multiple applicator approaches for radiofrequency and microwave ablation. *International Journal of Hyperthermia* 2005;21(2):93–106.

151. Simon CJ, Dupuy DE, Mayo-Smith WW. Microwave ablation: Principles and applications. *Radiographics* 2005;25:S69–S84.

152. Wright AS, Sampson LA, Warner TF, Mahvi DM, Lee FT. Radiofrequency versus microwave ablation in a hepatic porcine model. *Radiology* 2005;236(1):132–139.

153. Vogl TJ, Straub R, Eichler K, Woitaschek D, Mack MG. Malignant liver tumors treated with MR imaging-guided laser-induced thermotherapy: Experience with complications in 899 patients (2,520 lesions). *Radiology* 2002;225(2):367–377.

154. Gangi A, Alizadeh H, Wong L, Buy X, Dietemann J, Roy C. Osteoid osteoma: Percutaneous laser ablation and follow-up in 114 patients. *Radiology* 2007;242(1):293–301.

155. Lindner U, Weersink RA, Haider MA, Gertner MR, Davidson SRH, Atri M et al. Image guided photothermal focal therapy for localized prostate cancer: Phase I trial. *Journal of Urology* 2009;182(4):1371–1377.

156. Carpentier A, McNichols RJ, Stafford RJ, Itzcovitz J, Guichard J, Reizine D et al. Real-time magnetic resonance-guided laser thermal therapy for focal metastatic brain tumors. *Neurosurgery* 2008;63(1):21–28.

157. Hawasli AH, Ray WZ, Murphy RKJ, Dacey RG, Jr., Leuthardt EC. Magnetic resonance imaging-guided focused laser interstitial thermal therapy for subinsular metastatic adenocarcinoma: Technical case report. *Neurosurgery* 2012;70(6):onsE332.

158. Lafon C, Melodelima D, Salomir R, Chapelon JY. Interstitial devices for minimally invasive thermal ablation by high-intensity ultrasound. *International Journal of Hyperthermia* 2007;23(2):153–163.

159. Kangasniemi M, Diederich CJ, Price RE, Stafford RJ, Schomer DF, Olsson LE et al. Multiplanar MR temperature-sensitive imaging of cerebral thermal treatment using interstitial ultrasound applicators in a canine model. *Journal of Magnetic Resonance Imaging* 2002;16(5):522–531.

160. Nau WH, Diederich CJ, Shu R, Kinsey A, Bass E, Lotz J et al. Intradiscal thermal therapy using interstitial ultrasound—An in vivo investigation in ovine cervical spine. *Spine* 2007;32(5):503–511.

161. Makin IRS, Mast TD, Faidi W, Runk MM, Barthe PG, Slayton MH. Miniaturized ultrasound arrays for interstitial ablation and imaging. *Ultrasound in Medicine and Biology* 2005;31(11):1539–1550.

162. Delabrousse E, Salomir R, Birer A, Paquet C, Mithieux F, Chapelon J et al. Automatic temperature control for MR-guided interstitial ultrasound ablation in liver using a percutaneous applicator: Ex vivo and in vivo initial studies. *Magnetic Resonance in Medicine* 2010;63(3):667–679.

163. Melodelima D, Prat F, Fritsch J, Theillere Y, Cathignol D. Treatment of esophageal tumors using high intensity intraluminal ultrasound: First clinical results. *Journal of Translational Medicine* 2008;6:28.

164. Zimmer JE, Hynynen K, He DS, Marcus F. The feasibility of using ultrasound for cardiac ablation. *IEEE Transactions on Biomedical Engineering* 1995;42(9):891–897.

165. Gentry KL, Smith SW. Integrated catheter for 3-D intracardiac echo cardiography and ultrasound ablation. *IEEE Transactions on Ultrasonics Ferroelectrics and Frequency Control* 2004;51(7):800–808.

166. Ninet J, Roques X, Seitelberger R, Deville C, Pomar JL, Robin J et al. Surgical ablation of atrial fibrillation with off-pump, epicardial, high-intensity focused ultrasound: Results of a multicenter trial. *Journal of Thoracic and Cardiovascular Surgery* 2005;130(3):803–809.

167. Wootton JH, Prakash P, Hsu IJ, Diederich CJ. Implant strategies for endocervical and interstitial ultrasound hyperthermia adjunct to HDR brachytherapy for the treatment of cervical cancer. *Physics in Medicine and Biology* 2011;56(13):3967–3984.

168. Hynynen K. The feasibility of interstitial ultrasound hyperthermia. *Medical Physics* 1992;19(4):979–987.
169. Nau WH, Diederich CJ, Burdette EC. Evaluation of multielement catheter-cooled interstitial ultrasound applicators for high-temperature thermal therapy. *Medical Physics* 2001;28(7):1525–1534.
170. Nau WH, Diederich CJ, Stauffer PR. Directional power deposition from direct-coupled and catheter-cooled interstitial ultrasound applicators. *International Journal of Hyperthermia* 2000; 16(2):129–144.
171. Lafon C, Chapelon JY, Prat F, Gorry F, Margonari J, Theillere Y et al. Design and preliminary results of an ultrasound applicator for interstitial thermal coagulation. *Ultrasound in Medicine and Biology* 1998;24(1):113–122.
172. Chopra R, Luginbuhl C, Foster FS, Bronskill MJ. Multifrequency ultrasound transducers for conformal interstitial thermal therapy. *IEEE Transactionson the Ultrasonics Ferroelectrics Frequency Control* 2003;50(7):881–889.
173. Ross AB, Diederich CJ, Nau WH, Gill H, Bouley DM, Daniel B et al. Highly directional transurethral ultrasound applicators with rotational control for MRI-guided prostatic thermal therapy. *Physics in Medicine and Biology* 2004;49(2):189–204.
174. Chopra R, Baker N, Choy V, Boyes A, Tang K, Bradwell D et al. MRI-compatible transurethral ultrasound system for the treatment of localized prostate cancer using rotational control. *Medical Physics* 2008;35(4):1346–1357.
175. Vigen KK, Jarrard J, Rieke V, Frisoli J, Daniel BL, Pauly KB. in vivo porcine liver radiofrequency ablation with simultaneous MR temperature imaging. *Journal of Magnetic Resonance and Imaging* 2006;23(4):578–584.
176. Lepetit-Coiffé M, Quesson B, Seror O, Dumont E, Le Bail B, Moonen CTW et al. Real-time monitoring of radiofrequency ablation of rabbit liver by respiratory-gated quantitative temperature MRI. *Journal of Magnetic Resonance and Imaging* 2006;24(1):152–159.
177. Sherar MD, Moriarty JA, Kolios MC, Chen JC, Peters RD, Ang LC et al. Comparison of thermal damage calculated using magnetic resonance thermometry, with magnetic resonance imaging post-treatment and histology, after interstitial microwave thermal therapy of rabbit brain. *Physics in Medicine and Biology* 2000;45(12):3563–3576.
178. Lepetit-Coiffé M, Laumonier H, Seror O, Quesson B, Sesay M, Moonen C et al. Real-time monitoring of radiofrequency ablation of liver tumors using thermal-dose calculation by MR temperature imaging: Initial results in nine patients, including follow-up. *European Radiology* 2010;20(1):193–201.
179. Jolesz FA, Bleier AR, Jakab P, Ruenzel PW, Huttl K, Jako GJ. MR imaging of laser–tissue interactions. *Radiology* 1988;168(1):249–253.
180. Mack M, Vogl T, Eichler K, Muller P, Straub R, Roggan A et al. Laser-induced thermoablation of tumours of the head and neck under MR tomographic control. *Minimum Invasive Therapeutic Allied Technology* 1998;7(6):573–579.
181. McNichols RJ, Gowda A, Kangasniemi M, Bankson JA, Price RE, Hazle JD. MR thermometry-based feedback control of laser interstitial thermal therapy at 980 nm. *Lasers in Surgery and Medicine* 2004;34(1):48–55.
182. Carpentier A, Chauvet D, Reina V, Beccaria K, Leclerq D, McNichols RJ et al. MR-guided laser-induced thermal therapy (LITT) for recurrent glioblastomas. *Lasers in Surgery and Medicine* 2012;44(5):361–368.
183. Chopra R, Burtnyk M, N'Djin WA, Bronskill M. MRI-controlled transurethral ultrasound therapy for localised prostate cancer. *International Journal of Hyperthermia* 2010;26(8):804–821.
184. Wust P, Hildebrandt B, Sreenivasa G, Rau B, Gellermann J, Riess H et al. Hyperthermia in combined treatment of cancer. *Lancet Oncology* 2002;3(8):487–497.
185. Van Der Zee J, De Bruijne M, Mens JWM, Ameziane A, Broekmeyer-Reurink MP, Drizdal T et al. Reirradiation combined with hyperthermia in breast cancer recurrences: Overview of experience in Erasmus MC. *International Journal of Hyperthermia* 2010;26(7):638–648.

186. Stauffer PR, Maccarini P, Arunachalam K, Craciunescu O, Diederich C, Juang T et al. Conformal microwave array (CMA) applicators for hyperthermia of diffuse chest wall recurrence. *International Journal of Hyperthermia* 2010;26(7):686–698.

187. Franckena M, Fatehi D, de Bruijne M, Canters RAM, van Norden Y, Mens JW et al. Hyperthermia dose-effect relationship in 420 patients with cervical cancer treated with combined radiotherapy and hyperthermia. *European Journal of Cancer* 2009;45(11):1969–1978.

188. Westermann AM, Jones EL, Schem B, van der Steen-Banasik EM, Koper P, Mella O et al. First results of triple-modality treatment combining radiotherapy, chemotherapy, and hyperthermia for the treatment of patients with Stage IIB, III, and IVA cervical carcinoma. *Cancer* 2005;104(4):763–770.

189. van der Zee J. Heating the patient: A promising approach? *Annals of Oncology* 2002;13(8):1173–1184.

190. Gellermann J, Faehling H, Mielec M, Cho CH, Budach V, Wust P. Image artifacts during MRT hybrid hyperthermia–Causes and elimination. *International Journal of Hyperthermia* 2008;24(4): 327–335.

191. Gellermann J, Wlodarczyk W, Feussner A, Fahling H, Nadobny J, Hildebrandt B et al. Methods and potentials of magnetic resonance imaging for monitoring radiofrequency hyperthermia in a hybrid system. *International Journal of Hyperthermia* 2005;21(6):497–513.

192. Craciunescua OI, Stauffer PR, Soher BJ, Wyatt CR, Arabe O, Maccarini P et al. Accuracy of real time noninvasive temperature measurements using magnetic resonance thermal imaging in patients treated for high grade extremity soft tissue sarcomas. *Medical Physics* 2009;36(11):4848–4858.

193. Gellermann J, Hildebrandt B, Issels R, Ganter H, Wlodarczyk W, Budach V et al. Noninvasive magnetic resonance thermography of soft tissue sarcomas during regional hyperthermia—Correlation with response and direct thermometry. *Cancer* 2006;107(6):1373–1382.

194. Cheng KS, Stakhursky V, Stauffer P, Dewhirst M, Das SK. Online feedback focusing algorithm for hyperthermia cancer treatment. *International Journal of Hyperthermia* 2007;23(7):539–554.

195. Stakhursky VL, Arabe O, Cheng K, MacFall J, Maccarini P, Craciunescu O et al. Real-time MRI-guided hyperthermia treatment using a fast adaptive algorithm. *Physics in Medicine and Biology* 2009;54(7):2131–2145.

196. Sapozhnikov OA, Maxwell AD, MacConaghy B, Bailey MR. A mechanistic analysis of stone fracture in lithotripsy. *Journal of Acoustic Society of America* 2007;121(2):1190–1202.

197. Roberts WW, Hall TL, Ives K, Wolf JS, Fowlkes JB, Cain CA. Pulsed cavitational ultrasound: A noninvasive technology for controlled tissue ablation (histotripsy) in the rabbit kidney. *Journal of Urology* 2006;175(2):734–738.

198. Nyborg WL. Heat-generation by ultrasound in a relaxing medium. *Journal of the Acoustic Society of America* 1981;70(2):310–312.

199. Pauly H, Schwan HP. Mechanism of absorption of ultrasound in liver tissue. *Journal of the Acoustic Society of America* 1971;50(2):692–699.

200. Goss SA, Frizzell LA, Dunn F. Frequency dependence of ultrasonic absorption in mammalian testis. *Journal of the Acoustic Society of America* 1978;63(4):1226–1229.

201. Kennedy JE. High-intensity focused ultrasound in the treatment of solid tumours. *Nature Reviews Cancer* 2005;5(4):321–327.

202. Billard BE, Hynynen K, Roemer RB. Effects of physical parameters on high temperature ultrasound hyperthermia. *Ultrasound Medicine and Biology* 1990;16(4):409–20.

203. Fan X, Hynynen K. Ultrasound surgery using multiple sonications—Treatment time considerations. *Ultrasound Medicine and Biology* 1996;22(4):471–482.

204. Ebbini ES, Cain CA. Optimization of the intensity gain of multiple-focus phased-array heating patterns. *International Journal of Hyperthermia* 1991;7(6):953–973.

205. Fjield T, Hynynen K. The combined concentric-ring and sector-vortex phased array for MRI guided ultrasound surgery. *IEEE Transactions Ultrasonic, Ferroelectric, Frequency Control* 1997;44(5): 1157–1167.

206. Hynynen K, Watmough DJ, Mallard JR. The effects of some physical factors on the production of hyperthermia by ultrasound in neoplastic tissues. *Radiation and Environmental Biophysics* 1981;19:215–226.

207. Lele PP. Physical aspects and clinical studies with ultrasound hyperthermia. In: Storm FC, editor. *Hyperthermia in Cancer Therapy*. Boston, MA: Hall Medical Publishers; 1983. pp. 333–367.

208. Moros EG, Roemer RB, Hynynen K. Simulations of scanned focused ultrasound hyperthermia: The effects of scanning speed and pattern on the temperature fluctuations at the focal depth. *IEEE Transactions Ultrasonic Ferroelectric Frequency Control* 1988;35(5):552–560.

209. Daum DR, Hynynen K. Thermal dose optimization via temporal switching in ultrasound surgery. *IEEE Transactions Ultrasonic Ferroelectric Frequency Control* 1998;45(1):208–215.

210. Hynynen K, DeYoung D, Kundrat M, Moros E. The effect of blood perfusion rate on the temperature distributions induced by multiple, scanned and focused ultrasonic beams in dogs' kidneys in vivo. *International Journal of Hyperthermia* 1989;5(4):485–97.

211. Hynynen K, Roemer R, Moros E, Johnson C, Anhalt D. The effect of scanning speed on temperature and equivalent thermal exposure distributions during ultrasound hyperthermia in vivo. *IEEE Transactions Microwave Theory Techniques* 1986;34(5):552–559.

212. Hynynen K, Roemer R, Anhalt D, Johnson C, Xu ZX, Swindell W et al. A scanned, focused, multiple transducer ultrasonic system for localized hyperthermia treatments. *International Journal of Hyperthermia* 1987;3(1):21–35.

213. Lagendijk JJ, Crezee J, Hand JW. Dose uniformity in scanned focused ultrasound hyperthermia. *International Journal of Hyperthermia* 1994;10(6):775–784.

214. Hynynen K, Freund WR, Cline HE, Chung AH, Watkins RD, Vetro JP et al. A clinical, noninvasive, MR imaging-monitored ultrasound surgery method. *Radiographics* 1996;16(1):185–195.

215. McDannold N, King RL, Jolesz FA, Hynynen K. The use of quantitative temperature images to predict the optimal power for focused ultrasound surgery: In vivo verification in rabbit muscle and brain. *Medical Physics* 2002;29(3):356–365.

216. Cline HE, Schenck JF, Hynynen K, Watkins RD, Souza SP, Jolesz FA. MR-guided focused ultrasound surgery. *Journal of Computer Assisted Tomography* 1992;16(6):956–965.

217. Hynynen K, Darkazanli A, Unger E, Schenck JF. MRI-guided noninvasive ultrasound surgery. *Medical Physics* 1993;20(1):107–115.

218. Chopra R, Curiel L, Staruch R, Morrison L, Hynynen K. An MRI-compatible system for focused ultrasound experiments in small animal models. *Medical Physics* 2009;36(5):1867–1874.

219. Krafft AJ, Jenne JW, Maier F, Stafford RJ, Huber PE, Semmler W et al. A long arm for ultrasound: A combined robotic focused ultrasound setup for magnetic resonance-guided focused ultrasound surgery. *Medical Physics* 2010;37(5):2380–2393.

220. Tempany CMC, McDannold NJ, Hynynen K, Jolesz FA. Focused ultrasound surgery in oncology: Overview and principles. *Radiology* 2011;259(1):39–56.

221. Tempany CMC, Stewart EA, McDannold N, Quade BJ, Jolesz FA, Hynynen K. MR imaging-guided focused ultrasound surgery of uterine leiomyomas: A feasibility study. *Radiology* 2003;226(3):897–905.

222. Hynynen K, Clement G. Clinical applications of focused ultrasound—The brain. *International Journal of Hyperthermia* 2007;23(2):193–202.

223. Schmidt B, Antz M, Ernst S, Ouyang F, Falk P, Chun JKR et al. Pulmonary vein isolation by high-intensity focused ultrasound: First-in-man study with a steerable balloon catheter. *Heart Rhythm* 2007;4(5):575–584.

224. Gliklich RE, White M, Slayton MH, Barthe PG, Makin IRS. Clinical pilot study of intense ultrasound therapy to deep dermal facial skin and subcutaneous tissues. *Archives of Facial Plastic Surgery* 2007;9(2):88–95.

225. Fatemi A. High-intensity focused ultrasound effectively reduces adipose tissue. *Seminars Cutaneous Medicine and Surgery* 2009;28(4):257–262.

226. Lele PP, Parker KJ. Temperature distributions in tissues during local hyperthermia by stationary or steered beams of unfocused or focused ultrasound. *British Journal of Cancer* 1982;45:108–121.

227. Kratzik C, Schatzl G, Lackner J, Marberger M. Transcutaneous high-intensity focused ultrasonography can cure testicular cancer in solitary testis. *Urology* 2006;67(6):1269–1273.

228. Leon-Villapalos J, Kaniorou-Larai M, Dziewulski P. Full thickness abdominal burn following magnetic resonance guided focused ultrasound therapy. *Burns* 2005;31(8):1054–1055.

229. Mougenot C, Kohler MO, Enholm J, Quesson B, Moonen C. Quantification of near-field heating during volumetric MR-HIFU ablation. *Medical Physics* 2011;38(1):272–282.

230. Uterine leiomyomas: MR imaging-guided focused ultrasound surgery—Results of different treatment protocols. *Radiology; 91st Scientific Assembly and Annual Meeting of the Radiological-Society-of-North-America*; Nov. 27–Dec. 02, 2005; Jun.; Oak Brook; 820 Jorie Blvd, Oak Brook, IL 60523 USA: Radiological Soc North America; 2007.

231. Kim YS, Keserci B, Partanen A, Rhim H, Lim HK, Park MJ et al. Volumetric MR-HIFU ablation of uterine fibroids: Role of treatment cell size in the improvement of energy efficiency. *European Journal of Radiology* 2011.

232. Gianfelice D, Khiat A, Amara M, Belblidia A, Boulanger Y. MR imaging-guided focused US ablation of breast cancer: Histopathologic assessment of effectiveness—Initial experience. *Radiology* 2003;227(3):849–855.

233. Furusawa H, Namba K, Nakahara H, Tanaka C, Yasuda Y, Hirabara E et al. The evolving nonsurgical ablation of breast cancer: MR guided focused ultrasound (MRgFUS). *Breast Cancer* 2007;14(1):55–58.

234. Schmitz AC, Gianfelice D, Daniel BL, Mali WPTM, van den Bosch MAAJ. Image-guided focused ultrasound ablation of breast cancer: Current status, challenges, and future directions. *European Radiology* 2008;18(7):1431–1441.

235. Mougenot C, Tillander M, Koskela J, Kohler MO, Moonen C, Ries M. High intensity focused ultrasound with large aperture transducers: A MRI based focal point correction for tissue heterogeneity. *Medical Physics* 2012;39(4):1936–1945.

236. Payne A, Merrill R, Minalga E, Vyas U, de Bever J, Todd N et al. Design and characterization of a laterally mounted phased-array transducer breast-specific MRgHIFU device with integrated 11-channel receiver array. *Medical Physics* 2012;39(3):1552–1560.

237. McDannold N, Barnes AS, Rybicki FJ, Oshio K, Chen N, Hynynen K et al. Temperature mapping considerations in the breast with line scan echo planar spectroscopic imaging. *Magnetic Resonance in Medicine* 2007;58(6):1117–1123.

238. Taylor BA, Hwang KP, Elliott AM, Shetty A, Hazle JD, Stafford RJ. Dynamic chemical shift imaging for image-guided thermal therapy: Analysis of feasibility and potential. *Medical Physics* 2008;35(2):793–803.

239. Hey S, Maclair G, de Senneville BD, Lepetit-Coiffe M, Berber Y, Kohler MO et al. Online correction of respiratory-induced field disturbances for continuous MR-thermometry in the breast. *Magnetic Resonance in Medicine* 2009;61(6):1494–1499.

240. Sprinkhuizen SM, Bakker CJG, Bartels LW. Absolute MR thermometry using time-domain analysis of multi-gradient-echo magnitude images. *Magnetic Resonance in Medicine* 2010;64(1):239–248.

241. Crouzet S, Rebillard X, Chevallier D, Rischmann P, Pasticier G, Garcia G et al. Multicentric oncologic outcomes of high-intensity focused ultrasound for localized prostate cancer in 803 patients. *European Urology* 2010;58(4):559–566.

242. Uchida T, Shoji S, Nakano M, Hongo S, Nitta M, Murota A et al. Transrectal high-intensity focused ultrasound for the treatment of localized prostate cancer: Eight-year experience. *International Journal of Urology* 2009;16(11):881–886.

243. Madersbacher S, Pedevilla M, Vingers L, Susani M, Marberger M. Effect of high-intensity focused ultrasound on human prostate-cancer in-vivo. *Cancer Research* 1995;55(15):3346–3351.

244. Rouviere O, Souchon R, Salomir R, Gelet A, Chapelon JY, Lyonnet D. Transrectal high-intensity focused ultrasound ablation of prostate cancer: Effective treatment requiring accurate imaging. *European Journal of Radiology* 2007;63(3):317–327.

245. Chopra R, Tang K, Burtnyk M, Boyes A, Sugar L, Appu S et al. Analysis of the spatial and temporal accuracy of heating in the prostate gland using transurethral ultrasound therapy and active MR temperature feedback. *Physics in Medicine and Biology* 2009;54(9):2615–2633.

246. Sokka S, Hynynen K. The feasibility of MRI-guided whole prostate ablation with a linear aperiodic intracavitary ultrasound phased array. *Physics in Medicine and Biology* 2000;45(11):3373–3383.

247. Wharton IP, Rivens IH, ter Haar GR, Gilderdale DJ, Collins DJ, Hand JW et al. Design and development of a prototype endocavitary probe for high-intensity focused ultrasound delivery with integrated magnetic resonance imaging. *Journal of Magnetic Resonance Imaging* 2007;25(3):548–556.

248. Petrusca L, Salomir R, Brasset L, Chavrier F, Cotton F, Chapelon J-. Sector-switching sonication strategy for accelerated HIFU treatment of prostate cancer: In vitro experimental validation. *IEEE Transactions in Biomedical Engineering* 2010;57(1):17–23.

249. Gianfelice D, Gupta C, Kucharczyk W, Bret P, Havill D, Clemons M. Palliative treatment of painful bone metastases with MR imaging-guided focused ultrasound. *Radiology* 2008;249(1):355–363.

250. Liberman B, Gianfelice D, Inbar Y, Beck A, Rabin T, Shabshin N et al. Pain palliation in patients with bone metastases using MR-guided focused ultrasound surgery: A multicenter study. *Annals of Surgical Oncology* 2009;16(1):140–146.

251. Callstrom MR, Atwell TD, Charboneau JW, Farrell MA, Goetz MP, Rubin J et al. Painful metastases involving bone: Percutaneous image-guided cryoablation—Prospective trial interim analysis. *Radiology* 2006;241(2).

252. Hynynen K, Jolesz FA. Demonstration of potential noninvasive ultrasound brain therapy through an intact skull. *Ultrasound in Medicine and Biology* 1998;24(2):275–283.

253. Sun J, Hynynen K. Focusing of therapeutic ultrasound through a human skull: A numerical study. *Journal of Acoustic Society of America* 1998;104(3):1705–1715.

254. Clement GT, Sun J, Giesecke T, Hynynen K. A hemisphere array for non-invasive ultrasound brain therapy and surgery. *Physics in Medicine and Biology* 2000;45(12):3707–3719.

255. Clement GT, Hynynen K. A non-invasive method for focusing ultrasound through the human skull. *Physics in Medicine and Biology* 2002;47(8):1219–1236.

256. Aubry JF, Tanter M, Pernot M, Thomas JL, Fink M. Experimental demonstration of noninvasive transskull adaptive focusing based on prior computed tomography scans. *Journal of the Acoustic Society America* 2003;113(1):84–93.

257. Hynynen K, McDannold N, Clement G, Jolesz FA, Zadicario E, Killiany R et al. Pre-clinical testing of a phased array ultrasound system for MRI-guided noninvasive surgery of the brain—A primate study. *European Journal of Radiology* 2006;59(2):149–156.

258. Aubry J-, Marsac L, Pernot M, Robert B, Boch A-, Chauvet D et al. High intensity focused ultrasound for transcranial therapy of brain lesions and disorders. *IRBM* 2010;31(2):87–91.

259. Elias J, Monteith S, Kassell N, Khaled M, Huss D, Wintermark M. A phase 1 study of magnetic resonance guided focused ultrasound thalamotomy for the treatment of medication-refractory essential tremor. *12th International Symposium on Therapeutic Ultrasound*; Heidelberg, Germany, June 10–13, 2012.

260. Zhang L, Zhu H, Jin C, Zhou K, Li K, Su H et al. High-intensity focused ultrasound (HIFU): Effective and safe therapy for hepatocellular carcinoma adjacent to major hepatic veins. *European Radiology* 2009;19(2):437–445.

261. Illing RO, Kennedy JE, Wu F, ter Haar GR, Protheroe AS, Friend PJ et al. The safety and feasibility of extracorporeal high-intensity focused ultrasound (HIFU) for the treatment of liver and kidney tumours in a Western population. *British Journal of Cancer* 2005;93(8):890–895.

262. Zhu H, Zhou K, Zhang L, Jin C, Peng S, Yang W et al. High intensity focused ultrasound (HIFU) therapy for local treatment of hepatocellular carcinoma: Role of partial rib resection. *European Journal of Radiology* 2009;72(1):160–166.

263. Leslie TA, Kennedy JE, Illing RO, Ter Haar GR, Wili F, Phillips RR et al. High-intensity focused ultrasound ablation of liver tumours: can radiological assessment predict the histological response? *British Journal of Radiology* 2008;81(967):564–571.

264. Okada A, Murakami T, Mikami K, Onishi H, Tanigawa N, Marukawa T et al. A case of hepatocellular carcinoma treated by MR-guided focused ultrasound ablation with respiratory gating. *Magnetic Resonance in Medical Science* 2006;5(3):167–171.

265. Fischer K, Gedroyc W, Jolesz FA. Focused ultrasound as a local therapy for liver cancer. *Cancer Journal* 2010;16(2):118–124.

266. Lehmann JF, Delateur BJ, Warren CD, Stonebridge JS. Heating produced by ultrasound in bone and soft tissue. *Archives of Physical Medicine and Rehabilitation* 1967;48(8):397–402.

267. Smith NB, Temkin JM, Shapiro F, Hynynen K. Thermal effects of focused ultrasound energy on bone tissue. *Ultrasound Medicine and Biology* 2001;27(10):1427–1433.

268. Jung SE, Cho SH, Jang JH, Han J. High-intensity focused ultrasound ablation in hepatic and pancreatic cancer: Complications. *Abdominal Imaging* 2011;36(2):185–195.

269. Civale J, Clarke R, Rivens I, Ter Haar G. The use of a segmented transducer for rib sparing in HIFU treatments. *Ultrasound Medicine and Biology* 2006;32(11):1753–1761.

270. Cochard E, Prada C, Aubry JF, Fink M. Ultrasonic focusing through the ribs using the DORT method. *Medical Physics* 2009;36(8):3495–3503.

271. Quesson B, Merle M, Kohler MO, Mougenot C, Roujol S, de Senneville BD et al. A method for MRI guidance of intercostal high intensity focused ultrasound ablation in the liver. *Medical Physics* 2010;37(6):2533–2540.

272. Liu H, Hsu C, Huang S, Hsi Y. Focal beam distortion and treatment planning for transrib focused ultrasound thermal therapy: A feasibility study using a two-dimensional ultrasound phased array. *Medical Physics* 2010;37(2):848–860.

273. Grissom WA, Rieke V, Holbrook AB, Medan Y, Lustig M, Santos J et al. Hybrid referenceless and multibaseline subtraction MR thermometry for monitoring thermal therapies in moving organs. *Medical Physics* 2010;37(9):5014–5026.

274. Mougenot C, Quesson B, de Senneville BD, de Oliveira PL, Sprinkhuizen S, Palussiere J et al. Three-dimensional spatial and temporal temperature control with MR thermometry-guided focused ultrasound (MRgHIFU). *Magnetic Resonance in Medicine* 2009;61(3):603–614.

275. Ries M, de Senneville BD, Roujol S, Berber Y, Quesson B, Moonen C. Real-time 3D target tracking in MRI guided focused ultrasound ablations in moving tissues. *Magnetic Resonance in Medicine* 2010;64(6):1704–1712.

276. Holbrook AB, Santos JM, Kaye E, Rieke V, Pauly KB. Real-time MR thermometry for monitoring HIFU ablations of the liver. *Magnetic Resonance in Medicine* 2010;63(2):365–373.

277. Jacco A. de Zwart. Fast magnetic resonance temperature imaging for control of localized hyperthermia in medicine. Theis. University of Bordeaux; 2000.

278. Chopra R, Wachsmuth J, Burtnyk M, Haider MA, Bronskill MJ. Analysis of factors important for transurethral ultrasound prostate heating using MR temperature feedback. *Physics in Medicine and Biology* 2006;51(4):827–844.

279. Enholm JK, Kohler MO, Quesson B, Mougenot C, Moonen CT, Sokka SD. Improved volumetric MR-HIFU ablation by robust binary feedback control. *IEEE Transactions in Biomedical Engineering* 2010;57(1):103–113.

280. Staruch R, Chopra R, Hynynen K. Localised drug release using MRI-controlled focused ultrasound hyperthermia. *International Journal of Hyperthermia* 2011;27(2):156–171.

281. Vanne A, Hynynen K. MRI feedback temperature control for focused ultrasound surgery. *Physics in Medicine and Biology* 2003;48(1):31–43.

282. Arora D, Cooley D, Perry T, Guo J, Richardson A, Moellmer J et al. MR thermometry-based feedback control of efficacy and safety in minimum-time thermal therapies: Phantom and in-vivo evaluations. *International Journal of Hyperthermia* 2006;22(1):29–42.

283. Salomir R, Palussiere J, Vimeux FC, de Zwart JA, Quesson B, Gauchet M et al. Local hyperthermia with MR-guided focused ultrasound: Spiral trajectory of the focal point optimized for temperature uniformity in the target region. *Journal of Magnetic Resonance in Imaging* 2000;12(4):571–583.

284. Quesson B, Laurent C, Maclair G, de Senneville BD, Mougenot C, Ries M et al. Real-time volumetric MRI thermometry of focused ultrasound ablation in vivo: A feasibility study in pig liver and kidney. *NMR Biomedicine* 2011;24(2):145–153.

285. Ranjan A, Jacobs G, Woods DL, Negussie AH, Partanen A, Yarmolenko PS et al. Image-guided drug delivery with magnetic resonance guided high intensity focused ultrasound and temperature sensitive liposomes in a rabbit Vx2 tumor model. *Journal of Control Release* 2012;158(3):487–494.

286. Duck FA. *Physical Properties of Tissue: A Comprehensive Reference book*. London: Academic Press; 1990.

30

Medical Robotics

Kevin Cleary
Georgetown University
Medical center

Charles C. Nguyen
Catholic University of
America

30.1 Introduction

Medical robotics is an interdisciplinary field that focuses on developing electromechanical devices for clinical applications. The goal of this field is to enable new medical techniques by providing new capabilities to the physician or by providing assistance during surgical procedures. Medical robotics is a relatively young field, as the first recorded medical application occurred in 1985 for a brain biopsy. The field is still emerging. However, medical robotics has tremendous potential for improving the precision and capabilities of physicians when performing surgical procedures and it is believed that the field will continue to grow as improved systems become available.

Although several commercial companies sell medical robots, the total number of robots sold is still very small, and the market will most likely continue to grow slowly. Unlike the area of factory robotics, which grew rapidly during the 1970s and 1980s, medical robotics has not yet gained widespread acceptance. However, the authors feel that the benefits of medical robotics will become increasingly clear, leading to a continued rise in their use in medicine.

This chapter will begin with an introduction to robotics, followed by a historical review of their use in medicine. Clinical applications in several different medical specialties will then be presented. The chapter concludes with a discussion of technology challenges and areas for future research.

30.2 Robotics Review

The word "robot" originated from a 1922 play called *Rossum's Universal Robots (R.U.R.)* by Karel Capek.[1] The play was about a future in which all workers are automatons, who then revolt when they acquire souls.

The initial application of robotics was for handling of nuclear materials during the 1950s. These systems were teleoperated and controlled by a human operator. The first programmable robots were developed for industrial applications. The UNIMATE robot was used by General Motors in the early 1960s for stacking hot pieces of metal. The robot could be programmed using a teach pendant to move the system through the desired motions. The robot was hydraulically powered and available with five or six degrees of freedom (DOF).

DOF is a common term in the robotics field and indicates the ability of the robot to position and orient an object in space. To be able to carry out general tasks, a robot must have at least six DOFs: three for translations (such as x, y, and z translations in Cartesian space) and three for rotations (such as rotations about x, y, and z Cartesian axes).

The field of industrial robotics grew rapidly throughout the 1960s and 1970s, when many factories became automated, particularly in the electronics and automotive areas. A trade group, the Robotics Industries Association, was founded in 1974 (www.robotics.org). This group sponsors conferences aimed at industrial applications and publishes a magazine titled *Robotics World*. Their definition of a robot is often quoted, which is "a reprogrammable, multifunctional manipulator designed to move material, parts, tools, or specialized devices through variable programmed motions to perform a variety of tasks." While this definition may seem somewhat narrow, we will discuss the medical robotics devices in this chapter that fall mainly in this category. These robots typically consist of rigid links that allow relative motion from one link to another.[2] Attached to the end of the links is the robot hand, usually referred to as the end effector. The robot is controlled by a computer system that is used to move the end effector to any desired point and orientation within its workspace.

On the academic side, robotics became an increasingly popular subject in the engineering school and computer science departments in the 1980s. Several of the leading U.S. schools such as Stanford, MIT, and Carnegie Mellon developed robotics groups during this time. Carnegie Mellon formed a Robotics Institute in 1980 and began offering the first PhD in robotics in 1989. While many early students in the field started with the classical text by Richard Paul,[3] several other textbooks began to appear.[2-5] International conferences such as the *IEEE International Conference on Robotics and Automation*, which was first held in 1984, began to occur. The *IEEE Transactions on Robotics and Automation* also began publishing in that year, originally as the *Journal of Robotics and Automation*. In the late 1980s and continuing in the 1990s, the field of robotics began to branch out from factory robotics into other applications such as service robotics and medical robotics. Mobile robotics became a popular topic and medical applications, including surgical robots and robots for rehabilitation, began to appear. In this chapter of medical robotics, we will focus on the robots that play an active role during a surgical intervention. These systems are not meant to replace the physician, but rather, to augment the capabilities of the physician. There are other categories of medical robotics, such as robotics for rehabilitation or miniature robots that might be placed inside the body, but these will not be discussed here. This chapter is not intended to be comprehensive, but rather, to give an overview of the field of medical robotics.

Several other medical robotics review articles with a focus on surgical procedures have also been written. Davies[6] describes the history of surgical robotics and gives one classification for the types of robot systems studied by researchers. Taylor[7] discusses several taxonomies for surgical robotics and presents a different classification. Troccaz and Delnondedieu[8] give a historical review and describe passive, semiactive, and active robotic systems. Howe and Matsuoka[9] overview applications in image-based procedures, orthopedic surgery, and neurosurgery, among other specialties. Specialized reviews also exist, such as the article by Cadeddu et al.[10] on urology robotics.

30.3 Medical Robotics History

Medical robotics is a relatively new field, with the first recorded medical application of a robot occurring in 1985.[11] Here, the robot was a simple positioning device to orient a needle for biopsy of the brain. A 52-year-old man was put on a computed tomography (CT) scanner table and a series of CT images were obtained. The target within the brain was then identified on these images, and a PUMA 560 industrial robot was used to orient a guide tube through which the needle was inserted. Unfortunately, safety issues concerning the operation of the robot in close proximity to people prevented this work from continuing.[6]

About the same time, research groups in Asia, Europe, and the United States began investigating other medical applications of robotics. In Asia, Dohi and colleagues at Tokyo University developed a prototype of a CT-guided needle insertion manipulator.[12] In Europe, a group at Imperial College in London under the direction of Davies began developing a robot for resection of the prostate.[13] At Grenoble University Hospital in France, Benabid, Lavallee, and colleagues started work on neurosurgical applications.[14] In the United States, Taylor and associates at IBM began developing a system for hip implants later known as ROBODOC.[15]

Currently, there are several commercial ventures and a number of research laboratories active in the field of medical robotics. Interest in the topic has grown rapidly in the last few years in particular, and dedicated sessions on medical robotics can now be found at medically related conferences. Early research efforts have led to some commercial products. For example, the work at Grenoble University Hospital led to the NeuroMate robot of Integrated Surgical Systems (ISS), as described later.

30.4 Clinical Applications

Robots have been used in a number of clinical applications in medicine as shown in Table 30.1. This table is not exhaustive, but representative research groups and commercial vendors in several areas have been selected to give the reader an overview of the field. The column labeled "Studies" refers to whether human trials, animal studies, cadaver studies, or phantom studies have been done. Each of these areas will be discussed in the following sections.

TABLE 30.1   Clinical Application Areas and Representative Robotic Developments

Clinical Area	Country	Institution/Company	System	Studies	Reference
Neurosurgery	Switzerland	University of Lausanne	Minerva	Human	16,17
	United States	Integrated Surgical/Grenoble University Hospital	Neuro Mate	Human	18
	Japan	University of Tokyo	MRI compatible	Tissue samples	19
Orthopedic	United States	Integrated Surgical	ROBODOC	Human	15
	United States	University of Tokyo/Hopkins	PAKY/RCM	Phantom	19
	United States	Marconi	Kawasaki	Pig	20
	United Kingdom	Imperial College	Acrobot	Human	21
Interventional	United States	Georgetown/Hopkins	PAKY/RCM	Cadaver	22
	Austria	ARC Seibersdorf Research/Vienna UniversityHospital	Custom built	Phantom	23
Urology	United Kingdom	Imperial College	Probot	Human	24
	United States	Hopkins	PAKY/RCM	Human	25
Maxillofacial	Germany	Charite	SurgiScope	Pig	26
	Germany	Karlsruhe/Heidelberg	RX 90	Pig	27
Radiosurgery	United States	Accuray	Cyberknife	Human	28
Ophthalmology	United States	Hopkins	Steady Hand	Phantom	29
Cardiac	United States	Intuitive Surgical	da Vinci	Human	30
	United States	Computer Motion	Zeus	Human	31
	France	Grenoble	PADyC	Phantom	32

30.4.1 Neurosurgery

As mentioned in the historical review, neurosurgery was the first clinical application of robotics and continues to be a topic of current interest. Neurosurgical stereotactic applications require spatial accuracy and precision targeting to reach the anatomy of interest while minimizing collateral damage. This section presents three neurosurgical robotic systems:

1. Minerva from the University of Lausanne in Switzerland
2. NeuroMate from ISS in the United States
3. A magnetic resonance imaging (MRI)-compatible robot developed by Dohi and colleagues in Japan

30.4.1.1 Minerva

One of the earliest robotic systems developed for precise needle placement was the neurosurgical robot Minerva,[16] designed for stereotactic brain biopsy. A special-purpose robot was constructed that was designed to work within the CT scanner so that the surgeon could follow the position of the instruments on successive CT scans. This constraint ensured that CT images would be available throughout the procedure, keeping all procedures under the surgeon's supervision and control. A diagram of the system and associated components is shown in Figure 30.1.

The system consists of a five-DOF structure with two linear axes (vertical and lateral), two rotary axes (moving in a horizontal and vertical plane), and another linear axis (to move the tool to and from the patient's head).[17] The robot is mounted on a horizontal carrier that moves on rails. A stereotactic frame, the Brown–Roberts–Wells (BRW) reference frame, is attached to the robot gantry and coupled to the motorized CT table by two ball-and-socket joints arranged in series. The system was used for two operations on patients in September 1993 at the CHUV Hospital in Switzerland, but the project has since been discontinued.

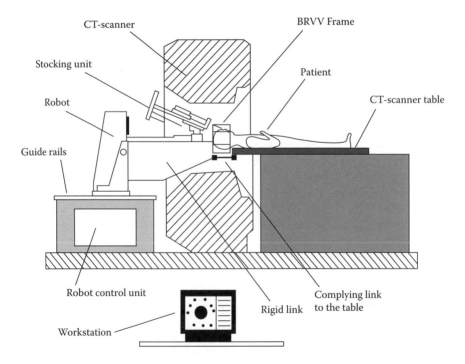

FIGURE 30.1 Minerva components and system overview. (From Burckhart, C.W., Flury, P., and Glauser, D., Stereotactic brain surgery, *IEEE Eng. Med. Biol.*, 14(3), 314–317, 1995. ©1995 IEEE. With permission.)

30.4.1.2 NeuroMate

The NeuroMate is a six-axis robot for neurosurgical applications that evolved from work done by Benabid, Lavallee, and colleagues at Grenoble University Hospital in France.[14,18,19] The original system was subsequently redesigned to fulfill specific stereotactic requirements and particular attention was paid to safety issues.[20] The current version (Figure 30.2) is a commercial product that has been licensed by ISS (Davis, CA) and is Food and Drug Administration (FDA) approved.

The system has been used in over 1600 procedures since 1989, covering a range of neurosurgical procedures. The major clinical applications include:

- Tumor biopsies (1100 cases)
- Stereoelectroencephalographic investigations of patients with epilepsy (200 cases)
- Midline stereotactic neurosurgery and functional neurosurgery of the basal ganglia (200 cases)

A typical clinical procedure consists of an initial data-acquisition step, followed by data transfer to the control computer, and then the procedure itself. Data acquisition involves obtaining images of the brain from which path planning from the skin entry point to the target point can be done using a specially developed software program. The images can be in digital form (DSA, CT, or MRI images) or can be digitized (e.g., radiographs) using a digitizing table or a scanner. Once the path is planned, the images are transferred directly from the planning workstation to the control workstation in the operating room over an Ethernet link.

To carry out the procedure, the robot must know where it is located relative to the patient's anatomy. This is typically done using a calibration cage, which is placed on the end effector of the robot around the patient's head (Figure 30.3). This cage looks like an open cubic box. The four sides are implanted with nine x-ray opaque beads, the positions of which have been precisely measured. Two x-rays are taken that show the position of these beads along with the fiducial markers of the patient's frame. This information is used to determine the transformation matrix between the robot and the patient. The defined trajectory is used to command the robot to position a mechanical guide, which is aligned with this trajectory. The robot is then fixed in this position, and the physician uses this guide to introduce the surgical tool such as a drill, probe, or electrode.

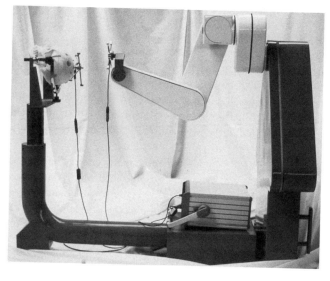

FIGURE 30.2 NeuroMate neurosurgical robot. (Courtesy of Integrated Surgical Systems, Davis, CA.)

FIGURE 30.3 Calibration cage held by the robot. (From Lavallee, S et al., Image-guided operating robot: A clinical application in stereotactic neurosurgery, In *Computer-Integrated Surgery*, Taylor, R.H. et al., eds., MIT Press, Cambridge, 1995, pp. 343–351. With permission from MIT Press.)

30.4.1.3 MRI-Compatible Robot

While several robots have been developed for stereotactic neurosurgery, including those mentioned above, almost all these systems used CT images for guidance. However, many structures in the brain are best visualized using MRI. The robotic systems described so far are not suitable for use in an MRI scanner because the strong magnetic fields generated dictate that only nonmagnetic materials can be used. In Japan, in the Mechatronics Laboratory at the University of Tokyo, Dohi, Masamune, and colleagues developed an MRI-compatible needle insertion manipulator intended for use in stereotactic neurosurgery.[21] The manipulator frame was manufactured using polyethylene terephthalate (PET), and ultrasonic motors were used for the actuators. Other parts such as bearings, feed screws, and gear that must be strong and precisely fabricated are made of nonmagnetic materials including brass, aluminum, delrin, and ceramics. In phantom tests using watermelons, the robot performed satisfactorily with a positioning error of less than 3.3 mm from the desired target. The unit was small enough at 491 mm in maximum height to fit inside the MRI gantry of 600 mm in diameter.

Rather than retrofitting an industrial robot, Masamune developed a completely new design based on the clinical requirements for safety, MRI compatibility, and compactness. As shown in Figure 30.4, the system includes an *X–Y–Z* base stage. An arch mechanism is mounted on the base stage along with a linear needle carriage. This isocentric design was adopted for its mechanical safety and simplicity. The system was controlled by a personal computer. The control computer and motor–driver boards were remotely located in the MRI control room and were connected by shielded cables to the robot.

In a related development, a new MRI-compatible robot has been developed to work within the interventional MRI unit at the Brigham and Women's Hospital in Boston, Massachusetts.[22] The interventional MRI has a pair of parallel-facing donut-shaped magnets, with an air gap of 560 mm. The robot sits between the magnets and is mounted on at the top of the unit as shown in Figure 30.5. The system is currently undergoing testing. One potential clinical application is needle placement for prostate brachytherapy.

Finally, researchers in Germany have developed an MRI-compatible robotic biopsy system, focusing on breast cancer as an initial application *in vitro* experiments using pig livers in a 1.5-T magnet and 4 mm targets resulted in all eight targets being successfully hit.[23]

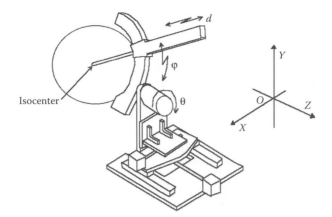

FIGURE 30.4 MRI-compatible robot design. (Courtesy of Ken Masamune, Tokyo Denki University, Tokyo, Japan.)

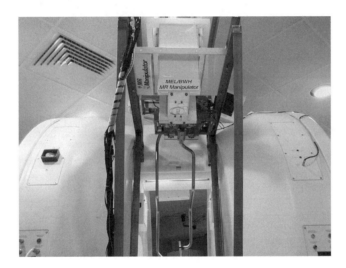

FIGURE 30.5 MRI-compatible robot in an interventional MRI system. (Courtesy of Kiyoyuki Chinzei, AIST, Tsukuba, Japan, and Ron Kikinis, BWH, Boston, MA.)

30.4.2 Orthopedic

Orthopedics is well suited for robotic assistance due to the rigid nature of the bone. Since bone does not deform significantly when it is drilled or cut, it is possible to intraoperatively apply preoperative imaging and planning information more easily than for soft tissues such as the brain or abdominal organs.[24] Orthopedics was also an early adopter of robotics, as the ROBODOC system described next was used to assist surgeons in performing part of a total hip replacement in 1992. The robot was used to mill out the hole for the hip implant. This marked the first use of an active robot for hip surgery.

30.4.2.1 ROBODOC

The ROBODOC system was developed clinically by ISS for total hip replacement procedures from a prototype created at IBM Research. The system was used in over 1000 cases at a hospital in Frankfurt, Germany from 1994 until 1998.[25] The system consists of three major components: a planning workstation, the robot itself that does the cutting, and the workstation that guides and controls the robot.

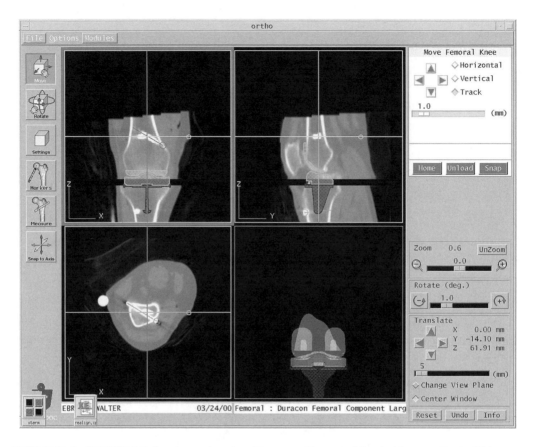

FIGURE 30.6   ORTHODOC planning workstation. (Courtesy of Integrated Surgical Systems, Davis, CA.)

A typical hip replacement procedure using ROBODOC is carried out as follows.[26] The procedure starts with the surgeon implanting three locator pins into the hip. These pins are later used as fiducial points for registering the patient anatomy with the robot. A CT scan is then obtained and the CT data are transferred to the planning workstation (ORTHODOC). The surgeon can then choose a suitable implant from a library of possible implants. The surgeon can virtually position the implant on the planning workstation, check different positions, and assess the impact on anteversion, neck length, and stress loading (Figure 30.6). When the planning session is finished, the data are transferred to the computer that controls ROBODOC.

In the operating room, the hip joint is exposed and the robotic system is moved into position to mill out the femoral cavity. The locator pins are used to register the hip joint with the robot. Cutting time is between 20 and 35 min. The surgeon monitors this process by watching a computer screen that shows the progress of the cutting operation. The robot can also be stopped at any time. When the milling is complete, the robot is removed and the rest of the operation is completed by hand in the conventional manner. A photograph of ROBODOC milling the cavity for the implant is shown in Figure 30.7.

30.4.2.2 University of Tokyo/Johns Hopkins Collaboration

An integrated robotic system for percutaneous placement of needles under CT guidance was developed by Masamune at the University of Tokyo in collaboration with Johns Hopkins.[27] Single image-based coregistration of the percutaneous access of the kidney (PAKY)/remote center of motion (RCM) robot and image space was achieved by stereotactic localization using a miniature version of the BRW head

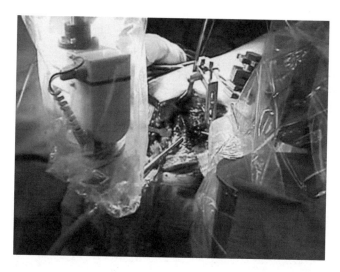

FIGURE 30.7 ROBODOC milling implant cavity for hip replacement surgery. (Courtesy of Integrated Surgical Systems, Davis, CA.)

frame built into the radiolucent needle driver. A phantom study was done with an orientation accuracy of 0.6° and a needle tip to target distance of 1.04 mm. The system is applicable to orthopedic (spine) and many other percutaneous procedures.

30.4.2.3 Marconi Medical Systems

An active robot has been integrated with a CT scanner for interventional procedures by Yanof and colleagues at Marconi Medical Systems.[28] The advantage of this approach is that the coordinate system of the robot can be registered with the coordinate system of the CT during the integration phase. A separate registration step is thus not required for clinical use. Animal experiments using pigs were completed to investigate needle placement in the abdomen. The path was planned based on the CT scans and this information was sent to the robot, which automatically moved to the skin entry point and then oriented and drove the needle.

30.4.2.4 Imperial College

A special-purpose robot called Acrobot (for active constraint robot) has been developed for safe use in the operating room for total knee replacement surgery.[29] The surgeon guides the robot using a handle attached to a force sensor attached to the robot tip. The concept is called "active constraint control" and allows for a synergy between the surgeon and the robot. The robot provides geometric accuracy and increases safety by providing motion constraint outside a predefined region. Following two preliminary clinical trials, the first clinical trial was conducted in which the Acrobot was used to register and cut the knee bones.

30.4.3 Interventional

30.4.3.1 Georgetown University/Johns Hopkins Collaboration

At Georgetown University Medical Center, our research group has been focusing on the use of robots for precision placement of instruments in minimally invasive spine procedures.[30,31] This work is a collaboration with the Urology Robotics Laboratory of the Johns Hopkins Medical Institutions and the Computer-Integrated Surgical Systems and Technology (CISST) Engineering Research Center at Johns Hopkins University.

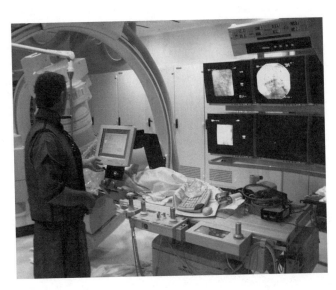

FIGURE 30.8 Needle driver robot cadaver study. The physician uses touch screen/joystick interface to control the robot and monitors the intervention on the fluoroscopy screens.

Low back pain is a common medical problem, and minimally invasive procedures such as nerve blocks are rapidly growing in popularity as a potential method of treatment. To assist the physician in needle placement during these procedures, we have begun to use a newly developed version of the PAKY/RCM needle driver robot developed at the Urology Robotics Laboratory. Robotic systems such as these have great potential as physician-assist devices for improving the precision of needle placement and enabling the development of the next generation of precision guidance systems for interventional techniques.

The newly developed needle driver robot consists of a three-DOF translational stage, a seven-DOF passive positioning stage, and a three-DOF orientation/driving stage. The robot is mounted on the interventional table and the physician controls the system through a touch screen/joystick interface as shown in Figure 30.8. A cadaver study using the robot to place a 22-gauge needle for the nerve and facet blocks has been completed (Figure 30.9).[32] Small-metal BB nipple markers (1 mm in diameter) were inserted percutaneously into the paraspinal region of an embalmed cadaver. Six BBs were placed near the nerve root and six BBs were placed near the facet joint. Using the touch screen/joystick interface, the physician controlled the robot to drive the needle toward each target BB. All needles were placed within 3 mm of the target, and the average distance was 1.44 ± 0.66 mm (standard deviation).

After the cadaver study was completed, approval was obtained from the FDA for a randomized clinical trial of 20 patients for the nerve and facet blocks using the robot under an investigational device exemption (IDE) protocol. This protocol was also approved by the Georgetown Institutional Review Board (IRB) and the U.S. Army Human Subjects Research Review Board. The procedure is done in the standard manner except the robot is used to position, orient, and drive the needle under physician control. Anterior/posterior (A/P) fluoroscopy is used to position and orient the needle, and lateral fluoroscopy is used to monitor the depth of insertion. To date, 11 patients have been enrolled and preliminary results show that it is feasible to use the robot for this clinical procedure. A photograph of one of the cases is shown in Figure 30.10.

30.4.3.2 ARC Seibersdorf Research Biopsy Robot

A mobile robotic system to assist interventional radiologists in ultrasound or CT-guided biopsies has been developed by ARC Seibersdorf Research in Austria in cooperation with the Vienna University Hospital.[33,55] The system includes a four-DOF gross positioning system consisting of three linear axes and one rotational axis for initial positioning of the biopsy needle (Figures 30.11 and 30.12). The final

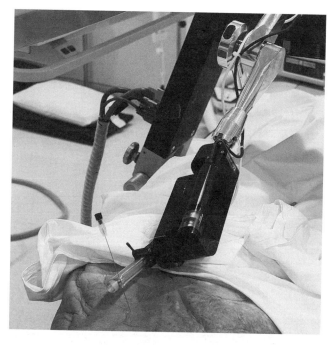

FIGURE 30.9 Close-up of the cadaver study showing the needle driver and passive positioning arm.

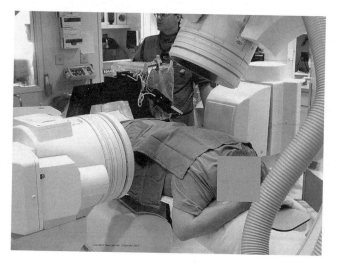

FIGURE 30.10 Clinical trial of the joystick-controlled robot to drive the needle for spinal block under fluoroscopy.

pose of the needle is controlled by a three-DOF needle positioning stage that includes two DOFs for orienting the needle at any angle, followed by a linear DOF with a stroke of 50 mm for moving the needle along the desired path to the skin entry point. Once the needle is positioned, the physician will then drive the needle to the target point by hand.

The system is controlled by two personal computers (PCs). A Windows 2000-based industrial PC is used for the high-level control of the robot. A Linux-based industrial PC is employed for the user interface and for monitoring of the intervention. The Linux PC includes a video capture card that receives the video input from the ultrasound or CT monitor and displays it on the screen for planning of the biopsy

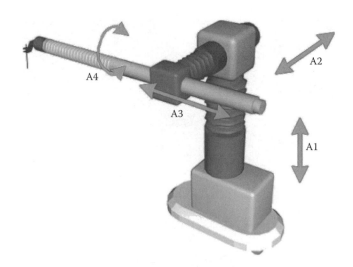

FIGURE 30.11 DOF of the gross positioning system. A1–A3 are linear axes and A4 is a rotational motion. (Courtesy of Gernot Kronreif, ARC Seibersdorf Research, Seibersdorf, Austria.)

FIGURE 30.12 The prototype gross positioning system includes the mobile cart and control electronics. (Courtesy of Gemot Kronreif, ARC Seibersdorf Research, Seibersdorf, Austria.)

path (Figure 30.13). Once the path has been planned, the robot can then be commanded to move to the desired location and orient the needle.

To register the coordinate system of the robot with that of the ultrasound or CT machine, an optical tracking system (Polaris, Northern Digital, Waterloo, Canada) is used to track the three-dimensional position of all the components. This scheme was tested in a phantom study where peas with a mean diameter of 9.4 mm were embedded within a custom-made gel. The ultrasound transducer was tracked using the Polaris and the path toward each target was placed on the user interface. The robot arm was then commanded to move to the planned insertion point and to orient itself toward the center of the target. A 17-gauge coaxial puncture needle and an 18-gauge biopsy needle were then manually driven

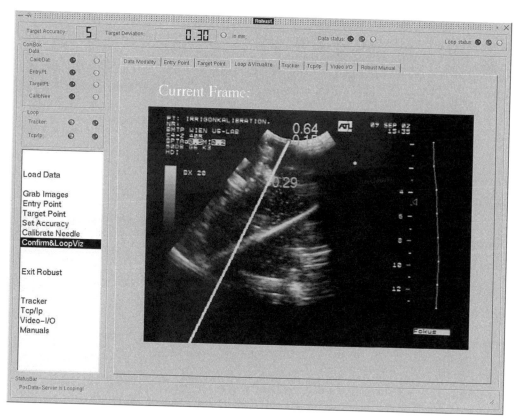

FIGURE 30.13 User interface showing planned and actual biopsy path overlaid with ultrasound scan. (Courtesy of Gernot Kronreif, ARC Seibersdorf Research, Seibersdorf, Austria.)

toward the center of the target by the radiologist. The goal was successfully achieved for all 20 targets with only a single needle pass. The average radial distance between the needle tip and the center of the target was 1.9 ± 1.1 mm (standard deviation).

30.4.4 Urology

30.4.4.1 Prostate Resection

One of the pioneering research groups in medical robotics is the Mechantronics in Medicine Laboratory at Imperial College in London. Starting in 1988, the group began developing a robotic system named the *Probot* to aid in transurethral resection of the prostate.[34] While an initial feasibility study was carried out using a standard six-axis PUMA industrial robot, such a system was determined not to be practical for medical purposes as these robots were not designed to work in close proximity with humans. Therefore, a special-purpose robotic frame was designed to hold the surgical instrument. The first patient was treated in April 1991. This was the first use of a robot to remove the substantial quantities of tissue from a human patient.[6]

The robotic frame shown in Figure 30.14 consists of three axes of movement. An additional axis is provided by the resectoscope, which is the surgical instrument used to remove the tissue. The geometry of the system is designed to allow a cavity to be hollowed out from within the prostate and to restrict movements outside an allowable range. This restriction provides an additional margin of safety.

The clinical application consists of four stages: (1) measurement, (2) imaging, (3) cavity design, and (4) cutting. To begin the procedure, the patient is positioned on the operating table and the Probot is

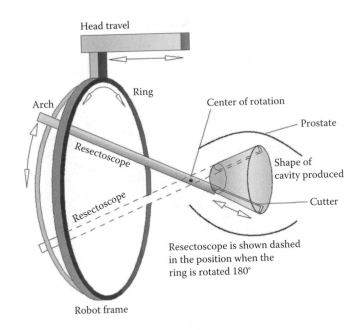

Head travel

Ring

Arch   Center of rotation

Prostate

Resectoscope

Shape of
cavity produced

Resectoscope

Cutter

Resectoscope is shown dashed
in the position when the
ring is rotated 180°

Robot frame

FIGURE 30.14   Prostate robot frame. (Courtesy of Brian Davies, Imperial College, London.)

positioned at the bladder neck. The user interface allows the surgeon to view the internal anatomy from a video camera within the resectoscope. An ultrasound probe is then passed down the resectoscope and the robot is set to acquire a series of scans at 5-mm intervals to build up a three-dimensional image of the prostate. The surgeon can then outline the cavity to cut on each slice of the ultrasound image using a light pen. The final step is the actual cutting operation. A picture of the operating room and the Probot in clinical use is shown in Figure 30.15. The surgeon is sitting to the left and can observe the progress of the cutting on a video monitor as shown in Figure 30.16. The real-time image of the prostate is at the top left of the monitor and an overlay of the cuts on an ultrasound image is shown at the bottom right.

30.4.4.2 Urology Robotics Laboratory

The Urology Robotics (URobotics) Laboratory, part of the Urology Department at Johns Hopkins Medical Institutions, is dedicated to the development of new technology for urological surgery.[35] The program combines engineering and medical personnel in close cooperation and is the only academic engineering program devoted exclusively to urology. This group and colleagues at the Engineering Research Center at Johns Hopkins University have developed the PAKY needle driver[36] and RCM robot[37] that has been applied to minimally invasive kidney procedures.

30.4.5 Maxillofacial

Maxillofacial surgery is a branch of surgery that is concerned primarily with operations on the jaws and surrounding soft tissues. Maxillofacial surgery often requires the manipulation of the skull bone, including drilling, cutting, shaping, and repositioning operations. Accuracy is at the premium since the shape of the bone and the aesthetic appearance of the skull and face are extremely important to patients. The current procedures are done manually using tools such as pliers, chisels, and electric saws and drills. Maxillofacial surgery may be a good application area for robotics since primarily, bony structures are involved and accuracy is at a premium.[38]

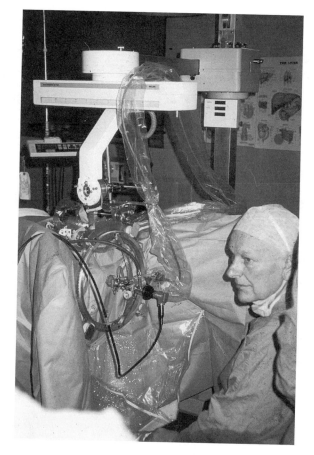

FIGURE 30.15   Probot in clinical use. (Courtesy of Brian Davies, Imperial College, London.)

For example, the following clinical tasks must be supported by a robot in maxillofacial surgery:[39]

- Guidance for nonflexible catheter implantation (brachytherapy)
- Handling of electrical drills, taps, and screwdrivers for fixing bones and implants (anaplastology)
- Handling of the electric saw and retractor hooks

30.4.5.1 Experimental Operating Room

For developing an interactive robot system for maxillofacial surgery, an experimental operating room has been set up at the Charite Hospital of Humbolt University in Berlin, Germany as shown in Figure 30.17. This operating room includes a unique robotic system, the SurgiScope. While most robotic systems described in this chapter are based on a serial kinematic structure in which the links are attached one after the other as in the human arm, at least one company has developed a medical robot based on a parallel kinematic structure. The SurgiScope is a general-purpose six-DOF robotic device consisting of a fixed base, three parallel links, and a movable end effector. The system is designed to be fixed on the ceiling and provides a large workspace while not cluttering the operating room floor. The parallel kinematic structure also provides a very stable structure for precision operations. The robot was originally sold by Elekta, but is now being marketed by Jojumarie Intelligente Instrumente in Berlin. The system has been demonstrated in animal studies for placement of the radiation source in brachytherapy.[40]

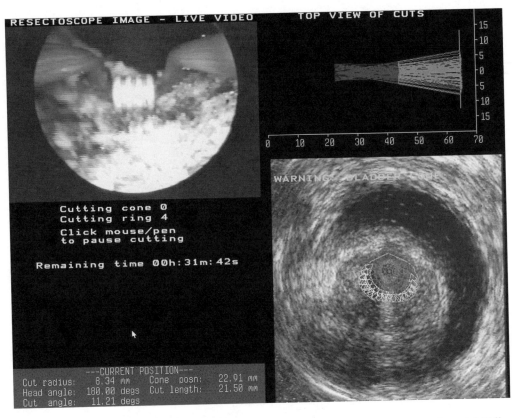

FIGURE 30.16 Video monitor display during the procedure. (Courtesy of Brian Davies, Imperial College, London.)

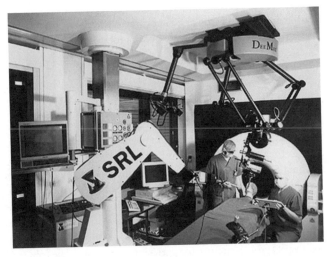

FIGURE 30.17 Surgical robotics laboratory showing the parallel robot (ceiling mounted), mobile CT (back of photo), and serial robotics arm (foreground). (Courtesy of Tim Lueth, Charite Berlin.)

30.4.5.2 Craniofacial Osteotomy

Another system for maxillofacial surgery has been developed at the Institute of Process Control and Robotics in Karlsruhe, Germany, in cooperation with the Clinic of Craniofacial Surgery at the University of Heidelberg. Animal studies were carried out to perform osteotomies where an RX 90 surgical robot (Orto Maquet, Staubli) was used to guide a surgical cutting saw.[41] The studies were carried out as follows. Twelve titanium screws were implanted into the head of a pig to be used as landmarks. A CT scan with 1.5 mm slice spacing was done, and the resulting images were used to create a surface model for surgical planning. A haptic interface was used to trace the cutting lines on the surface of the skull (Figure 30.18). Once the planning was completed, the robot was registered with the pig in the operating room (Figure 30.19), and the surgeon manually guided the robot arm along the trajectory where his movements perpendicular to the cutting line were restricted. This system has also been evaluated using sheep for the autonomous milling of a cavity in the skull needed for a customized titanium implant.

30.4.6 Radiosurgery

Radiation is a common means of treatment for tumors. Radiosurgery is the delivery of radiation to a tumor while attempting to spare the adjacent normal tissue. In the brain, radiosurgery has typically been carried out using stereotactic frames that are rigidly fixed to the patient's skull. A novel method for precision irradiation called image-guided radiosurgery has been developed by Adler and associates at Stanford University (Stanford, CA).[42] The system consists of a lightweight linear accelerator, a Kuka robot, paired orthogonal x-ray imagers, and a treatment couch as shown in Figure 30.20. During a radiosurgery treatment session, the x-ray imaging system determines the location of the lesion. These coordinates are sent to the robot, which adjusts the pointing of the accelerator beam toward the lesion. The robot arm moves the beam through a series of preset positions to maximize the dose to the lesion while minimizing the dose to the surrounding normal tissue.

30.4.7 Ophthalmology

Many surgical operations on the eye, ear, brain, nerves, and blood vessels require extremely precise positioning and manipulation of surgical instruments. It is not uncommon for a microsurgeon to perform

FIGURE 30.18 Planning of the bone cuts using a haptic interface. (Courtesy of Catherina Burghart, University of Karlsruhe, Germany.)

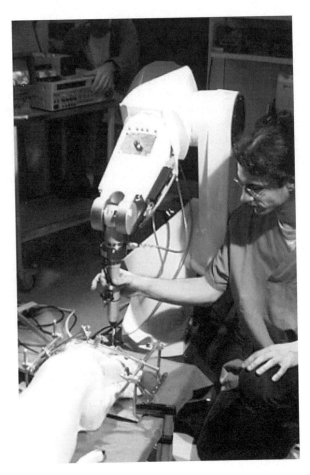

FIGURE 30.19 Registration by force-controlled manual guiding of the robot arm. (Courtesy of Catherina Burghart, University of Karlsruhe, Germany.)

150–200 μm movements during an operation but smaller movements would be desirable.[43] One representative microsurgical application is eye surgery and prototype systems for this purpose have been developed by Das et al.[44] and Hunter et al.[45]

Taylor and colleagues at Johns Hopkins University recently developed a "Steady-Hand" robot for microsurgical augmentation as shown in Figure 30.21.[46] While the initial target application is eye surgery, the system is applicable to numerous clinical specialties. The system consists of four modular subassemblies:

1. An off-the-shelf XYZ translation assembly (only the Z-axis can be seen here)
2. An orientation assembly
3. An end-of-arm motion and guiding assembly including a force/torque sensor
4. Specialized instruments

The major difference between this robotic device and the other robotic systems described in this chapter is that the Steady-Hand robot is designed to work cooperatively with the physician. In operation, the physician will grasp the tool held by the robot and manipulate the tool with the aid of the robot. The control system of the robot senses the forces exerted by the physician on the tool and by the tool on the environment and responds accordingly. The robot can thus provide smooth, tremor free, precise positioning, and force scaling.

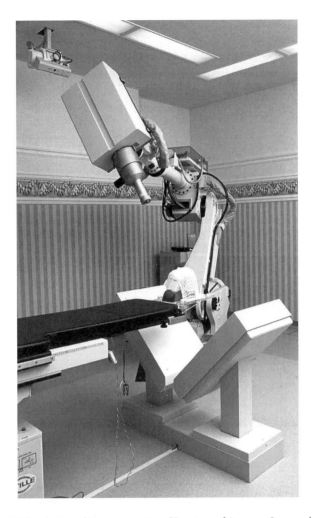

FIGURE 30.20 CyberKnife robotic radiosurgery system. (Courtesy of Accuray, Sunnyvale, CA.)

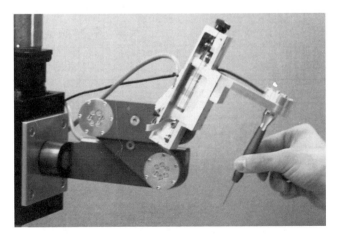

FIGURE 30.21 Steady-Hand robot for microsurgical augmentation. (Courtesy of Russell Taylor, Johns Hopkins University, Baltimore, MD.)

The Steady-Hand robot was employed in a series of experiments to test the ability of a human to position a 10-0 microsurgical needle to 250-, 200-, and 150-μm accuracy.[47] A datum surface was fabricated consisting of two metallic sheets separated by an insulating surface. Three different versions of the experiments were performed: (1) unassisted series (only human), (2) handheld (human plus Steady Hand), and (3) autonomous (Steady Hand was registered to the plates). The use of the Steady-Hand robot was found to significantly improve the ability of the human to position the needle, as success rates improved from 43% unassisted to 79% handheld for the 150-μm holes (autonomous performance was even better at 96.5%).

30.4.8 Cardiology

Two companies have recently developed master–slave systems for minimally invasive surgery that are aimed at restoring the dexterity that is lost when using the traditional laproscopic instruments. The introduction of these systems is a paradigm shift for surgical applications; in that the physician is no longer directly manipulating the surgical tool, but rather controlling the device from a remote interface. While these systems might be used for remote telesurgery in the future, in current practice, the master and slave devices are in the same operating room. The initial clinical applications of these systems have been in cardiac surgery, although other applications are beginning to appear as well.

30.4.8.1 Intuitive Surgical: da Vinci

The Intuitive Surgical system (Figure 30.22), called da Vinci, consists of the surgeon's viewing and control console, a control unit, and a three-arm surgical manipulator.[48] The system is designed to combine the freehand movements used in open surgery with the less traumatic methods of minimally invasive surgery. The surgeon sits at the console and sees a high-resolution three-dimensional image of the surgical field. The surgeon's hands grasp the instrument handles that control the remote endoscopic

FIGURE 30.22   da Vinci surgeon's console. (Courtesy of Intuitive Surgical, Sunnyvale, CA.)

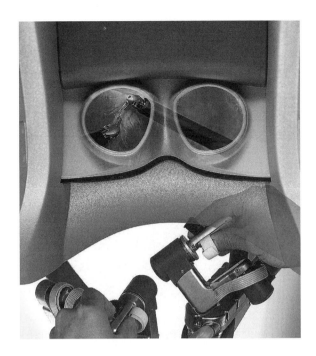

FIGURE 30.23   da Vinci instrument handles and remote manipulators. (Courtesy of Intuitive Surgical, Sunnyvale, CA.)

manipulators and end effectors. The surgeon's console is shown in Figure 30.22 and a view of the instrument handles along with the remote manipulators is shown in Figure 30.23. The manipulators provide three DOFs (pitch, yaw, and insertion). The end effector consists of a miniature wrist that adds three more DOFs (pitch, yaw, and roll) and one motion for tool actuation (such as grip). The system allows increased precision by providing motion scaling whereby large motions of the input devices can be scaled down proportionally to produce small motions at the end effector. Finally, unintended movements caused by tremor, which typically occur with a frequency of 6–10 Hz, are filtered by applying a 6 Hz motion filter.[49] Design issues associated with these types of systems are described by Madhani et al.[50]

The da Vinci system had been used to perform over 500 procedures as of October 1999.[48] The system has not only been used for cardiac procedures such as fully endoscopic coronary artery bypass grafts (CABG), but also for a wide variety of other procedures, including Nissen fundoplication, cholecystecomy, and lumbar sympathectomy.

30.4.8.2 Computer Motion: Zeus

A similar telesurgical system, called Zeus, has been developed by Computer Motion. A picture of the surgeon's console is shown in Figure 30.24. The Zeus slave system consists of three interactive robotic arms (two endoscopic instrument arms and one endoscopic camera arm) that are mounted on the operating room table. However, while the Intuitive Surgical system is a six-DOF system (plus grip motion), only the Computer Motion system has four DOFs, and therefore is not dexterous. Still, the performance of these systems in clinical applications is just beginning to be investigated, and it is difficult to draw conclusions about their efficacy at this point.

The clinical use of the system for endoscopic coronary artery bypass on 25 patients has been described by Boehm et al.[51] This study showed that endoscopic coronary artery bypass on the beating heart is possible, but further development of the technology and techniques is required to minimize the procedure time.

FIGURE 30.24 Zeus surgeon's console. (From Copyright 2001 Computer Motion, Santa Barbara, CA. Photograph by Bobbi Bennett.)

30.4.8.3 Grenoble: Pericardial Puncture

A prototype robot for pericardial puncture has been developed by Troccaz and colleagues at the TIMC/IMAG laboratory of Grenoble University Hospital. The robot is a six-DOF SCARA design consisting of a vertical translational axis, three vertical rotational axes, a rotation about a horizontal axis, and a last modular joint that can be a rotational or a translational axis. Similar to the Acrobot system for knee replacement surgery described earlier, this robot is designed as a synergistic device that is to be used in cooperation with a human operator. The design of the prototype robot and preliminary experiments are presented in an article by Schneider and Troccaz.[52]

30.5 Technology Challenges/Research Areas

While a number of different clinical areas are being explored as noted in Section 30.4, the field of medical robotics is still in its infancy and we are just at the beginning of this era. Only a handful of commercial companies exist and the number of medical robots sold each year is very small. Part of the reason for this is the difficulty of introducing new technology into the complex medical environment. In addition, the completion of a medical robotics project requires a partnership between engineers and clinicians that is not easy to establish.

Technology challenges and research areas for medical robotics include both the development of system components and the development of systems as a whole. In terms of system components, research is needed in:

1. System architecture
2. Software design
3. Mechanical design
4. Imaging compatible systems
5. User interface
6. Safety issues

For medical robotics systems, the development of application testbeds is critical to move the field forward. These testbeds can also serve to improve the dialog between engineers and clinicians. However, at least in the United States, it is difficult to obtain funding to develop these testbeds. Governmental funding agencies such as NIH or NSF will usually not fund such efforts as they are geared more toward basic research rather than applied research and development. Manufacturers are usually not interested because the environment and investment payback for medical robotics is uncertain. The regulatory issues for medical robotics have not been fully explored, although several systems have been approved by the FDA. These factors remain obstacles to advancing the field.

In the following sections, each of the six system components listed above are briefly discussed.

30.5.1 System Architecture

For medical robotics to evolve as its own field and for the cost and difficulty of developing prototype systems to decrease, the establishment of a system architecture would be an enabling step. The system architecture should emphasize modularity, as noted by Taylor in the design of the Steady-Hand robot, which emphasizes modularity in mechanical design, control system electronics, and software.[46] A modular approach has also been emphasized in the Urology Robotics Laboratory of Stoianovici, where a number of mechanical modules have been developed for precision interventional procedures.[35]

30.5.2 Software Design

The development of a software environment for medical robotics, possibly including an appropriate real-time operating system, is a significant challenge. Many researchers developing medical robotics systems base their software development on commercially available software that may not be suitable for the surgical environment. However, the low cost and widespread availability of these software packages makes their use attractive and there are steps that can be taken (such as watchdog timers, backup systems, and error recovery procedures) to make these systems more reliable. Still, it is believed that along with the system architecture mentioned above, a robust software environment geared to the medical environment would be a substantial contribution. While this software environment would still need to be customized for different surgical procedures, researchers would at least have a starting point for their development work.

30.5.3 Mechanical Design

In addition to better software design, novel mechanical designs are needed to improve the utility of robotics in medical procedures. As noted in the historical review in this chapter, the first recorded medical application of a robot was for biopsy of the brain, using a standard PUMA industrial robot. While some other researchers have described the use of the industrial robot for medical tasks, it is the belief of these authors and others[53] that special-purpose mechanical designs are more appropriate for most applications. In particular, these designs should be safer, as they can be designed specifically for the medical environment and can be customized for different medical procedures. The novel mechanical designs presented in this chapter include the Probot[34] and the Steady-Hand robot.[46] However, special-purpose designs will not enjoy the same economies of scale as more general designs, and one other solution may be to develop more general-purpose medical robots with specialized end effectors.

30.5.4 Imaging Compatible Systems

With the increasing popularity of image-guided interventions, robotic systems are required that can work within the constraints of various imaging modalities such as CT and MRI. While these systems are for the most part still under the direct control of the physician, in the future, they will be increasingly linked to these imaging modalities. In this chapter, some systems were noted that fall within this category, such as the MRI-compatible manipulator of Masamune et al.[21] and the CT-integrated robot Minerva.[16]

30.5.5 User Interface

One question that arises in the development of all medical robotics systems concerns the user interface. What is a suitable user interface for a medical robot? Should the robot be given a commanded path or volume and then autonomously carry out the task? Is a joystick or a push-button interface appropriate? Or would the physician rather manipulate the tool directly with the assistance of the robot? Is force feedback required for a high-fidelity user interface?

These are all questions that require further investigation by the medical robotics community. The answer certainly will vary depending on the medical task for which the robot is designed. It seems that medical robots will at least initially be more accepted by physicians if the physicians feel that they are still in control of the entire procedure.

30.5.6 Safety Issues

Safety is a paramount concern in the application of these systems and must be addressed to move the field forward. Safety issues have been discussed by Davies[53] and Elder and Knight.[54] According to Davies, medical robotics is a completely different application from industrial robotics in that medical robots must operate in cooperation with people to be fully effective. Therefore, appropriate safety levels should be defined and discussed by the community at large. Safety measures that can be taken include the use of redundant sensors, the design of special-purpose robots with capabilities tailored to the task at hand, and the use of fail–safe techniques so that if the robot does fail, it can be removed and the procedure can be completed by hand. One other safety issue for medical robotics is the need for sterilization and infection control in the operating room and interventional suite.

Davies also presents a hierarchical scheme for the host of tools available to surgeons, ranging from hand-held tools to a fully powered autonomous robot. As the hierarchy moves toward autonomous robots, the surgeon is less and less in control, and more dependent on the mechanical and software systems of the robot. Davies contends that until a consensus is developed on what level of safety is acceptable for what level of autonomy, the medical manufacturers will be slow to develop robotic systems.

While mechanical constraints are one means of assuring safety, programmable constraints, while inherently not as safe, are more flexible. The idea is to dynamically constrain the range of possible motions.[52] Four programming modes can be envisioned: free mode, position mode, trajectory mode, and region mode. As an example, region mode is particularly suited to resection operations such as total knee replacement in that the surgical tool is constrained to remain within a predefined region. This mode could also be valuable in the training of residents and fellows.

30.6 Conclusions

This chapter has reviewed the state of the art in surgical robotics. Several prototype and commercial medical robotics systems were described. Technology challenges and areas for future research were discussed. The use of robots in medicine clearly offers great promise.

We are just in the initial stages of the application of robotics to medicine, and much more work remains to be done. In particular, the development of more testbeds is required for different medical procedures so that more experience can be gained with the technology and how it can be integrated into clinical practice. The issues of cost, safety, and patient outcomes also need to be considered. While there have been some modestly successful commercial medical robots such as ROBODOC and da Vinci, they still are not completely accepted by the medical community.

It may be that the full benefits of robots in medicine will not be realized until more integrated systems are developed, in which the robots are linked to the imaging modalities or to the patient anatomy directly. This link will highlight the potential advantages of robots such as the ability to follow respiratory motion, and enable physicians to successfully complete procedures that can only be imagined today.

Acknowledgments

This is a revised version of an article titled *State of the Art in Surgical Robotics: Clinical Applications and Technology Challenges* that appeared in *Computer Aided Surgery*, Vol. 6, Number 6, 2001, pp. 312–328, Copyright 2002 by Wiley-Liss. The authors would like to thank Wiley for permission to republish this material. The authors would also like to thank all their colleagues who generously donated photographs. This work was funded in part by U.S. Army Grants DAMD 17-96-2-6004 and DAMD 17-99-1-9022. The content of this manuscript does not necessarily reflect the position or policy of the U.S. Government.

References

1. Capek, K., *R.U.R. (Rossum's Universal Robots)*, Dover Publications, Mineola, New York, 2001.
2. Craig, J.J., *Introduction to Robotics*, 2nd ed., Addison-Wesley, Reading, PA, 1989.
3. Paul, R.P., *Robot Manipulators: Mathematics, Programming, and Control*, MIT Press, Cambridge, 1981.
4. Fu, K.S., Lee, C.S.G., and Gonzales, R.C., *Robotics: Control, Sensing, Vision & Intelligence*, McGraw-Hill, New York, 1987.
5. Spong, M.W. and Vidyasagar, M., *Robot Dynamics and Control*, Wiley, New York, 1989.
6. Davies, B., A review of robotics in surgery, *Proc. Inst. Mech. Eng.* [H], 214, 129–140, 2000.
7. Taylor, R.H., Robots as surgical assistants: Where we are, wither we are tending, and how to get there, in AIME '97, Grenoble, France, 1997, pp. 3–11.
8. Troccaz, J. and Delnondedieu, Y., Robots in surgery, in *IARP Workshop on Medical Robots*, Vienna, 1996, pp. 161–168.
9. Howe, R.D. and Matsuoka, Y., Robotics for surgery, *Annu. Rev. Biomed. Eng.*, 1, 211–240, 1999.
10. Cadeddu, J.A., Stoianovici, D., and Kavoussi, L.R., Robotic surgery in urology, *Urol. Clin. North Am.*, 25(1), 75–85, 1998.
11. Kwoh, Y.S. et al., A robot with improved absolute positioning accuracy for CT guided stereotactic brain surgery, *IEEE Trans. Biomed. Eng.*, 35(2), 153–160, 1988.
12. Yamauchi, Y. et al., A needle insertion manipulator for x-ray CT image-guided neurosurgery, *Proc. LST*, 5, 814–821, 1993.
13. Davies, B.L. et al., A surgeon robot for prostatectomies, in *5th International Conference on Advanced Robotics (ICAR '91)*, IEEE, 1991, pp. 871–875.
14. Benabid, A.L. et al., Computer-driven robot for stereotactic surgery connected to CT scan and magnetic resonance imaging. Technological design and preliminary results, *Appl. Neurophysiol.*, 50 (1–6), 153–154, 1987.
15. Taylor, R.H. et al., An image-directed robotic system for precise orthopaedic surgery, *IEEE Trans. Robot. Autom.*, 10(3), 261–273, 1994.
16. Burckhart, C.W., Flury, P., and Glauser, D., Stereotactic brain surgery, *IEEE Eng. Med. Biol.*, 14(3), 314–317, 1995.
17. Glauser, D. et al., Mechanical concept of the neurosurgical robot Minerva, *Robotica*, 11(6), 567–575, 1993.
18. Lavallee, S. et al., Image-guided operating robot: A clinical application in stereotactic neurosurgery, in *Computer-Integrated Surgery*, Taylor, R.H. et al., eds., MIT Press, Cambridge, 1995, pp. 343–351.
19. Lavallee, S., A new system for computer assisted neurosurgery, in *Proceedings of the 11th IEEE Engineering Medical and Biological Conference*, 1989, 926–927.

20. Benabid, A.L. et al., Robotic guidance in advanced imaging environments, in *Advanced Neurosurgical Navigation*, E.A. III and Maciunas, R.J., eds., Thieme Medical Publishers, Inc., New York, 1999, pp. 571–583.

21. Masamune, K. et al., Development of an MRI-compatible needle insertion manipulator for stereotactic neurosurgery, *J. Image Guid. Surg.*, 1(4), 242–248, 1995.

22. Chinzei, K. et al., MR compatible surgical assist robot: System integration and preliminary feasibility study, in *Medical Image Computing and Computer Assisted Intervention*, Springer, 2000, pp. 921–930.

23. Kaiser, W.A. et al., Robotic system for biopsy and therapy of breast lesions in a high-field whole-body magnetic resonance tomography unit, *Invest. Radiol.*, 35(8), 513–519, 2000.

24. DiGioia, A.M., What is computer assisted orthopaedic surgery? *Clin. Orthop.*, 3542–4, 1998.

25. Borner, M. et al., Experiences with the ROBODOC system in more than 1000 cases, in *Computer Aided Radiology and Surgery*, Elsevier, Amsterdam, 1998, pp. 689–693.

26. Bauer, A., Borner, M., and Lahmer, A., Clinical experience with a medical robotic system for total hip replacement, in *Computer Assisted Orthopedic Surgery*, Nolte, L.P. and Ganz, R., eds., Hogrefe & Huber, Bern, 1999, pp. 128–133.

27. Masamune, K. et al., System for robotically assisted percutaneous procedures with computed tomography guidance, *Comput. Aided Surg.*, 6(6), 370–383, 2001.

28. Yanof, J. et al., CT-integrated robot for interventional procedures: Preliminary experiment and computer–human interfaces, *Comput. Aided Surg.*, 6(6), 352–359, 2001.

29. Jakopec, M. et al., The first clinical application of a "hands-on" robotic knee surgery system, *Comput. Aided Surg.*, 6(6), 329–339, 2001.

30. Cleary, K. et al., Robotics for percutaneous spinal procedures: Initial report, in *Computer Assisted Radiology and Surgery (CARS)*, Elsevier, Amsterdam, 2000, pp. 128–133.

31. Cleary, K. et al., Robotically assisted spine needle placement: Program plan and cadaver study, in *Computer Based Medical Systems (CBMS) 14th IEEE International Symposium*, 2001, pp. 339–342.

32. Cleary, K. et al., Robotically assisted nerve and facet blocks: A cadaveric study, *Acad. Radiol.*, 9(7), 821–825, 2002.

33. Kronreif, G. and Fürst, M., Robotic guidance for percutaneous interventions, in *Proceedings of the 11th International Workshop on Robotics in Alpe-Adria-Danube Region (RAAD 2002)*, Balatonfüred, Budapest Polytechnic, Hungary, 2002, pp. 277–281.

34. Davies, B.L. et al., A clinically applied robot for prostatectomies, in *Computer-Integrated Surgery*, Taylor, R.H. et al., eds., MIT Press, Cambridge, 1995, pp. 593–601.

35. Stoianovici, D., URobotics—Urology robotics at Johns Hopkins, *Comput. Aided Surg.*, 6(6), 360–369, 2001.

36. Stoianovici, D. et al., An efficient needle injection technique and radiological guidance method for percutaneous procedures, in *CVRMed-MRCAS*, Troccaz, J., Grimson, E., and Mosges, R., eds., Springer, Grenoble, 1997, pp. 295–298.

37. Stoianovici, D. et al., A modular surgical robotic system for image guided percutaneous procedures, in *MICCAI 98*, Wells, W.M., Colchester, A., and Delp, S., eds., Springer, Cambridge, MA, 1998, pp. 404–410.

38. Lueth, T.C. et al., A surgical robotic system for maxillofacial surgery, in *Proceedings of the 24th Annual Conference of the IEEE Industrial Electronics Society (IECON)*, 1998, pp. 2470–2475.

39. Lueth, T. and Bier, J., Robot assisted intervention in surgery, in *Neuronavigation—Neurosurgical and Computer Scientific Aspects*, Gilsbach, J.M. and Stiehl, H.S., eds., Springer-Verlag, Heidelberg, 1999.

40. Heissler, E. et al., Robot supported insertion of catheters for hyperthermia and brachytherapy, in *Computer Assisted Radiology and Surgery*, Elsevier, Amsterdam, 1998, pp. 660–663.

41. Burghart, C. et al., Robot assisted craniofacial surgery: First clinical evaluation, in *Computer Assisted Radiology and Surgery*, Elsevier, New York, 1999, pp. 828–833.

42. Adler, J.R., Jr. et al., Image-guided robotic radiosurgery, *Neurosurgery*, 44(6), 1299–1306; 1306–1307, 1999.

43. Charles, S., Williams, R.E., and Hamel, B., Design of a surgeon – machine interface for teleoperated microsurgery, in *Proceedings of the Annual International Conference of IEEE Engineering Medical and Biological Society*, IEEE, 1989, pp. 883–884.

44. Das, H. et al., Evaluation of a telerobotic system to assist surgeons in microsurgery, *Comput. Aided Surg.*, 4, 15–25, 1999.

45. Hunter, I.W. et al., A teleoperated microsurgical robot and associated virtual environment for eye surgery, *Presence*, 2(4), 265–280, 1993.

46. Taylor, R.H. et al., A steady-hand robotic system for microsurgical augmentation, *Int. J. Robotics Res.*, 18(12), 1201–1210, 1999.

47. Kumar, R. et al. Performance of robotic augmentation in microsurgery-scale motions, in *Second International Symposium of Medical Image Computing and Computer Assisted Surgery*, 1999, pp. 1108–1115.

48. Guthart, G.S. and Salisbury, J.J.K., The intuitive telesurgery system: Overview and application, in *IEEE International Conference on Robotics and Automation*, 2000, pp. 618–621.

49. Falk, V. et al., Quality of computer enhanced totally endoscopic coronary bypass graft anastomosis— Comparison to conventional technique, *Eur. J. Cardiothorac. Surg.*, 15(3), 260–264; 264–265, 1999.

50. Madhani, A.J., Niemeyer, G., and J.K.S. Jr., The Black Falcon: A teleoperated surgical instrument for minimally invasive surgery, in *International Conference of Intelligent Robots and Systems*, IEEE, Victoria, BC, Canada, 1998, pp. 936–944.

51. Boehm, D.H. et al., Clinical use of a computer-enhanced surgical robotic system for endoscopic coronary artery bypass grafting on the beating heart (In Process Citation), *Thorac. Cardiovasc. Surg.*, 48(4), 198–202, 2000.

52. Schneider, O. and Troccaz, J., A six-degree-of-freedom passive arm with dynamic constraints (PADyC) for cardiac surgery application: Preliminary experiments, *Comput. Aided Surg.*, 6(6), 340–351, 2001.

53. Davies, B.L., A discussion of safety issues for medical robotics, in *Computer-Integrated Surgery*, Taylor, R.H. et al., eds., MIT Press, Cambridge, 1996, pp. 287–296.

54. Elder, M.C. and Knight, J.C., Specifying user interfaces for safety-critical medical systems, in *Medical Robotics and Computer Assisted Surgery*, Wiley-Liss, Baltimore, MD, 1995, pp.148–155.

55. Kronreif, G., Fürst, M., Kettenbach, J., Figl, M., and Hanel, R., Robotic guidance for percutaneous interventions, *Adv. Robot*, 17(6), 541–560, 2003.

Index

O